10th Anniversary International Conference on Nanomaterials – Research and Application (NANOCON 2018)

Brno, Czech Republic
17 - 19 October 2018

ISBN: 978-1-5108-8183-9

Printed from e-media with permission by:

Curran Associates, Inc.
57 Morehouse Lane
Red Hook, NY 12571

Some format issues inherent in the e-media version may also appear in this print version.

Printed by Curran Associates, Inc. (2019)

For permission requests, please contact Tanger Ltd
at the address below.

Tanger Ltd
Keltickova 62
710 00 Slezka Ostrava
Czech Republic

www.tanger.cz

Additional copies of this publication are available from:

Curran Associates, Inc.
57 Morehouse Lane
Red Hook, NY 12571 USA
Phone: 845-758-0400
Fax: 845-758-2633
Email: curran@proceedings.com
Web: www.proceedings.com

TANGER Ltd.
The Czech Society for New Materials and Technologies
Regional Centre of Advanced Technologies and Materials

2018 NANOCON®

10th Anniversary International Conference on Nanomaterials - Research & Application

CONFERENCE PROCEEDINGS

Peer Reviewed

October 17th - 19th 2018
Hotel Voronez I, Brno, Czech Republic, EU

SESSION A
NANOMATERIALS FOR ELECTRONIC, MAGNETIC AND OPTIC APPLICATIONS. CARBON NANOSTRUCTURES, QUANTUM DOTS

Chairmen

Prof. Ing. Eduard HULICIUS, CSc.	Institute of Physics of the CAS, Prague, Czech Republic, EU
RNDr. Ing. Martin KALBAC, Ph.D.	J. Heyrovsky Institute of Physical Chemistry of the CAS, Prague, Czech Republic, EU
Ing. Oldrich SCHNEEWEISS, DrSc.	Institute of the Physics of Materials of the CAS, Brno, Czech Republic, EU
Assoc. Prof. Dr. Bohuslav REZEK	Czech Technical University in Prague, Czech Republic, EU

SELF-ORGANIZED AGGREGATION OF A TRIPLE OF COLLOIDAL QUANTUM DOTS INTO STABLE STRUCTURES WITH VARIOUS SHAPES CONTROLLED BY A LASER FIELD

[1]Victoria S. KORNIENKO, [2]Aleksei S. TSIPOTAN, [2,3]Aleksandr S. ALEKSANDROVSKY, [4]Vitaliy V. SLABKO

[1]Siberian Federal University, Department of Computational Mathematics, Institute of Computational Modeling of Federal Research Center KSC SB RAS, Krasnoyarsk, Russian Federation, vika-svetlakova@yandex.ru
[2]Siberian Federal University, Krasnoyarsk, Russian Federation, acipotan@sfu-kras.ru
[3]Laboratory of Coherent Optics, Kirensky Institute of Physics Federal Research Center KSC SB RAS, Krasnoyarsk, Russian Federation, aleksandrovsky@kirensky.ru
[4]Siberian Federal University, Krasnoyarsk, Russian Federation, vslbko49@mail.ru

Abstract

Dynamical model of self-assembly of nanoparticles in the field of laser radiation is developed and applied to the investigation of the possibility of the assembly of variously shaped structures comprised of the particles. Specifically, the computer model of the process of formation of structures with pre-defined geometry from a set of three initially isolated nanoparticles. The possibility of control of the geometry of the formed structure via the choice of the wavelength of external field.

Keywords: Nanoparticles, nanostructure, self-assembly, quantum dots

1. INTRODUCTION

In recent decades, nanostructures with unique properties other than properties of a bulk sample and dependent on both composition and shape have been studying actively. Therefore, the problem of developing a universal method of nanostructures formation is a subject of interest of many scientists. The least expensive method of obtaining colloidal crystals, which doesn't require a local physical impact on the system, is based on the ability of nanoparticles to self-organize in the process of random Brownian collisions in real disperse systems [1-4]. However, in such case it is impossible to control the processes of formation of nanostructures with predetermined shapes. Therefore, one of the possible solutions of this problem is the physical impact on the ensembles of nanoparticles that allows the formation of complex nanostructures without additional surface modifications [5-8]. However, it is not always possible to achieve the selectivity of the self-assembly process.

Earlier in [9-11], it was shown that the interaction of resonant nanoparticles in the field of laser radiation makes possible the formation of predetermined structures, since the energy of the induced dipole-dipole interaction is sufficient to overcome the barrier that prevents spontaneous aggregation (which makes possible the formation a stable structure), and their geometry depends on the wavelength and the polarization of the external field. In this case, the formation of structures, which scale is much smaller than the wavelength of the laser radiation caused this polarization, becomes possible. The presence of optical resonances in the particle leads to an increase of the interparticle interaction and is the basis for the selective formation of different structures with a predetermined position of the particles in the aggregate. It becomes possible because the energy of the interparticle interaction in the laser radiation field depends on its frequency, the resonance frequencies of the particles, and on the orientation of their group relatively to the plane of polarization.

The experiments [10] on the formation of pairs of colloidal quantum dots in the field of laser radiation have shown the possibility of realizing this method. The further formation of more complex structures (three or more particles) can be realized in the way of a step-by-step process, when a third particle is added to an already formed pair of particles, via selecting the wavelength of the external field.

In this study, a dynamic self-assembly model for a triple of particles in a laser field using Brownian dynamics is proposed. The possibility of forming a three-particle structure with a predetermined geometry from triple of isolated particles is studied. The obtained results allows to make significant progress in the study of the method of structures self-assembly in the field of laser radiation, which can be use as universal method to form structures with specified properties that will find application as sensors, photodiode elements and solar cells.

2. RESULTS AND DISCUSSIONS

In the present study, we used the computational model described in [11]. Using this model, the possibility of formation the structures under the action of laser field via a two-stage process was earlier shown [11], when a third particle was adjoined to a previously formed pair of particles. Such approach, however, requires certain complications of the experimental procedures. Present study the possibility of formation of structures with pre-defined geometry in the course of self-assembly of three separated particles under the action of the field of short pulse laser radiation. The parameters of the medium and of the particles are analogous to those considered in [11]. The nanoparticle's parameters correspond to the parameters of CdTe quantum dots with average diameter 3 nm (the wavelength of the first exciton transition λ=525 nm, Δλ=3 nm, electric dipole moment of the transition $|d_{12}|^2 = 1.91 \cdot 10^{-44}$ J·m^3). The choices of the material and of the size of QDs are the factors that determine both the resonant wavelength of exciton transition and the stability of QDs under the action of laser radiation.

As the result of computer modeling, the dependences of the probabilities of structures' formation on the external laser radiation wavelength were obtained. As it was shown earlier in [9,10], the interaction of dipole polarizations induced on QD by laser field in the given structure leads to the shift of resonances of QDs in comparison to those of isolated particles, and this shift is dependent on the position of QDs with respect to the polarization plane of laser field. Therefore, interparticle interaction energy is determined by the laser wavelength, intensity and polarization, which leads to the spectral selectivity of the structure obtained. **Figure 1** depicts such dependence of maximum probability of linear structure formation. In general, it was found that at certain wavelengths choices the formation of three types of stable structures: "line" (**Figure 1**, θ=0°), "angle" (θ=90°) and "pyramid" (θ=120°).

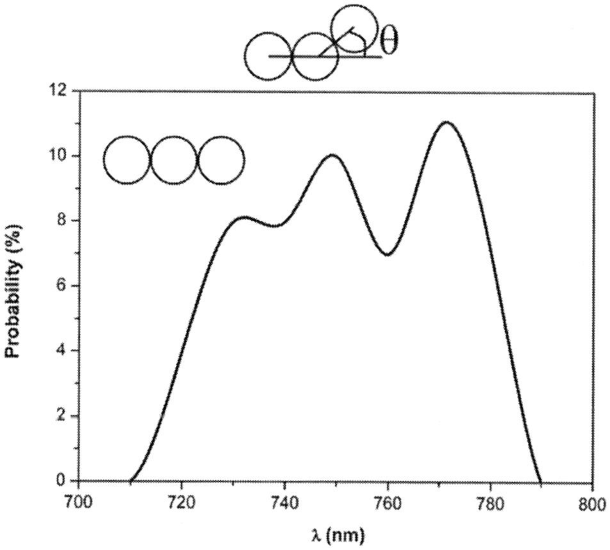

Figure 1 Dependence of the maximum of the probability of formation of the structure of "line" type on the external field wavelength

As one can see from **Figure 1**, the probability of "line" type structure formation reaches 12 % per a single pulse with 10 ns duration; in case when laser irradiation is performed in the long wavelength part of the spectrum, only line type structures are formed, evidencing high selectivity of self-assembly process. Irradiation in the short wavelengths part of the spectrum are shown to be favorable for the formation of other types of stable structures as can be seen in **Figure 2**.

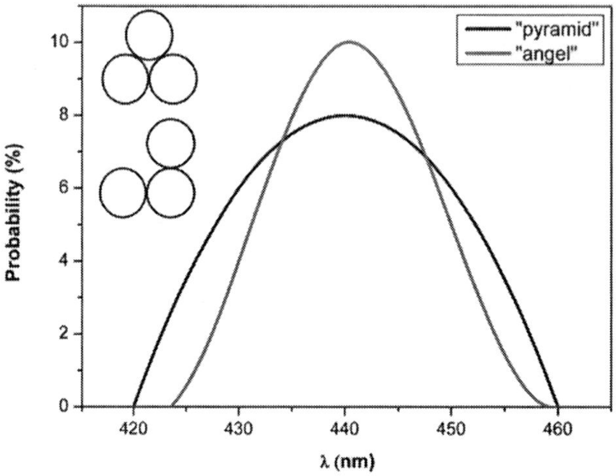

Figure 2 Dependence of maximum probability of formation of structures of a pyramid and angle types on the wavelength of external field

As one can see in **Figure 2**, radiation in the short wavelength range allows formation of two types of stable structures ("pyramid" and "angle") with a good selectivity on the wavelength. However, since for three particles unfixed in space it is impossible to use the selection on the laser field polarization, in these case the formation of both structures is expected to happen with approximately equal probability (10 % и 7.5 %).

Therefore, formation of structures is shown to be possible with the help of a laser with 10 ns pulse duration. Using of longer pulses as well as higher intensity is not unambiguously useful since they will lead not only increase of formation probability but decrease of the selectivity formation of pre-determined structure on the laser frequency. Increase of repetition rate of laser pulses will lead to the increase of overall probability of structures' formation per the unit of time, however, it can produce the radiation heating of a solution.

3. CONCLUSION

The possibility of formation of structures with pre-defined configuration in the field of laser radiation is shown via computer modeling. Using laser pulse with short duration of order of 10 ns, three types of stable structures can be obtained. The geometry of structures depends on the external field wavelength that allows performing their selective formation. Linear structures will be formed for the laser wavelengths in the 720-780 nm range, and the mixture of structures of two types ("pyramid" and "angle" will be formed under irradiation in the 420-460 nm range.

ACKNOWLEDGEMENTS

The reported study was funded by Russian Science Foundation (Grant 18-72-00003). V.V. Slabko is grateful for the support from the Ministry of Education and Science of the Russian Federation (Grant 3.6341.2017/VU).

REFERENCES

[1] KARPOV, S.V., ISAEV, I.L., GAVRILYUK, A.P. General principles of the crystallization of nanostructured disperse systems. *Colloid Journal.* 2009. Vol. 71, no. 3, pp. 313-328.

[2] WANG, T., LAMONTAGNE, D., LYNCH, J., ZHUANGC, J., CAO, Y.C. Colloidal superparticles from nanoparticle assembly. *Chem Soc Rev.* 2013. vol. 42, no7, pp. 2804-2823.

[3] QINGQING, L., BAOLIANG, C., BAOSHAN, X. Aggregation Kinetics and Self-Assembly Mechanisms of Graphene Quantum Dots in Aqueous Solutions: Cooperative Effects of pH and Electrolytes. *ENVIRONMENTAL SCIENCE & TECHNOLOGY.* 2017. vol. 51, no. 3, pp. 1364-1376.

[4] AL-MILAJI, K. N., RADHAKRISHNAN, V., KAMERKAR, P., ZHAO, H. pH-modulated self-assembly of colloidal nanoparticles in a dual-droplet inkjet printing process. *JOURNAL of COLLOID AND INTERFACE SCIENCE.* 2018. vol. 529, pp. 234-242.

[5] GONG, J., LI, G., TANG, Z. Self-assembly of noble metal nanocrystals: Fabrication, optical property, and application. *Nano Today.* 2012. vol. 7, no. 6, pp. 564-585.

[6] SRIVASTAVA, S., SANTOS, A., CRITCHLEY, K. Light-Controlled Self-Assembly of Semiconductor Nanoparticles into Twisted Ribbons. *Science.* 2010. vol. 327, no. 5971, pp. 1355-1359. 2.

[7] BENDIX, P. M., JAUFFRED, L., NORREGAARD, K. and ODDERSHEDE, L.B. Optical trapping of nanoparticles and quantum dots. *IEEE journal of selected topics in quantum electronics.* 2014. vol. 20, no. 3, 4800112.

[8] LEHMUSKERO, A., JOHANSSON, P., RUBINSZTEIN-DUNLOP, H., TONG, L. and KÄLL, M. Laser trapping of colloidal metal nanoparticles. *ACS Nano.* 2015. vol. 9, no. 4, pp. 3453-3469.

[9] SLABKO, V. V., KHACHATRYAN, G. G., ALEKSANDROVSKY, A. S. Self-organized aggregation of small metal particles controlled by an external light field. *JETP LETTERS.* 2006. vol. 84, is. 6, pp. 300-304.

[10] TSIPOTAN, A.S., GERASIMOVA, M.A., SLABKO, V.V., ALEKSANDROVSKY, A.S. Laser-Induced Wavelength-Controlled Self-Assembly of Colloidal Quasi-Resonant Quantum Dots. *Opt. Express.* 2016. vol. 24, no. 10, pp. 11145-11150.

[11] KORNIENKO, V.S., TSIPOTAN, A.S., SLABKO, V.V. Self-assembly of three nanoparticles in structures with a predetermined geometry controlled by quasiresonant laser field. *Journal of Siberian Federal University. Mathematics & Physics.* 2018. vol. 11, no. 6, pp. 1-10.

THE EFFECT OF ZIRCONIA NANOPOWDERS' STRUCTURE MODIFICATION BY GeO_2 ADDITION ON THE KINETICS OF THE INITIAL SINTERING STAGE

Marharyta LAKUSTA, Igor DANILENKO, Galina VOLKOVA, Larisa LOLADZE, Tetyana KONSTANTINOVA

Donetsk Institute for Physics and Engineering named after O.O. Galkin of the NAS of Ukraine, Kiev, Ukraine, marharyta.lakusta@outlook.com

Abstract

The impact of GeO_2 on the sintering kinetics and mass transfer mechanisms in tetragonal zirconia nanopowders has been investigated. The shrinkage kinetics of 3Y-TZP was investigated by the dilatometry analyses. The constant rate of heating method was used to determine the dominant mass transfer mechanisms at the initial sintering stage. It was found that GeO_2 in depends on the synthesis method and its amount caused the viscous flow mechanism participation at the initial sintering stage. The sintering mechanism in co-precipitated 3Y-TZP-GeO_2 powders was changed from volume diffusion mechanism to the volume diffusion with viscous flow mechanism. In mixed 3Y-TZP-GeO_2 powders, the sintering mechanism changed from grain boundary diffusion to the volume diffusion with viscous flow mechanism participation. In both types of nanopowders such sintering mechanism changing leads to the activation energy of sintering decreasing that reduces a sintering temperature and activates the densification process in 3Y-TZP.

Keywords: Zirconia, nanopowders, nanocomposites, sintering, additives

1. INTRODUCTION

Tetragonal zirconia (Y-TZP) ceramics is commercially very significant material, which used as an anode material in SOFCs, as a catalyst oxygen sensor and structural, medicine materials because of its physical and mechanical properties such as: high strength and fracture toughness, chemical and radiation resistance, ionic conductivity, biocompatibility, etc. [1]. It is known that ceramics properties strongly depend on the powders structure, thus, it is necessary to control the characteristics of nanoparticles' structure such as particle size, porosity, density, agglomeration degree, surface properties etc. [2-3]. Ceramic materials with improved physical and mechanical properties can be obtained by doping of 3Y-TZP with different additives. Matsui, Sakka and Suarez have reported that low amounts of Al_2O_3 lead to an increase in the strength and hardness of the zirconia ceramic [4-6]. GeO_2 and SiO_2 cause a super plasticity of the sintered material, the addition of NiO is used to create of solid oxide fuel cells [1]. Regarding the production of nanoparticles with given properties, not only doping by additives but also synthesis conditions have a great effect which leads to the powders structure modifying as was shown in our previous studies [7-9]. Additives can form solid solutions in zirconia lattice [1,3]. In addition, dopants create different defects and active centers on the particles' surface, which caused significant changes in nanopowders characteristics [2-3]. Zirconia nanopowders structure modifying by dopants influences on the sintering and can effect on the mass transfer mechanisms at the initial sintering stage [1, 4-6]. Different additives influence on the zirconia densification as well as a synthesis conditions in a different way. The sintering can be activated and slow down by dopants addition in depends on synthesis methods and additives amount [2-3]. One of the topical problems of materials science today is zirconia based ceramics with improved properties and activation of its sintering process. The main goal of the present study is to study the effect of GeO_2 addition on the 3Y-TZP nanopowders' structure and investigate influence of GeO_2 on the densification and mass transfer mechanisms at the initial sintering stage.

2. METHODS

For the investigation the nanopowders ZrO_2+3 mol% Y_2O_3 (3Y-TZP), 3Y-TZP+ n·GeO_2, PMM8-3Y-TZP+ n·GeO_2 obtained by co-precipitation and mechanical mixing methods were used. Where n is the additive amount, n = 0.2, 1, 2, 3, 5 wt% and PMM8 marks of mixed and milled for 8 h powders. 3Y-TZP was synthesized using of $ZrOCl_2$·nH_2O, $Y(NO_3)_3$·nH_2O salts. The preparation technique is described in studies [7-9]. After precipitation, the synthesized hydrogels were dried in a microwave furnace with output power of 700 W and at a frequency of 2.45 GHz. The dried hydroxides were calcined in a resistive furnace at 1000 °C with for 2 h. Calcined at 1000 °C for 2 h 3Y-TZP and GeO_2 nanopowders were mixed in the planetary mill (MSK-SFM-1 (MTI Corp., USA) at 400 rpm for 8 h using YSZ milling balls).Thereafter, all nanopowders were pressed at 300 MPa and sintered up to the 1500 °C with different heating rates of 2.5, 5, 10, 20 °C /min in the dilatometer (NETZSCH DIL 402 PC). The shrinkages data of specimens during sintering were measured by dilatometry. Thermal expansion of specimens was corrected with the cooling curve [4-5]. The final density of sintered specimens was measured by the Archimedes method. The lattice parameters, phase compositions and crystallites size of all nanopowders were investigated by X-ray diffraction (XRD) employing the Dron-3 diffractometer with Cu-K α radiation using the Debay-Scherrer equation. Fitting and analysis of the XRD curves were made by Powder Cell. The particles size and powders structures were studied by the transmission electron microscope TEM (Jem 200A, JEOL, Japan). The microstructures of the sintered ceramics were studied by scanning electron microscopy (JSM 6490LV JEOL) after polishing and thermally etching of specimens' surfaces (at 1450 °C for 0.5 h). The standard constant rate of heating (CRH) technique was used to analyze the obtained dilatometric data [4-6]. Using the CRH technique let to identify the sintering parameter *n* and activation energy of sintering Q. Considering that if n = 1, this means that the viscous flow mechanism dominates (VF). If n =0.4-0.5, the volume diffusion mechanism (VD) dominates and if n = 0.3-0.4, the grain boundary diffusion mechanism (GBD) dominates [4-6].

3. RESULTS AND DISCUSSION

The characteristics of the 3Y-TZP-GeO_2 nanopowders are shown in the **Table 1**. The XRD spectra of nanopowders 3Y-TZP, 3Y-TZP+2 wt% GeO_2, PMM8-3Y-TZP+2 wt% GeO_2 synthesized by mixing and co-precipitation techniques are present in the **Figure 1**. GeO_2 additive were not detected as a separate phase in 3Y-TZP as can be seen from the **Figure 1a** (the additive' peak should be situated within 25-27 degree in the XRD pattern). The GeO_2 does not effect on the phase composition of zirconia but 8 hours of milling leads to the monoclinic phase amount increasing up to 15.5 % in the initial PMM8-3Y-TZP and to 25 % in PMM8-3Y-TZP-5 wt%GeO_2. Crystallites size of all investigated nanopowders decrease and its specific surface area increasing at the same time (**Figure 1b, Table 1**).

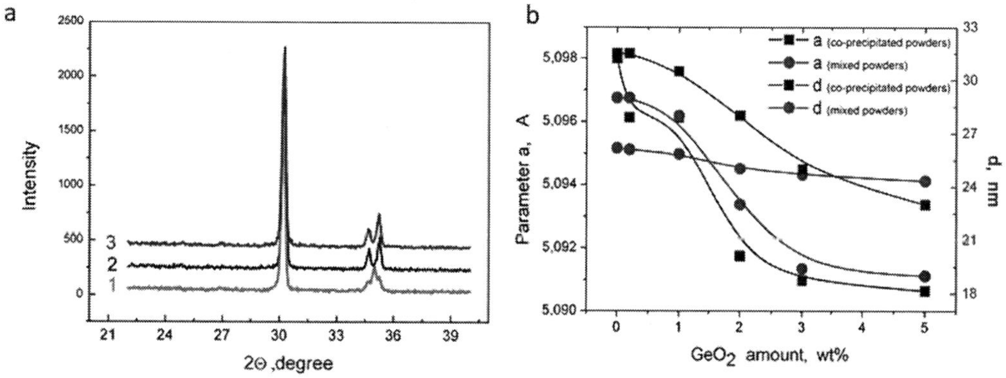

Figure 1 The XRD patterns of nanopowders (a): 1- 3Y-TZP; 2-3Y-TZP-2 wt% GeO_2; 3-PMM8-3Y-TZP-2 wt% GeO_2 and (b) the lattice parameters crystallites sizes changing of the investigated nanopowders

Figure 1b shows that GeO_2 has a greater effect on the particles' size and the lattice parameters of zirconia in co-precipitated(black curves) powders than in mechanically mixed powders (blue curves). In the **Figure 2** the nanopowders' structure of co-precipitated and mixed nanopowders are shown. The separate GeO_2 phase was observed in mixed 3Y-TZP-GeO_2 powders (**Figure 2e, f**). Moreover, as was mentioned above the GeO_2 was not detected as a separate phase by XRD in contrast to the mixed powders (**Table 1, Figure 2**). Thus, it can be concluded that in 3Y-TZP-GeO_2 system obtained by co-precipitation was the solid solution GeO_2 with zirconia matrix created because of precipitation. Mixed PMM8-3Y-TZP-GeO_2 powders systems create a two-phase structure (separate phases of GeO_2 and 3Y-TZP). The decrease in the crystallites sizes, for example, from 29 nm to 19 nm in case of PMM8-3Y-TZP-5 wt % GeO_2 nanopowder was caused by milling because the solid solutions were not created in mixed nanopowders (**Figure 1b, Table 1**).

Table 1 The XRD and BET analysis results

Nanopowders composition	% M-phase	Particle size, nm	Lattice parameters, Å	S_{BET}, g/m³
3Y-TZP	4	31.5	a=5.0980, c=5.17396	14.2
3Y-TZP-0.2 wt % GeO_2	4	31.5	a=5.09613, c=5.17186	10.2
3Y-TZP-1 wt % GeO_2	2	30.5	a=5.09613, c=5.16893	11.0
3Y-TZP-2 wt % GeO_2	1	28	a=5.09175, c=5.1677	14.0
3Y-TZP-3 wt % GeO_2	1	25	a=5.09065, c=5.16695	16.0
3Y-TZP-5 wt % GeO_2	0	23	a=5.09065, c=5.16695	19.0
PMM8-3Y-TZP	15.5	29	a=5.095163, c=5.16670	19.9
PMM8-3Y-TZP-0.2 wt % GeO_2	12	29	a=5.095163, c=5.16670	19.9
PMM8-3Y-TZP-1 wt % GeO_2	13	28	a=5.095163, c=5.16670	19.7
PMM8-3Y-TZP-2 wt % GeO_2	22.5	23	a=5.095163, c=5.16670	19.5
PMM8-3Y-TZP-3 wt % GeO_2	23	19.4	a=5.095163, c=5.16670	19.2
PMM8-3Y-TZP-5 wt % GeO_2	25	19	a=5.095163, c=5.16670	18.9

M-the monoclinic phase amount, standard deviation of the lattice parameters determination is $\Delta a = \pm 00006$, $\Delta c = \pm 0.0005$.

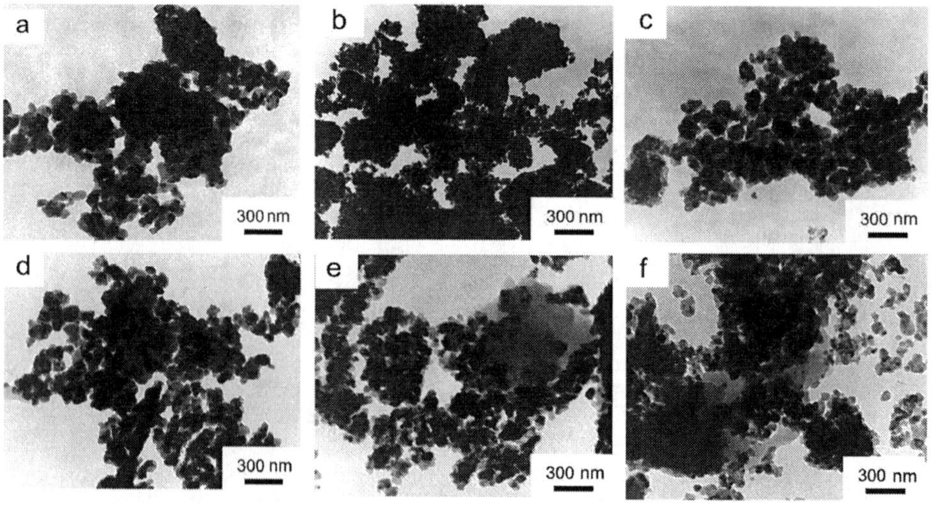

Figure 2 TEM images of nanopowders (a) 3Y-TZP, (b) 3Y-TZP-2 wt% GeO_2, (c) 3Y-TZP-5 wt% GeO_2, (d) PMM8-3Y-TZP, (e) PMM8-3Y-TZP-2 wt% GeO_2, (f) PMM8-3Y-TZP-5 wt% GeO_2

The densification kinetics of 3Y-TZP with 2 wt% GeO_2 specimens obtained by both methods is shown in the **Figure 3a**. As can be seen GeO_2 additive causes the densification rate changing in 3Y-TZP nanopowders. Namely, the temperature value of maximum densification rate (dp/dT) in co-precipitated and mixed 3Y-TZP powders with GeO_2 shifted to smaller values of temperatures (**Figure 3a**). This means that all specimens with GeO_2 were sintered faster than initial 3Y-TZP and PMM8-3Y-TZP powders without additive. Thus, can be concluded that GeO_2 accelerates the sintering of 3Y-TZP.

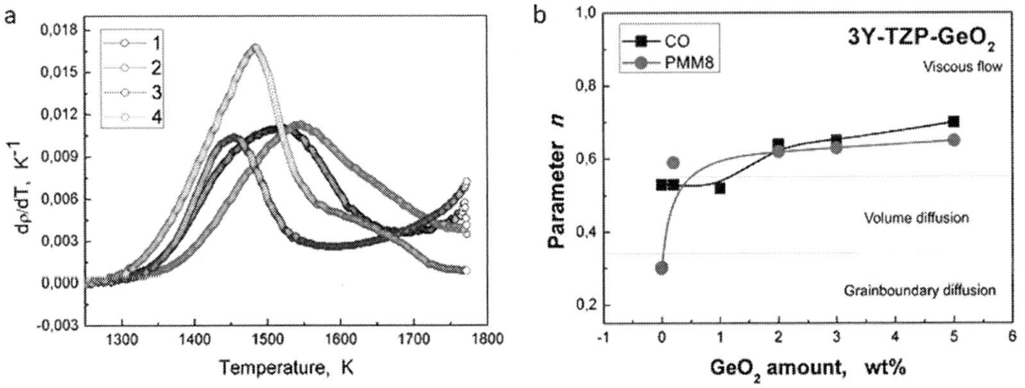

Figure 3 Temperature dependence of densification rate of nanopowders (a): (1) - 3Y- TZP; (2) - PMM8-3Y; (3) - 3Y- TZP -2 wt% GeO_2, (4) - PMM8-3Y- TZP -2 wt% GeO_2 at 10 ^{0}C/min and (b) the sintering mechanism changing in 3Y-TZP- GeO_2

The dominant mass transfer mechanism in all investigated specimens at the initial sintering stage has been determined using the CRH technique [4-6]. The **Figure 3b** shows the sintering mechanism changing in all investigated specimens according to the additive amounts and synthesis methods. As can be seen from the **Figure 3b** the co-precipitated 3Y-TZP-GeO_2 specimens were sintered due to the VD mechanism domination. But with additive amount increasing (1-5 wt%) the parameter n tends to 1 and this means that the sintering occurred with the participation of the viscous flow mechanism (VF). The liquid phase effected on the densification of 3Y-TZP-GeO_2 and caused the activation energy of sintering decreasing that led to the sintering process activation.

In mixed PMM8-3Y-TZP-GeO_2 specimens the sintering mechanism changing from GBD to VD with a small amount (0.2 wt%) of GeO_2 addition was observed. Similarly, the parameter n tends to 1 with additive amount increasing (1-5 wt%) and this means that the VF mechanism participates in sintering process too.

The VF mechanism was caused by the low melting point of GeO_2 (T=1116 °C) which coincides to the temperatures of the initial stage of sintering. The boiling point of GeO_2 is 1200 °C. Due to these facts, GeO_2 leaves the solid solution with temperature increasing during sintering, activates the VF mechanism and evaporates in co-precipitated powders. The same GeO_2 behavior was observed in mixed nanopowders but the melting and evaporation started earlier since at the beginning of sintering GeO_2 was a separate phase in 3Y-TZP. Based on XRD, TEM, SEM, dilatometry data the scheme of the GeO_2 behavior was build. The scheme and SEM images are shown in the **Figure 4** to confirm the results of investigations and our conclusions.

As can be seen from the **Figure 4** the initial nanopowders states of obtained by different methods were the solid solution (1.1) and two-component mix of 3Y-TZP and GeO_2 (2.1). At the initial sintering stage the solid solution decomposition (1.2, 2.2), melting and evaporation of GeO_2 occur in both types of powders. Because of sintering of different nanopowders types (1.3 and 2.3) the ceramic materials with the single-phase 3Y-TZP structure without GeO_2 was obtained (this confirmed by SEM images at 1200 and 1500 ^{0}C as shown in the **Figure 4**).

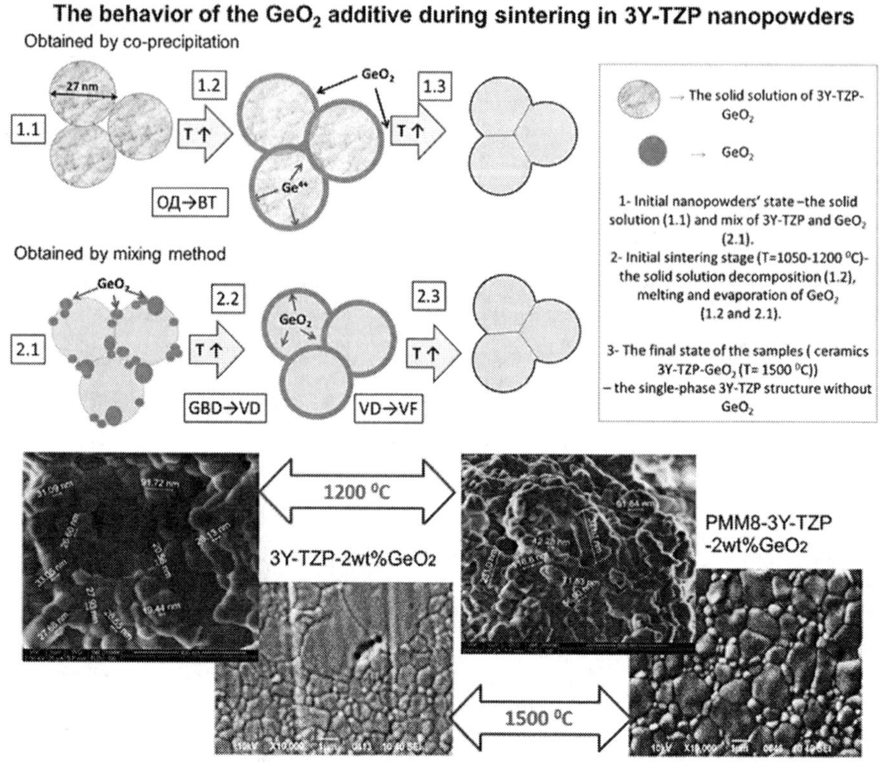

Figure 4 The scheme of GeO_2 behavior during 3Y-TZP sintering and SEM images of 3Y-TZP-2 wt% GeO_2, PMM8-3Y-TZP-2 wt% GeO_2

CONCLUSION

The impact of GeO_2 addition on sintering kinetics of tetragonal zirconia nanopowders has been investigated. The following results were obtained: i) The predominant sintering mechanism in the initial 3Y-TZP prepared by co-precipitation was VD mechanism and GBD in mixed PMM8-3Y-TZP nanopowders; ii) The small amount of the GeO_2 in mixed nanopowders led to the sintering mechanism changing from GBD to VD; iii) The additive amount increasing (1-5 wt%) in both types of powders (3Y-TZP-GeO_2, PMM8-3Y-TZP-GeO_2) caused the participation of the VF mechanism in sintering process. It was found that the GeO_2 accelerated the sintering process and helped create a single-phase 3Y-TZP ceramics with a narrow grain size distribution, which can have the most favorable effect on the properties of obtained ceramic materials for the functional application.

ACKNOWLEDGEMENTS

The authors are thankful the H2020-MSCA-RISE-2015. Programme, project N 690968 NANOGUARD2AR and project AZ 90355 Volkswagen foundation for support of this work.

REFERENCES

[1] RAGULYA, Andrey V. and SKOROCHOD Volodimir V. *Scientific book: Consolidated nanostructured materials.* Kiev: Naukova Dumka, 2007. p. 499.

[2] GEGUZIN Jakob E. *Scientific book: Physics of sintering.* Moscow: Science, 1967. p.321.

[3] KINGERY WD. *Scientific book: Introduction to ceramics.* 3rd ed. Moscow: Stroiizdat, 1967. p.500.

[4] LAKUSTA, Marharyta, DANILENKO, Igor, KONSTANTINOVA, Tetyana and VOLKOVA, Galina. Influence of Obtaining Conditions on Kinetics of the Initial Sintering Stage of Zirconia Nanopowders. *Nanoscale Research Letters*. 2016. Vol. 11, pp.238-245.

[5] LAKUSTA, Marharyta, DANILENKO, Igor, KONSTANTINOVA, Tetyana and VOLKOVA, Galina, GORBAN, Oksana. The Effect of a Small Amount SiO_2 on Sintering Kinetics of Tetragonal Zirconia Nanopowders. *Nanoscale Research Letters*. 2017. Vol.12, pp.398-407.

[6] LAKUSTA, Marharyta, DANILENKO, Igor, KONSTANTINOVA, Tetyana and VOLKOVA, Galina, LOLADZE Larisa, BURCHOVETSKIY, Valeriy. Comparative analyses of the IV group oxides additives influence on the sintering kinetics of zirconia nanopowders. *PLOS ONE*. 2018, pp.1-17. Available from: doi.org/10.1371/journal.pone.0200869.

[7] MATSUI, Koji. Sintering kinetics at constant rates of heating: effect of GeO_2 addition on the initial sintering stage of 3 mol% Y_2O_3-doped zirconia powder. *J Mater Sci*. 2008. Vol. 43, pp. 852-859.

[8] MATSUI, Koji. Sintering Kinetics at Constant Rates of Heating: Mechanism of Silica-Enhanced Sintering of Fine Zirconia Powder. *J. Am. Ceram. Soc*. 2008. Vol. 91,8, pp. 2534-2539.

[9] SUARREZ, Gustavo, SAKKA Yoshio, Effect of alumina addition on initial sintering of cubic ZrO2 (8YSZ). *Ceramics International*. 2010, Vol. 36. pp. 879-885.

OPTICAL PROPERTIES OF PbS-GO AND PbS-rGO SYSTEM FOR SOLAR CELLS FABRICATION

[1]Anton BABAEV, [1]Aliaksei DUBAVIK, [1]Sergei CHEREVKOV, [1]Peter PARFENOV, [1]Elena USHAKOVA, [1]Mikhail BARANOV, [2]Igor NABIEV, [1]Alexandr BARANOV, [1]Aleksandr LITVIN

[1]ITMO University, Saint-Petersburg, Russian Federation, a.a.babaev@corp.ifmo.ru
[2]Research Nuclear University MEPhI, Moscow, Russian Federation

Abstract

Quantum dots (QDs) are of great interest for creating optoelectronic devices due to their tunable bandgap, high stability and simple solution processability for device fabrication. One of the ways to increase the efficiency of such devices is to combine QDs with different carbon nanostructures. Lead sulfide PbS QDs with a ligand shell of oleic acid (OA), methylammonium iodide (MAI) and PbI2 were employed to study the process of charge transfer process between QDs and sheets of reduced graphene oxide (rGO) and graphene oxide (GO). QDs have been linked to graphene sheets by the (3-mercaptopropyl) trimethoxysilane in colloidal solution. As a result, we observed a change in intensity and kinetics of QDs PL, that allowed to estimate the efficiency of the charge transfer. The transfer efficiency for the QD-rGO system was found to be higher than that of the QD-GO system, ~ 80 % vs. 66 % respectively. Modification of the QDs surface with MAI and PbI_2 allowed to achieve a charge transfer efficiency of up to 83 % and 89 %, respectively.

Keywords: Colloidal quantum dots, reduced graphene oxide, graphene oxide, PbS

1. INTRODUCTION

Lead sulfide colloidal quantum dots (CQDs) have widely spread as a material for optoelectronic devices including photodetectors [1,2], light-emitting diodes [3,4], and solar cells [5-7] due to their huge bandgap tunability, multiple-exciton generation effect and facile solution-processing. Most common tools to improve the properties of such devices are CQDs surface passivation and optimization of the device structure. For example, outstanding stability was achieved in solution-processed CQDs solar cells even in ambient air conditions through surface passivation [8]. Recent advances in QD solar cells fabrication by device architecture engineering, surface passivation and layers optimization led to reported certified power conversion efficiency (PCE) of 11.6 % [9]. However, a lot of challenges is still to go. One of the main problems for PbS based optoelectronic devices is their rapid degradation under the high relative humidity of ambient air. Recently, it was shown that creating hybrid materials with graphene for active layer fabrication results in the reduction of moisture sensitivity in non PbS based devises [10-14] as well, as in PbS based solar cells [15]. Beside this, combing PbS CQDs with graphene materials is a promising way to fabricate highly effective photodetectors [16,17] and phototransistors [18,19]. In such devices, the charge carriers are generated in the CQDs and after this electrons are transported to the graphene layer according to the energy structure of CQDs-graphene/rGO/GO interface that schematically depicted in **Figure 1** for rGO/PbS case. The charge transfer efficiency of this interface limits the efficiency of this type of devices.

In our study we involved time-resolved PL characterization to evaluate the speed of charge carrier transfer speed in the blend of PbS CQDs and rGO/GO sheets. The PbS CQDs were attached to the rGO and CO sheets using the organic linker molecule. Moreover, the surface of QDs was modified by PbI_2 and MAI to improve the CQDs quality and optimize the charge transfer efficiency.

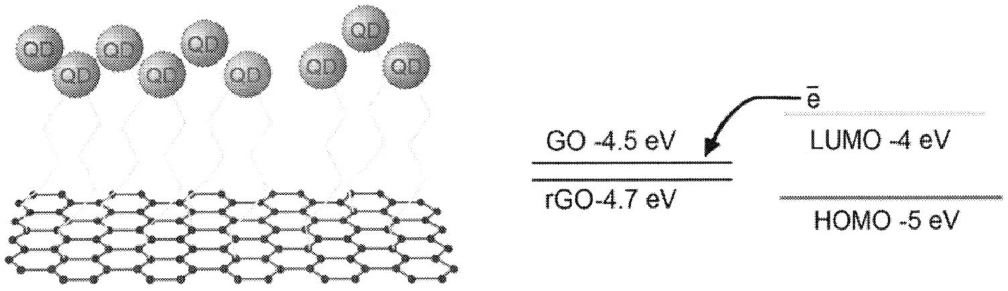

Figure 1 Schematic energy diagram of PbS-rGO

2. MATERIALS AND METHODS

Stabilized with poly(sodium 4-styrenesulfonate) rGO and GO water dispersions were purchased in Sigma-Aldrich (900197 and 777676 aldrich). Moreover, we prepare thin films from both dispersions to ensure the parameters of materials using SEM and AFM. **Figure 2** illustrates a typical single rGO sheet. The thickness we found trough AFM measurements was approximately 1 nm for both materials.

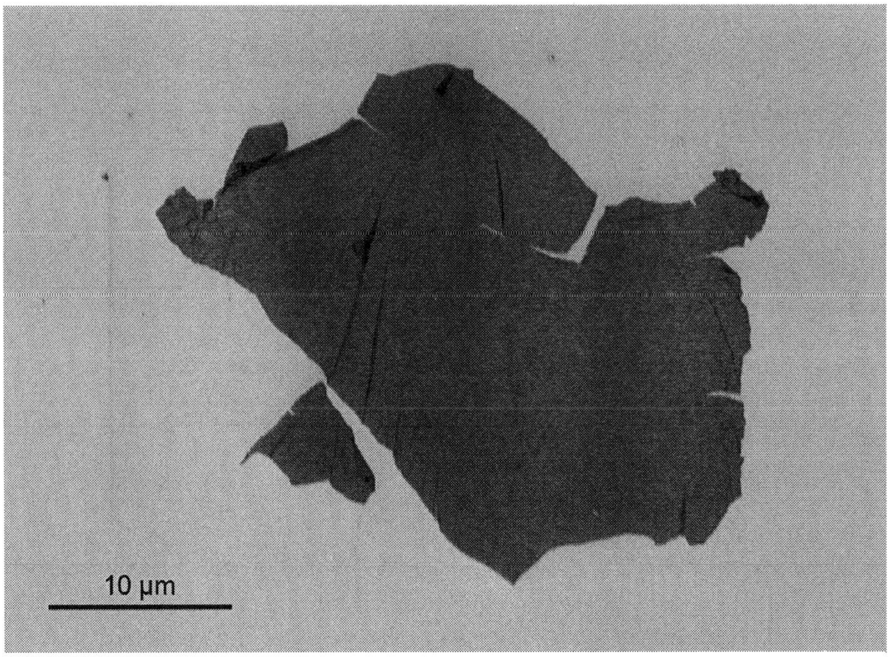

Figure 2 SEM image of rGO sheet

The oleic acid (OA)-capped PbS CQDs were synthesized via well-known protocol from [20]. Briefly, 1 mmol PbO (99.99 %, Aldrich), 4 mmol OA (90 %, Fisher) and 10 ml octadecene (ODE) (90 %, Acros) were mixed in the three-neck 25 ml for Pb precursor preparation. The mixture was heated up to 170 °C under vacuum for approximately 30 minutes until clarification and flushed with Ar. At T=134 °C the mixture of 0.2 mmol hexamethyldisilathiane ($[TMSi]_2S$) in 0.5ml ODE was swiftly injected and heated for 10 min. The reaction was quenched by cooling down in Ar atmosphere and by the addition of acetone to precipitate the PbS nanocrystals. The solution was centrifuged (6000 rpm, 10 min), redispersed in hexane (Sigma-Aldrich) and precipitated 2 more times with acetone. The final nanocrystals were redissolved in toluene for further use and measurements.

The iodide-capped PbS was obtained by modifying OA-capped PbS via protocol from [21]. In short, 0.02 M PbI_2 solution in 12 ml of dimethylformamide (DMF) and 6 ml of methanol was injected into a suspension of as-prepared PbS nanocrystals (2 mL, 50 mg mL^{-1}). The resulting solution was shaken for 1−2 min, leading to a precipitation of the PbS nanocrystals. After centrifugation, the PbS nanocrystals were redispersed in a mixture of 1,2-dichlorobenzene (DHBZ)(1 mL) and butylamine (BA) (0.2 mL). MAI treatment as well was provided using OA-capped PbS in the prolonged protocol described in [22].

The CQDs were attached with the rGO/GO sheets using 3-mercaptopropyltrimethoxysilane(MPTS), as described in [23]. For MPTS treatment of CQDs, MPTS was added in toluene/DHBZ:BA/DMF OA/iodide/MAI-capped CQDs solution with vortexing for 2 minutes. The solution was precipitated by acetone with centrifugation (6000 rpm, 10 min) and redispersed in the initial solvent.

3. RESULTS AND DISCUSSION

We prepared a series of samples to make our calculations precise. In the beginning, we simply spin-cast GO and OA-capped PbS CQDs layers one after another and observed no decrease in PL lifetime compared to the OA-capped PbS CQDs on the glass. In this case, OA acts like isolator and further, we believe that the change in PL kinetics was induced by the charge transfer process. Otherwise, when the iodide-capped PbS CQDs were spin-coated on GO layer we observe the 30 % decrease of the PL lifetimes because iodide shell is conductive. Then, we mix CQDs by the MPTS. On **Figure 3** one can observe the Pl spectra from OA-capped PbS CQDs in toluene, MPTS-treated OA-capped PbS CQDs in DMF and OA-capped PbS CQDs - MPTS - GO blend in DMF. MPTS treatment results in redshift in spectra accompanied by an increase of PL lifetimes. Such behavior is in contrast to the [23], where 11-mercaptoundecanoic acid treatment was provided to obtain reference PL lifetimes and results in a decrease of PbS CQDs PL lifetimes. We believe the change in surface passivation and medium permittivity affect our CQDs in this way. However, the difference in synthetic protocols and in the applied chemical treatment can explain such discrepancy in the results.

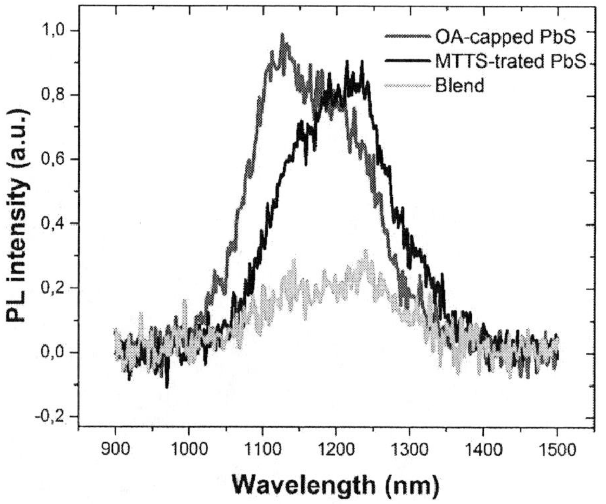

Figure 3 PL spectra of the OA-capped PbS, OA-capped PbS treated with MPTS and OA-capped PbS - GO blend

For all of our blend samples we provide our calculations via the next two equations from [24] and [23]:

$$k_{et} = \frac{1}{\tau_{blend}} - \frac{1}{\tau_{ref}} \quad (1)$$

where:

k_{et} - the rate of electron transfer

τ_{blend} - the decay time of QD-GO/rGO blend

τ_{ref} - the decay time of MPTS treated CQDs

$$\eta = \frac{k_{et}}{k_{et}+k_{rad}+k_{trap}} = 1 - \frac{k_{ref}}{k_{blend}} \quad (2)$$

where:

k_{rad}- the radiative decay rate

k_{trap} - the rate of trapping carriers on CQDs surface

k_{ref} - the decay rate of MPTS treated CQDs

k_{blend} - the decay rate of QD-GO/rGO blend

Our calculations for the samples are given in **Table 1**. This is noticeable that transfer speed calculations give opposite results to the efficiency calculations in the case of OA-capped CQDs and MAI-capped CQDs. In this case the different shell compositions result in different k_{trap} and k_{rad} rates, what counts like (Δk_{trap} +Δk_{rad}) linear mistake for decay rates calculations and like a nonlinear mistake in efficiency calculations. Both parameters for iodide-capped CQDs case are outstanding in comparison to all other structure types. It is obvious that the iodide shell around the CQDs core provides easier charge transfer because of better conductivity.

Table1 The charge transfer rate and efficiency of the charge transfer rate for the obtained samples

Sample	k_{et}, $s^{-1}\cdot 10^6$	η, %
OA capped PbS+rGO	6.8	80
Iodide capped PbS+rGO	7.4	89
MAI capped PbS+rGO	6.6	83
OA capped PbS+GO	4.3	66

Figure 4 illustrates the formation of a spin-coated film from our solution. There are rGO sheets densely occupied by the CQDs. It should be emphasized, that the values of charge transport efficiency strongly depend on the CQDs-rGO proportion. For our study, we employ actual proportions that result in effective device following [6]. The enlargement of rGO content results in the further decay of PL lifetime. For example, 10-time amplification of the rGO content in MAI capped PbS+rGO blend leads to an increment of decay rate to k_{et}=17.9 and transfer efficiency to η=93 %. However, such blend proportions do not result in high performance in despite better charge transfer.

Figure 4 SEM image of the spin-coated film of OA-capper PbS-GO blend

4. CONCLUSIONS

Our group produced a series of experiments that clarify the PbS-rGO/GO charge transfer mechanism. MPTS organic linker forms a well-defined CQDs-rGO/GO system providing both a mechanical connection between system components and better charge transport. The shell also plays an important role in charge process; the iodide shell affords 89 % charge transport efficiency against 80 % in case of basic OA-capped CQDs. The efficiency and transfer rate values were calculated for real device components proportions. The SEM measurements show that CQDs-rGO/GO system forms layers that similar to the rGO/GO layers. The results of our findings reveal details that could important for engineering of solution-processed photovoltaic devices.

ACKNOWLEDGEMENTS

This study was supported by the Ministry of Education and Science of the Russian Federation through the grant No. 14.584.21.0032 (ID RFMEFI58417X0032)

REFERENCES

[1] KUFER, Dominik, NIKITSKIY, Ivan, LASANTA, Tania, NAVICKAITE, Gabriele, KOPPENS, Frank H. L., and KONSTANTATOS, Gerasimos. Hybrid 2D-0D MoS2-PbS quantum dot photodetectors. *Adv. Mater.* 2015. vol. 27, no. 1, pp. 176-180.

[2] SARAN, Rinku and CURRY, Richard J. Lead sulphide nanocrystal photodetector technologies. *Nat. Photonics.* 2016. vol. 10, no. 2, pp. 81-92.

[3] LITVIN, Aleksandr P., MARTYNENKO, Irina V., PURCELL-MILTON, Finn, BARANOV, Alexander V., FEDOROV, Anatoiy V., and GUN'KO, Yurii K. Colloidal quantum dots for optoelectronics. *J. Mater. Chem. A.* 2017. vol. 5, no. 26, pp. 13252-13275.

[4] GONG, Xiwen *et al.* Highly efficient quantum dot near-infrared light-emitting diodes. *Nat. Photonics.* 2016. vol. 10, no. 4, pp. 253-257.

[5] LIU, Mengxia *et al.* Hybrid organic-inorganic inks flatten the energy landscape in colloidal quantum dot solids. *Nat. Mater.* 2017. vol. 16, pp. 258-263.

[6] LU, Kunyuan *et al.* High-Efficiency PbS Quantum-Dot Solar Cells with Greatly Simplified Fabrication Processing via "Solvent-Curing". *Adv. Mater.* 2018. vol. 30, no. 25, pp. 1-9.

[7] CAREY, Graham H., ABDELHADY, Ahmed L., NING, Zhijun, THON, Susanna M., BAKR, Osman M., and SARGENT, Edward H. Colloidal Quantum Dot Solar Cells. *Chem. Rev.* 2015. vol. 115, no. 23, pp. 12732-12763.

[8] CAO, Yiming, STAVRINADIS, Alexandros, LASANTA, Tania, SO, David, and KONSTANTATOS, Gerasimos. The role of surface passivation for efficient and photostable PbS quantum dot solar cells. *Nat. Energy.* 2016. vol. 1, no. 4, p. 16035.

[9] CHOI, Jongmin *et al.* Chloride Passivation of ZnO Electrodes Improves Charge Extraction in Colloidal Quantum Dot Photovoltaics. *Adv. Mater.* 2017. vol. 29, no. 33, p. 1702350.

[10] LIU, Zhike, LAU, Shu Ping, and YAN, Feng. Functionalized graphene and other two-dimensional materials for photovoltaic devices: device design and processing. *Chem. Soc. Rev.* 2015. vol. 44, no. 15, pp. 5638-5679.

[11] GUO, Yunlong, LIU, Chao, TANAKA, Hideyuki, and NAKAMURA, Eiichi. Air-Stable and Solution-Processable Perovskite Photodetectors for Solar-Blind UV and Visible Light. *J. Phys. Chem. Lett.* 2015. vol. 6, no. 3, pp. 535-539.

[12] LEE, Seungae, KIM, Yun Ki, and JANG, Jyongsik. Long-term stability improvement of light-emitting diode using highly transparent graphene oxide paste . 2016. vol. 8, p. 17551.

[13] GLEN, Tom S. *et al.* Dependence on material choice of degradation of organic solar cells following exposure to humid air. *J. Polym. Sci. Part B Polym. Phys.* 2016. vol. 54, no. 2, pp. 216-224.

[14] BAI, Yang *et al.* Enhancing stability and efficiency of perovskite solar cells with crosslinkable silane-functionalized and doped fullerene, *Nat. Commun.* 2016. vol. 7, p. 12806.

[15] MARTÍN-GARCÍA, Beatriz *et al.* Reduction of moisture sensitivity of PbS quantum dot solar cells by incorporation of reduced graphene oxide. *Sol. Energy Mater. Sol. Cells*. 2018. vol. 183, no. March, pp. 1-7.

[16] NIAN, Qiong, GAO, Liang, HU, Yaowu, DENG, Biwei, TANG, Jiang, and CHENG, Gary J. Graphene/PbS-Quantum Dots/Graphene Sandwich Structures Enabled by Laser Shock Imprinting for High Performance Photodetectors. *ACS Appl. Mater. Interfaces*. 2017. vol. 9, no. 51, pp. 44715-44723.

[17] GONG, Maogang *et al.* Printable Nanocomposite FeS_2-PbS Nanocrystals/Graphene Heterojunction Photodetectors for Broadband Photodetection. *ACS Appl. Mater. Interfaces*. 2017. vol. 9, no. 33, pp. 27801-27808.

[18] SONG, Xiaoxian *et al.* Graphene and PbS quantum dot hybrid vertical phototransistor, *Nanotechnology*. 2017. vol. 28, no. 14, p. 145201.

[19] CHE, Yongli *et al.* High-performance PbS quantum dot vertical field-effect phototransistor using graphene as a transparent electrode. *Appl. Phys. Lett.* 2016. vol. 109, no. 26, p. 263101.

[20] USHAKOVA, Elena V. *et al.* Anomalous size-dependent decay of low-energy luminescence from PbS quantum dots in colloidal solution. *ACS Nano*. 2012. vol. 6, no. 10, pp. 8913-8921.

[21] LU, Haipeng, JOY, Jimmy, GASPAR, Rachel L., BRADFORTH, Stephen E., and BRUTCHEY, Richard L. Iodide-Passivated Colloidal PbS Nanocrystals Leading to Highly Efficient Polymer:Nanocrystal Hybrid Solar Cells. *Chem. Mater.* 2016. vol. 28, no. 6, pp. 1897-1906.

[22] LAN, Xinzheng *et al.* 10.6 % Certified Colloidal Quantum Dot Solar Cells Via Solvent-Polarity-Engineered Halide Passivation. *Nano Lett.* 2016. vol. 16, no. 7, pp. 4630-4634.

[23] MARTÍ N-GARCÍ, Beatriz, POLOVITSYN, Anatolii, PRATO, Mirko, and MOREELS, Iwan. Efficient charge transfer in solution-processed PbS quantum dot-reduced graphene oxide hybrid materials. *J. Mater. Chem. C*. 2015. vol. 7088, p. 7088.

[24] JARZAB, Dorota, SZENDREI, Krisztina, YAREMA, Maksym, PICHLER, Stefan, HEISS, Wolfgang, and LOI, Maria A. Charge-separation dynamics in inorganic-organic ternary blends for efficient infrared photodiodes. *Adv. Funct. Mater.* 2011. vol. 21, no. 11, pp. 1988-1992.

MOVPE GaN/AlGaN HEMT NANO-STRUCTURES

[1]Eduard HULICIUS, [1]Karla KULDOVÁ, [1]Alice HOSPODKOVÁ, [1]Jiří PANGRÁC, [1]Filip DOMINEC, [2]Josef HUMLÍČEK, [1]Ivan PELANT, [1]Ondřej CIBULKA, [1]Kateřina HERYNKOVÁ

[1]*Institute of Physics of the CAS, Prague, Czech Republic, EU, hulicius@fzu.cz*
[2]*Masaryk University, Central European Institute of Technology, Brno, Czech Republic, EU*

Abstract

GaN/AlGaN-based high electron mobility transistors (HEMTs) attain better performance than their state-of-the-art full silicon-based counterparts, offering higher power, higher frequency as well as higher temperature of operation and stability, although their voltage and current limits are somewhat lower than for the SiC-based HEMTs. GaN/AlGaN-based HEMTs are a potential choice for electric-powered vehicles, for which they are approved not only for their power parameters, but also for their good temperature stability, lifetime and reliability. It is important to optimize HEMT structures and their growth parameters to reach the optimum function for the real-world applications.

HEMT structures described and discussed here were grown by MOVPE technology in AIXTRON apparatus on (111)-oriented single-surface polished 6" Si substrates.

Structural, optical and transport properties of the structures were measured by X-ray diffraction, optical reflectivity, time-resolved photoluminescence and μ-Raman spectroscopy.

Keywords: HEMT, MOVPE, nitrides, quantum well

1. INTRODUCTION

There are many approaches how to prepare HEMT structures which can be used for many potential application namely for electric-powered vehicles, for which they are approved not only for their power parameters, but also for their good temperature stability, lifetime and reliability. GaN can be the semiconductor of choice for power converters throughout vehicle electronics with devices capable of switching as much as 100 A at 900 V.

Silicon-based insulated gate bipolar transistors (IGBTs), superjunction MOSFETs and HEMTs are fundamental components of current power electronic systems for the conversion, control and conditioning of electrical energy. If silicon devices were to be replaced by a more efficient semiconductor such as GaN, compact converters with ultra-high density could be designed only because the breakdown strength and electron mobility in GaN are much higher [1,2]. GaN based HEMTs are very hot topic. There are about 5 000 papers at Web of Science and more than 50 review articles. We can recommend topical review "The 2018 GaN power electronics roadmap" [3] and the literature cited there. Industrial companies (eg. Transphorm, EPC, Panasonic, Infineon, GaN Systems, Dialog and Navitas) made much investments of a wide variety of markets exploring the use of the technology in new circuit topologies, packaging solutions and system architectures that are required to achieve and optimize the system advantages offered by GaN transistors, nevertheless it is important to optimize HEMT structures and their growth parameters to reach the optimum function for the real-world applications. A small part of this effort is presented in the article [3].

2. EXPERIMENTAL

MetalOrganic Vapor Phase Epitaxy (MOVPE) is the leading technology currently; Molecular Beam Epitaxy (MBE) covers only few percent of structure growing, mainly for the research.

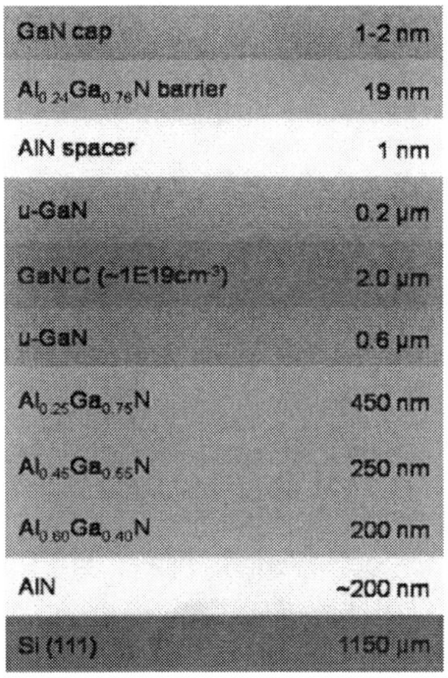

Figure 1 HEMT structure No. 6516 grown on 1.15 mm thick Si substrates with diameter of 200 mm

The structure No. 6516 presented in this article was grown with MOVPE AIXTRON apparatus on (111)-oriented single-surface polished 8" Si substrates. GaN/AlGaN structure is shown in **Figure 1**.

The second structure No. 6508 used in 2.1.2. and 2.1.4. sections as a reference sample was grown on (111)-oriented single-surface polished 6" Si substrates with slightly different Al and Ga composition in the AlGaN layers (see **Figure 8**).

Room-temperature time resolved photoluminescence (TR-PL) was excited by the four harmonic wavelength of a femtosecond laser Pharos (257 nm, 4.8 eV) with pulse duration of 160 fs, energy 17 µJ and repetition rate of 1 kHz. PL detection was carried out by a Hamamatsu streak camera.

Raman and steady-state photoluminescence (PL) spectra were measured at RT using LabRAM HR Evolution Raman microscope in backscattering configuration. The PL spectra were excited with a 325 nm laser line and the Raman spectra with a 532 nm line, both with a circular spot of about 1 µm in diameter. The signal was collected from the top of the wafer in z(y,-)-z configuration in Porto´s notation. The Raman measurements were performed with spectral resolution better than 1 cm^{-1}.

2.1. Measurements and discussion

2.1.1. X-ray diffraction measurement

The sample exhibits rather good crystallographic quality, as can be seen in **Figure 2**.

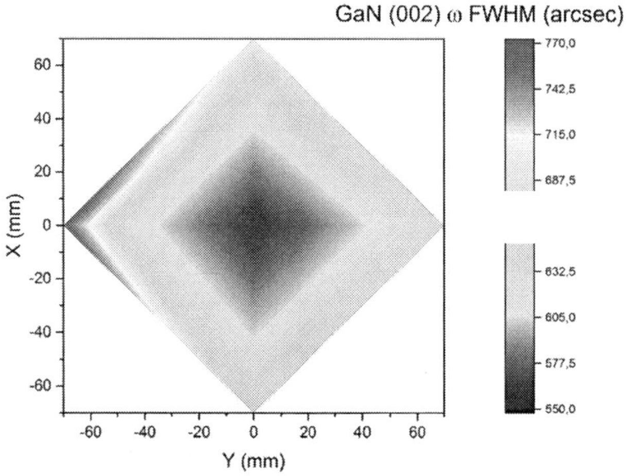

Figure 2 X-ray diffraction measured of GaN (002) ω FWHM - 9 points cross-map. GaN (102) ω FWHM in center point = 1234 arcsec

2.1.2. Optical reflectivity

Reflectivity of two different HEMT structures and a bare Si substrate was measured from 1.2 to 5.1 eV - see **Figure 3**. The lateral homogeneity of the wafers was tested by taking the spectra at several positions, the diameter of the circular measurement area being about 1 mm. Interference (Fabry-Perot-like) pattern below the gap of GaN (~3.4 eV) is due to the coherent reflections in the buffer and channel layers. Above ~3.5 eV, the spectra show reflectivity of the opaque GaN channel layer, modified by the presence of a thin AlGaN barrier layer. The latter leads to pronounced spectral changes around the energy of the (exciton) bandgap, moving to higher photon energies with increasing Al content (the arrows in **Figure 3**).

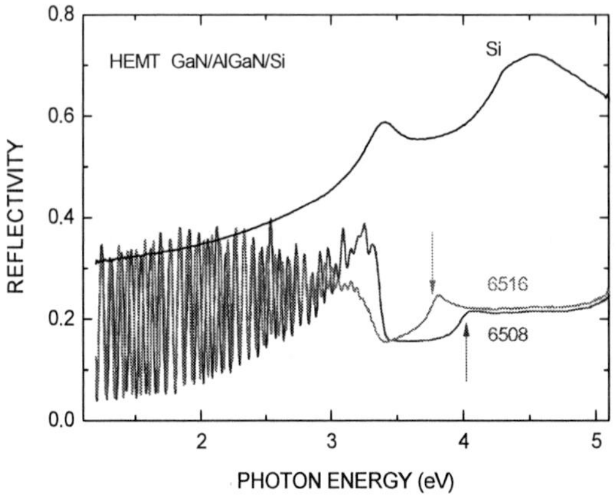

Figure 3 Reflectivity of two HEMT samples 6516 and 6508, measured at their center and 1 cm off center - red and blue colors

A closer look at the spectral positions, using the second derivatives of the reflectance in order to localize the transition energies, leads the Al contents of 15 and 22 at. % for the 6516 and 6508 samples, respectively. These estimates are rather different from the nominal values of 24 and 25 at. % Al.

2.1.3. Steady-state photoluminescence

The micro-PL spectra excited with three different intensities of 325 nm line are presented in **Figure 4**. We can distinguish in these spectra three maxima: the first one at 360 nm is due to emission of GaN exciton, the second one between 500 nm and 720 nm is the well-known yellow band (YB) of GaN defects and, between them, a low intensity emission from 400 nm to 500 nm corresponding to unidentified defects which are still under discussion. The spectra are normalized to the exciton peak intensity, and strong saturation of the YB for higher excitation intensities is evident.

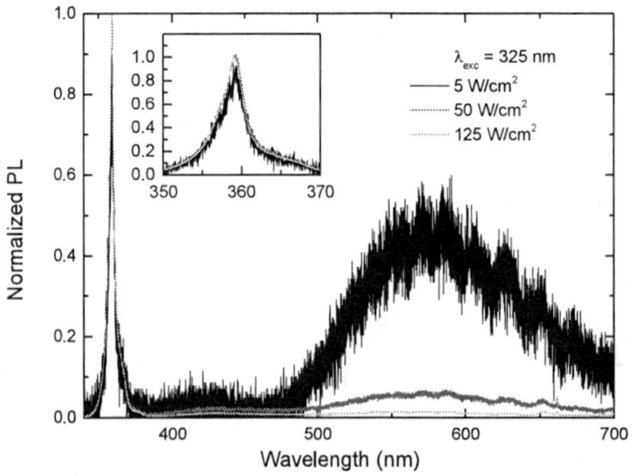

Figure 4 Comparison of PL spectra normalized to exciton peak intensity excited by three different excitation intensities. Relative contribution of YB luminescence is decreasing with excitation intensity due to a long decay time and saturation of this maximum

2.1.4. Time resolved photoluminescence

TR-PL and PL decay time at 555 nm of the GaN YB measured at our HEMT sample are shown in **Figure 5**. This emission band is modulated due to the interferences in some of thin layers, most probably that of GaN 400 nm thick. Decay time for 1/e of intensity is 34 µs.

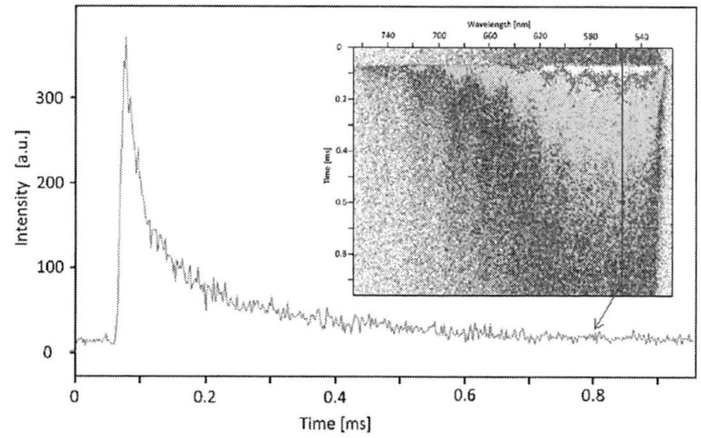

Figure 5 PL decay at 555 nm: decay time (1/e) = 34 µs. Insert: Two-dimensional map of TR-PL of the complete YB at 520-760 nm

Rather fast TR-PL emissions at around 360 nm (3.44 eV) - probably from the 400 nm thick GaN layer -, and at 315 nm (3.93 eV) - probably exciton from a 25 nm thick AlGaN barrier, are shown in **Figure 6**. The sample was excited as described above, only the repetition frequency was of 3 kHz. Detection was carried out again by the Hamamatsu streak camera; notice, however, that the time scale here is by 6 orders of magnitude faster.

The PL peak at 315 nm (3.93 eV) cannot be excited by the 325 nm laser, but its spectral position is probably indicated in the reflectivity spectrum at around 3.8-4.0 eV - see **Figure 3**. PL dynamic properties of this weak peak are shown in **Figure 6**. Decay time is expected to be faster than the measured 22 ps because deconvolution of the excitation pulse was not executed.

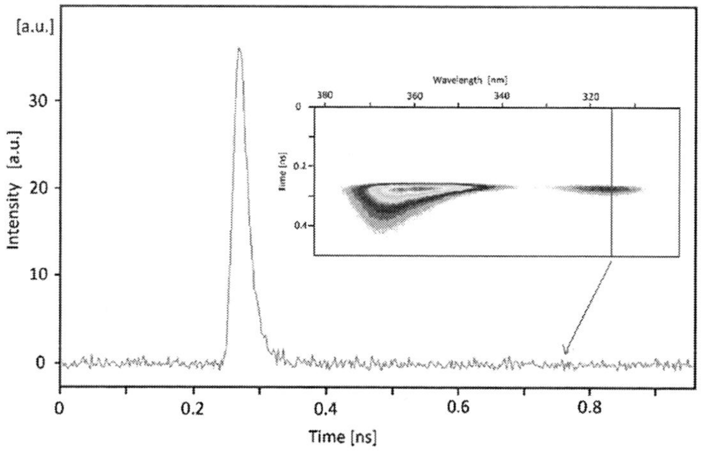

Figure 6 TR-PL of AlGaN emission at 315 nm, decay time (1/e) = 22 ps, dynamic of PL is much faster than for YB. Insert: Two-dimensional map of TR-PL of the GaN and AlGaN emission, the spectral line of AlGaN emission at 315 nm has FWHM of 14 nm

TR-PL of the GaN emission at around 360 nm is shown in **Figure 7**. Dynamic properties of PL are much faster than for the YB, decay times are again in the order of ten ps.

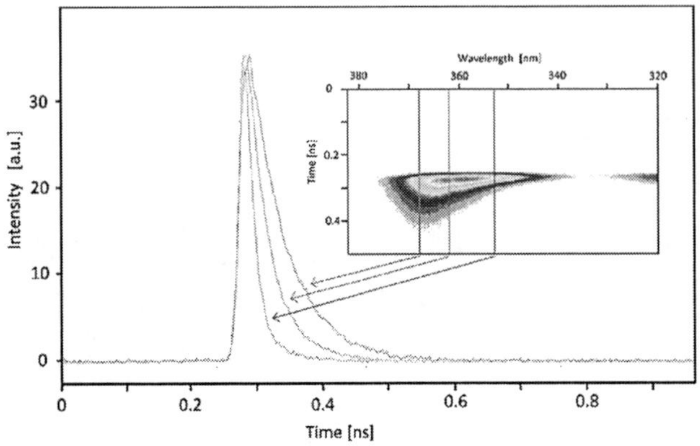

Figure 7 PL decay of GaN emission at 353, 362 and 368 nm: decay times (1/e) = 20, 38 and 56 ps, resp. Insert: Map of time resolved GaN PL, the spectral FWHM of the line is about 17 nm

2.1.5. Micro-Raman spectra

In Raman spectroscopy, we focused on the longitudinal optical mode A1(LO) because it is sensitive to Al composition in $Al_xGa_{1-x}N$ heterostructure but rather insensitive to strain [4]. The A1(LO) mode exhibits a single-mode behavior and its energy in $Al_xGa_{1-x}N$ as a function of x was several times refined by various authors. We have used the calibration curve for the A1(LO) phonon mode from [4].

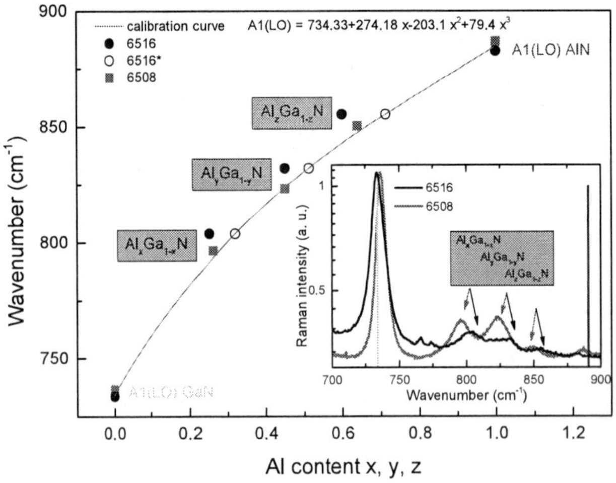

Figure 8 Calibration curve (red line) equation is taken from [4]. Insert shows Raman spectra of samples No. 6516 and 6508. The reference sample No. 6508 has x = 0.26, y = 0.45 and z = 0.64

Shown in the insert of the **Figure 8** are Raman spectra of sample No. 6516 and reference sample No. 6508 in the spectral region between A1(LO) mode of GaN and that of AlN. Green and blue lines mark theirs spectral positions in bulk materials, respectively. Between this two positions, one can see three small peaks corresponding to three $Al_xGa_{1-x}N$ layers with Al composition x = 0.25, y = 0.45 and z = 0.60; x = 0.26, y = 0.45 and z = 0.64 for the reference sample. Raman signal from a thin $Al_mGa_{1-m}N$ layer under the GaN cap with m =

0.24 (6516) or m = 0.25 (6508) of Al composition cannot be distinguished from the first thick $Al_xGa_{1-x}N$ layer with x = 0.25 (6516) or 0.26 (6508).

Positions of maxima of all peaks corresponding to A1(LO) mode frequencies are shown by full symbols in **Figure 8**.

As we can see, the positions of the A1(LO) peak of heterostructure layers of the reference sample are in good agreement with the values of the calibration curve. This is no more true for the sample No. 6516. There is a systematic shift towards a lower composition of Al in the structure. The composition values corresponding better to the measured frequencies, according to the calibration curve, are marked by open circles. The values of Al composition corresponding to those points are x = 0.30, y = 0.50 and z = 0.70.

3. CONCLUSIONS

GaN/AlGaN HEMT structures were prepared by MOVPE. X-ray diffraction measurement shows a good crystallographic quality of the structure.

Reflectivity measurement was used to establish thickness and Al composition. Measurements at different positions show good structure homogeneity.

PL, and especially TR-PL, have proved suitable tools for structure speed response and defect type and composition estimation.

Al composition in $Al_xGa_{1-x}N$ heterostructure as measured by Raman spectroscopy supplies feedback for grower intentions and aims.

ACKNOWLEDGEMENTS

Authors acknowledge support from the Technology Agency of the Czech Republic, project No. TH02010014 and MSMT NPU project no. LO1603 - ASTRANIT.

REFERENCES

[1] YANG, L., CHENG, C.W., BULSARA, M.T. and FITZGERALD, E.A., High mobility $In_{0.53}Ga_{0.47}As$ quantum-well metal oxide semiconductor field effect transistor structures. *J. Appl. Phys.* 2012. vol. 111, 104511.

[2] BRENNAN, B., ZHERNOKLETOV, D.M., DONG, H., HINKLE, C.L., KIM, J. and WALLACE, R.M., In situ surface pre-treatment study of GaAs and $In_{0.53}Ga_{0.47}As$. *Appl. Phys. Lett.* 2012. vol. 100, 151603.

[3] AMANO H., BAINES Y., BEAM E., et al, The 2018 GaN power electronics roadmap. *J. Phys. D: Appl. Phys.* 2018. vol. 51, 163001.

[4] WANG, Ch., CAHA, O., MUNZ, F., KOSTELNÍK, P., NOVÁK, T. and HUMLÍČEK, J., Mid-infrared ellipsometry, Raman and X-ray diffraction studies of $Al_xGa_{1-x}N$/AlN/Si structures. *Applied Surf. Sci. 2017.* vol. 421, 859-865.

RESONANT Z-SCAN MEASUREMENT OF NON-LINEAR PROPERTIES OF COLLOIDAL LEAD SULFIDE QUANTUM DOTS

Ivan SKURLOV, Evgenia PONOMAREVA, Sergey PUTILIN, Aleksei POPOV, Aliaksei DUBAVIK, Anton TSYPKIN, Aleksandr LITVIN

ITMO University, Saint-Petersburg, Russian Federation, ivan.skurlov.23@gmail.com

Abstract

Studying the nonlinearities in the resonant mode one may face thermo-optical effects due to sample overheating, occurring upon exposure to high-frequency repetition rate probing pulse. If present, thermo-optical contribution to the refractive index are extremely hard to separate from the electronic contribution. In order to obtain purely electronic non-linear change in refractive index and to avoid thermal effects we modify classic Z-scan method to operate with low probing laser pulse repetition rate of 1 Hz. In our work, we present results of a Z-scan measured non-linear refraction coefficient of colloidal PbS QDs of different sizes free from thermal contribution. Our measured values of nonlinear refractive index is in order of 10^{-16} cm^2/W and independent on the QD concentration in the solution.

Keywords: Nanocrystals, quantum dots, lead sulfide, z-scan, non-linear refraction

1. INTRODUCTION

Due to low energy bandgap, high absorption coefficient in near infrared region, size-tunable properties and suitability for wet processing, colloidal lead sulfide (PbS) quantum dots (QD) are widely studied material for applications in the near-infrared region [1]. This, combined with potentially high resonant non-linear optical response [2] and tunable non-linear absorption and reflection, make PbS QDs possible alternative material for use in the non-linear optics.

Common technique to study optical nonlinearities was introduced in 1990 by Sheik-Bahae et. al. [3] and ever since providing the simple and sensitive way to measure absolute values of nonlinear refraction and absorption coefficient of thin samples.

Previous Z-scan studies on the third-order nonlinear susceptibility of PbS quantum dots report either saturable [4] or mostly multiphoton absorption [5-7]. The common issue lies in the varying values of the nonlinear refractive index from different authors, e.g. $-(10 \div 60)\ 10^{-12}$ cm^2/W in C_2Cl_4 [6], or $-8.2 \cdot 10^{-16}$ cm^2/W in toluene, $-2.3 \cdot 10^{-15}$ cm^2/W in hexane, $-1.2 \cdot 10^{-15}$ cm^2/W in CCl_4 [8]. Mentioned problems arising from differences in the sample preparation method and, more important, in the origins of nonlinearity measured. We can split nonlinear contributions to one, reflecting purely nonlinear electron response (with response time of 10^{-15} - 10^{-10} s) to the applied electromagnetic field, and thermal (with response time of 10^{-8} s or more), arising when absorbed excitation energy is converted to heat due to thermo-optical coefficient *dn/dT* is higher than zero. Which in turn leads to thermal lensing effect, changing the resulting z-scan trace, thus acting as a parasite source of nonlinearity if one have to determine the electronic nonlinearity.

There is a way to reduce, if not entirely remove, thermal counterpart from the result. Electronic and thermal nonlinearity response time are different in magnitude by a few orders, so utilizing ultra-fast optical pulses can reduce the thermal counterpart. Although, using high repetition rate pulsed lasers can lead to formation of steady-state temperature profile hence accumulating thermal lens effect [9] so it is necessary to increase the time spacing between pulses to a minimum of $t_c = w^2/4D$, where D is thermal diffusion coefficient and w - beam diameter). Commonly used polar solvents have D in order of 10^{-3} cm^2/s thus the maximum repetition rate is in order of tens of kHz for typical z-scan setup. In our work, we demonstrate that the absence of the thermal counterpart to the nonlinear refractive index of PbS QD changes the behavior of the nonlinear refraction coefficient.

2. METHODS

Lead sulfide quantum dots (sizes ranging from 3.4 nm to 5.2 nm) were synthesized according to [10]. To remove excess of organic ligands QD were precipitated by acetone and then redispersed in the non-polar solvent. For power, concentration and size dependent measurements we chose hexane as a solvent for its lowest non-linear response (n_2 in order of 10^{-17} cm^2/W) among other common non-polar solvents (such as toluene, chloroform and tetrachloromethane). Absorption coefficient was calculated from the absorption spectra measured with Shimadzu UV3600 UV-VIS-NIR spectrophotometer. To measure the nonlinear refractive index we used standard z-scan setup with femtosecond Ti:sapphire laser (Avesta) with center wavelength 800 nm, pulse width >35 fs, acting as an excitation light source. In order to get rid of the thermal counterpart of the refractive index we lowered the pulse repetition rate to 1 Hz. Z-scan layout is schematically shown in **Figure 1.**

Typical open aperture Z-scan trace of pure hexane and colloidal QD solution are shown in the **Figure 2**. Presuming the additivity of nonlinear indexes of the QDs and the solvent we can subtract z-scan curve of the solvent to obtain the z-scan trace, attributed only to QDs [11]. n_2 values were calculated using formula (1) from [3]

$$n_2 = \frac{\Delta T}{0.406} \cdot \frac{\lambda}{2\pi I_{in} L_\alpha} \cdot \frac{\sqrt{2}}{(1-S)^{0.25}}, \quad (1)$$

here ΔT is the difference between the normalized peak and valley transmittance; I_{in} - light intensity within the sample; $L_\alpha = \left(1-e^{-\alpha L}\right)/\alpha$ - effective length, where α is the linear absorption coefficient and L - sample thickness; λ - wavelength of light, S - aperture.

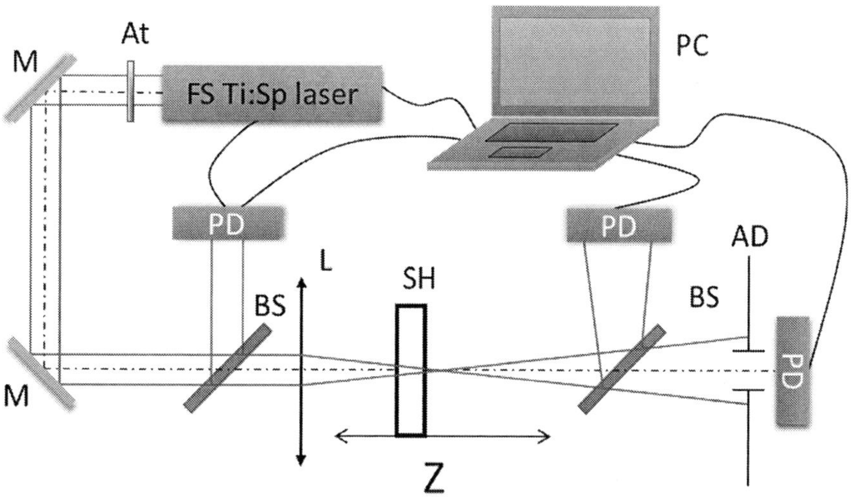

Figure 1 Scheme of the z-scan setup: **At** - laser output power attenuator, **M** - mirrors, **PD** - photodetectors; **BS** - beam splitters; **L** - condensing lens (f = 20 cm), **SH** - sample holder on a positioning table, **AD** - aperture diaphragm

3. RESULTS AND DISCUSSION

First, we have performed Z-scan measurement using the described setup with a pulse repetition rate of 70 MHz, MHz repetition rates are commonly used in works, studying PbS QD with z-scan technique [4-7]. Z-scan measurements were performed for the 4.5 nm QD colloidal solutions with different concentrations ranging from $3 \cdot 10^{-6}$ M to $14 \cdot 10^{-6}$ M.·The calculated n_2 values are $-(25 \div 90)\ 10^{-12}$ cm²/W and close to the values reported in the literature [5-7]. Negative sign of the nonlinear refraction index indicates the self-defocusing process. From **Figure 2(a)** it is clearly seen that non-linear refractive index changes linearly with the concentration.

Considering isolated QDs in colloidal solution, change the concentration leads only to change in the absorption. However, absorption change is taken into account in the calculations of n_2 via absorption-dependent effective length L_{eff} in equation (1). Thus, purely electronic nonlinear refractive index should remain constant. High absorption in the resonant (i.e. above-bandgap) excitation mode leads to the diffusion heating of the colloidal sample and consequent formation of the thermal lens, altering the n_2 values. Long-term thermal diffusion arises when sample cannot return to the equilibrium temperature because the time spacing between pulses are shorter than $t_c = w^2/4D$, where D is thermal diffusion coefficient and w - beam diameter, for most z-scan setups t_c>40 μs [9]. Hence, to get rid of the thermal diffusion (and thermal lens) it is necessary to reduce pulse repetition rate lower than few kilohertz.

Lowering the pulse repetition rate, we observed that n_2 concentration dependence vanishes when the repetition rate approaches ~10 Hz, and are completely gone at 1 Hz in the concentration range $2 \cdot 10^{-7}$ M to $3 \cdot 10^{-5}$ M, see **Figure 2(b)**. Resulting n_2 values are still negative, but significantly changed from ~10^{-12} cm²/W to ~10^{-16} cm²/W. Hence, we assume that colloidal QD n_2 concentration dependence indicates the presence of the thermal lens effect. To prove that our results obtained from equation (1) are correct, we have performed z-scan measurements with four different pulse powers (pulse repetition rate 1 Hz). Results for 4.5 nm diameter QD colloidal solution (concentration $1.1 \cdot 10^{-5}$ M) are shown on **Figure 2(c)**. As it can be seen, n_2 values are independent from the intensity in the range of 110-235 GW/cm².

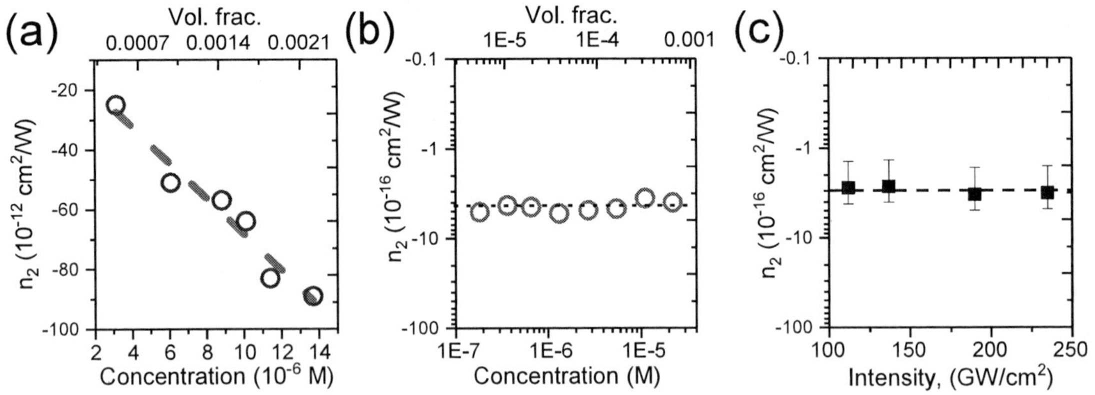

Figure 2 (a) n_2 dependence on QD (d = 4.5 nm) concentration pulse repetition rate - 70 MHz; **(b)** n_2 dependence on QD (d = 4.5 nm) concentration (or volume fraction) pulse repetition rate - 1 Hz; **(c)** n_2 dependence on input power I_{in} for lead sulfide QDs (d = 4.5 nm)

To estimate QD nonlinearity size dependence we measured a set of QD with sizes ranging from d = 3.2 nm to d = 5.2 nm, with the same excitation wavelength, concentration and incident beam power. The absorption spectra of the studied batch of PbS QD solution are shown on **Figure 3(a)**. Arrow notes red shift of the QD first excitonic peak with increasing QD size, red line marks the excitation wavelength. As shown in **Figure 3(b)**, nonlinear refraction index increases (from $-0.5 \cdot 10^{-16}$ cm²/W to $-4 \cdot 10^{-16}$ cm²/W) with increasing QD size, and consequently, reducing with increasing bandgap energy. n_2 dependence on QD radius was approximated with power function and found to be close to cubic. However, we measured only limited range of PbS QD sizes and further research has to be done to thoroughly analyze quantum confinement effect on the nonlinear properties of the QDs.

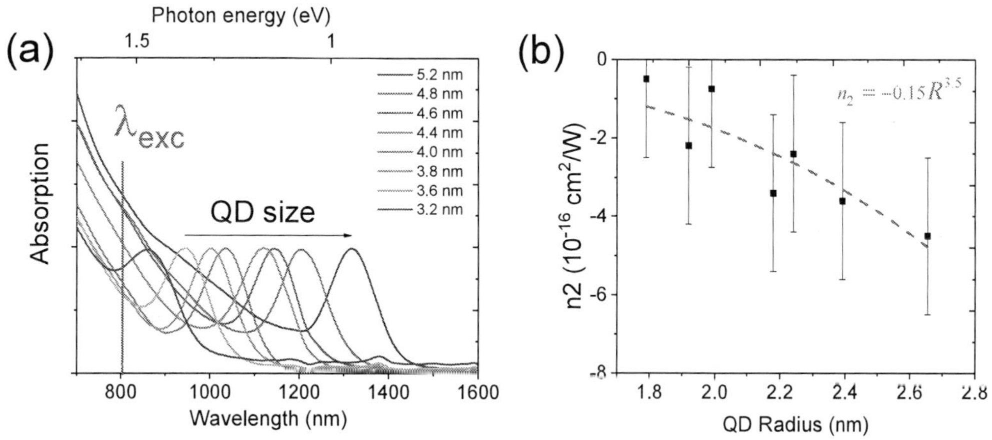

Figure 3 **(a)** absorption spectra for QD with different sizes, from d = 3.2 nm to d = 5.2 nm, λ_{exc} = 800 nm - excitation wavelength; **(b)** nonlinear refractive index size dependence, red dashed line stands for power function fit ($n_2 = -0.15R^{3.5}$)

4. CONCLUSION

We report that colloidal QD nonlinear refractive index concentration dependence is a sign of a thermal nonlinearity arising from the thermal lens effect. Performing z-scan measurement with low pulse repetition rate laser, thus free from thermal lens effect, we observed that nonlinear refractive index remain constant for different QD solution concentrations, and is in order of 10^{-16} cm²/W, which is by four order lower than previously reported results. QD nonlinear refractive index are expectedly size dependent and we found it to be close to cubic dependence

ACKNOWLEDGEMENTS

This work was supported by Russian Foundation of Basic Research project № 18-32-00027.

REFERENCES

[1] KERSHAW, Stephen V. SUSHA, Andrei S. and ROGACH Andrey L., Narrow bandgap colloidal metal chalcogenide quantum dots: synthetic methods, heterostructures, assemblies, electronic and infrared optical properties. *Chem. Soc. Rev.* 2013. vol. 42, no. 7, pp. 3033-3087.

[2] BANFI, GianPiero, DEGIORGIO, Vittorio and RICARD, Daniel. Nonlinear optical properties of semiconductor nanocrystals. *Adv. Phys.* 1998. vol. 47, no. 3, pp. 447-510.

[3] SHEIK-BAHAE, Mansoor, SAID, Ali A. WEI, Tai-Huei, HAGAN, David J. and VAN STRYLAND, E. W. Sensitive measurement of optical nonlinearities using a single beam. *IEEE Journal Quantum Electroics.* 1990. vol. 26, no. 4, pp. 760-769.

[4] ASUNSKIS, Daniel J.BOLOTIN, Igor L. and HANLEY, Luke. Nonlinear Optical Properties of PbS Nanocrystals Grown in Polymer Solutions. *J. Phys. Chem. C.* 2008. vol. 112, no. 26, pp. 9555-9558.

[5] OMARI, Abdoulghafar, MOREELS, Iwan, MASIA, Francesco, LANGBEIN, Wolfgang, BORRI, Paola, VAN THOURHOUT, Dries, KOCKAERT, Pascal and HENS, Zeger. Role of interband and photoinduced absorption in the nonlinear refraction and absorption of resonantly excited PbS quantum dots around 1550 nm. *Phys. Rev. B.* 2012 vol. 85, no. 11, p. 115318.

[6] KIM, Hyun S and YOON, Kyung B. Increase of 3rd-Order Nonlinear Optical Activity of PbS Quantum Dots in Zeolite Y by Increasing Cation Size. *J. Am. Chem. Soc.* 2012. vol. 134, no 5, pp. 2539-2542.

[7] WAWRZYNCZYK, Dominika, SZEREMETA, Janusz, SAMOC, Marek and NYK, Marcin. Third-Order Nonlinear Optical Properties of Infrared Emitting PbS and PbSe Quantum Dots. *J. Phys. Chem. C.* 2016. vol. 120, no. 38, pp. 21939-21945.

[8] CHENG, Hui, WANG, Yuhua, DAI, Hongwei, HAN, Jun-Bo and LI, Xianchang. Nonlinear Optical Properties of PbS Colloidal Quantum Dots Fabricated via Solvothermal Method. 2015. *J. Phys. Chem. C.* vol. 119, no. 6, pp. 3288-3292.

[9] FALCONIERI, Mauro. Thermo-optical effects in Z -scan measurements using high-repetition-rate lasers. *J. Opt. A: Pure Appl. Opt. 1999.* vol. 1, no. 6, pp. 662-667.

[10] USHAKOVA, Elena V. LITVIN, Aleksandr P. PARFENOV, Peter S. FEDOROV, Anatoly V. ARTEMYEV, Mikhail, PRUDNIKAU, Anatoly V, RUKHLENKO, Ivan D. and BARANOV, Alexander V. Anomalous size-dependent decay of low-energy luminescence from PbS quantum dots in colloidal solution. *ACS Nano.* 2012. vol. 6, no. 10, pp. 8913-8921.

[11] KHOROSHIKH, D. A. TSYPKIN, Anton N. MEL'NIK, M. V. CZINGEL, Kristian and KOKENYESI, Sandor. A study of the nonlinear refractive index in amorphous Ge-As-S layers and nanocomposites with gold nanoparticles. 2016. *Opt. Spectrosc.* 2016. vol. 121, no. 5, pp. 736-738

MODEL OF CARRIER MULTIPLICATION DUE TO IMPURITY IMPACT IONIZATION IN BORON-DOPED DIAMOND

[1,2]Vincent MORTET, [1,3]Nicolas LAMBERT, [4]Pavel HUBÍK, [5,6]Ali SOLTANI

[1]Institute of Physics, Czech Academy of Sciences, v.v.i., Prague, Czech Republic, EU, mortetv@fzu.cz
[2]Faculty of Biomedical Engineering, Czech Technical University in Prague, Kladno, Czech Republic, EU
[3]Faculty of Electrical Engineering, Czech Technical University, Prague, Czech Republic, EU, lambenic@fel.cvut.cz
[4]Institute of Physics, Czech Academy of Sciences, Prague, Czech Republic, EU, hubik@fzu.cz
[5]Laboratoire Nanotechnologies & Nanosystèmes, Université de Sherbrooke, Sherbrooke, Canada, Ali.Soltani@usherbrooke.ca
[6]Institut d'Electronique, Microélectronique et Nanotechnologie, Villeneuve d'Ascq, France, EU

Abstract

Boron-doped diamond exhibits a characteristic S-shaped I-V curve at room temperature [1] with two electrical conductivity states, i.e., low and high conductivity, at high electric fields (50 - 250 kVcm^{-1}) due to the carrier freeze-out and impurity impact ionization avalanche effect. To our knowledge, the carrier multiplication during the change of the conductivity state has not been studied. In this article, we investigate theoretically the effect of acceptor concentration and compensation level on the carrier multiplication coefficient at room temperature to determine the optimal dopants concentration of maximum carrier multiplication. The room temperature hole concentration of boron-doped diamond has been calculated for various acceptor concentration and compensation ratio by solving numerically the charge neutrality equation within the Boltzmann approximation of the Fermi-Dirac statistic. These values were used to determine theoretical carrier concentration multiplication coefficients as a function of the acceptor concentration and compensation ratio. The calculated multiplication coefficient is maximum for an acceptor concentration of ca. 2 10^{18} cm^{-3} and it increases with the compensation above 0.2 %. These theoretical values are consistent with the ratio of the carrier concentration at room temperature and the acceptor concentration available in the literature as well as the current multiplication observed in boron-doped diamond due to impurity impact ionization avalanche [1].

Keywords: Boron-doped diamond, semiconductor, carrier multiplication coefficient, impurity impact ionization

1. INTRODUCTION

The high ionization energy of electrically active dopants is one of the specificities of diamond as a semiconductor. As a result, the carrier freeze-out regime, i.e, the region of incomplete ionization of the dopants, extends towards temperatures well above "normal" operation temperature. The complete ionization of the dopant can be achieved within this temperature range by impurity impact ionization avalanche. This effect, first studied in germanium at cryogenic temperature, has been recently observed in boron-doped diamond in a wide range of temperature and in a wide range of acceptor concentration [1,2]. In this article, we investigate theoretically the effect of acceptor concentration and compensation level on the room temperature's carrier multiplication coefficient to determine the optimal theoretical dopants concentration to obtain maximum carrier multiplication attainable by impurity impact ionization avalanche effect.

2. MODEL

In this work, the carrier multiplication coefficient is defined as the ratio between the carrier concentration at complete ionization of acceptors, i.e., N_a-N_d with N_a the acceptor concentration and N_d the compensating donor concentration, and the carrier concentration at a given operating temperature, room temperature in this work. The carrier concentration is determined by solving numerically the charge neutrality equation within the Boltzmann approximation of the Fermi-Dirac statistic, i.e., assuming a non-degenerated semiconductor. The key parameters in the determination of the hole concentration (p) in boron-doped diamond are the effective density of states of the conduction (N_c*) and valence (N_v*) bands, the acceptor degeneracy factor (g_a), and the acceptor activation energy (E_a) [3]. In this work, the density of states and the degeneracy factor values are assumed to be independent of the acceptor concentration. The activation energy of dopants in semiconductors varies with their concentration [4]. In this work, we use the model proposed by Pearson and Bardeen [4] to model the activation energy dependence using equation (1):

$$E_a = E_I - \alpha . N_a^{1/3} \tag{1}$$

Where:

E_I -the ionization energy of an isolated boron acceptor (eV).

E_a - the acceptor activation energy (eV)

N_a - the acceptor concentration (cm^{-3})

The parameter α was determined assuming the Mott transition occurring at an acceptor concentration of 2 10^{20} cm^{-3}. All parameters used in this work are reported in **Table 1**.

Table 1 List of parameters used to numerically determine carrier concentration

N_c* (300 K)	N_v* (300 K)	g_a	E_I	α
10^{20} cm^{-3}	10^{19} cm^{-3}	6	0.370 eV	6.3 10^{-8} eV.cm

3. RESULTS

Figure 1 shows the Arrhenius plot of the calculated hole concentration in a boron-doped diamond with an acceptor concentration of 3 10^{18} cm^{-3} for various concentration of compensation impurities. This figure clearly shows that diamond is in the carrier freeze-out regime for temperature as high as 1000 K and above. The carrier concentration at room temperature is nearly independent of the concentration of compensation impurities below 3 10^{15} cm^{-3}, but it significantly decreases above this threshold value. Hence, the compensation can significantly increase the carrier multiplication coefficient without noticeably reducing the carrier concentration in complete acceptor ionization conditions. **Figure 2** shows the calculated room temperature carrier concentration multiplication coefficient for various boron concentration and percentage of compensation impurities. This calculation has been limited to acceptor concentration below 10^{20} cm^{-3} to remain within the Boltzmann approximation of the Fermi-Dirac statistic. This figure shows there is a maximum of the carrier multiplication coefficient for an acceptor concentration of ca. 2 10^{18} cm^{-3}. The multiplication coefficient is low, below 200, and it is nearly independent of the compensation impurities for percentages below 0.2 %, and it increases to value up to 10^4 and above in highly compensated boron-doped diamond. The decrease of the multiplication coefficient is attributed to the increase in the number of the carrier at room temperature with the steep decrease of the activation energy above 2 10^{18} cm^{-3} (see **Figure 3**). To our knowledge, there are no available measurements of the carrier multiplication coefficient in boron-doped diamond in the literature. One might estimate the value of the multiplication coefficient based on the ratio of the reported room-temperature carrier concentration determined by Hall effect characterization and acceptor concentration or the boron concentration determined from temperature dependent Hall effect characterization and secondary ions mass spectrometry (SIMS), respectively. The multiplication coefficient determined from the data available in the literature are compared to theoretical room temperature multiplication coefficient on **Figure 2**. These data points, that span over three decades, are within the range of theoretical multiplication coefficient. This figure also indicates that most of the boron-doped diamond layers have a compensation larger than 1 %. On this figure is also reported the current multiplication coefficient due to impurity impact ionization avalanche determined from data in ref. [1]. Here, we assume the current multiplication coefficient is comparable to the carrier concentration multiplication coefficient as the two current values are measured, on the S-shaped I-V curve, at the same voltage, i.e., same field. This calculated value is consistent with the theoretical value of low compensated and highly boron-doped diamond.

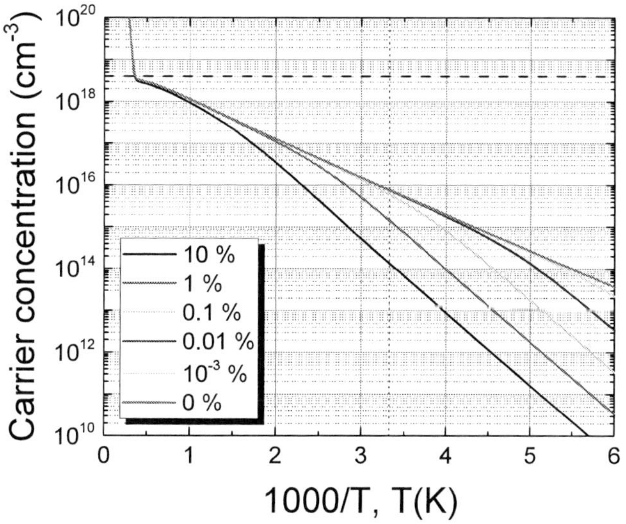

Figure 1 Arrhenius plot of the hole concentration in doped diamond with a boron concentration of 3 10^{18} cm^{-3} for various percentages of compensation impurities

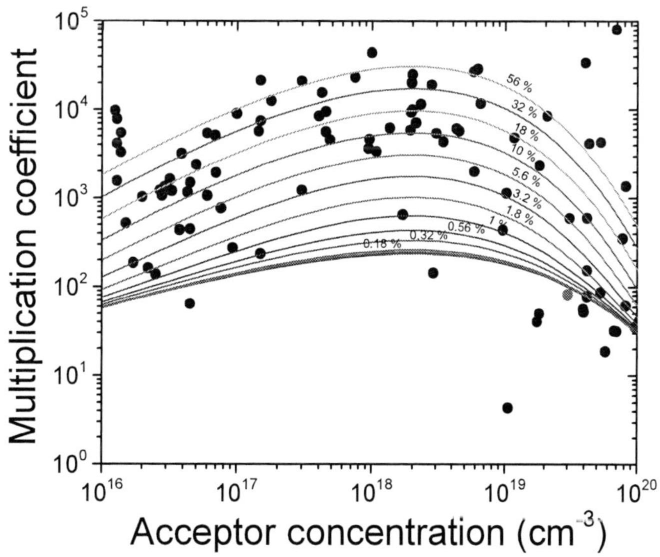

Figure 2 Room temperature carrier concentration multiplication coefficient for various boron concentration and compensation impurities percentages (from 0.01 % to 56 %) and its comparison with acceptor to carrier concentration ratios calculated from data available from the literature (black points) [5-17] and the current multiplication factor (red point) due to impurity impact ionization avalanche in boron-doped diamond determined from ref. [1]

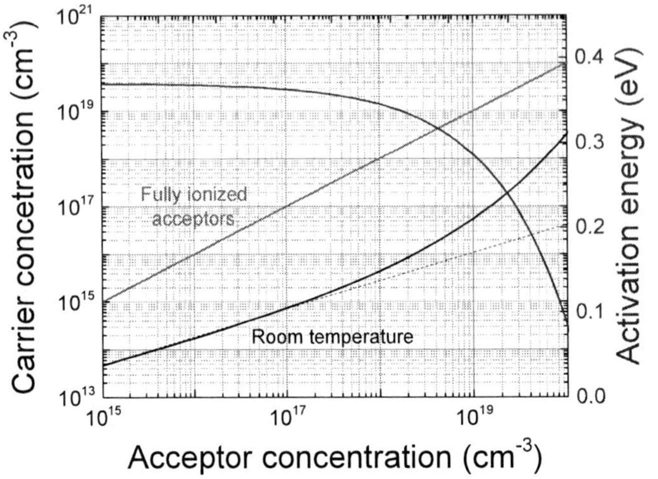

Figure 3 Comparison of the room temperature carrier concentration and the fully ionized acceptor carrier concentration as a function of the acceptor concentration in low compensated boron-doped diamond and variation of the modeled activation energy as a function of the acceptor concentration. Dotted line displays hypothetical carrier concentrations corresponding to the boron activation energy independent of acceptor concentration.

4. CONCLUSION

The theoretical room-temperature hole concentration of boron-doped diamond has been calculated for various acceptor concentration and compensation ratio by solving numerically the charge neutrality equation within the Boltzmann approximation of the Fermi-Dirac statistic and assuming a variation of the activation energy of the dopant in a -1/3 power function. These values were used to determine theoretical carrier concentration multiplication coefficients. The multiplication coefficient is the largest for an acceptor concentration of ca. 2 10^{18} cm^{-3} and it increases with the compensation above 0.2 %. These theoretical values are consistent with 1- the ratio of the carrier concentration at room temperature and the acceptor concentration available in the literature and 2- the current multiplication coefficient due to impurity impact ionization avalanche in boron-doped diamond.

ACKNOWLEDGEMENTS

The financial support from the following sources is gratefully acknowledged: the Czech Science Foundation grant ID: GACR 17-05259 S and J. E. Purkyne fellowship awarded to V. Mortet by Academy of Sciences of the Czech Republic.

REFERENCES

[1] MORTET, V., TRÉMOUILLES, D., BULÍŘ, J., HUBÍK, P., HELLER, L., BEDEL-PEREIRA, E. and SOLTANI, A. Peculiarities of high electric field conduction in p-type diamond. *Applied Physics Letters*. 2016. vol. 108, p. 152106.

[2] MORTET, V. and SOLTANI, A. Impurity impact ionization avalanche in p-type diamond. *Applied Physics Letters*. 2011. vol. 99, p. 202105.

[3] FONTAINE, F. Calculation of the hole concentration in boron-doped diamond. *Journal of Applied Physics*. 1999. vol. 85, pp. 1409-1422.

[4] PEARSON, G.L. and BARDEEN, J. Electrical Properties of Pure Silicon and Silicon Alloys Containing Boron and Phosphorus. *Physical Review*. 1949. vol. 75, pp. 865-883.

[5] BUSTARRET, E., GHEERAERT, E. and WATANABE, K. Optical and electronic properties of heavily boron-doped homo-epitaxial diamond. *Physica Status Solidi (A)*. 2003. vol. 199, pp. 9-18.

[6] BORST, T.H. and WEIS, O. Electrical characterization of homoepitaxial diamond films doped with B, P, Li and Na during crystal growth. *Diamond and Related Materials*. 1995. vol. 4, pp. 948-953.

[7] NISHIMURA, K., DAS, K. and GLASS, J.T. Material and electrical characterization of polycrystalline boron-doped diamond films grown by microwave plasma chemical vapor deposition. *Journal of Applied Physics*. 1991. vol. 69. pp. 3142-3148.

[8] WERNER, M., DORSCH, O., BAERWIND, H.U., OBERMEIER, E., HAASE, L., SEIFERT, W., RINGHANDT, A., JOHNSTON, C., ROMANI, S., BISHOP, H. and CHALKER, P.R. Charge transport in heavily B-doped polycrystalline diamond films. *Applied Physics Letters*. 1994. vol. 64, pp. 595-597.

[9] HATTA, A., SONODA, S. and ITO, T. Electrical Properties of B-doped homoepitaxial diamond films grown from UHP gas sources. *Diamond and Related Materials*. 1999. vol. 8, pp. 1470-1475.

[10] HUANG, J.T., GUO, W.H., HWANG, J. and CHANG, H. Reliability of Hall effect measurements on chemical vapor deposited polycrystalline B-doped diamond films. *Applied Physics Letters*. 1996. vol. 68, pp. 3784-3786

[11] WERNER, M., LOCHER, R., KOHLY, W., HOLMES, D.S., KLOSE, S. and FECHT, H.J. The diamond Irvin curve. *Diamond and Related Materials*. 1997. vol. 6, pp. 308-313.

[12] MORTET, V., PERNOT, J., JOMARD, F., SOLTANI, A., REMES, Z., BARJON, J., D'HAEN, J. and HAENEN, K. Properties of boron-doped epitaxial diamond layers grown on (110) oriented single crystal substrates. *Diamond and Related Materials*. 2015. vol. 53, pp. 29-34.

[13] BARJON, J., CHIKOIDZE, E., JOMARD, F., DUMONT, Y., PINAULT-THAURY, M.-A., ISSAOUI, R., BRINZA, O., ACHARD, J. and SILVA, F. Homoepitaxial boron-doped diamond with very low compensation. *Physica Status Solidi (A)*. 2012. vol. 209, pp. 1750-1753.

[14] VOLPE, P.-N., ARNAULT, J.-C., TRANCHANT, N., CHICOT, G., PERNOT, J., JOMARD, F. and BERGONZO, P. Boron incorporation issues in diamond when TMB is used as precursor: Toward extreme doping levels. *Diamond and Related Materials*. 2012. vol. 22, pp. 136-141.

[15] VOLPE, P.-N., PERNOT, J., MURET, P. and OMNES, F. High hole mobility in boron doped diamond for power device applications. *Applied Physics Letters*. 2009. vol. 94, p. 092102.

[16] TAVARES, C., OMNES, F., PERNOT, J. and BUSTARRET, E. Electronic properties of boron-doped {111}-oriented homoepitaxial diamond layers. *Diamond and Related Materials*. 2006. vol. 15 pp. 582-585.

[17] RI, S.-G., KATO, H., OGURA, M., WANATABE, H., MAKINO, T., YAMASAKI, S. and OKUSHI, H. Electrical and optical characterization of boron-doped (111) homoepitaxial diamond films. *Diamond and Related Materials*. 2005. vol. 14, pp. 1964-1968.

PROPERTIES OF PRESSURELESS SINTERED SILICON NITRIDE

Olga LUKIANOVA[1], Dmitrii KOLESNIKOV[1], Vladimir KRASILNIKOV[1], Alexander KHMARA[1], Sergey PEREVISLOV[2]

[1]*Belgorod National Research University, Belgorod, Russian Federation, sokos100@mail.ru*
[2]*Institute of Silicate Chemistry, Russian Academy of Sciences, Saint Petersburg (ex Leningrad), Russian Federation, perevislov@mail.ru*

Abstract

Silicon nitride find its application in quality of high-temperature ceramics and can be considered as one of the most promising dielectrics. Different properties and microstructure of pressureless sintered silicon nitride have been described a lot. The effect of various oxide additives and sintering regimes has also been often shown and discussed. However, there is not so much information about the influence of complex triple additives on properties of silicon nitride. In our work, we described the manufacture of high-dense Al_2O_3-MgO-Y_2O_3, Al_2O_3-MgO-La_2O_3 and Al_2O_3-La_2O_3-Y_2O_3 doped presureless sintered silicon nitride with high microhardness. The effect of different oxide compositions of yttrium, magnesium and lanthanum with Al_2O_3 on such important properties as shrinkage, microhardness and both green body density and density of sintered ceramics have been characterized.

Keywords: Materials science, ceramics, properties, oxide additives, powders

1. INTRODUCTION

Obviously, morphology of initial powders is one of the key factors affecting both microstructure and mechanical properties of pressureless sintered silicon nitride. Also, phase transformation of silicon nitride and formation of β-silicon nitride is characterized by a higher fracture toughness due to the ability to stop the propagation of cracks by elongated grains. Thus, it is obvious that investigation of mechanical properties and microstructure depending on the type of additives is very important at the moment. Rare earth oxides are usually used as additives. But the high cost is the main disadvantage of these oxides. Influence of such oxides as La_2O_3 has recently been of great interest. In particular, the separate and combined effect of Y_2O_3, La_2O_3, MgO and Al_2O_3 on microhardness, oxidation resistance, microstructure and phase composition of silicon nitride have partially been analyzed [1]. In particular, Guedes-Silva et al. showed that Gd_2O_3 leads to an intensification of the phase transformation and formation of elongated β grains of silicon nitride, an increase in crack resistance, while La_2O_3 led to an increase in microhardness [1]. Tatarko et al. showed decreasing of fracture toughness of silicon nitride with an increasing of the ionic radius of rare-earth element [3]. At the same time decreasing of the ionic radius of rare-earth element leads to increasing of the aspect ratio of Si_3N_4 composites [3]. Choi et al. reported hot pressed Y, Yb, Dy, Er, Sm, Ce, Lu, La, Pr, and Gd doped silicon nitride [4]. So, in described paper authors showed increasing of flexural strength and oxidation resistance with decrease in cation radius. Hong et al. reported that the rare-earth oxides affect such key properties and regularaties as the densification process and the shrinkage temperature [5]. Kitayama et al. described hot pressed fully dense Si_3N_4 with such rare-earth oxides as La, Nd, Gd, and Yb [6]. These authors showed the linear dependence of the phase transformation kinetics from the ionic radius of rare earth element atoms. Satet et al. investigated influence of various rare-earth and related elements (R = Lu, Sc, Yb, Y, Sm, La) on mechanical properties and grain growth anisotropy of β-silicon nitride ceramics [7]. They showed increasing of grain anisotropy with increasing of ionic radius because of non-linear growth kinetics.

In our work morphology of the starting powders of such oxide additives as La_2O_3,Y_2O_3, Al_2O_3, and MgO are demonstrated and characterized. Hirosaki et al. described the effect of rare earth oxides on densification of silicon nitride produced by pressureless sintering at 1600-1700°C and silicon nitride manufactured by gas pressure sintering at 10 MPa and 1800-2000 °C. Such single-component oxides as CeO_2, Nd_2O_3, La_2O_3, Sm_2O_3, and Y_2O_3 were used as an additive. It was shown that in this case the sintering temperature required to reach approximate theoretical density became higher as the melting temperature of the oxide increased. It was also clearly shown that the higher densification was achieved below 2000°C because of a lower liquid formation temperature when a mixed of such oxide additive as Y_2O_3-Ln_2O_3 (Ln=Ce, Nd, La, Sm) was used [2].

Short-term pressureless sintering is a relatively inexpensive and promising method for silicon nitride manufacture.

The main aim of the present study is to investigate features of the effect of different triple Al_2O_3, MgO, Y_2O_3 and La_2O_3 compositions on the linear shrinkage, microhardness and density of both green-compacts and pressureless sintered silicon nitride.

2. FIGURES, EQUATIONS, TABLES

The ground and homogenized powder mixture was cold isostatically pressed at 180 MPa (EPSI CIP 400B-9140 press) during 30 s. Conventional sintering of samples was carried out in a high temperature furnace (Nabertherm VHT8/22-GR) in nitrogen atmosphere at 1780 °C for 60 min. All reported compositions are listed in **Table 1**.

The microstructure of initial powders was characterized by scanning electron microscopy (SEM). Structural characterization was performed using Quanta 600 FEG (FEI company, Hillsboro, OR) scanning electron microscope. The density of the test samples was measured by the Archimedes method. The microhardness was measured by an automated Shimadzu DUH-211/DUH-211S.

There are three main criteria affecting properties and microstructure of ceramics. These criteria are: morphology, type and microstructure of initial powder, sintering conditions and type and content oxide additives. They play an important role in both equiaxial pressing and liquid phase sintering. It is expected that globular morphology of powder is more preferable. Morphology of initial powders is shown in **Figure 1**. Aluminum oxide grade Granolox (**Figure 1b**) demonstrated the most preferable morphology in comparison with the calcined aluminum oxide. Agglomerates of powder are shown in **Figure 1b**. Lanthanum oxide and magnesium oxide are distinguished by small particles of powder, as can be seen in **Figure 1c** and **Figure 1e**, respectively. **Table 1** shows described compositions.

Table 1 Described powder compositions

№	Si3N4 (wt%)	Al_2O_3 (wt%)	MgO (wt%)	Y_2O_3 (wt%)	La_2O_3 (wt%)
1	90	4.3	-	5.7	-
2	90	5.0	2.0	-	3.0
3	92	6.0	2.0	-	-
4	90	5.0	2.0	3.0	-
5	-	4.0	-	3.5	2.5

There is a large number of papers devoted to manufacturing of silicon nitride with such typical additives as Al_2O_3-Y_2O_3 and Al_2O_3-MgO. In our work, we considered such triple additive systems as Al_2O_3-MgO-Y_2O_3, Al_2O_3-MgO-La_2O_3 and Al_2O_3-La_2O_3-Y_2O_3, respectively. On the one hand, the ternary systems have low-melting eutectics allowing lowering the sintering temperature. On the other hand, there are not so many works describing the complex effect of lanthanum oxide with other more popular oxide additives on properties and microstructure. The density of green bodies, microhardness and the shrinkage of the produced ceramics are shown in **Table 2**. The maximum shrinkage equal to 19 % and 18 % was measured for Al_2O_3-MgO-Y_2O_3 and Al_2O_3-La_2O_3-Y_2O_3 doped silicon nitride, respectively. Maximum density was demonstrated for Al_2O_3-MgO-Y_2O_3. The growth of linear shrinkage of manufactured silicon nitride linearly depends on the density of the sintered body, however, there is no linear correlation with the green body density. Al_2O_3-La_2O_3-Y_2O_3 doped silicon nitride showed maximum microhardness equal to 1798 HV (see **Table 2**). Detailed information on the properties of pressureless sintered silicon nitride can be found in our earlier articles [8-15].

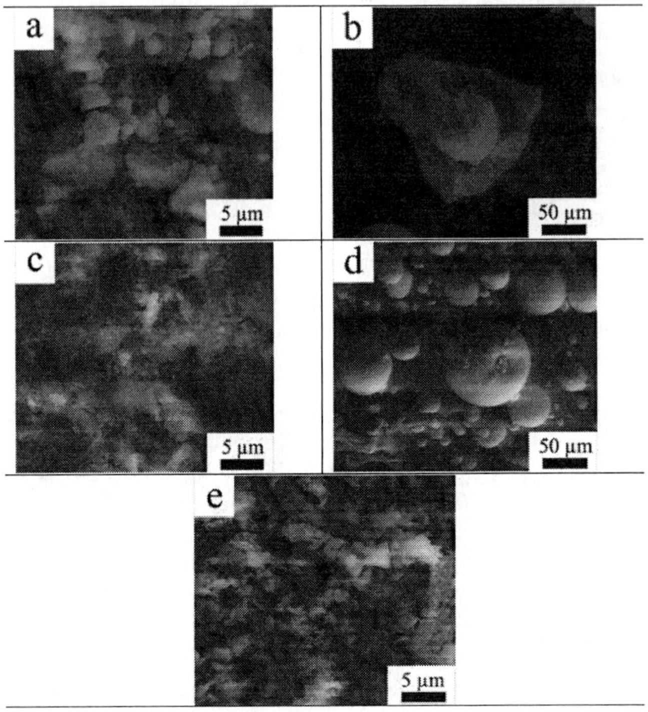

Figure 1 Microstructure of initial powders a) Al_2O_3 calcined b) Al_2O_3 Granolox c) MgO d) Y_2O_3 e) La_2O_3

Table 2 Density, microhardness and shrinkage of the obtained materials

№	$\rho_{green\ body}$, g/cm^3	$\rho_{sintered}$, g/cm^3	Shrinkage, %	HV
1	2.23	96	15	1506
2	1.81	96	17	1784
3	1.93	93	16	1710
4	1.98	97	19	1665
5	1.94	94	18	1798

For comparison, we briefly review the results of other authors. The density of the green body is an important indicator in ceramics processing. In particular, Almeida et al. described manufacturing of low temperature pressureless sintered silicon nitride with high content of Al_2O_3-SiO_2-Y_2O_3. Green body density varied from 1.74 g/cm^3 to 1.94 g/cm^3 in comparison with our materials1.81-2.23 g/cm^3 [16]. Huang et al. described AlN-La_2O_3-Y_2O_3 doped high-strength pressureless sintered silicon nitride ceramics [17]. However, Ling et. al described MgO-Y_2O_3 doped high-strength pressureless sintered silicon nitride ceramics [18]. Terwilliger et al. described the fundamental processes of the liquid-phase sintering of pressureless sintered silicon nitride with magnesium oxide. In particular, he showed pore growth due to decomposition leading to a decrease in the driving force for sintering and leads to the cessation of shrinkage [19]. Also Yang et al. established efficiency of MgO-CeO_2 additive for pressureless sintered silicon nitride compared to a single MgO or CeO_2. The optimal ratio of reported oxides was equal. Linear shrinkage of the considered ceramics was approximately 16-18 % [20]. Yonezawa et al. reported about slip casted and pressureless sintered silicon nitride ceramics with whiskers. A high density, good mechanical and thermal shock resistance properties, characterizes this ceramics. However, the shrinkage of such materials was not uniform [21].

3. CONCLUSION

Such important properties as linear shrinkage, green body density and density of sintered materials have been described. This study is devoted to show the influence of various oxides as lanthanum, yttrium, magnesium and aluminum on the linear shrinkage, microhardness, green body and sintered density of silicon nitride obtained by pressureless sintering. Suitable powder morphology of such reported oxide additives as Al_2O_3, MgO, La_2O_3 and Y_2O_3 have been shown. Pressureless sintered silicon nitride with Al_2O_3-MgO-Y_2O_3, Al_2O_3-MgO-La_2O_3 and Al_2O_3-La_2O_3-Y_2O_3 with high density and microhardness have been obtained, discussed and analyzed. It has been shown that ceramics with triple Al_2O_3-MgO-Y_2O_3, Al_2O_3-MgO-La_2O_3 and Al_2O_3-La_2O_3-Y_2O_3 additives have higher microhardness compared to ceramics with traditional double Al_2O_3-Y_2O_3 and Al_2O_3-MgO additives equal to 1506 HV and 1710 HV, respectively. Also, considered ternary additives led to a higher linear shrinkage 17-19 % of reported silicon nitride in contrast with the 15-16 % linear shrinkage of Si_3N_4 with traditional binary additives. Similarly, density of manufactured silicon nitride with ternary additives was even higher compared to reported Al_2O_3-Y_2O_3 and Al_2O_3-MgO silicon nitride density. The growth of linear shrinkage was coexisted by increasing the density of the sintered body. However, the density of the sintered material did not depend linearly on the density of the green body.

ACKNOWLEDGEMENTS

O.L. acknowledge with gratitude the financial support received through the Russian Science Foundation, under Grant No.18-79-00246.

REFERENCES

[1] GUEDES-SILVA, Cecilia C., CARVALHO, Flávio M. S. and BRESSIANIA, José C. Effect of rare-earth oxides on properties of silicon nitride obtained by normal sintering and sinter-HIP. Journal of rare earths. 2012. vol. 30, pp. 1177-1183.

[2] HIROSAKI Naoto, OKADA Akira, MATOBA Kazuo. Sintering of Si3N4 with the addition of rare-earth oxides. Journal of the American Ceramic Society. 1988. vol. 71, pp. 144-147.

[3] TATARKO P., LOJANOVÁ Š., DUSZA J., ŠAJGALÍK P. Influence of rare-earth sintering additives in Si3N4 based nanocomposites. NANOCON 2009: 1st Conference with International Participation. Roznov pod Radhostem. TANGER, 2009.

[4] CHOI, Heon J., LEE, June G. and KIM, Young W. High temperature strength and oxidation behaviour of hot pressed silicon nitride-disilicate ceramics. Journal of Materials Science. 1997. vol. 32, pp. 1937-1942.

[5] HONG Z L, YOSHIDA H, IKUHARA Y, SAKUMA T, NISHIMURA T, MITOMO M. The effect of additives on sintering behaviour and strength retention in silicon nitride with RE-disilicate. Journal of the European Ceramic Society. 2002. vol. 22, pp. 527-534.

[6] KITAYAMA, Mikito, HIRAO, Kiyoshi, KANZAKI, Shuzo. Effect of Rare Earth Oxide Additives on the Phase Transformation Rates of Si3N4. Journal of the American Ceramic Society. 2006. vol. 89, pp. 2612-2618.

[7] SATET, Raphaëlle L., HOFFMANN,Michael J. and CANNON,Rowland M. Experimental evidence of the impact of rare-earth elements on particle growth and mechanical behaviour of silicon nitride. Materials Science and Engineering: A. 2006. vol. 422, pp. 66-76.

[8] LUKIANOVA, O.A., NOVIKOV, V.Y., PARKHOMENKO, A.A., SIROTA, V.V., KRASILNIKOV, V.V. Microstructure of spark plasma-sintered silicon nitride ceramics. Nanoscale Research Letters. 2017 vol. 12, pp. 223.

[9] LUKIANOVA, O.A., SIROTA, V.V., KRASIL'NIKOV, V.V., PARKHOMENKO, A.A. Mechanical properties and microstructure of silicon nitride fabricated by pressureless sintering. 6th International Conference on Nanomaterials: Applications and Properties (NAP). Lviv: IEEE, 2016, pp. 02NNSA09.

[10] LUKIANOVA, O.A., IVANOV, O.N. The effect of Al2O3-MgO additives on the microstructure of spark plasma sintered silicon nitride. Ceramics International. 2018. vol. 44, pp. 390-393.

[11] LUKIANOVA, O.A., SIROTA, V.V. Dielectric properties of silicon nitride ceramics produced by free sintering. Ceramics International. 2017. vol. 43, pp. 8284-8288.

[12] LUKIANOVA, O.A., KHMARA, A.N., KRASIL'NIKOV, V.V., PARKHOMENKO, A.A., SAVOTCHENKO, S.E. Thermophysical properties of pressureless sintered silicon nitride., 2017 IEEE 7th International Conference on Nanomaterials: Applications and Properties (NAP). Odessa: IEEE, 2017, pp. 02MAN14.

[13] PARKHOMENKO, A.A., LUKIANOVA, O.A., KHMARA, A.N., KRASIL'NIKOV, V.V. Effect of the various oxide additives on the microhardness of silicon nitride ceramics. 2017 IEEE 7th International Conference on Nanomaterials: Applications and Properties (NAP). Odessa: IEEE, 2017, pp. 02MAN15.

[14] LUKIANOVA, O.A., KRASIL'NIKOV, V.V., KHMARA, A.N., PARKHOMENKO, A.A. Finite element model of mechanical tests of pressureless sintered silicon nitride. 2017 IEEE 7th International Conference on Nanomaterials: Applications and Properties (NAP). Odessa: IEEE, 2017, pp. 02MAN13.

[15] LUKIANOVA, O.A., KHMARA, A.N., PEREVISLOV, S.N., KOLESNIKOV, D.A., KRASILNIKOV V.V. Electrical resistivity of silicon nitride produced by various methods. Ceramics International. 2018. In Press. DOI: 10.1016/j.ceramint.2018.09.198.

[16] ALMEIDA, J.C., FONSECA, A.T., CORREIRA, R.N., BAPTISTA, J.L. Pressureless sintering of silicon nitride with additives of the Y2O3-Al2O3-SiO2 system. Materials Science and Engineering, A. 1989. vol. 109, pp. 395-400.

[17] HUANG, Zhen-Kun, ROSENFLANZ, Anatoly, CHEN, I-Wei. Pressureless sintering of Si3N4 ceramic using AlN and rare-earth oxides. Journal of the American Ceramic Society. 1997. vol. 80, pp. 1256-1262.

[18] LING, Gao, YANG, Haitao. Pressureless sintering of silicon nitride with magnesia and yttria. Materials Chemistry and Physics. 2005. vol. 90, pp. 31-34.

[19] TERWILLIGER, G.R., LANGE, F.F. Pressureless sintering of Si3N4. Journal of Materials Science. 1975. vol. 10, pp. 1169-1174.

[20] YANG, H., YANG, G., YUAN, R. Pressureless sintering of silicon nitride with magnesia and ceria. Materials Research Bulletin. 1998. vol. 33, no. 10, pp. 1467-1473.

[21] YONEZAVA, Takao, SAITOH, Shin-ichi, MINAMIZAWA, Masatoshi, MATSUDA, Toshitsugu. Pressureless sintering of silicon-nitride composites. Composites Science and Technology. 1994. vol. 51, pp. 265-269.

MOLECULAR DYNAMICS SIMULATIONS OF INTERACTION OF VOLATILE ORGANIC COMPOUNDS WITH GRAPHENE SURFACE

Michal LANGER, Martin PYKAL, Michal OTYEPKA

Regional Centre of Advanced Technologies and Materials (RCPTM), Department of Physical Chemistry, Faculty of Science, Palacký University, Olomouc, Czech Republic, EU, michal.otyepka@upol.cz

Abstract

Graphene is unarguably very promising material for countless applications including also gas sensing. For this particular application, it is very beneficial to understand the nature and strength of interaction of graphene with the adsorbed molecules. Molecular dynamics (MD) simulations matured to a very useful technique providing insights into complex molecular systems on the atomistic scale. However, the prediction potential of classical MD is given by quality of used parameters, including parameters for Lennard-Jones (LJ) term describing dispersion and repulsion part of intermolecular interaction potential. So far, LJ parameters used for graphene have not been thoroughly tested. Using MD we quantified the adsorption enthalpies of several organic volatile molecules (dichloromethane, nitromethane, ethanol, acetone, acetonitrile, ethyl acetate, hexane, cyclohexane, dioxane, benzene, and toluene) to a few layered graphene surface. The calculated adsorption enthalpies were compared with those acquired experimentally using an inverse gas chromatography technique. For each molecule, seven set of simulations were performed with altered LJ parameters and the accuracy of derived adsorption enthalpies was assessed. The average error of $\Delta\langle H^{FF}\rangle$ in respect to ΔH^{exp} suggests that OPLS-AA parameters delivered the best agreement with experiments. Needless to say that all tested LJ parameters were suitable for semi-quantitative estimates of the interaction energies of the molecules with the graphene. This implies that MD simulation can provide correct order of adsorption enthalpies of adsorbates and OPLS-AA in addition provides reliable quantitative estimates.

Keywords: Graphene, nanomaterials, molecular dynamics, force field, adsorption enthalpies

1. INTRODUCTION

Graphene is a two-dimensional material having many extraordinary properties like enormous surface area, high mechanical strength, high transmittance of light, great thermal conductivity and especially outstanding electronics properties [1-5]. These properties predestine graphene for many promising applications in very diverse areas. Graphene is an exceptionally low-noise material electronically, which makes it a potentially very promising candidate for chemical sensing [6]. There are many molecular dynamics (MD) simulations in the field of graphene, other carbon nanoallotropes and their derivatives [7-9]. Although reliability of classical MD simulations is conditioned by quality of used force field, including Lennard-Jones parameters, these LJ parameters used for graphene have not been thoroughly tested, yet. So far, authors either adopted Lennard-Jones parameters from biomolecular force fields considering validity of a transferability principle [8] or they parameterized their own LJ parameters for their specific case [10]. Because the graphene is predominantly formed by carbon atoms, the majority of interaction with approaching molecules is mediated by them. Therefore, proper Lennard-Jones parameters of carbon atoms are needed for getting reliable results. Here, we tried to compare different LJ parameters, which were either specifically developed for the carbon nano-allotropes or adopted from common force fields. In order to conclude which Lennard-Jones parameters are better for utilizing in the MD simulations of the graphene, adsorption enthalpies of small organic volatile molecules on the graphene surface were calculated via MD simulations $\Delta\langle H^{FF}\rangle$ and the results were compared with experimentally determined adsorption enthalpies ΔH^{exp} from inverse gas chromatography measurements [11,12]. Our critical assessment of Lennard-Jones parameters against experimental data can provide useful information for future MD studies of adsorption properties of graphene and its derivatives.

2. MAIN TEXT

2.1. Methods

The adsorption enthalpies were calculated using MD simulations as follows (equation (1):

$$\Delta\langle H^{FF}\rangle = \Delta\langle E^{FF}\rangle - RT \quad (1)$$

where $\Delta\langle H^{FF}\rangle$, $\Delta\langle E^{FF}\rangle$, $-RT$ correspond to adsorption enthalpy from MD, average interaction energy and enthalpy correction, respectively.

A value of $\Delta\langle H^{FF}\rangle$ was calculated for the adsorption of single molecule from a representative set of small organic molecules on a few-layered graphene surface, while varying the non-bonded LJ parameters σ and ε of the aromatic carbons of the graphene. The molecules were chosen in order to cover diverse set, ranging from nonpolar to polar, from aliphatic across cyclic to aromatic, containing different heteroatoms, namely dichloromethane, nitromethane, ethanol, acetone, acetonitrile, ethyl acetate, hexane, cyclohexane, dioxane, benzene, and toluene.

Table 1 LJ parameters of graphene carbons for different force fields, which are commonly used in the field of MD simulations of graphene

Force field	**σ (Å)**	**ε (kcal·mol^{-1})**
Girifalco (gra-gra) [10]	3.41214	0.0551
Cheng&Steele [13]	3.39967	0.0557
Girifalco C_{60}-C_{60} [14-16]	3.46900	0.0661
OPLS-AA [17]	3.55000	0.0700
Parm99 [18]	3.39967	0.0860

Periodic model of graphene was used in all MD simulations. As the experiments were carried out using a few-layered graphene, the graphene surface was modeled as six-layered; each layer consisted of 3936 atoms, which were kept rigid on a planar hexagonal lattice with a bond distance set to 0.14 nm. Periodic boundary conditions were applied in all three dimensions of the simulation box with the size of approximately 10 nm × 10 nm × 13 nm. The Newtonian equations of motion were integrated using leap-frog algorithm using 2 fs time step, hydrogen atoms were constrained with LINCS algorithm [19]. All simulations were carried out using GROMACS simulation package [20] in vacuum in NVT ensemble with a temperature of 323 K held with Berendsen thermostat [21]. The electrostatic interaction was treated with cut-off method with a radius of 1.0 nm. Intermolecular vdW interactions were calculated with a cut-off radius of 1.5 nm. Bonded and non-bonded parameters of the organic molecules were described using OPLS-AA force field [22]. The graphene was also described using OPLS-AA force field, but the Lennard-Jones parameters σ, ε of the graphene carbons were varied (see **Table 1**). As the graphene was modelled being periodic, each carbon atom was approximated as an uncharged LJ sphere. At first, each molecular dynamics simulation was equilibrated for 0.2 ns and then a productive 1 ns long simulation was performed. The interaction energy $\Delta\langle E^{FF}\rangle$ of a graphene-molecule system was calculated as an average over 5000 values. Some simulations were extended to 30 ns with the purpose of checking and ensuring that all the simulations successfully converged. In each simulation one molecule was placed in the center of the few-layered graphene surface approximately ~0.35 nm above its surface. For each particular graphene-molecule system of particular LJ parameters 5 independent simulations were performed, which differed in random generations of starting velocities.

2.2. Results and Discussion

The adsorption enthalpies $\Delta\langle H^{FF}\rangle$, as described by equation (1), of the interaction of the graphene-one molecule systems for all studied LJ parameters, together with experimentally determined ΔH^{exp}, are displayed in **Table 2**. Confidence intervals were obtained based on the statistical analysis equations for small samples containing 5 values (see equation (2)).

$$L = \bar{X} \mp K_n R \tag{2}$$

where L, $\bar{X}$, R and K_n correspond to confidence interval, arithmetic mean, a tabulated value and statistics range, respectively.

Table 2 Adsorption enthalpies $\Delta\langle H^{FF}\rangle$ calculated with different LJ parameters for graphene carbons and also ΔH^{exp}, both with their confidence intervals (at α = 0.05) for the adsorption of one molecule on the few-layered graphene surface. All values are in kJ·mol^{-1}.

	Ch&St	Ch&St 2	Girifalco gra-gra	Girifalco C_{60}-C_{60}	OPLS-AA	Parm99	Exp
Dich	-29.5±0.1	-29.6±0.2	-29.5±0.1	-32.8±0.1	-34.8±0.0	-36.6±0.2	-24.7±2.1
Nitr	-28.2±0.2	-28.7±0.6	-28.1±0.2	-31.5±0.2	-33.3±0.1	-35.1±0.0	-26.4±0.4
Etha	-23.0±0.2	-23.2±0.2	-22.9±0.4	-25.5±0.2	-27.5±0.1	-28.5±0.2	-30.5±2.9
Acet	-30.5±0.1	-30.9±0.2	-30.6±0.1	-34.0±0.1	-36.0±0.1	-38.0±0.1	-34.3±1.3
Acni	-22.2±0.5	-22.3±0.6	-21.8±0.6	-24.3±0.5	-25.3±0.1	-27.1±0.7	-31.8±1.3
Etat	-44.5±0.1	-44.8±0.3	-44.5±0.2	-49.8±0.1	-52.9±0.2	-55.7±0.2	-48.1±0.8
Hexa	-46.2±0.2	-46.5±0.3	-46.1±0.1	-52.1±0.3	-55.4±0.7	-57.8±0.3	-56.5±0.8
Cyc	-32.7±0.3	-33.0±0.8	-32.6±0.4	-37.1±0.3	-39.4±0.8	-40.9±0.4	-47.7±1.3
Diox	-35.6±0.4	-35.8±0.5	-35.5±0.3	-40.2±0.4	-43.2±0.3	-45.5±0.5	-45.2±0.4
Benz	-47.1±0.3	-47.6±0.2	-46.9±0.4	-52.0±0.3	-54.6±0.2	-58.8±0.1	-49.8±1.3
Tolu	-49.0±0.5	-49.8±0.3	-49.0±0.2	-54.6±0.4	-57.8±0.3	-61.7±0.3	-57.7±1.7
AE	7.0	6.8	7.1	4.8	4.5	5.5	
SDE	709.9	671.7	723.7	348.5	323.9	459.6	

Ch&St ... Cheng&Steele parameters; Ch&St 2 ... Cheng&Steele parameters where combining rules were changed (arithmetic mean for σ_{ij}); Girifalco gra-gra ... parameters suggested by Girifalco for interaction between two graphene sheets; Girifalco C_{60}-C_{60} ... parameters suggested by Girifalco for interaction between two fullerene molecules; Parm99 ... arithmetic mean for σ_{ij}; AE ... Average error; SDE ... Squared difference from experiment

The results from **Table 2** demonstrate that comparing to experiments $\Delta\langle H^{FF}\rangle$ for all studied LJ parameters were underestimated in case of some molecules, e.g. for ethanol, acetonitrile, cyclohexane. On the other hand, $\Delta\langle H^{FF}\rangle$ for all studied LJ parameters were overestimated, e.g. for dichloromethane, nitromethane. For the rest of the molecules, there was a particular LJ parameter which gave $\Delta\langle H^{FF}\rangle$ very close to the experimental value. In general, the difference between the calculated and experimental values may be partly explained by the neglect of an electrostatic interaction between an uncharged graphene surface and an approaching molecule or by the absence of a polarization in a pair additive force field. So, the absolute differences of $\Delta\langle H^{FF}\rangle$ and ΔH^{exp} can be accounted to the fact that in reality, there might be other contributions to the total interaction energy of the graphene-molecule interface [11]. The average error (AE) of $\Delta\langle H^{FF}\rangle$ in respect to ΔH^{exp}, averaged through all molecules, suggests that the OPLS-AA parameters, followed by Girifalco C_{60}-C_{60}, provided the best agreement with experiments. The values of squared difference from experiments (SDE) also confirmed the best agreement of $\Delta\langle H^{FF}\rangle$ with ΔH^{exp} when calculated using OPLS-AA parameters for graphene carbons, followed by Girifalco C_{60}-C_{60}.

The adsorption enthalpies calculated using molecular dynamics simulations are also graphically displayed in **Figure 1** for acetonitrile and ethyl acetate. For all molecules the same trend of $\Delta\langle H^{FF}\rangle$ was observed and these values correlated with epsilon LJ parameter.

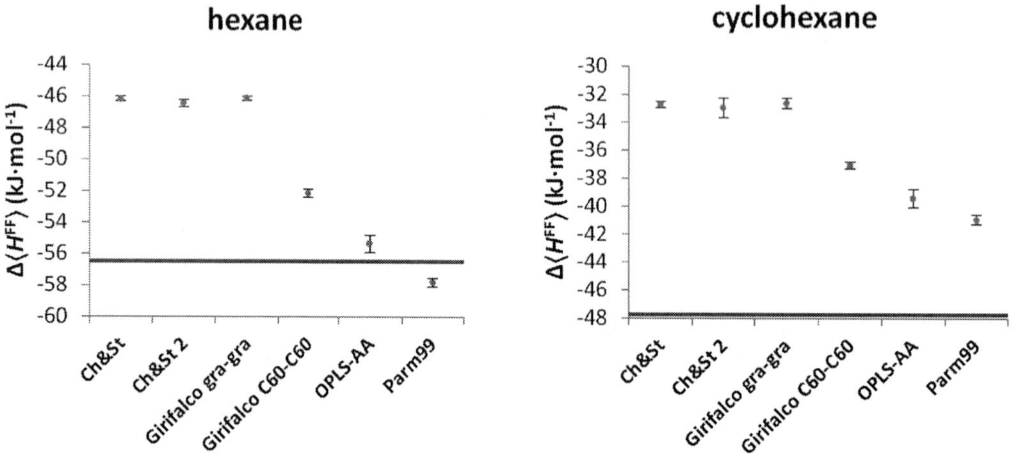

Figure 1 Adsorption enthalpies $\Delta\langle H^{FF}\rangle$ from MD simulations of adsorption of single molecule on the few-layered graphene surface calculated utilizing different studied LJ parameters of graphene carbons. Red line represents experimental value, error bars represent standard deviation determined by five independent simulations.

Many approximations were introduced in our MD simulations; therefore the absolution differences of calculated and experimental enthalpies might not be the most important observable quantity. More important is that if particular LJ parameter is able to differentiate on strongly and weakly bound molecules to the graphene surface, if it is able to order the molecules according to their affinity to the graphene surface. The data showed that all LJ parameters ordered the molecules in a same manner, more importantly with a good agreement with the experimental data. This observance was also statistically tested with Spearman's rank correlation coefficient [23]. On our chosen significance level $\alpha = 0.01$ the calculated correlation coefficient was higher than a critical value (0.709) for all studied LJ parameters, which means that a strong correlation between the order of ΔH^{exp} and $\Delta\langle H^{FF}\rangle$ was noticed. Therefore, it could be concluded that regardless of used LJ parameters of graphene carbons, MD simulations in our developed protocol may be suitable for semi-quantitative estimates of the interaction energies of the organic volatile molecules with graphene.

Systematically, a biggest deviation of our results from experiments can be seen especially for dichloromethane and acetonitrile. This could be possibly explained such that electronic effects or a cluster formation may contribute to the total interaction energy, which was already published in literature [11,24]. SAPT analysis [11] proposes that that polarization and electrostatics contribute to the total interaction energy more in case of acetone, acetonitrile and dichloromethane than in case of other molecules form our set. This SAPT analysis also completes underestimation of our calculated adsorption enthalpies in comparison with the experiments and advises that if polarization and electrostatic contributions were considered in our MD simulations, other studied LJ parameters would very likely give better agreement with experiments than OPLS-AA parameters.

In the last place, our results were discussed regarding Hansen solubility parameters [25], which are based on mutual miscibility/affinity. Three Hansen solubility parameters δ_d, δ_p, δ_h, relative polarity, calculated and experimental adsorption enthalpies, distance in Hansen space R_a, and molar volume V_m are listed in **Table 3**. Although a correlation of $\Delta\langle H^{FF}\rangle$ with Hansen dispersive solubility parameters was anticipated, unfortunately, the collected data demonstrated no such correlation. Also the rule "like dissolves like" was not confirmed because the distance in Hansen space did not correlate with our calculated enthalpies. On the other hand, some connection between calculated enthalpies and molar volume of the molecules could be observed, where bigger molecules adsorbed on the few-layered graphene surface more.

Table 3 Hansen solubility parameters of the molecules [25] and graphene [26] and other observable quantities

Molecule	rel. polar.	ΔH^{exp} (kJ·mol^{-1})	$\Delta\langle H^{FF}\rangle$ (kJ·mol^{-1})	δ_d (Mpa$^{0.5}$)	δ_p (Mpa$^{0.5}$)	δ_h (Mpa$^{0.5}$)	R$_a$	V_m (cm^3·mol^{-1})
Dich	0.309	-24.7	-34.8	18.2	6.3	6.1	6.8	63.9
Nitr	0.481	-26.4	-33.3	15.8	18.8	5.1	20.2	54.3
Etha	0.654	-30.5	-27.5	15.8	8.8	19.4	23.8	58.5
Acet	0.355	-34.3	-36.0	15.5	10.4	7.0	5.6	74.0
Acni	0.460	-31.8	-25.3	15.3	18.0	6.1	18.5	52.6
Etat	0.228	-48.1	-52.9	15.8	5.3	7.2	9.2	98.5
Hexa	0.009	-56.5	-55.4	14.9	0.0	0.0	24.9	131.6
Cyc	0.006	-47.7	-39.4	16.8	0.0	0.2	24.0	108.7
Diox	0.164	-45.2	-43.2	19.0	1.8	7.4	15.1	85.7
Benz	0.111	-49.8	-54.6	18.4	0.0	2.0	21.8	89.4
Tolu	0.099	-57.7	-57.8	18.0	1.4	2.0	19.5	106.8
Graphene [26]				18.0	9.3	7.7		

3. CONCLUSION

The prediction potential of classical MD on the graphene and the suitability of tested LJ parameters for the simulations of the graphene were verified. The adsorption enthalpies for eleven organic molecules calculated with different LJ parameters using MD simulations were assessed and all considered LJ parameters provided adsorption enthalpies in a qualitatively good agreement with experimental data. OPLS-AA parameters provided the adsorption enthalpies in the best agreement with experimental data with an average error less than 4.5 kJ·mol^{-1}, which is close to an experimental error of inverse gas chromatography. Needless to say that all tested LJ parameters were suitable for semi-quantitative estimates of the interaction energies of the molecules with the graphene. This implies that MD simulation is able to correctly order the adsorbates in respect to their adsorption enthalpies and to differentiate on strongly and weakly bound molecules. Therefore, our protocol is robust enough with respect to the level of approximation we introduced and may serve to other studies of sensing capacity of the graphene. In conclusion, both LJ parameters adopted from biomolecular force field and LJ parameters derived for their specific issue served well for the purpose of studying the adsorption of molecules on the graphene surface.

4. ACKNOWLEDGEMENTS

Financial support was provided by the project LO1305 of the Ministry of Education, Youth, and Sports of the Czech Republic.

REFERENCES

[1] GEIM, A. K. and NOVOSELOV, K. S. The Rise of Graphene. *Nat. Mater.* 2007. vol. 6, no. 3, pp. 183-191.

[2] STOLLER, M. D., PARK, S., ZHU, Y., An, J. and RUOFF, R. S. Graphene-Based Ultracapacitors. *Nano Lett.* 2008. vol. 8, no. 10, pp. 3498-3502.

[3] LEE, C., WEI, X., KYSAR, J. W. and HONE, J. Measurement of the Elastic Properties and Intrinsic Strength of Monolayer Graphene. *Science*. 2008. vol. 321, no. 5887, pp. 385-388.

[4] BALANDIN, A. A., GHOSH, S., BAO, W., CALIZO, I., TEWELDEBRHAN, D., MIAO, F. and LAU, C. N. Superior Thermal Conductivity of Single-Layer Graphene. *Nano Lett.* 2008. vol. 8, no. 3, pp. 902-907.

[5] ZHU, Y., MURALI, S., CAI, W., LI, X., SUK, J. W., POTTS, J. R. and RUOFF, R. S. Graphene and Graphene Oxide: Synthesis, Properties, and Applications. *Adv. Mater.* 2010. vol. 22, no. 35, pp. 3906-3924.

[6] SCHEDIN, F., GEIM, A. K., MOROZOV, S. V., HILL, E. W., BLAKE, P., KATSNELSON, M. I. and NOVOSELOV, K. S. Detection of Individual Gas Molecules Adsorbed on Graphene. *Nat. Mater.* 2007. vol. 6, no. 9, pp. 652-655.

[7] CHEN, J., WANG, X., DAI, C., CHEN, S. and TU, Y. Adsorption of GA Module onto Graphene and Graphene Oxide: A Molecular Dynamics Simulation Study. *Physica. E.* 2014. vol. 62, pp. 59-63.

[8] CHENG, Y., ZHANG, Z. and TEO, Z. Deformation of Graphene Induced By Adsorption of Peptides: A Molecular Dynamics Study. *Int. J. Appl. Mech.* 2013. vol. 5, no. 1, p. 1350007.

[9] PEI, Q. X., ZHANG, Y. W. and SHENOY, V. B. A Molecular Dynamics Study of the Mechanical Properties of Hydrogen Functionalized Graphene. *Carbon N. Y.* 2010. vol. 48, no. 3, pp. 898-904.

[10] GIRIFALCO, L. A., HODAK, M. and LEE, R. S. Carbon Nanotubes, Buckyballs, Ropes, and a Universal Graphitic Potential. *Phys. Rev. B - Condens. Matter Mater. Phys.* 2000. vol. 62, no. 19, pp. 13104-13110.

[11] LAZAR, P., KARLICKÝ, F., JUREČKA, P., KOCMAN, M., OTYEPKOVÁ, E., ŠAFÁŘOVÁ, K. and OTYEPKA, M. Adsorption of Small Organic Molecules on Graphene Adsorption of Small Organic Molecules on Graphene. *J. Am. Chem. Soc.* 2013. vol. 135, no. 16, pp. 6372-6377.

[12] OTYEPKOVÁ, E., LAZAR, P., ČÉPE, K., TOMANEC, O. and OTYEPKA, M. Organic Adsorbates Have Higher Affinities to Fluorographene than to Graphene. *Appl. Mater. Today.* 2016. vol. 5, pp. 142-149, 2016.

[13] CHENG, A. and STEELE, W. a. Computer Simulation of Ammonia on Graphite. I. Low Temperature Structure of Monolayer and Bilayer Films. *J. Chem. Phys.* 1990. vol. 92, no. 6, p. 3858.

[14] GIRIFALCO, L. A. Interaction Potential for Carbon (C60) Molecules. *J. Phys. Chem.* 1991. vol. 95, no. 14, pp. 5370-5371.

[15] GIRIFALCO, L. A. Molecular Properties of Fullerene in the Gas and Solid Phases. *J. Phys. Chem.* 1992. vol. 96, no. 2, pp. 858-861.

[16] MONTICELLI, L. On Atomistic and Coarse-Grained Models for C60 Fullerene. *J. Chem. Theory Comput.* 2012. vol. 8, no. 4, pp. 1370-1378.

[17] JORGENSEN, W. L., MAXWELL, D. S. and TIRADO-RIVES, J. Development and Testing of the OPLS All-Atom Force Field on Conformational Energetics and Properties of Organic Liquids. *J. Am. Chem. Soc.* 1996. vol. 118, no. 45, pp. 11225-11236.

[18] WANG, J., CIEPLAK, P. and KOLLMAN, P. A. How Well Does a Restrained Electrostatic Potential (RESP) Model Perform in Calculating Conformational Energies of Organic and Biological Molecules? *J. Comput. Chem.* 2000. vol. 21, no. 12, pp. 1049-1074.

[19] HESS, B., BEKKER, H., BERENDSEN, H. J. C. and Fraaije, J. G. E. M. LINCS: A Linear Constraint Solver for Molecular Simulations. *J. Comput. Chem.* 1997. vol. 18, no. 12, pp. 1463-1472.

[20] van der SPOEL, D., LINDAHL, E., HESS, B., GROENHOF, G., MARK, A. E. and BERENDSEN, H. J. C. GROMACS: Fast, Flexible, and Free. *J. Comput. Chem.* 2005. vol. 26, no. 16, pp. 1701-1718.

[21] BERENDSEN, H. J. C., POSTMA, J. P. M., van GUNSTERSEN, W. F., DiNOLA, A. and Haak, J. R. Molecular Dynamics with Coupling to an External Bath. *J. Chem. Phys.* 1984. vol. 81, no. 8, pp. 3684-3690.

[22] van der SPOEL, D., van MAAREN, P. J. and CALEMAN, C. GROMACS Molecule & Liquid Database. *Bioinformatics.* 2012. vol. 28, no. 5, pp. 752-753.

[23] SPEARMAN, C. The Proof and Measurement of Association between Two Things. *Am. J. Psychol.* 1904. vol. 15, no. 1, p. 72.

[24] LAZAR, P., OTYEPKOVÁ, E., PYKAL, M., ČÉPE, K. and OTYEPKA, M. Role of the Puckered Anisotropic Surface in the Surface and Adsorption Properties of Black Phosphorus. *Nanoscale.* 2018. vol. 10, no. 19, pp. 8979-8988.

[25] HANSEN, Ch. M. *Hansen Solubility Parameters: A User´s Handbook.* 2nd ed. Boca Roca: CRC Press LLC, 1999. p. 544.

[26] HERNANDEZ, Y., LOTYA, M.; RICKARD, D.; BERGIN, S. D. and COLEMAN, J. N. Measurement of Multicomponent Solubility Parameters for Graphene Facilitates Solvent Discovery. *Langmuir.* 2010. vol. 26, no. 5, pp. 3208-3213.

THE METHOD OF CARBON NANOTUBES ADDITION IN TO Al - Cu DISSIMILAR FRICTION STIR WELDED JOINT

Oleg PANCHENKO[1], Ke CHEN[2,3], Pedro VILACA[4], Fedor ISUPOV[1], Aleksey SHAMSHURIN[1]

[1] *National Technology Initiative Center of Excellence in Advanced Manufacturing Technologies at Peter the Great St. Petersburg Polytechnic University St. Petersburg, Russian Federation, panchenko_ov@spbstu.ru*
[2] *School of Materials Science and Engineering, Shanghai Jiao Tong University, Shanghai, China*
[3] *Shanghai Key Laboratory of Material Laser Processing and Modification, Shanghai Jiao Tong University, Shanghai, China*
[4] *Aalto University, Finland, EU*

Abstract

The method of carbon nanotubes (CNT) addition to a friction stir welded dissimilar joint has been developed. The CNT agglomerates are produced in shape of paper sheets. CNT sheets added to the gap of butt joint during fixture before friction stir welding. Use of CNT sheets provide ability to measure the weight ratio of CNT in welds. Experimental addition of CNT in Al to Cu dissimilar joint for obtaining 7 % CNT in weld was done. XRD study of weld metal shows no formation of carbides as CNT degradation products. SEM study of different weld join parts shows evidence of carbon presence. EDX mapping of carbon in the joint showed that both Al and Cu weld matrix parts have carbon inside.

Keywords: Carbon nanotubes, metal matrix composites, friction stir welding, dissimilar joint

1. INTRODUCTION

Electrical conductivity of dissimilar Al-Cu joints in bus-bars, foil conductors in transformers, connectors, etc. usually made with use of bolts, rivets. Due to behavior under the thermal cyclic load (expanding/contracting) driven by the Joule effect [1] the connection between solid bodies might be lost or became poor and electrical resistance increases. In some cases, energy loses in such joints may be very high even with good connection [2].

Those problems might be solved by welding of Al to Cu, but standard welding approaches of fusion processes (arc, laser) are not possible to use due to difference in physical properties and formation of intermetallic phases between Al and Cu. But solid state welding processes can overcome those limitations and sound weld joint with decreased amount of intermetallics can be obtained by friction stir welding [2,3].

One of possible way to modify the electrical conductivity and mechanical properties of such joints is addition of carbon nanotubes (CNT), nevertheless no methods of addition are proposed yet.

The aim of this work is development of the method for CNT addition to the friction stir welded dissimilar joint, and evaluation of CNT distribution and degradation after welding.

2. EXPERIMENTAL APPROACH

2.1. Materials

Commercially available multi-walled carbon nanotubes were used as a raw material for addition in to the welding joint.

Alloys are: Al - of electro chemical purity, Cu- of electro chemical purity (ETP), chemical composition give in **Table 1** and **2** respectively.

Table 1 Chemical composition of Al-ETP

Element	**Fe**	**Si**	**Mn**	**Ti**	**Al**
Concentration, (wt.%)	0.3	0.3	0.025	0.1	Bal.

Table 2 Chemical composition of Cu-ETP

Element	**Fe**	**Ni**	**S**	**As**	**Sn**
Concentration, (wt.%)	0.005	0.002	0.004	0.002	0.002
Element	Pb	Zn	O	Sb	Cu
Concentration, (wt.%)	0.005	0.004	0.05	0.002	Bal.

2.2. Methodology

So, called buckypaper was made of purchased CNTs. There are various types of processing for preparation buckypaper described in the literature [4-6]. **Figure 1** shows the prepared buckypaper sample and **Figure 2** - SEM images with different magnification.

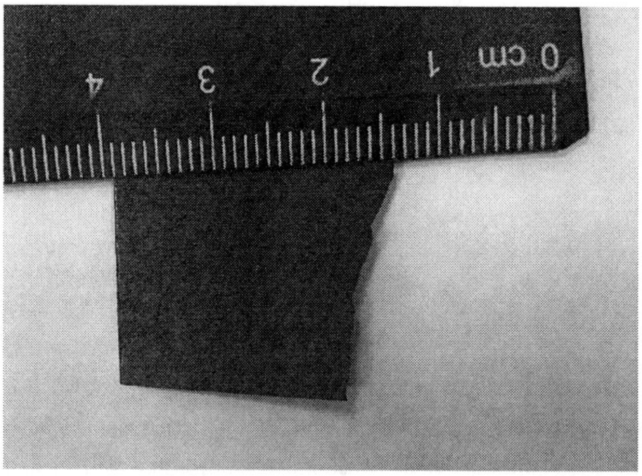

Figure 1 Sample of buckypaper

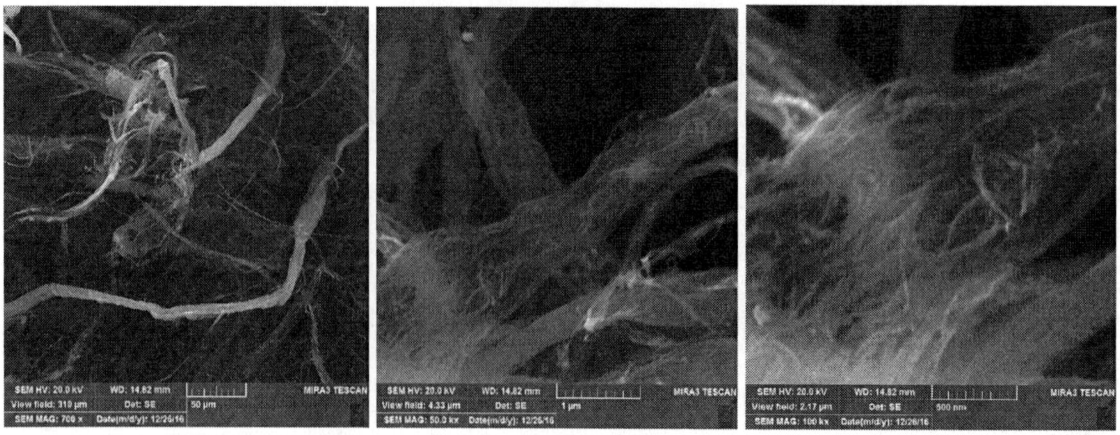

Figure 2 SEM images of prepared buckypaper

Friction stir welding parameters were found with a several tryouts of Cu to Al butt welding. Parameters are given in **Table 3** it is important to notice that during welding the tool has offset to aluminium side with respect to good welding practice [7]. Buckypaper was sliced to 3 mm wide strips and placed in the butt joint between 3 mm thick Al and Cu sheets without visible gap.

Mass of the added CNT buckypaper was measured and it is corresponded to 7 % of CNT in weld metal.

XRD of polished upper surface was evaluated (**Figure 4**) as well as SEM of stir zone with EDX mapping in cross section (**Figures 5 - 9**).

Table 3 Friction stir welding parameters

Rotation speed, rpm	**Travel speed, mm/min**	**Force/Impulse Force, kN**	**Tool probe offset to Al side, mm**
1200	100	7	2.0

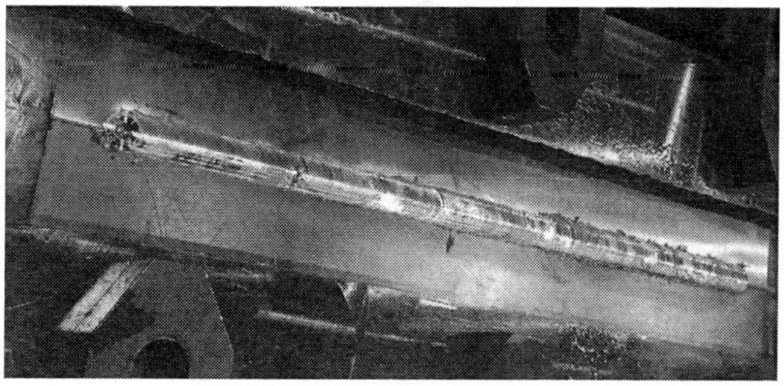

Figure 3 Overview of the joint

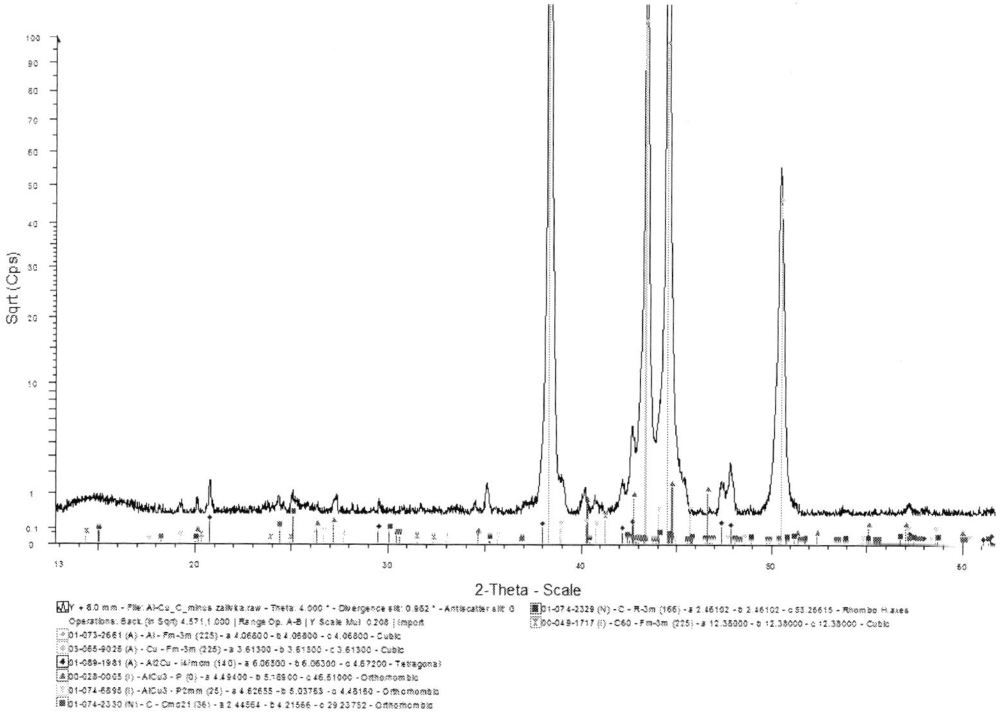

Figure 4 XRD patterns

Figure 5 SEM image of stir zone near boundary with copper

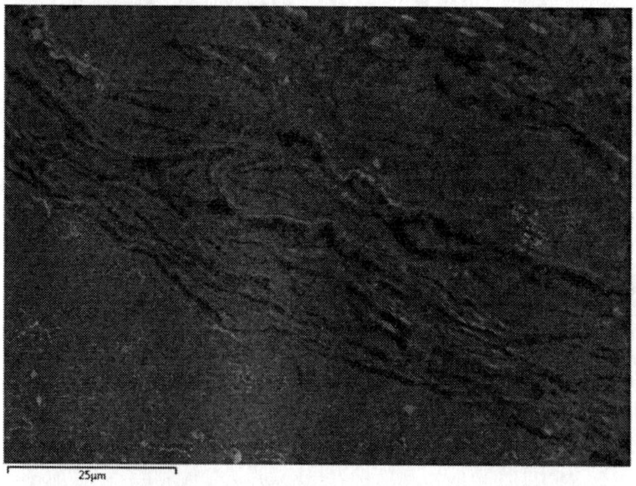

Figure 6 SEM image of stir zone near boundary with copper

Figure 7 EDX mapping of copper and carbon in stir zone near boundary with copper

Figure 8 EDX mapping of carbon in stir zone near boundary with copper

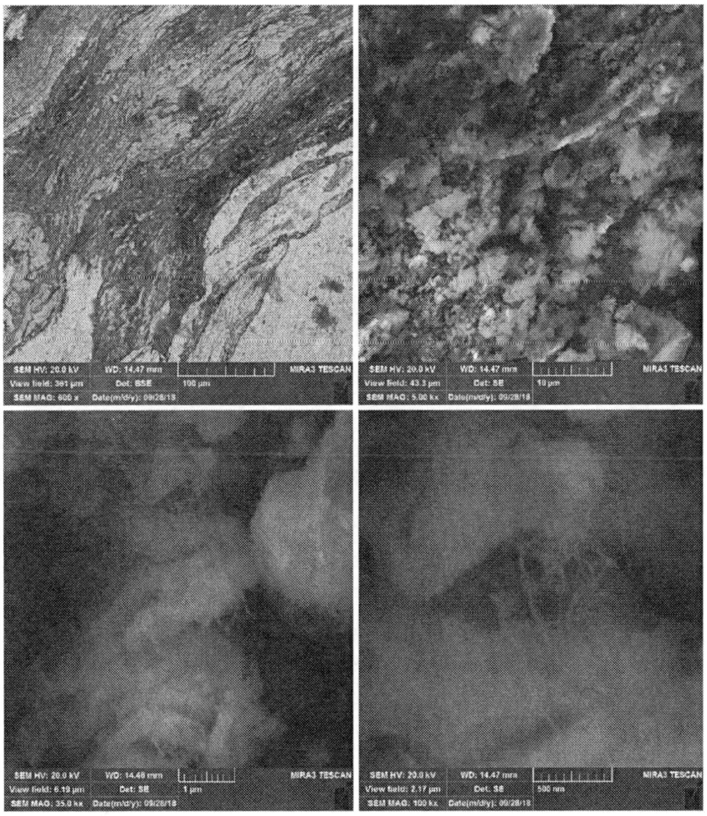

Figure 9 SEM image of stir zone between material clusters

3. RESULTS AND DISCUSSION

Analysis of XRD patterns shows evidence of Cu_xAl_y - intermetallic phase formation but no carbides are found as any other possible chemicals with carbon (**Figure 5**). So due to temperature and stir processes during welding intermetallics formation was stated in several papers [2,7]. As well as carbides, carbonates and many other carbon containing chemicals are stated to be the CNT degradation products [8-11]. At that point, no CNT degradation during friction stir welding of Al to Cu was found using XRD.

SEM mapping analysis of carbon distribution near Cu boundary (**Figures 7, 8**) is evidence that CNT distributed both in Al and Cu metal matrix. Nevertheless, it is well seen that carbon distributed not evenly. The reason for that might be the metal flow during stir process. With the same reason Al and Cu metal forms a sort of material clusters of complex shape. It is well seen form **Figures 5 - 7** that metals also distributed not evenly. Analysis of SEM images of boundary between clusters with high magnification (**Figure 9**) shows CNT in metal matrix.

4. CONCLUSION

Method for the addition of carbon nanotubes to friction stir welded joint has been developed. Carbon nanotubes were found in stir zone of weld metal between clusters of Cu and Al. No evidence of CNT degradation during friction stir welding was found. Distribution of CNT in both copper and aluminium matrix of weld metal has achieved.

ACKNOWLEDGEMENTS

The research was performed with financial support of the Ministry of Science and Education of Russian Federation in terms of the project No 14.Y30.18.3670-MK.

REFERENCES

[1] NIX F. and MACNAIR D. The thermal expansion of pure metals: copper, gold, aluminum, nickel, and iron. *Physical Review*, 1941, vol. 60, (8), pp. 597

[2] OLAFSSON D., VILACA P. and VESANKO J. Characterization of FSW of Aluminium to Copper for Electrical Busbars, *IIW*, Doc. III-1859-18

[3] DEBROY T. and BHADESHIA H. Friction stir welding of dissimilar alloys-a perspective, *Science and Technology of Welding and Joining*, 2010, vol. 15, (4), pp. 266-270

[4] WANG Ding, SONG Pengcheng, LIU Changhong, WU Wei and FAN Shoushan. Highly oriented carbon nanotube papers made of aligned carbon nanotubes, *Nanotechnology*. 2008, 19, 7, pp. 1-6, doi:10.1088/0957-4484/19/7/075609

[5] http://www.nano-lab.com/buckypaper.html

[6] Edited by HUANG Wenyi. *Nanopapers: From Nanochemistry and Nanomanufacturing to Advanced Applications*, Amsterdam, Netherlands: Elsevier, 2018, p. 256.

[7] GALVAO I., LOUREIRO A., VERDERA D., GESTO D. and RODRIGUES DM. Influence of tool offsetting on the structure and morphology of dissimilar aluminum to copper friction stir welds, *Met and Mater Trans*: 2012, A, 43 (13), pp. 5096-5105.

[8] LIU, Q., KE, L., LIU, F., HUANG, C. and XING, L. Microstructure and mechanical properties of multi-walled carbon nanotubes reinforced aluminum matrix composites fabricated by friction stir processing. *Materials and Design*, 2013, no. 45, pp. 343-348.

[9] IZADI, H. and GERLICH, A. Distribution and stability of carbon nanotubes during multi-pass friction stir processing of carbon nanotube/aluminum composites. *Carbon*, 2012, no. 50, pp. 4744-4749.

[10] PANCHENKO O., IVANOV S., KOL'TSOVA T. and MICHAILOV V., Survivability of carbon nanotubes during arc welding.*Conference Proceedings, 8th International Conference on Nanomaterials - Research and Application*, pp. 59-63

[11] KOL'TSOVA, T.S., SHAKHOV, F.M., VOZNYAKOVSKII, A.A., LYASHKOV, A.I.,TOLOCHKO, O.V, NASIBULIN, A.G., RUDSKOI, A.I. and MICHAILOV, V.G. Fabrication of a compact aluminum -carbon nanofiber material by hot pressing. *Technical Physics*, 2014, vol. 59, no. 11, pp. 1626-1630.

GRAPHENE NANOSHEETS SYNTHESIZED IN MICROWAVE PLASMA AND LIQUID EXFOLIATED GRAPHENE: STRUCTURAL CHARACTERIZATION STUDY

Jana JURMANOVÁ, Ondřej JAŠEK, Jozef TOMAN, Miroslav Šnírer, Vít KUDRLE

Department of Physical Electronics, Faculty of Science,
Masaryk University, Brno, Czech Republic, EU,
janar@physics.muni.cz

Abstract

Graphene based nanomaterials produced by liquid exfoliation techniques and microwave plasma decomposition of ethanol were studied by scanning electron microscopy (SEM), Raman spectroscopy and X-Ray powder diffraction (XRD). Graphene oxide (GO) and reduced graphene oxide (r-GO) prepared by liquid exfoliation technique consisted of several types of structures - thin flakes with thickness of 10-15 nm and lateral dimensions of several micrometers and graphite like structures, 100 µm in size, with not fully exfoliated graphene layers. The graphene nanosheets prepared in microwave plasma torch (MWGNs) were several nm thick and rectangular in shape with size of several hundred nanometers. Contamination by elements S, Mn and K was detected by EDX analysis in both GO and r-GO samples. Similarly, Fe was detected in MWGNs. This contamination was caused by synthesis process in all cases. The Raman spectroscopy analysis showed presence of D (1360 cm^{-1}] and G (1580 cm^{-1}) band and second order 2D (2690 cm^{-1}) and D+G (2920 cm^{-1}) in the measured spectra. The I_D/I_G intensity ratio was 2.2 and 2.5 for GO and r-GO samples, respectively. The graphene nanosheets Raman spectra exhibited I_D/I_G ratio of 0.5 and high intensity 2D peak at 2690 cm^{-1}. XRD analysis revealed graphite like structure, graphite peak (002), of all samples and presence of contamination (Fe, intercalation compounds) in agreement with EDX analysis results. XPS analysis confirmed results of XRD and Raman analysis and sp^2/sp^3 ratio of carbon C1s peak was increasing with decreasing I_D/I_G intensity ratio.

Keywords: Graphene, microwave plasma, liquid exfoliation, material analysis

1. INTRODUCTION

There exists wide range of methods to prepare advanced carbon materials like carbon nanotubes (CNT), nanofibers, graphene, nanoribbons or nanoflakes [1]. Liquid exfoliation belongs to most frequently used techniques for preparation of graphene and graphene oxide. The method is based on separation of individual graphene layers from graphite flakes by means of ultrasonication, intercalation or shear forces. The method generally involves three steps a) dispersion in a solvent, b) exfoliation and c) purification. During the preparation the graphite source and graphene based output material are exposed to number of chemicals (acids, reduction compounds) and processing steps. To avoid these elaborate procedures, an alternative to liquid exfoliation was developed in a form of synthesis of graphene nanosheets in microwave plasma. It was shown that ethanol decomposition in microwave plasma at atmospheric pressure leads to successful formation of graphene nanosheets [2]. This method was further investigated by Tatarova and Tsyganov [3] using microwave plasma surface wave discharge at atmospheric pressure in Ar/ethanol gas mixture. By further analysis and modelling they found that C_2 molecule promotes nucleation and growth of graphene and atomic C leads to deposition of amorphous and sp^3 phase. We have showed recently that microwave plasma torch operated at atmospheric pressure can also be used for such synthesis [4].

In this work we compare structure and composition of product of both methods: liquid exfoliation by Hummers method and decomposition of ethanol in microwave plasma torch at atmospheric pressure.

2. EXPERIMENTAL

Graphene oxide (GO) and reduced graphene oxide (r-GO) powder was prepared by Hummers method [5]. High purity artificial graphite dissolved in concentrated H_2SO_4 and $NaNO_3$ was added to the mixture. The mixture was magnetically stirred for 10 min and kept at the temperature of 0 °C in ice bath. Then anhydrous $KMnO_4$ was added to the mixture and kept for 20 min at 0 °C followed by the heating to 35 °C for 30 min. The sample was then hydrolyzed in water and H_2O_2 was added. Finally the sample was washed by deionized water and dried in air. To prepare the r-GO, the prepared GO was reduced at 350°C for 1 hour. More details about the preparation technique could be found in [6, 7].

As prepared powder was dispersed in ethanol and ultrasonicated for 10 minutes. The prepared suspension was then transferred by drop casting onto Si wafer and left in the fume hood under heating for 5 minutes to evaporate the solvent. Second set of samples was prepared by direct transfer of nanopowder to double-sided carbon adhesive tape.

The graphene nanosheets were synthesized by ethanol decomposition in microwave plasma torch at atmospheric pressure. The setup consisted of a microwave (MW) generator (2.45 GHz, 2 kW) with a standard rectangular waveguide, transmitting the MW power through a coaxial line to a hollow nozzle electrode. The gases were supplied to the discharge chamber through the carbon or iron nozzle which has two gas channels. The central channel (0.8 mm diameter), in the nozzle axis, was used for introduction of working gas - argon (360 - 920 sccm) and subsequent ignition of plasma. The secondary channel (annulus with outer radius 8.4 mm and inner radius 7.7 mm) was used for introduction of carrying gas - argon (500 - 1400 sccm) with precursor (ethanol 2-25 sccm) vapours into the plasma environment. The gas flows were controlled by Bronkhorst electronic flow controllers. Reactor chamber consists of a quartz tube (8 cm diameter, 20 cm long) terminated by dural flanges. Synthesized nanopowder was collected from the reactor wall or on the Si/SiO_2(92 nm) substrates fixed in the holder, quartz tube with fastening mechanism, 10 cm from the plasma nozzle.

Raman spectroscopy was carried out using HORIBA LabRAM HR Evolution system with 532 nm laser, using 100x objective and 25 % ND filter in the range from 1000 to 3200 cm^{-1}. Samples were imaged with TESCAN scanning electron microscope (SEM) MIRA3 with Schottky field emission electron gun equipped with secondary electron (SE) and back-scattered electron (BSE) detectors as well as characteristic X-ray detector Oxford Instruments EDX analyzer. XRD analysis was carried out with X-Ray Diffractometer SmartLab Type F from Rigaku capable of small angle X-ray scattering (SAXS) technique and equipped with with copper anode and vertical theta-theta goniometer with horizontal sample site with independent source and detector movement. XPS analysis was performed using ESCALAB 250Xi (Thermo Scientific, UK) X-ray photoelectron spectrometer equipped with a conventional hemispherical analyzer. Monochromatized Al K_α (1486.6 eV) X-ray source was focused into elliptical spot size at 650 µm.

3. RESULTS AND DISCUSSION

3.1. SEM Structural Characterization

Graphene oxide and reduced graphene oxide were observed in two ways: in the natural state (pieces of material adhered to a double-sided adhesive tape) and in a more coherent layer (dissolved in pure ethanol, shaken by an ultrasonic cleaner and dried on a silicon wafer) using secondary electron (SE) and back-scattered electron (BSE) mode. There are differences between the two preparation techniques which are expressed in the alignment of graphite flakes on the substrate. The drop casted sample from liquid formed thin layer showing individual pieces of (reduced) graphene oxide (**Figure 3**). The adhesive tape technique formed thicker layer and was better suited to obtaining information on the total elemental composition by EDX (Energy Dispersive X-Ray (Spectroscopy)) analysis, while the first one better depicts the topology of the resulting graphene oxide. The graphene nanosheets obtained by microwave plasma synthesis were imaged as collected directly on the substrate. A typical shape of the produced nanopowders was captured on the following micrographs **Figure 1**.

Figure 1 Topological structure of the studied nanomaterials. r-GO on the left, GO in center and MWGNs on the right

One can see that the GO crystals are partially exfoliated on the edges and one can observe large pieces, hundreds of μms, as well as small thinner and smaller flakes of material, several μms. The reduction process of GO formed its reduced form r-GO and changed nanopowder structure as well. The exfoliation on the edges was more pronounced and the large graphite flakes started to disintegrate. The structure of MWGNs was completely different in comparison to GO structure. The bottom-up approach of synthesis resulted in foam like ordering of individual curved graphene nanosheets with thickness of several nanometers and lateral dimensions of hundreds of nanometers.

3.2. Contamination characterization by EDX

The GO and r-GO materials, due to the way they were prepared, are expected to be contaminated with the chemicals used in preparation process. Sample contamination may be already revealed by the images obtained with the BSE detector, where brighter spots represented heavier elements than carbon and oxygen. Precise chemical analysis of such a site was performed using an EDX analyzer. In the following figures, **Figures 2 - 4**, contamination maps are given without carbon and oxygen (EDX cannot detect hydrogen), which are main constituents of the material. Because EDX can only determine elemental composition of contaminants but not its crystalline form, we cannot determine by this method chemical state of the used compounds after the exfoliation process. Not including oxygen bonded to the foreign elements, the r-GO sample contained 1.5 atomic percent of contamination (N, Si, S, Al, K, Mn, Fe in order of concentration) and the GO sample contained about 4 atomic percent of contamination (S, N, Si, Al, Fe, Mn, K in order of concentration). It is necessary to note that due to the particle form of studied samples and its chemical inhomogeneity, the information obtained by local EDX analysis needed to be complemented by more global technique such as XRD. EDX analysis clearly demonstrated a different ratio of carbon and oxygen in the reduced graphene oxide, about 90:10, and GO about 75:25. The detail of the topological structure of graphene oxide in SE mode, the chemical compounds contamination imaged by BSE mode with visible places of chemical inhomogeneities (the brightness is determined by the atomic number of the elements) can be seen in **Figure 2** for large flake of r-GO, in **Figure 3** for small flake of r-GO and in **Figure 4** for GO flakes. Contamination of graphene nanosheets was very small, below 1 percent. Iron nanoparticles, 10 - 200 nm in diameter, formed by sputtering of the nozzle electrode could be found in the deposit and were clearly visible on the images in BSE mode **Figure 5**. The composition of bright spots was determined by EDX analysis as iron coming from plasma nozzle. This contamination could be almost completely eliminated by using carbon as nozzle material. The assumption that the plasma discharge could cause breakage of carbon nozzle and contaminate the sample with large graphite particles was not confirmed and such objects could not be found in the deposit.

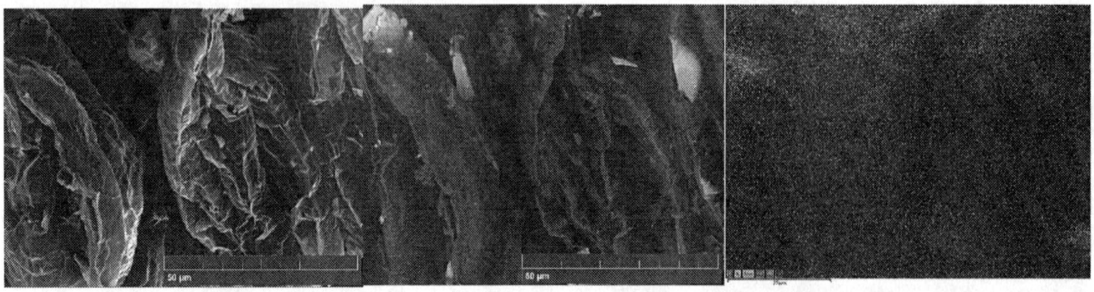

Figure 2 Reduced graphene oxide r-GO - a typical sample structure. Images taken by SE detector (on the left), BSE detector (in the center) and EDX detector (on the right), HV=8kV. Elements color key: S - violet (0.4 at%), K+Mn - green (0.2+0.2 at%), Fe - red (0.1 at%), Si,Al - blue (0.4, 0.2 at%).

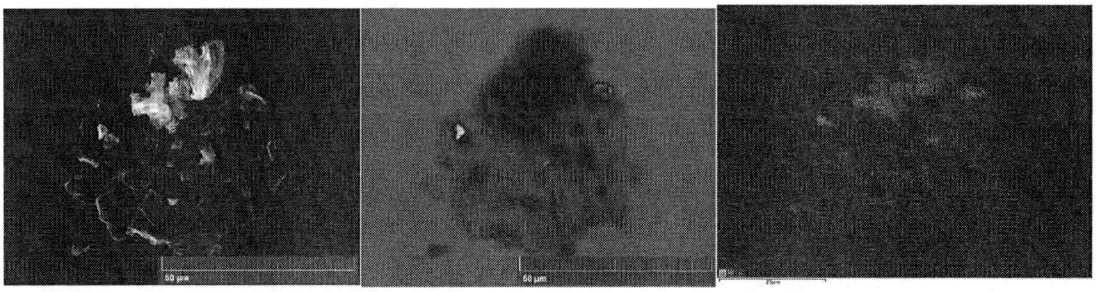

Figure 3 Detail of individual reduced graphene oxide flake. Images taken by SE detector (on the left), BSE detector (in the center) and EDX detector (on the right), HV= 15 kV. Elements color key: C - red (50.1 at%), O - green (1.0 % at), Si substrate - blue (48.9 at%).

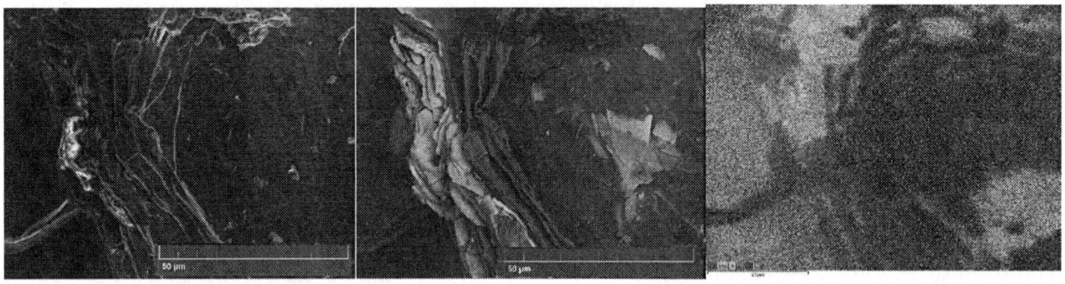

Figure 4 Graphene oxide GO - a typical sample structure. Images taken by SE detector (on the left), BSE detector (in the center) and EDX detector (on the right), HV=8kV. Elements color key: Fe - red (0.5 at%), K+Mn - green (0.1+0.2 at%), S - violet (2.0 at%), Si,Al - blue (2.1, 1.5 at%).

3.3. Material characterization by Raman spectroscopy and XRD

Further information about material structure and composition was obtained by Raman spectroscopy and XRD. XRD analysis of graphene nanomaterial powder was carried out by Bragg-Brentano technique from 10 to 100 2θ angle (**Figure 6**). In all types of materials strong graphite (002) diffraction at 26.7° (0.335nm) could be observed, but this peak appears to be slightly shifted to lower angle values in MWGNs sample. Weaker graphite diffraction peaks (100), (101) and (004) could be observed in GO and r-GO samples as well.

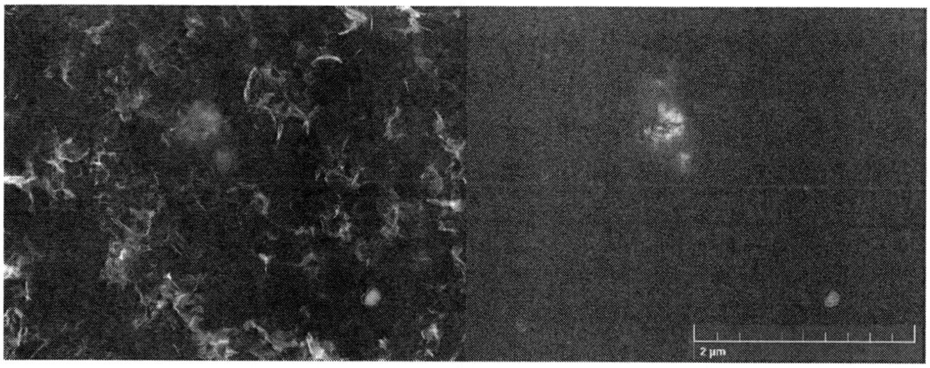

Figure 5 Detail of graphene nanosheets and its contamination by iron nanoparticles. Images taken by SE detector (on the left) and BSE detector (on the right), HV=15kV.

GO powder analysis also showed presence of strong GO peak at 10.4° (0.855nm) typical for intercalated graphite structure of graphite oxide [8]. This peak was not found in r-GO powder nor microwave plasma graphene nanosheets sample. Further wide peaks between 20-25° and at 40° could be assigned to contamination by intercalation compounds such as $KMnO_2$ etc. MWGNs sample also exhibited presence of diffractions at 44.3, 64.5, 81.8 and 98.5° assigned to Si substrates and some small intensity peaks could be assigned to Fe. The amount of contaminants was consistent with results of EDX analysis.

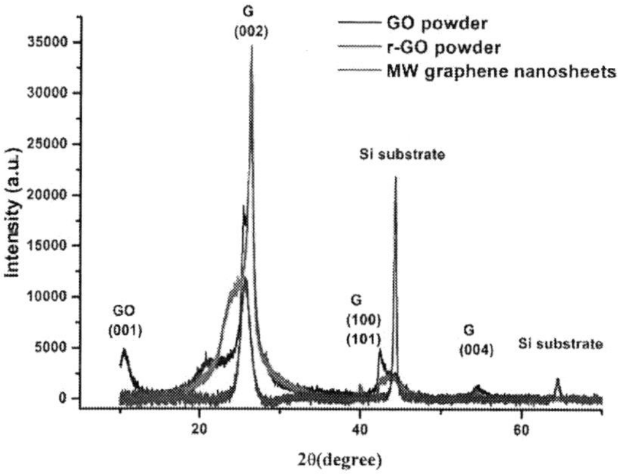

Figure 6 XRD pattern of prepared nanopowders

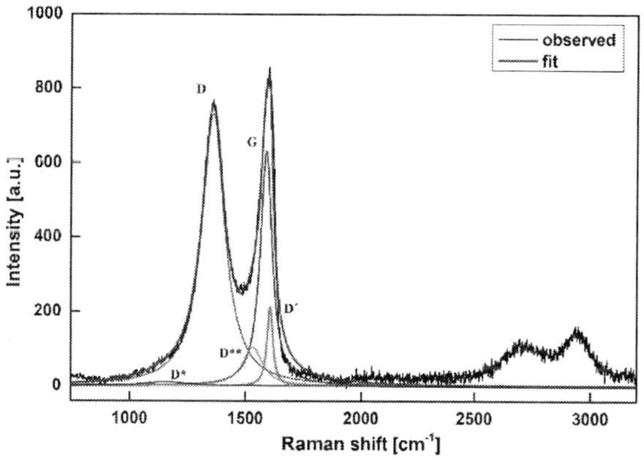

Figure 7 Raman spectra of prepared GO sample

Raman spectroscopy analysis showed presence of D, G, 2D and D+G bands [9]. **Figure 7** represented Raman spectra of both GO and r-GO samples which were very similar. The main feature of the spectra corresponded to the conventional D and G bands with peaks centered at ~1360 cm^{-1} and ~1590 cm^{-1}. Additional three peaks appeared during deconvolution of the region described as D*, D** and D´ centered at 1130 cm^{-1},1520 cm^{-1} and 1610 cm^{-1}. The occurrence of broad and intensive D peak in combination with D* and D** peaks were a sign of defective and disordered structures caused by exfoliation process. On the contrary, Raman spectra of MWGNs exhibited sharp D peak of low relative intensity (0.5) to G peak and intensive 2D peak (FWHM 65 cm^{-1}) as expected from few layer graphene.

Results of previous analyses were supported by XPS analysis of MWGNs. The results showed 98 % content of carbon with small addition of oxygen (2 %) in the sample. The sp^2 bonded structure was confirmed by fitting C1s peak profile with 95 % sp^2 (284.4 eV) and small content of sp^3 (285.4 eV) and carbon - oxygen bonds (286-287 eV) [10]. This result confirmed higher quality of MWGNs in comparison to GO and r-GO structures.

4. CONCLUSIONS

We have compared structure and composition of two kinds of graphene related materials - graphene oxide and its reduced form prepared by Hummers methods and graphene nanosheets prepared by microwave plasma decomposition of ethanol. The structural study by SEM proved that top-down preparation method of GO and r-GO led to only partial exfoliation of large portion of graphite flakes. This exfoliation on the flakes edges further increased after reduction process and formation of r-GO. The amount of oxygen substantially decreased from 25 to 10 % in r-GO structure as shown by EDX and XRD analysis, but the structure still exhibited very similar XRD diffraction pattern to graphite. Therefore, it would be better to call this structure reduced graphite oxide. On the contrary, bottom-up approach of microwave plasma synthesis was successfully used for synthesis of graphene nanosheets only several atomic layers thick, and therefore, much closer to graphene structure, but their lateral size was order of magnitude smaller than GO structures. As a consequence of preparation process both types of material contained impurities from processing chemicals (K, Mn, S, N) or plasma torch nozzle material (Fe). The structural and compositional characteristics were confirmed by Raman spectroscopy (I_D/I_G intensity ratio), XRD (graphite (002) crystal diffraction) and XPS (C1s energy spectra) measurements. In the future the amount of contamination could be reduced by modifying preparation process of both types of nanomaterial.

ACKNOWLEDGEMENTS

This work was supported by the Czech Science Foundation under project 18-08520S and in part by the project LO1411 (NPU I) funded by Ministry of Education, Youth and Sports of Czech Republic. We would like to thank Alexander G. Bannov for preparation of GO and r-GO powders.

REFERENCES

[1] A.C. FERRARI, F. BONACCORSO, V. FAL'KO, ET AL. Science and technology roadmap for graphene, related two-dimensional crystals, and hybrid systems. *Nanoscale*. 2015, vol. 7, no. 11, pp. 4598-4810.

[2] A. DATO, V. RADMILOVIC, Z. LEE, ET AL. Substrate-free gas-phase synthesis of graphene sheets. *Nano Lett.* 2008, vol. 8, no. 7, pp. 2012-2016.

[3] D. TSYGANOV, N. BUNDALESKA, E. TATAROVA, ET AL. On the plasma-based growth of 'flowing' graphene sheets at atmospheric pressure conditions. *Plasma Sources Sci. Technol.* 2015, vol. 25, no. 1, pp. 015013.

[4] J. TOMAN, O. JAŠEK, and J. JURMANOVÁ. The Influence of Gas Admixtures on The Synthesis of Graphene Nanosheets In Argon Microwave Plasma Torch Discharge. 8th International Conference on Nanomaterials - Research & Application, Nanocon 2016. pp. 122-126. *TANGER Ltd.*

[5] W.S. HUMMERS and R.E. OFFEMAN. Preparation of Graphitic Oxide. *J. Am. Chem. Soc.* 1958, vol. 80, no. 6, pp. 1339-1339.

[6] A.G. BANNOV, J. PRÁŠEK, O. JAŠEK, et al. Investigation of Pristine Graphite Oxide as Room-Temperature Chemiresistive Ammonia Gas Sensing Material. *Sensors* . 2017, vol. 17, no. 2, pp 320.

[7] A.G. BANNOV, A.A. TIMOFEEVA, V.V. SHINKAREV, et al. Synthesis and studies of properties of graphite oxide and thermally expanded graphite. *Prot. Met. Phys. Chem. Surf.* 2014, vol. 50, no. 2, pp. 183-190.

[8] S.-J. MU, Y.-C. SU, L.-H. XIAO, et al. X-Ray Diffraction Pattern of Graphite Oxide. *Chin. Physics Lett.* 2013, vol. 30, no. 9, pp. 096101.

[9] A.C. FERRARI: Raman spectroscopy of graphene and graphite: Disorder, electron-phonon coupling, doping and nonadiabatic effects. *Solid State Commun.* 2007, vol. 143, no. 1, no. 2, pp. 47-57.

[10] *Adv. Eng. Mater.* 2003, vol. 5, no. 7, pp. 519-522.

GOLD NANOPARTICLES-CYSTEINE-MULTIWALL CARBON NANOTUBE MODIFIED ELECTRODE FOR HEAVY METAL IONS DETECTION

[1]Zuzana KOUDELKOVA, [1,2]Lukas RICHTERA, [1]Pavel SVEC, [1,2]Vojtech ADAM

[1]Department of Chemistry and Biochemistry, Mendel University in Brno, Czech Republic, EU
[2]Central European Institute of Technology, Brno University of Technology, Brno, Czech Republic, EU

Abstract

The environmental contamination by heavy metals is one of the global problems that disturb the ecological balance. A platform for the simultaneous electrochemical detection of trace heavy metal ions has been developed based on the gold nanoparticles-cysteine-multiwall carbon nanotubes (AuNPs-Cys-MWCNTs) modified carbon paste electrode using square wave anodic stripping voltammetry. The voltammetric method was optimized with respect to accumulation time, accumulation potential and other parameters. The syntheses of AuNPs-Cys-MWCNTs composite with different AuNPs diameters were described and characterization using Dynamic light scattering and Scanning electron microscopy was performed. The AuNPs-Cys-MWCNTs composite significantly increases the electrode sensitivity towards Cd(II), Pb(II), Cu(II) and Zn(II). The AuNPs-Cys-MWCNTs composite has promising significance for future using, for example as a modification of the screen printed electrodes (SPEs), which would be used for low-cost and fast *in situ* measurement.

Keywords: Gold nanoparticles-cysteine-multiwall carbon nanotube composite, electrochemistry, paste electrode, screen printed electrode, ecological problems, heavy metals

1. INTRODUCTION

One of the global problems that disturb the ecological balance is the environmental contamination by heavy metals. Heavy metals, is a term for a group of toxic substances, which are dangerous for the environment. The group includes for example lead, mercury, cadmium, copper, zinc, chromium. The most common sources of these substances come from anthropogenic activities of different sectors (heavy industry, agriculture, and pharmaceutical industry) [1,2]. The accumulation of these substances in nature, watercourses and the soil, is a major problem. In fish and agricultural crops there is accumulation of these hazardous substances, through the food chain, heavy metals get into the human body [3,4]. Most of these heavy metals are carcinogens. For this reason it is necessary to check the concentrations of heavy metals in waste water [5,6].

Electrochemical methods are a suitable tool for detecting heavy metals in an aqueous environment. The multiwall carbon nanotubes (MWCNTs) and gold nanoparticles (AuNPs), both materials have recently received increased attention as materials suitable for use in electrochemical detection [7,8]. Their advantage is that they increase the conductivity of the electrode, facilitate the transfer of electrons, and have good chemical stability and MWCNTs increase the active surface area of the electrode. The aim of this study was to make use of nanotube, gold and cysteine connections to obtain excellent material for the formation of a paste electrode that would be sensitive to the determination of heavy metals [9]. The next step was to use the properties of this composite material and modify the surface of the SPEs and then use them for in situ analyzes.

2. MATERIALS AND METHODS

2.1. Chemicals

All chemicals were obtained from Sigma-Aldrich (St. Louis, MO, USA). High purity deionized water (Milli-Q Millipore 18.20 MΩ/cm, Bedford, MA, USA) was used throughout the study.

2.2. Preparation of gold nanoparticles-cysteine-multiwall carbon nanotubes composite (AuNPs-Cys-MWCNTs)

0.1 g MWCNTs were suspended in 30 mL 2.2 mM trisodium citrate. The solution was heated to boiling under vigorous stirring. Upon reaching the boiling point, 25 mM $H[AuCl_4]$ added. After 10 min of interaction, the temperature from 100 °C was lowered to 90 °C. After setting the temperature to 90 °C, 200 µL 60 mM sodium citrate solution was added with stirring and 200 µL 25 mM $H[AuCl_4]$ was added after an additional 2 min. After 10 min of interaction, the process with adding 60 mM sodium citrate solution and 25 mM $H[AuCl_4]$ was repeated 12 times. Thereafter, heating and stirring were stopped and the resulting product was cooled to room temperature and then 9 times washed. 30 mL of the resulting solution was dried.

2.3. Characterization of AuNPs-Cys-MWCNTs

The size of the presented AuNPs was measured by the Zetasizer Nano ZS Instrument (Malvern Instruments Ltd., Worcestershire, UK). The Scanning Electron Microscopy (SEM) MAIA3 (Tescan, a.s., Brno, Czech Republic) was used for visualization of presented composite material (AuNPs-Cys-MWCNTs).

2.4. Preparation of carbon paste electrode modified by AuNPs-Cys-MWCNTs (CPE-AuNPs-Cys-MWCNTs)

To prepare the modified carbon paste, 300 mg of expanded graphite and 75 mg of composite material (AuNPs-Cys-MWCNTs) were used. These materials were homogenized in a mortar with a pestle for 30 min. Subsequently, 260 µL of mineral oil was added. Again homogenization was carried out for 30 min. This mixture was transferred with a spatula into electrode body with 2.5 mm diameter.

2.5. Electrochemical determination of for Cd^{2+}, Pb^{2+}, Cu^{2+} and Zn^{2+}

Determination of Cd(II), Pb(II), Cu(II) and Zn(II) by square wave anodic stripping voltammetry (SqWASV) were performed with 663 VA Autolab (Metrohm, Herisau, Switzerland), using a standard cell with three electrodes. The CPE-AuNPs-Cys-MWCNTs was the working electrode. An Ag/AgCl/3M KCl as a reference electrode and platinum electrode was auxiliary. Software NOVA 1.8 (Metrohm, Herisau, Switzerland) was used for data evaluation. Prior to each measurement, approximately 0.1 mm of CPE-AuNPs-Cys-MWCNTs was wiped on a filter paper to obtain a new surface. The parameters of the measurement were as follows: initial potential of −1.3 V, end potential +0.3 V, deposition time 90 s, deposition potential −1.3 V, voltage step 5 mV, amplitude 25 mV, frequency 200 Hz and equilibration time 5 s. As supporting electrolyte was used acetate buffer pH 5.5.

3. RESULT AND DISCUSSION

In the **Figure 1** shows the structure of composite material AuNPs-Cys-MWCNTs. There is a detailed view of the structure of MWCNTs on which AuNPs are attached. The average particle size of AuNPs was estimated to be 60 nm.

An optimal ratio of the individual components in the paste electrode was determined. The electrochemical sensor (CPE-AuNPs-Cys-MWCNTs) was optimized for simultaneous detection of Zn(II), Cd(II), Pb(II) and Cu(II). The electrochemical method SqWASV was optimalized with respect to deposition time and potential, amplitude and frequency. The calibration curves for metals and the equation coefficient of the calibration curves with the coefficients of determination r^2 are shown in **Figure 2.** The detection limits were found to be 4.5 µg/L for Zn(II), 9.8 µg/L for Cd(II), 8.0 µg/L and 4.0 µg/L for Pb(II) and Cu(II) respectively. In addition, solution of AuNPs-Cys-MWCNTs were applied to the surface of the SPEs. The modified SPEs were successfully tested for Zn(II), Cd(II), Pb(II) and Cu(II) detection.

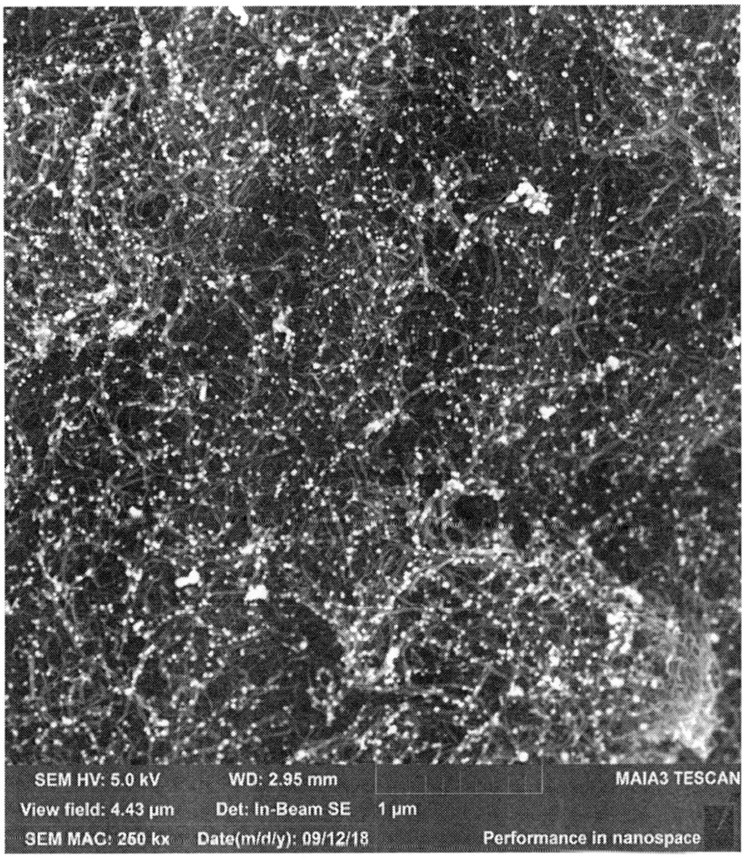

Figure 1 SEM characterization of AuNPs-Cys-MWCNTs

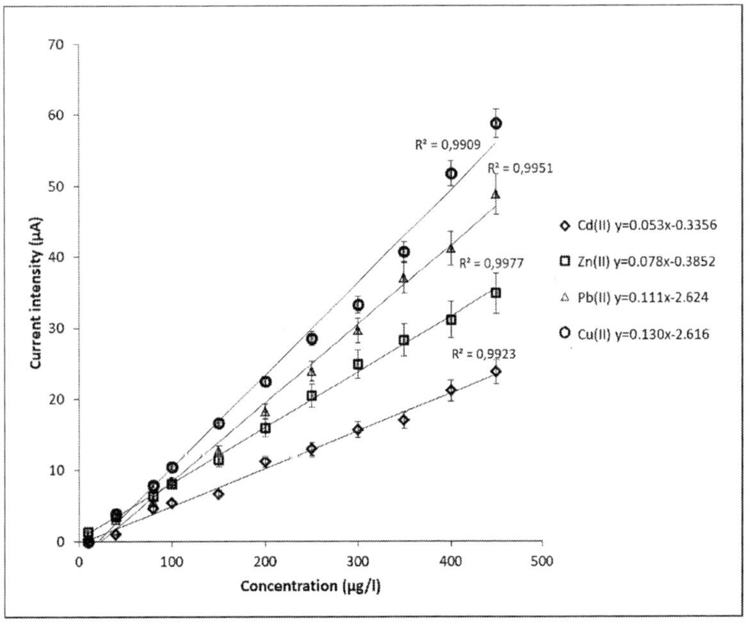

Figure 2 The determined calibration curves for Cd(II), Pb(II), Cu(II) and Zn(II)

4. CONCLUSION

In conclusion, we have developed a new modification for the CPE, which could potentially be used to determine the trace amounts of heavy metals in the aquatic environment. The materials used in this study (AuNPs, MWCNTs, cystein) are nontoxic, environmentally friendly substances, so they are not other potential threat to the environment. The electrochemical potential of this composite material has been demonstrated for the use of SPEs surface modification.

ACKNOWLEDGEMENTS

This work was supported from EFRR " Multidisciplinary research to increase application potential of nanomaterials in agricultural practice " (No. CZ.02.1.01/0.0/0.0/16_025/0007314).

REFERENCES

[1] FURNESS, R.W., *Heavy Metals in the Marine Environment: 0*2017: CRC press.

[2] FASANO, E., ARNESE, A., ESPOSITO, F., ALBANO, L., MASUCCI, A., CAPELLI, C., CIRILLO, T., NARDONE, A, *Evaluation of the impact of anthropogenic activities on arsenic, cadmium, chromium, mercury, lead, and polycyclic aromatic hydrocarbon levels in seafood from the Gulf of Naples, Italy.* Journal of Environmental Science and Health, Part A, 2018: p. 1-7.

[3] LANGSTON, W., *Toxic effects of metals and the incidence of metal pollution in marine ecosystems*, in *Heavy metals in the marine environment*2017, CRC Press. p. 101-120.

[4] AYANGBENRO, A., S., BABALOLA, O., *A new strategy for heavy metal polluted environments: a review of microbial biosorbents.* International journal of environmental research and public health, 2017. **14**(1): p. 94.

[5] DURUIBE, J., O., OGWUEGBU, M., O., C., EGWURUGWU, J., N., *Heavy metal pollution and human biotoxic effects.* International Journal of physical sciences, 2007. **2**(5): p. 112-118.

[6] BANSOD, B., KUMAR, T., THAKUR, R., RANA, S., SINGH, I., *A review on various electrochemical techniques for heavy metal ions detection with different sensing platforms.* Biosensors and Bioelectronics, 2017. **94**: p. 443-455.

[7] ZHU, L., XU, L., HUANG, B., JIA, N., TAN, L., YAO, S., *Simultaneous determination of Cd (II) and Pb (II) using square wave anodic stripping voltammetry at a gold nanoparticle-graphene-cysteine composite modified bismuth film electrode.* Electrochimica Acta, 2014. **115**: p. 471-477.

[8] SONG, Y., BIAN, Ch., TONG, J., LI, Y., XIA, S., *The graphene/l-cysteine/gold-modified electrode for the differential pulse stripping voltammetry detection of trace levels of cadmium.* Micromachines, 2016. **7**(6): p. 103.

[9] MORTON, J., HAVENS, N., MUGWERU, A., WANEKAYA, A., *Detection of trace heavy metal ions using carbon nanotube-modified electrodes.* Electroanalysis: An International Journal Devoted to Fundamental and Practical Aspects of Electroanalysis, 2009. **21**(14): p. 1597-1603.

PLASMON-EXCITON COUPLING IN A HYBRID FILMS OF METALLIC NANOPARTICLES AND J-AGGREGATES OF CYANINE DYES

Rezida NABIULLINA, Anton STAROVOYTOV, Igor GLADSKIKH

ITMO University, St. Petersburg, Russian Federation, anton.starovoytov@gmail.com

Abstract

The effect of the resonance overlapping of pseudoisocyanine J aggregates exciton and plasmon of the inhomogeneous ensemble of metallic nanoparticles was studied for thin films. The absorption of a hybrid structure was not the sum of the absorption of the cyanine dye layer and nanoparticle extinction. Regardless the increase of the molecular absorption due to near field of metal nanoparticles, the spectral dip at the absorption maximum of the J aggregates was observed for hybrid films. The magnitude of the increase in absorption and the shape of the dip were different for silver and gold nanoparticles, and depend on the equivalent thickness of the metal island films. The dip was explained by the strong coupling of the exciton transition in J aggregate with the plasmon resonance of nanoparticles.

Keywords: Plasmon, exciton, film, nanoparticle, J-aggregate

1. INTRODUCTION

Interaction of localized surface plasmons in metallic nanostructures and excitons in quantum-sized objects such as organic chromophores or quantum dots has been a subject of intense research on both the theoretical [1,2] and experimental [3,4] level recently. The strong interaction of J-aggregate excitons and plasmons of metallic nanostructures leads to the formation of a hybrid state, which called plexiton. The study of such plasmon−exciton systems improves the understanding of the interaction of light with matter and the development of novel sensors [5], harvesting solar cells [6] and other photonic devices [7]. There are many studies of plasmon-exciton structures in colloidal solutions [8,9], which are simple and good models for basic research. However, the dry solid samples, such as thin films, with thermal and photostability at room conditions are needed for practical application.

The optical properties of thin films obtained by coating of dye layer on an island metal film formed by the heterogeneous assembly of plasmonic nanoparticles were studied. Pseudoisocyanine (PIC) was used as dye to form self-organized J-aggregate with the maximum of exciton absorption of 583 nm. To adjust the overlap of the exciton and plasmon resonances, the properties of the island film such as deposited metal or the equivalent thickness. In the first case, it is possible to provide a shift of 100 nm by switching from silver to gold. The second is responsible for the characteristic sizes of nanoparticles in the ensemble.

2. SAMPLES AND METHODS

The sapphire substrates, transparent in the visible range, was used to obtain the dry thin-film samples. Metal island films were obtained by thermal deposition of metal (Ag or Au) with deposit rate of 0.01 nm/s on sapphire substrates heated to 200 ° C in a vacuum chamber Kurt Lesker PVD 75. The pressure of residual gases in the chamber at deposition was maintained at a level of ~ $3 \cdot 10^{-7}$ Torr. The equivalent thickness of films (3.5 or 5 nm) was controlled by quartz microbalances [10]. Growth island film on sapphire substrates occurs according to the mechanism of Volmer-Weber [11]. The absorption spectrum of the island film is inhomogeneously broadened due to a size variation of the obtained nanoparticles (**Figure 1**).

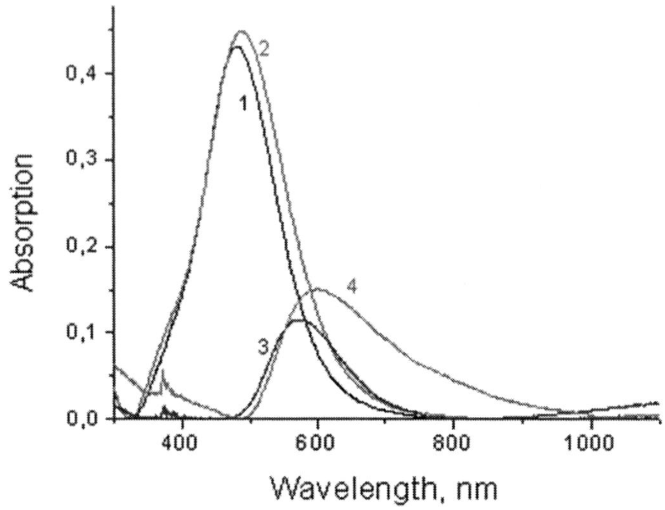

Figure 1 Absorption spectra of: Ag films with thickness of 3.5 nm (1) and 5 nm (2), Au films with thickness of 3.5 nm (3) and 5 nm (4)

1,1'-diethyl-2,2'-cyanine iodide (**Figure 2**), pseudoisocyanine dye (PIC), was used to form J-aggregates by self-organization. The standard method for obtaining J-aggregates is the dissolution of the dye in water with addition of salt to activate aggregation [12]. We have developed an original method of obtaining J-aggregates [13]. The saturated ethanol solution of PIC without the addition of salt was heated up to 60 °C to provide self-aggregation. After cooling for 10 minutes, 4 drops of the dye solution (total volume is 0.006 ml) was spin-coated upon substrate with island film. After rotation for 2 minutes with 1000 rpm the sample was dried under room conditions for 30 minutes. The thickness of the resulting dye layer varied from 30 to 50 nm.

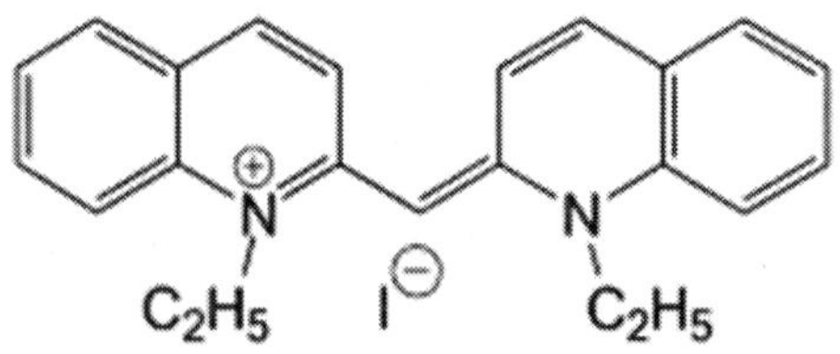

Figure 2 Chemical structure of PIC

The optical density spectra were measured in the range 200-1100 nm with a step of 1 nm by the spectrophotometer SF-56 (OKB "Spectrum" - LOMO).

3. EXPERIMENTAL RESULTS AND DISCUSSION

The absorption spectrum of the PIC ethanol solution, prepared by original method, is predominantly determined by the monomer, and the concentration of aggregates is insignificant [13]. The spectrum of the organic film, obtained from the solution, is broadened due to presence of different molecular forms (**Figure 3**), such as all-trans- (A) and cis-isomers (C), dimer (D), and J-aggregate (J) [14,15]. The most intense absorption band at 583 nm is demonstrated by the J-aggregate.

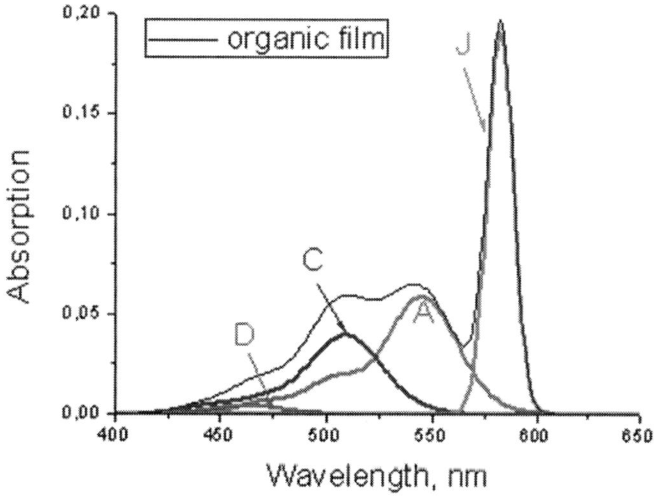

Figure 3 Absorption spectrum of PIC layer on sapphire substrate and its separation into the spectra of the molecular forms: all-trans- (A) and cis-isomers (C), dimer (D) and J-aggregates (J)

The absorption of a hybrid film was not the sum of the absorption of the cyanine dye layer and nanoparticle extinction. The increase of the absorption of all molecular forms is observed for hybrid film with silver nanoparticles in contrast to gold nanoparticles (**Figure 4**), assuming that the surface concentration of molecules is the same. Nevertheless, the spectral dip at a wavelength of 583 nm is noticed for the hybrid films with nanoparticles of both metals. The increase of the absorption of all molecular forms is observed for hybrid film with silver nanoparticles in contrast to gold nanoparticles (**Figure 4**), assuming that the surface concentration of molecules is the same. Nevertheless, the spectral dip at a wavelength of 583 nm is noticed for the hybrid films with nanoparticles of both metals. This is the wavelength of the exciton absorption maximum of the J-aggregate [12], which can overlap with the plasmon frequency of some nanoparticles of inhomogeneous ensemble of island film. This can lead to the strong interaction of the plasmon and the exciton, as well as, the formation of energy level splitting, which shows itself as a spectral dip [1,2,8].

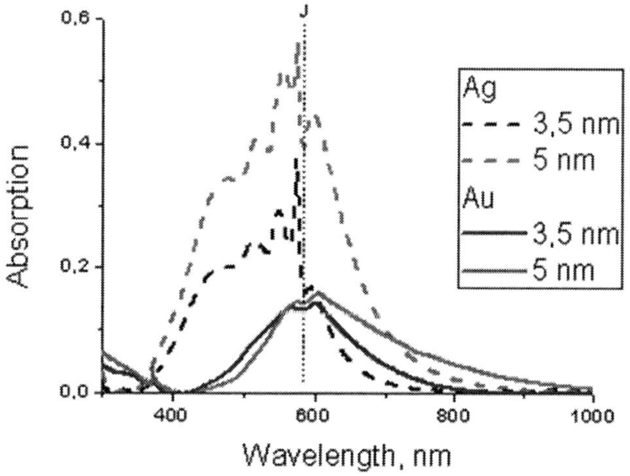

Figure 4 Absorption spectra of the organometallic films depending on the thickness of the island film

The ratio of the peaks near the spectral dip in the absorption maximum of J-aggregates changes depending on equivalent thickness (3.5 or 5 nm) of the island film (**Table 1**). This reflects the change in the formation and concentration of nanoparticles with plasmon resonance at the wavelength of the J-aggregates.

Table 1 Dependence of the ratio of the absorption intensities of the right and left peaks on the equivalent thickness of the island film for different metals

Equivalent thickness, nm	Ratio of the peaks, I_{right}/I_{left}	
	Ag film	Au film
3.5	0.45	1.05
5	0.80	1.09

4. CONCLUSION

The optical properties of thin films performed the resonance overlapping of pseudoisocyanine J aggregates exciton and plasmon of the inhomogeneous ensemble of metallic nanoparticles was studied. The absorption of a hybrid film was not the sum of the absorption of the cyanine dye layer and nanoparticle extinction. The increase of the absorption of all molecular forms is observed for hybrid film with silver nanoparticles in contrast to gold nanoparticles. The spectral dip in the absorption spectrum of the hybrid film at the maximum of the J-aggregate band at 583 nm was observed. The dip can be explained by the strong coupling of the exciton transition in J aggregate with the plasmon resonance of nanoparticles.

REFERENCES

[1] FOFANG, N.T., GRADY, N.K., FAN, Z., GOVOROV, A.O. and HALAS, N.J. Plexciton dynamics: exciton-plasmon coupling in a J-aggregate-Au nanoshell complex provides a mechanism for nonlinearity. *Nano Letters*. 2011. vol. 11, no. 4, pp. 1556-1560.

[2] CAOA, E., LINA, W., SUN, M., LIANG, W. and SONG, Y. Exciton-plasmon coupling interactions: from principle to applications. *Nanophotonics*. 2018. vol. 7, no. 1, pp. 145-167.

[3] BAUDRION, A.L., PERRON, A., VELTRI, A., BOUHELIER, A., ADAM, P.M. and BACHELOT, R. Reversible strong coupling in silver nanoparticle arrays using photochromic molecules. *Nano Letters*. 2013. vol. 13, no. 1, pp. 282-286.

[4] SCHLATHER A, E., LARGE, N., URBAN, A.S., NORDLANDER, P. and HALAS, N.J. Near-field mediated plexcitonic coupling and giant Rabi splitting in individual metallic dimmers. *Nano Letters*. 2013. vol. 13, no. 7, pp. 3281-3286.

[5] TANG, Y., YU, X., PAN, H., Chen, J., AUDIT, B., ARGOUL, F., ZHANG S. and XU J. Numerical study of novel ratiometric sensors based on plasmon-exciton coupling. *Applied Spectroscopy*. 2017. vol. 71, no. 10, pp. 2377-2384.

[6] NAN, F., DING, S., MA, L., CHENG, Z., ZHONG Y., ZHANG, Y., QIU, Y., LI, X., ZHOU L. and WANG Q. Plasmon resonance energy transfer and plexcitonic solar cell. *Nanoscale*. 2016. vol. 8, no. 32, pp. 15071-15078.

[7] FEDELE, S., HAKAMI, M., MURPHY, A., POLLARD, R. ans RICE, J. Strong coupling in molecular exciton plasmon Au nanorod array systems. *Applied Physics Letters*. 2016. vol. 108. pp. 053102:1-4.

[8] DELACY, B.G., MILLER, O.D., HSU, C.W., ZANDER, Z., LACEY, S., YAGLOSKI, R., FOUNTAIN, A.W., VALDES, E., ANQUILLARE, E., SOLJAČIĆ, M., JOHNSON, S.G. and JOANNOPOULOS, J.D. Coherent plasmon-exciton coupling in silver platelet-J-aggregate nanocomposites. *Nano Letters*. 2015, vol. 15, pp. 2588-2593.

[9] SIMON, T., MELNIKAU, D., SÁNCHEZ-IGLESIAS, A., GRZELCZAK, M., LIZ-MARZAN, L.M., RAKOVICH, Y.P., FELDMANN, J. and URBAN, A.S. Exploring the optical nonlinearities of plasmon-exciton hybrid resonances in coupled colloidal nanostructures. *The Journal of Physical Chemistry C*. 2016. vol. 120, no. 22, pp. 12226-12233.

[10] LEONOV, N.B., GLADSKIKH, I.A., POLISHCHUK, V.A. and VARTANYAN, T.A. Evolution of the optical properties and morphology of thin metal films during growth and annealing. *Optics and spectroscopy.* 2015, vol. 119, no. 3, pp. 450-455.

[11] OURA, K., LIFSHITS, V.G., SARANIN, A.A., ZOTOV, A.V. AND KATAYAMA M. *Surface Science: An Introduction.* Berlin: Springer, 2003, p.490. ISBN 3-540-00545-5.

[12] BRICKS, J.L., SLOMINSKII, Y.L., PANAS, I.D. and DEMCHENKO, A.P. Fluorescent J-aggregates of cyanine dyes: basic research and applications review. *Methods and Applications in Fluorescence.* 2017, vol. 6, no. 1, pp. 54:1-43.

[13] STAROVOYTOV, A.A., NABIULLINA, R.D., GLADSKIH, I.A., POLISCHUK, V.A., PARFENOV, P.S. and KAMALIEVA, A.N. Plasmon-exciton interaction in the thin film of inhomogeneous ensemble of silver nanoparticles and cyanine J aggregates. *Proceedings of SPIE.* 2018, vol. 10672, pp. 106723O.

[14] KALITEEVSKAYA, E. N., KRUTYAKOVA, V. P., RAZUMOVA, T. K. and STAROVOYTOV, A. A. The Influence of the Molecular Structure of Cyanine Dye on the Component Composition of Molecular Layers. *Optics and Spectroscopy.* 2016, vol. 120, no. 3, pp. 482-491.

[15] KALITEEVSKAYA, E. N., KRUTYAKOVA, V. P., RAZUMOVA, T. K. and STAROVOYTOV, A. A. Optical properties and component composition of layers of cyanine dyes on dielectric supports: influence of asymmetry of the molecular electron density distribution. Optical and Quantum Electronics. 2017, vol. 49, no. 1, pp. 32.

THE PHOTOLUMINESCENCE AND OPTICAL ABSORPTANCE OF PLASMA HYDROGENATED NANOCRYSTALLINE ZNO THIN FILMS

[1]Zdenek REMES, [1,2]Yu Ying CHANG, [1]Jiri STUCHLIK, [3]Julia MICOVA

[1]Czech Academy of Sciences, Institute of Physics, Prague, Czech Republic, EU,
remes@fzu.cz, chang@fzu.cz, stuj@fzu.cz
[2]Czech Technical University, Faculty of Biomedical Engineering, Prague, Czech Republic, EU,
changyuy@fbmi.cvut.cz
[3]Slovak Academy of Sciences, Institute of Chemistry, Bratislava, Slovak Republic, EU,
chemjumi@savba.sk

Abstract

We have developed the technology of the deposition of the nominally undoped ZnO nanocrystalline thin films by DC reactive magnetron sputtering of Zn target in the gas mixture of argon and oxygen plasma. We have optimized the photoluminescence spectroscopy for measuring optically scattering thin layers with the high sensitivity, precise sample positioning and very low influence of the scattered excitation light. Here we present the latest results on the enhancement of the photoluminescence of the nanocrystalline ZnO thin films after plasma hydrogenation. The photoluminescence in near UV region has been enhanced whereas the deep defect related photoluminescence has been significantly decreased. We found good room temperature stability of the plasma hydrogenated ZnO nanocrystals in air, but fast degradation at elevated temperature.

Keywords: Nanomaterials, ZnO, photoluminescence, magnetron sputtering

1. INTRODUCTION

The high electron mobility, high thermal conductivity and tunable electrical conductivity make ZnO suitable for a wide range of devices, including transparent thin-film transistors, photodetectors, light-emitting diodes and laser diodes that operate in the blue and ultraviolet region of the spectrum. ZnO is a semi-conductor with optical transparency in the visible spectral range, large exciton binding energy and strong room temperature photoluminescence [1]. The native point defects [2] and oxygen vacancies [3] have been studied including their sensitivity to hydrogen. The experimental results indicate that the OH and H bonds play the dominant role in facilitating surface recombination [4]. Moreover, hydrogen works as a shallow donor as shown by photoluminescence and photoconductivity measurements [5]. The hydrogen plasma treatment passivates deep defects [6] and it influences the relative luminescence intensities of bound excitons, in particular, the PL intensity at 3.363 eV (369 nm) [7]. The annealing in H_2 atmosphere at temperature above 800°C enhanced significantly excitonic emission in ZnO nanoparticles [8].

The RF sputtering technique has drawn considerable attention since the films properties can be controlled by changing the sputtering conditions. In our previous paper we have shown the effect of hydrogen plasma treatment on the nominally undoped ZnO thin film deposited by DC reactive magnetron sputtering of Zn target in the gas mixture of argon and oxygen plasma [9]. After oxidation due to thermal annealing in air, the optical absorption was significantly reduced in the infrared region and the electrical resistivity increased. After hydrogen plasma treatment, the increase of the infrared optical absorption, related to free carrier concentration, was detected below the optical absorption edge [10]. The increase of the optical absorption goes with the increase of the electrical conductivity related to the increase of the free carrier concentration. Moreover, we have developed the software based on the contactless reflectance interferometry as the fast method of the thin film thickness evaluation from the optical interference fringes [11].

In this paper we show how the photoluminescence of the plasma hydrogenated nanocrystalline ZnO layer correlates with optical absorption edge. Moreover, we study the stability of the plasma hydrogenated nanocrystalline ZnO layer upon exposition to air at room and elevated temperature.

2. EXPERIMENTAL

2.1. Reactive magnetron sputtering

The 450 nm thick ZnO layer was deposited on fused silica glass substrate by reactive magnetron sputtering in the stainless-steel vacuum chamber using a Zn target and O_2 plasma. The Zn target was sputtered in a DC capacitive couple glow discharge plasma (grounded substrate holder at temperature 400°C, constant potential of +400 V on the target, DC current of 0.13 A) of a reactive mixture of argon (purity 99.99 %, flow rate 2.0 sccm) and oxygen (purity 99.95 %, flow rate 0.5 sccm) under pressure of 1 Pa [12].

2.2. Plasma hydrogenation

The plasma hydrogenation was done in the stainless-steel vacuum chamber at room temperature using 9 W rf discharge at 13.56 MHz, hydrogen flow 50 sccm and pressure 70 Pa. Prior the plasma hydrogenation, the vacuum chamber had been evacuated down to the pressure vacuum 10^{-7} Pa.

2.3. Photothermal deflection spectroscopy

The photothermal deflection spectroscopy (PDS) measures directly the optical absorption of thin films with sensitivity of four orders of magnitude [13]. The measured sample is immersed in transparent liquid with the probe laser beam passing parallel to the sample surface. The sample is illuminated by the monochromatic light and the heat absorbed in the sample deflects the probe laser beam with the amplitude of the deflection being proportional to the optical absorption. Thus, the optical absorptance is detected by the position detector normalized on the deflection measured using black sample [14].

2.4. Photoluminescence spectroscopy

The steady state photoluminescence (PL) spectra of the highly scattering ZnO thin films were measured in the 360–660 nm spectral range using the f/4 double gratings monochromator SPEX 1672 equipped with two 1200 grooves/mm gratings blazed at 500 nm with less than 10-9 scattered light. The photoexcitation was provided by the 1 mW LEDs with maximum intensity centered at the wavelength 340 nm and optically filtered by narrow band pass optical filter featuring 90 % transmission and OD 6 blocking outside of the 330–350 nm passband. The sample holder was positioned by two perpendicularly oriented translation stages manually driven by adjuster screws for precision motion, see **Figure 1**. The scattered and emitted light was collected and focused onto the monochromator input slit by two 90° off-axis mirrors coated by UV enhanced aluminium. The PL signal was

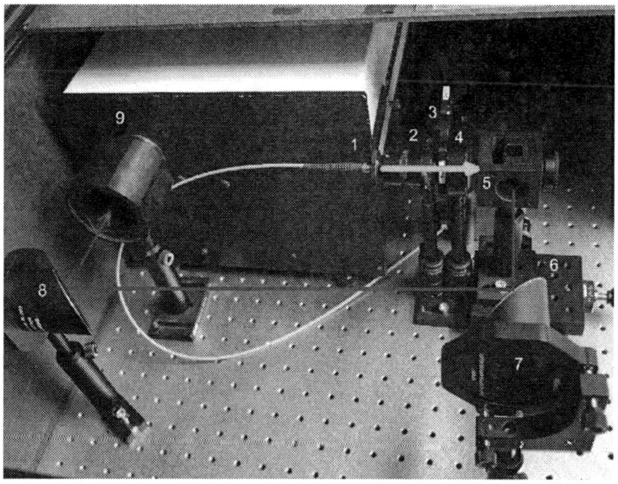

Figure 1 The detail of the photoluminescence setup: (1) optical fiber, (2) collimator, (3) bandpass filter, (4) focusing lens, (5) sample holder, (6) XY stage, (7-8) 90° off axis mirrors, (9) monochromator entrance slit

detected in dc mode by the cooled multi-dynode multi-alkali red sensitive photomultiplier (Photonis XP2203B) and the electrometer Keithley 6517A providing a high voltage for photomultiplier. The dark anode current was 2 pA, maximum photocurrent 1 μA and the noise 0.5 pA

3. RESULTS AND DISCUSSION

Figure 2 compares the photoluminescence (PL) emission and optical absorptance (A) spectra of the as grown and plasma hydrogenated ZnO thin film grown on fused silica glass substrate by reactive magnetron sputtering. Both the PL as well as A spectra were measured at room temperature. The A spectra show the optical absorption edge below 400 nm. The band gap of a single crystal ZnO 3.37 eV at room temperature corresponds to the wavelength 368 nm and the free-exciton binding energy in ZnO is 60 meV [15]. This large exciton binding energy indicates that efficient excitonic emission in ZnO can persist at room temperature. Indeed, **Figure 2** shows the exciton-related peak at 372 nm (3.33 eV) in the A spectrum of the hydrogenated nanocrystalline ZnO [16]. It should be noted that various authors report lower band gap values for the polycrystalline ZnO depending on the deposition conditions [17]. **Figure 2** shows that the PL signal is very low in the as grown sample. The PL signal strongly increases after plasma hydrogenation. The emission peak at 380 nm has been contributed to the indirect annihilation of intrinsic excitons with the simultaneous emission of one LO phonon [18].

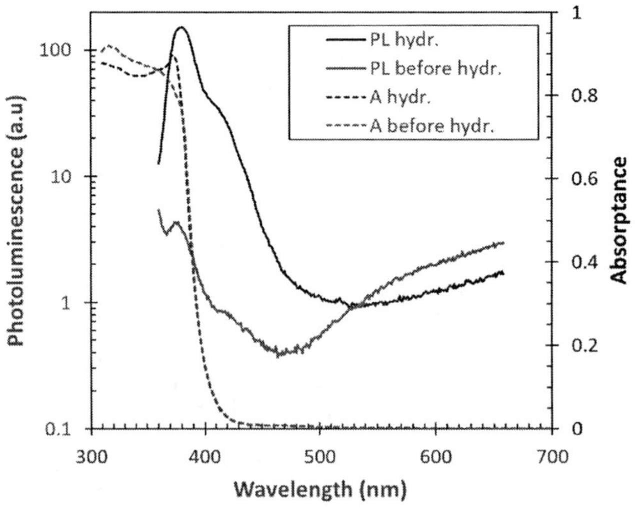

Figure 2 The comparison of the photoluminescence (PL) emission and optical absorptance (A) spectra of the as grown (PL before hydr.) and plasma hydrogenated 450 nm thick nanocrystalline ZnO layer grown on fused silica glass substrate by reactive magnetron sputtering

The photoluminescence of the hydrogenated nanocrystalline ZnO layer slightly decreased after several weeks in air, see **Figure 3**. This indicate good room temperature stability of the plasma hydrogenated ZnO nanocrystals in air. However, the degradation of the UV PL intensity is much faster at elevated temperatures. **Figure 3** shows that the significant deterioration occurs at 150°C in air within 1 h. Since our sample has been deposited in O_2 rich environment at 400°C, the ZnO nanocrystals should be stable in air at 150 °C. In general, the annealing in air atmosphere improves the crystal structure and optical properties of ZnO nanocrystals [19]. Thus, the observed degradation of the exciton-related PL intensity indicates that the involved processes are related to the surface oxidation. This hypothesis needs to be proved by further XPS and FTIR measurements that we plan in near future. However, our previous results on ZnO nanorods have shown that the surface composition drastically changes upon the exposure to O-plasma. The plasma treatments increased the presence of non-lattice oxygen in the form of Zn-O-H and Zn-O–Zn [20].

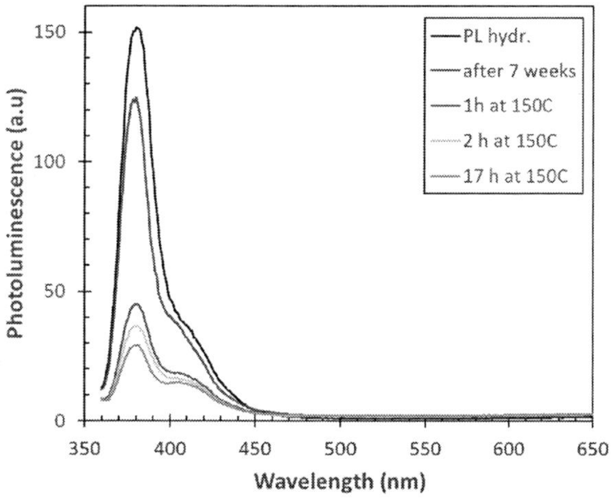

Figure 3 The degradation of the photoluminescence intensity of the plasma hydrogenated nanocrystalline ZnO layer after 7 weeks in air at room temperature followed by annealing in air for 1, 2 and 17 hours at 150°C

4. CONCLUSION

The nanocrystalline ZnO thin films were deposited by DC reactive magnetron sputtering of Zn target in the gas mixture of argon and oxygen plasma followed by hydrogen plasma treatment. The optical properties of the optically scattering were measured using the optimized spectrophotometer based on the double grating monochromator equipped with the fiber coupled LEDs as the excitation light sources. We have observed that the exciton related emission band centered at the wavelength 380 nm is influenced significantly by the plasma treatment and it correlates with the peak in the optical absorption spectra. In general, hydrogenation enhances the UV luminescence and passivates the visible emission. Our results indicate good room temperature stability of the exciton-related photoluminescence of the plasma hydrogenated ZnO nanocrystals in air, but the fast degradation at elevated temperatures.

ACKNOWLEDGEMENTS

This work was supported by the CSF project 16-10429J, the MEYS project LTC17029 INTER-COST Action MP1406 and by Operational Programme Research, Development and Education financed by European Structural and Investment Funds and the Czech Ministry of Education, Youth and Sports (Project No. SOLID21 - CZ.02.1.01/0.0/0.0/16_019/0000760.

REFERENCES

[1] JANOTTI, Anderson and VAN DE WALLE, Chris G. Fundamentals of zinc oxide as a semiconductor. *Reports on Progress in Physics.* 1 December 2009. Vol. 72, no. 12, p. 126501. DOI 10.1088/0034-4885/72/12/126501.

[2] JANOTTI, Anderson and VAN DE WALLE, Chris G. Native point defects in ZnO. *Physical Review B.* 4 October 2007. Vol. 76, no. 16. [Accessed 8 May 2017]. DOI 10.1103/PhysRevB.76.165202.

[3] JANOTTI, Anderson and VAN DE WALLE, Chris G. Oxygen vacancies in ZnO. *Applied Physics Letters.* 19 September 2005. Vol. 87, no. 12, p. 122102. DOI 10.1063/1.2053360.

[4] YANG, L.L., ZHAO, Q.X., WILLANDER, M., LIU, X.J., FAHLMAN, M. and YANG, J.H. Origin of the surface recombination centers in ZnO nanorods arrays by X-ray photoelectron spectroscopy. *Applied Surface Science.* March 2010. Vol. 256, no. 11, p. 3592-3597. DOI 10.1016/j.apsusc.2009.12.160.

[5] LI, Q. H., GAO, T., WANG, Y. G. and WANG, T. H. Adsorption and desorption of oxygen probed from ZnO nanowire films by photocurrent measurements. *Applied Physics Letters*. 21 March 2005. Vol. 86, no. 12, p. 123117. DOI 10.1063/1.1883711.

[6] SEKIGUCHI, Takashi, OHASHI, Naoki and TERADA, Yoshihiro. Effect of Hydrogenation on ZnO Luminescence. *Japanese Journal of Applied Physics, Part 2: Letters*. 1997. Vol. 36, p. L289-L291.

[7] ZHANG, Y., MA, Y., ZHANG, B. and DU, G. Effect of hydrogenation on luminescence properties of ZnO crystals. *Pan Tao Ti Hsueh Pao/Chinese Journal of Semiconductors*. 2008. Vol. 29, no. 3, p. 526-529. Scopus

[8] PROCHÁZKOVÁ, Lenka, GBUR, Tomáš, ČUBA, Václav, JARÝ, Vítězslav and NIKL, Martin. Fabrication of highly efficient ZnO nanoscintillators. *Optical Materials*. September 2015. Vol. 47, p. 67-71. DOI 10.1016/j.optmat.2015.07.001.

[9] CHANG, Yu-Ying, NEYKOVA, Neda, STUCHLIK, Jiri, PURKRT, Adam and REMES, Zdenek. Hydrogen plasma treatment of ZnO thin films. In: *NANOCON 2016 - 8th International Conference on Nanomaterials - Research and Application, Conference Proceedings*. Hotel Voronez I, Brno, Czech Republic: TANGER Ltd., October 2016. p. 161-165

[10] REMES, Z., STUCHLIK, J., PURKRT, A., CHANG, Y.-Y., JIRASEK, V., STENCLOVA, P., PRAJZLER, V. and NEKVINDOVA, P. The intrinsic submicron ZnO thin films prepared by reactive magnetron sputtering. In: *NANOCON 2016 - Conference Proceedings, 8th International Conference on Nanomaterials - Research and Application*. Hotel Voronez I, Brno, Czech Republic: TANGER Ltd., October 2016. p. 36-41

[11] CHANG, Yu-Ying, STUCHLÍK, Jiří, NEYKOVA, Neda, SOUČEK, Josef and REMEŠ, Zdeněk. Optical properties of the plasma hydrogenated ZnO thin films. *Journal of Electrical Engineering* [online]. 1 December 2017. Vol. 68, no. 7. [Accessed 23 March 2018]. DOI 10.1515/jee-2017-0060.

[12] MUSIL, J., BAROCH, P., VLČEK, J., NAM, K.H. and HAN, J.G. Reactive magnetron sputtering of thin films: present status and trends. *Thin Solid Films*. March 2005. Vol. 475, no. 1-2, p. 208-218. DOI 10.1016/j.tsf.2004.07.041.

[13] JACKSON, W.B., AMER, N.M., BOCCARA, A.C. and FOURNIER, D. Photothermal deflection spectroscopy and detection. *Applied Optics*. 1981. Vol. 20, no. 8, p. 1333-1344. DOI 10.1364/AO.20.001333.

[14] REMES, Z., PHAM, Tuan T., VARGA, M., KROMKA, A. and MAO, H. B. Carbon Coatings Prepared by Magnetron Sputtering and Microwave Plasma Enhanced Chemical Vapor Deposition Measured by the Photothermal Deflection Spectroscopy. *Advanced Science, Engineering and Medicine*. 1 April 2015. Vol. 7, no. 4, p. 321-324. DOI 10.1166/asem.2015.1697.

[15] MANG, A., REIMANN, K. and RÜBENACKE, St. Band gaps, crystal-field splitting, spin-orbit coupling, and exciton binding energies in ZnO under hydrostatic pressure. *Solid State Communications*. April 1995. Vol. 94, no. 4, p. 251-254. DOI 10.1016/0038-1098(95)00054-2.

[16] PANKOVE, Jacques I. *Optical processes in semiconductors*. Unabridged republication, with slight corr. Mineola [NY]: Dover, 1975. ISBN 978-0-486-60275-2.

[17] AL-HARDAN, N.H., ABDULLAH, M.J., ABDUL AZIZ, A., AHMAD, H. and RASHID, M. The effect of oxygen ratio on the crystallography and optical emission properties of reactive RF sputtered ZnO films. *Physica B: Condensed Matter*. February 2010. Vol. 405, no. 4, p. 1081-1085. DOI 10.1016/j.physb.2009.11.006.

[18] WEIHER, R. L. and TAIT, W. C. Contribution of Excitons to the Edge Luminescence in Zinc Oxide. *Physical Review*. 15 February 1968. Vol. 166, no. 3, p. 791-796. DOI 10.1103/PhysRev.166.791.

[19] YANG, L. L., ZHAO, Q. X., WILLANDER, M., YANG, J. H. and IVANOV, I. Annealing effects on optical properties of low temperature grown ZnO nanorod arrays. *Journal of Applied Physics*. March 2009. Vol. 105, no. 5, p. 053503. DOI 10.1063/1.3073993.

[20] NEYKOVA, Neda, STUCHLIK, Jiri, HRUSKA, Karel, PORUBA, Ales, REMES, Zdenek and POP-GEORGIEVSKI, Ognen. Study of the surface properties of ZnO nanocolumns used for thin-film solar cells. *Beilstein Journal of Nanotechnology*. 16 February 2017. Vol. 8, p. 446-451. DOI 10.3762/bjnano.8.48.

RELATION BETWEEN OPTICAL AND MICROSCOPIC PROPERTIES OF HYDROGENATED SILICON THIN FILMS WITH INTEGRATED GERMANIUM AND TIN NANOPARTICLES

[1]Jiri STUCHLIK, [1]The Ha STUCHLIKOVA, [1]Jan CERMAK, [2]Jaroslav KUPCIK, [2]Radek FAJGAR, [1]Zdenek REMES

[1]Czech Academy of Sciences, Institute of Physics, Prague, Czech Republic, EU,
stuj@fzu.cz, hotheha@fzu.cz, cermakj@fzu.cz, remes@fzu.cz
[2]Institute of Chemical Process Fundamentals of the CAS, Prague, Czech Republic, EU,
kupcik@icpf.cas.cz, fajgar@icpf.cas.cz

Abstract

The hydrogenated amorphous silicon layers (a-Si:H) were deposited by PECVD method on quartz substrates. During Interruption of PECVD process the vacuum chamber was pumped up to 10^{-5} Pa and 1 nm thin films of Germanium or Tin were evaporated on the surface. The materials form isolated nanoparticles (NPs) on the a-Si:H surface. Then the deposited NPs were covered and stabilized by a-Si:H layer by PECVD. Those two deposition processes were alternated 5 times. The a-Si:H thin films with integrated Ge or Sn NPs were characterized optically by PDS and CPM methods, and microscopically by SEM and AFM microscopies. Optical and microscopic properties of the structures are correlated and discussed considering their application in photovoltaics.

Keywords: Thin films, a-Si:H, nanoparticles, Germanium, Tin

1. INTRODUCTION

Single solar cells based on hydrogenated silicon thin films show the highest efficiency of photovoltaic conversion for the mixed-phase structure (partly amorphous and partly microcrystalline) [1]. The amorphous phase is characterized by higher optical absorption coefficient in near infrared region, while the crystalline phase shows better electrical properties and high stability. However, its band gap is indirect, the absorption coefficient is lower and the final photovoltaic structure has to be thicker [2]. A way to increase the optical absorption coefficient is to integrate another kind of nanoparticles (NPs) with direct or near direct band gap to silicon structures. The Plasma Enhanced Chemical Vapor Deposition (PECVD) in combination with Vacuum Evaporation (VE) and Plasma Treatment (PT) was already demonstrated as convenient solution for the deposition of hydrogenated amorphous silicon (a-Si:H) thin films with integrated silicide NPs [3]. We have shown that the NPs embedded in a-Si:H thin film significantly enhance the near infrared optical absorption. The a-Si:H based diodes with embedded magnesium silicide NPs showed under illumination similar open circuit voltage and fill factor as diodes without NPs, but the defects induced by NPs deteriorated the short circuit current. In this contribution we present the combined technique as a suitable process for in situ depositions of diode structures based on a-Si:H with embedded Ge or Sn NPs. The possible application of such NPs to increase absorption in very thin a-Si:H films is discussed with a view of improving the efficiency of the resulting photovoltaic structure.

2. EXPERIMENTAL

A special setup with a home made vacuum chamber (**Figure 1**) allows to alternate both deposition processes, PECVD and VE from a Tungsten boat. For the deposition of a-Si:H thin films we used these parameters: flow rates of gasses: hydrogen 50 sccm, silane 2.36 sccm, (4.5 % of silane in mixture), pressure of gasses during the deposition 30 Pa, RF power 18 W for glow discharge excitation at the industrial frequency 13.56 MHz (size of electrode: 62 x 62 mm^2, distance of electrodes: 32 mm) and deposition temperature 220 °C. The thickness of evaporated Ge or Sn was controlled by the Standard Quartz Crystal Sensor, typical thickness was 1 nm. Thin film structures were deposited on a fused silica glass substrates. Always two substrates were used for deposition of a-Si:H, one for the thin film with Sn or Ge NPs and one for the thin film without NPs which was used as reference. The deposition processes were alternated 5 times, the last deposition of NPs was not covered by a-Si:H. Final deposition process was evaporation of Titanium stripes with 2 mm distance as electrodes. **Figure 1** (on the right) shows a schematic of the final structure. This method was used to prepare samples 3K77, 3K78, 3K79, 3K80 with Sn NPs with different regime of hydrogen PT and sample 3K93 with Ge NPs. The deposited layers were characterized by scanning electron microscope (SEM) Maia3 FE-SEM (Tescan), high resolution transmission electron microscope (HRTEM) JEOL JEM 3010 (Japan) equipped with a thermo-emission LaB6 cathode and operated at 300kV. Surface topography was scanned also by an atomic force microscope (AFM) NTEGRA Prima (NT-MDT) by a Multi75E-G probe (BudgetSensors) in semicontact mode. The optical absorption coefficients of a-Si:H with embedded Ge NPs were measured by CPM - Constant Photocurrent Method [4] and with embedded Sn NPs measured by PDS - Photothermal Deflection Spectroscopy [5].

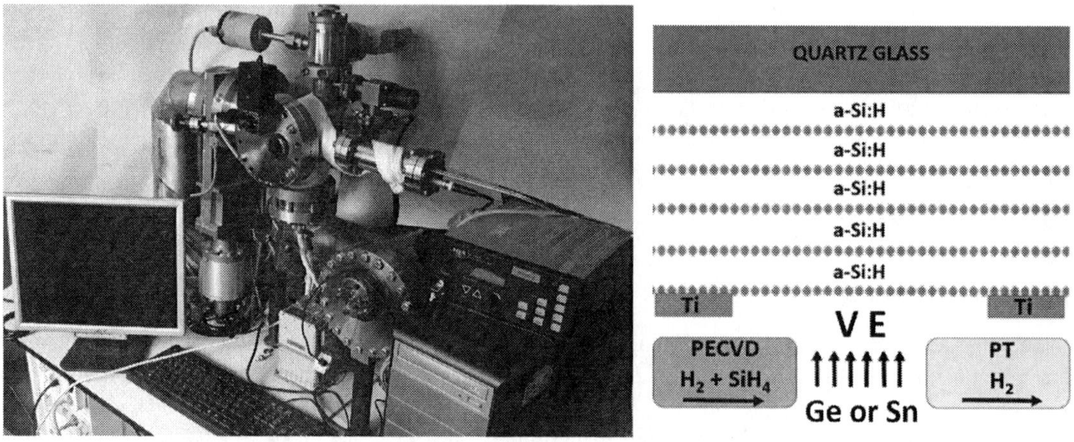

Figure 1 Deposition setup capable of alternation of two different deposition processes - PECVD and VE (on the left) and a schematic picture of requested structures of a-Si:H thin films with embedded Ge or Sn NPs (on the right).

3. RESULTS AND DISCUSSION

While a control structure of a-Si:H without NPs (**Figure 2a**) shows only fluffy 40 nm structures belonging to hydrogenated silicon, the picture of a-Si:H with NPs (**Figue 2b**) proves presence of Sn NPs with maximal size about 20 nm. HRTEM analysis of prepared films reveals that Sn is present in form of round nanoparticles with sizes in the range of approx. 4-10 nm. In the bright field (BF) images (**Figure 2c**) the Sn particles are displayed as darker spots on brighter a-Si:H film. Diffraction patterns (**Figure 2c**, in corner) obtained from the film show only week diffuse circles typical for amorphous material. The dark field (DF) image (**Figure 2d**) shows crystal origins with size 1-2 nm embedded in mostly amorphous material. Although for the recrystallization of the amorphous structures Si and evaporated metallic Sn we used high RF power for hydrogen PT the crystal NPs (like the one in **Figure 2b**) were observed rarely. That is why we could not prove any creation of the diamond respectively α-crystalline structure of the Sn NPs.

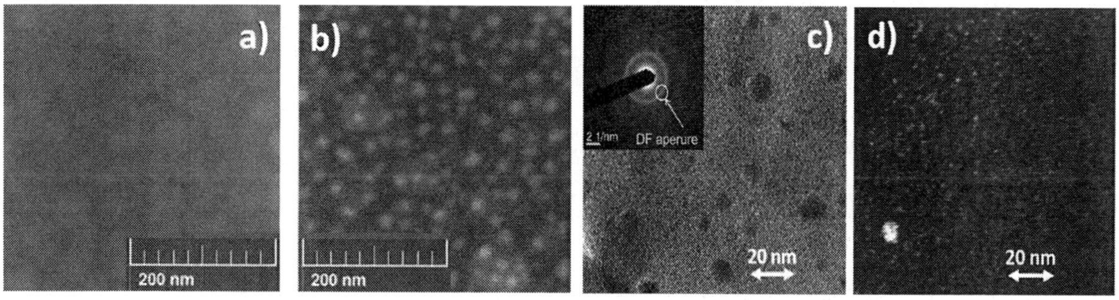

Figure 2 Typical SEM images of a) surface of a control a-Si:H film, b) a-Si:H thin film covered by Sn NPs, c) typical HRTEM BF image of a-Si:H with Sn particles on its surface, with corresponding diffraction pattern (inserted in corner), d) HRTEM DF image of the film. DF aperture used for d) is marked in the diffraction pattern in c).

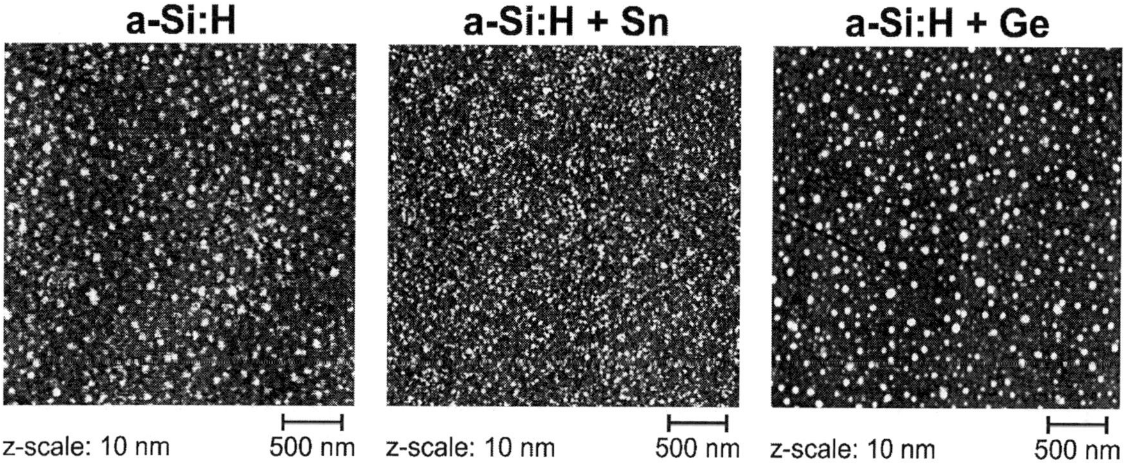

Figure 3 AFM surface topography of a) control a-Si:H,
b) a-Si:H with Sn nanoparticles, and c) a-Si:H with Ge nanoparticles

AFM surface topography of the control a-Si:H structure (**Figure 3**) corresponds well to the SEM image. The structure with Sn NPs is visually finer, yet the RMS roughness is slightly higher (3.4 nm) than in the case of the control structure (2.8 nm). Features on the surface of the structure with Ge NPs look similarly to the control sample and the RMS roughness is also the same.

Figure 4 shows the spectra of the optical absorption coefficients of a-Si:H with embedded Ge NPs (measured by CPM) and with embedded Sn NPs (measured by PDS). The absorption coefficient of a reference a-Si:H sample without NPs and of a crystalline silicon (c-Si) is displayed for comparison [6]. In the near infrared region the measurements show the optical absorption edge of a-Si:H that is for all thin films with encapsulated NPs shifted to lower energies together with an enhanced absorption. For proper comparison both samples (with and without NPs) deposited at the same time were treated by high energy RF Hydrogen plasma immediately after evaporation of NPs. For whole set of samples this process was repeated 5 times during deposition of non-doped a-Si:H thin film. The basic requirement for successful usage of embedded NPs is an increased absorption. However, NPs must not increase the number of defect levels in the amorphous structure. According to the measured results this requirement was successfully fulfilled for the sample 3K93 - for embedded Ge NPs.

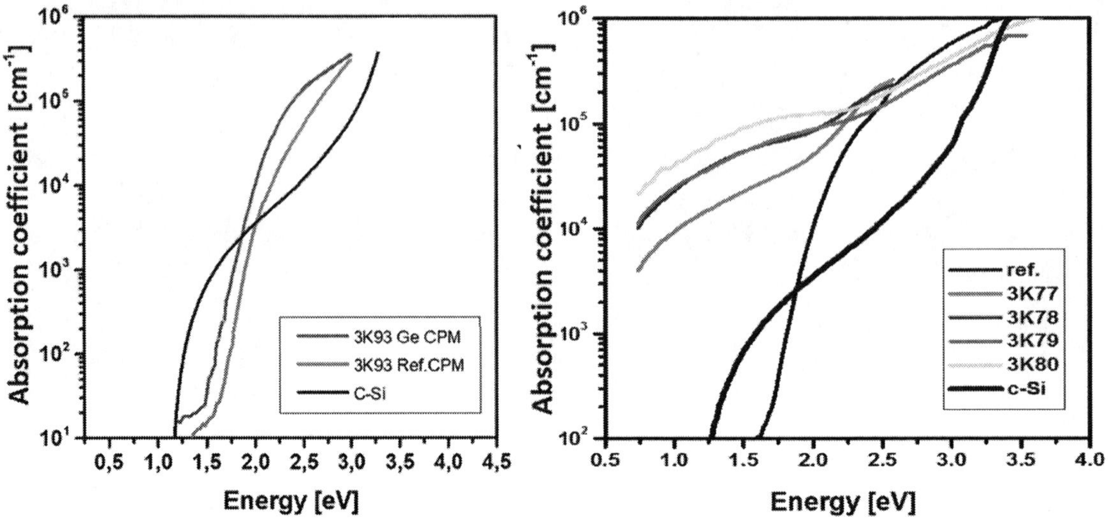

Figure 4 Spectra of the optical absorption coefficient a-Si:H thin films with and without embedded NPs. a-Si:H with and without Ge NPs (see the left picture) were measured by CPM. The right picture demonstrates absorption coefficient for thin films with and without embedded Sn NPs measured by PDS. Differences in dependences are given by different plasma treatment of deposited Sn NPs - see text. The absorption coefficient of c-Si in both pictures has been added for comparison.

For the set of samples with embedded Sn NPs the PDS measurements are significantly more sensitive than optical transmittance spectroscopy. The a-Si:H generally shows enhanced optical absorption in the visible light spectral region 1.9 - 3.4 eV compared to c-Si [6]. The exponential decrease of the optical absorption coefficient in a-Si:H at photon energies below 1.9 eV is related to tails of band-to-band electronic transition extended to band gap (Urbach edge). In the near infrared region below 1.5 eV the optical absorption coefficient is very low. Since there is the negligible optical absorption of the near infrared light, the quantum efficiency of the light conversion is also very low. Therefore, we have introduced the embedded Sn nanoparticles to enhance the optical absorption. **Figure 4** shows that the Sn NPs embedded in a-Si:H increase the optical absorption coefficient above 10^4 times. This means that the light is strongly absorbed already in thin film with thickness of 35 nm. As both SEM and AFM characterization revealed changes in surface features of sizes below 20 nm (far lower than visible or even near infrared light wavelengths), it is highly unlikely that such huge change in optical absorption originates in different surface scattering and is a result of the presence of NPs. The differences in the absorption in this spectral region between different structures with Sn NPs reflect a different hydrogen plasma treatment. The a-Si:H surfaces after interruption of PECVD deposition were exposed to different power of the PT: 100 W for sample 3k77, 33 W for 3k79, 66 W for 3K80. In the case of samples from 3K78 deposition the hydrogen PT was not applied.

4. CONCLUSION

In this contribution we demonstrate a possibility to integrate Ge or Sn NPs to intrinsic of a-Si:H thin films and to use this approach to increase the optical absorption in the films. The formation of Ge NPs in the amorphous structures, in the case of Sn NPs in metallic form, was confirmed by TEM measurements. The α-crystalline structure of the Sn NPs was not confirmed. We present the increase of absorption coefficient in visible part of light spectrum as well as a shift of the absorption edge toward lower energies. In the case of Ge NPs it is a needed prerequisite which can lead toward the requested increased efficiency of very thin PV diode structures of thicknesses only a few tens of nm.

ACKNOWLEDGEMENTS

This work was supported by the CSF project 16-10429J, the MEYS project LTC17029 INTER-COST Action MP1406 and by the Operational Programme Research, Development and Education financed by European Structural and Investment Funds and the Czech Ministry of Education, Youth and Sports (Project No. SOLID21 - CZ.02.1.01/0.0/0.0/16_019/0000760).

REFERENCES

[1] GREEN, M.A. Potential for low dimensional structures in photovoltaics. *Mater. Sci Eng. B.*2000. vol. 74, pp 118-124.

[2] PORUBA, A., FEJFAR, A., REMES, Z., SPRINGER, J., VANECEK, M., KOCKA, J., MEIER, J., TORRES, P. and SHAH, A. Optical absorption and light scattering in microcrystalline silicon thin films and solar cells. *J. Appl. Phys.* 2000. vol. 88, pp. 148-160.

[3] STUCHLIKOVA, T.H., PIC, V., LEDINSKÝ, M., PURKRT, A., REMEŠ, Z. and STUCHLIK, J. In-situ formation of magnesium silicide nanoparticles on the surface of the hydrogenated silicon films. In *JJAP Conf. Proc.* 2017, pp. 011303(1)-011303(5).

[4] VANĚČEK, M., KOČKA, J., STUCHLÍK, J., KOŽÍŠEK, Z., ŠTIKA, O. and TŘÍSKA, A. Density of the gap states in undoped and doped glow discharge a-Si:H. *Solar Energy Materials.* 1983, vol. 8, pp. 411-423.

[5] REMES, Z., BABCHENKO, O., VARGA, M., STUCHLIK, J., JIRASEK, V., PRAJZLER, V., NEKVINDOVA, P. and KROMKA, A. Preparation and optical properties of nanocrystalline diamond coatings for infrared planar waveguides. *Thin Solid Films.* 2016, vol. 618, pp. 130-133.

[6] GREEN, M.A. and KEEVERS, M.J. Optical properties of intrinsic silicon at 300K. Progress in photovoltaics: research and applications. 1995, vol. 3, pp. 189-192.

EFFECT OF Eu DOPING ON THE OPTICAL PROPERTIES OF ZnO NANORODS PREPARED BY CHEMICAL BATH DEPOSITION

Roman YATSKIV, Nikola BAŠINOVÁ, Jan GRYM, Jan VANIŠ

Institute of Photonics and Electronics of the CAS, Prague, Czech Republic, EU, yatskiv@ufe.cz

Abstract

We report on the properties of Eu-doped ZnO nanorods prepared by chemical bath deposition from aqueous solutions in an autoclave. The autoclave enables to grow the nanorods above the boiling point of water. The nanorods were characterized by scanning electron microscopy, secondary ion mass spectroscopy, and low temperature photoluminescence spectroscopy. Under the UV light excitation, the nanorods show a strong blue emission at 450 nm, which is associated with the $4f^65d$ - $4f^4$ transitions of Eu^{2+} ions. We discuss the mechanisms responsible for the blue luminescence. We further clarify how the growth parameters affect the morphology and the optical properties of the nanorods. Growth temperatures above the boiling point of water significantly influence the intensity ratio of the near band edge emission to the deep level emission and greatly enhance optical quality of the nanorods.

Keywords: ZnO nanorods, rare earth elements, europium, energy transfer, luminescence

1. INTRODUCTION

The low growth temperature of chemical bath deposition generally results in a high density of native point defects, which strongly affect electrical and optical properties of ZnO nanorods (NRs) [1]. Defect control in ZnO is one of the major issues to be solved for successful application in optoelectronic devices. While the growth temperature in conventional reactors is limited by the boiling point of the chemical bath (water solution), polypropiolactone (PPL) lined autoclaves enable to increase the temperature to 280 °C.

ZnO is also attractive as a host material for visible luminescence centers. This is because ZnO or its nanostructures have a potential to be used as highly efficient and environmentally friendly phosphors [2-6]. In recent years, doping of ZnO with Eu ions has attracted a great attention. Eu can be incorporated either as Eu^{3+} ions, which show red luminescence, or as Eu^{2+} ions, which show blue luminescence. The emission from the Eu state under UV excitation is associated with the energy transfer from the host ZnO to Eu ions. The direct energy transfer from the ZnO exciton level to Eu excited state is unlikely, because the radiative/nonradiative decay of ZnO excitons is much faster than the energy transfer to the rare earth [5]. However, deep defect levels have been shown to facilitate energy transfer, resulting in efficient luminescence from Eu ions [6]. In this work, we demonstrate how the growth in an autoclave can improve optical properties of ZnO NRs. We further report on blue luminescence from Eu doped ZnO NRs.

2. EXPERIMENTAL

The nanostructured ZnO seed layers were prepared by the sol-gel synthesis on n-type Si substrates. Detailed information on the preparation of c-axis oriented seed layers was given in our previous work [7], where the oriented seed layer was shown to be a prerequisite for the vertical alignment of ZnO NRs. The ZnO NRs were grown on the seed layers by chemical bath deposition from aqueous solutions consisting of zinc nitrate hexahydrate and hexamethylenetetramine (HMTA) in PPL lined autoclave placed in a muffle furnace at 95 - 280 °C for 1 - 6 h. Europium nitrate hydrate was used for the doping of ZnO NRs with Eu. The surface morphology of undoped and Eu-doped ZnO NRs was observed by scanning electron microscopy (SEM). The seed layer thickness was evaluated from the cross section SEM images. PL spectra were measured with a set-up comprising a HeCd laser (325 nm) as an excitation source, a grating monochromator Jobin Yvon THR 1000, a closed cycle He optical cryostat, and a photomultiplier detection system (S1 operated at 77 K). The PL spectra for all samples were measured at identical condition (slits, power density, and set-up configuration). Secondary-ion mass spectrometry (SIMS) was used to analyze the composition of ZnO:Eu NRs and to determine the Eu dopant depth concentration using the time of flight SIMS spectrometer installed in a dual beam (FIB-SEM) chamber. All TOF-SIMS spectra were measured using the Ga^+ ion beam with the energy of 30 keV and current of 200 pA.

3. RESULT AND DISCUSSION

Figure 1 shows top-view and cross-sectional SEM images of ZnO NRs prepared in both a conventional open-system reactor (batch reactor) [8] and in the autoclave.

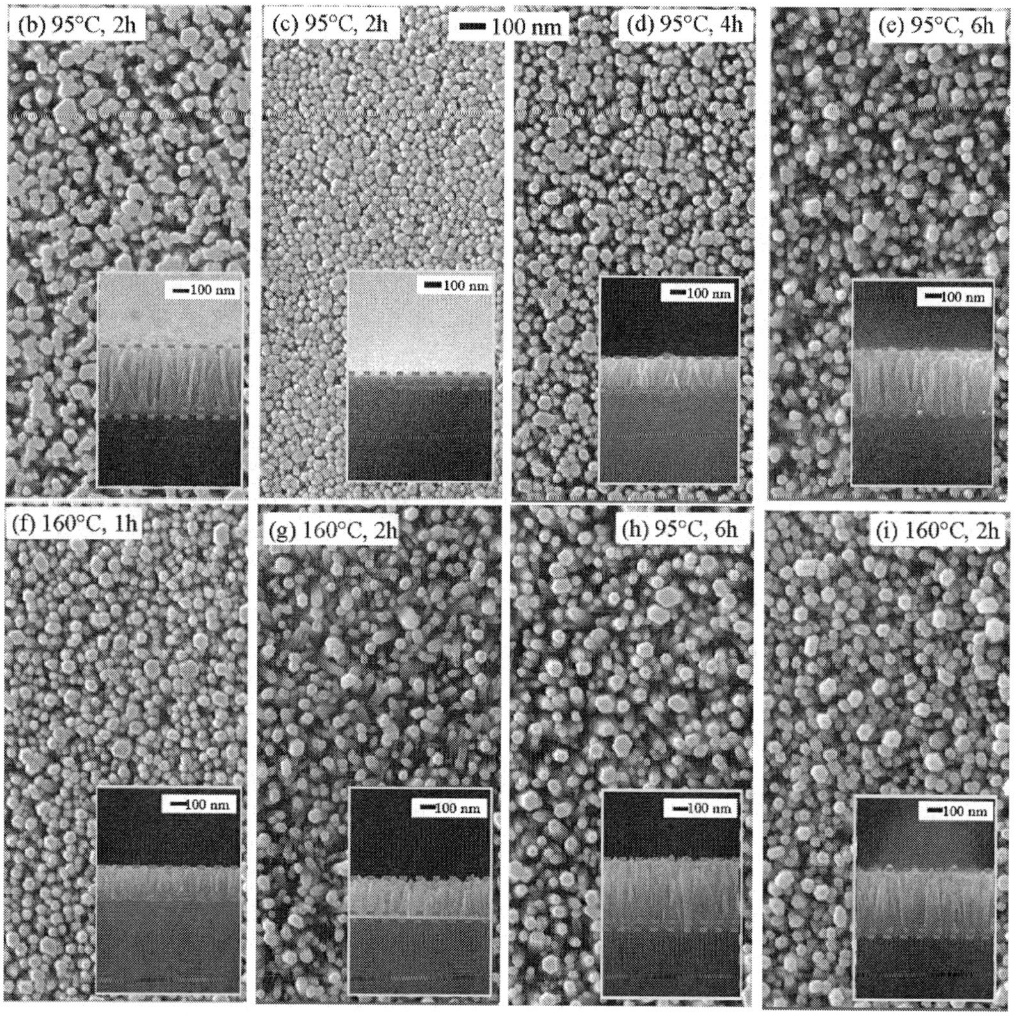

Figure 1 Top-view and cross-sectional SEM images of: (b) undoped ZnO NRs grown in the batch reactor and (c-g) in the autoclave; and (h,i) Eu doped ZnO NRs grown in the autoclave

Vertically oriented NRs with the diameters corresponding to the size of the nanocrystals of the seed layer were grown in both reactors when identical growth conditions (95 °C and 2 h) were employed (**Figure 1b,c**).

Nevertheless, in the autoclave, the length of the ZnO NRs was significantly lower, while their density was larger, which is related to higher nucleation density of the NRs in the autoclave. This difference in the NR length is related to the fact that the growth is diffusion limited; the growth units are quickly incorporated into the c-planes of the NRs and the diffusion is the limiting factor. The larger is the integral area of c-planes, the lower is the vertical growth rate. The NR length can be easily tuned by increasing the growth time (**Figure 1c-e**). At elevated temperature of 160 °C, the top surface of the nanorods changes form flat to pyramid-like morphology (**Figure 1g**). This change in morphology can be ascribed to the etching of ZnO nanorods by the undersaturated growth solution. At higher temperatures, the growth solution becomes quickly depleted of the growth units by both homogeneous and heterogeneous nucleation and growth. Growth times increased above 2 h result in the etching of the seed layer. **Figures 1h** and **1i** show SEM images for Eu-doped ZnO NRs (5 % in the growth solution) prepared under optimum growth conditions.

Figure 2 shows positive ion mass spectra of undoped ZnO NRs (a) and Eu-doped ZnO NRs (b) grown under the same conditions (160 °C, 2 h) with well pronounced peaks of Ga, Zn, Si and Eu ion isotopes. Corresponding depth distribution of ^{151}Eu and ^{64}Zn ion isotopes is shown in **Figure 3**. Depth of the interface corresponds to the NR height distribution observed by SEM. The higher ion yield close to the ZnO/Si interface is related to the unique geometry enhancing the generation of secondary ions.

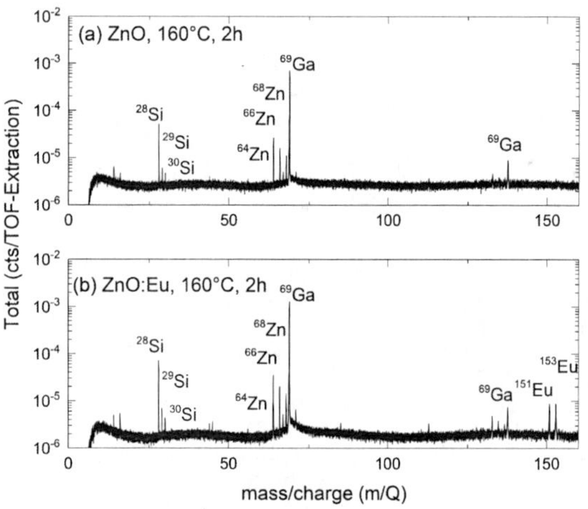

Figure 2 Mass spectra for undoped ZnO NRs (a) and Eu-doped ZnO NRs (b)

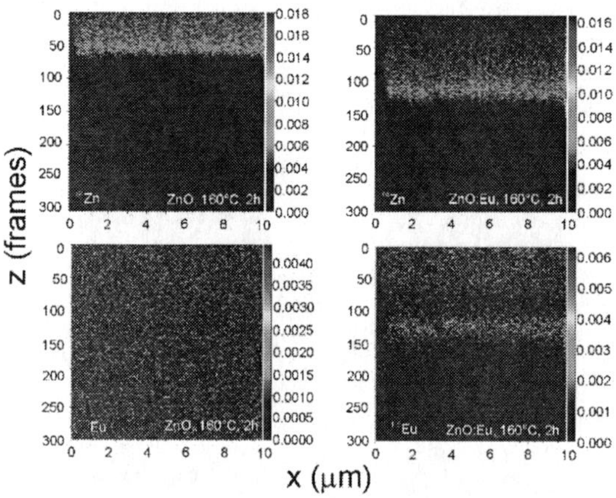

Figure 3 Depth distribution of the ^{64}Zn (top) and ^{151}Eu (bottom) ion isotopes for ZnO NRs and Eu-doped ZnO NRs

Figure 4i shows photoluminescence (PL) spectra of ZnO nanorods from **Figure 1**. The measured spectra are typical for the nanorods grown from solutions with the near band edge emission (NBE) in the UV region and the broad deep level emission (DLE) in the visible region. At room temperature the DLE is composed of the red luminescence at 1.82 eV (RL), orange luminescence at 2.05 eV (OL), and yellow luminescence at 2.2 eV (YL) [7,8]. The quantitative picture of the PL spectra for different growth temperatures and different growth times does not change (**Figure 4i-b,c,d,e**); however, when the growth temperature increases, the NBE/DLE intensity ratio also increases. This indicates substantial improvement of the optical and structural properties of the ZnO NRs (inset **Figure 2i**). The low temperature PL spectra in **Figure 4ii-c,d,e**, show that when the length of the ZnO NRs is higher than 500 nm, the near band edge emission is dominated by the donor bound exciton at 3.36 eV (DBE) and donor-acceptor transition 3.22 eV (DAP) [9]. The lines at 3.31 eV labeled as A are associated with the bound exciton to structural defect transition and were observed only for the seed layers or for the NRs with the length lower than 500 nm [8,10]. The presence of the luminescence at 3.31 eV for the ZnO NRs ($l < 500$nm) is due to superimposed defect luminescence from the seed layer.

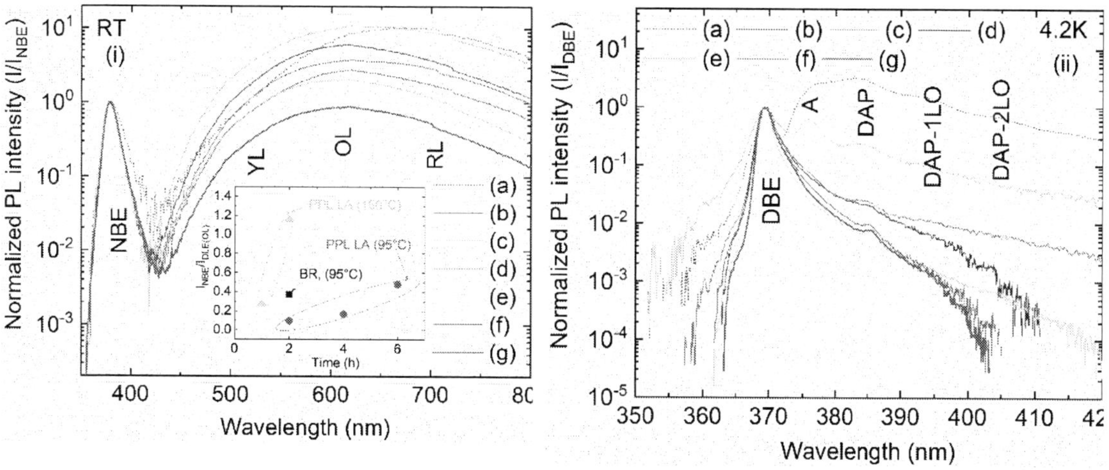

Figure 4 PL spectra measured at (i) room temperature and (ii) at 4.2 K. The samples (b) - (g) correspond to nomenclature from Figure 1. (a) PL spectra for the seed layer prepared by the sol-gel method.The inset shows the I_{NBE}/I_{DLE} ratio as a function of the growth time. BR - batch reactor, PPL LA - autoclave.

The Influence of Eu-doping on the emission spectra of the ZnO NRs is presented in **Figure 5**. Under 325 nm excitation, a new blue emission band (BL) peaked at 450 nm was observed for the samples with the growth temperature higher than 160 ° C. The BL emission can be associated with the $4f^6 5d$ - $4f^4$ transitions of Eu^{2+} [11]. The diagram of the energy transfer process from ZnO to Eu^{2+} is shown in **Figure 6**. Two mechanisms can be responsible for the BL; the Eu^{2+} 5d state is either excited directly by the 325 nm laser wavelength or the energy transfer is mediated by the deep levels.

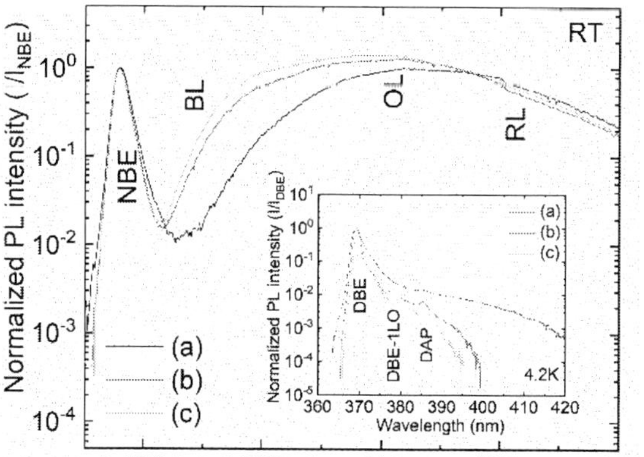

Figure 5 PL spectra measured at room temperature from ZnO NRs grown in the autoclave at 160 °C for 2 h. (a) undoped, (b,c) ZnO:Eu (5 %). The inset shows the PL spectra measured at 4.2 K.

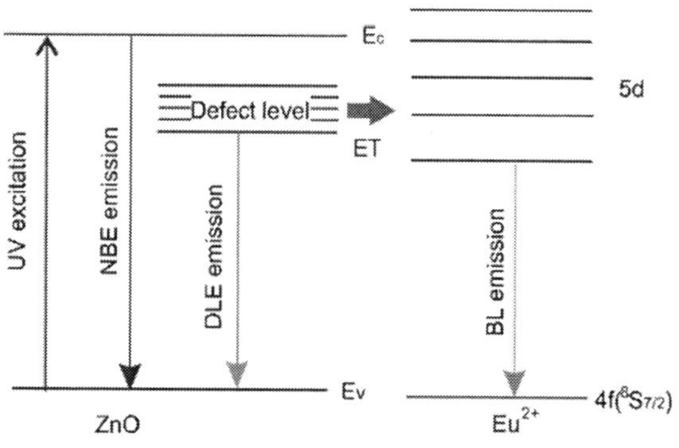

Figure 6 Energy level scheme: excitation and emission mechanisms of Eu^{2+} ions in ZnO

4. CONCLUSION

We demonstrated that the ZnO NRs grown in the autoclave above the boiling point of water possess improved optical properties in comparison with conventionally grown NRs. Moreover, these elevated temperatures are essential for the incorporation of Eu ions into ZnO crystal lattice. Under the 325 nm laser excitation, we did not observe the typically reported red emission associated with intra - 4d transition of Eu^{3+} ions. However, we did observe a blue emission peaked at 450 nm associated with the $4f^6 5d$ - $4f^4$ transitions of Eu^{2+}. We ascribed the origin of the blue luminescence to the energy transfer between the deep level defects in ZnO NRs and the Eu^{2+} ions.

ACKNOWLEDGEMENTS

This work was supported by the Czech Science Foundation projects 17-00546S and 17-00355S.

REFERENCES

[1] JANOTTI A. and VAN DE WALLE C.G., Fundamentals of zinc oxide as a semiconductor. *Rep Prog Phys.* 2009. vol. 72, no. 126501.

[2] SENAPATI S. and NANDA K.K., Red emitting Eu:ZnO nanorods for highly sensitive fluorescence intensity ratio based optical thermometry. *J Mater Chem C.* 2017. vol. 5, pp.1074-82.

[3] GEBURT S., LORKE M., DA ROSA A.L., FRAUENHEIM T., RODER R. and VOSS T., Intense Intrashell Luminescence of Eu-Doped Single ZnO Nanowires at Room Temperature by Implantation Created Eu-O-i Complexes. *Nano Lett.* 2014. vol. 14. no. 4523-8.

[4] HAYASHI Y., NARAHARA H., UCHIDA T., NOGUCHI T. and IBUKI S., Photoluminescence of Eu-Doped Zno Phosphors. *Jpn J Appl Phys 1* 1995. vol. 34. pp. 1878-82.

[5] AHMED S.M., SZYMANSKI P., EL-NADI L.M. and EL-SAYED M.A., Energy-Transfer Efficiency in Eu-Doped ZnO Thin Films: The Effects of Oxidative Annealing on the Dynamics and the Intermediate Defect States. *Acs Appl Mater Inter.* 2014. vol. 6. pp. 1765-72.

[6] WANG D.D., XING G.Z., GAO M., YANG L.L., YANG J.H. and WU T., Defects-Mediated Energy Transfer in Red-Light-Emitting Eu-Doped ZnO Nanowire Arrays. *J Phys Chem C.* 2011. vol. 115. pp. 22729-35.

[7] YATSKIV R., TIAGULSKYI S., GRYM J. and CERNOHORSKY O., Electrical and Optical Properties of Rectifying ZnO Homojunctions Fabricated by Wet Chemistry Methods. *Phys Status Solidi A.* 2018. vol. 215. no. 2.

[8] YATSKIV R. and GRYM J., Luminescence properties of hydrothermally grown ZnO nanorods. *Superlattice Microst.* 2016. vol. 99. pp. 214-20.

[9] TEKE A., OZGUR U., DOGAN S., GU X., MORKOC H., NEMETH B., NAUSE J. and EVERITT H.O., Excitonic fine structure and recombination dynamics in singlecrystalline ZnO. *Phys. Rev. B.* 2004. vol. 70. no. 195207.

[10] MEYER B.K., ALVES H., HOFMANN D.M., KRIEGSEIS W., FORSTER D., BERTRAM F., CHRISTEN J., HOFFMANN A., STRASSBURG M., DWORZAK M., HABOECK U. and RODINA A.V., Bound exciton and donor-acceptor pair recombinations in ZnO. *Phys. Status Solidi B Basic Solid State Phys.*2004. vol. 241. pp. 231-260.

[11] CHEN J., LIU Y.-G., MEI L., WANG Z., FANG M. and HUANG Z., Emission red shift and energy transfer behavior of color-tunable $KMg_4(PO_4)_3$:Eu^{2+},Mn^{2+} phosphors. *J Mater Chem C.* 2015. vol. 3, no. 21. pp. 5516-23.

STOICHIOMETRY ISSUE OF IRON(III) OXIDE NANOMATERIALS BY MÖSSBAUER SPECTROSCOPY

Jana HAVLÁKOVÁ, Jiří TUČEK, Radek ZBOŘIL

Regional Centre of Advanced Technologies and Materials, Departments of Experimental Physics and Physical Chemistry, Faculty of Science, Palacký University Olomouc, Czech Republic, EU
jana.havlakova@upol.cz

Abstract

Iron oxides, especially the Fe_3O_4 and γ-Fe_2O_3 phases, play a significant role in potential applications of nanomaterials due to their suitable features such as biocompatibility, magnetic and optoelectronic properties. Moreover, it is convenient to use Fe_3O_4 for some applications, such as Li-ion batteries, because of its combined valence states of iron. Therefore, it is highly desired to distinguish the two, otherwise similar, iron oxide phases and to evaluate the degree of Fe_3O_4 stoichiometry. Genuinely, a proper characterization of these compounds can be problematic due to their similar structures and the common occurrence of non-stoichiometric iron oxide forms and their mixtures in the nanosystems. In this respect, ^{57}Fe Mössbauer spectroscopy is regarded as a powerful tool for the identification of different forms of iron oxides as well as for the distinguishing between core-shell structures and mixtures. The present contribution is focused on theoretical ^{57}Fe Mössbauer spectra of Fe_3O_4, γ-Fe_2O_3 and their different degrees of stoichiometry, with regards to the correct evaluation of the spectra. ^{57}Fe Mössbauer spectra of two distinct levels of non-stoichiometric γ-Fe_2O_3 and Fe_3O_4 were modelled and discussed. Furthermore, theoretical examples of ^{57}Fe Mössbauer spectra of Fe_3O_4/γ-Fe_2O_3 mixture of 1/1 particle ratio and a core-shell structure, where Fe_3O_4 core takes up 75 % of the particle's volume, are presented.

Keywords: γ-Fe_2O_3, Fe_3O_4, stoichiometry, mixtures, ^{57}Fe Mössbauer spectroscopy

1. INTRODUCTION

Iron oxides are considered as intriguing nanomaterials due to their importance in fundamental research as well as their convenient physicochemical properties such as nontoxicity, biocompatibility, outstanding magnetic, electronic and optoelectronic features together with relatively low-cost production possibilities, that offer a wide application potential [1-4]. They are frequently classified into two main groups - hydrated (i.e., hydroxides and oxide-hydroxides) and non-hydrated phases; in non-hydrated forms, iron ions are present in divalent (FeO), trivalent (Fe_2O_3) or both states (Fe_3O_4). Additionally, ferric oxide exhibits polymorphism; five crystalline (α-Fe_2O_3, β-Fe_2O_3, γ-Fe_2O_3, ε-Fe_2O_3, and ζ-Fe_2O_3) structures and amorphous Fe_2O_3 are known to date, from which Fe_3O_4 (mineralogically known as magnetite) and γ-Fe_2O_3 (mineralogically known as maghemite) are the two leading phases regarding applications [4,5].

When the particle size decreases, it generally becomes difficult to distinguish between Fe_3O_4 and γ-Fe_2O_3, because of the presence of their mixtures and nonstoichiometric structures [6]. Moreover, the degree of stoichiometry can affect particle's physical and chemical properties such as reduction potential, conductivity, coercivity and saturation magnetization, etc. [3]. For some applications, e.g., in lithium-ion batteries, it is crucial to use materials with mixed valence states, and therefore to know and evaluate the degree of their stoichiometry. These factors could raise an issue - how to sufficiently characterize these oxides and their stoichiometry? The solution can be to employ ^{57}Fe Mössbauer spectroscopy; a powerful tool which can provide the information about the present phases as well as the degree of their stoichiometry, i.e., the quantification of Fe^{2+}/Fe^{3+} ratio.

In the present contribution, we focus on theoretical simulation of nanoparticulate assembly composed of nonstoichiometric Fe_3O_4, γ-Fe_2O_3, Fe_3O_4/γ-Fe_2O_3 mixture and a core-shell structure. Furthermore, we predict their Mössbauer spectral features which can be related to the experimental data of nanosystems with particle size not smaller than ~10 nm (below such threshold size limit, ultrafine nanoparticles adopt amorphous nature). The ^{57}Fe Mössbauer spectra were simulated with the help of *MossWinn* programme with the input data from our vast knowledge on iron oxide nanoparticle systems [1,7]. The only variables in modelling of the spectra involved the degree of stoichiometry and the phase volume ratio for mixture and core-shell structure while the rest of the parameters (e.g., particle size distribution, morphology, temperature fluctuations, etc.) were not considered. Therefore, we accept possible slight differences between theoretical spectra modelled here and those experimentally observed.

2. γ-Fe_2O_3, Fe_3O_4 AND THE EVALUATION OF THEIR STOICHIOMETRY

The two phases of non-hydrated iron oxide, γ-Fe_2O_3 and Fe_3O_4, are not trivial to distinguish as mentioned earlier, which is due to their nearly identical crystal structures [8]. γ-Fe_2O_3 has a cubic crystal structure of an inverse spinel with a lattice parameter a = 8.35 Å and two non-equivalent cation positions. Tetrahedral (A) sites are occupied solely by Fe^{3+} ions and octahedral (B) sites are divided between ferric ions and vacancies. Ferric ions occupy A- and B-sites in stoichiometric γ-Fe_2O_3 in the ratio of 1/1.67 and its formula can be written as $(Fe^{3+})^A(Fe^{3+}_{5/3}\square_{1/3})^B O_3$, where the symbol □ means vacancies [9-11]. The space group of γ-Fe_2O_3 is $Fd\overline{3}m$ if the vacancies are distributed over B-sites randomly, $P4_132$ with partially ordered vacancies and tetragonal $P4_32_12$ when the vacancies are fully ordered [11]. Magnetic moments of ferric ions in A- and B-sites are not equal and are in antiparallel alignment with respect to each other, therefore γ-Fe_2O_3 is a collinear ferrimagnet. Additionally, the experimental measurement of the Curie temperature is not possible due to the transformation to thermodynamically more stable α-Fe_2O_3 (hematite), in nanomaterials possibly via a rare phase ε-Fe_2O_3.

Fe_3O_4 has a cubic crystal structure of an inverse spinel as well as γ-Fe_2O_3 but the lattice parameter a is slightly increased, i.e., a = 8.39 Å. Stoichiometric Fe_3O_4 has tetrahedral (A) sites and octahedral (B) sites occupied in the ratio of 1/2 since the B-sites are split between both Fe^{2+} and Fe^{3+} ions equally [3,8]. Nonetheless, Fe_3O_4 often exists in non-stoichiometric form due to oxidization, which causes formation of vacancies and therefore shifting of the A/B ratio towards γ-Fe_2O_3. Its formula can be therefore written as $Fe_{3-\delta}O_4$, where the degree of stoichiometry δ can vary from 0 to 1/3, while zero value represents stoichiometric Fe_3O_4 and δ = 0.33 implies γ-Fe_2O_3 [2]. Moreover, if we account for the distribution of vacancies and their ratio to Fe^{2+} and Fe^{3+} ions, then the structural non-stoichiometric formula can be written as $(Fe^{3+})^A(Fe^{2+}_{1-2\delta}Fe^{3+}_{1+2\delta}\square_{\delta})^B O_4$, where the symbol □ stands for vacancies [2,3]. Fe_3O_4 is a collinear ferrimagnet as well as γ-Fe_2O_3 with the Curie temperature of ~850 K. At room temperature, Fe_3O_4 exhibits a conductive behaviour due to the fast electron hopping between octahedral cations. As the temperature decreases to ~124 K, Fe_3O_4 undergoes the Verwey transition; a structural distortion from cubic to orthorhombic symmetry accompanied with a rapid drop in conductivity [8,12-14].

Crystal structures of stoichiometric γ-Fe_2O_3 and Fe_3O_4 are shown in **Figure 1a,b** together with their typical powder X-ray diffraction (XRD) patterns (see **Figure 1c,d**) and typical ^{57}Fe Mössbauer spectra, with isomer shift values referred to δ-Fe at room temperature and zero applied field (see **Figure 1e,f**), while their Mössbauer hyperfine parameters are listed in **Table 1**. The ^{57}Fe Mössbauer spectrum of stoichiometric γ-Fe_2O_3 consists of two spectral components corresponding to ferric ions in the A-sites and in the B-sites. The stoichiometric form of Fe_3O_4 is in Mössbauer spectrum manifested by two spectral components, one originating from Fe^{3+} ions in the A-positions and the other from Fe^{2+} and Fe^{3+} ions in the B-positions with effective valence state of 2.5+ due to the fast electron hopping above the Verwey temperature [3,12].

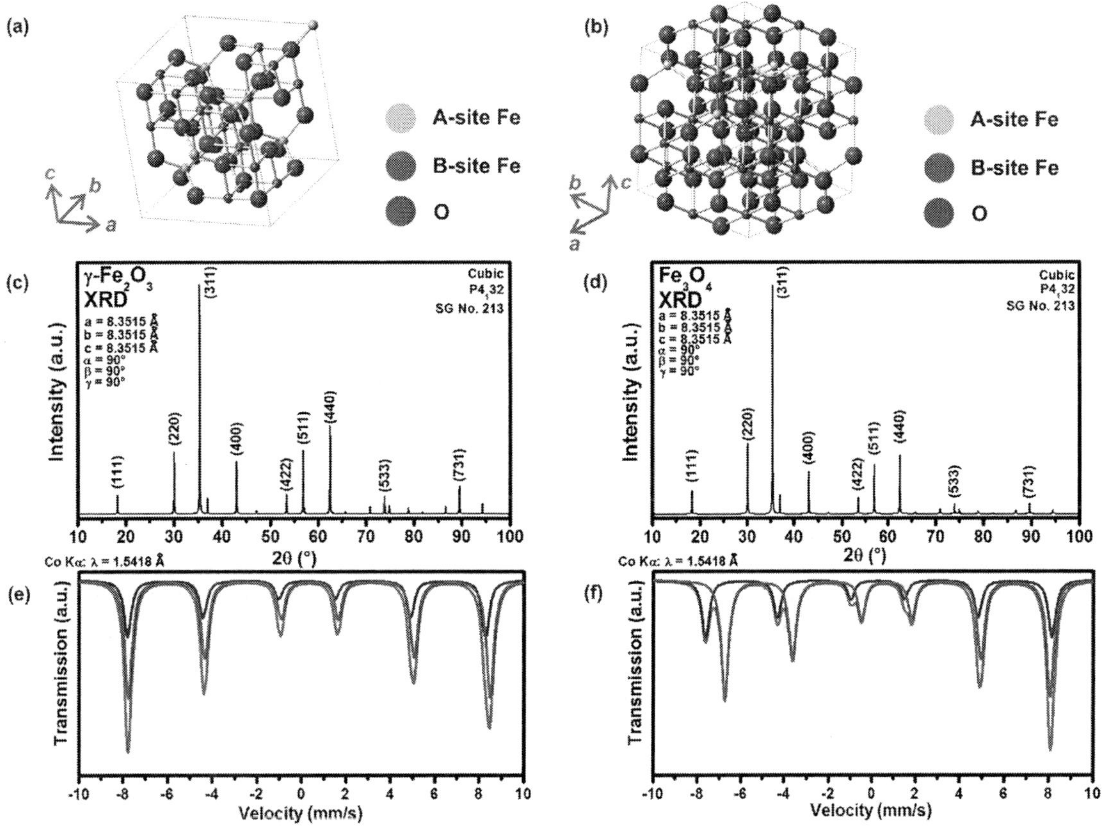

Figure 1 Crystal structures and typical XRD patterns and ^{57}Fe Mössbauer spectra of (a, c, e) γ-Fe_2O_3 and (b, d, f) Fe_3O_4, on the right-hand side. In the Mössbauer spectra, the red line is corresponding to the overall fit, the blue line corresponds to ions in the A-sites and the green line refers to ions in the B-sites. Panel (a) is reprinted with permission from [1].

Table 1 ^{57}Fe Mössbauer hyperfine parameters typical for ideally stoichiometric forms of γ-Fe_2O_3 and Fe_3O_4, where δ_{Fe} is the isomer shift, ΔE_Q is the quadrupole splitting, B_{hf} is the hyperfine magnetic field and RA is the relative spectral area. The isomer shift values are referred to metallic δ-Fe at room temperature

phase	δ_{Fe} (mm/s)	ΔE_Q (mm/s)	B_{hf} (T)	RA (%)	Assignment
γ-Fe_2O_3	0.25	0.00	50.0	37.5	Fe^{3+} A-site
	0.37	0.00	50.5	62.5	Fe^{3+} B-site
Fe_3O_4	0.25	0.00	48.9	33.3	Fe^{3+} A-site
	0.65	0.00	45.7	66.7	$Fe^{2.5+}$ B-site

Additionally, in the non-stoichiometric $Fe_{3-\delta}O_4$, Fe^{3+} ions also occur as isolated in the B-positions, which are not involved in the electron hopping effect and are manifested in the Mössbauer spectra by the presence of the additional third sextet. The degree of stoichiometry of $Fe_{3-\delta}O_4$ can be described also by the ratio x_m of structural amount of Fe^{2+} and Fe^{3+} ions, given by the following relationship stated by Gorski and Scherer [3]:

$$x_m = \frac{Fe^{2+}}{Fe^{3+}} = \frac{1-3\delta_m}{2+2\delta_m}. \quad (1)$$

The stoichiometry x_m can vary from 1/2 for stoichiometric Fe_3O_4 to zero corresponding to completely oxidized form (i.e., γ-Fe_2O_3). The value of x_m is obtained from ^{57}Fe Mössbauer spectra as a ratio of relative areas of spectral components corresponding to Fe^{2+} and Fe^{3+} ions by the following equation [3]:

$$x_m = \frac{Fe^{2+}}{Fe^{3+}} = \frac{\frac{1}{2}{}^{B}Fe^{2.5+}}{\frac{1}{2}{}^{B}Fe^{2.5+} + {}^{A+B}Fe^{3+}}. \quad (2)$$

As the degree of stoichiometry shifts towards γ-Fe_2O_3, the amount of Fe^{2+} decreases; at some point, there is not enough Fe^{2+} ions to be detected by ^{57}Fe Mössbauer spectroscopy, therefore the $Fe^{2.5+}$ sextet is not resolved in the spectrum and x_m cannot be evaluated. Nevertheless, these assemblies are rather closer to the structure of γ-Fe_2O_3 and their degree of stoichiometry can be estimated as the occupancy ratio of B/A sites, since we know that for stoichiometric γ-Fe_2O_3, the spectral ratio is equal to 1.67 [9].

We addressed this issue by modelling ^{57}Fe Mössbauer spectra of two systems with different degrees of stoichiometry; one closer to the Fe_3O_4 and the other one closer to γ-Fe_2O_3, as can be seen in **Figure 2**. The theoretical ^{57}Fe Mössbauer spectrum of non-stoichiometric Fe_3O_4 consists of three spectral components, corresponding to Fe^{3+} in A-sites, isolated Fe^{3+} in B-positions and $Fe^{2.5+}$ in B-sites. The spectrum of the non-stoichiometric γ-Fe_2O_3 consists of two spectral components corresponding to Fe^{3+} ions in A-sites and Fe^{3+} ions in B-sites. Their predicted hyperfine parameters are listed in **Table 2**. We estimated the degree of stoichiometry according to the B/A relative spectral areas ratio as $\delta_1 = 0.14$ and $\delta_2 = 0.26$. On the other hand, when we follow equation (2) and then equation (1) to obtain the value of δ_m (this procedure can be employed only for the non-stoichiometric Fe_3O_4), then $\delta_{m1} = 0.22$. The vast difference in values δ_1 and δ_{m1} can be caused by the applicability of equations (1) and (2) to structures close to stoichiometric Fe_3O_4. This brings us to several issues and questions: (i) What is the threshold value of δ (or x_m) when the structure cannot be considered Fe_3O_4-like and hence becomes non-stoichiometric γ-Fe_2O_3 illustrated by the grey area in **Figure 2**?; (ii) How to correctly evaluate the degree of stoichiometry in the close proximity of this threshold value?

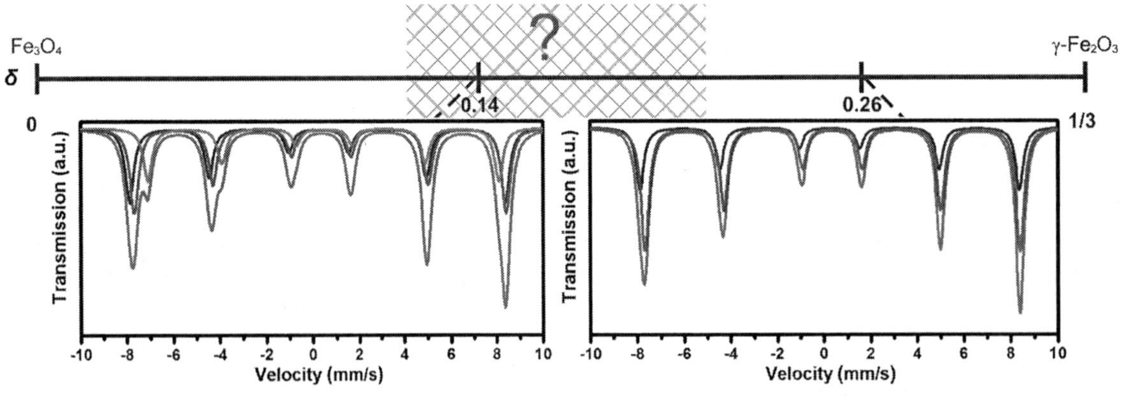

Figure 2 Theoretical ^{57}Fe Mössbauer spectra of Fe_3O_4 particles with different degrees of stoichiometry δ; the blue lines correspond to Fe^{3+} in A-sites, the green lines refer to Fe^{3+} isolated in B-sites, the pink line corresponds to $Fe^{2.5+}$ in B-sites and the red lines represent the overall fits

The second modelled situation is the difference between Mössbauer spectra of Fe_3O_4/γ-Fe_2O_3 mixtures and core-shell assemblies, which is shown in **Figure 2**. The theoretical Mössbauer spectrum of a 1/1 mixture consists of four spectral components of which two are assigned to Fe_3O_4 particles, hence Fe^{3+} ions in A-sites and $Fe^{2.5+}$ ions in B-sites. The other two components refer to Fe^{3+} in A-sites and Fe^{3+} in B-sites in γ-Fe_2O_3. The theoretical spectrum of a core-shell assembly, on the other hand, consists of three spectral components. The first and second component is assigned to the Fe_3O_4 core and its Fe^{3+} ions in A-sites and $Fe^{2.5+}$ ions in B-sites, respectively. Since the core takes-up 75 % of the particle's volume, the γ-Fe_2O_3 surface layer is extremely thin, hence the spectral components referring to γ-Fe_2O_3 cannot be resolved separately and are seen as one component.

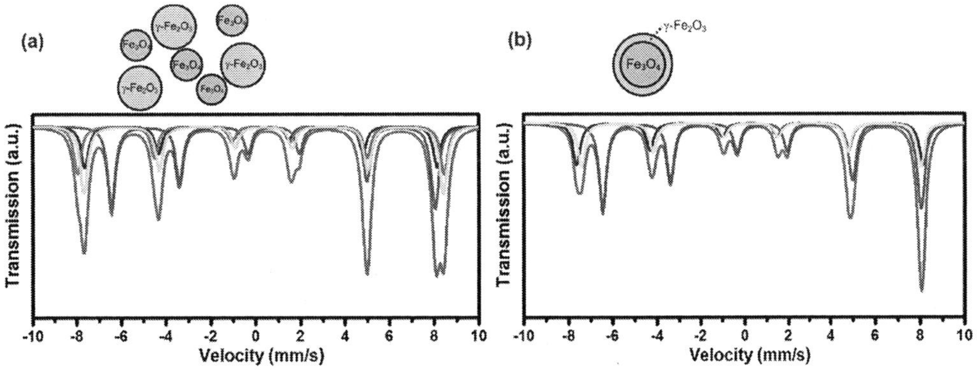

Figure 3 Theoretical ^{57}Fe Mössbauer spectra of (a) a 1/1 mixture Fe_3O_4/ γ-Fe_2O_3; (b) core-shell particles Fe_3O_4/ γ-Fe_2O_3 of 3/1 ratio, where the red lines correspond to the overall fit, the blue lines refer to the Fe^{3+} ions in A-sites, the dark green lines represent $Fe^{2.5+}$ in B-sites, the light green line refers to isolated Fe^{3+} in B-sites and the blue line represents the sum of Fe^{3+} in A- and B-sites.

Table 2 ^{57}Fe Mössbauer hyperfine parameters modelled for non-stoichiometric forms of γ-Fe_2O_3, Fe_3O_4, γ-Fe_2O_3/ Fe_3O_4 and of a core-shell Fe_3O_4/γ-Fe_2O_3 structure, where δ_{Fe} is the isomer shift, ΔE_Q is the quadrupole splitting, B_{hf} is the hyperfine magnetic field and RA is the relative spectral area. The isomer shift values are referred to metallic δ-Fe at room temperature.

phase	δ_{Fe} (mm/s)	ΔE_Q (mm/s)	B_{hf} (T)	RA (%)	Assignment
non-stoichiometric Fe_3O_4	0.25	0.00	50.0	35	Fe^{3+} A-site
	0.37	0.00	50.0	40	Fe^{3+} B-site
	0.50	0.00	47.2	25	$Fe^{2.5+}$ B-site
non-stoichiometric γ-Fe_2O_3	0.25	0.00	48.9	33.3	Fe^{3+} A-site
	0.65	0.00	45.7	66.7	Fe^{3+} B-site
Fe_3O_4/γ-Fe_2O_3 1/1 mixture	0.25	0.00	49	17.5	Fe^{3+} A-site
	0.80	0.00	45	34.5	$Fe^{2.5+}$ B-site
	0.25	0.00	51	20	Fe^{3+} A-site
	0.37	0.00	50	28	Fe^{3+} B-site
Fe_3O_4/γ-Fe_2O_3 core-shell	0.25	0.00	49	25	Fe^{3+} A-site
	0.80	0.00	45	50	$Fe^{2.5+}$ B-site
	0.31	0.00	48	25	Fe^{3+} A-sites + B-sites

3. CONCLUSION

In the present contribution, we theoretically addressed the topic of distinguishing between non-stoichiometric forms of γ-Fe_2O_3, Fe_3O_4 and between Fe_3O_4/γ-Fe_2O_3 mixtures and core-shell assemblies by ^{57}Fe Mössbauer spectroscopy. Some approaches to determine and quantify the degree of $Fe_{3-\delta}O_4$ stoichiometry have been proposed. Nevertheless, non-stoichiometric $Fe_{3-\delta}O_4$ materials can exist in a wide range of stoichiometry degrees δ varying from zero (Fe_3O_4) to 1/3 (γ-Fe_2O_3), when, in some cases, the procedures of stoichiometry determination, suggested in the literature so far, fail to some extent in terms of identifying the correct phase nature. In other words, here we highlight a peculiar situation frequently occurring in experimental samples, when the non-stoichiometry of the two iron oxide forms is in between or close to the middle δ value and it is hard to state if this phase carries the structural features of Fe_3O_4 or γ-Fe_2O_3. Therefore, it is an issue of further research to find the threshold value of stoichiometry degree δ (or x_m) between the two structures Fe_3O_4 and γ-Fe_2O_3. A possible solution is viewed in employing ^{57}Fe Mössbauer spectroscopy with an applied magnetic field, which would provide better resolution of individual spectral components and, hence, more precise determination of the stoichiometry parameter. Such information could then be related with data from XRD technique in terms of changing the lattice parameters values with vacancy ordering upon oxidation from Fe_3O_4 to γ-Fe_2O_3 as frequently observed in the literature [1].

ACKNOWLEDGEMENTS

The authors gratefully acknowledge support from the Ministry of Education, Youth and Sports of the Czech Republic under project No. LO1305.

REFERENCES

[1] TUČEK, J., MACHALA, L., FRYDRYCH, J., PECHOUŠEK, J. AND ZBOŘIL, R. Mössbauer Spectroscopy in Study of Nanocrystalline Iron Oxides from Thermal Processes. SHARMA, V.K., KLINGELHOEFER, G. and NISHIDA, T. *Mössbauer Spectroscopy: Applications in Chemistry, Biology, and Nanotechnology*. Hoboken: JWS, 2013, Part V: Iron Oxides, chapter 18, pp. 349-392.

[2] VANDENBERGHE, R.E., BARRERO, C.A., DA COSTA, G.M., VAN SAN, E. and DE GRAVE, E. Mössbauer characterization of iron oxides and (oxy)hydroxides: the present state of the art. *Hyperfine Interact.*, 2000. vol. 126, iss. 1-4, pp. 247-259.

[3] GORSKI, C.A. and SCHERER, M.M. Determination of nanoparticulate magnetite stoichiometry by Mössbauer spectroscopy, acidic dissolution, and powder X-ray diffraction: A critical review. *Am. Mineral.* 2010. vol. 95, iss. 7, pp. 1017-1026.

[4] JOHNSON, C.E., JOHNSON, J.A., HAH, H.Y., COLE, M., GRAY, S., KOLESNICHENKO, V., KUCHERYAVY and P., GOLOVERDA, G. Mössbauer studies of stoichiometry of Fe_3O_4: characterization of nanoparticles for biomedical applications. *Hyperfine Interact.* [online] 2016. vol. 237, no. 27, pp. 1-10 [viewed 2018-04-26]. Available from: DOI: https://doi.org/10.1007/s10751-016-1277-6.

[5] TUCEK, J., MACHALA, L., ONO, S., NAMAI, A., ZOSHIKIYO, M., IMOTO, K., TOKORO, H., OHKOSHI, S. and ZBORIL, R. Zeta-Fe_2O_3 - A new stable polymorph in iron(III) oxide family. *Sci. Rep.* [online] vol. 5, no. 15091, pp. 1-11 [viewed 2015-12-09]. Available from: DOI: 10.1038/srep15091.

[6] SINGH, L.H., PATI, S.S., GUIMARAES, E.M., RODRIGUES, P.A.M., OLIVEIRA, A.C. and GARG, V.K. Synthesis, structure, morphology and stoichiometry characterization of cluster and nano magnetite. *Mater. Chem. Phys.* 2016. vol. 178, pp. 182-189.

[7] KLENCSAR, Z., KUZMANN. E. and VERTES, A. User-friendly software for Mössbauer spectrum analysis. *J. Radioanal. Nucl. Chem.* 1996. vol. 210, iss. 1, pp. 105-118.

[8] CORNELL, R.M., SCHWERMANN, U. The Iron Oxides. 2nd ed. Weinheim: Wiley, 2000, p. 703

[9] ZBORIL, R., MASHLAN, M. and PETRIDIS, D. Iron(III) oxides from thermal processes-synthesis, structural and magnetic properties, Mössbauer spectroscopy characterization, and applications. *Chem. Mater.* 2002. vol. 14, iss. 3, pp. 969-982

[10] DA COSTA, G.M., DEGRAVE, E., DEBAKKER, P.M.A. and VANDENBERGHE, R.E. Influence of nonstoichiometry and the presence of maghemite on the Mössbauer spectrum of magnetite. *Clays Clay Miner.* 1995. vol. 43, iss. 6, pp. 656-668

[11] FOCK. J., BOGART, L.K., GONZALEZ-ALONSO, D., ESPESO, J.I., HANSEN, M.F., VARON, M., FRANDSEN, C. and PANKHURST, Q.A. On the 'centre of gravity' method for measuring the composition of magnetite/maghemite mixtures, or the stoichiometry of magnetite-maghemite solid solutions, via Fe-57 Mössbauer spectroscopy. *J. Phys. D.* [online] 2017. vol. 50, iss. 26, pp. 1-17 [viewed 2018-04-26]. Available from: DOI: 10.1088/1361-6463/aa73fa.

[12] DANIELS, J.M., ROSENCWAIG, A. Mössbauer spectroscopy of stoichiometric and non-stoichiometric magnetite. *J. Phys. Chem. Solids.* 1969. vol. 30, iss. 6, pp. 1561-1571.

[13] HAGGSTROM, L., ANNERSTEN, H., ERICSSON, T., WAPPLING, R., KARNER, W. and BJARMAN, S. Magnetic dipolar and electric quadrupolar effects on Mössbauer spectra of magnetite above Verwey transition. *Hyperfine Interact.* 1978. vol. 5, iss. 3, pp. 201-214.

[14] HONIG, J.M. Analysis of the Verwey transition in magnetite. *J. Alloy. Compd.* 1995. vol. 229, iss. 1, pp. 24-39.

COMPUTER SIMULATION OF ELECTRONIC DEVICE FOR GENERATION OF ELECTRIC OSCILLATIONS BY NEGATIVE DIFFERENTIAL CONDUCTIVITY OF SUPERCOOLED NANOSTRUCTURED SUPERCONDUCTORS IN ELECTRIC FIELD

Todor MISHONOV, Victor DANCHEV, Albert VARONOV

Sofia University St. Kliment Ohridski, Sofia, Bulgaria, EU,
mishonov@bgphysics.eu, spectrefdx@gmail.com, akofmg@gmail.com

Abstract

We use the formerly derived explicit analytical expressions for the conductivity of nanostructured superconductors supercooled below the critical temperature in electric field. Computer simulations reveal that the negative differential conductivity region of the current-voltage characteristic leads to excitation of electric oscillations. We simulate a circuit with distributed elements as a first step to obtain megahertz oscillations. This gives a hint that a hybrid device of nanostructured superconductors will work in terahertz frequencies. If the projected setup is successful, we consider the possibility to be put in a nanosatellite (such as a CubeSat). A study of high temperature superconductor layers in space vacuum and radiation would be an important technological task.

Keywords: Thin nanostructured superconductor film, negative differential conductivity, oscillations

1. INTRODUCTION

The purpose of the present work is to present a numerical simulation, which shows the generation of high-frequency oscillations created by a thin superconducting film supercooled bellow the critical temperature T_c in external electric field E. The idea was first proposed more than 10 years ago [1] and now we are giving details on how to start the experimental work using radio frequency oscillations with distributed elements (resistors, capacitors and inductors). The main idea is to insert a supercooled superconductor with negative differential conductivity as an active element in a resonance circuit. The simulation presented in this work aims to trigger the analogous development of the terahertz region where the resonance circuit will be implemented on the same hybrid nanodevice which will emit coherent terahertz waves.

2. FLUCTUATION CONDUCTIVITY IN ELECTRIC FIELD

Our starting point is the result for the two-dimensional fluctuation current density j_{2D} in external electric field [2-3] see also [4]

$$j_{2D} = \sigma_0 \,[\, S_n + S(\epsilon, \beta)]E, \tag{1}$$

where β is the dimensionless electric field and ϵ is the reduced temperature

$$\beta = \frac{\pi e\, \xi E}{16\, k_B T_c} = \frac{e \xi E}{\hbar/\tau_0}, \qquad \epsilon = \ln\frac{T}{T_c} \approx \frac{T - T_c}{T_c} \tag{2}$$

defined by the coherence length ξ

$$-T_c \left.\frac{\mathrm{d}B_{c2}}{\mathrm{d}T}\right|_{T_c} = \frac{\Phi_0}{2\pi\xi^2}, \qquad \Phi_0 = \frac{\pi\hbar}{|e|}, \tag{3}$$

the critical temperature T_c or a time constant τ_0, related to the relaxation of the superconducting order parameter

$$\frac{\hbar}{\tau_0} = \frac{16\, k_B T_c}{\pi}, \tag{4}$$

where k_B is the Boltzmann constant. The constant $\sigma_0 = e^2/16\hbar$ with dimension of conductivity represents the Aslamazov-Larkin conductivity above T_c for evanescent electric field $E \to 0$ $\sigma_{AL} = e^2/16\hbar\epsilon$ for $\epsilon>0$. The normal conductivity σ_n is parameterized by the dimensionless $S_n = \sigma_n/\sigma_0$.

The dimensionless function $S(\epsilon, \beta)$ of reduced temperature ϵ and electric field β

$$S(\epsilon,\beta) = \int_0^\infty \exp\left(-\epsilon v - \frac{\beta}{3} v^3\right) \mathrm{d}v, \qquad (5)$$

describing the fluctuation conductivity is depicted in **Figure 1**.

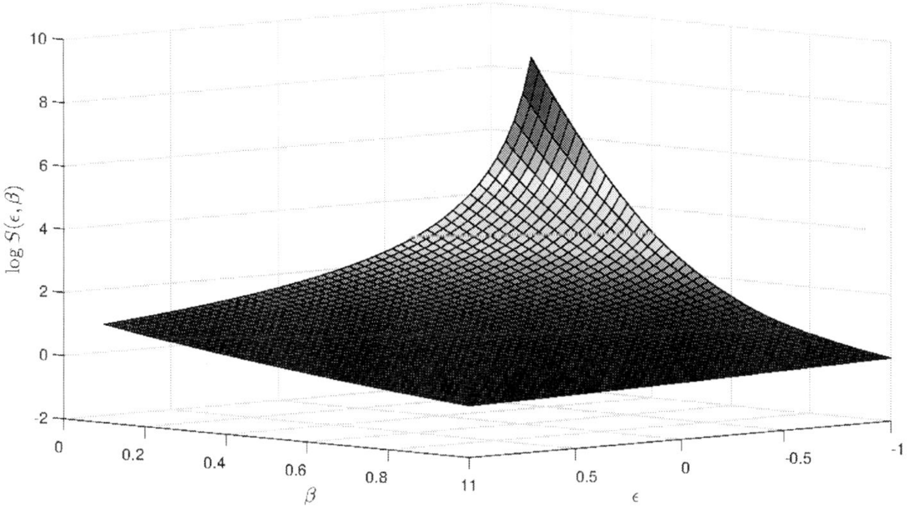

Figure 1 The logarithm of the dimensionless conductivity $S(\epsilon, \beta)$ as a function of reduced temperature ϵ and dimensionless electric field β, equation (5).

This formula for the current equation (5) is a consequence of time-dependent Ginzburg-Landau equation, which is applicable below T_c if the coherent superconducting order parameter is destroyed by a constant electric field. For $\epsilon = 0$ the integral is elementary, while in other cases the substitution $u = |\epsilon| v$ reduces the integral to well-known special functions.

For low frequencies where the heat capacity of the superconductor is negligible, the temperature of the superconducting layer is determined by the Ohmic heating per unit area $E\, j_{2\mathrm{D}}(E)$ and interface boundary resistance R_θ

$$E\, j_{2\mathrm{D}}(E,T) = \frac{T-T_\mathrm{s}}{R_\theta}, \qquad \rho_\theta = \frac{16\, k_\mathrm{B}^2 T_c R_\theta}{\pi^2 \hbar \xi^2}, \qquad (6)$$

where ρ_θ is a convenient dimensionless parameter useful for the further analysis and T_s is the substrate temperature. Until the critical temperature T_c can be found by fitting the temperature dependence of the resistivity above T_c, the coherence length ξ can be extracted without applying magnetic field B from the non-linear fluctuation conductivity [5] at T_c

$$j_{2\mathrm{D}} = \frac{3^{1/3}\Gamma\left(\frac{4}{3}\right)}{2^{4/3}\hbar} \left(\frac{k_\mathrm{B} T_c}{\pi e\, \xi}\right)^{2/3} E^{1/3} + \sigma_n E, \qquad T = T_c\,. \qquad (7)$$

Since the fluctuational current formally tends to infinity at evanescent electric field bellow T_c, one can easily derive the asymptotics

$$j(E)E|_{\mathrm{E}\to 0} = (T_c - T_\mathrm{s})/R_\theta = \frac{1}{lw} U I(U)|_{U\to 0}\,, \qquad (8)$$

which allows us to determine the interface conductance R_θ. Or else the current

$$j(E) = \frac{T_c - T_s}{R_\theta E} + \sigma_n E, \tag{9}$$

meaning that we have hyperbolic approximation for the current-voltage characteristic. Here the voltage U and the current I are given through the strip length l and its width w. They are related to $j(E)$ and E through $I = j_{2D}\, w$, $U = E\, l$. Below the critical temperature $T<T_c$ at small electric fields the current is significant and the temperature of the superconductor has to be calculated in a self-consistent way solving simultaneously equation (6) and equation (1) shown in **Figure 2**.

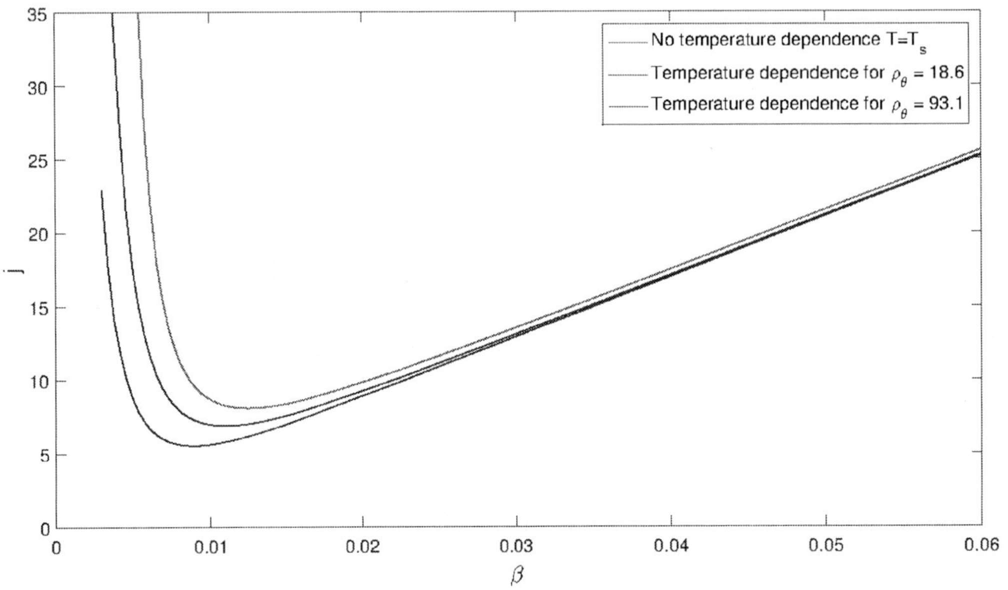

Figure 2 Dimensionless current-voltage characteristics $j_{2D}(E)$ or rescaled $I(U)$ based on equation (1) and equation (6) for different interfacial thermal resistances

The shape of the thin film is not necessarily a strip. For example it can be in Corbino geometry, a narrow ring between radii R_1 and R_2 with $l=R_2-R_1<<R_2$ and $w=2\pi R_1$. Specifically for high T_c superconductors, the frequency $1/\tau_0$ is in the terahertz range and for them the above-derived static formulas for the current are valid practically for the whole radio frequency range. In the next section, we present a circuit working within the framework of the theory presented so far.

3. CIRCUIT GENERATING OSCILLATIONS

The circuit for generation of stable oscillations by using negative differential conductivity in supercooled superconductors is shown in **Figure 3**.

One can easily see the parallel LC resonance circuit generating the oscillations. In short, the negative differential resistivity R_{SC} amplifies the generated oscillations, while the two oppositely connected diodes D_1 and D_2 limit the amplitude of the oscillations. Applying Kirchoff's laws, the dynamical laws governing the currents and voltages, to the circuit in **Figure 3** the current and voltage equations are easily found to be

$$\mathcal{E} - R_{\text{pot}}\, I_1 - U - \mathcal{L}\dot{I}_1 = 0, \tag{10}$$

$$\dot{U} - \frac{2}{C_0} I_2 - \frac{I_6}{C} = 0, \tag{11}$$

$$\frac{I_6}{C} - r\dot{I}_7 - L\ddot{I}_7 = 0, \tag{12}$$

$$I_5 - F(U) = 0, \tag{13}$$

$$I_1 - I_2 - I_5 = 0, \tag{14}$$

$$I_2 - I_3 - I_6 = 0, \tag{15}$$

$$I_3 - I_4 - I_7 = 0, \tag{16}$$

This system of equations governs the circuit operation.

The current through the diodes D_1 and D_2 (I_4) is governed by the volt-ampere characteristic

$$I_4 = \sigma_d\left[(\mathcal{E}_L - U_0)\theta(\mathcal{E}_L - U_0) + (\mathcal{E}_L + U_0)\left(1 - \theta(\mathcal{E}_L + U_0)\right)\right], \tag{17}$$

where $\mathcal{E}_L$ is the voltage between the two ends of the inductance L, σ_d is the diode's conductance when it is "open" and U_0 is its opening voltage. Obviously I_4 is entirely determined by the inductance voltage, which from equation (17) is completely determined by I_7 and $\dot{I}_7$.

Eliminating I_2, I_3, I_5 and I_6 and defining the additional functions $Y(t) = \dot{I}_7(t)$ and $D(x) = \sigma_d\left[(x - U_0)\theta(x - U_0) + (x + U_0)\left(1 - \theta(x + U_0)\right)\right]$, we can rewrite this system as

$$\dot{I}_1 = \frac{\mathcal{E}}{\mathcal{L}} - \frac{R_{\text{pot}}}{\mathcal{L}} I_1 - \frac{U}{\mathcal{L}}, \tag{18}$$

$$\dot{I}_7 = Y, \tag{19}$$

$$\dot{U} = \left(\frac{2}{C_0} + \frac{1}{C}\right)(I_1 - F(U)) - \frac{1}{C}\left(D(rI_7 + LY) + I_7\right), \tag{20}$$

$$\dot{Y} = \frac{1}{LC}\left(I_1 - I_7 - F(U) - D(rI_7 + LY)\right) - \frac{r}{L}Y. \tag{21}$$

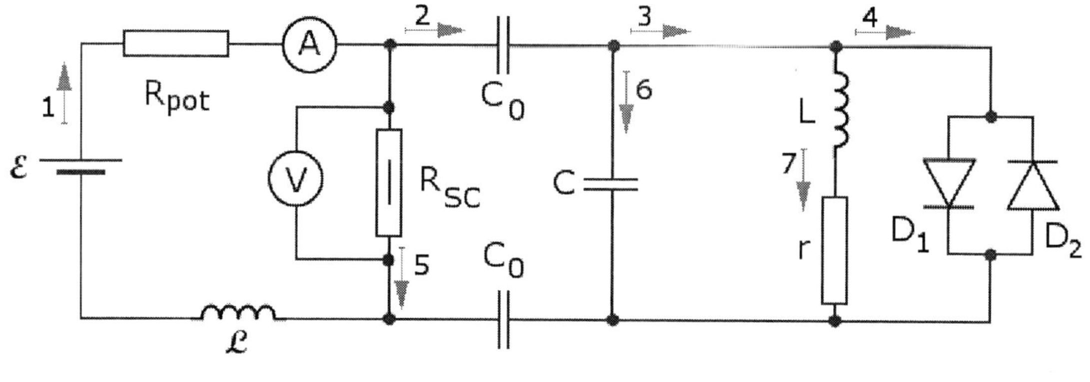

Figure 3 Circuit with negative differential conductivity R_{SC}^{-1} of a superconductor supercooled below T_c in an external electric field. The direct current is created by a battery $\mathcal{E}$ and the current I_1 (denoted by the arrow with number 1 in the figure) measured by the ammeter A is regulated by the potentiometer R_{pot}. High-frequency current through the battery is stopped by the large inductance $\mathcal{L}$ and voltmeter V measures the dc voltage of the thin superconductor film. The dc current supply is galvanically separated from the parallel LC resonator by the large capacitors C_0. The resonance frequency $\omega_0 = 1/\sqrt{LC}$ is fixed by the values of the small capacitor C and the small inductance L with internal resistance r. The amplitude of the electric oscillations is limited by diodes D_1 and D_2. The different currents in the circuit denoted in the figure by their numbers only, participate in the circuit equations (10)-(16).

4. RESULTS

The calculated oscillations generated by the thin supercooled below T_c superconductor from the circuit in **Figure 3** are shown in **Figure 4** with the numerical values in **Table 1**.

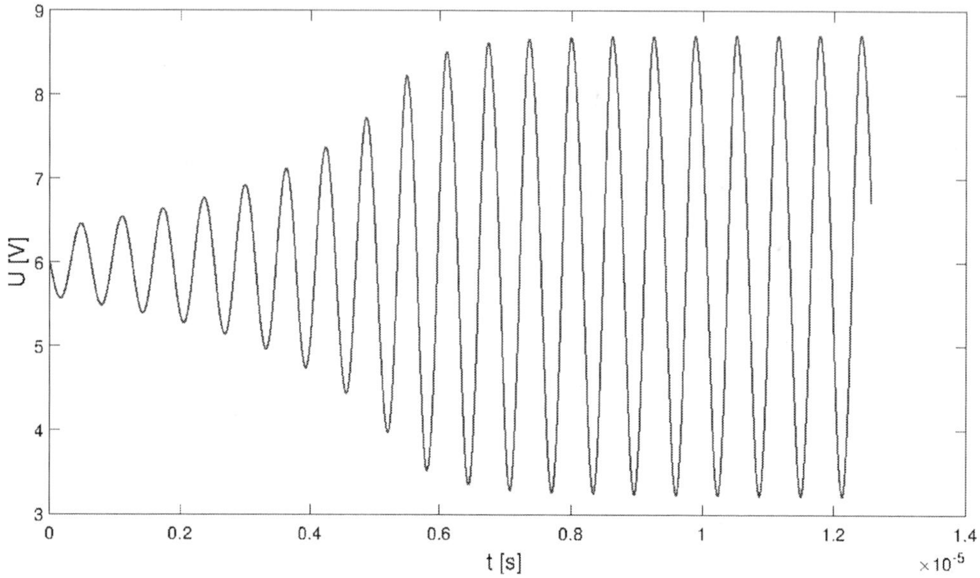

Figure 4 Stable electric oscillations generated by the circuit in **Figure 3** using the values in **Table 1**

Table 1 Numerical values of the circuit elements and parameters from Figure 3 and the equations (10)-(16)

Element	Value	Parameter	Value
$\mathcal{E}$	6 V	l	2 mm
R_{pot}	1 kΩ	w	2 mm
$\mathcal{L}$	100 μH	ξ	2 nm
C_0	100 μF	S_n	137
C	10 nF	T_s	80 K
L	1 μH	T_c	90 K
r	0.1 Ω		
U_0	2 V		
σ_d	0.1 Sm		

The oscillations are in the megahertz region as it is expected from the values of L and C from **Table 1**. Both diodes limit the oscillations amplitude preventing exponential grow. These results give an optimistic estimate that it is worth starting an experimental research with the available nanostructured samples.

5. CONLUSIONS

We demonstrate megahertz range of electric oscillations created by the negative differential conductivity in external electric field of supercooled below T_c superconductor. Using distributed elements, the frequency of generations is fixed but using as a resonator 2D plasmons [6] the resonance frequency can easily reach terahertz range. The theory of damping rate of 2D plasmon will be given elsewhere.

The development work should start with investigation of current voltage characteristics of a nanostructured superconductor similar to a field effect transistor. The area of source and drain should be maximal in order to ensure high currents. If the gate is implemented by de-pairing magnetic material, the described sample will be a new type superconducting field effect transistor operating in the terahertz range. Crucial starting point will be observation of annulation of differential conductivity at decaying electric field below T_c. It was suggested by Gor'kov [7] which tried to interpret the generation of high-frequency electric oscillations in thin superconducting films observed half a century ago [8]. Now the time has come to develop technical applications based on hybrid nanostructured superconductors.

ACKNOWLEDGEMENTS

This research was presented in COST school on quantum materials and workshop on vortex behavior in unconventional superconductors, 7-12 October, 2018 in Braga, Portugal. Authors are thankful to Anna Palau and Andrey Varlamov for the stimulating discussions during the COST school on quantum materials.

Authors will greatly appreciate any correspondence related to an experimental realization of the proposed generator.

REFERENCES

[1] MISHONOV, T.M. and MISHONOV, M.T. Theory of terahertz electric oscillations by supercooled superconductors. *Supercond. Sci. Technol.* 2005. vol. 18, pp. 1506--1512.

[2] MISHONOV, T.M., PACHOV, G.V., GENCHEV, I.N., ATANASOVA, L.A. and DAMIANOV, D.Ch. Kinetics and Boltzmann kinetic equation for fluctuation Cooper pairs. *Phys. Rev. B.* 2003. vol. 68, p. 054525.

[3] MISHONOV, T.M., POSAZHENNIKOVA, A. and INDEKEU, J. Fluctuation conductivity in superconductors in strong electric fields. *Phys. Rev. B.* 2002. vol. 65, p. 064519.

[4] VARLAMOV, A.A. and REGGIANI, L. Nonlinear fluctuation conductivity of a layered superconductor: Crossover in strong electric fields. *Phys. Rev. B.* 1992. vol. 45, p. 1060.

[5] HURAULT, J.P. Nonlinear Effects on the Conductivity of a Superconductor above Its Transition Temperature. *Phys. Rev.* 1969. vol. 179, p. 494.

[6] MISHONOV, T. and GROSHEV, A. Plasmon excitations in Josephson arrays and thin superconducting layers. *Phys. Rev. Let.* 1990. vol. 64, p. 2199.

[7] GORKOV, L.P. Singularities of the resistive state with current in thin superconducting films. *JETP Let.* 1970. vol. 11, p. 32.

[8] CHURILOV, G.E., DMITRIEV, V.M. and BESKORSIY, A.P. Generation of high-frequency oscillations in thin superconducting thin films. *JETP Let.* 1969. vol. 10, pp. 146-148.

SESSION B
INDUSTRIAL AND ENVIRONMENTAL APPLICATIONS OF NANOMATERIALS

Chairmen

Prof. RNDr. Radek ZBORIL, Ph.D.	Palacky University Olomouc, Czech Republic, EU
Prof. Dr. Ing. Miroslav CERNIK, CSc.	Technical University of Liberec, Czech Republic, EU
Prof. Ing. Miloslav PEKAR, CSc.	Brno University of Technology, Czech Republic, EU

GRAPHENE EMBEDDED BITUMINOUS ROADS: A NEW LOOK INTO THE DESIGN OF COMPLEX NANOCOMPOSITES

Govindan INDUCHOODAN[1], Bijan ADL-ZARRABI[1], Jan SWENSON[2], Luping TANG[1]

[1]*Chalmers University of Technology, Dept. of Architecture and Civil Engineering, Göteborg, Sweden, EU*
govindan@chalmers.se, bijan.adl-zarrabi@chalmers.se, tang.luping@chalmers.se
[2]*Chalmers University of Technology, Dept. of Physics, Göteborg, Sweden, EU, jan.swenson@chalmers.se*

Abstract

The chemistry of bitumen and the percentage of asphaltene can vary significantly depending on external and internal factors such as the source and method of bitumen extraction. This makes graphene embedded bitumen a complex ternary phase system. Within the framework of this paper, we focus on the phase stability and particle interaction between graphene and asphaltene in order to optimize the process of tailoring the microstructure of graphene embedded bitumen to improve the gas barrier and mechanical properties. We use characterization techniques such as rotational-rheometer, DSC, chemical titration and road engineering standards such as RTFOT and penetration-grade to study the critical-clustering-concentration, phase behavior and the impact of embedding graphene to manufacturing process of asphalt. Phase separation and agglomeration of dissimilar nanoparticles was observed when graphene embedded bitumen (1 % vol) was subjected to RTFOT, increasing the change in viscosity by around 13 % when compared to 10 % in std. bitumen. Hence this paper aspires to provide a brief look into tailoring of nanocomposites with dissimilar nanoparticles and embedding nanoparticles such as graphene into complex matrices.

Keywords: Graphene, material design, bitumen, material characterization, dissimilar nanoparticles

1. INTRODUCTION

Carbon based nanoparticles have been of interest to researchers for decades. Carbon allotropes such as graphene have been embedded in various matrices to improve the material property such as gas-barrier, thermal-conductivity and tensile-strength of the parent material. One such material is bitumen, for the reason that nearly 70 % of all global road is made of asphalt (bitumen + aggregate) [1]. Bitumen is a petroleum byproduct with complex chemistry, contains naturally occurring nanoparticles known as asphaltene [2]. There are global attempts by researchers from various fields to modify the properties of bitumen using nanoparticles. They most often aim to improve material properties such as thermal conductivity, electrical conductivity, tensile strength, rutting resistance, chemical aging and fatigue.

The two key strategies widely used are to either directly incorporate the additive to bitumen or enhance the polymer that is used in the polymer modified bitumen (PMBs). This paper does not intend to review existing works by other researchers, because Yang et al. and Fang et al. have reviewed near 100 such works by researchers in this field [3][4]. For the current study the assumptions formed by other authors, for their publications, are of interest.

Firstly, before attempting to study the mechanism of incorporating nanomaterial into bitumen it is important to define bitumen, its microstructure and properties. The true microstructure of bitumen is still widely debated but it is well established that bitumen can be divided based on various factors such as acidity, polarity, solubility and molecular weight. Separating bitumen fractions based on its solubility is the most widely used and well-established method [2]. Using this method bitumen is divided into Saturates, Aromatics, Resins and Asphaltenes. The most common and simple model that is followed to explain the arrangement of these 4 fractions is the sol-type or gel-type [2].

Yen et al. derived from x-ray diffraction that asphaltene exists as nanosheets about 8 layers [5] thick. Some well cited works, such as the one by Jahromi et al., assume bitumen to be a complex polymer with asphaltene existing as particles that are larger than 10 mm [6]. In their paper, Jahromi et al. illustrate a diagram in which nanoparticles adhere perpendicular to the asphaltene surface. This is contradictory to the assumption that asphaltene is nanometers in length and a few layers thick [2]. In another work, Yu et al. provide a model where nanoparticles are dispersed in a bitumen matrix with random orientation [7].

Arguably, in the two afro mentioned papers and similar works by other researchers, the authors have assumed the behavior of bitumen as that of a viscoelastic solid-matrix rather than a viscoelastic liquid, hence assuming nanoparticle embedded bitumen to be a nanocomposite. Nanoparticles embedded bitumen should not be treated as a nanocomposite in the same way as incorporating materials such as fillers or Portland cement into bitumen, during the design of asphalt, does not categorize the newly modified bitumen as composite. Thus, modified bitumen cannot be categorized as a composite.

This paper aspires to study the first strategy and understand the underlying mechanism and limitations of directly incorporating nanoparticles into bitumen. Two critical assumptions are made during this study. Firstly, Yen's interpretation of asphaltene is considered and hence asphaltene is assumed to exist as nanoparticles and bitumen is treated as colloidal system [5]. It was shown by Induchoodan et al. that there exists a critical concentration above which nanoparticles began to agglomerate in polymer matrix [8]. Hence, the second assumption is that there must exist a critical concentration that restricts further addition of nanoparticles into bitumen and this concentration should be below the limitation predicted by Induchoodan et al for a homogenous matrix.

2. METHOD

2.1. Materials

M-grade XGnP graphene nanoplatelets (GnP) was used for the study. The GnP has a surface area of 120-160 m^2g^{-1}, thickness of 6-8 nm, density of 2.2 gcm^{-1} and diameter of 25 µm. XGnP was purchased from XGScience. 70/100 grade bitumen was supplied by Skanska for the studies. Analytical grade N-Heptane was purchased from Sigma Aldrich.

2.2. Sample Preparation

XGnP was further exfoliated using shear-mixer but no chemical modification was performed on the material. Graphene and bitumen were blended using a shear mixer with 4-blade setup. Graphene was sputter coated with gold for a duration of 60 seconds while asphaltene precipitate was sputter coated with palladium for a duration of 30 seconds.

2.3. Experimental Setup

Anton paar MSR302 with a 25 mm plate-plate setup with a gap of 1 mm and temperature controller was used for the rheological studies. Oscillation test was performed on all samples with a shear strain from 0.01 to 0.16 at a constant frequency of 10 rad/s and the experiments were repeated from a temperature range of 40 °C to 70°C with a step increment of 6°C. For the thermal oxidation of bitumen, a rolling thin film oven test (RTFOT) was performed. Two 35 g samples for each type of bitumen was prepared and subjected to 163 °C for 85 mins in an aerobic environment with a constant air supply of 4000 ml/min and the entire setup was in constant rotation at a rate of 15 rpm.

Chemical extraction of asphaltene from bitumen was performed by dissolving bitumen in N-Heptane. The solution was stirred at 1000 rpm for 1 hour and then centrifuged using a Sigma 4-16 at 4500 rpm for 15 mins in two 30 ml sample holders. The solution was decanted after the centrifugation. This was repeated 3 times and the precipitate was air dried. After which, Graphene and asphaltene were observed under a FEG-SEM.

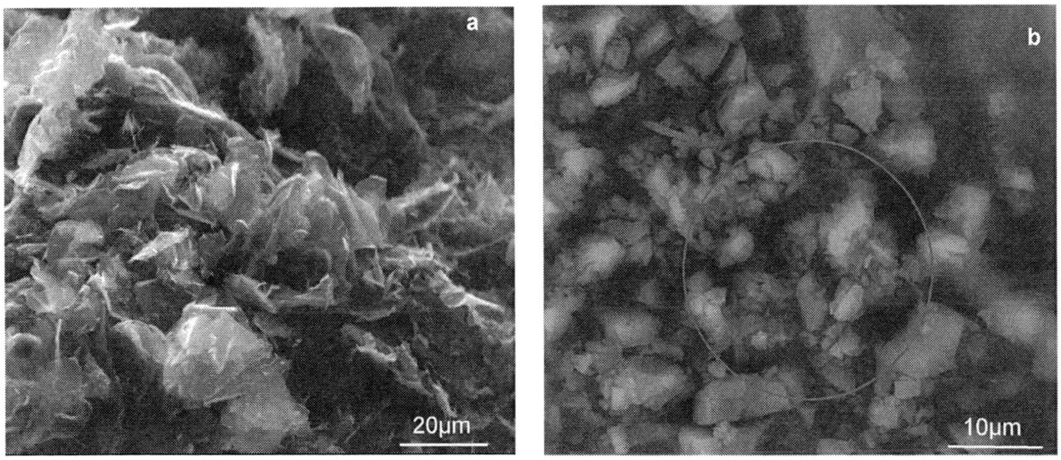

Figure 1 SEM image of a) exfoliated graphene flakes.
b) agglomerated asphaltene particles after chemical extraction from bitumen

3. HYPOTHESIS

Asphaltene should in theory have a diameter of 1-2 nm [2]. This goes to show that there are pre-existing nanoparticles in bitumen and many papers that were reviewed by Yang et al. and Fang et al. [3,4] do not account for these nanoparticles. This further implies that bitumen must not be considered as a uniform binder, polymer i.e. matrix in a conventional sense, but rather as a complex colloidal system with existing nanoparticles. Hence, adding unmodified graphene to bitumen makes the mixture an n+1 phase system, where n>0. For simplification, bitumen is assumed to be a simple two-phase colloidal system, i.e. n=2, we assign bitumen with a hard phase and liquid phase. The hard phase being n-heptane insoluble asphaltene, and the rest of the fractions are considered as a homogeneous liquid phase.

$$\phi_{tot} = \phi_{lp} + \phi_{hp} \tag{1}$$

Where ϕ_{tot} is the total volume fraction of the system and ϕ_{tot} = 1. ϕ_{lp} and ϕ_{hp} are the volume fraction of liquid and solid phase respectively. By adding graphene to this system (ϕ_{gr}, volume fraction of graphene), it has gone to being n=3. Due to the similar dimensions and interactive forces that both graphene and asphaltene exert, there must be a common upper limit to the amount of graphene that can be added, because the two families of nanoparticles could form agglomerates with each other. This upper limit is represented by ϕ ', where ϕ ' is the volume fraction beyond which nanoparticles in bitumen start to form agglomerates. Hence, there must exist an optimal ratio of graphene to asphaltene for a nanomodified bituminous system.

i.e. $$\phi_{tot} = \phi_{lp} + \phi_{hp} + \phi_{gr} \tag{2}$$

where $\phi_{hp} + \phi_{gr} < \phi'$

Furthermore, the volume fraction of asphaltene in bitumen increases during processing of asphalt and this increase has to be accounted during calculation. When bitumen is subjected to elevated temperatures, i.e. above 130°C, the rate of thermal oxidation of bitumen increases, this phenomenon is called short term aging. During short term aging, asphaltene fraction in bitumen increases, due to the chemical process that leads to the production of asphaltene from resins [2].

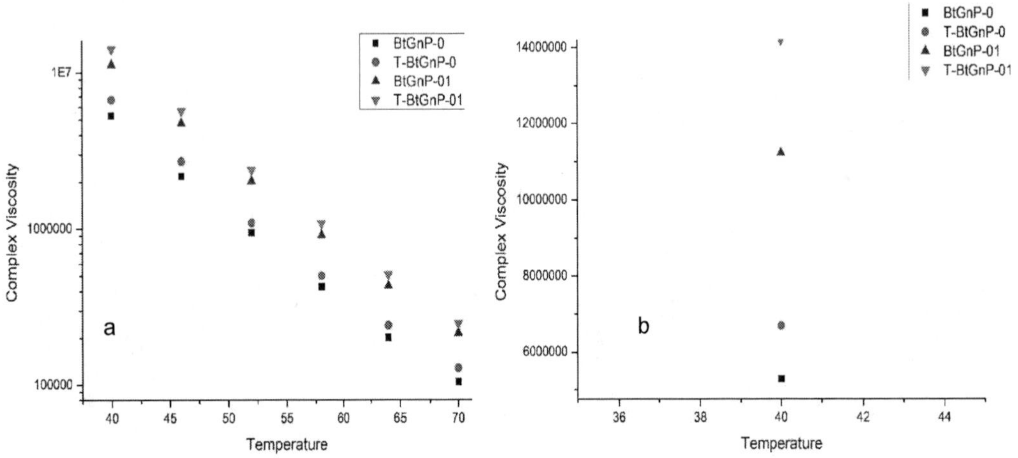

Figure 2 a) Log of complex viscosity vs temperature for neat binder and graphene embedded bitumen, before and after thermal oxidation. $\mathrm{BtGnP-0}$ and $\mathrm{T-BtGnP-0}$ are neat binder before and after thermal oxidation and $\mathrm{T-BtGnP-0}$ and $\mathrm{T-BtGnP-01}$ are graphene embedded bitumen, before and after thermal oxidation. b) Complex viscosity vs temperature of neat binder and graphene embedded bitumen, before and after thermal oxidation, at 40°C.

4. RESULTS AND DISCUSSION

(**Figure 1**) shows graphene sheets and asphaltene particles. It is evident from **Figure 1b** that asphaltene appears to be an independent hard phase, ϕ_{hp}. Asphaltene precipitates are circled red to differentiate it from the background. Asphaltene starts to precipitate when N-heptane introduced to bitumen. As the ratio of N-heptane to bitumen is increased the amount of asphaltene precipitated increases until it reaches a saturation point [2]. The precipitate is often obtained as an agglomerate and this is observed in **Figure 1b**.

During the rheological study bitumen was tested within its linear viscoelastic region and the properties were studies. During this study the complex viscosity (η*) was measured instead of the shear viscosity (η), because at low frequency the viscous viscosity (η') component of the complex viscosity is approximately equal to the shear viscosity.

$$\eta^* = \eta' + i\eta'' \tag{3}$$

$$\eta'_{\omega\to0} = \eta \tag{4}$$

This is true because at lower temperatures the contribution of the imaginary viscosity (η'') is neglectable and η* is similar to the η' , as shown by Li et al. [9]. Since the working frequency is 10 radians the condition is applied and η* is studied to understand the hypothesis.

$$\eta^*_{\omega\to0} = \eta \tag{5}$$

In this study the nomenclature BtGnP shall stand for bitumen embedded with graphene and following the digits represent the volume of graphene in bitumen. The prefix 'T' shall be used to describe any material that has been thermally oxidized and should not be confused with temperature. Neat binder, which is bitumen with no graphene ($\mathrm{BtGnP-0}$), and graphene embedded bitumen with 1 % volume of graphene ($\mathrm{BtGnP-01}$) were thermally oxidized, compared and studied.

Graphene, asphaltene and graphene embedded bitumen were successfully characterized. $\mathrm{BtGnP-01}$ exhibited immediate signs of agglomerations. Considering $\eta^*_{\mathrm{BtGnP-01}}$ and $\eta^*_{\mathrm{T-BtGnP-01}}$ as the complex viscosities before and after RTFOT of $BtGnP-01$ and similarly $\eta^*_{BtGnP-0}$ and $\eta^*_{T-BtGnP-0}$ as the complex viscosities before and after RTFOT for $BtGnP-0$, we can plot a graph with log of η^* as abscissa and temperature (T) as the ordinate, as shown in (**Figure 2 a**). It is immediately observed that there is a larger increase in complex viscosity of GnP-01 when compared to $BtGnP-0$ after thermal oxidation by RTFOT. Upon further observation, it is evident that η^* for $BtGnP-01$ increased by 12.21-14.65 % while η^* for $BtGnP-0$ increased by only 10.31 %, (**Figure 2b**).

Rheological studies of polymers have shown that the size of the molecules/particles has a direct relationship to the viscosity of the material. The larger the particle size the higher is the viscosity [10] and Induchoodan et al. showed that increasing the volume fraction of the nanoparticles leads to an increase in the viscosity of the nanocomposite [8]. So, both neat binder and graphene embedded bitumen should show an increase in the viscosity after RTFOT.

In both $BtGnP-0$ and $BtGnP-01$, there is an increase in asphaltene concentration. Thus, the contribution of ϕ_{hp} must be similar and ϕ_{gr} is fixed, and its contribution is observed in the difference between $\eta^*_{BtGnP-01}$ and $\eta^*_{BtGnP-0}$. Thus, the increase in the complex viscosity of $BtGnP-01$ after RTFOT can be related to the agglomeration of graphene and asphaltene. Therefore, the total volume of nanoparticles $\phi_{hp}+\phi_{gr}$ has exceeded ϕ', and hence

$$\eta^*_{T-BtGnP-01} - \eta^*_{BtGnP-01} > \eta^*_{T-BtGnP-0} - \eta^*_{BtGnP-0} \quad (6)$$

This goes to validate the hypothesis proposed in this paper. Thus, while directly incorporating nanofillers such as graphene into bitumen the total volume fraction ϕ', for graphene and asphaltene, must be accounted rather than ϕ_{gr} only.

5. CONCLUSION

Bitumen is considered to be a colloidal solution. Asphaltene and graphene were shown to be dissimilar nanoparticles. An upper limit ϕ' was predicted for bitumen, beyond which graphene and asphaltene start to agglomerate with each other. This was experimentally determined by thermally oxidizing bitumen, both with and without graphene, and comparing their respective complex viscosities.

ACKNOWLEDGEMENTS

The research is funded by the Norwegian Public Road Administration. The authors are grateful to Skanska for providing their lab facility for research.

REFERENCE

[1] Introduction, in: HUNTER, Robert, SELF, Andy and READ, John. *Shell Bitumen Handbook (6th Edition) - 4.2.2 More Complex Models,* pp. 05-10.

[2] Constitution and Structure of Bitumens, in: HUNTER, Robert, SELF, Andy and READ, John. *Shell Bitumen Handbook (6th Edition) - 4.2.2 More Complex Models,* pp. 54-57.

[3] YANG, Jun and TIGHE, Susan A Review of Advances of Nanotechnology in Asphalt Mixtures. *Procedia - Social and Behavioural Sciences*. 2013. vol. 96, no. 2, pp. 1269-1276.

[4] FANG, Changqing, RUIEN, Liu, and SHAOLONG, Yan, Nanomaterials Applied in Asphalt Modification: A Review, *Journal of Materials Science & Technology*. 2013. vol. 29, pp. 589-594.

[5] DICKIE, John P and YEN, Teh Fu, Macrostructures of the Asphaltic Fractions by Various Instrumental Methods, *Analytical Chemistry*. 1967, vol.39, pp. 1847-1852

[6] JAHROMI, Saeed and KHODAII, Ali, Effects of Nanoclay on Rheological Properties of Bitumen Binder, *Construction and Building Materials*. 2009, vol. 23, pp. 2894-2904

[7] YU, Jian-Ying, FENG, Peng-Cheng, ZHANG, Heng-Long, and WU, Shao-Peng, Effect of Organo-Montmorillonite on Aging Properties of Asphalt, *Construction and Building Materials.* 2009, Vol. 23, pp. 2636-2640

[8] INDUCHOODAN, Govindan and KÁDÁR, Roland, Tailoring Polymer Nanocomposite Microstructure by Controlling Orientation, Dispersion and Exfoliation of GnP in LDPE via Extrusion Flow, *Annual Transactions of the Nordic Rheology Society*. 2016, vol. 26, pp. 187-191

[9] LI, Shu-Ping, ZHAO, Ge and CHEN, Hong-Yuan, The Relationship between Steady Shear Viscosity and Complex Viscosity, *Journal of Dispersion Science and technology*. 2006, vol.26, pp. 415-419

[10] RICHARDS, John, Viscosity and the Shapes of Macromolecules, *Journal of Chemical Education*. 1993, vol. 70, pp. 685-689

POSSIBILITIES OF REMOVING THE CRUST FROM THE SURFACE OF HISTORIC STONE WITH NANOPARTICLES I TiO_2

[1]Klára KROFTOVÁ, [2]David ŠKODA, [2]Ivo KUŘITKA

[1] *Department of Architecture, Faculty of Civil Engineering, CTU in Prague, Czech Republic, EU,* *Klara.Kroftova@seznam.cz*

[2] *Centre of Polymer Systems, Tomas Bata University in Zlin, Zlín, Czech Republic, EU,* *dskoda@utb.cz, kuritka@utb.cz*

Abstract

The most common tasks, we encounter during restoration stone historical monument buildings, include mainly cleaning and removing of the crust from their surface. As part of the experimental research of nanomaterials and the widespread use of these materials for the preservation and restoration of stone monuments, the use of TiO_2 is currently being addressed. Applications have shown that this oxide can be used to protect against microorganisms and UV degradation due to its antifungal and antibacterial effects. At the same time, the photocatalytic properties of TiO_2 represent the possibility of chemical decomposition and the creation of a self-cleaning environment and self-sterilization area. Applying lime nanosuspensions with TiO_2 nanoparticles will reduce the rate of deposition of pollutants as a result of photocatalytic phenomena and thus slow the development of degradation processes.

Keywords: Lime nanosuspension, TiO_2, titanium dioxide, self-cleaning surfaces, stone crust, historical material

1. INTRODUCTION

The aging of historical objects and their materials is mainly influenced by degradation processes caused by long term effects of cyclical weather changes and harmful substances contained in the atmosphere. Among the most important negative factors are, in addition to repeated changes in temperature and humidity, especially carbon dioxide, sulfur dioxide, acid peat and solid particles contained in the atmosphere [1]. The settling of impurities and chemical processes occurring between the components of the external environment and the building material itself lead to the formation of a crust firmly embedded in the surface structure of the material. The formation of crusts damages historical objects not only aesthetically, but above all by the development of surface tensions between the crust and the core of material with different material properties, and chemical processes that influence the composition of the material and, on the other hand, lead to the development of tension in the structure of the material. Last but not least, it provides uneven surface crusts for depositing impurities from the air (resulting in its gradual darkening) and subsequently the broth for the development of microorganisms. The nature and development of degradation processes and their manifestations are influenced in this context also by the type of material (so called lythotype: marble, limestone, travertine, sandstone, granite, etc.), surface grain, material humidity, chemical composition, porosity, structure and texture of materials.

As a result of these negative effects, degradation processes occur in surface layers, which in the case of historical building materials lead to structural breakdown, loss of coherence of individual layers and consequent loss of surface layers [1].

Within the development of the use of nanomaterials in the care of cultural heritage, it has been demonstrated that metal oxides, especially ZnO and TiO_2, can be used as protection against microorganisms and UV degradation due to their antifungal and antibacterial effects [2-4]. Studies show, on one hand, the possibility of incorporating nanoparticles into building materials [5-8], the next route to new applications leads through the incorporation of nanoparticles which either functionalize suspensions themselves or can be functionalized.

Titanium dioxide is one of the most widely known photocatalytic materials used not only in the building industry, but also in other industries, eg for water and air purification, and for its antibacterial properties it is also used as an ingredient in eg tiles, PVC, etc. [9]. It is also known from the studies that nano-TiO_2 coating provides excellent corrosion protection for marine submerged materials [10]. The photocatalytic properties of TiO_2 were discovered by A. Fujishima and K. Honda, who first used titanium dioxide to decompose chemicals [11-13]. The photocatalytic properties of TiO_2 are stimulated by UV radiation and can suppress surface layers by algae and lichens. The intensity of UV radiation with a peak below 385 nm [17-20] is significantly influenced by the effect of TiO_2 [14-17] so that photocatalytic activity is triggered which produces reactive particles capable of oxidative degradation of pollutants.

2. MATERIALS AND METHODS

The research project "Development and Research of Materials, Methods and Technologies for the Restoration, Preservation and Strengthening of Historic Masonry Constructions and Surfaces and Systems of Preventive Conservation of Cultural Heritage Buildings Threatened by Anthropogenic and Natural Hazards" addresses besides the issues of the care of the historic materials and its surfaces by the different types of lime nanosuspensions. The institutions currently participating in the research project are the Faculty of Civil Engineering, CTU in Prague and the Centre of Polymer Materials, TBU in Zlín.

2.1. The Lime nanosuspension with titanium dioxide

For the purpose of experimental research, lime nanosuspensions were prepared to further prepare and subsequently test suspensions doped with the active substances for a specific use, ie to clean the surfaces of building materials. Limestone nanosuspensions form nanoparticles of calcium hydroxide scattered in alcoholic environments, which are primarily used to consolidate the surface layers of lime-based building materials. By supplementing these nanosuspensions with other effective nanoparticles (eg TiO_2), there are means that can provide multiple efficiencies - consolidation, self-purification and biocide protection [18]. For this purpose, the synthesis of $Ca(OH)_2$ (Ca4, **Figure 1a**), which was subsequently enriched with titanium dioxide (Ca4-Ti, **Figure 1b**), was developed within the research project.

The basic nanosuspension of Ca4 was made from the weight of calcium methoxide $Ca(OCH_3)_2$ (7.18 g, 70.2 mmol). The charge of calcium methoxide ($Ca(OCH_3)_2$) was dispersed in isopropyl alcohol (150 ml). To this suspension was added 100 ml of distilled water. The reaction time, constant stirring, was 90 minutes. After the reaction, the product was separated by centrifugation and washed with distilled water and isopropyl alcohol. The yield was 4.16 g [19].

To test the self-cleaning effect of titanium dioxide, the TIO_2 nanoparticles were prepared by a procedure known from the literature, an acid-assisted sol-gel method [20]. For the purpose of the research project, the synthesis was modified: Ti_2 was applied as an addition of isopropyl alcohol. Hydrolysis of $Ti(OiPr)_4$ in the presence of acetic acid was carried out by mixing 150 ml of H_2O, 5 ml of $Ti(OiPr)_4$ and 50 ml of CH_3COOH. The reaction was stirred at 70 °C for 3 hours on a magnetic stirrer. The product was separated by centrifugation and dried. To Ca4-Ti nanosuspension 1.0 g of Ti_2 sample was added. After addition of TiO_2, the slurry was dispersed in an ultrasonic bath.

The XRD powder analysis on a Rigaku MiniFlex 600 diffractometer equipped with a CoKα cathode (λ = 1.7903 Å, 40 kV, 15 mA) was used to determine the crystalline structure of the product. The observed diffraction of the TiO_2 nanoparticle sample corresponds to the JCPDS card card for TiO_2 (anatase) 01-071-1169 (**Figure 2**). Diffraction record of $Ca(OH)_2$ forming the initial slurry of Ca4 is also shown in this figure. The TiO2 nanoparticle electron microscope image is shown in **Figure 1b**.

Figure 1 Electron microscope images: a) nanosuspension CA4 ($Ca(OCH_3)_2$ precursor); b) CA4-Ti nanosuspension (Photo CPS UTB)

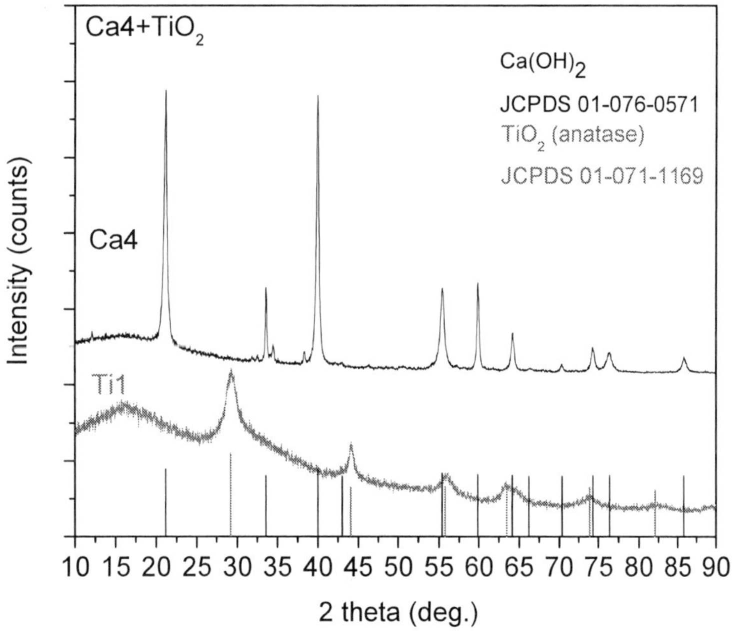

Figure 2 XRD diffractogram of prepared TiO_2 nanoparticles and Ca4 nanosuspension

2.2. Application of nanomaterials on stone samples

The application of Ca4-Ti nanosuspension was performed in situ on stone (marlite) walls in Prague, the surface of which is degraded by surface crust and massively contaminated by microorganisms (**Figure 3**). In a place exposed to sunlight, dry stone blocks were used to apply nanosuspension so that a photocatalytic reaction could take place. On their surface, the tested nanosuspension (Ca4-Ti) was applied sequentially using the sprayer. The application of nanosuspension Ca4-Ti with a total volume of 20 ml was repeated with the sprayer, where the individual applications followed successively after application of the applied composition, without its significant leakage on the surface.

Figure 3 Detail of the stone block before application of the nanosuspension of Ca4-Ti (stone - marlite - wall in Prague): a) before application; b) after application

Skins and scraps for laboratory bioassay were also performed. From the samples taken, the microorganisms were transferred to nutrient agar (Czapek Dox agar), which was cultured for 7 days at 23 ± 3 °C. The molds were isolated to individual species from which microscopic native preparations were prepared. Mold identification was performed on the Olympus BX41 microscope according to significant characters (in conformity with Fassati (1979), Singha, Frisvalda, Thrane Mathura (1991) and Samson, Houbraken, and Thrane (2010) microscopes). To determine the presence of algae, a 50 µl culture medium was inoculated with stone granulate (weight approximately 0.5 gram). Cultivation was carried out on a Ika vibrax shaker (150 rpm) for 14 days at 25 °C ± 5 °C. The illumination intensity was 1.5 kLux with a cycle of 18 : 6 (day : night). At the end of the 1st week of cultivation, bacteria were identified on the samples on 50 % of the area and mold on 30-40 % of the area.

2.3. Efficacy assessment of nano - titanium dioxide preparations

After 3 weeks of application, control documentation, skins and scraps were performed. The collected samples were transferred to Czapek Dox agar broth and allowed to cultivate for 7 days at 23 ± 3 °C. After a week of cultivation, significant algae and mold loss (only 10-15 % of the sample area), increased bacteria (up to 50 % of the sample area) were observed in all the samples studied. By visual comparison and by laboratory analysis, no change in the extent, character and color of the surface crust has been demonstrated.

3. CONCLUSION

The titanium dioxide suspensions were partially proven to be directly applied to areas with a massive presence of microorganisms (algae, bacteria, fungi). TiO_2-doped nanosuspension results show that mold and algae growth at Czapek Dox agar has decreased by about 20-30 %. However, experimental tests of nanosuspension doped with TiO_2 nanoparticles did not clearly show a significant results of the selfcleaning effect. The possibility of removing gypsum crusts using prepared lime nanosuspension with TiO_2 nanoparticles has not been demonstrated.

The results obtained have been influenced, among other things, by the particle size, concentration and stability of the suspension, the carrier media, the presence of ballast substances which suppress the self-cleaning and biocidal effect, etc. As mentioned above, the mineralogical composition also has a significant effect on the activity of the antifungicide and antibacterial agent material, its pH value and porosity.

These results were also limited to a relatively small number of in situ and laboratory tests. Obtaining fully objective results requires more extensive experimental research so that the results are statistically evaluable. At the same time, the need for further research stems from the need to eliminate the potential risk of failure and damage, especially on listed buildings. In addition to verifying any modified functional nanomaterials, it will be necessary to address the requirements of durability, long-term reliability, and last but not least reversibility requirements.

ACKNOWLEDGEMENTS

This article was written as part of the NAKI DG16P02M055 research project "Development and Research of Materials, Methods and Technologies for the Restoration, Preservation and Strengthening of Historic Masonry Constructions and Surfaces and Systems of Preventive Conservation of Cultural Heritage Buildings Threatened by Anthropogenic and Natural Hazards (2016 - 2022, MKO/DG)".

REFERENCES

[1] HOŠEK, J. and MUK, J., *Omítky historických staveb*, 1989, ISBN 80-04=23349-x, p. 117-120

[2] GRAZIANI, L., QUAGLIARINI, E., BODIOLI, F. and D´ORAZIO, M. *Durability of self-cleaning TiO2 coatings on fired clay brick façades: Effects of UV exposure and wet & dry cycles.* Building and Environment [online], 2014. vol. 71, pp. 193-203, ISSN 0360-1323, [viewed 2016-02-19] Available from: DOI: 10.1016/j.buildenv.2013.10.005.

[3] DE FILPO, G. et al *Preventing fungal growth in wood by titanium dioxide nanoparticles*, International Biodeterioration & Biodegradation, 2013. vol. 85, pp. 217-222, [viewed 2017-05-14] Available from: DOI: 10.1016/j.ibiod.2013.07.007

[4] PINHO, L., ELHADDAD, F., FACIO, D.S. and MOSQUERA, M. *A novel TiO2-SiO2 nanocomposite converts a very friable stone into a self-cleaning building material.* Applied Surface Science. 2013, vol. 275, pp. 389-396, ISSN 0169-4332 [viewed 2016-12-18] DOI: 10.1016/j.apsusc.2012.10.142.

[5] LI, H., XIAO, H., YUAN, J. and OU, J. *Microstructure of cement mortar with nano-particles.* Composites - Part B - Engineering, 2004, vol. 35, no. 2, pp. 185-189 [viewed 2018-02-10] Available from: DOI:10.1016/S1359-8368(03)00052-0

[6] OLAR, R. *Nanomaterials and nanotechnologies for Civil Engineering*, Buletinul Institutului Politehnic din Iasi. Sectia Constructii, Architectura; Iasi, 2011. vol. pp. 109-117. [viewed 2018-12-10] Available from: http://www.bipcons.ce.tuiasi.ro/Archive/266.pdf

[7] SOBOLEV, K., VIVIAN, I.F., TORRES-MARTÍNEZ, L.M., VALDEZ, P., ZARAZUA, E., CUÉLLAR, E.L. *Engineering of SiO2 Nanoparticles for Optimal Performance in Nano Cement-Based Materials*, Nanotechnology in Construction, 2009. vol. 3, pp 139-148 [viewed 2018-09-02] Available from: DOI: 10.1007/978-3-642-00980-8_18

[8] GUPTA, S.M., TRIPATHI, M. *A Review of TiO2 Nanoparticles*, Chinese Science Bulletin, 2011, vol. 56, no. 16, pp. 1639-1357, [viewed 2017-09-26] Available from: DOI: 10.1007/s11434-011-4476-1

[9] HOCHMANNOVÁ, L. and VYTŘASOVÁ, J. *Vliv nanočástic TiO2 a ZnO na fotokatalytické a antimikrobiální účinky silikátových nátěrů*, Chemické listy, 2010. vol. 104, pp. 940-944 [viewed 2017-08-21] Available from: https://www.researchgate.net/profile/Libuse_Hochmannova/publication/288291269_Effect_of_TiO2_and_ZnO_nanoparticles_on_photocatalytic_and_antimicrobial_activities_of_silicate_coatings/links/56c2fcef08ae8a6fab59e2ff/Effect-of-TiO2-and-ZnO-nanoparticles-on-photocatalytic-and-antimicrobial-activities-of-silicate-coatings.pdf

[10] LIU, T., Zhang, F., XUE, Ch., LI, L. and YIN, Y. *Structure stability and corrosion resistance of nano-TiO2 coatings on aluminum in seawater by a vacuum dip-coating method*, Surface and Coatings Technology. 2010, vol. 205, no. 7, pp. 2335-2339 [viewed 2017-08-19] Available from: DOI: 10.1016/j.surfcoat.2010.09.028

[11] FUJISHIMA, A.: *Discovery and applications of photocatalysis - Creating a comfortable future by making use of light energy.* In: Japan Nanonet Bulletin, 44th Issue (2005). [viewed 2018-02-10] Available from: http://archive.li/29CtP

[12] HASHIMOTO, K., IRIE, H. and FUJISHIMA, A. *TiO2 Photocatalysis: A Historical Overview and Future Prospects, Japanese journal of applied physics*, 2005. vol. 44, no. 12, pp. 8269 - 8285, [viewed 2018-02-10] Available from: DOI: 10.1143/JJAP.44.8269

[13] BÍLÝ, P. and FLÁDR, J. *Experimentální studie vlivu nanomateriálů na fotokatalytické vlastnosti betonu*, [viewed 2017-10-10] Available from: http://stavba.tzb-info.cz/beton-malty-omitky/14178-experimentalni-studie-vlivu-nanomaterialu-na-fotokatalyticke-vlastnosti-betonu

[14] QUAGLIARINI, E., BONDIOLI, F., GOFFREDO, G.B., LICCIULLI, A. and MUNAFO, P. *Smart surfaces for architectural heritage: Preliminary results about the application of TiO2 based coating on travertine*, Journal of Cultural Heritage, 2012. vol. 13, no. 2, pp. 204-209 [viewed 2015-06-22] Available from: DOI: 10.1016/j.culher.2011.10.002

[15] QUAGLIARINI, E., BONDIOLI, F., GOFFREDO, G.B., LICCIULLI, A. and MUNAFO, P. *Self-cleaning materials on Architectural Heritage: Compatibility of photo-induced hydrophibilicity of TiO2 coatingy on stone surfaces*, Journal of Cultural Heritage, 2013, vol.14, no. 1, pp.1-7 [viewed 2015-06-22] Available from: DOI: 10.1016/j.culher.2012.02.006

[16] GRAZIANI, L., QUAGLIARINI, E., OSIMANI, A., AQUILANTI, L. et al: *Evaluation of inhibitory effect of TiO2 nanocoatings against microalgal growth on clay brick facades under weak UV exposure conditions*, Building and Enviroment, 2013, vol. 64, ppp.38-45 [viewed 2016-02-10] Available from: DOI: 10.1016/j.buidenv.2013.03.003

[17] GRAZIANI, L., QUAGLIARINI, E., BONDIOLI, F. and D´ORAZIO, M. *Durability of self-cleaning TiO2 coatings on fired clay brick facades: Effect of UV exposure and wet &dry cycles*, Building and Enviroment, 2014. vol.71, pp.193-203 [viewed 2016-02-10] Available from: DOI: 10.1016/ j.buidenv.2013.10.005

[18] KROFTOVÁ, K., ŠMIDTOVÁ, M., KUŘITKA I. and ŠKODA, D. *Nanotechnology in the cultural heritage - influence of nanosuspensions adopted by nanoparticles od TiO2 for cleaning the surface of historical plasters*, The Civil Engineering Journal, No. 3, 2017, [viewed 2017-02-10] Available from: DOI: 10.14311/CEJ.2017.03.0019

[19] KROFTOVÁ, K., KUŘITKA, I., ŠKODA, D., ŠMIDTOVÁ, M. and MASAŘ, M. *Synthesis of nanolime suspension and their potential use in cultural heritage preservation* In. Proceedings of EASEC 15 - The Fifteenth East Asia - Pacific Conference on Structural Engineering and Construction, 2017

[20] LEYVA-PORRAS, A., TOXQUI-TERAN, O., VEGA-BECERRA, M., MIKI-YOSHIDA, M., ROJAS-VILLALOBOS, M., GARCÍA-GUADERRAMA, J.A. and AGUILAR-MARTÍNEZ; *Low-temperature synthesis and characterization of anatase TiO2 nanoparticles by an acid assisted sol-gel method*; Journal of Alloys and Compounds, 2015. vol. 647, pp. 627-636. [viewed 2018-03-15] Available from: DOI: 10.1016/j.jallcom.2015.06.041

NANOCELLULOSE AS POTENTIAL FILLER MATERIAL, PROTECTIVE SHELL AND ADHESION PROMOTER OF MICROBIAL AGROFORMULATIONS

Claudia PREININGER[1], Ursula SAUER[1], Sara DOPPLER[1], Einav KULBAK[2], Sems YONSEL[3], Carmen Michaela POPESCU[4], Cristina POPESCU[4]

[1]AIT Austrian Institute of Technology, Tulln, Austria, EU, claudia.preininger@ait.ac.at
[2]Melodea ltd, Rehovot, Israel
[3]Simbiyotek, Tuzla/Istanbul, Turkey
[4]PPIMC Petru Poni Institute of Macromolecular Chemistry, Iasi, Romania, EU

Abstract

Nanocellulose has emerged as attractive new biobased reinforcement and gas barrier material in for example sustainable food packaging. Herein we report on Cellulose Nano Crystals (CNC) for application in agriculture, as potential filler and adhesion promoter of microbial biofertilizer formulations. Using different types of bacteria we elucidate the physicochemical characteristics of CNC-based microbial seed coatings and spray formulations and show that CNC impacts mechanical strength of seed coatings, but not microbial viability or germination of seeds. CNC is not UV-absorbant except in combination with gelatine, it´s not a good film former, but due to its color it might help to visualize microbial spray pattern.

Keywords: Nanocellulose, microbial formulation, filler, adhesion promoter

1. INTRODUCTION

Most of the biofertilizers and biopesticides are of microbial and biochemical origin (bacteria, fungi, plant extracts) promoting plant growth and resilience or allowing for efficient plant pest and disease management. Compared to conventional pesticides biocontrol agents are usually inherently less toxic and generally affect only the target pest and closely related organisms. Though sustainable and environment-friendly shelf life of microbial inoculants, on-field stability, and active dosage remain critical issues.

Cellulose Nano Crystals (CNC) is a primary building block of the cell wall of plants. CNC has high mechanical strength modulus of 150 GPa and tensile strength of 10 GPA in line with supreme synthetic material and can be used as a nano building block for the enhancement of existing materials and for the production of novel eco-friendly materials [1].

Formulation granting prolonged survival on storage and both survival and sufficient level of bio-activity in field use represents the bottleneck in the broad scale implementation of biological means in plant protection urged by the EU directive on Integrated Pest Management 2009/128/EG. Concerning spray formulations adhesion to the leaves, protection against sunlight (UV), wind and rain and precision of delivery (spraying pattern, spraying volume, time and frequency) are important issues.

In order to tackle these challenges we evaluate nanocellulose as filler in microbial seed coatings for maize and protective shell and adhesion promoter in spray formulations.

2. EXPERIMENTAL

Nanocellulose (Cellulose Nano Crystals - CNC) from pulp (Melodea (**Figure 1**) was added to biopolymer solutions of Gelatine (Sigma Aldrich, Austria), whey protein (Springfield Nutraceuticals BV, Netherlands), milk powder (Carl Roth GmbH and Co. KG, Germany), polyvinylpyrrolidone (PVP), poly vinyl alcohol (PVA) and carboxymethylcellulose (CMC), the latter ones all from Sigma Aldrich, Austria and either used as spray formulations or coated onto maize seeds kindly provided by Saatbau Linz.

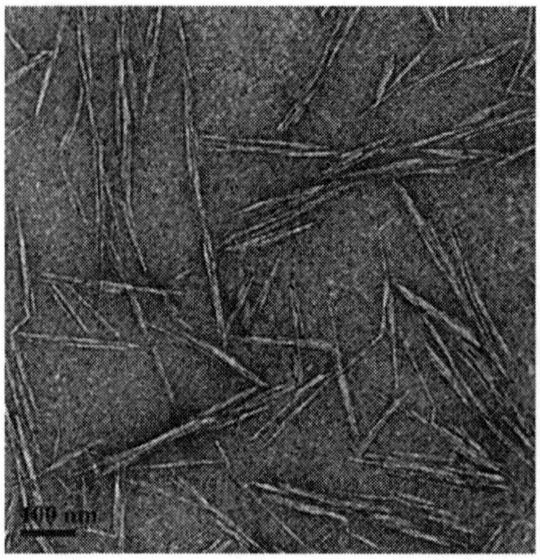

Figure 1 TEM Image of CNC

Bacteria for CNC-based formulations were grown in liquid Luria-Bertani (LB) medium to stationary phase for 72 h as shake culture incubating at 200 rpm at 28 °C. Bacteria were harvested by centrifugation at 4500 rpm and 4 °C during 10 min and resuspended in sterile 0.9 % NaCl before mixed with other formulation components. Number of microbial cells in seed coatings and spray formulations was estimated as colony forming units (CFU) using the serial dilution plate method with nutrient agar medium. Scanning electron microscopy (SEM) scans were imaged with an QUANTA FEG 250 operated at environmental mode (chamber pressure between 30 and 200 Pa), using a GSED Detector and cooling stage.

3. RESULTS AND DISCUSSION

3.1. Nanocellulose as filler in microbial seed coatings

Nanocellulose (CNC) is supposed to increase mechanical strength of biopolymer-based microbial seed coatings and protect the bacteria from adverse environmental conditions during storage. We assume that nanocellulose - similar to nanoclay - attaches to the bacterial cell (1-10 μm) forming a protective shell. To investigate potential benefits of such nanocellulose composite coatings maize seeds were coated with microbial formulations and the different compositions were evaluated with respect to the viability of the bacteria. The bacteria selected include desiccation-sensitive Gram negative *Paraburkholderia phytofirmans*, PsJN (DSM17436) and *Pseudomonas putida* (DSM6125), furthermore robust Gram positive spore-forming strains (*Paenibacillus* sp.). CNC was tested up to 3 % in the coating formulations and at pHs ranging from pH 3 to 5. The coating slurry also contained biobased binder materials like gelatine, whey protein, polyvinylpyrrolidone (PVP), poly vinyl alcohol (PVA), carboxymethylcellulose (CMC) that were tested in both distilled water and 0.9 % NaCl to evaluate the effect of osmolarity on microbial viability. No matter whether spore-forming or non-sporulating bacteria were used microbial viability was significantly (by 2 logs CFU/mL) improved in NaCl-based (about 300 mOsm $^{\times}$ L-1) compared to water-based CNC formulations (between 13 and 22 mOsm $^{\times}$ L-1). Addition of salts also leads to thicker CNC gels and this increased viscosity might also contribute to the enhanced viability. However, film forming capability was improved with aqueous CNC suspensions, the coatings were also more homogeneous, most probably due to better dispersion of CNC.

The seed coating process imposes stress on the bacteria which typically leads to loss of microbial viability during drying. Depending on the type of bacteria - sporulating bacteria are more robust - viability loss might be up to 3 log CFU/mL. To investigate a potentially positive effect of nanocellulose on the desiccation tolerance of bacteria different coating slurries were air-dried in an oven at 25°C under ventilation, rehydrated and the viability determined using agar plate counting. Highest viability was obtained for whey protein-based formulations with and without CNC independently from the type of bacteria used.

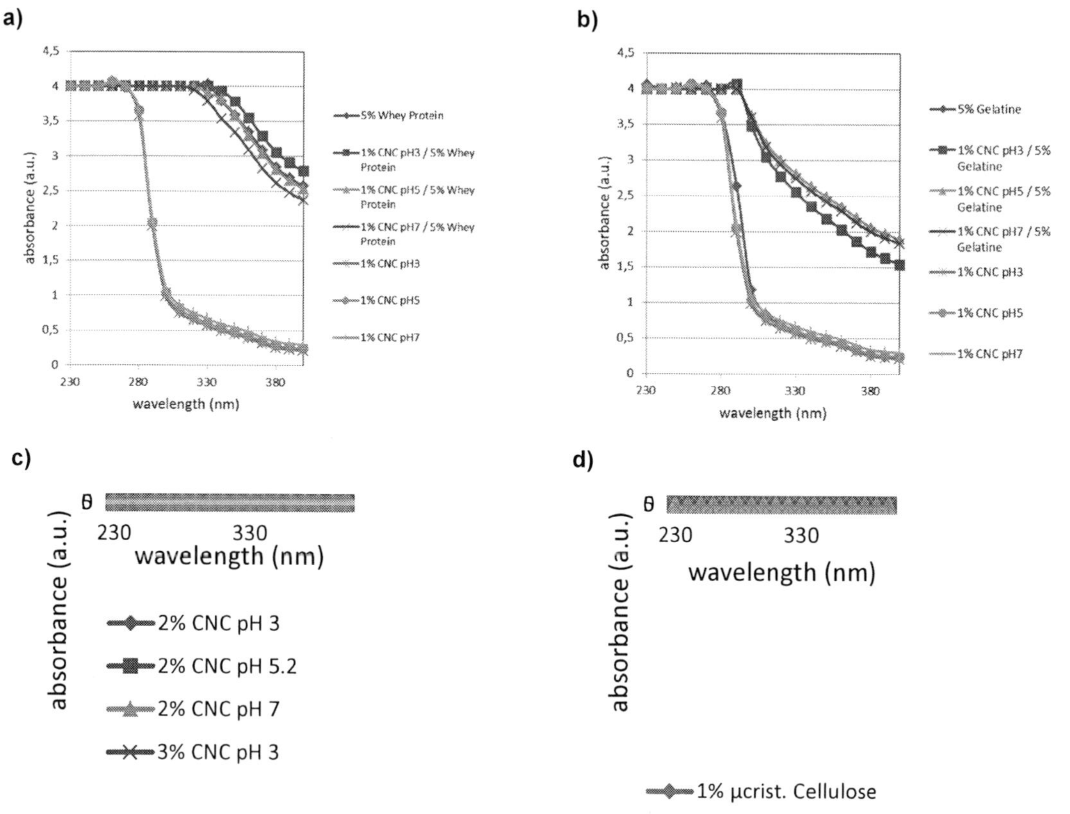

Figure 2 UV spectrum of CNC with a) whey protein and b) gelatine and respective plain solutions, c) the UV spectrum in dependence of pH and CNC concentration and d) different types of celluloses.

Thirdly, the absorbance of seed coating slurries at UV light (280-380 nm) was measured. Protection of bacteria from sunlight to prevent UV damage is generally important, in seed coatings as well as in spray formulations. For this purpose either indicators like congo red, reflectants, such as TiO_2 or lignin-based materials are typically used [2]. Herein we investigate, if nanoparticulate matter like nanocellulose could achieve similar effects. **Figure 2** shows that CNC only weakly absorbs UV-light, independently from the pH (**Figure 2a**), but its absorbance increases with increasing CNC concentration (**Figure 2c**). Compared with other types of celluloses (**Figure 2d**) CNC shows similar UV absorbance than carboxymethylcellulose (CMC), a conventional binder in agroformulations, but less absorbance compared to µcristalline cellulose, often used as absorbent or anti-caking agent.

In combination with stickers a synergistic effect was observed with gelatine: both pure materials do not absorb UV light, while combinations thereof strongly absorb at the relevant wavelengths. This behaviour is in contrast to that of a suspension of CNC in whey protein, where UV absorbance is determined by the properties of the single components only. FTIR studies of nanocellulose, gelatin and their bionanocomposites by Mondragon G. *et al.* [3] reveal that there is no strong intermolecular interaction between gelatin molecules and nanocellulose (no spectral changes in characteristic bands for gelatin and its nanocomposite). Thus it might be more likely that the enhanced UV absorbance of the CNC/gelatin composite is due to the fact that gelatine as the better solubilizer is capable of more uniformly dispersing CNC.

In addition, the seed coating must not prevent the seed from germination. A cold test was performed to evaluate seed vigour under stress conditions and to examine whether the CNC-composite coatings negatively influence germination. The outcome clearly indicated that seed germination was not prevented by any of the CNC-based coatings.

3.2. NANOCELLULOSE AS ADHESION PROMOTER IN MICROBIAL SPRAY FORMULATIONS

In foliar spray development & application not only the microbe-material interaction, but also the interaction of the formulation with the plant leaf surface plays an important role. Adhesion and spreadability of formulations on leaves is of utmost importance, especially at early growth stages where the plant surfaces are very waxy and water-repellent; furthermore for effective spraying good adhesion must be guaranteed to prevent wash-off during rainfall or wind. In order to meet these challenges we aim to use nanocellulose as adhesion promoter, and also for visualization of the spraying pattern on the leaf. **Figure 3** shows the contact angles of 9 formulations including the reference treatments. In **Table 1** the composition of the spraying suspensions and reference treatments (no. 1, 2) are listed. Formulation 3 to 7 were all CNC-based, formulations 6-9 contain (bio)surfactants. A contact angle below 90 °C was defined as a measure for good spreadability of the drops. Results depicted in **Figure 3** show that contact angles of formulation 4-9 were significantly reduced compared to the non sprayed leaves (no. 1). This clearly demonstrates that films of formulation 4-9 are still present on the maize leaves several days after spraying.

Table 1 Composition of selected formulations

No.	Spray formulation
1	no spraying
2	water
3	0.1 % CNC
4	0.1 % CNC, 5 % whey protein
5	0.1 % CNC, 5 % milkpowder
6	0.1 % CNC, 0.1 % biosurfactant
7	0.1 % CNC, 0.1 % commercial surfactant
8	0.1 % biosurfactant
9	0.1 % biosurfactant, 0.1 % talc

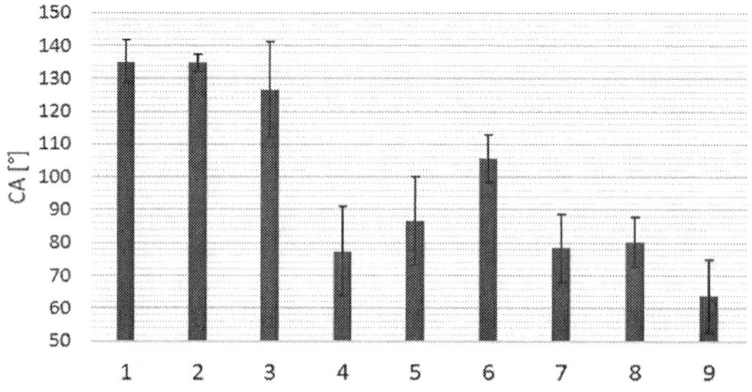

Figure 3 Contact angle of leaves 5 days after spraying with formulations 3-9, and the two reference leaves (no. 1, 2) (mean +/- SD, n=6)

SEM was used to characterize the film sprayed onto the maize leaves (**Figure 4**). High resolution scans of the control leaves show the distinct wax features on young maize leaves causing the high water contact angle of the plant material. With aging the leaves become more hairy and the wax features less prominent.

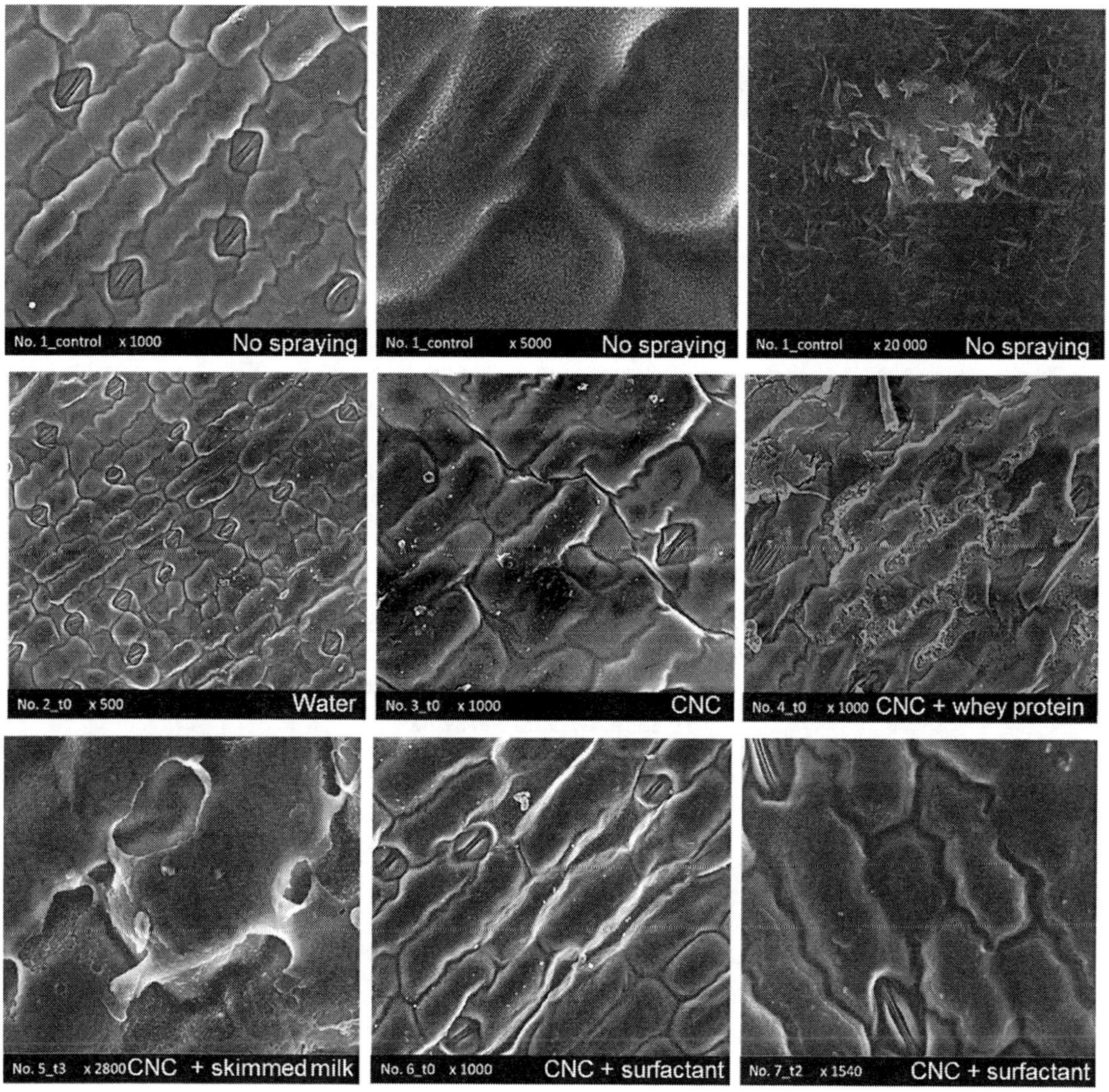

Figure 4 SEM images of plant leaves after spraying, and reference leaves (no. 1, 2)

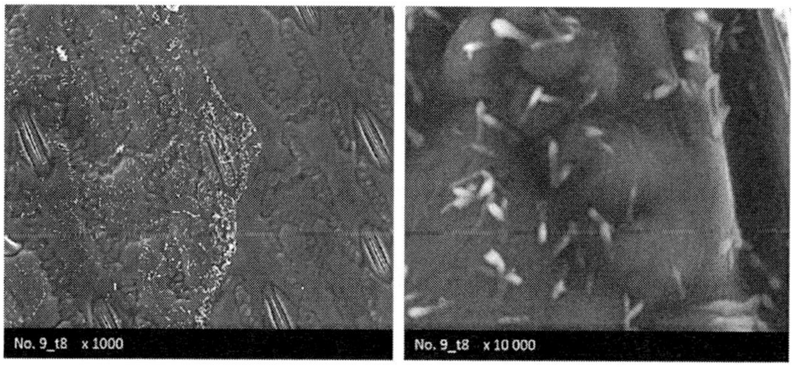

Figure 5 Film of formulation no. 9 (left) with bacterial lawn of rod-shaped bacteria PsJN (right)

Reference leaves in **Figure 4**, show the maize leaf surface compartments including the stomata at 500x to 20000x magnification; as expected there is no film. Leaves sprayed with formulation no. 3 display an incomplete film, only distinct areas of the leaf are covered with spray, but no homogeneous film was obtained. With formulation 4 some crystalline zones are observed. Formulation 5 appears as a thick film on the leaf, while formulation 7 and 8 homogeneously cover the whole leaf surface. Similar to formulations 7 and 8, formulation 6 leads to homogeneous film formation. This clearly indicates that nanocellulose (no. 3) *per se* has no good film forming capability, but with the help of a surfactant fully covers the leaf (no. 6, 7). **Figure 5** shows the film of formulation no. 9 with PsJN after spraying.

4. CONCLUSION

Cellulose Nano Crystals (CNC) used in microbial formulations for agricultural application were shown to improve mechanical strength of seed coatings, but had no protective effect on the selected microbes (microbial viability was not improved). CNC was furthermore tested as additive in spray formulations displaying good spreadability and film forming capability in the presence of a surfactant. UV absorbance was improved in combination with gelatine. Nevertheless lower concentrations of CNC might be more suitable for use with bacteria.

ACKNOWLEDGEMENTS

This work was partly supported by European Commission, Austrian Promotion Agency (FFG), Tubitak, PNCDI III and MATIMOP-ISERD within the m-era.net project COMPIO (www.compio-formulation.eu).

REFERENCES

[1] http://www.melodea.eu/, accessed on 1st October 2018

[2] PREININGER, C., SAUER, U., BEJARANO, A., BERNINGER, T. Concepts and applications of foliar spray for microbial inoculants. Applied Microbiology and Biotechnology. 2018. vol. 102, pp. 7265-7282.

[3] MONDRAGON, G., PENA- RODRIGUEZ, C., GONZALEZ, A., ECEIZA , A., ARBELEIZ, A. Bionanocomposites based on gelatin matrix and nanocellulose. European Polymer Journal. 2015. vol. 62, pp. 1-9.

CYCLIC PHOTOCATALYTIC REGENERATION OF SORBENT, BASED ON CARBON NITRIDE ($g\text{-}C_3N_4$) INCORPORATED IN SILICA NET STRUCTURE

Pavel MANČÍK[1,2], Jiří BEDNÁŘ[1,2], Ladislav SVOBODA[2], Richard DVORSKÝ[1,2,3]

[1]Nanotechnology Centre, VSB - Technical University of Ostrava, Czech Republic, EU
[2]IT4Innovations, VSB - Technical University of Ostrava, Czech Republic, EU
[3]Regional Materials Science and Technology Centre, Ostrava, Czech Republic, EU

Abstract

In this work, we present an advanced preparation method, of photoactive layered graphitic carbon nitride ($g\text{-}C_3N_4$) incorporated into the silica structure containing zinc ions. This new nanocomposite has enhanced sorption properties and photocatalytic activity due to presence of $g\text{-}C_3N_4$. Samples with 20 wt.% and 10 wt.% of $g\text{-}C_3N_4$ were prepared and successfully tested. Specific surface area of nanocomposites was increased from 36 m^2/g of pure carbon nitride $g\text{-}C_3N_4$ to 366 and 320 m^2/g of functional composites respectively. The material was prepared by vigorous stirring of aqueous solution of zinc acetate and g C_3N_4 within ultrasonic field of power density of 1 kW/L. During the sonication, sodium water glass was added at once to synthesize $ZnO\text{-}m\cdot SiO_2$ nanoparticles by precipitation reaction. The dispersion was then rapidly frozen and subjected to vacuum freeze-drying process to obtain powder material, which is easier to handle and more suitable for tests. Prepared nanocomposites were subjected to tests of sorption and photocatalytic degradation of methylene blue dye. Furthermore, samples did undergo four successful cycles of sorption of methylene blue with starting concentration of 1 mg/L followed by photocatalytic regeneration. The sorption capacity of sample with 20 wt.% of $g\text{-}C_3N_4$ was 0.21 mg/g and after the fourth cycle of regeneration it was restored to 75 % of its original value. The sorption capacity of sample with 10 wt.% of $g\text{-}C_3N_4$ was 0.18 mg/g and after fourth cycle of regeneration it was restored to 90 % of its original value.

Keywords: Photocatalysis, photocatalytic regeneration, sorption, nanoparticles, graphitic carbon nitride

1. INTRODUCTION

An environmental pollution is one of the major threats nowadays. People are turning to more ecological solutions and the industry is under pressure to reduce the amount of released pollutants. For these reasons industry companies search for new economic and environmentally friendly solutions. An effective and environmentally friendly removal and disposal of pollutants is therefore the goal of many studies [1,2]. Photocatalytic and sorption processes are considered as a good way how to ecologically remove pollutants. Photocatalytic process would be ideally powered by solar energy, making it energy-saving. Beside decomposition of organic pollutants [3], the photocatalysis has many other uses such as production of hydrogen by water splitting [4], self-cleaning materials [5], water disinfection using antibacterial effects [6] and reduction of carbon dioxide [7] and nitrogen oxides [8].

Sorption processes are another way how to remove pollutants. They are based on the binding of substances on their surface, thereby reducing their concentration in the environment. Sorption processes are used for sanitization of oil spills [9], removal of heavy metals from water [10], removal of organic pollutants [11], carbon capture and storage [12] or storage of other substances, such as hydrogen [13]. After its saturation, the sorbent must be disposed or subjected to a regeneration process. Simply disposal of saturated sorbents could be expensive. For example, it is the most expensive and time-consuming part of removal of oil spills [9].

On the other hand, there is a regeneration process that restores sorption capacity of the sorbent and allows its further use. Regeneration of the saturated sorbent is performed mostly thermally and chemically [14,15]. Thermal processes are energy-intensive and therefore economically inappropriate. For chemical regeneration, toxic or otherwise dangerous substances are often used. Therefore, there is an effort to develop and improve eco-friendly methods such as biological or photocatalytic regeneration [16,17].

If the sorbent also exhibits photocatalytic activity, it can be regenerated using electromagnetic radiation. The following photocatalytic reaction results in sorbate decomposition. Ideally the pollutants would be decomposed into substances with less sorbent affinity and washed away. Pardo et al. [17] prepared hydrogel containing titanium dioxide. The material was tested on sorption of methylene blue and antibiotics Ofloxacin and Ciprofloxacin and was subjected to photocatalytic regeneration under UV light irradiation. Sorption capacities were 96 mg/g for methylene blue, 84 mg/g for Ofloxacin and 63 mg/g for Ciprofloxacin when the starting concentrations of tested pollutants was 100 mg/L. Material preserved its sorption capacity for ten cycles of regeneration. Several other materials, such as zeolite [18], were modified by titanium dioxide and subjected to successful photocatalytic regeneration under UV irradiation. For better separability of saturated sorbent, it is very convenient to modify it with magnetic material [19]. The regeneration of Fe_3O_4/Bi_2S_3 sorbent with magnetic properties took place under simulated solar radiation (300 W xenon lamp). Magnetic properties then provided better handling of the material but after five regeneration cycles the sorption capacity has fallen to 25 % of the original value.

This work continues in the study of regeneration of graphitic carbon nitride incorporated into the silica structure containing zinc [20]. Following material has greater specific surface area, greater sorption capacities than ZnO-m·SiO_2/g-C_3N_4 nanocomposite [20] and did undergo four successful photocatalytic regeneration under UV light irradiation. Prepared sorbent was characterized by EDX, BET analysis, SEM and was subjected to test of sorption and photocatalytic regeneration.

2. EXPERIMENTAL

2.1. Materials and chemicals

Chemicals used for experiments are zinc acetate of 99 % purity purchased from PENTA s.r.o., sodium water glass 38-40 °Bé (Na_2SiO_3) with module of 3 purchased from company Vodní sklo, a.s. and melamine of 99 % purity purchased from Sigma-Aldrich spol. s.r.o. For all aqueous solutions, demineralized water was used.

2.2. Preparation of graphitic carbon nitride g-C_3N_4

Bulk graphitic carbon nitride was prepared the same way as in our previous work [21]. Melamine was heated in laboratory furnace in air from 20 °C to 550 °C with step of 2 °C per minute and lasted there for 4 hours. Material was then comminuted by grinding in agate mortar and further exfoliated by heating up to temperature of 500 °C with step of 2 °C per minute for next two hours. The specific surface area of prepared exfoliated graphitic carbon nitride was 157 m^2/g.

2.3. Preparation of modified graphitic carbon nitride g-C_3N_4

To prepare a sample with 20 wt.% of g-C_3N_4, 27 g of zinc acetate and 0.295 g of g-C_3N_4 were added to 75 ml of demineralized water and homogenized in the ultrasonic field. To prepare a sample with 10 wt.% of g-C_3N_4, 0.131 g of g-C_3N_4 was used. 225 ml of demineralized water was then mixed with 2.57 g of water glass (the fraction of solids was 39.67 wt.%). The mixture was homogenized in an ultrasonic field with power density of 1 kW/L and it was continuously stirred at 500 RPM. After five minutes, a dispersion of g-C_3N_4 and zinc acetate was added at once. The mixture was further stirred at 500 RPM in the ultrasonic field for fifteen minutes. The residue zinc acetate was then removed by washing the material. The mixture was transferred to a two-liter cylindrical container which was filled by demineralized water to its maximum. After two hours of sedimentation, the clear liquid above the sediment was carefully removed and the cylinder was again filled to the maximum by demineralized water. This washing cycle was repeated five times. The sedimented dispersion was then quickly frozen and subjected to vacuum sublimation.

2.4. EDAX and SEM analysis

The composition of nanocomposite was determined by EDX analysis on scanning electron microscope SEM FEI Quanta 650 FEG. This scanning electron microscope was used to capture SEM images of powdered sample.

2.5. Specific surface area

Dynamic BET analysis was performed on Qsurf HORIBA SA9601 to measure specific surface area (SSA). Powder material was degassed for 5 hours at 150 °C and then subjected to six-point analysis.

2.6. Continual measuring of adsorption of simulant dye

We used reactor for continuous dye adsorption measurement. This method applies the principle of dye concentration measurement from patent application "Continuous measurement of photocatalysis of dye simulants" [22]. The advantage is a fast in situ measurement, without the need of taking samples and therefore compromising the system. Setup was the same as described in [20]. The decrement of the dye concentration was calculated from measurement of diffused and backscattered light intensity from a laser located at the top of the measuring vessel. Ocean Optics USB4000-UV-VIS was used as a detector. Sorption process took place at the temperature of 20 °C, in the dark environment and under continuous stirring until the sorbent was saturated. 1 mg/L of methylene blue was used as starting concentration of simulant dye.

3. RESULTS AND DISCUSSION

Nanocomposite sorbent based on graphitic carbon nitride incorporated into silicate structure containing zinc was successfully prepared and tested on sorption of methylene blue and photocatalytic regeneration. The samples were prepared in two variants: 1. ECN20 with 20 wt.% of g-C_3N_4 and 2. ECN10 with 10 wt.% of g-C_3N_4. The SSA, which has an important role in sorption and photocatalytic processes, was determined to be 366 m^2/g for ECN20 and 320 m^2/g for ECN10. The SSA of prepared samples is higher than the SSA of previously prepared sorbent capable of photocatalytical regeneration [20] with value of 238 m^2/g.

Figure 1 SEM images of ECN20 (left) and ECN10 (right)

Images (**Figure 1**) from scanning electron microscope showed ragged morphology of the samples causing high values of the SSA. The EDX analysis (**Figure 2**) confirmed final content of both samples to be 18.14 wt.%

for ECN20 and 8.23 wt.% for ECN10. Lower content than the estimated was probably caused by imperfect incorporation of g-C_3N_4 during synthesis.

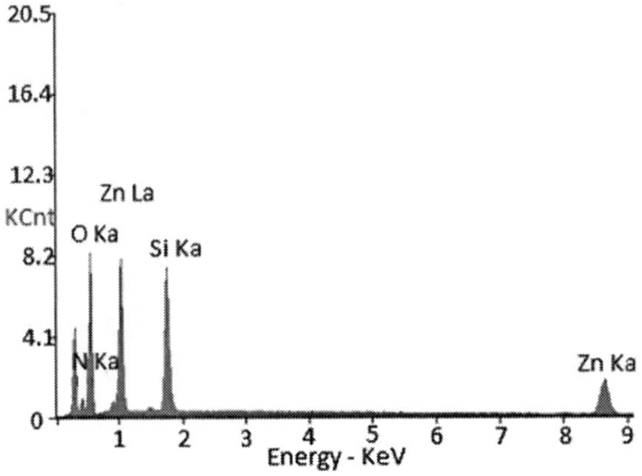

Figure 2 EDX element analysis of ECN20 used for calculation of g-C_3N_4 content (18.14 wt.%)

Samples were subjected to repeated sorption of methylene blue with starting concentration of 1 mg/L and photocatalytic regeneration process. The regeneration process consisted of several steps. At first, the saturated sorbent was separated from the methylene blue residue by centrifugation. The sediment was then moved into beaker and filled up to 50 ml with deionized water and subjected to photocatalytical decomposition of sorbed methylene blue. Photocatalytic regeneration was performed under LED UV irradiation source with the maximum of light emission located at 365 nm. After decomposition, the sample was centrifugated again to remove by-products and then it was ready for next sorption process. Sorption characteristics are shown on **Figure 3** and **Figure 4**. Sorption capacity of ECN10 was 0.18 mg/g and 90 % of it was regenerated after fourth cycle. The decrement of sorption capacity was due to loses of sorbent during manipulation or stronger interactions were created between sorbent and methylene blue due to deformation and enclosing of some pores.

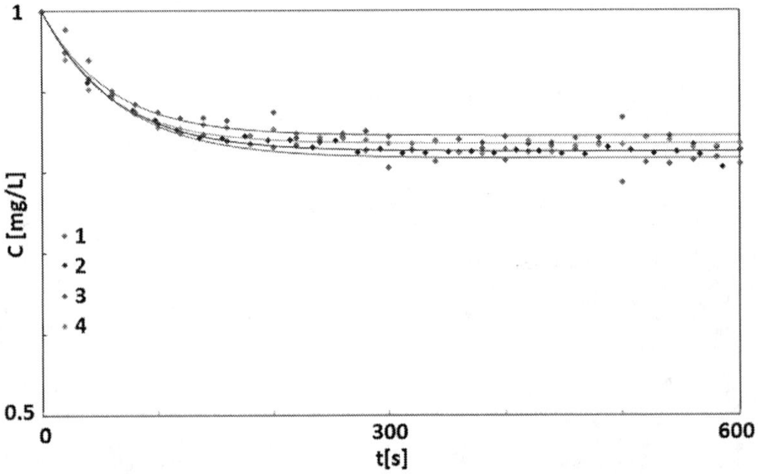

Figure 3 Sorption kinetics of methylene blue sorption process on the ECN10 for four cycles (different colours) of photocatalytic regeneration. Numbers 1 to 4 indicate the order of sorption process.

Sorption capacity of ECN20 was 0.21 mg/g and about 75 % of it was available after fourth cycle. The decrement of sorption capacity of ECN20 is greater than of ECN10 due to greater content of g-C_3N_4 which probably serves as centre for stronger interactions.

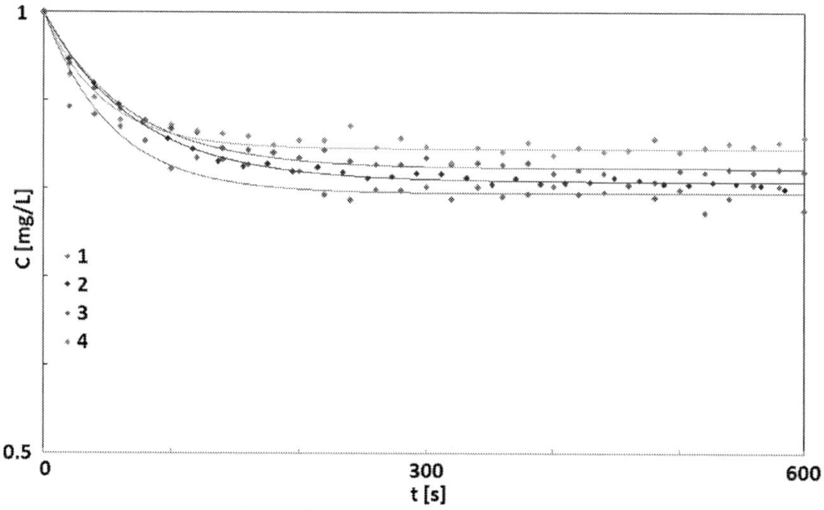

Figure 4 Sorption kinetics of methylene blue sorption on the ECN20 for four cycles (different colours) of photocatalytic regeneration. Numbers 1 to 4 indicate the order of sorption process.

Rate constants of adsorption tests were calculated from linear parts of acquired data for both samples. The rate constants were 0.015 s^{-1} for ECN10 and 0.019 s^{-1} for ECN20. These rate constants are greater than previously tested material [20] with the value of 0.004 s^{-1}. Modified nanocomposite showed better performance in photocatalytic regeneration and sorption process of methylene blue than previous nanocomposite ÚZnO-m·SiO_2/g-C_3N_4 [20] and was tested with four regeneration cycles.

4. CONCLUSION

Nanocomposite sorbent based on graphitic carbon nitride incorporated into silicate structure containing zinc was successfully prepared and tested for sorption of methylene blue and photocatalytic regeneration. The EDX analysis confirmed estimated composition and SEM images showed ragged morphology of both prepared samples, which caused the great values of the specific surface area. The specific surface area of sample with 20 wt.% of g-C_3N_4 was 366 m^2/g and was greater than at the sample with 10 wt.% of g-C_3N_4 with the value of 320 m^2/g.

Both samples were successfully regenerated four times by UV LED light source. Greater content of g-C_3N_4 led to greater decrease in sorption capacity. The sorption capacity of ECN20 was 0.21 mg/g and after the fourth cycle of regeneration it was restored to 75 % of its original value. The sorption capacity of ECN10 was 0.18 mg/g and after fourth cycle of regeneration it was restored to 90 % of its original value. Although both samples were successfully regenerated, the sample ECN10 performed better in the photocatalytic regeneration process under the UV light irradiation.

ACKNOWLEDGEMENTS

This work was supported by the European Regional Development Fund in the IT4Innovations national supercomputing center - path to exascale project, project number CZ.02.1.01/0.0/0.0/16_013/0001791 within the Operational Programme Research, Development and Education., the project Double degree CZ.02.2.69/0.0/0.0/16_018/0002708, the student projects SP2018/79, SP2018/75 and the project "Regional Materials Science and Technology Centre" - LO1203.

REFERENCES

[1] NATH, R. K., ZAIN, M. F. M. and JAMIL M. An environment-friendly solution for indoor air purification by using renewable photocatalysts in concrete: A review. *Renew. Sustain. Energy Rev.* 2016. vol. 62, pp. 1184-1194.

[2] MOLINARI, A., SARTI, E., MARCHETTI, N. and PASTI L. Degradation of emerging concern contaminants in water by heterogeneous photocatalysis with $Na_4W_{10}O_{32}$. *Appl. Catal. B Environ.* 2017. vol. 203, pp. 9-17.

[3] SVOBODA, L., DVORSKÝ, R., PRAUS, P., MATÝSEK, D. and BEDNÁŘ, J. Synthesis of ZnO nanocoatings by decomposition of zinc acetate induced by electrons emitted by indium. *Applied Surface Science*. 2016. vol. 388, p. 491-496.

[4] ISMAIL, A. A. and BAHNEMANN, D. W. Photochemical splitting of water for hydrogen production by photocatalysis: A review. *Sol. Energy Mater. Sol. Cells.* 2014. vol. 128, pp. 85-10.

[5] FAN, Y. Photocatalysis and self-cleaning from g-C_3N_4 coated cotton fabrics under sunlight irradiation. *Chem. Phys. Lett.* 2018. vol. 699, pp. 146-154.

[6] HELALI, S. Solar photocatalysis: A green technology for E. coli contaminated water disinfection. Effect of concentration and different types of suspended catalyst. *J. Photochem. Photobiol. A Chem.* 2014. vol. 276, pp. 31-40.

[7] XIONG, Z., LEI, Z., LI, Y., DONG, L., ZHAO, Y. and ZHANG, J. A review on modification of facet-engineered TiO2 for photocatalytic CO2 reduction. *J. Photochem. Photobiol. C Photochem. Rev.* 2018. vol. 36, pp. 24-47.

[8] ÂNGELO, J., ANDRADE, L., MADEIRA, L. M. and MENDES A. An overview of photocatalysis phenomena applied to NOx abatement, *J. Environ. Manage.* 2013. vol. 129, pp. 522-539.

[9] BHARDWAJ, N. and BHASKARWAR, A. N.. A review on sorbent devices for oil-spill control. *Environ. Pollut.* 2018.

[10] LÓPEZ-DELADO, A., PÉREZ, C. and LÓPEZ, F. A. Sorption of heavy metals on blast furnace sludge. *Water Res.* 1998 vol. 32, no. 4, pp. 989-996.

[11] ZHU, R., ZHU, GE, J., F. and YUAN, P. Regeneration of spent organoclays after the sorption of organic pollutants: A review. *J. Environ. Manage.* 2009 vol. 90, no. 11, pp. 3212-3216.

[12] THEO, W. L., LIM, J. S., HASHIM, H., MUSATFFA, A. A. and HO, W. S.. Review of pre-combustion capture and ionic liquid in carbon capture and storage. *Appl. Energ.* 2016 vol. 183, pp. 1633-1663.

[13] HUANG, J. Revealing contribution of pore size to high hydrogen storage capacity. *Int. J. Hydrogen Energy.* 2018 vol. 43, no. 39, pp. 18077-18082.

[14] SALVADOR, F., MARTIN-SANCHEZ, N., SANCHEZ-HERNANDEZ, R., SANCHEZ-MONTERO, M. J. and IZQUIERDO, C. Regeneration of carbonaceous adsorbents. Part I: Thermal Regeneration. *Microporous Mesoporous Mater.* 2015. vol. 202, pp. 259-276.

[15] LIU, D. Electrochemical regeneration of a graphite adsorbent loaded with Acid Violet 17 in a spouted bed reactor. *Chem. Eng. J.* 2016. vol. 304, pp. 1-9.

[16] KLIMENKO, N., SMOLIN, S., GRECHANYK, S., KOFANOV, V., NEVYNNA, L. and SAMOYLENKO, L. Bioregeneration of activated carbons by bacterial degraders after adsorption of surfactants from aqueous solutions. *Colloids Surfaces A Physicochem. Eng. Asp.* 2003 vol. 230, no. 1-3, pp. 141-158.

[17] PARDO, A. Self-regenerating photocatalytic hydrogel for the adsorption and decomposition of methylene blue and antibiotics in water. *Environ. Technol. Innov.* 2018. vol. 11, pp. 321-327.

[18] LIU, S., LIM, M. and AMAL, R..TiO_2-coated natural zeolite: Rapid humic acid adsorption and effective photocatalytic regeneration. *Chem. Eng. Sci.* 2014. vol. 105, pp. 46-52.

[19] ZHU, H., JIANG, R., LI, J., FU, Y., JIANG, S. and YAO, J. Magnetically recyclable Fe_3O_4/Bi_2S_3 microspheres for effective removal of Congo red dye by simultaneous adsorption and photocatalytic regeneration. *Sep. Purif. Technol.* 2017. vol. 179, pp. 184-193.

[20] MANČÍK, P., BEDNÁŘ, J., SVOBODA, L., DVORSKÝ, R. and MATÝSEK, D. Photocataylytic Reactivation of g-C_3N_4 Based Nanosorbent. In *NANOCON 2017 - 9th International Conference on Nanomaterials - Research & Application*. Ostrava: TANGER, 2017, pp. 289-294.

[21] SVOBODA, L., PRAUS, P., and LIMA, M. J. Graphitic carbon nitride nanosheets as highly efficient photocatalysts for phenol degradation under high-power visible LED irradiation. *Materials Research Bulletin*. 2018. vol. 100, pp. 322-332.

[22] DVORSKÝ, R., BEDNÁŘ, J., SVOBODA, L. and MANČÍK, P. Přihláška vynálezu: "Způsob kontinuálního měření fotokatalýzy barvivových simulantů, PV 2018-421, 2018.

DESIGN OF EXPERIMENT METHOD FOR OPTIMIZATION OF SYNTHESIS OF g-C_3N_4 PHOTOCATALYST

[1]Jiří ZELINKA, [2]Petra HALFAROVÁ, [2]David VYKYDAL, [2]Vojtěch LIŠKA, [1,3]Petr PRAUS, [1,3]Vlastimil MATĚJKA

[1]VSB - Technical University of Ostrava, Faculty of metallurgy and materials engineering, Department of Chemistry, Ostrava, Czech Republic, EU, zelinkaj@seznam.cz
[2]VSB - Technical University of Ostrava, Faculty of metallurgy and materials engineering, Department of Quality Management, Ostrava, Czech Republic, EU, petra.halfarova@vsb.cz
[3]VSB - Technical University of Ostrava, Institute of Environmental Technology, Ostrava, Czech Republic, EU, petr.praus@vsb.cz

Abstract

G-C_3N_4 is an intensively studied material for photocatalytic applications, mainly for the degradation of pollutants in water and air. The thermal polymerization of suitable precursors (urea, melamine, dicyanamide etc.) is the most often used approach for the g-C_3N_4 synthesis. In this research, g-C_3N_4 was prepared by the thermal polymerization of melamine and calcination conditions were tuned using Design of Experiment approach (DOE). Five factors, such as the temperature ramp rate, final temperature, holding time, cooling rate and amount of initial melamine, were preselected and tested on two levels: 3 and 10 °C/min, 500 and 550°C, 0 and 1h, shock and slow cooling, 2.5 and 10 g. A half testing plan including 16 calcination experiments was set using fractional factorial design. The yield of final products of calcination was used as the measured output of the experiments. The photogeradation activity against the model dye acid orange 7 (AO7) was evaluated for selected samples. The results indicated the weight loss of melamine for a given temperature depended mainly on its initial weight and heating rate. The photodegradation activity was higher for the samples calcined at the temperature of 500 °C.

Keywords: Graphitic carbon nitride, melamine, polymerization, DOE, photocatalysis

1. INTRODUCTION

The photocatalysts are materials studied for their potential to degrade organic, but also inorganic compounds and this feature can be utilized, for example, for cleaning of the air and water [1]. The principle of functionality of these materials is the light assisted generation of electron - hole pairs. The energy of incident light has to be equal or lower that the band gap energy of a photocatalyst. For example, widely studied TiO_2 and ZnO are being intensively studied as the efficient photocatalysts [2,3]. The band gap energy of both TiO_2 and ZnO reachs the values close to 3.2 eV what means their photoactivity can be evoked by the light with wavelength equal or lower to 386 nm (UVA region). Practically, it means that less than 10 % of the sun light can be used for activation of these materials and thus the artificial UV light has to be used for the efficient photocatalytic process over these materials.

Although the many approaches for shifting the photoactivity of these materials to the VIS part of spectra already exist (mainly metal and non-metal doping). The efforts for searching of photocatalysts active in the visible range of spectra led, for example, to indication of g-C_3N_4 as the promising material for this purpose [4]. The band gap energy of g-C_3N_4 reaches the value 2.5 eV what means that it can be activated with light of the wavelength equal or lower than 495 nm. The preparation of g-C_3N_4 is based on the thermal treatment of suitable precursors, for example, melamine [5]. During heating the precursors of g-C_3N_4 melam, melem and melon as the by-products of the melamine thermal polymerization were found [6]. Interestingly, there is the number of papers dealing with the calcination of melamine with the aim to prepare g-C_3N_4 with high photoactivity. The reported calcination temperature ranges from low temperatures below 500°C to high temperatures above 600 °C [7,8]. Undoubtedly, there are other factors which could affect the final photoactivity of synthesized g-C_3N_4 as well as the yield of final products, for example, the heating rate and holding time at a given temperature.

Due to the variety of factors, which are considered to have the effect on the melamine calcination, DOE represents a systematic approach for the creation of a testing plan and the statistical evaluation of relations between factors affecting a studied process and the outputs of experiments [9]. DOE is a systematic method to determine the relationship between factors affecting a process and the output of that process. In other words, DOE allows to find factors that have an influence on the production process and determine their optimal values of the factors. When experimenting, we tried to find the relationship "cause-effect" when we used the targeted input changes to achieve changes in output. We tried to capture and describe this relationship using mathematical methods. From these outputs, it is possible to analyse many links and patterns about the process.

Typical the DOE methodology is based on several steps. The first step is the analysis of studied process and indication of relevant factors influencing the studied process, setting their levels and defining the output(-s) of the process. The number of factors and the number of levels give the total number of experiments which need to be performed according to the equation (1).

$$n = 2^k \quad (1)$$

where n is the number of experiments, k is the number of factors and *2* is the number of levels.

The further steps of DOE include i) preparation of a testing plan, iii) performing of tests, collection of outputs and iv) evaluation of the outputs. Based on the number of the factors and levels, the number of experiments, which are necessary to be performed, could be high and thus the methods of reduced testing plan are also adopted. Half testing plans based on the fractional factorial design [10] or the Taguchi approach [11] are the typical examples.

In our research we focused on the evaluation of applicability of the half testing plan for tuning the calcination of melamine with the aim to identify the best combination of selected factors with respect to the yield of final products. The defined five factors, such as temperature, heating rate, hold time, cooling rate and the amount of melamine, were tested at two levels (minimum, maximum) resulted in 16 calcination experiments. The yield of the calcination product was used as the main output of the DOE experiments. The photodegradation of AO7 was measured for the prepared materials.

2. EXPERIMENTAL

2.1. Design of experiment

The half testing plan and its individual experiments were planned using the statistical software Minitab 17. Five factors, which should have the major influence on the calcination of melamine, were identified and are summarized in **Table 1**.

Table 1 Selected factors and defined levels

Factor	Identification	Level	
		Low	High
A	Temperature (°C)	500	550
B	Heating rate (°C·min^{-1})	3	15
C	Hold time (h)	0	1
D	Cooling rate	F	S
E	Weight of melamine (g)	2.5	10

Note: The cooling rate levels were fast (F) and slow (S).

Using the equation (1) the number of the experiments for full factorial testing plan for the combination of the factors and levels listed in **Table 1** is equal to 32. The half testing plan included 16 experiments selected by Minitab 17 and is shown in **Table 2**. Variable "-1" represents the low factor level and "+1" is the high factor level.

Table 2 The definition of the half testing plan, the average yield of the final product

No.	Factors (Inputs)					Output
	A	B	C	D	E	Yield (%)
1	-1	-1	-1	-1	1	64
2	1	-1	-1	-1	-1	53
3	-1	1	-1	-1	-1	47
4	1	1	-1	-1	1	50
5	-1	-1	1	-1	-1	57
6	1	-1	1	-1	1	57
7	-1	1	1	-1	1	49
8	1	1	1	-1	-1	42
9	-1	-1	-1	1	-1	60
10	1	-1	-1	1	1	63
11	-1	1	-1	1	1	55
12	1	1	-1	1	-1	47
13	-1	-1	1	1	1	64
14	1	-1	1	1	-1	54
15	-1	1	1	1	-1	47
16	1	1	1	1	1	47

2.2. Calcination experiments and photodegradation activity measurement

Melamine (Karl Roth GmbH, Germany) was used for all calcination experiments. The calcination was performed in a muffle furnace (LAC, Czech Republic). Glazed ceramic crucibles with glazed ceramic covers were for the calcination experiments. The calcination was performed in static air atmosphere. Two replicas for each experiment defined in **Table 2** were performed. TiO_2 P25 (Degussa) was used for the comparison of the photodegradation activity of the prepared samples.

The photodegradation activity of the prepared samples was tested by the photocatalytic decomposition of AO7 (Synthesia,a.s., Czech Republic). In the typical experiment, the tested samples were weighted (0.0500 g) into a glass bowl filled with 150 ml of distilled water mixed using an electromagnetic stirrer (300 rpm). After 5 min of stirring 15 ml of the AO7 solution ($c = 7.136 \cdot 10^{-4}$ mol·dm^{-3}) was added. The resulting suspension was further stirred in dark for 1 h to achieve adsorption equilibrium. After this period, 5 ml of the suspension was taken and filtered under vacuum using a membrane filter with porosity of 0.4 μm (Pragopor, Czech Republic). The absorbance of filtered solutions was measured at 480 nm using a HELIOS Σ spectrometer (ThermoSpectronic, USA) and as A_0. In the same time a source of UV light (BLB 36 W, NARVA) was turned on and the sample was subjected to irradiation for 120 min. After this step, 10 ml of sample was taken out of the stirred suspension and filtered using a membrane filter. The absorbance of the filtered solution was measured at 480 nm and assigned A_{120}.

UV-VIS diffuse reflectance spectroscopy measurements (UV-VIS DRS) were performed using Shimadzu UV-2600 spectrometer (Shimadzu, Japan) equipped with integrating sphere IRS-2600Plus (Shimadzu, Japan).

The samples were pressed inside the holder and DRS spectra were acquired in range 300 - 500 nm. BaSO4 powder (Nacalai Tesque, Inc., extra pure reagent grade) was used for the setting of the baseline. The Tauc's plots were used for the evaluation of indirect band gap values.

3. RESULTS AND DISCUSSION

The calcination yields obtained during the experiments defined by the half testing plan are shown in **Table 2**. The effect of the factor levels on the yields of the final product is expressed in **Figure 1**. Based on **Figure 1** the positive effect of the low temperature (500 °C), low heating rate (3 $°C \cdot min^{-1}$) and low duration of the calcination on the yield was observed. However, an opposite effect of the level of cooling rate and weight of the sample was observed. Practically, **Figure 2** shows us that for the maximum yield we should perform the calcination with the low heating rate, at the lower temperature, with the low holding time but the cooling rate should be faster and the higher amount of melamine should be used.

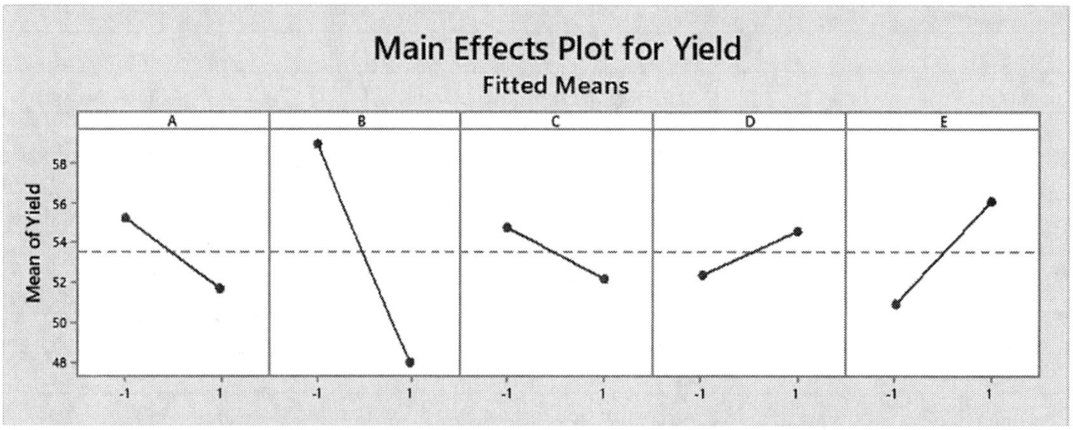

Figure 2 The effect of the level of given factor on the process of melamine calcination

The experiments ranking based on the yields is shown in **Figure 2**. **Figure 2** indicated the sample 8 as the sample with the lowest yield, whereas the samples 1 and 13 as that with the highest product yield.

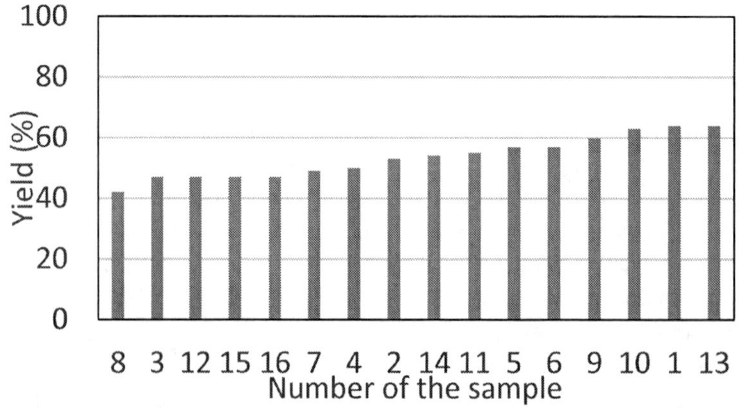

Figure 2 The ranking of the samples based on the yield of the melamine calcination

For the samples 1, 8 and 13 the photodegradation activity against AO7 was measured. After 2 h of the UVA irradiation the sample 8 showed 43 % of the AO7 degradation, whereas the sample 1 showed 80 % and sample 13 showed 87 % of the AO7 degradation. The reference sample TiO_2 P25, which is usually used as the reference material for evaluation of the photodegradation efficiency, showed 71 % of the AO7 degradation after the same time.

The morphology of the sample 13 was studied using SEM technique and the image is shown in **Figure 3a**). The stacked particles of the platy character formed the irregularly shaped agglomerates. The UV-VIS DRS spectra of the samples 1, 8 and 13 are shown in **Figure 3b**). The band gap energies of the samples 1 and 13 are similar and reached the values 2.81 and 2.79 eV, respectively; the calculated band gap energy of sample 8 reached value 2.73 eV.

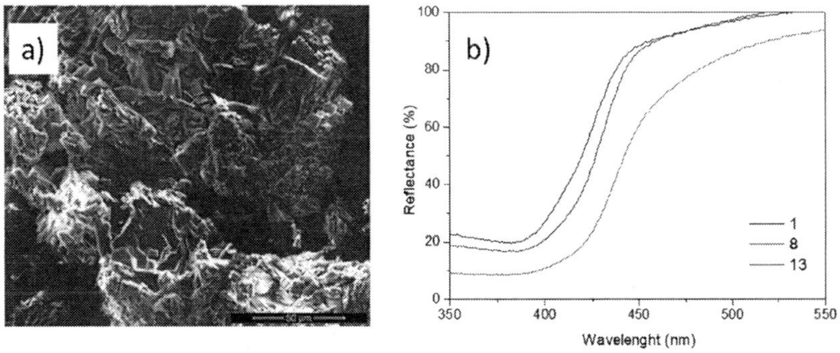

Figure 3 SEM image of the sample 13 (a), reflectance spectra of the samples 1, 8 and 13 (b)

4. CONCLUSION

The half testing plan approach as one of the DOE methods was used for the optimization of the melamine calcination with respect to the yield of the final product. The results indicated the optimal calcination conditions were: i) the calcination temperature of 500 °C, ii) the heating rate of 3 $°C \cdot min^{-1}$ and iii) the weight of melamine 10 g as the optimal conditions with respect to the yield of the calcination. The measured photoactivity indicated the samples 1 and 13 as more efficient in comparison to the sample 8. The photodegradation activity of the samples 1 and 13 was higher than the reference sample TiO_2 P25.

The next step of our research will be the finalization of the photodegradation activity according to the half testing plan and using the DOE approach for the indication of the factor(-s) of the melamine calcination with significant effects on the photodegradation activity of final products. In parallel the full testing plan including next 16 calcination experiments will be performed and the same outputs (yield and photodegradation activity) will be measured. The results of the full testing plan will be compared with the results of the half testing plan. The selected samples will be characterized using the methods of chemical and phase analysis. The relations between observed characteristics of the final samples, yield and photodegradation activity will be described.

ACKNOWLEDGEMENTS

This work was supported by the Grant Agency of the Czech Republic (reg. No. 16-10527S) and by the student projects SP2018/79 and SP2018/97 of VSB-Technical University of Ostrava.

REFERENCES

[1] REDDY, P.A.K., REDDY, P.V.L., KWON, E., KIM, K.H, AKTER, T. and KALAGARA, S. Recent advances in photocatalytic treatment of pollutants in aqueous media. *Environment International*. 2016, vol. 91, pp. 94-103.

[2] LEE, S.-Y. and PARK S.-J. TiO2 photocatalyst for water treatment applications. *Journal of Industrial and Engineering Chemistry*. 2013, vol. 19, no. 6, pp. 1761-1769.

[3] DI MAURO, A., FRAGALA, M.E., PRIVITERA, V. and IMPELIZZERI, G. ZnO for application in photocatalysis: From thin films to nanostructures. *Materials Science in Semiconductor Processing*. 2017, vol. 69, pp. 44-51.

[4] MAMBA, G. and MISHRA, A.K. Graphitic carbon nitride (g-C3N4) nanocomposites: A new and exciting generation of visible light driven photocatalysts for environmental pollution remediation. *Applied Catalysis B: Environmental.* 2016, vol. 198, pp. 347-377.

[5] DEVTHADE, V., KULHARI, D. and UMARE, S.S. Role of precursors on photocatalytic behavior of graphitic carbon nitride. *Materials Today: Proceedings.* 2018, vol. 5, no. 3, pp. 9203-9210.

[6] KESSLER, F.K., ZHENG, Y., SCHWARZ, D., MERSCHJANN, C., SCHNICK, W., WANG, X. and BOJDYS, M.J. Functional carbon nitride materials - design strategies for electrochemical devices. *Nature Reviews Materials.* 2017, vol. 2, no. 6, pp. 1 - 17.

[7] ZENG, Y., LIU, X., LIU, C., WANG, L., XIA, Y., ZHANG, S., LUO, S., PEI, Y. Scalable one-step production of porous oxygen-doped g-C3N4 nanorods with effective electron separation for excellent visible-light photocatalytic activity. *Applied Catalysis B: Environmental.* 2018, vol. 224, pp. 1-9.

[8] KOČÍ, K., RELI, M., TROPPOVÁ, I., ŠIHOR, M., KUPKOVÁ, J., KUSTROWSKI, P. and PRAUS P. Photocatalytic decomposition of N2O over TiO2/g-C3N4 photocatalysts heterojunction. *Applied Surface Science*. 2017, vol. 396, pp. 1685-1695.

[9] JIJU A. *Design of Experiments for Engineers and Scientists.* 1st ed. Elsevier, 2003. pp. 109.

[10] MONTGOMERY, D.C. *Design and analysis of experiments.* 6th ed. Hoboken, NJ: John Wiley, 2005. pp 752.

[11] ROY, R.K. *Design of experiments using the Taguchi approach: 16 steps to product and process improvement.* New York: Wiley, 2001. pp 560.

ELECTROCHEMICAL ANODIZATION METHOD TOWRDS MULTI-LEG TIO_2 NANOTUBES FOR PEC APPLICATIONS

[1]Mahdi SHAHREZAEI, [1]Rambabu YALAVARTHI, [1]Alberto NALDONI, [1]Stepan KMENT, [1]Patrik SCHMUKI, [1]Radek Zboril

[1]Regional Centre of Advanced Technologies and Materials, Faculty of Science, Palacky University, Olomouc, Czech Republic, EU

stepan.kment@upol.cz, radek.zboril@upol.cz

Abstract

Nanostructuring and morphological engineering are the two important aspects of the control of the performance of functional materials. In the present work, we report on the fabrication of high aspect discrete TiO_2 nanotubes possessing two or multi-legs *via* electrochemical anodization of titanium foil. The effect of applied voltage and anodization time with appropriate optimization of electrolyte containing (NH_4) HF_2 in diethylene glycol-based electrolyte were investigated. Scanning electron microscopy was employed to examine the morphology of the synthesized TiO_2 nanotubes. We observed that prolonged anodization times from 2 hrs to 6 hours, results in the formation of dense and high aspect nanotubes with improved lengths compared to the sample synthesized with 2 hours of anodization time.

Keywords: TiO2, nanotubes, anodization, photoelectrochemistry

1. INTRODUCTION

Nanostructure materials have attracted enormous attention due to their interesting properties which differ from those of bulk materials. Among various types of ceramics, titanium dioxide as a significant inorganic functional material has extensively been investigated because of its useful chemical, optical and electronical properties. TiO_2 nanotubes with dictinct size and morphologies have been synthesized using different method such as hydrothermal method, template synthesis and anodization method [1]. Highly ordered and vertical TiO_2 nanotubes with controllable pore size and length can be obtained by anodization method for use in applications incluing dye-sentisized solar cells, photocatalysis, gas sensor, drug release systems and photoelectrochemical water splitting [2]. Synthesis of anodic hollow nanostructure of TiO_2 by anodizing titanium in chromic acid electrolyte solution was reported by Zwilling et al [2]. Gong et al. [3], reported the prepration of self-organized titanium nanotubes arrays by direct anodization of Ti foil in HF-H_2O electrolyte at room tempreture in which the length of the synthesized TiO_2 nanotubes were about 500 nm. Several anodization parameters have been considered, including potential, composition of electrolyte and time duration which can affect the morphology of nanotubes.

Shakar et al [4] obtained multi-leg TiO_2 nanotubes by a modified electrochemical bath solution consist of NH_4F and formamide and HF contining di-ehylene glycol (DEG) electrolytes. Such nanotubes possess wider diameter for the sake of joining together towards the top. It is well known that greater interface can be offerred proportionally by multi-legs TiO_2 nanotubes, which improves their photo-electrochemical ability than conventional TiO_2 nanotube array [5]. Furthermore, the graded refractive index presented by these nanotubes makes the effective scattering of light and thereby harnesses the available light appreciably.

In this present work, we report highly ordered, discrete (*i.e.* complately seperated and unconnected) multi-leg nanotubes with high aspect ratio which are effective candidates for light harvesting applications due to their branching nature, inter-tube spacing and wall thickness.

2. EXPERIMENTAL

The TiO_2 nanotubes were fabricated by the electrochemical anodisation method using Ti foil 0.25 mm thick (99.7 % purity, Sigma Aldrich) as an anode and Pt foil as a cathode. Prior to the anodisation the Ti foil was cleaned with low concentration HCl and subsequently ultrasonicated in acetone, isopropyl alcohol and de-ionised water, for 10 min each. The Ti foil was dried with nitrogen gas before mounting onto the anodisation set-up. The anodisation was carried out at 60. 80 and 100 V for 2, 4 and 6 h in an electrolyte mixture containing 96 ml DEG, 0.5 ml HF (48 % solution, Rankem) and 3.5 ml (1.5 M) NH_4F aqueous solution (see **Table 1**). After the anodization, the samples were rinsed thoroughly with isopropyl alcohol and deionised water and kept at room temperature for 1 h to dry. Subsequently, as-prepared samples were annealed at 400 °C for 2 h to achieve the anatase phase.

Field emission scanning electron microscopy (Hitachi SU6600- Tokyo, Japan) was used to investigate the surface morphology of the prepared samples. X-ray diffraction (PANalytical, The Netherlands) was employed to analyze the crystalline structure of the synthesized TiO_2; the analyses were performed using a Co-Kα as radiation (40 Kv, 30 Ma, λ = 0.1789 nm) in the range of $10^{\circ} \leq 2\theta \leq 80^{\circ}$.

Table 1 Experimental conditions of TiO_2 nanostubes arrays synthesized by anodization method

Sample name	Anodization time (h)	Voltage (V)	Temprature (°C)
NT1	2	60	Room Tempreture
NT2	4	60	35
NT3	6	60	35
NT4	2	80	35
NT5	4	80	35
NT6	6	80	35
NT7	2	100	35
NT8	4	100	35
NT9	6	100	35

3. RESULTS AND DISCISSION

Mohammadpour et al. [6] reported the synthesis of multi-leg TiO_2 nanotubes in HF containing DEG-based electrolyte at large anodization time (>40 h) and high voltage (120V). **Figure 1**. shows cross sectionals and top view images of synthesized TiO_2 nanotubes via anodization method at 60 V for 2, 4 and 6 h. SEM image of sample NT1 with using lower distilled water (**Figure 1a**) indicates a not compact structure. However, with increasing the temperature of solution up to 35 °C, and decreasing distilled water to 3.5 ml in the solution the results show a clearly seperated nanotubes with approximately 1 µm long and pore diameter of ~200 nm (sample NT2, **Figure 1b**). Prolonging the anodization time to 4 and 6 h (**Figure 1c** and **1d** respectively), could be used to increase the length of synthesized TiO_2 nanotubes up to 4 µm, while the tubes' diameter remains the same of approximately 220 nm. Further observation indicated that more dense nanotubes can be clearly seen in **Figure 1d** comparing to that in **Figure 1c**. It can be concluded that an increase in the anodization time can decrease the distance between the synthesized nanotubes.

A furture increase of the anodization voltage to 80 V for 2 h decreased the compression of multi-leg nanotube arrays with the same electrolyte solution (**Figure 2a**) comparing to the sample at 60 V for 2h (**Figure 1b**). However, by incresing the anodization time to 4 and 6 h, more nanotubes arrays can be observed (**Figures 2b** and **2c**). Moreover, from the SEM images, the diameter and the length of the nanotubes for the NT7 sample

(sample with 6 h anodization time at 80 V) were about 400 nm and 4-4.5 µm, respectively. In order to corroborate the TiO_2 nanotube arrays dispersion, SEM image of the sample NT7 with low magnification were depicted in **Figure 2d**.

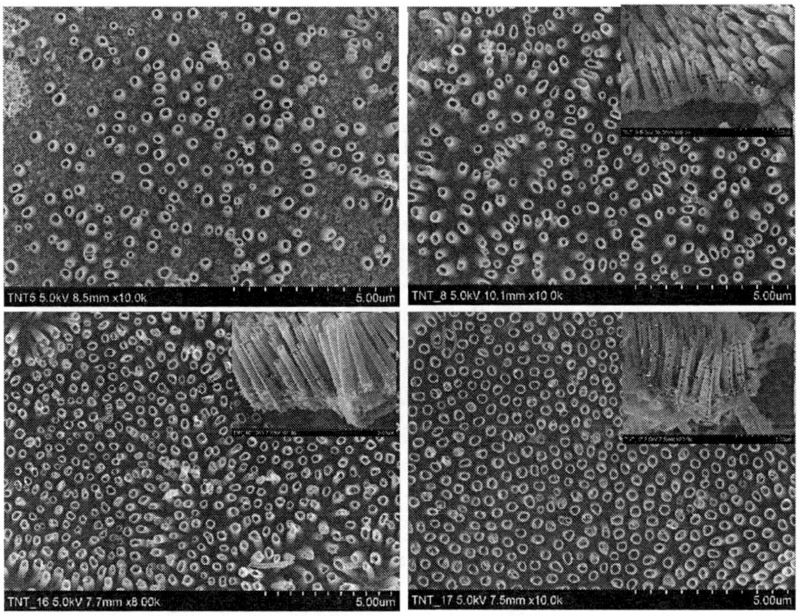

Figure 1 SEM images of the synthesized TiO_2 nanotubes arrays at 60 V for a) 2 h with 4 ml of distilled water, b) 2h with 3.5 ml of distilled water, c) 4 h and d) 6 h.

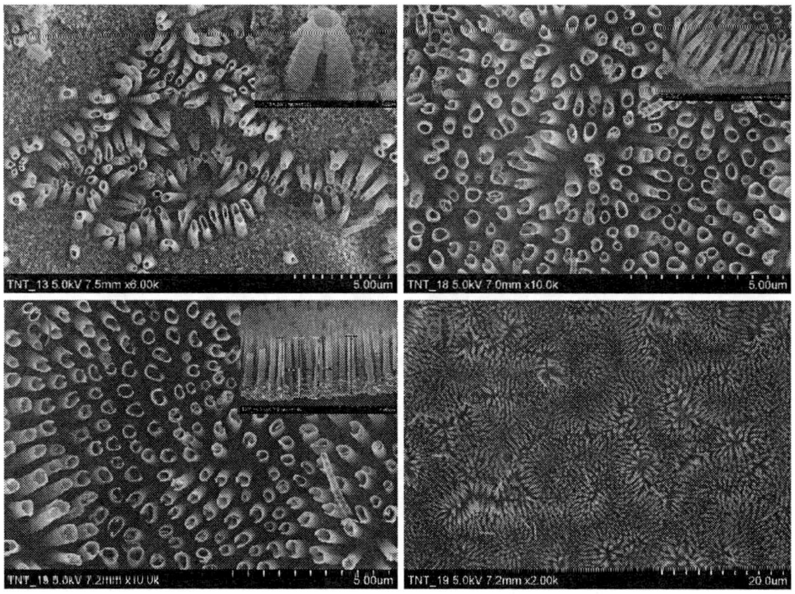

Figure 2 SEM images of the synthesized TiO_2 nanostubes arrays at 80 V for a) 2 h, b) 4h, c) and d) 6 h with different magnifications

The SEM images in **Figure 3a, 3b and 3d** confirm that prolonging the anodization time at 100 V for 2, 4 and 6 h respectively, results in considerable change in the obtained TiO_2 nanotube arrays density. Furthermore, by increasing the voltage under same conditions, the formation of a sponage oxide layer is observed, showing only little self-organized ordering of TiO_2 nanotubes.

Figure 3 SEM images of the synthesized TiO_2 nanostubes arrays at 100 V for a) 2 h, b) 4h, and c) 6 h

X-Ray diffractions (XRD) were used to demonstrate phase structure of the synthesized TiO_2 nanotubes after calcination at 400 °C for a 2 h in air. XRD patterns of calcined NT2 indicate anatase phase (see **Figure 4**).

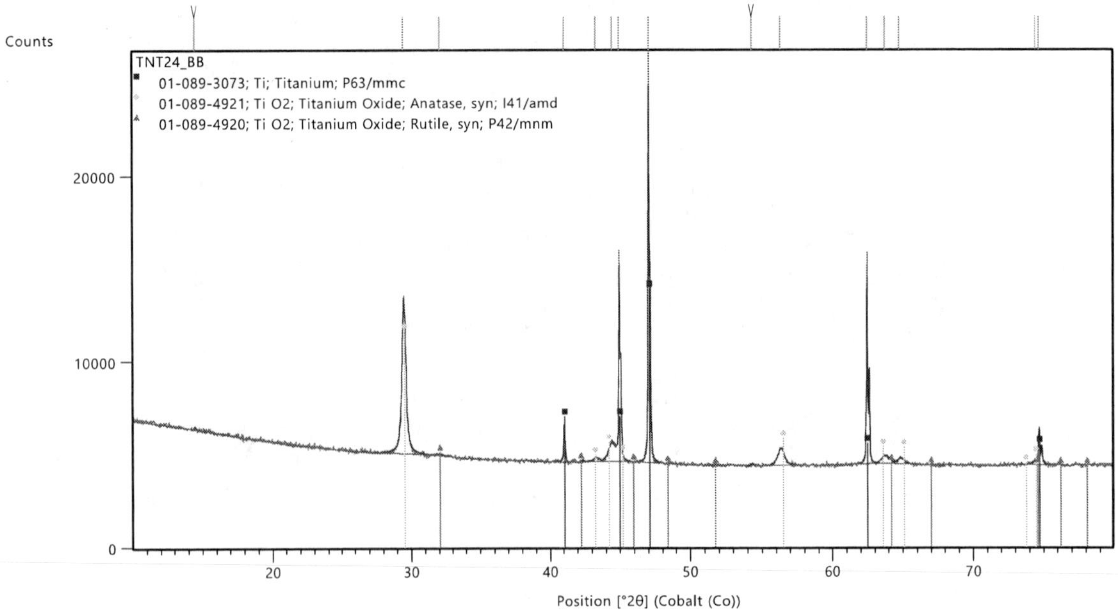

Figure 4 XRD patterns of sample NT2

4. CONCLUSION

In conclusion, we report on synthesis of discrete multi-leg TiO_2 arrays using the electrochemical anodization method in diethylene glycol (DEG) containing fluoride ions. An increase in the anodization time from 2 up to 6 h leads to synthesis longer nanotubes and at anodization potential of 80 V for 6 h, the longest nanotubes with wider diameter were obtained. These multi-leg nanotubes are suitable for enhanced light harvesting, leading to higher photocurrents in photoelectrochemical water splitting experiments.

ACKNOWLEDGEMENTS

The authors gratefully acknowledge the support by the project LO1305 of the Ministry of Education, Youth, and Sports of the Czech Republic.

REFERENCES

[1] KMENT, Stepan, RIBONI, Francesca, PAUSOVA, Sarka, WANG, Lei, WANG, Lingyun, HAN, Hyungkyu, HUBICKA, Zdenek, KRYSA, JOSE, SCHMUKI, Patrik, ZBORIL, Radek. Photoanodes based on TiO_2 and α-Fe_2O_3 for solar water splitting-superior role of 1D nanoarchitectures and of combined heterostructures. *Chemical Society Reviews.* 2017, vol. 46, pp. 3716-3769

[2] TAHMASEBPUOR, Roghayeh, BABALUO, Aliakbar, SHAHROUZI, Javad Rahbar, TAHMASEPOOR, Maryam and SHAHREZAEI, Mahdi. Theoretical and experimental studies on the anodic oxidation process for synthesis of self-ordering TiO_2 nanotubes: effect of TiO_2 nanotube lengths on photocatalytic activity. *Journal of environmental chemical engineering.* 2017, vol. 5, pp. 1227-1237.

[3] GONG, Dawei, GRIMES, Craig, VARGHESE, Oomman C, HU, Wenchong, SINGH, R.S, CHEN, Zhi, DICKEY, Elizabeth C. Titanium oxide nanotube arrays prepared by anodic oxidation. *Journal of material research*, 2011, Vol. 16, no.12, pp. 3331-3334.

[4] RAMBABU, Yalavarthi, JAISWAL, Manu, ROY, Somnath. Enhanced photoelectrochemical performance of multi-leg TiO_2 nanotubes through efficient light harvesting. *Journal of Physic D: Applied Physics.* 2015. vol. 48, pp. 1-9.

[5] BODA, Muzaffar Ahma, AFSHAR SHAH and Mohammad. Nhancd phoo-electrochemical potential of Fe_2O_3 by nanotube arra wit multiple legs. *Journal of material science*, 2018, vol. 29, pp. 4569-4601.

[6] MOHAMMADPOUR, Arash, WAGHMARE, Prashant, MITRA, Sushanta and SHANKAR, Karthik. Anodic groeth of large -diameter multipodal TiO_2 nanotubs. *ACS Nano*, 2010, vol. 4, no. 12, pp. 7421-7430.

INACTIVATION OF *ASPERGILLUS NIGER* ON PAINTS CONTAINING ZnO

[1]Zuzana RÁCOVÁ, [2]Michal BAUDYS, [2]Josef KRÝSA, [1]Petra TICHÁ, [5]Pavel DEMO

[1]CzechTechnical University in Prague, Prague, Czech Republic, EU,
zuzana.racova@fsv.cvut.cz, petra.ticha@fsv.cvut.cz
[2]University of Chemistry and Technology, Prague, Czech Republic, EU,
baudysm@vscht.cz, krysaj@vscht.cz
[5]Academy of Sciences of the Czech Republic v.v.i., Institute of Physics, Prague, Czech Republic, EU,
demo@fzu.cz

Abstract

Many paints contain special additives such as biocides which act as a prevention against biodeterioration. Presence of nanoparticles of silver and copper may inhibit the grow of microorganism. Self cleaning facade paints containing photocatalytic particles (mostly TiO_2) appear also on the market. Such paints are after irradiation of sunlight no to susceptible to soiling and thus help to keep exterior of building clean. Another type of paints is prepared with aim to improve air quality in exterior via photocatalytic oxidation organics compounds such as acetaldehyde, formaldehyde etc. This contribution deals with the application of ZnO particles to water based acrylic paints. The objective is to determined antifungal capability against A. niger. Attention will be also paid to determination of photocatalytic activity. Photocatalytic activity will be determined using indicator ink containing Resazurin dye. This method is based on observation of color change from pink to blue which occurs on photocatalytic surface.

Keywords: Self-cleaning, ZnO, *Aspergillus niger*, photocatalysis

1. INTRODUCTION

Self-cleaning paints contains besides usual pigment and fillers also particles of photocatalytic materials mostly TiO_2 of anatase modification but also paints containing ZnO [1] are not unusual. Such surfaces are after irradiation by UV light not to susceptible to soiling [2] and also exhibit superhydrophilic properties, which means that water do not formed droplets but it flows out from the surface in the form of thin film due to the decrease of water contact angle [3]. Such paints are also potential application to prevent biodeterioration of buildings material [4].

The resistance of building materials to micro-organisms is one of their monitored properties [5]. It is used to be tested the resistance to bacteria, mould and fungi. This paper deals with the resistance of the paint containing ZnO to fungi.The deal of this contribution is preparation and characterization of laboratory paints containing ZnO in term its photocatalytic and antifungal properties using *Aspergillus niger.*

2. EXPERIMENTAL

As a photocatalyst was used commercial ZnO with crystall size about 66 ± 3 nm (Sheerrer equation) and specific area 6 m^2/g (BET isotherm). Applying of ZnO in aqueous acrylic dispersion with non-photoactive rutile pigment, calcium carbonate and special additives were prepared samples of interior paints containing 8 % ZnO (in dry matter) Beside photoactive paint also reference paint without photocatalytic pigment containing only inactive rutile was prepared. Samples of paints were coated on glass (5x10 cm) using wire wound rot, the thickness of wet film was 120μm.

Photocatalytic activity was measured by method published in [6] which is based on reduction of Resazurin ink. Samples with photoactive coating were covered by Resazurin ink using 24 micron K bar, after drying samples were periodically irradiated by UV A light (λ = 351 nm, intensity 2 mW/cm^2). During UV exposition, it comes to reduction of Resazurin (blue) to Resorufin which is pink. Photocatalytic activity is than expressed as time of UV exposition which is needed to reach 90 % of overall color change.

Antifungal properties were studied using inactivation of *Aspergillus niger* [7]. Petri dishes with malt extract agar (Oxoid, Great Britain) containing samples of paints (25 x 25 mm) were inoculated by spraying of suspension of physiological solution containing spores of *Aspergillus niger* (Czech Collection of Microorganisms). Reference plates without paint samples were used as a control of growth of *Aspergillus niger*. Each test was performed with samples in five replicates. Samples were cultivated at temperature 23 ± 3 °C in light/dark cycle (16/8 hours). The experiment was evaluated as antifungal activity of tested paint visually after 28 days. The growth and development of mould mycelium on the dish and surface of the coating on a glass mat was monitored during the time of experiment to obtain comparison for paint with and without additive of 8 % ZnO. The scale used to evaluate mold growth is based on the standard ČSN EN ISO 846 (0 - no growth, 1 - growth visible only under the microscope, 2 - growth visible to the naked eye, covering less than 25 % of the surface, 3 - growth visible to the naked eye, covering less than 50 % of the surface, 4 - significant growth covering more than 50 %, 5 - surface covered with 100 %).

3. RESULTS AND DISCUSION

In **Figure 1**, there is illustrated dependence of red channel depending on time of UV irradiation for 6 paints samples containing 8 % of ZnO (in dry matter). We can see that during UV irradiation it comes to increase of Red channel, which corresponds to gradual color change of blue Resazurin to pink Resorufin. After 85 seconds of irradiation we can see in dependence in **Figure 1** typical elbow which corresponds to the time when the color of ink stop changing. With further UV irradiation the values of red channel didn't change and Photocatalytic activity is than expressed as time which is needed to reach 90 % of overall color change (time to bleach - Ttb_{90}). Calculated values are calculated by red squares and corresponds to the 83 ± 3 s. Reference paint without photocatalytic ZnO do not who measurable photocatalytic activity as results from **Table1**.

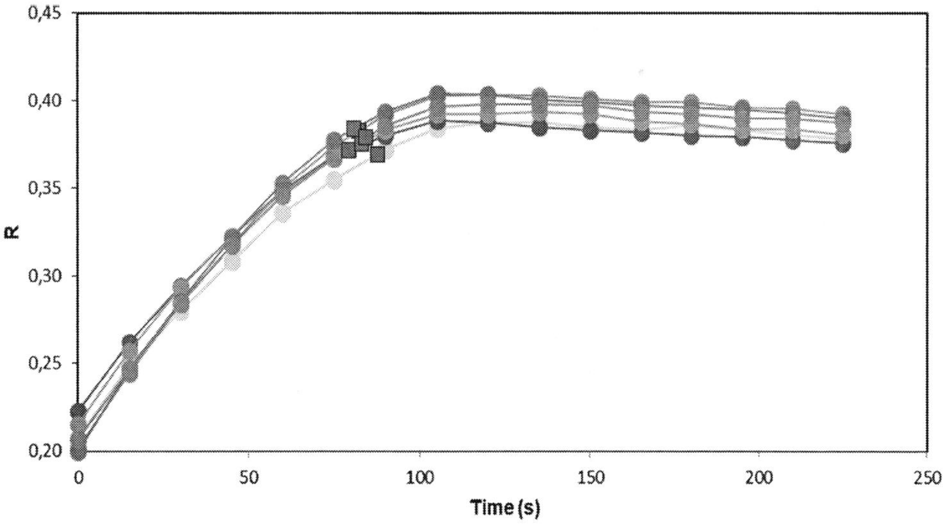

Figure 1 Dependence of photocatalytic experiment - paint containing 8 % of ZnO (experiment was proved for 6 identical samples, calculated TTb_{90} values express as red full squares)

Similar results of photocatalytic activity using Resazurin ink was observed in paper [8] in which photocatalytic activity of water based acrylic paint containing 8 % of TiO_2 was studied. On the other hand in paper published by Hochmanová et al. [1] was observed higher activity for ZnO compared to TiO_2 (measured using degradation of azo dye AO7).

Table 1 Comparison of color change of Resazurin ink on paint containing 8 % of ZnO (up) and reference sample

Sample/Time of UV exposition (seconds)	**0**	**15**	**30**	**60**
Active paint 8 % of ZnO				
Reference paint without ZnO			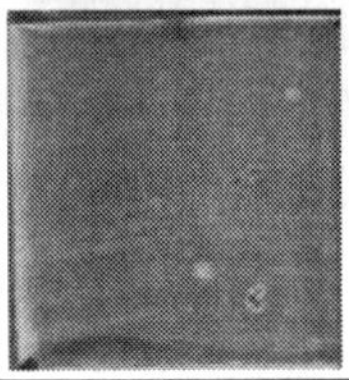	

Table 2 *Aspergillus niger* growth on the surface of paint samples during experiment

Sample/day of cultivation	**6th**	**8th**	**12th**	**28th**
Active paint 8 % of ZnO				
Reference paint without ZnO				
Reference paint without ZnO without inoculation		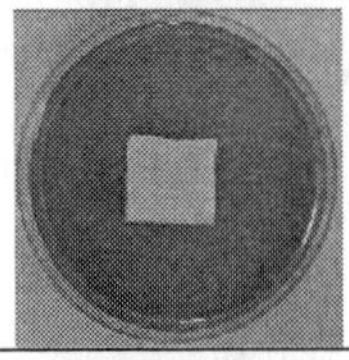	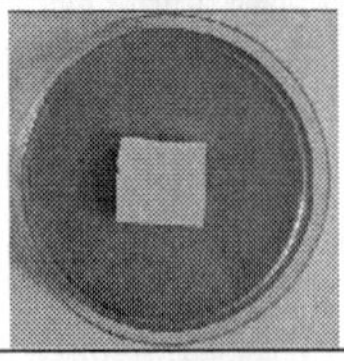	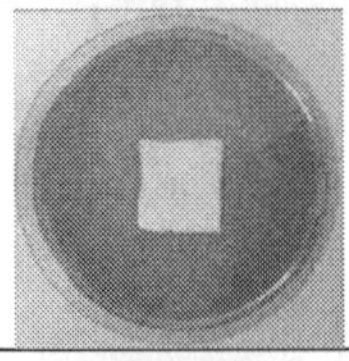
Control				

Aspergillus niger growth was monitored and noted for one month. The growth of mold is presented in the photos taken during experiment in **Table 2**. In **Table 3** there are shown results of the selected time intervals as an average value obtained from five repetition of samples. The results did not show significant difference between tested paint during the first six days. At eighth day first variation in mold growth was observed by kind of paint samples with inoculation. Growth of Aspergillus niger was more massive in the case of a sample without the addition of ZnO. It should be noted that even on the sample with the addition was the growth observed (growth visible only under the microscope - **Table 3**). Growth was registered up to the 28th day after inoculation. Even after the basic monitoring (28 days) has not acquired growth tendencies such as the sample without the addition of ZnO (meaning 50 days after inoculation). Those results are in the good agreement with paper [9] mentioned antifungal properties of ZnO.

Table 3 Evaluation of mold growth on the surface of paint samples

Sample/day of cultivation	6th	8th	12th	28th
Active paint 8 % of ZnO	0	0	0	1
Reference paint without ZnO	0	1	2	3
Active paint 8 % of ZnO without inoculation	0	0	0	0
Reference paint without ZnO without inoculation	0	0	0	0
Control	4	5	5	5

4. CONCLUSION

Photocatalytic and antifungal activity of water based acrylic paints containing 8 % of ZnO was determined. Prepared paints exhibit measurable photocatalytic activity which was quantified as Ttb_{90} (83 ± 3 s). Obtained results were similar to Ttb_{90} values determined for paints containing TiO_2. Antifungal properties was proved using *Aspergillus niger* as a model organism. After 8 days of the test apparent difference of mold grow was recorded. In sample containing ZnO, the growth of model organism was much lower which indicate antifungal properties of ZnO under presented setup of experiment. Given the obtained results it can be classified that the prepared coating has antifungal effect.

ACKNOWLEDGEMENTS

This work was supported by the Technical University in Prague, grant No. SGS17/118/OHK1/2T/11 and SGS16/197/OHK1/3T/11.

REFERENCES

[1] HOCHMANNOVA, L., VYTRASOVA, J. Photocatalytic and antimicrobial effects of interior paints In *Progress in Organics Coatings.* 2010. vol. 67, pp. 1-5.

[2] CHEN, J., POON C. Photocatalytic construction and building materials: From fundamentals to applications. In *Building and Enviroment.* 2009. vol. 44, pp. 1899-1906.

[3] FUJISCHIMA, A., RAO, T.N., TRYK, D.A. Titanium dioxide photocatalysis. In *J. Photochem and Photobiol A: Chemistry.* 2000. vol. 1, pp. 1-21.

[4] GAYLARDE, C.C., MORTON, L.H.G., LOH K., SHIRAKAWA, M.A. Biodeterioration of external architectural paint films - A review. In *International Biodeterioration & Biodegradation.* 2011. vol 65, pp. 1189-1198.

[5] TICHÁ, P. et al. Hydrophobic and biocidal properties of electrospun SiO2 nanofibers. In *9th International Conference on Materials - Research and Application (NANOCON).* Brno, 2017, pp. 206-211.

[6] MILLS, A. et al. A simple, inexpensive method for the rapid testing of the photocatalytic activity of self-cleaning surfaces. In *J. Photochem and Photobiol A: Chemistry.* 2013. vol. 272, pp. 18-20.

[7] CHEN, F. et al. Antifungal capability of TiO2 coated film on moist wood. *Building and Enviroment.* 2009. vol. 44, pp. 1088-1093.

[8] BAUDYS, M., KRÝSA, J., MILLS, A. *Smart inks as photocatalytic activity indicators of self-cleaning paints.* In Catalysis Today. 2017. vol. 280, pp. 8-13.

[9] SEVEN, O., et al. Solar photocatalytic disinfection of a group of bacterial and fungi aqueous suspensions with TiO2, Zno and Sahara desert dust. *Journal of Photochemistry and Photobiology A: Chemistry.* 2004. vol. 165, pp. 103-7.

SIMPLE ONE-POT HYDROTHERMAL SYNTHESIS AND PHOTOCATALYTIC ACTIVITY OF ZnS AND ZnS/KAOLINITE NANOCOMPOSITE

Kateřina MAMULOVÁ KUTLÁKOVÁ[1], Jonáš TOKARSKÝ[1,2]

[1]*Nanotechnology Centre, VSB - Technical University of Ostrava, Ostrava, Czech Republic, EU,*
katerina.mamulova.kutlakova@vsb.cz, jonas.tokarsky@vsb.cz
[2]*IT4Innovations, VSB - Technical University of Ostrava, Ostrava, Czech Republic, EU,*
jonas.tokarsky@vsb.cz

Abstract

This study is focused on preparation of pristine zinc sulphide (ZnS) and zinc sulphide/kaolinite (ZnS/K) nanocomposite using one-pot hydrothermal synthesis under atmospheric pressure and without any surfactant. Together with finding a simple preparation method, the aim was to achieve a high photocatalytic activity, and also reduce the environmental risk of ZnS (in the case of ZnS/K). Pristine ZnS was synthesized from aqueous solutions of zinc chloride and sodium sulphide in various ratios. ZnS/K nanocomposite was prepared similarly by addition of kaolin KKAF into the mixture of zinc chloride and sodium sulfide solutions. Reaction mixtures were continuously stirred at 100°C for several hours, resulting white solids were separated by centrifugation, washed with distilled water, and dried at 105°C overnight. X-ray powder diffraction analysis revealed the presence of cubic modification sphalerite both in ZnS and ZnS/K samples. Amount of ZnS in ZnS/K (~32 wt.%) was determined by Rietveld method which also confirmed the sphalerite structure. Morphology of the samples was observed using scanning electron microscopy. In the case of ZnS/K, the ZnS nanoparticles anchored on the kaolinite surface were observed. Photocatalytic activity evaluated using discoloration of Acid Orange 7 (AO7) in a liquid phase was found very high for ZnS samples. After 1 h of UV irradiation, ~99 % of AO7 was degraded by ZnS. Photodegradation efficiency reached 96 % in the case of ZnS/K. Taking into account that only ~32 wt.% of ZnS is present in ZnS/K, the photocatalytic activity of the nanocomposite can be considered three times higher compared to pristine ZnS.

Keywords: ZnS, kaolinite, hydrothermal synthesis, photoactivity, nanocomposite

1. INTRODUCTION

Zinc sulfide (ZnS), a direct wide band gap semiconductor crystallizing either in cubic sphalerite structure (E_g = 3.3-3.7 eV) or hexagonal wurtzite structure (E_g = 3.7-3.9 eV) [1], has found its application in many practical applications. In addition to medical diagnostics, senzoring, and optoelectronic devices [2], the ZnS plays an important role also in the field of photocatalysis, and ZnS quantum dots [3], nanoparticles [4], or microparticles [5] were reported as suitable for this purpose. ZnS can be prepared from a range of zinc and sulphur precursors (zinc acetate [3,6], zinc chloride [7], zinc nitrate [4,5], thioacetamide [3], sodium sulphide [4,5], etc.) via various preparation methods including sonochemical synthesis [3], melting the precursors in autoclave at 210°C for 24 h [4], hydrothermal synthesis in autoclave at 180°C for 12 h [5], or low-temperature solid-state chemical synthesis [7]. Surfactants for controlling the size and preventing an agglomeration of ZnS are quite often used [6,7]. Since the free nanoparticles can be hazardous [8], it is reasonable to prepare nanocomposites in which the mobility of nanoparticles is limited. Earlier studies have shown clay minerals as suitable matrices for anchoring the nanoparticles. Moreover, these nanocomposites exhibit higher photocatalytic activity compared to pure nanoparticles [6,9]. Nanocomposites of clay/ZnS type are not reported as commonly as clay-based nanocomposites containing TiO_2 or ZnO. Only few works on montmorillonite/ZnS can be found, e.g., Kozák et al. [6] who reported a two-step synthesis in which pure ZnS nanoparticles were prepared and then anchored to pure MMT by shaking both components in aqueous solution. The aim of this work is to test the preparation of ZnS nanoparticles and ZnS containing nanocomposite by a simplified one-step method without surfactants, without temperature higher than 100°C, and without complicated instrumentation. With respect to our earlier experiments [9], a similar preparation process and the same clay mineral, i.e., kaolinite KKAF, have been used in this work. X-ray powder diffraction, Rietveld analysis, and scanning electron microscopy were used to characterize the prepared samples. Finally, photoactivity of pure ZnS nanoparticles and the kaolinite/ZnS nanocomposite was determined and compared.

2. MATERIALS AND METHODS

2.1. Preparation of the samples

Zinc chloride ($ZnCl_2$; anhydrous; Lach-Ner) and sodium sulphide (Na_2S; anhydrous, Fisher Scientific) were used as received. For the preparation of pure ZnS nanoparticles, aqueous solutions of precursors having various concentrations (**Table 1**) were mixed in a beaker and continuously stirred at constant temperature 100°C. The kaolinite/ZnS nanocomposite was prepared similarly by mixing aqueous solutions of precursors in presence of 0.5 g of kaolinite KKAF (LB MINERALS). Size fraction < 40 µm was used. The resulting precipitates were centrifuged, washed with distilled water and dried at 100°C overnight. While the samples of pure ZnS nanoparticles were consecutively numbered, the nanocomposite was denoted as K/ZnS (**Table 1**).

Table 1 Reaction conditions for the synthesis of ZnS nanoparticles and K/ZnS nanocomposite

sample	$c(ZnCl_2)$ (mol/dm^3)	$c(Na_2S)$ (mol/dm^3)	molar ratio	$m_{kaolinite}$ (g)	reaction time (h)
ZnS_1	1.00	1.00	1:1	-	5
ZnS_2	0.50	1.00	1:2	-	5
ZnS_3	0.25	1.00	1:4	-	5
K/ZnS	0.50	1.00	1:2	0.5	4

2.2. Characterization methods

X-ray powder diffraction (XRPD) patterns were recorded (range 5-80°2θ) in reflection mode in symmetrical Bragg-Brentano arrangement using Bruker D8 Advance diffractometer equipped with fast position sensitive detector VÅNTEC 1. Radiation $Co_{K\alpha}$ was used (λ = 0.1789 nm). Sizes of ZnS crystallites (L_c) were calculated from ZnS(111) reflection according to Scherrer formula [10]:

$$L_c = \frac{K\lambda}{\beta \cos\theta} \tag{1}$$

where K is the dimensionless shape factor (0.9 was used), λ is the radiation wavelength (nm), β is the full width at half maximum of the most intensive ZnS reflection, i.e., the ZnS(111) reflection (°), and θ is the position of the ZnS(111) reflection (°). Lanthanum hexaboride (LaB_6) was used as a standard. Rietveld analysis (RA) of XRPD data was performed in Diffrac.Suite Topas software. Conditions of the RA were set up according to the experimental setting of the diffractometer, i.e., slit width 0.6 mm, Soller slits angle 2.3°, Lorentz polarization factor 0° (no monochromator). Chebyshev fifth-order polynomial was used for modeling the background. Goodness of fit was evaluated using χ^2 parameter [11] calculated via the least-square refinement of model XRPD pattern according to the experimentally obtained XRPD pattern. Scanning electron microscope (SEM) Hitachi SU6600 equipped with secondary electron detector was used to observe the samples. Prior the analysis, each sample was sputtered with Au/Pd. Accelerating voltage 5.0 kV was used. Photocatalytic experiment was performed in liquid phase using Acid Orange 7 (AO7). For each sample, two suspensions containing AO7 aqueous solution (V = 5 ml, c = $6.259 \cdot 10^{-4}$ mol/dm^3), 65 ml of demineralized water, and 50 mg of the sample were stirred in the dark for 1 h in order to obtain the adsorption equilibrium. One suspension was subsequently stirred under UV irradiation (λ = 254 nm) for 1 h, the second one was stored in the dark.

Photocatalytic activity (PA) was evaluated using CINTRA 303 UV-VIS spectrometer according to the following equation:

$$PA(\%) = \left(1 - \left(\frac{A_i}{A_c}\right)\right) \cdot 100 \tag{2}$$

where A_i and A_c is the intensity of AO7 absorption maximum (480 nm) for irradiated and control suspension, respectively.

3. RESULTS AND DISCUSSION

3.1. Phase composition of the samples

XRPD analysis showed that all three samples ZnS_1, ZnS_2, and ZnS_3 contain the same ZnS phase (**Figure 1a**), however, two options - sphalerite (PDF no. 05-0566) and wurtzite 8H (PDF no. 39-1363) - offered by ICDD PDF 2 Release 2014 database were very similar. Therefore, the RA was involved and two different models were used. Starting from unit cell parameters a = b = c = 5.409 Å, α = β = δ = 90° (sphalerite [12]) and a = b = 3.829 Å, c = 25.041 Å, α = β = 90°, δ = 120° (wurtzite 8H polytype [13]), the RA (performed on ZnS_3 pattern) revealed that the ZnS is present as the face centered cubic phase sphalerite. The χ^2 values 1.54 and 3.74 for sphalerite and wurtzite 8H, respectively, were obtained (compare also **Figures 1b and 1c**). For ZnS_1, ZnS_2, ZnS_3 samples, the L_c values (Eq. 1) were 3.89 nm, 7.70 nm, 7.81 nm, respectively. Consistency of the increase in L_c values and the increase in Na_2S : $ZnCl_2$ ratios (**Table 1**) is evident.

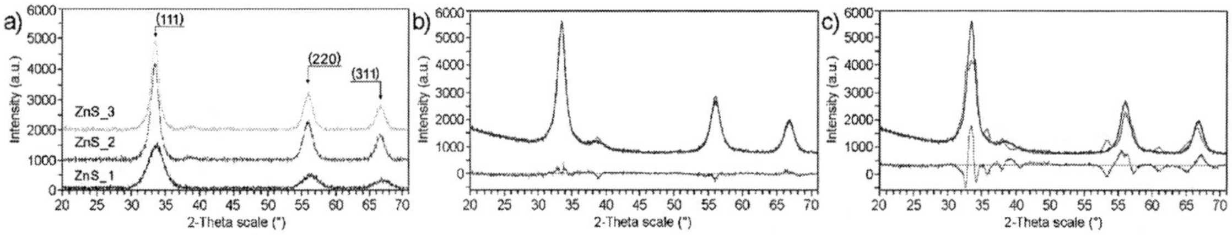

Figure 1 (a) XRPD patterns of ZnS_1, ZnS_2, and ZnS_3 sample in the range 20-70°2θ. (b) Experimentally observed (ZnS_3, blue), Rietveld calculated (red), and difference (grey bottom line) profiles for (b) sphalerite and (c) wurtzite 8H model applied to ZnS_3 XRPD data

XRPD pattern of the K/ZnS nanocomposite (**Figure 2**) revealed four phases: kaolinite (PDF no. 58-2028), muscovite (PDF no. 58-2037), quartz (PDF no. 46-1045, and sphalerite (PDF no. 05-0566). The latter phase was confirmed again by the RA which was used also for quantitative determination of the K/ZnS composition (**Figure 2**). Starting from the cell parameters a = 5.155 Å, b = 8.945 Å, c = 7.405 Å, α = 91.70°, β = 104.86°, δ = 89.82° (kaolinite [14]), a = 5.189 Å, b = 8.995 Å, c = 20.097 Å, α = δ = 90°, β = 95.11° (muscovite [15]), a = b = 4.913 Å, c = 5.405 Å, α = β = 90°, δ = 120° (quartz [16]), and the abovementioned cell parameters for sphalerite, the χ^2 value 2.26 for the XRPD pattern of K/ZnS sample was obtained. According to the quantitative RA, 49.16 wt.%, 11.85 wt.%, 7.26 wt.%, and 31.73 wt.% of kaolinite, muscovite, quartz, and sphalerite, respectively, is present in the K/ZnS nanocomposite. Good agreement between XRPD pattern and theoretical fit (**Figure 2)** indicates accuracy of the calculated weight percentages of individual phases in the K/ZnS sample. For ZnS in K/ZnS sample, value L_c = 4.83 nm was obtained using Eq. (1). The size of crystallites is, therefore, ~ 1.6 times lower compared to the ZnS_2 sample which was prepared using the same Na_2S : $ZnCl_2$ ratio as the K/ZnS (**Table 1**).

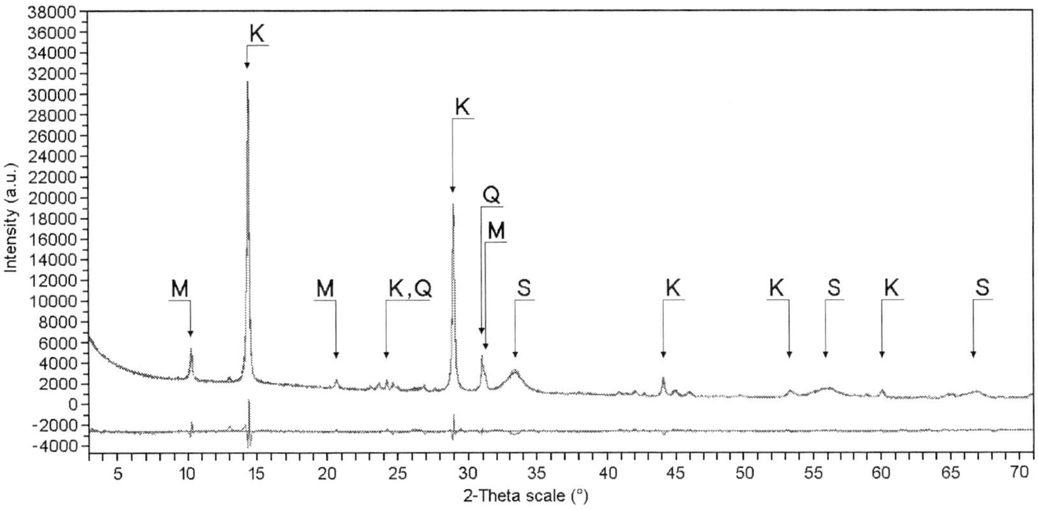

Figure 2 Experimentally observed (blue), Rietveld calculated (red), and difference (grey bottom line) profiles for the K/ZnS nanocomposite. M - muscovite, K - kaolinite, Q - quartz, S - sphalerite.

3.2. Scanning electron microscopy

Morphology of pure ZnS (**Figures 3a-c**) and K/ZnS (**Figures 3d-f**), was studied using SEM analysis.

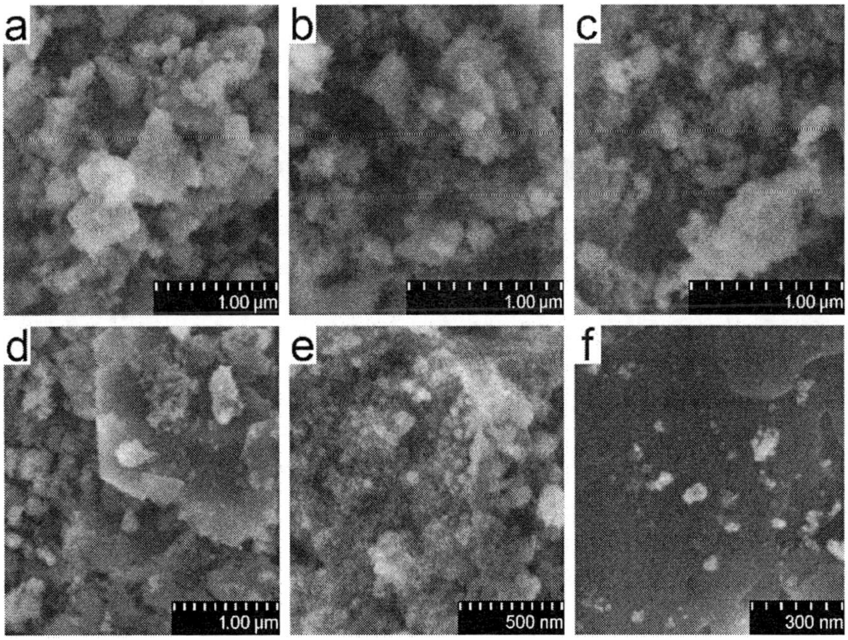

Figure 3 SEM images of samples (a) ZnS_1, (b) ZnS_2, (c) ZnS_3, and (d) K/ZnS

No significant differences between the pure ZnS samples were observed (**Figures 3a-c**), i.e., no influence of the Na_2S : $ZnCl_2$ ratio on the morphology was found. A common feature is the high degree of agglomeration, however, although agglomerates reach sizes in the micrometer range, round shaped nanoparticles having size < 100 nm can be recognized (**Figures 3a-c**). In the case of K/ZnS nanocomposite, SEM analysis confirmed the XRPD results in the meaning of ZnS growing on the clay matrix (**Figures 3d-f**). **Figure 3d** shows typical hexagonally shaped particle of kaolinite covered by ZnS nanoparticles. Further observation of the K/ZnS nanocomposite revealed uneven coverage of the clay matrix by the ZnS nanoparticles. In addition to completely covered areas (**Figure 3e**), also the relatively empty areas with only few nanoparticles (more or less isolated) were found (**Figure 3f**). The nanoparticles observed on the surface at the greatest magnification (**Figure 3f**) have the size from units to tens of nanometers.

3.3. Photocatalytic activity

Photodegradation of AO7 revealed almost identical and very high PA values for all three samples ZnS_1 (99.8 %), ZnS_2 (100 %), and ZnS_3 (99.2 %) (**Figure 4**). These results suggest negligible correlation between Na_2S : $ZnCl_2$ ratio used and the resulting PA value. For K/ZnS nanocomposite, the PA value is slightly lower (95.6 %), however, significantly lower ZnS amount in this sample (31.73 wt.%; see section 3.1) compared to samples ZnS_1, ZnS_2, and ZnS_3 (containing all 100 wt.% of ZnS) need to be kept in mind when evaluating the results. Thus, the ZnS in the K/ZnS nanocomposite exhibit nearly three times higher PA than pure ZnS nanoparticles when the ZnS amount in tested samples is taken into account. The observed increase in PA can be attributed to a lower degree of agglomeration of the ZnS nanoparticles in K/ZnS nanocomposite (see **Figures 3d,f**) due to their anchoring to the kaolinite matrix. The anchoring, along with a smaller size of nanoparticles, results in a higher area directly accessible to AO7 dye molecules.

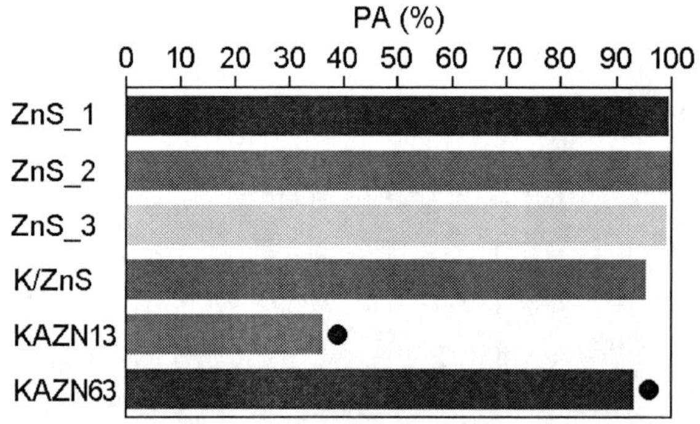

Figure 4 Photocatalytic activity (PA) against AO7 after 1 h of UV irradiation. Calculated according to Eq. 2. Samples and PA values marked with • were taken from [9] for comparison.

The K/ZnS nanocomposite was also compared with kaolinite-based nanocomposite containing 30 wt.% of ZnO [9] (see KAZN13 in **Figure 4**). For the KAZN13 sample tested under the same conditions as in this work, PA = 36 % (**Figure 4**) was found [9], while to reach PA similar to the K/ZnS, the KAZN13 had to be calcined at 600 °C for 1 h (see KAZN63 in **Figure 4)** [9]. Unnecessity of the calcination step to obtain the same PA value is an important economic factor reducing the cost of K/ZnS preparation.

4. CONCLUSION

Pure ZnS nanoparticles and K/ZnS nanocomposite were successfully prepared using simple one-pot hydrothermal method under atmospheric pressure and without any surfactant. Qualitative XRPD analysis supported by RA confirmed sphalerite phase in prepared samples, quantitative analysis revealed 31.73 wt.% of the sphalerite in the K/ZnS nanocomposite. Morphology of the samples and anchoring of nanoparticles on the clay matrix was observed using SEM. For the pure ZnS samples, the photocatalytic experiment performed in liquid phase for 1 h under UV irradiation showed nearly complete degradation of AO7 dye. Slightly lower PA value 95.6 % was obtained for the K/ZnS sample, however, with respect to the ZnS amount (31.73 wt.%), the photoactivity of this nanocomposite can be considered very high. Unlike the previously prepared kaolinite-based nanocomposite containing 30 wt.% of ZnO, no calcination is needed in the case of K/ZnS to reach such high PA value. The K/ZnS nanocomposite seems to be a very promising material for potential practical use in photocatalysis.

ACKNOWLEDGEMENTS

This work was supported by The Ministry of Education, Youth and Sports from the National Programme of Sustainability (NPS II) project „IT4Innovations excellence in science -LQ1602" and by the ERDF in the IT4Innovations national supercomputing center - path to exascale project (CZ.02.1.01/0.0/0.0/16_013/0001791) within the OPRDE. Financial support from The Ministry of Education, Youth and Sports project no. SP2018/95 is also gratefully acknowledged. Authors thank Dalibor Hroch and Lukáš Herman for preparation of samples and Rietveld analysis.

REFERENCES

[1] GAYOU, V. L., SALAZAR-HERNANDEZ, B. and CONSTANTINO, M.E. Structural studies of ZnS thin films grown on GaAs by RF magnetron sputtering. *Vacuum*. 2010. vol. 84, pp. 1191-1194.

[2] UMMARTYOTIN, S. and INFAHSAENG, Y. A comprehensive review on ZnS: From synthesis to an approach on solar cell. *Renewable and Sustainable Energy Reviews*. 2016. vol. 55, pp. 17-24.

[3] GOHARSHADI, E. K., HADADIAN, M., KARIMI, M. and AZIZI-TOUPKANLOO, H. Photocatalytic degradation of reactive black 5 azo dye by zinc sulfide quantum dots prepared by a sonochemical method. *Materials Science in Semiconductor Processing*. 2013. vol. 16, no. 4, pp. 1109-1116.

[4] XIANG, D., ZHU, Y., HE, Z., LIU, Z., and LUO, J. A simple one-step synthesis of ZnS nanoparticles via salt-alkali-composited-mediated method and investigation on their comparative photocatalytic activity. *Materials Research Bulletin*. 2013, vol. 48, no. 2, pp. 188-193.

[5] ZHANG, S. Preparation of controlled-shape ZnS microcrystals and photocatalytic property. *Ceramics International*. 2014. vol. 40, no. 3, pp. 4553-4557.

[6] KOZÁK, O., PRAUS, P., KOČÍ, K. and KLEMENTOVÁ, M. Preparation and characterization of ZnS nanoparticles deposited on montmorillonite. *Journal of Colloid and Interface Science*. 2010. vol. 352, no. 2, pp. 244-251.

[7] CHEN, F., CAO, Y. and JIA, D. Facile synthesis of ZnS nanoparticles and their excellent photocatalytic performance. *Ceramics International*. 2015. vol. 41, pp. 6645-6652.

[8] WU, T. and TANG, M. Review of the effects of manufactured nanoparticles on mammalian target organs. *Journal of Applied Toxicology*. 2018. vol. 38, no. 1, pp. 25-40.

[9] MAMULOVÁ KUTLÁKOVÁ, K., TOKARSKÝ, J. and PEIKERTOVÁ, P. Functional and eco-friendly nanocomposite kaolinite/ZnO with high photocatalytic activity. *Applied Catalysis B: Environmental*. 2015. vol. 162, pp. 392-400.

[10] SCHERRER, P. Estimation of size and internal structure of colloidal particles by means of Röntgen rays. *Nachrichten von der Gesellschaft der Wissenschaften zu Göttingen*. 1918. vol. 2, pp. 96-100.

[11] TOBY, B. H. R factors in Rietveld analysis: How good is good enough? *Powder Diffraction*. 2006. vol. 21, no. 01, pp. 67-70.

[12] WYCKOFF, R. W. G. *Crystal Structures, Volume 1.* 2nd ed. New York: John Wiley & Sons, p. 110.

[13] CHAO, G. Y. and GAULT, R. A. The occurence of two rare polytypes of wurtzite, 4H and 8H, at Mont Saint-Hilarie, Quebec. *The Canadian Mineralogist*. 1998. vol. 36, no. 3, pp. 175-177.

[14] BISH, D. L. and VON DREELE, R. B. Rietveld refinement of non-hydrogen atomic positions in kaolinite. *Clays and Clay Minerals*. 1989. vol. 37, no. 4, pp. 289-296.

[15] WYCKOFF, R. W. G. *Crystal Structures, Volume 4.* 2nd ed. New York: John Wiley & Sons, p. 346

[16] WYCKOFF, R. W. G. *Crystal Structures, Volume 1.* 2nd ed. New York: John Wiley & Sons, p. 312.

PREPARATION AND CHARACTERIZATION OF PHOTOCATALYTICALLY ACTIVE BISMUTH IODATE

[1,2]Vlastimil MATĚJKA, [2]Petr MARŠOLEK, [2]Alexandr MARTAUS, [3]Lucie GEMBALOVÁ, [1,2]Petr PRAUS

[1]VSB - Technical University of Ostrava, Faculty of metallurgy and materials engineering, Department of Chemistry, Ostrava, Czech Republic, EU, vlastimil.matejka@vsb.cz
[2]VSB - Technical University of Ostrava, Institute of Environmental Technology, Ostrava, Czech Republic, EU, petr.marsolek@vsb.cz
[3]VSB - Technical University of Ostrava, Faculty of Mining and Geology, Faculty of Mining and Geology, Institute of Clean Technologies for Mining and Utilization of Raw Materials for Energy Use, Ostrava, Czech Republic, EU, lucie.gembalova@vsb.cz

Abstract

Bismuth iodate was prepared using a hydrothermal synthesis of $Bi(NO_3)_3 \cdot 5H_2O$ and I_2O_5. The time effect of the hydrothermal synthesis (4, 24, 48 and 72 h) on the structure, specific surface area (SSA) and morphology of bismuth iodate particles, their optical properties and photodegradation activity were studied. A major phase presented in the prepared samples was $BiIO_4$ of which crystallite size increased with the duration of the synthesis as revealed with X-ray powder diffraction method. The values of specific surface area decreased with increasing time of the thermal treatment. The particles size and their morphology changed from very fine particles stacked into fluffy agglomerates observed after 4 h to agglomerates of tabular particles observed after 72 h long synthesis. The photocatalytic performance of the prepared samples irradiated with UVA light was evaluated by the photodegradation of model organic dye acid orange 7. It was observed that with the increasing duration of the $BiIO_4$ synthesis the photodegradation activity of resulting samples decreased. The photodegradation activity of $BiIO_4$ prepared at 4 and 24 h was higher than that of TiO_2 P25 (Degussa).

Keywords: Bismuth iodate, hydrothermal synthesis, characterization, photocatalysis

1. INTRODUCTION

The photocatalytic degradation of hazardous substances in water and air is part of perspective and widely studied methods mainly due to the fact that only an appropriate photocatalyst and light source of suitable wavelength is necessary. Titanium dioxide especially in its anatase form is undoubtedly the most studied material used for this purpose [1,2]. The band gap energy of anatase is approximately 3.20 eV what means this photocatalyst can be activated with light having wavelength equal to or lower than 387 nm what means it can be activated by UVA irradiation. Nowadays, many efforts are focused on the preparation of photocatalysts active in visible region of spectra, for example, intensively studied graphitic carbon nitride (g-C_3N_4) [3,4]. Efficient photocatalysts activated also by artificial UV light are still on high demand. The searching for new photocatalysts mainly reflects the demand for higher photodegradation efficiency, especially shifted towards the visible region [5].

It was already proved that the different bismuth based compounds, such as $BiVO_4$ [6], Bi_2WO_6 [7], Bi_2O_3 [8], exhibit high photodegradation efficiency, probably ascribed to lone pair electrons of Bi^{3+} [9]. $BiIO_4$ with layered structure is one of promising photocatalytically active semiconductors with the band gap energy close to that of TiO_2 [10,11]. Its layered structure is favorable for the efficient separation of electrons and holes and in general, this feature predetermines the layered photocatalytically active materials as the efficient photocatalysts.

In this research we focused on the synthesis of the $BiIO_4$ photocatalysts using the hydrothermal synthesis of aqueous solutions of $Bi(NO_3)_3{\cdot}5H_2O$ and I_2O_5. The prepared samples were characterized using X-ray powder diffraction method. Specific surface area was measured using the BET method; morphology of the particles was studied using scanning electron microscopy and band gap energy was determined using UV-VIS diffuse reflectance spectroscopy. The photodegradation activity of the prepared samples was evaluated by the UV induced photodegradation of azo dye acid orange 7.

2. MATERIALS AND METHODS

2.1. BiIO4 synthesis

$Bi(NO_3)_3{\cdot}H_2O$ was obtained from Penta s.r.o., I_2O_5 was obtained from Sigma Aldrich, distilled water was used for all of the experiments. The weighted amounts of $Bi(NO_3)_3{\cdot}5H_2O$ and I_2O_5 were dissolved separately in distilled water with assistance of ultrasound. The resulting solutions were mixed together to reach a weight ratio of 1:3 of pure $Bi(NO_3)_3{\cdot}5H_2O$ and I_2O_5. The resulting mixture was further mixed using an electromagnetic stirrer for 1 h. After this step, the reaction mixture was treated in an autoclave at 150 °C for 4, 24, 48 and 72 h. The resulting white slurry was filtered and washed with distilled water and dried at 60 °C. Resulting coarse particles were pulverized using an agate mortar to give the final samples assigned as $BiIO_4$(4), $BiIO_4$(24), $BiIO_4$(48) and $BiIO_4$(72), respectively.

2.2. Characterization methods

The X-ray powder diffraction (XRPD) examination of the samples was performed on a SmartLab diffractometer (Rigaku, Japan) equipped with a detector D/teX Ultra 250. Cobalt tube (CoKα, λ_1 = 0.178892 nm, λ_2 = 0.179278 nm) operated at 40 kV and 40 mA was used as an irradiation source. The powder samples pressed in a silicon holder were measured in a reflection mode with the Bragg-Brentano geometry in a 2θ range of 5° – 90°, with a step size of 0.02° and speed of 1.5 deg·min^{-1}.

Specific surface area (SSA) of the samples was measured using an automated volumetric apparatus 3Flex (Micromeritics Instruments, USA) after sample degassing at 60 °C for 24 h at vacuum higher than 1 Pa. Nitrogen was used as an adsorbate and the adsorption-desorption isotherms of nitrogen were measured at −196 °C. The values of SSA were calculated according the Brunauer-Emmett-Teller (BET) theory for the range of relative pressure of 0.05-0.25.

UV-VIS diffuse reflectance spectroscopy (UV-VIS DRS) measurements were performed using a Shimadzu UV-2600 spectrometer (Shimadzu, Japan) equipped with an integrating sphere IRS-2600Plus (Shimadzu, Japan). The samples were pressed inside a holder and UV-VIS DRS spectra were acquired in the range of 300 - 500 nm. $BaSO_4$ powder (Nacalai Tesque, Inc., extra pure reagent grade) was used for a baseline setting. The Tauc's plots were used for the evaluation of indirect band gap energies [12].

Scanning electron microscopy (SEM) observations were conducted using a FEI Quanta 650 FEG microscope (FEI, USA). The powders were spread over a conductive tape attached to a sample holder and sputtered with chromium using Quorum Q150T ES (Quorum Technologies, Great Britain) to reach its layer thickness of approximately 35 nm. The particles were observed using a back scattered electron detector (BSE).

The photodegradation activity of prepared samples was tested by the UV induced photodegradation of acid orange 7 (AO7) (Synthesia, a.s.). In the typical experiment, the tested samples were weighted (0.0500 g) into a glass bowl filled with 150 ml of distilled water and mixed using an electromagnetic stirrer (300 rpm). After 5 min of stirring 15 ml of the AO7 solution (c = 7.136 10^{-4} mol dm^{-3}) was added. The resulting suspension was further stirred in dark for 1 h to achieve the adsorption equilibrium. After this period, 5 ml of suspension was taken and filtered through a syringe filter Chromafil PP/GF/RC-20/25 (Macherey-Nagel, Germany). The absorbance of filtered solutions was measured at 480 nm using a Shimadzu UV-2600 spectrometer and assigned as A_0. In the same time a source of UV light (BLB 36 W, NARVA) was turned on and the sample was subjected to irradiation for 150 min. 5 ml of sample was taken out of the stirred suspension in selected time intervals (30, 60, 90 and 150 min) and filtered using a syringe filter. The absorbance of the filtered solution was measured at 480 nm; the values were assigned as A_{30}, A_{60}, A_{90} and A_{150}, respectively.

3. RESULTS AND DISCUSSION

X-ray powder diffraction analysis

The XRPD patterns of the synthesized samples $BiIO_4$(4), $BiIO_4$(24), $BiIO_4$(48) and $BiIO_4$(72) are compared in **Figure 1**. The phase $BiIO_4$ (01-080-7870) was identified as a major phase in all of the prepared samples. It is evident the intensity of the diffraction lines (0 1 0) and (0 4 2) are growing with duration of the hydrothermal synthesis.

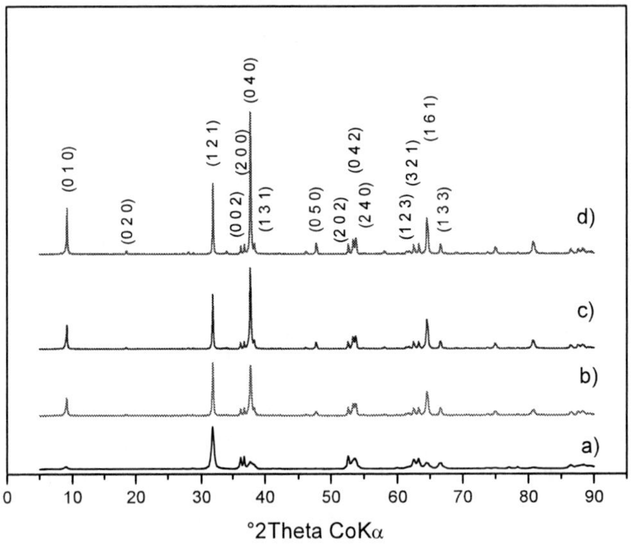

Figure 1 XRPD patterns of $BiIO_4$(4) (a), $BiIO_4$(24) (b), $BiIO_4$(48) (c) and $BiIO_4$(72) (d) samples

The sizes of the crystallites were calculated according to the Halder-Wagner equation implemented in the evaluation software PDXL2 (Rigaku, Japan) and are listed in **Table 1**.

Table 1 The values of crystalite size (L_D), specific surface area (SSA), band gap energy (Eg)

Sample	L_D (nm)	SSA ($m^2 \times g^{-1}$)	Eg (eV)
BiIO4(4)	20	16	3.20
$BiIO_4$(24)	30	11	3.20
$BiIO_4$(48)	39	6	3.17
BIO_4(72)	52	7	3.17
TiO_2 P25	Anatase - 20; Rutile - 29	50	3.20

The intensities of the diffraction lines obtained for the samples $BiIO_4$(4) and $BiIO_4$(24) are closely similar with that reported for $BiIO_4$ (JCPDS 01-080-7870). For the samples $BiIO_4$(48) and $BiIO_4$(72) the significantly higher intensity was registered for the diffraction peak (0 4 0) in comparison to that reported in its PDF card (JCPDS 01-080-7870) which could be attributed to preferential growing of the $BiIO_4$ crystals in this direction. The XRPD diffraction pattern of TiO_2 P25 (not shown in **Figure 1**) revealed the presence of anatase (A) and rutile (R) with the crystallite sizes of 20 and 29 nm, respectively. The calculated weight proportion of anatase and rutile in the P25 sample was 80 : 20.

The dependency of optical reflectance of the sample powders on wavelength was measured using UV-VIS DRS and is shown in **Figure 2**. The appropriate Tauc´s plots for indirect transitions are shown in **Figure 3** and the determined band gap energies are listed in **Table 1**.

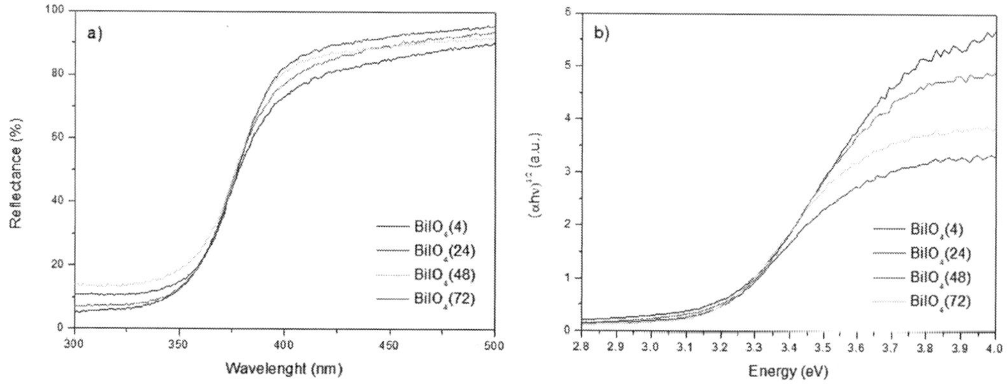

Figure 2 UV-VIS DRS spectra (a) and Tauc´s plots (b) of $BiIO_4(4)$, $BiIO_4(24)$, $BiIO_4(48)$ and $BiIO_4(72)$

The obtained values of E_g for the samples $BiIO_4(4)$ and $BiIO_4(24)$ are the same as that obtained for P25 and indicate the photocatalytic activity could be activated with irradiation having the wavelength ≤ 387 nm. The E_g values obtained for the samples $BiIO_4(48)$ and $BiIO_4(72)$ reached slightly lower values and the photocatalytic properties of both materials could be activated with UV irradiation having the wavelength ≤ 391 nm. The E_g values obtained for the synthesized samples were higher than that (2.94 eV) published by Huang et al [11].

The SEM images of the samples BiIO4(4) and BiIO4(72) obtained using SEM operated in a back scattered electron mode are shown in **Figure 3**. The $BiIO_4(4)$ sample (**Figure 3a**) occurred as fluffy aggregates of very fine particles contrary to the aggregates of particles of the BiIO4(72) sample which appeared as needles and also small plates which is evident from **Figure 3b**. The bigger particle size was evidenced for the sample $BiIO_4(72)$ (see **Figure 3b**). Growing intensity of the diffraction peaks (0 X 0) with the duration of the hydrothermal synthesis was identified by the XRPD method (**Figure 1**) and correlates with the observations achieved by the SEM method where the needle like shaped particles were observed.

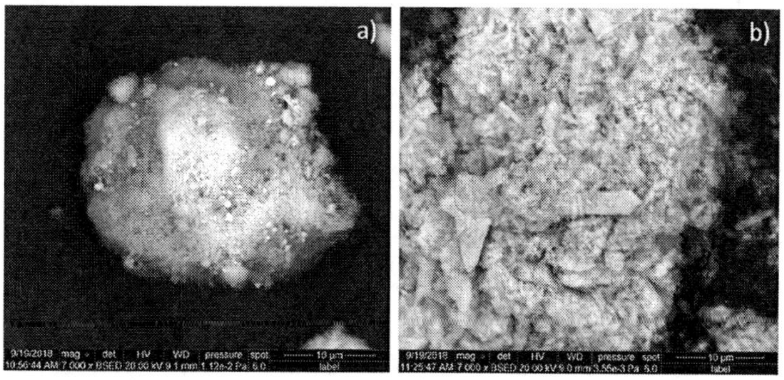

Figure 3 SEM images of $BiIO_4(4)$ (a) and $BiIO_4(72)$ (b) samples

The UV induced photodegradation activity of the prepared samples against AO7 is graphically presented in **Figure 4** and in the same figure it is compared with the photodegradation activity of TiO_2.

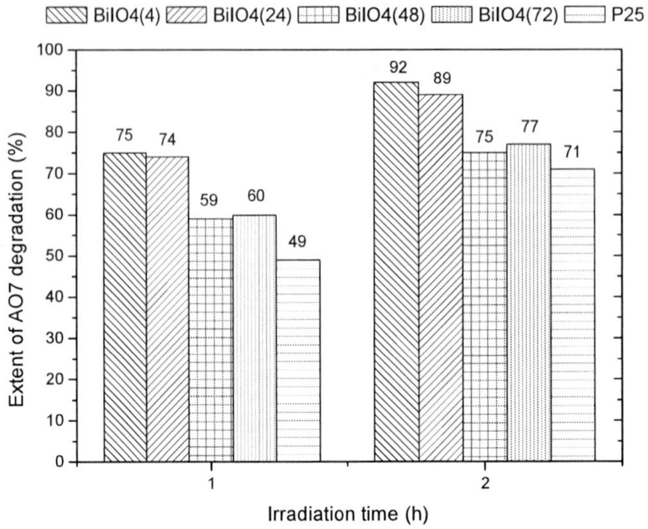

Figure 4 Photodegradation bars of AO7 for $BiIO_4(4)$, $BiIO_4(24)$, $BiIO_4(48)$, $BiIO_4(72)$ and TiO_2

Figure 4 indicates that more than 70 % of AO7 was degraded with the samples $BiIO_4(4)$ and $BiIO_4(24)$ after 1 h of irradiation. Approximately 60 % of AO7 was degraded with $BiIO_4(48)$ and $BiIO_4(72)$ and only 50 % of AO7 was degraded with TiO_2 after 1 h of irradiation. After 2 h of irradiation approximately 90 % of AO7 was photodegraded in case of the samples $BiIO_4(4)$ and $BiIO_4(24)$. Less than 80 % of AO7 was removed with the samples $BiIO_4(48)$, $BiIO_4(72)$ and TiO_2. Looking at the **Table 1** there is the correlation between crystallite size, SSA and the photodegradation activity of $BiIO_4$ samples: the lower the value of L_D, was the higher the value of SSA was and the photodegradation activity increased. The results pictured in **Figure 4** indicate the $BiIO_4(4)$ and $BiIO_4(24)$ samples as perspective photocatalysts for the photocatalytic degradation of azo dyes.

4. CONCLUSION

The samples of $BiIO_4$ were successfully prepared using the hydrothermal synthesis of the aqueous solutions of $Bi(NO_3)_3{\cdot}5H_2O$ and I_2O_5. It was observed that the different synthesis time had the effect on the structure, texture, morphology and photodegradation activity of the resulting samples. The results of photodegradation experiments indicated the samples synthesized for 4 and 24h were of the best photocatalytic degradation efficiency of AO7 reaching approximately 90 % after 2 h of UV irradiation. It is obviously higher than the performance of TiO_2 of which the extent of AO7 degradation reached 71 %.

Bismuth iodates are promising nanomaterials for photocatalytic applications and further research will be focused on shifting their photodegradation activity towards visible region by preparing nanocomposites with g-C_3N_4. The photocatalytic activity of both $BiIO_4$ and its nanocomposites with g-C_3N_4 will be also tested for degradation of various substances.

ACKNOWLEDGEMENTS

This work was supported by the Grant Agency of the Czech Republic (reg. No. 16-10527S) and by the student projects SP2018/79 of VSB-Technical University of Ostrava.

REFERENCES

[1] LEE, S.-Y. and PARK S.-J. TiO2 photocatalyst for water treatment applications. *Journal of Industrial and Engineering Chemistry*. 2013, vol. 19, no. 6, pp. 1761-1769.

[2] REN, H., KOSHY, P., CHEN, W.-F., QI, S. and SORRELL, C.C. Photocatalytic materials and technologies for air purification, *Journal of Hazardous Materials.* 2017, vol 325, pp. 340-366.

[3] XU, B., BOSHIR AHMED, M., ZHOU, J.L., ALTAEE, A., XU, G. and WU, M. Graphitic carbon nitride based nanocomposites for the photocatalysis of organic contaminants under visible irradiation: Progress, limitations and future directions. *Science of The Total Environment.* 2018, vol. 633, pp. 546-559.

[4] MASIH D., MA Y. and ROHANI, S. Graphitic C3N4 based noble-metal-free photocatalyst systems: A review. *Applied Catalysis B: Environmental.* 2017, vol. 206, pp. 556-588.

[5] SHAHAM-WALDMANN, N. and PAZ, Y. Away from TiO2: A critical minireview on the developing of new photocatalysts for degradation of contaminants in water. *Materials Science in Semiconductor Processing.* 2016, vol.48 Part 1, pp. 72-80.

[6] VENKATESAN, R., VELUMANI, S., ORDON, K., MALOWSKA-JANUSIK, M., CORBEL, G. and KASSIBA, A. Nanostructured bismuth vanadate (BiVO4) thin films for efficient visible light photocatalysis, *Materials Chemistry and Physics.* 2018, vol. 205, pp. 325-333.

[7] SELVI, M.H., VANGA, P.R. and ASHOK M. Photocatalytic application of Bi2WO6 nanoplates structure for effective degradation of methylene blue. *Optik.* 2018, vol. 173, pp. 227-234.

[8] RAZA, W., HAQUE M.M., MUNEER, M., HARADA, T. and MATSUMA, M. Synthesis, characterization and photocatalytic performance of visible light induced bismuth oxide nanoparticle. *Journal of Alloys and Compounds.* 2015, vol. 648, pp. 641-650.

[9] LI, F., HOU, X., PAN, S. and WANG, X. Growth, Structure, and Optical Properties of a Congruent Melting Oxyborate, Bi2ZnOB2O6. *Chemistry of Materials.* 2009, vol. 21, no. 13, pp. 2846-2850.

[10] HUANG, H., LIU, L., ZHANG, Y. and TIAN, N. One pot hydrothermal synthesis of a novel BiIO4/Bi2MoO6 heterojunction photocatalyst with enhanced visible-light-driven photocatalytic activity for rhodamine B degradation and photocurrent generation. *Journal of Alloys and Compounds.* 2015, vol. 619, pp. 807-811.

[11] HUANG, H., HE, Y., HE, R., JIANG X., LIN, Z., ZHANG, Y. and WANG S. Novel Bi-based iodate photocatalysts with high photocatalytic activity. *Inorganic Chemistry Communications.* 2014, vol. 40, pp. 215-219.

[12] TAUC, J., GRIGOROVICI, R. and VANCU, A. Optical Properties and Electronic Structure of Amorphous Germanium. *Physica Status Solidi (b).* 1966, vol. 15, no. 2, pp. 627-637.

ELECTRICAL AND ANTIBACTERIAL PROPERTIES OF POLYESTER NONWOVEN WITH CONDUCTIVE NANOMETALLIC SURFACE PREPARED USING TANNIC ACID

Hana KŘÍŽOVÁ[1], Jakub WIENER[2]

[1]*Technical University of Liberec, Institute for Nanomaterials, Advanced Technologies and Innovation, Liberec, Czech Republic, EU, hana.krizova@tul.cz*
[2]*Technical University of Liberec, Faculty of Textile Engineering, Department of Material Engineering, Liberec, Czech Republic, EU, jakub.wiener@tul.cz*

Abstract

The surface of the polyester nonwoven fabric was covered with a layer of silver and copper-based nanoparticles obtained from their salts and reduced using tannic acid. Polyester nonwoven MILIFE® without surface modification was used as a carrier of the metallic layer. The repetitive technological steps of preparing the conductive layer led to an increase of color intensity of fabric with bronzing effect, which also manifested in their decreased electric resistance. Simultaneously, all the samples were tested for resistance to selected G+ and G- bacteria.

Keywords: Tannic acid, silver, copper oxide, nanoparticles, electric resistance, antibacterial

1. INTRODUCTION

With the development and miniaturization of electronics, textiles are at the forefront of developers' interest due to their flexibility, breathability, comfort, low weight and considerable mechanical resistance. At the same time, the connection to electronics shifts the textiles to a qualitatively higher level. Therefore, in recent years, considerable attention has been paid to the electrical properties of textiles.

Most of the textiles act as very strong natural insulators. Electrically conductive fabrics can be made in several ways. Generally, conductive fabrics can be prepared by two methods. Firstly, conductive materials (e.g. metal nanoparticles, carbon nanotubes, graphene, conductive polymers) are added during the spinning process to obtain conductive fibers, and the fibers are then woven into conductive fabrics. [1,2] The disadvantage and limitation of this procedure are reduction of fiber spinnability, flexibility and many other mechanical properties. The second method of preparation of a conductive fabric is an additional covering of the fabric surface using above-mentioned conductive materials or their composites. [3] The disadvantages of this process are, for example, the unevenness of the conductive layer, the reduction of the breathability of the fabric, and the danger of exfoliation and delamination of the coating. Despite these risks, the second procedure is technologically simpler and cheaper.

One of the possible and sustainable paths of the development of new methods and technologies for eco-friendly advancements in preparation of conductive textiles is to use the green chemistry, particularly, its application in the deposition of metallic layers and reduction of metal nanoparticles on the surface of the fibers. Reduction of metal salts by plant derived compound like tannins fulfil the requirement of a chemical reductant. Hence, it renders the characteristics of green synthesis. Tannins as natural reductants (antioxidants) have potential to reduce some of their salts, which has been experimentally proven in the past research studies. Tannic acid has been used experimentally in the past for example to prepare nanoparticles of silver, gold, palladium and zinc. [4] Also, bimetallic nanoparticles such as Au-Pt were prepared using tannic acid. [5]

In this work, the bath method has been used for metal nanolayer coating of fibers surface when metal salts were reduced by natural reducing agent (tannic acid) in an alkaline environment. Thus, two aspects of polyester nonwoven with nanometallic layer were investigated, namely electrical resistivity and the ability to resist the action of selected G- and G+ bacteria (*Escherichia coli* and *Staphylococcus epidermidis).*

1.1. Mechanism of interaction of tannic acid with metal salts

Tannic acid is rich in electrons and embodies the capability of liberating freely reactive hydrogen atom. Presence of a hydrophobic "core," a hydrophilic "shell," and above all the polyphenolic nature of tannic acid make it an effective reducing agent. Numerous polyphenols bear antioxidant nature owing to the relative facileness of donating hydroxyl group to a free radical and the potency of the aromatic ring to carry an unpaired electron. Carboxylic acid groups (COOH) present in the tannic acid lose their hydrogen atom to become carboxylate ion (COO-) during the reduction process. The COO- so formed attaches to the surface of metal nanoparticles along with the remaining part of the polymer to act as surfactant and stabilize metal nanoparticles. [6]

1.2. Antibacterial properties of metal nanoparticles

Silver nanoparticles and nanoparticles of copper or copper oxides have been examined many times as biocides. The mechanism how nanoparticles can penetrate bacteria is not understood completely but studies have suggested that when bacteria are treated with copper nanoparticles, changes take place in its cell membrane morphology. Copper ions have the capacity to kill bacteria by destroying their cell walls and membranes because they have a strong reduction ability, which can extract electrons from the bacteria, causing their cytoplasm to escape and oxidizing the cell nucleus. [7] The opposite charges of bacteria and copper ions released from nanoparticles are thought to cause adhesion and bioactivity due to electrostatic forces. Copper ions can combine with the plasma membrane by electrostatic attraction and then penetrate the cell membrane through opening or closing of membrane channels. This alters the permeability of cellular membranes, which causes leakage of intracellular ions and low molecular weight metabolites. At the same time, copper ions entering cells combine strongly with intracellular amino acids and proteases, resulting in degeneration that leads ultimately to denaturation of proteins. The overall charge of bacterial cells at biological pH-values is negative due to the excess of carboxylic groups present in the lipoproteins at the bacterial surface, which, upon dissociation, makes the cell surface negative. [8] Also, the actual bactericide mechanism of silver nanoparticles is not well known. Some researchers support the idea that silver species release Ag^+ ions and they interact with the thiol groups in bacteria proteins, affecting the replication of DNA. It has also been reported that Ag^+ ions uncouple the respiratory chain from oxidative phosphorylation or collapse the proton-motive force across the cytoplasmic membrane. [9]

2. EXPERIMENTAL PART

2.1. Material and chemicals

Silver nitrate, copper sulfate pentahydrate, sodium hydroxide and sodium carbonate (all chemicals in analytical grade, Lach-Ner, Czech Rep.), tannic acid (Sigma Aldrich), thermally bonded polyester nonwoven MILIFE® of area weight 10 g/m^2 and fiber diameter 20 μm (JX Nippon, Japan), TSA (tryptone-soya agar) in Petri dishes (Oxoid, Czech Rep.), *Escherichia coli* and *Staphylococcus epidermidis* in suspensions.

2.2. Devices

Digital insulation tester (Chauvin Arnoux CA 6543 1 kV Megohmmeter) was used to determine the electrical conductivity or electrical resistance of textile samples with metallic layers, and the values were compared with the value of MILIFE® without the nanometallic layer. The measurement was carried out 3 times in different areas of the samples, each fabric in 1, 2 and 4 layers lying on each other. The basic statistical analysis was carried out again. Scanning electron microscope (UHR SEM Ultra Plus, Carl Zeiss, Germany) was used to display and measure the metallic nanolayer on the fibre surface.

2.3. Antibacterial tests

The ability to resist G- and G+ bacteria was based on standard [10] and tested using suspension of *Escherichia coli* and *Staphylococcus epidermidis*, each at a concentration of 10^6 CFU/ml. The test samples were placed on agar inoculated with the bacterial strain. Cultivation took 24 hours at 37 ˚C. Subsequently, the growth of the bacteria under the sample and the size of the inhibition zone around the sample were evaluated.

Second antibacterial test was based on modified standard [11]. This dynamic shake flask test is designed to evaluate the antimicrobial activity of non-leaching, antimicrobial-treated specimens under dynamic contact conditions. Its first phase is the dynamic contact of the antimicrobial treated surface with a bacterial suspension. The contact of samples with bacterial suspension took 1 hour. In the subsequent phase, shaking of the textile samples in water is carried out, shake ratio of 1:50 (weight of the fabric in grams: volume of water in ml). These extracts were further diluted 100x and 1 ml of this eluate was plated on agar. After 24 hours at 37 ˚C, colonies of bacteria were counted. This second test was provided with *E. coli* only.

2.4. Preparation of samples

The following stock solutions were prepared: tannic acid (3g/l), $AgNO_3$ (2 g/l), $CuSO_4{\cdot}5H_2O$ (2 g/l), Na_2CO_3 (2 g/l), NaOH (4 g/l). The $CuSO_4$ solution was converted to $Cu(OH)_2$ by addition of NaOH. The solutions of both metals were heated to 75 °C. A sample of the clean fabric MILIFE® was placed in an $AgNO_3$ solution for 5 minutes. Subsequently, it was gently dried and immersed into tannic acid solution for 1 minute and further alkalized by brief immersion in Na_2CO_3 solution. Since 2-3 repetitions were not sufficient to create a conductive layer on the surface of the fabric, this procedure was repeated 8 times until the visual effect of the so-called „bronzing“ (change in the color of the surface giving the appearance of aged bronze). This is a condition where multiple metallic layers with a typical optical effect are formed on the fabric surface. Preparation of the copper nanolayer on the fabric was carried out similarly but without additional alkalization of Na_2CO_3. The procedure was repeated 8 times, too.

3. RESULTS AND DISCUSSION

3.1. SEM images

Figure 1 and **Figure 2** show the structure of the metallic layers formed on the surface of the PES fibers by the process of repeated application of metal ions from the solution and reduction by tannic acid.

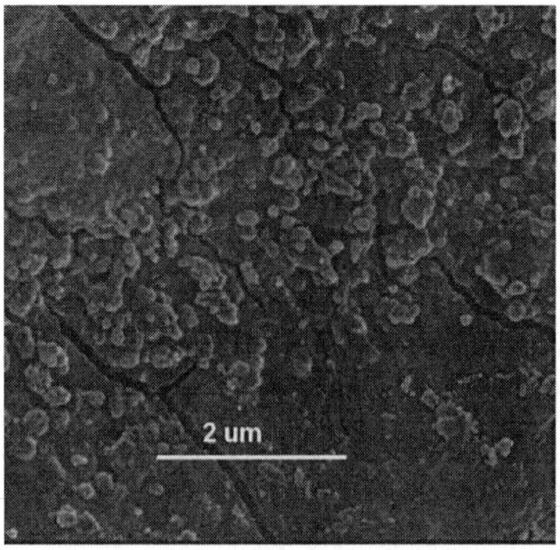

Figure 1 Copper-based nanolayer (10 000x)

Figure 2 Bronzing silver (multi) nanolayer (500x)

In **Figure 1**, the amorphous particles are visible, but it can hardly be recognized whether they are copper particles, copper oxide or copper tannates. The orange color indicates that it could be copper (I) oxide, but at the same time, it is impossible to recognize the cubic structure typical for spontaneous crystallization of Cu_2O, even at high magnification. It is very likely that a mixture of oxides and copper tannates is present. **Figure 2** confirms that the cause of the optical bronzing effect was actually the presence of multiple metallic layers of reduced silver. This layer was not homogeneous, and its disadvantage would therefore be that it would be delaminated and exfoliated under mechanical abrasion.

3.2. Electrical resistance

Table 1 contains an overview of measuring the electrical resistivity of the samples in a simple, double and quadruple-layer fabric to each other. The measurement was performed at a set value of 500 V. It is obvious that the resistivity decreased with the number of multiple layers because of increased conductivity due to a growing number of contacts fibers with metallic layers. Silver layer values show much better conductivity compared to copper-based coating. This result supports the assumption that, in the case of copper, a layer of Cu_2O was formed on the surface of PES fibers. In general, the oxides are non-conductive as they have a full valence band. Copper (I) oxide does not complete the valence rings and this is a metal deficient semi-conductor. Cu_2O is known to have a high electrical resistivity which varies with the method of preparation. [12] Whereas the resistivity of the carrier fabric was in the hundreds of GΩ, the fabric with copper-based layer had an electric resistivity of 1-2 orders of magnitude lower and the fabric with silver surface of 4 orders of magnitude lower compared to the non-metallic fabric.

Table 1 Mean values and 95 % confidence intervals of means for electric resistivity of samples

	MILIFE® (blank)	**MILIFE® (copper)**	**MILIFE® (silver)**
1 layer	450 ± 25.1 GΩ	28.5 ± 8.6 GΩ	117 ± 14.1 MΩ
2 layers	470 ± 32.2 GΩ	14.2 ± 4.8 GΩ	36 ± 5.6 MΩ
4 layers	550 ± 37.4 GΩ	7.7 ± 1.9 GΩ	25 ± 0.9 MΩ

3.3. Antibacterial properties

Figures 3a and b show the resulting interaction of samples with bacteria after 24 hours of cultivation. The samples are placed from left to right in the order of: MILIFE® (blank), MILIFE® with copper and MILIFE® with silver. Samples without surface treatment are completely covered by bacteria in both cases, the copper sample having about 1 mm zone of resistance. The best antibacterial effect shows the silver sample. Bacteria have not grown under it and, in the case of *E. coli*, a significant halo zone of 2.5 mm is visible.

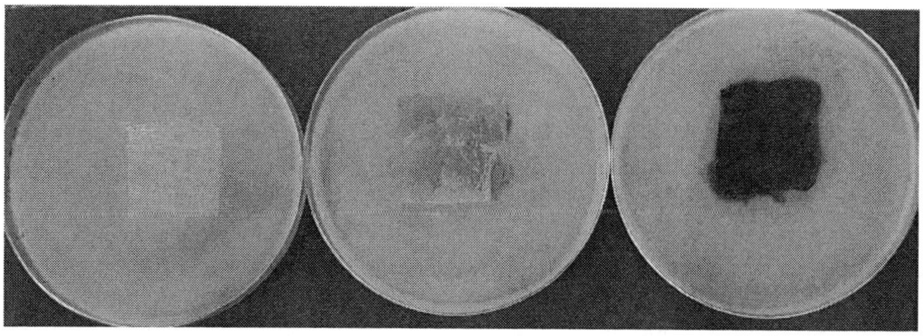

Figure 3a Samples on agar plates with *Staphylococcus epidermidis*, bacteria 10^6 CFU/ml

Figure 3b Samples on agar plates with *Escherichia coli*, both bacteria 10^6 CFU/ml

Dynamic shake flask test (**Figure 4**) provided some surprising results. The number of colonies in the resulting diluted eluates was 310 (blank), 470 (copper), 24 (silver) CFU/ml. A similar result, when the number of colonies from the copper sample exceeded the number of colonies from the blank, was reached even with repeated tests. One explanation for the cause could be that the layer did not contain copper or copper oxide, but rather a complex copper tannate whose antibacterial effects were not sufficient. The rugged surface of copper-based nanolayer was the reason that they caught larger number of bacteria than untreated samples.

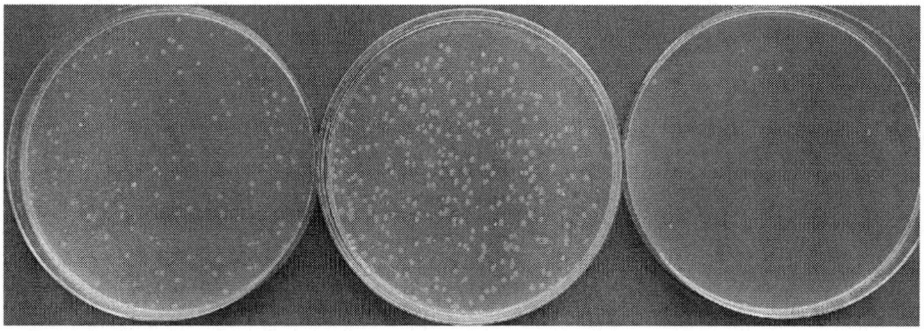

Figure 4 Cultivation of 100 times diluted eluates following dynamic contact with *Escherichia coli* (10^6 CFU/ml). From left to right: leach samples of MILIFE® (blank), MILIFE® (copper) and MILIFE® (silver)

In any case, even in this test, excellent antibacterial properties of silver nanoparticles were confirmed again, as the eluate from the silver surface sample was almost sterile.

4. CONCLUSION

In this work, we studied the preparation of metallic nanolayers on the surface of polyester nonwoven MILIFE® via green chemistry and the verification of the antibacterial and electrical properties of these samples. Thanks to the silver layer, MILIFE® obtained up to 10,000 times lower electrical resistivity than the original untreated fabric. Resistivity has decreased in the electrical measurement of multiple layers of fabric by creating a plurality of conductive joints. However, to achieve significant conductivity, it was necessary to repeat the application process and reduction of silver on the surface of the fibers up to eight times, but this resulted in non-homogeneous coating and formation of multiple metallic layers and plates, which optically manifested itself in the so-called bronzing. On the other hand, the simple layer of silver fabric also exhibited good antibacterial properties, especially against G- bacteria *Escherichia coli*. To prepare a copper-based conductive layer, a different procedure must be chosen because the layer resulting from the tannic acid reduction appears to have a high resistivity due to content of copper (I) oxide. The process of preparing the silver layer by means of green chemistry is promising. However, it is necessary to develop a process for production a homogeneous silver layer on the surface of the fibers.

ACKNOWLEDGEMENTS

This study is a result of the research funded by the project Nr. LO1201 with a financial support from the Ministry of Education, Youth and Sport of the Czech Republic under the NPU I program.

REFERENCES

[1] DURAN, Deniz and KADOĞLU, Hüseyin. Electromagnetic shielding characterization of conductive woven fabrics produced with silver-containing yarns. *Textile Research Journal*. 2015. vol. 85, no. 10, pp. 1009-1021.

[2] BHAT, Narendra V., et al. Development of conductive cotton fabrics for heating devices. *Journal of Applied Polymer Science*. 2006. vol. 102, no. 5, pp. 4690-4695.

[3] GREGORY, R. V. KIMBRELL, W. C. KUHN, H. H. Electrically conductive non-metallic textile coatings. *Journal of Coated Fabrics*. 1991. vol. 20, no. 3, pp. 167-175.

[4] SANCHEZ-MENDIETA, Victor and VILCHIS-NESTOR, Alfredo Rafael. Green synthesis of noble metal (Au, Ag, Pt) nanoparticles, assisted by plant-extracts. In: Noble Metals. *InTechOpen*. 2012. pp. 391-408.

[5] DANG-BAO, Trung, et al. Bimetallic Nanoparticles in Alternative Solvents for Catalytic Purposes. *Catalysts*. 2017. vol. 7, no. 207, pp. 1-33.

[6] AHMAD, Tufail. Reviewing the tannic acid mediated synthesis of metal nanoparticles. *Journal of Nanotechnology*. 2014. vol. 2014, no. 1, pp. 1-11.

[7] STOIMENOV, Peter K., et al. Metal oxide nanoparticles as bactericidal agents. *Langmuir*. 2002. vol. 18, no. 17, pp. 6679-6686.

[8] TONG, Guo, et al. Antibacterial effects of the Cu (II)-exchanged montmorillonite on Escherichia coli K88 and Salmonella choleraesuis. *Veterinary mikrobiology*. 2005. vol. 105, no. 2, pp. 113-122.

[9] HOLT, Katherine B. and BARD, Allen J. Interaction of silver (I) ions with the respiratory chain of Escherichia coli: an electrochemical and scanning electrochemical microscopy study of the antimicrobial mechanism of micromolar Ag+. *Biochemistry*. 2005. vol. 44, no. 39, pp. 13214-13223.

[10] ČSN EN ISO 20645. Fabrics - Determination of antibacterial activity - Agar plate spread test. 2005.

[11] ASTM E2149-13a. Standard Test Method for Determining the Antimicrobial Activity of Antimicrobial Agents Under Dynamic Contact Conditions. ASTM International, West Conshohocken, PA, 2013.

[12] OGWU, A. A., DARMA, T. H. and BOUQUEREL, E. Electrical resistivity of copper oxide thin films prepared by reactive magnetron sputtering. *Journal of achievements in materials and manufacturing engineering*. 2007. vol. 24, no. 1, pp. 172-177.

STUDY OF PHOTOVOLTAGE STABILITY ON NANODIAMOND-POLYPYRROLE COMPOSITES BY KELVIN PROBE METHOD

[1,2]Yen-Chen CHANG, [1,2]Daria MILIAIEVA, [1,2]Bohuslav REZEK

[1]*Institute of Physics of the Czech Academy of Science, Prague, Czech Republic, EU,*
changyen@fzu.cz
[2]*Faculty of Electrical Engineering, Czech Technical University, Prague, Czech Republic, EU*

Abstract

Diamond thin films and nanodiamond particles are considered promising for addressing the long-term challenge of organic photovoltaics: efficiency and stability. In this work, we characterize stability of surface potential and photovoltage on polyfunctional detonation nanodiamonds merged with polypyrrole oligomers. The polyfunctional-detonation nanodiamond (poly-DND), polypyrrole (PPy), and poly-DND/PPy composites were prepared on p-type silicon wafer substrates. In order to investigate illumination induced effects, surface potential and photovoltage of the samples were characterized by macroscopic Kelvin Probe method as a function of time. The poly-DND/PPy composite exhibits better stability of surface potential and surface photovoltage in short-term as well as in long-term (up to 8 weeks) compared to individual materials. The nanodiamond composites thus appear advantageous for use in organic photovoltaics.

Keywords: Diamond, nanoparticles, polypyrrole, scanning Kelvin probe

1. INTRODUCTION

Organic-based photovoltaic (OPV) devices are not so much limited in output under low-light or diffuse light conditions compared with silicon PV modules. OPV production and installation is also easy on large areas and in various shapes as they can be made on polymer foils. However, the main drawback of OPV is their long-term instability due to various chemical degradation processes and susceptibility to humidity. Many solutions including non-fullerene inorganic acceptors are being investigated. Diamond may provide interesting properties in this respect. The diamond interface with polypyrrole (PPy) was already shown to promote exciton generation and dissociation [1]. This property of the hybrid diamond/PPy composite may be thus beneficial to generate free charge. For this purpose it may be beneficial to use diamond in form of nanoparticles, so called nanodiamonds, which are commercially available in large quantities. Similarly as bulk diamond, nanodiamonds are reported as non-toxic [2,3] and generally chemically stable [4]. Synthesis process and structural interactions in nanodiamond/PPy composite were recently reported in our previous article [5]. In this work, opto-electronic properties of such composite are studied by Kelvin Probe. We present time-resolved measurements [6] of surface potential in dark and under illumination and we assess how stable the photovoltage of the nanodiamond-dye composite is compared with individual materials.

2. MATERIALS AND METHODS

We employed commercially available polyfunctional detonation nanodiamonds (NanoAmando) with oxygen and hydrogen surface chemical groups (poly-DND). The poly-DND particles, polypyrrole (PPy), or poly-DND/PPy composite were deposited on p-type silicon wafers according to the established protocol [5]. Briefly, the substrates were covered with poly-DND particles using 1 hour ultrasonication in aqueous colloidal poly-DND solution with concentration of 20 mg/mL [7] and then rinsed with deionized water. The process leads to a dense coverage by poly-DND [5]. The substrate with poly-DND layer was immersed in Py monomer solution (98 %) for 1 hour in dark to adsorb Py monomers on the nanodiamonds. The substrate with adsorbed Py was immersed into HCl acid (0.3 M) for 16 hours under indoor illumination to induce polymerization. Finally, the sample was taken out, rinsed with deionized water, and blown dry with nitrogen.

Scanning Kelvin probe (SKP) method was used to determine the surface potentials [6,8] of the samples in dark and under illumination with light intensity of about 200 W/m^2 (white cold light source with halogen lamp). SKP setup is shown in **Figure 1 (a)**. **Figure 1 (b)** is the detailed photograph of the sample in Kelvin probe setup. Measurement of each sample was conducted with a gold-plated tip electrode of 2 mm in diameter. The surface potential was detected at similar location on the sample in each measurement. At first, we measured surface potential in dark and waiting till it was stable for 100 seconds. Then we turned the light on quickly within 5 seconds and kept measuring the potential for 100 seconds. The procedure was continuously repeated during 12 min. Finally the measurement was stopped in the dark. Then the surface photovoltage was calculated from the surface potential data.

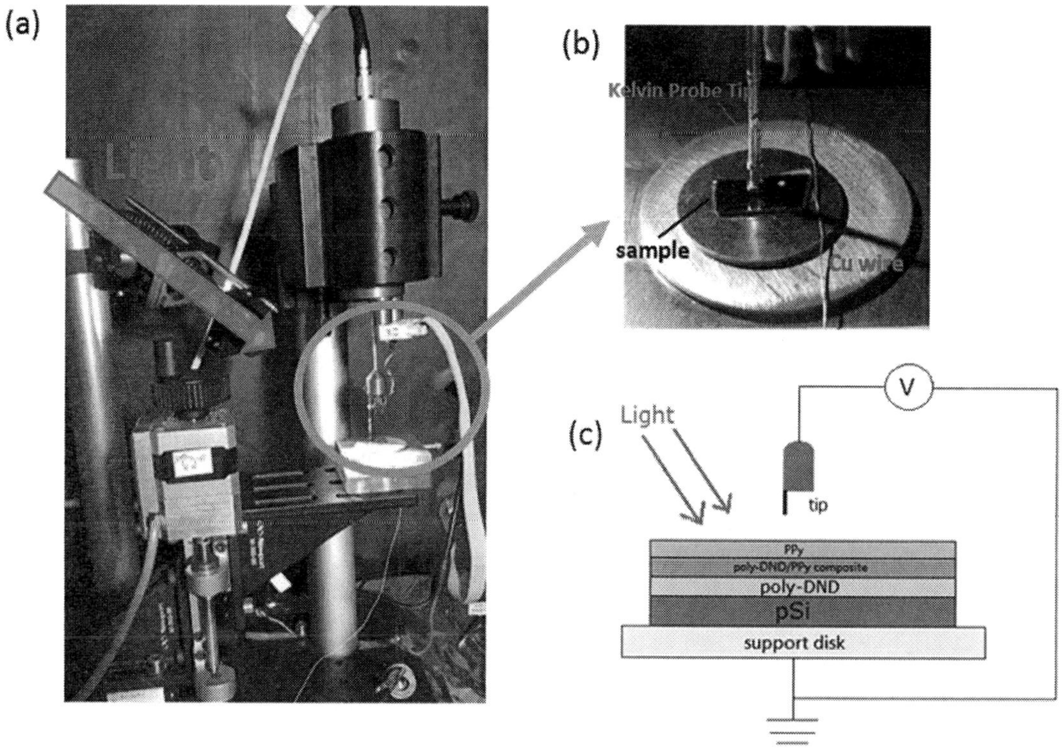

Figure 1 (a) Photograph of Scanning Kelvin probe setup. (b) Detailed photograph of the sample in Kelvin probe setup. (c) The scheme of macroscopic Scanning Kelvin Probe measurement with poly-DND/PPy composite.

3. RESULTS AND DISCUSSION

Figure 2 shows the results of surface potential (SP) measurement in the short-term. We can see that all surface potentials are stable (within 6 mV of standard deviation) and regularly reproducible during light switching in each measurement. Thus we averaged the whole surface potential measurement data in dark and under illumination for each material. The averaged surface potential with standard deviation in dark and under illumination was: pSi (-101 ± 6 mV and -272 ± 5 mV), PPy (-189 ± 3 mV and -311 ± 5 mV), poly-DND (-34 ± 4 mV and -230 ± 6 mV), and poly-DND/PPy composite (-202 ± 4 mV and -69 ± 3 mV).

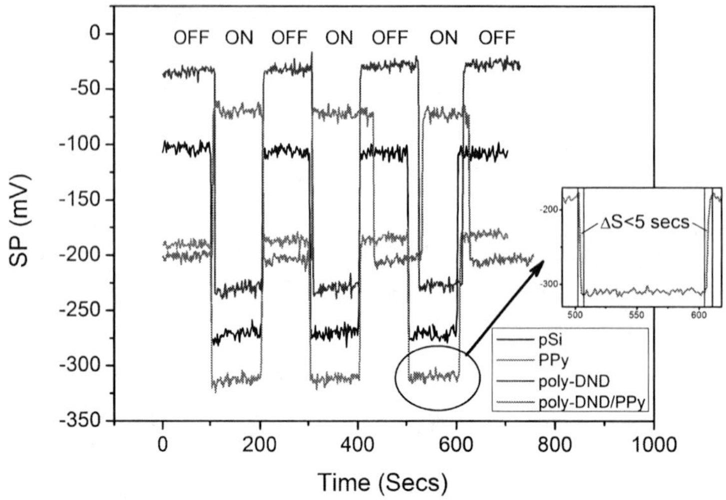

Figure 2 Surface potential of pSi, PPy, poly-DND, poly-DND/PPy as a function of time and visible light illumination in the short-term (0 - 12 min). The potential profiles correspond to measurements after 7 weeks. The inset shows that the rise and fall edges when illumination changes.

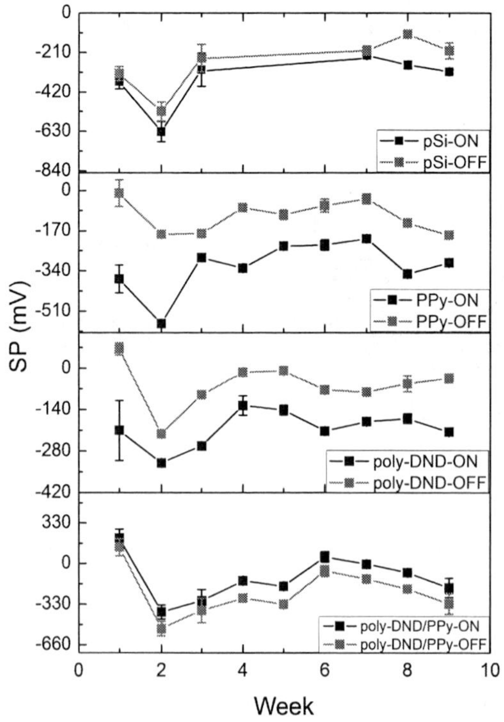

Figure 3 Average value of surface potential in dark and under illumination plotted as a function of time for the long-term stability evaluation

Then we repeated the short-term measurements every week for 8 subsequent weeks to investigate the long-term stability. Shape of the short-term profiles with the fast rise and fall edges was the same for the whole 8 weeks. Thus we averaged Kelvin Probe data in dark and under illumination each week. The averaged potential in dark and under illumination is plotted in **Figure 3**. Standard deviation of the average value is used as the error bar. We can see that in the long-term (during 8 weeks), the potentials in dark and under illumination are slowly fluctuating. The potential fluctuations are qualitatively very similar for all the studied materials, hence they are most likely related with some external factors.

The surface potentials exhibit similar fluctuation in dark and under illumination. Evaluation of surface photovoltage was thus use to observe actual material (in)stability in spite of the overall potential fluctuations. In **Figure 4 (a)** presents surface photovoltage (SPV) that was calculated as ($SP_{illumination}$ - SP_{dark}) each week for 8 weeks.

SPV exhibits the most pronounced changes during the first 3 weeks (area I), than during the remaining 6 weeks it seems fluctuating around a stabilized value (area II). To characterize this (in)stability, we calculated average SPV and its standard deviation in each area (values at the boundary were accounted for in both areas). The averaged SPV values with standard deviation in area I and area II are: pSi (-85 ± 33 mV and -91 ± 59mV), PPy (-281 ± 155 mV and -166 ± 55 mV), poly-DND (-182 ± 90 mV and -137 ± 30 mV), and poly-DND/PPy (95 ± 33 mV and 124 ± 22 mV).

Figure 4 (b) shows in the form of bar graph the standard deviation as a measure of SPV (in)stability. We can see the standard deviation difference of the averaged SPV between area I and area II on pSi (33 mV and 59 mV), PPy (155 mV and 55 mV), poly-DND (90 mV and 30 mV), and poly-DND/PPy (33 mV and 22 mV). The minimum standard deviation is observed on poly-DND/PPy composite during the whole 8 weeks.

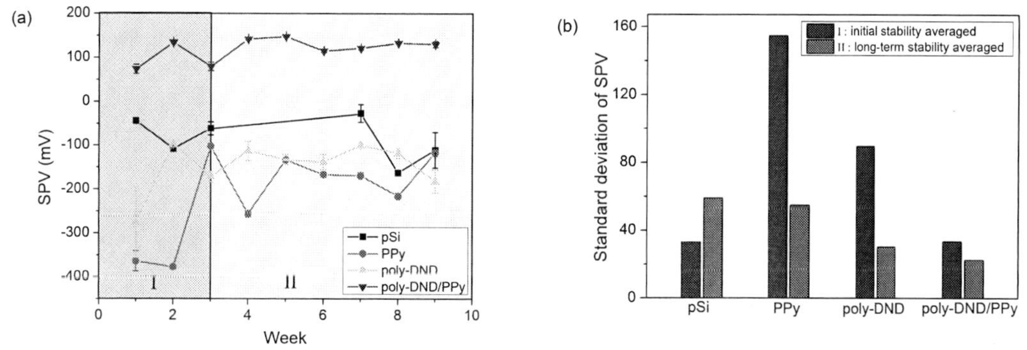

Figure 4 (a) Surface photovoltage (SPV) of all the materials for 8 weeks. Area I is defined as initial stability and area II as long-term stability. (b) The standard deviation of the averaged SPV value.

Thus the SPV of the poly-DND/PPy composite was the most stable from the very beginning after fabrication till the last measurement after 8 weeks. The material interaction in poly-DND/PPy composites thus contributes to photovoltage stability [6,9]. Note in this respect that surface photovoltage of of poly-DND/PPy composite has opposite (positive) sign compared the SP of pSi, PPy and poly-DND. Such positive charging of the surface, i.e. polypyrrole, is in agreement with theoretical atomic scale calculations [10] and indicates role of nanodiamonds as electron acceptors. This effect together with polypyrrole tight binding to nanodiamond may explain the improved stability of the composite.

4. CONCLUSION

In the range of minutes, surface potential (SP) under dark and illumination was stable and well reproducible for all the materials. In the long-term (weeks), the surface potentials of all materials were slowly fluctuating in similar way, most likely due to external factors. The generated photovoltage revealed actual material stability. The photovoltage was the most stable in the case of the poly-DND/PPy composite compared with individual materials. The improved stability is most likely due to the tight material interaction in poly-DND/PPy composites and nanodiamond acting as an electron acceptor. Nanodiamond composites may be thus advantageous for improving organic photovoltaics.

ACKNOWLEDGEMENTS

The work was financially supported by the GACR project 15-01809S and MSMT project "Centre of Advanced Photovoltaics" CZ.02.1.01/0.0/0.0/15_003/0000464.

REFERENCES

[1] REZEK, B., ČERMÁK, J., KROMKA, A., LEDINSKÝ, M., and KOČKA, J. Photovoltage Effects in Polypyrrole-Diamond Nanosystem. *Diamond and Related Materials.* 2009. vol. 18, no. 2-3, pp. 249-252. Available from: DOI: 10.1016/j.diamond.2008.07.019.

[2] VAIJAYANTHIMALA, V., TZENG, Y.K., CHANG, H.C., and LI, C.L. The Biocompatibility of Fluorescent Nanodiamonds and Their Mechanism of Cellular Uptake. *Nanotechnology.* 2009. vol. 20, no. 42, p. 425103. Available from: DOI: 10.1088/0957-4484/20/42/425103.

[3] SCHRAND, A.M., HENS, S.A.C., and SHENDEROVA, O.A. Nanodiamond Particles: Properties and Perspectives for Bioapplications. *Critical Reviews in Solid State and Materials Sciences.* 2009. vol. 34, no. 1-2, pp. 18-74. Available from: DOI: 10.1080/10408430902831987.

[4] KOZAK, H., REMES, Z., HOUDKOVA, J., STEHLIK, S., KROMKA, A., and REZEK, B. Chemical Modifications and Stability of Diamond Nanoparticles Resolved by Infrared Spectroscopy and Kelvin Force Microscopy. *Journal of Nanoparticle Research.* 2013. vol. 15, no. 4, p. 1568. Available from: DOI: 10.1007/s11051-013-1568-7.

[5] MILIAIEVA, D., STEHLIK, S., STENCLOVA, P., and REZEK, B. Synthesis of Polypyrrole on Nanodiamonds with Hydrogenated and Oxidized Surfaces. *Physica Status Solidi (A).* 2016. vol. 213, no. 10, pp. 2687-2692. Available from: DOI: 10.1002/pssa.201600278.

[6] REZEK, B., ČERMÁK, J., KROMKA, A., LEDINSKÝ, M., HUBÍK, P., MAREŠ, J.J., PURKRT, A., CIMROVÁ, V., FEJFAR, A., and KOČKA, J. Synthesis, Structure, and Opto-Electronic Properties of Organic-Based Nanoscale Heterojunctions. *Nanoscale Research Letters.* 2011. vol. 6, no. 1, p. 238. Available from: DOI: 10.1186/1556-276X-6-238.

[7] STEHLIK, S., VARGA, M., LEDINSKY, M., JIRASEK, V., ARTEMENKO, A., KOZAK, H., ONDIC, L., SKAKALOVA, V., ARGENTERO, G., PENNYCOOK, T., MEYER, J.C., FEJFAR, A., KROMKA, A., and REZEK, B. Size and Purity Control of HPHT Nanodiamonds down to 1 nm. *The Journal of Physical Chemistry C.* 2015. vol. 119, no. 49, pp. 27708-27720. Available from: DOI: 10.1021/acs.jpcc.5b05259.

[8] ČERMÁK, J., KOIDE, Y., TAKEUCHI, D., and REZEK, B. Spectrally Dependent Photovoltages in Schottky Photodiode Based on (100) B-Doped Diamond. *Journal of Applied Physics.* 2014. vol. 115, no. 5, p. 053105. Available from: DOI: 10.1063/1.4864420.

[9] ZHONG, Y.L., MIDYA, A., NG, Z., CHEN, Z.K., DAENEN, M., NESLADEK, M., and LOH, K.P. Diamond-Based Molecular Platform for Photoelectrochemistry. *Journal of the American Chemical Society.* 2008. vol. 130, no. 51, pp. 17218-17219. Available from: DOI: 10.1021/ja805977f.

[10] MATUNOVÁ, P., JIRÁSEK, V., and REZEK, B. Computational Study of Physisorption and Chemisorption of Polypyrrole on H-Terminated (111) and (100) Nanodiamond Facets. *Physica Status Solidi (A).* 2016. vol. 213, no. 10, pp. 2672-2679. Available from: DOI: 10.1002/pssa.201600228.

ATOMIC LAYER DEPOSITION OF LITHIUM OXIDE, TIN OXIDE, AND LITHIATED TIN OXIDE NANOFILMS FOR HIGH PERFORMANCE THIN FILM BATTERIES ANODES

Maxim MAXIMOV[1], Denis NAZAROV[1,2], ILya MITROFANOV[1], Yury KOSHTYAL[1], Aleksander RYMYANTSEV[1], Anatoly POPOVICH[1]

[1]*Peter the Great Saint-Petersburg Polytechnic University, Saint-Petersburg, Russian Federation,* maximspbstu@mail.ru, carlemeros@gmail.com, yury.koshtyal@gmail.com, rumyantsev.amr@gmail.com popovicha@mail.ru

[2]*Saint Petersburg State University, Saint Petersburg, Russian Federation,* dennazar1@yandex.ru

Abstract

In recent years atomic layer deposition has proved to be a successful method of fabrication and modification of electrodes for lithium-ion batteries and thin film power sources. Tin oxide demonstrated considerable excellence over currently used materials for thin film batteries anodes, however, lithiated tin oxides are expected to show even better performance. In this research nanofilms of LiOx, SnO_2 and lithiated tin oxide were deposited by atomic layer deposition (ALD) using lithium hexamethyldisilylazide ($LiN[Si(CH_3)_2]_2$, LiHMDS), tetraethyl tin ($Sn(C_2H_5)_4$, TET) as the metal-containing reagents and water or ozone or oxygen plasma as co-reagents. The monocrystalline silicon (100) and stainless steel (316SS) were used as supports. The thickness of the films was measured with spectral ellipsometry. The composition was studied by X-ray photoelectron spectroscopy and time-of-flight secondary ion mass spectrometry. Regardless of the type of co-reagent, the resulting films are unstable under long-term storage in air, but their stability is significantly improved by ALD of aluminum oxide with a thickness of several nanometers. The optimal temperature range was about 250-300 °C for deposition of lithium oxide, tin oxide, and lithium-tin oxide nanofilms. It was observed the remarkable influence of the type of co-reactant on the properties of the prepared nanofilms.

Keywords: Li-ion batteries, lithium oxide, anodes, lithiated tin oxide, thin film electrodes, atomic layer deposition

1. INTRODUCTION

Development of new materials for high-performance power sources is one of the most actual tasks for modern material science. Li-ion batteries (LIBs) are the most successful and perspective type of power sources which are actively used in hybrid electric vehicles and portable electronics [1,2]. Nowadays one of the most promising and interesting trends in the development of materials for lithium-ion batteries is thin film technologies [3,4]. Thin film power sources are the most progressive and sometimes the only possible ones for nanodevices such as smart cards, microchips with integrated power supplies, for various portable devices, medical implants and etc. [5,6]. To create such nanodevices, it is necessary to develop a reliable and successful technology for their production. In this case, the most significant is the method of producing thin-film electrodes.

Among the variety of methods it should be noted the atomic layer deposition (ALD) which is actively used in microelectronics [7,8] and also has broad prospects for producing of catalysts [9], biomaterials [10], photovoltaic devices [11] etc. ALD deserves specific attention due to its unique features, such as good conformity and absence of pinholes and defects, high precision of thickness and composition controllability and a capability to deposit high-quality coating both on planar, high aspect ratio and complex 3-dimensional substrates [12,13]. The ALD is based on a sequence of chemical reactions of gaseous reagents with the substrate surface separated in time by inert gas purges, which makes the self-limited growth of the films is possible [12,13]. Due to this features the ALD is ideal for preparation of thin film electrodes [14]. Today tin oxide thin film anodes are regarded as the most perspective for LIBs due to safety, cycleability and specific charge capacity (theoretically - 1491 mAh/g) [15-17]. But lithiated tin oxide probably is more perspective due to its better stability during many charge/discharge cycles [18,19].

In this work, we are studying the processes of ALD lithium oxide, tin oxide and lithium-tin oxide prepared using lithium hexamethyldisilylazide (LiHMDS), tetraethyl tin (TET) as the metal-containing reagents and water or ozone or oxygen plasma as co-reagents and also, we are trying to find the optimal parameters for ALD of lithium oxide and lithium-tin oxide thin films.

2. MATERIALS AND METHODS

Monocrystalline silicon (100) substrates (20x20 mm) and 316SS stainless steel plates, 16 mm in diameter were used as support for tin oxide, lithium oxide and lithium-tin oxide deposition. Before deposition silicon and stainless-steel supports were cleaned in an ultrasonic bath in acetone and deionized water for 10 min. After that, the substrates were dried in the inert gas atmosphere.

Atomic layer deposition was carried out with commercial "Picosun R-150" setup. Tetraethyltin ($Sn(C_2H_5)_4$ - TET) and remote oxygen plasma were used for tin oxide deposition. The more detailed the procedure is described in [20]. For lithium oxide ALD lithium hexadimethyldisilazide ($LiN[(CH_3)_3Si]_2$ - LiHMDS) and different co-reagent (ozone, water and remote oxygen plasma) were used. The power of plasma was 2500-3000 W. To find optimal regimes for deposition we varied the reactor temperature in the range of 250-300 °C. Stabilization of lithium oxide was carried out by additional of Al_2O_3 ALD coating (50 cycles) using trimethylaluminium (TMA) and water. All depositions were performed as follows. After pulsing of reagent (TET - 0.1s, LiHMDS - 0.1s, TMA - 0.1s) precursor had been injected for a certain pulse time, the excess of the precursor was purged (4-6s) with nitrogen (99.999 %) before the counter-reactant was introduced. Pulse times for co-reagents were H_2O - 0.1s, ozone - 4s, oxygen plasma - 10s. The temperatures of reagent containers were TET - 65 °C, LiHMDS - 150 °C, TMA-25 °C.

The thickness of the films was determined by spectral ellipsometry (SE) and Scanning Electron Microscopy (SEM). The ellipsometry setup "Ellips-1891 SAG" (Novosibirsk, the Russian Federation) ensured the accuracy in measuring the thickness equal to 0.3 nm. The measurements were carried out at 5-9 points of the support surface. Scanning electron microphotographs (SEM) showing plan-views and cross-sectional views were obtained using the SEM Zeiss Merlin. The microscope spatial resolution was around 1 nm. X-ray photoelectron spectra (XPS) were registered with a "Thermo Fisher Scientific Escalab 250Xi" spectrometer. The samples were excited by Al Kα (1486.7 eV) X-rays in the vacuum of 7×10^{-8} Pa. The binding energy scale was referenced using the C1s carbon line (284.8 eV). The depth profiles of elemental concentrations were examined by time-of-flight - secondary ion mass spectroscopy (TOF.SIMS 5 instrument, ION-TOF GmbH, Germany). Depth profiles of the elements were recorded in dynamic SIMS mode using the primary ion gun (Bi+ at 30 keV energy and 3.1 pA measured sample current, area 70x70 μm) for analysis and Cs (0.5 keV, area 120x120 μm) for sputtering. Both positive and negative ion regimes were used.

3. RESULTS AND DISCUSSION

The initial results showed that films of lithium oxide obtained using water were hydrolyzed very quickly when stored in air condition (discoloration and the presence of a significant gradient of thickness after 1 day of storage). The samples deposited using of plasma and ozone was less prone to hydrolysis. However, they also hydrolyzed during prolonged storage in air. In this regard, the samples of lithium oxide were future deposited using ozone, water and oxygen plasma as co-reactants and stabilized with alumina. To this end, immediately after the synthesis, films were coated by ALD with aluminum oxide using trimethylaluminum and water (**Table 1**).

Table 1 Characteristics of the samples

Short name	Reagents	Temperature (°C)	Number of cycles	Passivation (number of cycles)	Thickness, nm	Growth per cycle (nm)
Li-O_3-300	LiHMDS+ozone	300	500	Al_2O_3 (50)	26.5±0.8	0.048±0.001
Li-H_2O-250	LiHMDS+water	250	400	Al_2O_3 (50)	290±25	0.64±0.06
Li-H_2O-300	LiHMDS+water	300	400	Al_2O_3 (50)	500±100	1,11±0.22
Li-O_2-250	LiHMDS+O_2 plasma	250	400	Al_2O_3 (50)	165±25	0.37±0.06
Li-O_2-300	LiHMDS+O_2 plasma	300	400	Al_2O_3 (50)	70±22	0.16±0.05
Li-O_3-Sn-1-1	LiHMDS+O_3+TET+O_3	300	(1+1)*500	no	28±0.8	0.028±0.001
Li-H_2O-Sn-1-3	LiHMDS+H_2O+TET+O_2 plasma	250	(1+3)*200	no	67±5	0.084±0.006
Li-O_2-Sn-1-3	LiHMDS+O_2 plasma +TET+O_2 plasma	300	(1+3)*380	no	170±25	0.11±0.02

The samples obtained using H_2O showed a growth rate of more than 1 nm (**Table 1**), which is much larger than the thickness of the monolayer. In addition, they are quickly hydrolyzed in the air, forming crystals (presumably lithium hydroxide) despite the stabilization of the surface with Al_2O_3 (**Figure 1a, c**). The use of O_2 plasma and O_3 at 300 °C gives a growth per cycle of 0.16 and 0.048 nm, respectively. These samples are less prone to hydrolysis (**Figure 1b, d**). Uncontrolled growth of films deposited with LiHMDS and H_2O was observed earlier [21]. It appears that the use of water leads to uncontrolled growth due to the reservoir effect, where the bulk of the film absorbs water providing a large supply of hydroxide groups in addition to the self-limiting surface reaction. Any LiOH thereby formed in the bulk of the film may react with the subsequent pulses of LiHMDS leading to a growth rather limited by the thickness of the deposited film [21].

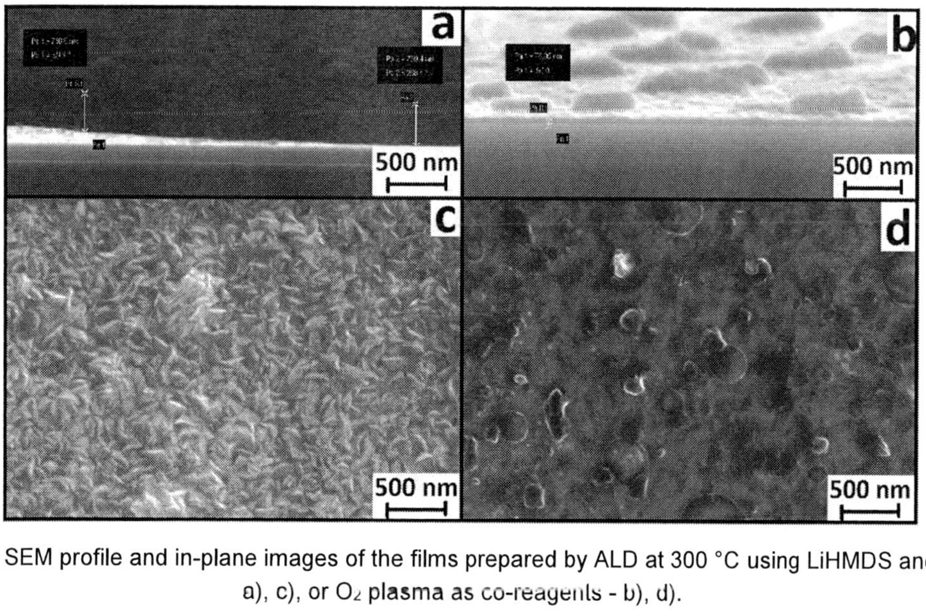

Figure 1 SEM profile and in-plane images of the films prepared by ALD at 300 °C using LiHMDS and water - a), c), or O_2 plasma as co-reagents - b), d).

The films of tin oxide were deposited using TET and oxygen plasma. A detailed description of the process was described earlier [20, 22]. Despite the fact that it is possible to obtain films at low temperatures (up to 200 °C) [23], the temperature range 250-300 °C is the optimal [20]. The use of this temperature range is applicable to the deposition of tin oxide using ozone and oxygen plasma [20]. A similar temperature range is also applicable to lithium oxide. Thus, the range of 250-300 °C is optimal for the deposition of lithium-tin oxides.

Then we obtained thin films of complex oxides using the sequential pulses of LiHMDS+O_3+TET+O_3. This sequence represents one supercycle ALD. The number of supercycles was 500. The XPS data showed that only Li, O and C are present on the surface of the samples. Moreover, Sn is either absent or its concentration is much less than 1 %.

In connection with this fact, for further work, we increased the number of TET pulses, and also oxygen plasma was used as a co-reagent, which is more active then ozone. The samples were obtained using supercycles: LiHMDS+H_2O+(TET+O_2 plasma)*3 - sample Li-H_2O-Sn-1-3 and LiHMDS+O_2 plasma+(TET+O_2 plasma)*3 - sample Li-O_2-Sn-1-3. According to XPS, these samples contained Sn on the surface, but its atomic concentration does not exceed 1 % (**Figure 2a**). In the bulk of the film, the Sn concentration exceeds 4 at. % (**Figure 2b**), however, the concentration of Si (> at. 12 %) is significantly increased. The presence of silicon in the films was observed earlier and is associated with the low reactivity of silicon containing a precursor (LiHMDS) under the ongoing ALD reactions [24,25].

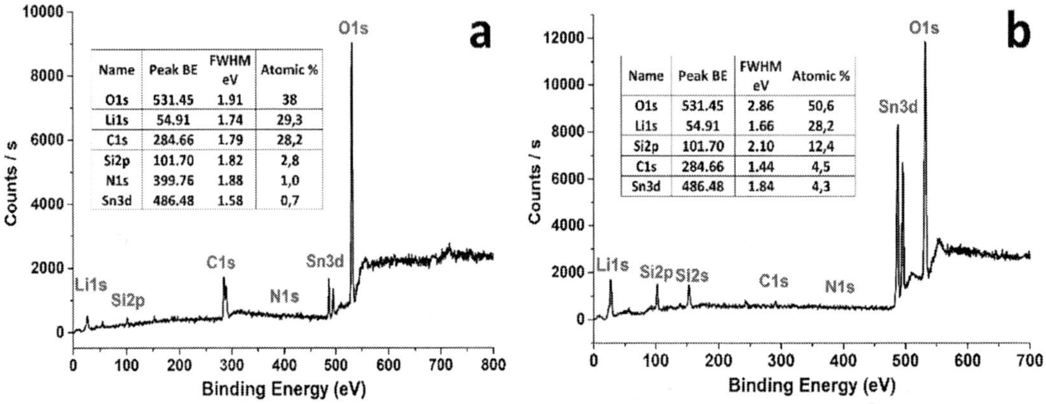

Name	Peak BE	FWHM eV	Atomic %
O1s	531.45	1.91	38
Li1s	54.91	1.74	29,3
C1s	284.66	1.79	28,2
Si2p	101.70	1.82	2,8
N1s	399.76	1.88	1,0
Sn3d	486.48	1.58	0,7

Name	Peak BE	FWHM eV	Atomic %
O1s	531.45	2.86	50,6
Li1s	54.91	1.66	28,2
Si2p	101.70	2.10	12,4
C1s	284.66	1.44	4,5
Sn3d	486.48	1.84	4,3

Figure 2 XPS survey spectra of the sample Li-H_2O-Sn-1-3 before - a) and after ion-etching - b)

TOF-SIMS study of lithium oxides coated with aluminum oxide revealed that aluminium is presented presumably on the surface. Secondary ions were detected both in the form of AlO^+, and $LiAl^+$, $LiAlO_2^+$ (**Figure 3a**). Thus, aluminum on the surface is present both in the form of an oxide (Al_2O_3) and as a complex oxide ($LiAlO_2$). However, the concentration of aluminum on the surface is a very small. The concentration of silicon in the surface layer is much lower than in the bulk of the film. Probably, the source of silicon in the bulk is the incompletely reacted precursor (LiHMDS) [24,25].

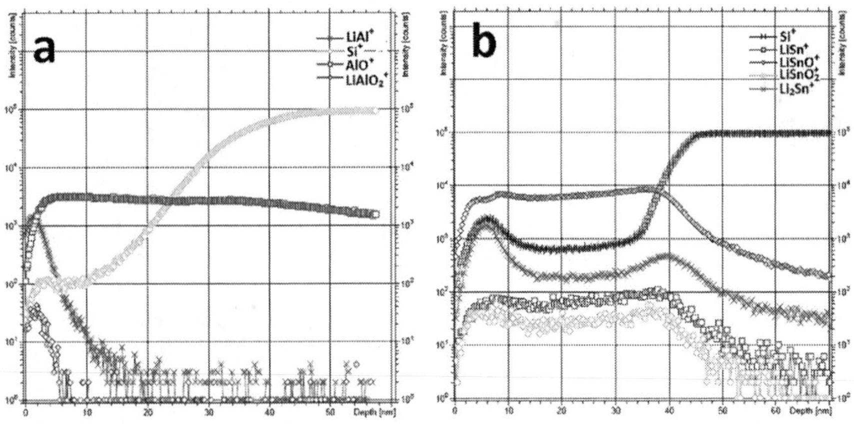

Figure 3 TOF-SIMS profiles of the samples Li-O_2-300 - a) and Li-H_2O-Sn-1-3 - b)

TOF-SIMS study of lithium-tin oxides revealed that tin concentration is very low, especially on the surface (**Figure 3b**). This result is in good agreement with the XPS data. In the sub-surface layer, the concentration of Li_2Sn^+ and $LiSnO^+$, as well as of Si^+, increases, while $LiSnO_2^+$, $LiSn^+$ practically does not change. The XPS and TOF-SIMS data show that the coating contains a lot of silicon but little nitrogen and carbon, which could also remain in the coating from the precursor. In addition, it can be assumed that lithium, silicon and tin form a mixture of complex oxides and the sub-surface layer rather thick (about 10 nm) and significantly different in composition from bulk.

4. CONCLUSION

Nanofilms of lithium oxide, tin oxide and lithiated tin oxide were successfully deposited by ALD using lithium LiHMDS or/and TET as the precursors. It was revealed that lithium oxide is nonstable at storage in air conditions but can be stabilized by ALD alumina coatings. The lithiated tin oxide can be prepared by ALD using LiHMDS, TET and O_2 plasma as co-reactant, but the resulting films have a relatively low concentration of tin (~4.3 at. %) and a remarkable concentration of silicon (>12 % at. %). Nevertheless, the results showed that the increase in the number of TET pulses in each supercycle of ALD can significantly increase the concentration of tin and probably lead to the producing of lithiated tin oxide which is promising material for high-performance Li-ion anodes.

ACKNOWLEDGEMENTS

The research was conducted under the financial support from the grant of the President of the Russian Federation, Grant No. MK-2961.2018.3.

REFERENCES

[1] GOODENOUGH, J.B. and PARK, K.S. The Li-ion rechargeable battery: a perspective. *J. Am. Chem. Soc.*, 2013. vol. 135, no. 4, pp. 1167-1176. DOI: 10.1021/ja3091438

[2] BLOMGREN, G.E. The Development and Future of Lithium Ion Batteries. *Journal of the Electrochemical Society*, 2017. vol. 164, no.1, pp. A5019-A5025. DOI: 10.1149/2.0251701jes

[3] CHEN, J. Recent Progress in Advanced Materials for Lithium Ion Batteries. *Materials,* 2013. vol. 6, no. 1, pp. 156-183. DOI: 10.3390/ma6010156

[4] GUAN, C. and WANG, J. Recent Development of Advanced Electrode Materials by Atomic Layer Deposition for Electrochemical Energy Storage. *Adv. Sci.*, 2016. vol. 3, no. 10, pp. 1500405:1-23. DOI: 10.1002/advs.201500405

[5] DUDNEY, N.J. Thin film micro-batteries. *Electrochem. Soc. Interface,* 2008. vol. 17 no. 3, pp. 44-48.

[6] WANG, Y., LIU, B., LI, Q., CARTMELL, S., FERRARA, S., DENG, Z.D. and XIAO, J. Lithium and lithium-ion batteries for applications in microelectronic devices: review. *J. Power Sources,* 2015. vol. 286, pp. 330-345. DOI: 10.1016/j.jpowsour.2015.03.164

[7] HWANG, C.S. and YOO, C.Y. *Atomic Layer Deposition for Semiconductors.* New York: Springer, 2013. p. 263.

[8] KIM, H., LEE, H.-B.-R. and MAENG W.-J. Applications of atomic layer deposition to nanofabrication and emerging nanodevices. *Thin Solid Films,* 2009. vol. 517, no. 8, pp. 2563-2580 DOI: 10.1016/j.tsf.2008.09.007

[9] LU, J., ELAM, J.W. and STAIR, P.C. Atomic layer deposition - Sequential self-limiting surface reactions for advanced catalyst "bottom-up" synthesis. *Surface Science Reports,* 2016. vol. 71, no. 2, pp. 410-472. DOI: 10.1016/j.surfrep.2016.03.003

[10] NAZAROV, D.V., SMIRNOV, V.M., ZEMTSOVA, E.G., YUDINNTCEVA, N.M., SHEVTSOV, M.A. and VALIEV, R.Z. Enhanced Osseointegrative Properties of Ultra-Fine-Grained Titanium Implants Modified by Chemical Etching and Atomic Layer Deposition. *ACS Biomaterials Science and Engineering,* 2018. vol. 4, no. 9, pp. 3268-3281. DOI: 10.1021/acsbiomaterials.8b00342

[11] PALMSTROM, A.F., SANTRA, P.K. and BENT, S.F. Atomic layer deposition in nanostructured photovoltaics: tuning optical, electronic and surface properties. *Nanoscale,* 2015. vol. 7, no.29, pp. 12266-12283. DOI: 10.1039/c5nr02080h

[12] GEORGE, S.M. Atomic layer deposition: an overview. *Chem. Rev.*, 2010. vol. 110, no. 1, pp. 111-131. DOI: 10.1021/cr900056b

[13] MIIKKULAINEN, V., LESKELÄ, M., RITALA, M. and PUURUNEN, R.L. Crystallinity of inorganic films grown by atomic layer deposition: Overview and general trends. *J. Appl. Phys.*, 2013. vol. 113, no. 2, pp. 021301:1-021301:101. DOI: 10.1063/1.4757907

[14] NILSEN, O., MIIKKULAINEN, V., GANDRUD, K.B., ØSTRENG, E., RUUD, A. and FJELLVAG, H. Atomic layer deposition of functional films for Li-ion microbatteries. *Phys. Status Solidi, A,* 2014. vol. 211, no. 2, pp. 357-367. DOI: 10.1002/pssa.201330130

[15] AHMED, B., ANJUM, D.H., GOGOTSI, Y. and ALSHAREEF, H.N. Atomic layer deposition of SnO_2 on MXene for Li-ion battery anodes. *Nano Energy,* 2017. vol. 34, pp. 249-256. DOI: 10.1016/j.nanoen.2017.02.043

[16] CHEN, J.S. and LOU, X.W. SnO_2-based nanomaterials: synthesis and application in lithium-ion batteries. *Small,* 2013. vol. 9, no. 11, pp. 1877-1893. DOI: 10.1002/smll.201202601

[17] POPOVICH, A.A., MAXIMOV, M.Yu., NOVIKOV, P.A., SILIN, A.O., NAZAROV, D.V. and RUMYANTSEV, A.M. Cyclic stability of the anode material based on tin(IV) oxide for thin-film current sources. *Russ. J. Appl. Chem.*, 2016. vol. 89, no. 4, pp. 679-681. DOI: 10.1134/S1070427216040236

[18] WINTER, M. and BESENHARD, J. O. Electrochemical lithiation of tin and tin-based intermetallics and composites. *Electrochim. Acta,* 1999. vol. 45, no. 1, pp. 31−50. DOI: 10.1016/S0013-4686(99)00191-7

[19] ZHANG, H., CHEN, Z., HU, R., LIU, J., CUI, J., ZHOU, W. and YANG, C. Enabling a highly reversible conversion reaction in a lithiated nano-SnO_2 film coated with Al_2O_3 by atomic layer deposition. *J. Mater. Chem. A,* 2018. vol. 6, pp. 4374-4385. DOI:10.1039/C8TA00290H

[20] NAZAROV, D.V., MAXIMOV, M.Y., NOVIKOV, P.A., POPOVICH, A.A., SILIN, A.O., SMIRNOV, V.M., BOBRYSHEVA, N.P., OSMOLOVSKAYA, O.M., OSMOLOVSKY, M.G. and RUMYANTSEV, A.M. Atomic layer deposition of tin oxide using tetraethyltin to produce high-capacity Li-ion batteries. *J. Vac. Sci. Technol. A,* 2017. vol. 35, no.1, art. No. 01B137. DOI: 10.1116/1.4972554

[21] ØSTRENG, E., VAJEESTON, P., NILSEN, O. and FJELLVAG, H. Atomic layer deposition of lithium nitride and carbonate using lithium silylamide. *RSC Advances,* 2012. vol. 2, pp. 6315-6322. DOI: 10.1039/c2ra20731a

[22] MAXIMOV, M.Y., NOVIKOV, P.A., NAZAROV, D.V., RYMYANTSEV, A.M., SILIN, A.O., ZHANG, Y. and POPOVICH, A.A. Characterization and Electrochemical Performance at High Discharge Rates of Tin Dioxide Thin Films Synthesized by Atomic Layer Deposition. *Journal of Electronic Materials,* 2017. vol. 46, no. 11, pp. 6571-6577. DOI: 10.1007/s11664-017-5701-8.

[23] POPOVICH, A.A., MAXIMOV, M.Yu., NAZAROV, D.V., NOVIKOV, P.A., SILIN, A.O. and SHAMSHURIN, A.I. Low-temperature deposition of tin(IV) oxide films for thin-film power sources. *Russ. J. Appl. Chem.,* 2016. vol. 89, no. 5, pp. 805-808. DOI: 10.1134/S1070427216050190

[24] HÄMÄLÄINEN, J., MUNNIK, F., HATANPÄÄ, T., HOLOPAINEN, J., RITALA, M. and LESKELÄ, M. Study of amorphous lithium silicate thin films grown by atomic layer deposition. *J. Vac. Sci. Technol. A,* 2012. vol. 30, no. 1, pp. 01A106:1-5. DOI: 10.1116/1.3643349

[25] TOMCZAK, Y., KNAPAS, K., SUNDBERG, M., LESKELA, M. and RITALA, M. In Situ Reaction Mechanism Studies on Lithium Hexadimethyldisilazide and Ozone Atomic Layer Deposition Process for Lithium Silicate. *J. Phys. Chem. C,* 2013. vol. 117, no. 27, pp. 14241−14246. DOI: 10.1021/jp312309g

OPEN TOP AND SELF-ORGANIZED LONG TITANIUM DIOXIDE NANO TUBES PREPARED BY ANODIZATION OF SPUTTERED TITANIUM LAYER

[1]S.M. Hossein HEJAZI, [1]Tapan BARMAN, [1]Alberto NALDONI, [1]Zdeněk HUBIČKA, [1]Štěpán KMENT, [1]Patrik SCHMUKI [1], Radek ZBOŘIL

[1]*Regional Centre of Advanced Technologies and Materials, Faculty of Science, Palacky University, Olomouc, Czech Republic, EU,* radek.zboril@upol.cz

Abstract

Anodization is a simple and effective method to synthesize self-organized titanium dioxide nanotubes. But depending on the electrolyte used for anodization, there is a disorder "nano grass" or an "initiation layer" on top of the nano tubes. Additionally peeling off the nanotubes from substrate might happen during the anodization of titanium which is sputtered on the fluorine doped tin oxide (FTO). Removing this unwanted layer (without delamination), especially in high length TiO_2 nanotubes could increase the photo electrochemical performance. Here we report a novel method to anodize the titanium layer homogenously and to produce open top nano tubes. Results show that the photo current density of samples will be increase up to 15 % after opening the top surface of nanotubes.

Keywords: Open top TiO_2 nano tubes, anodization, nano grass, initiation layer, sputtering

1. INTRODUCTION

Titanium dioxide (TiO_2) is a well-known semiconductor for its photoelectrochemical properties and it is used for instance in perovskite solar cells (PSCs), dye sensitized solar cells (DSSCs) [1,2], water splitting [3] and photocatalysis [4]. One dimensional nanostructures such as nanotubes, provide a straight pathway for photogenerated electrons and could increase the efficiency of such devices [5]. Anodization is a simple method for producing self-organized TiO_2 nanotubes. Fluoride contained electrolyte like HF or NH_4F solution is necessary for growth of TiO_2 nanotubes. However, in both cases long TiO_2 nanotube films grown by these solutions are always covered by a grassy top layer [6] or initiation layer [7]. This disorder top layer can significantly reduce the efficiency of charge transfer [7-9]. Therefore, long TiO_2 nanotubes with open and clean top surface are highly desirable for electron-transport applications such as solar water splitting, PSCs and/or DSSCs. In addition, in these applications it is desirable to illuminate TiO_2 nanotubes from both back and front sides [5]. Therefore, growing TiO_2 directly on a transparent conductive charge collecting layer such as fluorine doped tin oxide (FTO) by the electrochemical anodization of magnetron sputtered titanium on FTO is highly favorable. Preparation of TNTs from the sputtered titanium films without peeling off requires homogenous anodization.

In order to obtain the open top TiO_2 nanotubes several approaches have been proposed, including producing different polymorph of TiO_2 at the top and removing it after anodization [8], resistant layer technique [7] and multi-step anodization [10]. All these methods produce open-top TiO_2 nanotubes. But all of these methods produce open top TiO_2 nanotube layer which were grown on pure Ti or Ti alloy films.

Here we proposed two step anodization with optimized conditions to remove the as-anodized disorder top layer easily and effectively in long length TiO_2 nanotubes which were grown directly on FTO glass. These technique consist of growing a weak and non-sticky layer at top surface of nanotubes as sacrificial layer and then removing this layer by sonication. Furthermore, we proposed a simple technique to produce homogenous and highly adherent TiO_2 nanotubes directly grown on FTO substrate. By using this technique, it is possible to synthesize TiO_2 nanotubes with length up to 20 μm.

2. EXPERIMENTAL

2.1. Deposition of Ti-Ta thin film on ultrasonically cleaned FTO

The stainless steel plasma reactor used in this work is equipped with a 4 inch magnetron sputtering unit and a stepper motor controlled rotatory substrate holder with a heater. For the deposition of the titanium tantalum composite, a titanium 4 inch target of 99.99 % purity with a tantalum wire of diameter 0.05 mm at the center of the Ti target is fitted in the magnetron source. Ultrasonically cleaned (subsequently in acetone, ethanol and deionized water for 5 minutes each) and nitrogen dried FTO glass is used as substrates for titanium tantalum alloy film deposition. The plasma reactor is pumped down to the base pressure of 5×10^{-5} Pa by means of turbo molecular pump in combination with a rotary pump. A RF plasma cleaning of the FTO glass substrates is carried out at pressure 15 Pa in argon plasma (100 sccm) at -120 V self-bias by applying 13.56 MHz RF 50 Watt. Afterwards, an adhesion treatment is performed by a separate plasma treatment in presence of forming gas (mixture hydrogen and nitrogen) and argon (flow 10 sccm each) at pressure 15 Pa and self-bias -120 V by applying RF 13.56 MHz power 45 Watt. Then, working pressure 0.2 Pa is achieved by introducing 15 sccm argon gas and controlling the gate valve. At the working pressure sputtering is carried out by pulsed DC power supply at frequency 50 kHz with duty cycle 50 %, power ~ 700 watt, discharge current ~ 630 mA and discharge voltage ~ 1120 V for 145 minutes without intentional substrate heating.

2.2. TiO_2 synthesize

TiO_2 nanotubes were synthesized at room temperature via electrochemical anodization. **Figure 1** shows the setup that used for anodization. The thickness of sputtered layer varied from 2.2 μm up to 7 μm. Two different solutions were used as electrolyte. The first one was HF (0.2 M, Sigma Aldrich), ethylene glycol (Sigma Aldrich) and distilled water (4 wt. %) and the second one was NH_4F (0.15 M, Sigma Aldrich), ethylene glycol (Sigma Aldrich) and distilled water (3 wt. %). The surface of the sample was connected to the positive terminal of power supply as anode and 1 cm^2 platinum sheet with 1 mm thickness was connected to the negative terminal as the cathode. In case of HF electrolyte, the voltage was set to 120V (D.C.) and in case of NH_4F electrolyte the voltage was set to 20V (D.C.) for the first step anodization (10 min) and 70V (D.C.) for second step anodization. The vertical distance between anode and cathode was kept constant in about 1.5 cm. In all cases the anodization continued until the consumption of all Ti-Ta layer. After anodization the samples was rinsed with ethanol and dried in nitrogen stream immediately. The samples was finally annealed at 450°C for 3h in air atmosphere with heating and cooling rate of 2°C/min.

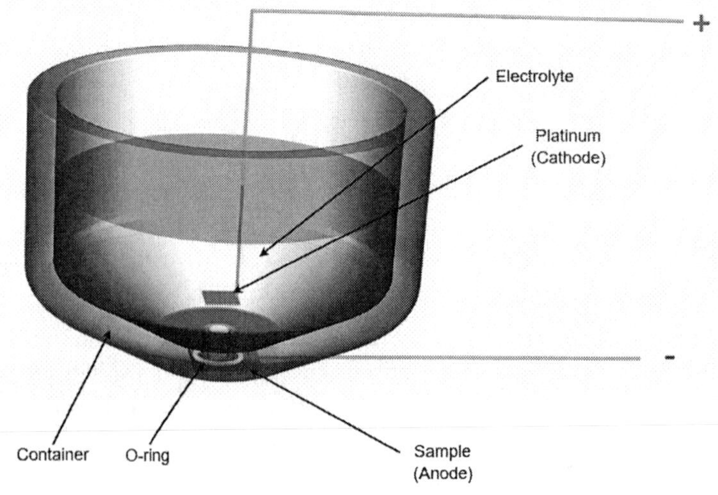

Figure 1 Anodization setup

2.3. Sample characterization and measurement

The morphology of the top surface and thickness of the TiO_2 layer after anodization were measured using scanning electron microscope Hitachi SU6600 (Hitachi, Tokyo, Japan). The thickness of the films before anodization were measured with the help of profilometry. X-ray diffraction patterns were recorded at room temperature with an Empyrean (PANalytical, The Netherlands) diffractometer in the Bragg-Brentano geometry, Co-K_α radiation (40 kV, 30 mA, λ = 0.1789 nm). The photoelectrochemical data were collected using a standard three-electrode electrochemical cell with a Gamry Series G 300 Potentiostat (Warminster, PA, USA). The TiO_2 nanotubes served as working electrode (photoanode), the Ag/AgCl (3M KCl) as the reference electrode and the Pt wire was used as the counter electrode. A 150 W Xenon lamp coupled with an AM1.5G filter was used as a light source. The power intensity was kept at 1 sun (100 mW/cm^2) which was calibrated through a silicon reference solar cell (Newport Corporation, Irvine, CA, USA). The photoelectrochemical behavior of the prepared electrodes was investigated by means of sweep voltammetry measurement in 1M NaOH electrolyte (pH= 13.5).

3. RESULTS AND DISCUSSION

Figure 2 shows the surface of samples anodized with HF and NH_4F solutions. In samples anodized with HF solution, some nanograsses are obvious in the top surface of nanotubes. These nanograsses are formed due to etching of the walls of TiO_2 nano tubes during long duration anodization [11]. The wall thickness of top part of nanotubes become too thin to withstand against their own weight and during drying they will bend and collapse. In the case of NH_4F solution, at the initial step of anodization, a typical initial current spike is present that results in a compact oxide layer on the Ti surface according to a high field oxide formation mechanism [7].

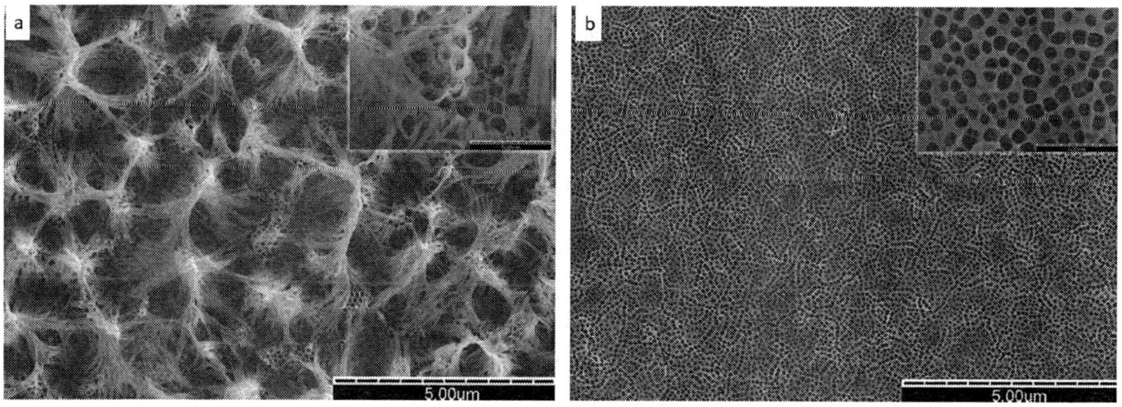

Figure 2 SEM images of top view of samples anodized with a. HF solution (nanograss) and b. NH_4F solution (initiation layer)

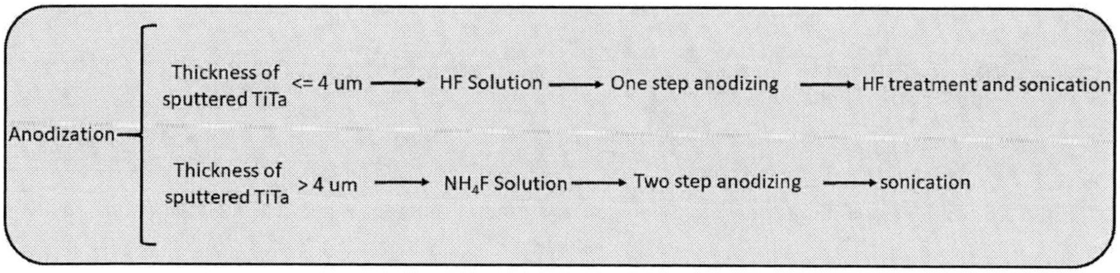

Figure 3 Anodization protocol for samples with various initial thickness

To clean these disorder top layers, we proposed a protocol for anodization of sputtered Ti layer on FTO glass with thicknesses from 2.2 µm up to 7 µm (**Figure 3**). Since TiO_2 easily dissolves in HF, for samples with initial Ti thickness more than 4 µm, the final thickness of TiO_2 does not increase more than 10 µm. Therefore we find that in order to synthesize TiO_2 nanotube with length more than 10 µm, it is better to use NH_4F solution as electrolyte for anodization.

For samples with thickness less than 4 µm, HF solution produce grassy top layer that can be removed by sonication in dilute HF solution in water for maximum 4 minutes (**Figure 3**). Removing the initiation layer is not as easy as in the case of nanograsses. Generally, double step anodization is an effective way to eliminate this layer. During the first step a sacrificial layer grows at the top and by second step, the main layer grows at the bottom. The initiation layer vanishes by removing the sacrificial layer with sonication in water for 4 min (**Figure 4**).

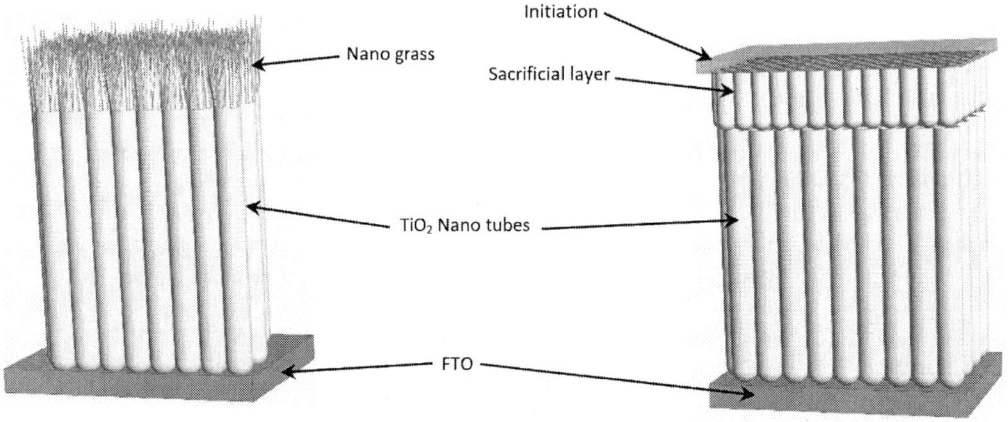

Figure 4 Schematic of nano grass, sacrificial layer and initiation layer

Figure 5 shows the top view of samples after surface treatment with open top nanotube. These open top nanotubes could increase the dye loading in case of dye sensitized solar cell applications. The maximum length of nanotubes that can produce with this technique is around 20 µm.

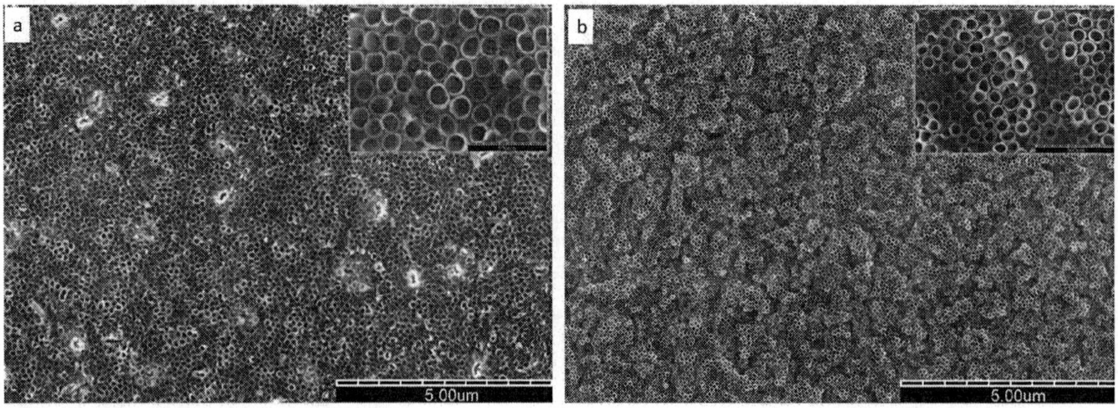

Figure 5 SEM images of top view of samples after surface treatment with a. HF solution and b. NH_4F solution

Photocurrent measurements for sample with about 10 µm thickness before and after surface treatment are shown in **Figure 6**. Cleaning the top surface results in about 15 % increase in photocurrent at 0.6 V (from 40 μA/cm² to 46 μA/cm²). The nano-grasses inhibit smooth electron transport in one direction and decrease the photocurrent by increasing the electron recombination [12].

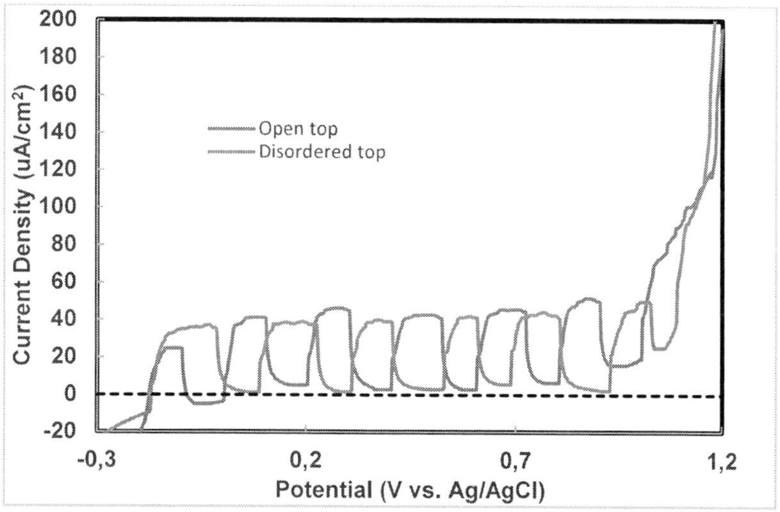

Figure 6 Chopped light polarization curve for TiO_2 nano tubes on FTO glass with open and disorder top, electrolyte 1M NaOH

In order to anodize the Ti-Ta layer homogenously and without delamination, it is necessary to produce completely homogenous conditions for anodization in every parts of the sample. To keep the conditions constant it is necessary to change periodically the position of the cathode (sample) against the anode (platinum) during the anodization without changing the vertical distance between them. By changing the position, every unwanted effects due to inhomogeneity in surface and distance will be distribute in whole part of sample and production of a homogenous anodized surface will be guaranteed. **Figure 7** shows the effect of changing the position of anode during anodization on the homogeneity of anodized sample.

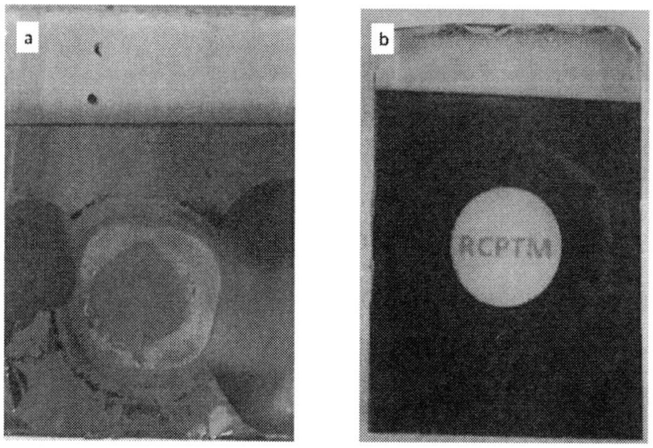

Figure 7 a. Heterogeneous and b. homogenous anodized sample

4. CONCLUSION

Producing open top and long length TiO_2 nanotubes directly on FTO glass is a key factor to fabricate efficient devices like dye sensitized solar cells. In present work we used HF and NH_4F solution to anodize Ti-Ta thin film directly sputtered on FTO. To open the top of synthesized nanotubes with HF solution, we used a simple surface treatment including HF treatment and sonication in ultrasonic bath. Using NH_4F solution results in initiation layer that could be removed by producing a sacrificial layer at the top of the samples and sonication. These surface treatments result in better photocurrent and will be promising for TiO_2 nanotubes applications.

ACKNOWLEDGEMENTS

The authors gratefully acknowledge the support by the project LO1305 of the Ministry of Education, Youth, and Sports of the Czech Republic.

REFERENCES

[1] LEE, Kiyoung, and SCHMUKI, Patrik. Ta doping for an enhanced efficiency of TiO2nanotube based dye-sensitized solar cells. *Electrochemistry Communications*. 2012. vol. 25, no. 1, pp. 11-14.

[2] PANG, Qi et al. Dye sensitized solar cells using freestanding TiO2nanotube arrays on FTO substrate as photoanode. *Materials Chemistry and Physics*. Elsevier B.V., 2011. vol. 125, no. 3, pp. 612-616.

[3] ROY, Poulomi et al. Oxide nanotubes on Ti-Ru alloys: Strongly enhanced and stable photoelectrochemical activity for water splitting. *Journal of the American Chemical Society*. 2011. vol. 133, no. 15, pp. 5629-5631.

[4] ALSALKA, Yamen et al. Co-catalyst-free photocatalytic hydrogen evolution on TiO2: Synthesis of optimized photocatalyst through statistical material science. *Applied Catalysis B: Environmental*. Elsevier, 2018. vol. 238, no. July, pp. 422-433.

[5] KATHIRVEL, Sasipriya et al. The growth of TiO<inf>2</inf> nanotubes from sputter-deposited Ti film on transparent conducting glass for photovoltaic applications. *Vacuum*. Elsevier Ltd, 2015. vol. 118pp. 17-25.

[6] KIM, Doohun, GHICOV, Andrei, and SCHMUKI, Patrik. TiO2 Nanotube arrays: Elimination of disordered top layers (“nanograss”) for improved photoconversion efficiency in dye-sensitized solar cells. *Electrochemistry Communications*. Elsevier B.V., 2008. vol. 10, no. 12, pp. 1835-1838.

[7] ROY, Poulomi, ALBU, Sergiu P., and SCHMUKI, Patrik. TiO2 nanotubes in dye-sensitized solar cells: Higher efficiencies by well-defined tube tops. *Electrochemistry Communications*. 2010. vol. 12, no. 7, pp. 949-951.

[8] SONG, Yan-Yan et al. TiO[sub 2] Nanotubes: Efficient Suppression of Top Etching during Anodic Growth. *Electrochemical and Solid-State Letters*. 2009. vol. 12, no. 7, pp. C17.

[9] WANG, Daoai et al. TiO Nanotubes with Tunable Morphology , Diameter , and Length : Synthesis and Photo-Electrical / Catalytic Performance TiO 2 Nanotubes with Tunable Morphology , Diameter , and Length : Synthesis and Photo-Electrical / Catalytic Performance. *Chemistry of Materials*. 2009. no. 6, pp. 1198-1206.

[10] CHANMANEE, Wilaiwan et al. Formation and characterization of self-organized TiO2 nanotube arrays by pulse anodization. *Journal of the American Chemical Society*. 2008. vol. 130, no. 3, pp. 965-974.

[11] ROY, Poulomi, BERGER, Steffen, and SCHMUKI, Patrik. TiO2nanotubes: Synthesis and applications. *Angewandte Chemie - International Edition*. 2011. vol. 50, no. 13, pp. 2904-2939.

[12] ROY, Poulomi et al. TiO2 nanotubes and their application in dye-sensitized solar cells. *Nanoscale*. 2010. vol. 2, no. 1, pp. 45-59.

ELECTRICAL CONDUCTIVE AND GAS SENSITIVE TRANSPARENT TIN OXIDE THIN FILMS PREPARED BY ATMOSPHERIC PRESSURE PECVD

Sven GERULLIS, Oliver BEIER, Andreas PFUCH, Marc RAUGUST, Sten Mark KRETZSCHMAR, Bernd GRUENLER

INNOVENT e.V. Technology Development Jena, Jena, Germany, EU,
ob@innovent-jena.de

Abstract

In this contribution, the deposition of tin oxide thin films and the resulting layer properties are investigated. An atmospheric pressure PECVD process using a non-thermal plasma jet was applied, creating the coatings on silicon and glass slides. For a highly reactive oxygen atmosphere, the plasma process was driven with air as working gas. The precursor substance tetra-n-butyltin was provided as an aerosol towards the active plasma zone and is further converted to form the ~100 nm thin layers. Structural, electrical and gas sensing properties were determined of the undoped SnO_x layers, deposited at substrate temperatures between 20°C and 500°C.

With increasing substrate temperature a decrease in film thickness and roughness of the films could be observed. In general, the films consists of a dense part near the substrate surface and a more porous structure on top. At 300 °C the lowest electrical resistivity with about 7.2×10^{-2} Ω·cm occurs in 4-point probe measurements. Experiments for the gas sensing characteristics revealed an electrical response to different atmospheres (water vapour, aqueous ammonia solution) already at elevated measuring temperatures of 150°C. Ammonia concentrations until 25 % were tested. The thin films are showing distinct properties in gas selectivity between the changing atmospheres, gas sensitivity during ammonia interaction as well as recurring reversibility after a surface recovery step with pressurized air. So, tin oxide coatings created by atmospheric pressure PECVD are showing already in its undoped state and without further layer post-treatments high application potential.

Keywords: Atmospheric pressure PECVD, plasma jet, tin oxide, transparent conductive oxide (TCO), gas sensor

1. INTRODUCTION

Semiconductors, including SnO_2, TiO_2, ZnO, WO_3 and ZnS have been intensively investigated for diverse applications [1]. Tin oxide in form of polycrystalline SnO_2 and amorphous SnO_x and their various material properties are of high interest. Transparent highly conducting oxides are used as electrodes for electronic devices such as flat displays [2], solar cells [3] or gas sensors [4]. Moreover, such films can be deposited and applied in form of undoped, In-, F-, or Sb-doped SnO_2 on a wide variety of substrates. Often used methods to create thin tin oxide films include physical vapor deposition (PVD) such as sputtering [5], chemical vapor deposition (CVD) [6], spray pyrolysis [7] and sol-gel technology [8]. Most PVD methods require high vacuum and will be performed in batch-processes but can be performed advantageously at low temperatures. The energy to pyrolyse tin precursors in CVD or spray pyrolysis methods is mostly provided by high substrate temperatures (> 500 °C). Hence, these methods are often inappropriate to deposit films on more temperature-sensitive materials. Beneficial is the possibility to treat larger substrate areas under atmospheric pressure conditions (in-line ability). In this study an atmospheric pressure plasma jet (APPJ) working with pressurized air as process gas was used. Film properties were characterized in dependence of the applied substrate temperature (20 °C - 500 °C) during film deposition. An additional heat post-treatment was not implemented.

2. MATERIALS AND METHODS

2.1. Film deposition

Films were deposited on monocrystalline Si wafer pieces (Si-Mat Silicon Materials) and soda lime glass slides (VWR International), latter coated with a 20 nm thick silicon dioxide film (deposited by Pyrosil® procedure). These silicon dioxide films were created to prevent Na^+-diffusion during substrate heating into the SnO_x films. The tin-organic precursor tetra-butyl-tin(IV) (TBT) was used as starting substance for film deposition and was purchased from Sigma Aldrich in 98 % purity. This precursor was further diluted with isopropanol to a concentration of 50 % (v/v). Experimental setup, the atmospheric pressure plasma jet and detailed information on the film deposition process are described in previous publications [9]. Shortly: An APPJ based on the Plasma MEF (Tigres GmbH) technology was used. This jet system was driven with pressurized air as working gas and creates an electrically potential-free cold plasma (DC pulsed discharge). The nozzle of this plasma source was modified to spray liquids or liquid dispersed particulate materials into the active plasma discharge. TBT and the TBT/isopropanol mixture is liquid at room temperature, the corresponding flow rates were kept constant at 50 µl/min. Substrates were placed on a hot plate (temperature: 20 °C to 500 °C) fixed on a x-y-table. This x-y-positioning system performed a meandering movement of the samples relatively to the plasma nozzle exit with a velocity of 100 mm/s. 30 successive runs were performed to create thin SnO_x films.

2.2. Characterization

Films were extensively examined and results are presented in literature [9]. Supplementary results will be presented in the following sections with focus on gas sensing properties. The morphology of the deposited tin films were characterized by scanning electron microscopy (SEM) with a Supra 55 VP (Carl Zeiss NTS GmbH) using the in-lens detector and an operating voltage of 5 kV. Applying 4-point probe measurements using the Sheet Resistivity Meter SD-600 (Nagy Instruments) the sheet resistance was measured. To demonstrate the response of the deposited SnO_x to oxidizing and reducing gases a custom-made test system was used (see **Figure 1**).

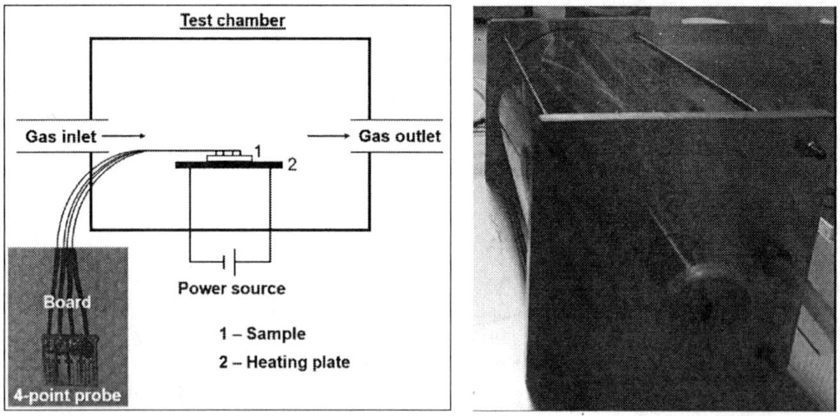

Figure 1 Setup investigating the gas sensing properties: scheme (left) and the build test chamber (right)

This setup is composed of a sealable glass cylinder (length = 250 mm, inner diameter = 130 mm) with a gas inlet and a gas outlet. Furthermore, a hot plate was used and placed inside the chamber. The samples (coated glass slides) were mounted on top of this hot plate and during the experiments the sample temperatures were controlled between 20 °C and 150 °C. In order to measure sheet resistance during the gas interaction four copper wires were mechanically fixed to the sample. Outside of the chamber the opposite parts of the wires were soldered to a board containing four copper stripes. 4-point probe measurements were conducted on top of these copper stripes every 30 s during the test procedure. During the experiment ammonia was used as test gas. Therefore 10 ml aqueous ammonia solutions with different ammonia concentrations (1.56 %, 3.12 %, 6.25 %, 12.5 %, 25.0 %) were filled in a bubbling system and heated up to 50 °C by use of a water bath. Pressurized air (flow rate 1.5 l/min) was used as carrier gas and directed through this bubbling system in order to transport gaseous ammonia into the test chamber. After each test run the sample surfaces were recovered using pressurized air (flow rate = 30 l/min) for 20 min.

3. RESULTS

3.1. Morphology

Films deposited at optimized working conditions appear homogenously on Si and glass substrates and clear interference pattern can be observed. To demonstrate this, **Figure 2** presents a coated stainless steel substrate. Furthermore, the related cross-sectional SEM investigation (SnO_x on Si) is also shown in this figure. SEM investigations reveal that created films compose of a thin and dense layer close to the substrate and a more porous layer with an aligned columnar structure on top of it. A change from dense film growth to columnar film growth mechanism led to the formation of this structure. Such an effect is well-known in CVD processes: The formation of a thermal barrier film on a thermally conductive substrate could be a reason for this, whereby such thermal barrier lead to a modification in the heat flux toward the substrate [10].

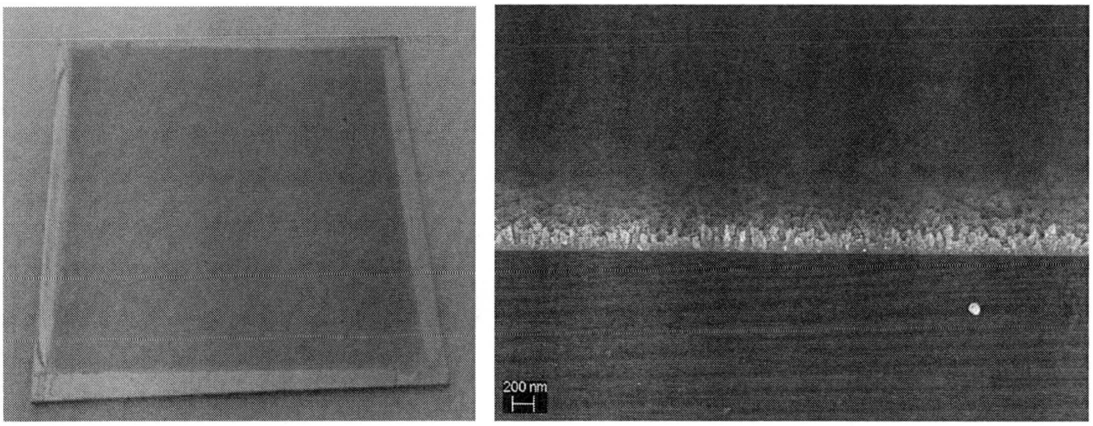

Figure 2 Stainless steel (size: 5 cm × 5 cm) coated with SnO_x (left) and cross-sectional SEM image of SnO_x coated silicon wafer (right); both at substrate temperature 300°C

3.2. Electrical properties

Figure 3 presents the calculated resistivities of SnO_x thin films deposited on glass. By 4-point probe measurement no electrical conductivity could be determined (conductivity to low) for films deposited at substrate temperatures lower than 250 °C. Films produced at 300 °C substrate temperature show the lowest sheet resistivity (~ 0.07 Ω·cm) and therewith the highest electrical conductivity. With increasing the substrate temperature to 400 °C the sheet resistivity increases to ~ 0.25 Ω·cm and a further increase to 500 °C leads to ~ 0.5 Ω·cm sheet resistivity. A commonly used method to create SnO_x is spray pyrolysis. In literature reported values for the resistivity of undoped films ranges from ~ 0.008 Ω·cm to 0.7 Ω·cm [11, 12]. To obtain these low resistivities elevated substrate temperatures between 450 °C and 500 °C were necessary. Lowest resistivities obtained in our work were at substrate temperatures of 300 °C. Therewith a significant reduction of substrate temperatures down to 300 °C was possible and resistivities of SnO_x films comparable to films deposited by spray pyrolysis using > 450 °C could be achieved. One reason for this improvement could be the creation of a reactive and oxygen-rich environment in the plasma and hence, a promotion of oxidation processes and the reduction of organic residuals within the films.

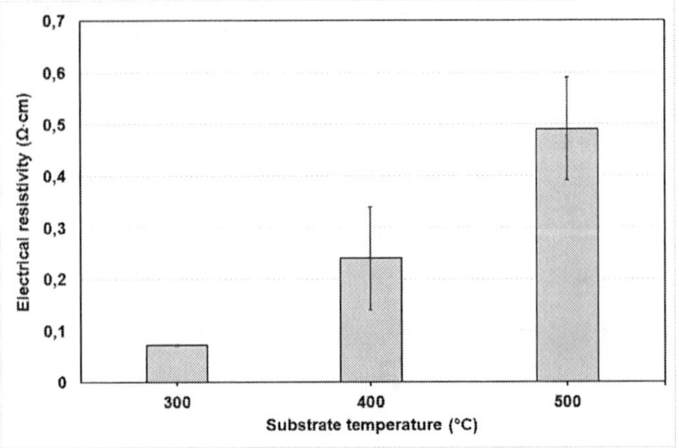

Figure 3 Electrical resistivity of plasma deposited SnO_x thin films on glass as a function of substrate temperature

3.3. Gas sensing

The presented results in this chapter were obtained for films deposited using 400 °C substrate temperature. During the gas sensing measurements the films temperature was kept at 150 °C. Tests were also conducted at lower measuring temperatures, however, first changes of the sheet resistance during surface/ammonia-interaction was monitored at 150 °C. These findings are in good accordance to literature data: surface temperatures of 200 °C were often necessary for SnO_2 films [13]. **Figure 4** shows the relative sheet resistance ($R_{ammonia}/R_{air}$ × 100 (%)) in dependence of time at different ammonia solution concentrations.

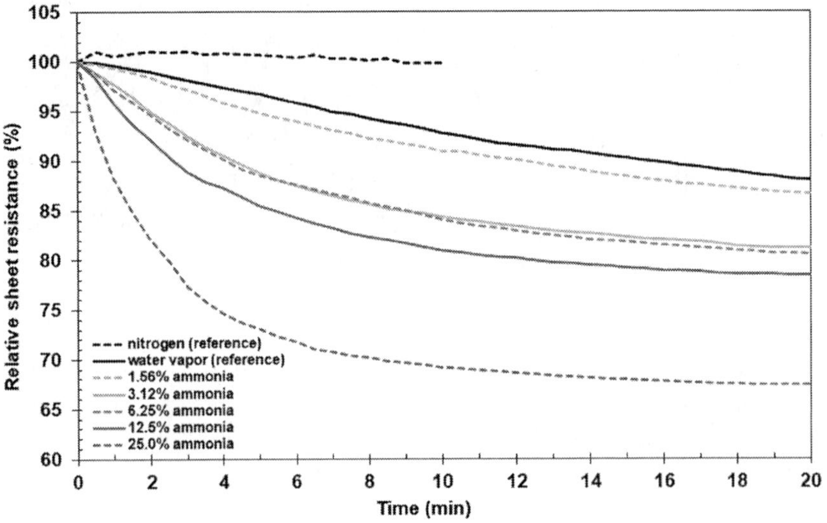

Figure 4 Electrical response of SnO_x films (T_{Sub} = 400 °C) to different atmospheres: Relative sheet resistance vs. time at ammonia-containing atmospheres and references [9].

The solvent (pure deionized water) and nitrogen as an inert gas medium were used as references and fed into the test chamber. The water vapor itself causes a reduction of the sheet resistance and a relative sheet resistance of ~88 % was achieved following 20 min treatment. In comparison to water vapor, nitrogen did not influence the sheet resistance. By introducing ammonia into the test chamber the relative sheet resistance decrease stronger and values of approximately 86 %, 81 %, 80 %, 78 % and 67 % were achieved using 1.56 %,

3.12 %, 6.25 %, 12.5 % and 25.0 % ammonia concentration, respectively. Therewith a clear decrease of relative sheet resistance with increasing ammonia concentration was observed. Hence, films show distinctive gas sensitivity and selectivity. This behavior is typical for SnO_2 gas sensors and described in literature [13,14].

Furthermore, measurements could be repeated several times consecutively with similar results (see **Figure 5**). Following an ammonia test step, the chamber was purged by air and a recovery of the sheet resistance was observed. Thus, one test cycle consists of one ammonia exposure step and one air-purging step. By applying this routine, several cycles were conducted consecutively. The sheet resistance at the start (before introducing ammonia) and the end of each cycle (saturation of sheet resistance and start of air-purging) were comparable among several test cycles.

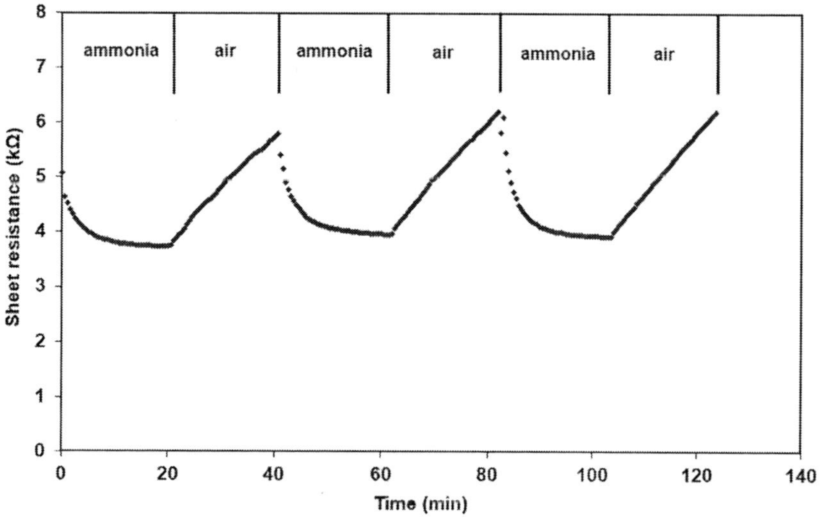

Figure 5 Reversibility of SnO_x films (T_{Sub} = 400 °C): Sheet resistance vs. time for recurring changing atmospheres; ammonia (25.0 %) and pressurized air.

4. CONCLUSION

By using a modified APPJ system, nanostructured SnO_x thin films were successfully deposited in one single operation step. TBT was introduced in form of an aerosol into the reactive plasma, chemically converted and transported to the substrate were film formation took place. Good electrical conductivity was achieved using substrate temperatures between 250 °C and 500 °C. The deposition procedure itself can be described as a combination of plasma assisted CVD and spray pyrolysis. However, compared to spray pyrolysis, lowest sheet resistivity could be achieved at lower substrate temperatures. Hence, this technology could be a more suitable method also for temperature sensitive materials. First experiments clearly indicate the gas (ammonia) sensing properties of the deposited films. So, tin oxide coatings created by atmospheric pressure PECVD using the APPJ technology are showing already in its undoped state and without further layer post-treatments high application potential.

REFERENCES

[1] MEYER, J., HAMWI, S., KRÖGER, M., KOWALSKY, W., RIEDL, T., KAHN, A. Transition metal oxides for organic electronics: energetics, device physics and applications. *Advanced Materials*, 2012, vol. 24, no. 20, pp. 5408-5427.

[2] PARK, S.K., HAN, J.I., KIM, W.K., KWAK, M.G. Deposition of indium-tin-oxide films on polymer substrates for application in plastic-based flat panel displays. *Thin Solid Films*, 2001, vol. 397, pp. 49-55.

[3] MORALES, C., JUAREZ, H., DIAZ, T., MATSUMOTO, Y., ROSENDO, E., GARCIA, G., RUBIN, M., MORA, F., PACIO, A., GARCIA, A. Low temperature SnO_2 films deposited by APCVD. *Microelectron*, 2001, vol. 39, pp. 586-588.

[4] WAN, Q., WANG, T.H. Single-crystalline Sb-doped SnO_2 nanowires: synthesis and gas sensor application. *Chemical Communication*, 2005, vol. 30, pp. 3841-3843.

[5] MICHEL, H.J., LEISTE, H., HALBRITTER, J. Structural and electrical characterization of PVD-deposited SnO_2 films for gas-sensor application. *Sensors and Actuators B: Chemical*, 1995, vol. 25, pp. 568-572.

[6] KARSLIOGLU, R., UYSAL, M., AKBULUT, H. The effect of substrate temperature on the electrical and optic properties of nanocrystalline tin oxide coatings produced by APCVD. *Journal of Crystal Growth*, 2011, vol. 327, pp, 22-26.

[7] OKUYA, M., KANEKO, S., HIROSHIMA, K., YAGI, I., MURAKAMI, K. Low temperature deposition of SnO_2 thin films as transparent electrodes by spray pyrolysis of tetra-n-butyltin(IV). *Journal of the European Ceramic Society*, 2001, vol. 21, pp. 2099-2102.

[8] BISWAS, P.K., DE, A., PRAMANIK, N.C., CHAKRABORTY, P.K., ORTNER, K., HOCK, V., KORDER, S. Effects of tin on IR reflectivity, thermal emissivity, Hall mobility and plasma wavelength of sol-gel indium tin oxide films on glass. *Materials Letters*, 2003, vol. 57, pp. 2326-2332.

[9] GERULLIS, S., POHLE, L., PFUCH, A., BEIER, O., KRETZSCHMAR, B.S.M., RAUGUST, M., RÄDLEIN, E., GRÜNLER, B., SCHIMANSKI, A. Structural, electrical and optical properties of SnO_x films deposited by use of atmospheric pressure plasma jet. *Thin Solid Films*, 2018, vol. 649, pp. 97-105.

[10] OJEDA, F., CUEMO, R., SALVAREZZA, R., VAZQUEZ, L. Dynamics of rough interfaces in chemical vapor deposition: experiments and a model for silica films. *Physical Review Letters*, 2000, vol. 84, pp. 3125-3128.

[11] PATIL, P.S., KAWAR, R.K., SETH, T., AMALNERKAR, D.P., CHIGARE, P.S. Effect of substrate temperature on structural, electrical and optical properties of sprayed tin oxide (SnO_2) thin films. *Ceramics International*, 2003, vol. 29, pp. 725-734.

[12] TURGUT, G., KESKENLER, E.Y., AYDIN, S., SÖNMEZ, E., DOGAN, S., DÜZGÜN, B., ERTUGRUL, M. Effect of Nb doping on structural, electrical and optical properties of spray deposited SnO_2 thin films. *Superlattices and Microstructures,* 2013, vol. 56, pp. 107-116.

[13] LASSESSON, M., SCHULZE, M., VAN LITH, J., BROWN, S.A. Tin oxide nanocluster hydrogen and ammonia sensors. *Nanotechnology*, 2007, vol. 19.

[14] WANG, G.X., PARK, J.S., PARK, M.S., GOU, X.L. Synthesis and high gas sensitivity of tin oxide nanotubes. *Sensors and Actuators B*, 2008, vol. 131, pp. 313-317.

SURFACE MODIFICATION OF POLYMERS TO IMPART ANTISTATIC PROPERTIES

Alexandra ISAKOVA[1], Vladimir TVERSKOY[2], Andrey INDENBOM[1,3], Victor IVANOV[1], Oxana GRIBKOVA[1], Aleksandr NEKRASOV[1]

[1]*Russian Academy of Sciences A.N. Frumkin Institute of Physical Chemistry and Electrochemistry RAS, Moscow, Russian Federation,*
Isakova_Aleks@list.ru, oxgribkova@gmail.com
[2]*Moscow Technological University, Institute of Fine Chemical Technologies of M. V. Lomonosov, Moscow, Russian Federation*
[3] *Moscow Institute of Physics and Technology (State University), Dolgoprudny, Moscow Region, Russian Federation*

Abstract

Static electricity on the surface of dielectric polymers can be dangerous in various areas of human life. Antistatic agents reduce or eliminate accumulation of static electricity by creation a conductive layer on the polymer surface. The most known antistatic agents are surfactants. They increase surface conductivity by absorbing moisture from the air. But this effect is short-lived and depends on the temperature and air humidity. Modification of a thin surface layer of polymers by conductive polymer, e.g., polyaniline (PANi), is a promising method. In this work fragments of polyacrylic and polystyrenesulfonic acids were grafted onto the surface of polyethylene films before the synthesis of PANi. This technique allows to create long lasting antistatic effect. Synthesis of PANi on the modified polyethylene films was studied by optical spectroscopy. The autocatalytic character of this process was proved. Optimal synthesis conditions for obtaining homogeneous films were defined. The electrical characteristics of the obtained materials were measured. Thus, the possibility of synthesizing PANi on polyethylene films with incorporated fragments of sulfonated polystyrene or with its mixture with acrylic acid was demonstrated. The minimum sheet resistance of the modified PE film was achieved with the maximum content of styrene sulfonic acid. It was shown that the proposed technique is promising for the creation of antistatic coatings of polymers. The proposed method is also suitable for application of antistatic coatings on the inner surface of polymer containers of complex shape.

Keywords: Polyaniline, polyethylene, antistatic properties, conductive polymers

1. INTRODUCTION

The occurrence of static electricity on the surface of dielectric polymers can be dangerous in various areas of human life. The resulting electrical discharge can cause combustion or explosion of flammable substances, lead to breakdown of various electronic devices, etc. Static charge on polymer surface can be significantly reduced. Depending on the conditions and the environment of use, two methods to reduce or eliminate accumulation of static electricity can be distinguished. The first method is to create a surface hydrophilic layer that promotes the emergence of a thin film of water under natural moisture conditions and temperatures above 0°C. Various kinds of surfactants are commonly used as such substances, including nitrogen-containing compounds (long-chain amines, amides or quaternary ammonium salts), sulfonic acids, polyols and their derivatives, etc. The second method consists in introducing into the surface layer various electroconductive additives (nanoparticles, carbon materials, etc.) that provide the electrical conductivity of the surface layer irrespective of atmospheric humidity and temperature. Despite the wide choice of antistatic additives, their introduction can lead to a deterioration in the physico-mechanical properties of the polymers being modified. Many additives can not be used to modify industrially important polyolefins (thermoplastics - polyethylene, polypropylene, polybutylene), and the antistatic effect persists for a short time. In addition, some antistatic agents are difficult to use at low temperature and low relative humidity (for example, in the conditions of the

Far North and the Arctic). In our opinion, the most promising method of imparting antistatic properties to polymers is their modification with special additives that, after chemical interaction with the base polymer, create a thin layer of new electrically conductive polymeric material. In this case, the electrostatic properties not only do not depend on the environmental conditions, but are also most resistant to various external influences.

A promising polymer for use as an antistatic agent is polyaniline (PANi). It is easily synthesized, resistant to oxidation in air, and has sufficient electrical conductivity. A method for synthesizing PANi in hydrochloric acid in a matrix of porous polyethylene was described [1]. The authors showed that this method leads to the production of conductive composites. The disadvantage of this synthesis is the influence of the polyethylene film relief on the uniformity of the PANi distribution. We have previously shown that the use of polysulfonic acids together with polyaniline allows to obtain coatings with improved properties [2,3]. Complexes of PANi with polysulfonic acids contain hydrophilic fragments, which gives the modified material additional electrostatic properties. However, such complexes are soluble in water and form easily removable films on the surface of polymers (for example, polyethylene). More stable coatings of this type could be obtained by synthesizing PANI on polyethylene with grafted fragments of polysulfonic acids.

2. MATERIALS AND METHODS

We used polyethylene (PE) films with grafted polystyrene sulfonic acid fragments (PE-PSSA) and PE films with grafted fragments of copolymer of acrylic and styrene sulfonic acids (PE-ASA). The main characteristics of the films are shown in **Table 1**.

Table 1 Physico-chemical properties of PE films

Film thickness, μm	Fraction of grafting, Δp(%)	
	[St]	[acrylic acid]
50	93	-
50	39	-
50	28	-
22	17.8	58.7
16	13.6	57.0
14	16.8	56.9

Ammonium persulfate (reagent grade) employed as an oxidizer was used as received. Aniline (reagent grade) was distilled in vacuum in nitrogen atmosphere, and a fraction with a boiling temperature of 82-84°C (20 mmHg) was collected. Polymerization experiments were performed with the as-distilled product (n_d^{20}= 1.583)

Polymerization of aniline on PE-PSSA and PE-ASA films was carried out at a temperature of 25 ° C. Few days before the synthesis, the PE-ASA was transferred from Na + to the H + -form by placing in a 1M HCl solution, and then washed with water and dried in air.

For PANi synthesis, PE-PSSA and PE-ASA films were left for a few days in 0.1 M aniline solution, washed with water and air-dried. Thus, the aniline-treated film was transferred to a PSA solution for the aniline polymerization. After polymerization, the films were washed with water and dried.

For comparison, the oxidative polymerization of aniline in an aqueous solution of polystyrene sulfonic acid (PSSA, Aldrich Mw ~ 7 × 10^4) was carried out. The molar ratio of aniline to oxidizer was 1:1 mol/mol, the ratio of aniline to a sulfonic group of PSSA was 1:2 mol/g-eq. sulfonic groups. Aniline concentration was equal to 0.003 M. [3].

Spectral measurements in the visible region of the spectrum during the matrix synthesis were carried out using a high-speed scanning single-beam spectrophotometer Avantes 2048, equipped with flexible light guides. Instrument control, data acquisition and processing were done by a computer equipped with the appropriate software. The spectral range of was 300 to 900 nm, the recording time was 100-150 ms, which made it possible to neglect the change in the shape of the spectrum during its recording. The spectra of the modified films in the 250-3200 nm region were recorded using a two-beam spectrophotometer "Shimadzu UV-3101PC".

The sheet resistance was measured using digital ohmmeter by applying two narrow contacts made of silver conductive paste onto 2 opposite edges of a rectangular sample (1 cm X 1 cm) of the modified films.

3. RESULTS AND DISCUSSION

3.1. Synthesis of polyaniline in polyethylene films with grafted polystyrene sulfonic acid

The changes in the optical absorption spectra during the synthesis of PANi on the films of PE-PSSA (a) and in the solution of polystyrene sulfonic acid-PSSA (b) are shown in **Figure 1**. Synthesis of PANi on PE-PSSA reveals some similarity to the synthesis of PANi in the PSSA solution. Thus, during the synthesis of PANi on PE-PSSA, as well as in the case of PANi synthesis in the PSSA solution, the growth of absorption bands attributed to cation-radical (430 nm) and quinoneimine (600 nm) forms is observed. In the future, a shift of the absorption maximum to the long-wavelength region around 750 nm, which is caused by the formation of localized polarons is observed. Changes in the absorption spectra during the synthesis on PE-PSSA films are less pronounced, in comparison with the PANi synthesis in the PSSA solution. This may be due to several factors, including the spatial distribution of aniline and its orientation over the surface/inside the film, the presence of a crystalline phase of polyethylene, the formation of a surface layer of PANi contributing to delocalization of oxidized fragments, and others.

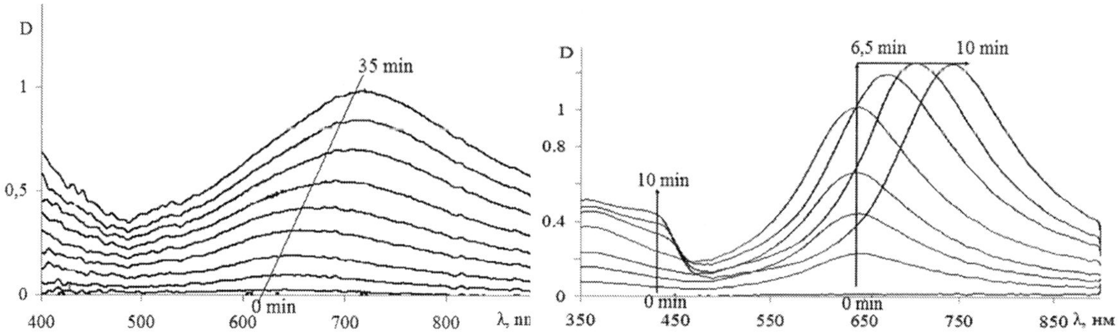

a b

Figure 1 Spectrum changes during PANi synthesis on PE-PSSA films (a) and in the presence of PSSA (b). The proportion of grafted PS in case (a) is 39 %. The reagent ratio for (a): nAn: n PSSA = 1: 2, nAn: nOx = 1: 1; for (b): nAn: nOx = 1: 1

The polymerization of PANi on PE-PSSA has an autocatalytic character, as evidenced by the S-shaped kinetic curves (**Figure 2**). It should be noted that the rate of synthesis is influenced by the ratio of reagents. Thus, rapid oxidation of aniline (the case with an excess of oxidizer, nAn / nOx = 1/2) leads to the acceleration of polymerization process by approximately a factor of 2.

Figure 3 shows the PANi polaron absorbance changes depending on the proportion of grafted polystyrene. It can be seen that the growth rate (slope of the linear part of the curve) reaches maximum in the interval 28-39 % and then does not change up to 93 %. The induction period of polymerization is hard to be determined at the polystyrene concentration of Δp =28 %. With a further increase in the proportion of grafted polystyrene from 39 to 93 % (i.e. 2.38 times), the induction period increases from about 20 to 47 min (2.35 times), i.e. approximately proportionally to the increase of fraction of grafted polystyrene.

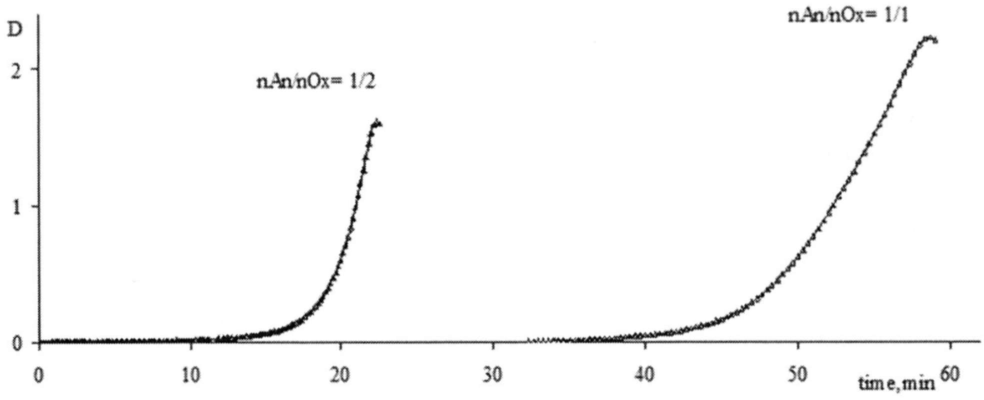

Figure 2 Absorbance changes at 750 nm (polaron absorption) during PANi synthesis on PE-PSSA (Δp=39 %), depending on the ratio of reagents

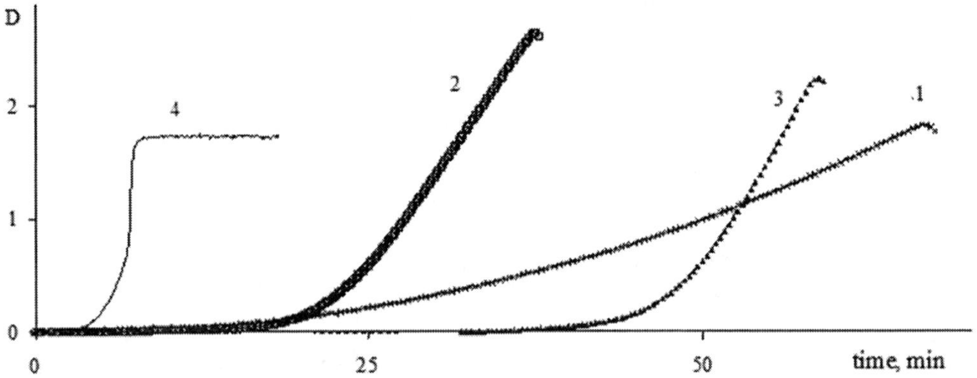

Figure 3 Change in the absorption at 750 nm (polaron form of PANi) during synthesis in PE-PSSA films depending on the proportion of grafted polystyrene. Curve 1 - PANi-PE-PSSA (Δp=28 %); 2- PANi-PE-PSSA (Δp =39 %), 3- PANi - PE- PSSA (Δp=93 %), 4- PANi-PSSA (synthesis of PANi in the solution of PSSA for comparison).

For PE films with grafted fragments of the copolymer of acrylic (AA) and styrene sulfonic (SSA) acids the effect was studied of total proportion of grafted acids and the ratio of the AA and SSA on the shape of final electronic absorption spectra of the films at a given PSA concentration.

The change in the composition of the film with a partial replacement of SSA by AA groups led to a significant slowdown in the synthesis of PANi (the synthesis occurred for a day or more). Analysis of the spectra of optical absorption of grafted copolymers (**Figure 4**) shows the presence of characteristic absorption bands of PANI.

The position of the maximum in the region of localized polarons (750 nm) shows that the optimum is PE-ASA with a grafting degree of 70.6 %, including 57.0 % AA and 13.6 % SSA. Differences in the characteristic absorption spectra for PE-ASA films with Δp = 70.6 and Δp = 76.5 (58.7 % AA+ 17.8 % SSA) indicate that it is necessary to select a specific concentration of PSA for each film to avoid the formation of short oligomers, as in the case of a film with Δp = 73.7 % (56.9 % AA+ 16.8 % SSA)

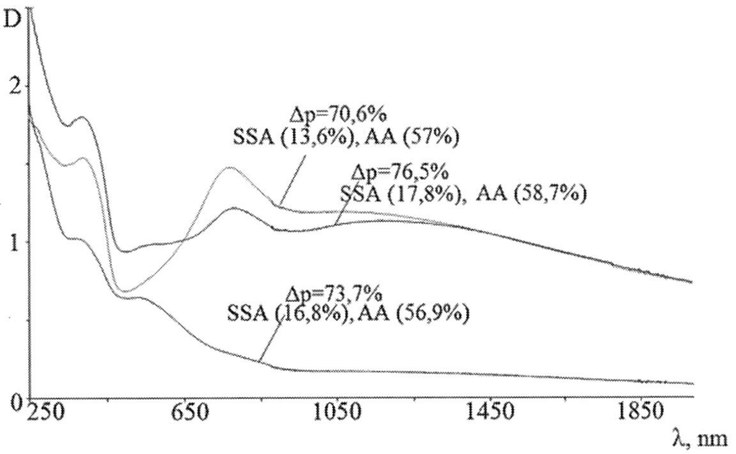

Figure 4 Electronic absorption spectra of modified PE-ASA films with different ratios of AA and SSA groups in the graft copolymer at a total PSA concentration of 5.7 mM.

The sheet resistance of polyaniline-modified PE-PSSA films was measured. The results, presented in **Table 2**, indicate successful modification of these films by PANi. It can be seen that the film with the maximum degree of polystyrene grafting has the lowest sheet resistance. Unfortunately, the sheet resistance of PANi-PE-ASA films could not be measured despite the electronic absorption spectrum proper to PANi. Apparently, this is due to irregular distribution of the conductive areas of PANi phase in the polyethylene film. In particular, the frequency and regularity of the arrangement of the sulfonic groups, the distribution of the crystalline and amorphous phase in polyethylene can affect the uniformity of the distribution.

Table 2 Surface resistance of modified PE-PSSA*

Films	Sheet resistance, MOм/□
PANi-PE-PSSA 93 %	0.08
PANi-PE-PSSA 39 %	3.6
PANi-PE-PSSA 28 %	5

*The measurement temperature is 22 °C, relative air humidity is 45 %.

Thus, the possibility of synthesizing PANi on polyethylene films with incorporated fragments of sulfonated polystyrene or with its mixture with polyacrylic acid was demonstrated. The minimum sheet resistance of the modified PE film was achieved with the maximum content of styrene sulfonic acid. It was shown that the proposed technique is promising for the creation of antistatic coatings of polymers.

REFERENCES

[1] ELYASHEVICH Galina K., SIDOROVICH Askold V., SMIRNOV Michail A., KURYNDIN Ivan S., BOBROVA Natalia V., TRCHOVA Miroslava, STEJSKAL Jaroslav. Thermal and structural stability of composite systems based on polyaniline deposited on porous polyethylene films. *Polymer Degradation and Stability*.2006. vol. 91, pp.2786-2792

[2] GUSEVA Marina A., ISAKOVA Alexandra A., GRIBKOVA Oxana L., TVERSKOI Vladimir A., IVANOV Victor F., VANNIKOV Anatolyi V., FEDOTOV Yuri A. Matrix polymerization of aniline in the presence of polyamides containing sulfo acid groups. *Polymer Science - Series A*.2007. vol. 49, no. 1. pp. 4-11.

[3] GRIBKOVA Oxana L., NEKRASOV Alexandr A., TRCHOVA Miroslava, IVANOV Victor F., SAZIKOV Vadiml., RAZOVA Anna B., TVERSKOY Vladimir A. VANNIKOV Anatolyi V. *Polymer*. 2011. vol. 52. pp. 2474-2484.

ADHESION IMPROVEMENT BETWEEN BITUMEN AND MINERAL AGGREGATE: THEORETICAL PRINCIPLES AND THEIR IMPACT ON PAVEMENT STRUCTURE LIFETIME

Jan TREJBAL[1,2,a], Pavel DEMO[1,2,b], Václav NEŽERKA[1,c], Alexey SVESHNIKOV[1,2,d], Tereza VALENTOVÁ[1,e], Jan VALENTIN[1,f]

[1]Czech Technical University in Prague, Faculty of Civil Engineering, Prague, Czech Republic, EU
[2]Czech Academy of Science, Institute of Physic, Prague, Czech Republic, EU
[a]jan.trejbal@fsv.cvut.cz, [b]demo@fzu.cz, [c]vaclav.nezerka@fsv.cvut.cz, [d]alexey.sveshnikov@fsv.cvut.cz, [e]tereza.valentova@fsv.cvut.cz, [f]jan.valentin@fsv.cvut.cz

Abstract

Pavement structures often fail due to insufficient interphase interaction between bituminous binder and mineral aggregate. A thin bituminous film covering aggregate particles is often replaced by water, resulting in bitumen-water-aggregate system, which is firmly linked to adhesion failure between these materials. This paper describes how to avoid this phenomenon using nanomaterial based admixtures - adhesion promoters maximizing the formation of intermolecular interactions between bitumen and aggregate surfaces. Physicochemical and topological principles about interaction between the two materials were described. Regular paving grade bitumen 50/70 was modified by additives based on: (i) alkylsilanes; (ii) phosphorus compounds; and (iii) amines. Mineral aggregate grains (maximal size of 16 mm, pre-heated on 160±5 °C) were covered by the thin film of reference and such modified binders (170±5 °C). In this way prepared samples were exposed to water (60 °C) for 60 minutes. After samples removing from water, their digital images were obtained and then an area of aggregate grains covered by the bituminous thin film was assessed using: (i) bitumen adhesion test; following relevant technical standards; (ii) digital image analysis based on both gray level thresholding and entropy-based segmentation. It was shown that adhesion of the bitumen to aggregate, expressed by aggregate particle coating area, was increased by almost 18 %, if adhesion promoters were used.

Keywords: Asphalt mixture, adhesion promoters, interface interaction, coated area

1. INTRODUCTION

Asphalt mixtures, consisting of bituminous binder, mineral aggregate, and other admixtures, dominate during construction of road pavements worldwide. Their mechanical performance as well as pavement lifetime significantly depends on an interaction between individual phases, namely between bituminous binder and mineral aggregate. If adhesion between the bitumen film and aggregate fails, aggregate stripping usually occurs, resulting in subsequent damage of pavement structure and its integrity lost. Such estimated damages, known as potholes or just stripping, have a strong negative impact on asphalt pavements from the both the operational, traffic safety, and economical point of view [1].

Interphase interactions between bitumen, aggregate and other substances (e.g. dust, water) is complex process resulting in formation of adhesive bonds. Bitumen is usually considered to be a colloid system of intricate chemical composition and of complicated internal arrangement of molecules. Bituminous adhesive forces are promoted by polar components (such as polynuclear aromatics, phenolics, pyridinics, ketones, sulfides, etc.) lowering the interfacial energy between asphalt and aggregate [2]. Similarly, the temperature dependent viscosity reflects the capacity of bituminous material to be absorbed onto the aggregate surface. Furthermore, adhesive bonds are considerably influenced also by physico-chemical properties of the present mineral aggregates (surface morphology - roughness, porosity, surface area, dislocations, geometrical shape, etc. [3]) and the final adhesion is then the result of combination of the above mechanisms. It has to be stressed out that aggregates exhibit a strong affinity for water, in particular, the so-called acidic (hydrophilic) rocks with a high content of silicon-based substances (granite, quartz). Consequently, bitumen is frequently replaced by residual water, followed by formation of interfacial system bitumen-water-aggregate and this stripping effect may be firmly linked to adhesion failure between bituminous material and aggregate surface. This problem can be optimized by addition of adhesion promoters maximizing the formation of intermolecular interactions between bitumen and aggregate surface (apolar Lifshitz-van der Waals, resp. Lewis acid- based components) resulting in plausible values of the final surface free energy. Even a small amount of these promoters, usually between 0.01 and 0.5 wt.% to bitumen, causes significant enhancement in bituminous binder to aggregate adhesion [4,5].

In order to reveal interphase interaction between the two materials, several methods can be employed. Contact angle measurement of hot bitumen droplet applied on the planar surfaces of cut aggregate is able to provide an information about the interphase interaction [6,7]. However, this method cannot take into account a shape factor and the surface roughness of the aggregate. The same problem is typical also for peel-tests, when steel plates, bonded to aggregate by thin layer of the binder, are removed from planar aggregate surfaces [8]. Empirical methods based on visual observations of aggregate coated by the binder are frequently used throughout technical praxis; however, these cannot provide accurate and consistent results due to subjective assessment by evaluators [9,10]. Instead of evaluators, digital and computational technic can be employed to assess the rate of interaction between two materials of different state [11,12].

In this research study, we modified standard paving grade bitumen by several types of adhesion promoters. Such modified binders were used to cover mineral aggregate within test specimens which are defined by a selected aggregate fraction coated by hot bitumen. The area of coated aggregate by binder, pointing to adhesion between the two materials, was assessed using both conventional and primarily digital image analysis based methods.

2. ASPHALT MIXTURES

2.1. Bituminous binder

Paving grade bitumen 50/70 (EN 12591 and ČSN 657204) was used to cover aggregate grains. Reference mix marked as "MR" was produced without any additives. Compared to that, mixtures M1-M5 were made from the same type of bitumen, but these contained selected types of adhesion promoters.

2.2. Mineral aggregate

Crushed aggregate the Brant quarry, where granite porphyry is the key mineral, was chosen with respect to the availability of natural aggregate sources. Moreover, it was targeted to use a less suitable aggregate (hydrophilic) more susceptible to water damage, with poor adhesiveness to the binder and elevated potential of stripping failures shown in the asphalt mix composite. The criterion of less favorite adhesion was fully met by granite porphyry, or porphyry microgranite. A considerable proportion of the rock consists of quartz, potassium feldspar (orthoclase) and mica. The fraction was equal to 8-16 mm.

2.3. Adhesion promoters

To improve adhesion between the bituminous binder and aggregate, liquid-form chemical-based additives were used. In sum, five different products were selected: M1, alkylsilanes-based; M2, phosphorus-based; M3, based on chemical surfactants comprising amines; M4 and M5, based on unsaturated fatty acids combined with diethanolamine. Their amount differed from 0.2 to 0.4 wt. % to binder. Their specification and amount are summarized in **Table 1**.

Table 1 Summarization of asphalt mixture composition

Mixture	Type of mineral aggregate	Bituminous binder	Adhesion promoter	Adhesion promoter (wt. % to binder)
MR	Granite porphyry fraction 8/16 originated from the Brant quarry	50/70	-	0
M1			Additive based on alkylsilanes	0.2
M2			Additive based on phosphorus	0.3
M3			Additive based on chemical surfactants comprising amines	0.3
M4			Additive based on unsaturated fatty acids combined with diethanolamine	0.2
M5				0.4

3. DETERMINATION FOR ADHESION OF BITUMINOUS BINDER TO AGGREGATE

3.1. Conventional visual analysis

Following ČSN EN 73 6161 technical standard, test specimens in a defined glass bowl were prepared. These composed of several individual pieces of aggregate with fraction of 8/16 mm (in sum ca. 300 g, temperature 160±5°C) coated by the bituminous binder (12±0.3 g, temperature 170±5°C). Such prepared samples were stored for 24 hours, then submerged to water (60 °C) and stored additional 60±5 minutes. Then, they were subjected to conventional visual assessment according to cited standard. The conventional analysis is based on subjective assessment of two evaluators, who classify the rate of stripped areas of aggregate.

It is clear from the procedures mentioned above that the subjective assessment cannot exclude "human factor" and provide consistent results. In order to overcome this drawback, the test specimens placed in glass bowls were photographed using digital camera (Canon EOS 70D, focus length 50 mm, Av 8). Such obtained images were subjected to testing by other two assessment methods, as described below.

3.2. Gray level thresholding

The percentage of the aggregate surface coated by the bitumen was assessed using "in-house" developed Matlab software Binder. The main principle is based on identification of areas brighter than the dark bituminous binder. These areas are then considered to be parts of stripped aggregate particles. Such an approach introduced inaccuracies in the case of dark aggregates or light reflections on the bitumen surface. The reflected light causes local over-exposition within images, which the software evaluates as uncovered.

3.3. Entropy-based image segmentation

To overcome the issues connected with dark aggregate or local over-exposition, an alternative approach was addressed. Namely, we employed local entropy calculation in order to assess roughness of the texture. Such a procedure is computationally more demanding but still feasible and with much more accurate outcomes. The areas with high entropy are considered to belong to aggregates while low entropy areas represent the bituminous matrix.

4. RESULTS

Conventional visual analysis of asphalt mixture, executed following ČSN EN 73 6161 standard, revealed that the adhesion between aggregate and reference bitumen (mixture MR) reached only to "unsatisfactory" classification, i.e., the area of aggregate particles covered by bitumen was less than 80 %. According to gray-level and entropy-based analysis, the area was equal to 80.3 % (satisfactory) and 73.9 % (unsatisfactory), respectively. The results acquired using the entropy-based analysis were considered to be more accurate than those obtained from gray-level-based analysis due to problems with proper photographing caused by light reflection from the surface of shiny bituminous binder surfaces (this held true for all results comparison between these two methods).

Covered aggregate surface area of mixture M1 (containing adhesion promoters at the amount of 0.2 wt. %) reached on 91.9 % and 85.9 % (good / satisfactory), depending on the evaluating method (gray-level-based and entropy-based).

If whatever other adhesion promoters were used (M2-M5), the rate of aggregate covered area increased significantly up to ca. 95-99 % (good / excellent) depending on the evaluating method. The results are summarized in **Table 2** and graphically illustrated in **Figure 3.** These finding were comparable to research by Rosi et al. [3,6]. For illustration, the extreme cases (MR and M4) are shown in **Figure 1** and **Figure 2.**

Based on these results, it can be concluded that adhesion promoters significantly contribute to bituminous binder adhesion to mineral aggregate. Moreover, it was found that entropy-based assessment of interphase interaction between the two materials is a potential technique from the practical point of view, providing proper results and regarding shape factor of the aggregate.

Table 2 Assessment of adhesion of bituminous mixtures to mineral aggregate

Mixture	Visual assessment	Gray-level-based assessment (% / classification)	Entropy-based assessment (% / classification)
MR	unsatisfactory	80.3 / satisfactory	73.9 / unsatisfactory
M1	satisfactory	91.9 / good	85.9 / satisfactory
M2	excellent	98.8 / excellent	95.6 / good
M3	good	97.5 / excellent	93.7 / good
M4	good	97.8 / excellent	95.8 / good
M5	good	97.9 / excellent	94.6 / good

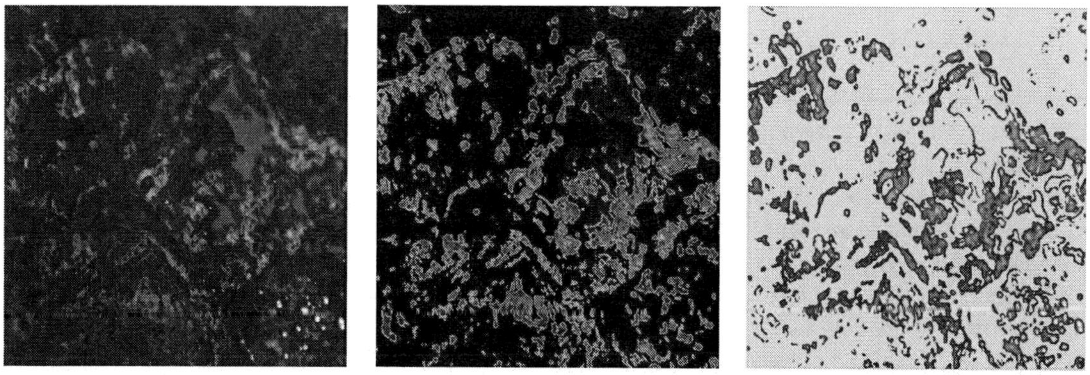

Figure 1 Images of poorly covered aggregate by binder (MR mixture), from left: original photo, gray level thresholding-, and entropy-based image segmentation

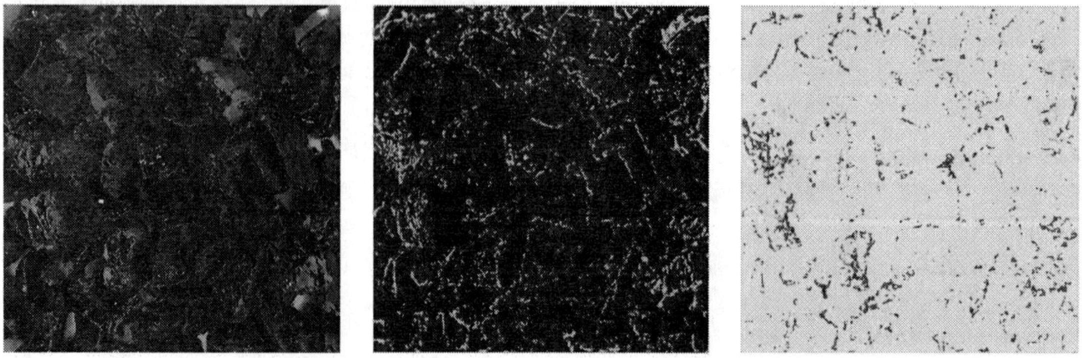

Figure 2 Images of good covered aggregate by binder (M4), from left: from left: original photo, gray level thresholding-, and entropy-based image segmentation

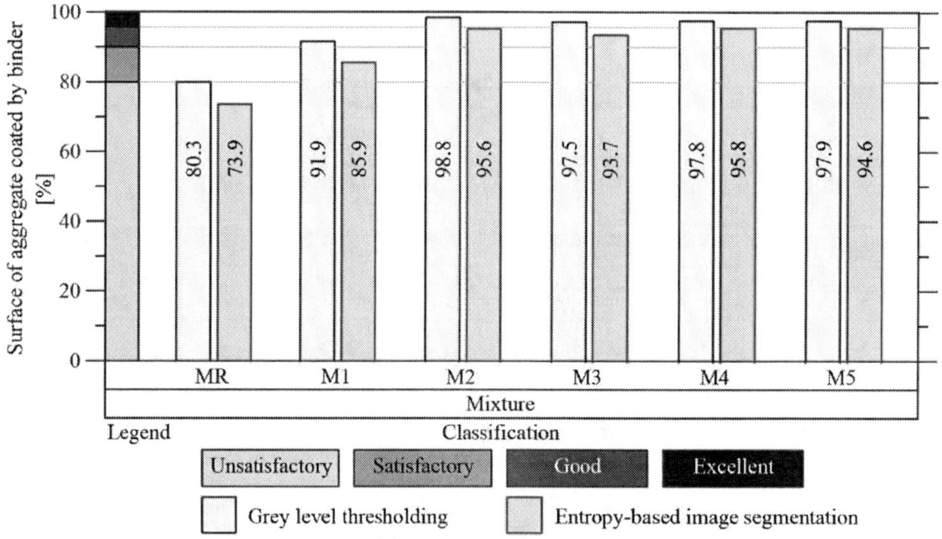

Figure 3 Assessment of adhesion of bituminous mixtures to mineral aggregate

5. CONCLUSION

Interphase interaction between straight-run distilled bituminous binder (50/70) thin film and mineral hydrophilic aggregate (8-16 mm, granite porphyry) from Brant quarry in Central Bohemian region was assessed using conventional and newly developed digital image-based methods. The binder was modified with nanomaterial chemical-based adhesion promoters in order to increase the adhesion between these two materials and thus to ensure creation of aggregate-bitumen interphase system only, with no other materials. For these purposes, chemical-based additives were used at the amount of 0.2-0.4 wt. % (to binder). Asphalt mixtures composed of the aggregate particles coated by the thin film of the bitumen were prepared, photographed and then subjected to the adhesion assessment. Focusing on the surface of these particles, an area coated by the binder was evaluated subjectively, following relevant technical standard and automatically, using grey level thresholding and entropy-based image segmentation analysis. It was shown that the coated area was approx. 74-80 % in the case of reference bitumen (without any nanomaterial based additives, depending on evaluating method. If adhesion promoters were used, the examined area was increased up to almost 100 %. Moreover, it was found that the evaluating method based on entropy-based image segmentation was considered to be more accurate, because it overcame problems relating with light reflection from shiny bituminous binder surfaces, resulting in local over-exposed areas within the photography.

ACKNOWLEDGEMENTS

This work was financially supported by the Czech Science Foundation research project 18-13830S and the support is gratefully acknowledged. The author would also like to thank the Center for Nanotechnology in Civil Engineering at the Faculty of Civil Engineering, CTU in Prague and the Joint Laboratory of Polymer Nanofiber Technologies of the Institute of Physics Academy of Science of the Czech Republic.

REFERENCES

[1] AHMAD, Naveed. *Asphalt Mixture Moisture Sensitivity Evaluation Using Surface Energy Parameters.* Nottingham: Nottingham ePrints service, 2011. p. 286.

[2] LESUEUR, Didier. The colloidal structure of bitumen: Consequences on the rheology and on the mechanisms of bitumen modification. *Advances in Colloid and Interface Science.* 2009. vol. 145, no. 1-2, pp. 42-82.

[3] ROSSI, Cesare Oliviero, TELTAYEV, Bagdat, ANGELICO, Ruggero. Adhesion Promoters in Bituminous Road Materials: A Review. *Applied Sciences.* 2017. vol. 7, no. 524, pp. 1-10.

[4] GAO, Yangming, ZHANG, Yuqing, GU, Fan, XU, Tao, WANG, Hao. Impact on Mineral and Water on Bitumen-mineral Adhesion and Debonding Behaviors Using Molecular Dynamic Systems. *Construction and Building Materials.* 2018. vol. 171, pp. 214-222.

[5] BALDI-SEVILLA, Alejandra, MONTERO, L. Mavis, AGUIAR-MOYA, P. José, LORIA-SALAZAR, G. Luis, BHASIN, Amit. Influence of Bitumen and Aggregate Polarity on Interfacial Adhesion. *Road Materials and Pavement Design.* 2017. vol. 18, pp. 304-317.

[6] ROSI, Oliviero Cesare, CAPUTO, Paolino, BALDINO, Noemi, ROMANA LUPI, Francesca, MIRIELO, Domenico, ANGELICO, Ruggero. Effect of Adhesion Promoters on the Contact Angle of Bitumen-aggregate Interface. *International Journal of Adhesion and Adhesives.* 2016. vol. 70, pp. 297-303.

[7] GRÖNNIGER, Jens, WISTUBA, Michael, RENKEN, Peter. Adhesion in Bitumen-aggregate-systems. *Materials and Pavement Design.* 2010. vol. 10, pp. 881-898.

[8] CUI, Shuang, BLACKMAN, R.K. Bamber, KINLOCH, J. Anthony, TAYLOR, C. Ambrose. Durability of Asphalt Mixtures: Effect of Aggregate Type and Adhesion Promoters. *International Journal of Adhesion and Adhesives.* 2014. vol. 54, pp. 100-111.

[9] VALENTOVÁ, T., VALENTIN, J., ALTMAN, J. Development and verification of a suitable methodology for stability assessment of bitumen adhesion promoters. In *Proceedings of 6th Eurosphalt & Eurobitume Congress.* Prague: E&E, 2016. Prague.

[10] VALENTOVÁ, Tereza, VALENTIN, Jan, ALTMAN, Jan. Impact of Asphalt Ageing on the Activity of Adhesion Promoters and the Moisture Susceptibility. *Transportation Research Procedia.* 2016. vol. 14, pp. 768-777.

[11] LAMPERTI, R., Semi-automatic Evaluation of the degree of Bitumen Coverage on Bitumen-coated Aggregates. In *8th RILEM International Symposium on Testing and Characterization of Sustainable and Innovate Bituminous Materials.* Springer Netherlands, 2016. p. 1042.

[12] NEŽERKA, Václav, SOMR, Michael, TREJBAL, Jan. Contact Angle Measurement Tool Based on Image Analysis. *Experimental Techniques.* 2018. vol. 42, no. 3, pp. 271-278.

COMPARISON OF MAGNETICALLY MODIFIED MONTMORILLONITE AND SMECTA AS HEAVY METALS SORBENTS

Michaela TOKARČÍKOVÁ, Jana SEIDLEROVÁ, Oldřich MOTYKA

Nanotechnology Centre, VSB - Technical University of Ostrava, Ostrava, Czech Republic, EU,
michaela.tokarcikova@vsb.cz, jana.seidlerova@vsb.cz, oldrich.motyka@vsb.cz

Abstract

The magnetically modified composites for Cd(II), Pb(II) and Zn(II) ions removal from aqueous solution were synthetized. Montmorillonite/Fe_xO_y composites were successfully prepared using ferrofluid. The presence of Fe_xO_y nanoparticles (NPs) on the clay surface did not influence its chemical composition and did not lead to the release of clay elements to the solution. Moreover, the Fe_xO_y NPs enhanced the sorption properties of the clays and - due to the magnetic properties of composites - they could be easily removed by external magnetic field after the sorption. The influence of initial pH of the aqueous solution containing Cd(II), Pb(II) and Zn(II) ions on sorption capacity of montmorillonite/Fe_xO_y composites was studied. The accompanying elements released from clays to final extracts were observed as well. Results show that the initial pH of aqueous solution affected the sorption capacity of the observed elements and the release of the elements from montmorillonite originated from Ivančice (M) and magnetically montmorillonite (MM). The maximum sorption capacity was determined in the acid area of pH (3-4). After the sorption, Fe_xO_y NPs were determined in final extracts at low concentrations and are, therefore, fixed onto MM strongly. However, Fe_xO_y NPs are not fixed onto magnetically modified Smecta (SM) as strongly as onto MM because they leached to the final extracts more than from MM. Due to the aforementioned results of Fe_xO_y leaching, both studied composites were still magnetic after sorption experiments. The prepared composites are functional sorbents and are stable in pH 3-7.

Keywords: Montmorillonite, magnetically composite, iron oxide nanoparticles, stability, sorption

1. INTRODUCTION

Heavy metal pollution of the environment has remained a serious problem. Its efficient removal from wastewaters eliminates the contamination and cumulation in soils, surface and underground water. Cleaning technics for heavy metals removal such as ion exchange, reverse osmosis, and electrodialysis are effective, although costly [1]. Chemical precipitation is simple and low-cost method, however, it produces large volumes of sludge and does not reduce the heavy metal content sufficiently [2]. Adsorption is, considering the criteria aforementioned, simply, low cost and quick. Biochar [3], soybean hull biosorbent [4], rice and peanut husks [5,6], zeolite, bentonite [7], montmorillonite [8] are one of the most commonly biosorbents studied for the elimination of Pb, Zn, Cd and other elements from aqueous solution. Especially clays and clay mineral have been studied as sorbents due to their large specific area, low cost, chemical stability, cation-exchange capacity, environmental safety and excellent sorption properties [9]. Potencial sorbents has been modified by various technics including treatments with steam, acids, bases, metal oxides, carbonaceous materials, clay minerals, organic compounds, and biofilms to enhance their surface area and sorption properties [10-12]. One of the best treatments is magnetic modification of sorbents using iron oxides nanoparticles (NPs). Fe_3O_4 shows paramagnetic properties therefore, it can be influenced by magnet and loses its magnetic properties after its removal from the magnetic field.

There are several methods of clay magnetization: calcination of pre-treated clay minerals [13,14], mixing clay with iron oxide particles and prepared in a microwave [15], precipitation of magnetic iron oxides from a mixture of ferrous and ferric salts by a strong hydroxide in the presence of clay [15]. The magnetization of sorbents can increase the sorption capacity and the sorbent can be removed from solution by magnet.

Smecta (S) and montmorillonite (M) were magnetized using ferrofluid. Magnetically modified Smecta (MS) and montmorillonite (MM) composites were tested as sorbents of Cd, Pb and Zn ions from aqueous solution with various initial concentration and pH (3-7). pH and elements (Fe, Si, Al, Na, K, Mg and Ca) released from composites to final extracts were determined by atomic emission spectroscopy with inductively coupled plasma (AES-ICP)

2. EXPERIMENTAL

Chemical composition (**Table 1**) of Smecta (denoted as S) and montmorillonite originated from Ivančice (M) was determined using an energy dispersive fluorescence spectrometer (XRFS) SPECTRO XLAB (SPECTRO Analytical Instruments GmbH). For this measurement, samples in powder form were pressed into tablets using wax as a binder. Magnetically modified Smecta (MS) and montmorillonite (MM) composites prepared prof. Ivo Šafařík in the Institute of Nanobiology and Structural Biology of GCRC, Academy of Sciences of the Czech Republic.

Table 1 The chemical composition of Smecta (S) and montmorillonite (M)

Element	S (wt.%)	M (wt.%)
Na_2O	< 0.005	0.093 ± 0.047
MgO	3.6 ± 1.2	3.07 ± 0.62
Al_2O_3	14.2 ± 0.5	13.9 ± 0.84
SiO_2	49.8 ± 1.0	57.47 ± 4.03
K_2O	0.168 ± 0.009	0.925 ± 0.084
CaO	1.48 ± 0.08	2.04 ± 0.63
TiO_2	0.17 ± 0.012	0.296 ± 0.030
Fe_2O_3	2.77 ± 0.09	3.40 ± 0.14
LOI	27.3 ± 0.3	17.75 ± 1.07

2.1. Adsorption experiment

$Cd(NO_3)_2.4H_2O$ (SIGMA-ALDRICH, p.a. ≥ 99.0 %), $Pb(NO_3)_2$ (LACHEMA, p.a.) and $ZnCl_2$ (Lach-Ner, s.r.o., M_r 136.29 g/mol) were used to prepare aqueous solutions with Cd(II), Pb(II) or Zn(II) ions, respectively. The initial pH was adjusted by NaOH (MACH CHEMIKÁLIE, s.r.o., p.a. M_r 40 g/mol) or HNO_3 (MACH CHEMIKÁLIE, s.r.o., p.p. 65 %) solution to range 3 - 7 pH with the initial concentration 10 mg/l of Cd(II), Pb(II) and Zn(II) ions. 0.2 g of sample was mixed with 50 ml of initial solution and was shaken in a rotating shaker at a constant speed (4.5 rpm) at laboratory temperature. The suspensions were, after 1 hour of shaking, filtered through 0.45 µm filter paper immediately after 30 minutes of centrifugation (3000 rpm). Elements leached from S, MS, M and MM samples (Al, Si, Fe, Na, K, Mg, Ca) and concentration of Cd(II), Pb(II) and Zn(II) ions were determined by AES-ICP (CIROS SPECTRO VISION) in final extracts.

The differences between the initial and final concentrations of adsorbed elements (Cd, Pb and Zn) were used to calculate the amount of sorbed metal ions. The final metal uptake capacity was calculated according to Eq. 1 [16]:

$$q = \frac{V.(c_0 - c_r)}{m}.100 \tag{1}$$

where:

q - amount of contaminant removed from solution (%)
V - volume of the solution (l)
c_0 - concentrations of the metal ions in the initial solution (mg/l)
c_r - concentrations of the metal ions in the final solution (mg/l)
m - sorbent dose in the mixture (g)

3. RESULTS AND DISCUSSION

3.1. Effect of initial pH

Figure 1 illustrates the effect of initial pH on adsorption of metals ions from solution onto S, MS, M and MM samples. Initial concentration of metal ions in solution was 10 mg/L.

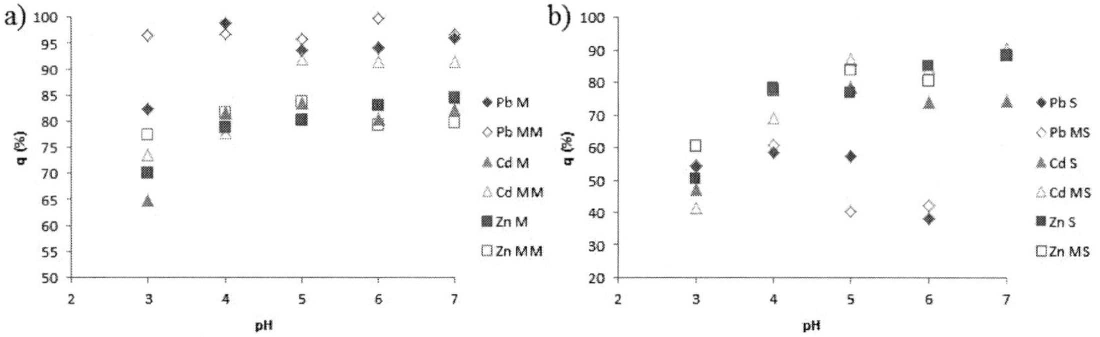

Figure 1 The amount of Pb, Cd and Zn ions removed from solution by a) original montmorillonite (M) and magnetically modified montmorillonite (MM), b) Smecta (S) and magnetically modified Smecta (MS)

The adsorbed amounts of monitoring ions increased with increasing pH of initial solution. The Cd(II), Pb(II) and Zn(II) ions uptake increased at pH ≥ 4 for original M and MM composite. Pb(II) ions show the best affinity to M and MM composite (> 95 %). The adsorption of Cd(II) and Zn(II) onto original S and MS composite increased at pH ≥ 4 as well. The removal of Cd(II), Pb(II) and Zn(II) ions from solution by the original M and MM composite was more efficient than by S or MS composite. M and MM composite removed more than 80 % of Cd(II), Pb(II) and Zn(II) from solution with pH ≥ 4. Moreover, magnetization of both clays increased their sorption capacity.

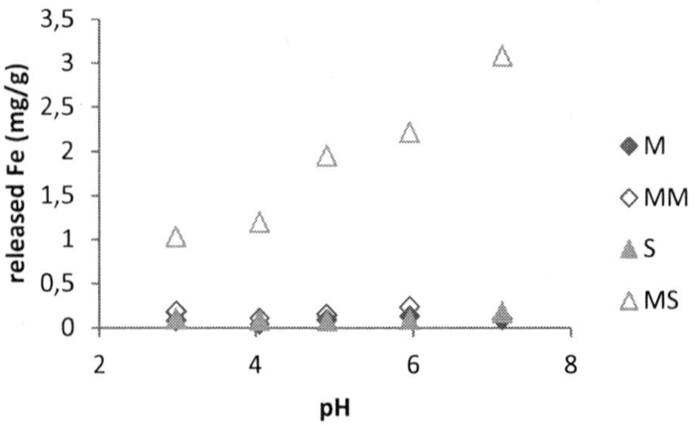

Figure 2 The amount of Fe released from original montmorillonite (M) and Smecta (S) and magnetically modified montmorillonite (MM) and Smecta (MS) to extraction agents with various initial pH

The concentration of Fe released from original M and S and from MM and MS composites after leaching test was determined as well (**Figure 2**). Fe released from MM to extraction agent is negligible in comparison to the original M. Therefore, Fe_xO_y NPs are strongly fixed onto the composite and MM composite is stable in a wide pH range. Concentration of Fe determined in final extract obtained from MS increased with increasing pH of initial solution. Amount of Fe released from MS to the final extracts did not exceed 3.5 mg/g. The after-sorption magnetic properties of MM and MS were successfully verified by magnet.

The determined concentration of Al, Si, Fe, Na, K, Mg and Ca were also evaluated using factorial analysis of mixed data (**Figure 3**).

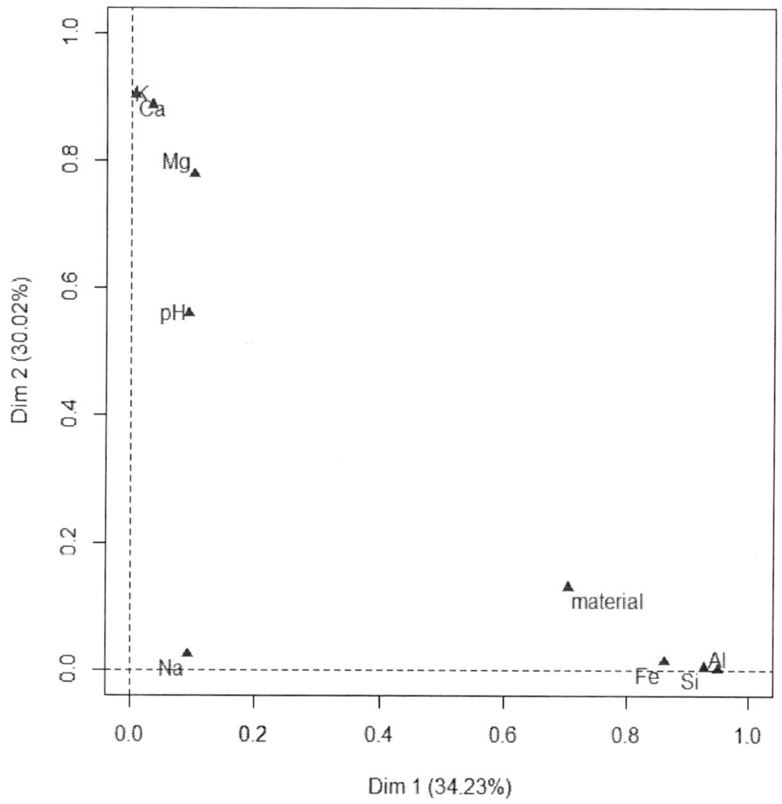

Figure 3 Factorial analysis of mixed data (FAMD)

Figure 3 shows the relation between the concentration of leached elements from samples and initial pH of extraction agents. The determined concentration of Al, Si and Fe in final extracts are display association with with materials, however seem not to be affected by the initial pH. The concentrations of K, Ca and Mg in the final extracts are not affected with tested materials, but they are related to the initial pH of extraction agents. Aforementioned relationships determined by FAMD were evaluated in more details by analysis of variance (ANOVA) and consequent Tukey HSD test.

The results of ANOVA show that the initial pH of extraction agents significantly influences the concentration of leached elements (Na, K, Mg and Ca) in final extracts (α = 0.05). Tukey HSD test shows significant difference in released amount of Na between pH 5 and 6 as well as between pH 5 and 7. Significant difference between pH and released amount of K or Ca were observed in acidic pH area. The influence of initial pH on Mg release from materials was insignificant. Tukey HSD test proved significant difference in released amount of Al and Si between MM and MS composites and between M and MS. Significant difference to released amount of Fe was proved between S and MS, MM and MS, M and MS. This suggests that the release of Fe from composite is influenced by matrix, mainly by Smecta.

4. CONCLUSION

Magnetically modified montmorillonite (MM) and Smecta (MS) composites were successfully prepared and tested as sorbents of heavy metals ions from aqueous solution with various initial pH. Original montmorillonite and Smecta were tested as sorbents for the comparison with composites as well. The sorption properties of original montmorillonite and Smecta increased after magnetically modification. Magnetically modified Smecta removed 35-90 % of tested ions. Magnetically modified montmorillonite adsorbed up to 99 % of Pb(II), 92 % of Cd(II) and 85 % of Zn(II). Composites show the best sorption properties at pH ≥ 4. The concentration of released Fe from original clays and composites were determined in solution after sorption as well. The amount of released Fe from original montmorillonite and montmorillonite composite were similar. Therefore, the Fe_xO_y NPs are strongly fixed in montmorillonite composite in wide range of pH 3-7. The amount of released Fe from Smecta composite increased with increasing pH of solution. The released amount of Fe from magnetically modified Smecta did not exceed 3.5 mg/g and the magnetic properties were successfully verified by magnet. From the statistical point of view, the release of Fe from magnetically modified montmorillonite and magnetically modified Smecta composites was significantly influenced by the substrate.

ACKNOWLEDGEMENTS

The work was supported from ERDF/ESF New Composite Materials for Environmental Applications (No. CZ.02.1.01/0.0/0.0/17_048/0007399).

Authors also wish to thank prof. Ivo. Safarik for providing the magnetically samples and all colleagues, who participated in the measurements.

REFERENCES

[1] BHATTACHARYA, A. K., MANDAL, S. N., DAS, S. K. Adsorption of Zn(II) from aqueous solution by using different adsorbents. *Chemical Engineering Journal*. 2006. vol. 123, pp. 43-51.

[2] KANDAH, Munther Issa. Zinc and cadmium adsorption on low-grade phosphate. *Separation and Purification Technology*. 2004. vol. 35, pp. 61-70.

[3] WONGROD, Suchanya, SIMON, Stéphane, GUIBAUD, Gilles, LENS, Piet N. L., PECHAUD, Yoan, HUGUENOTA, David, D. VAN HULLEBUSCH, Eric. Lead sorption by biochar produced from digestates: Consequences of chemical modification and washing. Journal of Environmental Management. 2018. vol. 219, pp. 277-284.

[4] MÓDENES, Aparecido N., ESPINOZA-QUIÑONES, Fernando R., COLOMBO, Andréia, GERALDI, Claudinéia L., TRIGUEROS, Daniela E. G. Inhibitory effect on the uptake and diffusion of Cd^{2+} onto soybean hull sorbent in Cd-Pb binary sorption systems. *Journal of Environmental Management*. vol. 154, pp. 22-32.

[5] SAWASDEE, Suchada, JANKERD, Hathairat, WACHARABUNDIT, Prachart. Adsorption of dyestuff in household-scale dyeing onto rice husk. *Energy Procedia*. vol. 138, pp. 1159-1164.

[6] HAN, Runping, HAN, Pan, CAI, Zhaohui, ZHAO, Zhenhui, TANG, Mingsheng. Kinetics and isotherms of Neutral Red adsorption on peanut husk. *Journal of Environmental Sciences*. 2008. vol. 20, pp. 1035-1041.

[7] HAMIDPOUR, Mohsen, KALBASI, Mahmoud, AFYUNI, Majid, SHARIATMADARI, Hossein, HOLM, Peter E., HANSEN, Hans *Christian Brunn. Sorption hysteresis of Cd(II) and Pb(II) on natural zeolite and bentonite. Journal* of Hazardous Materials. 2010. vol. 181, pp. 686-691.

[8] ABDELLAOUMI, Youness, OLGUÍN, María Teresa, ABATAL, Mohamed, ALI, Bassam, MÉNDEZ, Sosimo Emmanuel Díaz, SANTIAGO, Arlette A. Comparison of the divalent heavy metals (Pb, Cu and Cd) adsorption behavior by montmorillonite-KSF and their calcium- and sodium-forms. *Superlattices and Microstructures*. 2017. In press.

[9] JIANG, Liying, YE, Qichao, CHEN, Jianmeng, CHEN, Zhong and GU, Youli. Preparation of magnetically recoverable bentonite-Fe_3O_4-MnO_2 composite particles for Cd(II) removal from aqueous solutions. *Journal of Colloid and Interface Science.* 2018. vol. 513, pp. 748-759.

[10] SIMUZ, Tom, FRESNO, Teresa, AKGUL, Gökçen, FROST, Harrison, MORENO-JIMÉNEZ, Eduardo. Biochar modification to enhance sorption of inorganics from water. *Bioresource Technology.* 2017. vol. 246, pp. 34-47.

[11] ROZUMOVÁ, Lucia, ŽIVOTSKÝ, Ondřej, SEIDLEROVÁ, Jana, MOTYKA, Oldřich, ŠAFAŘÍK, Ivo, ŠAFAŘÍKOVÁ, Mirka. Magnetically modified peanut husks as an effective sorbent of heavy metals. *Journal of Environmental Chemical Engineering.* 2016. vol. 4, pp. 549-555.

[12] NADERI, Arman, DELAVAR, Mohmmad Amir, GHORBANI, Yousef, KABOUDIN, Babak, HOSSEINI, Mehdi. Modification of nano-clays with ionic liquids for the removal of Cd (II) ion from aqueous phase. Applied Clay Science. 2018. vol. 158, pp. 236-245.

[13] BOURLINOS, A.B., KARAKASSIDES, M.A., SIMOPOULOS, A., PETRIDIS, D. Synthesis and characterization of magnetically modified clay composites. *Chemistry of Materials.* 2000. vol. 12, pp. 2640-2645

[14] GAO, Z., LI, X., WU, H., ZHAO, S., DELIGEER, W., ASUHA, S. Magnetic modification of acid-activated kaolin: Synthesis, characterization, and adsorptive properties. *Microporous and Mesoporous Materials.* 2015, pp. 1-7.

[15] SAFARIK, Ivo, HORSKA, Katerina, POSPISKOVA, Kristyna, SAFARIKOVA, Mirka. One-step magnetic modification of non-magnetic solid materials. *Powder Technology.* 2012, pp. 285-289.

[16] CORRAMI, Alessia, MIGNARDI, Silvano, FERRINI, Vincenzo. Corami, A., Mignardi, S., Ferrini, V., 2008. Cadmium removal from single- and multi-metal (Cd + Pb + Zn + Cu) solutions by sorption on hydroxyapatite. *Journal of Colloid and Interface Science.* 2008, pp. 402-408.

MICROSCOPIC PERFORMANCE OF CEMENTITIOUS COMPOSITE WITH RECYCLED CONCRETE ADMIXTURE

[1]Vladimír HRBEK, [2]Zdeněk PROŠEK, [2]Václav NEŽERKA, [2]Pavel TESÁREK

[1]Czech Technical University in Prague, Faculty of Civil Engineering, Institute of Theoretical and Applied Mechanics, CAS, Centre of Excellence Telč, Czech Republic, EU vladimir.hrbek@fsv.cvut.cz
[2]Czech Technical University in Prague, Faculty of Civil Engineering, Prague, Czech Republic, EU, zdenek.prosek@fsv.cvut.cz, vaclav.nezrka@fsv.cvut.cz, pavel.tesarek@fsv.cvut.cz

Abstract

The topic of emission reduction and recycling in civil engineering is highly discussed topic over last years. With respect to construction waste production, it is obvious to reuse materials such as concrete. Coarse aggregate recycling and reinforcement rebars separation are the most used procedure in this field so far. Recently, methods including recycling of concrete cementitious matrix were introduced to the market. As such, the concrete matrix consists of matured hydration products of the cement (calcium hydroxide, calcium silica hydrate, etc.) as well as non-hydrated Portland cement particles (clinker), which can represent up to 5 % of the matrix. The focus of this article is placed on microscopic level of cementitious composite with admixture of finely grounded recycled concrete. The impact of recycled addition on micro-structure of the composite, namely the percentage representation of individual material phases, is investigated by image analysis of micro-graphs obtained with scanning electron microscopy. The micro-mechanical performance of each composite is examined with displacement driven grid indentation on several levels of the sample. In consideration of the complexity of the cementitious composite with addition of finely grounded concrete, additional testing of input materials and referential cement paste sample is indispensable to properly distinguish mechanical properties of each composite phase. Young's modulus of individual phases is thus determined from spectral deconvolution of obtained indentation data with respect to the composite structure.

Keywords: Recycled concrete, phase, nano-indentation, SEM, spectral deconvolution, image analysis

1. INTRODUCTION

The construction and demolition waste (CDW), as the largest portion of stored waste material, became worldwide problem in both environmental and economic way. For example, according to data collected by Czech Statistical Office over the year 2016, 14.3 million tons of construction and demolition waste was produced in Czech Republic. Even though the long-term monitoring indicates decrease of CDW production, the amount of land-filled construction and demolition waste increases approx. 4.5 % annually. This projected trend underlines the potential development of much efficient and thrifty CDW re-cycling methods, that could reduce waste deposition while saving natural sources and environment. One of economic problems of CDW recycled materials (compare to traditional raw materials), can be paradoxically higher price due to the transportation costs [1, 2]. In general, concrete is 30 to 60 % made of coarse aggregate which is typically reused. In other words, 70 to 40 % of the concrete is considered as secondary waste product. Absolute recycling of ordinary concrete, when high-speed milling is used, transforms the waste material into very fine powder. Such recycling processing opens material ability for potential use as micro-filler (for non-reactive hydration products and aggregate) and partial cement substitute in case of non-hydrated clinker in cement paste [3].

The potential of recycled finely grinded concrete as substitute for pulverized limestone in asphalt mixture was shown in M. Chen study [4]. Compare to pulverized limestone (based on XRD SEM analysis), recycled concrete powder particles have rougher surface and show higher content of silicon dioxide (SiO_2). Mechanical properties of this component depend on the environment thermal conditions, as shown in this study. The results indicate increased strength and fatigue persistence in higher temperatures. Features of cement paste with partial replacement of a binder by fine grinded recycled concrete are discussed in [5]. As corroborated by the results, the replacement impacts the mixture in fresh state due to decreased workability and prolonged hardening time (delay up to 2 hours). The study recommendation inclines to 15 % limitation of the replacement, as mechanical results of samples with higher fine grinded recycled concrete content (45 % cement replacement) show reduction of compressive strength up to 30 %. The potential pozzolanic activity of pulverized recycled concrete from railway sleepers due to presence of non-hydrated clinkers, likely unfolded by fine milling [6,7]. The recycled material,as the level of non-hydrated clinker in raw concrete is low, is thus feasible for subsoil under railway structures strengthening [8]. Other potential small scale utilization can be found in cementitious composite material.

This study is focused on micro-mechanical performance of the recycled concrete drain gutter, in form of finely grinded powder (0 - 180 μm particles), used as partial replacement in cementitious composite. To ensure adequate interpretation of measured data, both referential cement paste and original concrete drain gutter were also subjected to micro-mechanical and micro-structural testing.

2. METHODOLOGY AND DATA ANALYSIS

For purpose of this study, referential cement paste and cementitious composite with grinded recycled drain gutter inclusion were prepared and left to mature for 28 days. The referential specimen (REF) was made of CEM II 42.5 R cement with water to cement ratio of 0.35. In case of second specimen (REC30), 30 % of cement weight was replaced with finely grinded recycled drain gutter. In addition, third sample of original concrete drain gutter (10 years old, further denoted as ORG) was added for proper identification of the recycled powder inclusion effect on the paste. All three samples were embedded in epoxy-resin and their surface grinded and polished, avoiding further hydration and selective abrasion, for purposes of nano-indentation and scanning electron microscopy.

2.1. Indentation and spectral deconvolution

The micro-scale elastic properties of individual phases present in each material were evaluated based on nano-indentation results (Ti 750 series, Hysitron Inc.). In general, indentation technique reflects th dependency of material resistance to the probe propagation into the indented half-space. Indentation modulus (and/or other elastic mechanical properties) of indented region are determined from unloading part of the record. Incorporated errors of the measurement (creep, visco-elasticity etc.) can be avoided by appropriate setup of measurement [9 - 13]. In all cases of testing, the displacement driven load function was selected, where both loading and unloading lasted 5 seconds each with an in-between 60 seconds segment with constantly held maximum displacement of 150 nm. The measurement was performed on different levels of the sample, based on specimen heterogeneity and expected number of present phases. Histograms of measured data were further processed with spectral deconvolution, where micro-mechanical features of material individual phases are determined with respect to possible interference of main, soft, phase with other, tougher, phases. The maximum allowed error between indentation and processed data and structural representation did not exceed 10 %.

The referential specimen (cement paste) indentation was performed in three separate locations. In each position, rectangular grid constructed of 21 by 21 indents with mutual separation of 10 μm was implemented. In case of original concrete drain gutter, two material levels were selected for indentation - area of cementitious matrix (without aggregate present) and inter-facial transition zone (ITZ) between matrix and aggregate (presence of clinker not expected). On both material levels, two indentation grids (similar to grids imposed on referential specimen) were implemented. The composite with 30 % recycled pulverized gutter replacement was also tested via indentation on two material levels. 12 locations indicating presence of inclusion were subjected to grid indentation similar to previous samples (REF and ORG). 4 random positions, capturing mostly cement matrix, were indented in 51 by 51 indents pattern, mutually separated by 15 μm. The exponentially increased number of measurement data inevitable for proper mechanical features identification of most individual phases present in the composite.

2.2. SEM and image analysis

The scanning electron microscopy (SEM) investigation was performed in MIRA II LMU (Tescan corp., Brno) on polished specimens coated with thin layer of carbon necessary to ensure proper conductivity of the surface. Chemical composition (in weight percentage) of each phase was determined using EPMA based on the detection of the X-rays by energy dispersive X-ray detector (EDX, Bruker corp., Berlin) on several point of each position. Due to the time-demands and statistical insufficiency of EPMA phase determination, image analysis of SEM BSE micrographs was used to obtain phase volumetric representations in material. In general, energy of back-scattered electron is dependent on atomic number of the element which is relevant to brightness of the phase in BSE diagram. Based on estimated number of phases and their gray-scale range from EDX measurements of each sample, it is possible numerically quantify percentage representation of each phase from several BSE SEM graphs by image analysis.

The EDX SEM investigation of cement paste (REF sample) was applied in 20 points of interest from which basic phases of the matrix were identify, including their brightness range in gray-scale. Furthermore, 20 SEM BSE diagrams were analyzed providing average description specimen structure. In case of original concrete drain gutter sample, image analyses of two magnification levels was performed in order to adequately and effectively determine all present phases. Close magnification BSE diagrams image analyses (identical to REF specimen) determined average representation of cement matrix phases. Image analysis of SEM BSE diagrams (12 images) capturing outlying structure of the specimen served for aggregate - matrix ratio evaluation. Unfortunately, identical procedure of micro-structure investigation of composite specimen (REC30) gave not conclusive information of phases representation due to very close ranges of brightness in gray-scale of chemically equal phases.

3. RESULTS AND DISCUSSION

The results of spectral deconvolution and histograms, from which individual phases features were derived, are summarized in **Table 1** and **Figure 1**. Direct indentation modulus comparison between can be made due to incorporation of mutual influence of phases with different stiffness. For example, integration of recycled LD CSH into the matrix of new composite seams to increase its modulus by approx. 12 %, compare to features of same phase present in original concrete gutter. On the other hand, the incorporation and interaction of recycled HD CSH phase with matrix with significantly lower stiffness results in decrease of its indentation modulus by approx. 6 %. In case of portlandite evaluation in REC30 sample, newly formed and recycled inclusion of this phase indentation modulus are too close and could not be separated. Presented results can be described as average (homogenized) value of both. Non-hydrated clinker and aggregate inclusion, given their stiffness and possible measurement error, remain intact, i.e. with same value of indentation modulus.

Table 1 Results of individual phases micro-mechanical features (indentation modulus E_r [GPa])

Sample	ITZ	LD CSH	R-LD CSH	HD CSH	R-HD CSH	Portlandite	Clinker	Aggreg.
REF	nonex.	17.1 ± 5.6		40.1 ± 6.1		77.7 ± 18.6	125.7 ± 6.6	nonex.
ORG	16.7 ± 3.6		22.0 ± 2.1		46.2 ± 4.6	72.6 ± 9.3	127.5 ± 3.8	106.1 ± 8.5
REC30	not det.	16.2 ± 3.4	24.7 ± 6.2	39.7 ± 2.4	43.7 ± 4.0	75.5 ± 12.2	122.6 ± 7.2	103.7 ± 7.7

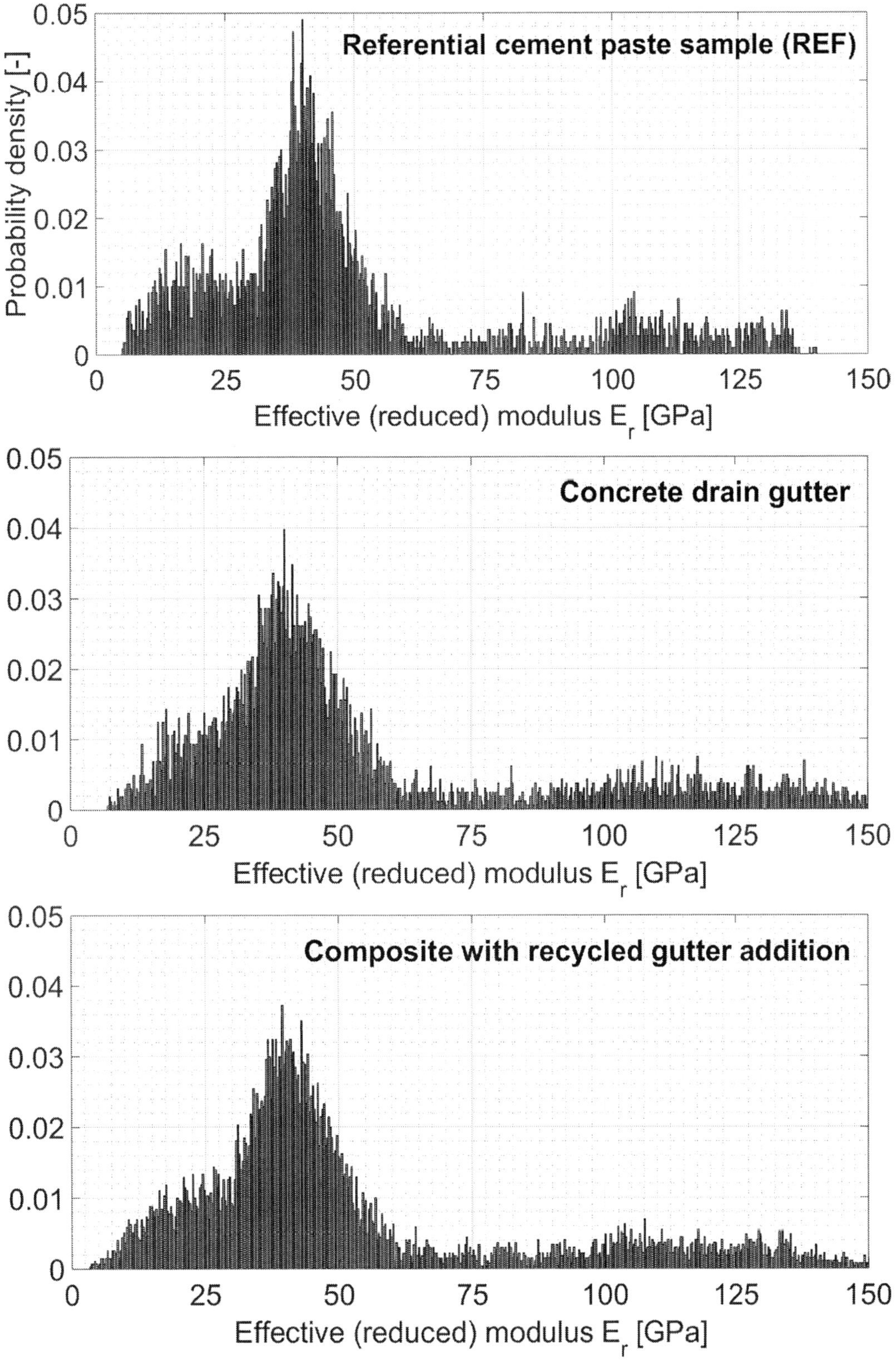

Figure 1 Histograms of indentation results

Micro-structural results of referential and original drain gutter samples are shown in **Table 2.** The REC30 sample, due to its complex structure (possible 9 present phases) and almost identical gray-scale brightness range of several phases, could not be properly evaluated with respect to number of measurements (data, diagrams). Even so, results of phase percentage representation were used for in spectral deconvolution procedure (with respect to used pulverized recycled gutter amount and its individual phases representation) as an error indicator.

Table 2 Results of average individual phases percentage representation [%]

	Pores	LD CSH	HD CSH	Portlandite	Clinker	Aggregate
Cement paste	10.39 ± 0.82	44.33 ± 3.73	18.20 ± 1.46	18.50 ± 1.73	9.72 ± 0.93	nonexisting
Concrete drain gutter	1.83 ± 0.12	13.92 ± 0.97	20.45 ± 1.22	6.53 ± 0.30	4.91 ± 0.18	52.36 ± 3.54

4. CONCLUSION

The nano-indentation technique in combination with SEM structure observation was use in the study of cementitious composite with 30 % cement replacement with pulverized recycled concrete gutter. Based on the results of post-processed measured data, following can be stated:

- The addition of recycled concrete, in newly produced cementitious composite, does not impact the micro-mechanical features of newly formed matrix phases.
- Incorporated particles of recycled pulverized concrete drain gutter function in the composite as micro- and nano- filler, i.e. impact mainly mechanical features of the matrix on the meso- and macro-scale.
- The combination of indentation technique and SEM proved essential in complex composite investigation.
- Detailed investigation of composite's inclusions was essential for proper identification of micro-mechanical features of individual phases.
- The number of indentation data necessary for investigation exponentially increases with number of expected phases in the material, leading to very time-demanding testing and measurement evaluation.
- The micro-structural investigation of cementitious composite with 30 % cement replacement with pulverized recycled concrete gutter via SEM and image analysis was not conclusive. Closer investigation and higher number of diagrams is recommended for in order to describe composite's structure.
- The presumption of reactive clinker involvement in hydration process of cement can be also corroborated with further comparison of composite structure and clinker phase representation of referential and original material.

ACKNOWLEDGEMENTS

This was financially supported by the Czech Science Foundation research projects 17-06771S and by the Czech Technical University in Prague project SGS17/168/OHK1/3T/11.

REFERENCES

[1] MARINKOVIC, S., et al., Comparative environmental assessment of natural and recycled aggregate concrete, *Waste Management*, 2010, vol. 30, pp. 2255 - 2264.

[2] POON, C. S., et al., Effect of microstructure of ITZ on compressive strength of concrete prepared with recycled aggregates, *Construction and Building Materials*, 2004, vol: 18, pp. 461 - 468.

[3] TOPIČ, J., PROŠEK, Z., Properties and microstructure of cement paste including recycled concrete powder, *Acta Polytechnica*, 2017, vol. 57, pp. 49 - 57.

[4] CHEN, M., et al., Potential of recycled fine aggregates powder as filler in asphalt mixture, *Construction and Building Materials*, 2011, vol. 25, pp. 3909 - 3914.

[5] KIM, Y. J., et al., Utilization of waste concrete powder as a substitution material for cement, *Construction and Building Materials*, 2012, vol. 30, pp. 500 - 504.

[6] LIDMILA, M., et al., Utilization of Recycled Fine-ground Concrete from Railway Sleepers for Production of Cement-based Binder, *Applied Mechanics and Materials*, 2014, vol. 486, pp. 323 - 326.

[7] ŠEPS, K., et al., Mechanical Properties of Mixtures with Crushed Sleepers Stiffened by Synthetic Fibers, *In Proceedings of the 4th Conference Nano and Macro Mechanics, 2013, pp.* 193 - 198.

[8] LIDMILA, M., et al., Utilization of Recycled Binder for Improvement of Subsoil under Railway Sleepers, *Applied Mechanics and Materials*, 2016, vol. 825, pp. 53 - 56.

[9] OLIVER, Warren C., PHARR, George M. An improved technique for determining hardness and elastic modulus using load and displacement sensing indentation experiments. *Journal of materials research*. 1992. vol. 7, no. 6, pp. 1564-1583.

[10] CONSTANTINIDES, Georgios, et al. Grid indentation analysis of composite microstructure and mechanics: Principles and validation. *Materials Science and Engineering: A*. 2006. vol. 430, no.1-2, pp. 189-202.

[11] HARDING, J. W., SNEDDON, I. N., The Elastic Stresses Produced by The Indentation of the Plane Surface of a Semi-Infinite Elastic Solid by a Rigid Punch, Mathematical Proceedings of the Cambridge Philosophical Society, 1945, vol. 41, no. 1, pp. 16 - 2.

[12] PHARR, G. M., BOLSHAKOV, A. Understanding nanoindentation unloading curves. *Journal of Materials Research*. 2002. vol. 17, no. 10, pp. 2660-2671.

[13] OLIVER, Warren C., PHARR, Georges M. Measurement of hardness and elastic modulus by instrumented indentation: Advances in understanding and refinements to methodology. Journal of materials research. 2004. vol. 19, no.1, pp. 3-20.

ISOLATION OF ARGININE USING MAGHEMITE PARTICLES

Natalia CERNEI[1,2], Pavel KOPEL[1,2], Pavel HORKY[3], Ondrej ZITKA[1,2], Vojtech ADAM[1,2]

[1]*Department of Chemistry and Biochemistry, Faculty of Agronomy, Mendel University in Brno, Brno, Czech Republic, EU, vojtech.adam@mendelu.cz*
[2]*Central European Institute of Technology, Brno University of Technology, Brno, Czech Republic, EU*
[3]*Department of Animal Nutrition and Forage Production, Faculty of Agronomy, Mendel University in Brno, Brno, Czech Republic, EU*

Abstract

In this paper, we report the optimal conditions for the isolation of amino acid arginine from different types of matrix, such as blood, plasma or tissue and their characterization using ion-exchange liquid chromatography with VIS detector. Paramagnetic microparticles able to isolate and immobilize amino acid arginine were prepared and used to preconcentrate arginine for subsequent analysis on ion-exchange liquid chromatography with VIS detector. Commercially available Dowex was covered by maghemite nanoparticles and the composite showed excellent binding and superparamagnetic properties. It was found that the paramagnetic microparticles were able to bind specifically amino acid arginine. Therefore, these paramagnetic microparticles have potential for better isolation of arginine and in future can use for application as a platform of delivery system.

Keywords: Ion exchange chromatography, maghemite microparticles, amino acids, arginine

1. INTRODUCTION

Arginine is semiessential amino acid and plays a various roles in biosynthetic pathways in tumor biology [1]. Major cellular functions of arginine were nitric oxide production [2], creatine production and polyamine synthesis [3]. Arginine influences in tumor cells has been shown to inhibit metastatic tumor growth [4]. Long-term and significant of arginine deprivation induces free radical formation and peroxynitrite stress especially in connection to arginine-dependent cancer cells [5]. Based on intensive study, it was shown the effect of arginine is to reduce nitric oxide (NO) production and the significant increase of oxidative stress [6]. On the other hand, L-arginine regulates blood flow, is the main cause of blood pressure and increased of the oxidative stress conditions of patients suffering from cardiomyopathy [7]. L-arginine, shows various functions in human health and may play a major role in age-related degenerative diseases such as Alzheimer's disease (AD) [8]. Nowadays, multiple methods are developed for semiessential amino acid arginine determination. Most of them are based on chromatography (GC, LC, UHPLC) [9] with tandem of mass spectrometry (MS) [10], IEC [11] or CE [12]. In our study we decided to use ion exchange chromatography with post-column ninhydrin derivatization and VIS detector for arginine determination [13].

2. METHODOLOGICAL BASES AND EXPERIMENTAL PART

2.1. Chemicals

Arginine of purity 99 % was obtained from Sigma Aldrich (St. Louis, Missouri, USA). Solution of arginine for preparing of calibration curve was prepared in the dilution buffer Na: TDG (NaN_3 - 0.10 g, NaCl -11.5g, $C_6H_8O_7$-14 per 1L H_2O). For experiment were used Citric acid, NaCl, NaN_3, TDG, HCl 35 % from Sigma - Aldrich, Ninhydrine from Sigma - Aldrich, Methylcellosolve (Ingos, Prague, Czech Republic), $SnCl_2$ (Ingos).

2.2. Ionex chromatography

An AAA 400 (Ingos, Czech Republic) liquid chromatography apparatus was used for determination of amino acids. The system consists of a glassy filling chromatographic column and steel precolumn, two chromatographic pumps for transport of elution buffers and derivatization reagent, cooled carousel for 25 test tubes of 1.5-2.0 mL volume, dosing valve, heat reactor, VIS detector and cooled chamber for derivatization reagent. Chromatographic columns for transfer of elution buffers and derivatization reagent are able to work at flow 0.01-10 mL•min^{-1} under a maximum pressure of 40 MPa. Volume of injected sample was 100 µL with an accuracy of application RSD of about 1 %. A two-channel VIS detector with a 5 µL flow volume cuvette was operated at wavelengths of 440 and 570 nm.

2.3. SECM characterization of DOWEX modified maghemite particles

Structure and elemental composition of paramagnetic microparticles were characterised by electron microscope. For documentation of the selected nanomaterials a FEG-SEM MIRA XMU (Tescan, a.s., Brno, Czech Republic) was used. This model is equipped with a high brightness Schottky field emitter for low noise imaging at fast scanning rates. The SEM was fitted with Everhart-Thronley type of SE detector, high speed YAG scintillator based BSE detector, panchromatic CL Detector and EDX spectrometer. The MIRA 3 XMU system is based on a large specimen chamber with motorized stage movements 130×130 mm. Samples were coated by 10 nm of carbon to prevent sample charging. A carbon coater K950X (Quorum Technologies, Grinstead, United Kingdom) was used. For automated acquisition of selected areas a TESCAN proprietary software tool called Image Snapper was used. The Accepted Article software enables automatic acquisition of selected areas with defined resolution. Different conditions were optimized in order to reach either minimum analysis time or maximum detail during overnight automated analysis. An accelerating voltage of 15 kV and beam currents about 1 nA gives satisfactory results regarding maximum throughput.

2.4. X-ray fluorescence analysis of PMPs

XRF elemental analysis of PMPs was carried out on Xepos (SPECTRO analytical instruments GmbH, Kleve, Germany) fitted with three detectors: Barkla scatter - aluminium oxide, Barkla scatter - HOPG and Compton/secondary molybdenum respectively. Analyses were conducted in Turbo Quant cuvette method of measuring. Analysis parameters were set to - measurement duration: 300 seconds, tube voltage from 24.81 to 47.72 kV, tube current from 0.55 to 1.0 mA, with zero peak at 5000 cps and vacuum switched off.

2.5. Preparation of microparticles

Maghemite nanoparticles were prepared by sodium borohydride ($NaBH_4$) reduction of iron chloride ($FeCl_3.6H_2O$). The product was separated by a magnet, washed with water and dried at 40 °C. Product was heated at 400 °C for approximately 1 h in the muffle furnace (Verkon, Prague, Czech Republic). Product has a red color and was dispersed in water. To this solution was added Dowex with stirring on Biosan OS-10 overnight. Then the product was separated by a magnet, washed with water and dried at 40 °C.

3. RESULTS AND DISCUSSION

Aim of our study was an mainly isolation and 2D separation of the semiessential amino acid arginine, able to serve as a possible marker of metastatic tumor growth inhibition, based on the adsorption on the paramagnetic microparticles on commercially available Dowex and subsequent determination using ion-exchange liquid chromatography (IEC). We modified Dowex with nanomaghemite and NH_2 functional groups, and these PMPs showed excellent properties for binding of arginine (recovery 42.48 %) (**Figure 1.A**). Britton-Robinson buffer with pH 2 causes semiessential amino acid arginine protonation which leads to a positive charging of molecules due to its pI = 10.76. Interaction between surface of our magnetic microparticles and positively charged molecule provides the binding between them. These interactions depend on isoelectric point of amino acid, which are in this mode behaving as the ion-exchangers. Other amino acids and tripeptides such as GSH and GSSG, showed also relatively good recoveries (serine recovery 1.1 %, leucine 10 %, sarcosine 1.3 %, GSSG 0.5 %,GSH 0.45 %, glutamic acid 6.7 %, glycine 0.8 %, valine 4.35 % and arginine 42.48 %, respectively); however, their quantity in real sample of blood are not so high to interfere during arginine binding to PMPs (**Figure 1D**). Furthermore, our second separation step - analysis using Ion Exchange Chromatography (IEC) - eliminates the influence of interferents to a minimum.

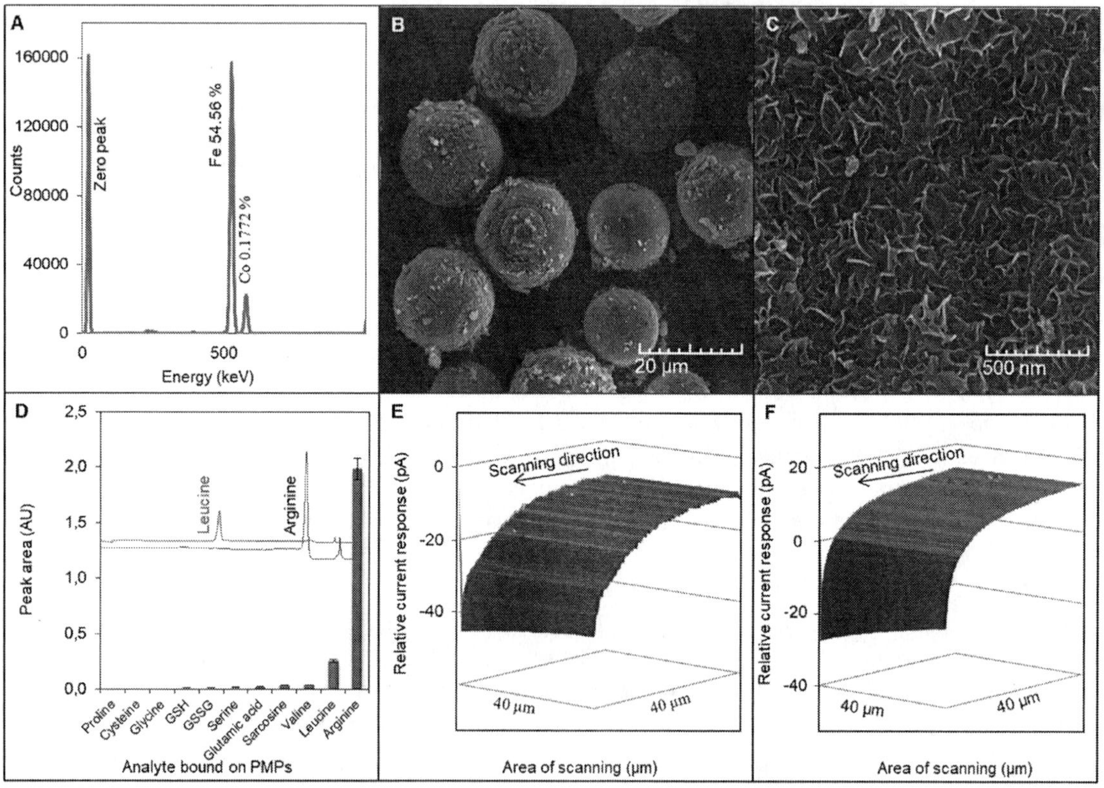

Figure 1 Characterization of PMPs. (**A**) XRF showing elemental composition of paramagnetic microparticles. (**B**) SEM image in resolution of 20 µm. (**C**) SEM scan in resolution of 500 nm. (**D**) IEC results showing ability of PMPs bound required substances specifically. (**E**) SECM scan showing relative current response of PMPs without arginine bound. (**F**) SECM scan of PMPs with arginine bound showing increased relative current response.

After confirmation of capability to bind amino acid arginine, we carried out a various characterization of PMPs. Primarily characterization was XRF analysis providing information about elemental composition of microparticles. In the **Figure 1A** can be seen that iron was determined to be the most abundant element (Fe represented in 54.56 %) and cobalt (Co represented 0.1772 %). This information was not surprising, because we carried out SECM analysis for recognition of PMPs surface relative current response changes in dependence on arginine binding. In Fig 1B can be seen a micrograph expressing microparticles surface and morphology in resolution of 20 µm for PMPs and in **Figure 1C** is SEM micrograph of PMP surface in resolution of 500 nm. In **Figure 1E, F** can be seen 3D images, expressing a relative current response of PMP surface without arginine and with arginine bound.

4. CONCLUSION

In this study, we synthesized paramagnetic microparticles able to bind semiessential amino acid arginine - as a possible marker of metastatic tumor growth inhibition. The paramagnetic microparticles have potential to better isolation of arginine from the samples of plasma, cells or tissue and in future can serve for application as a platform for delivery system.

ACKNOWLEDGEMENTS

The financial support from NAZV QK1720349 is highly acknowledged.

REFERENCES

[1] ASSELIN, B.L. The three asparaginases - comparative pharmacology and optimal use in childhood leukemia. In Drug resistance in leukemia and lymphoma iii, Kaspers, G.J.L.; Pieters, R.; Veerman, A.J.P., Eds. Kluwer Academic / Plenum Publ: New York, 1999; Vol. 457, pp 621-629.

[2] WU, G.Y.; MORRIS, S.M. Arginine metabolism: Nitric oxide and beyond. Biochem. J. 1998, 336, 1-17.

[3] MORRIS, S.M. Arginine metabolism revisited. J. Nutr. 2016, 146, 2579-2586.

[4] LIND, D.S. Arginine and cancer. J Nutr 2004, 134.

[5] XIA, Y.; ZWEIER, J.L. Superoxide and peroxynitrite generation from inducible nitric oxide synthase in macrophages. Proc Natl Acad Sci U S A 1997, 94, 6954-6958.

[6] DASGUPTA, T.; HEBBEL, R.P.; KAUL, D.K. Protective effect of arginine on oxidative stress in transgenic sickle mouse models. Free Radical Biology and Medicine 2006, 41, 1771-1780.

[7] TRIPATHI, P.; PANDEY, S. L-arginine attenuates oxidative stress condition during cardiomyopathy. Indian J Biochem Biophys 2013, 50, 99-104.

[8] YI, J.; HORKY, L.L.; FRIEDLICH, A.L.; SHI, Y.; ROGERS, J.T.; HUANG, X. L-arginine and alzheimer's disease. International Journal of Clinical and Experimental Pathology 2009, 2, 211-238.

[9] AL-TANNAK, N.F. Uhplc-uv method for simultaneous determination of perindopril arginine and indapamide hemihydrate in combined dosage form: A stability-indicating assay method. Sci. Pharm. 2018, 86.

[10] MESONZHNIK, N.V.; POSTNIKOV, P.V.; APPOLONOVA, S.A.; KROTOV, G.I. Characterization and detection of erythropoietin fc fusion proteins using liquid chromatography-mass spectrometry. J. Proteome Res. 2018, 17, 689-697.

[11] DUSKOVA, K.; VESELY, S.; SILVA, J.D.; CERNEI, N.; ZITKA, O.; HEGER, Z.; ADAM, V.; HAVLOVA, K.; BABJUK, M. Differences in urinary amino acid patterns in individuals with different types of urological tumor urinary amino acid patterns as markers of urological tumors. In Vivo 2018, 32, 425-429.

[12] DE CARVALHO, R.C.; MATHIAS, T.R.D.; NETTO, A.D.P.; MARQUES, F.F.D. Direct determination of amino acids in brewery worts produced by different processes by capillary zone electrophoresis. Electrophoresis 2018, 39, 1613-1620.

[13] CERNEI, N.; ZITKA, O.; RYVOLOVA, M.; ADAM, V.; MASARIK, M.; HUBALEK, J.; KIZEK, R. Spectrometric and electrochemical analysis of sarcosine as a potential prostate carcinoma marker. Int. J. Electrochem. Sci. 2012, 7, 4286-4301.

STUDY OF ANTIMONY REMOVAL EFFICIENCY USING FERRATE (VI)

Robert PRUCEK, Martin SOLÁR, Aleš PANÁČEK, Jiří TUČEK, Jan FILIP, Radek ZBOŘIL

Palacký University in Olomouc, RCPTM, Olomouc, Czech Republic, EU, robert.prucek@upol.cz

Abstract

Heavy metals are widely used in a range of materials including paints, motor vehicles, guns, and electrical components. After discarding such materials, the metal ions after dissolution can penetrate to the environment and are toxic to plants, organisms, and humans. Among such dangerous heavy metals belongs antimony. The aim of presented work was to study removal efficiency of antimony from waters using ferrate (VI). The experiments were performed for antimony in oxidation state III and V and the influence of ferrate(VI) to antimony ratio on antimony removal efficiency was studied. The experiments were performed under neutral pH value of the reaction mixture. The antimony concentration was determined using AAS method and residual concentration of antimony in reaction mixture was traced within two weeks. With increasing ferrate (VI) to antimony ratio from 10:1 to 60:1, the removal antimony efficiency slightly increased. However, even for ferrate(VI)to antimony equal to 10:1, the removal efficiency was after 30 minutes and after two weeks was even 80 %. This means that the effectiveness of removal increase with prolonged contact time of components and no leaching of antimony back to aqueous solution proceed. The impact of concentration of some inorganic ions (nitrates, phosphates, sulfates, carbonates, and chlorides) usually present in real waters on antimony removal efficiency was also studied.

Keywords: Antimony, removal, ferrate(VI), ions influence

1. INTRODUCTION

Heavy metals are highly resistant to environmental degradation and tend to exhibit accumulative tendencies. In addition, metals are mobile, depending on the surrounding water chemistry. US Environmental Protection Agency (EPA) regulates heavy metals and, therefore, many methods have been studied for their removal. Among the toxic elements belongs the also antimony. Antimony is exploited in semiconductors, infrared detectors and diodes [1]. Among the most famous antimony alloys are the so-called "hard lead", found especially in ammunition. The lead itself without addition of antimony is too soft and unsuitable for this purpose and the antimony itself is too fragile [2]. The antimony is also involved in lead-acid-based automotive batteries electrodes or in corrosion-resistant pipes and alloys used to bearings metals which is also closely related to automotive industry [2], [3]. For another example, Sb_2S_3 is utilized in photoelectric devices or electrophotographic recording media and Sb_2O_3 is exploited in a self-extinguishing additive or as a moisturizing agent in brake pads [3]. Antimony oxides are used to fabrication of dyes, adhesives, plastics, rubber, ceramics, glass enamels, and fireworks [4] or fire protection of materials such as plastics, coatings, electronics or other polybrominated flame retardants [3]. Sb_2O_3 is also used as a catalyst in the production of polyethylene terephthalate (PET) [5]. Sodium antimony is used as a bleaching component, for example, for optical glasses in cameras or telescopes [3]. From all of the above mentioned materials or products, antimony can enter the aqueous and subsequently soil environment. Antimony can be found in four oxidation states: -III, 0, + III and + V. The predominant oxidation states in the environment are + III and + V and these oxidation states can be found in natural waters. Antimony is practically almost completely present as Sb^{+3} ions only in strongly acidic solutions or under anaerobic conditions [6]. In recent years, however, interest in antimony has increased considerably and has been included in the list of priority environmental pollutants. The Council of the European Union has set this limit value under Directive 98/83/EC to a maximum of 5 $\mu g \cdot dm^{-3}$. Although in the natural waters the concentration of antimony ranged usually from 0.01 to 5 μg·dm^{-3}, its concentration may be significantly higher. The toxic properties of antimony are dependent to its form and the oxidation state. The antimony in oxidation state +III is about 10 times more toxic than the oxidation state +V, also the antimony in the oxidation state +III is more mobile in the environment [7]. Inorganic forms of antimony is more toxic than its compounds, organic forms, such as methylated Sb. Trivalent antimony has a high affinity for red blood cells and for thiol groups, but antimony in oxidation state +V is almost impermeable through red blood cells. Heavy metal cations generally tend to bind to the carboxyl or to the amino group, which are characteristic for genetic information related substances. Antimony has also been described to inhibit DNA replication [8]. The mutagenic or carcinogenic properties was reported, in particular antimony oxide Sb_2O_3 was included among potential carcinogens [9]. Long-term inhalation of antimony and its compounds has a dangerous impact on the heart, blood pressure, lung, liver, and kidneys functions [10,11]. Antimony compounds have a very harmful effect on pregnancy and can lead also to loss of fertility [1].

Adsorption, chemical precipitation, ion flotation, membrane filtration, and reverse osmosis have been used to remove metals from water. Among several methods, sorption/adsorption, based on iron containing compounds, has received greater attention because of the environmentally friendly nature and magnetically separable attributes of iron-containing compounds. In recent years, potassium salt of tetraoxy iron(VI) (K_2FeO_4, Fe(VI)) has emerged as an effective agent for the oxidation, disinfection, and coagulation/coprecipitation to remove a wide range of contaminants. Ferrate(VI) represents an environmentally friendly oxidant; its redox potential ranges from 2.20 V in acidic environment to 0.72 V in basic environment [12]. It has been reported to be effective for the treatment of various contaminants, for example, ibuprofen, sulfamethoxazole, phenol and cyanides, and a variety of inorganic compounds. [13,14] The rate of ferrate(VI) reaction with various compounds mostly varies from several milliseconds to several minutes. [15] The additional benefit of using ferrate(VI) lies in the fact that resulting secondary iron oxide nanoparticles can further serve as an efficient adsorbent for oxidized pollutants. [16]

The aim of this work is to study of antimony (III, V) removal using ferrate(VI) treatment, investigate the influence of ferrate(VI) to antimony ratio and impact of the presence of inorganic ions usually present in real water on efficiency of antimony removal.

2. EXPERIMENTAL

Stock solutions of antimony at a concentration of 1000 mg L^{-1} were prepared using $SbCl_3$ or $SbCl_5$ (Sigma-Aldrich). Inorganic Salts (NaCl, $NaNO_3$, $NaHCO_3$, and Na_2SO_4) were purchased from Sigma-Aldrich. Standard solutions of Sb and Fe for atomic absorption spectroscopy (AAS) measurements were prepared using commercial standards TraceCERT (concentration 1000 mg/L of element in nitric acid), which were purchased from Fluka Inc. All chemicals were used as received without any further purification. Potassium ferrate (K_2FeO_4) (>90 %, Sigma-Aldrich) was used throughout the experiments. Solutions of ferrate(VI) were prepared with deionized water (18 MΩ cm^{-1}, Millipore system).

The influence of concentration of ferrate(VI) on the removal of antimony was tested at neutral pH value. In the experiments, the initial concentration of antimony (III, V) was 10 mg L^{-1} (as elemental Sb). The concentrations of ferrate(VI) as Fe were varied from 100 to 600 mg L^{-1}. In all these experiments, solutions of ferrate(VI) were added to the solutions of antimony (IIII, V) water in the Erlenmeyer flasks (the final volume was 100 mL). Mixed solutions were shaken on a conventional end-over-end shaker for 30 min, 1 week, and 2 weeks and thereafter filtered through 0.2 μm syringe filters. The residual concentration of antimony in the filtrate were immediately determined using an AAS-flame ionization technique.

In the investigation of the effect of ions on antimony (III, V) removal efficiency using ferrate(VI), chlorides, nitrates, carbonates, sulfates, and phosphates were employed. After adding ferrate(VI), the pH of the solutions was adjusted and solution was shaken for 2 weeks. Samples for AAS analyses were taken 30 min, 1 week, and 2 weeks after the mixing of the components. The concentrations of ions were in the range from 25 to 200 mg L^{-1} of element in ion (i.e., chlorides ion as Cl, nitrates as N, carbonates as C, sulfate as S, and phosphates as P). The experiments were carried out in triplicates.

Antimony concentrations were determined by the AAS-flame technique using a ContrAA 300 (Analytik Jena AG, Germany) equipped with a high-resolution Echelle double monochromator (spectral band width of 2 pm at 200 nm) and with a continuum radiation source (xenon lamp). The absorption line used for Sb analyses was 206.8300 nm. TEM images of formed nanoparticles were performed by transmission electron microscopy (TEM) on a JEOL JEM-2010 transmission electron microscope equipped with a LaB_6 cathode (accelerating voltage of 160 kV; point-to-point resolution of 0.194 nm). A drop of high-purity distilled water was placed onto a holey carbon film supported by a copper-mesh TEM grid (SPI Supplies, USA) and air-dried at room temperature.

3. RESULTS AND DISCUSSION

The K_2FeO_4 material was tested for the removal of antimony(III) and antimony(V) ions. The ferrate(VI) anions react with water which leads to in-situ formation of ferric oxides. The capacity of antimony species, as well as arsenic species sorption by iron oxides is generally influenced by pH value of the solution. Based on our previous studies with arsenic removal using ferrate(VI), we focused on experiments performed at neutral pH value (approx. 7). These experiments were aimed to the examination of the influence of K_2FeO_4 concentration on the efficiency of antimony removal. The initial concentration of antimony (III, V) was 10 mg L^{-1} (as Sb) and the concentrations of ferrate(VI) (referred as Fe) were varied from 100 to 600 mg L^{-1}, i.e. that ratio of Fe:Sb was ranged from 10:1 to 60:1. The limit of antimony in drinking waters is 5 mg L^{-1}. However, in some case, for example in mining area in China, concentrations up to 30 mg L^{-1} were reported.[17] The kinetics of antimony (III, V) removal was monitored by time-dependent changes of arsenic concentration in the solution. From the **Figure 1** it is clear that with increasing ferrate (VI) to antimony(III) ratio from 10:1 to 60:1, the removal antimony efficiency slightly increased. For ferrate(VI) to antimony(III) equal to 10:1, the removal efficiency of antimony(III) was approximately 65 % after 30 minutes and for higher ferrate (VI) to antimony ratio, the removal effectiveness reached up to almost 90 %. In the case of Sb(V), the efficiency was very similar just in the case of Sb(III) **(Figure 2)**. The more important is the fact, that after 1 week and 2 weeks after mixing of ferrate(VI) and the antimony(III, V) the antimony concentration gradually decreased and no leaching of antimony back to the aqueous environment was observed (**Figures 1 and 2**). So, the using the ferrate(VI) can be an effective tool for antimony treatment in waters. The TEM image of iron nanoparticles formed after the reaction of ferrate(VI) in the presence of antimony(V). Concentration of ferrate(VI) was 200 mg·dm^{-3} (as Fe) and the concentration of antimony was 10 mg·dm^{-3} (as Sb) is shown at **Figure 3** . Just in the case arsenic (III), also in the case of antimony(III) the fast oxidation to antimony(V) and possible incorporation of antimony to the structure of the formed iron oxide can be assumed.[18,19] This hypothesis needs to be confirmed by in-field ^{57}Fe Mössbauer spectroscopy and XPS measurement. The ionic radius of antimony is 0.076 nm, so this ion can be embedded into the structure of formed iron oxide just in the case of arsenic because the limit of ionic radii for possible incorporation is 0.092 nm.[18] In the subsequent experiment, the influence of selected ions (chlorides, nitrates, sulfates, carbonates, phosphates) concentrations were examined. In this type of experiments, the ratio of ferrate(VI) and antimony(III, V) was set up to 20:1. The concentrations of the above mentioned ions were ranged between 25 to 200 mg L^{-1} of element in ions In case of the presence of sulfates, nitrates, and chlorides, the efficiency of antimony (III, V) remained unchanged. Residual concentrations of antimony was very similar, just in the case when only antimony (III, V) was present in the solution together ferrate(VI). Also of antimony concentrations decreasing with time show the same trend. The situations were very different in the case of presence of phosphates and carbonates. The presence of phosphates even in the lowest tested concentration (25 mgL) eliminates completely removal of antimony from aqueous environment. The carbonates influence the removal of antimony at the higher concentrations (100 and 200 mg L).

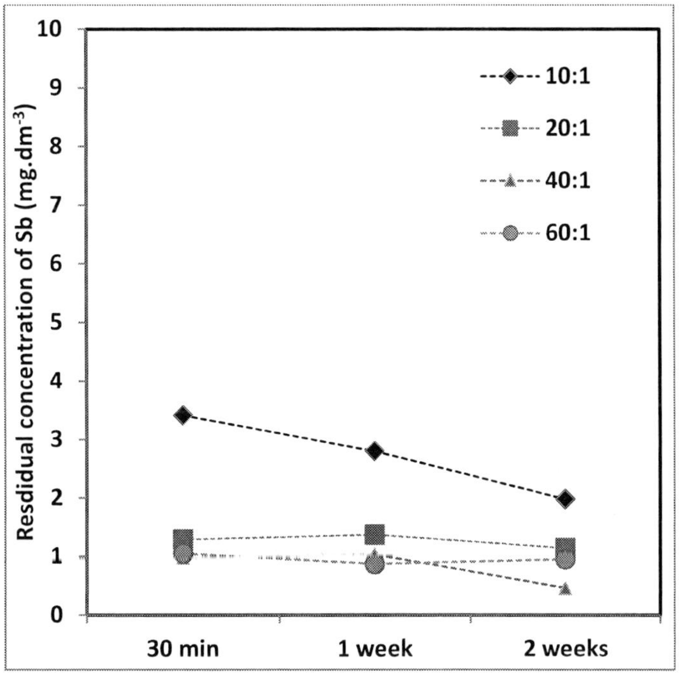

Figure 1 Residual concentrations of Sb for different antimony(III) to ferrate(VI) weight ratios. The initial concentrations of antimony(III) was 10 mg·dm^{-3}.

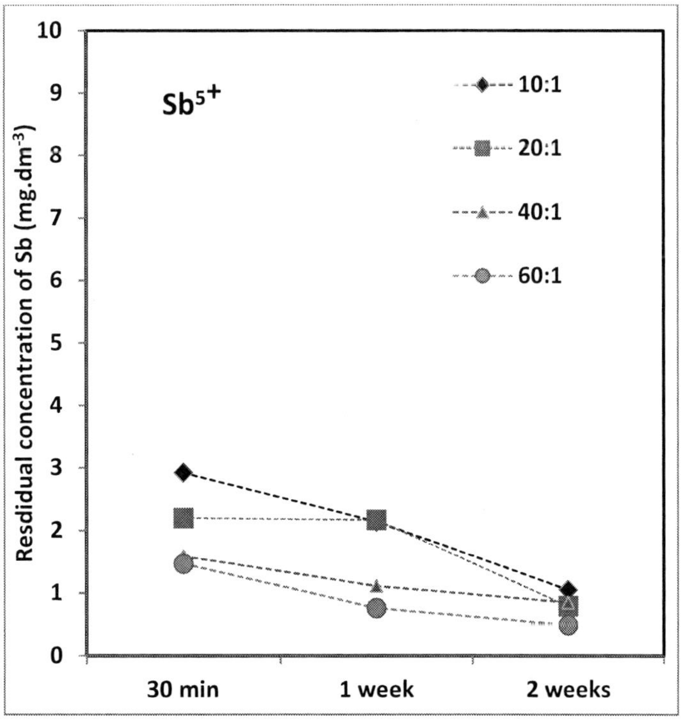

Figure 2 Residual concentrations of Sb for different antimony(V) to ferrate(VI) weight ratios. The initial concentrations of antimony(V) was 10 mg·dm^{-3}.

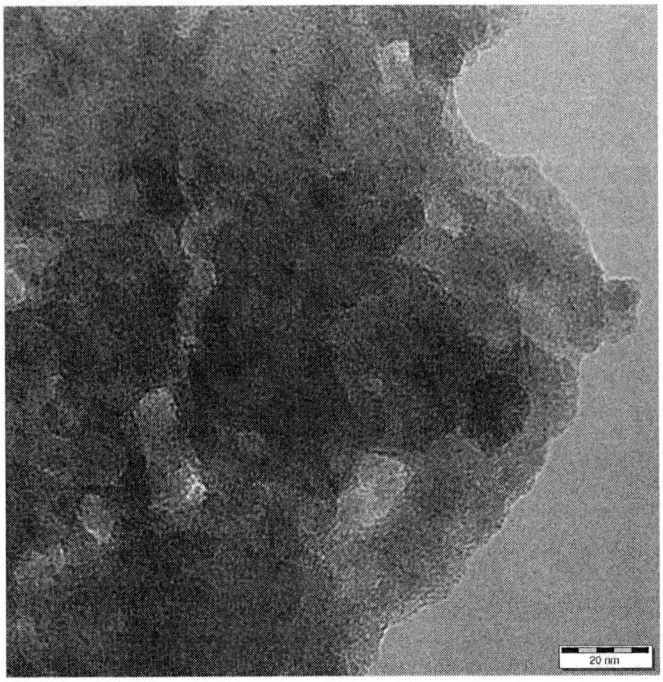

Figure 3 TEM images of iron based particles after the reaction of ferrate(VI) in the presence of antimony(V). Concentration of ferrate(VI) was 200 mg·dm^{-3} (as Fe) and the concentration of antimony was 10 mg·dm^{-3}(as Sb) .

4. CONCLUSION

The reported study reported possibilities of antimony treatment using ferrate(VI) The achieved results have demonstrated that selected inorganic ions, typically present in natural waters, have different impact on removal of antimony(III, V) by ferrate(VI) from aqueous environment. Interestingly, ions such as chlorides, nitrates, and sulfates had marginal on removal efficiency of antimony by ferrate(VI). On the contrary, presence of phosphates and carbonates had negative influence on efficiency of antimony removal by ferrate(VI). With the increased concentration of phosphates, the efficiency of antimony removal was decreased.

ACKNOWLEDGEMENTS

The authors are grateful for the support by the projects LO1305 of the Ministry of Education, Youth and Sports of the Czech Republic.

REFERENCES

[1] CHOMISTEKOVÁ, Z., CULKOVÁ, E., BELLOVÁ, R., MELICHERČÍKOVÁ, D., DURDIAK, J., BEINROHR, E., RIEVAJ, M., TOMČÍK, P.: Metódy a postupy stanovenia antimónu ako environmentálne významného analytu. Chem. Listy, 2016, vol. 110, no.10, pp. 671-677.

[2] ANDERSON, C. G.: The Metallurgy of Antimony. Chemie der Erde - Geochemistry. 2012, vol. 72, no. 4, pp. 3-8,.

[3] DUPONT, D., ARNOUT, S., JONES, P. T., BINNEMANS, K.: Antimony Recovery from End-of-Life Products and Industrial Process Residues: A Critical Review. J. Sustain. Metall., 2016, 2 (1), s. 79-103, ISSN 2199-3831.

[4] HOUSECROFT, C. E., SHARPE, A. G.: Anorganická chemie. Prekl.: Ondřej BENEŠ. Praha: Vysoká škola chemicko-technologická. 2014, pp. 389, 472, 474-476, 481-482, 500-501, 527, 694, 730-732.

[5] UNGUREANU, G., SANTOS, S., BOAVENTURA, R., BOTELHO, C. Arsenic and antimony in water and wastewater: Overview of removal techniques with special reference to latest advances in adsorption. Journal of Environmental Management. 2015. vol. 151, pp. 326-342.

[6] MUBARAK, H., CHAI, L.-Y., MIRZA, N., YANG, Z.-H., PERVEZ, A., TARIQ, M., SHAHEEN, S., MAHMOOD, Q.: Antimony (Sb)-Pollution and Removal Techniques-Critical Assessment of Technologies. Toxicological & Environmental Chemistry. 2015. vol. 97, no. 10, pp. 1296-1318.

[7] SHARMA, M., PATEL, K. S.: Determination and Speciation of Antimony in Waters. International Journal of Environmental Analytical Chemistry. 1993. vol. 50, no. 1, pp. 63-71.

[8] OBIAKOR, M. O., TIGHE, M., PEREG, L., WILSON, S. C.: Bioaccumulation, trophodynamics and ecotoxicity of antimony in environmental freshwater food webs. Critical Reviews in Environmental Science and Technology, 2017, 47 (22), s. 2208-2258, ISSN: 1064-3389.

[9] FOWLER, B. A., YAMAUCHI, H., CONNER, E. A., AKKERMAN, M.: Cancer risks for humans from exposure to the semiconductor metals. Scand J Work Environ Health. 1993. Vol. 19, no. 1, pp. 101-103.

[10] JOHNSON, M., LORENZ, B.: Antimony Remediation Using Ferrate(VI). Separation Science and Technology. 2015. vol. 50, no. 11, pp. 1611-1615.

[11] BRUNE, D., NORDBERG, G., WESTER, P. O.: Distribution of 23 Elements in the Kidney, Liver and Lungs of Workers from a Smeltery and Refinery in North Sweden Exposed to a Number of Elements and of a Control Group. The Science of the Total Environment. 1980. vol. 16, no. 1, pp. 13-35.

[12] WOOD, R. H. The heat, free energy and entropy of ferrate(VI) ion. J. Am. Chem. Soc. 1958. vol. 80, 2038−2041.

[13] SHARMA, V. K.; MISHRA, S. K. Ferrate(VI) oxidation of ibuprofen: A kinetic study. Environ. Chem. Lett. 2006, vol. 3, 182−185.

[14] SHARMA, V. K.; MISHRA, S. K.: Oxidation of inorganic contaminants by ferrates (VI, V, and IV)-kinetics and mechanisms: A review. J. Environ. Manage. 2011, vol. 92, 1051−1073.

[15] SHARMA, V. K.; MISHRA, S. K.; NESNAS: Oxidation of sulfonamide antimicrobials by ferrate(VI) [$Fe^{VI}O_4^{2-}$]. Environ. Sci. Technol. 2006, vol. 40, 7222−7227.

[16] MURMANN, R. K.; ROBINSHON, P. R. Experiments utilizing FeO_4^{2-} for purifying water. Water Res. 1974, vol. 8, pp. 543−547.

[17] HE, M.C., WANG, X.Q., WU, F. C., Fu Z. Y. Antimony pollution i n China. Sci. Total. Environ. 2012, vol. 421, pp. 41-50.

[18] PRUCEK, R., TUČEK, J., KOLAŘÍK, J., FILIP, J., MARUŠÁK, Z., SHARMA, V. K., ZBOŘIL, R. Ferrate(VI)-Induced Arsenite and Arsenate Removal by In Situ Structural Incorporation into Magnetic Iron(III) Oxide Nanoparticles. Environ. Sci. Technol. 2013. vol. 47, no. 7, pp. 3283-3292.

[19] PRUCEK, R., TUČEK, J., KOLAŘÍK, J., HUŠKOVÁ, I., FILIP, J., VARMA, R. S., SHARMA, V. K., ZBOŘIL, R. Ferrate(VI)-Prompted Removal of Metals in Aqueous Media: Mechanistic Delineation of Enhanced Efficiency via Metal Entrenchment in Magnetic Oxides. *Environ. Sci. Technol.* **2015,** vol. *49, no. 4*, pp. 2319-2327.

ENHANCEMENT OF ELECTRICAL CONDUCTIVITY AND STABILITY OF POLYPYRROLE BY INTERCALATION INTO MONTMORILLONITE

Jonáš TOKARSKÝ[1,2], Pavlína PEIKERTOVÁ[1,2], Kateřina MAMULOVÁ KUTLÁKOVÁ[1], Lenka KULHÁNKOVÁ[3]

[1]*Nanotechnology Centre, VSB - Technical University of Ostrava, Ostrava, Czech Republic, EU,* jonas.tokarsky@vsb.cz, pavlina.peikertova@vsb.cz, katerina.mamulova.kutlakova@vsb.cz

[2]*IT4Innovations, VSB - Technical University of Ostrava, Ostrava, Czech Republic, EU,* jonas.tokarsky@vsb.cz, pavlina.peikertova@vsb.cz

[3]*Faculty of Metallurgy and Materials Engineering, VSB - Technical University of Ostrava, Ostrava, Czech Republic, EU,* lenka.kulhankova@vsb.cz

Abstract

Natural phyllosilicate montmorillonite (MMT) was intercalated by electrically conducting polymer polypyrrole in order to obtain nanocomposites with enhanced electrical conductivity and stability in comparison with pure polypyrrole (PPYR). Samples in powder form were prepared via oxidative polymerization of pyrrole in aqueous solution using $FeCl_3$ as oxidizing agent. Two nanocomposites MMT/PPYR_1 and MMT/PPYR_2 containing 32 wt.% and 45 wt.% of PPYR, respectively, were prepared and characterized. Composition of the samples was studied using elemental analysis and X-ray fluorescence spectroscopy, and amount of PPYR in each nanocomposite was calculated from results of thermogravimetric analyses. Intercalation rate was monitored by X-ray powder diffraction. Raman spectroscopy was used to characterize the polypyrrole in the samples. Electrical measurements performed over 700 days revealed the course of changes in conductivity. Comparison of conductivities of fresh samples showed enhancement of the conductivity and its preservation in the case of nanocomposites. The enhancement was observed throughout the whole testing period, the nanocomposites constantly exhibited higher conductivity compared to PPYR.

Keywords: Polypyrrole, montmorillonite, intercalation, conductivity, stability

1. INTRODUCTION

Polypyrrole (PPYR) belongs to the family of electrically conductive polymers with π-conjugated structure. Although the pure PPYR itself is suitable for various practical applications, e.g. sensors [1], attention is paid to tuning the conductivity by preparation of PPYR-based nanocomposites. Natural layered phyllosilicate montmorillonite (MMT) whose interlayer cations are exchangeable via intercalation proces is very suitable for this purpose. These nanocomposites exhibit increased conductivity in comparison with pure PPYR [2,3]. Stable conductivity is an important factor for practical applications, and, in the case of pure PPYR samples, temperature and time dependence of conductivity has been studied [4,5]. However, these experiments are still rare in papers devoted to MMT/PPYR nanocomposites. Effect of thermal ageing on AC conductivity of chemically prepared Fe-MMT/PPYR was studied by Zidi et al. [6], and effect of ageing under ambient conditions (60 days) on DC conductivity of electrochemically prepared PPYR mixed with caprolactam-modified MMT was reported by Liu and Tsai [7]. Present work describes chemical synthesis of pure PPYR and two MMT/PPYR nanocomposites and their characterization. Pure unmodified Na-MMT was also characterized for comparison. Attention was paid to determination of PPYR amount in the nanocomposites, but the main goal of the experiment was to monitor the changes in conductivity of the nanocomposites and the PPYR over a long-time horizon of 700 days.

2. MATERIALS AND METHODS

Pyrrole (C_4H_5N; ≥ 98 %, reagent grade) and ferric chloride ($FeCl_3$) were used as received. Crystallochemical formula of the Na-MMT Portaclay® is ($Al_{2.85}Mg_{0.71}Ti_{0.02}Fe^{3+}_{0.42}$) ($Si_8$)$O_{20}(OH)_4$ with negative layer charge -0.69 el. per unit cell. Size fraction < 40 μm was used. Pure PPYR was prepared via oxidative polymerization of 0.05 mol of pyrrole by 0.1 mol $FeCl_3$ in 250 ml of water. MMT/PPYR_1 and MMT/PPYR_2 nanocomposites were prepared similarly via oxidative polymerization of pyrrole (n = 0.025 or 0.05 mol) by $FeCl_3$ (n = 0.05 or 0.1 mol) in 250 ml of MMT aqueous dispersion (m = 4 g). After 6 h, black solids were collected on filters by rinsing with distilled water and dried for 24 hours at 40 °C. Samples were stored in closable plastic vessels in the dark at an average room temperature of 25 °C.

Thermogravimetric (TG) analysis was performed using Netzsch STA 409 EP thermogravimeter. Samples were heated up to 800 °C (10 °C/min) in dry air with a flow rate 100 cm^3/min. Elemental analysis was performed using Vario EL Cube Elementar Analyzer with sulfanilic acid as a standard. Chlorine was determined by titration. Chemical composition was also determined by X-ray fluorescence spectroscopy (XRFS) using SPECTRO XEPOS spectrometer equipped with 50 W PdX-ray tube. Intercalation was controlled by Bruker D8 Advance diffractometer equipped with detector VÅNTEC 1. X-ray powder diffraction (XRPD) patterns were obtained in reflection mode under CoKα radiation (λ = 1.7889 Å). Raman microspectroscopy was performed using Raman microscope XploRA™, Horiba Jobin Yvon, equipped with 785 nm excitation laser source, with 50× objective and using 600 gr/mm grating. Acquisition time was set to 30 s with ten repetitions. Electrical measurements of powder samples (V ~ 60 mm^3; height of column ~ 5 mm) were performed using home made device equipped with measuring card PCI-6221. Electric current passing through the sample between two polished Cu electrodes under the constant voltage U = 1 V (DC voltage source HY3003D) was measured for 600 s and the mean value was used to calculate the conductivity. Each measurement was repeated five times in order to obtain average conductivity value.

3. RESULTS AND DISCUSSION

3.1. Composition of samples

XRFS analysis revealed increasing Cl content in MMT/PPYR nanocomposites compared to pure MMT (**Table 1**). This finding together with decrease in amount of other analytes and increase in LOI values suggests expected increase in PPYR amount in the nanocomposites.

Table 1 Composition of samples according to XRFS analysis. LOI - loss on ignition.

sample	Al_2O_3 (wt.%)	Cl (wt.%)	Fe_2O_3 (wt.%)	MgO (wt.%)	SiO_2 (wt.%)	SO_3 (wt.%)	LOI (wt.%)
MMT	19.4±2.3	0.02±0.01	3.81±0.75	3.22±0.54	56.7±1.4	0.68±0.12	11.9
MMT/PPYR_1	13.0±2.0	3.39±1.13	3.18±0.07	2.47±0.37	41.5±6.3	0.23±0.07	34.2
MMT/PPYR_2	10.2±1.6	5.71±0.86	2.56±0.39	2.32±0.35	30.5±4.6	0.08±0.01	43.7

TG analysis in temperature range 30-800 °C (**Table 2**) showed four temperature intervals of weight loss for pure MMT. The first two intervals can be attributed to loss of water adsorbed on MMT surface and loss of water in the MMT interlayer space, respectively. No weight loss was detected in the third interval. Weight loss in the last interval is caused by dehydroxylation of the MMT [8]. Total weight loss of the MMT was 7.8 wt.%. In the case of PPYR sample, two temperature intervals of weight loss were observed (**Table 2**). The first one corresponds to loss of water, and the second one can be attributed to PPYR degradation. This observation agrees with previously reported experiments [2]. Total weight loss of the PPYR was 90.6 wt.%. Weight losses of the nanocomposites in the first two intervals were similar to the MMT. However, unlike the MMT, significant weight loss was observed in the third interval (**Table 2**). This can be attributed to the decomposition of PPYR in the nanonocomposite. Since the weight loss in the last interval is higher in comparison with pure MMT, simultaneous dehydroxylation of MMT and additional decomposition of PPYR can be expected.

Table 2 Weight losses of MMT, PPYR, and MMT/PPYR nanocomposites according to TG analysis. Only two intervals of weight loss were observed for PPYR.

sample	$\Delta m_{30\text{-}110°C}$ (wt.%)	$\Delta m_{110\text{-}200°C}$ (wt.%)	$\Delta m_{200\text{-}580°C}$ (wt.%)	$\Delta m_{580\text{-}800°C}$ (wt.%)
MMT	3.8	1.0	0.0	3.0
PPYR	4.4		86.2	
MMT/PPYR_1	2.1	1.4	32.1	8.1
MMT/PPYR_2	3.1	1.5	44.9	3.2

Rough estimation of the PPYR amount in the nanocomposites according to the third interval in **Table 2** was refined by calculation using the following formula [9]:

$$\Delta m_{MMT/PPYR} = w_{PPYR} \cdot \Delta m_{PPYR} + (1 - w_{PPYR}) \cdot \Delta m_{MMT} \tag{1}$$

where $\Delta m_{MMT/PPYR}$ is the residue of MMT/PPYR nanocomposite after heating (wt.%), w_{PPYR} is the proportion of PPYR (wt.%), Δm_{PPYR} is the residue of PPYR after heating (wt.%), and Δm_{MMT} is the residue of MMT after heating (wt.%). Calculated amount of PPYR, i.e., 31.9 wt.% and 44.7 wt.% for MMT/PPYR_1 and MMT/PPYR_2, respectively, is not only very close to the initial estimation but is also in good agreement with data obtained from elemental analysis (**Table 3**).

Table 3 Elemental analysis of MMT/PPYR nanocomposites

sample	C (wt.%)	H (wt.%)	N (wt.%)	S (wt.%)	Cl (wt.%)
MMT/PPYR_1	18.49±0.06	2.226±0.014	5.318±0.008	0.156±0.005	4.17±0.14
MMT/PPYR_2	28.22±0.03	2.710±0.002	8.157±0.019	0.120±0.002	5.59±0.08

Elemental analysis confirmed presence of all elements contained in PPYR (**Table 3**). Small amount of S originates from the MMT as revealed by XRFS (**Table 1**). Total amount of C, H, N, and Cl for MMT/PPYR_1 and MMT/PPYR_2, i.e., 30.20 wt.% and 44.67 wt.%, respectively, agrees well with PPYR amounts obtained from TG analysis using Eq. (1). While the C/N ratio in PPYR is ideally equal to 3.43, values 3.48 and 3.46 were found for MMT/PPYR_1 and MMT/PPYR_2, respectively. This result shows successful polymerization.

3.2. Monitoring the intercalation

Original MMT sample exhibits asymmetric shape of *(001)* basal reflection having FWHM = 0.94 °2θ. Quite regular arragement of part of the MMT layers with d_{001} = 12.569 Å (**Figure 2a**) corresponds with presence of hydrated Na^{+} cations in the MMT interlayer space [10], however, another part of the MMT layers exhibits less regular arrangement (6.0-7.5 °2θ in **Figure 2a**). Significant increase in basal distances observed for nanocomposites suggests that the intercalated structure was successfully prepared (**Figure 2a**). Finding that basal distance of the MMT/PPYR_1 (d_{001} = 17.631 Å; FWHM = 1.17 °2θ) is higher than basal distance of the MMT/PPYR_2 (d_{001} = 15.315 Å; FWHM = 1.50 °2θ) led to repeated preparation of both nanocomposites, however, the same result was obtained. It can be explained by rapid reaction of pyrrole with $FeCl_3$, which in the case of MMT/PPYR_2 with higher content of these compounds in the reaction mixture leads to disorder in arrangement of MMT layers and to lower intercalation. The explanation is supported by both lower intensity of *(001)* basal reflection and higher FWHM value of the MMT/PPYR_2. XRPD analyses of prepared nanocomposites were performed three times during the experiment: 1 day, 303 days, and 700 days after preparation. No significant differences were found, d_{001} values varied only in tenths of angstrom. For illustration, basal reflections of the MMT/PPYR_1 nanocomposite are shown in **Figure 2b**.

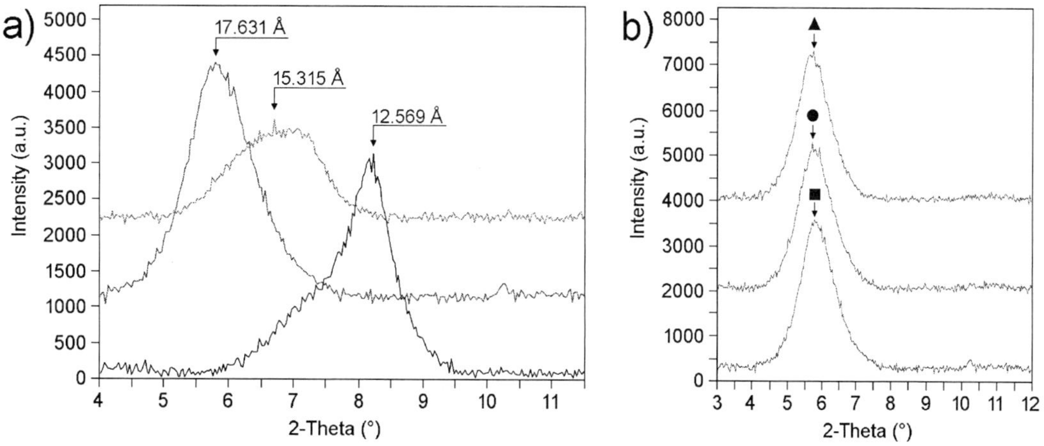

Figure 1 (a) XRPD profiles of *(001)* basal reflections for pure MMT (black), MMT/PPYR_1 (blue), and MMT/PPYR_2 (red) measured 1 day after preparation. (b) Comparison of *(001)* basal reflections for MMT/PPYR_1 over time. ■ 1 day, d_{001} = 17.631 Å; ● 303 days, d_{001} = 17.801 Å; ▲ 700 days, d_{001} = 17.714 Å.

3.3. Electrical conductivity and Raman spectroscopy

Electrical measurements of fresh samples performed one day after preparation showed enhancement of the conductivity in the case of nanocomposites (σ_{PPYR} = 21.7±1.5 S/m, $\sigma_{MMT/PPYR\_1}$ = 49.1±6.5 S/m, $\sigma_{MMT/PPYR\_2}$ = 41.1±5.2 S/m; see **Figure 2**). Although the decrease of initial conductivities is obvious for all samples throughout the whole testing period, MMT/PPYR_1 and MMT/PPYR_2 nanocomposites constantly exhibited higher conductivity compared to pure PPYR. After 700 days, conductivities σ_{PPYR} = 0.7±0.03 S/m, $\sigma_{MMT/PPYR\_1}$ = 21.2±1.7 S/m, and $\sigma_{MMT/PPYR\_2}$ = 24.1±4.7 S/m were obtained (**Figure 2**). Concerning preservation of the conductivity, the PPYR exhibits only 3 % of the original value after 700 days, while the MMT/PPYR_1 and MMT/PPYR_2 nanocomposites exhibit 45 % and 54 %, respectively, after the same time period.

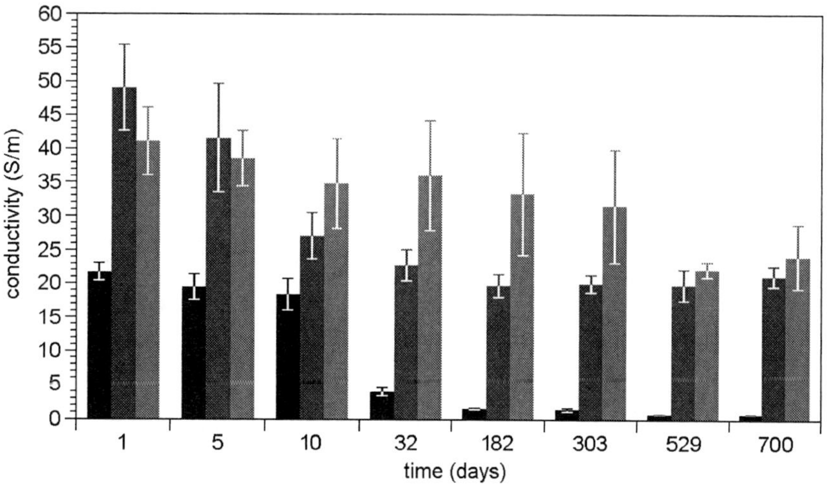

Figure 2 Changes in electrical conductivity of pure PPYR (black), MMT/PPYR_1 (blue), and MMT/PPYR_2 (red) monitored over a long time horizon of 700 days

A drop below 50 % of the original conductivity was recorded on both PPYR and MMT/PPYR_1 samples between the 1st and the 32nd day. While the conductivity of the PPYR further decreased, the conductivity of the MMT/PPYR_1 sample was preserved until the end of the experiment (32nd - 700th day; **Figure 2**). Average value calculated from five mean conductivities during this period is 20.7±1.2 S/m. Conductivity of the MMT/PPYR_2 sample was relatively stable from the 10th to the 303rd day (**Figure 2**), and the average value calculated from four mean conductivities of the MMT/PPYR_2 during this period is 33.9±1.7 S/m.

Raman spectra, measured in the 1st and the 700th day, are shown in **Figure 3**. All characteristic bands of PPYR are present [11,12], however, MMT bands are not visible due to the MMT fluorescence. The spectra of fresh samples (the 1st day; **Figure 3a**) showed that the pure PPYR sample is less conductive compared to MMT/PPYR_1 and MMT/PPYR_2 samples, because band of C=C and C-C stretching related to oxidation state of the polymer (marked with ▼in **Figure 3**) is shifted to the highest wavenumbers [11,12]. The lowest wavenumber of this band detected for the sample MMT/PPYR_1 indicate the highest conductivity. Also, the more intensive band of C-H in-plane deformation vibrations in radical cation units in protonated polypyrrole (marked with ★) in comparison with band of C-H in-plane deformations in the neutral units of polypyrrole base (marked with ✦) is connected to the better protonation of the sample MMT/PPYR_1 (see **Figure 3a**) [11,12]. This is in good agreement with electrical measurements (**Figure 2**).

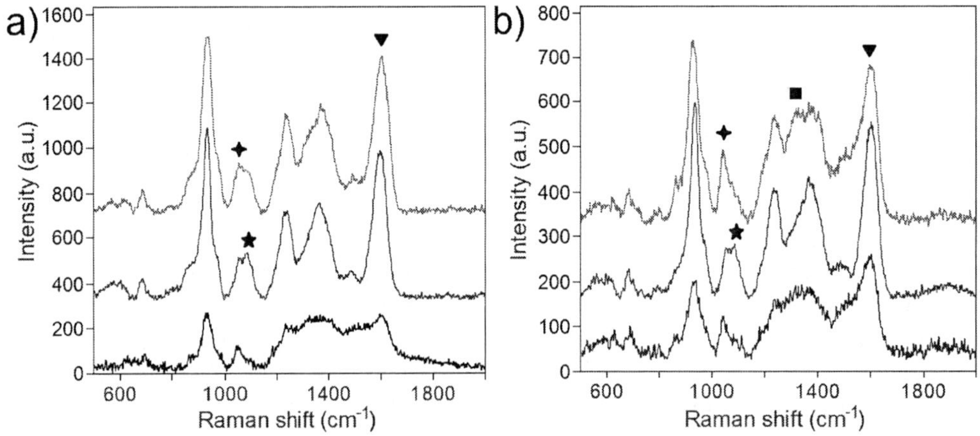

Figure 3 Raman spectra of pure PPYR (black), MMT/PPYR_1 (blue), and MMT/PPYR_2 (red) measured 1 day after preparation (a) and 700 days after preparation (b)

Raman spectra measured after 700 days (**Figure 3b**) are noisier and their overall intensities are lower compared to spectra from the 1st day. The band ★ is still more intensive than the band ✦ in spectrum of the MMT/PPYR_1 (**Figure 3b**), and the band of neutral C-C ring stretching vibrations related to of reduced units (denoted as ■ in **Figure 3b)** appeared in spectrum of the MMT/PPYR_2. This should indicate the highest conductivity of the MMT/PPYR_1 sample [11,12], however, the wavenumber of the band ▼ is higher compared to spectrum of the MMT/PPYR_2 sample (**Figure 3b**) suggesting lower conductivity. Conductivity can be, therefore, considered equal, which agrees with electrical measurements (**Figure 2**).

4. CONCLUSION

Two nanocomposites MMT/PPYR_1 and MMT/PPYR_2 containing 32 wt.% and 45 wt.% of PPYR, respectively, were succesfully prepared by intercalation of MMT. Composition and structure of the nanocomposites and pure components, i.e., PPYR and MMT, were studied using XRFS, TGA, elemental analysis, XRPD, and Raman spectroscopy. Electrical conductivity of the samples was monitored over 700 days, and higher conductivity of the nanocomposites and significantly lower decrease in the conductivity compared to pure PPYR were observed. After 700 days, the conductivity of MMT/PPYR_1 and MMT/PPYR_2 decreased by 55 % and 46 %, respectively, while the conductivity of pure PPYR decreased by 97 %. Stable conductivity is important for practical applications. Conductivity of the MMT/PPYR_1 was stable between 32nd and 700th day, i.e., for 668 days, and the average conductivity during this period was $\sigma_{average}$ = 20.7±1.2 S/m. For the MMT/PPYR_2, the stability in conductivity was observed between 10th and 303rd day day, i.e., for 293 days, and the average conductivity during this period was $\sigma_{average}$ = 33.9±1.7 S/m.

ACKNOWLEDGEMENTS

This work was supported by The Ministry of Education, Youth and Sports from the National Programme of Sustainability (NPS II) project „IT4Innovations excellence in science -LQ1602" and by the ERDF in the IT4Innovations national supercomputing center - path to exascale project (CZ.02.1.01/0.0/0.0/16_013/0001791) within the OPRDE. Financial support from The Ministry of Education, Youth and Sports project no. SP2018/95 is also gratefully acknowledged. Authors thank Silva Vallová for thermogravimetric analyses.

REFERENCES

[1] QIN, H., KULKARNI, A., ZHANG, H., KIM. H., JIANG, D. and KIM, T. Polypyrrole thin film fiber optic chemical sensor for detection of VOCs. *Sensors and Actuators B*. 2011. vol. 158, pp. 223-228.

[2] RAMÔA, S.D.A.S., BARRA, G.M.O., MERLINI, C., SCHREINER, W.H., LIVI, S. and SOARES, B.G. Production of montmorillonite/polypyrrole nanocomposites through oxidative polymerization of pyrrole: Effect of anionic and cationic surfactants on structure and properties. *Applied Clay Science*. 2015. vol. 104, pp. 160-167.

[3] LIU, Y.-C. and GER, M.-D. Modification of electropolymerized polypyrrole with Na-montmorillonite. *Chemical Physics Letters*. 2002. vol. 362, pp. 491-496.

[4] TRUONG, V.T. Thermal degradation of polypyrrole: effect of temperature and film thickness. *Synthetic Metals*. 1992. vol. 52, pp. 33-44.

[5] RICKS-LASKOSKI, H.L. and BUCKLEY, L.J. Twenty-year aging study of electrically conductive polypyrrole films. *Synthetic Metals*. 2006. vol. 156, pp. 417-419.

[6] ZIDI, R., BEKRI-ABBES, I., SDIRI, N., VIMALANANDAN, A., ROHWERDER, M. and SRASRA, E. Electrical and dielectric investigation of intercalated polypyrrole montmorillonite nanocomposite prepared by spontaneous polymerization of pyrrole into Fe(III)-montmorillonite. *Materials Science and Engineering B*. 2016. vol. 212, pp. 14-23.

[7] LIU, Y.C. and TSAI, C.J. Enhancements in conductivity and thermal and conductive stabilities of electropolymerized polypyrrole with caprolactam-modified clay. *Chemistry of Materials*. 2003. vol. 15, pp. 320-326.

[8] HATAKEYAMA, T. and LIU, Z. *Handbook of Thermal Analysis*. John Wiley&Sons, 1998. p. 325.

[9] MARINS, J.A., SOARES, B.G., DAHMOUCHE, K., RIBEIRO, S.J.L., BARUD, H. and BONEMER, D. Structure and properties of conducting bacterial cellulose-polyaniline nanocomposites. *Cellulose*. 2011. vol. 18, pp. 1285-1294.

[10] FONSECA, C.G., VAISS, V.S., WYPYCH, F., DINIZ, R., LEITÃO, A.A. Structural and thermodynamic investigation of the hydration-dehydration process of Na^+-Montmorillonite using DFT calculations. *Applied Clay Science*. 2017. vol. 143, pp. 212-219.

[11] VARGA, M., KOPECKÝ, D., KOPECKÁ, J., KŘIVKA, I., HANUŠ, J., ZHINGUNOV, A., TRCHOVÁ, M., VRŇATA, M. PROKEŠ, J. The ageing of polypyrrole nanotubes synthesized with methyl orange. *European Polymer Journal*. 2017. vol. 96, pp. 176-189.

[12] STEJSKAL, J., TRCHOVÁ, M., BOBER, P., MORÁVKOVÁ, Z., KOPECKÝ, D., VRŇATA, M., PROKEŠ, J., VARGA, M., WATZLOVÁ, E. Polypyrrole salts and bases: superior conductivity of nanotubes and their stability towards the loss of conductivity by deprotonation. *RSC Advanced*. 2016. vol. 6, pp. 88382-88391.

GERMANIUM AND TIN NANOPARTICLES ENCAPSULATED IN AMORPHOUS SILICON MATRIX FOR OPTOELECTRONIC APPLICATIONS

The Ha STUCHLIKOVA, Zdenek REMES, Jiri STUCHLIK

Institute of Physics, Czech Academy of Sciences, Prague, Czech Republic, EU
hotheha@fzu.cz, remes@fzu.cz, stuj@fzu.cz

Abstract

The plasma enhanced chemical vapour deposition was combined with in situ deposition of Ge and Sn thin film by evaporation technique at surface temperature about 220 °C to form nanoparticles on the surface of hydrogenated silicon thin films to prepare diodes. Formation of nanoparticles was additionally stimulated by plasma treatment through a low pressure hydrogen glow discharge. The diodes based on PIN diode structures with and without the embedded Ge or Sn nanoparticles were characterized by temperature dependence of electrical conductivity, activation energy of conductivity, measurement of volt-ampere characteristics in dark and under solar illumination.

Keywords: a-Si:H, Ge NPs, Sn NPs, diode structures, I-V characteristics

1. INTRODUCTION

The efficiency and stability improvement of inexpensive solar cells based on amorphous hydrogenated silicon (a-Si:H) were very attractive are actual up to now. The question how to increase too weak photo- and electro-luminescence of amorphous hydrogenated silicon is open too. Our effort is focused on a possibility to deposit *in situ* amorphous or microcrystalline Si:H thin films with embedded nanoparticles (NPs). We choose the semiconductors which band gap is below the band gap of a-Si:H. For it we develop new deposition systems, which allow combine convenient deposition techniques. In this contribution we pay attention to germanium, but in the contrary with lot of published attempts to apply a-Si_xGe_y:H alloys in the photovoltaic [1,2], even the a-Si_xSn_y:H alloys were studied [3], we prefer form of Ge nanoparticles (Ge NPs) integrated into intrinsic a-Si:H structure. Special attention we paid to behaviour of tin. Ge NPs and tin nanoparticles (Sn NPs) were encapsulated in a-Si:H thin films to improve optoelectronic properties of this thin layer structures. Presented work has continuity on previous published results [4 - 8].

2. EXPERIMENTAL

The intrinsic a-Si:H film was deposited by plasma enhanced chemical vapour deposition (PECVD) technique on a fused silica glass substrates size 10x5 mm at surface deposition temperature 220 °C. During interruption of PECVD the vacuum chamber was pumped up to 10^{-5} Pa and on the a-Si:H surface the germanium or tin films were evaporated in the thickness of 1 nm. The elements formed isolated NPs on the a-Si:H surface with support of hydrogen plasma treatment. Then the deposited NPs were covered and stabilized by a-Si:H layer by PECVD. Both deposition processes were alternated 5 times. Finally on the surface were evaporated two titanium strips with the distance 2 mm. By this way were deposited samples with Ge NPs - 3K93 and samples with embedded Sn NPs - 3K78, 3K79 and 3K80. Those alloys with integrated Ge or Sn NPs with the co-planar configuration of electrodes were characterized by measurement of temperature dependence of electrical conductivity. On the Corning glasses covered by transparent conductive layer (ITO) after previous deposition of p-type of a-Si:H. were deposited a-Si:H films which were immediately in situ covered by Ge or Sn nanoparticles. The NPs were formed on a-Si:H surface during Vacuum Evaporation (VE) of Ge thin film with thickness 1 nm and Sn thin films with thicknesses 1 or 0.3 nm. By this way were deposited final PIN diode structures 3K95 with Ge NPs and 3K90 and 3K91 with embedded Sn NPs - see **Figure 1**.

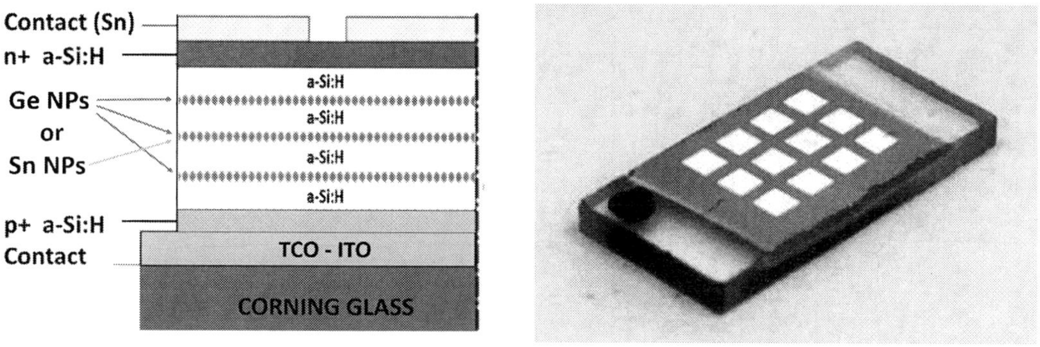

Figure 1 Schematic illustration of PIN diode structures with embedded NPs and photography of the size 10x5 mm sample with 12 diodes. Ge NPs were embedded in the intrinsic layer in 1/4, 1/2 and 3/4 of its thickness, the Sn NPs only in the 1/2 of its thickness.

3. RESULTS AND DISCUSSION

For characterization of semiconductors the temperature dependence of electrical conductivity i.e. Arrhenius plot is important measurement, see **Figure 2**. While the integration of Ge NPs increases the conductivity and decreases the activation energy of conductivity only little, the effect of Sn NPs is marked in changes of activation energy. The relatively light increases of conductivities at the room temperature 293 K for samples with Sn NPs proved the NPs are well mutually insulated by a-Si:H thin film.

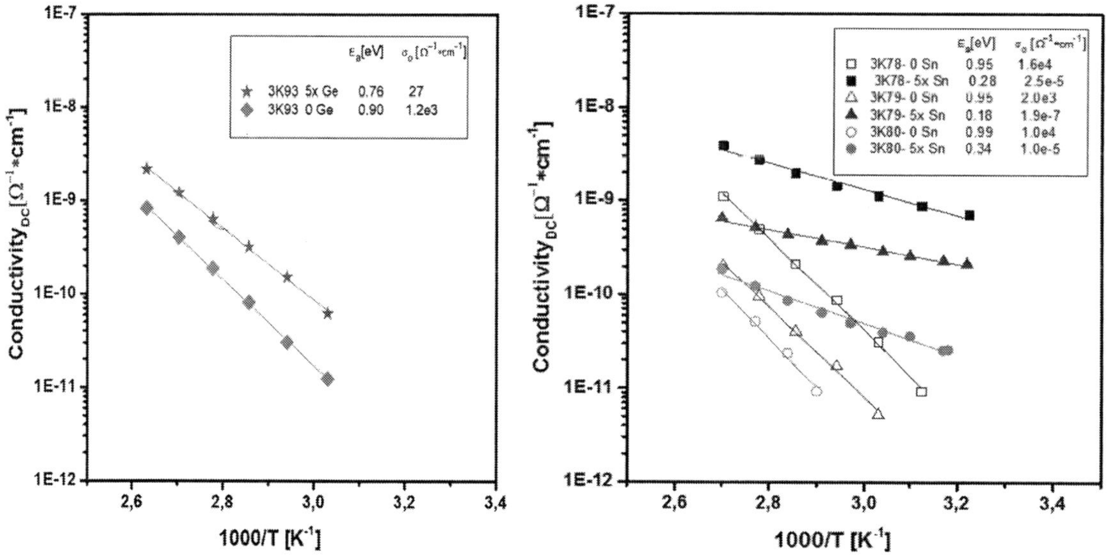

Figure 2 Temperature measurement of electrical conductivity (Arrhenius plot) of reference samples and samples with 5 times integrated monolayers of Ge (left picture) and Sn NPs (right picture)

In the case of measurement of I-V characteristics we compare two diode structures with and without Ge or Sn NPs. Here the deposition and creation of Ge NPs was repeated 3 times for one diode and at same time for second diode the deposition process of a-Si:H was interrupted only. The measurement of I-V characteristics in dark gives the first results, both currents (reverse and forward) increased, see **Figure 3**.

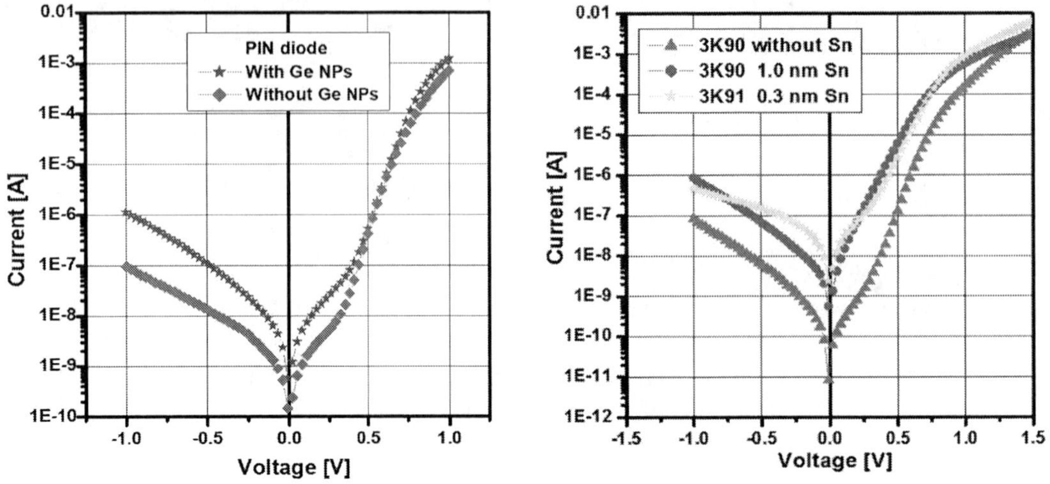

Figure 3 I-V characteristics of PIN diode measured in dark with 3 evaporated Ge monolayers of thickness 1.0 nm. Monolayers of formed Ge NPs are in the middle and at 1/4 and 3/4 of intrinsic a-Si:H thin film (3K95 - left picture), one monolayer of Sn NPs is only in the middle of a-Si:H (3K90 and 3K91 - right picture).

Most important is the measurement under illumination by solar simulator. From the results shown on the **Figure 4**. we conclude the Voltage open current (V_{oc} = 0.67 - 0.68 V) is same for both PIN diodes and as well as the Fill Factor (FF = 64) does not change too.

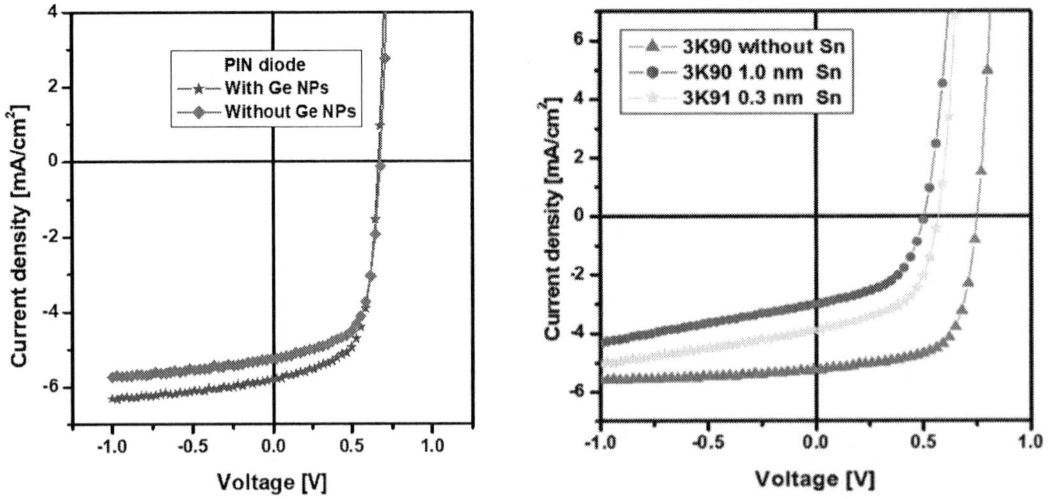

Figure 4 I-V characteristics of PIN diodes (3K95) measured under illumination by solar simulator. Requested effect of Ge NPs on the characteristic parameters of PV elements is demonstrated by increase of I_{sc} while V_{oc} and FF are without changes (see left picture) and as well as measurement of PIN diodes (3K90 and 3K91 - right picture) with embedded Sn NPs with different sizes.

Positive effect of Ge NPs we observe on short circle Current (Isc increases from 52 to 58 μA), for small photovoltaic element size 0.64 mm^2. On the contrary the integration of Sn NPs if the evaporated thickness was same i.e. 1 nm the effect on the basic parameters of diodes was negative while the smaller size of Sn NPs (evaporated thickness 0.3 eV) leads toward an increase of those characteristics values.

4. CONCLUSION

In this contribution we demonstrate possibility to integrate Ge or Sn NPs to intrinsic layer of PIN diodes on the base of a-Si:H thin films. Although the measurements in dark of I-V characteristics of the diode structures show negative effect on the reverse current the increase of forward current is positive. The most important result which we introduce here is the increase of I_{sc} if the Germanium is in the form of NPs embedded in diode structures while the U_{oc} and FF remain same. The integration of Ge NPs leads toward the requested increase efficiency of very thin film PV elements which are only a few tenths of nm. In the case of metallic Sn NPs we demonstrate surprising result - although the influence on diode parameters is negative, the character of diodes were preserved. The negative impact of NPs increases with the increase of NPs size.

ACKNOWLEDGEMENTS

This work was supported by the CSF project 16-10429J, the MEYS project LTC17029 INTER-COST Action MP1406 and by the Operational Programme Research, Development and Education financed by European Structural and Investment Funds and the Czech Ministry of Education, Youth and Sports (Project No. SOLID21 - CZ.02.1.01/0.0/0.0/16_019/0000760).

REFERENCES

[1] HAMAKAWA. Y. *Amorphous semiconductor technologies and devices silicon*. Japan Annual Reviews in Electronics Computers and Telecommunications, Vol.6, Tokyo: OHMSHA, LTD, 1983, p. 135

[2] FRITZSCHE. H. *Amorphous silicon and related materials*, Advances in Disordered Semiconductors - Vol. 1, Singapore: Word Scientific Publishing Co. Pte. Ltd., Vol 1B, 1989, p.887

[3] HAMAKAWA. Y. *Amorphous semiconductor technologies and devices silicon*. Japan Annual Reviews in Electronics Computers and Telecommunications, Vol.22, Tokyo: OHMSHA, LTD, 1987, p. 99

[4] KRIVYAKIN, G. K., VOLODIN, V. A., SHKLYAEV, A. A., MORTET, V., MORE-CHEVALIER, J., ASHCHEULOV, P., REMES, Z., STUCHLIKOVÁ, T. H., STUCHLIK, J., Formation and study of p-i-n structures based on two-phase hydrogenated silicon with a germanium layer in the i-type region, *Semiconductors 51* 2017, pp 1370 - 1376.

[5] CHERNEV, I. M., SHEVLYAGIN, A. V., GALKIN, K. N., STUCHLÍK, J., REMEŠ, Z., FAJGAR, GALKIN, N. G. On the way to enhance the optical absorption of a-Si in NIR by embedding Mg2Si thin film, *Appl. Phys. Lett. 109*, 2016, pp. 043902-1 - 043902-4.

[6] KRIVYAKIN, G. K., VOLODIN, V. A., KOCHUBEI, S. A., KAMAEV, G. N., PURKRT, A, REMEŠ, Z., FAJGAR, R., STUCHLIKOVÁ, T. H., STUCHLIK, J., Optical properties of p-i-n structures based on amorphous hydrogenated silicon with silicon nanocrystals formed via nanosecond laser annealing, *Semiconductors 50*, 2016, pp. 935 - 940.

[7] DVURECHENSKII, A.V., VOLODIN, V.A., KRIVYAKIN, G.K., SHKLYAEV, A.A., KOCHUBEI, S.A., NEIZVESTNY, I.G., STUCHLIK, J., Raman Studies of Phase and Atomic Compositions of GeSi Nanosystems after Pulsed Annealing, *Optoelectronics, Instrumentation and Data Processing 52*, 2016, pp. 496-500.

[8] STUCHLÍKOVA, T. H., FAJGAR, R., KOŠTEJN, M., DŘÍNEK, V., REMEŠ, Z., PURKRT, A., STUCHLÍK, J., The Laser Ablation as a perspective technique for the deposition of metal-silicide nanoparticles in situ embedded in PECVD of Si:H thin films, *JJAP Conf. Proc. 3*, 2015, pp. 011302-1 - 011302-5.

ANALYSIS OF ELECTRIC FIELD DEPENDING ON THE DIAMETER OF THE BALL ELECTRODE

Ondřej BAŤKA, Jaroslav BERAN, Josef SKŘIVÁNEK

Department of textile machine design - Technical University of Liberec, Liberec, Czech Republic, EU,
ondrej.batka@tul.cz

Abstract

This paper deals with the analysis of the distribution of the electric field around the spinning electrodes used for the production of nanofibres by the AC-electrospinning method. Understanding the behavior of the electrostatic field around the electrodes is an important part for design a spinning device. The performed analyzes were carried out depending on the previously performed optimization of the shape of the spherical electrode, which ensures a uniform distribution of the intensity of the electric field across the functional surface of the electrode. Analyzes were performed for multiple diameters of the spherical electrode and the dependence of the intensity of the electric field on the electrode diameter was monitored. Furthermore, the electrical voltage dependence on the electrode diameter and surface was monitored at the same electric field strength.

Keywords: Electric field, spinning electrode, spinning process, electrospinning

1. INTRODUCTION

This article builds on the research that was described in the paper "Analysis and optimization of the ball-shaped electrodes designed for AC-electrospinning" published at Nanocon 2016 [1]. The article describes the optimization of the shape of a spherical electrode designed for the production of nanofibers by the method AC-electrospinning, which was developed at the Technical University of Liberec [2]. The optimization was performed to achieve a uniform distribution of the electric field on the ball surface of the electrode. [1] **Figure 1** shows the result of the electric field simulation for an optimized electrode, taken from [1]. This shape was used as the default for further simulations. Simulations were performed for electrodes of different diameters and the electric field dependence on the size of the spherical head of the electrode was monitored. As described above, for the stability of the electrospinning process, it is important that the intensity be evenly distributed throughout the ball head, so that the electrode shapes for all the analyzed diameters are optimized.

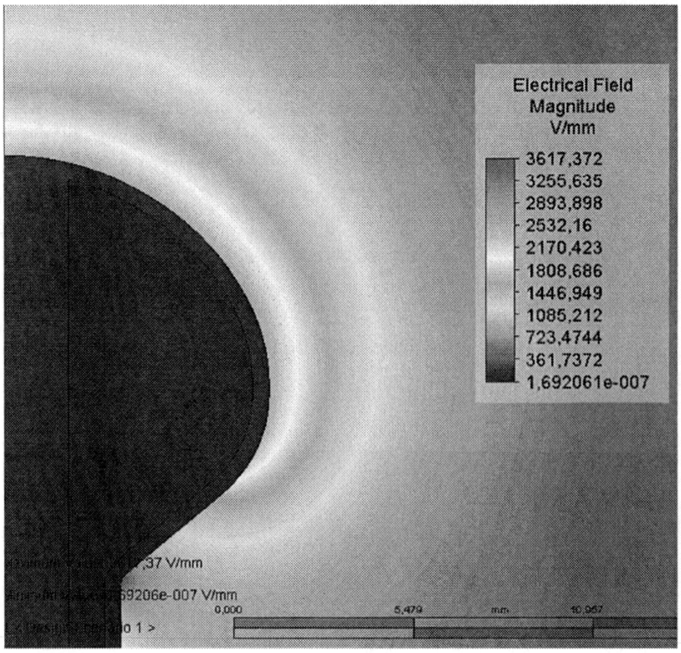

Figure 1 optimized electrode [1]

2. SIMULATION

Figure 2 shows a cross section of the electrode. The electrode is formed by a tube which is provided with a spherical spinning head at its upper end. As can be seen from the **Figure 2**, the head is not an accurate sphere, but it is made up of four radiuses. (R1, R2, R3, R4) Changing the magnitude of these radii was done by optimizing the shape so that the electric field was evenly distributed throughout the electrode head. Simulations were made for four different outer diameters of the electrode head (ϕD), specifically for diameters 11, 13, 15 and 17 mm. For all simulations performed, the same regional conditions were selected. An electrical voltage of 25 kV was applied to the electrode. In the distant surroundings of the spherical shape a voltage of 0 V was entered. Steel was chosen as the material of the electrode. Next, a thin layer of PVB polymer was imitated on the electrode surface. Surrounding of the electrode is air. **Figures 3** and **4** show the results of simulated for all optimized electrode shapes. The results show that the electric is evenly distributed in the ball head. The intensity at the point of transition of the spherical part to the pipe rapidly decreases. This part of the electrode is not functional for the spinning because the intensity values in this part are so small that there is no fiber production. Only the head of the electrode is therefore important

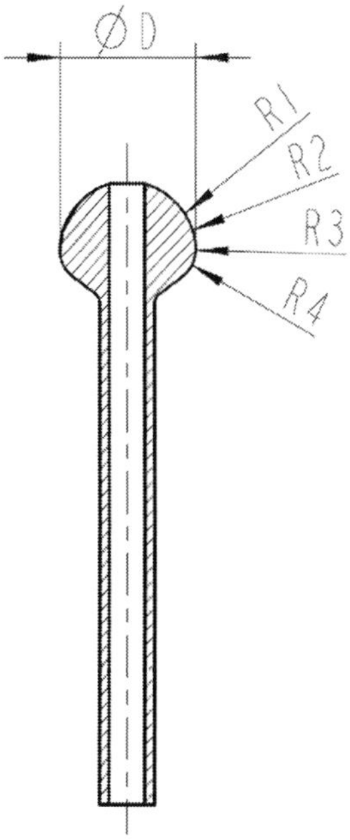

Figure 2 shape of electrode

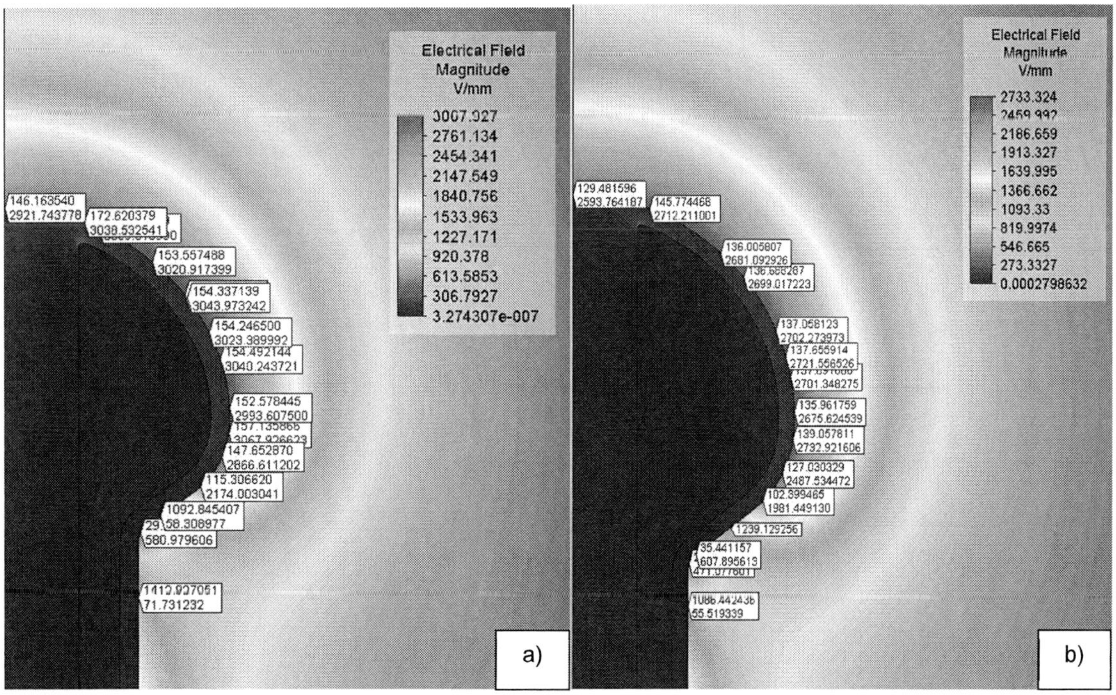

Figure 3 Simulation results, a) diameter 11 mm, b) diameter 13 mm

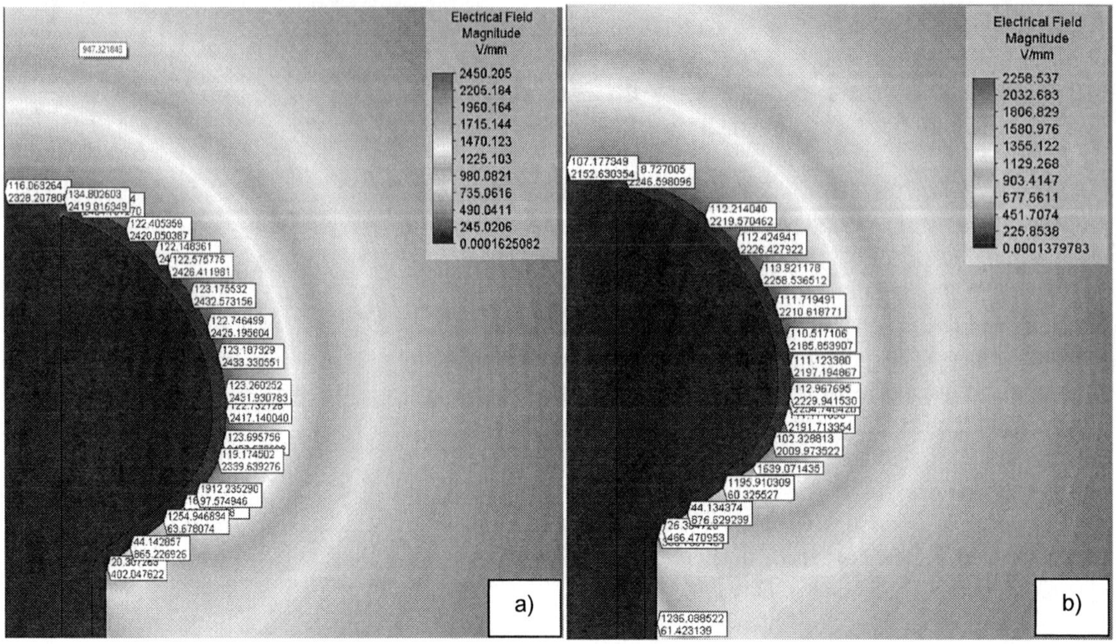

Figure 4 Simulation results, a) diameter 15 mm, b) diameter 17 mm

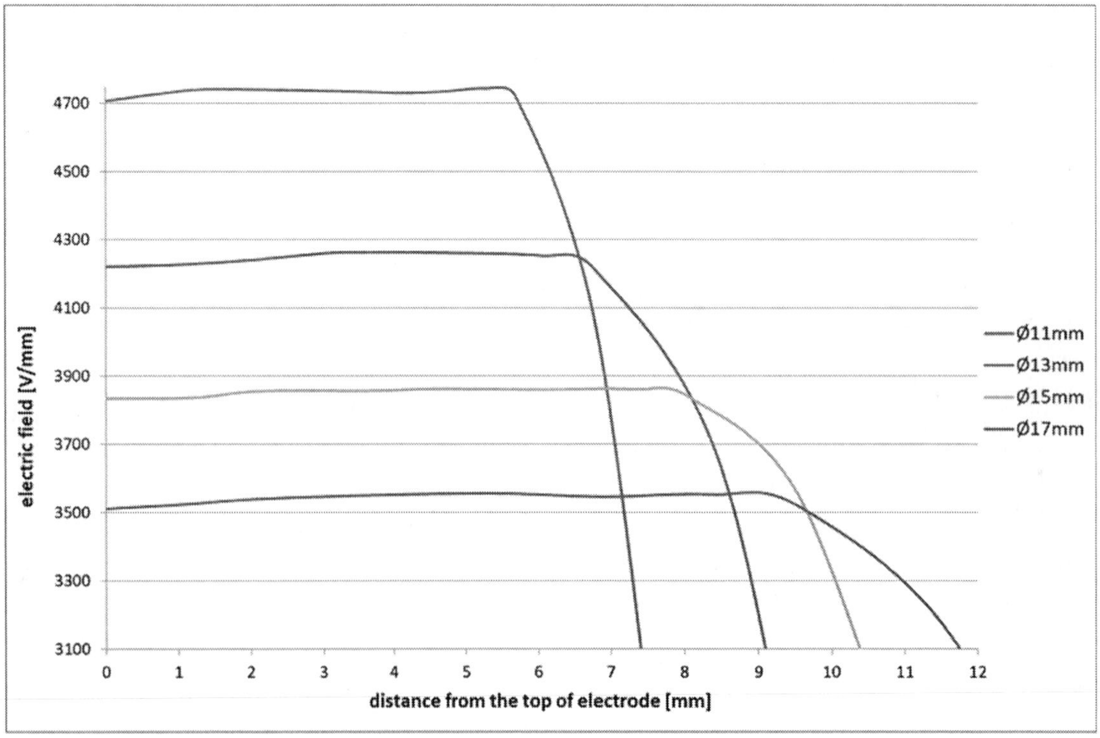

Figure 5 distribution of electric field

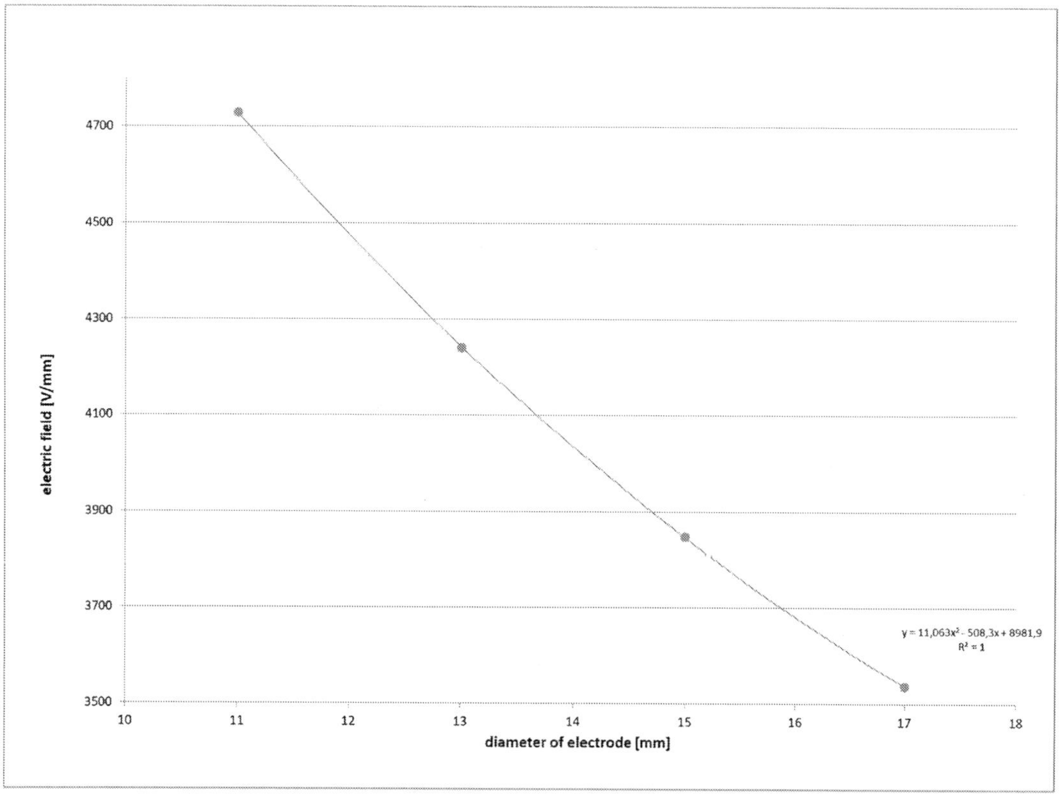

Figure 6 Electric field dependence on diameter of electrode

3. CONCLUSION

It follows from these simulations that the optimization of the shape of the electrode for the purpose of uniformity of the electric field can be done for more diameters of the electrodes. Further, it is observable from the graphs of **Figures 5** and **6** that with the increasing diameter the values of the electric field intensity on the surface of the electrode decrease. The values in the graph in **Figure 6** are obtained as the average of the electrode head values in the range where the electric field values are uniform. The electric field value is important for the stability and efficiency of the electrospinning process and also affects the resulting fibers. With an increasing electric field, the efficiency of the electrospinning process increases. Simulation results show that it is better to use smaller diameters of the electrodes for higher production efficiency. However, the surface area of the electrode also influence the efficiency of the process because more fibers are produced on the larger surface. Therefore, it would be advisable to devote this task in the future to laboratory measurements.

ACKNOWLEDGEMENTS

The research presented in this article was supported by the Ministry of Education, Youth and Sports in the framework of the targeted support of the "National Programme for Sustainability I" LO 1201

This publication was written at the Technical University of Liberec as part of the project " Research and development of devices for production of nanofibrous materials using AC-electrospnning process " with the support of the Specific University Research Grant, as provided by the Ministry of Education, Youth and Sports of the Czech Republic in the year 2016.

REFERENCES

[1] BATKA O., BERAN J., STRNAD M., SKRIVANEK J., Analysis and optimization of the ball shaped electrode designed for the AC-electrospinning, (2016) *NANOCON 2016 - Conference Proceedings, 8th International Conference on Nanomaterials - Research and Application*, pp. 358-361. ISBN: 978-808729471-0.

[2] LUKAS, D., et al.: Process for preparing polymeric nanofibers by spinning a solution of polymer melt in electric field and linear form of polymeric nanofibers prepared in such a manne *CZ 304173*.

[3] BAŤKA O., LE VAN, K., BÍLEK, M., SKŘIVÁNEK, J., VALTERA, J., Analysis of electric field around wire spinneret (2017) *NANOCON 2017* - Conference Proceedings, 9th International Conference on Nanomaterials - Research and ApplicationVolume 2017-October, 2018, pp 241-245. ISBN 978-80-87294-81-9.

JIRSAK, O., et al.: A method of nanofibers production from polymeric solution using electrostatic spinning and device for carrying out the method. Technical University of Liberec. WO2005024101 (A1). 2015.

[4] OLDRICH JIRSAK, PETR SYSEL, FILIP SANETRNIK, JAKUB HRUZA, and JIRI CHALOUPEK, "Polyamic Acid Nanofibers Produced by Needleless Electrospinning," Journal of Nanomaterials, vol. 2010, Article ID 842831, 6 pages, 2010. doi:10.1155/2010/84283

[5] XIN WANG, HAITAO NIU, XUNGAI WANG, and TONG LIN, "Needleless Electrospinning of Uniform Nanofibers Using Spiral Coil Spinnerets," Journal of Nanomaterials, vol. 2012, Article ID 785920, 9 pages, 2012. doi:10.1155/2012/785920.

[6] SKRIVANEK J., BATKA O., BILEK M., VALTERA J., LE VAN K., Effect analysis of the string cross-section on the electrostatic spinning proces. (2017) NANOCON 2017 - Conference Proceedings, 9th International Conference on Nanomaterials - Research and ApplicationVolume 2017-October, 2018, pp 224-228.

[7] YOUN-MOOK L., HUI-JEONG G, JOON PYO J, YOUNG-CHANG N. Preparation of Cellulose-based Nanofibers Using Electrospinning [online] URL: <http://www.intechopen.com/books/nanofibers/preparation-of-cellulose-based-nanofibers-using-electrospinning>

MÖSSBAUER STUDY OF REACTION MECHANISMS OF FERRATES (Fe^{VI}, Fe^{V}, and Fe^{IV}): DETERMINATION OF EFFICIENCY IN WATER DECONTAMINATION

Michaela TOMÍČKOVÁ, Radek ZBOŘIL, Jiří TUČEK

Regional Centre of Advanced Technologies and Materials, Departments of Experimental Physics and Physical Chemistry, Palacký University in Olomouc, Olomouc, Czech Republic, EU,
michaela.tomickova@upol.cz, radek.zboril@upol.cz, jiri.tucek@upol.cz

Abstract

At present, treatment of surface, underground and waste water is viewed as a high priority and many studies deal with this issue. Water can be contaminated by organic pollutants, inorganic pollutants, toxic, heavy metals and their compounds, and many other substances that do not degrade to non-toxic compounds in natural waters. Therefore, it is necessary to prevent accumulation of these substances in water and remove them from environment.

In this respect, iron compounds in the high oxidation state (IV, V and VI), called ferrates, are very promising materials to water treatment. Especially, ferrates are able to remove hormones, such as estrogen, or medicines, for example antibiotics, from water. Reactions of ferrates with pollutants occur very quickly and thus, it is difficult to monitor the course of the reaction.

This study brings a deeper insight into the mechanism of ferrate reactions with pollutants because, thanks to Mössbauer spectroscopy, we are able to *in-situ* observe the course of the reaction. Moreover, analyzing the input materials, intermediates and also final products of reaction is possible once this technique is employed. In this regard, the application of Mössbauer spectroscopy for monitoring the reaction of ferrate(VI) with arsenic for removal of arsenic from environment is mentioned.

Keywords: Water treatment, ferrates, Mössbauer spectroscopy

1. INTRODUCTION

Currently, many studies are focused on cleaning and treatment of water. There is no wonder as, a large increase in human and industrial activity (i.e., industrial production, traffic, agriculture, extraction, chemical disasters, etc.) leads to increased contamination of groundwater and surface water. In this context, water can be contaminated by toxic, heavy metals and their compounds, organic pollutants (e.g., insecticides, herbicides, petroleum derivatives, solvents, medicaments, etc.) or inorganic pollutants (e.g., acids, ammonia, DDT, polychlorinated biphenyls, nitrogen, phosphorus, mercury, etc.). Many of these pollutants do not naturally decompose to non-toxic compounds in water. For this reason, a significant attention is currently devoted to the search for new methods and materials for the rapid, efficient and, as far as possible, a universal way of decontamination of water. Therefore, iron-based technologies still capture an eminent interest in the scientific community, especially, because they are environmentally friendly and use an Earth-abundant material. In addition, some materials containing iron are magnetic and their removing or recycling after culmination of the treatment process from environment is easier and less problematic. [1,2]

Iron occurs in several valence states (i.e., 0, II, III, IV, V and VI). In other words, it exists as metallic iron (Fe(0)), ferrous (Fe(II)), and ferric (Fe(III)) forms in the nature. Moreover, higher oxidation states of iron, such as +IV, +V and +VI, can be synthesized. These compounds of iron, where the Fe atom is in a higher oxidation state, with oxidation number +IV, +V or +VI, are generally called ferrates. These compounds comprise FeO_4^{4-} (Fe^{+IV}), FeO_4^{3-} (Fe^{+V}), and FeO_4^{2-} (Fe^{+VI}) ions. Recent studies have shown that ferrates are very promising materials for decontamination of water. They have a strong oxidation properties; in this respect, ferrate(VI) is the most powerful and most used agent. The stability and reactivity of the ferrates is dependent on pH and initial ferrate concentration of the solution, as well as on the presence of foreign ions in solution. Ferrates are the most stable in range of pH from 9.4 to 9.7. More dilute solutions of the ferrate ions are more stable, which means, that the stability of ferrates decreases with increasing concentration of ferrates solutions. Currently, a significant scientific interest is devoted to the +VI oxidation state of iron (Fe(VI)) because of its potential use as a "greener" oxidant for organic synthesis, a water oxidation catalyst and in treatment contaminants and toxins in water and wastewater. [1,3,4,5]

As already mentioned, ferrate(VI) is the most widely used due to strong oxidizing properties. It exhibits a reduction potential of 2.2 V in acidic medium; the reduction potential decrease to 0.7 V in alkaline solution. Maximum performance is thus achieved in an acidic or a neutral environment. Ferrate(VI) can easily oxidize different organic contaminants and effectively inactive microorganisms or bacterial indicators. Ferrate(VI) can also remove metal and metalloid contaminants. Together with the reduction of Fe(VI) to Fe(III), coagulation and precipitation of certain pollutants is observed. The main advantage of the use of ferrates in water treatment is that ferrates have a disinfecting effect and the resulting reaction products are nontoxic. They have a versatile use, always combining several reaction processes. Ferrates remove also micropollutants, estrogens or antibiotics. Small doses of ferrates are enough for their application in water purification. [2,3,6]

Despite a very wide range of possible applications, the real utilization of ferrates in practice in greater amounts is hindered by their instability. Ferrate(VI) salts are comparatively stable for quite a long time in the case they are stored in a dry atmosphere. Otherwise, ferrates are very unstable in water, and ferrate(VI) is reduced to a compound containing iron in oxidation state +III (Fe^{+III}), which is much more stable. As already mentioned, ferrates are very promising materials, but for real application of these materials in water treatment and others, is necessary, at first, to understand the reaction mechanisms between these compounds and pollutants. Reactions between pollutants and ferrates are very fast and it is difficult to in-situ monitor their course. Fortunately, this obstacle can be overcome by ^{57}Fe Mössbauer spectroscopy. This contribution thus introduces ^{57}Fe Mössbauer spectroscopy as a powerful tool to describe and monitor the reaction mechanisms and also to analyze precursors and intermediate and final products. ^{57}Fe Mössbaeur spectroscopy is presented as a main technique for assessment of sample purity and identifying the intermediate states during transformation from Fe(VI) to Fe(III) as it provides distinguishing between various valence states of iron based on their Mössbauer hyperfine parameters. [3]

More specifically, the methodology of ^{57}Fe Mössbauer spectroscopy is described here for ferrate research. The Mössbauer parameters of oxidation states (+III, +IV, +V and +VI) of iron are also listed. Further, the preparation of ferrates and characterization of the precursors and reaction products are described. Finally, an example of ferrate(VI) application for removing arsenic from the environment is briefly outlined.

2. METODOLOGY OF MÖSSBAUER SPECTROSCOPY FOR INVESTIGATION OF FERRATES REACTION MECHANISM

2.1. Mössbauer spectrum and hyperfine parameters of ferrates

Mössbaeur spectroscopy represents a unique technique to distinguish different oxidation states of iron using the values of the isomer shift (δ). The δ values are very sensitive to the oxidation state of iron. Then, the differences in the position of the Mössbauer resonant lines of ferrates(IV, V, VI) and iron(III) in the velocity ranges are observed (**Figure 1a**).

The typical ^{57}Fe Mössbauer spectrum of iron(III), that is usually the final product of reactions of pollutants with ferrates, is shown in **Figure 1a)** on the top. The spectrum contains a doublet with the value of isomer shift δ = 0.35 ± 0.05 mm/s and the value of quadrupole splitting ΔE_Q = 0.70 ± 0.05 mm/s. For ferrate(IV), a doublet is typically observed with the value of isomer shift δ = - 0.25 ± 0.05 mm/s and the value of quadrupole splitting ΔE_Q = 0.20 ± 0.05 mm/s. Ferrate(V) also gives a doublet with the isomer shift δ = - 0.45 ± 0.05 mm/s and quadrupole splitting ΔE_Q = 0.80 mm/s and the ferrate(VI) exhibits singlet with the value of isomer shift δ = -0.90 ± 0.05 mm/s. In **Figure 1b)**, the diagram with ranges of the isomer shift values observed for iron-containing compounds is displayed. Information on the values of the Mössbauer hyperfine parameters for each oxidation state is necessary to know to subsequently determine a composition of a characterized compound.

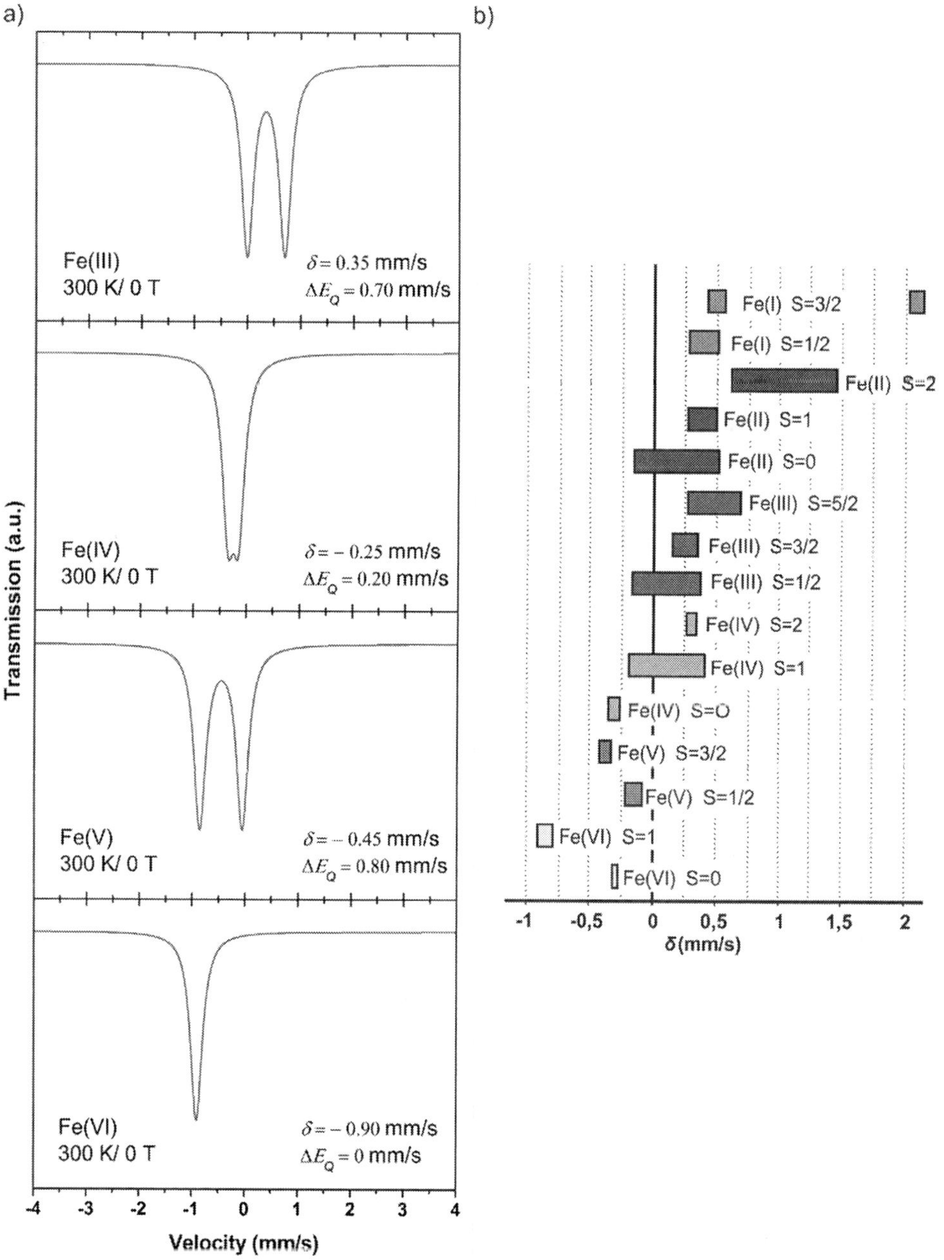

Figure 1 a) Representative [57]Fe Mössbauer spectrum of Fe(III), Fe(IV), Fe(V) and Fe(VI) at room temperature, without an external magnetic field. In the lower left corner, respective Mössbauer hyperfine parameters, such as isomer shift (δ) and quadrupole splitting (ΔE_Q) are displayed for each case. b) Diagram of isomer shift values observed for iron-containing compounds in various oxidation states at room temperature. The isomer shift values are referred to metallic α-Fe at room temperature. [8]

2.2. Preparation of ferrates

In general, three methods are often used to synthesize ferrates. The first method involves dry oxidation by heating a mixture of iron(III) oxides under strongly alkaline conditions and in the presence of oxygen. For example, adopting this strategy K_2FeO_4 (i.e., ferrate(VI)) can be prepared using heating of iron(III) oxides and potassium nitrate over 1100 °C. To synthesize Na_2FeO_4 (i.e., ferrate(VI)), a protocol involving heating of iron(III) oxides and sodium peroxide at 600 °C can be exploited. Then, the Na_4FeO_4 (i.e., ferrate(IV)) can be obtained by heating of iron(III) oxide and sodium peroxide at 370 °C. Generally, dry methods involve mixing ferric salt, often iron oxide, with alkali peroxide (Na_2O_2, K_2O_2) at high temperatures. Depending on the composition of the reaction mixture (Fe, Na, and O_2) and temperature, ferrate(IV), ferrate(V) or ferrate(IV) can be prepared. The second method to synthesize ferrates is based on wet oxidation, when iron(III) oxides or their salts are oxidized by an oxidant (e.g., potassium hypochlorite, sodium hypochlorite, ozone, etc.) in strong alkaline environment. Na_2FeO_4 can be synthesized using sodium hydroxide. K_2FeO_4 can be then precipitated from the solution by subsequent adding of potassium hydroxide to the Na_2FeO_4 solution. Using dry methods, various ferrates(VI) can be prepared, such as Cs_2FeO_4, $RbFeO_4$, $BaFeO_4$, $SrFeO_4$, and others. Third strategy for preparation of ferrates involves electro-chemical methods of anodic oxidation. Iron or iron salts are selected as an anode and sodium or potassium hydroxides are used as an electrolyte. The advantage of this approach for the synthesis of ferrates is often connected with the efficiency and quality, i.e., the yield of the reaction and the phase purity of the final products. The drawbacks include the complexity of the method and financial costs. [1,7]

Based on the study [9] (see the experimental section there for more details) Na_4FeO_4 - Fe(IV) and K_3FeO_4 - Fe(V) salts were prepared and the reproducibility of the solid state synthesis has been verified by ^{57}Fe Mössbauer spectroscopy and X-ray powder diffraction. The following values are referred to metallic α-Fe at room temperature. At first, ferrate(IV), Na_4FeO_4, was synthesized following the reaction **Equation 1**, i.e.,

$$8Na_2O_2 + 2Fe_2O_3 \rightarrow 4Na_4FeO_4 + 3O_2 \quad (1)$$

The reaction took place in a glovebox, where Na_2O_2 was grinded to a powder. Then, Fe_2O_3 was added to Na_2O_2 and homogenized and heated to 400 °C for hour. After cooling down to room temperature, analyzes were performed. The room-temperature ^{57}Fe Mössbauer spectrum of Na_4FeO_4 features a doublet with the isomer shift δ = - 0.23 ± 0.05 mm/s and the quadrupole splitting ΔE_Q = 0.41 ± 0.05 mm/s. Based on the results from ^{57}Fe Mössbauer spectroscopy, the amount of Fe(IV) was 94 at.% and Fe(III) 6 at.%. The X-ray powder diffraction determined the purity of the sample, i.e., the amount of Fe(IV), higher than 90 at.% in accordance with ^{57}Fe Mössbauer spectroscopy measurements. At the top of **Figure 2**, ^{57}Fe Mössbauer spectrum (a)) and X-ray diffraction pattern (b)) of synthesized Na_4FeO_4 are shown. To prepare Fe(V), the two-steps reaction process was performed (reaction **Equation 2** and **Equation 3**).

$$Fe^{III}O_x(OH)_y + KNO_3 \rightarrow KFeO_2 + NO_x \quad (2)$$

$$KFeO_2 + 2KNO_3 \rightarrow K_3FeO_4 + 2NO_2 \quad (3)$$

At first, homogenized mixture of KNO_3 and $Fe^{III}O_x(OH)_y$ was heated up to 950 °C for 30 minutes. After cooled down, the $KFeO_2$ was observed and then, in the second step, the $KFeO_2$ was added to KNO_3. This mixture was again homogenized and heated up to 1000 °C in presence of nitrogen. After that, the prepared K_3FeO_4 was frozen using liquid nitrogen. The room-temperature ^{57}Fe Mössbauer spectrum of K_3FeO_4 shows a doublet with the isomer shift δ = - 0.52 ± 0.05 mm/s and the quadrupole splitting ΔE_Q = 0.88 ± 0.05 mm/s. In the middle of **Figure 2**, the ^{57}Fe Mössbauer spectrum (c)) and X-ray diffraction pattern (d)) of synthesized K_3FeO_4 are displayed. The X-ray powder diffraction again determined the purity of the sample, i.e. the amount of Fe(V), higher than 90 at.%. Based on the analysis of the respective ^{57}Fe Mössbauer spectrum, the amount of Fe(V) was 93 at.% and the amount of Fe(III) was 7 at.%. Using ^{57}Fe Mössbauer spectroscopy, it is also feasible to check the quality of potassium ferrate(VI), with declared purity higher than 95 %, from Sigma-Aldrich. This ferrate is usually used in applications for water treatment. At the bottom of the **Figure 2**, the ^{57}Fe Mössbauer spectrum (e)) and X-ray diffraction pattern (f)) of K_2FeO_4 are shown. The respective ^{57}Fe Mössbauer spectrum of K_2FeO_4 features a singlet with the isomer shift δ = - 0.89 ± 0.05 mm/s. The X-ray powder diffraction and also ^{57}Fe Mössbauer spectroscopy determined 96 at.% of Fe(VI) and 4 at.% of Fe(III). [9]

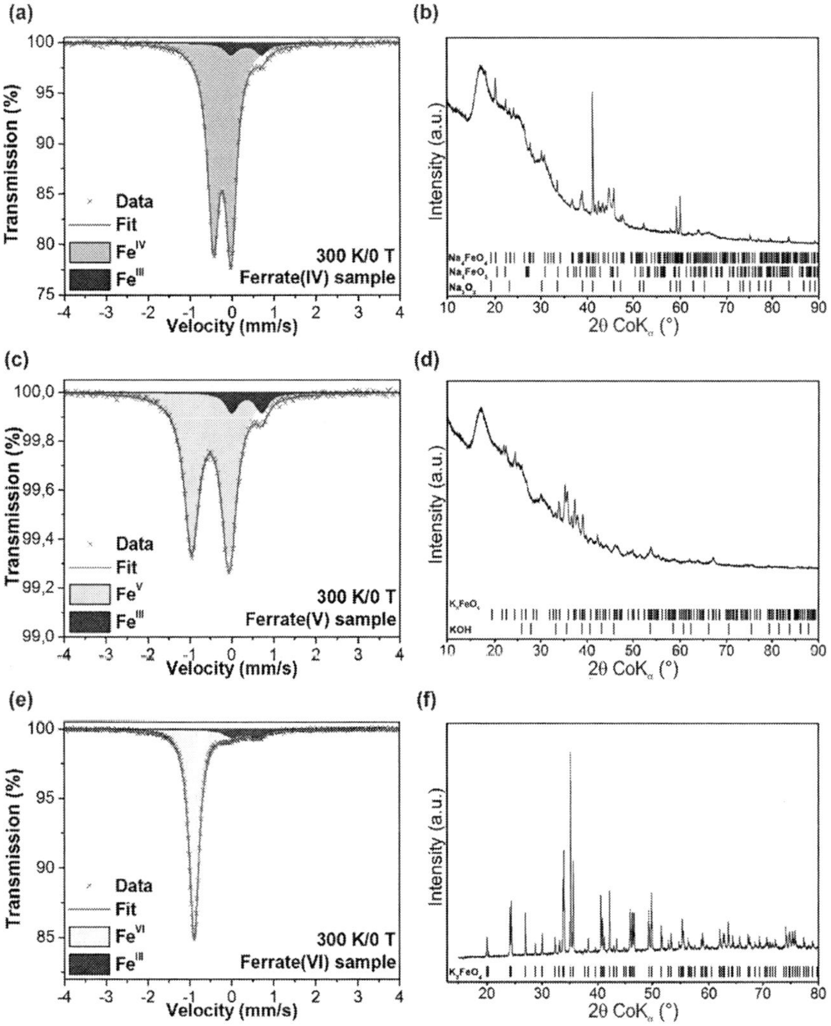

Figure 2 a) ^{57}Fe Mössbauer spectrum of Na_4FeO_4 - Fe(IV), b) X-ray diffraction pattern of synthesized Na_4FeO_4, c) ^{57}Fe Mössbauer spectrum of K_3FeO_4 - Fe(V), d) X-ray diffraction pattern of synthesized K_3FeO_4, e) ^{57}Fe Mössbauer spectrum of K_2FeO_4 - Fe(VI), f) X-ray diffraction pattern of synthesized K_2FeO_4. [9]

2.3. Reactions of ferrates

As already mentioned, ferrates have a high reduction potential. Thus, they are equipped with powerful oxidation properties; among them ferrate(VI) is the most strongest oxidant. Due to this feature, ferrate(VI) is the most used agent from group of ferrates. Ferrate(VI) has a reduction potential of 2.2 V in acidic medium and the reduction potential equals to 0.7 V in alkaline medium. Below, the equations of reactions of ferrate(VI) in the alkaline solution (**Equation 4**) and in the acidic solution (**Equation 5**) are listed. [1,6,7]

$$FeO_4^{2-} + 4H_2O + 3e^- \rightarrow Fe(OH)_3 + 5OH^- \quad (4)$$

$$FeO_4^{2-} + 8H^+ + 3e^- \rightarrow Fe^{3+} + 4H_2O \quad (5)$$

Thus, in water, ferrate(VI) immediately decomposes to stable oxidation state of Fe^{III}. When an electron is gained in water, ferrate(VI) is reduced to intermediate products, i.e., high valence states of iron, Fe(V) and Fe(IV), through 1-e^- and 2-e^- transfer processes. Simultaneously, the final iron-containing compound with iron in oxidation state Fe^{III} is also observed. Proof of the formation of intermediate iron species, Fe(V) and Fe(IV), during decomposition of Fe(VI), can be performed by fast freezing of reaction solution and subsequent characterization using ^{57}Fe Mössbauer spectroscopy (see **Figure 3**). The reaction rate is always based on pH, the concentration of the solution and the presence of foreign ions. The resulting products of ferrate(VI) decomposition in water are always iron(III) oxides or hydroxides (e.g., Fe_2O_3, Fe_3O_4, FeOOH, $Fe(OH)_3$). The decomposition is accompanied by a change of color from black-purple to sorrel. Once the color is changed, the substances are no longer usable. [1,10,11]

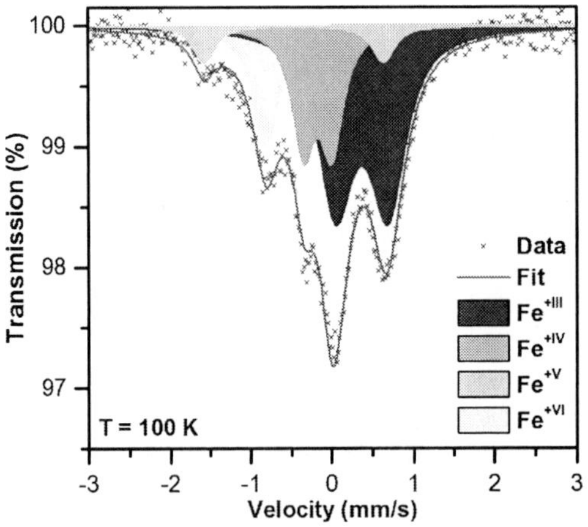

Figure 3 ^{57}Fe Mössbauer spectrum of the sample taken during decomposition of Fe(VI) to Fe(III) and characterized using ^{57}Fe Mössbauer spectroscopy. All possible oxidation states of iron are observed. [9]

2.4. Example of application ^{57}Fe Mössbauer spectroscopy in the removal of arsenic from environment using ferrate(VI)

In aerobic surface water, primarily arsenate ($As^VO_4^{3-}$, As(V)) occurs. Then, arsenite ($As^{III}O_3^{3-}$, As(III)) predominantly exists in anaerobic environment. However, arsenite (As(III)), that is more toxic than arsenate (As(V)), oxides to arsenate under aerobic conditions. One of the arsenic removal options is to use a sorbent, where arsenic is bonded either on the surface or into the sorbent pores. Current technologies are limited by a low sorption capacity of sorbents. The ratio between sorbent and arsenic is usually in the ranges from 1000:1 to 100:1 in weight fraction. The second restriction of available technologies bonding arsenic onto the surface sorbent lies in release of arsenic back to environment due to leaching effect. Thus, exploitation of ferrate(VI) combines the properties of an oxidant, used to oxidizing of As(III) to As(V), and efficient sorbent with a low sorbent/As ratio. More specifically, here we mention an example of removal of arsenic from environment using K_2FeO_4, with concentration of iron equal to 0.5 g/L (see more details in study [13]). As sources of As(III) and As(V), $NaAsO_2$ and $Na_2HAsO_4 \cdot 7H_2O$ were used, respectively, with a final concentration of arsenic of 100 mg/L. Then, the ratio between Fe:As is 5:1. Three studied samples are shown in **Figure 4**. The "blank" sample was prepared by adding of ferrate(VI) to deionized water, at the "ex-situ" sample was at first added ferrate(VI) to deionized water and subsequently As-containing compound. At the third "in-situ" sample was simultaneously added ferrate(VI) and also As-containing compound to deionized water.

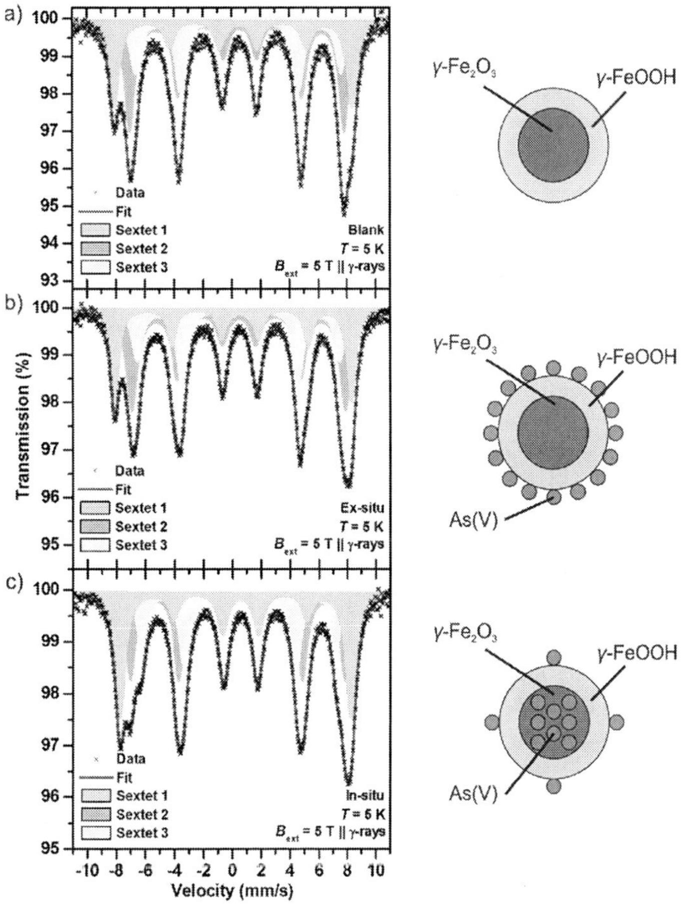

Figure 4 ^{57}Fe Mössbauer spectra collected at a temperature of 5 K and in a field of 5 T for a) "blank" sample (sample of ferrate(VI) added to deionized water), b) "ex-situ" sample (As(V)-containing compound was added to mixed deionized water with ferrate(VI)) and c) "in-situ" sample (simultaneously addition of ferrate(VI) and also As(V)-containing compound to deionized water). On the left side, the embedding of As(V) onto the surface nanoparticle (panel (b)), and primarily into the structure of nanoparticle and subsequently onto the surface (panel (c)) is illustrated. [13]

The experiment is based on forming γ-Fe_2O_3 nanoparticles, which can be after the completion of the treatment process, very easily removed from environment using an external magnet. As seen in **Figure 4 c)**, the uniqueness of using ferrate(VI) in water treatment is that arsenite is "in-situ" structurally embedded into the tetrahedral sites of γ-Fe_2O_3 nanoparticles formed during reaction with water and the pollutant. Thus, As(V) ions are not released back to the environment and they are firmly bonded into the structure of nanoparticles. Moreover, using ^{57}Fe Mössbauer spectroscopy, the amount of As(V), which are embedded in the nanoparticles (*x* in **Equation 6**) is possible to determine based on the evaluation of spectral ratio of the components belonging to the tetrahedral and octahedral sites in the γ-Fe_2O_3 crystal structure, i.e.,

$$\left[Fe^{3+}_{1-x}As^{5+}_{x}\right]^{T}\left[Fe^{3+}_{5/3-y}o_{1/3+y}\right]^{O}O_4 \qquad (6)$$

In the "blank" sample, the spectral ratio between tetrahedral (T) and octahedral (O) sites is T:O ~ 0.65, in the "ex-situ" sample, T:O ~ 0.7 and in the "in-situ" sample, T:O ~ 1.22. Based on these values, the amount of As(V) in the nanoparticles was estimated to be approximately 25 at.%. [13]

3. CONCLUSION

This study highlighted the significance of ^{57}Fe Mössbauer spectroscopy for studying reaction mechanism of ferrates with pollutants. Ferrates, particularly ferrate(VI) and their preparation and reaction mechanism were described in more details. Finally, the experiment presenting an unique properties of ferrate(VI) for removing of arsenic from environment was reviewed.

ACKNOWLEDGEMENTS

The authors acknowledge support from the Ministry of Education, Youth and Sports of the Czech Republic under project No. LO1305.

REFERENCES

[1] SHARMA, Virender K., ZBORIL, Radek and VARMA, Rajender S. *Ferrates: Greener Oxidants with Multimodal Action in Water Treatment Technologies.* Acc Chem Res., 2015. pp. 182-191.

[2] PRUCEK, Robert, TUČEK, Jiří, KOLAŘÍK, Jan, HUŠKOVÁ, Ivana, FILIP, Jan, VARMA, Rajender S., SHARMA, Virender K., ZBOŘIL, Radek. *Ferrate(VI)-Prompted Removal of Metals in Aqueous Media: Mechanistic Delineation of Enhanced Efficiency via Metal Entrenchment in Magnetic Oxides.* Environmental Science & Technology, 2015, pp. 2319-2327.

[3] MACHALA, Libor, ZBORIL, Radek, SHARMA, Virender K., FILIP, Jan, JANCIK, Dalibor and HOMONNAY, Zoltan. *Transformation of Solid Potassium Ferrate(VI) (K_2FeO_4): Mechanism and Kinetic Effect of Air Humidity.* Eur. J. Inorg. Chem., 2009. pp. 1060-1067.

[4] MACHALA, Libor, ZBORIL, Radek, SHARMA, Virender K., FILIP, Jan, SCHNEEWEISS, Oldrich and HOMONNAY, Zoltán. *Mössbauer Characterization and in Situ Monitoring of Thermal Decomposition of Potassium Ferrate(VI), K2FeO4 in Static Air Conditions.* J. Phys. Chem. B, 2007. pp. 4280-4286.

[5] SHARMA, Virender K., PERFILIEV, Yurii D., ZBOŘIL, Radek, MACHALA, Libor and WYNTER, Clive I.. *Mössbauer Spectroscopy, Applications in Chemistry, Biology, and Nanotechnology.* WILEY. 2013. FERRATES(IV, V AND VI): MÖSSBAUER SPECTROSCOPY CHARACTERIZATION, chapter 24, pp. 505-506.

[6] DARKO, Beatrice, JIANG, Jia-Qian, KIM, Hyunook, MACHALA, Libor, ZBORIL, Radek, SHARMA, Virender K. *Advances Made in Understanding the Interaction of Ferrate(VI) with Natural Organic Matter in Water.* Elsevier, 2014. pp.183-197.

[7] JIANG, Jia-Qian. *Advances in the development and application of ferrate(VI) for water and wastewater treatment.* J. Chem. Technol. Biotechnol. 2014. pp. 165-177.

[8] GÜTLICH, Philipp, BILL, Eckhard, TRAUTWEIN, Alfred X. *Mössbauer Spectroscopy and Transition Metal Chemistry,Fundamentals and Applications* .Springer: Springer-Verlag Berlin Heidelberg, 2011. p. 85.

[9] ŠIŠKOVÁ, MACHALOVÁ, Karolína, JANČULA, Daniel, DRAHOŠ, Bohuslav, MACHALA, Libor, BABICA, Pavel, ALONSO, Paula G., TRÁVNÍČEK, Zdeněk, TUČEK, Jiří, MARŠÁLEK, Blahoslav, SHARMA, Virender K., and ZBOŘIL, Radek. *High-valent iron (Fe^{VI}, Fe^{V}, and Fe^{IV}) species in water: characterization and oxidative transformation of estrogenic hormones.* Phys. Chem. Chem. Phys., 2016, **18**. pp.18802-18810.

[10] CUI, Junkui, ZHENG, Lei, and DENG, Yang. *Emergency water treatment with ferrate(VI) in response to natural disasters.* Environ. Sci.: Water Res. Technol., 2018, **4**. pp.359-368.

[11] DEDUSHENKO, S. K., PERFILIEV, Yu. D., KULIKOV, L. A. *Mössbauer spectroscopy and quality control in ferrate technology.* Hyperfine Interact, 2013. pp.59-65.

[12] PRUCEK, Robert, TUČEK, Jiří, KOLAŘÍK, Jan, MARUŠÁK, Zdeněk, SHARMA, Virender K., and ZBOŘIL, Radek. *Ferrate(VI)-Induced Arsenite and Arsenate Removal by In Situ Structural Incorporation into Magnetic Iron(III) Oxide Nanoparticles.* Environ. Sci. Technol., 2013. pp.3283-3292.

STRUCTURAL AND MAGNETIC PROPERTIES OF NANOMATERIALS PREPARED BY MODIFICATION OF IRON NANOPARTICLES AND THEIR EFFICIENCY IN Cr(VI) REMOVAL: A COMPARATIVE STUDY

Josef KAŠLÍK, Ivo MEDŘÍK, Jan FILIP, Eleni PETALA, Ondřej MALINA, Radek ZBOŘIL

Regional Centre of Advanced Technologies and Materials, Palacký University, Olomouc, Czech Republic, EU, josef.kaslik@upol.cz

Abstract

Technologies involving iron-based nanomaterials, such as nanoscale zero-valent iron (nZVI), are extensively used for elimination of various contaminants in the environment. In this study, we modified commercial nZVI particles NANOFER 25P (from NANO IRON company, CZ) via one-step thermal treatment under different conditions (including temperature, time, and atmosphere), in a way that has not been previously reported. The structural, magnetic and air-stability aspects of the as-prepared materials were investigated by means of X-ray powder diffraction, Mössbauer spectroscopy, and magnetic and surface area measurements. The characterization revealed that each of the materials is dominantly formed by different crystalline phase (i.e., iron nitrides Fe_4N and $Fe_{2.43}N$, FeO, and Fe_3C) and that the modification occurred in almost entire volume of the particles rather than only on the surface. Moreover, the as-prepared materials were evaluated for their efficiency of Cr(VI) removal and compared with bare nZVI used for their preparation.

Keywords: Solid state reaction, modification, iron nitride, iron carbide, characterization

1. INTRODUCTION

Nanoscale zero valent iron (nZVI) belongs to the group of widely studied materials with a great potential in various environmental applications. It can be effectively utilized in reductive remediation technologies for degradation of numerous inorganic and organic contaminants, including heavy metals, nitrates, chemical warfare agents, polychlorinated hydrocarbons, pesticides, organic dyes and antibiotics [1-3].

Although the high reactivity of bare nZVI is undoubtedly their great advantage, it also causes fast oxidation during their exposure to the air. Therefore such materials have to be stored and handled under inert conditions or in slurry which increase cost of the material. The other possibility is stabilization of the particles by creation of inorganic passivating shell on the particle surface, which can be formed either by iron oxide or by carbon and/or iron carbides [4,5]. Such shell protects the particles against further oxidation and the materials can therefore be stored and manipulated without special requirements.

In this paper, we present a different approach of nZVI modification. Four materials were prepared via one-step thermal treatment of bare nZVI in various conditions (namely temperature, time, composition and pressure of environment). The as-prepared materials were characterized from structural, magnetic and stability point of view. Moreover, their efficiency in Cr(VI) removal was evaluated.

2. EXPERIMENTAL SECTION

2.1. Materials preparation

All materials were synthesized via one step solid state-gas reaction using commercial nZVI NANOFER 25P (NANO IRON, CZ) as a precursor. The samples N1 and N2 were prepared by annealing of 35g of nZVI in a mixture of NH_3 and N_2 (in ratio 1:2) with relative overpressure 0.5 bar for 120 min at temperatures 200 and

250 °C, respectively. The samples W1 and C1 were prepared by annealing of 55 g of NANOFER 25P in the mixture of CO_2 and H_2 (ratio 4:1) and relative gas overpressure 0.25 bar at 580 °C for 300 min and 500 °C for 240 min, respectively. All prepared materials were cooled down to room temperature in the furnace, transported under inert conditions to the glove-box, and stored there prior use. The commercial nZVI NANOFER 25P (labeled as NF25) was used as a reference for evaluation of the efficiency of prepared materials for Cr(VI) removal.

2.2. Characterization techniques

X-ray powder diffraction was performed on X`Pert PRO MPD powder diffractometer (Malvern Panalytical) equipped with iron filtered Co$K_\alpha$ radiation source and fast X`Celerator detector. The samples were deposited (under nitrogen atmosphere) on zero background Si slide and covered with Mylar foil. The High Score Plus software in conjunction with PDF-4+ and ICSD databases were used for evaluation of captured data.

Transmission ^{57}Fe Mössbauer spectra were measured at room temperature employing a MS2007 Mössbauer spectrometer based on virtual instrumentation technique, using a constant acceleration drive spectrometer and a commercial ^{57}Co(Rh) radiation source. The acquired spectra were evaluated using the MossWinn software. The isomer shift values were referred to α-Fe foil at room temperature.

A superconducting quantum interference device (SQUID, MPMS XL-7, Quantum Design, USA) was used for the magnetization measurements. The hysteresis loop measurements were recorded at the temperature of 300 K and in external magnetic fields ranging from -5 to +5 T.

Specific surface area (SSA) was measured employing Autosorb iQ analyzer (Quantachrome) at temperature of liquid nitrogen (77.4 K). The materials were degassed before measurement at room temperature for 12 hours. SSA was calculated using the multipoint BET3 method in p/p_0 between 0.01 and 0.30. The analysis was performed with the AsiQwin software (Quantachrome).

2.3. Reactivity evaluation

The efficiency of the as-prepared materials towards Cr(VI) removal was evaluated under aerobic conditions. In a typical procedure, 10 mg of each sample were mixed with 100 mL of prepared solution with concentration of Cr(VI) 6.5 mg/L. The pH was adjusted to 5.5 by HCl and the bottle with solution was shaken on a rotational shaker at room temperature for 48 hours. After 48 hours, 0.3 mL of the Cr(VI) solution was withdrawn and used for colorimetric measurement. The Cr(VI) concentration was determined by the 1,5-diphenylcarbazide method, which is based on the reaction of Cr(VI) cations with 1,5-diphenylcarbazide molecules leading to the formation of a red−purple chromium 1,5-diphenylcarbazide complex that shows absorbance maximum at 541.5 nm. 0.3 mL of the Cr(VI) solution was mixed with 2.7 mL of distilled water, 120 µL of 1,5-diphenylcarbazide acetone solution (10.3 mM) and 60 µL of H_3PO_4 aqueous solution (0.86 M). The solution was left 10 min to allow color development, and then the concentration was determined by UV-Vis spectroscopy. A calibration curve (in the range 0 - 1 mg/L) has been previously determined.

3. RESULTS AND DISCUSSION

3.1. Structural and magnetic aspects of prepared materials

The structural and phase analysis revealed that each of the prepared samples is dominantly formed by different crystalline phase (see **Table 1** and **Figure 1**). The samples prepared in a mixture of ammonium with nitrogen are formed mostly by iron nitrides, namely Fe_4N for sample N1, and $Fe_{2.43}N$ for sample N2. The samples prepared in mixture of carbon dioxide with nitrogen are formed mostly by wüstite and iron carbide (samples W1, and C1, respectively). This indicates that the modification took place in almost full volume of the particles rather than just by creation of the surface shell on the particles. However, except sample W1, all other samples contained small amount of α-Fe (less than 11 wt%; for details see **Table 1**). Additionally, all samples contained small amount of magnetite/maghemite (less than 12 wt%). When the temperature applied during thermal treatment was equal to 500 °C or higher (i.e., during preparation of the samples W1 and C1), then the silicon, which is the trace element contained in nZVI precursor, formed separate Fe_2SiO_4 phase (fayalite). Its content increased with increasing temperature (i.e., from 6 wt% for sample W1 to 10 wt% for sample C1). The crystallinity (i.e., the mean X-ray coherence length (MCL)) of the dominant phases increased with increasing temperature from 35 nm for Fe_4N in N1 sample to 119 nm for Fe_3C in C1 sample.

The SSA was measured for all prepared materials, and additionally for precursor NF_25. While the samples N1 and N2 keep the SSA similar to the precursor (17.9, 18.0, m^2/g for N1 and N2, respectively, compared to 17.8 m^2/g for NF_25), the SSA of samples W1 and C1 significantly drop to 7.3 and 6.2 m^2/g, respectively. Such decrease is presumably caused by the temperature induced sintering of the particles during preparation of the materials, being thus consistent with the evolution of MCL.

Table 1 Quantification of relative content (RC) of crystalline phases in prepared materials and their mean X-ray coherence lengths (MCL)

Sample	α-Fe		FeO		Fe_3O_4/ γ-Fe_2O_3		Fe_2SiO_4	Fe_3C		$Fe_{2.43}N$		Fe_4N	
	RC (wt%)	MCL (nm)	RC (wt%)	MCL (nm)	RC (wt%)	MCL (nm)	RC (wt%)	RC (wt%)	MCL (nm)	RC (wt%)	MCL (nm)	RC (wt%)	MCL (nm)
N1	11	44	---	---	12	31	---	---	---	2	35	75	35
N2	1	43	---	---	4	32	---	---	---	69	54	26	34
W1	---	---	94	58	<1	41	6	---	---	---	---	---	---
C1	3	80	2	33	<1	65	10	85	119	---	---	---	---

The Mössbauer spectra of all samples are quite complicated (see **Figure 1** and **Table 2**). For the both samples containing nitrides (i.e., N1 and N2), each spectrum composes of 7 components: one sextet (*S1*) can be ascribed to α-Fe, two sextets (*S2* and *S3*) to Fe atoms in nonstoichiometric magnetite (where sextet *S3* represents octahedral iron atoms with effective valence 2,5+ observed because of electron hopping effect; i.e., paired Fe^{2+} and Fe^{3+}; and sextet *S2* represents tetrahedral Fe^{3+} and unpaired octahedral Fe^{3+} ions) [6], three sextets (*S4*, *S5*, and *S6*) to Fe atoms with nonequivalent neighborhood in Fe_4N [7] and doublet (*D1*), which in the case of N1 sample represents magnetically disordered atoms Fe^{3+}, and in the case of N2 sample Fe atoms belonging to $Fe_{2.x}N$ [8]. The Mössbauer spectrum of sample W1 contains three components: two doublets (*D1* and *D2*) which can be ascribed to wüstite in its nonstoichiometric form [9], and the third doublet (*D3*) representing Fe^{2+} ions in fayalite. For the last sample, C1, the spectrum is again composed of 7 components: one sextet (*S1*) represents α-Fe, two sextets (*S2* and *S3*) Fe atoms in nonstoichiometric magnetite, one sextet (S4) Fe atoms in iron carbide Fe_3C [10], two doublets (*D1* and *D2*) Fe^{2+} ions in nonstoichiometric wüstite, and last doublet (*D3*) Fe^{2+} ions in fayalite. Generally, the results of Mössbauer spectroscopy and XRD correlate well. The observed differences can be caused by presence of the low-crystalline to amorphous shell, which cannot be principally observed by XRD analysis.

The hysteresis loops measured at 300 K show the values of saturation magnetization for prepared materials (**Figure 1**). For sample W1 the hysteresis loop manifests the paramagnetic behavior of the material, which is for wüstite typical at room temperature. The magnetization of other samples is saturated at 5 T and reaches values 161.3, 93.9, and 102.6 Am^2/kg for samples N1, N2, and C1, respectively.

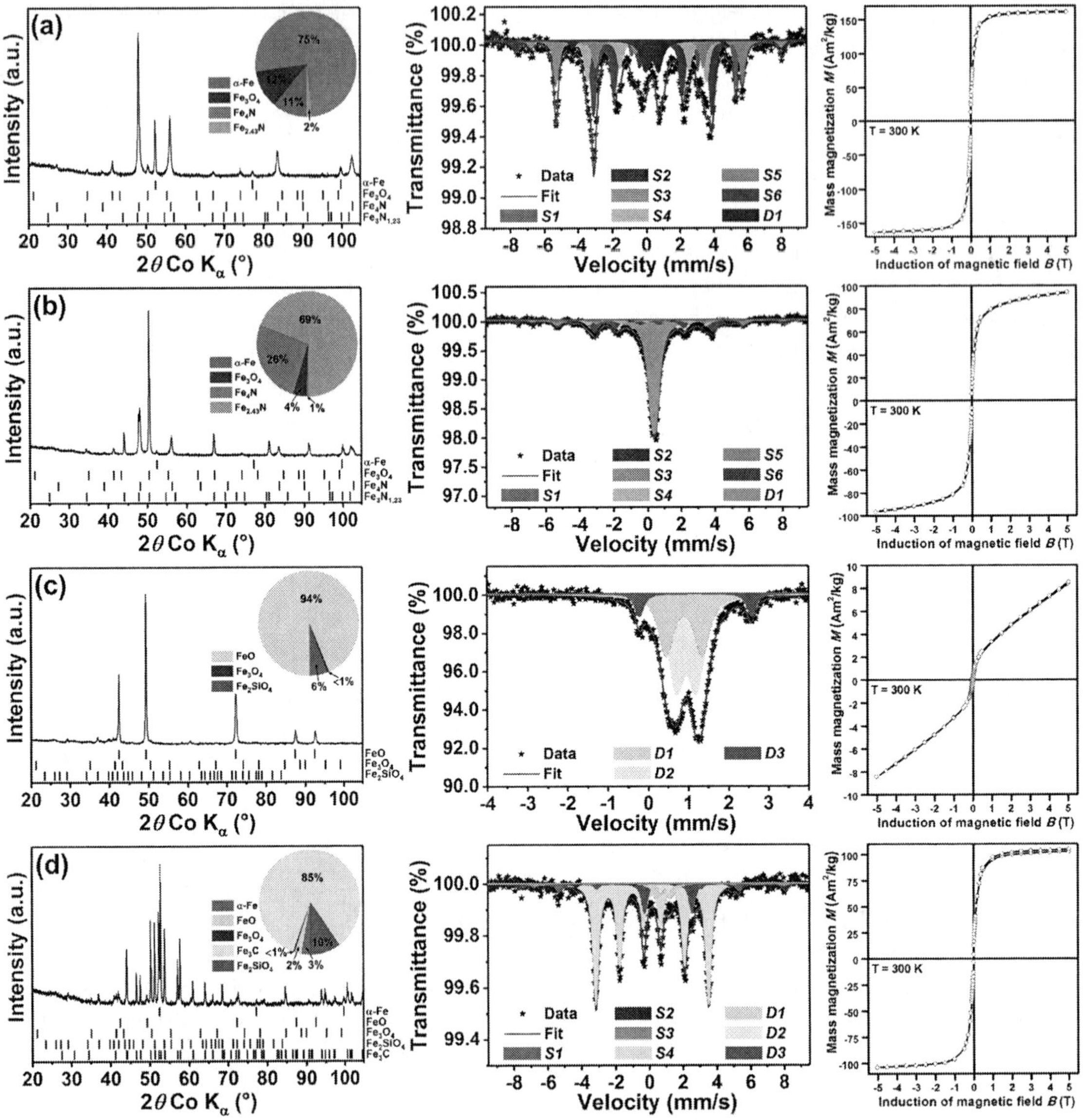

Figure 1 Diffraction patterns (left), Mössbauer spectra (middle), and hysteresis loops captured at 300 K (right) of prepared samples (a) N1, (b) N2, (c) W1, and (d) C1

3.2. Stability of prepared materials

The stability of prepared materials was evaluated after their exposure to the ambient atmosphere for 7 days. After their first contact with air, the materials evolved some heat as a result of their partial surface oxidation. Immediately performed structural and phase analysis (performed by XRD) revealed that the composition did not change significantly. However, the content of dominant phase decreased about approximately 2 or 3 wt% to the detriment of magnetite/maghemite. The additional storage did not initiate any further changes in composition of crystalline phases supporting the idea that the initial exothermic effect is a manifestation of creation of passivating shell on the surface of the particles.

Table 2 Hyperfine parameters of prepared samples derived from Mössbauer spectra. δ is the center shift, ΔE_Q is the quadrupole splitting, B_{hf} is hyperfine magnetic field, Γ is line width, and RA is relative subspectra area.

Sample	Component	δ ± 0.01 (mm/s)	ΔE_Q ± 0.01 (mm/s)	B_{hf} ± 0.3 (T)	Γ ± 0.01 (mm/s)	RA ± 1 (%)	Assignment
N1	*S1*	0.00	0.00	33.1	0.24	13	α-Fe
	S2	0.26	-0.04	49.1	0.38	5	^{tet}Fe (magnetite)
	S3	0.67	-0.11	46.0	0.38	3	^{oct}Fe (magnetite)
	S4	0.29	0.24	21.7	0.39	34	Fe_4N
	S5	0.22	0.00	34.1	0.38	18	Fe_4N
	S6	0.29	-0.45	21.9	0.38	18	Fe_4N
	D1	0.35	0.78	---	0.58	9	Fe^{3+}
N2	*S1*	0.00	0.00	33.0	0.33	1	α-Fe
	S2	0.26	-0.04	49.0	0.37	3	^{tet}Fe (magnetite)
	S3	0.67	-0.10	46.0	0.43	2	^{oct}Fe (magnetite)
	S4	0.29	0.24	21.7	0.50	18	Fe_4N
	S5	0.22	0.00	34.2	0.49	8	Fe_4N
	S6	0.29	-0.45	21.9	0.50	8	Fe_4N
	D1	0.39	0.26	---	0.58	60	$Fe_{2.x}N$
W1	*D1*	0.90	0.79	---	0.52	59	Fe^{2+} (wüstite)
	D2	0.95	0.44	---	0.45	33	Fe^{2+} (wüstite)
	D3	1.17	2.80	---	0.26	8	Fe^{2+} (Fayalite)
C1	*S1*	0.00	0.00	33.0	0.36	4	α-Fe
	S2	0.26	-0.04	49.1	0.38	3	^{tet}Fe (magnetite)
	S3	0.67	-0.11	46.0	0.38	2	^{oct}Fe (magnetite)
	S4	0.19	0.01	20.7	0.38	74	Fe_3C
	D1	0.90	0.79	---	0.58	6	Fe^{2+} (wustite)
	D2	0.95	0.44	---	0.26	2	Fe^{2+} (wustite)
	D3	1.14	2.84	---	0.36	9	Fe^{2+} (Fayalite)

3.3. Reactivity evaluation

The prepared materials were evaluated in terms of their efficiency towards Cr(VI) reduction (**Figure 2**). Moreover, their efficiency was comparable with unmodified nZVI (sample NF25). NF25 exhibited high removal efficiency, reaching 80 % of Cr(VI) removal after 48 hours. In comparison with that, the modified materials performed lower capacity but still adequate for such environmental applications. The most efficient were materials based on nitrides (i.e., the samples N1 and N2), which additionally contain a small residual portion of Fe^0. Their efficiencies were between 40 and 50 %. On the other hand, the material formed dominantly by wüstite (i.e., sample W1) exhibited relatively small efficiency, i.e., 15 % approximately. The iron carbide based material (sample C1) showed sufficient removal ability equal to 35 %.

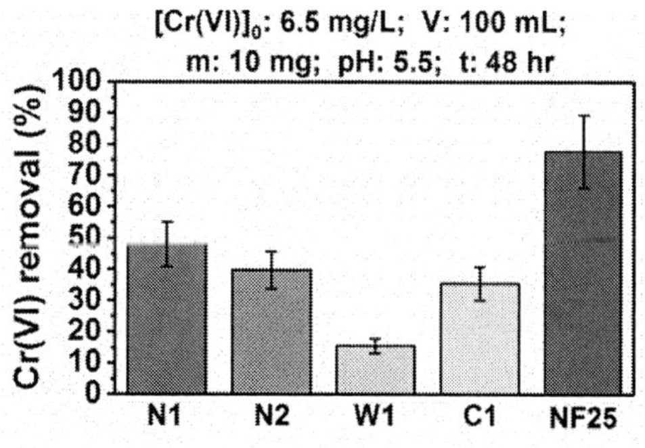

Figure 2 The efficiency of Cr(VI) removal by prepared materials after 48 hours

The mechanism of Cr(VI) removal was evaluated by XPS characterization of solid fractions obtained by filtering of the solution at the end of reactivity experiments. In all cases the Cr was identified only as Cr(III) without any traces of Cr(VI) (data not shown). Thus, the suggested removal mechanism involves reduction and subsequent precipitation of the reduced chromium species. It has to be mentioned, that the estimated lower removal capacity of the modified materials in comparison with the unmodified nZVI, is not a restrictive factor. In overall, these materials showed relatively high removal capacity reaching values of 35 mg of Cr(VI) per gram of material, values among the highest that involves nanoparticles for water remediation.

4. CONCLUSION

Four types of modification of bare nZVI were performed via one-step thermal treatment using different conditions. nZVI particles were modified in almost full volume of the particles rather than by only creation of a shell on the surface of the particles. Detailed structural and magnetic characterization revealed that each of the prepared materials is dominantly formed by different phase (i.e., by Fe_4N, $Fe_{2.43}N$, FeO, and Fe_3C). The prepared materials exhibited relatively high removal capacity which is comparable to conventional nanoparticles involved for water remediation. Moreover, these materials showed better stability and applicability in comparison to the bare nZVI, which is crucial for water treatment technologies.

ACKNOWLEDGEMENTS

The authors gratefully acknowledge support from project LO1305 of the Ministry of Education, Youth and Sports of the Czech Republic.

REFERENCES

[1] FU, F., DIONYSIOU, D.D. and LIU, H. The use of zero-valent iron for groundwater remediation and wastewater treatment: A review. *Journal of Hazardous Materials*. 2014. vol. 267, pp. 194-205.

[2] ZBORIL, R., ANDRLE, M., OPUSTIL, T., MACHALA, L., TUCEK, J., FILIP, J., MARUSAK, Z. and SHARMA, V.K. Treatment of chemical warfare agents by zero-valent iron nanoparticles and ferrate(VI)/(III) composite. *Journal of Hazardous Materials*. 2012. vol. 211, no. SI, pp.126-130.

[3] LI, X.-Q., ELLIOTT, D.W. and ZHANG, W.X. Zero-valent iron nanoparticles for abatement of environmental pollutants: materials and engineering aspects. *Critical Reviews in Solid State and Materials Sciences*. 2006. vol. 31, no. 4, pp. 111-122.

[4] KAŠLÍK, J., KOLAŘÍK, J., FILIP, J., MEDŘÍK, I., TOMANEC, O., PETR, M., MALINA, O., ZBOŘIL, R. and TRATNYEK, P.G. Nanoarchitecture of advanced core-shell zero-valent iron particles with controlled reactivity for contaminant removal. *Chemical Engineering Journal*. 2018. vol. 354, pp. 335-345.

[5] JOZWIAK, W.K., KACZMAREK, E., MANIECKI, T.P., IGNACZAK, W. and MANIUKIEWICZ, W. Reduction behavior of iron oxides in hydrogen and carbon monoxide atmospheres. *Applied Catalysis A: General*. 2007. vol. 326, no. 1, pp.17-27.

[6] GORSKI, C.A. and SCHERER, M.M. Determination of nanoparticulate magnetite stoichiometry by Mössbauer spectroscopy, acidic dissolution, and powder X-ray diffraction: A critical review. *American Mineralogist*. 2010. vol. 95, no. 7, pp. 1017-1026.

[7] BORSA, D.M. and BOERMA, D.O. Phase identification of iron nitrides and iron oxy-nitrides with Mössbauer spectroscopy. *Hyperfine Interactions*. 2003. vol. 151, no. 1, pp. 31-48.

[8] KSENOFONTOV, V., REIMAN, S., WALDECK, M. NEWA, R., KNIEP, R. and GUTLICH, P. In situ - High temperature Mossbauer spectroscopy of iron nitrides and nitridoferrates. *Zeitschrift Fur Anorganische Und Allgemeine Chemie*. 2003. vol. 629, no. 10, pp.1787-1794.

[9] CORNELL, R.M. and SCHWERTMANN, U. *The Iron Oxides: Structure, Properties, Reactions, Occurrences and Uses*. 2nd ed. Weinheim: WILEY-VCH Verlag GmbH & Co. KGaA. p. 664.

[10] BI, X.-X., GANGULY, B., HUFFMAN, G.P., HUGGINS, F.E., ENDO, M. and EKLUND, P.C. Nanocrystalline α-Fe, Fe_3C, and Fe_7C_3 produced by CO_2 laser pyrolysis. *Journal of Materials Research*. 1993. vol. 8, no. 7, pp.1666-1674.

PREPARATION AND CHARACTERIZATION OF FIBROUS NON-WOVEN TEXTILE DECORATED BY SILVER NANOPARTICLES FOR WATER FILTRATION

Petra PEER[1], Martina POLÁŠKOVÁ[2,3], Jaroslav MOSNÁČEK[2,4], Jana SEDLAŘIKOVÁ[3,5], Milan MASAŘ[3]

[1]Institute of Hydrodynamics of the Czech Academy of Sciences, v. v. i., Prague, Czech Republic, EU, peer@ih.cas.cz
[2]Department of Polymer Engineering, Faculty of Technology, Tomas Bata University in Zlín, Zlín, Czech Republic, EU
[3]Centre of Polymer Systems, Tomas Bata University in Zlín, Zlín, Czech Republic, EU
[4]Slovak Academy of Sciences, Polymer Institute, Bratislava, Slovak Republic, EU
[5]Department of Fat, Surfactant and Cosmetic Technology, Faculty of Technology, Tomas Bata University in Zlín, Zlín, Czech Republic, EU

Abstract

The environmentally friendly preparation of silver nanoparticles was proposed for the production of antibacterial non-woven textile used in water filtration. The silver nanoparticles were prepared by chemical reduction of silver nitrate using two environmentally friendly reducing agents, such as fructose or ascorbic acid. For comparison also commonly used reducing agent - sodium borohydride was used. The silver nanoparticles in various size and yield were produced by immersion of the plasma pre-treated polypropylene (PP) non-woven textile in the colloidal solutions for different periods (35, 45 and 135 min). The morphology of the silver nanoparticles was characterized by SEM and EDX analysis. Additionally, the antibacterial activity of the silver decorated PP non-woven textile was evaluated by an agar diffusion test using both Gram-positive *Staphylococcus Aureus* and Gram-negative *Escherichia coli*. The results suggest that the type of the reducing agent has major effect both on the morphology of silver nanoparticles and the antibacterial activity.

Keywords: Silver nanoparticles, reducing agent, non-woven textile, antibacterial activity

1. INTRODUCTION

Membranes for water filtration are capable of removing a wide variety of contaminants, ranging from large colloids, algae, bacteria up to individual ions, depending on pore dimensions [1-3]. Porous membranes accomplish separation essentially by size exclusion; solutes larger than pore size are rejected, while solutes smaller that membranes pore flow through pore structure. As the solutes build up on the surface or in the pores the productivity of membrane decreases, known as fouling. This process is often accompanied by accumulation and growth of microbial cells on surface and within the pores.

In principle there are two main approaches to the control of biofouling process. The first one is based on the modification of surface of the membranes by various methods, which makes the area less attractive for bacterial adhesion [4,5]. The second approach comprises membranes with the antibacterial agent, which either kills or inhibits the growth of microorganism and the quality and lifetime of the filter can be prolonged.

Among the large number of antibacterial agents available, silver nanoparticles attract the attention due to their low toxicity to human tissue and high activity against broad spectrum of bacteria, viruses and fungi [6,7]. In this context, several physical and chemical methods have been used for synthesizing and stabilizing silver nanoparticles (NPs), including chemical reduction using a variety of organic and inorganic reducing agents. Most of these methods are still in developmental stage and the stability and aggregation of NPs, control of crystal growth, morphology, size and size distribution represent the common experienced problems.

The aim of this study was production of silver/polyprolylene non-woven textile intended for the application as an antibacterial filter. For this purpose, the silver nanoparticles were prepared by chemical reduction of silver nitrate using two environmentally friendly reducing agents (fructose or ascorbic acid) and commonly used reducing agent (sodium borohydride). The effect of reducing agents and time of reduction was investigated in correlation with particle size and density of resulting silver nanoparticles. Besides, the antibacterial properties of non-woven textiles against the Gram-positive and Gram-negative bacteria were determined.

2. EXPERIMENTAL

2.1. Material and Preparation of Silver Nanoparticles

Silver nitrate ($AgNO_3$), fructose (F), ascorbic acid (AA) and sodium borohydride (SB) of analytical grade purity (Penta, Czech Republic) were utilized without further purification. *S. Aureus* (CCM 3953) and *E. Coli* (CCM 3954) were obtained from the Czech Collection of Microorganisms (CCM, Czech Republic). The bacterial cultures were grown on Mueller Hinton agar (HiMedia Laboratories, India).

In this experiment, 6.4 g of 1M $AgNO_3$ was dissolved in 750 mL of distilled water. Similarly, reducing agent, such as 0.5 g of fructose or 0.5g ascorbic acid or 19 mg of $Na(BH_4)$ was dissolved in 500 mL of distilled water. Then, the plasma pre-treated polypropylene (PP) non-woven textiles were immersed first in $AgNO_3$ solution and subsequently in solution of reducing agent for different time periods as described in **Table 1**. Before further analysis, the PP non-woven textiles were rinsed in distilled water three times to remove unfixed silver NPs. The experiments were proceeded at laboratory temperature.

Table 1 Preparation of silver nanoparticles

	Immersion time in solution (min)			
Sample code	**AgNO3**	Fructose	Ascorbic acid	$Na(BH_4)$
F 35	5	30		
F 45	15	30		
F 135	15	120		
AA 35	5		30	
AA 45	15		30	
AA 135	15		120	
SB 35	5			30
SB 45	15			30
SB 135	15			120

2.2. Characterizations

Morphology of the silver nanoparticles sputtered by a thin gold layer was observed using a Vega 3 high resolution scanning electron microscope (Tescan, Czech Republic). The mean particle diameter and particle density were determined by help of Adobe Creative Suite software.

Elemental microanalysis was performed by the Octane SSD (area 30 mm^2) EDX (energy dispersive X-ray) detector (AMETEC, Inc) integrated into the NOVA NanoSEM 450 (FEI company) operated at 15 kV. The qualitative and quantitative results were obtained by applying the experimental procedure several times under following conditions: distance between EDX detector and holder with sample - 5 mm, magnification of 10 000 or 30 000.

Antibacterial efficiency of silver NPs coated PP fibres against *S. aureus* and *E.* coli was tested by an agar diffusion method. Square (9 × 9 mm^2) samples were placed on agar plates pre-inoculated with 1 mL of 0.5 McF turbid bacterial suspension in sterile saline solution. The plates were then incubated at 37°C (24 hours) and after cultivation the diameters of inhibition zones around the samples were recorded in mm.

3. RESULTS AND DISCUSSION

Pre-treatment of polypropylene (PP) non-woven textiles was done in order to create the polar moieties on the PP fibres to anchor the silver NPs during the course of their preparation. This procedure should ensure the fixation of silver NPs on the surface of PP fibres and prevent their agglomeration at the same time. The plasma pre-treated PP non-woven textiles were immersed into the colloidal solutions for different periods of time (35, 45 and 135 min). The solutions turned light yellow immediately after immersion of PP non-woven textile to solutions of reducing agents, indicating the initial formation of silver nanoparticles. The SEM images of the PP non-woven textiles with silver NPs are shown in **Figure 1**. All reducing agents led to the formation of metallic silver (Ag^0) that was found on the surface of PP fibres and additionally confirmed by EDX analysis (**Figure 2**). A strong signal for elemental silver is noticed in the spectrum inferring the existence of metallic silver, in addition to the C peak originated from PP fibres.

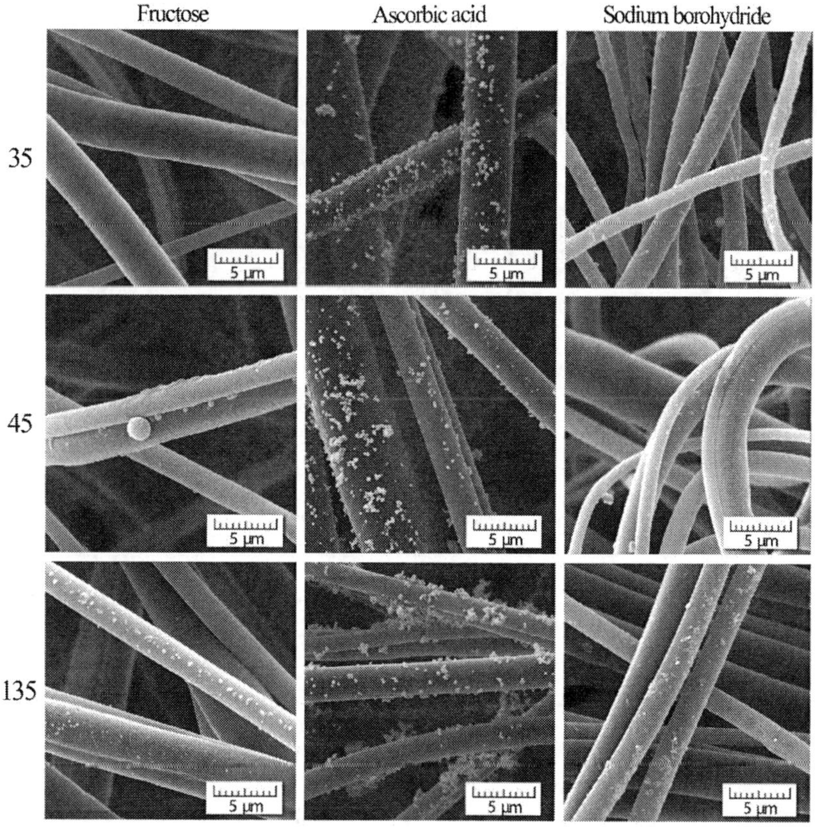

Figure 1 SEM images of silver nanoparticles prepared using various reducing agents

It is visible that changes in time of immersion provided nanoparticles with various average sizes (**Table 2**). In the shortest immersion time, the silver NPs were spherical in shape, well separated and size ranged between

91 to 166 nm depending on the type of reducing agent. When the immersion time was prolonged, the silver NPs started to conjugate and aggregate into bigger particles up to 263 nm. At the same time, the density of silver NPs was increased with longer immersion time.

The reducing agent has significant effect on the size and density of the silver NPs on PP fibres. The most distinctive difference was found between two environmentally friendly reducing agents, fructose and ascorbic acid. The silver NPs from ascorbic acid exhibited the highest size and density, at the longest immersion time the silver NPs created veil of crystals between fibres.

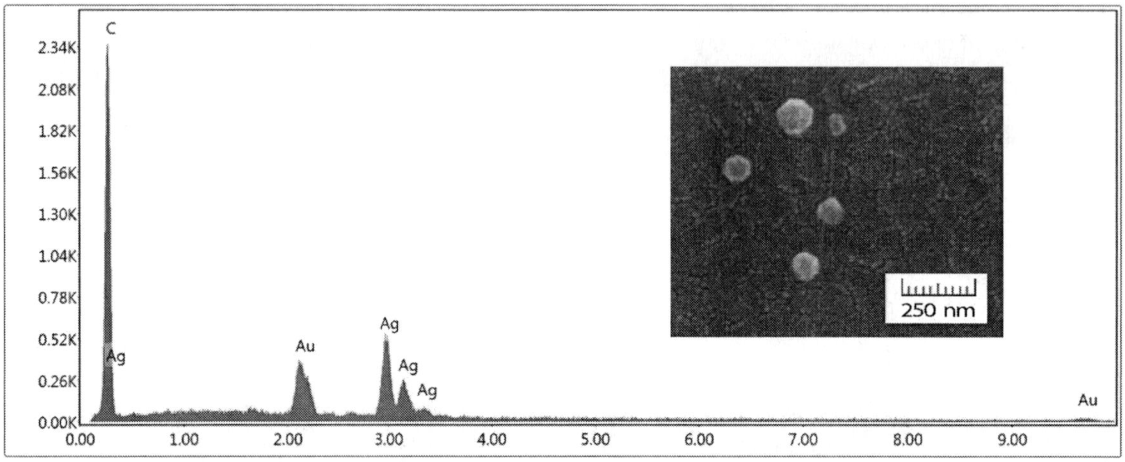

Figure 2 EDX pattern of silver nanoparticles prepared by ascorbic acid as reducing agent (AA 35)

Table 2 Size and density of silver NPs

Sample code	Diameter of NPs[a] (nm)	Density of NPs[a] (counts/100 µm²)
F 35	91 ± 20	4.4
F 45	138 ± 71	7.2
F 135	169 ± 96	31
AA 35	166 ± 55	62
AA 45	189 ± 54	92
AA 135	263 ± 75	101
SB 35	118 ± 54	22
SB 45	110 ± 37	45
SB 135	137 ± 69	70

[a] area of 430 μm^2 was analysed

Results from antibacterial tests proved the activity against both representatives of Gram-negative and Gram-positive bacteria depending on the immersion time and type of reducing agent. The huge discrepancies in measured inhibition zones can be ascribed to the non-homogenous surface covering of PP fibres by the silver NPs. It can be also concluded that the shortest immersion time of 35 min was not optimal for creation of the sufficient amount of silver NPs to inhibit the growth of bacteria in all tested cases. The samples prepared using longer immersion times (45 and 135 min) exhibited clear zones ranging from 5.5 to 12 mm regardless the applied reducing agent.

4. CONCLUSION

The silver nanoparticles were prepared by different reducing agents on the plasma pre-treated PP non-woven textile. It was proved that the size and density of the formed silver NPs were influenced by time of reduction as well as by reducing agent, from which the ascorbic acid showed the most positive results. The density of silver NPs was high even at the shorter reducing period. Besides, the PP non-woven textile covered by silver NPs showed antibacterial efficiency against both representatives of Gram-negative and Gram-positive bacteria.

ACKNOWLEDGEMENTS

The author (P.P.) wishes to acknowledge the financial support of the Institutional Research Project RVO 67985874. The others authors thank to the support of the Ministry of Education, Youth and Sports of the Czech Republic - Programme NPU I (LO1504).

REFERENCES

[1] MILLER, D. J., DREYER, D. R., BIELAWSKI, C. W., PAUL, D. R., FREEMAN, B. D. Surface modification of water purification membranes. *Angewandte Chemie International Edition*, 2017, vol. 56, pp. 4662-4711.

[2] SHAHKARAMIPOUR, N., TRAN, T. N., RAMANAN, S., HAIQING, L. Membranes with surface-enhanced antifouling properties for water purification. *Membranes*, 2017, vol. 7, 1-18.

[3] GEISE, G. M., LEE, H. S., MILLER, D. J., FREEMAN, B. D., MCGRATH, J. S., PAUL, D. R. Water purification by membranes: the role of polymer science. Journal of Polymer Science Part B: Polymer Physics, 2010, vol. 48, pp. 1685-1718.

[4] MCCLOSKEY, B. D., PARK, H. B., JU, H., ROWE, B. W., MILLER, D. J., CHUN, B. J., KIN, K., FREEMAN, B. D. Influence of polydopamine deposition conditions on pure water flux and foulant adhesion resistance of reverse osmosis, ultrafiltration, and microfiltration membranes. *Polymer*, 2010, vol. 51, pp.3472-3485.

[5] VLADKOVA, T. Surface engineering for non-toxic biofouling control (Review). *Journal of the University of Technical technology and Metallurgy*, 2007, vol. 42, pp. 239-256.

[6] GAO, X. Y., GUO, Z. G. Biomimetic superhydrophobic surfaces with transition metals and their oxides: A review. *Journal of Bionic Engineering*, 2017, vol. 14, pp. 401-429.

[7] DEHNAVI, A. S., AROUJALIAN, A., RAISI, A., FAZEL, S. Preparation and characterization of polyethylene/silver nanocomposite films with antibacterial activity. *Journal of Applied Polymer Science*, 2013, vol. 127, pp. 1180-1190.

ELECTROSPUN NANOFIBRES FOR SOLID-PHASE MICROEXTRACTION

Vojtěch ANTOŠ, Pavel HRABÁK, Klára LÍSKOVÁ

Technical University of Liberec, Liberec, Czech Republic, EU, vojtech.antos@tul.cz

Abstract

Solid phase microextraction has been established as modern analytic method, which provides several benefits such as sensitivity, rapid and solvent free technique for sample preparation in organic analytical chemistry. Nowadays there is a scientific effort dedicated to improve analytical methods based on new and nanostructured materials. Our study follows this trend and contributes to analytical chemistry approach. Explicitly, the investigation of using nanofibrous sorbents for solid phase microextraction.

Nanofibres of polyethersulfone (PES), polyetherimide (PEI) and polysulfone (PSU) were successfully prepared by needle electrospinning and fixed on a thin stainless steel wire SPME assmebly. The geometry of sorbents plays a crucial role in the extraction kinetics of the mass transfer in SPME. The intention of using electrospun fibres geometry is to enhance the sorbent sensitivity and capacity.

The basic properties of the prepared nanofibers were determined by thermogravimetry (TGA) and scanning electron microscope (SEM). The scanning electron microscopy images of prepared nanofibers showed a diameter range from 700 to 1400 nm for polyethersulfone, 300 to 600 nm for polyetherimide and 800 to 1400 nm for polysulfone. TGA analysis proved the thermal stability of all prepared nanofibers up to 250 °C. The extraction efficiency of new lab-made SPME fibres and selected common commercial SPME fibres (PDMS and PDMS/DVB) was investigated by the headspace SPME (HS-SPME) mode of gas chromatograph coupled with tandem mass spectrometer (GC/MSMS).

Hexachlorocyclohexanes isomers (HCH) were selected as model analytes in water matrix. With these pollutants, a set of experiments was conducted to proof the effect of key SPME method conditions extraction time on MSMS response.

Keywords: Solid phase microextraction, nanofibres, polyethersulfone, polysulfone, polyetherimide

1. INTRODUCTION

Solid phase microextraction can be defined as an extraction method (or sample preparation method), when the extraction agent (liquid or solid) had much smaller volume comparison to volume of sample. In equilibrium, the extract fraction of the analyte is often negligible [1]. SPME fibre is composed from carrier material (fused-silica core and stainless steel core), coating [2] and protective layer (if it´s needed, especially during the analysis of complex matrixes in direct mode of SPME [3]). A commercial available coating for GC application includes PDMS, PDMS/DVB, Carboxen/PDMS, DVB/Car/PDMS, polyacrylate amd Carbowax polyethylene glycol coatings. Geometry plays a crucial role in the extraction kinetics of the mass transfer in SPME [4]. The geometry of the fiber, thin film and a round particles was directly compared [5]. A syringe-like SPME fibre geometry appears to be most viable: polymeric coatings, sandwiches and mixtures are placed at the tip of the plunger, which is either hidden in a needle (during the transport phase) or exposed off the needle (during the extraction and thermal desorption phases of the analysis). The usual length of the plunger tip is 10 mm and its diameter is 0.1 mm. Polymeric layer thickness ranges from 7 to 100 μm. Amongst other parameters, the duration of the extractive sorption part of the sample introduction step (enrichment) seems to highly influence the overall time of analysis. This is especially the case if less volatile organic compounds are the target analytes [6]. Temperature is another parameter driving the analyte transfer velocity from the water phase towards the sorption fibre. However, water vapour competes with organic analytes vapours for sorption places at higher temperatures [7] therefore; temperatures exceeding 80 °C are rarely used even for organic compounds with a low Henry constant. Last but not least, sorption fibre polymeric material affinity for specific analytes and its surface preconditioning plays an important role [8].

Completed SPME assemblies were compared with commercial PDMS and DVB/Carboxene/PDMS fibres with an emphasis on the possible shortening of the duration of the enrichment step. As a matrix for enrichment time tests, water contaminated by hexachlorocyclohexanes was used. These compounds were selected for a few reasons. They belong to the group of persistent organochlorinated pesticides listed as POPs under the Stockholm Convention on persistent organic pollutants. Also HCH was former industrial production in the Czech Republic [9]. And, last but not least, these pesticides belong to widely studied groups of target analytes with SPME utilization [10,11,12].

2. MATERIALS AND METHODS

Lab-made SPME fibers were assembled from a stainless steel capillary and 304H wire supplied by Teseco and a RDG810 3D-printer polymer supplied by VeroClear (**Figures 1 and 2**). Polyetherimide, polyethersulfone and polysulfone pellets (supplied by Sigma-Aldrich) was dissolved overnight in a dimethylformamide into 22 % (PEI), 24 % (PSU) and 28 % (PES) w/w solution. These prepared solutions were pumped into the electrospinner needle with the flow of 3 ml per hour. The lab-made assembly used as a collector electrode was subjected to 20-25 kV voltage. The relative humidity 45 - 55 % and temperature 25 °C was kept during electrospinning of all fibers. SPME commercial fibre 100 µm PDMS for comparison was obtained from Restek. Helium of 5.0 purity and argon of 4.8 purity (Linde Gas) were used as a carrier gas and collision gas for GC-MS/MS measurements, respectively. Chromatographic standard of hexachlorocyclohexanes (HCH) - Mix 5 (100 $\mu g \cdot ml^{-1}$ in acetone) war purchased from Neochema.

Figure 1 Schema of SPME fibre

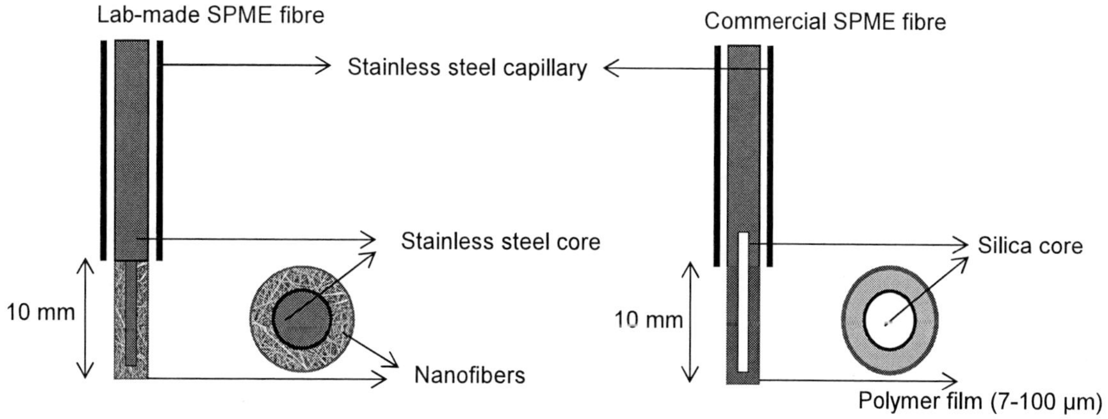

Figure 2 Schema of commercial and lab-made SPME fibre

3. APPARATUS AND EQUIPMENTS

Appearance of produced PEI, PES and PSU nanofibres on lab-made SPME fibers was documented by SEM microscope (UHR FE-SEM Carl Zeiss ULTRA Plus). Analytical performance of all SPME fibers (manufactured and commercial) was tested with gas chromatograph (Thermo Trace 1310) equipped with a mass spectrometer triple quadrupole detector (Thermo TSQTM 8000 EVO) and a programmed temperature vaporizing injector (PTV). For automatic handling of prepared samples was GC/MSMS equipped by an autosampler (CTC Analytics AG, PAL RTC), which was set to headspace operation mode (samples were agitated during enrichment time).

A gas chromatography column DB-5MS (30 m long, 0.25 µm thick and with 0.25 µm film thickness of stationary phase) was installed into GC oven. For the purpose of rapid testing, a very fast method was developed whereby the duration of one GC cycle was reduced to 10 minutes. Temperature program of the chromatographic oven started at 80 °C, graduating by 20 °C per min. to 260 °C. The carrier gas (helium 5.0) flow was adjusted to 1.5 ml per min.

For proper cleaning step of SPME fibres was set desorption step after injection to system. PTV injector was set to hold at 220 °C in splitless mode for 1 minute and for cleaning phase the temperature was set to 240 °C with the flow of carrier gas to 30 ml per min. All of the SPME fibers were tested for different extraction (enrichment) time in the headspace of the measured samples. Enrichment times for all tested fibers were set to: 2, 5, 10, 20, 30, 40 and 50 minutes. Temperatures during the agitations were set to 70 °C with constant speed 250 rpm. For each tested fiber 21 tap water samples spiked with chromatographic standard of hexachlorocyclohexanes (HCH) were prepared in 20 ml vials capped with PTFE/silicon septa and magnetic cap. Samples were spiked to reach the concentration of HCH 5 $\mu g \cdot l^{-1}$.

4. RESULTS AND DISCUSSION

Nanostructure surface of produced polyetherimide, polyethersulfone and polysulfone nanofibers were documented by SEM microscope (**Figures 3 - 8**). The scanning electron microscopy images of prepared nanofibers showed a very high number of pores presence at fibres surface. This effect was primarily observed at PES and PSU nanofibers. The diameter of the prepared nanofibers was in range from 700 to 1300 nm for polyethersulfone, 300 to 600 nm for polyetherimide and 800 to 1200 nm for polysulfone. All lab-made fibers with PEI, PES and PSU nanofibers were compared with commercially available SPME fibre - 100 µm PDMS. Enrichment time 2, 5, 10, 20, 30, 40 and 50 minutes was the parameter chosen for the comparison. The dependencies of GC-MS/MS peak areas on the enrichment time are shown on **Figures 9 - 13**. Obtained data indicate that the worst performance of all tested SPME fibers have polyetherimide fibers. PES fiber showed very similar responses as the commercial PDMS fiber. The highest response in all tested enrichment times of all tested fibers had showed PSU fiber. During the longest agitation time, the response of PSU fiber was 50 % bigger than the commercial PDMS fiber.

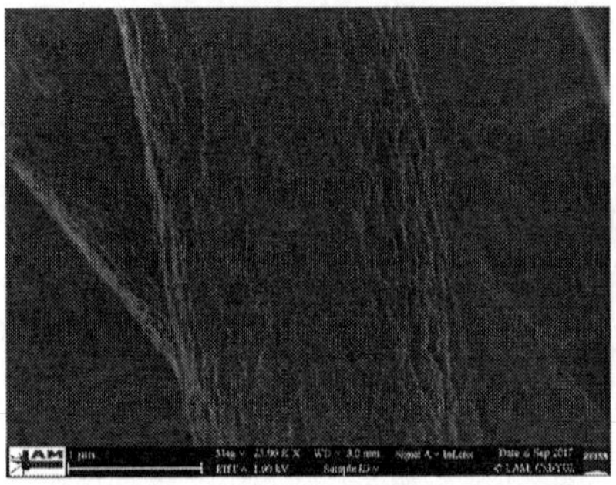

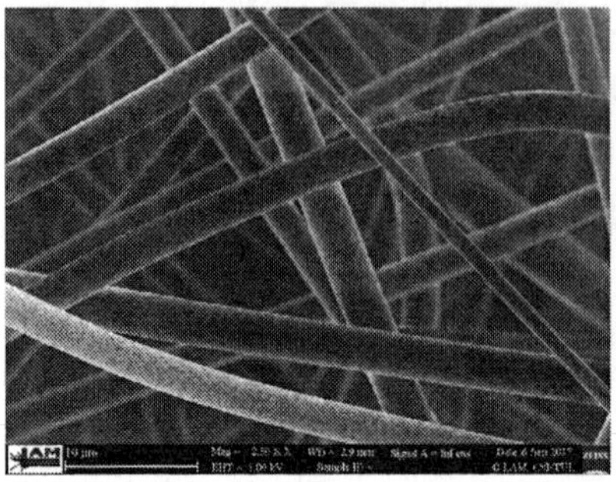

Figures 3 - 4 Electrospun polyetherimide (PEI) nanofibers

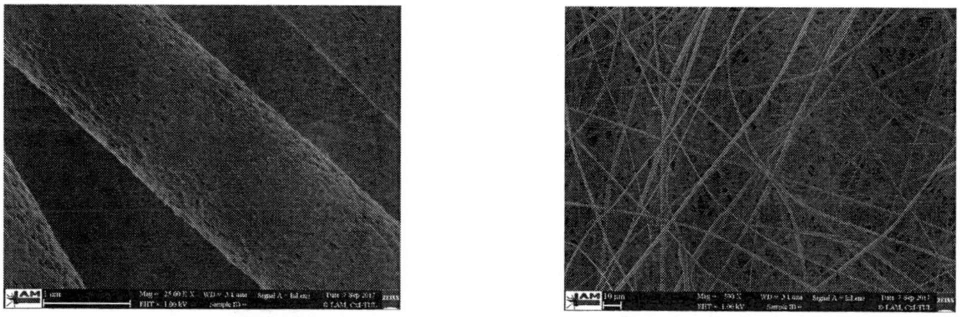

Figures 5 - 6 Electrospun polyethersulfone (PES) nanofibers

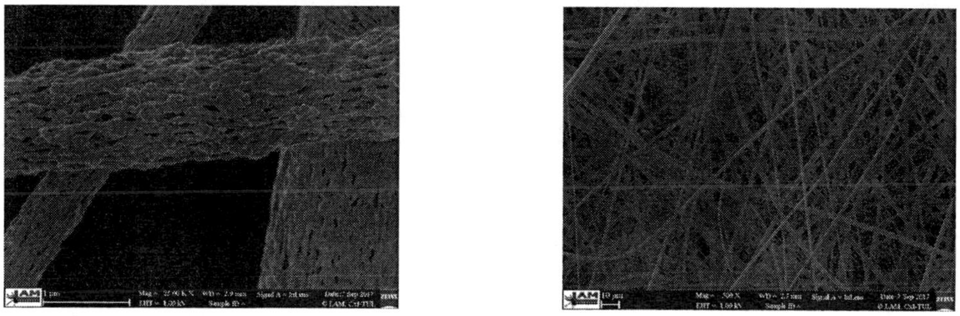

Figures 7 - 8 Electrospun polyethersulfone (PSU) nanofibers

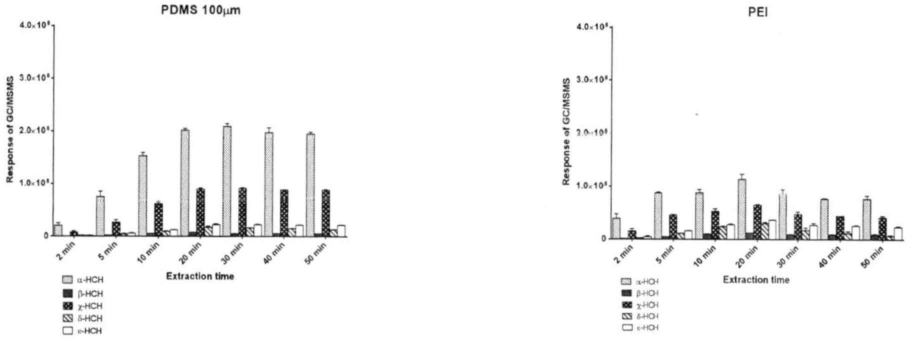

Figures 9 - 10 Comparison of PDMS 100 μm and PEI fibers, GC/MS response after SPME injection (error bars at 1 σ, n = 2)

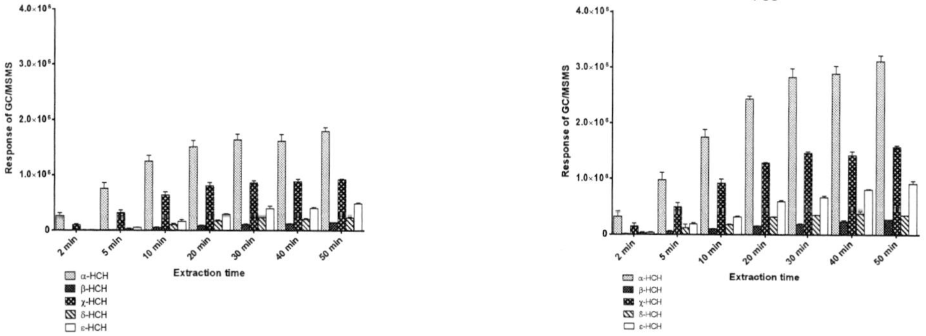

Figures 11 - 12 Comparison of PES and PSU fibers, GC/MS response after SPME injection (error bars at 1 σ, n = 2)

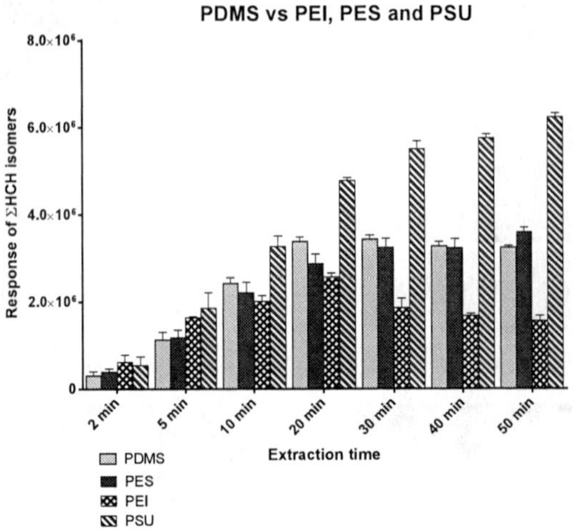

Figure 13 Comparison of commercial PDMS 100 µm fibre vs PEI, PES and PSU fibers, GC/MS response after SPME injection (error bars at 1 σ, n = 2)

5. CONCLUSION

Polyethersulfone, polyetherimide and polysulfone nanofibres were succesfully prepared from polymer solution by needle electrospinning. The scanning electron microscopy images of prepared nanofibers showed a diameter range from 700 to 1300 nm for polyethersulfone, 300 to 600 nm for polyetherimide and 800 to 1200 nm for polysulfone. Even though the achieved diameters for nanofibers wasn´t so small, the images of prepared nanofibres showed a very high number of pores presence at fibres surface, especially on PES and PSU nanofibres. TGA analysis proved the thermal stability of all prepared nanofibres (PEI, PES and PSU) up to 250 °C. BET analysis determined that specific surface was 57 $m^2 \cdot g^{-1}$ for PEI, 32 $m^2 \cdot g^{-1}$ for PES and 10 $m^2 \cdot g^{-1}$ for PSU of prepared nanofibres.

The parameter of the enrichment time had significantly positive influence on the response of MS detector. Obtained data indicate that the worst performance of all tested SPME fibers have polyetherimide fibers. PES fiber showed very similar responses as the commercial PDMS fiber. Saturation was achieved in case of PDMS, PES and PEI fibres. Just in case of polysulfone fibre the saturation wasn´t reached during extraction times 2 - 50 minutes. The highest response in all tested enrichment times of all tested fibers had showed PSU fiber. During the longest agitation time, the response of PSU fiber was 50 % bigger than the commercial PDMS fiber.

Handicap of using nanofibres as sorbent in analytical chemistry could be the homogeneity of producing nanofibres and their possible shorter lifetime during the longer usage. Obtained data indicated that used geometry in form of nanofibres can be successfully used as sorbent in analytical chemistry, especially in case of HS-SPME technique.

ACKNOWLEDGEMENTS

The research in this article was supported by the project OPR&DI Institute for Nanomaterials, Advanced Technologies and Innovation (CZ.1.05/2.1.00/01.0005) a National programme for sustainability I (L01201). The work of V.A. was supported by the Ministry of Education of the Czech Republic within the SGS project no. 21176/115 on the Technical University of Liberec. The authors acknowledge the assistance provided by the Research Infrastructure NanoEnviCz, supported by the Ministry of Education, Youth and Sports of the Czech Republic under Project No. LM2015073.

REFERENCES

[1] POOLE, C.; MESTER, Z.; MIRÓ, M.; Pedersen-Bjergaard, S.; Pawliszyn, J. Glossary of Terms Used in Extraction (IUPAC Recommendations 2016). Pure Appl. Chem. 2016, 88 (5), 517-558.

[2] LORD, H.; PAWLISZYN, J. Evolution of Solid-Phase Microextraction Technology. J. Chromatogr. A 2000, 885 (1), 153-193.

[3] SOUZA SILVA, É. A.; PAWLISZYN, J. Optimization of Fiber Coating Structure Enables Direct Immersion Solid Phase Microextraction and High-Throughput Determination of Complex Samples. Anal. Chem. 2012, 84 (16), 6933-6938.

[4] REYES-GARCÉS, N.; GIONFRIDDO, E.; GÓMEZ-RÍOS, G. A.; ALAM, M. N.; BOYAC, E.; BOJKO, B.; SINGH, V.; GRANDY, J.; PAWLISZYN, J. Advances in Solid Phase Microextraction and Perspective on Future Directions. Anal. Chem. 2018, 90 (1), 302-360.

[5] PIRI-MOGHADAM, H.; AHMADI, F.; GÓMEZ-RÍOS, G. A.; BOYAC, E.; REYES-GARCÉS, N.; AGHAKHANI, A.; BOJKO, B.; PAWLISZYN, J. Fast Quantitation of Target Analytes in Small Volumes of Complex Samples by Matrix-Compatible Solid-Phase Microextraction Devices. Angew. Chem. Int. Ed. 55 (26), 7510-7514.

[6] NIU, J. et al. A water resistant solid-phase microextraction fiber with high selectivity prepared by a metal organic framework with perfluorinated pores. J. Chromatogr. A doi:10.1016/j.chroma.2016.02.076

[7] BRUNTON, N. P., CRONIN, D. A. & MONAHAN, F. J. The effects of temperature and pressure on the performance of Carboxen/PDMS fibres during solid phase microextraction (SPME) of headspace volatiles from cooked and raw turkey breast. Flavour Fragr. J. 16, 294-302 (2001).

[8] CHONG, S. L., WANG, D., HAYES, J. D., WILHITE, B. W. & MALIK, A. Sol-gel coating technology for the preparation of solid-phase microextraction fibers of enhanced thermal stability. Anal. Chem. 69, 3889-3898 (1997).

[9] WACŁAWEK, S., ANTOŠ, V., HRABÁK, P., ČERNÍK, M. & ELLIOTT, D. Remediation of hexachlorocyclohexanes by electrochemically activated persulfates. Environ. Sci. Pollut. Res. Int. 23, 765-773 (2016).

[10] BOYD-BOLAND, A. A., MAGDIC, S. & PAWLISZYN, J. B. Simultaneous determination of 60 pesticides in water using solid-phase microextraction and gas chromatography-mass spectrometry. Analyst 121, 929-937 (1996).

[11] BOUSSAHEL, R., BOULAND, S., MOUSSAOUI, K. M., BAUDU, M. & MONTIEL, A. Determination of chlorinated pesticides in water by SPME/GC. Water Res. 36, 1909-1911 (2002).

[12] DONG, C., ZENG, Z. & YANG, M. Determination of organochlorine pesticides and their derivations in water after HS-SPME using polymethylphenylvinylsiloxane-coated fiber by GC-ECD. Water Res. 39, 4204 4210 (2005).

SURFACE MODIFICATION OF ZERO-VALENT IRON NANOPARTICLES BY BIOLOGICAL ACTIVE SUBSTANCE - COMPLEX TESTS

Kristýna PEŠKOVÁ, Kristýna MARKOVÁ, Ondřejka VOLOŠČUKOVÁ, Jaroslav NOSEK

Technical University of Liberec, Institute for Nanomaterials,
Advanced Technologies and Innovation (CXI), Liberec, Czech Republic;
kristyna.peskova@tul.cz, jaroslav.nosek1@tul.cz

Abstract

Chlorinated solvents such as trichloroethene (TCE) are widespread groundwater contaminants. These contaminants can be transformed by combination of abiotic and biotic methods under anaerobic conditions. Nowadays nanoscale zero-valent iron (nZVI) is used for the treatment of chlorinated compounds via its strong reducing property. Biological reductive dechlorination of chlorinated ethenes (CEs) is contributed by dehalorespiration. The influence of nZVI in combination with carboxymethyl cellulose (CMC) and molasses on the specific dehalorespiring microflora and besides the influence of biological surface modification on reactive properties of iron was tested within this study.

Groundwater contaminated with CEs was collected from the chemical factory Spolchemie a.s. Batch tests (reactors) with iron and various concentrations of CMC and molasses were performed. The samples for gas chromatography (GC) analyses were taken regularly after defined time period for determination of CEs concentration and physical-chemical parameters were monitored simultaneously. DNA was extracted after filtration of the tested water and used as a template for a real-time PCR amplification. 16S rDNA gene was used as a total bacterial community marker. Specific genes were used for detection of ongoing reductive dehalogenation (vcrA, bvcA, Dre, DHC-RT and Dsb) and to monitor denitrifying and sulphate reducing bacteria (nirK and apsA).

Keywords: Iron particles, chlorinated solvents, biological surface modification, biological reductive dechlorination, dehalorespiration, molecular genetic analysis

INTRODUCTION

The different surface modification or stabilization techniques were used for decrease of ZVI (zero valent iron) particles aggregation and for increase its mobility in the subsurface. The typical stabilization is coating of particles by polymer substances. These surface macromolecules serve as a barrier which protects the particles against agglomeration using electrostatic or spatial repulsion [4]. Due to the project is the research focused on group of substances, which can be biological decomposed and can serve as a source of nutrients for dehalorespiring microorganisms [5,6].

The aim of this study was to develop a biological active substance supporting natural dehalogenation processes. A suitable biological substance has to be non-toxic, with minimum passivation effect on ZVI surface and with minimum negative effect on its reactivity properties. It should also support the migration capability of ZVI particles [9], [10].

To develop this material the laboratory tests with biological active substances were performed. The influence of tested materials on the reactivity properties of ZVI and on the quantity of soil bacteria was monitored. To test the influence of different substances on the specific soil bacteria the molecular genetic analyses were used. These analyses were focused mainly on the organohalide-respiring microflora.

1. LABORATORY TEST METHODOLOGIES

Methodology of reactor tests for testing of reactivity properties

The reactor tests were designed as a closed mixing system, with contaminated water and a remedial reagent. The tests were carried out in 2.5 l glass bottles fitted with a joint [2]. The reactors were mixed with mechanical stirrers, since magnetic stirrers cannot be used for testing magnetic particles. In addition, the reactors were equipped with sampling ports, electrodes and other electrodes for continuous measurement of physical and chemical parameters. The special stirrer (according to utility model 25752 of the Technical University of Liberec) [3] was placed on the central joint. The system was sealed as tightly as possible, but a certain degree of leakage was expected during the experiment and was monitored using a control reactor without the reagent (or electric field).

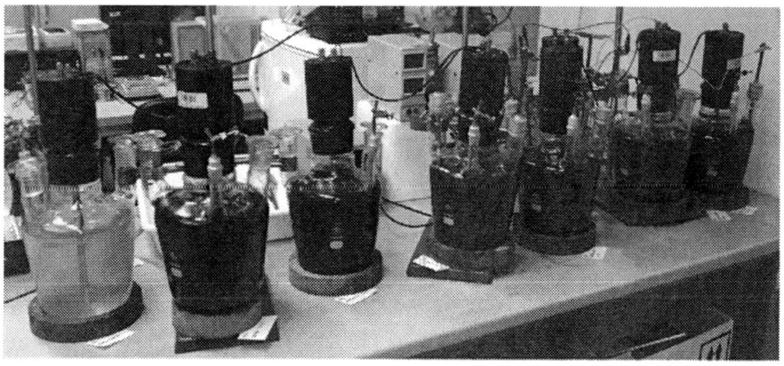

Figure 1 Photo of reactor tests

Reactor sampling took place in two modes. For control sampling, a Teflon tube was introduced into the sampling port and water samples were drawn through a glass syringe. In the first phase of the control sampling a water sample was taken for chlorinated hydrocarbons (CHC) analysis by GC/MS. Physical and chemical parameters (pH, redox potential and conductivity) were then measured. Samples were also regularly collected for the analysis of dissolved ions and metals. In the case of the enclosed electric field, the value of the electrical voltage and current in the system was deducted for each sample. The sampling frequency depended on the reagent used and the duration of the test. Sampling was usually more frequent at the beginning of the tests, and intervals were prolonged. These reactors were placed in laboratory temperature for periods ranging from 6 to 26 days. Direct current (DC) [1] was also applied in some reactors. **Figure 2** illustrates the possible reactor configurations for these types of test.

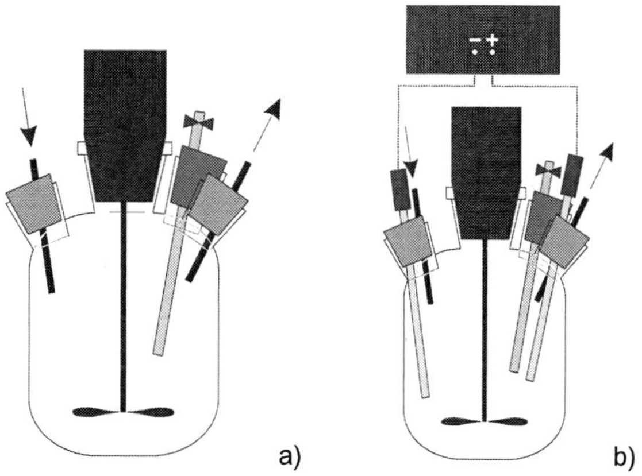

Figure 2 Possible reactor test configurations with a gas chromatogram connection: a) a simple reactor; (b) reactor with electrodes in a combined configuration

Methodology of molecular genetic analysis

0.5 L of water from each reactor was filtered through 0.22 µm pore membrane when reactor test was completed. DNA was isolated from filters with collected microorganisms by FastDNA SPIN Kit for Soil according to manufacturer's protocol.

Extracted DNA was used as a template for real-time PCR amplification. Real-time PCR assays were performed and analysed using LightCycler 480 SYBR Green system.

	Genes	Primers	Primer sequences (5'-3')
16S	16S rDNA broad-range	U16SRT-F	ACTCCTACGGGAGGCAGCAGT
		U16SRT-R	TATTACCGCGGCTGCTGGC
bvcA	vinyl chloride reductase	bvcA227F	TGGGGACCTGTACCTGAAAA
		bvcA523R	CAAGACGCATTGTGGACATC
vcrA	vinyl chloride reductase	vcrA880F	CCCTCCAGATGCTCCCTTTA
		vcrA1018R	ATCCCCTCTCCCGTGTAACC
Dre	Dehalobacter sp. 16S rDNA	Dre441F	GTTAGGGAAGAACGGCATCTGT
		Dre645R	CCTCTCCTGTCCTCAAGCCATA
DHC-RT	Dehalococcoides sp. 16S rDNA	DHC793F	GGGAGTATCGACCCTCTCTG
		DHC946R	CGTTYCCCTTTCRGTTCACT
Dsb	Desulfitobacterium spp. 16S rDNA	Dsb406F	GTACGACGAAGGCCTTCGGGT
		Dsb619R	CCCAGGGTTGAGCCCTAGGT
nirK	nitrite reductase	nirK876F	ATYGGCGGVCAYGGCGA
		nirK1040R	GCCTCGATCAGRTTRTGGTT
apsA	adenosine 5'-phosphosulfate reductase	RH1-aps-F	CGCGAAGACCTKATCTTCGAC
		RH2-aps-R	ATCATGATCTGCCAGCGGCCGGA

Figure 3 qPCR markers used in this study

16S rDNA gene was used as a total bacterial community marker. Specific genes were used for detection of ongoing reductive dehalogenation (vcrA, bvcA, Dre DHC-RT and Dsb) and to monitor denitrifying and sulphate reducing bacteria (nirK and apsA). All used markers are listed in **Figure 3**.

Cq value of each sample was compared with Cq value of feed water and relative quantity was given in the tables according to rules listed below:

- red colour - high increase of quantity (more than 80x),
- orange colour - lower increase of quantity,
- yellow colour - no changes,
- light blue colour - degrease of quantity,
- dark blue colour- no detection.

2. RESULTS

a) Carboxymethyl cellulose (CMC)

Iron particles (Fe) with diverse CMC concentration in combination with electric current (DC) were applied in this test.

Total bacterial abundance (16S rDNA) increased in the control reactor (contaminated water without treatment), in the reactor with CMC and in the reactor with Fe in combination with higher CMC concentrations (0.5 and 1 g/L). In the reactor with Fe in combination with the highest CMC concentration (1 g/l), the biggest increase of bacterial quantity was detected. In the reactors with Fe, the most of tested markers increased when higher CMC concentrations were used. These results indicated that CMC has probably a bacteria's protecting effect against Fe toxicity. (**Figure 4**).

The best growth conditions for dehalorespiring microflora were present in the control reactor and in the reactor with Fe particles in combination with high CMC concentration (1 g/l). According to these results, big decrease of DCE was also observed in this reactor. It appears that Fe could increase the amount of H_2 due to its strong reducing properties and H_2 serves as electron donor for dehalorespiring microflora. CMC seems to have a protecting effect on these bacteria [7], [8].

This test also demonstrated the devastating impact of electric current on all bacterial groups. It is visible on both total bacterial abundance (16S rDNA) and all specific markers. Although, this impact could be attenuated on contaminated sites due to bacterial recolonization of that site by groundwater flow.

			Fe + CMC			Fe + CMC + DC			
	Control	1 g/L CMC	0,25 g/L CMC	0,5 g/L CMC	1 g/L CMC	0,25 g/L CMC	0,5 g/L CMC	1 g/L CMC	Cq values
16S	6	1	0,04	10	83	0	0,0001	0,0005	12,25
bvcA	0,2	0	0	0,005	0,01	0	0	0	26,04
vcrA	0,6	0,01	0,001	0,01	0,1	0	0	0	19,4
Dre	3	1,4	0	0,1	0	0	0	0,8	33,38
DHC-RT	0,1	0	0	0,002	0,01	0	0	0	23,14
Dsb	0,5	0	0,0001	0,3	5	0,0004	0	0	25,36
nirK	70	0,18	0,01	1	44	0	0,01	0	25,03
apsA	0,3	0	0,2	10	6	0	0,001	0,01	19,95
	R1	R2	R3	R4	R5	R6	R7	R8	Feed water

Figure 4 Relative quantity of bacterial markers in reactors with CMC, ZVI composite and DC (5 V); Fe = 3 g/L; 21 period days

The following diagrams document the process of laboratory reactor tests. The **Figure 5** compares the normalized concentrations of DCE and TCE in time. The diagrams able easily to compare the influence of each system on dechlorination process.

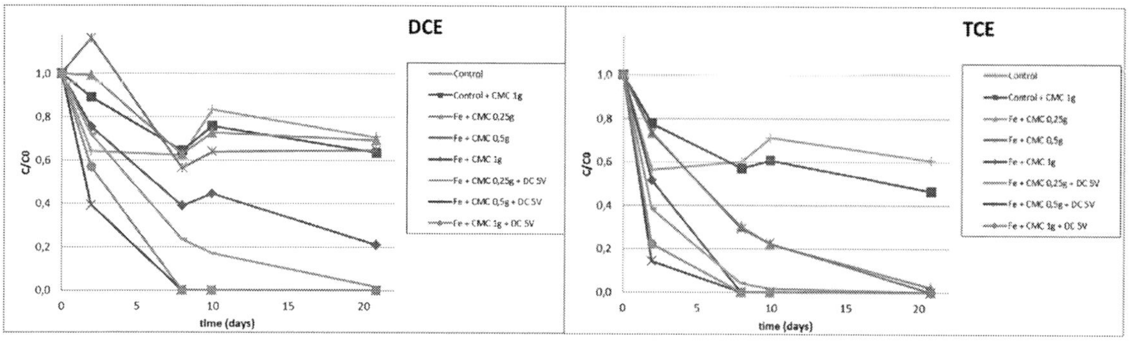

Figure 5 Degradation of TCE, DCE in reactors with CMC, ZVI composite and DC (5 V)

CMC had definitely positive effect on given microflora [7]. The surface modification by CMC passivated the surface of ZVI particles and so degreased its toxicity. However the surface passivation led to the rapid decrease of ZVI reactivity.

The systems with DC support achieved strongly better results from the point of view of dechlorination quickness and of total balance. The best results in CEs reduction were achieved in systems with high concentration of CMC and in combination with DC support.

b) Molasses

In this type of test the Molasses in combination with higher concentration of ZVI particles (10 g/l) were applied. The next configuration was the comparison of just ZVI influence and its combination with alternating current (AC) lower intensity. The sampling was performed twice after 11 and 19 days.

Total bacterial abundance (16S rDNA) increased in the control reactor (contaminated water without treatment), in the reactor with molasses and in the reactor with iron particles in combination with molasses. The highest increase of total bacterial quantity was observed in the reactor with molasses. In this reactor, the quantity of all tested markers increased, mostly denitrifying bacteria marker (nirK). Markers of dehalorespiring bacteria also increased in this reactor. It indicated that molasses served as the fermentable substrate and products of fermentation were utilised by dehalorespiring bacteria. The quantity of the most tested markers was lower in the second sampling due to ongoing substrate exhaustion.

In the reactor with the combination of Fe and molasses, the total bacterial abundance increased. The marker of denitrifying bacteria (nirK) was the only one of the specific markers whose amount increased in this reactor. This could be caused by toxicity of high Fe concentration used in this test (10 g/L).

	Control		Molasses		Molasses + Fe		Fe		Fe + AC	Cq values
16S	18,62	2,357	1298	512,7	824,2	88,03	3E-04	0	0,003	12,04
bvcA	0,039	0,064	3,029	0	0	0,216	0	0	0,012	27,11
vcrA	0,101	0,122	8,343	0,064	0,085	0,67	0,002	0	0,005	22,67
Dre	0,029	0,095	1,129	3,358	0	0,208	0	0	0,028	32,57
DHC-RT	0,006	0,03	1,788	0,005	0,002	0,177	0	0	0,001	23,90
Dsb	0,049	0,069	3,156	0,039	0,019	0,17	2E-04	0	0,000	24,82
nirK	20,63	5,994	114,3	1,44	1,334	5,871	0	0	0,005	28,29
apsA	0,161	0,179	2,034	0,39	0,164	0,201	0,003	0	0,006	22,71
	11 days	19 days	11 days	19 days	11 days	19 days	11 days	19 days	19 days	Feed water

Figure 6 Relative quantity of bacterial markers in reactors with molasses and ZVI composite; Fe = 10 g/L; Melasa = 10 g/L; two periods 11 and 19 days

This test also shoved lower impact of AC on bacteria in comparison with DC used in the previous test (above-mentioned). But the amount of all bacterial markers also significantly decreased when AC was used.

The following diagrams **Figure 7** compare the normalized concentrations of DCE and TCE in time. The diagrams able easily to compare the influence of each system on dechlorination process.

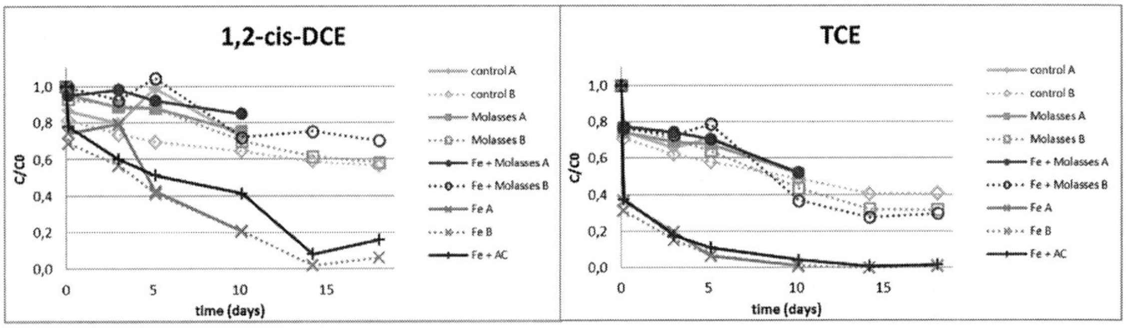

Figure 7 Degradation of TCE, DCE in reactors with molasses and ZVI composite

The development of monitored CEs concentration showed very well the influence of Fe surface modification by Molasses (Fe + Mel. A; Fe + Mel. B). Despite the using of high Fe concentration the combination with Molasses was not able to decompose the contamination. Applied Fe material served just for the equalizing of water pH, which had positive effect on biological processes. From the economical point of view this configuration does not make.

Separately applied Fe (Fe A; Fe B) achieved very good results because the CEs were during the experiment completely reduced. The reason was probably the high concentration of Fe - 10 g/L.

The results of the first testing of Fe application combinated with AC (Fe + AC) and its influence on dechlorination process were very interesting. Using AC showed lower impact on bacteria in comparison with DC but showed lower efficiency from the point of view of dechlorination rate of CEs.

3. CONCLUSIONS

Organic compounds used as ZVI modifiers could have large impact on the soil microflora of remediated sites.

Fe particles in combination with CMC increased the quantity of examined bacterial markers more than CMC by itself. This increasing was higher with higher CMC concentration. It confirmed positive influence of Fe particles on dehalorespiring bacteria when they are protected from its toxicity by CMC. The disadvantage of CMC usage could be its cost. Higher amount of ZVI would be necessary to apply because of the passivation effect of CMC on ZVI surface.

Molasses as carbon and electron source had positive effect on all studied groups of bacteria. Molasses is suggested to serve as the substrate for the fermentation which produces electrons utilised by dehalorespiration. Molasses supported the bacterial growth but had strong negative effect on ZVI reactivity due to its surface passivation.

Electricity (AC, DC) had devastating impact on tested bacteria in laboratory conditions. Remediated sites could be usually recolonised via ground water flow.

According to these results the use of organic compounds especially CMC as a surface modification of ZVI particles for the field applications does not represent a favourable product as was expected due to the literature sources. The CMC represents an expensive source of carbon and it passivates the surface of ZVI and decrease the reactivity of ZVI. The electricity prolongs the ZVI longevity but has a negative impact on given microflora. However at remediated sites is usually easy recolonised via ground water flow.

ACKNOWLEDGEMENTS

The presented research was supported by the project Nanomaterials for remediation of contaminated water No. TF02000064 supported by TAČR. The work of Kristýna Pešková was supported by the Ministry of Education of the Czech Republic within the SGS project no. 21176/115 on the Technical University of Liberec.

REFERENCES

[1] HRABAL, J.; ČERNÍK, M.; NOSEK, J.: The method of in situ remediation of subsurface contaminated by harmful chemical substances. Patent: 304152, accepted 9.10.2013, MEGA a.s., Technical University of Liberec

[2] NOSEK, J., CÁDROVÁ, L., ANTOŠ, V.: Reactor for laboratory modelling of contaminated water remediation. Utility model: 29308, accepted 22/03/2016, MEGA a.s., Technical University of Liberec.

[3] NOSEK, J., PLUHAŘ, T., ČERNÍK, M.: Special laboratory joint stirrer. Utility model: 25752. Technical University of Liberec, 08/08/2013.

[4] XUE, D., SETHI, R. (2012). Viscoelastic gels of guar and xanthan gum mixtures provide long-term stabilization of iron micro- and nanoparticles. J. Nanoparticle Res. *14*, 1239.

[5] HENN, K.W., WADDILL, D.W. (2006). Utilization of nanoscale zero-valent iron for source remediation—A case study. Remediat. J. *16*, 57-77.

[6] HE, F., ZHAO, D., PAUL, C. (2010). Field assessment of carboxymethyl cellulose stabilized iron nanoparticles for in situ destruction of chlorinated solvents in source zones. Water Res. *44*, 2360-2370.

[7] ZHOU, L., THANH, T.L., GONG, J., KIM, J.-H., KIM, E.-J., CHANG, Y.-S. (2014). Carboxymethyl cellulose coating decreases toxicity and oxidizing capacity of nanoscale zerovalent iron. Chemosphere *104*, 155-161.

[8] DONG, H., XIE, Y., ZENG, G., TANG, L., LIANG, J., HE, Q., ZHAO, F., ZENG, Y., WU, Y. (2016). The dual effects of carboxymethyl cellulose on the colloidal stability and toxicity of nanoscale zero-valent iron. Chemosphere *144*, 1682-1689.

[9] ZHAO, X., LIU, W., CAI, Z., HAN, B., QIAN, T., ZHAO, D. (2016). An overview of preparation and applications of stabilized zero-valent iron nanoparticles for soil and groundwater remediation. Water Res. *100*, 245-266.

[10] YAN, W., LIEN, H.-L., KOEL, B.E., ZHANG, W. (2013). Iron nanoparticles for environmental clean-up: recent developments and future outlook. Env. Sci Process. Impacts *15*, 63-77.

[11] KOCUR, C.M.D., LOMHEIM, L., BOPARAI, H.K., CHOWDHURY, A.I.A., WEBER, K.P., AUSTRINS, L.M., EDWARDS, E.A., SLEEP, B.E., O'CARROLL, D.M. (2015). Contributions of Abiotic and Biotic Dechlorination Following Carboxymethyl Cellulose Stabilized Nanoscale Zero Valent Iron Injection. Environ. Sci. Technol. *49*, 8648-8656.

A COMPARATIVE ASSESSMENT OF Au NANOPARTICLES SYNTHESIZED VIA GREENER PATHWAYS USING NATURAL HYDROCOLLOIDS

Mayza IBRAHIM, Vinod V.T. PADIL, Miroslav ČERNÍK

Institute for Nanomaterials, Advanced Technology and Innovation, Technical University of Liberec, Liberec, Czech Republic, EU, +420 485 353 017

Abstract

This paper presents the synthesis and stabilization of Gold nanoparticles (Au NPs) using five different hydrocolloids [Tragacanth (Tg), Karaya (Kg), Guar (Gg), Carrageenan (Cg), and Xanthan (Xg)] via greener pathways. The particle size, morphology, stabilization characteristics and catalytic efficiencies of the synthesized Au NPs were analyzed using spectroscopic and microscopic techniques. The Gg stabilized Au NPs was extremely stable (particle size 8.0 ± 1 nm) at 4 °C for a period up to three months as compared to other Au NPs synthesized and stabilized via other polysaccharides. However, zeta potential and UV-vis analysis of Au NPs produced by Tg, Kg, Gg, Cg and Xg did not show much variation in magnitudes during the tested periods (0 to 90 days). The catalytic performance of hydrocolloid stabilized Au NPs were also evaluated and compared their kinetics reaction rates, and stability for the degradation of 4-Nitrophenol (p-NP) to 4-Amino phenol (p-AP). The reduction of p-NP follows the pseudo-first order. The highest kinetic rate value is found to be 10.6 x 10^{-3} s^{-1} for Au Guar composites (5.1 x 10^{-6} mM).

Keywords: Hydrocolloids, Au NP, green synthesis, catalysis

1. INTRODUCTION

Synthesized metal nanoparticles have received considerable interest due to their unique chemical, physical, optical, electrical, and catalytic properties which totally differ from the bulk materials. Among the various nanomaterials, gold nanoparticles (Au NPs) are the most widespread materials in many applications especially in the fields of physics, chemistry, optics, catalysis, biology, electronics, and material science [1-5]. This is due to their stability, less toxicity, and biocompatibility [6]. There are several chemical agents used for the reduction of electropositive (Au^{+3}) to neutral atoms. The first reducing agent to prepare Au NPs was phosphorus that was used by Michael Faradey, after that sodium borohydride [7], citrate [8,9], hydrogen peroxide [10,11], and other organic compounds such as ethylene glycol [12]. However, these chemicals are highly reactive, and may be associated with environmental toxicity or biological risks.

In order to overcome these concerns, researchers agree to embrace the principle of green chemistry to achieve the safest possible future for advancing nanotechnology in sustainable world [13]. Green nanotechnology aims to eliminate or reduce the harmful polluting substances in the synthesis of nanomaterials or to utilize the produced nanomaterials in eliminating or reducing these pollutants. Recently there are several studies of green synthesis for Au NPs by using harmless alternative biocompatible molecules. There are many studies reported the efficient use of hydrocolloids (starch, chitosan, xanthan, cellulose, carrageenan, and tree gums) as reducing and stablizing agents [14,15].

4 Nitro-phenol (p-NP) is one of the hazardous and toxic pollutants [15] that are widely used in industry. It is known to be highly soluble and stable in water, hence traditional methods for water purification are not effective. There are many techniques that were developed for the removal of p-NP and include: microbial degradation, microwave-assisted catalytic oxidation, electron fenton method, electrochemical treatment, and electrocoagulation [14,16,17]. However, these techniques are energy-consuming and involve the use of organic solvents [18]. Therefore, it is important to employ aqueous phase degradation of p-NP under mild conditions by using catalytic reduction methods. Au NPs are known for their excellent catalytic behaviour, and

the catalytic behaviour of nanocrystals was generally recognised as dependant on their exposed facets and their structure [18-21].

In this study, we provide a green synthesis method for Au NPs and an assessment for the NPs produced by five different hydrocolloids. This assessment is based on the size, surface resonance plasmon band, and stability of these nanoparticles in aqueous solutions for nearly 3 months. These hydrocolloids are cost effective, non-toxic, and abundant in nature, which supports their sustainability. The hydrocolloids used in this study come from different resources such as: karaya (Kg) and tragacanth (Tg) from plant exudates, guar (Gg) from seed gums, xanthan (Xg) from microbial exudates, and carrageenan (Cg) from seaweed polysaccharides. In this study we also fulfilled the concept of green nanotechnology by providing a green synthesis of Au NPs and using these nanoparticles in the degradation of 4-Nitrophenol (p-NP) to 4-Amino phenol (p-AP). We have characterized these gold nanoparticles by centrifugal particle sedimentation (CPS), transmission electron microscopy (TEM), ultraviolent-visible (UV-Vis) spectroscopy, and Zetasizer.

2. MATERIALS AND METHODS

2.1. Materials

Guar, Xanthan, Karaya, Tragacanth, and Carrageenan and have been used without purification. Chloroauric acid ($HAuCl_4$), sodium hydroxide (NaOH) for adjusting pH, 4-nitrophenol, and Sodium borohydride were purchased from Sigma-Aldrich.

2.2. Synthesis of gold nanoparticles

5 ml of 0.1 % chloroauric acid ($H[AuCl_4]$) were added to 2 ml of 0.1 % hydrocolloids solutions and 3 ml of distilled water. The pH of the reactions were maintained at 8 by 0.1 % NaOH and the mixture was stirred at 70 °C for 15 minutes or until a colour change (the colour changed to wine red in the case of Gaur, and to purple with other hydrocolloids).

2.3. Characterization of Au NPs

UV-Vis spectroscopy (cintra 202 UV-VIS spectrophotometer, GBC, Australia), centrifugal particle sedimentation (DC24000UHR, CPS Instruments Inc., USA), zetasizer, ATR-FTIR spectrometry, transmission electron microscope (TEM), and selected area electron diffraction (SAED).

2.4. Catalytic properties of gold nanoparticles in reducing 4-nitro-phenols

For a catalytic reduction of 4-nitrophenol (p-NP), 23.4 µl of p-NP (3 mM), and 340 µl of sodium borohydride (63 mM) were added to a quartz cuvette containing 615 µl of deionized water. Thereafter, 21.6 µl of Au NPs (8×10^{-4}mM) were added to start the reaction. The intensity of the absorption peak at 400 nm was used to monitor the reduction of the 4-nitrophenol (p-NP) to 4-aminophenol (p-AP) in the UV-vis spectrograms.

3. RESULTS AND DISCUSSION

All the five aforementioned biopolymers have been used for the synthesis of Au NPs by chemical reduction. The formation of nanoparticles was visually observed by changing the colour of the solution from the colourless, at the beginning, to the wine red with guar bio-template or the purple with other hydrocolloids.

3.1. Stability of NPs

In order to study the NPs stability for a long term, UV-vis spectral were analyzed and the zeta potential of the solutions were recorded over time. The long term stability of the produced nanoparticles are also confirmed through the steady data of surface plasmon resonance (SPR) and zeta potential over time. **Figure 1** shows UV -Vis spectra for Au NPs colloids as a function of the ageing time.

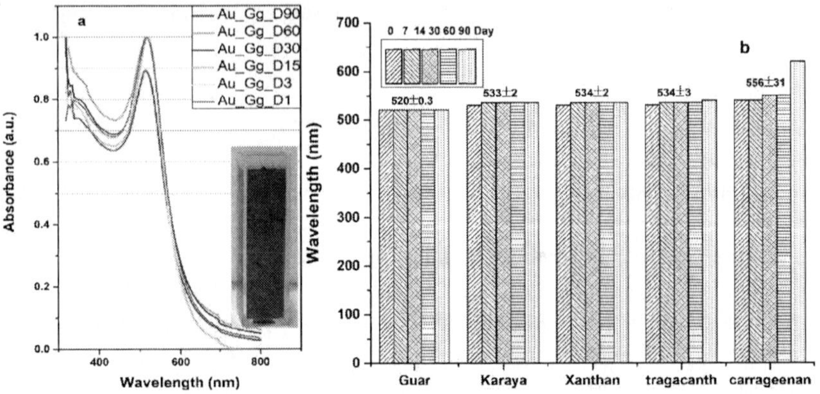

Figure 1 UV spectra of Au NP synthesized by Gg (a), all hydrocolloids (b)

Figure 1a shows Au NPs synthesized by guar where SPR band occurs at 520 ± 0.3 nm, without any shift in the peak for 90 days, which indicates the stability of gold nanoparticles synthesized by guar. The other hydrocolloids experience relatively small red shift over time as shown in **Figure 1b**. For more study about stability, the surface charge was measured at different time intervals. The values for zeta potential are shown in **Figure 2**.

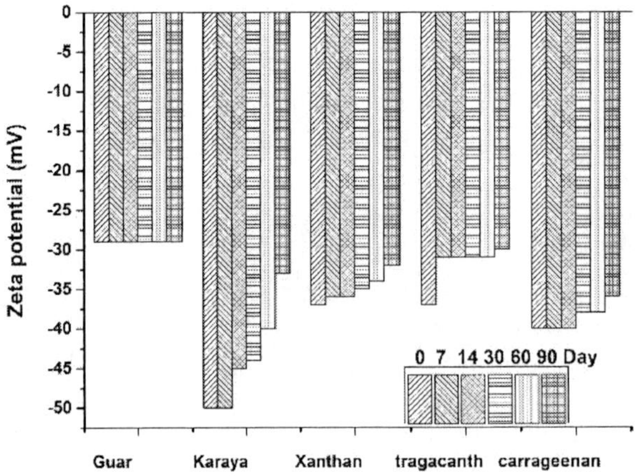

Figure 2 Zeta potential values for Au NPs composites at different time interval (0 to 90 days)

The surface charge for higher stable nanoparticles should be outside the range of -25 mV to 25 mV [22], [23], and the five used hydrocolloids were measured outside this range. Moreover, their values are quite stable even after 90 days, which indicates a good dispersion of the nanoparticles. For Au NPs guar composites, zeta potential values remained steady at -29 mV over the entire time of study (90 days). For other NPs composites, zeta potential were larger than -30 mV, which is high enough to cause repulsive forces between the nanoparticles, as long as the distances between these nanoparticles are not enough to overcome the Van der Waals attraction.

The nanoparticle have been analysed by CPS, as shown in **Figure 3**. The smallest particles size is 8 ± 1 nm for gold nanoparticles synthesized by guar. This was visually clear from its intense red colour shown in **Figure 1a** [24], and this value remains constant in agreement with both records from UV-Vis and zeta potential. For xanthan Au NPs, the measured diameter is 9 ± 1 nm. The largest size was for nanoparticles synthesized by carrageenan with a diameter of 17 ± 3 nm that tends to increase to 25 ± 4 nm at the end of the study (i.e. after 90 days).

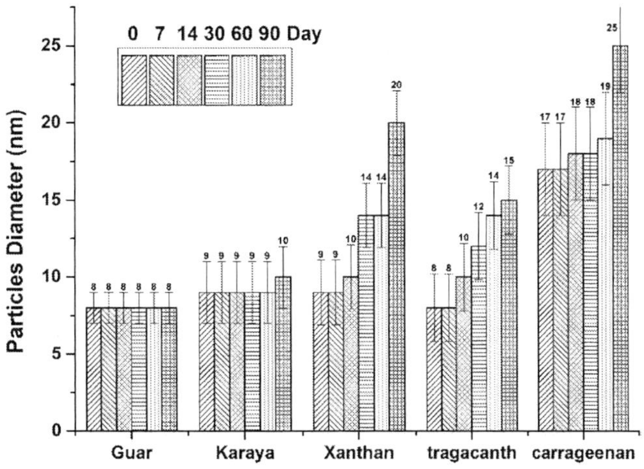

Figure 3 Particles size distribution by CPS at different time intervals.

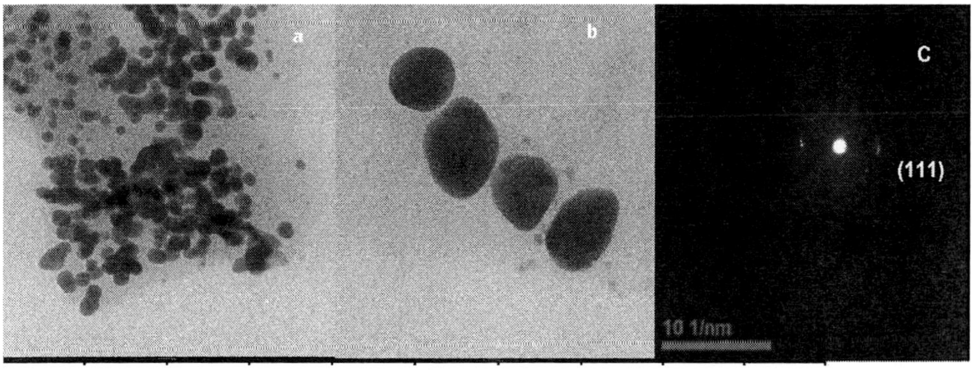

Figure 4 HRTEM images for Au NPs from Tg (a), Au NPs from Cg (b), and SAED for Au NPs from Cg (c)

Figure 4 shows images from high resolution TEM. These images show the well dispersity of nanoparticles in the solution. The average diameter is 11 ± 2 nm in the case of Au Tg and Au Cg. The representative selected area electron (SAED) in **Figure 4c** for Au Cg nanocomposites show a predominant (111) planes of face-centered cubic FCC structure.

3.2. Catalytic application of Au NPs for 4-nitrophenol reduction

The reduction of 4-NP was studied as a model reaction for the catalytic activity of the gold nanoparticles. 23.4 µL of 3 mM 4-NP, 340 µL of $NaBH_4$ of 63 mM were added to 615 µL distilled water at 10 pH. Before adding the Au NPs, the yellow colour of the solution remains the same even after one day. After adding 6.4 µL of 8×10^{-4} mM Au NPs (5.1 x 10-6 mM in the final reaction), the colour gradually disappeared which indicates the reduction of 4-NPs to 4-APs. The reaction was monitored by measuring the intensity of the absorption peak of 4-NPs at 400 nm and it was found to decrease over time. As the initial concentration of $NaBH_4$ was largely greater than the initial concentration of 4-NPs, we assumed the reaction follow the pseudo-first order.

The linear correlation shown in **Figure 5** suggests that, the reactions catalysed by Au NPs follow the first order kinetics. The kinetic rate constants (K) were determined by the slopes of the linear relationships. The highest value of 10.6×10^{-3} s^{-1} was found for Au guar composites, which was expected from its smallest size of 8 nm. The kinetic rate constants (K) for other nanocomposites are 10.6×10^{-3}, 9×10^{-3}, 7.9×10^{-3}, 7.9×10^{-3} s^{-1}, and 6×10^{-3} s^{-1} for Au Gg, Au Cg, Au Kg, Au Tg, and Au Xg, respectively.

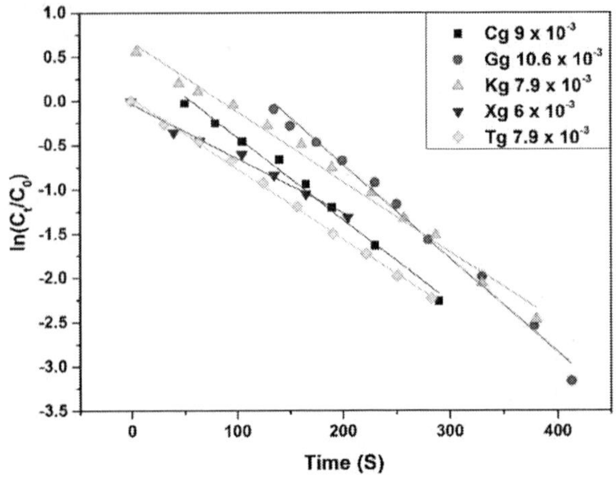

Figure 5 shows a plot of ln (Ct/Co) versus time for Au Nps synthesized by different hydrocolloids

4. CONCLUSION

A green synthesis of the gold nanoparticles has been reported herein with an assessment of their composites that were synthesized by green, biocompatible, sustainable, cost effective, and eco-friendly routes. The study shows that, guar Au NPs composite has the smallest particle size (8 nm) with utmost stability for 3 months and a high kinetic rate (10.6×10^{-3} s^{-1}) for the degradation of p-NP to p-AP. The zeta potential and the UV-vis analysis of Au NPs produced by Tg, Kg, Gg, Cg and Xg did not show much variation in magnitudes during the tested periods (0 to 90 days). Zeta potential values for all Au NPs composites are largely greater than 30 mV, which is a clear indication for their stability without any sign of aggregation during the study time which was three months.

ACKNOWLEDGEMENTS

This work was supported by the Ministry of Education, Youth and Sports of the Czech Republic and the European Union - European Structural and Investment Funds in the frames of Operational Programme Research, Development and Education - project Hybrid Materials for Hierarchical Structures (HyHi, Reg. No. CZ.02.1.01/0.0/0.0/16_019/0000843. The first author also acknowledges the Ministry of Education of the Czech Republic through the SGS project no. 21176/115 of the Technical University of Liberec.

REFERENCES

[1] C.-C. Wu and D.-H. Chen, "Facile green synthesis of gold nanoparticles with gum arabic as a stabilizing agent and reducing agent," Gold Bull., vol. 43, no. 4, pp. 234-240, 2010.

[2] T. Ishida and M. Haruta, "Gold catalysts: towards sustainable chemistry," Angew. Chemie Int. Ed., vol. 46, no. 38, pp. 7154-7156, 2007.

[3] M. T. Castaneda, A. Merkoci, M. Pumera, and S. Alegret, "Electrochemical genosensors for biomedical applications based on gold nanoparticles," Biosens. Bioelectron., vol. 22, no. 9-10, pp. 1961-1967, 2007.

[4] K. Kneipp, H. Kneipp, and J. Kneipp, "Surface-enhanced Raman scattering in local optical fields of silver and gold nanoaggregates from single-molecule Raman spectroscopy to ultrasensitive probing in live cells," Acc. Chem. Res., vol. 39, no. 7, pp. 443-450, 2006.

[5] M.-C. Daniel and D. Astruc, "Gold nanoparticles: assembly, supramolecular chemistry, quantum-size-related properties, and applications toward biology, catalysis, and nanotechnology," Chem. Rev., vol. 104, no. 1, pp. 293-346, 2004.

[6] R. Shukla, V. Bansal, M. Chaudhary, A. Basu, R. R. Bhonde, and M. Sastry, "Biocompatibility of gold nanoparticles and their endocytotic fate inside the cellular compartment: a microscopic overview," Langmuir, vol. 21, no. 23, pp. 10644-10654, 2005.

[7] I. A. Wani and T. Ahmad, "Size and shape dependant antifungal activity of gold nanoparticles: a case study of Candida," Colloids surfaces B Biointerfaces, vol. 101, pp. 162-170, 2013.

[8] G. L. Burygin, B. N. Khlebtsov, A. N. Shantrokha, L. A. Dykman, V. A. Bogatyrev, and N. G. Khlebtsov, "On the enhanced antibacterial activity of antibiotics mixed with gold nanoparticles," Nanoscale Res. Lett., vol. 4, no. 8, p. 794, 2009.

[9] L. Zhao, D. Jiang, Y. Cai, X. Ji, R. Xie, and W. Yang, "Tuning the size of gold nanoparticles in the citrate reduction by chloride ions," Nanoscale, vol. 4, no. 16, pp. 5071-5076, 2012.

[10] Q. Li, B. Lu, L. Zhang, and C. Lu, "Synthesis and stability evaluation of size-controlled gold nanoparticles via nonionic fluorosurfactant-assisted hydrogen peroxide reduction," J. Mater. Chem., vol. 22, no. 27, pp. 13564-13570, 2012.

[11] N. Vaklev, N. Vasileva, and C. Dushkin, "Synthesis of gold nanoparticles via hydrogen peroxide reduction enhanced by sonication," Nanosci. Nanotechnol., vol. 7, pp. 70-73, 2007.

[12] C. C. Li et al., "Mass synthesis of large, single-crystal Au nanosheets based on a polyol process," Adv. Funct. Mater., vol. 16, no. 1, pp. 83-90, 2006.

[13] G. B. Smith, "Green nanotechnology," in Nanostructured Thin Films IV, 2011, vol. 8104, p. 810402.

[14] S. Pandey and S. B. Mishra, "Catalytic reduction of p-nitrophenol by using platinum nanoparticles stabilised by guar gum," Carbohydr. Polym., vol. 113, pp. 525-531, 2014.

[15] M. Venkatesham, D. Ayodhya, A. Madhusudhan, and G. Veerabhadram, "Synthesis of Stable Silver Nanoparticles Using Gum Acacia as Reducing and Stabilizing Agent and Study of Its Microbial Properties: A Novel Green Approach," Int. J. Green Nanotechnol., vol. 4, no. 3, pp. 199-206, Jul. 2012.

[16] N. Modirshahla, M. A. Behnajady, and S. Mohammadi-Aghdam, "Investigation of the effect of different electrodes and their connections on the removal efficiency of 4-nitrophenol from aqueous solution by electrocoagulation," J. Hazard. Mater., vol. 154, no. 1-3, pp. 778-786, 2008.

[17] E. Marais and T. Nyokong, "Adsorption of 4-nitrophenol onto Amberlite® IRA-900 modified with metallophthalocyanines," J. Hazard. Mater., vol. 152, no. 1, pp. 293-301, 2008.

[18] C. Mahugo Santana, Z. Sosa Ferrera, M. Esther Torres Padrón, and J. Juan Santana Rodríguez, "Methodologies for the extraction of phenolic compounds from environmental samples: new approaches," Molecules, vol. 14, no. 1, pp. 298-320, 2009.

[19] K. S. Anseth, "Foreword A2 - Kumbar, Sangamesh G.," C. T. Laurencin and M. B. T.-N. and S. B. P. Deng, Eds. Oxford: Elsevier, 2014, p. xvii.

[20] G. Applerot, N. Perkas, G. Amirian, O. Girshevitz, and A. Gedanken, "Coating of glass with ZnO via ultrasonic irradiation and a study of its antibacterial properties," Appl. Surf. Sci., vol. 256, no. 3, pp. S3-S8, 2009.

[21] A. Aravamudhan et al., "Contributors BT - Natural and Synthetic Biomedical Polymers," Oxford: Elsevier, 2014, pp. xiii-xv.

[22] S. Sanyasi et al., "Polysaccharide-capped silver Nanoparticles inhibit biofilm formation and eliminate multi-drug-resistant bacteria by disrupting bacterial cytoskeleton with reduced cytotoxicity towards mammalian cells," Sci. Rep., vol. 6, p. 24929, 2016.

[23] M. F. Meléndrez, G. Cárdenas, and J. Arbiol, "Synthesis and characterization of gallium colloidal nanoparticles," J. Colloid Interface Sci., vol. 346, no. 2, pp. 279-287, 2010.

[24] Sigma-Aldrich, "Gold Nanoparticles: Properties and Applications." [Online]. Available: https://www.sigmaaldrich.com/technical-documents/articles/materials-science/nanomaterials/gold-nanoparticles.html. [Accessed: 25-Mar-2018].

FLEXIBLE AMMONIA GAS SENSOR BASED ON POLYANILINE

Alexandr POŠTA, Josef NÁHLÍK, Alexandr LAPOSA, Jan VOVES

CTU FEE- Czech Technical University in Prague Faculty of Electrical Engineering, Prague, Czech Republic, EU, postaale@fel.cvut.cz

Abstract

In this paper we prepared flexible ammonia (NH_3) gas sensor based on polyaniline (PAni) as an active sensing thin film layer. The sensors function was based on chemiresistive principle with thin film deposited on top of interdigitated electrodes. We measured the change in resistance of the thin film under gas exposure. The PET based foil has been used as the sensors substrate. The Interdigitated electrodes (IDE) has been printed with Fujifilm Dimatix DMP 2831 inkjet printer by nanocoloid silver ink. The active thin film has been prepared by drop casting method from polyaniline nanoparticle dispersion. The morphology of thin films surface was characterized by atomic force microscopy (AFM). The sensors response was investigated in a gas chamber, and the gas flow was controlled by mass flow meters. The sensitivity towards various gases under different concentration in synthetic air has been tested. The tested gases were NH_3, NO_2, CO_2 and CO. The measurement of the sensors response to the different levels of humidity was also conducted. All the measurements were performed at room temperature, which is important feature for the sensor to be able to operate in hazardous environment and it facilitates low-power operation of the sensor as well. We also observed the long-term stability of the sensors response, as well as the effect of bending of the sensing platform on the sensors response and stability.

Keywords: Ammonia, polyaniline, chemiresistive gas sensing, inkjet printing

1. INTRODUCTION

During the last decades, there has been growing interest in a gas sensing technology from many areas including industrial, environmental, domestic, military and many other industries. The ability to detect or monitor different gases became critical during some manufacture processes and in environmental issues, to avoid harmful effects caused by toxic gases. This demand accelerated research in gas sensor industry, and different gas sensing techniques were proposed. In this paper, we present a flexible chemiresistive ammonia gas sensor based on polyaniline as the sensing layer.

Ammonia is natural as well as manufactured compound on our planet. It can be found in our atmosphere, water, soil and is typically found in our body. Although ammonia can be found all around us, the exposure to the high concentration of ammonia gas causes chemical burns of the respiratory tract, skin, and eyes. The ammonia gas can be detected by the human nose since it has a strong, unpleasant and irritating smell [1]. The detection limit of the human nose towards the ammonia gas is around 50 ppm. However, even lower concentrations can be irritating for the humans. The maximal allowed workspace concentration of ammonia is set to be 20 ppm. Today, most of the ammonia in our atmosphere is emitted direct or indirect by human activity [2]. There are so far four major industry areas with a need for ammonia gas sensing application. It is environmental, automotive, chemical industry and medical diagnostics. The detection of ammonia gas in the environment is especially critical near farming areas, where the concentration of the gas can exceed allowed exposure limits. Ammonia from decomposing animal excreta probably accounts for the most significant proportion of the ammonia produced, with the decay of organic materials from plants, dead animals, etc. The presence of ammonia can cause unhealthy conditions for both farmers and the animals especially indoors. The usage of ammonia detection in the automotive industry can be divided into three categories. The first is monitoring of the pollution caused by vehicles exhaust system, the second is control of indoor air condition, while the third reason is the reduction of NOx in diesel engines. In the chemical industry, the vast majority of ammonia is produced for manufacturing of fertilizers. There are other processes using ammonia such as plastics, synthetic fibers, resins, explosives, and other uses. The detection of NH_3 in very low concentrations is very interesting for medical diagnostics. Ammonia is produced by the human body, and the amount of produced NH_3 can be used for diagnostics of certain diseases using breath analyzers. The detection limit in those systems needs to be in orders of tens to hundreds of ppb. Detection of such concentrations is so far possible by large and expensive optical sensors. The flexible sensor prepared in this paper belongs between solid state chemiresistive group of the gas sensors. As a sensing layer acted a thin film of polyaniline (PAni) and we observed the influence of light irradiation of the sensing layer on the response of the sensor as well as the influence of the sensors bending.

2. AMMONIA GAS SENSING MECHANISM IN POLYANILINE

The first report on chemiresistive properties of polyaniline towards ammonia gas was by Hirata et al. in 1994 [3]. In recent years the attention of the scientific community was dedicated to exploring different nanostructures, composites and flexible sensors based on polyaniline. [4], [5], [6], [7]. The polyaniline belongs to the group of conjugated polymers, although the molecular structure is a little bit different since its absence of strictly alternating single and double bonds. The PAni has unique ability since its doped state can be controlled by the acid/base reaction. The ability of PAni to undergo doping/dedoping under acidic/basic media suggested that this material could be applied as a chemiresistive sensor for the detection of a wide range of compounds. When exposed in ammonia gas, PAni undergoes dedoping by deprotonation [8], [9]. The illustration of the chemical process mentioned above can be seen in **Figure 1** The protons on -NH- groups, were transferred to NH_3 molecules to form ammonium ions while PAni itself turned into its base form. This process is reversible, and in fact, when ammonia atmosphere is removed the ammonium ion can be decomposed to ammonia gas and proton [8]

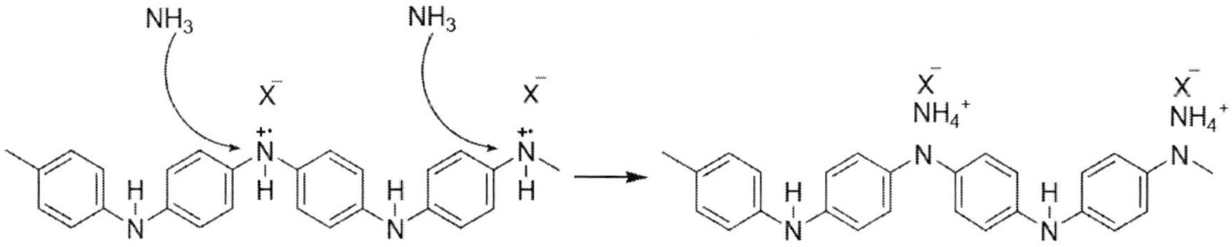

Figure 1 Schematic illustration of the dedoping process in polyaniline during exposure to ammonia gas [8]

3. GAS SENSOR PREPARATION AND CHARACTERIZATION

The gas sensors were prepared by an inkjet printer (IJP) on NoveleTM IJ-220 foil from Novacentrix. It is a PET-based printed electronics substrate for low-cost and low-temperature applications. The sensing array dimension was 16 x 7.7 mm and was connected to PCB by 8 PIN FPC connector. The sensing array consisted of four sensing IDE areas with dimensions approximately 2 x 2 mm each. The electrode width and spacing were designed to be 100 µm. The IDE was printed with Dimatix DMP 2831 drop on demand (DoD) IJP by silver nanoparticle ink Silver jet DGP-40LT-15C from Advanced Nano Products Co.Ltd. The masking layer was printed with ink of SU-8 photoresist XP PrlElex® SU-8 1.0 from MicroChem. The active sensing layer was prepared from doped polyaniline (PAni) powder in the form of emeraldine salt obtained from Sigma Aldrich. The PANI was dispersed in laboratory grade p-Xylene from Sigma Aldrich. The dispersion of 2wt% of PAni in p-Xylene was stirred at room temperature and sonicated at 60 °C for 120 min. The thin film of PAni was deposited by simple drop casting method by laboratory pipette The photo of the sensor array with AFM picture of the surface of the sensor in the inset, and the schematic illustration of the sensor can be seen in **Figure 2**

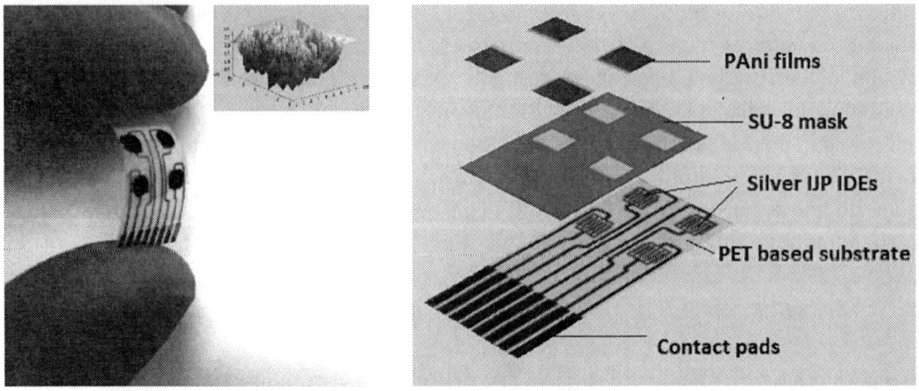

Figure 2 On the left photo of actual sensor (inset AFM picture of the PAni surface) and the schematic illustration of each layer of the fabricated sensors on the right

In **Figure 3** can be seen the experimental setup for measuring the responses and the picture of the test chamber with the attached sensor and LED used for irradiation. It consists of gas bottles with target analytes, reduction valves, mass flow meters, valve switch, source meters, and the test chamber. The data were acquired by the LabVIEW program through the PC. The detection and cleaning stage of the measurements were cycled using two-way our inputs/output valve driven by VICI universal valve actuator controlled by the same LabVIEW program via the serial port. The measurement setup was identical to the one used by Davydova et al. reporting the gas sensing behavior of ZnO/diamond nanostructures. [10]

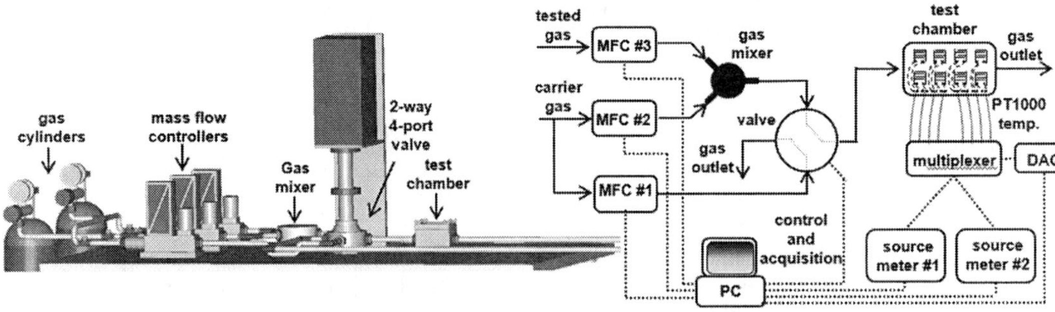

Figure 3 Schematic illustration of measurement setup

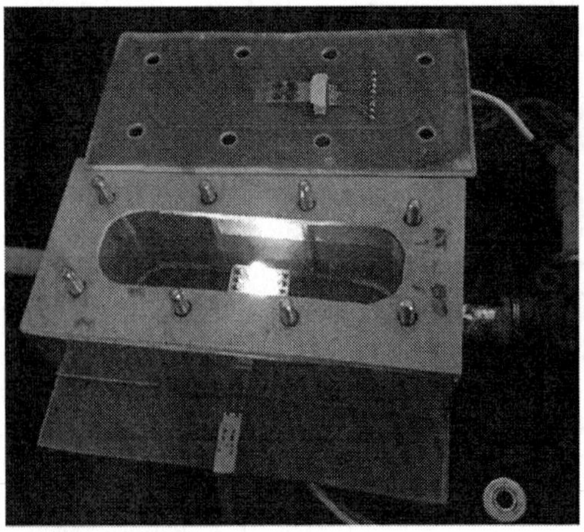

Figure 4 Photo of the test chamber with blue LED light source and the sensor array

The LEDs used for irradiation were powered by Keithley 2230G DC power supply. The LEDs used in experiment were blue, yellow and IR, with wavelength 470, 590 and 940 nm respectively. The picture of the test chamber with light irradiation setup is in **Figure 4**.

4. RESULTS

We measured the response of the sensor to different analytes in synthetic air (SA) (80 % N_2, 20 % O_2). The gas flow was set to 100 ml/min by mass flow meters. The measurement consisted of cleaning cycles (pure SA) and the testing cycles of target analyte mixed with SA. The tested gases were NH_3, CO, CO_2 and NO_2. The response to the 50 ppm of NH_3 analyte can be seen in **Figure 5**. The response to the step function of the NH_3 analyte concentration can be seen in **Figure 6**. In presence of humidity (50 % RH), the response of the sensor to the NH_3 was inhibited. The sensor exhibited negligible additive response towards CO_2 and NO_2 analytes. During exposure to the CO analyte we observed decrease in resistance. When exposed to the different levels of humidity, the resistance of the sensor decreases dramatically and after exposure to 25 % RH there was not any significant drop in the resistance upon exposure to higher RH levels. The bending of the sensor did not show any noticeable effect on the performance of the sensor. One of the goal of this paper was to research the influence of sensors surface by vis-IR light on the sensors response. The response did not show any significant changes upon the exposure to the light irradiation

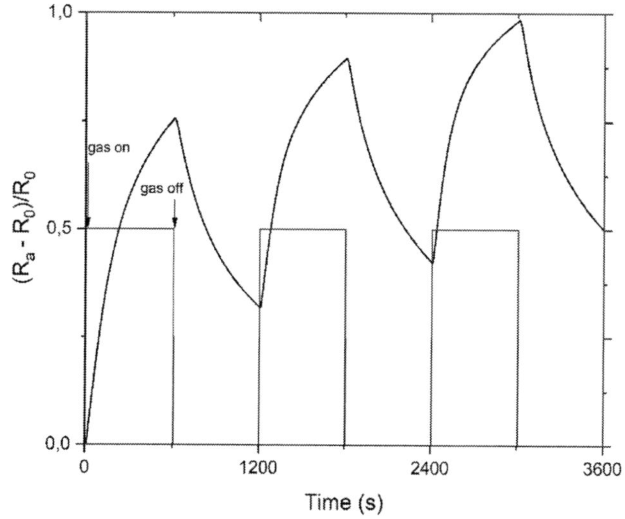

Figure 5 Response of the sensor to the 50 ppm of NH_3 at room temperature

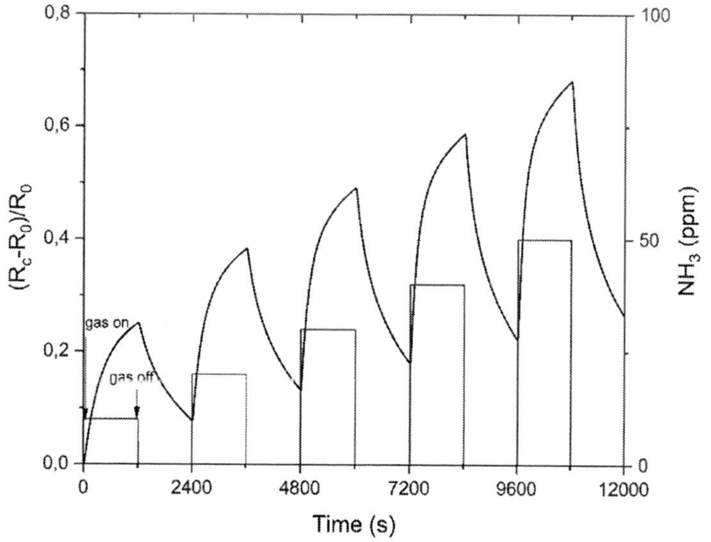

Figure 6 Response of the sensor to the step function of NH_3 concentration at room temperature

5. CONCLUSION

The sensor array was successfully printed by silver ink and then the platform masked by SU-8 photoresist ink utilizing inkjet printing. The drop casting method used for deposition of an active layer showed its limitation especially concerning the film uniformity. This lead to different baseline resistance of fabricated sensing films. The response of sensors array was tested towards different gases in synthetic air at room temperature. The sensors showed good sensitivity towards the NH_3 analyte. At room temperature, the sensor array did not fully recover during the cleaning cycles. The sensor had negligible response towards NO_2 and CO_2 analytes. When exposed to the CO analyte, the resistance of the sensor decreased possibly due to the reducing effect of the analyte and its reaction with adsorbed oxygen on the surface of the sensing layer. The response of the array was also measured under irradiation by vis-IR light. As light sources were used blue (470 nm) yellow (590 nm) and IR (940 nm) LEDs. There was not any noticeable influence caused by light irradiation on adsorption/desorption properties of the sensor. The possible interaction will also be investigated for other wavelengths. Repeated bending of the approximately by 45° did not show any significant decrease in performance.

ACKNOWLEDGEMENTS

This paper was supportedby the CTU student grant No. SGS17/096/OHK3/1T/13 and the project CAV CAAS CZ.02.1.01/0.0/0.0/16_019/0000737

REFERENCES

[1] U.S. Department of Health and Human Services - Agency for Toxic Substances and Disease Registry, "Toxicological Profile For Ammonia," 2004.

[2] W. OLTHUIS AND A. VAN DEN BERG, "Ammonia sensors and their applications - a review," vol. 107, pp. 666-677, 2005.

[3] M. HIRATA and L. SUN, "Characteristics of an organic semiconductor polyaniline film as a sensor for NH3 gas," *Sensors Actuators A. Phys.*, vol. 40, no. 2, pp. 159-163, 1994.

[4] L. ZHIHUA *et al.*, "Sensors and Actuators B : Chemical Fast response ammonia sensor based on porous thin film of polyaniline / sulfonated nickel phthalocyanine composites," *Sensors Actuators B. Chem.*, vol. 226, pp. 553-562, 2016.

[5] S. SRIVASTAVA, S. S. SHARMA, S. AGRAWAL, S. KUMAR, M. SINGH, and Y. K. VIJAY, "Study of chemiresistor type CNT doped polyaniline gas sensor," *Synth. Met.*, vol. 160, no. 5-6, pp. 529-534, 2010.

[6] L. KUMAR, I. RAWAL, A. KAUR, and S. ANNAPOORNI, "Flexible room temperature ammonia sensor based on polyaniline," *Sensors Actuators, B Chem.*, vol. 240, pp. 408-416, 2017.

[7] Z. F. LI, F. D. BLUM, M. F. BERTINO, and C. S. KIM, "Understanding the response of nanostructured polyaniline gas sensors," *Sensors Actuators, B Chem.*, vol. 183, pp. 419-427, 2013.

[8] H. BAI and G. SHI, "Gas Sensors Based on Conducting Polymers," *Sensors*, vol. 7, pp. 267-307, 2007.

[9] N. R. TANGUY, M. THOMPSON, and N. YAN, "Sensors and Actuators B : Chemical A review on advances in application of polyaniline for ammonia detection," *Sensors Actuators B. Chem.*, vol. 257, pp. 1044-1064, 2018.

[10] M. DAVYDOVA *et al.*, "Gas-sensing behaviour of ZnO/diamond nanostructures," *Beilstein J. Nanotechnol.*, vol. 9, no. 1, pp. 22-29, 2018.

INFLUENCE OF NANOPARTICLES ON MICROALGAE DETECTED BY IN VIVO IMAGING

Kristyna PAVELICOVA, Aneta STREJCKOVA, Ivan RANKIC, Tereza VANECKOVA, Jaroslava ZELNICKOVA, Dalibor HUSKA, Marketa VACULOVICOVA

Department of Chemistry and Biochemistry, Mendel University in Brno, Brno, Czech Republic, EU, xpavelic@node.mendelu.cz

Abstract

Fluorescence imaging is an efficient and powerful technique, which has been often utilized for tracking compounds in animals, however its application on plants or other organisms is not very common. In this work, the fluorescence imaging was used to investigate the effect of selected nanoparticles - NPs (ZnO, CdTe, etc.) on microalgae (*Scenedesmus quadricauda and Chlorella vulgaris*) growth. The growth and metabolic activity of microalgae was influenced by application of nanoparticles into the growing medium (liquid and solid) and therefore the intrinsic fluorescence was affected. Microalgae pigments (chlorophylls, carotenoids) are responsible for the autofluorescence of the species and therefore, the fluorescence can be used as a tool for quick and easy evaluation of expression these compounds. These pigments have main role in light harvesting and minimize photo-oxidative damage of the cells. Such protection mechanism can be disturbed by another stress related compounds like NPs that presumably damage this protection mechanism i) indirectly (reactive oxygen species) and ii) directly (disruption of biomembranes). Owing to broad fluorescence imaging and correlation with total amount of chlorophylls and carotenoids it was possible to evaluate the influence of NPs on individual microalgae just only by *in vivo* imaging.

Keywords: Nanoparticles, Scenedesmus quadricauda, Chlorella vulgaris, fluorescence, abiotic stress, metabolic

1. INTRODUCTION

Algae are a large group of organisms which including unicellular microalgae and more complex multicellular organisms (macroalgae). In recent years, large-scale production of microalgae has been very successful especially for use in many branches, such as food, feed and fuel products [1]. In microalgae, pigments with fluorescent activity, in particular, carotenoids and chlorophylls have an important role in photosynthesis. Greenish, fat-soluble chlorophylls are pigments with porphyrin ring responsible for conversion of solar energy into the chemical energy. Carotenoids (β-carotene, lycopen, astaxanthin, zeaxanthin, vialoxanthin a lutein) perform two important functions: light absorption in range of visible spectrum, where chlorophyll does not (400-530 nm [2]) and, thanks to their antioxidant activity, defence against formation of reactive oxygen species (ROS) [3].

Plant systems are subject abiotic stress, which can be caused by many factors (for example cold, heat, and drought). Therefore, it is important to study of metabolic pathways in plant cells. This can be accomplished by many methods such as quantitative polymerase chain reaction [4], liquid chromatography [5] or mass spectrometry [6]. Use of imaging technologies is more favorable than mentioned methods, mainly for their financial and time-saving benefits. Imaging methods allow imaging in living organisms at the cell (microscopy), organ and/or whole body level. The most commonly used ones include ultrasound, computed tomography and/or magnetic resonance. NMR spectroscopy has contributed to the study of plant primacy or specialized metabolism in very diverse ways [7].

On the other hand, fluorescence imaging is no very common yet even though compared to the above mentioned methods; it exhibits numerous advantages such as variability, multiplexing suitability and/or cost-effectivity. Fluorescence *in vivo* imaging has a great potential for a wide range of molecular diagnostic application. Fluorescence-based methods are increasingly used for their quantitative sensitivity, natural biological safety and fairly ease of use [8].

In this study, a fluorescence *in vivo* imaging was used and changes in growth microalgae exposed to nanoparticle effects were investigated.

2. MATERIALS AND METHODS

2.1. Preparation of nanoparticle solutions

Solutions of ZnO nanoparticles (NPs) were prepared by mixing a stock solution of the commercial NPs (Houston, USA) with distilled water at four different concentrations (0 mg/L, 10 mg/L, 50 mg/L a 100 mg/L). Solutions were sonicated for 10 minutes before use.

2.2. Preparation of bold´s basal medium (BBM)

Bold´s Basal Medium (BBM) has been prepared according to published method [9]. The final solutions were prepared by mixing 2 mL solutions of NPs (0 mg/L, 10 mg/L, 50 mg/L a 100 mg/L). Volumes were adjusted with BBM medium to 150 mL. One of these solutions (0 mg/L) was used as a control. All solutions were autoclaved and subsequent these media were added to petri dishes.

2.3. Spot test

The stock solution of microalgae (*Scenedesmus quadricauda and Chlorella vulgaris*) was prepared at absorbance of 0.1-0.2 AU. Then, four samples were prepared by subsequent dilution in ratio 1:1 with BBM without NPs. Individual samples (5 µL) were spotted onto petri dishes with NPs containing BBM (three spots per sample). In total, 4 petri dishes containing 12 spots (4 microalgae dilutions, in triplicates) were prepared for each microalgae specie *(Scenedesmus quadricauda and Chlorella vulgaris)*

2.4. *In vivo* fluorescence imaging

Monitoring of influence of nanoparticles on microalgae was performed using an In Vivo Xtreme Imaging System by Bruker (Massachusetts, USA). The parameters used for image acquisition were following: exposition time - 4 s, binning - 4x4 pixels, fStop - 1.1, field of view - 10x10 cm.

Monitoring of influence of nanoparticles on microalgae has been performed using excitation filter of 650 nm and emission filter of 700 nm. Individual samples were measured consecutively for 6 days.

For image post processing and analysis Bruker Molecular Imaging Software (Bruker, Massachusetts, USA) was used. In this software the spectral modeling was performed.

3. RESULTS AND DISCUSSION

In this study, the series of *in vivo* experiments has been done to evaluate the potential of influence of nanoparticles on microalgae. The samples of microalgae were monitored for 6 days and a change in fluorescence intensity was observed in pW/mm^2.

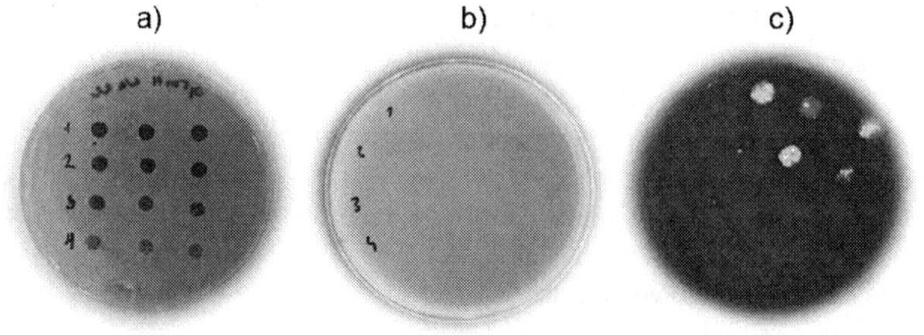

Figure 1 Microalgae *Chlamydomonas reinhardtii* (a) visible to the naked eye, *Chlorella vulgaris* including 100 mg/L NPs (b) invisible to the naked eye and *Chlorella vulgaris* including 100 mg/L NPs by fluorescence imaging (c).

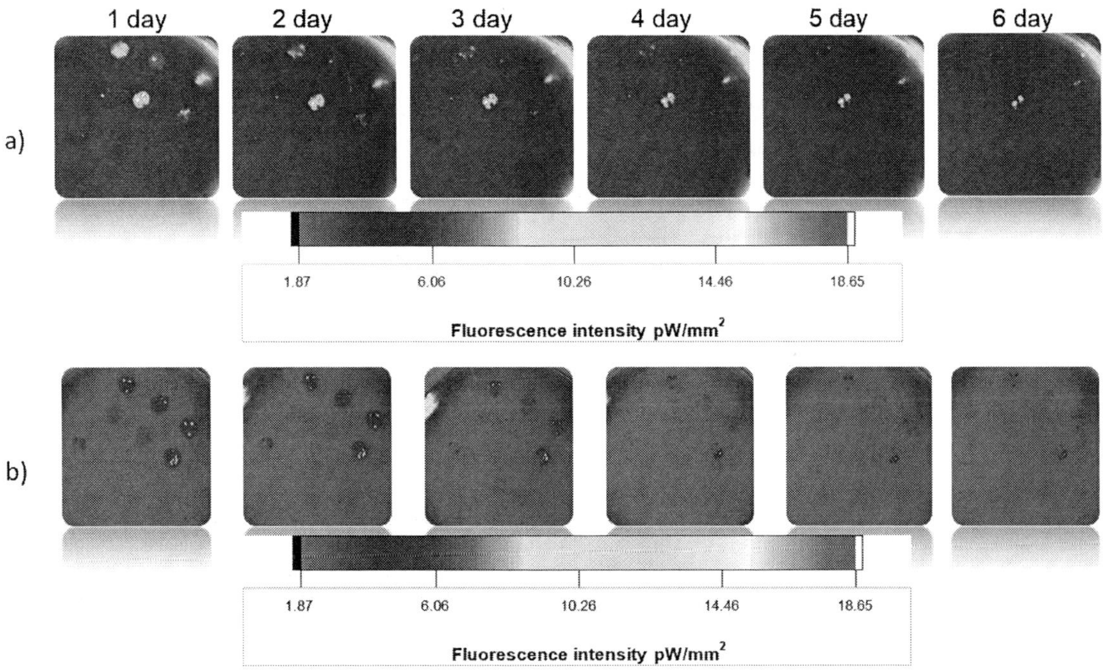

Figure 2 Microalgae *Chlorella vulgaris* including 100 mg/L NPs (a) and microalgae without NPs (b). Fluorescence images captured at time intervals as noted above picture respectively.

Fluorescence intensity was higher in microalgae including NPs (**Figure 2a**) than microalgae without NPs (**Figure 2b**). This could be due to the increased content of photosynthetic pigments, which could be explained by the balancing effect to overcome the toxicity induced by NPs [10]. But images show that fluorescence intensity was gradually fallen. Which could be the fact that metal NP and ions affecting metabolism in the form of excessive increasing levels of reactive oxygen species (ROS) within the microalgae [11].

The images were further analyzed using Bruker Molecular Imaging Software.

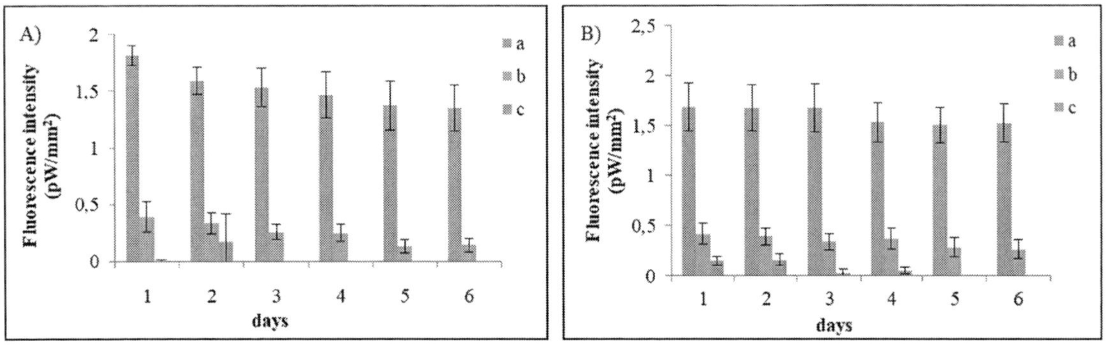

Figure 3 The graph A (*Chlorella vulgaris)* and the graph B (*Scenedesmus quadricauda)* show 3 concentrations of NPs (a-c). The graphs show pure fluorescence intensity with deducted control fluorescence intensity. All 3 NPs concentrations (a-c) gradually killed the cells, therefore fluorescence intensity decreased in time gradually. The highest fluorescence intensity was detected at the highest concentration of NPs (A and B, a).

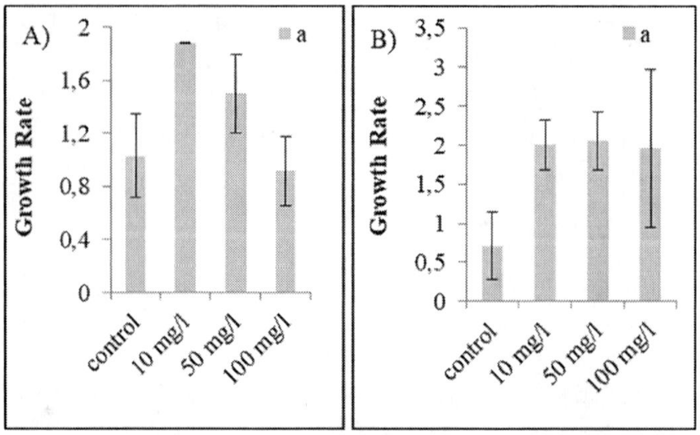

Figure 4 The growth rates of *Chlorella vulgaris* (A) and *Scenedesmus qudricauda* (B)

Based on our results we have come the following hypothesis. The highest concentrations of nanoparticles (100 mg/L) affect probably almost the biosynthesis of plant pigments instantly. For about 4 hours, fluorescence intensity was superior to the other concentrations (**Figure 3a-c**). Which may be caused by stimulated of photosynthesis. It is important to note that the literature is not available for assessment.

If we look at the growth rate (**Figure 4A**), we can see a concentration 10 mg/L the fluorescence intensity increased but concentrations 50 mg/L and 100 mg/L reduced it. We conclude it to reflect the amount of biomass. The smallest concentration of nanoparticles (10 mg/L) increased the amount of biomass, while concentration of NPs reduced it. The growth rate of *Scenedesmus quadricauda* (**Figure 4B**) was level-headed.

4. CONCLUSION

Microalgae are the first trophic level of the food chain, providing biological energy and oxygen for other organisms and playing very important roles in keeping the balance of aquatic ecosystems. Nanoparticles (NPs) are inevitably released into the aquatic environment for being widely used [12]. During this study the effect of nanoparticles on *Scenedesmus quadricauda* and *Chlorella vulgaris* was studied. The NPs can be a stress factor and promote the emergence of ROS. Mentioned changes in growth of microalgae were observed using modern *in vivo* imaging technology. Today, the *in vivo* imaging method is used primarily in animals, whereas in plants it is not so common. Thanks to this technology, we would like to optimize *in vivo* imaging methods in plant materials. The future should be, for example, recognizing the distribution of toxic substances in plants to preserve the safety of agriculture and food.

ACKNOWLEDGEMENTS

The research was financially supported by IGA grant, no. TP_1/2017.

REFERENCES

[1] ZHONG Y, JIN P, CHENG JJ. A comprehensive comparable study of the physiological properties of four microalgal species under different light wavelength conditions. *Planta* 2018; 248: 489-98.

[2] CHEN C, GONG N, QU F et al. Effects of carotenoids on the absorption and fluorescence spectral properties and fluorescence quenching of Chlorophyll a. *Spectrochimica Acta Part A: Molecular and Biomolecular Spectroscopy* 2018; 204: 440-5.

[3] D'ALESSANDRO EB, ANTONIOSI FILHO NR. Concepts and studies on lipid and pigments of microalgae: A review. *Renewable and Sustainable Energy Reviews* 2016; 58: 832-41.

[4] GAO D, KONG FN, SUN PP et al. Transcriptome-wide identification of optimal reference genes for expression analysis of Pyropia yezoensis responses to abiotic stress. *BMC Genomics* 2018; 19.

[5] CERÓN-GARCÍA MC, GONZÁLEZ-LÓPEZ CV, CAMACHO-RODRÍGUEZ J et al. Maximizing carotenoid extraction from microalgae used as food additives and determined by liquid chromatography (HPLC). *Food Chemistry* 2018; 257: 316-24.

[6] TUREMIS M, SILLETTI S, PEZZOTTI G et al. Optical biosensor based on the microalga-paramecium symbiosis for improved marine monitoring. *Sens Actuator B-Chem* 2018; 270: 424-32.

[7] DEBORDE C, MOING A, ROCH L et al. Plant metabolism as studied by NMR spectroscopy. *Progress in Nuclear Magnetic Resonance Spectroscopy* 2017; 102-103: 61-97.

[8] GHOROGHCHIAN PP, THERIEDN MJ, HAMMER DA. In vivo fluorescence imaging: a personal perspective. *WIREs Nanomedicine and Nanobiotechnology* 2009; 1: 156-67.

[9] DEMIREL Z, YILMAZ FF, OZDEMIR G, DALAY MC. Influence of Media and Temperature on the Growth and the Biological Activities of Desmodesmus protuberans (FE Fritsch & MF Rich) E. Hegewald. *Turk J Fish Quat Sci* 2018; 18: 1195-203.

[10] MIDDEPOGU A, HOU J, GAO X, LIN D. Effect and mechanism of TiO2 nanoparticles on the photosynthesis of Chlorella pyrenoidosa. *Ecotox Environ Safe* 2018; 161: 497-506.

[11] MORENO-GARRIDO I, PÉREZ S, BLASCO J. Toxicity of silver and gold nanoparticles on marine microalgae *Marine Environmental Research* 2015; 111: 60-73.

[12] YU Z, HAO R, ZHANG L, ZHU Y. Effects of TiO2, SiO2, Ag and CdTe/CdS quantum dots nanoparticles on toxicity of cadmium towards Chlamydomonas reinhardtii. *Ecotox Environ Safe* 2018; 156: 75-86.

DETECTION OF UV RADIATION IN VINEYARDS BY OPTICAL ANALYSIS OF CdTe QUANTUM DOTS

[1]Lenka SOCHOROVÁ, [1]Božena HOSNEDLOVÁ, [2]Michaela DOČEKALOVÁ, [2]Martina STAŇKOVÁ, [2]Dagmar UHLÍŘOVÁ, [3]Branislav RUTTKAY-NEDECKÝ, [4]Marta KEPINSKA, [4]Halina MILNEROWICZ, [1]Mojmír BAROŇ, [5]Hai Xuan NGUYEN, [5]Bach Ngoc DUONG, [5]Hoai Viet NGUYEN, [1]Jiří SOCHOR, [2]Josef RŮŽIČKA, [2,3,4]René KIZEK

[1] Mendel University in Brno, Faculty of Horticulture, Department of Viticulture and Enology, Lednice, Czech Republic, EU, sochor.jirik@seznam.cz
[2]Prevention Medicals, Studenka, Czech Republic, EU, docekalova@preventionmedicals.cz
[3]University of Veterinary and Pharmaceutical Sciences Brno, Pharmaceutical Faculty, Brno, Czech Republic, EU, kizek@sci.muni.cz
[4] Department of Biomedical and Environmental Analyses, Faculty of Pharmacy with Division of Laboratory Diagnostics, Wroclaw Medical University, Wroclaw, Poland, EU, zalewska.m@gmail.com
[5] Research Center for Environmental Monitoring and Modelling, VNU University of Science, Hanoi, Vietnam, nguyenviethoai@hus.edu.vn

Abstract

The type, number, and magnitude of environmental constraints are currently undergoing changes due to shifts in climate patterns already observed for the past and predicted for the future. These changes are already affecting grape composition with observed changes in sugar and acidity concentrations. UV sensor in vineyards is therefore needed. In this work, the ability of cadmium telluride quantum dots (CdTe QDs) to respond to UV radiation was examined. Green, yellow, orange, and red CdTe QDs were freshly prepared every day and exposed to UV daily radiation for 12 hours for seven consecutive summer days. Every day, solar irradiance was measured, reaching maximal values during the day of around 200-700 W/m^2 with a UV index of 2-8. The highest solar and UV irradiances were on the third, fourth, sixth, and seventh days (700 W/m^2, UV index 8). The lowest solar and UV irradiances were on the first and fifth days (600 W/m^2, UV index 6). On the second day, both irradiances were average (200-500 W/m^2; UV index 4-7). The physical properties of individual QDs (absorption and fluorescence maxima, electrochemical determination of cadmium concentration) were monitored. The rising dose of UV radiation led to the release of cadmium ions and the change in the size of individual QDs (a significant change of colour directed toward to the red area of the visible light spectrum). Yellow, orange, and green QDs responded most to UV radiation. This sensor will be used for detection of UV radiation in vineyards.

Keywords: UV radiation, quantum dots, electrochemical detection, fluorimetric detection, climate change

1. INTRODUCTION

The sun is a natural source of UV (ultraviolet light) radiation on Earth; however, only the UV-A and UV-B components of this radiation enter the troposphere. UV-A radiation has a wavelength range of 320-400 nm and accounts for/represents 99 % of the sun's UV rays that fall on the Earth's surface. UV-B radiation has been found to cause the release of Cd^{2+} ions from quantum dots (QDs) due to/as a result of cadmium telluride (CdTe) QDs surface oxidation [2].

Tsipotan et al. examined the effect of visible light and UV radiation on the aggregation stability of CdTe QDs [3]. In the case of ultraviolet light, photoinduced reduction of CdTe QDs size was confirmed. Typical CdTe QDs stabilised with mercaptosuccinic acid (MSA) after exposure to specific UV-B radiation at a given wavelength and energy resulted in excitation of a system capable of generating a whole group of molecularly reactive oxidation intermediates (ROI) and reactive oxygen species (ROS) [2,4]. Due to the photodynamic effect of CdTe QDs, free cadmium ions are released.

This mechanism has been used to construct a new, very simple, and completely non-demanding qualitative UV-B radiation sensor, even in places where technical equipment is not available. Such a sensor can be used to monitor the UV-B radiation in the cultivation of vine. In this study, for the very first time, a mercaptosuccinic acid-stabilised CdTe QDs sensor for monitoring of UV-B radiation was proposed, and the possibility of it using CdTe QDs has been investigated.

2. MATERIAL AND METHODS

2.1. Chemicals

All chemicals used in this study (Cd $(CH_3COO)_2 \cdot 2H_2O$, Na_2TeO_3, mercaptosuccinic acid, Trizma base, HCl) were purchased from Sigma-Aldrich (St. Louis, MO, USA), in ACS purity. Propanol, $NaBH_4$ was purchased from Merck (Darmstadt, Germany), and 25 % aqueous NH_4OH was purchased from Lach-Ner s.r.o. (Neratovice, Czech Republic). All chemicals that we used for gel electrophoresis were purchased from VWR (Germany). All plastic materials used (tubes, tips) in this study were purchased form Eppendorf (Hamburg, Germany).

2.2. CdTe quantum dots synthesis

The preparation of CdTe quantum dots (QDs) was as follows: 10 ml of solution $Cd(CH_3COO)_2 \cdot 2H_2O$ (0.02 M), 76 ml of H_2O, 1 ml of mercaptosuccinic acid (MSA) solution (0.4 M), 5 mL of Na_2TeO_3 (0.02 M), and 40 mg of $NaBH_4$ was stirred with a magnetic stirrer (VMS-C4, VWR International Ltd., Darmstadt, Germany) for at least two hours until bubbling stopped. Subsequently, the volume was adjusted to 100 ml. 2 ml of the prepared solution was pipetted into the glass vials (Sigma Aldrich, St. Louis, MO, USA) with a white cap (Anton Paar, Graz, Austria) and a Teflon cap (Anton Paar, Graz, Austria). The thus prepared vial was placed in a microwave, which was set to a power of 300 W, and the heating took place for two minutes for green QDs, four minutes for yellow QDs, six minutes for orange QDs, and eight minutes for red QDs. The final concentration of QDs was 2 mM.

2.3. Spectrometry and fluorescence measurement

Absorption and fluorescence spectra were obtained with the Tecan Infinite 200 PRO Multifunctional Reader (TECAN, Switzerland). For absorption spectra, the wavelength was in the range 400-800 nm. For fluorescence spectra, there was an excitation wavelength of 250 nm and an emission wavelength in the range 350 nm to 800 nm. The optical properties of nanoparticles were studied by spectral analysis. The absorbance spectra of nanoparticles were recorded within the range of 400 to 800 nm using a UV-3100PC UV-VIS spectrophotometer (VWR, Germany).

2.4. Electrochemical determination of cadmium ions

Determination of Cd^{2+} by difference pulse voltammetry (DPV) was performed at 663 VA Stand (Metrohm, Switzerland). A standard cell with three electrodes was used, and a hanging mercury drop electrode (HMDE) with a drop area of 0.4 mm^2 was employed as the working electrode. An Ag/AgCl/3M KCl electrode acted as the reference and a carbon electrode was auxiliary. For data processing, VA database software by Metrohm was employed. The analysed samples were deoxygenated prior to measurements by purging with argon (99.999 %). Acetate buffer (0.2 M sodium acetate and 0.2 M acetic acid, pH = 5) was used as a supporting electrolyte. The parameters of the measurement were as follows: initial potential -1.2 V, end potential 0 V, deoxygenating with argon 120 s, accumulation time 120 s, step potential 5 mV, modulation amplitude 25 mV, volume of injected sample: 50 µL, volume of measurement cell 10 mL. The samples, the electrolyte, and the measuring vessel were thermostated using the JULABO-200 circulation pump (Germany). The temperature was set to 20°C for all measurements.

2.5. Measurement of physical parameters

The location of the weather stations was an open non-industrial landscape in Boskovice, Czech Republic; for the monitoring, two independent measuring points with the frequency parameter of 60 s were selected. We used data from the Davis meteorological station (Hayward, USA). The station is located in Boritov, with 16° 58' 65" E longitude and 49° 43' 09" N latitude. Its altitude is 305 m a.s.l. Each weather station's temperature, humidity, and rain sensor are at 1.85 m, and solar and UV sensors are at 3.5 m. Individual parameters were collected at one-minute intervals. All data were automatically sent to the control unit and then to a computer database to store data.

2.6. Statistic and results evaluation

Experimental work was performed in at least three independent experiments. Each sample in the experiments was analysed at least five times. The obtained data presented in this paper are average values. No experimental points were excluded from the proposed experimental study. All the obtained data are stored in the Qinslab database. If possible, data were processed and evaluated mathematically and statistically in the Qinslab database. The results were expressed as mean ± standard deviation (S.D.). Photos were processed by the ColorTest program, which assigns intensity to the individual pixels of the studied image in a given colour area [5]. For preparing the publication, the data were processed using Microsoft software (USA).

3. RESULTS

3.1. Quantum dots as sensor for UV radiation

CdTe QDs were freshly prepared every day and exposed to UV radiation daily for 12 hours for seven consecutive summer days. After each exposure, their basic physicochemical characteristics were evaluated. The principle of our prepared sensor is shown in **Figure 1**.

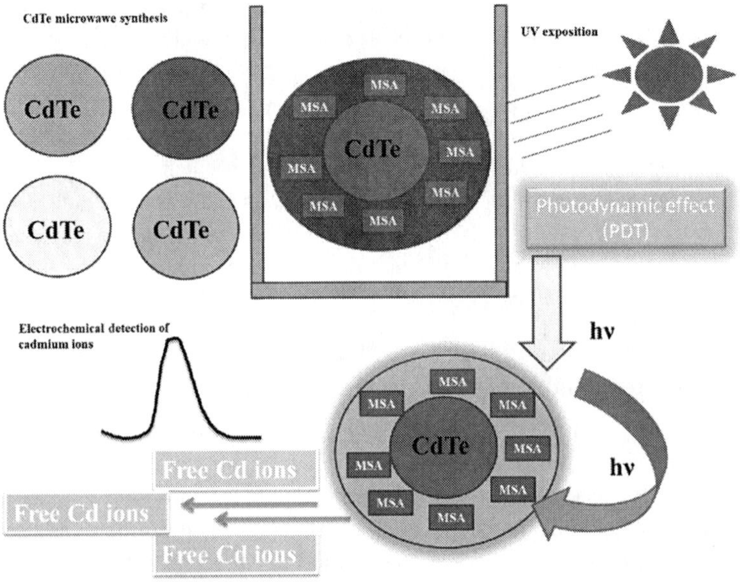

Figure 1 Schematic representation of the QDs function for sensor application. Typical CdTe QDs stabilized with mercaptosuccinic acid (MSA) after exposure to specific radiation at a given wavelength and energy leads to the excitation of a system that is capable of generating a whole group of molecular oxygen intermediates and relative oxygen species. In addition, due to the photodynamic effect of CdTe QDs, free cadmium ions are released.

3.2. Biophysical characteristics of prepared CdTe quantum dots

We decided to use CdTe QDs stabilised with mercaptosuccinic acid (MSA) in microwave synthesis. Four types of CdTe QDs: green, yellow, orange, and red (**Figure 2A**) were prepared by microwave synthesis (500 W) and under UV light (310 nm). A typical profile of the individual types of QDs on the PAGE gel after their separation is shown in (**Figure 2B**) and the intensities of the colour signals of the QDs electrophoretic bands can be seen in **Figure 2C**. Typical absorption spectra can be seen in **Figure 2D**. In the visible area of the spectrum, the absorption maxima were as follows: green QDs 519 nm, yellow QDs 544 nm, orange QDs 563 nm, and red QDs 564 nm. Typical fluorescence maxima are shown in **Figure 2E**. The fluorescence maxima were as follows: green QDs 532 nm, yellow QDs 560 nm, orange QDs 596 nm, and red QDs 654 nm. The cadmium concentration electrochemically determined was as follows: green QDs 1878 µM, yellow QDs 1739 µM, orange QDs 1636 µM, and red QDs 1181 µM (**Figure 2F,G**). The determined nanoparticle size ranged between 5 and 10 nm.

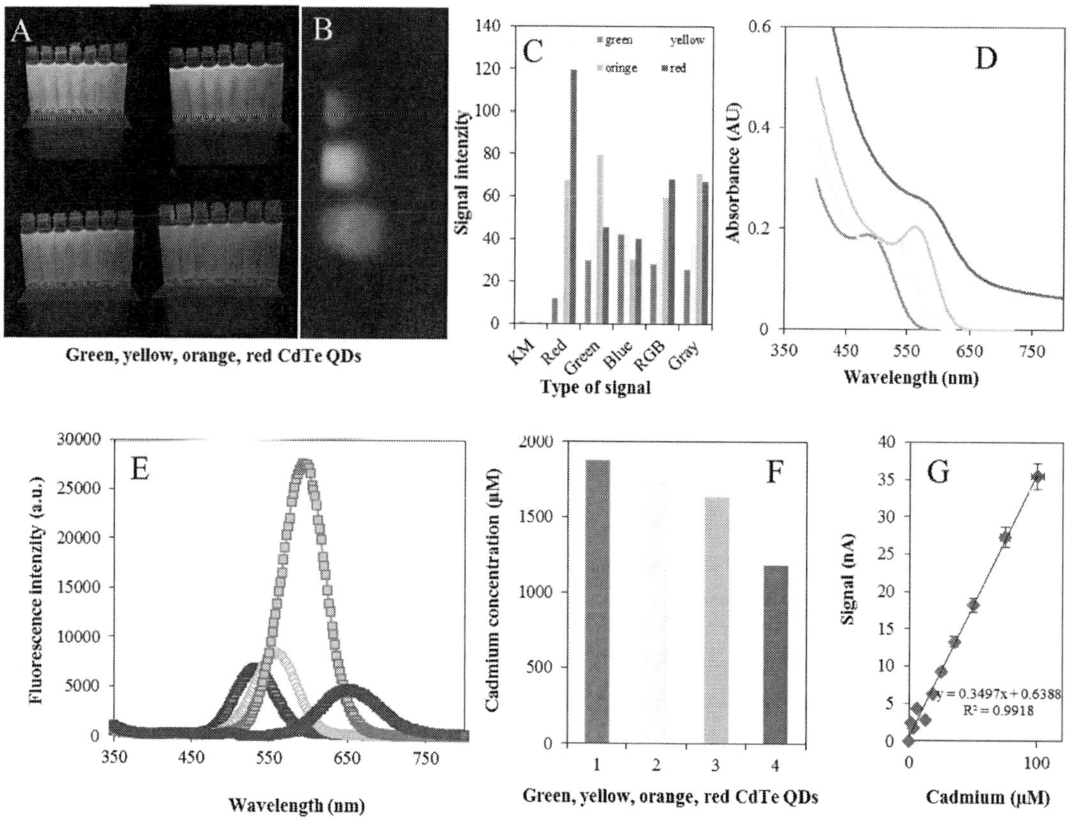

Figure 2 Biophysical characteristics of prepared CdTe quantum dots (QDs) used for environmental experiment. (A) Fluorescence of the prepared dots under UV light (310 nm) used in the experiment described above. (B) Typical profile of individual types of QDs on PAGE gel after separation (1.5h, 100V). (C) The created profile was analysed using the COLOR test program developed by us. The intensities of individual pixels are plotted in different colour profiles. (D) Spectrophotometric characterization of QDs, plastic cuvette 1 cm track, blank solution - water. (E) Fluorometric characteristics of QDs measured in a UV transparent plate at excitation of 250 nm. (F) The cadmium concentration in studied QDs. (G) Typical calibration curve for electrochemical cadmium analysis in organic form. All experiments were performed in five replicates, the displayed data being the average values. Other experimental details are described in Materials and Methods.

3.3. Summary of biophysical characteristics of QDs after their exposure to UV radiation

Four types of CdTe QDs (green, yellow, orange, and red) were exposed to UV radiation daily for 12 hours (QDs were freshly prepared each day) and the measurements were repeated for seven consecutive days in the summer. The course of the whole experiment, expressed in terms of average differential changes, is shown in **Figure 3**. The integral sum of the areas of the measured solar irradiance (W/m^2) and the UV index is shown in **Figure 3A** and **Figure 3B**. The highest solar and UV radiations were on the third, fourth, sixth, and seventh days, whereas the lowest solar and UV radiations were on the first and fifth days. On the second day, they were average. The cadmium concentration reached the highest values for non-irradiated (freshly prepared) QDs (day 0). The lowest Cd values were then reached on the sixth and seventh day (**Figure 3C**).

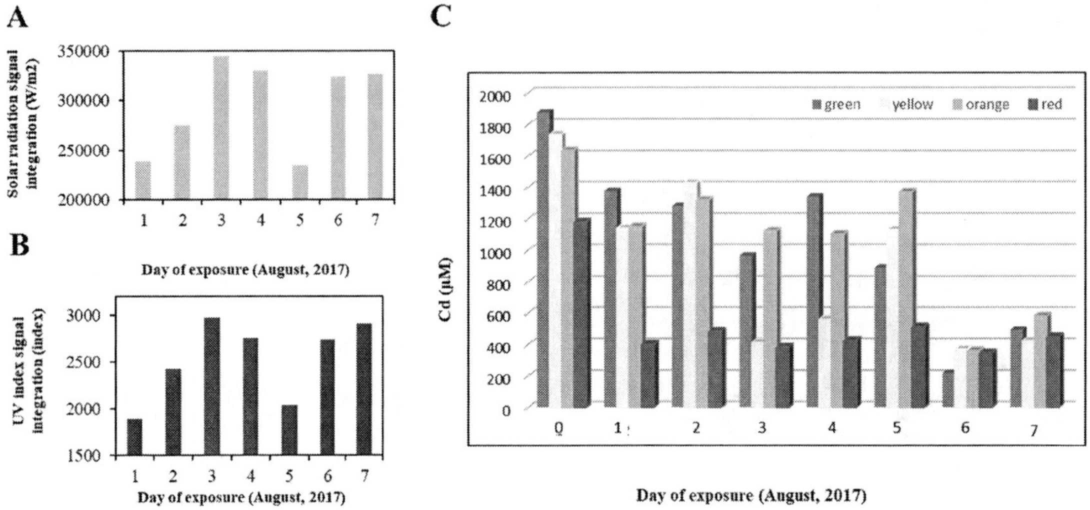

Figure 3 Changes in the observed biophysical parameters of CdTe QDs after exposure to the environment. The above figure shows the integral sum of the areas measured by solar irradiance (W/m^2) (A) and the UV index values (B) on the observed days in the experimental locality. On individual days, fluctuations in the integral radiation values are well traceable. The cadmium concentration (µM) of CdTe QDs on the observed days where in day 0 the initial cadmium concentration in CdTe QDs is given (C). The individual tested colour QDs are marked in the graphs with the appropriate colours. The results were evaluated as averages of five replicates. Other experimental details are described in Materials and Methods

4. CONCLUSION

This study provides an important basis for further understanding the relationship between the behaviour of CdTe QDs and UV radiation, and offers valuable information for their potential practical use as a sensor in the near future. The findings show that MSA-stabilised CdTe QDs are a suitable platform for the measurement of UV irradiance. This unique feature of the designed sensor seems to be a promising prospect for its versatile use in a variety of industrial and environmental applications, particularly in vineyards.

ACKNOWLEDGEMENTS

This work was supported by the Program of Applied Research and Development of National and Cultural Identity Project DG16P02R017 "Viticulture and enology for maintaining and restoring cultural identity wine regions in Moravia".

REFERENCES

[1] SCHULTZ, H. R. Global Climate Change, Sustainability, and Some Challenges for Grape and Wine Production. *Journal of Wine Economics,* 2016, vol. 11, no. 1, pp. 181-200.

[2] CHEN, H., GONG, Y.,HAN, R. Cadmium Telluride Quantum Dots (CdTe-QDs) and Enhanced Ultraviolet-B (UV-B) Radiation Trigger Antioxidant Enzyme Metabolism and Programmed Cell Death in Wheat Seedlings. *PLoS One,* 2014, vol. 9, no. 10, pp. e110400.

[3] TSIPOTAN, A. S., GERASIMOVA, M. A., ALEKSANDROVSKY, A. S., ZHARKOV, S. M.,SLABKO, V. V. Effect of visible and UV irradiation on the aggregation stability of CdTe quantum dots. *J Nanopart Res,* 2016, vol. 18, no. 11, pp. 324.

[4] DERFUS, A. M., CHAN, W. C.,BHATIA, S. N. Probing the cytotoxicity of semiconductor quantum dots. *Nano Lett,* 2004, vol. 4, no. 1, pp. 11-18.

[5] UHLIROVA, D., DOCEKALOVA, M., STANKOVA, M., MELICHAR, L., RUZICKA, J.,KIZEK, R. 2017. *A diagnostic strip for determining the amount of sarcosine, creatinine and hydrogen peroxide in a biological or environmental sample.* PUV Czech Republic patent application 2017-33695.

SESSION C
BIONANOTECHNOLOGY, NANOMATERIALS IN MEDICINE

Chairmen

Prof. Dr. Ing. Anton FOJTIK, CSc.	Czech Technical University in Prague, Czech Republic, EU
Prof. Jiri HOMOLA, CSc.	Institute of Photonics and Electronics of the CAS, Prague, Czech Republic, EU

PREPARATION OF COMPOSITE PERIODIC METAL-POLYMER NANOSTRUCTURES

Dominik FAJSTAVR, Markéta PIŠLOVÁ, Petr SLEPIČKA, Václav ŠVORČÍK

Department of Solid State Engineering, University of Chemistry and Technology Prague, Prague, Czech Republic, EU, fajstavd@vscht.cz

Abstract

This paper investigates the preparation of composite metal-polymer nanostructures formed on the surface of polyethersulfone (PES) by an excimer laser beam. Conditions for laser beam modification varied with the laser fluence of 4-28 mJ·cm^{-2} and number of pulses up to 6000. The samples were further deposited with a layer of metals Ag, Au, Pd and Pt with a thickness of 5-20 nm and their surface morphology was examined by atomic force microscopy (AFM). Electrical properties of layers were also investigated and for optical properties UV-Vis spectroscopy was used. Composites prepared by this approach were studied further for stability under laser modification. Surface morphology measurement revealed creation of laser induced periodic surface structures (LIPSS) in forms of nanoripples after modification with laser fluence value of 8 mJ·cm^{-2} and nanodots after modification with laser fluence value of 16 mJ·cm^{-2}. It has been found that metal layer deposition of the surface leads to formation of isolated clusters. Those metallized with the treatment of laser fluence 8 mJ·cm^{-2}, preferably aggregated on the ridge of the ripples and their electrical continuity was further examined. The nanostructured surfaces may be used as antibacterial material, also with potential to guide the specific types of cells, such are e.g. osteoblasts.

Keywords: Excimer laser, polymer, surface morphology, nanostructure, composite

1. INTRODUCTION

Nanostructuring of the surface brings a number of advantages and new material properties for the material under investigation. The process is based on the preparation of shapes of various patterns in the nanometer scale on a homogeneous surface of functional polymeric materials, which gives to the material modified physical and chemical properties [1]. Such an example of change in properties may be surface treatment associated with increased roughness. That also leads to an increase in reactivity of the material. Surface properties therefore play a significant role in interacting with other substrates, especially in the preparation of the metal-polymer composite material [2].

A suitable alternative for forming nanostructures on the polymer surface is excimer laser exposure. This technique provides the advantage that the morphology of the surface or chemical composition in the surface layer can be altered without changing the properties of the bulk material. Laser-induced Periodic Surface Structure (LIPSS) arises from a surface treatment technique that is governed by laser parameters such as wavelength, fluence, beam intensity, duration, and other radiation conditions [3-5]. The scientific goal of this paper is to introduce a simple method, how to prepare composite material merging both conductive and polymeric properties.

2. EXPERIMENTAL

2.1. Material and modification

For experiments we used polymer foils of polyethersulfone (PES, thickness 50 µm, density 1.37 g cm^{-3}, glass transition temperature T_g 226 °C) supplied by Goodfellow Ltd., Cambridge, Great Britain. The polymer was modified with a 248 nm laser beam using the Compex Pro50 excimer pulse KrF laser with the NovaTube. The value of laser fluence was set at 4-16 mJ·cm^{-2} with an attenuator, and the optical path of the beam was adjusted by dielectric mirrors from SiO_2 for combined reflection 248/633 nm. The modification time was determined by the number of pulses 1000-6000 at a frequency of 10 Hz. For the treatment of the laser beam, a 25x25x25 mm^3 prism polarizer was used with a polarizing layer and a polarizing circular Eksma Optics with a diameter of 25.4 mm. Subsequently, the beam passed through a metal aperture, which bounded the modification area to an area of 0.5 cm^2.

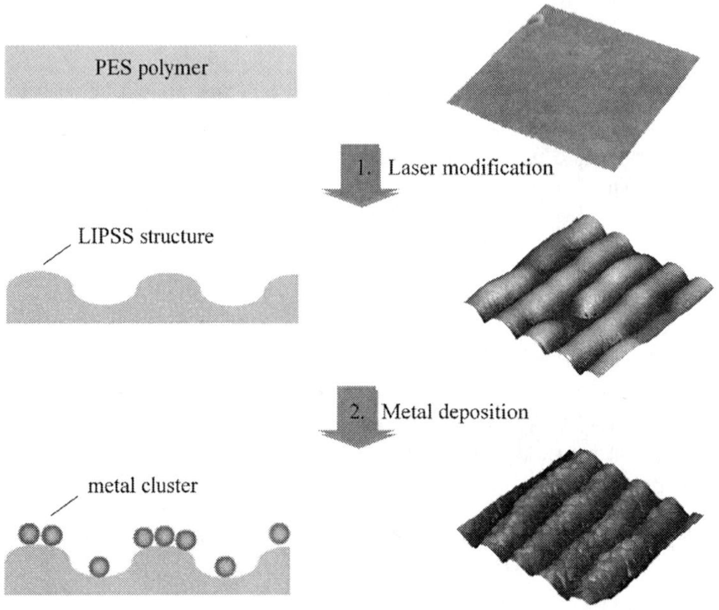

Figure 1 Schema of the preparation of metal-polymer nanostructures

2.2. Characterization techniques

To study morphology, a Dimension ICON (Bruker Corp., US) atomic force microscope was used in ScanAsyst mode. The silicon tip is mounted on a SCANASYST-AIR nitride lever with a spring constant of 0.4 N.m^{-1}. The NanoScope Analysis application was used to process the data, and the average roughness values (Ra) represent the mean of the deviations from the center plane of the sample. The image sizes ranged from 300x300 nm^2 to 10x10 μm^2.

Sheet electrical resistance of the samples was determined by a two-point method using KEITHLEY 487 picoammeter. For this measurement additional Au contacts, about 50 nm thick, were prepared by sputtering. The electrical measurements were performed at a pressure of about 10 Pa in a shielded chamber to minimize influence of atmospheric humidity and stray current. The UV-Vis spectra were measured using a PerkinElmer Lambda 25 spectrometer in the spectral range from 225 to 400 nm. The applicable range is 190-1100 nm with bandwidth of 1 nm (fixed).

3. RESULTS AND DISCUSSION

Substrates after modification with 6000 pulses and the linearly polarized laser beam at laser fluence 8 mJ·cm^{-2} showed the formation of a regular ripple nanostructure. However, the choice of a higher modifying fluence of 16 mJ·cm^{-2} led to formation of the regular nanodots structure and the substrates exhibited a lower surface roughness. After laser modification, samples were deposited with thickness of 5, 10 and 15 nm of Ag layer. The effect of the laser fluence and the thickness of the metal layer on the surface structure was studied.

3.1. Surface morphology and roughness

AFM images of sputtered nanostructures with metal nanolayers showed their similar surface morphology, but the most significant changes in surface roughness were achieved with the silver layers after modification by laser fluence of 8 mJ·cm^{-2}. In all the images, it was evident that metal sputtering did not result in the formation of a layer electrically continuous which would cover the surface of the polymer. The cluster formations of silver atoms, which has predominantly grown on the ridge of the ripples, were formed (**Figures 2A, 2B**). Longer metal deposition time caused cluster to grow, but there was still a significant ripple-like structure typical of the laser fluence value of 8 mJ·cm^{-2} and thus the prevailing influence of polymer on surface morphology. The surface roughness of the sample was 39 nm (**Figure 2A**).

Images of samples irradiated with laser fluence of 16 mJ·cm^{-2} had typical globular structure on the surface (**Figures 2C-D**). Metal sputtering in this case smoothens the surface, as metal clusters preferably fill the deepened formations in the polymer. The 15 nm layer thus appears to be completely planar and also exhibits the lowest surface roughness Ra = 2.5 nm (**Figure 2C**).

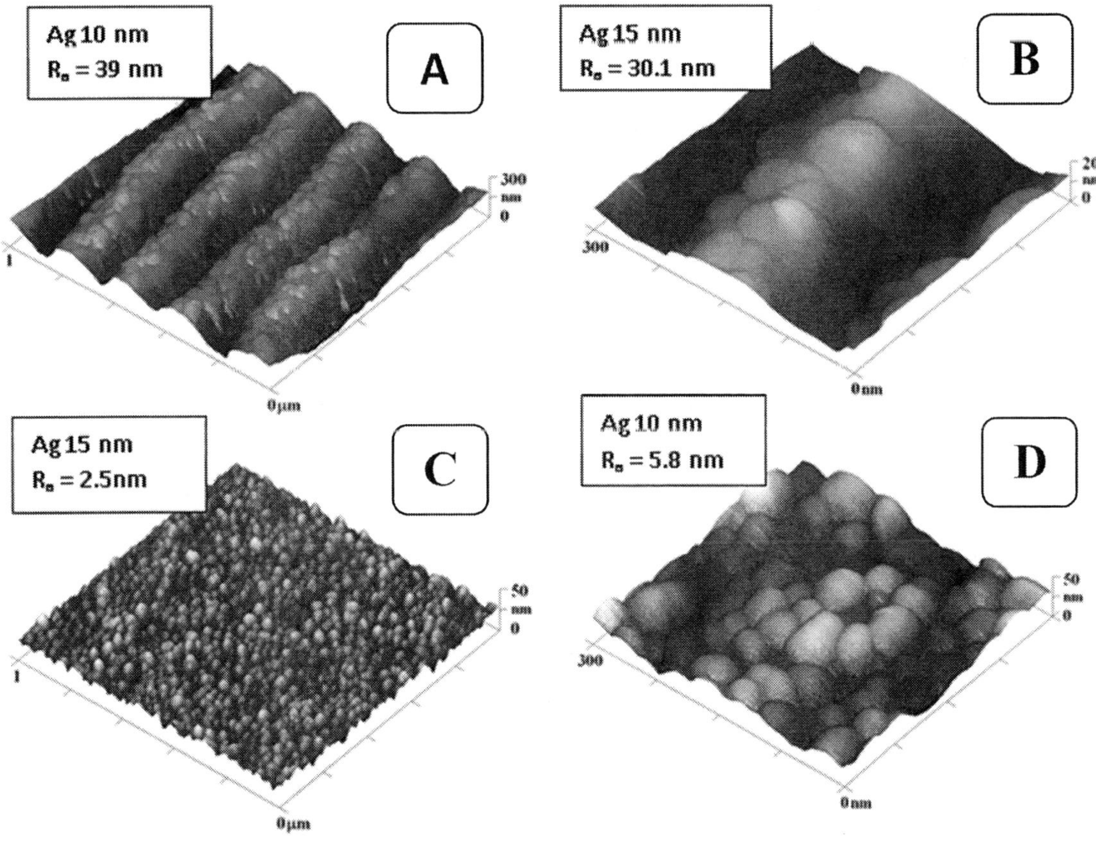

Figure 2 AFM images of samples after irradiation with 6000 pulses of laser fluence 8 mJ·cm^{-2} with a silver layer of 10 nm (A) and 15 nm (B, detailed 300x300 nm^2), and 16 mJ·cm^{-2} with a silver layer of 15 nm (C) and 10 nm (detailed 300x300 nm^2, D).

3.2. Electrical properties

A graph of resistance dependence on effective layer thickness typical for thin metal layers on isolating substrate can be seen (**Figure 3**). The sheet electrical resistance was examined in the direction corresponding to the ripple orientation at a laser fluence value of 8 mJ·cm^{-2}. It is shown that the 5 nm metal layers reach the highest resistances up to MΩ. This resistance values indicate the possible occurrence of conductive joints formed on the surface structure. Higher effective electrical continuity can be observed for metal layers of thickness 10 and 15 nm, where there is no significant decrease in resistance.

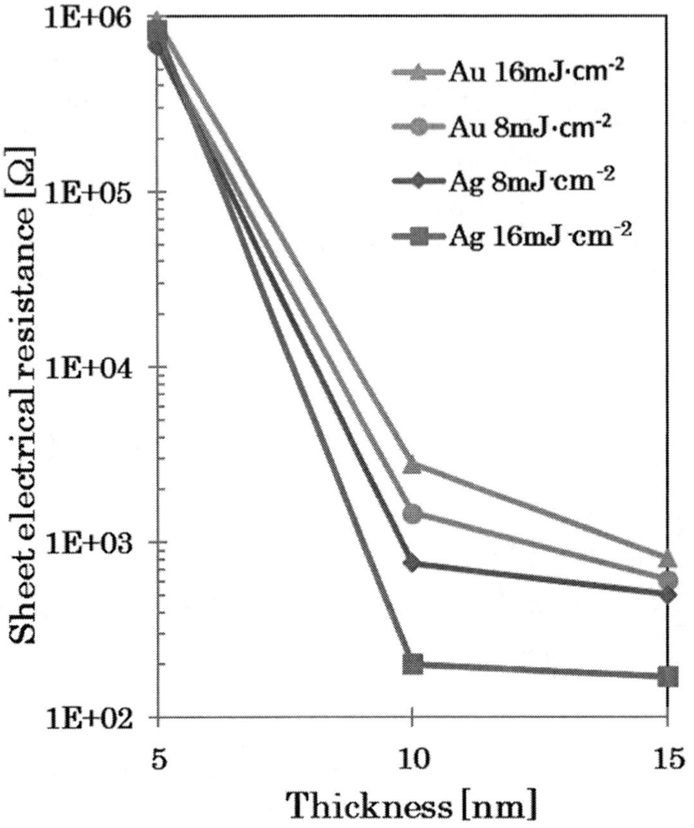

Figure 3 Sheet electrical resistance of thin Ag and Au layers after irradiation with 6000 pulses of laser fluence 8 mJ·cm^{-2} and 16 mJ·cm^{-2}

3.3. Optical properties

The UV-Vis spectroscopy method was subsequently used to characterize the samples. We investigated the absorbance of deposited metal layers in dependence on the wavelength. **Figure (4A)** shows the absorbance of deposited gold layers of 5, 10 and 15 nm thickness after modification of 6000 pulses by laser fluence value of 8 and 16 mJ·cm^{-2}. The graph shows the trend that samples modified by a higher laser fluence value reach higher absorbances across the wavelength range. Similarly, the increasing thickness of the layer resulted in higher absorbance for all wavelengths. Both of these trends were common to other studied samples of Ag, Pt and Pd.

The next **Figure (4B)** examines the effect of modifying pulses on the absorbance of the sample with a thin layer of 5 nm of gold. It is noteworthy that while the pristine PES polymer exhibited the lowest absorbance, the modified samples increased its absorbance with the increasing number of pulses.

A completely different situation occurred after the sample was sputtered with a 20 nm thick layer of metal, as illustrated in **Figure 4C**. Such a strong layer completely suppressed the influence of the substrate and its modification by different pulse counts. Further, the absorbance, unlike all previous cases, showed a stable absorbance value below 0.5 even in the shorter wavelength range.

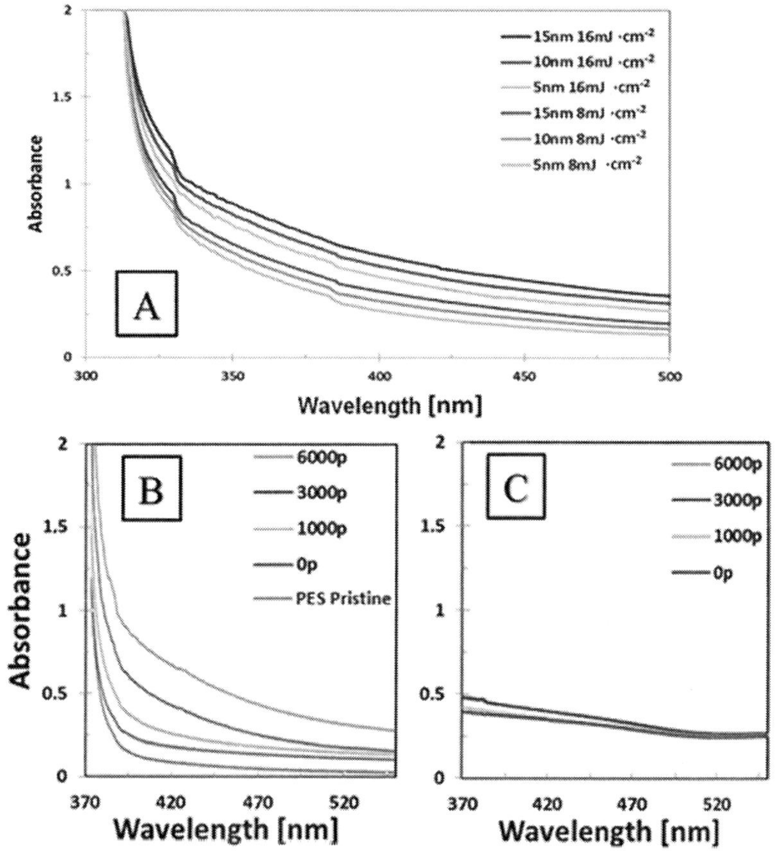

Figure 4 UV-Vis spectra of Au metal layers: 6000 pulses, 8 mJ·cm^{-2} and 16 mJ·cm^{-2}, 5-15 nm of Au (A); 0-6000 pulses, 8 mJ·cm^{-2}, 5 nm of Au (B), 0-6000 pulses, 8 mJ·cm^{-2}, 20 nm of Au (C)

CONCLUSION

This work focuses on the preparation of metal-polymer nanostructures achieved by the excimer laser beam. The PES substrate was modified by a laser beam with a pulse rate of 1000-6000 and a laser fluence in the range of 4 and 16 mJ·cm^{-2}. Modification by a laser fluence of 8 mJ·cm^{-2} led to the formation of nanoripples, whereas the higher value of the laser fluence 16mJ·cm^{-2} led to the formation dot periodic structures. Morphology was investigated on the surface of samples deposited with metal layers. Layer thicknesses of 5-15 nm deposited on the modified surface led to the formation of isolated metal clusters. Surface metallized after the treatment of laser fluence 8 mJ·cm^{-2}, preferably aggregated on the ridge of the ripples. All measurements of UV-Vis spectroscopy revealed that the lowest absorbance showed the original PES substrate, and the effect of increasing the absorbance is caused by the number of modifying pulses, the laser fluence value and the thickness of the layer itself. The rapid increase in absorbance occurred in wavelength ranges of 400 nm or less. If the layer thickness reached 20 nm, the modification of the samples no longer had an effect on the absorbance. Antibacterial properties of these metal layers deposited on nanostructured surface could be used for material supressing the bacteria biofilm formation, surface with regular nanostructures has also potential to guide the specific types of cells, such are e.g. osteoblasts.

ACKNOWLEDGEMENTS

This work was supported by the Ministry of Health of the CR under project 15-32497A.

REFERENCES

[1] CASTILLEJO, M., EZQUERRA, T. A., OUJJA, M., & REBOLLAR, E. Laser nanofabrication of soft matter. In: *Lasers in Materials Science*. Springer, Cham, 2014. p. 325-344.

[2] ZAPOROJTCHENKO, V., STRUNSKUS, T., BEHNKE, K., VON BECHTOLSHEIM, C., KIENE, M., & FAUPEL, F. Metal/polymer interfaces with designed morphologies. *Journal of Adhesion Science and Technology*, 2000, 14.3: 467-490.

[3] BÄUERLE, DIETER. Adsorbed Layers, Laser-MBE. In: *Laser Processing and Chemistry*. Springer, Berlin, Heidelberg, 2011. p. 457-475.

[4] REBOLLAR, ESTHER, MILDNER, J., GÖTTE, N., OTTO, D., SARPE, C., KÖHLER, J., ... & CASTILLEJO, M. Microstructuring of soft organic matter by temporally shaped femtosecond laser pulses. *Applied Surface Science*, 2014, 302: 231-235.

[5] FAJSTAVR, D., MICHALJANICOVA, I., SLEPICKA, P., NEDELA, O., SAJDL, P., KOLSKA, Z., & SVORCIK, V. Surface instability on polyethersulfone induced by dual laser treatment for husk nanostructure construction. *Reactive and Functional Polymers*, 2018, 125: 20-28.

[6] BASTING, DIRK; MAROWSKY, Gerd (ed.). *Excimer laser technology*. Springer Science & Business Media, 2005.

[7] BOHAC, V., D'ANNA, E., LEGGIERI, G., LUBY, S., LUCHES, A., MAJKOVA, E., & MARTINO, M. Tungsten silicide formation by XeCl excimer-laser irradiation of W/Si samples. *Applied Physics A*, 1993, 56.4: 391-396.

[8] REBOLLAR, ESTHER, DE ALDANA, J. R. V., MARTIN-FABIANI, I., HERNANDEZ, M., RUEDA, D. R., EZQUERRA, T. A., ... & CASTILLEJO, M. Assessment of femtosecond laser induced periodic surface structures on polymer films. *Physical Chemistry Chemical Physics*, 2013, 15.27: 11287-11298.

[9] KVITEK, O., FAJSTAVR, D., REZNICKOVA, A., KOLSKA, Z., SLEPICKA, P., & SVORCIK, V. Annealing of gold nanolayers sputtered on polyimide and polyetheretherketone. *Thin Solid Films*, 2016, 616: 188-196.

[10] THOMAS, B.; ALLONCLE, A. P. DELAPORTE PH., SENTIS M., SANAUR, BARRET M., COLLOT PH."Experimental investigations of laser induced forward transfer process of organic thin films" Appl. *Surf. Sci*, 2007, 254.4: 1206-1210.

[11] LAZARE, S., TOKAREV, V., SIONKOWSKA, A., & WISNIEWSKI, M. Surface foaming of collagen, chitosan and other biopolymer films by KrF excimer laser ablation in the photomechanical regime. *Applied Physics A*, 2005, 81.3: 465-470.

[12] REBOLLAR, ESTHER, VAZQUEZ DE ALDANA, J. R., PEREZ-HERNANDEZ, J. A., EZQUERRA, T. A., MORENO, P., & CASTILLEJO, M. Ultraviolet and infrared femtosecond laser induced periodic surface structures on thin polymer films. *Applied Physics Letters*, 2012, 100.4: 041106.

[13] CHAN, C.-M.; KO, T.-M.; HIRAOKA, HIROYUKI. Polymer surface modification by plasmas and photons. *Surface science reports*, 1996, 24.1-2: 1-54.

[14] REZNICKOVA, A., KVITEK, O., FAJSTAVR, D., SLAVIKOVA, N., & SVORCIK, V. Substrate Effects of Noble Metal Nanostructures Prepared by Sputtering. In: *Noble and Precious Metals-Properties, Nanoscale Effects and Applications*. IntechOpen, 2018.

POLYURETHANE MODIFIED BY PLASMA ION IMPLANTATION

[1,2]V.S. CHUDINOV, [2]I.N. SHARDAKOV, [2]A.L. SVISTKOV, [3]A.V. KONDYURIN

[1]Perm State University, Perm, Russian Federation, chudinovsl@mail.ru
[2]Institute of Continuous Media Mechanics of the Ural Branch of Russian Academy of Science, Perm, Russian Federation
[3]School of Physics, University of Sydney, Sydney, Australia

Abstract

Polyurethane was treated by nitrogen ions with energy of 20 keV and fluence of 10^{14} - 10^{16} ions/cm^2. Modified polyurethane was investigated by the IR FTIR, Raman and EPR spectroscopy, AFM microscopy. The contact angle and the surface energy of modified polyurethane are measured. Polyurethane after treatment has a carbonized layer of 70 nm thickness and partly depolymerized layer below the carbonized layer. Polyurethane after treatment have wave-like surface relief and cracks. Carbonized layer of the modified polyurethane contains highly active free radicals. It provides high hydrophilicity of modified polyurethane surface. High activity free radicals and hydrophilicity of modified polyurethane surface provides a saving of biological activity of the covalently attached proteins. As results, it provides to attachment and growing of endothelial cells on the polyurethane surface. The modified polyurethane can be applied for soft tissue medical implants.

Keywords: Carbonized layer, polyurethane, plasma ion implantation, implant, biocompatibility

1. INTRODUCTION

Inserting of synthetic medical implants induces an immune system complex of local cellular and responses of humoral immunity to a foreign body and inflammation [1,2]. The result of these reactions is the formation of a connective tissue capsule that isolates the implant, inflammatory complications, etc. The implants are rejected, become non-functional, and repeated surgical interventions are required. The risk of operations for patients in old age, patients with diabetes mellitus and other diseases is very significant. Patients receiving immunosuppressants create additional risks and are not always applicable due to side effects. For these reasons, the creation of synthetic implants without a foreign body reaction is very desirable. The recent studies on the creation of a continuous active protein layer on the implant surface show the possibility and prospects of such implants. The protein layer on the surface of the implant makes possible to exclude the contact of the cells of the body's immune system with the naked surface of the implant [3-5].

Earlier in our works, we show that the polyurethane modified by plasma-ion treatment on a device with a planar geometry of the electrode contains a surface carbonized layer about 50-70 thick and has activity, which is provided by the presence of free radicals. The high activity and hydrophilicity of the modified surface polymer layer ensures stable adsorption of the protein on the surface of the polyurethane with the formation of a covalent bond and with the preservation of the biological activity of the adsorbed protein [6,7].

2. MATERIALS AND METHODS

We synthesized polyurethane from a prepolymer based on 2,4-toluylene diisocyanate and polyether with a hardener methylene bis-orthochloroaniline. Components of polyurethane were mixed for 1 min. We get film of polyurethane in Petri dishes and anneal it in vacuum oven at a temperature of 100°C for 24 hours. The thickness of the final films was 0.5 mm. Films of polyurethane were washed in heptane and deionized water and dried in a dust-free environment.

Polyurethane was treated by nitrogen ions with energy of 20 keV at a different time of treatment, which correspond fluence of 10^{14} - 10^{16} ions/cm^2.

Modified polyurethane was investigated by the IR FTIR, Raman and EPR spectroscopy, AFM microscopy. Raman spectra were recorded on a Ramanshaw spectrometer with a microscope, a lens of x50. EPR spectra were recorded on electron paramagnetic resonance spectrometer Spinscan X. The topography of the polyurethane surface modified by plasma ion implantation was analyzed using the Park System atomic force microscope.

3. RESULTS

Raman spectra of polyurethane contain lines related to the vibrations of the polyurethane macromolecule (**Figure 1**). The most intense peaks are observed in the region of 1619 cm^{-1} (vibrations of the aromatic ring), 1453 and 1320 cm^{-1} (vibrations of the ether part of the macromolecule), as well as the less intense peaks of 1712 cm^{-1} and 1540 cm^{-1} oscillations of Amide I and II urethane groups, 1080, 1125 and 1267 cm^{-1}, the skeletal vibrations of the polyester chain. The Raman spectrum of polyurethane treated with a fluence of 10^{16} ions / cm^2 also showed polyurethane lines (1128, 1250, 1450, 1524 and 1690 cm^{-1}), but the intensity of these lines is much less than in the spectrum of untreated polyurethane. Two new intense lines, 1348 and 1595 cm^{-1}, related to the vibrations of carbon structures are also observed. The peak at 1595 cm^{-1} refers to the vibrations of E_{2g} graphite. The type of this oscillation is active in the Raman spectrum. The peak at 1348 cm^{-1} refers to the vibrations of A_{1g} graphite. This peak is not active in the Raman spectra. The appearance of this peak is associated with the presence of graphite structure defects and breaking of the selection rules for the Raman spectrum. The intensity and position of the peaks show that these lines refer to the resonance effect of Raman spectra when the absorption line of electronic levels of graphite structures coincides or is close to the laser excitation line (514 nm) in Raman spectra. According to the model of the Ferrari and Robertson carbon structures [8, 9], the position of the lines and the intensity ratio correspond to the a-C structure of amorphous carbon with graphite clusters, a characteristic size of about 2 nm. Thus, a graphite-like layer was created on the polyurethane surface as a result of plasma ion treatment.

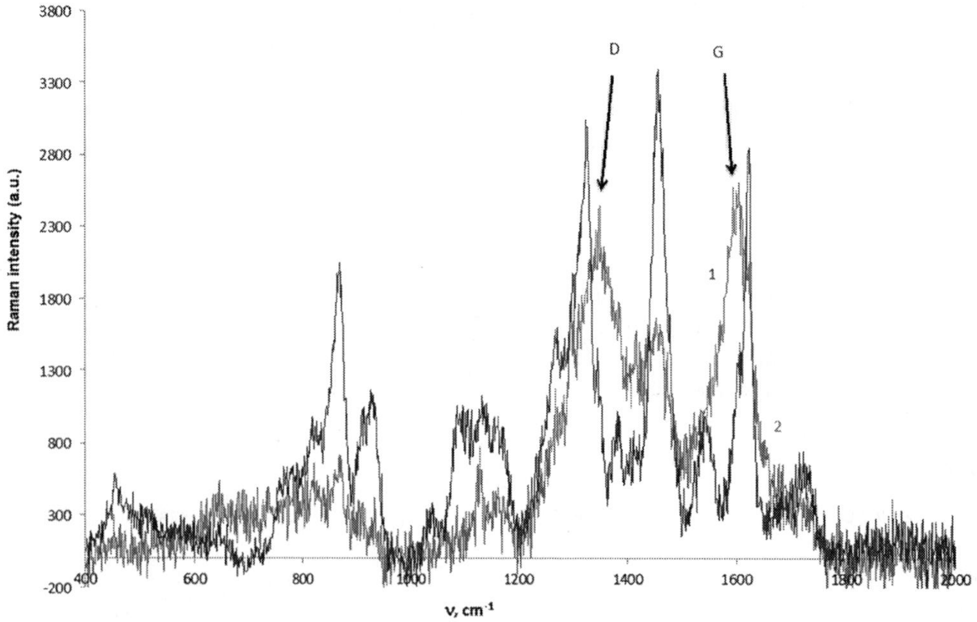

Figure 1 Raman spectra of polyurethane without the treatment (1, blue spectrum) and treated with nitrogen ions of 20 keV energy for 800 sec with a fluence of 10^{16} ions / cm^2 (2, red spectrum). Arrows show the lines D and G of carbon structures

EPR spectra show the presence of free radicals in the treated polyurethane (**Figure 2**).It supports our results of IR FTIR spectroscopy surface layer of treated polyurethane [7]. As result modified polyurethane surface is high hydrophilic. It provides to covalently attachment protein molecules on the carbonized layer of modified polyurethane and growing of endothelial cells on the polymer surface.

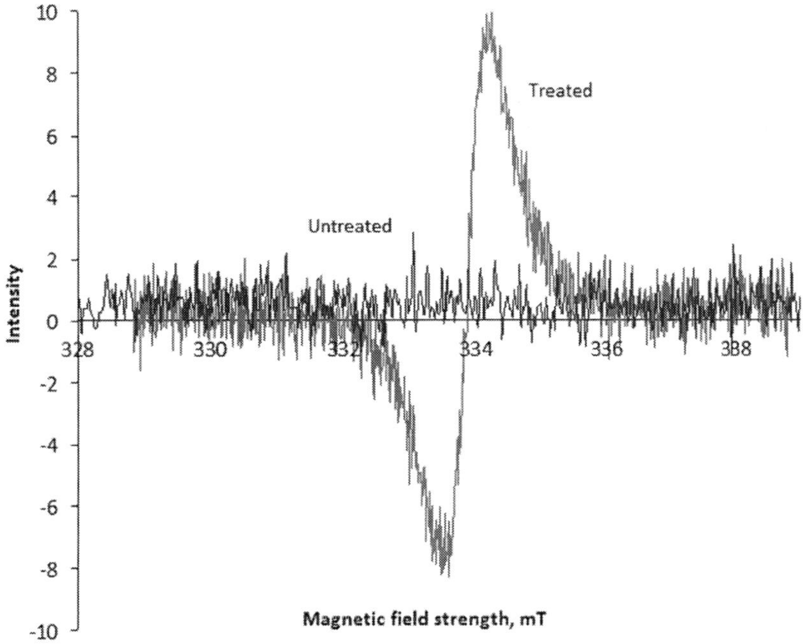

Figure 2 Electron paramagnetic resonance spectra of a polyurethane film (SCU-PFL), untreated and treated with nitrogen ions with an energy of 20 keV of 800 sec (corresponding to a fluences of 10^{16} ions / cm^2). The film was processed and stored 2.5 months in the laboratory conditions before recording the spectrum

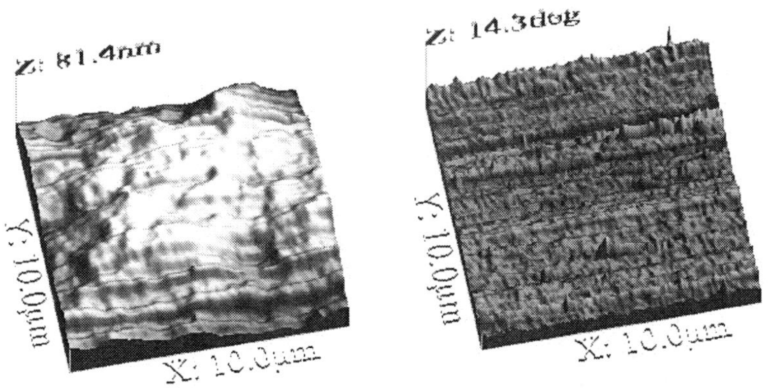

Figure 3 Topology of the surface and phase picture of interaction between cantilever and the surface for untreated polyurethane

The topology of the polyurethane film surface varies significantly after treatment (**Figures 3-4**). The surface of untreated polyurethane shows a roughness of 80 nm region for 10x10 um, but shows the homogeneity of interaction with the microscope cantilever by the phase image. The character of the surface roughness is random. The surface of the treated polyurethane is characterized by a large number of folds and cracks.

Analysis of the folds shows that the amplitude of the folds is much higher than the thickness of the modified layer (**Figure 5**). The direction of the folds is chaotic, not related to the direction of cracks or the dimensions of the sample.

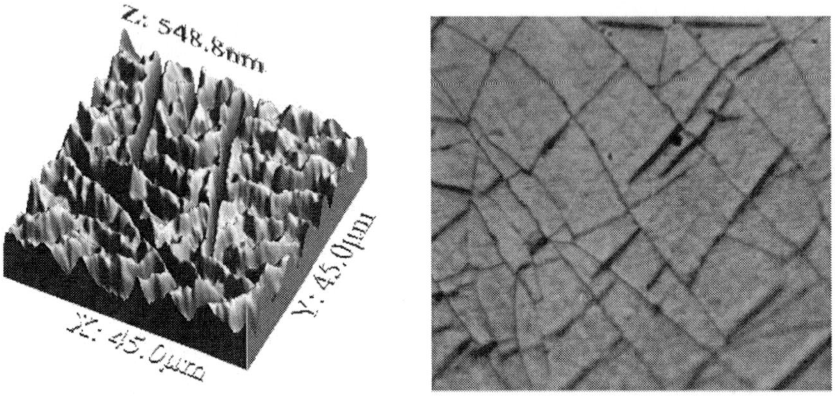

Figure 4 Topology according to the atomic force microscope and micrograph of the surface of the treated polyurethane

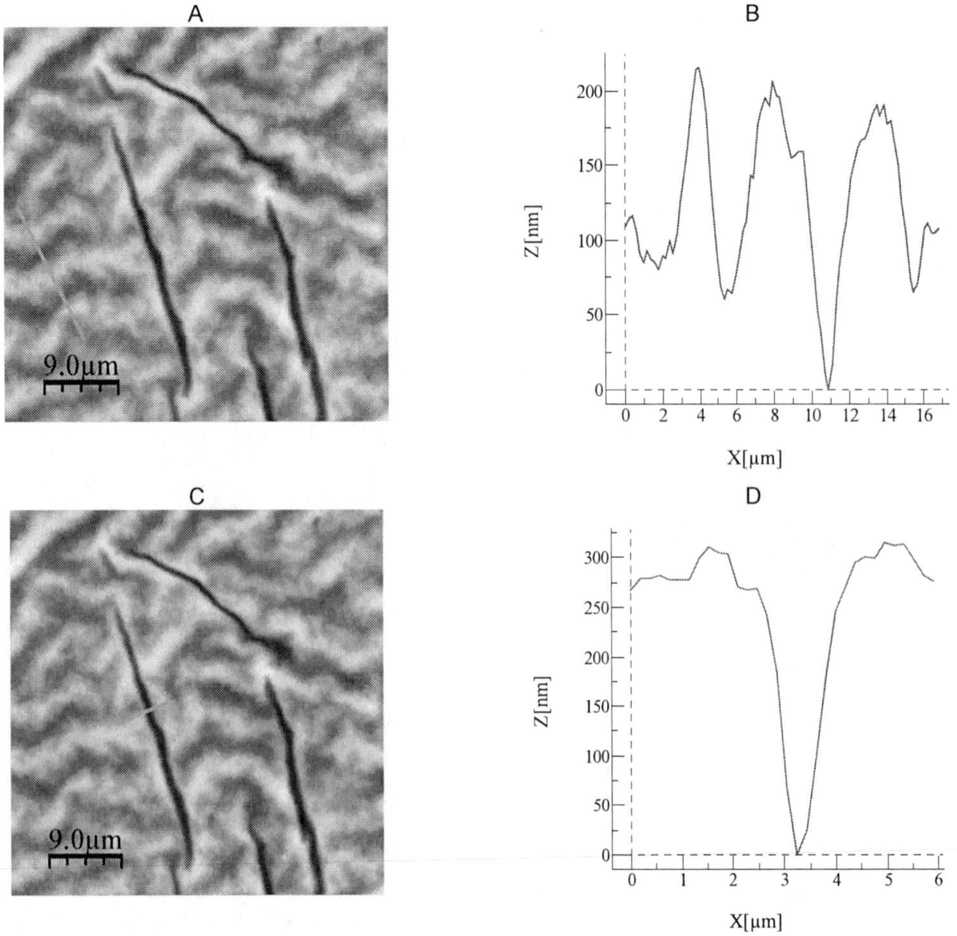

Figure 5 Surface analysis of treated polyurethane: amplitude of folds (A, B) and depth of cracks (C, D)

4. CONCLUSION

The treatment by 20 keV energy nitrogen ions creates on the surface of a polyurethane implant a carbonized layer consisting of amorphous carbon with inclusions of graphite clusters. Carbonized layer of modified polyurethane contains highly active free radicals. Treated polyurethane surface is highly hydrophilic. Polyurethane after treatment has a wave-like surface relief with cracks. High activity free radicals and hydrophilicity of the modified polyurethane surface provides a saving of biological activity of the covalently attached proteins. It provides to attachment and growing of endothelial cells on the polyurethane surface. The modified polyurethane can be applied for soft tissue medical implants.

ACKNOWLEDGEMENTS

This study was supported by the Russian Science Foundation (project no. 18-19-00574).

REFERENCES

[1] R.Y. KANNAN, H.J. SALACINSKI, P.E. BUTLER, G. HAMILTON, A.M. SEIFALIAN, Journal of Biomedical Materials Research B Applied Biomaterials, 74 (2005) 570-581.

[2] P.D. BALLYK, C. WALSH, J. BUTANY, M. OJHA, J Biomech, 31 (1998) 229-237.

[3] BEGISHEV V., GAVRILOV N., MESYATS G., KLYACHKIN YU., KONDYURINA I., KONDYURIN A., OSORGINA I., Modification of polyurethane endoprosthetics surface by pulse ion beam, Proc. of 12th Intern. Conf. on High-Power Particle Beams, Haifa, Israel, June 7-12, 1998. Ed. by M.Markovits and J. Shiloh, vol 2, p. 997-1000.

[4] KONDYURIN A., MAITZ M.F., Surface Modification of ePTFE and Implants using the same, US patent WO 2007/022174 A3, 2007.

[5] I. KONDYURINA, I. SHARDAKOV, G. NECHITAILO, V. TERPUGOV, A. KONDYURIN, Cell growing on ion implanted polytetrafluorethylene, Applied Surface Science 314 (2014) 670-678.

[6] I.V. KONDYURINA, V.S. CHUDINOV, V.N. TERPUGOV, A.V. KONDYURIN. Chemical attachment of albumin on the surface of a polyurethane implant modified by ion beam, Medical technology-Moscow, 2018. e. 4, pp. 19-21.

[7] V.S. CHUDINOV, I.V. KONDYURINA, I.N. SHARDAKOV, A.L. SVISTKOV, I.V. OSORGINA, A.V. KONDYURIN. Polyurethane Modified by Plasma Ion Implantation for Medical Application, Biophysics, 2018, Vol. 63, No. 3, pp. 330-339.

[8] A. C. FERRARI, J. ROBERTSON, Interpretation of Raman spectra of disordered and amorphous carbon, Phys. Rev. 2000, B 61, 14095.

[9] 13. A.C. FERRARI, D.M. BASKO, Raman spectroscopy as a versatile tool for studying the properties of graphene, Nature Nanotechnology 8, 235-246 (2013).

THE EFFECT OF SIMULATED BODY FLUID ON THE GROWTH AND DIFFERENTIATION OF HUMAN MESENCHYMAL STEM CELLS CULTURED ON NANOSTRUCTURED Ti6Al4V SURFACES

[1]Eva FILOVÁ, [1,2]Barbora VOLTROVÁ, [3]Petra JAROLÍMOVÁ, [4]Roman MATĚJKA, [1]Věra LUKÁŠOVÁ, [3]Jaroslav FOJT, [3]Luděk JOSKA, [5]Matej DANIEL

[1]Institute of Experimental Medicine of the Czech Academy of Sciences, Prague, Czech Republic, EU, eva.filova@iem.cas.cz
[2]Charles University in Prague, Faculty of Science, Czech Republic, EU, barbora.voltrova@iem.cas.cz
[3]University of Chemistry and Technology Prague, Faculty of Chemical Technology, Prague, Czech Republic, EU, petra.jarolimova@vscht.cz, jaroslav.fojt@vscht.cz, ludek.joska@vscht.cz
[4]Institute of Physiology of the Czech Academy of Sciences, Prague, Czech Republic, EU, roman.matejka@fgu.cas.cz
[5]Czech Technical University in Prague, Faculty of Mechanical Engineering, Czech Republic, EU, matej.daniel@fs.cvut.cz

Abstract

TI6Al4V is a standard medical titanium alloy. In this study, we compared ground Ti6Al4V discs with nanostructured Ti6Al4V discs prepared using the anodic oxidation method at 30V. Subsequently, both the nanostructured and ground discs were exposed to simulated body fluid (SBF) for 7 days in order to form a hydroxyapatite-like layer. Human mesenchymal stem cells (hMSC) were seeded on the discs in osteogenic differentiation medium and cultivated for 21 days. hMSCs were tested for metabolic activity (MTS assay) and rate of proliferation. The synthesis of type I collagen was evaluated using immunohistochemical staining and confocal microscopy analysis. All samples supported cell adhesion and proliferation. The highest amount of type I collagen was found on nanostructured (N) and ground Ti6Al4V exposed to SBF (TiH) surfaces. Thus, SBF and anodic oxidation positively influenced the growth and osteogenic differentiation of hMSCs.

Keywords: Anodic oxidation, hydroxyapatite, mesenchymal stem cell, titanium alloy

1. INTRODUCTION

Proper bone implants should promote fast adhesion of cells, stimulate cell differentiation and support the formation of new bone tissue. Currently, titanium alloys are used as biocompatible materials capable of bone tissue integration [1,2]. Titanium and its alloys are widely used in orthopaedic and dental surgery. Despite the benefits of titanium alloys for such applications, there are still questions which need to be answered It is possible to modify the titanium surface in nanoscale using anodic oxidation in fluorides containing electrolytes [3,4]. This process leads to the formation of self-organised TiO_2 nanotube layer on the titanium surface. Such a change in the nano-topography of the material can improve cell adhesion which is crucial for the implants osseointegration [5]. Incubation in the simulated body fluid (SBF) forms a hydroxyapatite-like layer on the exposed surfaces. The deposited layer of hydroxyapatite by immersing Ti6Al4V in the simulated body fluid (SBF) was investigated previously [6,7].

This study was focused on the comparison of nanostructured and ground Ti6Al4V titanium alloys, modified by hydroxyapatite coating, on the cytotoxicity and differentiation capacity of seeded human mesenchymal stem cells (hMSCs). Nanostructures were fabricated using anodic oxidation. Hydroxyapatite surface modification of the titanium samples was realized by exposure to SBF.

2. MATERIALS AND METHODS

2.1. Specimens preparation

Nanostructuring was realized using Potentiostat-Galvanostat IMP 88 PC-200V with PGU-AUTO Extern control unit (Jaissle). A silver/silver chloride reference electrode with 3 mol/L KCl (SSCE) was used and, glassy carbon bars were used as counter electrodes. The electrochemical measurement consisted of a potential ramp from the open circuit potential to the selected end potential 30V with a 100 mV/s sweep rate and potentiostatic exposure for 2200 s. An ammonium sulphate (1 mol/L) and ammonium fluoride (0.2 wt.%) solution was used as the electrolyte. All anodization experiments were carried out at room temperature. Following the electrochemical measurement, the samples were ultrasonicated in ethanol and deionised water for 10 minutes and then dried in air stream. Ground and nanostructured titanium (Ti6Al4V) discs were exposed in simulated body fluid (SBF) at temperature of 37 °C for 7 days in order to create a hydroxyapatite layer on the surface. In this study, 4 sets of flat round Ti6Al4V samples, with 16 mm diameter and height 4 mm, with different surface treatments were used (**Table 1**). Before treatment all samples were ground to the roughness of sandpaper P2500, degreased with ethanol and cleaned with distilled water in an ultrasonic bath. For the morphological characterization of samples, a Vega3 LMU scanning electron microscope (SEM, Tescan) with an Oxford Instruments EDS analyzer was used. The results of the elemental analysis by Energy-dispersive X-ray spectroscopy were evaluated using the Aztec software. The length and inner diameter of the nanotubes was evaluated from SEM images in ImageJ software. Data from four distant image fields from at least three samples were used for analysis.

Table 1 List of samples before and after exposition to simulated body fluid (SBF)

Samples	Name	SBF
Nanostructured Ti6Al4V	N	-
Ground Ti6Al4V	Ti	-
Ground Ti6Al4V exposed to SBF	TiH	+
Nanostructured Ti6Al4V exposed to SBF	NH	+

2.2. Cell culturing

hMSCs (ScienceCell, Carlsbad, CA), were cultivated in growth medium consisting of MEM Alpha medium (1X) (Gibco, Thermofisher Scientific, Waltham, MA, USA) supplemented with 10 % (v/v) fetal bovine serum (FBS), penicillin/streptomycin (100 IU/mL and 100 µg/mL, Sigma-Aldrich) in 5 % CO_2 and 37 °C. The hMSCs were seeded on Ti6Al4V scaffolds and control glass coverslip and control Tissue culture plast (TCP) at a density of 12,000 cells/cm^2. The samples were sterilized in collaboration with the Institute of Nuclear Physics of the Academy of Sciences of the Czech Republic using the Microtron MT25 cyclic particle accelerator with an electron beam of 9.8 MeV with a dose of 27 kGy. The cells were cultured in growth medium. After 3 days of cell culture, when the cells reached confluence, the cultivating medium was enriched by osteogenic differentiation additives: 40 µg/mL ascorbate-2-phosphate, 100 nM dexamethasone, and 10 mM glycerol-2-phosphate disodium salt. The osteogenic medium was changed twice a week.

2.3. Cell metabolic activity measurement

The metabolic activity of the hMSCs was determined using MTS assay (CellTiter96® AQueous One Solution Assay; Promega, Madison, WI, USA). To each scaffold, 800 µL of MTS solution in fresh medium was added. After 2 hours incubation, the absorbance of the metabolized solution (100 µL) was detected at 490 nm using Tecan Infinite® PRO 200 series reader (Tecan, Männedorf, SUI). The background absorbance (690 nm) was subtracted from the measured data, as well as the absorbance of media with MTS solution.

2.4. Cell proliferation

To determine the cell proliferation a fluorescence-based kit (Quant- iT™ PicoGreen® dsDNA Assay Kit; Invitrogen, Thermo Fisher Scientific, Waltham, MA, USA) was used. Media was aspirated from the samples after MTS assay and 1 mL of cell lysis buffer (10 mM L−1 Tris, 1 mM L−1 EDTA, 0.2 % v/v Triton X-100) was added, incubated 20 min in RT then frozen three times (-80°C) and thawed to facilitate the cell lysis and DNA release. Fluorescence intensity was detected using Tecan Infinite® PRO 200 series reader Tecan (λex = 485 nm, λem = 528 nm) and the DNA content was determined according to the λDNA calibration curve from the kit.

2.5. Immunohistochemical staining of type I collagen

On day 14, the samples were fixed with 70 % frozen methanol, washed with phosphate buffered saline (PBS) and incubated in 3 % FBS in PBS/0.1 % Triton-X for 30 min at RT. Primary monoclonal antibody against type I collagen (M-38-C, DSHB, USA, dilution 1:20) was added, and incubated overnight at 2°C-8°C. After 3 washes with PBS/0.05 % Tween, and one in PBS, the samples were incubated with secondary antibody Alexa Fluor 488-conjugated anti-mouse antibody for 45 min, RT (4408S, Cell Signalling Technologies, dilution 1:400). Cell nuclei were stained with 5 µg/mL propidium iodide in PBS for 5 min, RT (Sigma Aldrich, USA) and twice washed in PBS. The cells were visualized using a confocal microscope ZEISS LSM 5 DUO at λexc=488 nm and λem =505-550 nm for Alexa Fluor 488, and λex =560 nm, λem >575 nm for propidium iodide). The intensity of fluorescence was evaluated from 8 - 9 photomicrographs per sample group using MATLAB.

2.6. Statistical analysis

Quantitative data are presented as mean ± standard deviation. The results were evaluated statistically using one-way analysis of variance (ANOVA), normality of the data was tested, and Student-Newman-Keuls Method was used as a post-hoc test. Fluorescence intensity was evaluated using Multifactorial ANOVA. The level of significance was set at 0.05.

3. RESULTS

Prepared nanostructured Ti6Al4V discs were evaluated using SEM. According to obtained images, nanotubes of diameter of 70 nm (SD ± 16 nm) were fabricated on the surface of the discs (**Figure 1**). The elemental composition of the samples was measured using EDS (**Table 2**). The metabolic activity of seeded hMSC (**Figure 2A**) on four different titanium alloys was evaluated using MTS metabolic assay on days 1, 7, 14 and 21. On day 1, the highest metabolic activity was measured in groups Ti and TiH. In the following days the metabolic activity rose steadily and no differences were seen between the different titanium discs (**Figure 2**). Further differences were seen only on days 7 and 21 between all tested groups and the TCP control group that is specifically modified to support cell growth. These finding are in agreement with a study that confirmed biocompatibility of the titanium alloys [8]. However, a study by Saha et al. showed improved metabolic activity of MG63 cells on nanostructured Ti6Al4V discs [9]. The rate of hMSCs proliferation was evaluated using a PicoGreen kit on days 1, 7, 14, and 21. Both Ti and TiH samples showed a significantly higher amount of DNA (**Figure 2B**) compared to both nanostructured samples on day 1. This suggests that nanostructured surface did not improve cell adhesion regardless of the presence of hydroxyapatite layer.

Table 2 Samples composition

Samples	Ti [at. %]	Al [at. %]	V [at. %]	Ca [at. %]	P [at. %]
N	88.6	6.9	4.5	/	/
Ti	89.1	6.6	4.3	/	/
TiH	87.7	6.1	3.8	1.5	0.9
NH	78.1	4.2	1.9	9.7	6.1

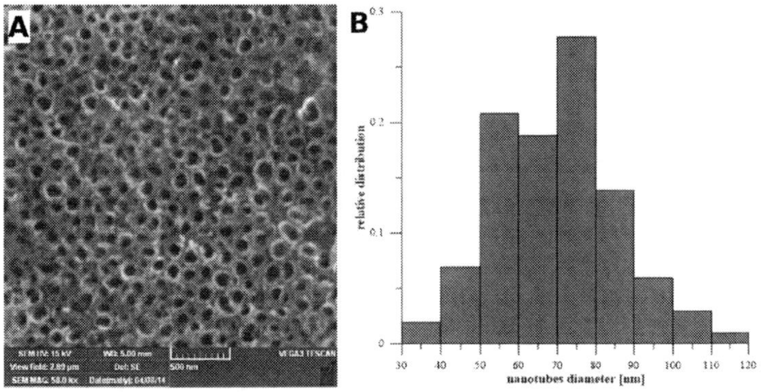

Figure 1 SEM image of nanostructure and histogram of nanotube inner diameters created on Ti6Al4V samples by anodic oxidation at 30V

On day 7, a significant difference in the amount of DNA was detected again on the samples Ti and TiH compared to sample N. No significant differences were detected on day 14. However, on day 21, a higher amount of DNA was found on Ti and TiH samples compared to N and TCP group. Thus, the results implied that nanostructures formed, by anodic oxidation, decreased the rate of cell proliferation in comparison to ground titanium alloys. However, the exposition of samples to SBF had no effect on the rate of cell proliferation. This is in contrast to several studies where it was proved that the nanostructured surface of the Ti6Al4V improved cell proliferation [9,10]. Similarly, Deligianni et al. reported that nanorough surfaces of Ti6Al4V showed increased adsorption of total proteins from FBS, and the amount of adsorbed fibronectin on surfaces with higher roughness compared to lower roughness. This finding correlated with number of cells adhered on the surface [1]. On titanium surfaces modified with nanotubes, the highest protein adsorption was found on samples containing nanotubes with the diameter of about 30 nm and the lowest with the diameter of 80 nm. Similarly, hMSCs proliferation correlated with the rate of protein adsorption.

Type I collagen, an extracellular protein, is a marker of osteogenic differentiation. The measurement of type I collagen (**Figure 3**) on day 14 revealed the highest amount of this extracellular protein on the samples N and TiH. Therefore, we conclude that formed nanostructures on the surface of Ti6Al4V discs had the same effect on collagen type I production as the exposition of the ground titanium alloy to SBF. Moreover, on all the samples, there was a statistically higher intensity detected in comparison to the glass control. Those results are in agreement with the study of Filova et al. [5]. Similarly, surfaces with higher roughness of Ti6Al4V showed higher alkaline phosphatase activity of hMSCs compared to surfaces with lower roughness [1]. Higher nanotube diameter on titanium samples improved osteogenic markers compared to smaller diameter [11].

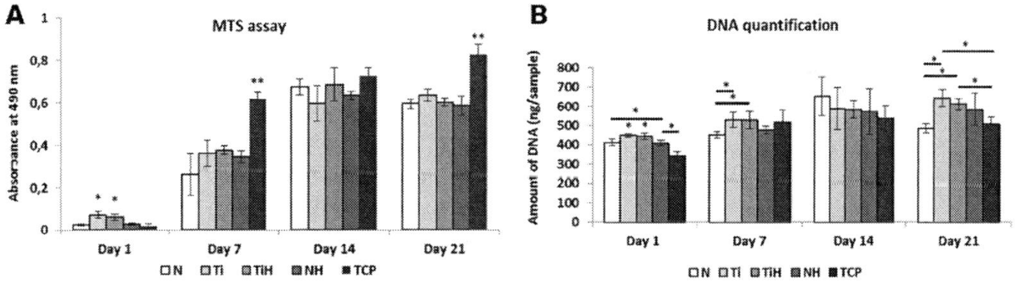

Figure 2 (A) Metabolic activity of the hMSCs measured by MTS assay (B), the proliferation of the hMSCs determined using the PicoGreen assay. Statistical difference with $P < 0.05$ marked as * **and P**< 0.001 as **.

In a human body, apatite layer is formed on the implant surfaces. Through this layer, bone is binding to the implant surface [7]. Thus, the properties of the formed layer may affect cell adhesion, growth and differentiation. Modification of titanium surfaces by hydroxyapatite layer (HAp) has been tested using different techniques, e.g. electrophoretic deposition, electrophoretic deposition, plasma spraying, phase-transited lysozyme coating etc. [12,13]. They provide HAp of slightly different composition, morphology, crystallinity, and different stability. HAp modified titanium surfaces showed higher osteoconductivity and osteoinductivity than non-modified surfaces [14]. In our study, we used sample incubation in SBF solution. The surfaces of TiH and NH were completely covered with HAp-like layer. TiH did not negatively affect hMSC proliferation compared to N samples. On the other hand, type I collagen significantly decreased in NH samples although it amount was higher than on Ti samples which was probably influenced by decreased roughness.

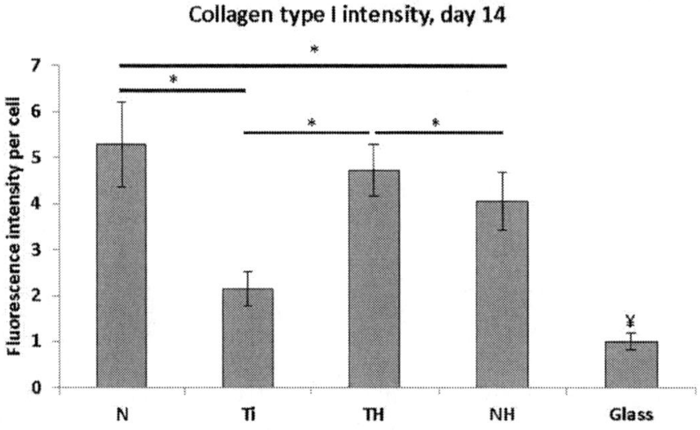

Figure 3 Type I collagen intensity on diverse Ti6Al4V discs. Statistical difference with P < 0.05 marked as *, ¥ mean this group showed the lowest fluorescence intensity statistically.

4. CONCLUSION

Following the surface modification of Ti6Al4V discs (Ti) by anodic oxidation, the nanotubes with a diameter of 70 nm ± 16 nm were formed (N). Subsequently, the hydroxyapatite-like layer was deposited on both the nanostructured and ground titanium surfaces (NH, TiH) by the discs exposure in SBF. Our results clearly demonstrated that all four kinds of Ti6Al4V discs supported hMSCs metabolic activity and growth. Significant differences in the fluorescence intensity of collagen type I revealed that both modifications, nanostructuring of the surfaces and exposition to the SBF, positively influenced the osteogenic differentiation of seeded hMSCS.

ACKNOWLEDGEMENTS

The project was supported by the Czech Science Foundation project Nr. 16-14758S and by the Ministry of Education, Youth and Sports of the Czech Republic under NPU I: LO1508.

REFERENCES

[1] DELIGIANNI, D.D., KATSALA, N., LADAS, S., SOTIROPOULOU, D., AMEDEE, J. and MISSIRLIS, Y.F. Effect of surface roughness of the titanium alloy Ti-6Al-4V on human bone marrow cell response and on protein adsorption. Biomaterials. 2001. vol. 22, no. 11, pp. 1241-1251.

[2] BORDJI, K., JOUZEAU, J.Y., MAINARD, D., PAYAN, E., NETTER, P., RIE, K.T., STUCKY, T. and HAGE-ALI, M. Cytocompatibility of Ti-6Al-4V and Ti-5Al-2.5Fe alloys according to three surface treatments, using human fibroblasts and osteoblasts. Biomaterials. 1996. vol. 17, no. 9, pp. 929-940.

[3] CHRZANOWSKI, W., SZEWCZENKO, J., TYRLIK-HELD, J., MARCINIAK, J. and ZAK, J. Influence of the anodic oxidation on the physicochemical properties of the Ti6Al4V ELI alloy. Journal of Materials Processing Technology. 2005. vol. 162-163, pp. 163-168.

[4] VELTEN, D., BIEHL, V., AUBERTIN, F., VALESKE, B., POSSART, W. and BREME, J. Preparation of TiO(2) layers on cp-Ti and Ti6Al4V by thermal and anodic oxidation and by sol-gel coating techniques and their characterization. Journal of Biomedical Materials Research. 2002. vol. 59, no. 1, pp. 18-28.

[5] FILOVA, E., FOJT, J., KRYSLOVA, M., MORAVEC, H., JOSKA, L. and BACAKOVA, L. The diameter of nanotubes formed on Ti-6Al-4V alloy controls the adhesion and differentiation of Saos-2 cells. Int J Nanomedicine [online]. 2015. vol. 10, p. 7145-7163 [viewed 2018-09-25]. Available from DOI: 10.2147/IJN.S87474.

[6] EL-WASSEFY, N.A., HAMMOUDA, I.M., HABIB, A.N., EL-AWADY, G.Y. and MARZOOK, H.A. Assessment of anodized titanium implants bioactivity. Clinical Oral Implants Research. 2014. vol. 25, no. 2, pp. e1-e9.

[7] TAKADAMA, H., KIM, H.-M., KOKUBO, T. and NAKAMURA, T. XPS study of the process of apatite formation on bioactive Ti-6Al-4V alloy in simulated body fluid. Science and Technology of Advanced Materials. 2001. vol. 2, no. 2, pp. 389-396.

[8] GALLARDO-MORENO, A.M., PACHA-OLIVENZA, M.A., SALDAÑA, L., PÉREZ-GIRALDO, C., BRUQUE, J.M., VILABOA, N. and GONZÁLEZ-MARTÍN, M.L. In vitro biocompatibility and bacterial adhesion of physico-chemically modified Ti6Al4V surface by means of UV irradiation. Acta Biomaterialia. 2009. vol. 5, no. 1, pp. 181-192.

[9] SAHA, S., KUMAR, R., K. PRAMANIK, K., and BISWAS, A. Interaction of osteoblast -TiO2 nanotubes in vitro: The combinatorial effect of surface topography and other physico-chemical factors governs the cell fate. Applied Surface Science. 2018, vol. 449, pp. 152-165.

[10] SHOKUHFAR, T., HAMLEKHAN, A., CHANG, J.-Y., CHOI, C.K., SUKOTJO, C. and FRIEDRICH, C. Biophysical evaluation of cells on nanotubular surfaces: the effects of atomic ordering and chemistry. Int J Nanomedicine [online]. 2014. vol. 9, pp. 3737-3748 [viewed 2018-09-20]. Available from DOI: 10.2147/IJN.S67344.

[11] DING, X., ZHOU, L., WANG, J., ZHAO, Q., LIN, X., GAO, Y., LI, S., WU, J., RONG, M., GUO, Z., LAI, C., LU, H. and JIA F. The effects of hierarchical micro/nanosurfaces decorated with TiO2 nanotubes on the bioactivity of titanium implants in vitro and in vivo. Int J Nanomedicine [online]. 2015. vol. 10, pp. 6955-6973 [viewed 2018-09-25]. Available from DOI: 10.2147/IJN.S87347.

[12] YANG, J., ZHANG, K., QUE, K., HOU, S., CHEN, Z., LI, Y., WANG, Y., SONG, Y., GUAN, B., ZHANG, W., ZHU, D., LI, C., WANG, D., GENG, P. and ZHANG, X. Surface modification of titanium with hydroxyapatite layer induced by phase-transited lysozyme coating. Materials Science and Engineering. C, Materials for Biological Applications. 2018. vol. 92, pp.206-215.

[13] MANO, T., UEYAMA, Y., ISHIKAWA, K., MATSUMURA, T. AND K. SUZUKI, K. Initial tissue response to a titanium implant coated with apatite at room temperature using a blast coating method. Biomaterials. 2002. vol. 23, no. 9, pp. 1931-1936.

[14] LIN, L., CHOW, K.L. and LENG, Y. Study of hydroxyapatite osteoinductivity with an osteogenic differentiation of mesenchymal stem cells. Journal of Biomedical Materials Research A. 2009. vol. 89, no. 2, pp. 326-335.

NATURAL HIERARCHICAL STRUCTURES AND THEIR REPLICATION

Dora KROISOVÁ

Technical University of Liberec, Liberec, Czech Republic, EU, dora.kroisova@tul.cz

Abstract

Hierarchical plant surface structures show many characteristics that are an inspiration for humans. They are associated with the process of photosynthesis and provide a hydrophobic or a hydrophilic behaviour, an absorption of solar radiation, etc. Due to evaluation of synthetic analogical structures we are able to obtain important information about their potential applicability in a technical practice. The easiest way to prepare them is a two-stage replication process. This process involves the creation of the so-called first replica, that is the mould for the second replica - the "copy" of natural surface. Replicas can be made from various types of polymers. The aim of the paper is to compare the influence of different types of polymers on the character and quality of synthetic structures - replicas. A hierarchical structure of the pansy flower *(Viola x wittrockiana)* was selected as a replication pattern. The silicone elastomer President Light Body (Coltene Wahledent, Hamburg, Germany) was used for the first (negative) replicas preparation. Thermoset (EP), thermoplastic (HDPE), biodegradable polymers (PCL, PLA), as well as aqueous (PVA) and alcoholic polymer solutions (PVB) were used to create the second (positive) replicas. Replicas quality was evaluated by scanning electron microscopy (Carl Zeiss ULTRA Plus) and by contact angle measuring (See System E). Different hierarchical surface structures and different contact angles were identified according to types of polymers (amorphous, semi-crystalline, chemical composition) and used technology of replication. Easy, fast and a low cost replication process can be used for copying structures up to hundreds/tens of nanometers.

Keywords: Nanostructures, biomimetics, hierarchical plant surfaces, replication, polymers

1. INTRODUCTION

The surfaces of natural objects show a number of interesting properties related to their structure and chemical composition. Hierarchical structures of plant/petal surfaces are very often built up by conically shaped epidermal cells at micrometre levels. Their surfaces are covered with nanostructured wax shapes in form of platelets, tubes or rods. The surfaces structured by the plant provide a protection against excessive wetting and excessive sunlight. They can act as sensors or light-catching systems, too [1,2]. Preparation of synthetic structures and their further testing is interesting from the technical practice point of view. At present, a number of experiments are being carried out to use of natural models. For example, these experiments are: increasing photovoltaic panels performance (pansy and rose petal surface), testing of self-cleaning surfaces (the lotus leaf surface) and a study of antibacterial surfaces based on physical behaviour (cicada and dragonfly wings) [3,4]. Preparation of synthetic structures for the selected properties testing is complicated due to their structural and dimensional hierarchy - nanometre structures are located on larger ones. A selection of the proper material for replica production as well as a low cost technological process is important. Laser, etching and even 3D printing are not suitable processes for the preparation of copies of the natural surface because it is not possible to copy the natural pattern in all its details. The two-step process of replication is an effective, fast and inexpensive way to prepare synthetic hierarchical surface samples. A silicone elastomer is used for the preparation of the mould (so-called the negative/the first replica), whereas an epoxy resin or a silicone elastomer is the most commonly used polymers for the preparation of synthetic samples (so-called the positive/the second replica) [2,3,5]. An evaluation of the influence of different polymers characteristics and quality of synthetic replicas is the aim of this article.

2. EXPERIMENTAL PART

A two-step replication process was used for the preparation of the petal surface polymer replicas. This process is called a soft lithography. Replica analysis was based on the evaluation of scanning electron microscopy micrographs and evaluation of contact angles of synthetic structures from different polymers

2.1. Materials

A pansy *(Viola x wittrockiana)* petal epidermal structure was selected for an evaluation of the replication process, the influence of polymer type on quality and properties of prepared replicas. The pansy petal surface is a typical hierarchical structure of epidermal cells at micrometre and sub-micrometre levels. The elastomeric material President Light Body (Coltene Wahledent, Hamburg) was selected for the preparation of moulds. Materials for the polymeric sample preparation were: epoxy resin EnviPOXY®, (Spolchemie, a. s.), high density polyethylene HDPE (Goodfellow GmbH), polylactic acid PLA (Goodfellow GmbH), polycaprolactone PCL (Merck KGaA), polyvinyl alcohol PVA (Merck KGaA) and polyvinyl butyral PVB (Merck KGaA). The selection includes thermosetting polymer, thermoplastics, biodegradable polymers, water and alcohol soluble polymers, amorphous and semi-crystalline polymers.

2.2. Technology of replication

Fresh and clean pansy petals were used for a surface structure replication. A sample was glued to a flat surface, was covered with a silicone elastomer and was loaded with a force up to 5 N. The crosslinking time of the silicone elastomer was 3 minutes at normal laboratory temperature. Thereafter the elastomer layer was removed from the plant surface, thereby creating the mould. This first replica was consequently and repeatedly used for the preparation of positive/the second replica from different type of polymers. A clean, dry mould was used for the casting of epoxy resin, PVA and PVB. The crosslinking process of epoxy resin samples runs at normal laboratory temperature during 24 hours, in a special vessel at a pressure of 0.7 MPa. Used epoxy resin filled the microscopic and sub-microscopic structure details of the mould well and minimized non-homogeneity of replicas due to an overpressure process. Evaporation of the solvent (water or alcohol) from PVA, respectively PVB solution took place in the air at normal laboratory temperature without overpressure. Samples of HDPE, PCL, and PLA were prepared by melting the polymer granules and subsequently compressing molten mass in the mould. In order to ensure a desired homogeneity of polymer replicas, namely a good filling of microscopic and sub-microscopic details of a mould, a force of 3 to 6 N was used. Thermoplastic replicas solidified during a few minutes.

The following materials and conditions were used to produce replicas: EnviPOXY®, curing agent Telatit 0492, modified cycloaliphatic amine (100 : 27 wt%), a curing time of 24 hours, a curing temperature of 22 °C, an overpressure of 0.5 to 0.8 MPa. 10 wt% aqueous solution of PVA, an evaporation of the solvent in the air for 12 hours, 10 wt% an alcoholic solution of PVB, an evaporation of the solvent in the air for 12 hours. HDPE granules, a melting temperature 190 °C, a melting time 5 minutes, a loading 3 to 6 N. PLA granules, a melting temperature 180 °C, a melting time 5 minutes, a loading 3 to 6 N. PCL granules, a melting temperature at 70 °C, a melting time 5 minutes, a loading 3 to 6 N.

2.3. Analytical methods

The surface structure of pansy petal as well as all type polymer replicas were investigated by scanning electron microscopy (FE SEM Zeiss ULTRA Plus). All samples were sputter coated with a 4 nm Pt-Pd layer (Quorum Q150R ES) prior to SEM investigation. Additional information was determined by static contact angle measurements (See System E). For this purpose 5 μl droplets of distilled water were applied on the replicas surfaces.

3. RESULTS AND DISCUSSION

The pansy flower conical cells show a diameter approximately 30 to 40 μm at the base, wrinkles on the top of epidermal cells and folding typical for all cultivar of pansy (**Figure 1**). The thin layer was retrieved from the surface of a flower petal. Due to the minimal volume of water in this layer only a slight deformation of the actual conical structure has occurred. Well visible folds show thickness from 200 to 300 nm. The epidermal cells height is about 40 μm.

Figure 1 The natural petal surface - epidermal conical cells of pansy *(Viola x wittrockiana)*

The silicone elastomer mould (**Figure 2**) is a good copy of the epidermal cell surface sample both at the micrometer and the sub-micrometer size. Foldings in the mould do not appear to be sufficiently deep in the images compar ed to the plant specimen.

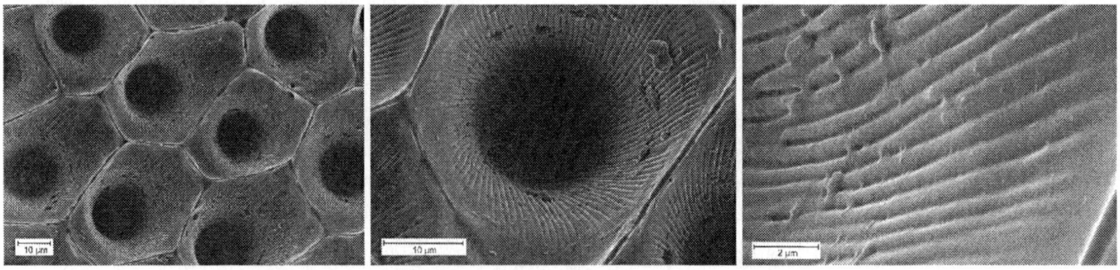

Figure 2 Silicone elastomer mould surface - negative replica

The epoxy resin (**Figure 3**) copies the pansy surface sample very well. Both epidermal cells of dimensions from 30 to 40 μm, as well as 200 to 300 nm folds are well visible.

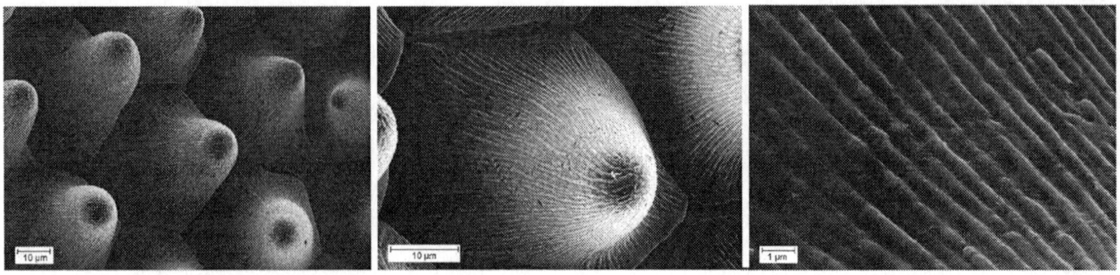

Figure 3 Synthetic replica of the plant surface - epoxy resin (EnviPOXY®)

The polyvinyl alcohol (**Figure 4**) copies the shape, the diametre and the folds of the pansy surface sample. Both epidermal cells of dimensions from 30 to 40 μm, wrinkles on the top of epidermal cells, as well as 200 to 300 nm folds are well visible. The surface of the folds is not smooth.

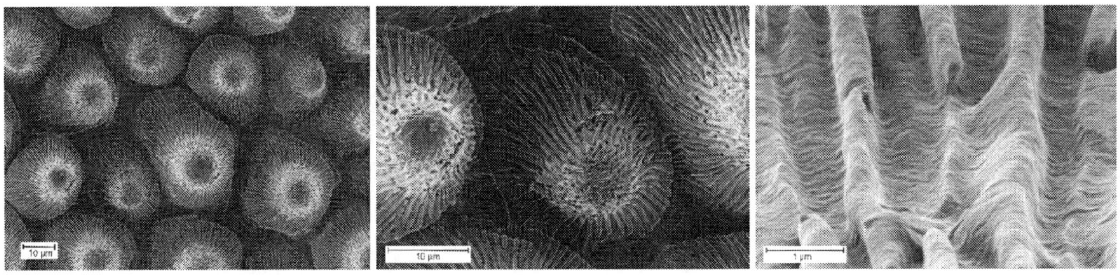

Figure 4 Synthetic replica of the plant surface - polyvinyl alcohol (PVA)

The polyvinyl butyral (**Figure 5**) copies the shape, the diametre and the folds of the pansy surface sample. The epidermal cells and the wrinkles on the top of epidermal cells are well visible. The folds are shallow compared to others polymer replicas. The surface of the folds is not smooth, visible particles were not studied in more detail.

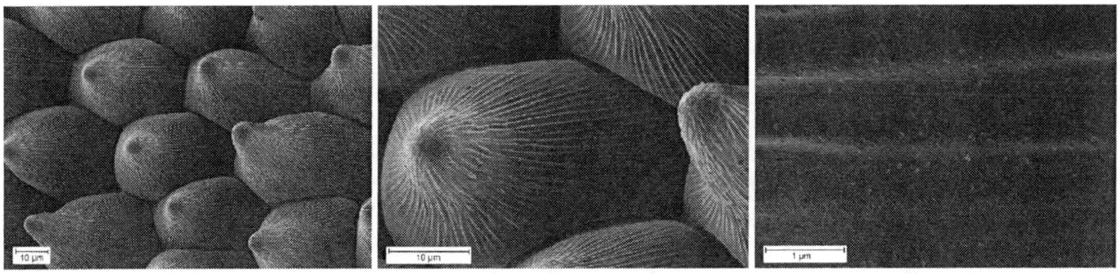

Figure 5 Synthetic replica of the plant surface - polyvinyl butyral (PVB)

The polylactic acid (**Figure 6**) copies the shape and the diametre of pansy surface sample similar as in previous examples. The folds are shallow, the surface of the folds is not smooth but it contains pores and visible particles. These inhomogeneities were not studied in more detail. The polylactic acid is a polymer with a low level of crystallization.

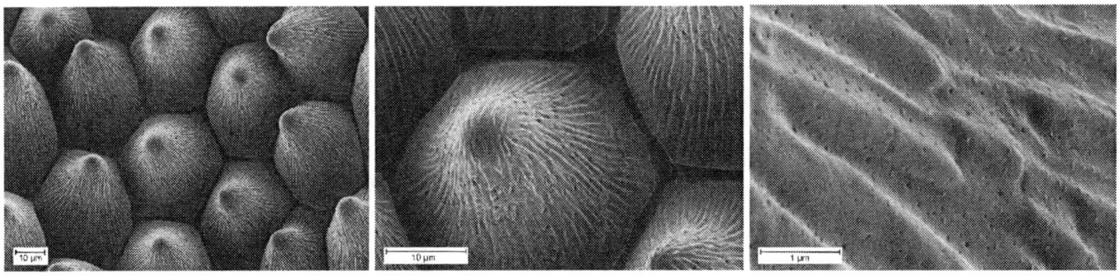

Figure 6 Synthetic replica of the plant surface - polylactic acid (PLA)

The polycaprolactone (**Figure 7**) copies the shape, the diametre, the wrinkles on the top of epidermal cells and the folds of the pansy surface sample. The folds are shallow, inhomogeneous with a slightly fibrous structure compared to other polymer replicas. The depth and dimensions of the folds do not reach the quality of the plant sample. The fibrous surface of the folds was not studied in more detail. The polycaprolactone is a semicrystalline polymer.

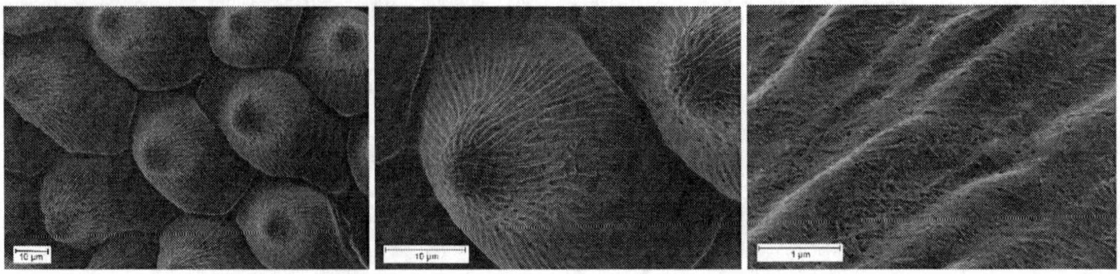

Figure 7 Synthetic replica of the plant surface - polycaprolactone (PCL)

The high density polyethylene (**Figure 8**) copies the shape, the diametre and the folds of the pansy surface sample. Both epidermal cells of dimensions from 30 to 40 μm, wrinkles on the top of epidermal cells, as well as 200-300 nm folds are well visible. The surface of the folds is not smooth but a new fibrous structure was created. The visible fibers are 10 to 20 nm in size and are transversally arranged to the folds. The high density polyethylene is a semicrystalline polymer.

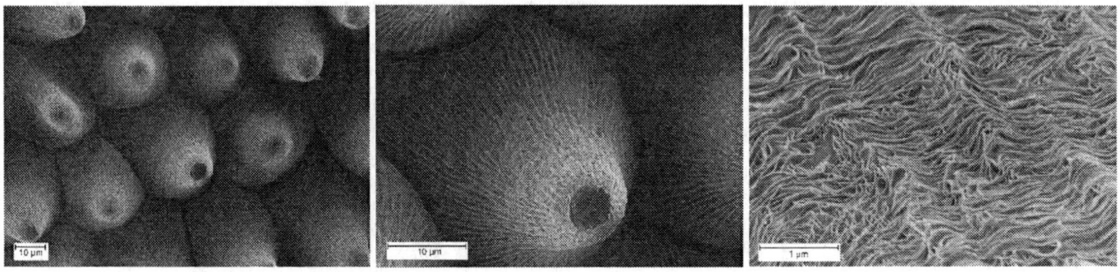

Figure 8 Synthetic replica of the plant surface - polyethylene (HDPE)

From the point of view of the replication process, all replicas prepared from the above mentioned polymers show good compliance with the pansy pattern on the epidermal cells level. The nanostructure built up from the tens of nanometres densely folded fibres was identified in the case of the high density polyethylene. This substructure does not appear on the plant surface or on replicas from the other used polymers. The described substructure was not found on the HDPE control samples which were made by a conventional injection moulding into a standard steel mould. This nanostructure was created due to the influence of the used polymer and the technology of replication.

Table 1 Contact angle of the plant pattern, the flat polymer surfaces and the structured polymer replicas

Polymer	Contact angle of flat surface (°)	Contact angle of replica surface (°)	Increase of contact angle (%)
Plant pattern	-	139.8	-
Epoxy resin	59.4	108.6	54.7
PVA*	-	-	-
PVB	72.6	119.9	60.6
PCL	77.8	113.2	68.7
PLA	74.9	110.3	67.9
HDPE	77.3	138.3	78.9

* *an impossible to measure due to a dissolution of PVA sample in water*

The contact angles were measured on the different polymer replicas. The difference in contact angle values depends on the type of the polymer used and the character of the surface. The flat surface of all the used polymers shows the hydrophilic behaviour because the contact angle is lower than 90°. Due to the structuring of the polymer sample surface, the contact angle increases and the polymer replicas surfaces show the hydrophobic behaviour (the contact angle is higher than 90°). The epoxy resin replica shows the lowest contact angle 108.6. On the contrary the HDPE as a semi-crystalline polymer shows the highest value of the contact angle 138.3 (**Table 1**). The difference in the contact angle values can be explained due to the dissimilar chemical composition of the used materials and the distinct surface structure of the replicas that is particularly due to the fibrous nanostructure that was analysed on the HDPE replicas samples. To confirm the pronounced opinion, further experiments will be required to verify the effect of the chemical composition and the influence of the nanofibrous structure.

4. CONCLUSION

The two step method of replication is a very efficient, fast and inexpensive method for the preparation of the polymer samples of the natural surfaces and their further testing. Due to this method it is possible to capture the actual state of the natural surface. The replicas could be created from amorphous and semi-crystalline polymers. The type of the selected polymer has the underlying effect on the resulting structure and contact angle values. The quality of the replication process is at a high level. The new nanostructure was identified in the case of the high density polyethylene. This nanostructure was created due to the influence of the used polymer and the technology of replication.

ACKNOWLEDGEMENTS

This work was supported by the Ministry of Education, Youth and Sports of the Czech Republic and the European Union - European Structural and Investment Funds in the frames of Operational Programme Research, Development and Education - project Hybrid Materials for Hierarchical Structures (HyHi, Reg. No. CZ.02.1.01/0.0/0.0/16_019/0000843).

My thanks belong to Pavel Kejzlar for microscopic analysis, Taťána Vacková and František Lednický for their point of view.

REFERENCES

[1] BHUSNAN, B., JUNG, Y.C. and KOCH, K. Micro-, nano- and hierarchical structures for superhydrophobicity, self-cleaning and low adhesion: *Philosophical Transactions of the Royal Society A: Mathematical, Physical and Engineering Sciences.* 2009. vol. 367, no. 1894, pp. 1631-1672.

[2] SCHULTE, A. J. *Light-trapping and Superhydrophobic Plant Surfaces - Optimized Multifunctional Biomimetic Surfaces for Solar Cells.* Bonn: Rheinischen Friedrich-Wilhelms-Universitat Bonn. 2012, p. 113.

[3] HUANG, Z., YANG, S., ZHANG, H., CAO, W. Replication of Leaf Surface for Light Harvesting. *Scientific Reports.* 2015. vol. 5, no. 14281.

[4] ELBOURNE, A., CRAWFORD, R. J., IVANOVA, E. P. Nano-structured antimicrobial surfaces: From nature to synthetic analogues. *Journal of Colloid and Interface Science.* 2017. vol. 508, pp. 603-616.

[5] HUANG, Z., CAI, C., KUAI, L., LI, T. HUTTULA, M., CAO, W. Leaf-structure patterning for antireflective and self-cleaning surfaces on Si-based solar cell. *Solar Energy.* 2018. vol. 159, pp. 733 741.

BACTERIAL RESISTANCE TO SILVER NANOPARTICLES

Aleš PANÁČEK, Lucie HOCHVALDOVÁ, David PANÁČEK, Monika SMÉKALOVÁ, Robert PRUCEK, Aristeidis BAKANDRITSOS, Michal OTYEPKA, Radek ZBOŘIL, Libor KVÍTEK

Regional Centre of Advanced Technologies and Materials, Department of Physical Chemistry, Faculty of Science, Palacký University Olomouc, Olomouc, Czech Republic, EU ales.panacek@upol.cz

Abstract

Silver nanoparticles (NPs) exhibit significant antimicrobial activity against a broad range of bacteria and fungi at concentrations ranging from a few ppm to tens of ppm that are not cytotoxic to human cells. Silver NPs also strongly enhance antibacterial activity against multiresistant, β-lactamase and carbapenemase-producing *Enterobacteriaceae* when combined with antibiotics. As a result, silver NPs have already been applied in various biomedical and antimicrobial technologies and products of every-day life as an alternative to conventional antibiotics. While antibiotic resistance has been discussed extensively in the literature, the possible development of resistance to silver NPs after repeated long-term exposure has not been fully explored. We report that bacteria can develop resistance to silver NPs after prolonged exposure. The observed resistance consists in aggregation of silver NPs which consequently eliminates their antibacterial effects. The resistance mechanism can be overcome by preventing the aggregation of silver NPs via their deposition on suitable inorganic substrate such as graphene oxide.

Keywords: Silver nanoparticle, graphene oxide, bacteria, antibacterial, antibiotic, resistance

1. INTRODUCTION

Bacterial resistance to antibiotics currently represents a very serious and complex problem transcending the borders of healthcare. The mounting resistance of pathogenic bacteria increasingly limits antibiotic effectiveness, significantly increasing the likelihood of failure of antibiotic therapy and the related morbidity and mortality of patients [1]. The development of entirely new antimicrobial drugs and their introduction into clinical practice is a lengthy and costly process. An option to overcome bacterial resistance is restoring antibacterial effect of antibiotics by their combination with other antibacterial substances. A promising way to overcome bacterial resistance is a combination of conventional antibiotics with nanomaterilas showing antibacterial effect such as silver based nanoparticles and nanocomposites. Silver-based nanostructured materials are considered to be promising nanomaterials in biomedical applications or biotechnology for their high antimicrobial activity [2]. Silver nanoparticles (NPs) exhibit high bactericidal activity at concentrations showing no cytotoxicity to mammalian cells and also strongly enhance the antibacterial activity of conventional antibiotics even against multiresistant bacteria through their synergistic effect [3-8]. Unlike the specific antibiotic's mode of action, the effects of silver NPs are not specific to a single cellular level, but influence many bacterial structures and metabolic processes concurrently [4,9-11]. The question whether silver NPs could be used in medicine to enhance the effectiveness of antibiotics or fully replace them in the treatment of local and systemic infections in the future is rather premature. However, it is no doubt too early now to deal with potential development of bacterial resistance to silver NPs. Silver in various forms has been used as an effective antibacterial agent for many centuries and its antibacterial effect is still maintained. Bacterial resistance to silver is mainly associated with the ionic form of silver [12,13]. Only few studies demonstrating possible resistance or decreased susceptibility to silver NPs have appeared recently, nevertheless resistance to silver NPs was driven by mechanisms eliminating ionic form of silver [14,15]. On the other hand, unlike the intensive, long-term and repeated administration of classical antibiotics or ionic forms of silver (silver sulfadiazine), silver NPs have not been commonly used in medical practice. It remains unclear whether bacteria may develop resistance to silver NPs, as is the case with long term and repeated exposure to antibiotics. Here we show that bacteria originally susceptible to silver NPs are able to develop a mechanism by which they can resist the antibacterial effect of silver NPs after long-term and repeated exposure.

2. MATERIALS AND METHODS

2.1. Chemicals and biological material

Silver nitrate (p.a., Fagron), ammonia (28-30 % [w/w], p.a., Sigma-Aldrich), sodium hydroxide (p.a., Lach-Ner), D-maltose (p.a., Sigma-Aldrich), and graphene oxide (powder, Sigma-Aldrich) were used for silver NPs synthesis and silver decorated graphene oxide nanocomposite, respectively. The bacterial strains used in this work were *E. coli* CCM 3954 obtained from the Czech Collection of Microorganisms (Masaryk University, Brno) and *S. epidermidis* 008 obtained from the culture collection of the Department of Microbiology, Faculty of Medicine and Dentistry, Palacký University Olomouc.

2.2. Synthesis and characterization of silver NPs and silver decorated graphene oxide (Ag@GO)

The modified Tollens process involving reduction of the complex cation $[Ag(NH_3)_2]^+$ by D-maltose was used to synthesize dispersions of silver NPs with a diameter of 28 nm and silver concentrations of 108 $mg{\cdot}L^{-1}$ or 432 $mg{\cdot}L^{-1}$. The concentrations of all the reaction components were as follows: silver nitrate 1×10^{-3} $mol{\cdot}dm^{-3}$; ammonia 5×10^{-3} $mol{\cdot}dm^{-3}$; sodium hydroxide 9.6×10^{-3} $mol{\cdot}dm^{-3}$, and D-maltose (as a reducing agent) 1×10^{-2} $mol{\cdot}dm^{-3}$. To prepare silver NP dispersions at a concentration of 432 mg L^{-1}, the concentrations of all the reaction components were increased fourfold except that of sodium hydroxide. The same procedure was applied for deposition of silver NPs on the surface of graphene oxide sheets. For this purpose, 10 mg of graphene oxide was dispersed in 10 ml of $AgNO_3$, NH_3 and NaOH solution followed by 30 min sonication. After that, maltose solution was added under vigorous stirring to reduce silver NPs on the surface of graphene oxide sheets. Silver NPs decorated graphene oxide was three times separated by centrifugation at 150 x g for 15 minutes and wash by distilled water. Silver amount in Ag@GO nanocomposite was determined using atomic absorption spectrophotometer (AAS) (ContrAA 300, Analytik Jena, Germany) using method of flame atomization. Before analysis, vacuum dried Ag@GO nanocomposites were dispersed in 2 % HNO_3 solution under 30 min of sonication to dissolve silver NPs followed by graphene oxide separation using syringe filter (Polyethersulfone membrane, 45 µm). The average size of the synthesized silver NPs were determined by the dynamic light scattering method using a Zetasizer Nano ZS instrument (Malvern, UK). Transmission electron microscopy measurements were conducted using a JEM 2010 instrument (Jeol, Japan).

2.3. Culture of bacteria in the presence of silver NPs

Bacteria were repeatedly exposed to subinhibitory concentrations of silver NPs by twenty repeated cultures in microplates. Dispersion of silver NPs was diluted in geometric progression with Mueller-Hinton Broth (Becton, Dickinson and Company) and inoculated with a bacterial strain at a concentration of 10^6 CFU/mL. Incubation was carried out at 37°C for 24 hours. After the incubation, the MICs of silver were read after each culture step as the lowest silver concentrations inhibiting the visible growth of microorganisms. Immediately after the 24-hour cultivation, 10 µL of Mueller-Hinton broth containing the surviving bacteria were taken from wells with subinhibitory concentrations of silver (concentrations below MIC) and consequently sub-cultured on blood agar (TRIOS) at 37°C for 24 hours. Bacteria grown on blood agar were used for inoculum preparation at a density of 10^6 CFU/mL and for the next culture step.

2.4. Proteins identification and detection bacterial biofilm formation

Silver aggregates were separated from 500 ml (10 × 50 ml) of bacterial suspension containing 27 mg of silver by centrifugation at 100 × g for 2 minutes. The pelleted aliquots with silver concentration 4.5 $mg{\cdot}mL^{-1}$ were each individually incubated with 1 mL of 2.5 % trifluoroacetic acid (TFA). Extracts obtained after overnight incubation at 37 °C with shaking at 800 rpm and the subsequent pelleting of the solid particles by a centrifugation at 20,000 g for 20 min, were vacuum-dried and the recovered proteins were dissolved in 50 µL of 50 mM NH_4HCO_3 (facilitated by a 5-min sonication). Then, 1.5 µL of 200 µM modified trypsin was added for in-solution protein digestion and the mixture was again incubated at 37 °C overnight. After vacuum drying, peptides from the digests were purified using ZipTip-C18 pipette tips (Merck-Millipore, Tullagreen-Carrigtwohill, Ireland) according to the manufacturer's instructions. Vacuum-dried peptide samples were dissolved in 10 µL of 0.1 % TFA and separated using a Dionex UltiMate3000 RSLCnano liquid chromatograph (Thermo Fisher Scientific, Germering, Germany), coupled to an amaZon speed ETD ion trap equipped with a CaptiveSpray ion source (Bruker, Bremen, Germany). The scan speed was 8,100 u s^{-1} for both mass spectrometry (MS) and tandem mass spectrometry (MS/MS); acquisition was performed by collision-induced fragmentation using helium as the collision gas.

Christensen method was applied for semi-quantitative detection of bacterial biofilm formation. *S. epidermidis* 008 biofilm was formed on sterilized 96-well microplates. A 10 µl of cell suspension was inoculated in 190 µl Mueller-Hinton Broth medium in each well to get 10^6 CFU/mL. Then microplate was incubated for 24 h at 37°C. After aspiration of planktonic cells biofilms were fixed with 99 % methanol. Plates were washed twice with phosphate buffer saline or sterile saline water and air-dried. Then, 200 µl of crystal violet solution (1 %) was added to all wells. After 10 min, the excess crystal violet was removed and plates were washed twice and air dried. Finally, the cell bound crystal violet was dissolved in 33 % acetic acid. Biofilm growth was monitored in terms of $O.D._{570}$ nm using micro plate reader (DYNEX MRX).

3. RESULTS AND DISCUSSION

The well-known and reliable modified Tollens process [5] was used for synthesis of silver NPs with a diameter of 28 nm with a very narrow size distribution. The silver NPs synthesized in this way exhibit good antibacterial activity as documented by their low minimum inhibitory concentrations (MICs) equal to 3.38 mg/L after the first culture step (**Table 1**). Nevertheless, it can be presumed that bacteria may resist the antibacterial effect of silver NPs after long-term exposure at subinhibitory silver concentrations thanks to their genetic material and ability to adapt to the effects of antimicrobials. This presumption has been confirmed by the gradually increasing MICs of silver NPs following repeated culture, as documented by **Table 1** summarizing MICs of silver NPs after each second cultivation step against tested bacterial strains. **Table 1** also includes MICs of silver NPs against reference (non-induced) strains after the 1st culture step in order to compare MICs of silver NPs against induced and non-induced bacteria. It can be clearly seen from the increasing MICs in consecutive culture steps that bacteria gradually developed resistance to silver NPs. In the case of *E. coli* CCM 3954 significantly higher MICs equal to 13.5 mg/L as compared to MIC of the reference strain equal to 3.38 mg/L was determined already in the sixth culture step and from the eighth step onwards, the MICs increased to as much as 54 mg/L. *S. epidermidis* 008 became gradually more resistant from the 14th to the 18th culture steps; in the 18th step, the MIC of silver NPs reached 54 mg/L. Dispersion of silver NPs with a silver concentration as high as 432 mg/L was used to determine the final MICs of silver NPs higher than 54 mg/L against all induced resistant strains. The final MICs of silver NPs against resistant strains increased to 108 mg/L after twenty culture steps (**Table 1**). Gradual development of bacterial resistance to silver NPs was also accompanied by precipitation of silver NPs at the bottom of microplate wells (**Figure 1a**). By contrast, in the presence of susceptible bacterial strains, the dispersion of silver NPs diluted in Mueller-Hinton broth retained the yellow color (**Figure 1b**). Because no precipitates were formed and the yellow color of the dispersion of silver NPs remained unchanged during culture with reference strains, aggregation stability of silver NPs as well as their antibacterial activity were unaffected. The good aggregation stability is also evidenced by transmission electron microscopy (TEM) images of silver NPs after culture with susceptible bacteria (**Figure 2a**). In the presence of resistant strains, however, the dispersion of silver NPs changed its color from yellow to grey- black due to the formation of a precipitate at the bottom of the well. The precipitate consists of large aggregates of silver NPs with sizes on the order of hundreds of nanometers as seen from TEM measurements (**Figure 2b**). Such large aggregates do not show antibacterial activity as the originally small silver NPs. It is evident that bacteria developed the ability to induce aggregation instability of silver NPs during repeated culture in order to minimize or even totally eliminate their antibacterial effects.

Table 1 Minimum inhibitory concentrations (MICs) of silver NPs after each second culture step against bacteria compared with MICs of silver NPs against reference strains after the first culture step.

bacteria	**MIC (mg/L) after appropriate culture step**										
	1st	2nd	4th	6th	8th	10th	12th	14th	16th	18th	20th
E.coli **CCM 3954**	3.38	6.75	6.75	13.5	54	>54	>54	>54	>54	>54	108
S. epidermidis **008**	3.38	6.75	3.38	1.69	1.69	6.75	6.75	13.5	27	54	108

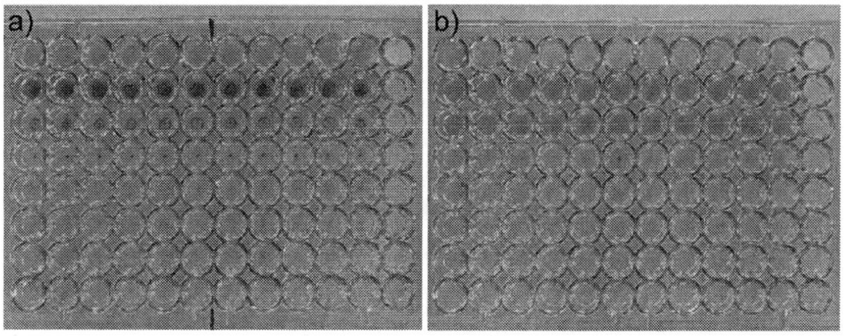

Figure 1 Microplates after culture of resistant (a) and sensitive (b) *E. coli* CCM 3954 with silver NPs diluted in Mueller-Hinton broth from the second to eighth rows to concentrations ranging from 54 mg/L to 0.84 mg/L. The upper row contains dispersion of silver NPs only.

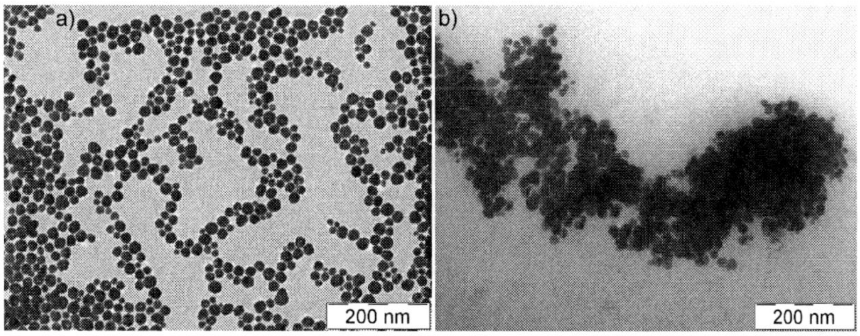

Figure 2 TEM images of stable non-aggregated silver NPs (a) and aggregated (b) silver NPs after culturing for 24 hours with "Ag-sensitive" and "Ag-resistant" *E. coli* CCM 3954, respectively.

It can be suggested that "Ag-resistant" bacteria induce aggregation of silver NPs by extracellular secretion of an agent that remains adsorbed on the aggregates' surfaces. This agent is likely to be a high molecular weight substance such as a protein. Therefore, proteins extracted from silver NPs precipitated by Ag-resistant bacteria were subjected to proteomic mass spectrometry analysis to identify proteins that might be responsible for the observed NP aggregation. In the case of *E. coli* CCM 3954, the most abundant proteins in the extract prepared with 2.5 % TFA were flagellin followed by type-1 fimbrial protein. To investigate the proposed role of flagellin in triggering the aggregation process, we conducted a simple experiment with a commercially available flagellin solution (1 mg/l), which was added to a dispersion of silver NPs in MH broth under conditions similar to those used in the culture experiments. The initially yellow dispersion of silver NPs immediately became grey upon adding flagellin, and a black silver precipitate was observed at the bottom of the Eppendorf tube after 24 hours at 37 °C. In the case of *S. epidermidis* 008, any proteins have not been analyzed on the surface of precipitated silver NPs. On the other hand, Ag-resistant *S. epidermidis* 008 produced bacterial biofilm much more than Ag-sensitive *S. epidermidis* 008 as it was confirmed by Christensen detection method. We therefore suppose that bacterial biofilm formation was responsible for aggregation of silver NPs and consequently for resistance of *S. epidermidis* 008 to silver NPs.

The mechanism of bacterial resistance based on aggregation of silver NPs can be overcome by preventing the particle aggregation using a strong deposition of silver NPs on the surface of appropriate nanostructured substrate. For example, graphene oxide can be considered as a suitable substrate for deposition of silver NPs. In order to overcome particle aggregation and bacterial resistance to silver NPs we synthesized silver NPs decorated graphene oxide (**Figure 3**). Diameter of silver NPs reduced on the GO sheets ranged from 15 to 30 nm, which is similar to diameter of silver NPs in the used water dispersion. Concentration of silver (determined using AAS) per 1 g of Ag@GO composite was 273 mg/g. Ag@GO nanocomposite showed a high antibacterial activity against both Ag-sensitive and Ag-resistant bacterial strains. Determined MIC of Ag@GO recalculated to silver content were as low as 2.5 mg/l for both Ag-sensitive *E. coli* CCM 3954 and *S. epidermidis* 008 and 5 mg/l for both Ag-resistant *E. coli* CCM 3954 and *S. epidermidis* 008. MICs of silver NPs equal to 5 mg/l against Ag-resistant *E. coli* CCM 3954 and Ag-resistant *S. epidermidis* 008 represent a very low concentration compared to MIC of water dispersion of silver NPs against sensitive bacterial strains. Bacterial resistance to uncoated silver NPs was overcome by preventing the particle aggregation through the deposition of silver NPs on the graphene oxide sheets.

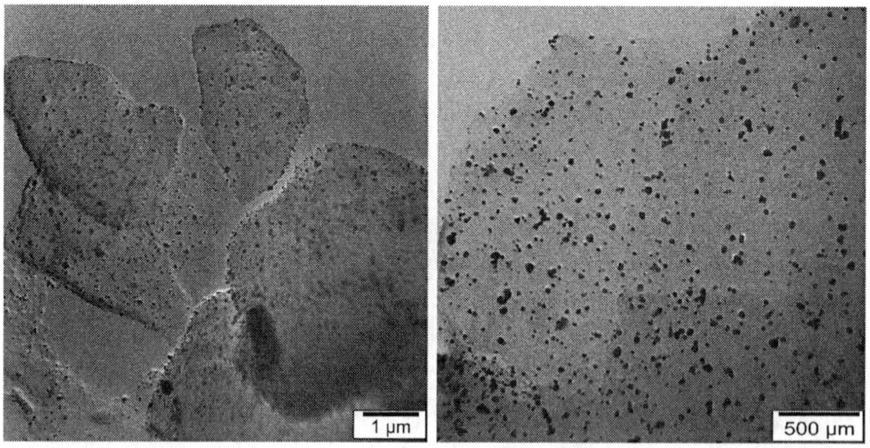

Figure 3 TEM images of Ag@GO nanocomposite at different magnifications

4. CONCLUSION

In summary, we have demonstrated that bacteria repeatedly exposed to subinhibitory silver concentrations are able to develop a resistance to silver NPs in relatively short time. The mechanism based on a physical effect by which bacteria can resist the antibacterial effect of silver NPs involves disruption of the colloidal stability of silver NPs through the flagellin (*E. coli* CCM 3954) or biofilm production (*S. epidermidis* 008). Bacterial resistance to silver NPs can be overcome by deposition of silver NPs of graphene oxide sheets preventing particle aggregation. These findings will certainly help to better understand the mechanisms that drive bacterial resistance to antibacterial agents and will be useful in prevention of bacterial drug resistance and fight against infectious bacteria.

ACKNOWLEDGEMENTS

The authors gratefully acknowledge the support provided by project LO1305 of the Ministry of Education, Youth and Sports of the Czech Republic

REFERENCES

[1] Livermore, D. M. Fourteen years in resistance. *Internatiuonal Journal of Antimicrobial Agents.* 2012. vol. 39, no. 4, pp. 283-294.

[2] Wong K. K. Y. and Liu X. L. Silver nanoparticles-the real "silver bullet" in clinical medicine? *MedChemComm.* 2010. vol. 1, pp. 125-131.

[3] Cui L., Chen P.Y, Chen S.D., Yuan Z.H., Yu C.P., Ren B. and Zhang K.S. In Situ Study of the Antibacterial Activity and Mechanism of Action of Silver Nanoparticles by Surface-Enhanced Raman Spectroscopy. *Analytical Chemistry.* 2013. vol. 85, no. 11, pp. 5436-5443.

[4] Lara H. H., Ayala-Nunez N. V., Turrent L. D. I. and Padilla C. R. Bactericidal effect of silver nanoparticles against multidrug-resistant bacteria. *World Journal of Microbiology & Biotechnology.* 2010. vol. 26, no.4, pp. 615-621.

[5] Panacek A., Kvitek L., Prucek R., Kolar M., Vecerova R., Pizurova N., Sharma V.K., Nevecna T. and Zboril R. Silver colloid nanoparticles: Synthesis, characterization, and their antibacterial activity. *Journal of Physichal Chemistry B.* 2006. vol. 110, no. 33, pp. 16248-16253.

[6] Panacek A.,Smekalova M., Kilianova M., Prucek R., Bogdanova K., Vecerova R., Kolar M.,Havrdova M., Plaza G.A., Chojniak J, Zboril R. and Kvitek L. Strong and Nonspecific Synergistic Antibacterial Efficiency of Antibiotics Combined with Silver Nanoparticles at Very Low Concentrations Showing No Cytotoxic Effect. *Molecules.* 2016. vol. 21., no. 1.

[7] Panacek A., Smekalova M., Vecerova R., Bogdanova K., Roderova M., Kolar M., Kilianova M., Hradilova S., Froning J.P., Havrdova M., Prucek R., Zboril R. and Kvitek L. Silver nanoparticles strongly enhance and restore bactericidal activity of inactive antibiotics against multiresistant Enterobacteriaceae. *Colloids and Surfaces B-Biointerfaces.* 2016. vol. 142, pp. 392-399.

[8] Brown A. N., Smith K., Samuels T.A., Lu J.R., Obare S.O. and Scott M.E. Nanoparticles Functionalized with Ampicillin Destroy Multiple-Antibiotic-Resistant Isolates of Pseudomonas aeruginosa and Enterobacter aerogenes and Methicillin-Resistant Staphylococcus aureus. *Applied and Environmental Microbiology.* 2012. vol. 78, no. 8, pp. 2768-2774.

[9] Xu H.Y., Qu F., Xu H., Lai W.H., Wang Y.A., Aguilar Z.P. and Wei H. Role of reactive oxygen species in the antibacterial mechanism of silver nanoparticles on Escherichia coli O157:H7. *Biometals.* 2012, vol. 25, no. 1, pp. 45-53.

[10] Lok C.N., Ho C.M., Chen R., He Q.Y., Yu W.Y., Sun H.Z., Tam P.K.H., Chiu J.F. and Che C.M. Proteomic analysis of the mode of antibacterial action of silver nanoparticles. *Journal of Proteome Research.* 2006. vol. 5, no. 4, pp. 916-924.

[11] Li W.R., Xie X.B., Shi Q.S., Zeng H.Y., Ou-Yang Y.S. and Chen Y.B. Antibacterial activity and mechanism of silver nanoparticles on Escherichia coli. *Applied and Environmental Microbiology.* 2010. vol. 85, no. 4, pp.1115-1122.

[12] Nies D. H. Efflux-mediated heavy metal resistance in prokaryotes. *Fems Microbiology Reviews.* 2003. vol. 27, no. 2-3, pp. 313-339.

[13] Silver S., Phung L. T. and Silver G. Silver as biocides in burn and wound dressings and bacterial resistance to silver compounds. *Journal of Industrial Microbiology & Biotechnology. 2006. vol.* 33, no. 7, pp. 627-634.

[14] Graves J. L., Jr., Tajkarimi M., Cunningham Q., Campbell A., Nonga H., Harrison S.H. and Barrick J.E. Rapid evolution of silver nanoparticle resistance in Escherichia coli. *Frontiers in Genetics.* 2015. vol. 6, 42.

[15] Losasso C., Belluco S., Cibin V., Zavagnin P., Micetic I., Gallocchio F., Zanella M., Bregoli L., Biancotto G. and Ricci A. Antibacterial activity of silver nanoparticles: sensitivity of different Salmonella serovars. *Frontiers in Microbioogy.* 2014. vol. 5, 227.

ON THE FEASIBILITY OF APPLICATION OF BIO-CONJUGATED MOS_2 BASED NANOPLATFORM FOR TARGETED CANCER TREATMENT

Anna KALOSI[1], Adriana ANNUSOVA[1,2], Martina LABUDOVA[3], Nikola BUGAROVA[4], Marianna ELIASOVA SOHOVA[2], Monika BENKOVICOVA[1], Michal BODIK[1], Matej JERGEL[1], Peter SIFFALOVIC[1], Eva MAJKOVA[1,2], Maria OMASTOVA[4], Silvia PASTOREKOVA[3]

[1] Department of Multilayers and Nanostructures, Institute of Physics, Slovak Academy of Sciences, Bratislava, Slovak Republik, EU
[2] Centre for Advanced Material Application, Slovak Academy of Sciences, Bratislava, Slovak Republik, EU
[3] Department of Cancer Biology, Institute of Virology, Biomedical Research Center, Slovak Academy of Sciences, Bratislava, Slovak Republik, EU
[4] Department of Composite Materials, Polymer Institute, Slovak Academy of Sciences, Bratislava, Slovak Republik, EU

Abstract

Surface functionalization of 2D MoS_2 materials offers a convenient way to tune their properties and facilitate their application in the field of bio-sensing and nanomedicine. In this study, three different liquid phase exfoliation (LPE) methods were utilized for MoS_2 exfoliation: LPE in 45 % ethanol solution, LPE in ultrapure water and amino acid intercalation assisted LPE. The three products were bio-conjugated with PEG molecules and through a biotin-avidin-biotin bridge they were functionalized with specific antibody M75 having exclusive affinity towards cancer cells expressing CAIX protein on their membrane. We are aiming to achieve high selectivity towards cancerous cells utilizing an antigen-antibody based cancer detection scheme. This study uses CAIX antigen tagged MDCK cells simulating the cancerous environment and pure MDCK cells as control. Cellular intake studies were performed by confocal Raman microscopy imaging and confocal laser scanning microscopy before and after binding the specific antibody. The preliminary results indicate higher density of internalized nanoplatforms in the case of MDCK cells expressing CAIX compared to the control cells. However, further studies are required to validate these results.

Keywords: MoS_2 based nanoplatform, bio-conjugation, antigen-antibody based detection

1. INTRODUCTION

The idea of producing a cancer detection platform based on 2D nanomaterials originates from the achievements of bio-nanotechnology over the last decade. Biocompatible graphene derivates [1] were used to produce nano-systems that can be utilized in biomedical applications. Several studies promoted systems for drug delivery mediated cancer treatment [2,3] or cancer treatment via photothermal therapy [4,5]. In 2013, Chou et al. [6] surveyed the photothermal performance of 2D TMD (transition metal dichalcogenide) materials and showed that MoS_2 nanosheets were better than graphene. Subsequently, different approaches of photothermal therapy were demonstrated [7-10]. Several studies described PEGylated (poly(ethylene glycol) terminated) MoS_2 nanosheets as nontoxic or with negligible toxicity. Teo et al. [11] showed that exfoliated MoS_2 and other TMDs have lower cytotoxicity than graphene and its analogues and they considered it as a biocompatible material. Another study by Appel et al. [12] stated both mechanically exfoliated and CVD grown MoS_2 as biocompatible, evaluating not only their cytotoxicity but also genotoxicity. Liu et al. [13] reported that lipoic acid (LA) conjugated PEG containing a disulfide group at the PEG terminal enables a strong binding to MoS_2 and the PEG greatly enhances the physiological stability. Furthermore, Liu et al. [8] reported that the PEG not only stabilizes the MoS_2 solution but also increases the cellular uptake of the PEG-MoS_2 sheets loaded with specific drug when compared to the case where the sheets were not PEGylated.

Taking into account these findings, the nanoplatform model that we are considering consists of a MoS_2 nanosheet base, which is covalently functionalized with LA terminated PEG. This PEGylated nanoplatform needs to be able to bind the proper antibodies that can attach to cancer cells expressing specific antigens. Therefore, this nanoplatform design uses a PEG molecule with LA at one end and biotin on the other. Biotin is a water-soluble vitamin which can be bonded to avidin (avidin is a glycoprotein which has very high affinity to bind biotin). The avidin-biotin non-covalent bond is widely used in biological sciences due to its high specificity and strength [14,15]. Moreover, an avidin molecule can bind up to four biotin molecules which can be conjugated to several different molecules. This feature allows to introduce biotinylated molecules, antibodies, cytotoxins, fluorescent dyes, etc. to the system.

Figure 1 Step by step scheme of the nanoplatform preparation process: exfoliation, PEGylation and attachment of the specific antibody

In this study, we describe the functionalization process of the MoS_2 nanoplatforms (**Figure 1**) and demonstrate the internalization process utilizing a Madin-Darby Canine Kidney (MDCK) mammalian cell line, which is frequently applied in various biochemical models as it provides an ideal representation for epithelial tissues. We discuss the successful Internalization within the MDCK cells expressing carbonic anhydrase protein (CAIX), which is associated with tumor cell growth and survival [16].

2. EXPERIMENTAL

2.1. Liquid phase exfoliation

LPE in 45% solution

The MoS_2 exfoliation starts from 1 % powder dispersion in ethanol/water in 7:9 ratio (ethanol purchased from Merck, grade: ACS, ISO, Reag. Ph Eur). The pristine MoS_2 powder was purchased from Alfa Aesar (99 % metals basis purity, approx. 3 µm of size). The selected ethanol/water ratio has the most favorable Hansen solubility parameters for an increased concentration of exfoliated product [17]. The MoS_2 dispersion was sonicated in ultrasonic bath (Sonorex RK 510 H, 35 kHz, Bandelin) at the temperature of 10°C for 48 hours. Subsequently the solution was centrifuged for 1 h with RCF 41.000 g (Model 3-30K, Sigma centrifuge) at 20°C, then it was carefully decanted to separate the exfoliated part from the sediment and stored at room temperature (sample A). The concentration of sample A is 100 µg/ml.

LPE in DI water

As in the previous case, the MoS_2 exfoliation starts from powder MoS_2 (Alfa Aesar), 50 mg was dispersed in 20 ml deionized (DI) ultrapure water. The MoS_2 dispersion was sonicated in ultrasonic bath (Sonorex RK 510 H, Bandelin) at the temperature of 10°C for 48 hours. Subsequently the solution was centrifuged for 30 min with RCF 21.000 g (Model 3-30K, Sigma centrifuge) at 20°C. The supernatant was collected by micropipette and stored at room temperature (sample B). The concentration of the sample B nanoflakes are 40 µg/ml.

Amino acid intercalation assisted LPE

In this case we used an amino acid - arginine (Sigma Aldrich) - as an exfoliating agent [18]. 100 mg of arginine was dissolved in DI and sonicated for 30 min. 50 mg of pristine bulk MoS_2 powder (Alfa Aesar) was added to arginine-DI water mixture and stirred at 560 rot/min for 24 h utilizing a laboratory shaker. Subsequently, 10 ml of DI water was added to the mixture and sonicated for 20h at 10°C (Sonorex RK510H, Bandelin). Then the mixture was centrifuged for 30 min at RCF 21.000 g (Model 3-30K, Sigma centrifuge). The precipitate is dissolved in 20 ml of DI water, sonicated again for 20 h and centrifuged in a similar manner as before. The solution is decanted to separate the exfoliated arginine functionalized part (sample C) from the sediment. The sample is then stored at 6°C in a refrigerator unit. The concentration of sample C is 20 µg/ml.

2.2. MoS_2 functionalization

First, all three types of exfoliated MoS_2 were functionalized by PEG. In the case of sample A, the nanoflakes were transferred into water after drying out 20 ml of the product in a desiccator. Subsequently, the dried-out nanoflakes were dissolved in 10 ml of DI water and sonicated for 30 min. From this step the PEGylation process was identical for sample A and B and took place in aqueous environment. We used PEG molecules terminated with LA at one end and with biotin at the other (LA-PEG-biotin, MW 2000, Nanocs). 10 mg of LA-PEG-biotin dissolved in 5 ml of DI water was added to 10 ml of the nanoflakes. The mixtures were stirred for 1 h on a laboratory shaker, then dialyzed for 72 h with a dialysis membrane of molecular weight cut-off (MWCO) 12-14 kDA (Standard RC Tubing Spectra/Por®2, SpectrumLabs). In the case of sample C, the PEGylation process required modified environment: 10 mg EDC (98+%, Acros Organics) and 10 mg of sulfo-NHS (98 %, Sigma-Aldrich) were dissolved in 5 ml of DI water. Additionally, 5 mg of poly(ethylene glycol) 2-aminoethyl ether biotin (NH_2-PEG-biotin, MW 5300, Sigma-Aldrich) was dissolved in the same amount of DI water. 5 ml of sample C was placed into sonicating bath. First, the EDC/sulfo-NHS was pipetted into the sample, and then the PEG solution was added. Subsequently, the mixture was left in the sonication bath for 2 h. Afterwards, it was dialyzed for 72 h with dialysis membrane of MWCO 25 kDA (Pre-wetted RC Tubing Spectra/Por®6, SpectrumLabs).

Second, a biotinylated anti-human carbonic anhydrase IX (M75 [19]) was attached to the nanoplatform through a biotin-avidin bridge. 1 mg of avidin (GBiosciences) was diluted in 1 ml of DI water. 1 µl of the biotinylated antibody solution (160 kDA, 0,98 µg/µl) was mixed into 10 ml of PEGylated nanoplatform solution. The mixture was placed into a laboratory shaker (180 rot/min) and 1 µl of the avidin solution was dropped into this mixture. In the case of the samples that were prepared for confocal laser scanning microscopy (CLSM) measurements without M75 antibody, fluorescently labelled streptavidin (ATTO532 streptavidin, ATTO-TEC) was added in the same concentration instead avidin.

2.3. Cell preparation

MDCK epithelial cell line (ATCC No. CCL-34) was cultivated using VLE Dulbecco's MEM medium containing 2 mM L-glutamine (Biochrom) with 10 % fetal bovine calf serum (FCS, Biochrom) and gentamicin 80 µg/ml (Lek). MDCK cells permanently transfected with full-length human carbonic anhydrase (CAIX) cDNA in the pSG5c plasmid (marked as MDCK-CAIX) were used in our experiments, mock-transfected cells (marked as MDCK-neo) being a negative control [16]. 200 000 cells were incubated in humidified atmosphere at 37°C in the presence of 5 % CO_2 attached to a glass coverslip (0.17 mm, Zeiss) in a 3.5 cm culture dish (Greiner Bio One). After 24 h, the solution of bio-conjugated MoS_2 nanoplatforms was added to the cells (500 µl to 1 ml of medium) and incubated further for 24 h. Viability (BioTek, Synergy H4 microplate reader) was tested on the MDCK epithelial cell line with 3 different concentrations (250 µl, 500 µl and 750 µl to 1 ml of the medium) of exfoliated nanoflakes after 24 h, 48 h and 72 h exposure. The results show no cytotoxicity, suggesting that the MoS_2 nanoflakes are not affecting the cell cycle of the examined samples.

In the case of samples that were prepared for CLSM measurements with M75 antibody, the MoS_2 nanoplatforms were first incubated with donkey anti-mouse IgG (H+L) secondary antibody conjugated with fluorescent tag Alexa Fluor 488 (Thermo Fisher Scientific) for 1 hour at 37°C and then they were added to the cells and incubated further. After the incubation phase the cells were fixed with methanol (5 min at -20°C), and then they were mounted on a glass slide with mounting medium Fluoroshield with DAPI (Sigma-Aldrich).

2.4. Cellular intake study

Confocal Raman Microscopy

Confocal Raman microscope (CRM, Alpha300 R+, WITec) complemented with an immersion objective (W Plan-Apochromat 63×, NA=1, Zeiss, Germany) was utilized on living cells in PBS (pH=7.4, Oxoid). The samples were excited at the wavelength of 532 nm and laser power of 5 mW (Spectra-Physics Excelsior 532-60 Multi Mode). The Raman spectra were acquired with a spectrometer (UHTS 300, WITec) equipped with a 600 gr/mm grating (blazed at 500 nm) and coupled to an EMCCD camera (Newton DU970N-BV-353, Andor).

Confocal Laser Scanning Microscopy

The CLSM (LSM 510 Meta, Zeiss) measurements were carried out on fixed cells using an oil immersion objective (Ec Plan Neofluar 40×/1.30 Oil DIC M27, Zeiss). The CLSM scanning unit was mounted on a Axiovert 200M inverted microscope (Zeiss). Images were scanned at the scan speed 6, 1024 x 1024 pixels, 12bit data depth with average mode 4x line.

3. RESULTS AND DISCUSSION

We examined the internalization of MoS_2 nanoplatforms with and without M75 antibody in MDCK-neo and MDCK-CAIX cell lines. All three samples A, B and C were terminated with either M75 or fluorescently tagged streptavidin.

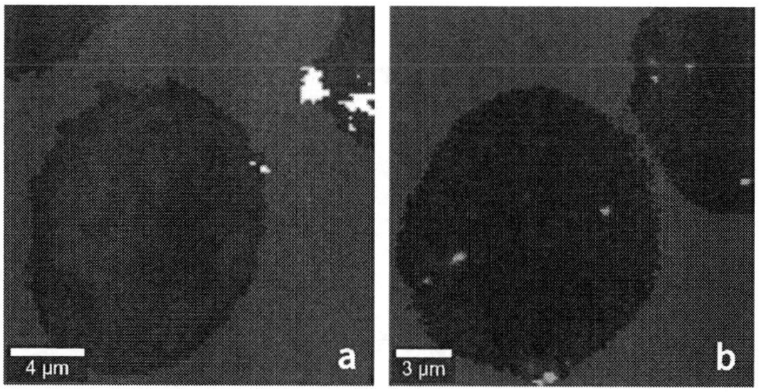

Figure 2 Localization of MoS_2 nanoplatforms (marked by white color) of sample A without M75 antibody on MDCK neo cells (a) and with M75 antibody internalized inside MDCK-CAIX cells (b) using CRM

By means of CRM imaging, we tracked the location of MoS_2 nanoplatforms. The MoS_2 Raman modes (E^1_{2g} and A_{1g}), the CH_2 stretch mode (2854 cm^{-1}) of lipids, fatty acids and proteins and the symmetric CH_3 stretch mode of proteins (2930 cm^{-1}) were used to determine whether the examined nanoplatform is localized inside the cell, on the cell membrane or outside the cell. For this purpose, we studied the z-stack composed of several cross sections through the inspected cell. **Figure 2** represents a typical CRM image showing a cell cross section with the MoS_2 nanoflakes highlighted in white. The nanoflakes shown in **Figure 2a** were found to be stopped on the cell membrane while the nanoflakes in **Figure 2b** passed through the cell membrane and were internalized. Our CRM measurements suggest that if the nanoplatforms are internalized (with or without M75), they are found deeper inside the cell rather than close to the membrane.

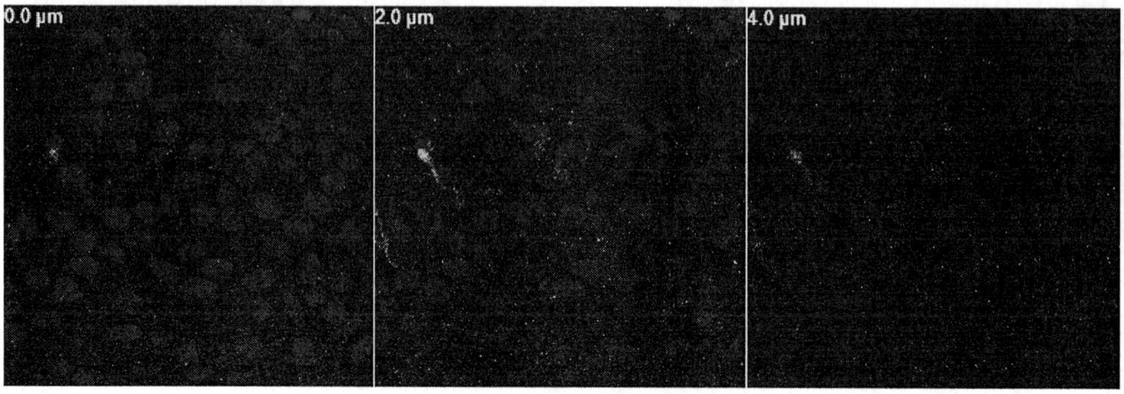

Figure 3 Selected CLSM z-stack images of PEGylated sample A scanned from the top (z=0 µm) of the cell to the glass slide (z=4 µm) with M75 in MDCK-CAIX cells. Cell nuclei and MoS_2 nanoplatforms are marked with blue and green color, respectively.

As mentioned in the introduction, other articles reported the successful internalization of PEGylated MoS_2 nanoflakes. The focus of our study is whether the specificity of the antigen-antibody bonding increases the number of internalized nanoflakes in cells expressing CAIX. Based on the CLSM z-stack images (**Figure 3**) we determined the ratio of the cells with internalized nanoplatforms to the cells without them. The internalization results are displayed in **Table 1**: for all three samples, we observed an increased number of internalized M75 bio-conjugated nanoplatforms when incubated with MDCK-CAIX compared to MDCK-neo. The results also show an increased probability of internalization in the latter for the ethanol and arginine assisted LPE produced samples without M75 (samples A, C).

Table 1 Evaluation of the nanoplatform internalization into MDCK-CAIX and MDCK-neo in case of nanoplatforms with and without M75 based on CLSM

Nanoplatform	CAIX	With M75	Without M75
Sample A	✓	12 %	0 %
	✗	3 %	5 %
Sample B	✓	7 %	3 %
	✗	3 %	3 %
Sample C	✓	12 %	5 %
	✗	4 %	9 %

4. CONCLUSION

In conclusion, the application of MoS_2 based nanoplatform for the next generation of highly selective anti-cancer drug nanocarrier was demonstrated. The nanoplatform internalization was tracked and validated on cellular level by the label-free Raman imaging and CLSM. The increased selectivity of the antibody conjugated nanoplatforms was confirmed by the increased internalization in CAIX expressing cells. Further studies are required to improve the selectivity to cancer cells and validate the observed internalization in different cell lines.

ACKNOWLEDGEMENTS

This research was supported by the Slovak Research and Development Agency under the projects APVV-15-0641 and APVV-14-0120.

REFERENCES

[1] GURUNATHAN, S., KIM, J. H. Synthesis, Toxicity, Biocompatibility, and Biomedical Applications of Graphene and Graphene-Related Materials. *Int J Nanomedicine* 2016, *11*, pp. 1927-1945.

[2] DONG, H., DONG, C., REN, T., LI, Y., SHI, D. Surface-Engineered Graphene-Based Nanomaterials for Drug Delivery. *J. Biomed. Nanotechnol.* 2014, *10*, pp. 2086-2106.

[3] BAO, H. Q., PAN, Y. Z., PING, Y., SAHOO, N. G., WU, T. F., LI, L., LI, J., GAN, L. H. Chitosan-Functionalized Graphene Oxide as a Nanocarrier for Drug and Gene Delivery. *Small* 2011, *7*, pp. 1569-1578.

[4] ROBINSON, J. T., TABAKMAN, S. M., LIANG, Y. Y., WANG, H. L. *et al.* Ultrasmall Reduced Graphene Oxide with High Near-Infrared Absorbance for Photothermal Therapy. *J. Am. Chem. Soc.* 2011, *133*,pp. 6825-6831.

[5] YANG, H. W., LU, Y. J., LIN, K. J., HSU, S. C. *et al.* EGRF Conjugated PEGylated Nanographene Oxide for Targeted Chemotherapy and Photothermal Therapy. *Biomaterials* 2013, *34*, pp. 7204-7214.

[6] CHOU, S. S., KAEHR, B., KIM, J., FOLEY, B. M., DE, M., HOPKINS, P. E. *et al.* Chemically Exfoliated MoS2 as Near-Infrared Photothermal Agents. *Angew Chem Int Ed Engl* 2013, *52*, pp. 4160-4164.

[7] YU, J., YIN, W. Y., *et al.* Smart MoS2/Fe3O4 Nanotheranostic for Magnetically Targeted Photothermal Therapy Guided by Magnetic Resonance/Photoacoustic Imaging. *Theranostics* 2015, *5*, pp. 931-945.

[8] LIU, T., WANG, C., CUI, W., GONG, H., LIANG, C., SHI, X. Z., LI, Z. W., SUN, B. Q., LIU, Z. Combined Photothermal and Photodynamic Therapy Delivered by PEGylated MoS2 Nanosheets. *Nanoscale* 2014, *6*, pp. 11219-11225.

[9] FENG, W., CHEN, L., QIN, M., ZHOU, X., ZHANG, Q., MIAO, Y., QIU, K., ZHANG, Y., HE, C. Flower-like PEGylated MoS2 Nanoflakes for near-Infrared Photothermal Cancer Therapy. *Sci Rep* 2015, *5*, pp. 17422.

[10] WANG, S. G., LI, K., CHEN, Y. *et al.* Biocompatible PEGylated MoS2 Nanosheets: Controllable Bottom-up Synthesis and Highly Efficient Photothermal Regression of Tumor. *Biomaterials* 2015, *39*, pp. 206-217.

[11] TEO, W. Z., CHNG, E. L., SOFER, Z., PUMERA, M. Cytotoxicity of Exfoliated Transition-Metal Dichalcogenides (MoS2 , WS2 , and WSe2) Is Lower than That of Graphene and Its Analogues. *Chemistry (Easton).* 2014, *20*, pp. 9627-9632.

[12] APPEL, J. H., LI, D. O., PODLEVSKY, J. D., DEBNATH, A., GREEN, A. A., WANG, Q. H., CHAE, J. Low Cytotoxicity and Genotoxicity of Two-Dimensional MoS2 and WS2. *ACS Biomater. Sci. Eng.* 2016, *2*, 361-367.

[13] LIU, T., WANG, C., GU, X., GONG, H., CHENG, L., *et al.* Drug Delivery with PEGylated MoS2 Nano-Sheets for Combined Photothermal and Chemotherapy of Cancer. *Adv. Mater.* 2014, *26*, pp. 3433-3440.

[14] WILCHEK, M., BAYER, E. A. The Avidin-Biotin Complex in Immunology. *Immunol. Today* 1984, *5*, pp. 39-43.

[15] WILCHEK, M., BAYER, E. A. The Avidin-Biotin Complex in Bioanalytical Applications. *Anal. Biochem.* 1988, *171*, pp. 1-32.

[16] PASTOREK, J., PASTOREKOVA, S., CALLEBAUT, I. *et al.* Cloning and Characterization of MN, a Human Tumor-Associated Protein with a Domain Homologous to Carbonic Anhydrase and a Putative Helix-Loop-Helix DNA Binding Segment. *Oncogene* 1994, *9*, pp. 2877-2888.

[17] ZHOU, K.-G., MAO, N.-N., WANG, H.-X., PENG, Y., ZHANG, H.-L. A Mixed-Solvent Strategy for Efficient Exfoliation of Inorganic Graphene Analogues. *Angew. Chemie Int. Ed.* 2011, *50*, pp. 10839-10842.

[18] SATHEESHKUMAR, E., BANDYOPADHYAY, A., *et al.* M. One-Step Simultaneous Exfoliation and Covalent Functionalization of MoS2by Amino Acid Induced Solution Processes. *ChemNanoMat* 2017, *3*, pp. 172-177.

[19] PASTOREKOVA, S., ZAVADOVA, Z., KOSTAL, M., BABUSIKOVA, O., ZAVADA, J. A Novel Quasi-Viral Agent, MaTu, Is a Two-Component System. Virology 1992, 187, pp. 620-626.

ADHESION AND GROWTH OF ADIPOSE TISSUE-DERIVED STEM CELLS ON FIBRIN ASSEMBLIES WITH ATTACHED GROWTH FACTORS FOR TISSUE ENGINEERING OF HEART VALVES

[1]Elena FILOVA, [1]Martina TRAVNICKOVA, [1]Jarmila KNITLOVA,[1,2]Roman MATEJKA, [3]Johanka KUCEROVA, [3]Zuzana RIEDELOVÁ, [3]Eduard BRYNDA, [1]Lucie BACAKOVA

[1]Department of Biomaterials and Tissue Engineering, Institute of Physiology of the Czech Academy of Sciences, Czech Republic, elena.filova@fgu.cas.cz, martina.travnickova@fgu.cas.cz, jarmila.knitlova@fgu.cas.cz, lucie.bacakova@fgu.cas.cz

[2]Faculty of Biomedical Engineering, Czech Technical University in Prague, Czech Republic, roman.matejka@fgu.cas.cz

[3]Department of Bioactive Polymers, Institute of Macromolecular Chemistry of the Czech Academy of Sciences, Czech Republic, kucerova@imc.cas.cz, riedelova@imc.cas.cz, brynda@imc.cas.cz

Abstract

Currently used xenogeneic biological heart valve prostheses are decellularized and crosslinked with glutaraldehyde. These grafts usually undergo degeneration and calcification. Pericardium-based heart valve prostheses, re-seeded with autologous cells, i.e. adipose tissue-derived cells (ASCs) and endothelial cells, could have longer durability and biocompatibility. In order to improve the adhesion of cells and their ingrowth into decellularized pericardium, various fibrin (Fb) layers were developed, i.e. Fb, Fb with covalently bound heparin (H), Fb with either vascular endothelial growth factor (VEGF) or fibroblast growth factor 2 (FGF) in various concentrations (1 ng/ml, 10 ng/ml, 100 ng/ml) or with both VEGF and FGF (100 ng/ml). Growth factors were attached onto Fb *via* heparin or were adsorbed. ASCs were seeded on theses layers in a DMEM medium supplemented with 2 % of fetal bovine serum, TGFβ1 and BMP-4 (both 2.5 ng/ml), and with ascorbic acid. Cell adhesion and growth/viability was assessed by counted cell number/MTS evaluation. ASCs were stained for differentiation markers of smooth muscle cells, such as alpha-actin, calponin, and myosin heavy chain. On day 7, ASCs on Fb_H_VEGF layers produced both calponin and alpha-actin. An increased FGF concentration caused reduced calponin staining of ASCs. Lack of heparin in fibrin assemblies with growth factors inhibited the production of both alpha-actin and calponin in ASCs. The highest ASCs density/viability was found on Fb_H_VEGF_FGF layer. The proper formulation of fibrin coatings could be favorable for ASCs growth and differentiation and could subsequently support endothelialization of cardiovascular prostheses with endothelial cells.

Keywords: Adipose tissue-derived stem cells, heart valves, fibrin assemblies, growth factors

1. INTRODUCTION

Currently used bovine pericardial prostheses or patches are glutaraldehyde-treated or decellularized. Glutaraldehyde (GA) treatment reduces host immune response, and improves the graft durability and mechanical stability. On the other hand, a long-term release of GA from the tissue results in the graft toxicity, macrophage activation, calcification, fibrotic deposition, chronic inflammation, and degeneration [1,2,3]. Decellularization improves pericardium biocompatibility by removing immunogenic and pro-calcific elements and, concurrently, it preserves extracellular matrix. Decellularized pericardial CardioCel patches (Admedus Regen Pty Ltd, Perth, Western Australia) [4], which were rapidly crosslinked with 0.6 % GA, allowed intima formation, neovascularization, and remodelling after their implantation in patients. Controlled recellullarization and endothelialization could improve biological properties of the grafts, reduce their degeneration, prolong their durability and could allow the graft growth in children. The choice of available cell type, optimizing cell culture conditions, composition of cell culture medium, and modifications of the scaffold structure are important factors for rapid, homogeneous recellularization and directed cell differentiation. Valve interstitial cells, present in heart valves have a phenotype partially similar to fibroblasts and smooth muscle cells (SMCs). Adipose tissue-derived stem cells (ASCs) are able to differentiate into SMCs, which make them a suitable candidate for recellullarization of pericardium scaffolds [5]. The way of growth factor (GF) attachment onto the graft surface, their concentration or sustained release, or even their combination, are important factors for recellularization. In the present study, various fibrin layers with adsorbed/attached FGF and/or VEGF have been developed; heparin was used for the long-term GF binding. We evaluated modified fibrin layers in term of ASCs adhesion, proliferation, and their differentiation towards SMC phenotype.

2. MATERIALS AND METHODS

2.1. Material preparation

Glass bottom wells in a cell culture plate (24- Well Glass Bottom Plate, Cellvis) were coated with a fibrin layer using technique described earlier [6]. The wells were incubated at 4°C with fibrinogen (5 µg/ml in Tris-HCl buffer, pH 7.4 (TB), (Sigma-Aldrich) overnight. Then, thrombin (2 U/ml in TB) (Sigma-Aldrich) was added for 30 min at room temperature. Subsequently, the wells were incubated with a mixture of antithrombin III (0.5 U/ml in TB) (Chromogenix) and fibrinogen (200 µg/ml in TB) for 2 hours. Heparin was activated by a procedure published earlier [7]. Heparin sodium salt from porcine intestinal mucosa (14 mg/ml in PBS) was oxidized by NaIO4 (2.4 mg/ml) (Sigma-Aldrich) in the dark for 90 minutes. The periodate was removed by ultrafiltration using Amicon Ultra Centrifugal filter unit with a molecular weight cut-off 3,000 Da. Activated heparin solution was diluted in the citrate buffer and immediately added to the wells and incubated overnight. In the last step, different concentration, i.e. 1 ng/ml, 10 ng/ml, and 100 ng/ml of human FGF (FGF-basic 154aa) and human VEGF 165 (GenScript) in PBS were added to the wells coated with only Fb or Fb with attached heparin) for 2 hours. Finally, the wells were three times washed with PBS, UV irradiated for 15 min, and seeded with cells. Glass cover slips, pure Fb and Fb_H were used as control samples.

2.2. Cell culture conditions

Human adipose tissue-derived stem cells (ASCs) were isolated from human lipoaspirates in compliance with the tenets of the Declaration of Helsinki for experiments involving human tissues and under ethical approval issued by the Ethics Committee in the Bulovka Hospital in Prague (August 28, 2014) [8,9]. Flow cytometry analysis revealed that the ASCs were positive for cell surface antigens CD73, CD90, CD105, and CD29, they were negative for CD31, CD34, and CD45, and about 3 % of the cells were positive for CD146.

The ASCs (passages 2-3) were seeded onto Fb layers at the density of 25,000 cells per well in DMEM supplemented with 2 % FBS, transforming growth factor β1 (ab50036, Abcam), and bone morphogenetic protein 4 (SRP6156-10UG, Sigma), both 2.5 ng/ml, 2-phospho - L - ascorbic acid trisodium salt (50 µg/ml 49752, Sigma), and cultured for 7 days.

2.3. Evaluation of cell adhesion, growth and differentiation, statistical evaluation

On day 1, the phase contrast images were taken using Olympus IX 51 microscope, equipped with DP70 digital camera from 30 microscopic fields (from 3 individual samples), and cells numbers were counted from the micrographs. On day 7, cell viability/metabolic activity was assessed by MTS assay (The CellTiter 96® AQueous One Solution Cell Proliferation Assay, Promega). 100 µl of MTS reagent was added into 500 µl of cell culture medium and cells were incubated for 45 min. Fluorescence intensity of 150 µl of the samples was measured at 490 nm in triplicates from 3 individual samples.

For evaluation cell number, ANOVA on ranks, Dunn's Method was used. Data were presented as median ± quartiles; p value ≤ 0.05 was considered significant. For MTS analysis, One way ANOVA, the Student-Newman Keuls method was used, data are expressed as mean + standard deviation.

Immunofluorescence staining of alpha-actin, calponin, and myosin heavy chain was performed. Briefly, the samples were incubated with solution containing 1 % bovine serum albumin (A9418, Sigma) and 0.1 % Triton X-100 (Sigma) for 20 min, washed with phosphate-buffered saline (PBS), and incubated with 1 % Tween 20 for 20 min. After washing with PBS, the primary antibodies (dilution 1:200) were applied overnight at 4°C, i.e. mouse monoclonal α-smooth muscle actin antibody (99A2547, Sigma), rabbit monoclonal anti-calponin antibody (EP798Y, ab46794, Abcam), mouse monoclonal anti-MYH11 (G-4) antibody (sc-6956, Santa Cruz Biotechnology, Inc.). After washing samples twice with PBS, secondary antibodies were applied for 1 hour at RT, i.e. Alexa Fluor 488 F(ab`)2 fragment of goat anti-rabbit IgG (H+L) (A11070), Alexa Fluor 546 goat anti-mouse IgG (H+L) (A11003), Alexa Fluor 488 F(ab`)2 fragment of goat anti-mouse IgG (H+L) (A11017) (dilution 1:400, Thermo Fisher Scientific). Cell nuclei were counterstained with Hoechst 33258 (Sigma). Finally, the samples were washed twice with PBS and the images were taken under IX71 Olympus epifluorescence microscope, equipped with DP71 digital camera.

3. RESULTS AND DISCUSSION

The morphology of ASCs attached to Fb layers was polygonal with long and numerous protrusions - filopodia (**Figure 1**), which was probably caused by low stiffness of the fibrin layers. On hard glass surface, the cells seemed to be better spread with smaller filopodia. Similarly, in a study by Wong et al [10], performed on a micropatterned substrate composed of rigid micro-islets surrounded by a soft gel, fibroblasts attached to the stiff islets used filopodia to test rigidity of the surrounding gel. On the rigid islets, the filopodia were stable and allowed cell spreading. On the contrary, on soft surfaces, the filopodia were less stable and tended to retract. In another study, NIH 3T3 cells evolved higher traction forces, higher migration speed, and they had higher spreading area on stiffer surfaces; Young's modulus was 300 and 140 kdyn/cm^2 for stiff and soft surface, respectively [11].

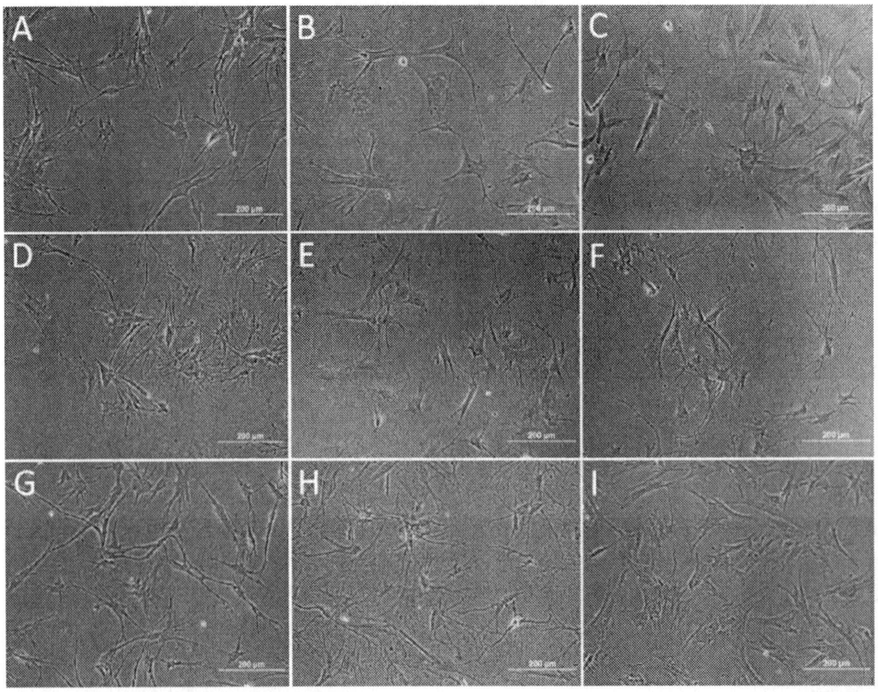

Figure 1 Phase contrast images of ASCs on Fb (A), Fb_VEGF (B), Fb_FGF (C), Fb_H (D), Fb_H_VEGF (E), Fb_H_FGF (F), Fb_VEGF_FGF (G), Fb_H_VEGF-FGF (H), and on glass coverslips (I) on day 1 after seeding. Olympus IX 51 microscope, DP70 digital camera, objective ×10

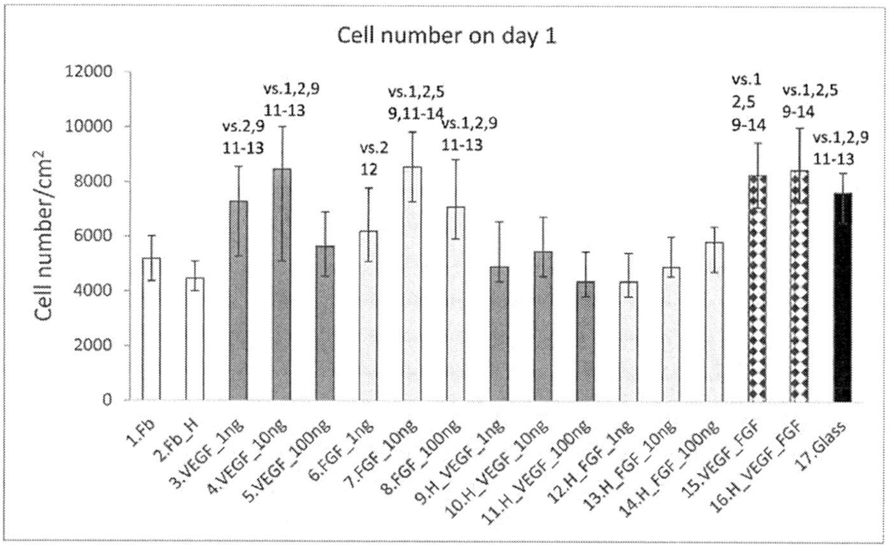

Figure 2 Number of ASCs on Fb layers modified by the attachment of H, VEGF, and FGF on day 1, measured from 30-31 microscopic fields. Data expressed as median ± quartiles. Statistical significance compared to the samples of the same number is displayed above the columns. P value ≤ 0.05 is considered significant.

High densities of ASCs were observed after one day on Fb modified with adsorbed VEGF and FGF (both 10 ng/ml) without heparin and on Fb modified with a combination of both growth factors regardless heparin presence (**Figure 2**). Heparin generally decreased cell adhesion, except for Fb_H_VEGF_FGF layer.

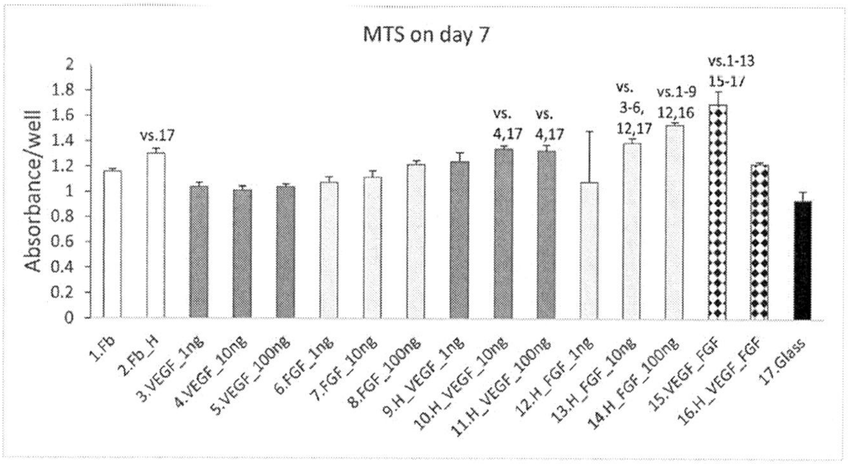

Figure 3 Viability/metabolic activity of ASCs on fibrin Fb layers modified by the attachment of H, VEGF, and FGF on day 7, measured from 3 samples. Data expressed as mean + standard deviation. Statistical significance compared to the samples of the same number is displayed above the columns.

Cell viability/proliferation was assessed by MTS assay (**Figure 3**) on day 7. The highest values were found on Fb_VEGF_FGF, slightly lower on Fb with either heparin-bound FGF or heparin-bound VEGF at the concentration of 100 ng/ml and 10 ng/ml. On glass and on Fb with adsorbed growth factors, ASCs proliferation was significantly lower. On Fb_H, cell viability was higher than on glass. On another study, heparin immobilized on chitosan films decreased smooth muscle cells proliferation, while heparin immobilized on chitosan-collagen films improved their proliferation increased [12]. We suppose that attachment of growth factors via heparin decreased their degradation and prolonged their activity compared to the Fb samples with just adsorbed GFs.

ASCs differentiation towards smooth muscle cells (SMCs) can be evaluated by the presence of early SMC markers, i.e. SM alpha-actin, myocardin, and SM22-α, medium-term markers, such as calponin and H-caldesmon, and late markers, such as meta-vinculin, smoothelin-B, myosin heavy chain 11, and desmin [5,8,9]. We assessed the presence of SM alpha-actin, calponin (**Figure 4**) and myosin heavy chain 11. The cells were apparently positively stained for SM alpha actin on all H-containing samples, on Fb, Fb_VEGF_FGF, and on glass. The presence of adsorbed growth factors reduced the production of alpha-actin in ASCs. This might be caused by different kinetics of GF release and degradation. Similarly, staining of calponin showed its higher production in ASCs growing on Fb with covalently-bound heparin, and among them on Fb_H_VEGF_10ng. The Fb_H_FGF surfaces, especially those with the highest concentration of FGF, are less convenient for stimulation of ASCs differentiation towards SMCs. The staining for myosin heavy chain 11 was negative for cells on all samples.

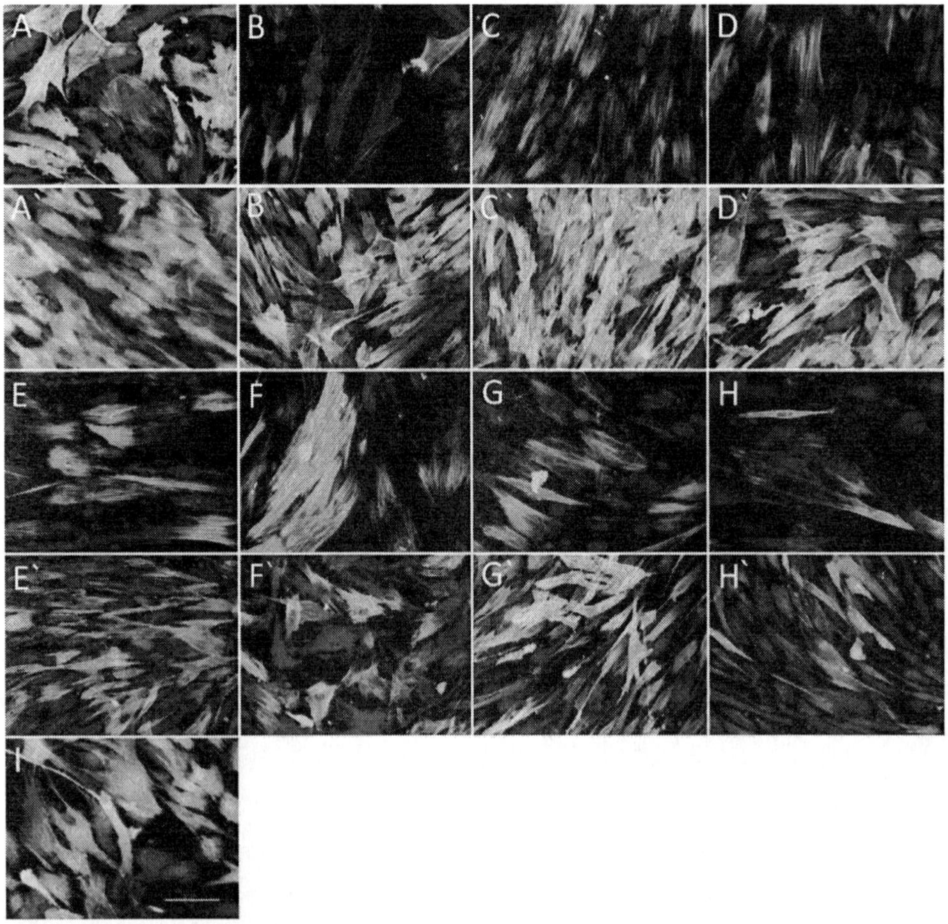

Figure 4 Differentiation markers of smooth muscle alpha actin (red) and calponin (green) in ASCs on various modified Fb layers as follows: Fb (A), Fb_H (A`), Fb_VEGF_1ng (B), Fb_H_VEGF_1ng (`B), Fb_VEGF_10ng (C), Fb_H_VEGF_10ng (C`), Fb_VEGF_100ng (D), Fb_H_VEGF_100ng (`D), Fb_VEGF_FGF (E), Fb_H_VEGF_FGF (E`), Fb_FGF_1ng (F), Fb_H_VEGF_FGF_1ng (F`), Fb_FGF_10ng (G), Fb_H_FGF_10ng (G`), Fb_FGF_100ng (H), Fb_H_FGF_100ng (H`), and glass (I), obj. × 20, scale bar = 100 μm. Cell nuclei counterstained with Hoechst 33258 (blue).

4. CONCLUSIONS

Various fibrin layers with either adsorbed or heparin-bound VEGF and/or FGF were prepared and their effect on ASCs adhesion, proliferation, and differentiation was assessed. We have observed that Fb samples with attached growth factors *via* heparin mostly decreased initial cell adhesion of ASCs. However, these samples supported cell viability/metabolic activity on day 7, and early differentiation towards smooth muscle cells. Heparin-bound VEGF or combination of both VEGF and FGF seems to be more effective than heparin-bound FGF for improving cell differentiation towards SMC phenotype.

ACKNOWLEDGEMENTS

Supported by the Grant Agency of the Czech Republic (Centre of Excelence P108/12/G108) and Ministry of Health of the Czech Republic (Grant No: 15-29153A).

REFERENCES

[1] Umashankar, P.R., Mohanan, P., Kumari, T. Glutaraldehyde treatment elicits toxic response compared to decellularization in bovine pericardium. Toxicology International 2012, vol.19, no.1, pp.51-58.

[2] Iop, L., Palmosi, T., Sasso, E.D., Gerosa, G. Bioengineered tissue solutions for repair, correction and reconstruction in cardiovascular surgery. Journal of Thoracic Disease. J Thorac Dis 2018, vol.10, no. Suppl 20, pp. S2390-S2411

[3] Aguiari, P., Iop, L., Favaretto, F., Fidalgo, C.M., Naso, F., Milan, G., Vindigni, V., Spina, M., Bassetto, F., Bagno, A., Vettor, R., Gerosa, G. In vitro comparative assessment of decellularized bovine pericardial patches and commercial bioprosthetic heart valves. Biomed Mater [online] 2017. vol. 12, iss. 1. pp. 015021, [viewed 2018-09-28]. From: doi: 10.1088/1748-605X/aa5644.

[4] Prabhu, S., Armes, J.E., Bell., D., Justo, R., Venugopal, P., Karl, T., Alphonso, N. Histologic Evaluation of Explanted Tissue-Engineered Bovine Pericardium (CardioCel). Seminars in Thoracic and Cardiovascular Surgery 2017, vol. 29, no.3, pp. 356-363.

[5] Bacakova, L., Travnickova, M., Filova, E., Matejka, R., Stepanovska, J., Musilkova, J., Zarubova, J., Molitor, M. In: Muscle Cell and Tissue, Kunihiro Sakuma, ed. IntechOpen, London, United Kingdom, ISBN 978-953-51-6092-2, 2018, Vascular smooth muscle cells (VSMCs) in blood vessel tissue engineering: the use of differentiated cells or stem cells as VSMC precursors, in press.

[6] RIEDEL, T., BRYNDA, E., DYR, J.E., HOUSKA M. Controlled preparation of thin fibrin films immobilized at solid surfaces. Journal of Biomedical Materials Research Part A. 2009, vol. 88, no. 2, pp. 437-447

[7] Kaplan, O., Hierlemann, T., Krajewski, S., Kurz, J., Nevoralová, M., Houska, M., Riedel, T., Riedelová, Z., Zárubová, J., Wendel, H.P., Brynda, E. Low-thrombogenic fibrin-heparin coating promotes in vitro endothelialization. Journal of Biomedical Materials Research Part A. 2017, vol. 105, No. 11, pp. 2995-3005.

[8] Przekora, A., Vandrovcova M., Travnickova, M., Pajorova, J., Molitor, M., Ginalska, G., Bacakova, L. Evaluation of the potential of chitosan/β-1,3-glucan/hydroxyapatite material as a scaffold for living bone graft production in vitro by comparison of ADSC and BMDSC behaviour on its surface. Biomed Mater [online] 2017. vol 12, no.1, 015030, [viewed 2018-09-28]. From: doi: 10.1088/1748-605X/aa56f9.

[9] Bacakova, L., Zarubova, J., Travnickova, M., Musilkova, J., Pajorova, J., Slepicka, P., Slepickova Kasalkova, N., Svorcik, V., Kolska, Z., Motarjemi, H., Molitor, M. Stem cells: their source, potency and use in regenerative therapies with focus on adipose-derived stem cells - a review. Biotechnol Adv [online] 2018. vol. 36, iss. 4, pp. 1111-1126 [viewed 2018-09-28]. From: doi: 10.1016/j.biotechadv.2018.03.011.

[10] Wong, S., Guo, W.H., Wang, Y.L. Fibroblasts probe substrate rigidity with filopodia extensions before occupying an area. Proceedings of the National Academy of Sciences of the United States of America. 2014, vol.111, no. 48, pp. 17176-17181.

[11] Lo, C.M., Wang, H.B., Dembo, M., Wang, Y.L. Cell movement is guided by the rigidity of the substrate. Biophysical Journal 2000, Vol. 79, No. 1, pp. 144-152

[12] Robu, I.S., Walters, H.L. 3rd, Matthew, H.W.T. Morphological and growth responses of vascular smooth muscle and endothelial cells cultured on immobilized heparin and dextran sulfate surfaces. Journal of Biomedical Materials Research Part A 2017, vol.105, no. 6, pp. 1725-1735.

RELATION BETWEEN RHEOLOGICAL, STRUCTURAL AND DISSOLUTION PROPERTIES OF COVALENTLY AND IONICALLY MODIFIED HYALURONIC ACID-BASED DRUG CARRIERS

[1]Ádám JUHÁSZ, [1]Norbert VARGA, [1]Árpád TURCSÁNYI, [2]Edit CSAPÓ

[1]University of Szeged, Faculty of Science and Informatics, Department of Physical Chemistry and Material Sciences, Szeged, Hungary, EU, Juhasz.Adam@stud.u-szeged.hu
[2]University of Szeged, Faculty of Medicine, MTA-SZTE Biomimetic Systems Research Group, Szeged, Hungary, EU, juhaszne.csapo.edit@med.u-szeged.hu

Abstract

This paper presents some findings from the rheological and drug dissolution characterizations of hyaluronic acid (HyA)-based colloidal drug delivery systems. During this investigations the widely applied, non-steroidal, anti-inflammatory ketoprofen (KP) has been used as model drug. For the moderation of the hydrophilic character of HyA and encapsulation of the hydrophobic drug the cross-linked derivatives at different cross-linking ratio have been synthesized. To achieve the increasing encapsulation efficiency, the hydrophobized variants of HyA have also been prepared in aqueous medium by modifying the structure with positively charged cetyltrimethylammonium bromide (CTAB) at different polymer/surfactant mass ratios. It was established that the original coherent gel-like structure of the hydrophilic and well-solvated polymer chains change into an incoherent colloidal system due to modifications that were verified by rheological investigations. Taking into account the dissolution profiles of covalent and ionic systems it can be concluded that nearly 70 % of the encapsulated KP molecules dissolve from the totally cross-linked HyA systems after 8h, while the dissolved amount of KP is about 20 % from the surfactant modified colloidal system at near 1:1 molar ratio. It was also established that increasing the amount of surfactant the drug release turns to diffusion and erosion controlled way. It has been confirmed that the ionically modified HyA may be a potential candidate for controlled drug release of KP.

Keywords: Hyaluronic acid, ketoprofen, encapsulation, controlled drug release

1. INTRODUCTION

Novel colloidal drug delivery systems have become one of the most fascinating research areas in modern pharmaceutical developments. Numerous biocompatible macromolecules such as polyesters, proteins, polysaccharides, polyelectrolites, lipids *etc.* are used as drug carriers in order to achieve a controlled drug delivery and drug release process.[1] The hyaluronic acid is a well-known linear polysaccharide of alternating units of β-1,4-D-glucuronic acid and β-1,3-N-acetyl-D-glucosamine. [2] Thanks to the biocompatible, biodegradable, non-toxic, non-immunogenic and non-inflammatory features this polymer is a perfect candidate for a number of medical and pharmaceutical applications. At physiological conditions the mentioned monomer units have a negative charge (hyaluronate form) which results in a water soluble nature. Because of this hydrophilic properties of HyA it is present in several organic fluids, the highest mass being found in the extracellular matrix of the soft connective tissues. The central drawback of this hydrophilic character is that HyA molecules, without chemical modification, cannot be applied as colloidal drug delivery systems. Numerous techniques have been developed for the construction of cross-linked HyA, but one of the frequently used method is the carbodiimde technique.[2] An alternative technique is the neutralization of HyA by the creation of electrostatic interactions using cationic surfactants like cetyltrimethylammonium bromide (CTAB), which also results in associated colloid structure having hydrophobic nature.[3,4] In this work the hydrophobic KP, as model drug molecule, has been used to synthetize and characterize diverse types of HyA-based colloids for controlled drug release. Depending on the drug dissolution results the possible application of a biopolymer founded and nano sized drug delivery systems for encapsulation of a NSAID drug have been reported.

2. MATERIALS AND METHODS

2.1. Materials

Hyaluronic acid sodium salt (HyA, 4000 kDa) was obtained from Gedeon Richter Plc. Ketoprofen (KP; $C_{16}H_{14}O_3$; ≥98 %) and hexadecyltrimethylammonium bromide, sodium phosphate dibasic dodecahydrate and sodium phosphate monobasic monohydrate were obtained from Sigma-Aldrich. Sodium chloride from Molar Chemicals was used to prepare isotonic (150 mM) NaCl solution. For the cross-linking reaction 2,2'(ethylenedioxy)bis(ethylamine) and 1-[3-(dimethylamino)propyl]-3-ethylcarbodiimid methiodide (CDI) were purchased from Sigma-Aldrich. Highly purified water was obtained by deionization and filtration with a Millipore purification apparatus (18.2 MΩ·cm at 25 °C). All solvents and reagents used were of analytical grade and no further purification were made.

2.2. Experimental procedures of the covalent and ionic modifications

Cross-linked (cl) HyA derivatives were prepared according to the previously published procedure. [2] The synthesis was performed at room temperature. 200 mg HyA was dissolved in water to produce 1 mg/mL solution and then adjusted the pH to pH = 5.5. The stochiometric ratio of cross-linking was 50 %, 75 % and 100 % resulting in cl-HyA/50 %, cl-HyA/75 % and cl-HyA/100 % samples. Accordingly, 1.88 mL, 2.82 mL and 3.76 mL 2,2'(ethylenedioxy)bis(ethylamine) solution (1 v/v%, pH = 5.5) was added to the HyA solution and mixed for 30 min. 80 mg, 120 mg and 160 mg CDI was dissolved in water and added to the mixture drop by drop, respectively. After an overnight stirring the product was purified by dialysis for 7 days against distilled water and the aqueous solution of the final product was freeze-dried. For CTAB modification different calculated amount of surfactant was added to the aqueous solution of HyA to change the phobicity. The mixture was stirred for 30 min before further use.

2.3. Preparation of drug containing systems

Because of the low solubility of KP in pure Milli-Q water (c_{max} = 0.051 mg/mL) all drug containing samples were prepared in phosphate buffer solution (PBS) at pH = 7.4 at 25 °C using constant ionic strength (0.9 % NaCl). In all cases constant KP (c_{KP} = 20 mg/mL) and constant HyA concentrations (100.0-100.0 mg lyophilized cl-HyA/ml) have been used. According to the preparation procedures the aqueous KP solutions were added to the different individual cross-linked HyA and HyA/CTAB samples which resulted in the formation of gel-like structure after 24 h.

2.4. Structural characterizations

The particle size and zeta potential of the prepared systems were determined by Dynamic Light Scattering (DLS) with a Zetasizer Nano ZS ZEN 4003 apparatus (Malvern Ins., UK) equipped with a He-Ne laser (λ = 633 nm). The measurements were performed at 25 ± 0.1 °C with an angle detection of 173° in clear disposable zeta cell. In order to determine the maximum amount of CTAB to be added prior to precipitation 0.02 M of CTAB was added stepwise (20-20 µl/ step) to 0.2 mg/mL concentration of HyA in PBS and the electrokinetic potential values were registered by DLS and parallel with this measurement the turbidity data were also registered. Turbidity measurements were performed by a Precision Bench Turbidity Meter LP2000 (Hanna Ins.), while the conductivity was measured by a Radelkis OK- 114 conduktometer equipped with an electrode with sheet plates.

2.5. Rheological studies

All rheological measurements were performed using an Anton Paar Physica MCR 301 Rheometer (Anton Paar, GmbH, Germany) at 25 °C. The measuring system equipped with a 25 mm diameter parallel cone-plate geometry (CP25-1-SN12204), a double-gap- (DG26.7-SN12740) and a concentric cylinder geometry (CC27-SN12793). The rheometer utilized a temperature controlled water bath in combination with a Peltier heating system for accurate control.

2.6. In vitro drug release experiments

The spectrophotometric properties of KP was studied by using an UV-1800 (Shimadzu) double beam spectrophotometer, using 1 cm quartz cuvette in the range of 200-500 nm. The concentration of KP was determined at 260 nm. The *in vitro* drug release experiments were carried out in phosphate buffer (PBS, pH = 7.4) at 25 °C. A cellulose membrane (Sigma-Aldrich) was used as a dialysis membrane. The release process was followed for 480 minutes (8 hours). Samples were taken every 10 minutes in the first hour and then once per hour.

3. RESULTS AND DISCUSSION

Both the cross-linked (cross-linking ratio of 50, 75 and 100 %) and hydrophobized derivatives (chemical modification of the HyA polymer chains and neutralization with positively-charged amines) have been successfully prepared by different methods.

3.1. Structural characterization of the different HyA-based carriers

We established that increasing cross-linking ratio in the formation of nanosized particles with increasing average particle diameter (ca. 45 nm (50 %), ca. 100 nm (110 %). However the particle size increases with the cross-linking ratio, the presence of more polydispersed system can be confirmed by DLS. Besides the syntheses of cross-linked variants of HyA the hydrophobization of the polymer has been performed by using positively-charged CTAB surfactant and aminodecane primer amine as well. Firstly, conductometric studies have been carried out in aqueous solution at 25 °C in order to determine the critical micelle concentration (cmc) of CTAB. According to the parallel measurements 0.94 ± 0.01 mM of *cmc* value is obtained, while in presence of 0.1 mg/ml of HyA the *cmc* of CTAB shifts to 1.16 ± 0.02 mM. Besides, the zeta potential of the negative charged polymer reaches the zero value at 0.5 mM surfactant concentration which is corresponds to 1:1 surfactant/monomer molar ratio.[5] Both the turbidity measurements and the change of the average particle diameter of this system confirm the aggregation of the polymer chains at and above that surfactant concentration (0.5 mM in the case of 0.2 mg/ml HyA in PBS) wich necessary for the charge compensation of the carboxylate groups of HyA monomers.

3.2. Rheological characterization of aquatic HyA solutions and hydrogels

Based on the rheological characterization of aquatic hyaluronic acid solution we found that, the increase in concentration of the polymer solution resulted in dramatically change of the rheological behavior. Namely, the Hya solutions shows Newtonian, shear-thinning and viscoelastic behavior as the polymer concentration in aquatic solutions and hydrogels rises up from 0.05 to 100 mg mL^{-1}. In the case of very dilute solutions the linear flow curves prove the validity of the Newton law, in this way these solutions were characterized by constant (shear rate-independent or Newtonian) viscosity. Viscosities of moderately concentrated solutions of high-molecular weight polymers are known to be variable quantities which decreases with increasing shear rate. This shear-thinning effect has been known for a long time and same behavior was observed above 1.00 mg mL^{-1} HyA concertation (**Figure 1A**).

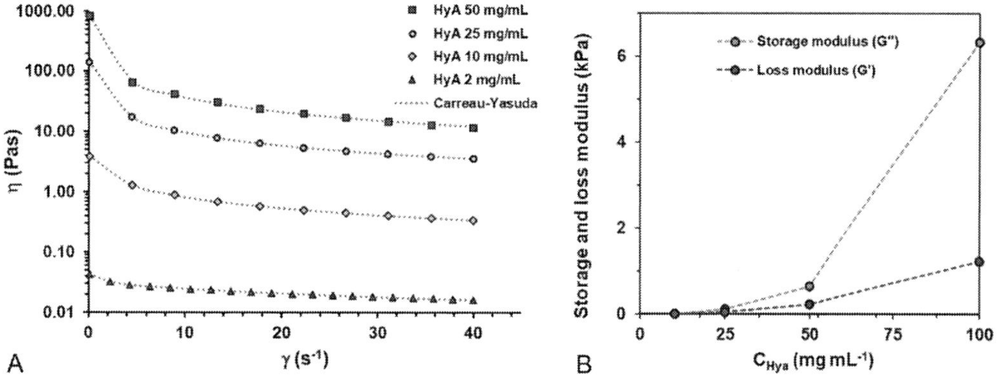

Figure 1 A: Measured and Carreau-Yasuda model fitted viscosity curves of HyA solutions (c_{HyA} = 2, 10, 25 and 50 mg mL^{-1}), B: concentration dependence of the viscoelastic of HyA hydrogels

Instead of the viscosity/shear rate profile (such as that shown in **Figure 1A**) is more suitable the determination of both loss and storage modulus (*G'* and *G"*) as a function of amplitude, under 10 s^{-1} oscillation frequency via amplitude sweep measurements. Amplitude scans of the linear HyA gels shown that the *G"/G'* ratio does not change as strain increases from 0.1 to 10.0 %, it means that the internal friction is independent of strain in this region in this way the viscoelastic behavior can be described by constant G' and G" values as presented in **Figure 1B** against the polymer concentration.

3.3. Rheological characterization of ionically- and covalently modified HyA solutions and hydrogels

To characterize the ionically modified form of polymer, a constant (0.1 mg/mL) initial HyA concentration was chosen for steady shear rate measurements of CTAB solution diluted HyA samples and a larger concentration (50 mg/mL) was applied in the case of amplitude sweep measurements. As it can be seen in **Figure 2A.** the steady shear rate tests determined apparent viscosity of polymer solution continuously decreased due to the added surfactant and a visible breakpoint can be observed. Above the 1:1 surfactant/monomer molar ratio the viscosity of the neutralized and thus hydrophpbized polymer containing solution shows a rising trend. After this observation it can be stated that in addition to conductivity and ITC measurement a modified rheological investigation also suitable for detection the structural change of the polymer - surfactant colloid system.

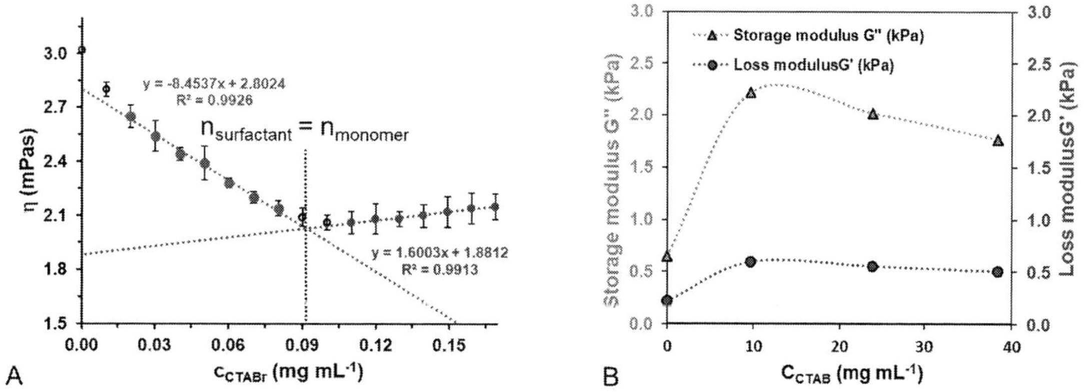

Figure 2 A: Steady shear rate determined apparent viscosity of linear HyA solution (0.1 mg mL^{-1}) titrated by CTAB (c = 25.0 mM) solution, B: Amplitude sweep determined rheological parameters of linear HyA hydrogels (c_{HyA} = 50 mg mL^{-1}) as a function of CTAB concentration

As illustrated in the **Figure 2B** the varying degrees hydrophobozed polymer based hydrogels are show elastic character at low strain, while at high strain range the loss modulus dominates (*G" > G'*) as the above reported linear hyaluronic acid hydrogels. Same trend was observed in the case of the varying degrees cross-linked polymer hydrogels (**Figuer 3A**) where the storage modulus almost reaches zero value due to the structure modification of polymer chains. The latter outlined two observations confirm the assumption that both the added CTAB and cross-linking agent break the coherent structure which spontaneous formed from the solvated biopolymer.

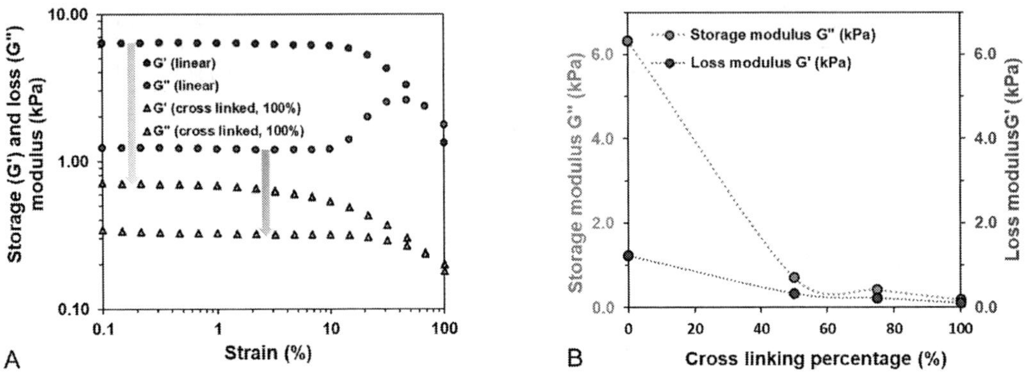

Figure 3 Changing of the amplitude sweep determined rheological parameters (storage and loss modulus) of linear and cross-linked hyaluronic acid hydrogels (c_{HyA} = 50 mg mL^{-1}) as a function of applied strain (A) and the degree of cross-linking (B)

3.4. Evaluation of in vitro drug release experiments

The evolution of KP dissolution from the HyA based carriers was investigated using first-order kinetic model.[6] Drug release curves of KP from varying degrees hydrophobized or crosslinked polymer based carrires are depicted in **Figure 3**. As it can be seen, all the formulations are able to sustain KP release for more than 6 h, but the release of the drug from the cross-linked polymer is almost complete within 7 h. By adding the linker from 50 % to 75 and 100 % into the polymer it was possible to achieve a more accurate control over drug release. However, it should be noted that nearly 80 % is released from the total drug content in the case of completely cross-linked polymer at the end of the seventh hour (**Figure 4B**). On the other hand, using the least amount of surfactant (**Figure 8A**) is causing that only 60 % of the active ingredient dissolved during the experiment.

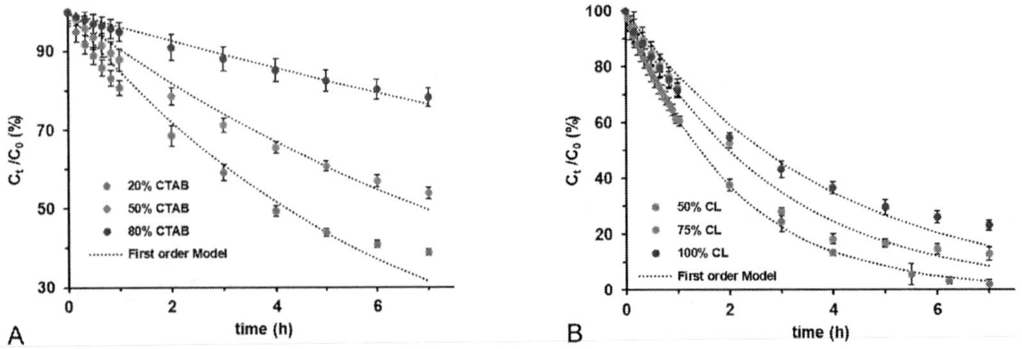

Figure 4 Photometric measured release profiles and first-order kinetic model predicted release curves of KP from varying degrees neutralized (A) or cross-linked (B) HyA-based drug carriers

Based on the observation of previous parts (change of the zeta potential, turbidity and particle size values as a function CTAB amount), it can also be established that the release of KP from the carrier faster and rather diffusion controlled when relatively small amount electrostatic adsorbed surfactant molecules are present in the system. As the polymer-ionic surfactant interactions lead to changes in polymer structure the dissolution of drug becomes slower and the release turns to diffusion and erosion controlled way.

4. CONCLUSION

During this study we have demonstrated the preparation and characterization of a biopolymer based colloidal drug delivery systems. Because of the hydrophilic character of the polymer, cross-linked derivatives have been synthesized to moderate the water soluble nature and produce polymer nanoparticles. The neutralized variants of polymer have also been produced by modifying with ionic surfactant to reach an incoherent „particle like" colloid system. Due to modifications the coherent structure of the polymer changes into an incoherent colloidal system that were verified by dynamic light scattering and rheological investigations. In the case of the cross-linked carrier, we could not significantly reduce the rate of drug dissolution but in the surfactant modified system the release rate has increased near four times larger in proportion to the amount of surfactant. It has been verified that the modified hyaluronic acid may be a potential candidate for controlled drug release of hydrophobic molecules.

ACKNOWLEDGEMENTS

This work was supported by the UNKP-18-3 New National Excellence Program of the Ministry of Human Capacities and the János Bolyai Research Fellowship of the Hungarian Academy of Sciences (Edit Csapó). The authors thanks the financial support of the National Research, Development and Innovation Office-NKFIH through the project "Synthesis, structural and thermodynamic characterization of nanohybrid systems at solid-liquid interfaces" K116323 and the project GINOP-2.3.2-15-2016-00060.

REFERENCES

[1] DANHIER, Fabinne, ANSORENA, Eduardo. JOANA, Silva. M. RÉGIS, Coco. AUDE, Le Breton and RRÉAT, Veronique. PLGA-based naoparticles: An owerview of biomedical applications. *Journal of Controlled Release*. 2012. vol. 161, no. 2, pp. 505-522.

[2] BODNÁR, Magdolna. DARÓCZI, Lajos. BATTA, Gyula. BAKÓ, József. HARTMANN, John. F. and BORBÉLY, János. Preparation and characterization of cross-linked hyaluronan nanoparticles. *Colloid and Polymer Science*. 2009. vol. 287, no. 8, pp. 991-1000.

[3] KARGEROVA, Andrea. PEKAR, Miroslav. High-resolution ultrasonic spectroscopy study of interaction between hyaluronan and cationic surfactants. *Langmuir*. 2014. vol. 30, no. 40, pp. 11866-11872.

[4] OUESLANTI, Nadia. LEBANC, Pierrick. HARSCOAT-SCHIAVO, Christelle. RONDAGS, Emmanuel. MEUNIER, Stéphane. KAPEL, Romain. MARC, Ivan. CTAB turbidimetric method for assaying hyaluronic acid in complex environments and under cross-linked form. *Carbohydrate Polymers*. 2014. vol. 112, pp. 102-108.

[5] CSAPÓ, Edit. SZOKOLAI, Hajnalka. JUHÁSZ, Ádám. VARGA, Norbert. JANOVÁK, László. DÉKÁNY, Imre. Cross-linked and hydrophobized hyaluronic acid-based controlled drug delivery systems. *Carbohydrate Polymers*. 2018. vol. 195, no. 1, pp. 99-106.

[6] COSTA, Paulo. SOUSA LOBO, José Manuel. Modelling and comparison of dissolution profiles. *European Journal of Pharmaceutical Sciences*. 2001. vol. 13, pp. 123-133.

OECT MULTIELECTRODE ARRAY FOR REAL-TIME CELL CULTURE MONITORING

Ota SALYK[1], Jan VÍTEČEK[2], Eva ŠAFAŘÍKOVÁ[2], Lukáš OMASTA[1], Stanislav STŘÍTESKÝ[1], Jiří EHLICH[1], Martin VALA[1], Martin WEITER[1]

[1]*Materials Research Centre, Faculty of Chemistry, Brno University of Technology, Brno, Czech Republic, EU, salyk@fch.vut.cz, xcomastal@fch.vut.cz, xcstritesky@fch.vutbr.cz, xcehlich@fch.vut.cz, vala@fch.vutbr.cz, weiter@fch.vutbr.cz*

[2] *Institute of Biophysics of the Czech Academy of Sciences, Brno, Czech Republic, EU, jan.vitecek@ibp.cz, safarikova.eva@ibp.cz*

Abstract

The significant cost of newly developed pharmacological products is due to the toxicity examination performed on animals in pre-clinical trials. Our ultimate goal is to develop a biosensor array to determine acute cardiotoxicity. It capitalizes on the fact that heart muscle cells (cardiomyocytes) show beating *in vitro*. Based on the beating characteristics their pathophysiological status can be deducted. In addition to microscopic imaging, the method of monitoring the action and field potential of synchronously beating cardiomyocyte syncytia on multielectrode arrays (MEA) appears to be a suitable method for the investigation of the condition of cardiac cells. For this purpose, MEA on the basis of organic electrochemical transistors (OECTs) has been developed. The organic polymer PEDOT:PSS was utilized to meet the strict surface biocompatibility demand for a surface in contact with media and living cardiac cells. The MEA consisted of a microplate of a 12x8 chimney - well array with transistors on the bottom for cell cultivation. For the fabrication (a screen printing method) primarily a contact field and an organic polymer was used. The device was tested in a simulation mode for the response of the drain current to the voltage pulses applied to the gate electrode up to 100 μV. The response to such gate voltage renders the device sensitive enough to detect the electrical signal of beating cardiomyocytes. Further development is directed towards higher time resolution by means of miniaturization of the transistors' channel.

Keywords: OECT, screen printing, organic electrochemical transistor, PEDOT:PSS, microplate, multi-electrode array, cell culture

1. INTRODUCTION

The high electrical conductivity of polyacetylene and other organic compounds with conjugated bonds, discovered by Hideki Shirakawa, Alan Heeger, and Alan MacDiarmid led to intense interest in the use of organics in electronics. This discovery was recognized with the Nobel Prize in Chemistry in 2000 [1]. Especially the following branches saw considerable effort put into related research: photovoltaics, photoluminescence, and bioelectronics as a brand-new field of science. In 1989 the first polymer-based light emitting diode (LED) was discovered using PPV as the emissive layer [2]. However, also the printable properties of organic semiconductors have led to the development of so-called printed electronics.

The biocompatibility of organic semiconductors, predominantly polymers, has made them an appealing choice for application in biosensors and bioactuators in biology and medicine as an ideal biointerface [3]. In 1984 White et al. [4] fabricated the first organic electrochemical transistor (OECT) based on polyaniline (PANI).Heywang and Jonas [5] (1992) reported the electrochemical polymerization of a new conducting polymer, poly(3,4-ethylenedioxythiphene) (PEDOT). Yamato et al. [6] (1995) showed the superiority of Poly(3,4-ethylenedioxythiophene)-poly(styrenesulfonate) (PEDOT :PSS) in terms of stability against pH and applied potential in comparison with PANI. This polymer became widespread in bioelectronics as active transistor channels and other fields as a conductive thin-film semi-transparent wiring.

Aqueous solution of PEDOT:PSS with various additives such as methanol, ethylene glycol, sorbitol serve as solvents for thin film deposition by spin coating and inks for inkjet printing and screen printing. The additives enhance the resulting conductivity by expanding the coil conformation of the macromolecule backbone into a linear conformation and reorganization of the molecule from benzoid form to the quinoid form [7]. The temporary conductivity of PEDOT:PSS reaches 300 S/cm [8].

Bernards and Malliaras (2007) [9] created a model of OECT electronic function. The electrodes are analogues to inorganic FET, i.e. a gate, source and drain, but it works in aqueous solutions of salts, frequently PBS (Phosphate buffered saline) physiological solution for cultivation of biological materials. The charge transport is mediated by holes moving along the PEDOT backbone. The charge is compensated by dangling bonds in some missing hydrogens on SO_3H groups on parallel PSS - see **Figure 1**. Instead of field-caused accumulation or depletion of charge carriers in the channel, in this case the high mobile holes are substituted by low mobile cations and reversely. Thus, the channel conductance is not modulated by the charge concentration but by the mobility, or we might say by the doping - dedoping process.

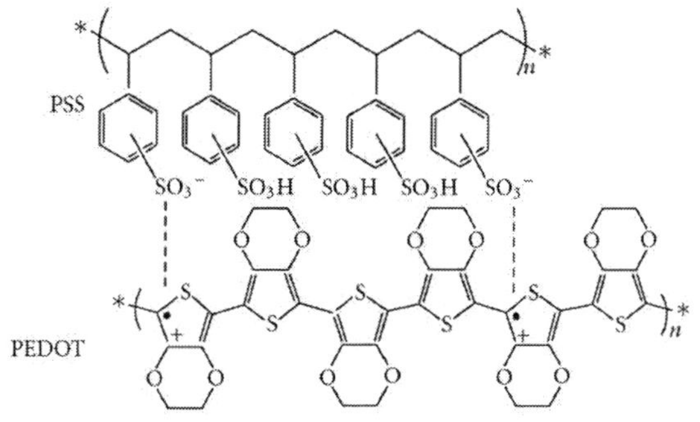

Figure 1 Chemical structure of PEDOT:PSS. The "dot" and "plus" represent the unpaired electron and positive charge on the PEDOT chain, respectively

Piro et al. (2018) [10] presented in their review a broader insight into the biological application of OECTs. Yao et al. (2014) [11] presented the first study of using OECT arrays for monitoring action potentials from cardiomyocyte-like (i.e. cardiac muscle cell-like) HL-1 cells on both rigid and flexible substrates with improved signal qualities compared to those recorded from grapheme-based electrolyte-gated FETs. Hempel et al. (2016) [12] developed OECT devices in a wafer-scale process and used them as electrophysiological biosensors measuring the electrophysiological activity of the cardiac cell line HL-1. The topography of OECT was further optimized by Rivnay et al. [13] (2015) with respect to enhancing its amplification (transconductance) and frequency range. Proctor et al. (2016) [14] improved the theoretical model of OECTs by introducing a new quantity, namely volumetric capacitance. Tymbrant et al. (2017) [15] completed the theoretical model for OECT and included the ionic double layer at the electrolyte interface and the effects of contact barriers. The biological part of the research is based on research into biocompatibility and cell proliferation [16,17]. The device development is based on our previous experimental work [18].

2. EXPERIMENT

The goal of this study is to investigate the possibility of applying OECTs based on PEDOT:PSS and fabricating a planar printed 12x8 microplate array of transistors. The device (see **Figure 1a**) was constructed on a 96-well polystyrene microplate (chimney-well, no bottom) from Greiner Bio-One GmbH for automated analysis of a multitude of substances. We provided the bottom with the all-printed 12x8 array of Organic Electrochemical Transistors (OECT) based on Poly(3,4-ethylenedioxythiophene) polystyrene sulfonate (PEDOT:PSS) polymer (see **Figure 1c**).

In the first stage, the screen-printing method was used. Printed organic electronic technology has brought the possibility of printing biosensor patterns on the biocompatible polymer foils PET (Poly(Ethylene Terephthalate)) or PEN (Poly(Ethylene 2,6-Naphthalate) (Goodfellow Cambridge Ltd), the latter enabling processing at a temperature of up to 170 °C. The bottom substrate was a 0.25 mm thick PEN foil due to better thermal stability during the subsequent annealing processing of the prints. The first printed pattern (with a screen mesh count of 77 threads/cm) of silver paste CB115v2 (©DuPont) Photopolymer and Electronic Materials) creates the contact and conductive field and paths that are 0.18 mm in width (see **Figure 1d**). The foil with the silver pattern was then annealed at 120 °C for fixation.

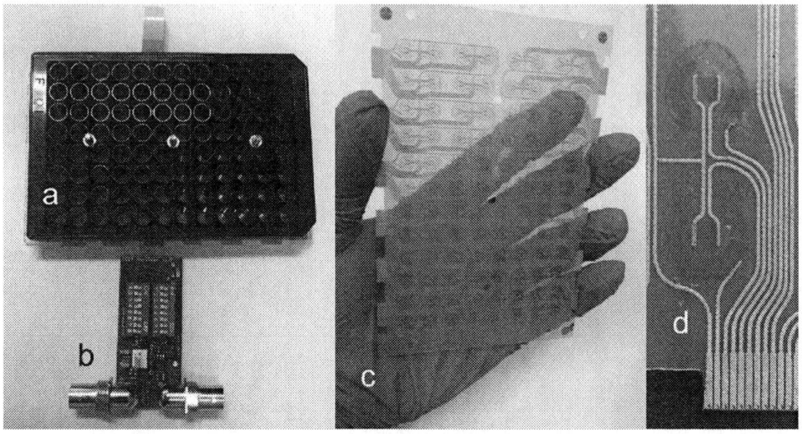

Figure 2 Encapsulated OECT 96 well microplate for electrogenic cell cultivation and investigation (a) with power source-signal amplifier and current/voltage converter (b), foil with functional printed patterns of all OECTs 12x8 array (c) and detail of a couple of OECTs with PEDOT:PSS printed channel and gate electrode (d).

The second pattern of PEDOT:PSS paste Clevios™ S V3 (screen mesh count 140 threads/cm) created a functional OECT gate and channel. The paste was first stirred intensively before printing in order to obtain a lower viscosity and better homogeneity for the resulting film. The silver conductive paste and the polymer patterns were annealed using a hot plate at 150 °C for 15 and 30 minutes respectively in the air to stabilize the film and improve adhesion.

The stirring slightly decreased the final thickness, improved the thickness homogeneity in roughness and waviness measured and the sheet resistance decreased significantly from about 800 Ω/□ to 400 Ω/□. It is considered that intensive stirring acts similarly to alcohol treatment. The system is masked and sealed by the third pattern of Sylgard® 184 (Dow Corning) silicone elastomer (mesh count 120 threads/cm), which insulates the silver conducting paths from the electrolyte environment and also prevents the tested biomaterial from coming into contact with the non-biocompatible parts and layers of the printed device. Other parts and materials were tested for biocompatibility [18]. A coating to increase biocompatibility was done with murine collagen type IV (BD Biosciences, cat. No. 354233) at 10 $\mu g \cdot cm^{-2}$ according to manufacturer instructions.

Patterning and completing of the microplate followed. The OECT's channel visible in **Figure 2d**, created by a screen-printed rectangular PEDOT:PSS layer on PEN foil, is surrounded by a planar circular gate electrode 6 mm in outer diameter. The rotary symmetry is expected to improve the field distribution. The electrodes are contacted by printed silver conductors 0.2 mm wide. The printed transparent silicone layer covers the surface of the sensing array with the exception of the functional PEDOT:PSS interfacing to the biomaterial and physiological solution. The exposed channel area is of a length of L = 1 mm and a width of W = 1.5 mm, the thickness of the PEDOT:PSS layer was 250 nm on average, and the typical resistance of the channel was

500-700 Ω. The foil was firmly fixed to the polystyrene microplate and bolted down from the bottom by means of a Plexiglas® plate in order to enable easier handling and transparency needed for microscope observations. The second pattern of PEDOT:PSS paste Clevios™ S V3 (screen mesh count 140 threads/cm) created a functional OECT gate and channel. The paste was first stirred.

In the second state the printed circuit technology with a PEN foil of 0.25 mm coated with copper (AKAFLEX® from KREMPEL GmbH) for the substrate was used due to its better conductance and resolution of the lithographically patterned board in comparison with a printed silver paste pattern (see **Figure 3b**). The biocompatibility was provided by the gilding of the conductor paths, and silicon masking was substituted by the SU-8 photoresist patterning with enhanced resolution. The single transistor is depicted in **Figure 3a**. The entire device is assembled as in the previous case. Eight transistors are connected via one Molex connector and sixteen switches to the current/voltage converter - amplifier (see **Figure 2b**), a diagram of which can be found in **Figure 3c**, for further digital processing in a scope or another recording facility.

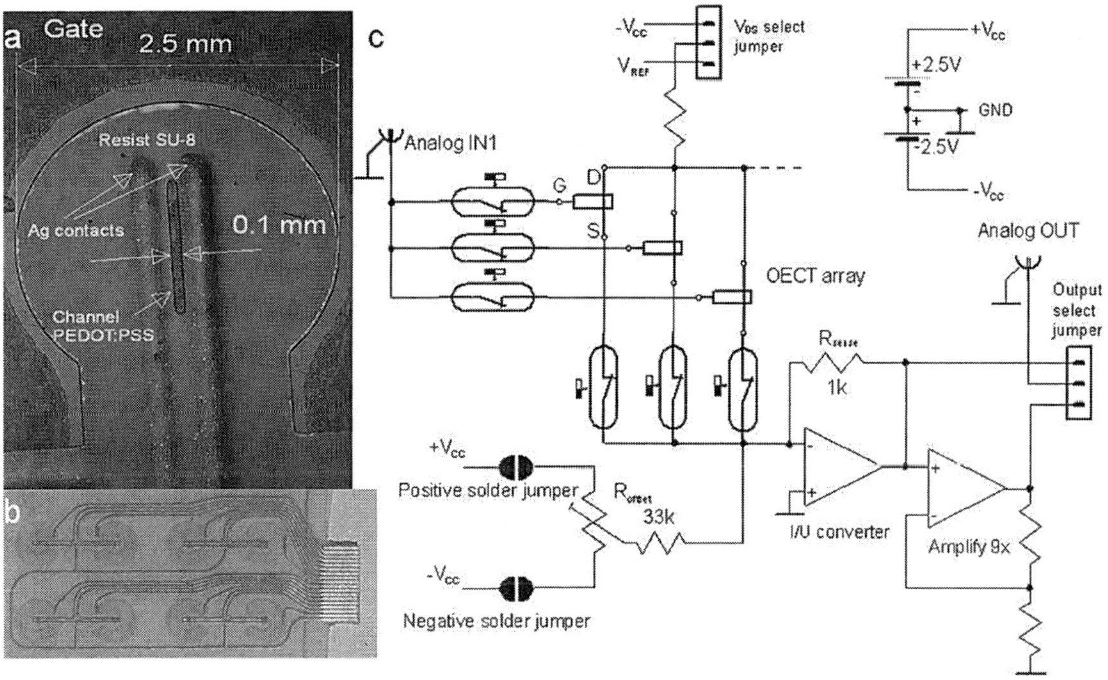

Figure 3 The amplifier - current/voltage converter for testing of the OECTs

3. RESULTS AND DISCUSSION

The device was tested for the OECT output (**Figure 4a**) and transfer (**Figure 4b**) characteristics, where the optimum setpoint with respect to high transconductance and safety against redox reactions was found. Also the frequency/time constant limits were tested by modulating the gate voltage by rectangular pulses 1 mV_{pp} and the frequency 1Hz and the responding gate induced drain current I_{Dpp} = 2.5 μA, which results in the transconductance g = 2.5 mS (**Figure 5**). The gate pulses of 250 μV were also reliably detected using FFT analysis. The time constant τ = 0.15 s is given by the serial combination of electrolyte resistance and volumetric capacity of the PEDOT:PSS channel. Other capacities and resistances (gate, electrical double layer - EDL) are neglected.

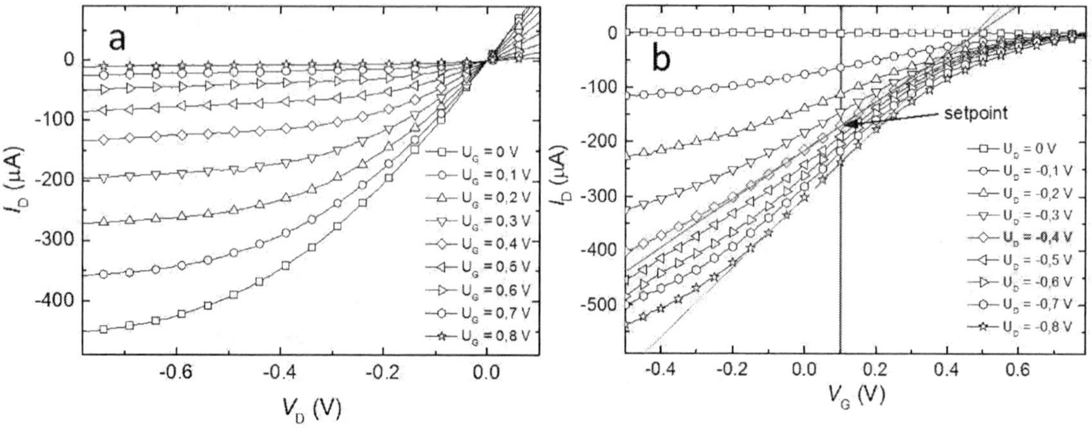

Figure 4 Output and transfer characteristics with the illustration of setpoint and the slope dependent on the drain potential

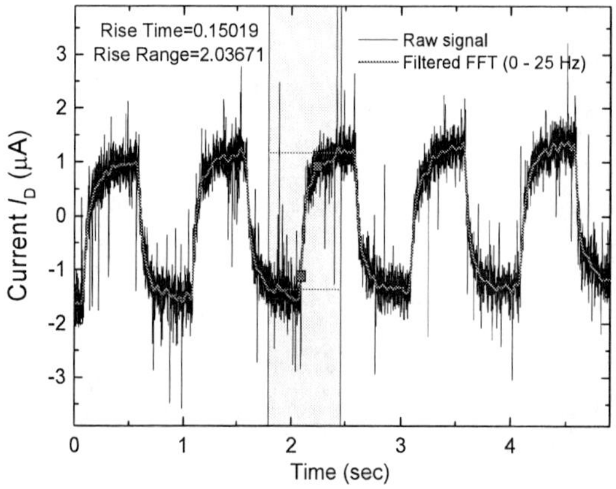

Figure 5 Response of OECT to 1 mV_{pp} gate rectangular signal at setting point V_{DS} = -0.7 V V_{GS} = 0.2 V. The derived transconductance gives g = 2.5 mS and time constant τ = 0.15 s.

4. CONCLUSIONS

The improved screen printing process of photoresist SU-8 masking was optimized and successfully implemented into previously adapted OECT sensors based on PEDOT:PSS and silicon masking [19]. The photolithographic process enabled shortening the channel length down to 100 µm and further and the OECT geometry to adapt on zero gate bias setpoint for maximum signal amplification. Simulating the pulsation of cardiomyocyte cells, the transconductance g = 2.5 mS was achieved with the S/N ratio about 10. The upper-frequency limit 7 Hz was concluded from the OECT gate circuit time constant of 0.15 s. Additional improvement of the device speed and amplification could be obtained by reducing the OECT channel dimensions, mainly by the reducing the PEDOT:PSS volume capacitance by decreasing the channel area and its thickness.

ACKNOWLEDGEMENTS

This work was supported by Czech Science Foundation via project No. 17-24707S, research infrastructure was supported by projects MŠMT No. LO1211 and J. Ehlich was supported by the Faculty of Chemistry at BUT under project No. FCH-S-17-4685.

REFERENCES

[1] HEEGER, Alan (2001). "Nobel Lecture: Semiconducting and metallic polymers: The fourth generation of polymeric materials". *Reviews of Modern Physics* (free download). 73 (3): 681.

[2] BURROUGHES, J. H. et al. Light-Emitting Diodes Based on Conjugated Polymers. *Nature.* 1990. 347, 539.

[3] STRAKOSAS, Xenofon.; BONGO, Manuelle.; OWENS, Roisin M. The organic electrochemical transistor for biological applications. *J. Appl. Polym. Sci.* 2015, 132, doi:10.1002/app.41735.

[4] WHITE, Henry S., KITTLESEN, Greg P., & WRIGHTON, Mark S. (1984). Chemical Derivatization of an Array of 3 Gold Microelectrodes with Polypyrrole - Fabrication of a Molecule-Based Transistor. *Journal of the American Chemical Society,* 106(18), 5375-5377. doi: 10.1021/ja00330a070

[5] HEYWANG, G.; JONAS, F., Poly(alkylenedioxythiophene)s - New , Very Stable Conducting Polymers. *Adv. Mater.* 1992, 4 (2), 116-118.

[6] YAMATO, H.; OHWA, M.; WERNET, W., Stability of Polypyrrole and Poly(3,4 ethylenedioxythiophene) for Biosensor Application. *Journal of Electroanalytical Chemistry* 1995, 397 (1-2), 163-170.

[7] LI, J., J.C. LIU, and GAO C.J., On the mechanism of conductivity enhancement in PEDOT/PSS film doped with sorbitol. *E-Polymers*, 2011

[8] https://www.sigmaaldrich.com/catalog/product/aldrich/739316?lang=en&region=CZ

[9] BERNARDS, D.A. and G.G. MALLIARAS, Steady-state and transient behavior of organic electrochemical transistors. *Advanced Functional Materials*, 2007. 17(17): p. 3538-3544.

[10] PIRO, B.; MATTANA, G.; ZRIG, S.; ANQUETIN, G.; BATTAGLINI, N.; CAPITAO, D.; MAURIN, A.; REISBERG, S., Fabrication and Use of Organic Electrochemical Transistors for Sensing of Metabolites in Aqueous Media. *Applied Sciences-Basel* 2018, 8 (6).

[11] YAO, C., LI, Q., GUO, J., YAN, F., & HSING, I. M. (2015). Rigid and Flexible Organic Electrochemical Transistor Arrays for Monitoring Action Potentials from Electrogenic Cells. *Advanced Healthcare Materials*, 4(4), 528-533. doi: 10.1002/adhm.201400406

[12] HEMPEL, F.; LAW, J.K.-Y.; NGUYEN, T.C.; MUNIEF, W.; LU, X.; PACHAURI, V.; SUSLOPAROVA, A.; VU, X.T.; INGEBRANDT, S. PEDOT:PSS organic electrochemical transistor arrays for extracellular electrophysiological sensing of cardiac cells. *Biosens. Bioelectron.* 2016, doi:http://dx.doi.org/10.1016/j.bios.2016.09.047

[13] RIVNAY, J.; LELEUX, P.; FERRO, M.; SESSOLO, M.; WILLIAMSON, A.; KOUTSOURAS, D. A.; KHODAGHOLY, D.; RAMUZ, M.; STRAKOSAS, X.; OWENS, R.M.; BENAR, C.; BADIER, J.M.; BERNARD, C.; MALLIARAS, G.G., High-performance transistors for bioelectronics through tuning of channel thickness. *Science Advances* 2015, 1 (4).

[14] PROCTOR, C. M.; RIVNAY, J.; MALLIARAS, G. G., Understanding Volumetric Capacitance in Conducting Polymers. *Journal of Polymer Science Part B-Polymer Physics* 2016, 54 (15), 1433-1436.

[15] TYBRANDT, K.; ZOZOULENKO, I. V.; BERGGREN, M., Chemical potential-electric double layer coupling in conjugated polymer-polyelectrolyte blends. *Science Advances* 2017, 3 (12).

[16] STŘÍTESKÝ, S. - MARKOVÁ, A. - VÍTEČEK, J. - ŠAFAŘÍKOVÁ, E. - HRABAL, M. - KUBÁČ, L. - KUBALA, LUKÁŠ - WEITER, M. - VALA, M. Printing inks of electroactive polymer PEDOT:PSS: The study of biocompatibility, stability, and electrical properties. *Journal of Biomedical Materials Research. Part A.* ROČ. 106, Č. 4 (2018), S. 1121-1128 ISSN 1549-3296

[17] VISTEJNOVA, L.; DVORAKOVA, J.; HASOVA, M.; MUTHNY, T.; VELEBNY, V.; SOUCEK, K.; KUBALA, L. The comparison of impedance-based method of cell proliferation monitoring with commonly used metabolic-based techniques. *Neuroendocrinol. Lett.* 2009, 30, 121-127

[18] SAFARIKOVA E., SINDLEROVA SVIHALKOVA L., STRITESKY S., KUBALA L., VALA M., WEITER M & VITECEK, J (2018) Evaluation and improvement of organic semiconductors' biocompatibility towards fibroblasts and cardiomyocytes. *Sensors and Actuators B: Chemical* 260, 418-425.

[19] SALYK, O.; VITECEK, J.; OMASTA, L.; SAFARIKOVA, E.; STRITESKY, S.; VALA, M.; WEITER, M., Organic Electrochemical Transistor Microplate for Real-Time Cell Culture Monitoring. *Applied Sciences-Basel* 2017, 7 (10).

CONJUGATED METALLOTHIONEIN-CARBON-DOXORUBICIN NANOTRANSPORTER FOR TARGETED BREAST CANCER THERAPY

[1]Sylvie SKALIČKOVÁ, [2]Marta KEPINSKA, [1]Martin LOFFELMANN, [3]Dagmar UHLÍŘOVÁ, [1]Michael GARGULÁK, [3]Michaela DOČEKALOVÁ, [1]Božena HOSNEDLOVÁ, [1]Branislav RUTTKAY-NEDECKÝ, [3]Halina MILNEROWICZ, [4]Carlos FERNANDEZ, [1,2,3]René KIZEK

[1]University of Veterinary and Pharmaceutical Sciences Brno, Pharmaceutical Faculty, Brno, Czech Republic, EU, kizek@sci.muni.cz
[2] Wroclaw Medical University, Faculty of Pharmacy with Division of Laboratory Diagnostics, Department of Biomedical and Environmental Analyses, Wroclaw, Poland, EU, zalewska.m@gmail.com
[3]Prevention Medicals, Studenka-Butovice, Czech Republic, EU, uhlirova@preventionmedicals.cz
[4]Robert Gordon University, School of Pharmacy and Life Sciences, Garthdee Road, Aberdeen, Scotland, United Kingdom, c.fernandez@rgu.ac.uk

Abstract

Metalothionein (MT) is a polypeptide of molecular weight in the range of 6-10 kDa. MT typically contains 60 to 68 amino acid residues. MT is characterized by its unique content of metal ions as well as its sulfur content. Higher MT levels were observed in proliferating cells. This fact demonstrates the importance of MT in the process of cellular regulation (relationship to cancer). The most widely used drug for patients with breast cancer metastases is an anthracycline antitumor antibiotic doxorubicin (DOX). However, the clinical use of DOX is limited by dose-related heart muscle damage (cardiomyopathy), more prevalent with increasing cumulative doses. For this reason, creation of novel pharmaceutical formulations based on using alternative methods as nanocarriers for targeted drug delivery to tumour cells is a crucial task in modern pharmacology. The aim of this work was to design a nanotechnological construct. The construct is designed as two separate nanotransporters. The nanotransporter (A) is formed by an antibody-modified AgNPs particle and a carbon nanotube with encapsulated DOX (AgNPs/Ab1/MWCNT/DOX/ODN1). The nanotransporter (B) is engineered with SPION particle modified with antibody and with bound MT (SPION / Ab2 / MT / ODN2). Construct AgNPs/Ab1/MWCNT/DOX/ODN1-SPION/Ab2/MT/ODN2 is formed using an oligonucleotide anchor. Individual parts of the nanotransporter were studied using appropriate methods. The presence of MT was monitored electrochemically by Brdicka method in connection with the transfer technique (AdTSV). Characteristic MT signals RS2CO (-1.15 V), Cat1 (-1.25 V), Cat2 (-1.45 V), Cat3 (-1.75 V) were observed at accumulation time of 120s. SDS PAGE confirmed the presence of MT on SPION nanoparticles at sizes 7 to 15 kDa. The DOX signal was fluorometrically monitored (Em 590 nm, Ex 490 nm). AgNPs sizes ranged from 15 - 20 nm, and the SPION nanoparticles ranged from 20 - 50 nm. Additionally, used AgNPs nanoparticles exhibited significant antiproliferative activity (growth inhibition by 20 - 40 %) on a model culture *S. cerevisiae*. Created nanoconstruct A showed growth inhibition *for S. cerevisiae* by more than 50 %. The nanoconstruct after these various analysis shows a high potential as an anticancer drug and may be an innovative way how to deal with the breast cancer in a targeted therapy

Keywords: Nanomedicine, carbon nanomaterials; anticancer drugs, electrochemical analysis, biophysical analysis

1. INTRODUCTION

Cancer belongs worldwide to the leading cause of human death. The early detection and precise diagnosis of the onset of a disease is the most promising approach to accelerate healing processes or to improve survival of patients. There are different approaches for cancer treatment, including surgery, radiation therapy, phototherapy, chemotherapy and biotherapy. Although chemotherapy has advantage in its efficiency, it has important limitations such as toxic side effects, damage of healthy cells, arising of the multidrug resistance and tumor recurrence. Therefore, new strategies are developed to overcome this problem. Nanoparticles and especially carbon nanoparticles have excellent physical properties, like NIR absorption or Raman enhancement and can be used for different approaches, among others for photothermal or photodynamic therapy, effective drug or gene delivery, or for diagnostic. Carbon is the fourth most abundant element in the universe, after hydrogen, helium and oxygen. Carbon atoms can build complicated organic structures. It can exist in different forms, which range from zero-dimensional (0-D, e.g. fullerenes) to three-dimensional (3-D, e.g. graphite). An analysis of a protein by electrochemical methods is not so broaden as by others techniques such as mass spectrometry [1,2]. On the other hand there have been published a number of papers, where the authors proved that an electrochemical technique could be a suitable tool for a determination of proteins [3-7]. More than 80 years ago Prof. Rudolf Brdicka laid the foundations of an electrochemical protein determination. He discovered an electrochemical catalytic signal of proteins in the presence of cobalt solution [8]. This technique has been used as a sensitive and selective tool for distinguishing of tumour and healthy tissue in 50s of the last century [9,10]. After that the electrochemical techniques were replaced by electrophoretic methods. In the beginning of 80s of the last century the method of Prof. Brdicka has been modified [11] and used for the detection of MTs [12]. The most widely used drug for patients with breast cancer metastases is an anthracycline antitumor antibiotic doxorubicin (DOX). However, the clinical use of DOX is limited by dose-related heart muscle damage (cardiomyopathy), more prevalent with increasing cumulative doses. For this reason, creation of novel pharmaceutical formulations based on using alternative methods as nanocarriers for targeted drug delivery to tumour cells is a crucial task in modern pharmacology [13]. The aim of this work was to design a nanotechnological construct. The construct is designed as two separate nanotransporters.

2. MATERIAL AND METHODS

Carbon (5 mg) was dispersed in 4 mL of an HNO_3:H_2SO_4 (1:3) mixture. The obtained suspension was heated for 7 h at 70°C (Electrothermal, UK). Then, it was placed in an ultrasonic bath (Ultrasonic Cleaner T, VWR) for 30 min and centrifuged at 14 000 *g* for 30 min. (Centrifuge 5418 R, Eppendorf). The obtained supernatant was removed, and the sediment washed with 500 µL H_2O (18 mΩ) and centrifuged at 14 000 × *g* for 30 min. The last step was repeated three more times. After the last centrifugation, the supernatant was removed and the sediment was dispersed in 500 µL of a 200 µg/mL doxorubicin (DOX) solution. The obtained suspension was placed in an ultrasonic bath for 30 min and centrifuged at 14 000 × *g* for 30 min. The obtained supernatant was removed and the sediment (fullerene-DOX nanocomposites) was washed twice with 500 µL H_2O (18 mΩ). Beside, 250 µL of the suspension of fullerene-DOX prepared as above, before placement in an ultrasonic bath for 90 min, was additionally treated with 1 % acetic acid. Electrochemical measurements were performed with AUTOLAB Analyser connected to VA-Stand 663 (Metrohm, Switzerland), using a standard cell with three electrodes. The working electrode was a hanging mercury drop electrode (HMDE) with a drop area of 0.4 mm^2. The reference electrode was an Ag/AgCl/3M KCl electrode and the auxiliary electrode was a graphite stick electrode. The supporting electrolyte was prepared by mixing buffer components. For smoothing and baseline correction the software NOVA. Differential pulse voltammetry (DPV) of metallothionein: The amount of MT was measured using AdTS DPV. The samples of the MT were reduced before each measurement by 1 mM tris(2-carboxyethyl)phosphine addition according to [6, 14]. AdTS DPV parameters were as follows: an initial potential of -1.2 V, an end potential -0.3 V, a modulation time 0.057 s, a time interval 0.2 s, a step potential of 1.05 mV/s, a modulation amplitude of 250 mV, E_{ads} = 0 V. All experiments were carried out at room temperature (24 °C). The DPV samples analyzed were deoxygenated prior to measurements by purging with argon (99.999 %) saturated with water for 60 s. AdTS DPV Brdicka reaction of MT: In our studies, the Brdicka supporting electrolyte containing 1 mM $Co(NH_3)_6Cl_3$ and 1 M ammonia buffer (NH_3(aq) + NH_4Cl, pH = 9.6) was used; surface-active agent was not added. AdTS DPV Brdicka reaction parameters were as follows: an initial potential of -0.35 V, an end potential -1.8 V, a modulation time 0.057 s, a time interval 0.2 s, a step potential of 1.05 mV/s, a modulation amplitude of 250 mV, E_{ads} = 0 V. For the analysis of SPIONs MT concentration the SDS-PAGE with silver staining has been used. Gels (7.5 %) were prepared using acrylamide stock solution 30 % (m / V) with bisacrylamide 1 % (m / V). Separating gel contained: acrylamide 7.5 % (m / V), bisacrylamide 0.5 % (m / V), 0.4 M Tris/HCl, 0.1 % (m / V) sodium dodecyl sulfate (SDS), pH 8.8. Stacking gel contained: 4.5 % acrylamide (m / V), 0.15 % bisacrylamide (m / V), 0.1 % SDS (m / V), 0.1M Tris/HCl, pH 6.8. Nanoconstructs were diluted 2:1 with a loading buffer (PLB Max). Each well contained 15 µl of the diluted solutions. Electrophoretic measuring conditions were: 120 V, 1.5 hours in a running buffer (24 mM Tris, 0.2M glycine and 3 mM SDS). After measurement the gel was stained with silver, scanned and evaluated by Colortest in the laboratory system Qinslab.

3. RESULTS

In our work, a unique DOX nanotransporter was prepared to target the tumor cell. The nanotransporter (A) is formed by an antibody-modified AgNPs particle and a carbon nanotube with encapsulated DOX (AgNPs/Ab1/MWCNT/DOX/ODN1). The nanotransporter (B) is engineered with SPION particle modified with antibody and with bound MT (SPION/Ab2/MT/ODN2). Construct AgNPs/Ab1/MWCNT/DOX/ODN1-SPION/Ab2/MT/ODN2 is formed using an oligonucleotide anchor. Individual parts of the nanotransporter were studied using appropriate methods.

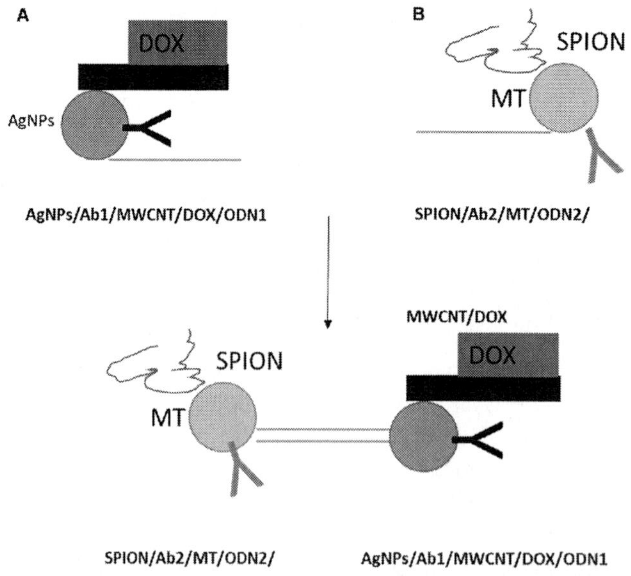

Figure 1 Schema of the proposed nanoconstruct for its detailed characterization. Part A is a dual antitumor nanotransporter. DOX is encapsulated within carbon nanomaterial (nanotube, fullerene), and then this structure is bound to AgNPs, ODN1 and model general antibody using EDC / NHS modification. Part B provides a model for the targeted protection of healthy tissue. SPION particle is modified by MT, ODN2 and antibody.

The presence of MT was monitored electrochemically by Brdicka method in connection with the transfer technique (AdTSV). It has been known more than 80 years that it is suitable to use the catalytic signal of hydrogen evolution in the presence of ammonium buffer (1 M NH_4Cl + NH_4OH) containing cobalt solution $Co(NH_3)_6Cl_3$ for the determination of the proteins containing cysteine amino acids, e. g. metallothionein (**Figure 2**).

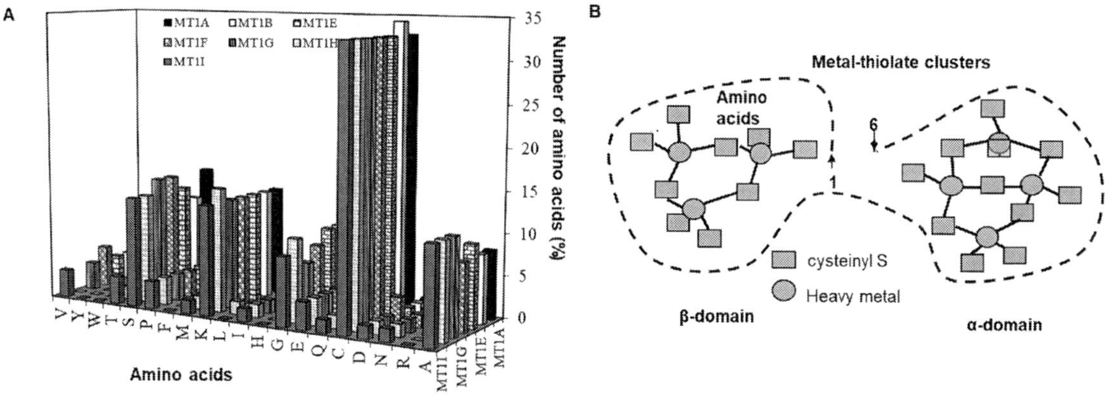

Figure 2 Content of aminoacids in human metallothioneins of MT1 class according to **Ex**pert **P**rotein **A**nalysis **Sy**stem - ExPASy (A). (**B**) Model of Metal-thiolate clusters, according to [6].

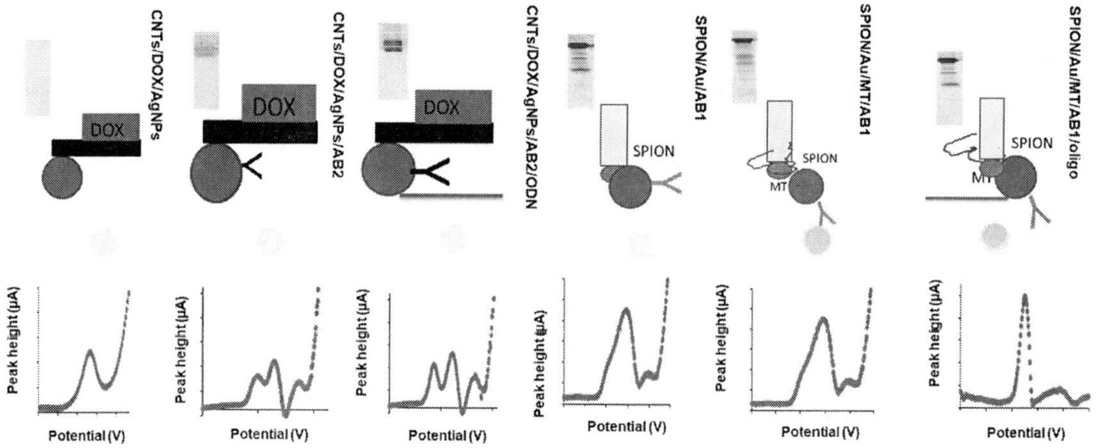

Figure 3 Detailed characterization of prepared nanoconstructs by means of electrophoresis, dot-blot analysis and electrochemical detection of proteins. On the NC45 nitrocellulose membrane (pore size 0.45µm, Serva, Germany) was pipetted 2 µl of antibody. The membrane was allowed to dry. Then blocked by 1 % BSA (bovine serum albumin) dissolve in PBS (Phosphate buffered saline) on a Multi-Rotator, Multi RS-60 (BioSan) for 40 minutes. Then the membrane was insered into a sample solution and again strirred on a rotator shaker for 1 hour. The membrane was then washed three times with PBS-T (Phosphate buffered saline with 0.05 % Tween 20) and then inserted into the gold-labeled antibody and stirred for 1 hour on rotator shaker. In the end the membrane was washed with PBS-T, allowed to dry and scaned for another evaluation. Other experimental details is in Material and Methods section.

A number of authors have concerned with explanations of the processes, which proceed on the surface of working mercury electrode during Brdicka reaction. We observed during modified Brdicka reaction by AdTS DPV analysis four MT signals - Co_1, RS_2Co, Cat1 and Cat2. Signals of Cat1 a Cat2 correspond to the reduction of hydrogen at the mercury electrode [15]. Another signal, which is appeared at the potential about -1.0 V, relates with the reduction of the RS_2Co complex [15]. In addition the signal called Co_1 could result from reduction of $[Co(H_2O)_6]^{2+}$ [15]. It clearly follows from the figure that character of the mentioned MT signals change with different MT amount. Signal Co_1 decreased and shifted to more negative potential with decreasing MT amount. The signal is almost un-detectable at MT amount under 0.6 pmol. The others mentioned MT signals of Brdicka reaction (RS_2Co, Cat1 and Cat2) are getting well-developed and separated with decreasing MT concentration. In addition RS_2Co and Cat signals decreased and slowly shifted to more positive potential according to decreasing MT concentration. Moreover, SDS PAGE confirmed the presence of MT on SPION nanoparticles at band sizes 7 to 15 kDa (see in **Figure 3**). Additionally, dot blot technique was used to confirm the presence of the bound antibody. Thus, it was clearly demonstrated that AgNPs were modified by antibody 1 and SPION (20-50 nm) were modified with antibody 2. MWCNTs were chemically modified for DOX binding according to established procedures. The DOX signal was fluorometrically monitored (Em 590 nm, Ex 490 nm). The amount of DOX bound to MWCNTs was around 50-60 %. Another part of the nanoconstruct were AgNPs (15-20 nm) that show antiproliferative effects. Additionally, used AgNPs nanoparticles exhibited significant antiproliferative activity (growth inhibition by 20-40 %) on a model culture *S. cerevisiae*. Created nanoconstruct A showed growth inhibition *for S. cerevisiae* by more than 50 %. The nanoconstruct after these various analysis shows a high potential as an anticancer drug and may be an innovative way how to deal with the breast cancer in a targeted therapy.

4. CONCLUSION

A unique universal nanotransporter, DOX and AgNPs, was developed to target the tumor cell. The nanotransporter (A) was formed by an antibody-modified AgNPs particle and a carbon nanotube with encapsulated DOX (AgNPs/Ab1/MWCNT/DOX/ODN1). The nanotransporter (B) was engineered with SPION particle modified with antibody and with bound MT (SPION / Ab2 / MT / ODN2). Construct AgNPs/Ab1/MWCNT/DOX/ODN1-SPION/Ab2/MT/ODN2 was formed using an oligonucleotide anchor. Individual parts of the nanotransporter were studied using appropriate methods. In the next work, the nanotransporter will be treated with suitable target antibodies and tested on suitable cell lines.

ACKNOWLEDGEMENTS

This work has been supported by the NANOCSDOXO (306/2018/FaF) and the Grant ST.D170 from Wroclaw Medical University and the international collaboration project of The European Technology Platform for Nanomedicine.

REFERENCES

[1] PETREK, J., VITECEK, J., VLASINOVA, H., KIZEK, R., KRAMER, K. J., ADAM, V., KLEJDUS, B.,HAVEL, L. Application of computer imaging, stripping voltammetry and mass spectrometry for study of the effect of lead (Pb-EDTA) on growth and viability of early somatic embryos of Norway spruce (Picea abies /L./ Karst.). *Anal. Bioanal. Chem.,* 2005, vol. in press, no. pp.

[2] KLEJDUS, B., ZEHNALEK, J., ADAM, V., PETREK, J., KIZEK, R., VACEK, J., TRNKOVA, L., ROZIK, R., HAVEL, L.,KUBAN, V. Sub-picomole high-performance liquid chromatographic/mass spectrometric determination of glutathione in the maize (Zea mays L.) kernels exposed to cadmium. *Anal. Chim. Acta,* 2004, vol. 520, no. 1-2, pp. 117-124.

[3] PETRLOVA, J., POTESIL, D., MIKELOVA, R., BLASTIK, O., ADAM, V., TRNKOVA, L., JELEN, F., PRUSA, R., KUKACKA, J.,KIZEK, R. Attomole voltammetric determination of metallothionein. *Electrochim. Acta,* 2005, vol. submitted, no. pp.

[4] PALECEK, E., MASARIK, M., KIZEK, R., KUHLMEIER, D., HASSMANN, J.,SCHULEIN, J. Sensitive electrochemical determination of unlabeled MutS protein and detection of point mutations in DNA. *Anal. Chem.,* 2004, vol. 76, no. 19, pp. 5930-5936.

[5] POTESIL, D., ZELENA, J., PETRLOVA, J., ADAM, V., VACEK, J., KLEJDUS, B., ZEHNALEK, J., TRNKOVA, L., HAVEL, L.,KIZEK, R. Simultaneous femtomole determination of cysteine, reduced and oxidized glutathione, and phytochelatin in maize (*Zea mays L.*) kernels using highe-performance liquid chromatography with electrochemical detection. *J. Chromatogr. A,* 2005, vol. 1084, no. pp. 134-144.

[6] KIZEK, R., VACEK, J., TRNKOVA, L.,JELEN, F. Cyclic voltammetric study of the redox system of glutathione using the disulfide bond reducant tris(2-carboxythhyl)phosphine. *Bioelectrochemistry,* 2004, vol. 63, no. pp. 19-24.

[7] BRAZDOVA, M., KIZEK, R., HAVRAN, L.,PALECEK, E. Determination of glutathione-S-transferase traces in preparations of p53 C-terminal domain (aa320-393). *Bioelectrochemistry,* 2002, vol. 55, no. 1-2, pp. 115-118.

[8] BRDICKA, R. Application of the polarographic effect of proteins in cancer diagnosis. *Nature,* 1937, vol. 139, no. 29, pp. 330-330.

[9] HEYROVSKY, J. Polarographic research on cancer. *Nature,* 1938, vol. 142, no. 3590, pp. 317-319.

[10] BRDICKA, R. Zur frage nach der natur der polarographisch feststellbaren serumverandrungen bei Krebs. *Klinische Wochenschrift,* 1938, vol. 18, no. pp. 305-308.

[11] PALECEK, E.,PECHAN, Z. Estimation of nanogram quantities of proteins by pulse polarographic techniques. *Anal. Biochem.,* 1971, vol. 42, no. pp. 59-71.

[12] OLAFSON, R. W.,SIM, R. G. Electrochemical approach to quantitation and characterization of metallothioneins. *Anal. Biochem.,* 1979, vol. 100, no. 2, pp. 343-351.

[13] PANCHUK, R. R., PRYLUTSKA, S. V., CHUMAK, V. V., SKOROKHYD, N. R., LEHKA, L. V., EYSTIGNEEV, M. P., PRYLUTSKYY, Y. I., BERGER, W., HEFFETER, P., SCHARFF, P., RITTER, U.,STOIKA, R. S. Application of C-60 Fullerene-Doxorubicin Complex for Tumor Cell Treatment In Vitro and In Vivo. *Journal of Biomedical Nanotechnology,* 2015, vol. 11, no. 7, pp. 1139-1152.

[14] ADAM, V., PETRLOVA, J., POTESIL, D., ZEHNALEK, J., SURES, B., TRNKOVA, L., JELEN, F.,KIZEK, R. Study of metallothionein modified electrode surface behaviour in the presence of heavy metal ions - biosensor. *Electroanalysis,* 2005, vol. 17, no. 18, pp. 1649-1657.

[15] RASPOR, B. Elucidation of the mechanism of the Brdièka reaction. *J. Electroanal. Chem.,* 2001, vol. 503, no. pp. 159-162.

NANODIAMONDS ENCAPSULATED IN NANOFIBERS AS A NEW STRATEGY FOR MICRORNA-BASED THERAPY OF PROSTATE CANCER

[1,2]Giuseppe BITTI, [2]Radek DIVÍN, [3]Veronika BENSON, [1]Eva FILOVA, [2]Evžen AMLER

[1]*Institute of Experimental Medicine, THE Czech Academy of Sciences, Prague, Czech Republic, EU,* giuseppe.bitti@iem.cas.cz, eva.filova@iem.cas.cz

[2]*University Center for Energy Efficient Buildings, Czech Technical University in Prague, Buštěhrad, Czech Republic, EU,* radek.divin@uceeb.cz, evzen.amler@cvut.cz

[3]*Institute of Microbiology, The Czech Academy of Sciences, Prague, Czech Republic, EU,* benson@biomed.cas

Abstract

Given the miRNA-34 family role as tumor suppressor in cancer, it was showed that using miRNA-34a as a new anticancer agent has a great therapeutic potential. Unfortunately, biopharmaceutical issues, rapid degradation in biological fluids, poor uptake into cells and not specific distribution into the body, prevent the use of wild-type non-coding RNA by systemic administration. To overcome these limits, in our previous study we developed a new therapeutic concept based on nanotechnology delivery of miRNA-34a into prostate cancer using fluorescent nanodiamond particles coated with polyethylenimine. In the current study, in order to achieve a more conservative, stable and controlled release of miRNA-34a, we designed a system based on nanodimaonds-miRNA34a complexes encapsulated into nanofibers. We successfully encapsulate nanodiamonds into fibers of polycaprolactione, a biodegradable polymer with a slow degradation rate and a high cellular adhesion, using an electropsinning process. We cultivated prostate cancer cells (PC3 and DU145) on nanofibers scaffolds and we analyzed the presence of nanodiamonds in cells using confocal microscopy. Moreover, in these prostate cancer cell lines we observed a replacement of miRNA-34a via real-time PCR and a decreased viability testing the metabolic activity. Our preliminar results suggest a higher increased and stable mid-term replacement of oncosuppressor miRNA-34a in prostate cancer when the nanodiamonds-miRNA complexes are delivered encapsulating them in nanofibers. Therefore we provided an effective strategy for cancer therapy ready to be tested for *in vivo* experiments.

Keywords: Nanodiamonds, nanofibers, gene therapy, prostate cancer

1. INTRODUCTION

1.1. The microRNA-34a

MicroRNAs are a class of short, single non-coding RNAs (19-22 nt), that have been identified in many eukaryotic organisms. They can prevent protein expression through cleavage of specific target mRNAs or through inhibition of their translation and may conceivably play a key role in tumorigenesis [1]. Some of miRNA molecules are associated with tumor proliferation and their level is increased inside cancer cells in comparison to healthy cells. Those miRNAs, which are contribute to carcinogenesis by inhibiting tumor suppressor genes, are considered oncogenic miRNAs (OncomiRs), while downregulated miRNAs, that normally prevent cancer development by inhibiting the expression of proto-oncogenes, are known as tumor suppressor miRNAs [2]. We focused on miRNA-34a, for which was proved and decreased level in samples from patients with prostate cancers [3]. MiR-34a may exert its tumor suppressor role via targeting several signaling molecules involved in various stages of prostate cancer progression. DNA damage and oncogenic stress strongly induced miR-34 expression in a p53-dependent pathway. The distinctive features of p53 activity are promoted by miR-34 activation: induction of cell cycle block and apoptosis through down-regulation of different proteins [4]. Given the miR-34 family role as tumor suppressor in cancer, we tried to increase the levels of miR-34a in prostate cancer cell lines through replacement with miRNA mimics transfected by *in vitro* experiments using nanodiamonds as a delivery system.

1.2. Nanodiamonds and nanofibers

The possible use of nanoparticles for diagnostic and therapeutic applications has always been very attractive. The increased interest about their usage with the increased presence of nanomaterials in commercial product has raised concerns about its potential environmental pollution and toxicity effects [5]. Nanodiamonds (ND) are carbon based nanostructures with a nanoscale size (tens to hunderds of nanometers). Given the small size of the individual nanodiamond crystals, the ratio of their surface to their volume si much greater than within the large diamonds. That means that they can be coated with large amount of functional chemicals. Their size is also suitable for transportation of molecules into the the cytoplasm of the cell. Moreover, biocompatibilty of nanodiamonds was studied under various conditions and so far nanodiamonds are considered to be non-toxic and highly biocompatible, even though no rigorous clinical trials have been done yet and the questions of expulsion of nanodiamonds or their possible accumulayion in organs remain to be answered. The use of so-called high-pressure high-temperature (HPHT) nanodiamonds has opened a possibility to track the transfection directly using engineered fluorescent nitrogen-vacancy (NV) centers. NV centers in nanodiamonds are widely used for fluorescence-based detection techniques because they provide photostability, emission in the near-infrared region and lack of photoblinking [6]. These properties have been utilized in a vast range of applications, from physics tyo biological applications such as single particle tracking inside cells or long-term *in vivo* particle tracking [7]. Finally, the HPHT fluoscencent nanodiamonds surface is negatively charged upon oxidation, and strongly binds the prototypical transfection reagent PEI for delivery of macromolecules as miRNAs. transfection reagent polyethylenimine (PEI) for delivery of biomolecules as miRNAs. Nanofibers (NF) formed by electrospinning have been shown to mimic the structure and biological function of the extracellular matrix (ECM) in terms of chemical and physical structure [8]. The diameter of electrospun fibers closely matches the size scale of the ECM, which is ideal for cell attachment, proliferation, and differentiation [9]. With respect to the stereological properties of nanofibers, fiber diameter, porosity, and pore size are the most important parameters. In addition to serving as scaffolds, nanofibers meshes can be designed to serve as delivery vehicles for bioactive factors. The main aspects for the development of successful delivery systems for tissue engineering are the bioactivity of the biomolecules incorporated within the scaffolds and the controlled release of these biomolecules according to the time frame of tissue regeneration [10]. Functionalized nanofibers have been produced directly by electrospinning polymer blends/mixtures with various chemical agents such as antibiotics, cytostatics, proteins, DNA, and small interfering RNA. Moreover, the combination of nanoparticles such liposomes with polymeric scaffolds could overcome the limitations of conventional liposomes and extend their use in drug delivery [11].

2. METHODS

PC3 and DU145 prostate cancer cell lines were maintained in RPMI1640 and DMEM media respectively supplemented with 10 % of fetal bovine serum and 1 % of antibiotics. MicroRNA-34a mimic and a scrambled microRNA (scr) as a negative control were delivered into the cells using NDs-PEI and RNAiMAx lipofectamine as a standard trasfection reagent (Tr.re.) in order to compere two different types of transfection methods. Prior to use, the nanodiamonds were dissolved in water and sonicated with a probe for 30 min. The resulting transparent colloid was filtered using a 0.2 µm PVDF microfilter. Nanodiamonds diluted in water were coated with 0.09 mg/mL of polyethylenimine (PEI 800) adding the same volume of both solutions. The mix was let vortex overnight and after the solution was centrifugated in 9000 g for 1 h to create a ND-PEI pellet. This pellet was diluted in RNAse/DNAse free water to reach a final concentration of 1 mg/µL. To link microRNA we mixed

12.5 μL of 1 mg/μL of ND-PEI with 5 μL of miRNA-34a and Scramble miRNA. Fibrous meshes were prepared using a centrifugal spinning device (Cyclone 1000 L/M Forcespinning® device; FibeRio, McAllen, TX, USA). Polycaprolactione (PCL) was dissolved in a mixture of chloroform and ethanol in a volume ratio of 9:1 to make 40 % solution, then it was miced with the nanodiamonds solution. An orifice G30 at a rotation speed of 6,000× g was used to prepare the fibrous meshes. Fibers were deposited on spunbond textile using vacuum-assisted deposition. During the electrospinning process we kept an environmental condition with 60 % of humidity. Moreover the voltage value of the intensity of the electric field between the electrode supplied by the polymer solution and the collector electrode was in range between 60 kV to 70 kV with a 60 % of humidity. Before cell seeding, PCL nanofibers were cut into round patches of 6 or 10 mm diameter and sterilized using ethylene oxide. The PC-3 and Du-145 cells were seeded for 48 and 72 h in a 96-well plate for confocal microscope visualization. Cell nuclei were stained with Hoechst 33342 and for visualization we used as excitation/emission parameters 405nm/461 (Hoechst) and 559/655-755 (NDs). Subsequently the images were recorded with an Olympus FV1000 SIM confocal microscope (objective 40×/0.95) and analyzed with Olympus FLUOVIEW 2.0a software. The cells were seeded in 96-well plates in order to evaluate cells proliferation using WST-1 assay after 48 h, 72 h and 7 days of incubation. Cells were seeded in 12-well plates and after 48 h, 72 h and 7 days incubation from the treatment total RNA was isolated. Those RNA samples were used to perform Real-time PCR in order to evaluate the expressions levels or miRNA-34 and RNU44 as a control housekeeping gene.

3. RESULTS AND DISCUSSION

3.1. Nanodiamonds-nanofibers visualization

Since fluorescent nanodiamond carriers are intended for monitoring, we performed confocal microscopy to see if it is possible to detect nanodiamonds in stained dead cells and if we can see any changes in cells seeded on nanofibers with nanodiamonds coated with miRNA-34a encapsulated. For visualization we used as excitation/emission parameters 405nm/461 (Hoechst) and 559/655-755 (ND) spectrum and any nonspecific fluorescence that interferes with nanodiamond luminescence was removed by bleach treatment. After 48 hours from the treatment, we were able to visualize nanodiamonds as clusters inside and outside nanofibers. Subsequently, at 72 hours from the incubation, we were able to visualize less nanodimonds and the nanofibers were showing a lower amount of fluorescence, probably due to the increased release of nanodiamonds. However, in the DU145 samples, it was difficult to find more clusters of nanodimonds due to the tendency that this cell line in growing as a layer of cells very close to each other's. Although, we noticed a decreased number of cells in the 72 hours samples in both cell lines, as a preliminary sign that the release of these nanodimonds with microRNA-34a was starting to affect the cell proliferation. (**Figure 1**).

3.2. Detection of miRNA-34a expression

After 48 h, 72 h and 7 days incubation from treatment with miRNA-34a and scramble miRNA transfected with both transfection methods and with encpasulation in nanofibers, we isolated total RNA and subsequently we performed Real-time PCR in order to evaluate gene expressions of miRNA-34a. We observed a significant increase of miRNA-34a expression levels in the samples treated with ND-PEI complexes with values higher than values from samples treated with the standard transfection reagent. However, the highest increase of levels of miRNA-34 was found in the PC3 cells seeded on nanofibers and after 48 h of incubation. In DU245 cell line, the replacement of miRNA-43 levels was also more successful after 48 h of incubation, however in this cell line we didn't find a significant increase of miRNA-34a levels comparing the incubation of nanodiamonds with the incubation in which the cells were seeded on nanofiers encpasulating nanodiamonds. The results were normalized using RNU44 as a control houskeeping gene (**Figure 2**).

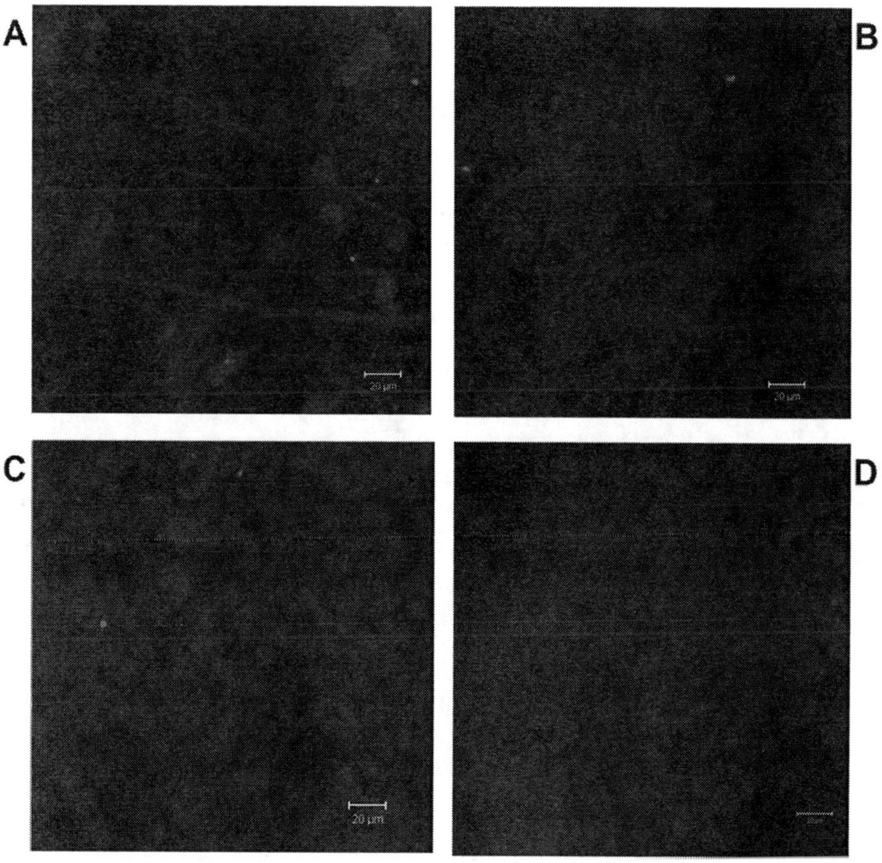

Figure 1 Images of cells seeded on nanofibers with nanodiamonds linked with microRNA encapsulated: PC3 48h (A), PC3 72h (B), DU154 48h (C), DU145 72h (D)

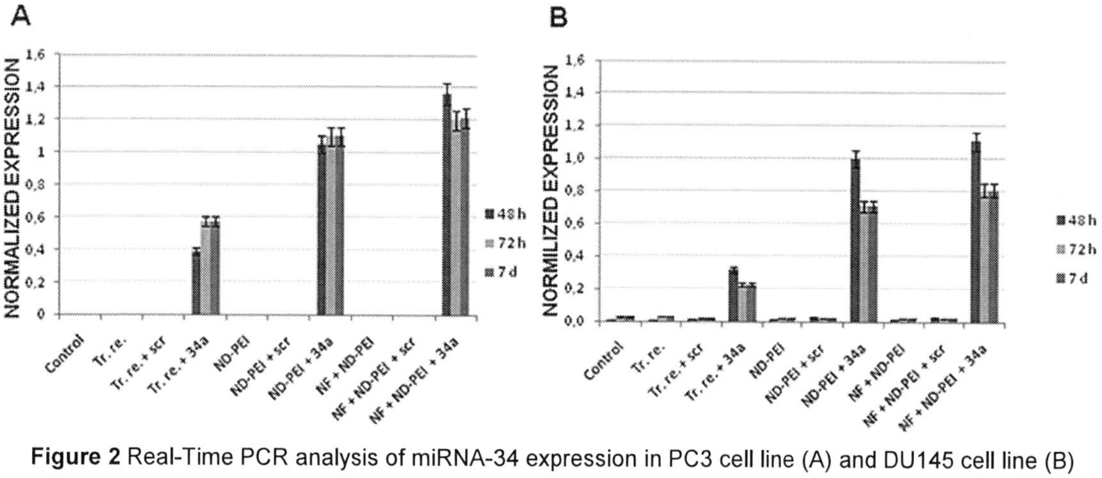

Figure 2 Real-Time PCR analysis of miRNA-34 expression in PC3 cell line (A) and DU145 cell line (B)

3.3. Evaluation of cell proliferation

The WST-1 assay was performed in order to evaluate cell proliferation in PC3 and DU145 treated with miRNA-34a and scramble miRNA transfected using ND-PEI complexes and standard transfection reagent .

Absorbance measurements were recorded after 48 and 72 h and 7 days of incubation from treatment and control samples were obteined from non-treated cells. Moreover we compared these results with the proliferation observed seeding these cell lines on nanofiber encapsulating our nanodiamonds complexes. We evaluated percentage of cell proliferation normalizing the absorbance results obteined from each samples with control samples (100 %). We observed a decrease of cell proliferation in both cell lines in samples treated with ND-PEI-miRNA34a with results similar to samples treated with the standard transfection reagent. However, the most significant difference was found when the cell lines were seeded on nanofibers after 7 days of incubation (**Figure 3**).

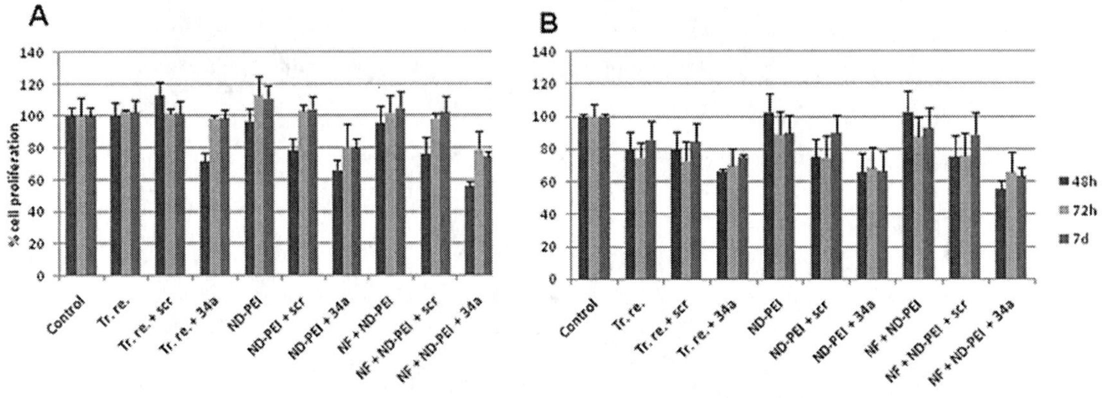

Figure 3 Cell proliferation analysis in PC3 cell line (A) and DU145 (B) cell line

4. CONCLUSION

In our previous study, we compared the transfection efficiency between a nanodiamons system and standard transfection reagent in order to delivery miRNA-34a, a well-known tumor suppressor miRNA, into prostate cancer cell lines. Our preliminary results suggested that the ND-based delivery systems seem to be a promising strategy to delivery miRNA-34a into prostate cancer in order to achieve a safe and effective targeting of the tumor. Given these promising results, in this study we observed the efficiency of such delivery system, encapsulating our nanodiamonds complexes in nanofibers of polycaprolactione made by electrospinning. We performed in vitro experiments treating PC3 and DU145 prostate cancer cell lines, and the first step was being able to visualize the fluorescence of these nanodiamonds inside the nanofibers. We noticed a decrease of fluorescence after 72 h, indicating the nanofibers' capability of releasing ND complexes coated with miRNA-34 mimic. Moreover, using WST-1 assay, we observed a significant decrease in cell proliferation in samples treated with ND-PEI-miRNA34a, especially when the cells were incubated seeding them on nanofibers and after 7 days after the treatment. However using Real-time PCR, we observed in both cell lines a very significant increase of miRNA-34a levels in samples treated with ND-PEI-miRNA34 encapsulated in nanofibers starting from the 48 h incubation time. These new results prove that seeding the cells on nanofibers encapsulating nanodiamonds can induce a constant release of these complexes even after 7 days. Overall, polycaprolactione nanofibers could be considered as a valid delivery system for a stable mid-term replacement of oncosuppressor miRNA-34a linked on fluorescent nanodiamonds.

ACKNOWLEDGEMENTS

The project was supported by the Ministry of Education, Youth and Sports of the Czech Republic under NPU: LO1309 and LO1508.

REFERENCES

[1] CALIN GA, Croce CM. MicroRNA signatures in human cancers. Nat Rev Cancer. 2006;6:857-866.

[2] ESQUELA-KERSCHER A, SLACK FJ. Oncomirs - microRNAs with a role in cancer. Nat Rev Cancer. 2006;6:259-269.

[3] LIU C, KELNAR K, LIU B, CHEN X, CALHOUN-DAVIS T, LI H, et al. The microRNA miR-34a inhibits prostate cancer stem cells and metastasis by directly repressing CD44. Nat Med. 2011;17:211-215.

[4] YAMAKUCHI M, LOWENSTEIN CJ. MiR-34, SIRT1 and p53: the feedback loop. Cell Cycle. 2009;8:712-715.

[5] V. L. COLVIN, The potential environmental impact of engineered nanomaterials, Na. Biotech, 2003, 21, 1166-1170

[6] V. VAIJAYANTHIMALA, H-C. CHANG, Functionalized fluorescent nanodiamonds for biomedical applications, Nanomedicine, 2009, 4,1,47-55

[7] L. RONDINI, G. DANTELLE, A. SLABLAB et al. Surface-induced charge state conversion of nitrogen-vacancy defects in nanodiamonds, Phys. Rev. B, 201, 82, 115499-115454.

[8] GARG K, BOWLIN GL. Electrospinning jets and nanofibrous structures.Biomicrofluidics 2011, 5, 13403.

[9] NATHAN A. HOTALING, KAPIL BHARTI et al. DiameterJ: A Validated Open Source Nanofiber Diameter Measurement Tool. Biomaterials 2005, 26, 3929–3939.

[10] JI W SU, Y, YANG, F; VAN DEN BEUCKEN J. et al. Biodegradable electrospun nanofibers coated with platelet-rich plasma for cell adhesion and proliferation. . Pharm. Res. 2011, 28, 1259–1272.

[11] SILL T. J, VON RECUM H. A.Electrospinning: applications in drug delivery and tissue engineering. Biomaterials 2008, 29, 1989–2006.

UNIVERSAL PLATFORM FOR SERS DETECTION OF "NORMAL" AND "TUMOR" CELLS CULTIVATION MEDIA

[1,2]Mariia ERZINA, [1]Andrii TRELIN, [1,2]Olga GUSELNIKOVA, [1,2]Yevgeniya KALACHYOVA, [1]Anastasia PERMINOVA, [2]Pavel POSTNIKOV, [1]Vaclac ŠVORČÍK, [1,2]Oleksiy LYUTAKOV

[1]VŠCHT - University of Chemistry and Technology Prague, Prague, Czech Republic, EU, lyutakoo@vscht.cz
[2]Tomsk Polytechnic University, Research School of Chemistry and Applied Biomedical Sciences, Tomsk, Rusian Federation

Abstract

Surface-enhanced Raman scattering (SERS) has become an attractive analytical tool for recognition of various biomolecules and bio-objects. In this work, we created a universal platform, based on surface modified gold multibranched nanoparticles (AuMs), plasmon coupling between surface plasmon polariton waves (SPP) and localized surface plasmon (LSP), and utilization of advanced statistical methods for analysis and estimation of SERS spectra, measured on the „tumor" cells cultivation media. At the first stage, we used the AuMs which have an extremely high number of plasmonic hot spots per single nanoparticle and provide excellent SERS signal intensification. The surface of AuMs was decorated with charged organic functional moieties to entrap various (bio) molecules from targeted solution. Chemical and morphological structures of the prepared nanoparticles were examined with different techniques such as SEM-EDX and Raman spectroscopy. The decorated AuMs were placed in the cultivation media of melanoma and fibroblast lines in vitro. After interaction with cultivation media the AuMs were drop-deposited on the periodical gold grating, to achieve SPP-LSP coupling and even more enhance the SERS signal from entrapped (bio)molecules. Raman spectra were collected from several kinds of melanoma cells lines and evaluated using the developed mathematical route.

Keywords: Gold nanoparticles, SERS detection, cultivation media, arenediazonium tosylates

1. INTRODUCTION

SERS spectroscopy is a well-known analytical tool [1-4], that offers a number of advantages with other methods of analysis such as Fourier transform infrared (IR) spectroscopy, UV-vis absorption, fluorescence, nuclear magnetic resonance (NMR) and others [5-7]. Also SERS has become an attractive analytical tool for recognition of various biomolecules especially in tandem with plasmonic gold multibranched nanoparticles (AuMs) [8].

AuMs comparing to other nanoparticles (NPs) are famous with their biocompatibility, chemical stability, and plasmon tunability [8,9]. Moreover, they have controllable synthesis, easy surface modification, high molar absorption coefficient and plasmon peaks that are tunable into the near-infrared (NIR) region. It means that AuMs are ideal analysis tools for biochemical assays [9].

However present of the art state of SERS does not allow performing the analyses of the "big" biomolecules, which represent one of central, but up-to-now unsolved challenge in this field. Many attempts to solve this question were performed through the creation of sophisticated SERS substrates (where the targeted molecule is sandwiched between the metal surface and metal nanoparticles) to enhance the SERS and prevent the Raman peak interference; specific functionalization of SERS-active surfaces (with the aim to reach the very definite "specific' entrapping of biomolecules and minimize unspecific sorption); or introduction of advanced algorithm of SERS data analysis. In this study, we aimed to decorate AuMs with charged organic functional moieties to entrap (bio) molecules from the cultivation media of melanoma and fibroblast lines and collect Raman spectra from several kinds of melanoma cells lines and evaluated using the developed non-linear mathematical route (DMR). To our knowledge this is the first report of detection melanoma and fibroblast lines with DMR. Our results indicate that discrimination of a complex biological systems is possible with the implementation of advanced data evaluation.

2. EXPERIMENTAL

2.1. Materials

Diethyl ether, deionized water, methanol (purist, p.a., absolute, ≥99.8 % (GC)), Chloroauric acid tetrahydrate (HAuCl4·4H2O, 99.9 %), silver nitrate (AgNO3,99.0 %), ascorbic acid (AA, 99.0 %), p-toluenesulfonic acid, acetic acid, terc-butyl nitrite, 4-nitroaniline, 4-aminobenzoic acid (+ 99 %), 4-aminophthalic acid (97.0 %), all were purchased from Sigma-Aldrich and used without further purification.

2.2. Gold grating preparation

Su-8 films were spin-coated onto freshly cleaned glass substrates and patterned using the linearly polarized excimer laser irradiation according to the procedure described in [10]. Gold thin films were deposited onto the patterned surface by vacuum sputtering.

2.3. Gold nanoparticles preparation

Gold nanoparticles (AuMs) were synthesized by the seed-mediated growth method with some modifications [11]. Briefly, seed solution was prepared by mixing boiling HAuCl4·4H2O solution (1 mM, 100 ml) with 1 % citrate solution (15 ml) under vigorous stirring. The solution was boiled 15 min, cooled, filtered by a 0.22 µm nitrocellulose membrane and kept in a dark, cold place. Nanostars were prepared by adding the seed solution (100 µl) to HAuCl4 solution (0.25 mM, 10 ml) with HCl (1 M, 10 µl). After that AgNO3 (20 mM, 100 µl) and AA (100 mM, 50 µl) were added simultaneously. After color change AuMs were immediately centrifugal washed [12].

2.4. Diazonium modification

4-Aminobenzenediazonium tosylate (ADT-NH2), 4-carboxybenzenediazonium tosylate (ADT-COOH) and 3,4-dicarboxybenzenediazonium tosylate ADT-(COOH)2 were prepared according to the published procedure [13,14]. AuMs were spontaneously modified due to the published procedure [12].

2.5. Entrapping of biomolecules expressed by the normal and tumor-associated cell lines

Decorated AuMs were placed in the 0.35 mkl of analyte (cultivation medium, cultivation medium from health cells, cultivation medium from melanoma associated cells) and put in a cold, dark place for 3 hours. After that there were taken drops of each analyte and placed on the gold grating. When drops were dried SERS spectra were measured immediately.

3. RESULTS AND DISCUSSION

A schematic representation of the multibranched gold nanoparticles (AuMs) synthesis and surface modification is showing in **Figure 1**.

Prepared AuMs were decorated by several arenediazonium tosylate salts (ADT-NH_2, ADT-COOH, ADT-$(COOH)_2$) with the aim of changing the surface charge of the NPs and introducing functional groups. The AuMs shape and size distribution were examined by TEM after preparation, modification and purification procedures.

The results are presented in **Figure 2**. It is apparent that modification of AuMs did not change the shape of nanoparticles, which contain a great number of plasmonic hot-spots - sharp edges.

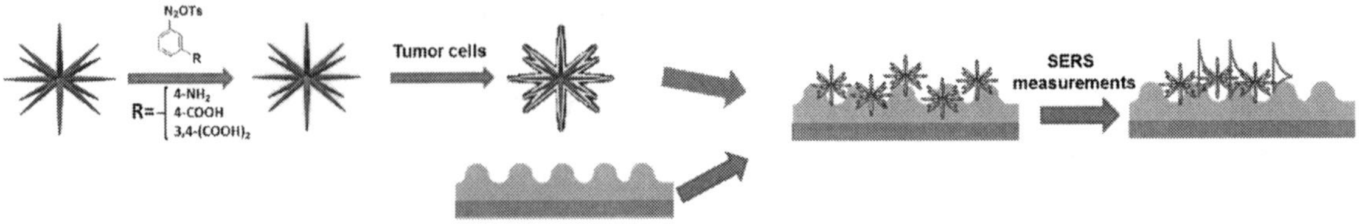

Figure 1 Schematic representation of AuMs synthesis and surface modification

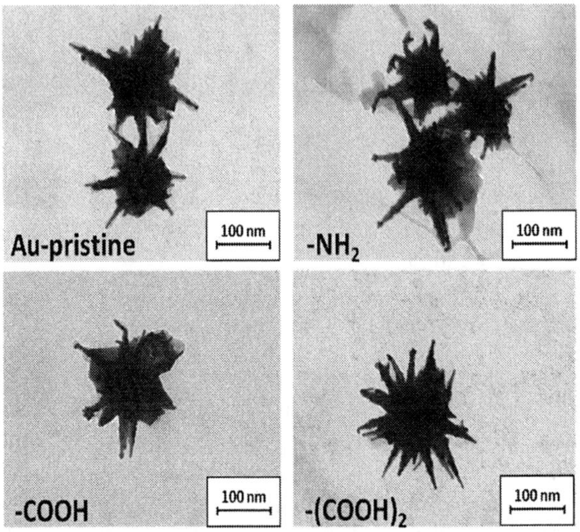

Figure 2 TEM images of modified gold multibranched nanoparticles. The scale bar is 100 nm.

The results of AuMs surface modification and organic functional group (OFG) grafting were also checked by SEM-EDS (**Figure 3**). EDS analysis confirmed the presence of OFGs characteristic elements on the NP surfaces with a spatial distribution correlating to the physical positions of AuMs on the SEM scans (**Figure 3**). The element ratio values (presented in the Table below **Figure 3**) correlate well with the composition of the attached OFGs.

Figure 4 presents the absorption spectra of the gold grating where the apparent plasmon-polariton absorption band at 780 nm is well visible (i.e. the effective excitation of SPP takes place). Additionally, **Figure 4** shows the absorption spectra of pristine and modified AuNPs. As evident, the main plasmon absorption peak of AuNPs is also located near the 780 nm. It must be also noted that used AuMs have an extremely high number of plasmonic hot spots per single nanoparticle that provide excellent SERS signal intensification, which can be further enhanced through the SPP wave, after the deposition of AuNPs on the gold grating surface.

After interaction with cultivation media the AuMs were drop-deposited on the periodical gold grating, to achieve surface plasmon polariton waves - localized surface plasmon (SPP-LSP) coupling and even more enhance the SERS signal from entrapped (bio)molecules. Raman spectra were collected from several kinds of melanoma cells lines and evaluated using the developed mathematical route. The SERS spectra of line A2058 tumor melanocytes are presented in **Figure 5**. As is evident, the as-measured SERS spectra do not contain any apparent Raman band and the evaluation of results seems to be complicated. However, utilization of non-linear mathematical algorithm provide an possibility to effectively process the Raman signal and clearly distinguish the spectral areas, attributed to the different kinds of (bio) molecules.

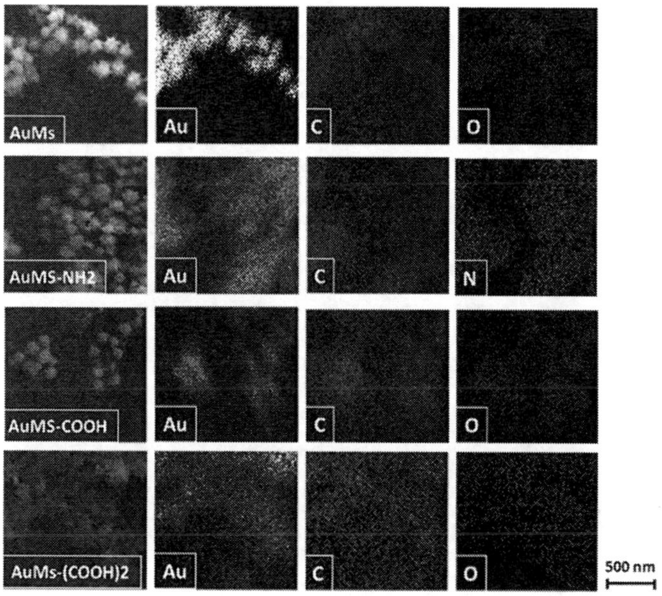

Element	Apparent Concentration, wt%			
	Au-Ms	AuMs-NH2	AuMs-COOH	AuMs-(COOH)2
C	8,8	19,1	21,5	23,8
N	-	3,7	-	-
O	0,9	-	3,2	3,5
Au	90,3	77,2	75,3	72,7

Figure 3 SEM-EDS analysis of pristine and modified (-NH2, -COOH, -(COOH)2) AuMS images of modified gold multibranched nanoparticles. The weight concentrations of organic elements

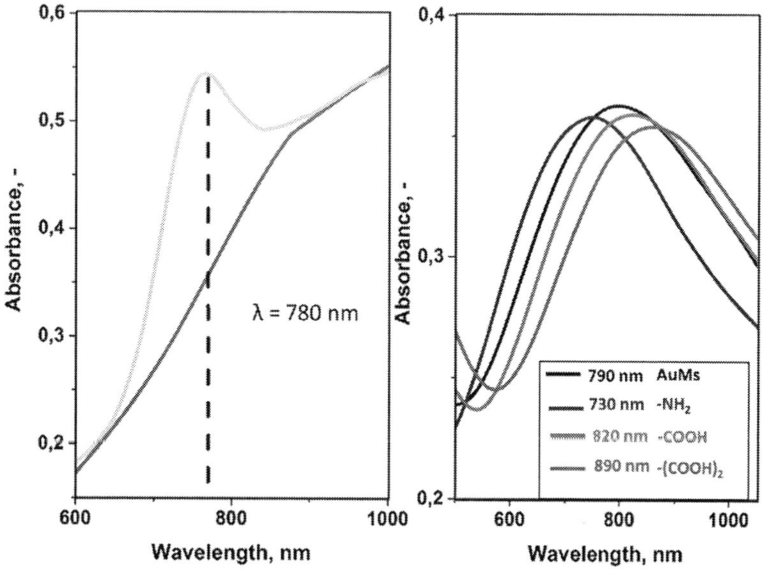

Figure 4 UV-Vis absorption spectra of the gold grating and pristine and modified (AuMs-NH2, AuMs-COOH, AuMs-COOH)$_2$) NPs

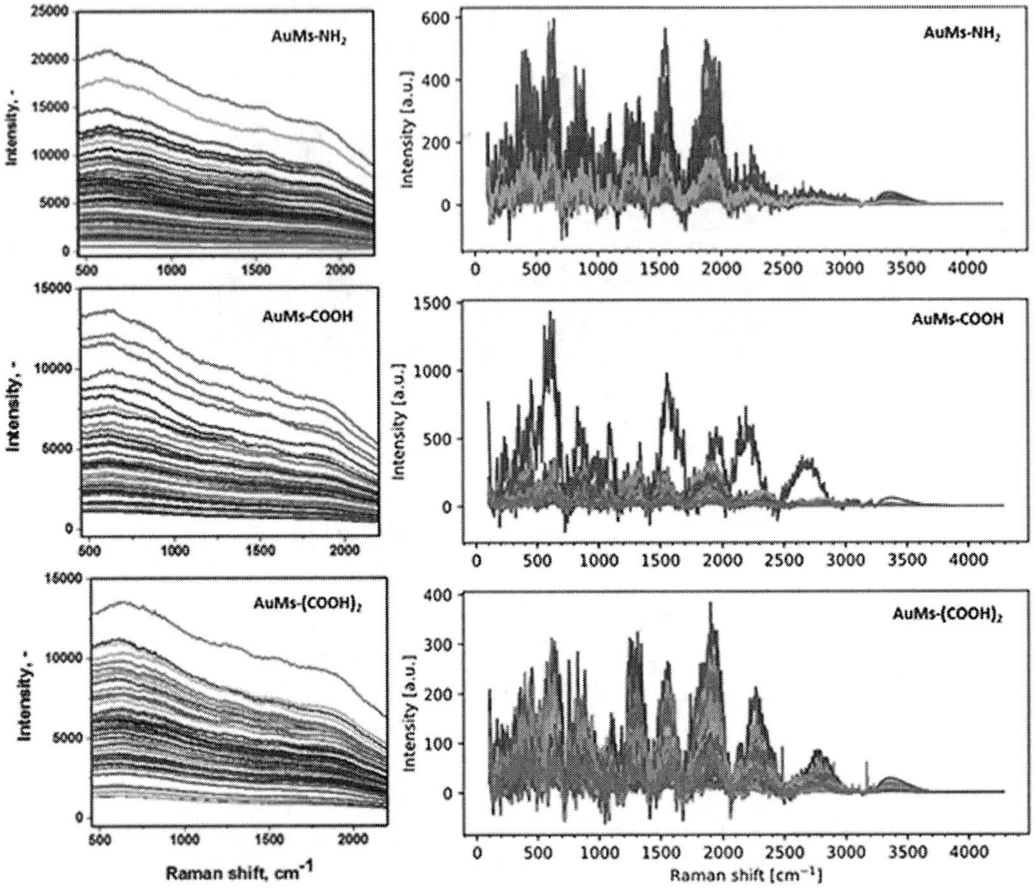

Figure 5 SERS spectra of line A2058 tumor melanocytes deposited on the AuMs decorated with -NH2, -COOH, -(COOH)2 organic groups to entrap (bio)molecules from solution (excitation wavelength - 785 nm)

4. CONCLUSION

We have developed and analytical route to apply the SERS spectroscopy for the analysis of sophisticated (bio)objects. Proposed procedure includes the combination of gold nanoparticles surface functionalization, utilization of plasmonic coupling and advanced Raman spectra evaluation using non-linear mathematical algorithms. In particular, we analysis the melanoma lines with DMR using Raman spectra of decorated AuMs with charged organic functional moieties and perform the transition from the initially complexed spectrum to the clearly distinguishable Raman information.

ACKNOWLEDGEMENTS

This work was supported by the Grant Agency of Health Ministry 15-33459A.

REFERENCES

[1] Martin. Surface-enhanced Raman Spectroscopy: A Brief Retrospective. Journal of raman spectroscopy journal of raman spectroscopy. 2005. vol. 36. pp. 485-496

[2] KNEIPP, Katrin. Surface-enhanced Raman scattering. Physics Today. 2007. vol. 60. pp. 40-46

[3] BRUS, Louis. Noble Metal Nanocrystals: Plasmon Electron Transfer Photochemistry and Single-Molecule Raman Spectroscopy. Accounts of Chemical Research. 2008. vol. 41. pp. 1742-1749

[4] STILES, Paul L., DIERINGER, Jon A., SHAH, Nilam C., and VAN DUYNE, Richard P. Surface-Enhanced Raman Spectroscopy. Annual Review of Analytical Chemistry. 2008. vol. 1. pp. 601-626

[5] PETRY, Renate, SCHMITT, Michael and POPP, Jürgen. Raman Spectroscopy--a Prospective Tool in the Life Sciences. *Chemphyschem: A European Journal of Chemical Physics and Physical Chemistry*. 2003. vol. 4. pp. 14-30

[6] BAENA, Josefa R. and LENDL, Bernhard. Raman Spectroscopy in Chemical Bioanalysis. Current Opinion in Chemical Biology. 2004. vol.8. pp. 534-539

[7] ABALDE-CELA, Sara, ALDEANUEVA-POTEL, Paula, MATEO-MATEO, Cintia, RODRÍGUEZ-LORENZO, Laura, ALVAREZ-PUEBLA, Ramón A. and LIZ-MARZÁN, Luis M. Surface-Enhanced Raman Scattering Biomedical Applications of Plasmonic Colloidal Particles. Journal of the Royal Society Interface. 2010. vol. 7, pp. 435-450

[8] YUAN, Hsiangkuo, KHOURY, Christopher G., HWANG, Hanjun, WILSON, Christy M., GRANT, Gerald A.and VO-DINH, Tuan. Gold Nanostars: Surfactant-Free Synthesis, 3D Modelling, and Two-Photon Photoluminescence Imaging. Nanotechnology. 2012. vol. 23. pp. 075102

[9] CHEN, Yiping, XIANYU, Yunlei and JIANG, Xingyu. Surface modification of gold nanoparticles with small molecules for biochemical analysis. Accounts of chemical research. 2017: vol. 50.2. pp. 310-319

[10] SVANDA, J., KALACHYOVA, Y., SLEPICKA, P., SVORCIK, V., & LYUTAKOV, O. Smart component for switching of plasmon resonance by external electric field. ACS applied materials & interfaces. 2015. vol. 8(1). pp. 225-231

[11] KALACHYOVA, Yevgeniya, MARES, David, JERABEK, Vitezslav, ULBRICH, Pavel, LAPCAK, Ladislav, SVORCIK, Vaclav and LYUTAKOV, Oleksiy. Ultrasensitive and reproducible SERS platform of coupled Ag grating with multibranched Au nanoparticles. Physical Chemistry Chemical Physics. 2017. vol. 19(22). pp. 14761-14769.

[12] KALACHYOVA, Yevgeniya, OLSHTREM, Anasiya, GUSELNIKOVA, Olga, POSTNIKOV, Pavel, ELASHNIKOV, Roman, ULBRICH, Pavel, RIMPELOVA, Silvie, ŠVORČÍK, Václav and Lyutakov, Oleksiy. Synthesis, Characterization, and Antimicrobial Activity of Near-IR Photoactive Functionalized Gold Multibranched Nanoparticles. 2017. ChemistryOpen. vol.6(2). pp. 254-260

[13] GUSELNIKOVA, Olga, POSTNIKOV, Pavel, ELASHNIKOV, Roman, TRUSOVA, Marina, KALACHYOVA, Yevgeniya, LIBANSKY, Milan, BAREK, Jiri, KOLSKA, Zdenka, ŠVORČÍK, Vaclav and LYUTAKOV, Oleksiy. Surface modification of Au and Ag plasmonic thin films via diazonium chemistry: Evaluation of structure and properties. 2017. Colloids and Surfaces A: Physicochemical and Engineering Aspects. vol. 516. pp.274-285

[14] GUSELNIKOVA, O., POSTNIKOV, P., ERZINA, M., KALACHYOVA, Y., ŠVORČÍK, V., and LYUTAKOV, O. Pretreatment-free selective and reproducible SERS-based detection of heavy metal ions on DTPA functionalized plasmonic platform. Sensors and Actuators B: Chemical. 2017. vol. 253. pp. 830-838

SUPERPARAMAGNETIC IRON OXIDE NANOPARTICLES (SPIONS) MODIFIED WITH SARCOSINE OXIDASE - ENZYMATIC ACTIVITY ANALYSIS BY SDS-PAGE

[1]Dagmar UHLÍŘOVÁ, [1]Martina STAŇKOVÁ, [1]Michaela DOČEKALOVÁ, [2]Božena HOSNEDLOVÁ, [3]Marta KEPINSKA, [2]Branislav RUTTKAY-NEDECKÝ, [1]Josef RŮŽIČKA, [4]Carlos FERNANDEZ, [3]Halina MILNEROWICZ, [1,2,3]René KIZEK

[1]Department of Research and Development, Prevention Medicals s.r.o., Studénka-Butovice, Czech Republic, EU, uhlirova@preventionmedicals.cz
[2]Department of Human Pharmacology and Toxicology, University of Veterinary and Pharmaceutical Sciences Brno, Brno, Czech Republic, EU, kizek@sci.muni.cz
[3]Department of Biomedical and Environmental Analyses, Faculty of Pharmacy with Division of Laboratory Diagnostics, Wroclaw Medical University, Wroclaw, Poland, EU zalewska.m@gmail.com
[4]School of Pharmacy and Life Sciences, Robert Gordon University, Scotland, United Kingdom, c.fernandez@rgu.ac.uk

Abstract

Sarcosine oxidase (SOX) is an enzyme that catalyzes the oxidative demethylation of sarcosine with the glycine as a product and is physiologically active in human body and other mammals. However prostate cancer cells have a high expression of sarcosine. In this study the superparamagnetic iron oxide nanoparticles (SPIONs) were prepared and their surface was modified with gold nanoparticles (AuNPs). These AuNPs were modified with chitosan (CS) and SOX. Obtained AuNPs were characterized by physicochemical methods, such as dynamic light scattering or spectrophotometry, where the pseudo-peroxidase activity of the AuNPs was used. Hydrogen peroxide decomposes because of the pseudo-peroxidase activity with the appearance of a blue coloration of 3,3',5,5'-tetramethylbenzidine (TMB). For the analysis of AuNPs enzymatic activity the SDS-PAGE with silver staining has been used. Gels (7.5 %) were prepared using acrylamide stock solution 30 % (m / V) with bisacrylamide 1 % (m / V). Separating gel contained: acrylamide 7.5 % (m / V), bisacrylamide 0.5 % (m / V), 0.4 M Tris/HCl, 0.1 % (m / V) sodium dodecyl sulfate (SDS), pH 8.8. Stacking gel contained: 4.5 % acrylamide (m / V), 0.15 % bisacrylamide (m / V), 0.1 % SDS (m / V), 0.1M Tris/HCl, pH 6.8. Nanoconstructs were diluted 2:1 with a loading buffer (PLB Max). Each well contained 15 µl of the diluted solutions. Electrophoretic measuring conditions were: 120 V, 1.5 hours in a running buffer (24mM Tris, 0.2M glycine and 3mM SDS). After measurement the gel was stained with silver, scanned and evaluated by Colortest in the laboratory system Qinslab. The SPIONs or AuNPs cannot be detected themselves alone using SDS-PAGE, therefore this method served as a confirmation, that all parts of the nanoconstruct are connected and we are able to analyze them and use them for other research or possible diagnostic purposes.

Keywords: Superparamagnetic iron oxide nanoparticles, gold nanoparticles, sarcosine oxidase, chitosan, enzymatic activity, SDS-PAGE

1. INTRODUCTION

Sarcosine is an amino acid that is formed by methylation of glycine and is present in trace amounts in the body. Increased sarcosine concentrations in blood plasma and urine are manifested in sarcosinemia and in some other diseases such as prostate cancer [1-3]. For this purpose, sarcosine detection using the nanomedicine approach was proposed. In this study, we have prepared superparamagnetic iron oxide nanoparticles (SPIONs) with different modified surface area. Gold nanoparticles (AuNPs) were modified by chitosan (CS), and sarcosine oxidase (SOX). SPIONs without any modification were taken as controls.

2. MATERIAL AND METHODS

All chemicals were purchased from Merck. All reagents were of analytical grade and used without any further purification. Ninhydrin, hydrindatin were purchased from Ingos, (Czech Republic). Aqueous solutions for size analysis were prepared using PURELAB® Ultra (Elga, High Wycombe, United Kingdom) resistivity 18 MΩ-cm. For other purposes deionized water was used. The obtained NPs were characterized by physicochemical methods. Structure of NPs was characterized by scanning electron microscopy (SEM). For documentation of the NPs structure, the MIRA3 LMU (Tescan, Brno, Czech Republic) was used. Absorbance scan was carried out in the range from 300-850 nm by 2 nm steps. All measurements were performed at 22 °C (V UV-3100PC, VWR, Germany). The absorbance or spectra in plate (Brand, Germany) were recorded by using a reader Infinite M200 (Tecan, Männedorf, Switzerland). The size distribution (i.e. the hydrodynamic diameter, DH) was determined by dynamic light scattering (DLS) using the Zetasizer Nano ZS ZEN3600 (Malvern Instruments, Malvern, UK) with the detection angle of 173° in optically homogeneous square polystyrene cells. For the analysis of SPIONs SOX concentration the SDS-PAGE with silver staining has been used. Gels (7.5 %) were prepared using acrylamide stock solution 30 % (m / V) with bisacrylamide 1 % (m / V). Separating gel contained: acrylamide 7.5 % (m / V), bisacrylamide 0.5 % (m / V), 0.4 M Tris/HCl, 0.1 % (m / V) sodium dodecyl sulfate (SDS), pH 8.8. Stacking gel contained: 4.5 % acrylamide (m / V), 0.15 % bisacrylamide (m / V), 0.1 % SDS (m / V), 0.1M Tris/HCl, pH 6.8. Nanoconstructs were diluted 2:1 with a loading buffer (PLB Max). Each well contained 15 µl of the diluted solutions. Electrophoretic measuring conditions were: 120 V, 1.5 hours in a running buffer (24mM Tris, 0.2M glycine and 3mM SDS). After measurement the gel was stained with silver, scanned and evaluated by Colortest in the laboratory system Qinslab. The SPIONs or AuNPs cannot be detected themselves alone using SDS-PAGE, therefore this method served as a confirmation, that all parts of the nanoconstruct are connected and we are able to analyze them and use them for other research or possible diagnostic purposes.

3. RESULTS

The obtained NPs were characterized by physicochemical methods. The size of the NPs determined by the dynamic light scattering method was as follows: SPIONs/Au/NPs (100-300 nm), SPIONs/Au/CS/NPs (300-700 nm), and SPIONs/Au/CS/SOX/NPs (600-1500 nm). The SPIONs surface was modified with CS. The amount of CS deposited on the NP surface was found to be 48 mg/mL and 39 mg/mL for SPIONs/Au/CS/NPs and SPIONs/Au/CS/SOX/NPs, respectively, and it repeatability varied around 10 %. Pseudo-peroxidase activity of NPs was verified using sarcosine, horseradish peroxidase (HRP) and 3,3' 5,5'-tetramethylbenzidine (TMB) substrate (**Figure 1**).

We used the following particle types for gel electrophoresis: SPIONs, chitosan modified SPIONs, SPIONs with SOX, SPIONs with SOX and horseradish peroxidase (HRP), SPIONs with SOX, HRP and dithiotreitol (DTT), SPIONs with SOX, HRP and zinc. The last type of nanoparticles, were SPION nanoparticles with SOX, HRP, DTT and zinc. In gel electrophoresis, SOX was shown as tetramer, therefore we followed 4 bands on a gel electrophoresis, which we evaluated individually. **Figure 2** shows the calibration curve of SOX at a concentration range, up to10 µg/mL. The calibration curve is evaluated using Colortest in Qinslab laboratory system. This is the dependence of the density (expressed by LOG (Io / I)) and the concentration of the enzyme.

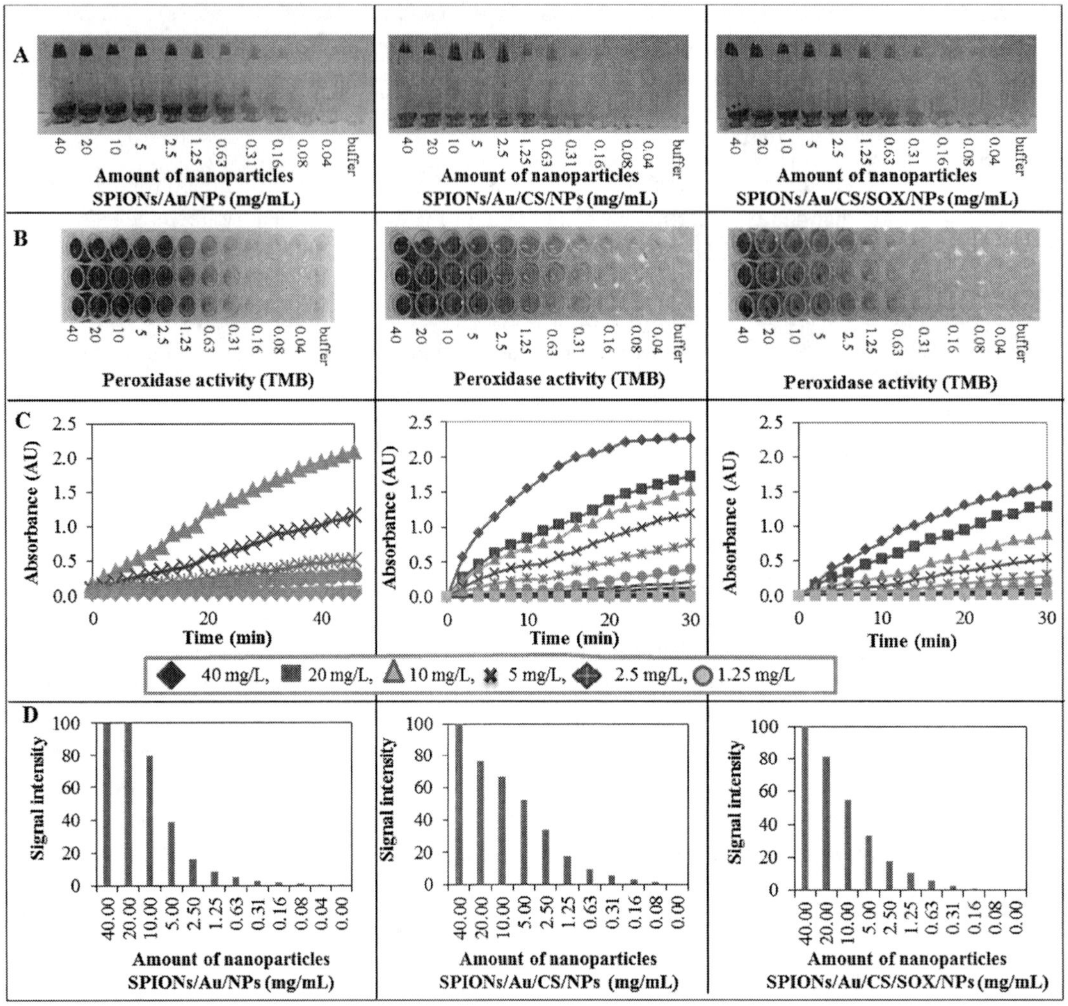

Figure 1 Pseudo-peroxidase activity of SPIONs/Au/NPs, SPIONs/Au/CS/NPs, SPIONs/Au/CS/SOX/NPs. Photographs (camera Canon PowerShot SX610 HS, 20.2 Mpx CMOS sensor, ultra wide-angle lens with 18x superzoom) showing (A) a typical appearance of individual used types of SPIONs in a range of concentrations from 0 to 40 mg/mL in 50 mM phosphate buffer (pH 8.0) and (B) a typical color course of pseudo-peroxidase reaction with SPIONs/Au/NPs in a range of concentrations from 0 to 40 mg/mL; representative image; (C) Time course of pseudo-peroxidase reaction (0 to 30 min) for SPIONs/Au/NPs in a range of concentrations from 0 to 10 mg/mL; SPIONs/Au/CS/NPs and SPIONs/Au/CS/SOX/NPs (0–40 mg/mL); (D) Signal intensity of SPIONs/Au/NPs in dependence on pseudo-peroxidase reaction signal in a range of concentrations from 0 to 40 mg/mL. Composition of reaction mixture for the measurement of pseudo-peroxidase activity was as follows: 5 mM TMB, hydrogen peroxide 30 %, 0.5 M acetate buffer pH 4. The measurement was carried out in the Brand plate after washing with 18 MΩ water, at 650 nm in 1 min reading interval, each value being the average of 5 repeated measurements.

SDS-PAGE electropherogram of SPIONs is shown in **Figure 3**. As negative controls SPIONs and SPIONs modified with CS alone were used, when no bands appeared on the gel. For the other nanoparticles, 4 bands of SOX appeared, the intensity of which was evaluated by Colortest. If DTT and zinc were coupled with the enzymes, the band intensity was low. Conversely, if we added FAD or magnesium chloride, the band intensity improved.

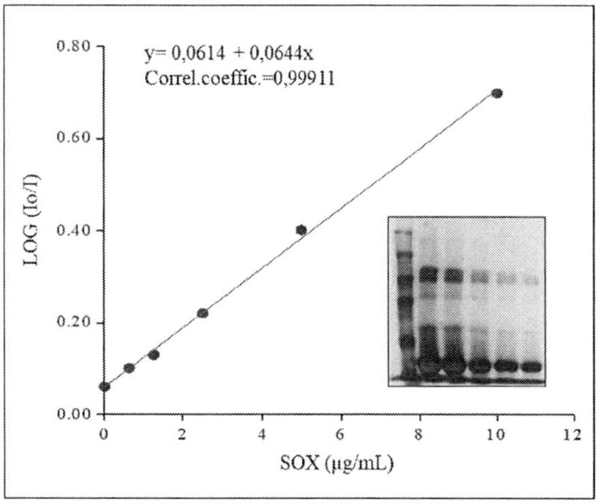

Figure 2 Calibration curve of sarcosine oxidase at concentrations of 0, 0.62, 1.25, 2.50, 5.00 and 10.00 µg / mL. The curve is evaluated using the Colortest in the Qinslab laboratory system. This is the dependence of the density (expressed by LOG (Io / I)) and the concentration of the enzyme. The inset shows different concentrations of sarcosine oxidase on gel electrophoresis from the highest concentration to the lowest.

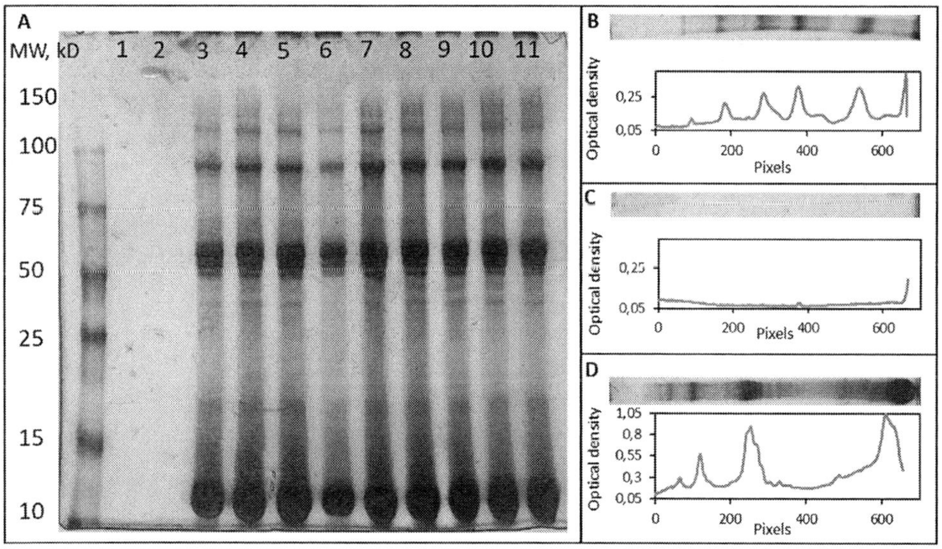

Figure 3 A) On gel electrophoresis, the order of samples behind the protein ladder is the following 1-SPIONs without modification; 2 - chitosan modified SPIONs; 3 - SPIONs modified with chitosan and sarcosine oxidase; 4 - SPIONs modified with chitosan, sarcosine oxidase and 1mM DTT; 5-Spions modified with chitosan, sarcosine oxidase, 1 mM DTT and 0.2 mg Zn; 6 - SPIONs modified with chitosan, sarcosine oxidase and 1 mg Zn; 7 - SPIONs modified with chitosan, sarcosine oxidase and 0.5 mg Zn; 8 - SPIONs modified with chitosan, sarcosine oxidase and 0.2 mg Zn; 9-SPIONs modified with chitosan, the enzyme sarcosine oxidase and 0.1 mg Zn; 10-SPIONs modified with chitosan, the enzyme sarcosine oxidase and 0.05 mg of Zn; 11 - SPIONs modified with chitosan, enzyme sarcosine oxidase and 0.025 mg Zn B) Evaluation of ladder by Colortest. The fragments are displayed by size. C) Evaluation of the first sample (SPION nanoparticles) using Colortest. In the picture it can be seen that no band was created. D) Evaluation of SPION specimen modified with chitosan, sarcosine oxidase enzyme and 1mM DTT again using Colortest. The individual peaks are beautifully visible on the graph, representing bands on the gel.

4. CONCLUSION

The different types of SPIONs particles with SOX were prepared. Pseudo-peroxidase activity of SPIONs was verified using sarcosine, horseradish peroxidase (HRP) and 3,3' 5,5'-tetramethylbenzidine (TMB) substrate. The SOX presence in SPIONs was verified by SDS-PAGE electrophoresis. We prepared calibration curve of SOX at a concentration range up to10 µg/mL. The calibration curve was evaluated using Colortest in Qinslab laboratory system. The limit of detection of this SOX determination in SPIONS using SDS-PAGE electrophoresis was 0,62 µg/mL.

ACKNOWLEDGEMENTS

The article has been funded by the grant SARCOTEST 165/2016 and the international collaboration project of The European Technology Platform for Nanomedicine. Zeta potential and particle size measurements were carried out in the Laboratory of Elemental Analysis and Structural Research, Faculty of Pharmacy with Division of Laboratory Diagnostics, Wroclaw Medical University. Carlos Fernandez would like to express also his gratitude to Robert Gordon University for its support.

REFERENCES

[1] BAR-JOSEPH, I., PRAS, E., REZNIK-WOLF, H., MAREK-YAGEL, D., ABU-HORVITZ, A., DUSHNITZKY, M., GOLDSTEIN, N., RIENSTEIN, S., DEKEL, M., PODE-SHAKKED, B., ZLOTNIK, J., BENARROSH, A., GILLERY, P., HOFLIGER, N., AURAY-BLAIS, C., GARNOTEL, R.,ANIKSTER, Y. Mutations in the sarcosine dehydrogenase gene in patients with sarcosinemia. *Human Genetics,* 2012, vol. 131, no. 11, pp. 1805-1810.

[2] SREEKUMAR, A., POISSON, L. M., RAJENDIRAN, T. M., KHAN, A. P., CAO, Q., YU, J. D., LAXMAN, B., MEHRA, R., LONIGRO, R. J., LI, Y., NYATI, M. K., AHSAN, A., KALYANA-SUNDARAM, S., HAN, B., CAO, X. H., BYUN, J., OMENN, G. S., GHOSH, D., PENNATHUR, S., ALEXANDER, D. C., BERGER, A., SHUSTER, J. R., WEI, J. T., VARAMBALLY, S., BEECHER, C.,CHINNAIYAN, A. M. Metabolomic profiles delineate potential role for sarcosine in prostate cancer progression. *Nature,* 2009, vol. 457, no. 7231, pp. 910-914.

[3] ALLEN, R. H., STABLER, S. P.,LINDENBAUM, J. Serum betaine, N,N-dimethylglycine and N-methylglycine levels in patients with cobalamin and folate-deficiency and relatned inborn-errors of metabolism *Metabolism-Clinical and Experimental,* 1993, vol. 42, no. 11, pp. 1448-1460.

γ-Fe_2O_3 NANOPARTICLES AS A POTENTIAL SENSOR FOR HISTAMINE ISOLATION

[1,2]Milica GAGIC, [1,2]Pavel KOPEL, [1,2]Natalia CERNEI, [1,2]Ondrej ZITKA, [1]Pavel SVEC, [3]Ewelina JAMROZ, [1,2]Vojtech ADAM

[1]Department of Chemistry and Biochemistry, Mendel University in Brno, Faculty of AgriSciences, Brno, Czech Republic, EU, gagic.milica@gmail.com
[2]Central European Institute of Technology, Brno University of Technology, Brno, Czech Republic, EU
[3]Insitute of Chemistry, University of Agriculture, Cracow, Poland, EU

Abstract

Histamine, biologically active amine, is normally present in the body and it is involved in a local regulation of physiological processes. It occurs in food as a product of microbial decarboxylation of amino acid histidine, and consumption of histamine-rich food can lead to intoxication. Hence, its identification, quantification, and awareness of this foodborne toxin are important for food safety as it can serve as an indicator of food spoilage. This study presents a synthesis of various γ-Fe_2O_3 nanoparticles, which differ in surface functionalization, application of particles for histamine detection and isolation, and subsequent reaction of desorbed histamine with ninhydrin for final ion exchange chromatography quantification. The aim of this study was to develop easy to use, cost-effective, and error-free procedure for quantification of histamine.

Keywords: Histamine, nanoparticles, food safety

1. INTRODUCTION

Histamine is normal constituent of many foods, such as fish, meat, cheese, and wines and is described as low molecular weight organic base with a heterocyclic structure. The formation of histamine in food is mainly derived from enzymatic decarboxylation of histidine and can have a toxicological effect especially in histamine sensitive consumers. Usually, histamine does not represent any health hazard but due to ingestion of large quantities, when food is spoiled, it causes anaphylactic reactions [1]. Therefore, the presence of histamine has been used as an indicator for quality control during food production [2].

Histamine poisoning also referred as scombroid poisoning, is the most common seafood intoxication in the world. It occurs after the consumption of food containing histamine at concentration higher than 500ppm [3]. To prevent potential risk, the European Legislation set a limit for histamine levels in fishery products, up to maximum of 10 mg/100 g [4].

Merits of the conventional methods for histamine quantifications are high sensitivity and accuracy and most of them are based on liquid chromatography where histamine must be converted to a derivative capable to be detected in the visible range of the spectrum [5]. Despite sensitivity and reproducibility, chromatographic methods are complex, expensive, and time-consuming. Therefore, there is an urgent need to develop simple, rapid and reliable method for the routine analysis of this potentially hazardous food compound.

The strategy described in this article used nanotechnology for separation of histamine, based on the adsorption of histamine on the paramagnetic particles (PMPs) with subsequent ion exchange chromatography (IEC) determination. The IEC based method used to determine histamine employed chemical derivatization of histamine using ninhydrin. A simple, sensitive and rapid method is described for histamine determination.

2. EXPERIMENTAL PART AND METHODOLOGY

2.1. Chemicals

The chemicals were purchased from Sigma-Aldrich (St. Louis, MO, USA) in ACS purity unless noted otherwise.

2.2. Synthesis of paramagnetic particles

Maghemite nanoparticles were prepared by sodium borohydride reduction of iron nitrate ($Fe(NO_3)_3{\bullet}9H_2O$) according to following procedure. 7.48g of $Fe(NO_3)_3{\bullet}9H_2O$ was dissolved in 400 mL of water. Under stirring 1g of $NaBH_4$ was added, which was previously dissolved in 50 mL of 3.5 % NH_3. The obtained solution was heated at boiling temperature for 2h. After cooling, dark product was separated by external magnetic field and washed several times with water. 100 mL of prepared nanoparticles were used as a core for surface modification.

2.3. Synthesis of modified paramagnetic particles

MAN181: 50 mL of methanol was added to a maghemite and mixture was shaken. Titanium(IV) butoxide was applied to modify the nanomaghemite surface Firstly 0.7 mL of titanium(IV) butoxide was applied to modify the nanomaghemite surface and subsequently 1 mL of 28 % ammonium (*w/v*) was mixed. The resulting mixture was stirred at Biosan OS-10 (Biosan, Riga, Latvia) for 2 h at room temperature. Resulting product was separated using magnet and washed with water.

MAN183: 50 mL of methanol was added to nanomaghemite and the mixture was shaken. Tetraethyl orthosilicate (TEOS) and titanium(IV) butoxide was applied to modify the nanomaghemite surface. Firsly 0.7 mL of TEOS was mixed with nanomaghemite, after 1oh 0.7 mL of titanium(IV) butoxide was added. After 1 h of constantly stirring 1 mL of 28 % ammonium (*w/v*) was mixed. The resulting mixture was stirred at Biosan OS-10 for 2 h at room temperature. Resulting product was separated using magnet and washed with water.

2.4. Ion exchange chromatography with visible light range detector

For the identification of paramagnetic particles binding capacity an IEC Model AAA - 400 (Ingos, Prague, Czech Republic) with post-column derivatization by ninhydrin and an absorbance detector in visible light range was used. Experimental conditions were applied according to our preliminary study [6]. A glass column with an inner diameter of 3.7 mm and length of 350 mm was filled manually with strong cation exchanger Ostion LG ANB (Ingos, Prague, Czech Republic) in sodium cycle with ~12 µm particles and 8 % porosity. The column was thermostated at 60 °C. Double channel VIS detector with an inner cell of 5 µL was set to two wavelengths: 440 and 570 nm. Prepared solution of ninhydrin was stored under a nitrogen atmosphere in the dark at 4 °C. Elution of histamine was carried out by a buffer containing 10.0 g of citric acid, 5.6 g of sodium citrate, and 8.4 g of sodium chloride per liter of solution (pH 2.7). The flow rate was 0.25 mL/min. The reactor temperature was set to 120 °C.

2.5. Scanning electron microscopy

Morphology of paramagnetic particles was revealed using scanning electron microscopy (SEM) FEG-SEM MIRA (Tescan, Brno, Czech Republic). The SEM was fitted with Everhart-Thronley type of SE detector, high speed YAG scintillator based BSE detector, panchromatic CL Detector, and EDX spectrometer.

3. RESULTS AND DISCUSSION

Purpose of our study was to synthesize the paramagnetic particles able to establish a binding with histamine, and thus to isolate and preconcentrate this analyte from complicate matrices for its subsequent determination using IEC. Nanomaghemite acts as an excellent superparamagnetic carrier due to its unique physic-chemical properties and ability to be simply functionalized with chemically active groups. We reported controlled procedure for preparation of stable PMPs which are responsive to magnetic field and can be used for histamine isolation. For particles coating, substances with potential to cover the nanomaghemite surface and to provide active binding sites for histamine were selected as it is mentioned in the previous chapter where the entire synthesis procedure is described. The surface coating is crucial for enhancement of the nanoparticles properties because it can regulate their stability and their selectivity.

3.1. Workflow process resulting in binding of histamine onto MAN181 and MAN183 surface

First, binding capacity of synthesized paramagnetic particles was tested by the described IEC method. Overall workflow process scheme can be seen in **Figure 1A**. Before PMPs application, it was necessary to include three washing steps with phosphate buffered saline (PBS) to remove undesired impurities, contained in the solid PMPs after their synthesis. Further, to achieve to highest yields of histamine, binding and elution conditions for histamine were optimized. The recoveries of histamine were 94.5 and 80.82 % for MAN181 and MAN183, respectively (**Figure 1C**). Synthesized PMPs showed high selectivity for binding of histamine which is probably based on the underlying principles of the proposed workflow process. Two pKa values of 5.8 and 9.4 have been reported for histamine [7]. Therefore, in strongly basic environments histamine will be predominantly present in its neutral form and upon lowering the pH below 5.8, histamine exists as a dication. In this experiment, pH of the sample played an important role in the adsorption of histamine and Briton-Robinson (BR) buffer pH 4 was used, causing histamine protonation and its interaction with the surface of PMPs (**Figure 1B**).

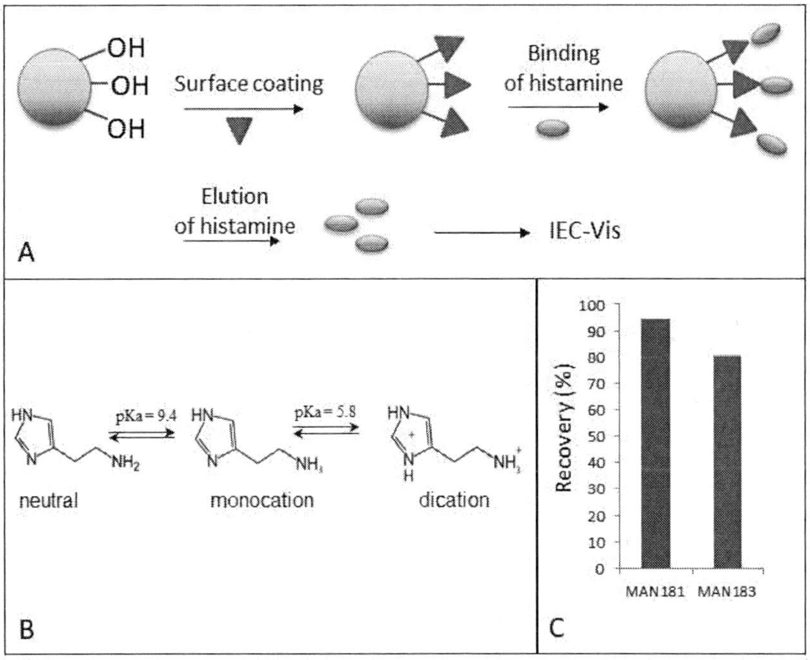

Figure 1 (A) The workflow process, providing the binding between paramagnetic particles and histamine molecules. Firstly, the paramagnetic beads are synthesized. Secondly, the particles are functionalized. Thirdly, beads are mixed with histamine. As a fourth step incubation is included, following elution step. Sample was prepared to be analyzed using IEC. (B) Effect of pH on the protonation of histamine. (C) Histamine recoveries, in percentage, for MAN181 and MAN183.

3.2. Morphological characterization

After confirmation of ability to bind histamine combination of complementary characterization techniques for comprehensive morphological and chemical characterization of PMPs were carried out. To reveal the particles morphology scanning electron microscopy (SEM) was employed. As it shown in **Figure 2**, paramagnetic particles are irregularly shaped, amorphic structures and form clusters very willingly, but it did not influence on their superparamagnetic ability.

3.3. Characterization of elemental composition of PMPs

The initially formed maghemite nanostructure changed into titanium modified particles evidenced by energy dispersive X-ray (EDX) spectra. EDX showed that the MAN181 and MAN183 are composed primarily of iron (Fe) and titanium (Ti) (**Figure 2**).

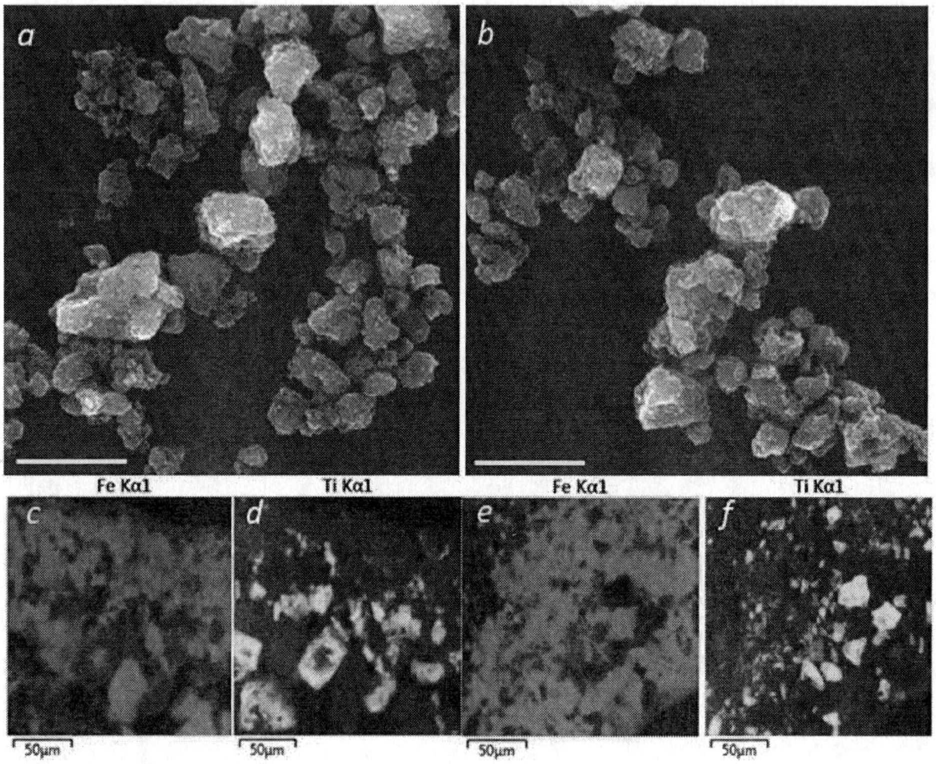

Figure 2 (a)(b) Micrographs of PMPs showing their morphology and size (Scale 2 μm) were obtained using SEM for MAN181 and MAN183, respectively. (c)(d) EDX results providing information about elemental composition of MAN181, Fe - iron, Ti - titanium (e)(f) EDX results for MAN183 Fe - iron, Ti - titanium.

4. CONCLUSION

In this study, we synthesized titanium(IV) oxide functionalized paramagnetic particles with high affinity and specificity towards histamine. The development of these potential sensors for histamine analysis appears to provide alternative methods to separation techniques. The main advantage of proposed protocol is simple and fast PMPs based isolation which makes it especially suitable for routine determination of histamine in order to control potential toxic effects. Further application of this method is quite possible and more research will be undertaken in our laboratory.

ACKNOWLEDGEMENTS

The research was financially supported by Internal Grant Agency of the Faculty of Agronomy, Mendel University in Brno no. AF-IGA-IP-2018/076.

REFERENCES

[1] LEHANE L. and OLLEY J., Histamine fish poisoning revisited. International Journal of Food Microbiology, Vol. 58, No. 1-2, 2000, pp. 1-37.

[2] GAO F., GRANT E., and LU X.N., Determination of histamine in canned tuna by molecularly imprinted polymers-surface enhanced Raman spectroscopy. Analytica Chimica Acta, Vol. 901, No., 2015, pp. 68-75.

[3] RUSSELL F.E. and MARETIC Z., SCOMBROID POISONING - MINIREVIEW WITH CASE-HISTORIES. Toxicon, Vol. 24, No. 10, 1986, pp. 967-973.

[4] SILVA C.C.G., DA PONTE D.J.B., and DAPKEVICIUS M., Storage temperature effect on histamine formation in big eye tuna and skipjack. Journal of Food Science, Vol. 63, No. 4, 1998, pp. 644-647.

[5] EEROLA S., HINKKANEN R., LINDFORS E., and HIRVI T., liquid-chromatographic determination of biogenic-amines in dry sausages. Journal of Aoac International, Vol. 76, No. 3, 1993, pp. 575-577.

[6] CERNEI N., LACKOVA Z., GURAN R., HYNEK D., SKLADANKA J., HORKY P., ZITKA O., and ADAM V., Determination of Histamine in Silages Using Nanomaghemite Core (-Fe2O3)-Titanium Dioxide Shell Nanoparticles Off-Line Coupled with Ion Exchange Chromatography. International Journal of Environmental Research and Public Health, Vol. 13, No. 9, 2016.

[7] COOPER DG Y.R., DURANT GJ, GANELLIN CR, Histamine receptors, in comprehensive medicinal chemistry. Comprehensive Medicinal Chemistry, Vol. 3, No., 1990, pp. 323-421.

THERANOSTIC TOOL FOR TARGETING OF BACTERIAL CELLS BASED ON SILVER NANOPARTICLES

[1]Michael GARGULÁK, [1]Matouš ČÍŽEK, [2,5]Michaela DOČEKALOVÁ, [2]Marta KEPINSKA, [1,3]Andre FERREIRA DA SILVA, [1]Aneta SUROVÍKOVÁ, [2]Dagmar UHLÍŘOVÁ, [2]Martina STAŇKOVÁ, [1,5]Branislav RUTTKAY-NEDECKÝ, [5]Aleš VAVŘINÍK, [5]Jiří SOCHOR, [6]Hoai Viet NGUYEN, [1,2,4,5]René KIZEK

[1]*University of Veterinary and Pharmaceutical Sciences Brno, Pharmaceutical Faculty, Brno, Czech Republic, EU, kizek@sci.muni.cz*

[2]*Prevention Medicals, Studenka-Butovice, Czech Republic, EU, uhlirova@preventionmedicals.cz*

[3]*University of Lisboa, Faculty of Pharmacy Avenida Professor Gama Pinto, Lisboa - Portugal, EU, andre_ferreira.silva@hotmail.com*

[4]*Department of Biomedical and Environmental Analyses, Faculty of Pharmacy, Wroclaw Medical University, Wroclaw, Poland, EU, zalewska.m@gmail.com*

[5]*Mendel University in Brno, Faculty of Horticulture, Department of Viticulture and Enology, Lednice, Czech Republic, EU, sochor.jirik@seznam.cz*

[6]*Research Center for Environmental Monitoring and Modeling, VNU University of Science, Hanoi, Vietnam, nguyenviethoai@hus.edu.vn*

Abstract

Infectious diseases are known to be a constant threat to human health also in the beginning of the 21st century. The rapid growth of the human population has led to the agricultural intensification, which has caused an increase in resistant bacterial species in the environment. The aim of the project was to design and verify a theranostic tool combining the diagnostic and therapeutic parts for targeting the bacterial cells. The construct was composed of: A) silver nanoparticles prepared by green synthesis (AgNPsGS); B) antibiotic encapsulated in apoferritin (APO); C) modified magnetic gold nanoparticles (AuNPs). The effect of theranostic nanotransporter on prokaryotic cell culture was model-monitored by altering their growth characteristics (OD monitoring). During the project, 20 various types of AgNPsGS were prepared. The best antibacterial effect (85 % growth inhibition of *S. aureus*) with the minimal inhibitory concentration (MIC50) of 0.4 µg/mL was observed by AgNPsGS4 (*Thymus serpyllum* L.). SPIONs (20 mg/mL, magnetic AuNPs) were modified with antibody (1 mg/mL IgY) and oxidized graphene sheets (1 mg/mL). The EDC/NHS polymer (2.2 mM EDC, 4 mM NHS, 3 h, 37 °C) was used for binding. The apoferritin (SPIONs/Au/APO) was attached to the SPION/AuNPs. In the SPIONs/Au/APO cavity (12 nm), the antibiotic doxorubicin (2 µM DOXO, the DOXO encapsulation efficiency was around 13 % of the applied concentration) was model-sealed. The created SPIONs/Au/AgNPsGS4/APO/DOXO construct was applied to the *S. aureus* model culture. In the presence of 70 µg of SPIONs/Au/AgNPsGS4/APO/DOXO, a dramatic growth inhibition of *S. aureus* (95 % growth inhibition and 13 mm wide inhibition zone on agar medium) was observed. The biological effect of the nanotransporter lies in ROS (reactive oxygen species) formation, DNA destruction and cell membrane damage.

Keywords: Nanomedicine, silver nanoparticles, nanoconstructs, targeted therapy

1. INTRODUCTION

Tumor diseases are one of the most common diseases worldwide [1]. Secondary infections are serious complications in immunocompromised patients who have been treated with cytostatics [2]. According to statistical analyzes, patients with neutropenia due to tumor diseases and their treatment have a 15-60 % higher mortality rate for this infectious disease [3]. The effectiveness of the fight against bacterial infectious disease is lower every year due to the phenomenon of bacterial resistance to the antibiotics used and antibacterial chemotherapeutics. The main problem is their huge consumption in livestock farming as well as in human medicine, or poorly chosen therapies for mistakenly determined FAO diagnoses [4]. In the Czech republic, almost 70 tons of antibiotics and antibacterial chemotherapeutics were consumed in 2010 [5].

A new and innovative option in the fight against resistant pathogens is the use of nanomaterials with antibacterial effect. The nanotechnology approach is an important strategy for obtaining new tools and procedures for overcoming bacterial infections and resistence [6]. Green synthesis of nanoparticles reduces chemical consumption and improves bioavailability, and consequently the surface modification of nanoparticles dramatically increases already known effects on target cells [7-10]. Green synthesis of nanoparticles uses the introduction of biological molecules from living organisms. Numerous studies have shown no significant antibacterial effects of silver nanoparticles (AgNPs) [7]. Surface modification of AgNPs provides further significant improvement of their antibacterial properties. In our concept we are coming with the use of modified gold nanoparticles, silver nanoparticles obtained by green synthesis in combination with a suitably chosen antibiotic enclosed in a nanoprotein construct [11]. Apoferritin is a protein that is known for its ability to self-organize in dependence on pH change [12]. This outstanding feature has been used for our experiments. The cavity of apoferritin is about the order of several nanometers, and therefore low-molecular compounds can be enclosed here in non-negligible concentrations.

2. MATERIAL AND METHODS

All used reagents were purchased in Merck-Sigma-Aldrich (USA) in analytical grade. Distilled water was prepared by the Aqual system (Tišnov, Czech Republic) and ultra-pure water was prepared by the ELGA (USA) system to the sterile 18 MΩ quality.

For the purpose of this project, 10 types of AgNPsGS from Vietnamese plants were prepared. Four of them were selected to have the greatest antimicrobial effect. All samples were transferred to a Petri dish and placed in a drier (HN101, Lang-Shan, China) for 24 hours at 60 °C. Then the samples were homogenized to a fine powder using a mixer (Mixer Grinder Mixer HL1643 / 06, Philips, The Netherlands). The material thus prepared was used to prepare the plant extract. 15 g of the prepared plant material was mixed with 500 ml of ultra-pure water at 80 ° C. Extraction was carried out for 1 hour with constant stirring (IKA RH basic, Malaysia) at 80 °C and 300 rpm. Each plant extract was filtered and allowed to cool to room temperature. After filtration, the plant extracts were mixed in a 1: 1 ratio with 0.1 M $AgNO_3$ and allowed to stir with a magnetic stirrer (IKA RH basic, Malaysia) at room temperature (23 °C) for 18 hours [13] By the green synthesis the following plants were prepared: AgNPsGS1 (*Lagerstroemia indica*), AgNPsGS2 (*Carica papaya*), AgNPsGS3 (*Polyalthia longifolia*), AgNPsGS4 (*Ficus bengalensis*).

Doxorubicin fluorescence (DOXO) measurements were performed by pipetting into the microtiter plate (Brand, Germany) 100 µl of sample with a concentration of 10 µM DOXO. The weight concentration of apoferritin in acid form in the solution was 3.0 mg / ml. For measurement, the Multimode microplate reader M200 PRO (Tecan, Japan) was used, fluorescence of doxorubicin (Ex 480 nm, Em 650 nm) with a typical peak at 560 nm.

For MIC testing in microtiter plates, the cultured sample of the selected microorganism was first diluted to OD 0.1 at 600 nm using a pure sterile LB medium. 250 µl of the prepared diluted micro-organism was pipetted into the well of the microtiter plate and 50 µl of the test sample was added to each well. Measurement conditions were set to 18 hours, absorbance reading for 30 min, shaking for 30 min, shaking at 50 Hz for 5 s, temperature 35 °C, wavelength 600 nm. The Multimode microplate reader M200 PRO (Tecan, Japan) was used for measurement.

3. RESULTS

3.1. Preparation of the theranostic SGA nanotransporter

In the experiment, a teranous nanotransporter was designed and studied by biophysical and microbiological methods for specific targeting of S. aureus. The proposed construct is composed of three separate parts (**Figure 1**). Part A of nanoconstruct is made of silver nanoparticle prepared by green synthesis with significant antibacterial effects. Part B of the nanocomposite consists of apoferritin encapsulated with an appropriate antibiotic.

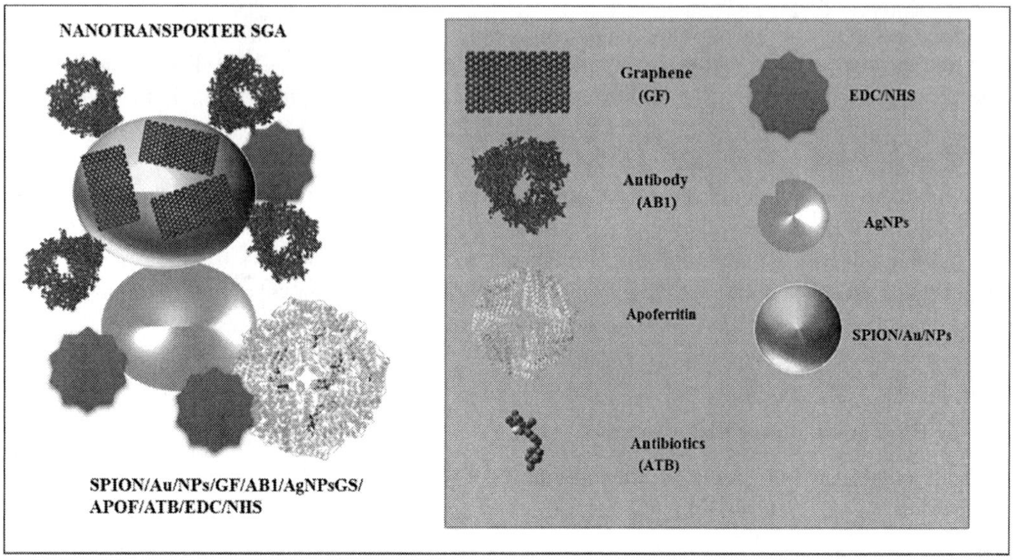

Figure 1 Proposed antibacterial nanotransporter scheme for sensitive and specific targeting on *S. aureus* bacterial cells. The nanotransporter consists of three separate units. Gold supraparamagnetic particles SPION modified by graphene sheet and antibody. (AND); silver nanoparticles formed by green synthesis of AgNPs (B); apoferritin cage with closed antibiotic (C). The individual parts are linked by the EDC / NHS binding polymer.

Part C of the nanoconstruct consists of a superparamagnetic gold nanoparticle (SPION/Au/NPs). On the surface of the nanoparticle, an anti-protein A antibody was then attached to target surface A protein *S. aureus*. Additionally, the SPION/Au/NPs surface was modified by graphene sheets to bind cytotoxic compounds produced by *S. aureus*. The individual parts of the construct are interconnected with the EDC / NHS binding polymer to form a complete SGA theranostic construct.

3.2. Monitoring of bacterial growth in the presence of AgNPs

Growth curves of S. aureus bacterial culture were recorded over 18 h. All AgNPsGS showed a high inhibitory activity (above 60 %). For all other experiments, AgNPsGS3 were selected because they exhibited the highest antibacterial effect on *S. aureus* culture (95.6 %).

3.3. Testing the SGA nanotransporter

The binding efficiency of AgNPsGS3 to the remaining parts of the construct was about 15 %. Various variants of the construct underwent biophysical and microbiological testing. DOT BLOT, as a biological immunoassay method, was used to demonstrate antibody binding to SPION / Au / NPs. DOT BLOT can detect anti-protein A at 40 ng / ml

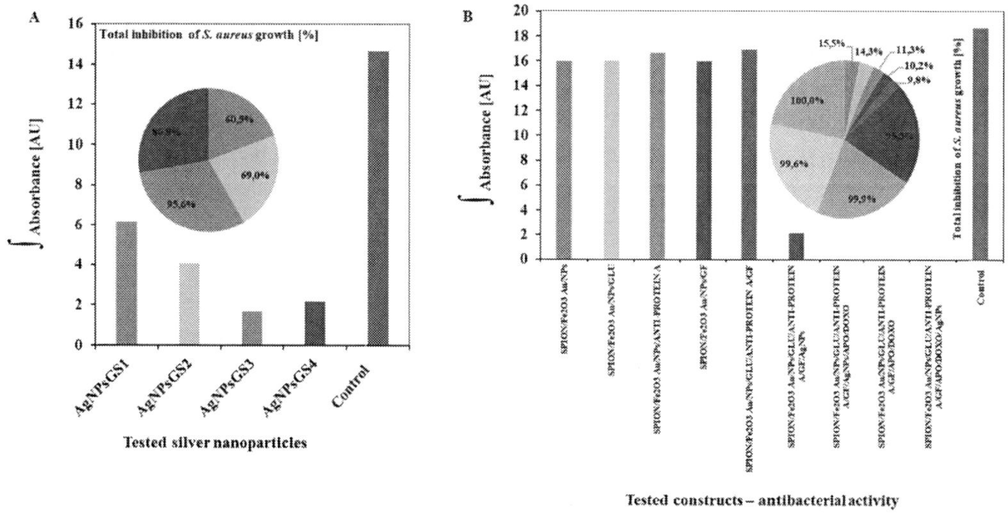

Figure 2 Integral absorbance of individual types of silver nanoparticles compared to control. In inset percentage inhibition of silver nanoparticles is given. The control is 100 %. Concentration 0.2 mg / ml, total volume 300 µl. Measured for 18 hours and readings were read at 30 ° C, at 37 ° C, shaking at 3 with 6 mm (A). Growth curves of S. aureus after application of various parts of the construct, total volume 300 µl. Measured for 18 hours and values were read at an interval of 30 minutes at 37 ° C (B).

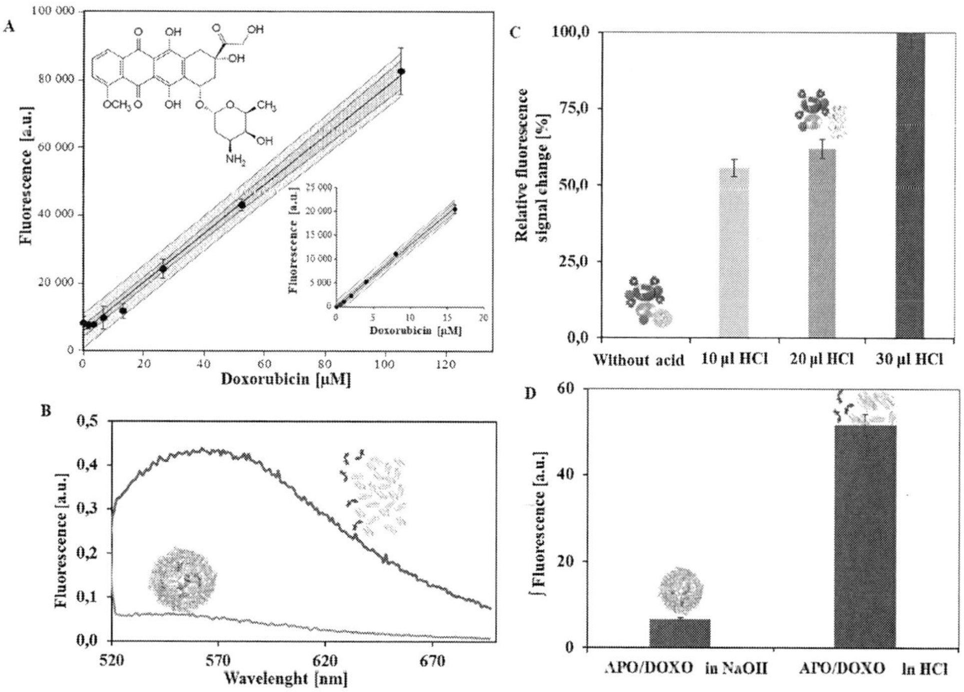

Figure 3 Doxorubicin Fluorescence Calibration Curve (Ex 480 nm, Em 650 nm) with typical peak at 560 nm. (A); Dependence of DOXO fluorescence on wavelength in open and closed apoferritin (B); Following the behavior of apoferritin in the acidic environment (different additions of 0.1 M HCl to a final volume of 30 µl), the release efficiency was determined as the fluorescence of doxorubicin. (C); Fluorescence integral of DOXO in apoferrin in acidic and basic media (D)

On the basis of all the experimental data obtained above, the nanoconstruct proposed by us was confirmed. The main aim of this work was to strengthen antibacterial properties. Therefore, the antibacterial effect of our nanoconstruct and its parts was monitored. (**Figure 2**). Part C composed of SPION / Au / NPs with bound antibody and graphene, coupling part C and A (AgNPs), linking part C and part B (apoferritin with encapsulated doxorubicin) and, of course, the entire final construct. **Figure 2** shows typical growth curves in the presence of the nanoconstructs tested. Part C showed no antibacterial effects, nor were they expected. In the presence of antibacterially active substances, however, dramatic inhibition has been observed.

Apoferritin is a protein that is known for its ability to self-regulate in dependence on pH change [12]. The apoferritin cavity is of the order of several nanometers, and thus low-molecular compounds can be enclosed at negligible concentrations at acidic pH, opening of the apoferrin structure, whereas basic environment for clustering. The anthracycline antibiotic doxorubicin at a concentration of 10 µM was stirred for 30 minutes while stirring. is sequentially closed (by changing the pH from acidic to alkaline) into the cavity of apoferritin. We found that doxorubicin (around 13 % of the DOXO concentration) was enclosed within the nanometric structure created (**Figure 3**). This fact can be used for the design of other nanoconstructions that will contain other antibiotics (penicillin G, vancomycin, etc.).

4. CONCLUSION

Advances in the treatment of bacterial infections are significant and, with the current antibiotic crisis, the importance of nanotechnology continues to grow. In recent years, researchers have been focusing on silver nanoparticles. In our work a nanoconstruct with multifunctional mechanisms of antibacterial activity was created - nanoantibiotic. The nanoconstruct binds to the bacterium by immunological binding via protein A with a specific antibody. Other parts of the SGA nanoconstruct are designed to damage the bacterial cell. The SGA nanoconstruct, designed by us, is functionalized by an antibiotic that is released from apoferithin at low pH. The antibiotic acts by a specific bacteriocidal mechanism. Another advantage of the prepared nanotransporter is the presence of graphene sheets. After bacterial lysis, bacterial toxins are released into the environment, from which are captured by graphene sheets with maximum sorption capacity to avoid damage to the host organism. Our SGA nanoconstruct showed very good overall stability in different environments. Its specific targeting to the bacterial cell was possible by binding to protein A. The synergistic effect of biogenic silver nanoparticles and antibiotic sealed in the apoferritin cage increased the inhibitory effect to more than 90 %. We expect the rapidly growing interest in theranostic applications based on metal nanoparticles that will not only encourage and stimulate research in this highly active region, but will also bring completely new fundamental knowledge.

ACKNOWLEDGEMENTS

This work has been supported by the project 3D ANTIMICROB (LC18002) and by the CA COST Action CA15114.

REFERENCES

[1] MCGUIRE, S. World Cancer Report 2014. Geneva, Switzerland: World Health Organization, International Agency for Research on Cancer, WHO Press, 2015. *Adv. Nutr. Int. Rev. J.,* 2016, vol. 7, no. 2, pp. 418-419.

[2] LANOIX, J.-P., PLUQUET, E., LESCURE, F. X., BENTAYEB, H., LECUYER, E., BOUTEMY, M., DUMONT, P., JOUNIEAUX, V., SCHMIT, J. L., DAYEN, C.,DOUADI, Y. Bacterial infection profiles in lung cancer patients with febrile neutropenia. *BMC Inf. Dis.,* 2011, vol. 11, no. 1, pp. 183.

[3] LYMAN, G. H., MICHELS, S. L., REYNOLDS, M. W., BARRON, R., TOMIC, K. S.,YU, J. Risk of mortality in patients with cancer who experience febrile neutropenia. *Cancer,* 2010, vol. 116, no. 23, pp. 5555-5563.

[4] WALL, B. A., MATEUS, A., MARSHALL, L., PFEIFFER, D. U., LUBROTH, J., ORMEL, H. J., OTTO, P.,PATRIARCHI, A. 2016. *Drivers, dynamics and epidemiology of antimicrobial resistance in animal production*, Food and Agriculture Organization of the United Nations.

[5] HERA, A., KOUTECKÁ, L., DORN, D.,POKLUDOVÁ, L. Spotreba antibiotik a antiparazitik ve veterinární medicíne v CR v letech 2003-2008. *Vestník ÚSKVBL,* 2009, vol. 3, no. pp. 33-8.

[6] NAIKA, H. R., LINGARAJU, K., MANJUNATH, K., KUMAR, D., NAGARAJU, G., SURESH, D.,NAGABHUSHANA, H. Green synthesis of CuO nanoparticles using Gloriosa superba L. extract and their antibacterial activity. *J. Taibah Univ. Sci.,* 2015, vol. 9, no. 1, pp. 7-12.

[7] SHARMA, V. K., YNGARD, R. A.,LIN, Y. Silver nanoparticles: Green synthesis and their antimicrobial activities. *Adv. Coll. Int. Sci.,* 2009, vol. 145, no. 1-2, pp. 83-96.

[8] IRAVANI, S. Green synthesis of metal nanoparticles using plants. *Green Chem.,* 2011, vol. 13, no. 10, pp. 2638-2650.

[9] CHALOUPKA, K., MALAM, Y.,SEIFALIAN, A. M. Nanosilver as a new generation of nanoproduct in biomedical applications. *Trends Biotechnol.,* 2010, vol. 28, no. 11, pp. 580-588.

[10] KUMAR, A., VEMULA, P. K., AJAYAN, P. M.,JOHN, G. Silver-nanoparticle-embedded antimicrobial paints based on vegetable oil. *Nat. Mater.,* 2008, vol. 7, no. 3, pp. 236-241.

[11] KIM, J. S., KUK, E., YU, K. N., KIM, J.-H., PARK, S. J., LEE, H. J., KIM, S. H., PARK, Y. K., PARK, Y. H., HWANG, C.-Y., KIM, Y.-K., LEE, Y.-S., JEONG, D. H.,CHO, M.-H. Antimicrobial effects of silver nanoparticles. *Nanomed.-Nanotechnol. Biol. Med.,* vol. 3, no. 1, pp. 95-101.

[12] RUOZI, B., VERATTI, P., VANDELLI, M. A., TOMBESI, A., TONELLI, M., FORNI, F., PEDERZOLI, F., BELLETTI, D.,TOSI, G. Apoferritin nanocage as streptomycin drug reservoir: Technological optimization of a new drug delivery system. *Int. J. Pharmacol.,* 2017, vol. 518, no. 1-2, pp. 281-288.

[13] RONAVARI, A., KOVACS, D., IGAZ, N., VAGVOLGYI, C., BOROS, I. M., KONYA, Z., PFEIFFER, I.,KIRICSI, M. Biological activity of green-synthesized silver nanoparticles depends on the applied natural extracts: a comprehensive study. *Int. J. Nanomed.,* 2017, vol. 12, no. pp. 871-883.

CIRCULATING TUMOR DNA AS A PROMISING BIOMARKER FOR BREAST CANCER DETECTION: PCR AMPLIFICATION IMPROVED BY USING GOLD NANOPARTICLES

[1,2]Andre FERREIRA DA SILVA, [3]Dagmar UHLÍŘOVÁ, [1,4]Christina KOLLIOPOULOU, [1,4]Anastasia SEIMENI, [5]Marta KEPINSKA, [1]Martin LOFFELMANN, [1]Michael GARGULÁK, [2]Michaela DOČEKALOVÁ, [1]Božena HOSNEDLOVÁ, [1]Branislav RUTTKAY-NEDECKÝ, [5]Halina MILNEROWICZ, [1,2,4]René KIZEK

[1]University of Veterinary and Pharmaceutical Sciences Brno, Pharmaceutical Faculty, Brno, Czech Republic, EU, kizek@sci.muni.cz
[2]University of Lisboa, Faculty of Pharmacy Avenida Professor Gama Pinto, Lisboa, Portugal, EU, andre_ferreira.silva@hotmail.com
[3]Prevention Medicals, Studenka-Butovice, Czech Republic, EU, uhlirova@preventionmedicals.cz
[4]Faculty of Pharmacy, University of Patras, Prefabricated Buildings Area, University Campus, Rio Patras, Greece, EU, Greecechristinakolliop@gmail.com
[5]Department of Biomedical and Environmental Analyses, Faculty of Pharmacy, Wroclaw Medical University, Wroclaw, Poland, EU, zalewska.m@gmail.com

Abstract

Breast cancer is the most commonly diagnosed cancer in women. Circulating tumor DNA (ctDNA) is related to breast cancer development and thus, screening of tumors using ctDNA blood detection is highly sensitive and may significantly improve early-stage breast tumor diagnosis. Recently, a nanomaterial-assisted PCR using gold nanoparticles (AuNPs) has been applied to dramatically improve the specificity and sensitivity of PCR, achieving better gene detection. In this study, we aimed to optimize AuNP-assisted PCR, using a new AuNPs-ODN-ThiC3 nanocomposite. Synthesized AuNPs-ODN-ThiC3 nanocomposites, obtained after 60 min of interaction between AuNPs and ODN-ThiC3, were characterized by differential UV/VIS spectroscopy (AuNPs 35 mg/mL), AuNP peroxidase-like activity (0.5 mM TMB, 60 min, 30 cycles), PAGE 25 % gel electrophoresis with silver staining visualization (V = 120 mV, I = 150 mA, P = 20 W) and adsorptive stripping voltammetry (HMDE, from start 0 to end -1.75 V, time of accumulation 240 s). AuNP-ODN-ThiC3-assisted PCR was performed using Elizyme HS Robust Polymerase (5 U/μL), denaturation: 95 °C/15 s, annealing: 59 °C/15 s, extension: 72 °C/15 s, 35 cycles. Analysis of AuNPs modification was achieved with ODN-ThiC3 (in different concentrations) and, therefore, new AuNPs-ODN-ThiC3 nanocomposites were obtained. We observed an increased absorbance signal of AuNPs at 527 nm, with the increase of ODN-ThiC3 concentration (at 260 nm) in the AuNPs-ODN-ThiC3 mixture ($y = -0.0397 + 0.0734x$, $r = 0.99$). Mixing ODNs-ThiC3 with AuNPs increased the peroxidase-like activity of AuNPs by 17 %. LS voltammograms of cytosine and adenine (CA) reduction (the average potential at -1.34 V) showed a linear dependence ($r = 0.99$) between peak current and concentration of ODN-ThiC3 in the AuNPs-ODN-ThiC3. A linear dependence ($r = 0.99$) between bands density and ODN-ThiC3 concentration was also achieved. Analysis of PCR protocols: AuNPs-ODN-ThiC3-assisted PCR (AuNPs 814 μg/mL, ODN-ThiC3 4 μg/mL) showed a better yield (140 % of conventional PCR and 120 % of AuNP-assisted PCR). Conventional PCR can be optimized by integrating ODN-ThiC3 on the AuNP surface during PCR. This improvement represents an innovative PCR method for breast cancer detection.

Keywords: Nanomedicine, gold nanoparticles, anticancer drugs, PCR, breast cancer

1. INTRODUCTION

Breast cancer is the most commonly diagnosed cancer in women, including 28 % of all types of cancers [1]. Circulating tumor DNA (ctDNA) is related to breast cancer development and, thus, screening of tumors using ctDNA blood detection is highly sensitive and may significantly improve early-stage breast tumor diagnosis [2]. Recently, a nanomaterial-assisted PCR using gold nanoparticles (AuNPs) has been applied to dramatically improve the specificity and sensitivity of PCR, achieving better gene detection [3]. In this study, we aim to optimize AuNP-assisted PCR, using a new AuNPs-ODN-ThiC3 nanocomposite.

2. MATERIAL AND METHODS

All used reagents were purchased in Merck-Sigma-Aldrich (USA) in analytical grade. Distilled water was prepared by the Aqual system (Tišnov, Czech Republic) and ultra-pure water was prepared by the ELGA (USA) system to the sterile 18 MΩ quality. Synthesized AuNPs-ODN-ThiC3 (5'-AGCCCTTGGGGASTTGAATTGCTG[ThiC3]-3') nanocomposites, obtained after 60 min of interaction between AuNPs and ODN-ThiC3, were characterized by differential UV/VIS spectroscopy (AuNPs 35 mg/ml), AuNP peroxidase-like activity (0.5 mM TMB, 60 min, 30 cycles), PAGE 25 % gel electrophoresis with silver staining visualization (V = 120 mV, I = 150 mA, P = 20 W) and adsorptive stripping voltammetry (HMDE, start 0 to end -1.75 V, time of accumulation 240 s). AuNP-ODN-ThiC3-assisted PCR was performed using Elizyme HS Robust Polymerase (5 U/µl), denaturation: 95 °C/15 s, annealing: 59 °C/15 s, extension: 72 °C/15 s, 35 cycles.

3. RESULTS

3.1. UV-visible spectrophotometry

Analysis of AuNPs' modification: was achieved with ODN-ThiC3 (different concentrations) and, therefore, new AuNPs-ODN-ThiC3 nanocomposites were obtained. We observed an increased absorbance signal of AuNPs at 527 nm, with the increase of ODN-ThiC3 concentration (at 260 nm) in the AuNPs-ODN-ThiC3 mixture (y = -0.0397 + 0.0734x, r = 0.99) (**Figure 1**), suggesting that ODN-ThiC3 interaction with AuNPs surface may change its absorption properties. It should be noted that coating of AuNPs with ODNs impact the surface density, hybridization with the target sequences, and the non-specific interactions. We hypothesize that chemisorption of ODN-ThiC3 on AuNPs surface increases AuNPs solubility, which may explain the increase of AuNPs'absorbance peaks when binding ODN-ThiC3.

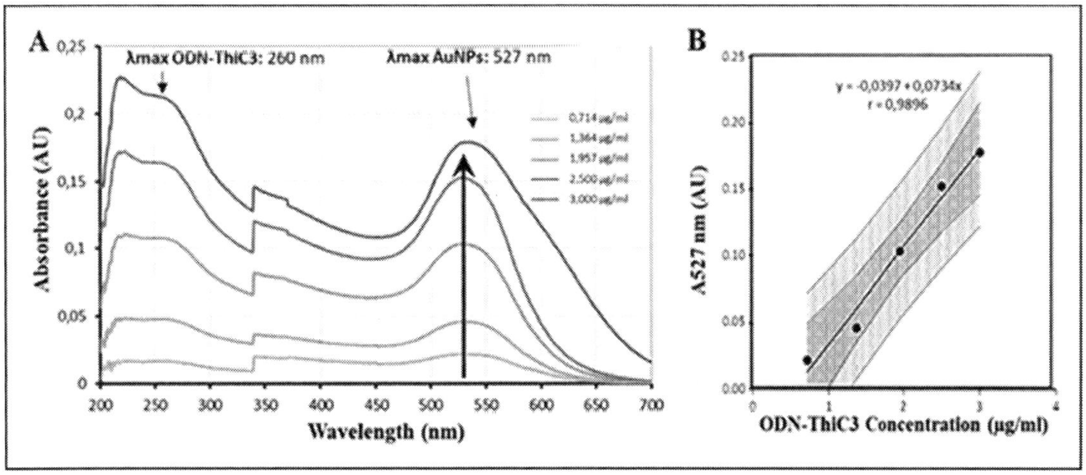

Figure 1 Effect of functionalization of AuNPs surface with increasing concentrations (0.714, 1.364, 1.957, 2.500 and 3.000 µg/mL) of ODN-ThiC3 on the AuNPs absorbance at 527 nm. (A) Increasing concentrations of ODN-ThiC3 leads to absorbance increase at 527 nm. (B) The occurrence of this effect was linear between 0.714 and 3.000 µg/mL.

3.2. Adsorptive Stripping Voltammetry (ASV) measurements

Mixing ODNs-ThiC3 with AuNPs increased by 17 % the peroxidase-like activity of AuNPs. Electrochemistry of DNA at electrodes provides valuable insights into the mechanisms involved in DNA-mediated charge transport at electrodes. Moreover, electrochemistry provides the opportunity for highly sensitive DNA detection. ASV voltammograms of cytosine and adenine (CA) reduction (average potential at -1.34 V) showed a linear dependence (r = 0.99) between peak current and concentration of ODN-ThiC3 in the AuNPs-ODN-ThiC3. A linear dependence (r = 0.99) between bands' density and ODN-ThiC3 concentration was also achieved. Analysis of PCR protocols: AuNPs-ODN-ThiC3-assisted PCR (AuNPs 814 µg/ml, ODN-ThiC3 4 µg/ml) showed a better yield (140 % of conventional PCR and 120 % of AuNP-assisted PCR). This data clearly demonstrates that interaction with AuNPs minimizes the reduction of ODN-ThiC3 on the electrode's surface, interfering with ODN-ThiC3 detection using electrochemical-based methodology (**Figure 2**).

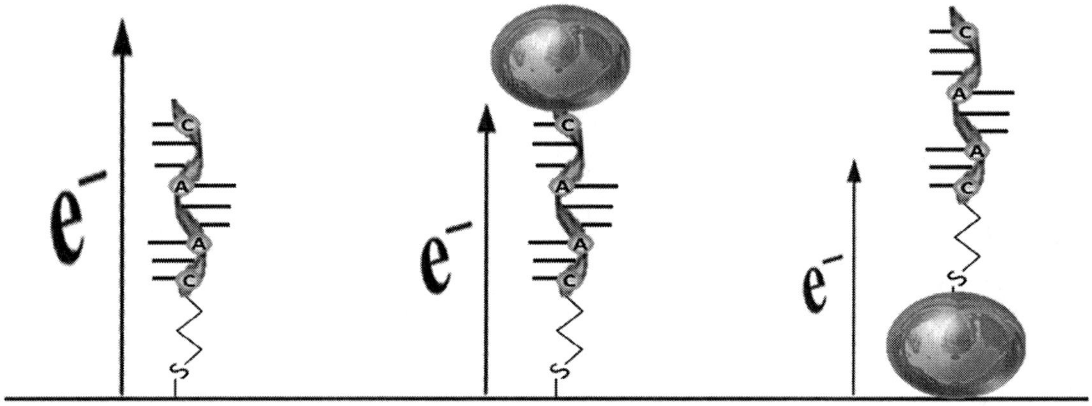

Figure 2 Schematic illustration of attenuation of ODN-ThiC3- mediated electrochemistry by gold nanoparticles (AuNPs)

3.3. AuNP-assisted PCR

The specificity of the primer pairs for the desired DNA region was analyzed first by single conventional PCR. The target gene was amplified specifically using its designed primers. The size of the amplified products was identical with the expected target fragments (237 bp). To perform AuNP-assisted PCR, different reaction mixtures were made by adding a different volume (0.55, 1.10, 2.20, 3.30, 4.40, 5.50 and 8.75 µL) of synthesized AuNPs (18.50 mg/mL) to conventional PCR mixtures, obtaining different AuNPs' final concentrations (407, 814, 1628, 2442, 3256, 4070 and 6475 µg/mL) in each mixture. The resulting PCR products were migrated on 15 % SDS-PAGE gel (**Figure 3A**). After executing silver staining protocol, the gel was scanned and processed by programmed ColorTest, which assigns intensity to the individual pixels of the studied image in the color area and allows densitometry quantifications of the bands as well as the drawing of its electrophoretic graphs. The enhancement of amplification yield (%) of the desired DNA sequence was defined as a ratio of the density value of the target DNA band obtained with AuNP-assisted PCR to the density value of the target DNA. band observed with conventional PCR, which is assigned to a value of 0 %. In AuNP-assisted PCR, the concentration of AuNPs that enabled the PCR to produce the large amount of DNA target band (maximum efficiency) on the gel was identified to be the optimum concentration. This study revealed that AuNP-assisted PCR using AuNP concentrations of 407 µg/m and 814 µg/mL showed improved yield of gene amplification when compared with conventional PCR (**Figure 3B**). Taken together, these data confirms that, at an appropriate concentration, presence of AuNPs may significantly improve the efficiency of PCR, which can be explained by three general effects of AuNPs in PCR already reported (**Figure 3C**)

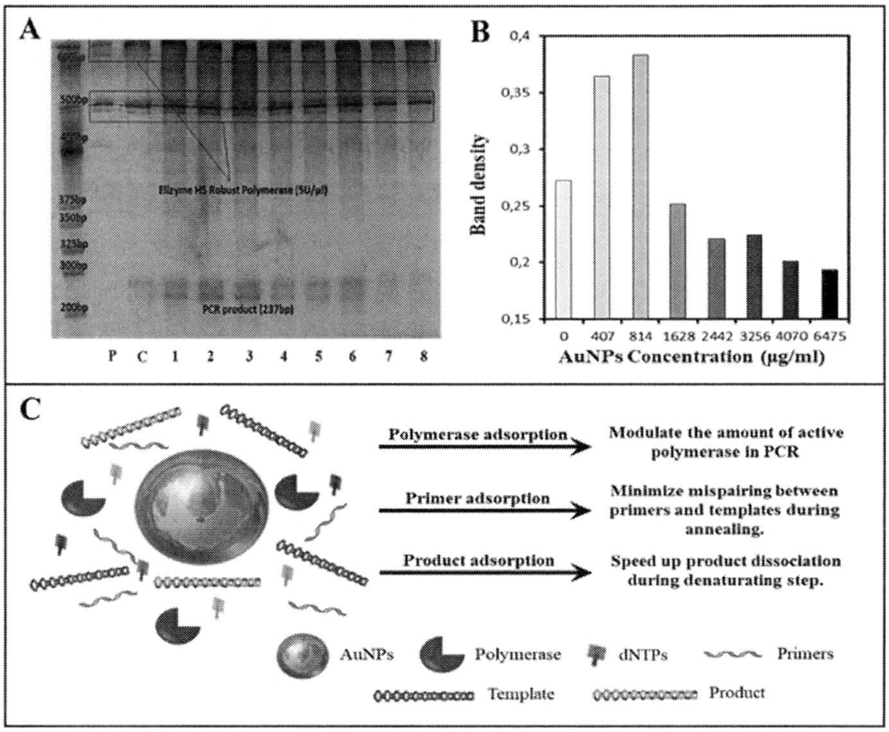

Figure 3 (A, B) The effect of the concentration (µg/mL) of gold nanoparticles on the yield of PCR: lane P (polymerase): 0; lane C (control): 0; lane 1: 0; lane 2: 407; lane 3: 814; lane 4: 1628; lane 5: 2442; lane 6: 3256; lane 7: 4070; lane 8: 6475. (C) Schematic illustration of potential effects of gold nanoparticles (AuNPs) in PCR

4. CONCLUSION

The AuNPs did indeed improve the yield of PCR. Conventional PCR can be optimized by integrating ODN-ThiC3 on the AuNP surface during PCR. Thus the use of AuNPs open new opportunities for improving PCR-based techniques, among which the improvement of PCR-based ctDNA detection may optimize breast cancer diagnosis, allowing earlier tumor detection and improved breast cancer management.

ACKNOWLEDGEMENTS

This work was supported by the H2020 ERASMUS+ and NANO LPR 2017 Liga proti rakovině Praha and international collaboration project of The European Technology Platform for Nanomedicine.

REFERENCES

[1] KARAMI, F.,MEHDIPOUR, P. A Comprehensive Focus on Global Spectrum of BRCA1 and BRCA2 Mutations in Breast Cancer. *Biomed Research International,* 2013, vol., no. pp.

[2] CHENG, F. F., SU, L.,QIAN, C. Circulating tumor DNA: a promising biomarker in the liquid biopsy of cancer. *Oncotarget,* 2016, vol. 7, no. 30, pp. 48832-48841.

[3] BOARD, R. E., WARDLEY, A. M., DIXON, J. M., ARMSTRONG, A. C., HOWELL, S., RENSHAW, L., DONALD, E., GREYSTOKE, A., RANSON, M., HUGHES, A.,DIVE, C. Detection of PIK3CA mutations in circulating free DNA in patients with breast cancer. *Breast Cancer Research and Treatment,* 2010, vol. 120, no. 2, pp. 461-467.

MICROSCOPIC STUDY OF MULTIFUNCTIONAL DRUG MOLECULE ADHESION TO ELECTRONIC BIOSENSORS COATED WITH DIAMOND AND GOLD NANOPARTICLES

[1]Tomáš FINSTERLE, [1]Ivana PILARČÍKOVÁ, [1]Ilona ALI BLÁHOVÁ, [2]Štěpán POTOCKÝ, [2]Alexander KROMKA, [2]Egor UKRAINTSEV, [3]Eugenie NEPOVIMOVÁ, [3]Kamil MUSÍLEK, [3]Kamil KUČA, [1]Bohuslav REZEK

[1]Faculty of Electrical Engineering, Czech Technical University in Prague, Czech Republic, EU
finsttom@fel.cvut.cz
[2]Institute of Physics, Czech Academy of Sciences, Prague, Czech Republic, EU
[3]Faculty of Science, University of Hradec Kralove, Czech Republic, EU

Abstract

The easy and fast detection of drug content and concentration levels is demanded in biological research as well as in clinical practice. Here we study on microscopic level how nanodiamonds and gold nanoparticles interact with a multifunctional drug molecule directly on a biosensor surface. The sensors are made of interdigitated Au electrodes coated by 5 nm hydrogenated or oxidized nanodiamonds and further combined with Au colloidal nanoparticles (size 20 nm) providing nanoscale composite (spacing 100 nm). Atomic force microscopy is employed to measure local tip-surface adhesion forces and surface topography. AFM adhesion maps show that the drug binds to all types of nanoparticles and the adhesion is also significantly influenced by the substrates on which the nanoparticles are deposited. Role of local AFM tip interaction with nanostructured surface is also discussed.

Keywords: Biosensors, nanoparticles, nanodiamonds, Alzheimer drugs

1. INTRODUCTION

The easy and fast detection of drug content and concentration levels is demanded in biological research as well as in clinical practice. Therefore, label-free platforms based on electrical detection are intensively developed. Nanomaterials and structures are often introduced to provide a fast and sensitive response. [1] Nanodiamonds have unique properties for use in biosensors.[2] Nanocrystalline diamond provides a convenient combination of physical, mechanical, chemical, biocompatible and electrical properties [3-5]. Based on analysis using SEM and AFM microscopy we study on microscopic level how nanodiamonds and gold nanoparticles interact with drug molecules. As a model system we use experimental multifunctional molecule for treatment of Alzheimer disease [6].

2. MATERIALS AND METHODS

For the preparation of sensors, commonly available sensors were used. Gold interdigital electrodes (IDE) are sputtered on the ceramic substrate (15-50 um gap, overall size 5 × 6 mm^2) coated by nanodiamonds. Suspensions of detonation nanodiamond particles (hydrogenated H-DND and air-annealed O-DND) were prepared by ultrasonication and centrifugation in water and then drop casted (10 uL) on the active sensor area. Both types of sensors were also further combined with Au colloidal nanoparticles (AuNP, size 20 nm) for providing nanoscale binding sites. For drug binding were then functionalized in MTAB (myristyltrimethylammonium bromide) and finally exposed to 1-EN-142 drug [6] in DMSO (10 % solution) and rinsed by water and methanol to remove all excess residues.

The electrode system allows the measurement of various sensitive layers. The used sensor consists of a layer structure with different roughness. The reason for using this sensor is that it is a commonly available platform.

The nanodiamond layer itself has different properties when it comes to O-DND or H-DND. Their properties are also strongly affected depending on the substrate used [7].

A set of sensor samples with different combinations of applied coatings (H-DND, MTAB, Au nanoparticles, drug 1-EN-142) were examined. Suspensions of detonation nanodiamond (DND) particles based on hydrogenated H-DND (as received, polyfunctional, positive zeta potential) and air-annealed O-DND (purified and oxidized by annealing in air at 450 °C, 30 min, negative zeta potential) were prepared by ultrasonication and centrifugation in water (20 mg of powder in 1 mL). Then 1 mL of supernatant was carefully extracted by a micropipette and diluted 10× by demi water prior to further use. Gold nanoparticles (AuNP) were obtained from a colloidal solution of citrate stabilized 20 nm Au particles (BBI Solutions, EM.GC 20, 0.06 wt%, particle density of 7×10^{11} particles per 1 mL).

The morphological differences of the prepared sensors were studied by scanning electron microscopy (SEM) using Tescan MAIA 3 in secondary electron mode. Atomic force microscopy (Bruker ICON, CF4 treated Multi75AL cantilevers) was employed to measure local tip-surface adhesion forces and surface topography. Characteristic sensor area of 1×1 um^2 was examined at a scan rate of 0.1 Hz and resolution of 512x512 pixels. Plane correction and correction of scars were applied on adhesion images using Gwyddion software [8,9] to improve the visibility of the features. Tip particle adhesion on individual particles were evaluated also Gwyddion program. The basic principle was to determine adhesions of the small area typical 20×20 nm^2 on individual particles. This area was always selected at the top of the particle. Determining adhesion from the area and not from the single top point on the particle allowed us to reduce the RMS of the adhesion signal.

3. RESULTS AND DISCUSSION

Figure 1a shows the schematic drawing of the sensor coated with DND and Au nanoparticles. **Figure 1b** shows typical SEM microscopic morphology of the sensor surface, in this case coated with O-DND/AuNP composite and with the adsorbed drug (EN). Fine nanostructured morphology is observed all over the electrode surface. This indicates dense coating of the electrodes by O-DND nanodiamonds. In addition, there are bright points noticeable. They most likely correspond to AuNP. This assumption is confirmed by further AFM adhesion analysis.

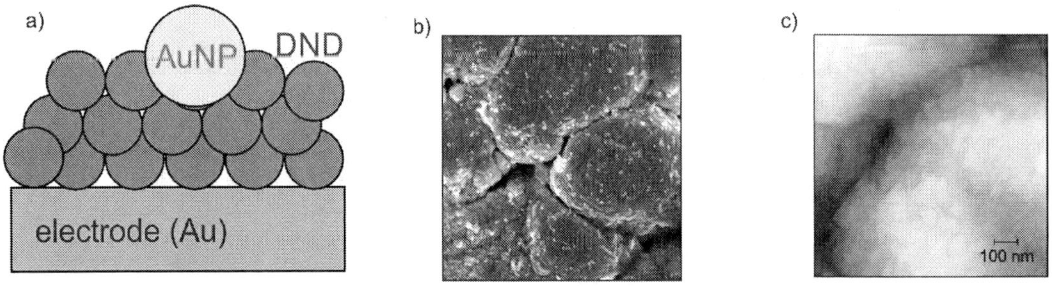

Figure 1a) Schematic cross-sectional drawing of the sensor coated with DND and AuNP, **b)** SEM morphology (2×2 um^2) of the sensor surface coated with O-DND/AuNP composite and with adsorbed drug, **c)** Detailed AFM topology (1×1 um^2) of the same surface

Figure 2 shows adhesion maps of the sensor coated with AuNP prior and after drug adhesion. Some rounded objects can be recognized on a smooth background in topography. Adhesion contrast reveals that they are different from the background and that they are composed of one to several rounded objects. Their height is about 20 nm. Their diameter is approximately 40 nm, however, that corresponds to 20 nm with AFM features being enlarged due to convolution with the tip of a typical diameter of 20 nm. Thus, these objects can be attributed to our 20 nm gold nanoparticles. Typical adhesion values are around 1.4 nN for these AuNP and 1.9 nN for the background electrode.

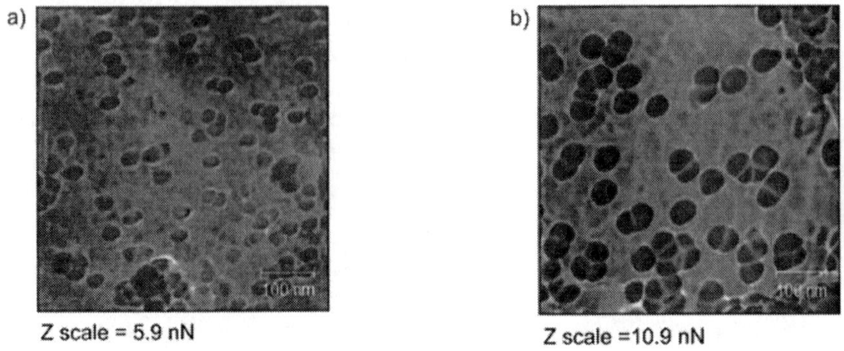

Figure 2 Detailed AFM surface adhesion maps of **a)** sensor coated with AuNP, **b)** sensor coated with AuNP after exposure to EN drug

Figure 2b shows that the drug absorption has a pronounced effect on the adhesion. Typical adhesion values are about 1.7 nN for AuNP and 3.2 nN for the electrode. This is a change of approximately 50 %. Importantly, adhesion on both AuNP and electrode change, hence the drug binds to both electrode and AuNP.

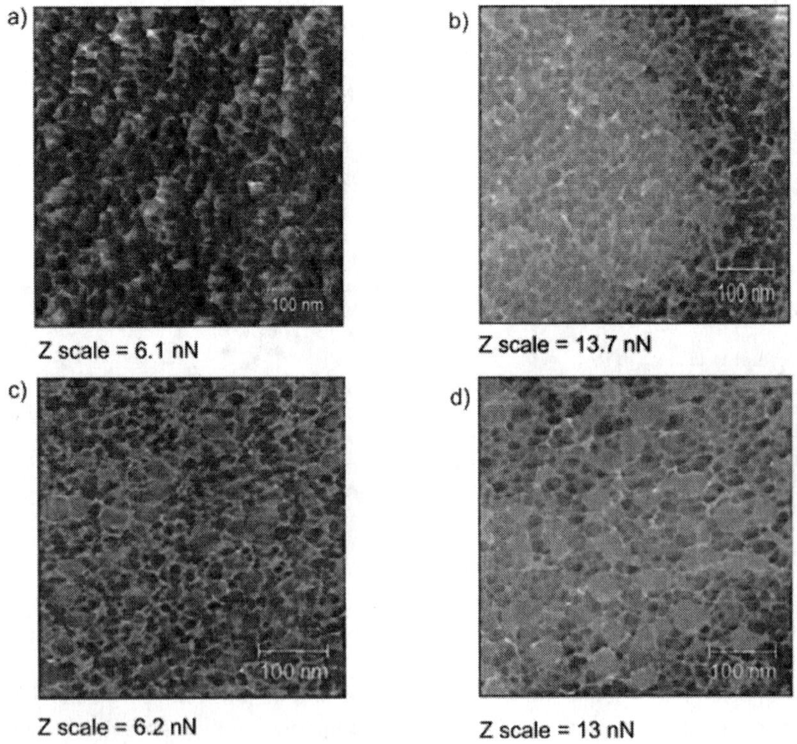

Figure 3 Detailed AFM surface adhesion maps of **a)** sensor coated with H-DND, **b)** sensor coated with H-DND after exposure to EN drug, **c)** sensor coated with H-DND/AuNP, **d)** sensor coated with H-DND/AuNP after exposure to EN drug

From adhesion map on the sensor coated with H-DND (**Figure 3a**) one can recognize a highly nanostructured surface morphology that is different to flat electrode background observed in **Figure 2**. There is a dense coverage by many small particles down to about 5 nm in size. It is thus obvious that the sensor surface is densely covered by H-DND. Typical adhesion values are around 1.8 nN on the H-DND nanoparticles and 3.3 nN on the electrode (measured in a small pinhole). After absorption of the drug (**Figure 3b**), the adhesion increased to 5.3 nN on the H-DND nanoparticles and to 7.4 nN on the electrode. The root mean square (RMS) of adhesion is around 0.5 nN in both cases, thus the difference is significant.

On the surface of a sensor where AuNP is added to H-DND (**Figure 3c**) one can recognize larger brighter particles among the smaller darker particles in the adhesion map. Considering the results on AuNP in **Figure 2**, these features can be attributed to the AuNP embedded in the H-DND layer. Typical adhesion values are about 1.4 nN on Au nanoparticles and 0.7 nN on the H-DND. Adhesion map on the H-DND/AuNP sensor after drug exposure (**Figure 3d**) shows an increase in adhesion. Typical adhesion on AuNP is 5.8 nN and on H-DND it is 3.3. nN. This is 320 % increase in adhesion. The drug again binds to both AuNPs and H-DNDs.

Figure 4 shows adhesion maps of the sensors coated with O-DND nanoparticles. Most of the surface is covered by O-DND nanoparticles. Typical adhesion on O-DND particles is 1.9 nN and on background electrode is 6.7 nN (measured in a pinhole). The drug adsorption reduces adhesion on O-DND. After drug exposure, typical adhesion value on O-DND particles is 0.8 nN and on the electrode it is 2.4 nN.

In the case of sensor coated with composite of O-DND/AuNP (**Figure 4 c-d**) AuNPs can be again recognized in the adhesion maps. Typical adhesion values on Au particles is 6.3 nN and on O-DND nanoparticles it is 9.9 nN. Adhesion after drug absorption decreases coated with composide O-DND/AuNP. Typical adhesion on Au particles is 4.2 nN and on O-DND nanoparticles is 2.4. nN.

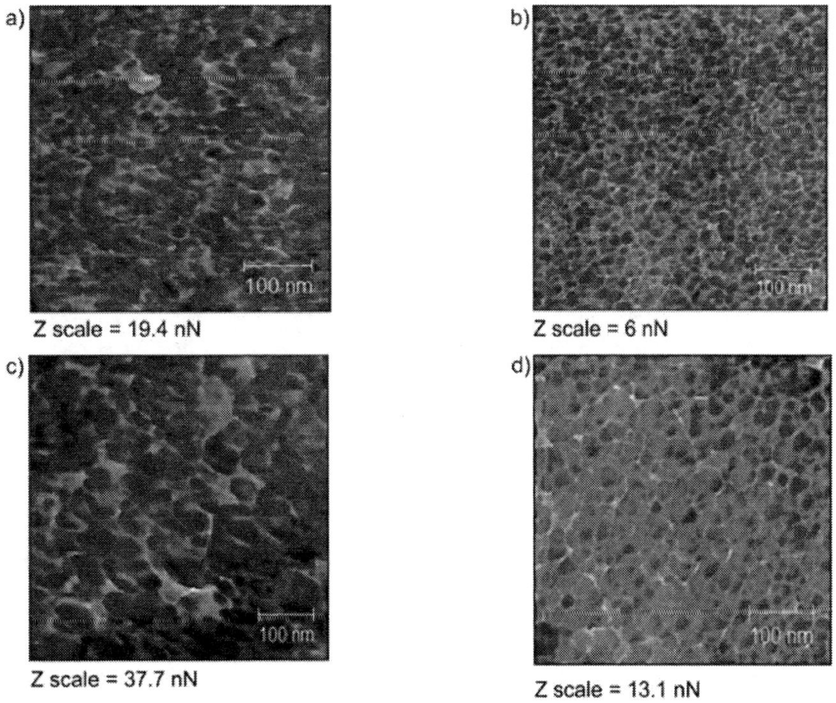

Figure 4 Detailed AFM surface adhesion maps of **a)** sensor coated with O-DND, **b)** sensor coated with O-DND after exposure to EN drug, **c)** sensor coated with O-DND/AuNP, **d)** sensor coated with O-DND/AuNP after exposure to EN drug

All values of adhesion, on original and drug exposed sensors are summarized in graph in **Figure 5**. It is evident that in the case of sensors coated with H-DND, drug 1-EN-142 increases adhesion, whereas on sensors coated with O-DND the drug decreases adhesion. The reproducible and consistent decrease of adhesion on O-DND based coatings compared to increase of adhesion on H-DND based coatings after drug exposure indicates different drug binding and conformation on these two nanomaterials.

In addition, H-DND and O-DND have also impact on the way how the drug binds to AuNP. Noticeably, after drug adsorption the adhesion on AuNP that is embedded in H-DND layer is 5× larger than on AuNP alone (see **Figure 2b**). Similarly this happens on AuNP that is embedded in O-DND, but the effect has opposite direction, the adhesion decreases after the drug adsorption. Noticeably, AuNP on O-DND exhibits much higher adhesion forces already before drug exposure.

The difference is most likely related to different AuNP charges when interacting with different substrates [7]. This is supported by the observation that the presence of H-DND also increased the density of deposited Au nanoparticles compared to other substrates. Different drug-adhesion responses on the same material can be thus observed with different substrates.

There is also different adhesion on H-DND or O-DND when alone or combined with AuNP. While the nanoparticles, coating process, and the sensor substrate are the same, the most likely explanation is change of nanodiamond surface due to exposure to AuNP colloidal solution containing citrates.

Different adhesion is observed also on background electrode when measured on different sensors. However, the free electrode area is small on coated sensors. Thus the adhesion is influenced not only by the interaction of the tip with the electrode but also with the nanoparticles on tip sides. The situation is schematically illustrated by the inset in **Figure 5**. Due to this effect the adhesion and its change after drug adsorption is the same as for given nanoparticle coating. In other words, due to sideway interaction of the tip with the nanoparticles the observed adhesion is not really controlled by the bottom electrode.

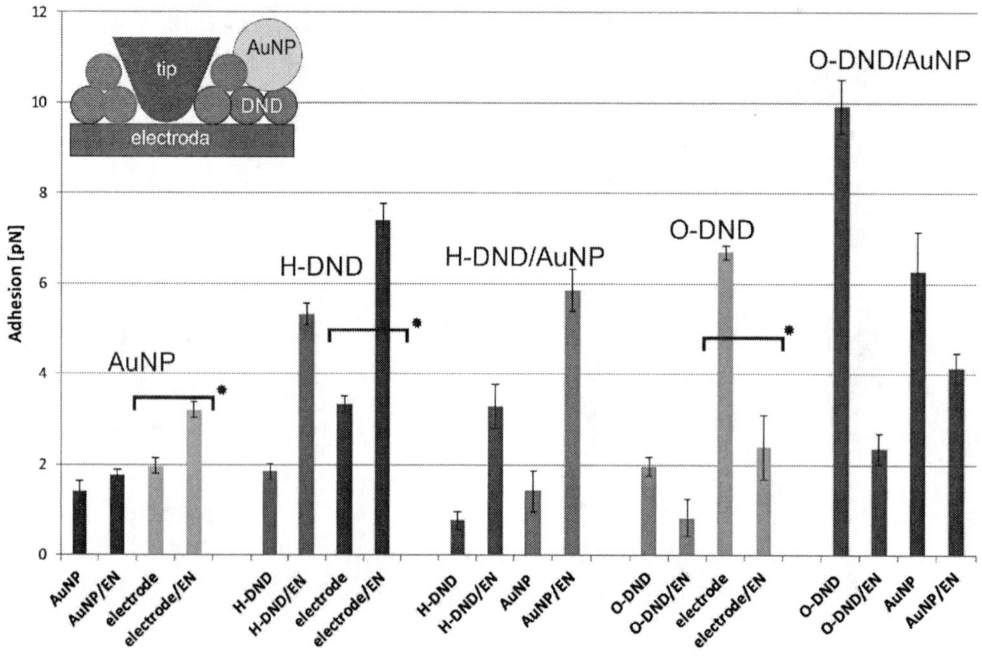

Figure 5 Bar graph of adhesion on individual particles on sensors with various nanoparticle coating.
* Adhesion when tip interacts with bottom electrode. The inset shows schematic drawing of such interaction.

4. CONCLUSION

We combined 5 nm nanodiamond coatings with 20 nm gold nanoparticles to create functional coatings on impedance sensor substrates. The microscopic morphology of the coatings was characterized by SEM and AFM, revealing nanoscale composite formation with uniformly distributed AuNP. The obtained adhesion maps from AFM showed that 1-EN-142 Alzheimer drug binds to all nanoparticle materials (DND-H, O-DND, AuNP). The drug binds to these nanoparticles whether they are standalone or combined in the composite. However, it binds in different configurations as indicated by opposite adhesion change on H-DND and O-DND based coatings. Interestingly the adhesion on AuNP depends significantly on surrounding materials, most likely due to different nanoparticle charging. The study thus may help understand biosensor response.

ACKNOWLEDGEMENTS

The work has been supported by the GACR project 17-19968S.

REFERENCES

[1] ZHANG, Wenli, Kush PATEL, Andrew SCHEXNIDER, Shirin BANU a Adarsh D. RADADIA. Nanostructuring of Biosensing Electrodes with Nanodiamonds for Antibody Immobilization. ACS Nano 2014, 8(2), 1419-1428 DOI: 10.1021/nn405240g. ISSN 1936-0851.

[2] CAI, Weibo. Applications of gold nanoparticles in cancer nanotechnology. Nanotechnology, Science and Applications [online]. 2008, 1, 17-32. DOI: 10.2147/NSA.S3788. ISSN 1177-8903.

[3] J.A. Carlisle, Diamond films: Precious biosensors, Nat. Mater. 3 (2004) 668-669. DOI:10.1038/nmat1225.

[4] STEINMULLERNETHL, D, F KLOSS, M NAJAMULHAQ, et al. Strong binding of bioactive BMP-2 to nanocrystalline diamond by physisorption. Biomaterials. 2006, 27(26), 4547-4556. DOI: 10.1016/j.biomaterials.2006.04.036. ISSN 01429612. .

[5] M. Krátká, A. Kromka, E. Ukraintsev, M. Ledinský, A. Brož, M. Kalbacova, B. Rezek, Function of thin film nanocrystalline diamond-protein SGFET independent of grain size, Sens. Actuators B Chem. 166-167 (2012) 239-245. doi:10.1016/j.snb.2012.02

[6] NEPOVIMOVA, Eugenie, Elisa ULIASSI, Jan KORABECNY, et al. Multitarget Drug Design Strategy: Quinone-Tacrine Hybrids Designed To Block Amyloid-β Aggregation and To Exert Anticholinesterase and Antioxidant Effects. Journal of Medicinal Chemistry. 2014, 57(20), 8576-8589 DOI: 10.1021/jm5010804. ISSN 0022-2623.

[7] STEHLIK, Stepan, Tristan PETIT, Hugues A. GIRARD, Alexander KROMKA, Jean-Charles ARNAULT a Bohuslav REZEK. Surface potential of diamond and gold nanoparticles can be locally switched by surrounding materials or applied voltage. Journal of Nanoparticle Research [online]. 2014, 16(4),. DOI: 10.1007/s11051-014-2364-8. ISSN 1388-0764.

[8] NEČAS, David a Petr KLAPETEK. Gwyddion: an open-source software for SPM data analysis. Open Physics [online]. 2012, 10(1) [cit. 2018-10-17]. DOI: 10.2478/s11534-011-0096-2. ISSN 2391-5471. Dostupné z: http://www.degruyter.com/view/j/phys.2012.10.issue-1/s11534-011-0096-2/s11534-011-0096-2.xml

[9] GAN, Yang. Atomic and subnanometer resolution in ambient conditions by atomic force microscopy. Surface Science Reports. 2009, 64(3), 99-121. DOI: 10.1016/j.surfrep.2008.12.001. ISSN 01675729.

THE USE OF BIOACTIVE COMPOUNDS AND THEIR ENCAPSULATION INTO LIPOSOMES TO INCREASE THE EFFECTIVENESS AND CONTROL RELEASE

Julie HOOVÁ, Marie VYSOKÁ, Lucia DZURICKÁ, Petra MATOUŠKOVÁ, Ivana MÁROVÁ

Brno University of Technology, Faculty of Chemistry, Materials Research Centre, Brno, Czech Republic, EU, xchoova@fch.vut.cz

Abstract

Extracts from naturally occurred sources, such as plants or microorganisms, have been recently tested for their interesting properties that can be used in food or cosmetic industry.

The aim of this work is to encapsulate microbial compounds, some plant extracts and also model hydrophilic and hydrophobic substances. Chosen components were encapsulated into liposomes.

Particle size and distributions of liposomes were analyzed by dynamic light scattering and their stability by using zeta potential. Spectrophotometric methods were used to measure encapsulation efficiency and to characterize the chosen plant extracts by its amount of total polyphenols, flavonoids and antioxidant activity.

Chosen samples were analyzed for their antimicrobial activity and cytotoxicity. *Serratia marcescens, Escherichia coli* as gram-negative bacterias, *Micrococcus luteus* as gram-positive bacteria and fungal strain (*Candida glabrata*) were used to test antimicrobial activity. Human keratinocytes and mouse melanoma cells were used to cytotoxicity determination.

It has been proved that encapsulation techniques help slowly release the active compounds from plant extracts and transport them to targeted location. Antimicrobial effect was also seen before and after encapsulation into liposomes. Moreover, none of the extracts showed evidence of cytotoxicity.

Keywords: Encapsulation, liposomes, polyphenols, antioxidant activity, antimicrobial activity, cytotoxicity

1. INTRODUCTION

Novel antimicrobial research and its results are very often used to combine with antibiotic therapy. This can be new way how to pursue in order to combat and delay resistance of pathogens. Many research groups are studying the effects of natural antimicrobial substances. These agents can be used as stand-alone or adjunctive therapies.[1] Mainly, phenolic compounds are increasingly attracting attention and in this context, the use of essential oils or other plant extract may exert synergistic antimicrobial activity. [1,2,3] Essentials oils and plant extracts are widely used in food, agricultural, cosmetics and pharmaceutical application. However, there are certain limitations, such as, low water solubility, strong organoleptic flavor and low stability. Using of encapsulation techniques might help increasing these components' chemical stability and solubility. Encapsulation into different type of particles also supports controlled and sustained release which enhances the bioavailability and efficiency against resistant pathogens. [2,3,4] So, the antimicrobial efficiency against microorganism can be enhanced by applying novel drug delivery systems such as nanoparticles or liposomes.[1] Liposomes are spherical lipid vesicles with a bilayered membrane structure consisting of amphiphilic lipid molecules. A lipid membrane surrounding an aqueous cavity enables to encapsulate both hydrophobic and hydrophilic compounds. One of the most commonly used lipids in liposome preparation is phosphotidylcholine where cholesterol is incorporated into the formulation to adjust membrane rigidity and stability. Currently, liposomes are the most widely used antimicrobial drug delivery system, thanks to direct fusing with bacterial membranes, the antimicrobial component of liposomes can be released to the cell membranes or the interior of the bacteria. [5]

2. MATERIALS AND METHODS

2.1. Chemicals

ABTS (2,2'azinobis-(3-ethylbenzthiazoline-6-sulfonic acid)), Gallic acid, (+)-catechin, Folin-Ciocalteu's reagent, cholesterol, soya bean lecithin, nisin and Trolox (6–hydroxy–2,5,7,8–tetramethylchroman)–2–carboxylic acid were purchased from Sigma-Aldrich Chemical Co. (St. Louis, MO, USA). All other solvents and reagents used in the analysis were of analytical grade.

2.2. Plant extracts and their characterization

Three different types of extract were prepared (water, ethanol and essential oils). Total phenolic and flavonoid contents, antioxidant activity and antimicrobial activity of these extracts were analyzed. Finally, the extracts were used for encapsulation into liposome. The total phenolic content was measured using the Folin–Ciocalteu colorimetric method. Readings were quantified using a standard curve of gallic acid and the results were expressed in mg of gallic acid equivalent (GAE) per g of dry sample. The total flavonoid content was measured by the aluminium chloride colorimetric method. Readings were quantified using a standard curve of catechin and the results were expressed is mg of catechin equivalent (CE) per gof dry sample. Total antioxidant activity was determined by ABTS radical cation decolorization assay and antioxidant activity was calculated as a change in the absorbance. Results were expressed in mg of Trolox equivalent (TE) per gram of dry sample. The measurements were performed using UV-Vis spectrophotometer (Thermo Spectronic Helios™ δ, Thermo Fisher UK Ltd., Hemel Hempstead, UK).

2.3. Microorganisms and their cultivation

The antibacterial activities of all extracts and liposome were determined against bacterial and yeast strains. Gram-positive *Micrococcus luteus* CCM 1569, Gram-negative bacteria *Serratia marcescens* CCM 8587, *Escherichia coli* CCM 7395 and yeast strain *Candida glabrata* CCM 8270 were used. All microorganism used in this study were supplied by the Czech Collection of Microorganisms in Brno. Bacterial cultures were grown in commercial Nutrient Broth medium, for *Candida glabrata* commercial medium YPD Broth was used. Temperature for cultivation of all tested microorganism was 37 °C.

2.4. Determination of antimicrobial activity

Agar diffusion method and broth dilution method were used to determine antimicrobial activity of extracts and liposomes. Mainly the microbroth dilution assay was used. In this method, microorganisms were inoculated into a liquid growth medium in the presence of different concentrations of testing samples. Specifically, 150 μL of 10^6 CFU/mL of microbial culture was added to sterile 96-well microtiter plates followed by 50 μL of sample. The negative control was prepared by adding 150 μL of microbial culture followed by 50 μL of solvent. Absorbance of the wells was measured using ELISA reader (BioTek Instruments GmbH, Bad Friedrichshall, Germany) at 630 nm, before and after 24 hour incubation at 37°C.

2.5. Preparation of liposomes and their characterization

In this work plant extracts were encapsulated into liposome nanoparticles prepared using sonication techniques from soya bean lecithin. Solution of lecithin 16 mg/mL with the addition of cholesterol 2 mg/mL was used. The lipid dispersion was sonicated (80 W, 20 kHz) using ultrasonic homogenizer (BANDELIN electronic GmbH & Co. KG, Berlin, GER) for a few minutes to form liposomes. Size of prepared liposomes and stability of liposomes were analyzed thanks to its zeta potential by dynamic light scattering using colloidal DLS analyzer Zetasizer Nano ZS (Malvern Instruments Ltd., Malvern, UK). Encapsulation efficiency was measured spectrophotometrically.

The content of encapsulated polyphenols components was measured before and after encapsulation. The percentage of encapsulation efficiency (EE) was then calculated according to the following equation:

$$\mathrm{EE} = \frac{m_t - m_f}{m_t} \cdot 100 \tag{1}$$

where:

m_t - total polyphenols before encapsulation (g)

m_f - free polyphenols after encapsulation (g)

3. RESULTS AND DISCUSSION

3.1. Characterization of plant extracts

In this study content of phenolics and flavonoids in prepared extracts was measured. Total antioxidant activity of tested extracts was determined too. The highest content of total phenolic components was determined in essential oil of clove, in ethanol extract and water water extract of clove and cinnamon (**Figure 1**). Aqueous extract sample of acerola was also characterized. Acerola extract showed even higher amount of total phenolics than tested spices - 23, 87 mg/g.

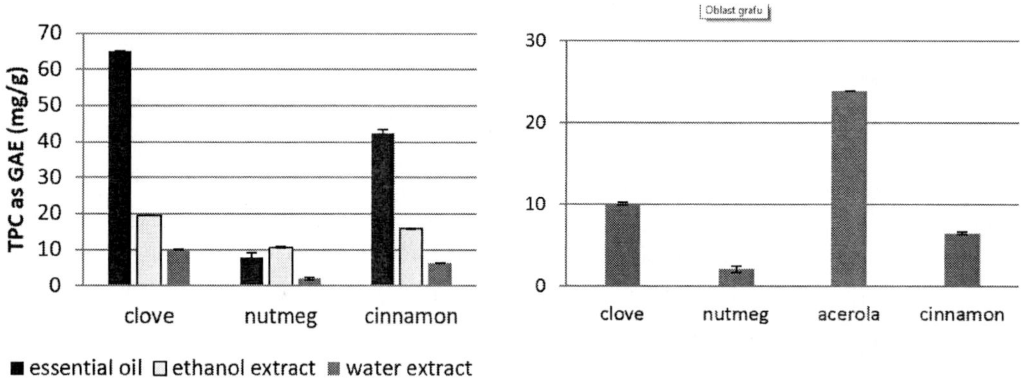

Figure 1 Content of total phenolic components in different type of extracts

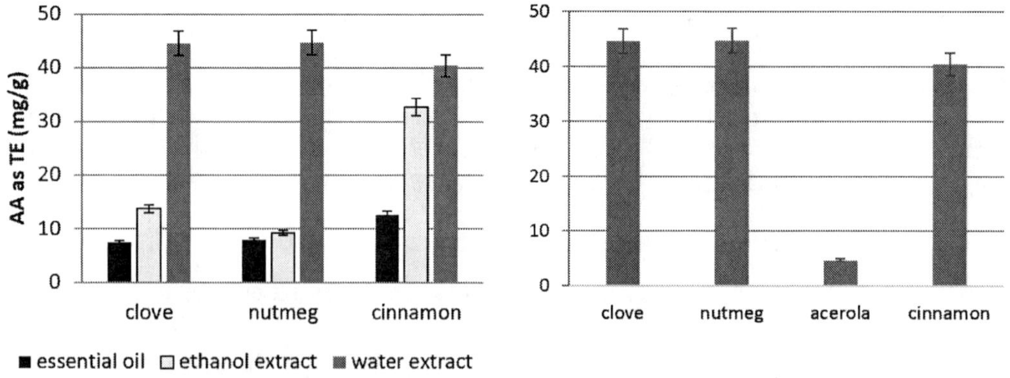

Figure 2 Antioxidant activity of different type of extracts

High antioxidant activity was observed mainly in water extracts (**Figure 2**) and the highest antioxidant activity was determined in nutmeg and clove. It is interesting that extracts of nutmeg had low concentration of total phenolic components, while exhibited high antioxidant activity. These results indicate that the phenolic components are not a sole source of antioxidant activity of these herbal extracts.

3.2. Encapsulation of plant extracts and determination of particle stability

Three different types of plant extracts were encapsulated into liposomes. Encapsulation was successful for all tested spice extract and encapsulation efficiency was in almost all case higher than 70 % (**Figure 3**).

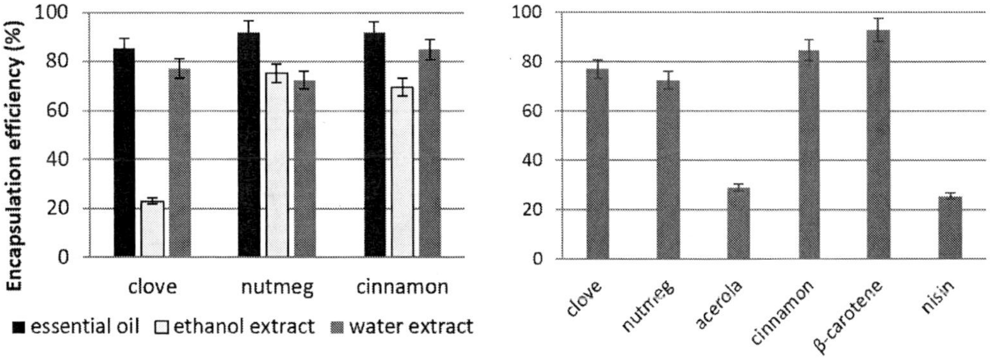

Figure 3 Encapsulation efficiency of different type of extracts

Size of particles was analyzed by dynamic light scattering and the size of liposomes was in ranged from 100 to 320 nm. Stability of the particles was then measured using zeta potential and all prepared particles exhibited very good stability (**Table 1**).

Table 1 Size and stability of liposomes measured by dynamic light scattering and zeta potential

source of encapsulated water extract	star anise	clove	nutmeg	cinnamon	acerola	β-carotene	nisin
Ø size [nm]	172.6 ± 2.1	146.7 ± 0.8	164.2 ± 1.0	159.1 ± 1.9	130.8 ± 1.6	132.3 ± 0.7	106.7 ± 0.9
ZP [mV]	- 49.7 ± 1.1	- 46.9 ± 0.8	- 45.1 ± 1.6	- 43.4 ± 0.6	- 38.9 ± 1.1	- 44.1 ± 1.2	- 42.4 ± 0.9

3.3. Determination of antimicrobial activity

In this study the antimicrobial activity of different types of plant extracts encapsulated into liposomes were measured. The best antimicrobial activity was found in liposomes with essential oils. These particles showed good antimicrobial activity against all tested strains (**Table 2**).

Table 2 Antimicrobial activity of spice extracts-load liposomes

microorganism	*M. luteus*			*S. marcescens*			*C. glabrata*		
type of encapsulated extract	essential oil	ethanol	water	essential oil	ethanol	water	essential oil	ethanol	water
clove	+++	+++	+	++	+++	+	+++	+	+
nutmeg	+++	+++	++	+++	+	+	+	-	+
cinnamon	+++	++	+	++	+	+	+++	-	+

0-35 % growth compared to the control (+++), 36-70 % growth compared to the control (++), 71-90 % growth compared to the control (+) 100 % growth compared to the control (-)

Liposomes with chosen aqueous extract were combined with nisin in specific ration according to their individual antimicrobial activity. Some samples showed synergism when antimicrobial activity was higher than activity of individual nisin sample. Acerola, cinnamon and clove extracts might be the best extracts to be combined with nisin thanks to high antimicrobial activity against *Micrococcus luteus, Escherichia coli and Candida glabrata* (**Table 3**).

Table 3 Antimicrobial activity of combined extracts of nisin and plant extract

microorganism	*M. luteus*				*S. marcescens*				*E. coli*				*C. glabrata*			
concentration [mg/mL] nisin / extract	0.1		20		0.1		10		0.2		50		0.1		10	
ratio (nisin:extract)	1:0	0:1	1:1	1:2	1:0	0:1	1:1	1:2	1:0	0:1	1:1	1:2	1:0	0:1	1:1	1:2
clove	-	++	+	+	+	+	-	+	+	-	-	+	-	+	+	+
cinnamon	-	+	-	+	+	+	-	-	+	-	-	-	-	+	+	+
acerola	-	+	+	++	+	+	-	+	+	++	+++	+++	-	+	+	+

0-20 % growth compared to the control (+++), 21-50 % growth compared to the control (++), 51-90 % growth compared to the control (+) 100 % growth compared to the control (-)

3.4. Long-term stability and safety of prepared liposomes

The long-term stability of liposomes containing plant extract was tested in model conditions. The liposomes were also added into different type of cosmetic product (gel and emulsion) and stability of liposomes and stability of cosmetic product during storage were measured. The prepared particles exhibited very good long-term stability in aqueous conditions. Relatively good stability of particles was determined in cosmetic emulsion. Nevertheless, particles stored in the presence of higher concentrations of fat were less stable.

Furthermore, it is necessary to test potential genotoxic and cytotoxic effects before application of prepared particles. Genotoxic effects of extract-loaded liposomes were analyzed by commercial SOS Chromotest kit and it was found, that neither particles show any genotoxic effect. Next, the cytotoxicity of liposomes was assessed by MTT assay on human epidermal keratinocytes (HEK). The MTT assay was used to assess the viability of HEC with different concentration of tested liposomes and it was confirmed that all tested liposomes particles with concentration from 2 to 14 % do not have toxicity effect.

4. CONCLUSION

Liposome particles with encapsulated hydrophilic or hydrophobic antimicrobial plant extract, especially spices, could be used in cosmetics and pharmacy applications as an alternative to antibiotics. Antimicrobial effect together with antioxidant activity of plant extracts even combined with animal origin substances, such as nisin, could be served as a very promising tool for food preservation or other antimicrobial products. Other potential applications of prepared particles are in disinfection or wound healing therapy.

ACKNOWLEDGEMENTS

This work was supported by project "Materials Research Centre - Sustainability and Development" No. LO1211 (Ministry of Education, Youth and Sport of the Czech Republic).

REFERENCES

[1] KHAMENEH, B., DIAB, R., GHAZVINIi, K., BAZZAZ, B.S.F. Breakthroughs in bacterial resistance mechanisms and the potential ways to combat them. *Microb Pathog*, 2016, vol. 95, pp. 32-42.

[2] RAJ M., PARALIKAR P., JOGEE P., AGARKAR G., INGLE A.P., DERITA M., ZACCHINO S. Synergistic antimicrobial potential of essential oils in combination with nanoparticles: Emerging trends and future perspectives. *International Journal of Pharmaceutics*, 2017, vol. 519, pp. 67-78.

[3] LU, W., KELLY, A. L., MIAO, S. Emulsion-based encapsulation and delivery systems for polyphenols. *Trends in Food Science &Technology*, 2016, vol. 47, pp. 1-9.

[4] El ASBAHAMI, A.et al. (2015) Essential oils: from extraction to encapsulation. *Int. J. Pharmacol.* 2015, vol. 483, pp. 220-243.

[5] ZHANG, L., POMPATTANANANGKU, D., HU, C.-M., HUANG, C.-M. Development of nanoparticles for antimicrobial drug delivery. *Curr. Med. Chem.* 2019, vol. 17, no. 6, pp. 585-594.

ANTIOXIDANT ACTIVITY OF SILVER NANOPARTICLES PREPARED BY GREEN SYNTHESIS

[1]Branislav RUTTKAY-NEDECKÝ, [2]Michaela DOČEKALOVÁ, [3]Božena HOSNEDLOVÁ, [2]Dagmar UHLÍŘOVÁ, [2]Martina STAŃKOVÁ, [4]Marta KEPINSKA, [4]Halina MILNEROWICZ, [5]Carlos FERNANDEZ, [3]Mojmír BAROŇ, [3]Jiří SOCHOR, [6]Hoai Viet NGUYEN, [1,2,4]Rene KIZEK

[1]University of Veterinary and Pharmaceutical Sciences Brno, Pharmaceutical Faculty, Brno, Czech Republic, EU, kizek@sci.muni.cz
[2]Prevention Medicals, Studenka-Butovice, Czech Republic, EU, uhlirova@preventionmedicals.cz
[3]Mendel University in Brno, Faculty of Horticulture, Department of Viticulture and Enology, Lednice, Czech Republic, EU, sochor.jirik@seznam.cz
[4]Department of Biomedical and Environmental Analyses, Faculty of Pharmacy with Division of Laboratory, Diagnostics, Wroclaw Medical University, Wroclaw, Poland, EU, zalewska.m@gmail.com
[5]Robert Gordon University, School of Pharmacy and Life Sciences Garthdee Road, Aberdeen, AB10 7QB, Scotland, United Kingdom; c.fernandez@rgu.ac.uk
[6]Research Center for Environmental Monitoring and Modeling, VNU University of Science, Hanoi, Vietnam, nguyenviethoai@hus.edu.vn

Abstract

At present, great attention is given to silver nanoparticles (AgNPs), which thanks to its unique properties, such as good electric conductivity, photoelectrochemical activity and antimicrobial activity are widely used. Green synthesis of nanoparticles uses biological molecules from living organisms. Biological extracts may contain molecules which exhibit significant antibacterial, antiviral and cytotoxic effects. The aim of this work was to study the diverse AgNPs synthesized from 10 different types of plant extracts. Extracts from dried plants (0.5 g/25 mL 18 MΩ water) were prepared at 70 °C by heating for 20 minutes. After filtration, the leachates were mixed in the 1:1 ratio with 0.1 M $AgNO_3$ and allowed to stir at room temperature for 18 hours. They were then mixed in the 1:1 ratio with methanol, shaken for 5 minutes on the rotary mixer, and centrifuged at 12,000 g for 30 minutes. After removing the supernatant, the pellet was dried at 60 °C for 24 hours. After weighing, the purified AgNPs were dissolved in 18 MΩ water. AgNPs were yellow, orange and brown. The extraction efficiency was monitored by organic solvents (methanol, ethanol, acetone, propanol). The yields of AgNPs ranged from 15 to 5 %, and the most suitable solvent was methanol with an average AgNPs yield of about 10 %.AgNPs were characterized spectrally (spectral maxima were in the range of 300-500 nm) by determining the zeta potential and the size of nanoparticles (30-80 nm). Furthermore, antioxidant activity was monitored using ABTS (670 nm), DPPH (517 nm), and FRAP (595 nm) methods. Two methods (ABTS and DPPH) based on the elimination of synthetic radicals and the FRAP method based on the reduction of iron complexes were used to monitor the antioxidant activity. Antioxidation assays were evaluated using calibration curve equations in which the standard was gallic acid. The results for the ABTS and DPPH methods were also expressed as percentage of inhibition of the radicals and in the FRAP method as percentage of reduction activity. The results were calculated using the ABTS method in the range (25.9 - - 84.9 %), in the DPPH method (19.2-86.6 %) and in the FRAP method (8.5 - 93.2 %). Most AgNPs prepared by green synthesis showed significant antioxidant activity.

Keywords: Nanomedicine, silver nanoparticles, green synthesis, antioxidant activity

1. INTRODUCTION

The synthesis of silver nanoparticles (AgNPs) attracts an increasing interest due to their new and different characteristics that allow applications in various fields of medicine and biotechnology such as antimicrobials, and anticancer agents [1]. The properties of AgNPs as a high surface area, very small size, and high dispersion make them one of the most commonly used nanomaterials. AgNPs are also known to have antioxidant properties [2]. Several techniques have demonstrated that AgNPs can be synthesized using chemical and physical methods, however due to the fact of usage of a huge amount of toxic chemicals and high temperature conditions, a search for alternative methods began. Synthesis of nanoparticles by biological methods, using microorganisms, enzyme and plant or plant extract, has been suggested as possible eco-friendly alternatives to chemical and physical methods [3]. The use of plant extracts to produce nanoparticles is one of environmental friendly green processes. Nanoparticles produced from plant extract, because of their medicinal properties, could be used in drugs, targeted drug delivery and cosmetic applications [4]. The various biomolecules present in the plant extract such as enzymes, proteins, flavonoids, terpenoids, and cofactors act as both reducing and capping agents [5]. The plant-mediated synthesis of nanoparticles is relatively fast as there is no need of maintaining specific media and culture conditions, unlike microbial synthesis [5].

Most of the oxidative diseases are due to oxidative stress resulting from free radicals [6]. Free radicals such as superoxide anion, hydroxyl radicals and non-radical species such as hydrogen peroxide and singlet oxygen are different varieties of activated oxygen constituting reactive oxygen species [7]. An active antioxidative defense system is needed to balance the output of free radicals. Antioxidant therapy by the curing of these diseases has an enormous importance. Nanotechnology is an interdisciplinary approach in biochemical applications and focusing on synthesis of nanoparticles have improved antimicrobial and antioxidant properties against the degenerative diseases and cancer [8]. Antioxidant activity in plant extract is due to the redox potential of phytochemicals [9], which can play an important role in quenching singlet and triplet oxygen, decomposing the peroxides or neutralizing the free radicals. Therefore, it is assumed that higher antioxidant activity of nanoparticles might be due to the preferential adsorption of the antioxidant material from the plant extract onto the surface of the nanoparticles. In our study we used ten types of greenly synthesized AgNPs from medicinal plants or food of plant origin and we tested them for their antioxidant activity using DPPH, ABTS and FRAP methods.

2. MATERIAL AND METHODS

2.1. Nanoparticle synthesis

For preparation of AgNPs extracts from 10 different plants were used. They were the following: AgNPs 1 - black tea (*Camelia sinensis*), AgNPs 2 - green tea (*Camelia sinensis)*, AgNPs 3 - coffee (*Coffea arabica)*, AgNPs 4 - common thyme (*Thymus vulgaris*), AgNPs 5 - red clover (*Trifolium pratense*), AgNPs 6 - red raspberry (*Rubus idaeus*), AgNPs 7 - absinthe wormwood (*Artemisia absinthium*), AgNPs 8 - common agrimony (*Agrimonia eupatoria*), AgNPs 9 - garden strawberry (*Fragaria ananassa*), AgNPs 10 - purple crownvetch (*Securigera varia)*. Plants except tea and coffee were harvested in the Boskovická brázda area in May 2018. After collection, they were washed in distilled water and dried in a 60 ° C oven for 48 hours followed by grinding to about 1 mm particle size. Extracts from dried plants (0.5 g/25 mL 18 MΩ water) were prepared at 70 ° C by heating for 20 minutes. After filtration, the leachates were mixed in the 1:1 ratio with 0.1 M $AgNO_3$ and allowed to stir at room temperature for 18 hours. The prepared AgNPs solution was stored at 4 ° C in closed containers and spectrally characterized. AgNPs solutions were also used for determination of antioxidant activity.

2.2. Characterization of nanoparticles

The absorbance spectra of nanoparticles were recorded within the range from 350 to 700 nm using an UV-3100PC UV-VIS spectrophotometer (VWR, Germany). The average nanoparticles' size and size distribution were determined by quasielastic laser light scattering with a Malvern Zetasizer (NANO-ZS, Malvern Instruments, Worcestershire, UK). Nanoparticle distilled water solution of 1.5 mL (1 $mg.mL^{-1}$) was put into a polystyrene latex cell to measure the following properties such as: detector angle 173°, wavelength 633 nm, refractive index 0.30, real refractive index 1.59, and temperature 25 °C.

2.3. Antioxidant properties of nanoparticles

Photometric measurements were carried out using an chemical analyser BS-300 (Mindray, China). For detection itself, the following range of wavelengths can be used - 340, 380, 412, 450, 505, 546, 570, 605, 660, 700, 740 and 800 nm. The DPPH test is based on the ability of the stable 2, 2-diphenyl-1-picrylhydrazyl (DPPH) free radical to react with hydrogen donors. A 150 µL volume of reagent (0.095 mM 2,2-diphenyl-1-picrylhydrazyl - DPPH•) was incubated with 15 µL of sample. Absorbance was measured at 505 nm for 12 minutes and output ratio was achieved by difference of absorbance at the last (12th) minute and second minute of the assay procedure. The ABTS radical method is one of the most used assays for the determination of the concentration of free radicals. It is based on the neutralization of a radical-cation arising from the one-electron oxidation of the synthetic chromophore 2,2'-azino-bis(3-ethylbenzothiazoline-6-sulfonic acid) (ABTS): ABTS• - e- ABTS•+. This reaction is monitored spectrophotometrically by the change of the absorption value. A 150 µL volume of reagent 7mM ABTS• (2,2′-azinobis 3-ethylbenzothiazoline-6-sulfonic acid and 4.95 mM potassium peroxodisulphate) is poured with 3 µL of sample. Absorbance is measured at 660 nm. For calculating of the antioxidant activity, difference between absorbance at the last (12th) minute and second minute of the assay procedure was used. The FRAP method (Ferric Reducing Antioxidant Power) is based on the reduction of complexes of 2,4,6-tripyridyl-s-triazine (TPTZ) with ferric chloride hexahydrate ($FeCl_3 \cdot 6H_2O$), which are almost colorless, or eventually slightly brownish. This chemical forms blue ferrous complexes after its reduction. Reagent preparation: Solution 1: 10 mmol.L^{-1} solution of TPTZ in 40 mmol.L^{-1} of hydrochloric acid. Solution 2: 20 mmol.L^{-1} solution of ferric chloride hexahydrate, in ACS water. Solution 3: 20 mmol.L^{-1} acetate,buffer pH 3.6. These three solutions (TPTZ, $FeCl_3$, acetate buffer) are mixed in a 1:1:10 ratio. A 150 µL volume of reagent is injected into a plastic cuvette with subsequent addition of a 3 µL sample. Absorbance is measured at 605 nm for 12 minutes. Difference between absorbance at the last (12th) minute and second minute of the assay procedure was used for calculating of the antioxidant activity.

3. RESULTS

3.1. Characterization of nanoparticles

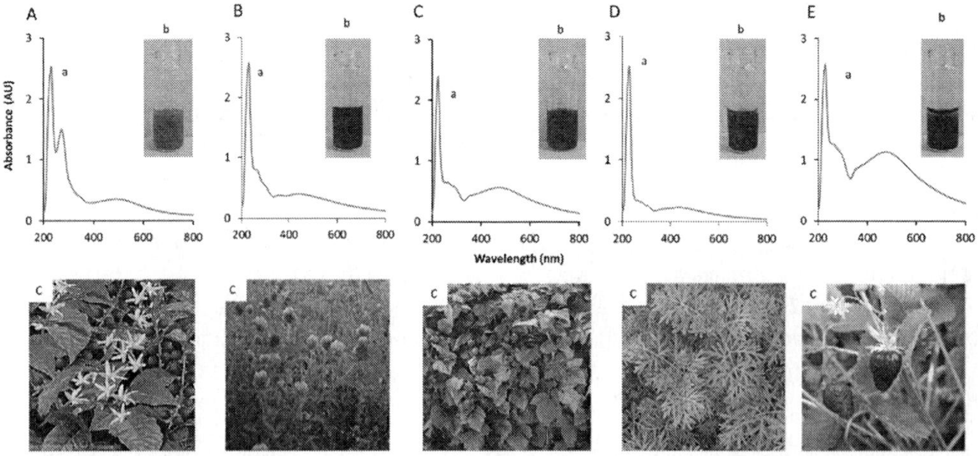

Figure 1 Characterization of AgNPs synthesized by green synthesis (A - AgNPs 3, B - AgNPs 5, C - AgNPs 6, D - AgNPs 7, E - AgNPs 9. AgNPs were prepared using following plant extracts: AgNPs 3 - coffee (*Coffea arabica,* AgNPs 5 - red clover (*Trifolium pratense*), AgNPs 6 - red raspberry (*Rubus idaeus*), AgNPs 7 - absinthe wormwood (*Artemisia absinthium*), AgNPs 9 - garden strawberry (*Fragaria ananassa*). AgNPs were characterized by absorption spectra (a). The relevant characteristic images of the nanoparticles are shown in the pictures marked with b. Pictures of plants used for the preparation of plant extracts are marked with c.

The AgNPs were characterized by absorbance spectra (**Figures 1, A, B, C, D, E, F, G, H, I, J-a**) with the two characteristic absorption maxima that ranged between 470-480 nm (AgNPs B) (**Figures 1, A-Ja**). **Figures 1, A-Jb** illustrates the relevant images of the AgNPs prepared by green synthesis. Size of nanoparticles ranged between 30 and 80 nm.

3.2. Antioxidant activity of nanoparticles

The antioxidant activity was analyzed by DPPH, ABTS and FRAP methods for AgNPs prepared by green synthesis and the results were expressed in both DPPH and ABTS methods as percentage of radical scavenging and in FRAP method as percentage of Fe^{3+} reduction as well as in all methods as gallic acid equivalent (GAE) in mg.L^{-1}. Percentage of DPPH and ABTS radical scavenging by AgNPs prepared by green synthesis is shown in **Figures 2A,B** and percentage of Fe^{3+} reduction measured using FRAP method is shown in **Figure 2C**. Results of antioxidant activity of AgNPs determined by DPPH, ABTS and FRAP method and expressed in GAE are shown in **Figures 2D,E,F**, respectively.

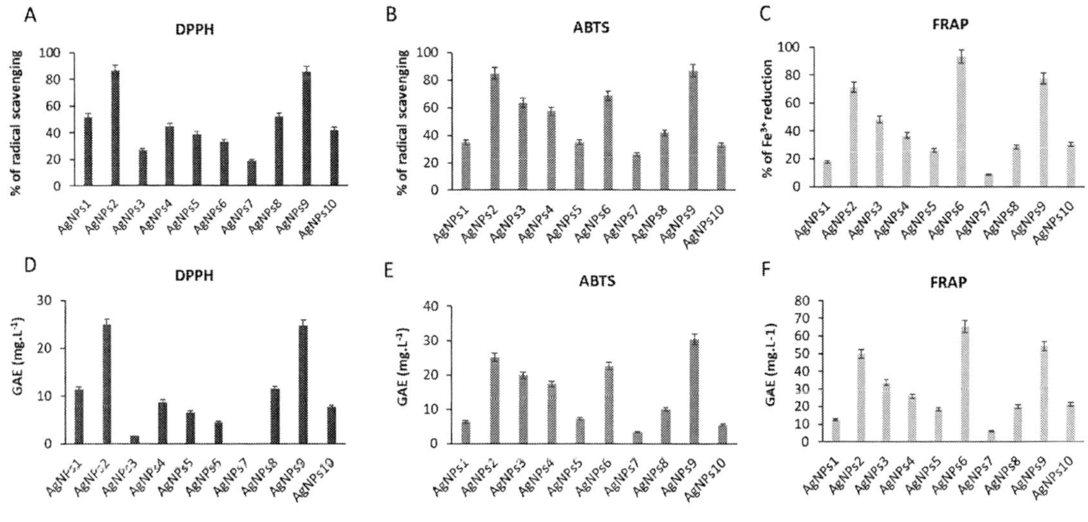

Figure 2 Antioxidant activity of AgNPs synthesized by green synthesis determined by DPPH method (A, D), ABTS method (B, E) and FRAP method (C, F). In Figures 2A and 2B, the antioxidant activity is expressed as percentage of radical extinction, in Figure 2C as a percentage of Fe^{3+} reduction, and in figures 2D, E and F, in GAE. AgNPs were prepared using following plant extracts: AgNPs 1 - black tea (*Camelia sinensis*), AgNPs 2 - green tea (*Camelia sinensis)*, AgNPs 3 - coffee (*Coffea Arabica)*, AgNPs 4 - common thyme (*Thymus vulgaris*), AgNPs 5 - red clover (*Trifolium pratense*), AgNPs 6 - red raspberry (*Rubus idaeus*), AgNPs 7 - absinthe wormwood (*Artemisia absinthium*), AgNPs 8 - common agrimony (*Agrimonia eupatoria*), AgNPs 9 - garden strawberry (*Fragaria ananassa*), AgNPs 10 - purple crownvetch *Securigera varia*.

The highest antioxidant activity was determined in AgNPs 2 (prepared using green tea extract) and in AgNPs 9 (prepared using garden strawberry extract) using DPPH and ABTS methods. Results of antioxidant activity of these samples ranged between 85-87 % of radical scavenging and 24-30 GAE (mg.L^{-1}). When FRAP method was used, the highest antioxidant activity was determined in AgNPs 6 (prepared using red raspberry extract). Result of antioxidant activity of the sample was 93 % of radical scavenging and 65 GAE (mg.L^{-1}).

4. CONCLUSION

Ten different AgNPs were prepared by green synthesis and spectrophotometrically characterized. The spectral maxima ranged between 470-480 nm. The determination of antioxidant activity using DPPH, ABTS and FRAP methods showed, that the highest antioxidant activity was determined in AgNPs 2 (prepared using green tea extract), AgNPs 9 (prepared using garden strawberry extract), and AgNPs 6 (prepared using red raspberry extract). The anticancer properties of these AgNPs will be investigated.

ACKNOWLEDGEMENTS

This work has been funded by the grant IGA GREEN-NANO 316/2018 /FaF.

REFERENCES

[1] ABOU EL-NOUR, K. M. M., EFTAIHA, A., AL-WARTHAN, A.,AMMAR, R. A. A. Synthesis and applications of silver nanoparticles. *Arabian Journal of Chemistry,* 2010, vol. 3, no. 3, pp. 135-140.

[2] ABDEL-AZIZ, M. S., SHAHEEN, M. S., EL-NEKEETY, A. A.,ABDEL-WAHHAB, M. A. Antioxidant and antibacterial activity of silver nanoparticles biosynthesized using Chenopodium murale leaf extract. *Journal of Saudi Chemical Society,* 2014, vol. 18, no. 4, pp. 356-363.

[3] SAXENA, A., TRIPATHI, R. M.,SINGH, R. P. Biological synthesis of silver nanoparticles by using onion (Allium cepa) extract and their antibacterial activity. *Digest Journal of Nanomaterials and Biostructures,* 2010, vol. 5, no. 2, pp. 427-432.

[4] MITTAL, A. K., CHISTI, Y.,BANERJEE, U. C. Synthesis of metallic nanoparticles using plant extracts. *Biotechnology Advances,* 2013, vol. 31, no. 2, pp. 346-356.

[5] RAJA, S., RAMESH, V.,THIVAHARAN, V. Green biosynthesis of silver nanoparticles using Calliandra haematocephala leaf extract, their antibacterial activity and hydrogen peroxide sensing capability. *Arabian Journal of Chemistry,* 2017, vol. 10, no. 2, pp. 253-261.

[6] GUTTERIDGE, J. M. C. Free-radicals in disease processes - a compilation of cause and consequence. *Free Radical Research Communications,* 1993, vol. 19, no. 3, pp. 141-158.

[7] HALLIWELL, B.,GUTTERIDGE, J. M. C. Role of free radicals and catalytic metal ions in human disease - an overview. *Methods in Enzymology,* 1990, vol. 186, no. pp. 1-85.

[8] NAZEM, A.,MANSOORI, G. A. Nanotechnology solutions for Alzheimer's disease: Advances in research tools, diagnostic methods and therapeutic agents. *Journal of Alzheimers Disease,* 2008, vol. 13, no. 2, pp. 199-223.

[9] ZHENG, W.,WANG, S. Y. Antioxidant activity and phenolic compounds in selected herbs. *Journal of Agricultural and Food Chemistry,* 2001, vol. 49, no. 11, pp. 5165-5170.

DESIGNING A COMPOSITE FILMS BASED ON CARBON QUANTUM DOTS AND FURCELLARAN

[1,2]Vedran MILOSAVLJEVIC, [3]Ewelina JAMROZ, [1]Milica GAGIC, [1,2]Pavel KOPEL, [1,2]Vojtech ADAM

[1]Department of Chemistry and Biochemistry, Faculty of Agronomy, Mendel University in Brno, Czech Republic, EU, vojtech.adam@mendelu.cz
[2]Central European Institute of Technology, Brno University of Technology, Brno, Czech Republic, EU
[3]Institute of Chemistry, University of Agriculture, Cracow, Poland, EU

Abstract

Composite films based on furcellaran (FUR) and carbon quantum dots (CQDs) has been obtained by casting method. The effects of CQDs films as nanofiller on optical, physical, mechanical and microstructural properties of furcellaran films were evaluated. The synthesized composites were characterized using Scanning Electron Microscopy (SEM), Fourier-transform infrared spectroscopy (FT-IR) and ultraviolet-visible spectroscopy (UV-Vis). The results from scanning electron microscopy (SEM) confirmed that the composite films have homogeneous and compact structure, with no fractures or cracks. Tensile strength (TS) and elongation at break (EAB) values increased with the increase of CQDs concentration. The UV-Vis results confirmed that composite films exhibited enhancement of photosensitivity. The FUR/CQDs composite films have high potential to be used in many applications.

Keywords: Furcellaran, carbon dots, composite film

1. INTRODUCTION

Recently, polysaccharides-, proteins- and lipids- based films have been examined as an alternative for plastic packaging materials [1]. These bio-based films and coating offer excellent biocompatibility and biodegradability. However, biopolymer-based films have limitations: lower mechanical and water properties. This type of films are also used as carriers of active agents, such as essential oils [2], plant extracts [3] or nanostructures. In this work, we focused on the formation of nanocomposite films based on furcellaran and carbon quantum dots. Furcellaran is a polysaccharide with negative charge, which is composed of a fragment of (1 → 3) β-D-galactopyranose with a sulphate group at C-4 and (1 → 4) -3,6-anhydro-α-D-galactopyranose [4]. Carbon quantum dots, as a new class of photoluminescence (PL) nanoparticles, have been studied due to its excellent photostability, biocompatibility and low toxicity. Recently, there has also been a number of studies which investigated various nanocomposite films with quantum dots, including chitosan, gelatin [5], starch [6] and carboxymethyl cellulose [7]. Moreover, there are some reports on the application of carbon quantum dots such as photo catalysts, nanomaterials in medicine [8], sensors [9] and so on. To the best of our knowledge, no research work has been reported on the formation of nanocomposite films using furcellaran and carbon quantum dots. In this study, we described preparation of furcellaran films (FUR) containing carbon quantum dots (CQDs). Furcellaran was used as a polymer matrix and CQDs as nanofiller. The FUR/CDQs nanocomposite films were characterized using SEM, FT-IR and UV-Vis analysis. Physical and mechanical properties of nanocomposite films were also tested.

2. METHODOLOGICAL BASES AND EXPERIMENTAL PART

2.1. Chemicals

Furcellaran was provided by Est-Agar AS (Karla village, Estonia). All other chemicals were of analytical grade and purchased from Sigma-Aldrich (St. Louis, MO, USA).

2.2. Preparation of carbon quantum dots (CQDs)

Citric acid monohydrate (2.1 g) was dissolved in 45 ml of MilliQ water. Then, 3,3´-diamino-N-methyl-dipropylamine (mdpta) (1.7 ml) was slowly dropped into the solution with stirring. Water was added to 50 ml. 2 ml of the solution was heated in glass vials in microwave (130 °C) (Multiwave3000, Anton-Paar GmbH, Graz, Austria).

2.3. Preparation of nanocomposite films

Furcellaran films (FUR or control film) were prepared using a solution casting method. 1 g of FUR powder was dispersed in 100 ml of deionized water. Following this, glycerol (0.5 ml), as a plasticizer, was added into film-forming solution and stirred at 50 °C for 30 min. For the preparation of nanocomposite films, the CQDs (1, 3, 5, and 10 wt. % based on furcellaran) were added to the furcellaran solution following the same procedure as above. Finally, each film-forming solution (50 ml) was poured into petri dish (dimeter 90 mm) and transferred under fume hood and dried for 24 h. The dried films were peeled off from the petri dishes and conditioned at 20 °C and 50 % relative humidity in desiccators before evaluating.

2.4. Characterization

Thickness The manual instrument Mitotuyo, no. 7327 (Kawasaki, Japan) was used to measure the film thickness. Five measurements were taken from each type of films and averaged.

Water behaviors Water behaviors (WC- water content and WS- solubility) were examined according to our previous work [10].

Mechanical properties Tensile strength (TS, MPa) and elongation at break (EAB, %) of films were examined using a texture analyzer TAxT2i Stable Micro System (Surrey, UK). Mechanical properties were calculated according to the following equations:

$$\mathrm{TS}\,[\mathrm{MPa}] = \frac{F_{max}}{S} \quad (1)$$

$$EAB[\%] = \frac{L_f}{L_0} x100 \quad (2)$$

where TS - tensile strength [MPa], F_{max} - breaking load [N] and S - an area of the film cross-section [mm]; EAB- elongation at break; L_f - final length of the film sample at the point of break, L_0 - initial length of the film sample.

Spectroscopic analysis Ultraviolet-visible (UV-Vis) spectroscopy was performed by using a UV-5500 spectrophotometer (Metash), in the wavelength range from 200 to 800 nm. FTIR spectra of film samples were determined using a MATTSON 3000 spectrophotometer (Madison, Wisconsin, USA). 32 scans were collected with 4 cm^{-1} resolution in the 600 - 4000 cm^{-1} wavelength range.

Scanning electron microscopy (SEM) For the documentation of the structure of films, a scanning electron microscope HITACHI S-3400 N (CA, USA) was used.

Statistical analysis Statistical analysis was performed with Statistica v12.0 software (StatSoft, Tulsa, USA). Differences were statistically significant at the $p < 0.05$ level.

3. RESULTS AND DISCUSSION

The furcellaran films (FUR, control) were transparent, flexible and homogenous. The nanocomposite films with CQDs (in concentration 1, 3, 5 and 10 %) were visually homogeneous with no bubbles, cracks or brittle areas (**Figure 1**).

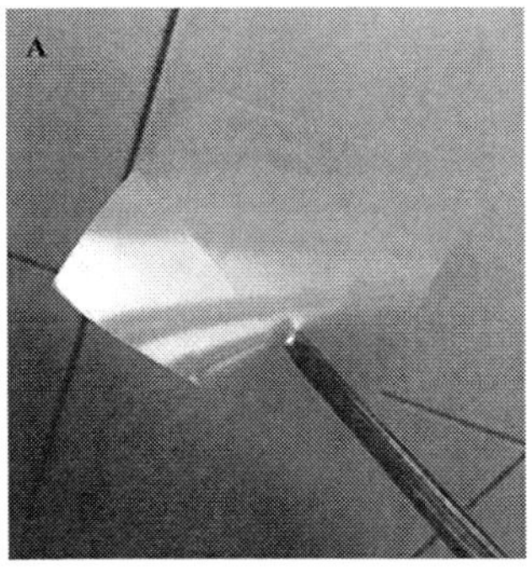

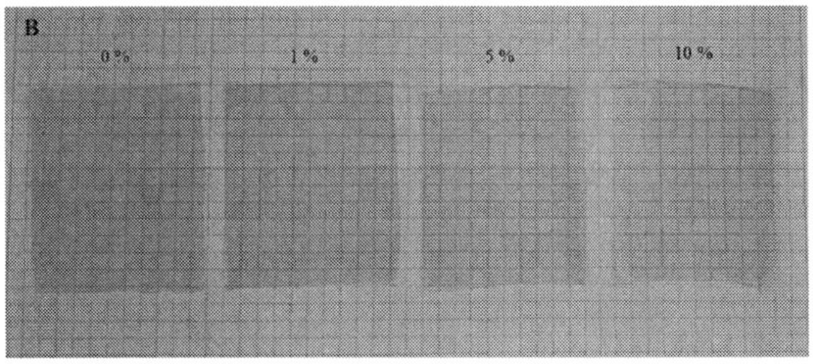

Figure 1 Appearance of nanocomposite FUR-based films with CDQs in different concentration (A) FUR films with 6 % CDQs, (B) FUR films with 0-10 % CDQs

Figure 2 shows SEM micrographs of the FUR/CQDs films. Uniform and smooth structure was observed for furcellaran films (**Figure 2A**), the presence of CQDs did not caused significant differences in the appearance of the films. SEM micrographs suggested the strong interaction between CQDs and furcellaran molecules.

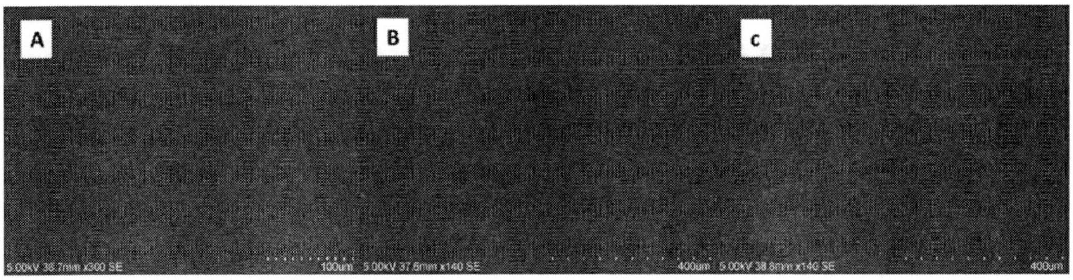

Figure 2 Scanning electron microscopy micrographs of the surface of films (A) control films, (B) FUR films with 5 % CDQs and (C) FUR films with 10 % CDQs

Table 1 Physical properties of furcellaran films with CQDs

Concentration of CQDs*	Physical properties of FUR/CQDs				
	Thickness (mm)	Water content (%)	Solubility (%)	Tensile strength (MPa)	Elongation at break (%)
0 %	0.10±0.01ª	24.54±1.66ª	100.00±0.00ª	4.44±0.74ª	54.77±2.75ª
1 %	0.10±0.01ª	21.48±0.57ᵇ	100.00±0.00ª	3.99±0.75ª	54.09±3.80ª
3 %	0.10±0.01ª	22.39±0.45ᵇ	100.00±0.00ª	5.28±0.76ᵇ	56.29±4.40ª
5 %	0.10±0.01ª	21.65±0.29ᵇ	94.18±1.97ᵇ	5.91±0.74ᵇ	56.16±4.02ª
10 %	0.10±0.01ª	22.00±0.65ᵇ	60.55±1.95ᶜ	8.94±0.37ᶜ	56.34±2.31ª

*Values are expressed as mean ± standard deviation. Different letters in the same rows indicate significant differences ($p < 0.05$).

Table 1 shows the effect of incorporation CQDs on physical properties of FUR films. The thickness of films was not affected significantly by addition of CQDs. Reduced solubility of nanocomposite films is related to strong interaction between polysaccharide and CQDs. The incorporation CQDs into FUR matrix caused reduction in water content. Tensile strength (TS) of nanocomposite films increase with increasing the CQDs concentration, and the values of elongation at break were at the same level. The increase in the TS value indicates the presence of increased stiffness of the film after the addition of CQDs.

The UV-Vis absorption spectra of FUR films, CQDs in the water, and FUR/CQDs films were shown in **Figure 3**. CQDs in the films absorbed radiation in the visible regions of 300-400 nm, which is related to strong fluorescence. The control film has no absorption above 300 nm, while increase in FUR films resulted in increase in peaks in the range 300-400 nm.

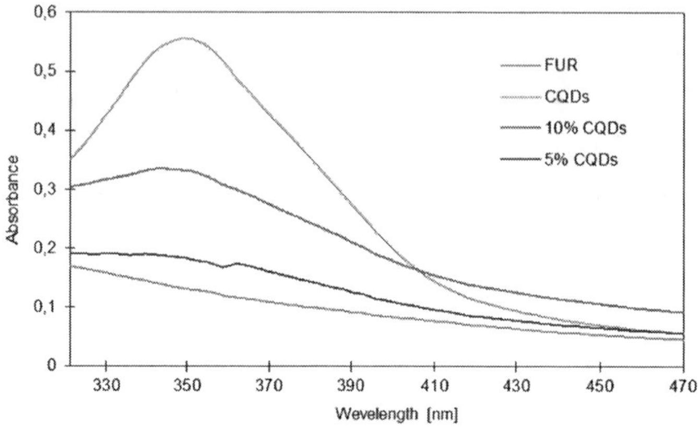

Figure 3 UV-Vis spectra of CQDs, FUR films and FUR films with different concentration of CQDs

The major bands in IR spectra of FUR films were observed at 841 cm^{-1} (-O-SO_3 stretching vibration at D-galactose-4-sulfate), at 920 cm^{-1} (C-O-O of 3,6-anhydro-D-galactose), at 1030 cm^{-1} (C-O stretch), at 1211 cm^{-1} (C-O bridge stretch), at 1632 cm^{-1} (water deformation), at 2923 cm^{-1} (C-H stretch), at 3267 cm^{-1} (O-H stretch). The spectrum of FUR-10 % CQDs film was similar to the bands of FUR, but there were new peaks: the band at 1734 cm^{-1} attributed to the C=O in ester group and 1629 cm^{-1} assigned to -COO- symmetric stretching. Thus, CQDs can interact with the functional group of FUR. According to You at al. [7], Javanakht and Namazi [6,11], the FT-IR analysis confirmed that incorporation of CQDs into polymer matrix modified intermolecular interaction.

4. CONCLUSION

The nanocomposite films were developed by adding CQDs in different concentration (1, 3, 5 and 10 %) to FUR matrix using the solution casting method. The incorporation CQDs into FUR matrix significantly influences water and mechanical properties of nanocomposite films. The furcellaran films with any concentration of CQDs were highly transparent and luminescent. The films with CQDs showed better UV-Vis light barrier properties, which make them good candidates as packaging materials. Excellent barrier properties against UV light were shown by FUR-10 % CQDs. FT-IR analysis confirmed the interaction between FUR and CQDs, by the presence of new bands in spectra. The results showed that CQDs can be used to improve the properties of furcellaran films. This study was preliminary research, so future studies on antimicrobial activity and toxicity test are required.

ACKNOWLEDGEMENTS

The study was financially supported from ERDF (No. CZ.02.1.01/0.0/0.0/16_025/0007314).

REFERENCES

[1] HASSAN, Bilal, CHATHA, Shazhad A.S., HUSSAIN, Abdullah I., ZIA, Khalid M. and AKHTAR, Nassem. Recent advances on polysaccharides, lipids and protein based edible films and coatings: A review. *International Journal of Biological Macromolecules*. 2018. vol. 109, pp. 1095-1107.

[2] ATERES, Lorena and CHIRALT, Amparo. Essential oils as additives in biodegradable films and coatings for active food packaging. *Trends in Food Science & Technology*. 2016. vol. 48, pp. 51-62.

[3] MIR, Shabir A., DAR, B.N., WANI, Ali A. and SHAH, Manzoor A. Effect of plant extracts on the techno-functional properties of biodegradable packaging films. *Trends in Food Science & Technology*. 2018. vol. 80, pp. 141-154.

[4] LAOS, Katrin, BROWNSEY, Geoffrey J. and RING, Stephen G. Interactions between furcellaran and the globular proteins bovine serum albumin and β-lactoglobulin. *Carbohydrate Polymers*. 2007. vol. 67, pp.116-123.

[5] BORKOVSKA, Lyudmyla, KORSUNSKA, N., STARA, Tetyana, GUDYMENKO, O., KLADKO, Vasyl, STROYUK, Oleksandr, RAEVSKAYA, Alexandra and KRYSHTAB, T. Photoluminescence and structural properties of cdse quantum dot-gelatin composite films. *Physica B: Condensed Matter*. 2014. vol. 453, pp. 86-91.

[6] JAVANBAKHT, Siamak and NAMAZI, Hassan. Solid state photoluminescence thermoplastic starch film containing graphene quantum dots. *Carbohydrate Polymers*. 2017. vol. 176, pp. 220-226.

[7] YOU, Yuqin, ZHANG, Haoran, LIU, Yingliang and LEI, Bingfu. Transparent sunlight conversion film based on carboxymethyl cellulose and carbon dots. *Carbohydrate Polymers* 201. vol., 151, pp. 245-250.

[8] MILOSAVLJEVIC, Vendran, JELINKOVA, Pavlina, JIMANEZ, Ana M.J., MOULICK, Amitava, HADDAD, Yazan, BUCHTELOVA, Hana, KRIZKOVA, Sona, HEGER, Zbynek, KALINA, Lukas, RICHTERA, Lukas, KOPEL, Pavel and ADAM, Vojtech. Alternative synthesis route of biocompatible polyvinylpyrrolidone nanoparticles and their effect on pathogenic microorganisms. *Molecular Pharmaceutics*. 2017. vol. 14, pp. 221-233.

[9] LIU, Yun, LIU, Chun-yan and ZHANG, Zhi-ying. Synthesis of highly luminescent graphitized carbon dots and the application in the hg2+ detection. *Applied Surface Science*. 2012. vol. 263, pp. 481-485.

[10] JAMRÓZ, Ewelina, KOPEL, Pavel, JUSZCZAK, Lesław, KAWECKA, Agnieszka, BYTESNIKOVA, Zuzanna, MILOSAVLJEVIC, Vedran, KUCHAREK, Mateusz, MAKAREWICZ, Małgorzata and ADAM, Vojtech. Development and characterisation of furcellaran-gelatin films containing senps and agnps that have antimicrobial activity. *Food Hydrocolloids*. 2018. vol. 83, pp. 9-16.

[11] JAVANBAKHT, Siamak and NAMAZI, Hassan. Doxorubicin loaded carboxymethyl cellulose/graphene quantum dot nanocomposite hydrogel films as a potential anticancer drug delivery system. *Materials Science and Engineering: C*. 2018. vol. 87, pp. 50-59.

CONTROLLED SYNTHESIS OF GD-DOPED SUPERPARAMAGNETIC IRON OXIDE NANOPARTICLES TOWARDS STABLE AND BIOCOMPATIBLE CONTRAST AGENT

Kateřina POLÁKOVÁ, Zdeňka MEDŘÍKOVÁ, Ondřej MALINA, Klára ČÉPE, Radek ZBOŘIL

Regional Centre of Advanced Technologies and Materials, Palacký University in Olomouc, Olomouc, Czech Republic, EU, katerina.polakova@upol.cz

Abstract

Superparamagnetic iron oxides (SPIO) nanoparticles (NPs) are used as negative contrast agents in magnetic resonance imaging (MRI). However, controlled synthesis of highly stable, biocompatible, reproducible SPIO NPs in mild conditions is still very rare which is demonstrated by the fact that many commercial contrast agents have been withdrawn from the market due to the problems with effective and reproducible synthesis. Here, we introduce a facile and controlled synthesis of Gd-doped SPIO NPs. We have found that different amount of Gadolinium (5, 10 and 20 %) replacing the Fe mols in the beginning of the synthesis influence the final physico-chemical properties of the later products. With increasing amount of Gd ions, the higher amount of organic versus inorganic content is calculated (from TGA analysis) which is mirrored in decreasing of saturation magnetization (from 78 Am^2/kg for SPIO without Gd to 15 Am^2/Kg for 20 % Gd-doped SPIO). All prepared samples are stable in PBS and till 10 % of Gd doping biocompatible at high concentrations used (500 μg/ml). Moreover, 10 % Gd-SPIO (cMNPs_Gd10) shows the most stable properties also in a high ionic strength (1 M of NaCl). This sample is promising for mesenchymal stem cells (MSC) labeling. After 24 hours, robust internalization into the MSC was confirmed by fluorescence optical microscopy and by strong negative signal of SPIO labeled MSC measured by clinical 1.5 T MRI.

Keywords: Controlled synthesis, superparamagnetic iron oxides (SPIO), in vitro characterization, mesenchymal stem cells (MSC), magnetic resonance imaging (MRI)

1. MATERIALS AND METHODS

1.1. Materials

Carboxymethylcellulose, sodium salt (CMC), iron (II) sulphate heptahydrate ($FeSO_4.7H_2O$), gadolinium (III) chloride ($GdCl_3.6H_2O$), N-(3-dimethylaminopropyl)-N′-ethylcarbodiimide hydrochloride (EDC) were from Sigma Aldrich (Germany) in analytical grade. Sodium hydroxide (p.a.), 99.8 % acetic acid, 37 % HCl, 28 % NH_4OH (p.a.), NaCl (p.a.) were obtained from P-LAB (Czech Republic). Alexa Fluor 488 Hydrazide (AlexaFluo) was obtained from Life Technologies, Czech Republic. All chemicals were used without further purification. For all experiments double, distilled water (ddH_2O) was used.

1.2. Preparation of carboxylated MNPs (cMNPs) and Gd-doped cMNPs (cMNPs_Gd)

Magnetite NPs stabilized by CMC (cMNPs) were prepared according to a protocol described earlier by Medříková et al. [1]. Briefly, CMC (1 g) was dissolved in ddH_2O (30 mL) and heated up to 303 K. $FeSO_4.7H_2O$ (720 mg) was dissolved in ddH_2O (10 mL), pre-acidified by 37 % HCl (total volume of 100 μL) and added to CMC solution under mechanical stirring (300 rpm). NH_4OH (28 %, 7 mL) was added to the mixture. Over a period of 15 min, the temperature was raised to 323 K followed by addition of 28 % NH_4OH (5 mL). After another 15 min, the reaction mixture was removed from the hot plate and the formed magnetic colloid was immediately centrifuged (20,000 g, 293 K) for 1 h. The magnetic pellet was resuspended in 60 mL ddH_2O and sonicated for 15 min. Subsequently, it was again centrifuged (20,000 g, 293 K, 1 h), the pellet resuspended in ddH_2O (40 mL) and sonicated.

Carboxylated Gd-doped MNPs (cMNPs_Gd) were prepared according the same protocol as cMNPs including replacing 5 %, 10 % or 20 % mols of Fe in the reaction mixture by the same amount of Gd mols. The prepared magnetic nanoparticles were marked cMNPs_Gd5, cMNPs_Gd10, and cMNPs_Gd20, respectively. Briefly, mixtures consisting of 0.684 g of $FeSO_4.7H_2O$ and 0.048 g of $GdCl_3.6H_2O$ (cMNPs_Gd5), 0.648 g of $FeSO_4.7H_2O$ and 0.096 g of $GdCl_3.6H_2O$ (cMNPs_Gd10), and 0.576 g of $FeSO_4.7H_2O$ and 0.193 g of $GdCl_3.6H_2O$ (cMNPs_Gd20) were used instead of 0.72 g of $FeSO_4.7H_2O$ in the protocol for the preparation of Gd-doped cMNPs.

The final magnetic colloids were centrifuged (1,000 g, 298 K) for 20 min to remove any aggregates. The supernatants containing carboxylated magnetic nanoparticles (cMNPs) or Gd-doped cMNPs were collected and stored at 4 °C.

1.3. Determination of stability

The stability of prepared magnetic nanoparticles (NPs) was tested in phosphate buffered saline (PBS) consisting of 0.05 M phosphate buffer (pH 7.4) and 0.15 M NaCl solution (final concentrations). For each experiment, 250 µL of NPs (1 g of Fe_3O_4/L) were taken into 0.5 mL of phosphate buffered saline (0.1 M phosphate buffer, 0.3 M NaCl, pH 7.4) adjusted to 1 mL by ddH_2O. The testing tubes were incubated at 310 K and at shaking speed 150 rpm. The hydrodynamic diameters of NPs (d_H) were determined before and after 24 hours of its incubation in PBS. The gadolinium release from NPs was determined after 24 hours of its incubation in PBS. The non-released and released Gd was determined in pellets and supernatants, respectively, after centrifugation (20 000 g, 30 min) by ICP-MS.

1.4. Preparation of AlexaFluo modified cMNPs or cMNPs-Gd10

Immobilizations of fluorescent mark onto NPs surface were carried out with the activations of carboxyl groups of NPs via EDC chemistry. In brief, 50 µL of EDC (1 mg) were added to 1 mL of NPs (1 g of Fe_3O_4/L) and the activation was performed for 30 min. Then, 0.03 mg of AlexaFluo was added and the immobilization was performed for 16 h, at 150 rpm, 10 °C, and in the dark. Unbounded agents were removed by centrifugation.

1.5. Techniques for physicochemical characterization

Microscopic images were obtained by transmission electron microscopy (TEM) JEM2100 microscope operated at 200 kV with a point-to-point resolution of 1.9 Å and high-resolution transmission electron microscopy (HRTEM) TITAN 60-300kV, operating at 300 kV. STEM - Energy Dispersive X-ray Spectroscopy (EDS) with acquisition time 20 min was used for chemical mapping. Thermogravimetric analysis was performed by a thermal analyzer STA 449 C Jupiter (Netzsch Instrument). Hydrodynamic diameter (d_h) and zeta potential values ($_p$) of cMNPs were determined by Dynamic Light Scattering (DLS) method using the instrument Malvern Zetasizer Nano (Malvern Instruments Ltd., Worcestershire, UK). The iron content in the solution was determined with a protocol of the ferrozine assay. The gadolinium concertation was determined by ICP-MS.

A superconducting quantum interference device (SQUID, MPMS XL-7, Quantum Design, U.S.A.) was employed for the magnetization measurements. The hysteresis loop of the studied sample was collected at a temperature of 300 K in external magnetic fields ranging from - 5 to + 5 T. The magnetization values were corrected assuming the response of the sample holder and respective Pascal constants.

1.6. In vitro cytotoxicity

The viability of NIH/3T3 cells (ATCC, USA) was determined with using the 3-(4,5-dimethylthiazol-2-yl)-2,5-diphenyltetrazolium bromide(MTT) assay (Sigma Aldrich). Cells were seeded in a 96-well plate (TPP, Biotech). After spreading (24h), cells were incubated with different concentration of various cMNPs nanoparticles in a fully-humidified incubator. After treatment with cMNPs (24 h), 20 µl of an MTT solution (concentration of 5 mg/ml) was added into the each well. After incubation (4 h), the MTT solution was gently removed and dimethylsulfoxide (100 ml) was added to dissolve formazan crystals. Absorbance was measured at 570 nm using an Infinite PRO M200 multiplate reader (Tecan, Austria). Results are presented in relative viability with respect of control untreated cells.

1.7. Stem cell labeling and MRI

hMSCs were isolated from the fatty tissue of three healthy donors who had undergone cosmetic liposuction (one male, two female). Isolation of hMSCs was based on the incubation of lipoaspirates with collagenase. MSCs were expanded in complete DMEM supplemented with 5 % platelet lysate. All animal experiments were performed in accordance with the law of Czech Republic (Law on Animal Protection and Decree of Ministry of Agriculture on Experimental Animal Use and Breeding) and were approved by the Committees for the Use of Experimental Animals at Masaryk University and Palacký University. All collection of waste lipoaspirates from liposuction was performed after approval of the Ethical Committee of Masaryk University and with signature-documented informed consent of patients.

For cellular internalization of cMNPs particles labeled by AlexaFluo and visual microscopic quantification of viable labeled cells Olympus IX 70 light fluorescence microscopes (Olympus Corporation) was used. The cells were grown on 24 well-plate. After spreading of cells (24hours), 100 μg/ml of cMNP_Gd10 conjugated with fluorescein dye was added and imaged after 4 hours, 1day and 5 days.

For MRI phantom study of labeled cells, cells were grown on 24well plate. After 24 hours, 100μg/ml of cMNPs_Gd10 particles were added into each well and incubated overnight. After that, medium was removed, adhered cells were trypsinized, detached from surface and centrifugated. The obtained pellet (500 thousand cells) was fixed in 1 % agar in 1.5ml eppendorf tube. The eppendorf with pellet containing labeled cells was measured on 1.5 T MRI clinical magnet using Sag T2 FrFSE sequences (TR:2600, TE:90, matrix 2X2)

2. INTRODUCTION

Magnetic resonance imaging (MRI) has become a powerful technique in the clinical diagnosis of disease and in animal imaging. MRI enables images of living subjects with high spatial resolution and no penetration limit only by using of external magnetic field on tissue water protons [2-4]. To increase a diagnostic accuracy and most recently for molecular targeting or stem cell monitoring superparamagnetic iron oxide nanoparticles (SPIO) and paramagnetic chelates are being used as negative and positive contrast agents [5-7]. SPIO NPs are consisted of maghemite (γ-Fe_2O_3) or magnetite (Fe_3O_4) spinel structure having two nonequivalent magnetic sublattices revealing very high magnetic susceptibility. Moreover, below a critical size (usually below 100 nm) superparamagnetic behavior occurs. This means that SPIO NPs show higher saturation magnetization than bulk material of the same chemical structure when using magnetic field, on the other hand, after switching off the external magnetic field the magnetization become zero and without remanence and coercivity. Therefore, during MR investigation SPIO can react very fast on MR magnetic field, causing a decrease of relaxation times of surrounding water protons which gives the negative signal of imaged tissue in T2-weighted images [8,9]. SPIO NPs of various sizes and polymer coatings has been clinically tested for imaging of bowel and abdomen (Lumirem, Abdoscan) [10,11], liver/spleen (Endorem, Feridex, Resovist), lymph node imaging (AMI-227 - Sinerem and Combidex), bone marrow imaging (AMI-227), perfusion imaging (NC100150 - Clariscan) or MR angiography (NC100150) [12] or cellular imaging [13].

However, almost all commercial contrast agents have been recently withdrawn from the market mainly due to the problems with effective and reproducible synthesis. Most of SPIO have been synthesized by coprecipitation of iron salts which leads to nonuniform particles with non-perfectly defined magnetic properties. Reproducibility is thus also difficult. Other types of synthetic procedures include microemulsion, hydro-thermal or thermal decomposition [14,15]. Generally, these syntheses are expensive with low quantities or hardly reproducible therefore the product in the clinics would be very expensive. Procedures leading to controlled synthesis of highly stable, biocompatible, reproducible SPIO NPs of high quantities are still very rare.

Here, we demonstrate a facile and controlled synthesis of SPIO nanoparticles coated with carboxymethylcellulose which are biocompatible and very stable in PBS and also in a high ionic strength only by using appropriate amount of Gd chelated in the carboxymethylcellulose polymer (CMC). SPIO sample with 10 % of Gd doped CMC shows promising results in mesenchymal stem cell imaging as negative MR contrast agent.

3. RESULTS AND DISCUSSION

The carboxymethylcellulose stabilized MNPs (cMNPs) were synthetized by the aqueous precipitation approach described earlier [1]. The morphology and structural organization of as-prepared nanoparticles were observed by TEM (**Figure 1**). The TEM image of colloidal cMNPs (**Figure 1a**) shows that nanoparticles were arranged into superclusters (densely packed nanoparticles) as described in the previous study [1]. **Figures 1b-d** show impact of Gd doping on structural organization of cMNPs. The increasing Gd content in the reaction leads to the destruction of superclusters. Whereas superclusters were partially recognizable in cMNPs_Gd5 (**Figure 1b**), the nanoparticles in samples cMNPs_Gd10 and cMNPs_Gd20 lacked supercluster arrangement (**Figure 1c** and **1d**, respectively).

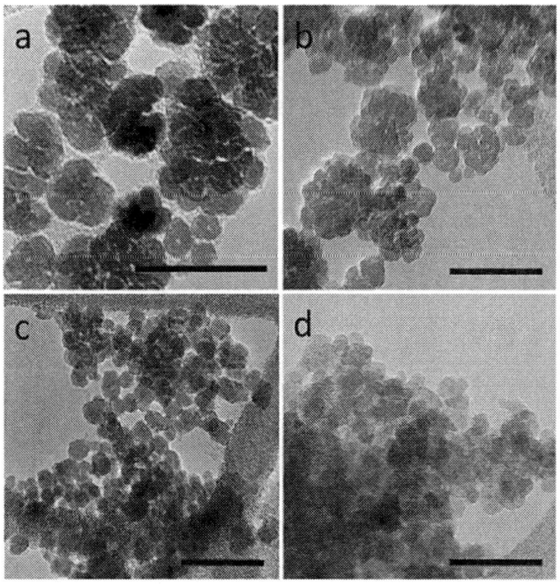

Figure 1 TEM images of (a) cMNPs, (b) cMNPs_Gd5, (c) cMNPs_Gd10, and (d) cMNPs_Gd20. Scale bar is 40 nm.

The accurate final contents of Gd in Gd-doped cMNPs (cMNPs_Gd5, cMNPS_Gd10, and cMNPs_Gd20) were measured by ICP-MS (**Table 1**) and were found to be 0.156, 0.332, and 0.712 g of Gd per 1 g of Fe, respectively. These concentrations are fully comparable with Gd:Fe ratios used in the reaction mixtures. The inorganic (Gd and Fe_2O_3) and organic (polymer) contents were determined thermogravimetrically (**Table 1**) and ranged from 80.4 to 48.5 for inorganic and from 19.6 to 51.5 for organic compounds, respectively. The polymer content increases by increasing of Gd amount in the sample which suggested that Gd play a key role as crosslinker agent for CMC. Due to the presence of Gd during the preparation of nanoparticles the higher amount of polymer is attached on the surface of nanoparticles. This effect is achieved by formation of complex between Gd^{3+} and carboxyl groups of CMC.

Table 1 Physico-chemical characterization of cMNPs and Gd-doped cMNPs

Sample	Inorganic /organic content[a] (w/w %)	Magnetization (emu/g)	Gd content[b] (g/g Fe)	Gd content[c] (g/g Fe)
cMNPs	80.4/19.6	78.1	NA	NA
cMNPs-Gd5	62.5/37.5	37.5	0.156 ± 0.004	0.157 ± 0.013
cMNPs-Gd10	51.8/48.2	32.8	0.332 ± 0.008	0.325 ± 0.011
cMNPs-Gd20	48.5/51.5	15.2	0.712 ± 0.009	0.694 ± 0.008

[a] Determined thermogravimetrically
[b] Determined by ICP-MS
[c] Determined by ICP-MS after 24 hours exposure to PBS

In order to investigate magnetic properties of the prepared materials, hysteresis loops were measured at 300 K (see **Figure 2**). The saturation magnetization values of cMNPs and Gd-doped cMNPs, which ranged between 78.1 and 15.2, are listed in **Table 1**. These values are lower than that reported for bulk Fe_3O_4 (saturation magnetization of 89 emu/g) [15]. The reduction in the saturation magnetization can be explained in terms of finite-size effects, presence of organic substance (having a diamagnetic nature), and in the case of Gd-doped cMNPs also presence of gadolinium complexed with carboxyl groups of the polymer (having a paramagnetic nature) contributes to lower values. In our case, the effects of organic substance and/or gadolinium chelated with organic substance are prevailing as the magnetization values are normalized to the weight content of the sample. Moreover, remanence and coercivity of all prepared materials are almost undetectable indicating that the cMNPs clusters and Gd-doped cMNPs possess superparamagnetic properties at room temperature.

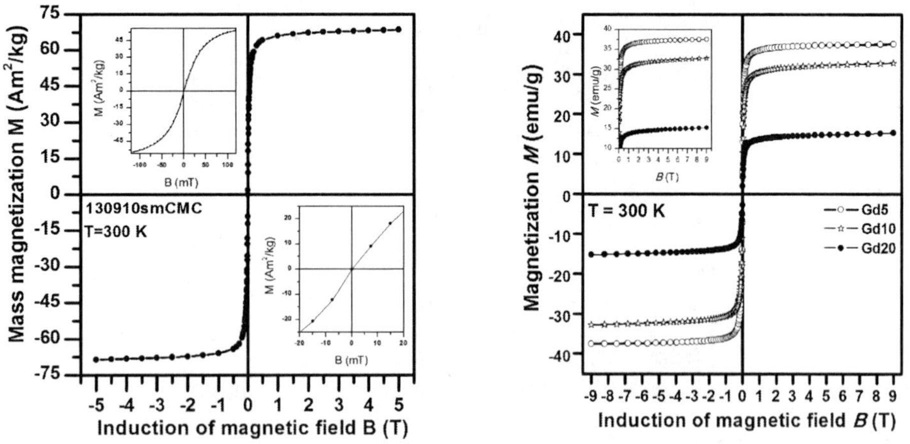

Figure 2 Hysteresis loops of the synthesized cMNPs (left) and Gd-doped cMNPs (right) measured at a temperature of 300 K

The stability of Gd-doped cMNPs (in terms of Gd release from polymer) were investigated using their incubation in PBS for 24 hours. It was measured that a content of Gd did not decrease in all Gd-doped cMNPs. Simultaneously, Gd was not detected in supernatants which indicate high stability of bonding between gadolinium and carboxyl groups of CMC. The mean hydrodynamic diameters (d_h) before and after exposure of nanoparticles to PBS (**Figure 3**) were determined by DLS measurements. The results indicate that the Gd doping led to increase in d_h values. However, Gd-doping did not decrease the stability of cMNPs because this increase of d_h values can be contributed to higher polymer content on the nanoparticles surface. This statement is supported by further lowering of d_h of Gd-doped cMNPs after their 24h exposure to PBS which again indicate a high stability of all prepared nanoparticles. These decreases of d_h values could be a result of interaction between Na ions and carboxyl groups.

We further confirmed that Gd is chelated in polymer (CMC) corona by using chemical mapping (**Figure 4a**). Moreover, the higher stability of cMNPs_Gd10 compared to bare cMNPs even at high ionic strange (**Figure 4b**) make this sample suitable and promising for in vivo bio-application. Results from viability measurements (**Figure 5**) show that only cMNPs_Gd20 sample is significantly more toxic in concentrations above 350 μg/ml (IC 50 about 500 μg/ml) probably due to the higher amount of Gd located on the outer shell of CMC. These Gd ions are thus in closer contact with the membrane of cells which can cause cytotoxicity. On the other hand, the rest three samples are biocompatible and viability of cells is not less than 80 % till concentration of 500 μg/ml.

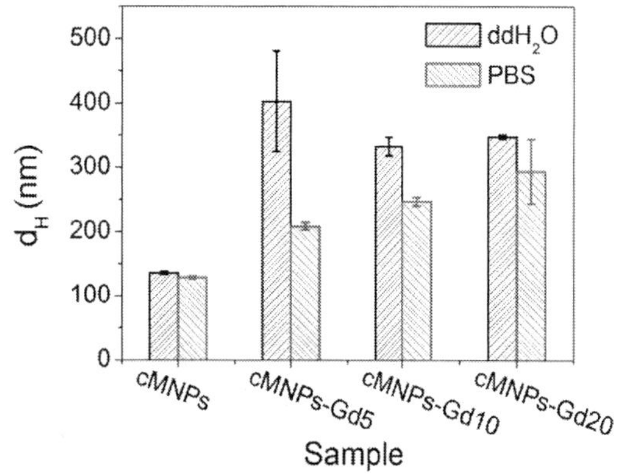

Figure 3 The mean hydrodynamic diameters (d_h) of cMNPs and Gd-doped cMNPs before and after 24 hour exposure of samples to PBS determined by DLS measurements

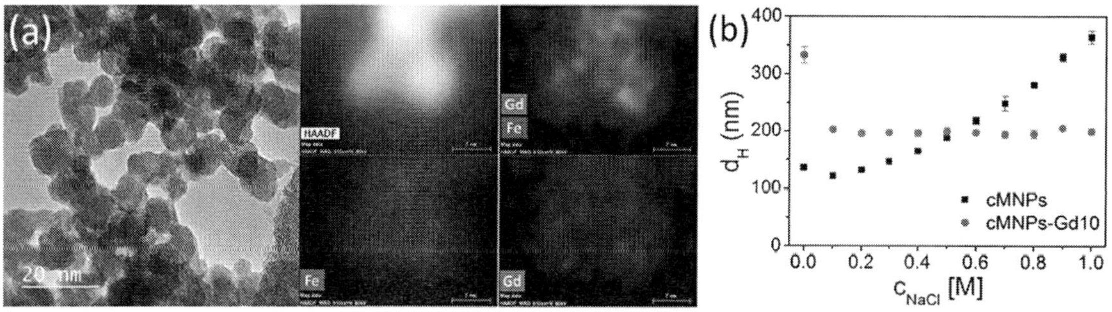

Figure 4 (a) HRTEM-STEM chemical mapping of cMNPs_Gd10 and (b) stability (hydradynamic diameter d_h) of cMNPs_Gd10 in diverse ionic strength.

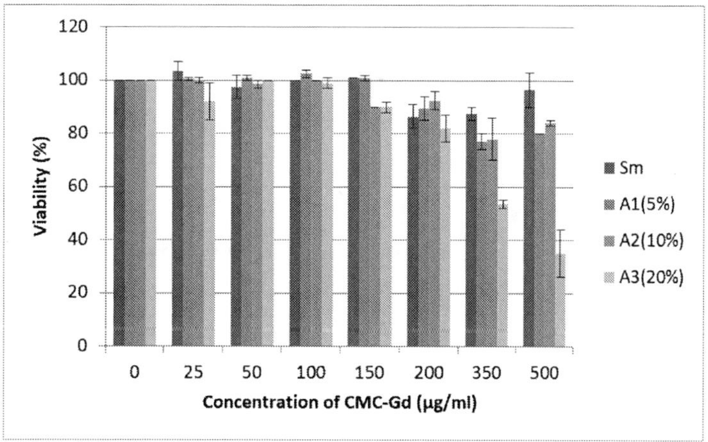

Figure 5 Viability of NIH3T3 cells after 24 hours of incubation with various cMNPs. Note: Sm (cMNPs), A1 (cMNPs_Gd5), A2 (cMNPs_Gd10), A3 (cMNPs_Gd20). Each error bar represents the average value from three independent measurements

Based on the above mentioned magnetization data, stability and viability results, we realized to perform further biological experiments on cMNPs_Gd10 sample.

First, we were curious if the labeled stem cells by our cMNPs_Gd10 can be visible by MRI technique. After 24 hours of incubation with 100 μg/ml NPs $5x10^5$ of MSCs were collected and centrifugated (see a reddish area in **Figure 6 left**). After that, this pellet of cells was imaged by 1.5 T MRI in T2-weighted images clearly visible as a black sphere in **Figure 6 (right)**. Highly efficient negative contrast effect of labeled MSCs when compared to surrounding medium make the cMNPs_Gd10 promising candidate for in vivo MSCs monitoring.

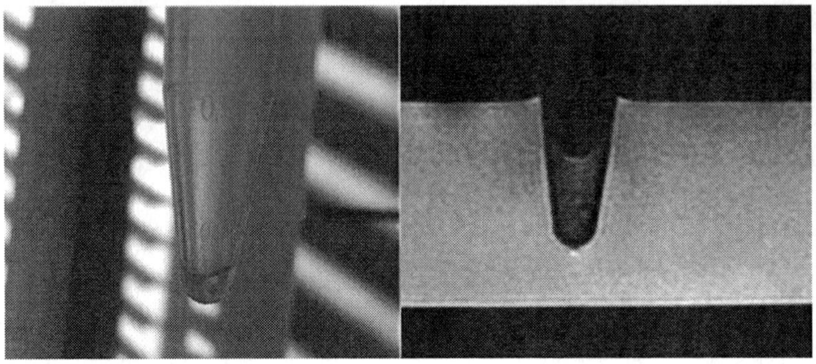

Figure 6 MRI contrast effect of MSC labeled by cMNPs_10. (left) reddish pellet of centrifugated labeled MSC, (right) black sphere of the same pellet measured by clinical 1.5 T MRI in T2 weighted images.

One of the important issues of the nanoparticles interacting with cells is their internalization mechanism, localization and degradation within the cells. For this purpose, we covalently bound green fluorescein dye on the surface of cMNPs_Gd10. After 24 hours, SPIO NPs, most probably internalized by endocytosis, are localized homogenously in the endosomes/lysosomes as green globular areas depicted in **Figure 7 (left)**. Moreover, MNPs are not found in nucleus of cells and cells keep their native morphology and proliferation rate (**Figure 7 left and right**) suggesting their high biocompatibility. Additionally, we can summarize that cMNPs_Gd10 sample could be used not only for MRI but also for optical imaging which in fact increase the sensitivity of the diagnostics [16].

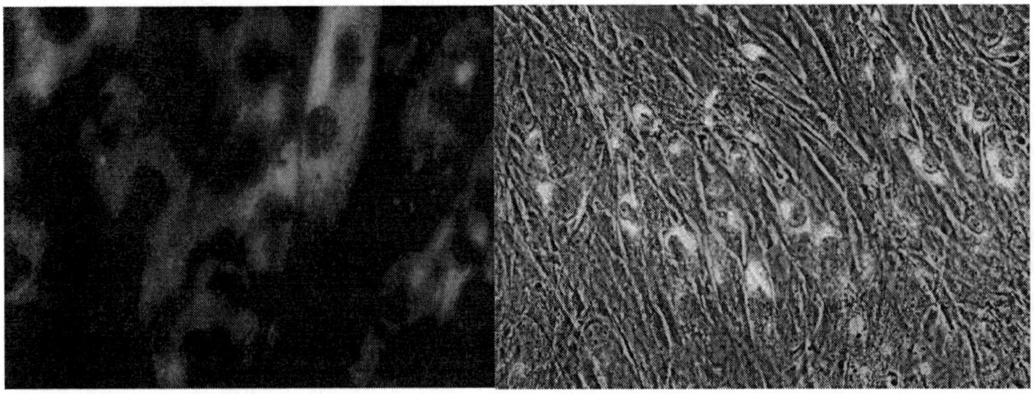

Figure 7 Localization of cMNPs_Gd10 nanoparticles in MSCs using optical fluorescence microscopy

CONCLUSION

In this work, we prepared easily highly stable superparamagnetic iron oxide nanoparticles coated with carboxymethylcellulose polymer containing different amount of chelated Gd. By controlling the amount of Gd in the synthesis the magnetization properties can be tuned. Moreover, we showed that the most stable cMNPs_Gd10 bound with fluorescent dye is appropriate for biomedical application as dual contrast agents for MCSs imaging.

ACKNOWLEDGEMENTS

The authors acknowledge the support from the Ministry of Education, Youth and Sports of the Czech Republic under Project No. LO1305.

REFERENCES

[1] MEDRIKOVA, Z., NOVOHRADSKY, V., ZAJAC, J. et. al. Enhancing Tumor Cell Response to Chemotherapy through the Targeted Delivery of Platinum Drugs Mediated by Highly Stable, Multifunctional Carboxymethylcellulose-Coated Magnetic Nanoparticles. *Chemistry-A European Journal.* 2016. vol. 22, iss 28, pp 9750-9759.

[2] WINTER, P. M., CARUTHERS, S. D., WICKLINE, S. A., LANZA, G. M. Molecular Imaging by MRI. *Curr. Cardiol. Rep.* 2006. vol. 8, pp 65-69.

[3] HU, X., NORRIS, D. G. Advances in High-Field Magnetic Resonance Imaging. *Annu. Rev. Biomed. Eng.* 2004. vol. 6, pp 157-184.

[4] CHEN, S., ZHANG, Q., HOU Y. et al. Nanomaterials in medicine and pharmaceuticals: nanoscale materials developed with less toxicity. *Eur. J. Nanomed.* 2013. vol. 5.

[5] SHOKROLLAHI, H. Contrast agents for MRI. *Materials Science and Engineering C.* 2013 vol. 33, pp 4485-4497.

[6] SKOPALIK, J., POLAKOVA, K., HAVRDOVA, M. et al.; Mesenchymal stromal cell labeling by new uncoated superparamagnetic maghemite nanoparticles in comparison with commercial Resovist - an initial in vitro study. *Int J Nanomed.* 2014, vol. 9, pp 5355-5372.

[7] LI, L., JIANG, W., LUO, K. et al. Superparamagnetic Iorn Oxide Nanoparticles as MRI contrast agents for Non-invasive stem cell labeling and tracking. *Theranostics* 2013, vol. 3, pp 595-615.

[8] KLUCHOVA, K., ZBORIL, R., TUCEK, J. et. al. Superparamagnetic maghemite nanoparticles from solid-state synthesis - Their functionalization towards peroral MRI contrast agent and magnetic carrier for trypsin immobilization. *Biomaterials* 2009. vol 30, pp 2855-2863.

[9] POLAKOVA, K., MOCIKOVA, I., PUROVA, D. et al. Magnetic resonance cholangiopancreatography (MRCP) using new negative per-oral contrast agent based on superparamagnetic iron oxide nanoparticles for extrahepatic biliary duct visualization in liver cirrhosis. *Biomed Pap Med Fac Univ Palacky Olomouc Czech Repub.* 2016. vol. 160, pp 512-517.

[10] MARKOVA, I., POLAKOVA, K., TUCEK, P. et al. MR enterography with a new negative oral contrast solution containing maghemite nanoparticles. *Biomed Pap Med Fac Univ Palacky Olomouc Czech Repub.* 2012, vol. 156, pp 229-235.

[11] WANG, Y.X., HUSSAIN, S.M., KRESTIN, G.P. et.al. Superparamagnetic iron oxide contrast agents: physicochemical characteristics and applications in MR imaging. *Eur. Radiol.* 2001. vol. 11, pp 2319-2331.

[12] GUPTA, A.K. and GUPTA, M. Synthesis and surface engineering of iron oxide nanoparticles for biomedical applications. *Biomaterials* 2005. vol. 26, pp 3995-4021.

[13] BULTE, J.W.M. and KRAITCHMAN, D.L. Iron oxide MR contrast agents for molecular and cellular imaging. *NMR in Biomedicine* 2004. vol. 17, pp 484-499.

[14] LAURENT, S., FORGE M., PORT, A. et.al. Magnetic Iron Oxide Nanoparticles: Synthesis, Stabilization, Vectorization, Physicochemical Characterizations, and Biological Applications. *Chem. Rev.* 2008. vol. 108, pp 2064-2110.

[15] CORNELL, R.M. and U. SCHWERTMANN, U. *The Iron Oxides*, 2nd Edition (Wiley-VCH). 2003.

[16] CMIEL, V., SKOPALIK, J., POLAKOVA, K. et al.; Rhodamine bound maghemite as a long-term dual imaging nanoprobe of adipose tissue-derived mesenchymal stromal cells, *Eur. Biophys. J.* 2017. vol. 46, pp 433-444.

TRANSPORT OF DYES IN POLYELECTROLYTE-SURFACTANT HYDROGELS

Martina KLUČÁKOVÁ, Sabina JARÁBKOVÁ, Miloslav PEKAŘ

Brno University of Technology, Faculty of Chemistry, Materials Research Centre, Brno, Czech Republic, EU, klucakova@fch.vutbr.cz

Abstract

Polyelectrolyte-surfactant mixtures are progressive systems studied for their potential use in numerous applications in daily life including delivery systems for pharmaceutics. Mixing polyelectrolytes with oppositely charged surfactant ions in aqueous media typically leads to a strong association providing a simple route for constructing ordered materials with a great scientific and application potential. In this study, the combination of cationized dextran with sodium dodecylsulphate as surfactant was used for the preparation of two types of hydrogels. The diffusion of Nile red and Atto 488 dyes from their solutions in surfactant or NaCl into the hydrogels was monitored in time. The time development of the diffusion flows was used for the determination of basic transport characteristics of hydrogels. The diffusion of the probe dyes in prepared hydrogels was compared and discussed.

Keywords: Hydrogel, surfactant, polyelectrolyte, dye, diffusion

1. INTRODUCTION

Hydrogels are widely used for drug delivery systems in medicine of its unique characteristics [1-4]. These materials can prolong drug action and reduce side effects [1,5,6]. One of the main problems is how to solubilize hydrophobic compounds in hydrogels. The incorporation of some hydrophobic domains in the hydrogel structure would provide its solution. Hydrogels with the domains can be prepared by the interactions of polyelectrolytes with oppositely charged surfactant ions [4,7,8]. In this work, cationized dextran was used as polyelectrolyte and sodium dodecyl sulfate (SDS) as surfactant. Dextran is a bacterial-derived polysaccharide consisting of α-1,6-linked D-glucopyranose residues, and generally produced by enzymes from certain strains [9]. It is non-toxic and offers high water solubility, biocompatibility, and biodegradability. The strong interactions between dextran and SDS were observed [10] and it was driven by the interaction between the bile acid moiety and the surfactant [11]. Therefore their combination can be suitable for the preparation of hydrogel with domains capable to solubilize hydrophobic substances. Recently published results showed that the adsorption capacity of hydrogel for murexide increased with increase in contact time and initial dye concentration [12]. Other work confirmed the electrostatic interaction and hydrogen bonding between ibuprofen and dextran which had positive effect on its flowability, tableting and dissolution characteristics [13].

In this work, the incorporation of two different dyes into hydrogels based on dextran and SDS was studied. The aim is to describe absorption capacity and capability to absorb dyes by the diffusion from their solutions into prepared hydrogel. The dyes were used for the incorporation into domains prepared in the hydrogels.

2. MATERIALS AND METHODS

In this work, cationized dextran was used as polyelectrolyte. Diethylaminoethyl-dextran hydrochloride (Sigma-Aldrich, Czech Republic; DEAED) was used in one molecular weight 729 ± 3 kDa. Sodium dodecyl sulfate (SDS, ≥ 99.0 %) was purchased from Sigma Aldrich (Czech Republic) and used as surfactant.

All solutions were prepared in 0.15M NaCl solution using purified water (Purelab ELGA system). Salt solution was used because preliminary experiments showed that a non-zero ionic strength of the aqueous medium is important for obtaining gel-like materials [14]. Samples were prepared by mixing polyelectrolyte and surfactant stock solutions in a 1:1 volume ratio. The concentrations of initial solutions of polyelectrolyte and surfactant are given in **Table 1**.

Table 1 Concentrations of initial solutions used to prepare hydrogels

Sample	Dextran (wt%)	SDS (mM)
D-I	4	400
D-II	4	100

Nile red and Atto 488 (both purchased from sigma Aldrich) were dissolved in SDS (400mM for D-I and 100mM for D-II) or physiological saline (0.15M NaCl). The dyes were used in ten different initial concentrations listed in **Table 2**. Hydrogels were equilibrated (24 h), separated from the liquid residue by centrifugation at 4000 rpm for 15 min and covered by 5 cm^3 of SDS or NaCl solution with dissolved dye. The concentration decrease of dyes in solution was monitored by means of UV/VIS spectrometry (Hitachi U-3300). The data were used for the determination of dye absorption in hydrogels [15,16] distribution coefficient [16,17] and their diffusivity [15-18].

Table 2 Initial concentrations of dyes in the solutions of SDS and NaCl used for diffusion experiments

Solution	a	b	c	d	e	f	g	h	i	j
Dye (μM)	0.1	0.3	0.5	0.7	0.8	0.9	1.0	3.0	5.0	7.0

3. RESULTS AND DISCUSSION

In this work, two different approaches for the investigation of transport of model dyes into hydrogels were applied. Nile red was prepared as its solution in SDS. The used concentrations of SDS were much higher than its critical micellar concentration (~8 mM) [19,20]. The aggregation number of SDS can be usually found between 40 and 70 [20,21], therefore the dye should be completely distributed in the micelles of SDS. In contrast, Atto 488 was dissolved in water and therefore we assumed that it diffused in hydrogels in the form of simple (partially dissolved) molecules. The theoretical ratio of charges between SDS and cationized dextran is ~4.6 for D-I and 1.2 for D-II. We assume that a part of SDS can be exhausted for the formation of hydrogel networks. In the case of D-I, this part is relatively small, therefore the pore structure in hydrogel contains solution with SDS micelles. In contrast, the large amount of SDS is consumed for the cross-linking of hydrogel D-II and the content of SDS in the pores is low. If we take account of these differences in the two used hydrogels, we can assume the faster diffusion into hydrogel D-II.

The examples of kinetic data are shown in **Figure 1**. We can see that the amount diffused into D-II hydrogels is higher for both used dyes. The dye content in hydrogels increased strongly mainly in first days. The rate of diffusion was strongly influenced by initial concentration of dye solution used as the source of diffusion particles. The concentration dependence is presented in detail in **Figure 2**. If we compare data obtained for Atto 488 and Nile red, we can say that the transport of Nile red into hydrogels is generally slower. Results published by Zhang et al. [22,23] showed that the diffusion of Atto 488 is faster also in water. The diffusion into hydrogels are more difficult as a result of their intrinsic structure constituted by the hydrogel network and pore structure. Due to various curvatures of pores, the diffusion is strongly influenced not only by the porosity (volume of pores) but also by tortuosity given by their shapes. Hydrogel D-I contains much more SDS than D-II and its structure should be more cross-linked. Therefore the diffusion into the hydrogel D-I is more difficult and the properties of hydrogel has the strong influence.

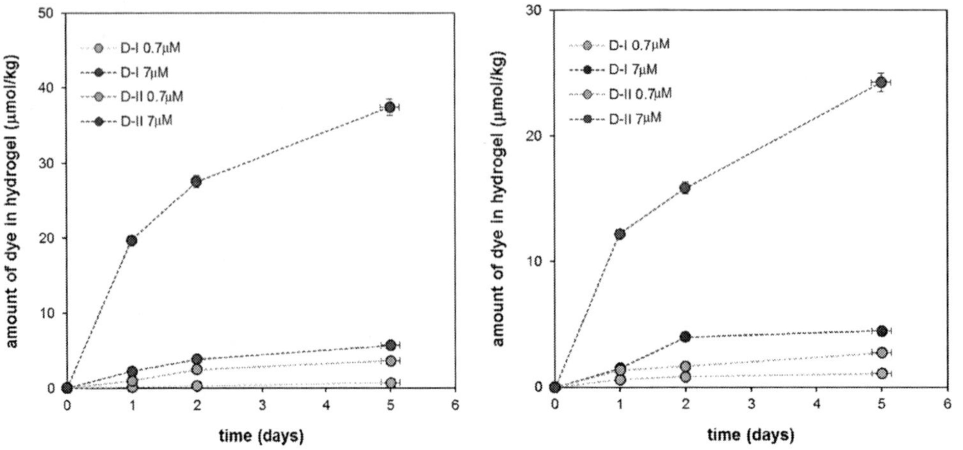

Figure 1 Kinetic data obtained for Atto 488 (left) and Nile red (right)

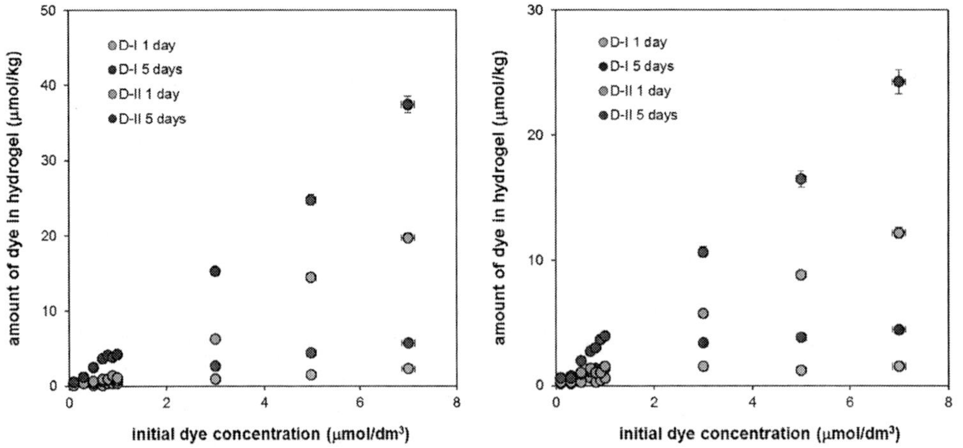

Figure 2 The amount of dyes diffused into hydrogels in the dependence on their initial concentrations: Atto 488 (left) and Nile red (right)

Maiti et al. [24] determined the diffusion coefficient of Nile red in SDS micelles. Its value (1.8×10^{-10} m^2/s) was similar to our results for the diffusion of Nile red into D-II (Table 3). The values of effective diffusion coefficients listed in **Table 3** were obtained on the basis of mathematical model developed in our previous work [17]. This model was developed for the diffusion couple with phase interface. In this work, the couple is comprised by the hydrogel (acceptor part) and the solution of dye above hydrogel (donor part). Both parts are placed in cylindrical vessel. We can see that diffusion coefficients of Atto 488 in water (value published in ref. [22]) and the hydrogels are higher in comparison with results obtained for Nile red. The differences in results between both used dyes obtained for the hydrogel D-I are given mainly by two factors. The first one is the fact that Nile red was diffused in the form of micelles and the second one is the denser hydrogel network less accessible for the bigger particles. We assume that the influence of hydrogel structure can predominate in this case as was described above. In the case of Atto 488, other circumstance can affect the diffusion. The presence of high amount of SDS in the hydrogel D-I can lead to the incorporation of dye into micelles which conduced to the strong decrease of the diffusion coefficients.

Table 3 Effective diffusion coefficients of dyes in water and hydrogels

Medium	Atto 488	Nile red
water[a)]	4.0×10^{-10} m²/s	3.3×10^{-10} m²/s
D-I	1.5×10^{-10} m²/s	1.1×10^{-10} m²/s
D-II	2.8×10^{-10} m²/s	1.5×10^{-10} m²/s

a) published values: Atto 488 ref. [22], Nile red ref. [23]

Results obtained for the diffusion of both dyes into the hydrogel D-II are more different. The most of SDS was consumed for the cross-linking and the pore structure in the hydrogel D-II should be more accessible for the diffusion of dyes. We assume that the diffusion of Nile red into hydrogel D-II was strongly influenced by the fact that Nile red was transported as the micelles in SDS. It can cause the slowdown of diffusion both in water [24] and in the hydrogel. The size of diffusion particles is probably the main influencing factor in the diffusion into the hydrogel D-II.

The structural factor of hydrogels defined in our previous work [22] is the ratio between porosity and tortuosity. If no chemical reaction proceeds between hydrogel and dye the factor can be calculated as the ratio between diffusion coefficient of dye in hydrogel and in water. These factors are lower for Atto 488 (0.38 for the hydrogel D-I and 0.70 for the hydrogel D-II). In the case of Nile red, we used the diffusion coefficient publushed for Nile red in SDS micelles [24] for the calculation of the structure factor. The values of 0.61 (D-I) and 0.63 (D-II) were obtained.

4. CONCLUSION

The diffusion of two different dyes into the hydrogels based on the combination of cationized dextran with sodium dodecylsulphate as surfactant was studied. It was found that the incorporation of dyes into hydrogels is influenced by several factors. Nile red was used in the form of its micelles in SDS thus its transport into hydrogel network was more difficult. The diffusion of Atto 488 was affected by the presence of SDS into the pore structure in hydrogel D-I and its micelles can be formed during its transport. The diffusion coefficient of Atto 488 in the hydrogel D-II (which contained only small amount of SDS non-consumed for the cross-linking) was much faster.

ACKNOWLEDGEMENTS

Materials Research Centre at FCH BUT-Sustainability and Development, REG LO1211, with financial support from National Programme for Sustainability I(Ministry of Education, Youth and Sports).

REFERENCES

[1] HURLER, J., BERG, O. A., SKAR, M., CONRADI, A. H., JOHNSEN, P. J. and SKALKO-BASNET, N. Improved burns therapy: Liposomes-in-hydrogel delivery system for mupirocin. *Journal of Pharmaceutical Sciences*. 2012. vol. 101, NO. 10, pp. 3906-3915.

[2] JU, C., SUN, J., ZI, P., JIN, X. and ZHANG, C. (2013). Thermosensitive micelles-hydrogel hybrid system based on Poloxamer 407 for localized delivery of Paclitaxel. *Journal of Pharmaceutical Sciences*. 2013. vol. 102, no. 8, pp 2707-2717.

[3] LU, C., YOGANATHAN, R. B., KOCIOLEK, M. and ALLEN, C. Hydrogel containing silica shell cross-linked micelles for ocular drug delivery. *Journal of Pharmaceutical Sciences*. 2013. vol. 102, no. 2, pp. 627-637.

[4] PEKAŘ, M. Hydrogels with micellar hydrophobic (nano)domains. *Frontiers in Materials*. 2015. vol. 1, no. 35, pp. 1-14.

[5] HUYNH, C. T., NGUYEN, M. K. and LEE, D. S. Injectable block copolymer hydrogels: achievements and future challenges for biomedical applications. *Macromolecules*. 2011. vol. 44, no. 17, pp. 6629-6636.

[6] SHI, W., JI, Y., ZHANG, X., SHU, S. and WU, Z. Characterization of pH- and thermosensitive hydrogel as a vehicle for controlled protein delivery. *Journal of Pharmaceutical Sciences*. 2011. vol. 100, no. 17, pp. 886-895.

[7] PICULELL, L. Understanding and exploiting the phase behavior of mixtures of oppositely charged polymers and surfactants in water. *Langmuir*. 2013. vol. 29, no. 33, pp. 10313-10329.

[8] LI, D. and WAGNER, N. J. Universal binding behavior for ionic alkyl surfactants with oppositely charged polyelectrolytes. *Journal of the American Chemical Society*. 2013. vol. 135, no. 46, pp. 17547-17555.

[9] LINDBLAD, M. S., SJOEBERG, J., ALBERTSSON, A.-C. and HARTMANN, J. Hydrogels from polysaccharides for biomedical applications. *ACS Symposium Series*. 2007. vol. 954, no. 10, pp. 153-167.

[10] DONG, J., CHEN, L., DING, Y. and HAN, W. Swelling and mechanical properties of a temperature-sensitive dextran hydrogel and its bioseparation applications. *Macromolecular Chemistry and Physics*. 2005. vol. 206, no. 19, pp. 1973-1980.

[11] BAI, G., CASTRO, V., NICHIFOR, M. and BASTOS, M. Calorimetric study of the interactions between surfactants and dextran modified with deoxycholic acid. *Journal of Thermal Analysis and Calorimetry*. 2010. vol. 100, no. 2, pp. 413-422.

[12] DERMIBELEK, C. and DINÇ, C. Ö. Diethylaminoethyl dextran/epichlorohydrin (DEAE-D/ECH) hydrogel as adsorbent for murexide. *Desalination and Water Treatment*. 2015. vol. 57, no. 15, pp. 1-10.

[13] ABIOYE, A. O. and KOLA-MUSTAPHA, A. Formulation studies on ibuprofen sodium-cationic dextran conjugate: effect on tableting and dissolution characteristics of ibuprofen. *Drug Development and Industrial Pharmacy*. 2016. vol. 42, no. 1, pp. 39-59.

[14] KROUSKÁ, J., PEKAŘ, M., KLUČÁKOVÁ, M., ŠARAC, B. and BEŠTER-ROGAČ, M. Study of interactions between hyaluronan and cationic surfactants by means of calorimetry, turbidimetry, potentiometry and conductometry. *Carbohydrate Polymers*. 2017. vol. 157, pp. 1837-1843.

[15] KLUČÁKOVÁ, M. Complexation of metal ions with solid humic acids, humic colloidal solutions, and humic hydrogel. *Environmental Engineering Science*. 2014. vol. 31, no. 11, pp. 612-620.

[16] KALINA, M., KLUČÁKOVÁ, M. and SEDLÁČEK, P. Utilization of fractional extraction for characterization of the interactions between humic acids and metals. *Geoderma*. 2013. vol. 207-208, pp. 92-98.

[17] KLUČÁKOVÁ, M. and PEKAŘ, M. Transport of copper(II) ions in humic gel - new results from diffusion couple. *Colloids and Surfaces, A: Physicochemical and Engineering Aspects*. 2009. vol. 349, no. 1-3, pp. 96-101.

[18] SMILEK, J., SEDLÁČEK, P., KALINA M. and KLUČÁKOVÁ, M. On the role of humic acids' carboxyl groups in the binding of charged organic compounds. *Chemosphere*. 2015. vol. 138, pp. 503-510.

[19] UMLONG, I. M. and ISMAIL, K. Micellization behaviour of sodium dodecyl sulfate in different electrolyte media. *Colloids and Surfaces, A: Physicochemical and Engineering Aspects*. 2007. vol. 299, no. 1-3, pp. 8-14.

[20] GRACIE, K., TURNER, D. and PALEPU R. Thermodynamic properties of micellization of sodium dodecyl sulfate in binary mixtures of ethylene glycol with water. *Canadian Journal of Chemistry*. 1996. vol. 74, no. 9, pp. 1616-1625.

[21] JOBE, D. J., REINSBOROUGH, V. C. and WETMORE, S. D. Sodium dodecyl sulfate micellar aggregation numbers in the presence of cyclodextrins. *Langmuir*. 1995. vol. 11, no. 7, pp. 2476-2479.

[22] ZHANG, X., PONIEWIERSKI, A., JELIŃSKA, A., ZAGOŻDŻON, A., WISNIEWSKA, A., HOU, S. and HOŁYST, R. Determination of equilibrium and rate constants for complex formation by fluorescence correlation spectroscopy supplemented by dynamic light scattering and Taylor dispersion analysis. *Soft Matter*. 2016. vol. 12, no. 39, pp. 8186-8194.

[23] ZHANG, L. X., CAO X. H., CAI, W. P. and LI Y. Q. Observations of the effect of confined space on fluorescence and diffusion properties of molecules in single conical nanopore channels. *Journal of Fluorescence*. 2011. vol. 21, no. 5, pp. 1865-1870.

[24] MAITI, N. C., KRISHNA, M. M. G., BRITTO, P. J. and PERIASAMY, N. Fluorescence dynamics of dye probes in micelles. *Journal of Physical Chemistry B*. 1997. vol. 101, no. 51, pp. 11051-11060.

AN APPROPRIATE METHOD FOR ASSESSING HYDROGEL PORE SIZES BY CRYO-SEM

[1]Katerina ADAMKOVA, [2]Monika TRUDICOVA, [1]Kamila HRUBANOVA, [2]Petr SEDLACEK, [1]Vladislav KRZYZANEK

[1]The Czech Academy of Sciences, Institute of Scientific Instruments, v. v. i., Brno, Czech Republic, EU, krzyzanek@isibrno.cz
[2]Brno university of technology, FCH, Brno, Czech Republic, EU, sedlacek-p@fch.vut.cz

Abstract

The aim of our work was to examine and describe ultrastructure of the agarose hydrogel and any possible structural concentration dependencies, and to assess the distribution and size of pores of agarose hydrogel in dependence on its concentration. Four concentrations were prepared (0.5 %, 1.0 %, 2.0 % and 4.0 % of dry weight content) and cryo-SEM and turbidimetry methods were executed on wet (original) samples in order to image the ultrastructure and measure the pore sizes within.

Reasonable results were obtained for the wet samples as they were closer to their native state they are usually used for applications in. Cryo-SEM and turbidimetry provided comparable results of pore diameters and allowed to compare pore diameters dependant on the concentrations; moreover, it showed more detailed and realistic structure.

Keywords: Hydrogel, agarose, cryo-SEM, freezing methods, image analysis

1. INTRODUCTION

Hydrogels are ubiquitous materials with three dimensional polymer network in which water is trapped due to its surface tension effect. Recently, there has been an increasing interest in hydrogels, since they can be used in a broad field of applications, e.g., human health and cosmetics (contact lenses, wound healing dressings, and artificial replacement tissues), drug delivery systems, bioengineering etc.

Structure differences between prepared hydrogel concentrations and its respective pore sizes were observed and analysed using freezing, freeze-fracture, freeze-etching, imaging in cryogenic scanning electron microscope (cryo-SEM) and image analysis. Simultaneously, it was necessary to find the best way to prepare hydrogel samples for freezing and the most suitable freezing method. Cryo-SEM imaging was applied because hydrogels, as their name implies, consist a large amount of water, and aqueous samples cannot be observed in high vacuum that is needed to be maintained in the SEM chamber without any preceding stabilization.

2. EXPERIMENTAL PART

2.1. Agarose hydrogel preparation

Agarose hydrogel was prepared and examined. Agarose is a hydrophilic [1] polysaccharide with gelling properties in solution that can be obtained by extraction from marine red algae. The agarose hydrogel is of high elasticity and comprises a thick [2] rigid [1] network of agarose chains with incorporated large water pores [2]. This polymer is similar to living tissue and extracellular matrix fluid, and therefore it is used for biophysical phenomena [1]. The agarose used within this experiment was Sigma Aldrich type II, medium EEO; CAS number 9012 36 6.

To prepare agarose hydrogel, the agarose powder was dissolved in deionized water; subsequently the mixture was stirred and heated for several minutes and left to cool down at the room temperature. There were four concentrations of this hydrogel prepared - 0.5 %, 1.0 %, 2.0 % and 4.0 %. Their density increased proportionally with the increase of the concentration.

2.2. Agarose hydrogel sample preparation and freezing

The agarose hydrogel samples were frozen by means of two freezing techniques that proved the most successful within preliminary testing. The techniques were: plunge freezing (technique based on immersing samples into a liquid cryogen [3, 4]) and high pressure freezing - HPF (hyperbaric freezing in hyperbaric device Leica EM HPM100 - the specimen "sandwiched" between two carriers is jet-freezed with liquid nitrogen under the pressure raised of 2100 bar, thus, ice crystal nucleation is limited [4]). The agarose hydrogel dries out when exposed to the air, which results in concomitant ultrastructure alterations, so there had to be fresh samples prepared prior each freezing process. Moreover, samples have to be frozen in very thin layers and very small volumes, otherwise they would not be frozen evenly - it would take too long for the deeper areas to get frozen and the ultrastructure would be damaged by the growth of ice crystals. The agarose solution density, however, increases rapidly at thin layers and small volumes at the room temperature (this process is called gelatinization), which makes the sample preparation more difficult.

For plunge freezing, small metal tubes filled with each concentration of the hydrogel were held in forceps and swiftly immersed in liquid nitrogen before the gelatinization took place. As for the HPF, the first attempt was to cut off slices using a scalpel and a special puncher. Due to their consistency, hydrogels are not easy to be sliced and the slices were too thick, however. In the end, it was decided to change the strategy completely and work with the mixture before it lost heat and thickened. From several alternative application methods that were tested, dropping a small amount of hot agarose hydrogel directly on the carrier by means of a pipette and covering it with a microscope slide until it thickened was chosen. The rapid density increase in small volumes had to be averted by quick manipulation with the pipette.

After having been frozen, freeze-fracture (scratching the surface superficially by a sharp blade at high vacuum and low temperature [3]) and freeze-etching (letting the surface ice sublime under the vacuum [3]) procedures were performed on all samples.

2.3. Agarose hydrogel imaging

All the frozen samples were imaged in the cryo-SEM FEI Magellan 400L at the temperature of -120 °C, then three steps of freeze-etching succeeded: the first and second steps meant to increase the temperature to -100 °C, and after it stabilized, decrease it back to -120 °C. Within the third step, the temperature was risen to -100 °C and lowered back to -120 °C after 15 minutes. Images were taken before the freeze-etching and after each step respectively.

The plunge frozen samples showed unwanted ice crystal growth effects altering the structure (e.g. enlarging the pores) - they are shown in the **Figure 1**. The most significant alterations are marked with red arrows - a) before and b) after the first step of freeze etching.

The **Figure 2** shows differences between the agarose hydrogel of the same concentration frozen using the plunging technique and the high pressure freezing a) and b) before the freeze etching, c) and d) after the third step of the etching. Again, structure alterations and enlarged perturbed pores can be seen in the plunge frozen samples images. The structure of high pressure frozen samples stays unaltered and undamaged, so the images could undergo further analysis as follows.

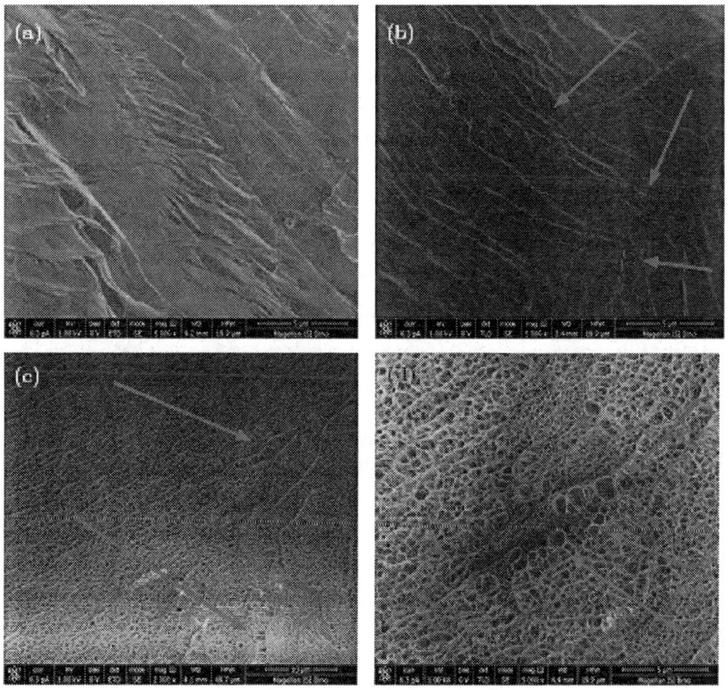

Figure 1 Unwanted structure alterations caused by ice crystal growth prior to the freeze etching (a, b) and after the first step of freeze etching (c, d) marked with red arrows

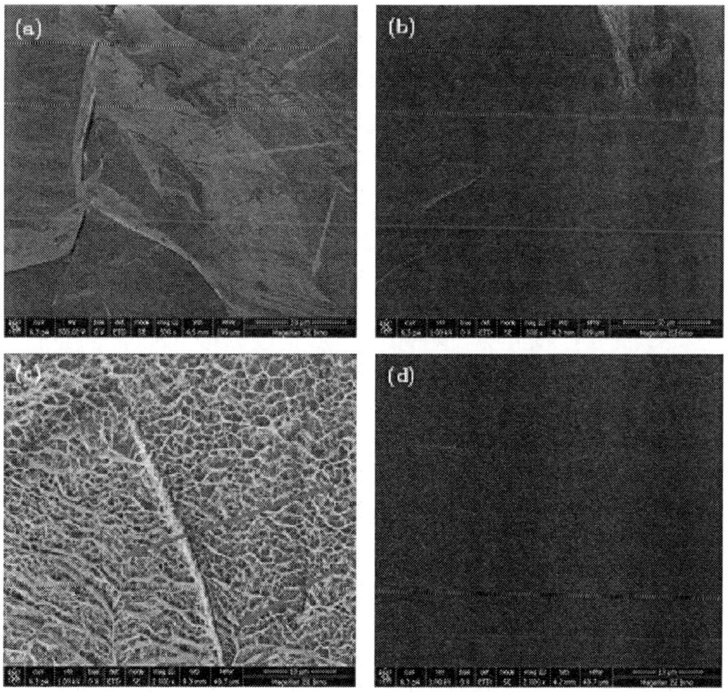

Figure 2 The comparison of agarose hydrogel of the same concentration frozen using the plunging technique (a, c) and the high pressure freezing (b, d), imaged before the freeze etching (a, b) and after the third step of the etching (c, d)

2.4. Image processing and analysis

Basically, the images obtained by the optimized cryo-SEM technique serve as a visual illustration of the qualitative characteristics of internal structure of the studied gel networks. Nevertheless, they can also be further processed in order to gain some quantitative structural parameters. For this purpose, we have applied two approaches of the image processing that are implemented in open-source scientific image processing toolbox ImageJ and that are proposed for the analysis of porous structures. Both analytical approaches were performed on the similar square sections (512×512 pixels) of the images recorded at the same level of magnification (see the first column in **Figure 3**). Before the very analysis, the processed images need to be converted from a grayscale to black & white projections with an effort to display only the uppermost layer of the 3D network structure. For this conversion, the image sections were thresholded using MaxEntropy algorithm. Subsequently, the actual analysis of the internal structure proceeded.

Firstly, the Particle analysis tool (an automatic particle segmentation algorithm implemented in ImageJ) was used to identify individual pores of the gel network. The outlines of the detected pores are shown in the third column in **Figure 3**. As a numerical result of this analysis, every pore which is detected in the structure is described by its area and perimeter. Wherever it is necessary to take care of the pores which are displayed in the binary picture touching one another, the Watershed algorithm can be used prior to the Particle analysis. This algorithm uses a density profile to determine if one object with a peninsula should actually be two objects. If it determines that they should, it will draw a line to separate them. From the Particle analysis, distribution of pore areas and perimeters is obtained and processed into statistical parameters, e.g. average or mean values.

Skeleton analysis represents another image processing technique, applicable in the morphological analysis of porous materials. In this technique, the network structure displayed in the analyzed image is skeletonized, i.e. replaced by the line skeleton using a topology-maintaining medial axis thinning algorithm (skeletonized representations of the analyzed images is shown in the second column in **Figure 3**). Using the Skeleton analysis tool, branches and junctions of such a skeleton are then classified, counted and measured. As an example of a quantitative parameter, useful in description of the porosity of the analyzed structure, average branch length is provided among the results of the Skeleton analysis.

Comparison of the quantitative results of the two image-analysis approaches is shown in **Figure 4**. For both methods, average value of a linear parameter which represents the pore size is provided as a function of concentration of agarose in the gel. In the case of Particle analysis, the average size of the pore was calculated from the average pore area (using the simplified assumption of the circular cross-section of the pores); from the Skeleton analysis, average branch length was used for the same purpose. It can be seen that results of both methods are in good agreement. Nevertheless, we propose the Skeleton analysis as a more suitable method because, unlike the Particle analysis approach, it does not suffer from the improper assumption of the circular pore shape.

Finally, the results of image analysis of cryo-SEM images were compared to the indirect pore size estimation by means of turbidimetry according to Aymard et al. [2]. In this approach, it is assumed that the pore size in agarose gels corresponds well with the correlation length, calculated from the wavelength dependence of the turbidity in spectral region of 800-900 nm. As can be seen in **Figure 4**, the absolute values of the correlation lengths as determined by turbidimetry are significantly lower (although of the same order of magnitude) than the linear parameters resulting from the cryo-SEM image analyses. Nevertheless, the decreasing trends of the correlation length with the agarose concentration is comparable with the concentration dependence if the cryo-SEM results.

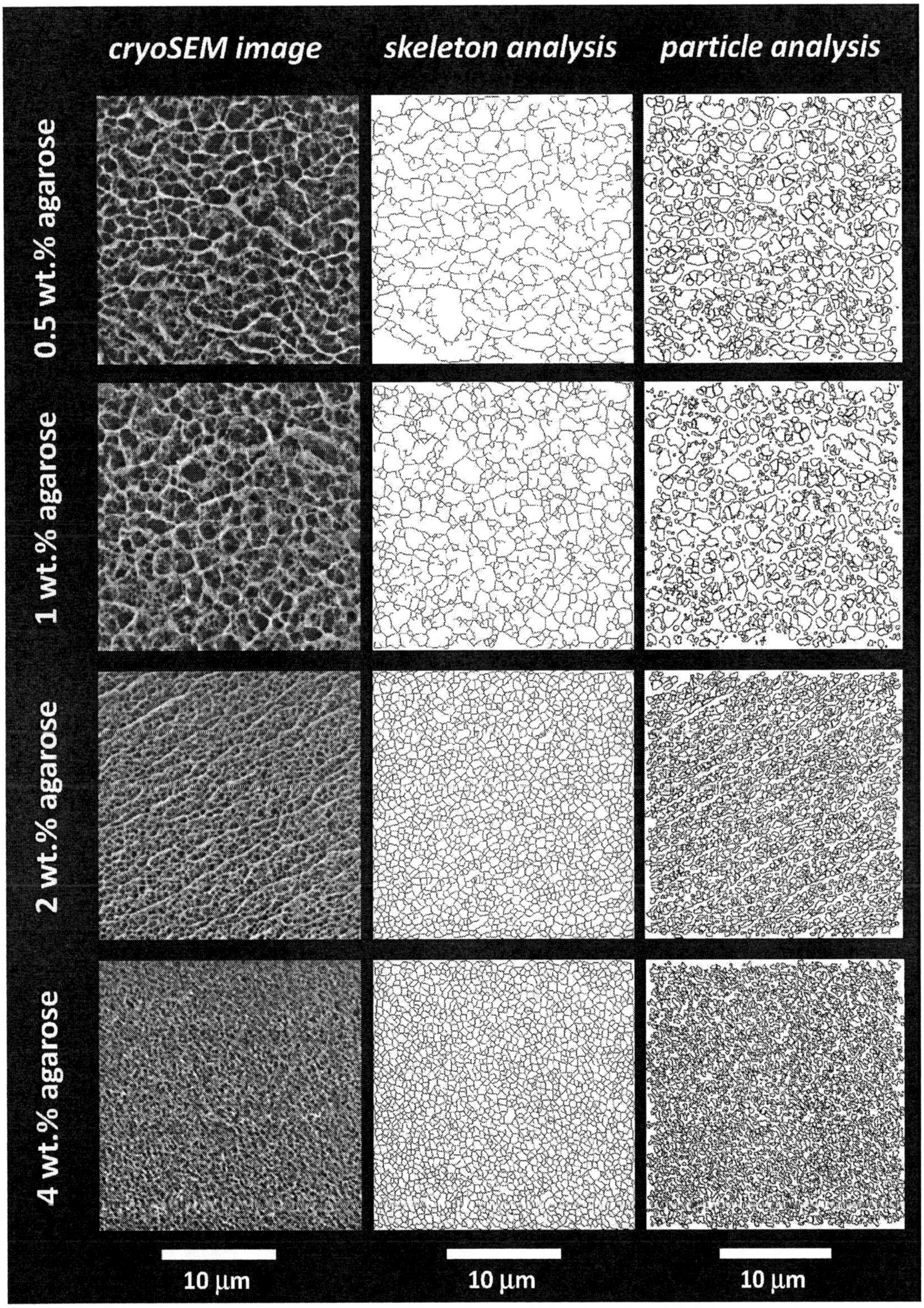

Figure 3 Image processing of the cryo-SEM images of various agarose gels. Sections of the micrographs which were subjected to further processing and analysis (left). Mask representation of the same sections processed for the Skeleton analysis (middle) and Particle analysis (right) in ImageJ.

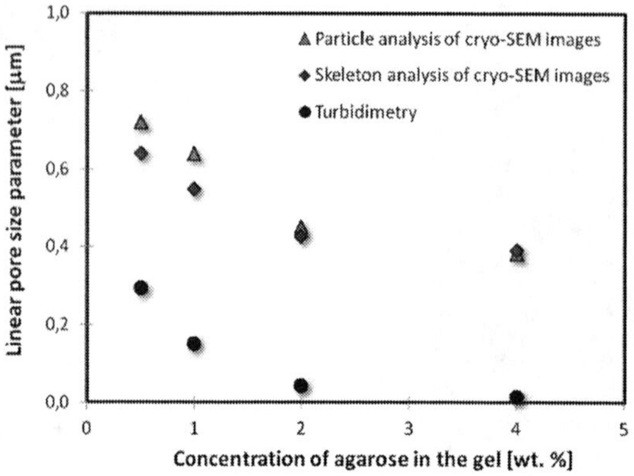

Figure 4 Linear parameters representing the average pore size of the agarose gels

3. CONCLUSION

Four concentrations of agarose hydrogel were prepared in order to observe and examine its structure and any concentration-dependant differences, as well as to perform image analysis. The samples were frozen by means of plunge-freezing and HPF. As for the plunge freezing, the hydrogel was filled into small metal tubes. For the HPF freezing, various sample preparation methods were tested because of rapid thickening of small volumes of the agarose hydrogel at the room temperature. From these preparation methods, only the direct application on the carrier was suitable for all the concentrations. All the frozen samples were then freeze fractured, freeze etched and imaged in the cryo-SEM. The plunge freezing technique proved insufficient, because ice crystal growth caused unwanted structure alterations and damage. HPF frozen samples showed no or minimum contamination, damage or alterations, so the structures of all four concentrations could be examined and compared. The cryo-SEM imaging provided a network like structures revealed by the progressive freeze etching. The agarose hydrogel of the lowest concentration was the most "airy" with the sparsest network and largest pores. The final output of this method, however, does not need to be mere image illustrating the structure - quantitative parameters can also be obtained, which was verified applying two approaches - particle analysis and skeleton analysis.

ACKNOWLEDGEMENTS

The research was supported by CSF (17-15451S), MEYS CR (LD15047 and LO1212), its infrastructure by MEYS CR and EC (CZ.1.05/2.1.00/01.0017).

REFERENCES

[1] MAALOUM, Mounir, PERNODET, Nadine and TINLAND, Bernard. Agarose gel structure using atomic force microscopy: Gel concentration and ionic strength effects. *Electrophoresis* [online]. 1998, vol. 19, iss. 10, pp. 1606-1610 [cit. 2018-04-22]. DOI:10.1002/elps.1150191015.

[2] AYMARD, Pierre, MARTIN, David R., PLUCKNETT, Kevin, FOSTER, Tim J., CLARK, Allan H. and NORTON, Ian T. Influence of thermal history on the structural and mechanical properties of agarose gels. *Biopolymers* [online]. 2001, vol. 59, iss. 3, pp. 131-144 [viewed 2018-04-22]. Available from: DOI: 10.1002/1097-0282(200109)59:3<131::AID-BIP1013>3.0.CO;2-8.

[3] HRUBANOVA, Kamila, NEBESAROVA, Jana, RUZICKA, Filip, KRZYZANEK, Vladislav. The innovation of cryo-SEM freeze-fracturing methodology demonstrated on high pressure frozen biofilm. *Micron*. 2018. vol. 110, pp. 28-35. DOI: 10.1016/j.micron.2018.04.006

[4] SEVERS, Nicholas J. and SHOTTON, David M. Rapid Freezing of Biological Specimens for Freeze Fracture and Deep Etching. In: Celis, Julio E. *Cell Biology*. 3rd ed. Burlington: Academic Press, 2006, pp. 249-255.

STUDY OF INTERACTION OF DYES WITH BIOPOLYMER CHITOSAN

[1]Pavel KOLESA, [1,2]Martina KLUČÁKOVÁ

[1]*Brno University of Technology, Faculty of Chemistry, Brno, Czech Republic, EU,* xckolesa@fch.vutbr.cz

[2]*Material Research Centre, Brno University of Technology, Brno, Czech Republic, EU,* klucakova@vutbr.cz

Abstract

The presented work was focused on the study of interactions of azo dyes (model diffusion probes) with cationic biopolymer chitosan. The interactions were realized via diffusion processes in hydrogel media based on thermoreversible agarose. The main aim was study of influence of solution pH on the diffusion process. Interactions of used dyes are based on electrostatic character. The amino group of chitosan interacts with the functional group of chosen anionic dyes (sulfonic group) and thus affects the process of diffusion. Organic anionic dyes Chicago Sky Blue 6B (C.I. 24 410), Sirius Red (C.I. 35 780) and Reactive Blue 49 (C.I. 621 526) were chosen for this experiment. Used diffusion technique was based on monitoring of the time progression of diffusion profile by UV-VIS spectrophotometry. The presented work follows the previous experiments and shows comprehensive view of the reactivity of chitosan and its behavior in different systems. Unsteady diffusion in cuvettes appears to be universal method for the study of reactivity of biopolymers and for the study of transport processes in hydrogel media.

Keywords: Diffusion, chitosan, hydrogel, agarose, organic dye, reactivity of biopolymers

1. INTRODUCTION

Each life organism is composed by a lot of complex macromolecules which are widely called biopolymers. There are compounds, such as proteins, nucleic acids or polysaccharides which are responsible for functioning of the organism. All these compounds are commonly used in industrial application, especially in pharmaceutical industry. One of the important biopolymers which is often used in pharmaceuticals applications is a linear polysaccharide chitosan. From a chemical point of view is chitosan partially deacetylated *N*-acetyl glucosamine and commercially is produced by deacetylation of chitin, which is the second most widespread polysaccharide. Different types of chitosan are distinguished especially by degree of deacetylation and length of chain, respectively molecular weight [1]. In the nature is chitosan most commonly occurs in the mixture with chitin. Due to its biodegradability and biocompatibility is chitosan significant material for use in many pharmaceutical applications. It also corresponds with high number of patents and articles focused on chitosan published in the last years [2].

Chitosan is a cationic biopolymer due to presence of amino groups which can be protonated in solutions. The presented experiment is based on reactivity of chitosan, especially its amino groups whose reactivity was studied by a simple diffusion technique called unsteady diffusion in cuvettes. The main aim of this work was study of influence of solution pH on diffusion process into hydrogel matrix. Used diffusion technique was based on periodic monitoring of the time progression of diffusion profile used model probes (azo dyes) by UV-VIS spectrophotometry. Interactions of azo dyes with chitosan were based on electrostatic character because amino group of chitosan interacted with sulfonic group of chosen azo dyes and thus affected the diffusion process. The observed parameter was effective diffusion coefficient.

2. EXPERIMENTAL PART

2.1. Materials

Agarose (≤ 10 % w/w humidity), chitosan (75-85 % degrees of deacetylation), 95 % w/w acetic acid, sodium hydroxide and used azo dyes Chicago Sky Blue 6B (C.I. 24 410), Sirius Red (C.I. 35 780) and Reactive Blue 49 (C.I. 621 526) were purchased from Sigma-Aldrich®, structures of dyes are shown in the figures below (**Figure 1-3**). Buffer solutions with different pH were prepared using: disodium hydrogen phosphate dihydrate, sodium dihydrogen phosphate and citric acid which were purchased from Penta®. These chemicals were used for preparation of buffer solutions with pH = 3, pH = 7 and pH = 11. Each of used organic dyes was dissolved in buffers. Final concentrations of organic dyes in buffers were different for each type of dye (Chicago Sky Blue 6B c = 0.01 g/dm^3, Sirius Red c = 0.04 g/dm^3 and Reactive Blue 49 c = 0.10 g/dm^3). Solution of chitosan was prepared by dissolving 0.001 g of chitosan in 50 ml 5 % w/w acetic acid and stirred for 24 hours. Then pH of solution was adjusted using sodium hydroxide to pH value pH = 7 and diluted by distilled water to volume 100 ml. Prior to diffusion experiments calibration curves of all dyes in agarose gel for each value of pH were measured. Hydrogels for diffusion experiments contained the same concentrations of agarose (1.0 % w/w) but different concentrations of chitosan in hydrogel matrix (0 % w/w and 0.001 % w/w). Diffusion experiments were realized in plastics UV-VIS cuvettes (PMMA, 10 x 10 x 45 mm) so this PMMA cuvettes were filled by hydrogel solutions prepared by heating. Cooling of the agarose solutions at room temperature led to the hydrogel formation and then were cuvettes placed for 72 hours at room temperature into glass diffusion container including 250 ml of dye buffer solution.

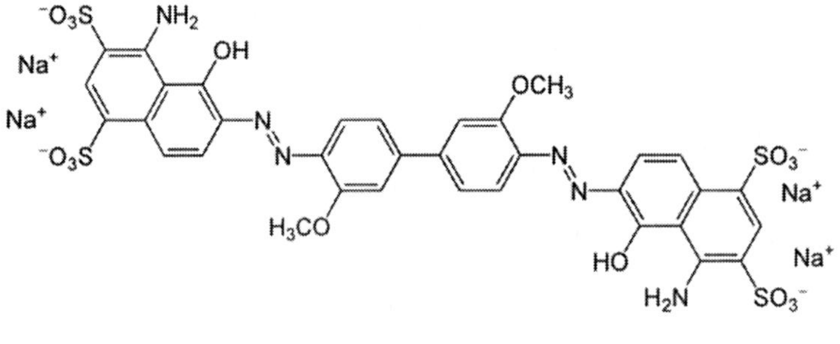

Figure 1 Structure of dye Chicago Sky Blue 6B C.I. 24 410 [3]

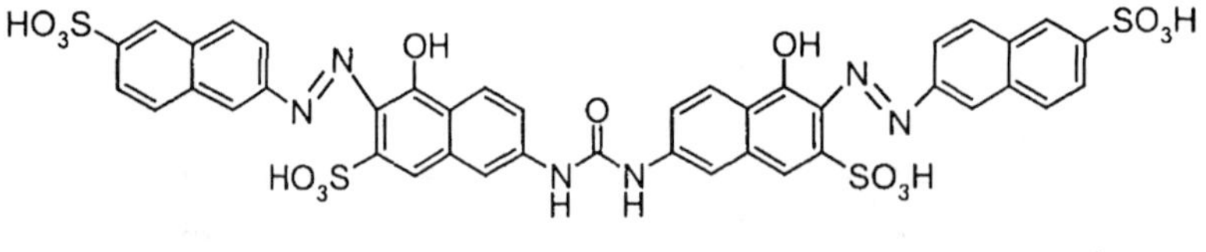

Figure 2 Structure of dye Sirius Red C.I. 35 780 [4]

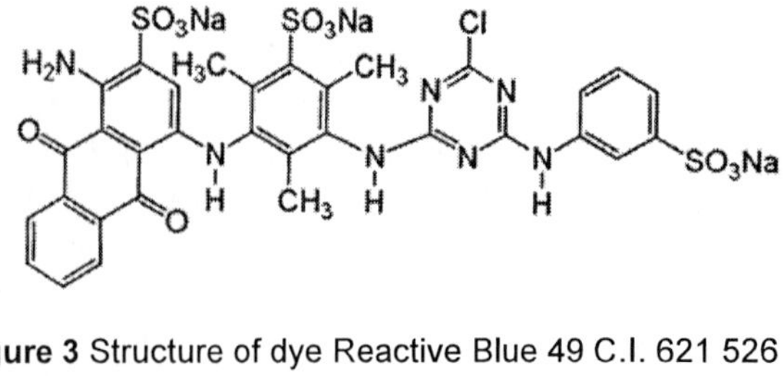

Figure 3 Structure of dye Reactive Blue 49 C.I. 621 526 [5]

2.2. Methods

PMMA cuvettes with hydrogel were measured by UV-VIS spectrophotometer at time interval 24, 48 and 72 hours. Vertically adjustable holder for cuvettes was made for this purpose which allowed to measure cuvettes at different distances from the interface hydrogel-solution. This procedure allowed to obtain complex UV-VIS spectra in different distances of cuvettes. All cuvettes placed in dye buffer solution (pH = 3, pH = 7 and pH = 11) were measured by this UV-VIS technique (**Figure 4**, **Figure 5**) and concentration profiles in each cuvette were obtained by using calibration curves.

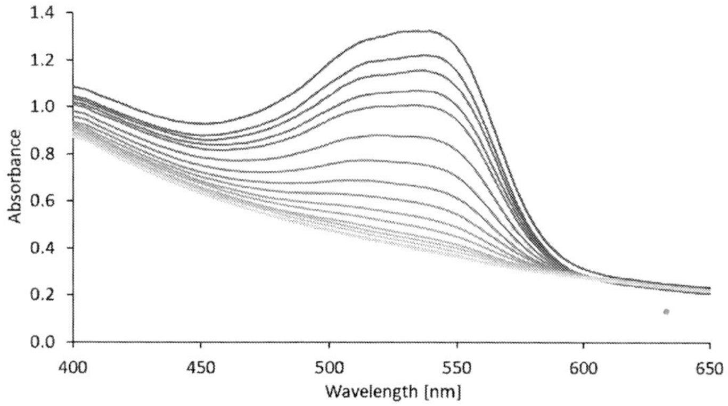

Figure 4 UV-VIS spectra correspond to different distances from interface hydrogel-solution in cuvette placed in dye Sirius Red after 72 hours

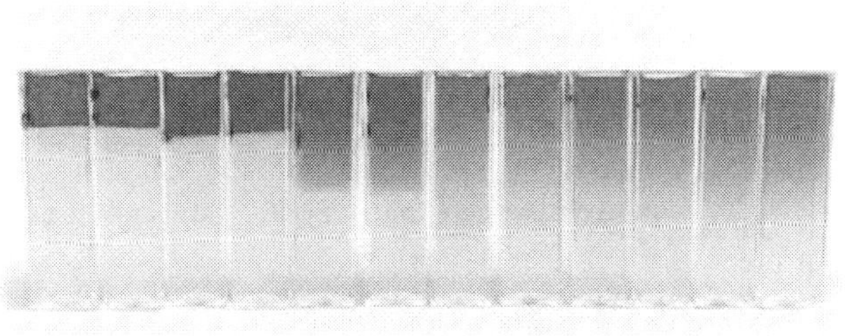

Figure 5 Cuvettes with hydrogel placed in dye buffers of Sirius Red after 72 hours at different pH (from left 6 cuvettes contained hydrogel with chitosan concentration 0.001 % w/w, the rest of hydrogel cuvettes without any addition of chitosan)

2.3. Results and discussion

The most important parameter which corresponds to diffusion process and its speed into hydrogel matrix is effective diffusion coefficient. This parameter was calculated for each cuvette. By evaluation of experimental data was found that pH of solution significantly affects speed of diffusion process in cuvettes with addition of chitosan.

As was already mention in the previous chapter chitosan is a cationic polymer due to presence of amino groups. Amino group of chitosan is protonated in solution and thus interacts with anionic sulfonic group of used dyes. These electrostatic interactions between opposite charged functional groups affect speed of penetration of dyes to the hydrogel matrix. Amino group of chitosan has a pK_a = 6.5. In alkaline solutions is small number of amino groups protonated so only few groups interacts with sulfonic groups of azo dyes. Otherwise acidic solutions lead to formation of $-NH_2^+SO_3^-$ respective $=NH^+SO_3^-$ complexes which affect diffusion process via electrostatic interactions of used dyes [6]. **Figure 6** below shows influence of decreasing pH value of solutions on effective diffusion coefficient for cuvettes with chitosan addition. Hydrogel cuvettes containing just 1 % w/w agarose gel did not show any influence of diffusion on different pH of solution. Organic azo dye Sirius red showed the most intensive interactions with chitosan in comparison with other used dyes (Chicago Sky Blue 6B and Reactive Blue 49).

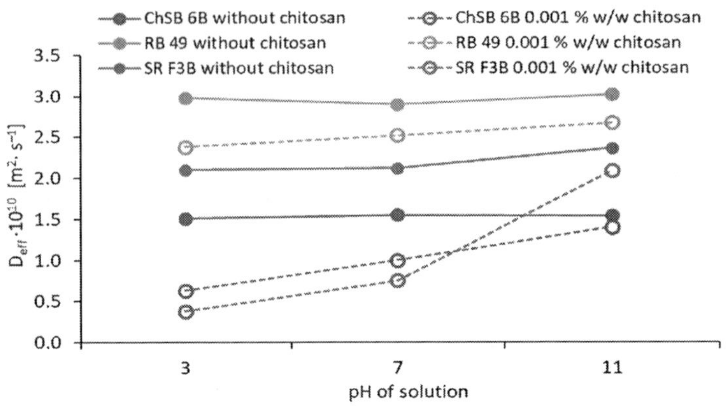

Figure 6 The influence of pH value of solution on effective diffusion coefficient

The obtained spectrophotometric data were evaluated using Origin 8 software. The baseline of each sample was corrected using preset function. Effective diffusion coefficients were obtained by fitting of experimental data to theoretic model based on Fick's laws. Mathematical model corresponds to:

$$c = c_0 \cdot \operatorname{erfc} \frac{x}{\sqrt{4D_{eff}t}},$$

where c is concentration of diffuse compound, c_0 is a concentration of compound in the phase interface, x is diffusion length, D diffusion coefficient of used compound, t is time of diffusion and erfc is an error function [7]. Fitting of theoretical model was performed by the smallest square method using Solver function in MS Excel. The table below (**Table 1**) shows summary of results for effective diffusion coefficients and concentrations of dyes in solution-hydrogel interface for all tested dyes.

Table 1 Effective diffusion coefficients for hydrogels with chitosan (0.001 % w/w) and without chitosan depending pH of solution

pH of solution	**$D_{eff}\cdot10^{10}$ [$m^2\ s^{-1}$]**	
Chicago Sky Blue 6B 0.01 g/dm^3		
	0.000 % w/w chitosan	0.001 % w/w chitosan
3	1.51 ± 0.05	0.63 ± 0.08
7	1.55 ± 0.07	1.00 ± 0.05
11	1.54 ± 0.10	1.40 ± 0.13
Sirius Red 0.04 g/dm^3		
	0.000 % w/w chitosan	0.001 % w/w chitosan
3	2.05 ± 0.12	0.38 ± 0.03
7	2.04 ± 0.10	0.74 ± 0.04
11	2.00 ± 0.12	2.01 ± 0.20
Reactive Blue 49 0.10 g/dm^3		
	0.000 % w/w chitosan	0.001 % w/w chitosan
3	2.98 ± 0.09	2.38 ± 0.13
7	2.90 ± 0.13	2.52 ± 0.05
11	3.02 ± 0.07	2.67 ± 0.08

3. CONCLUSION

Reactivity of biopolymers can be tested and measured by many experimental techniques. This study is focused on diffusion method Unsteady diffusion in cuvettes which seems to be a simple tool for reactivity testing of many compounds. This experiment follows the previous experiments which were focused on testing of different conditions, such as temperature or ionic strength and their influence on diffusion speed in hydrogel matrix with biopolymers. It helps to better understand to chitosan and other biopolymers and their behavior in different systems. The study of biopolymers using diffusion techniques has a lot of advantages, such as cheap or simplicity. But this technique is still relatively "new" and should be further tested and optimized. High amount of water in hydrogel matrix can simulate conditions suitable for use in e.g. pharmacy. The obtained results will be compared with other diffusion techniques (sorption experiments, diffusion in diaphragm cell...) in the future [8].

ACKNOWLEDGEMENTS

This research was supported by Materials Research Centre at FCH BUT-Sustainability and Development, REG LO1211, with financial support from National Programme for Sustainability (Ministry of Education, Youth and Sports).

REFERENCES

[1] KHOR, Eugene. Chitin: a biomaterial in waiting. *Current Opinion in Solid State and Materials Science*. 2002, Vol. 6, issue 4, p. 313-317. DOI: 10.1016/S1359-0286(02)00002-5.

[2] RAVI KUMAR, Majeti N.V. A review of chitin and chitosan applications. *Reactive and Functional Polymers*. 2000, Vol. 46, issue 1, p. 1-27. DOI: 10.1016/S1381-5148(00)00038-9.

[3] HE, Zongsheng, Lingdi YAN, Zheng YONG, Zhaoqi DONG, Huajin DONG a Zehui GONG. Chicago Sky Blue 6B, a vesicular glutamate transporters inhibitor, attenuates methamphetamine-induced hyperactivity and behavioral sensitization in mice. *Behavioural Brain Research* [online]. 2013, Vol. 239, issue 4, p. 172-176. DOI: 10.1016/j.bbr.2012.11.003. ISSN 01664328.

[4] DAPSON, RW, C FAGAN, JA KIERNAN a TW WICKERSHAM. Certification procedures for sirius red F3B (CI 35780, Direct red 80). *Biotechnic & Histochemistry*. 2011, Vol. 86, issue 3, p. 133-139. DOI: 10.3109/10520295.2011.570277. ISSN 1052-0295.

[5] WANG, Baoe a Xiu GUO. Reuse of waste beer yeast sludge for biosorptive decolorization of reactive blue 49 from aqueous solution. *World Journal of Microbiology and Biotechnology*. 2011, Vol. 27, issue 6, p. 1297-1302. DOI: 10.1007/s11274-010-0576-3. ISSN 0959-3993.

[6] ZHENG, Liuchun, Chenggang WANG, Yuehong SHU, Xiaomin YAN, Laisheng LI a Véronique PRÉAT. Utilization of diatomite/chitosan-Fe (III) composite for the removal of anionic azo dyes from wastewater: Equilibrium, kinetics and thermodynamics. *Colloids and Surfaces A: Physicochemical and Engineering Aspects*. 2015, Vol. 468, issue 244, p. 129-139. DOI: 10.1016/j.colsurfa.2014.12.015.

[7] CRANK, John. The mathematics of diffusion. 2nd ed., 1st publ. in paperback 1979. *Oxford: Oxford University Press*, c1975, Vol. 8, p. 414 Oxford science publications. ISBN 978-0-19-853411-2.

[8] SEDLÁČEK, P.; SMILEK, J.; KLUČÁKOVÁ, M. How the Interactions with Humic Acids Affect the Mobility of Ionic Dyes in Hydrogels - 2. Non-stationary Diffusion Experiments. *Reactive and Functional Polymers 2014*. Vol. 75, No. 1, p. 41-50.

SENSITIVE SERS PLATFORM OF BIMETALLIC NANOPARTICLES IMMOBILIZED ON THE GOLD GRATING SURFACE

Yevgeniya KALACHYOVA, Vaclav SVORCIK, Oleksiy LYUTAKOV

Department of Solid State Engineering, University of Chemistry and Technology Prague, Czech Republic, EU, kalachyy@vscht.cz

Abstract

Surface-enhanced Raman spectroscopy (SERS), stemming from the gigantic light energy focusing near the plasmon-active surface, provides the possibility to reveal the presence of targeted analytes under extremely dilute conditions. At present, SERS sensors have been making significant progress regarding practical applications in the analytical and bioanalytical fields. The main delaying factors in SERS development are related to the poor signal reproducibility. Proposed work is aimed at the achievement of highly-reproducible SERS results. The high sensitivity of SERS has been achieved through the immobilization of sharp-edges bimetallic specific nanoparticles - gold/platinum nanourchins (Au@PtNPs), on the gold grating surface, through the 1,4-benzendithiol (BDT) or 4-sulfobenzenediazonium tosylate (ADT-SH). Gold grating, able to support the surface plasmon polariton (SPP) excitation and propagation, was created on the polymer substrate using the excimer laser modification with further metal deposition. Bimetallic nanoparticles, able to efficiently excite the localized surface plasmons (LSPs), were synthesized using the wet chemical approach. Immobilization of Au@PtNPs on the gold grating allows to achieve SPP-LSP coupling and significantly SERS enhancement, with the maintaining of signal reproducibility, provided by the periodical surface. More detailed experiments show that in the case of ADT-SH, the better Au@PtNPs grafting density and SERS enhancement factor were achieved. Moreover, the produced structure allows SERS measurements with a portable Raman spectrophotometer.

Keywords: Multibranched nanoparticles, gold gratings, coupling, plamonics, SERS

1. INTRODUCTION

Surface-enhanced Raman spectroscopy (SERS) is a way to significantly increase the weak naturally weak Raman signal. Recently significant advancements in the field of SERS substrate design and preparation were reported [1-4]. However, despite the many methods, proposed to realize the uniform SERS response the construction and realization of large area, homogenious substrate is complicated. Among the common techniques, aimed to solve this challenge through achieving of the uniform distribution of plasmon hot spots, the self-assembling can be mentioned [5]. Alternatively, the significant SERS enhancement can be realized through the utilization of nanostructures with the high number of so-called hot-spots [6-8], for example nanoparticles with urchin-like symmetry [9]. The immobilization or growth of gold nanoparticles with shape edges on the different substrates was proposed to combine both mentioned above approaches: creation of hot spots homogenious distribution and high SERS enhancement factor [10].

In this work the immobilization of gold urchin-like particles on the patterned gold grating surface was proposed and studied. The immobilization of gold nanourchins was performed on the metal surface, using the thiol chemistry for monolayer formation [11-12]. The high SERS response has been achieved through the immobilization of specific nanoparticles Au@PtNPs on the gold grating surface, through BDT or ADT-SH [13]. Prepared structures were checked using the UV-Vis, AFM, and Raman spectroscopy techniques.

2. EXPERIMENTAL SECTION

2.1. Materials

Chloroauric acid tetrahydrate ($HAuCl_4 \cdot 4H_2O$, 99.9 %), silver nitrate ($AgNO_3$, 99.0 %), chloroplatinic acid hydrate ($H_2PtCl_6 \cdot xH_2O$, 99.0 %), ascorbic acid (AA, 99.0 %) and 1,4-benzendithiol (BDT, 99.0 %) were purchased from Sigma-Aldrich. 4-sulfobenzenediazonium tosylate (ADT-SH) were prepared according to the described procedure [4].

All chemical reagents were used as received without further purification. Deionized water was used throughout the experiments.

2.2. Preparation of Au@Pt Multibranched NPs

Au@Pt Multibranched NPs (Au@PtNPs) were synthesized by known procedure [11]. 100 µL of 10 mM aqueous $HAuCl_4$ solution was mixed with 3 mL of deionized water, and then 6 µL of 10 mM aqueous $AgNO_3$ solution was added under magnetic stirring for 30 s. After the solutions had been thoroughly mixed, 2 µL of 100 mM ascorbic acid was "quickly" added, and the solution was stirred vigorously for 10 s at room temperature (RT). When reducing agent was added the color of the mixture changed immediately from yellow to dark blue, indicating the formation of gold multibranched NPs (AuMs). Then the AuMs were dispersed in 3 mL of water and were purified by centrifugation at 5000 rpm for 10 min.

For further use the resultant precipitates were re-dissolved in 3 mL of deionized water. Then the 15 µL of 10 mM aqueous H_2PtCl_6 solution was mixed with 3 mL of prepared AuMs. After the solutions had been thoroughly mixed, 10 µL of 100 mM AA was "quickly" added and stirred for 30 s at room temperature (RT). The color of the prepared mixture changed from dark blue to grey, indicating the formation of Au@PtNPs. Then the Au@PtNPs were dispersed in 3 mL of water and were purified by centrifugation at 5000 rpm for 20 min. For further use the resultant precipitates were re-dissolved in 3 mL of deionized water.

2.3. Preparation of Au grating

Polymer films (Solution of epoxy resin - photoresist, Su-8, purchased from Microchem) were spin-coated (1000 rpm) from a solution onto freshly cleaned glass substrates (supplied by Glassbel Ltd, CR) during 10 min. The prepared samples were dried under 50 °C for 24 h and irradiated by UV-source for 30 min. After UV-irradiating samples were dried under 90 °C for 2 h.

Then the flat polymer surface was patterned by KrF excimer laser by known procedure [11]. As a result, the periodic surface structures were created on the Su-8 surface.

Gold was then deposited onto a patterned surface by vacuum sputtering (gold thickness 50 nm, DC Ar plasma, gas purity of 99.995 %, gas pressure of 4 Pa, discharge power of 7.5 W provided by Safina, Czech Republic).

2.4. Samples grafting with Au@Pt NPs

Freshly prepared samples of Au gratings were inserted into methanol solution of BDT (10^{-6} mol·L^{-1}) and water solution of ADT-SH (10^{-6} mol·L^{-1}) for 24 h. After incubation modified samples were washed with methanol and distilled water. Then prepared samples immersed into solutions of Au@PtNPs for 24 h.

Finally, the samples were cleaned in the ultrasonic bath (in distilled water) and dried under N_2 flow.

Procedures of sampel preparation is illustrated in **Figure 1**.

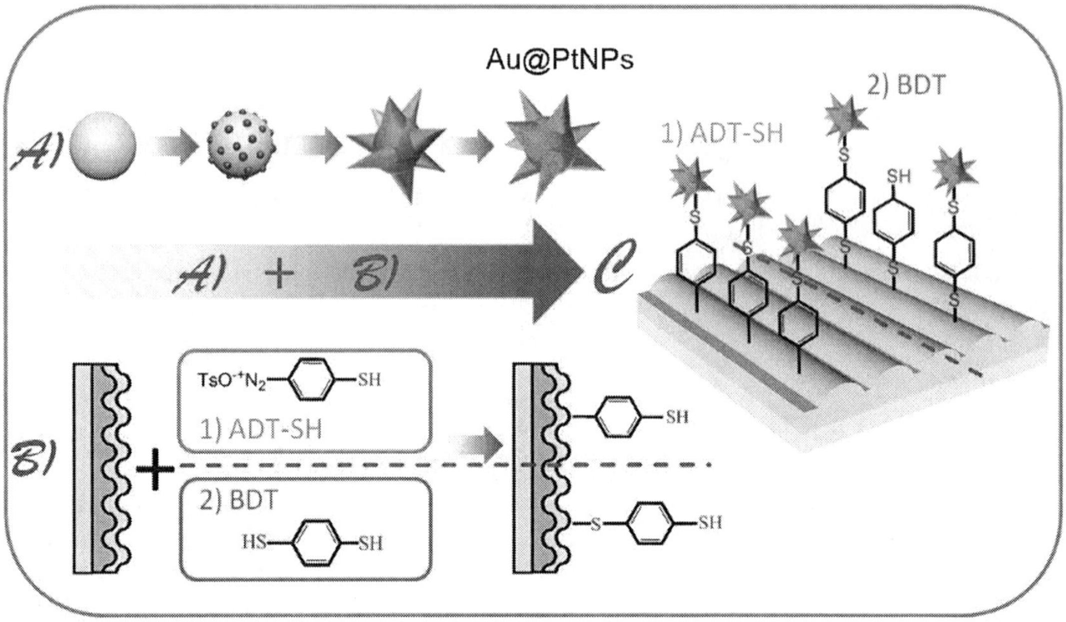

Figure 1 Schematic representation of the LSP-SPP coupled system preparation: (A) - preparation of Au@PtNPs; (B) - preparation of gold grating and grafting of ADT-SH and BDT linkers; (C) - grafting of Au@PtNPs through the ADT-SH and BDT linkers

3. RESULTS AND DISCUSSION

3.1. Characterization of solid films

3.1.1. Transmission electron microscopy (TEM)

TEM images of nanoparticles were obtained on JEOL JEM-1010 instrument operated at 80 kV (JEOL Ltd., Japan). This technique confirms the presence of shape edges in the AuMs structures, which is responsible for efficient photon-plasmon conversion and allows achieving huge SERS enhancement (**Figure 3a**). After deposition of Pt layer the AuMs conserve their shape (**Figure 3b**), providing the possibility for efficient SERS detection [14].

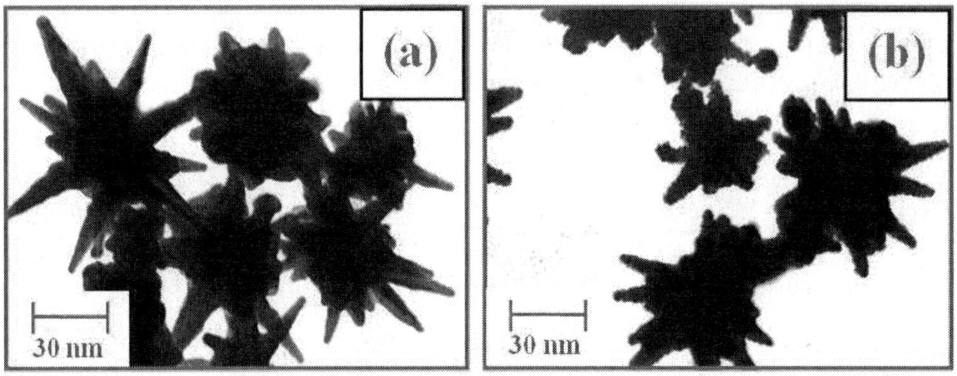

Figure 2 TEM images of (a) - AuMs; (b) - Au@PtNPs

3.1.2. X-ray photoelectron spectroscopy (XPS)

Concentration of elements on the surface of prepared samples was studied by X-ray photoelectron spectroscopy (XPS). Results are summarised in the **Table 1**. As expected, after thiol grafting the surface of samples contained elements typical for thiol grafting (sulfur), and elements of grafting Au@PtNPs (gold and platinum).

Table 1 Concentration of elements studied by XPS

Sample	Element concentration (At. %)				
	C (1s)	O (1s)	S (2p)	Au (4f)	Pt (4f)
Su-8/Au	56.1	24.5	-	-	-
Su-8/Au/ADT-SH	65.9	18.1	5.9	-	-
Su-8/Au/ADT-SH/Au@PtNPs	66.9	16.1	3.3	2.3	2.0
Su-8/Au/BDT	65.8	18.5	6.2	-	-
Su-8/Au/BDT/Au@PtNPs	67.1	16.3	4.3	1.9	1.5

3.1.3. Atomic force microscopy (AFM)

For characterization of the sample surface before and after surface AuMs immobilization, the peak force AFM technique was used. Surface mapping was performed with Icon (Bruker) set-up on 3×3 μm^2 samples area. In particular, **Figure 3** shows the different stage of the AuMs immobilization through the ADT-SH and BDT linkers. **Figure 3(a)** represents the grating structure before the samples immersing in the solution of Au@PtNPs. **Figure 3(b)** shows the surface morphology after the nanoparticles immobilizatio through the BDT spacer molecules (also during the 24 hours of entrapping procedure). Finally, **Figure 3(c)** shows the results of Au@PtNPs grafting after the 24 hours of sample immersing in the solution of ADT-SH, followed by the Au@PtNPs. As is evident, the entrapping of Au@PtNPs occurs, but the grating surface is not fully covered. The grating structure is fully screened by the attached Au@PtNPs, which indicate the complete covering of pristine surface during the reaction time.

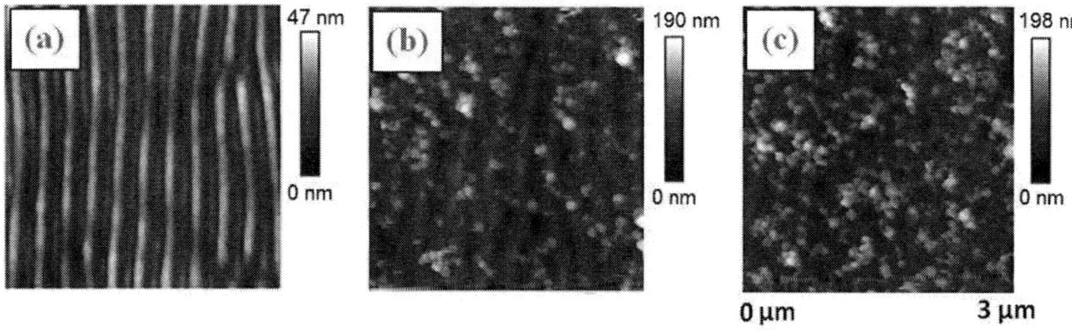

Figure 3 Surface morphology measured on the: (a) - patterned polymer surface covered by gold layer; (b) - patterned polymer surface covered by silver layer with grafted Au@Pt NPs after sample immersing in BDT solution during the 24 hours; (c) - patterned polymer surface covered by gold layer with grafted Au@Pt NPs after sample immersing in ADT-SH solution during the 24 hours

3.1.4. SERS

Raman scattering was measured on Nicolet Almega XR spectrometer (Laser power 30 mW) Raman spectrometers with 780 nm excitation wavelengths. Spectra were measured 10 times, each of them with 30 s accumulation time. As analytical standart the Rhodamine 6G was used (supplied from SigmaAldrich, dissolved in methanol (10^{-11} mol·L^{-1} solution) and added by dropping solution onto samples surface before SERS measurement).

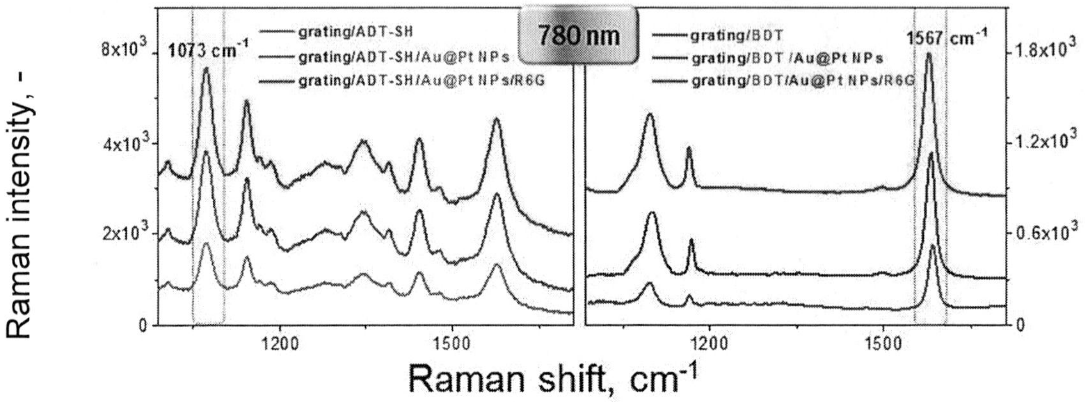

Figure 4 SERS spectra from the samples consisiting of gold gratings grafted through ADT-SH and BDT linkers with Au@PtNPs and additionally covered with R6G molecules

In the SERS experiments, the Au@PtNPs were grafted to the surface of silver gratings through ADT-SH and BDT linkers. Raman excitation was performed using 780 nm. It must be noted that in proposed system two molecules, able to give significant SERS response are presented: the BDT (or ADT-SH) and R6G. As is evident from the **Figure 4** the apparent SERS responses were observed from both molecules. But in the case of SERS excitation with 780 nm, the R6G peaks disappear (or becomes insignificant), and the BDT or ADT-SH response is dominant. This fact indicates that the plasmon intensity is distributed differently: when the excitation wavelength is closed to the plasmon maximum of the both kind of metal nanostructure the plasmon energy is focused in the gap between the Au@PtNPs and gold grating. In the case of ADT-SH, the better Au@PtNPs grafting density and high SERS enhancement factor were achieved. This approach implements the light focussing at the nanometer scale between the gold surface and grafted gold nanourchins.

4. CONCLUSION

The plasmonically coupled SERS substrates were prepared by the creation of gold grating (supported the SPP excitation) followed by the grafting of Au@PtNPs to the gold surface. Prepared structures were checked using the AFM and Raman spectroscopy techniques. Plasmon coupling and SPP-LSP interplay allows to achieve high SERS enhancement, compare to the array of Au@PtNPs, deposited in the dielectric environment or on the metal films. More detailed experiments show that in the case of ADT-SH, the better Au@PtNPs grafting density and high SERS enhancement factor were achieved. Moreover, the produced structure allows SERS measurements with a portable Raman spectrophotometer. From the viewpoint of sensitivity and reproducibility prepared SERS substrates to have key advantages for the various analytical applications.

ACKNOWLEDGEMENTS

This work was supported by the GACR under the project 15-33459A.

REFERENCES

[1] LAL, S., LINK, S., HALAS, N. J. Nano-optics from sensing to waveguiding. *Nature photonics*, 2007. vol. 11, 641.

[2] PEKSA, V., LEBRUSKOVA, P., SIPOVA, H., STEPANEK, J., BOK, J., HOMOLA, J., PROCHAZKA, M. Testing gold nanostructures fabricated by hole-mask colloidal lithography as potential substrates for SERS sensors: sensitivity, signal variability, and the aspect of adsorbate deposition. *Physical Chemistry Chemical Physics*. 2016. 18, no. 29, pp. 19613-19620.

[3] KALACHYOVA, Y., ERZINA, M., POSTNIKOV, P., SVORCIK, V., LYUTAKOV, O. Flexible SERS substrate for portable Raman analysis of biosamples. *Applied Surface Science*. 2018. vol. 458, pp. 95-99.

[4] GUSELNIKOVA, O., POSTNIKOV, P., ERZINA, M., KALACHYOVA, Y., SVORCIK, V., LYUTAKOV, O. Pretreatment-free selective and reproducible SERS-based detection of heavy metal ions on DTPA functionalized plasmonic platform. *Sensors and Actuators B: Chemical*. 2017. vol. 253, pp. 830-838.

[5] KHOURY, C. G., VO-DINH, T. Gold nanostars for surface-enhanced Raman scattering: synthesis, characterization and optimization. *The Journal of Physical Chemistry C*. 2008. vol. 112, no. 48, pp. 18849-18859.

[6] EVCIMEN, N. I., COSKUN, S., KOZANOGLU, D., ERTAS, G., UNALAN, H. E., ESENTURK, E. N. Growth of branched gold nanoparticles on solid surfaces and their use as surface-enhanced Raman scattering substrates. *RSC Advances*. 2015. vol. 5, no. 123, pp. 101656-101663.

[7] DONASONG, H., INAYANG, Y. Solution based, on chip direct growth of three-dimensionally wrinkled gold nanoparticles for a SERS active substrate. Chemical Communications, 2015. vol. 51, no. 1, pp. 213-216.

[8] DUAN, H. W., KUANG, M., WANG, D. Y., KURTH D. G., MOHWALD, H. Colloidally stable amphibious nanocrystals derived from poly {[2-(dimethylamino) ethyl] methacrylate} capping. Angewandte Chemie, 2005. vol. 44, pp. 1717-1720.

[9] XU, J., ZHANG, L., GONG, H., HOMOLA, J., YU, Q. Tailoring plasmonic nanostructures for optimal SERS sensing of small molecules and large microorganisms. *Small*. 2011. vol. 7, no. 3, pp. 371-376.

[10] ZHENG, P., LI, M., JUREVIC, R., CUSHING, S. K., LIU, Y., WU, N. A gold nanohole array based surface-enhanced Raman scattering biosensor for detection of silver (I) and mercury (II) in human saliva. *Nanoscale*, 2015. vol. 7, no. 25, pp. 11005-11012.

[11] KALACHYOVA, Y., MARES, D., JERABEK, V., ULBRICH, P., LAPCAK, L., SVORCIK, V., LYUTAKOV, O. Ultrasensitive and reproducible SERS platform of coupled Ag grating with multibranched Au nanoparticles. *Physical Chemistry Chemical Physics*. 2017. vol. 19, no. 22, pp. 14761-14769.

[12] GUO, Dao-Jun; MIRKHALAF, Fakhradin. *Modification of nano-objects by aryl diazonium salts*. Wiley-VCH: Weinheim, Germany, 2012. pp. 103-124.

[13] ALVAREZ-PUEBLA, R. L., LIZ-MARZAN M., GARCIA DE ABAJO, F. J. Light concentration at the nanometer scale. *The Journal of Physical Chemistry Letters*, 2010. vol. 1, pp. 2428-2434.

[14] RODRIGUES-LORENZO, L., ALVAREZ-PUEBLA, R. A, PASTORIZA-SANTOS, I., MAZZUCCO, S., STEPHAN, O., KOCIAK, M., GARCIA DE ABAJO, F. J. Zeptomol detection through controlled ultrasensitive surface-enhanced Raman scattering. *Journal of the American Chemical Society*. 2009. vol. 131, no. 13, pp. 4616-4618.

MAGNETIC MOLECULARLY IMPRINTED POLYMERS USED FOR STAPHYLOCOCCUS AUREUS ISOLATION AND DETECTION

Jaroslava BEZDEKOVA, Jitka HUTAROVA, Marketa VACULOVICOVA

Department of Chemistry and Biochemistry, Mendel University in Brno, Brno, Czech Republic, EU, jzelnickova@seznam.cz

Abstract

A sensitive detection of microorganisms plays an important role in prevention and identification of infectious and deadly diseases. However, conventional methods of bacterial detection are time-consuming, laborious or expensive. A promising way to overcome these problems is offered by the strategy of molecularly imprinted polymers (MIPs), which uses molecular templates to create selective binding sites in cross-linked polymer layers. By this approach, it is possible to prepare materials tailored for recognition of target analyte taking to account the detection method available. In this work, the novel sensitive method for *Staphylococcus aureus* isolation and detection was investigated. This method utilized the non-covalent molecular imprinting technique employing dopamine as a functional monomer. A layer of molecularly imprinted polymer was created on a surface of magnetic particles (MPs). MPs were chosen due to the ability to pre-concentrate bacteria from large sample volumes and thereby allow for the extremely sensitive detection.

Keywords: Molecularly imprinted polymers, bacteria, dopamine, fluorescence

1. INTRODUCTION

Bacteria belong to a large group of unicellular ubiquitous microorganisms. Due to of their ubiquity, size, and fast growth rates, bacteria rank to the most commonly encountered biological contaminants. Their presence must be monitored and controlled because they can cause serious public health problems [1]. Nowadays, the most common tools for bacterial detection are culture and colony counting methods, polymerase chain reaction, and immunology-based methods as enzyme-linked immunosorbent assay (ELISA) [2]. Unfortunately, these conventional microorganism identification methods have several disadvantages. Counting methods are time-consuming and methods as ELISA use antibodies that are expensive and have low stability [3]. Therefore, there is an effort to develop some alternative method of bacterial detection.

This work is concentrated to detection of bacteria by molecularly imprinted polymers (MIPs). Technique of molecular imprinting uses a template (in this case bacteria) to create selective binding sites in cross-linked polymer [4]. MIPs have a range of advantages, in comparison to other bio-macromolecules as antibody or enzyme, which include high chemical and physical stability, low production costs and possibility of preparation the complementary binding sites for variety types of molecules/particles [5]. Due to of these features, MIPs are ideal choice for a large number of biochemical applications.

For preparation of MIPs in this work, a non-covalent imprinting approach was used. Dopamine (DA) was chosen as a functional monomer because it undergoes an oxidative polymerization under alkaline conditions [6]. Other advantages of using DA are the fact that it is eco-friendly and contains a lot of functional groups that enable the interaction with imprinted bacteria [7]. Layer of molecularly imprinted polymer was prepared on the surface of magnetic particles (MPs). MPs were used because they can be employed for isolation of the analyte from large sample volumes due to high magnetic sensitivity [8]. Gram-positive facultative anaerobic bacterium *Staphylococcus aureus* (*S. aureus*) was chosen as a template. These bacteria cause range of serious infections and therefore, they are a critical threat to public health [9]. Scheme of preparation of *S. aureus* selective MPs is shown in **Figure 1**.

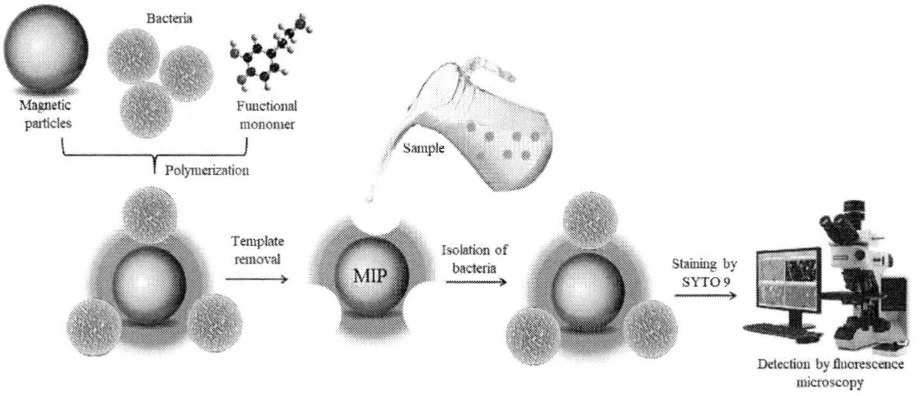

Figure 1 Scheme of preparation of *S. aureus* selective MIPs. The templates (bacteria) were mixed with magnetic particles and functional monomer (dopamine). After polymerization, the templates were removed. Prepared MIPs were used for isolation of analytes (bacteria) from sample. Bound bacteria were stained by fluorescent dye (SYTO 9) and determined by fluorescence microscopy/capillary electrophoresis with laser-inducted fluorescence detection

2. MATERIALS AND METHODS

2.1. Materials

Dopamine hydrochloride, Trizma base, sodium dodecyl sulfate and acetic acid were purchased from Sigma-Aldrich (St. Louis, MO, USA) in ACS purity. Dynabeads MyOne Silane, tetraborate decahydrate and SYTO 9 were obtained from Thermo Fisher Scientific (Waltham, MA, USA).

2.2. Preparation of bacteria

The bacterial strains of *S. aureus* NCTC 8511 and *Enterococcus faecalis* ATCC 11700 (Czech Collection of Microorganisms, Brno, Czech Republic) were cultivated in Muller-Hinton broth (MHB; Oxoid, Hampshire, UK) overnight at 37 °C and 150 rpm. The concentrations of bacterial solutions were determined by optical density at 600 nm and using the MHB the cultures were adjusted to a concentration of 1×10^6 CFU·ml^{-1}. The cells were centrifuged at 8000 rpm for 10 min at ambient temperature and the supernatant was discarded. The cells were re-suspended in the same volume of 20 mM TRIS pH 8.5.

2.3. Preparation of magnetic molecularly imprinted polymers (MPs-MIPs)

The MPs-MIPs were prepared according the method reported by Yang and co-workers with slight modification [10]. Briefly, 50 µl of Fe_3O_4 nanoparticles (40 mg/ml) were washed three times by 200 µl of 20 mM TRIS (pH 8.5). Subsequently, 600 µl of *S. aureus* (1×10^6 CFU/ml) suspended in 20 mM TRIS (pH 8.5) was added to the washed MPs. To prepare non-imprinted polymers (NIPs) that are used as control, 600 µl of 20 mM TRIS (pH 8.5) was added to the washed MPs. The mixtures were mechanically stirred for 2 hours until the MPs were well suspended. Then, 100 µl of dopamine (17.5 mg/ml) suspended in 20 mM TRIS (pH 8.5) was added and the reaction continued over night at laboratory temperature. Next, the product was collected with an external magnetic field and the template was washed out three times by 200 µl of solution containing mixture of 5 % acetic acid and 1 % SDS and once by 200 µl of MilliQ water.

2.4. Binding ability of the imprinted materials

200 µl of S. *aureus* (1×10^6-1×10^2 CFU·ml^{-1}) suspended in 20 mM TRIS (pH 8.5) was added to the MPs-MIPs and mixture was shook for 1 hour. Then, the unbound bacteria were removed and MPs-MIPs were

washed by 200 µl of MilliQ water. 18 µl of sample (MP-MIPs with bound analyte or MPs-NIPs) was mixed with 2 µl of 5 mM SYTO 9 (dissolved in DMSO) and the mixture reacted for 10 minutes. Next, the mixture was analyzed by capillary electrophoresis with laser-induced fluorescence detection (ex. 488 nm, em. 530 nm) or by fluorescence microscopy (ex. 545 nm, em. 610 nm, dichroic mirror 600 nm).

2.5. Detection of binding analytes

Capillary electrophoresis

Samples were analyzed by CE (Agilent technologies, Waldbronn, Germany) with laser-induced fluorescence detector (ZetaLIF, Picometrics, Toulouse, France) with solid-state laser (λ_{em} = 488 nm) as an excitation source. Fused silica capillary with internal diameter of 75 µm and with the total length of 64.5 cm and effective length of 56 cm was used. The sample was introduced hydrodynamically by 50 mbar for 5 s and separation voltage of 25 kV was applied, 20 mM borate buffer pH 9 was used as an electrolyte. Prior to the analysis, the capillary was washed for 60 s by the separation electrolyte.

Fluorescent microscopy

An Olympus IX71 inverted fluorescence microscope was used for imaging of MIPs/NIPs. The used objective was LUCPLFLN 20 X PH. Total magnification was 200×. Detector of emitted light was Hamamatsu CCD ORCA-HR (C4742-95-12HR) with pixels of 1600 × 1200 was used and exposure time was 4 s. Applied filter was TX Red (ex. 545 nm, em. 610 nm, dichroic mirror 600 nm).

3. RESULTS AND DISCUSSION

3.1. Binding properties of the imprinted materials

To assess the binding properties of created MIPs, the binding experiments were carried out. The binding experiments were accomplished with various initial concentrations of *S. aureus*, ranging from 1 × 10^6 to 1 × 10^2 CFU·ml^{-1} and the binding capacities of MIPs against NIPs were compared. Acquired data is shown in **Figure 2**. It was found that the adsorption of bacteria on the surface of MIPs rapidly increased with growing bacterial concentration. On the other hand, the use of higher concentration caused increased binding of analyte on surface of NIPs. NIPs are prepared under same conditions as MIPs. However, NIPs are prepared without presence of the template. NIPs were used as an indicator of non-specific binding. The functional groups in NIPs are randomly organized on the surface and can interact with analyte causing a non-specific interaction. Non-specific interactions are weaker than specific interaction caused by selective binding sites in MIPs created by imprinting of the template molecule. Therefore, NIPs are able to bind lower amount of analyte.

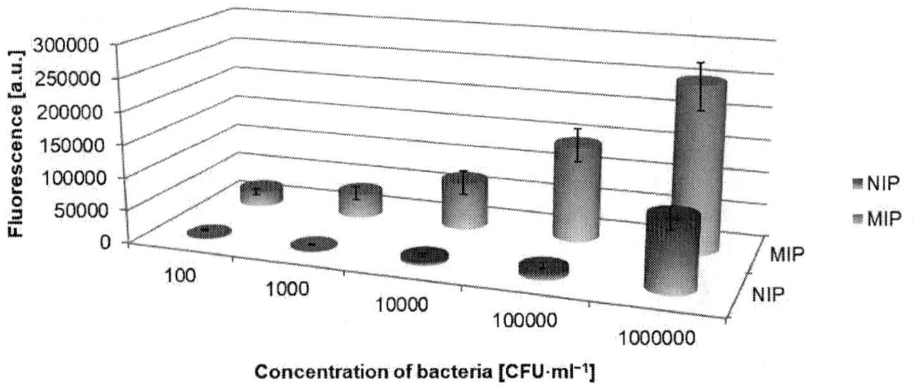

Figure 2 Adsorption isotherm of prepared magnetic MIPs and NIPs selective for *S. aureus*

The binding properties of prepared MPs-MIPs/NIPs were evaluated by two different methods. First of them was fluorescence microscopy. The images of MIPs/NIPs are shown in **Figure 3 A, B**. These images were subsequently evaluated using the BRUKER Molecular Imaging software. From the quantification of the fluorescence images follows that bacterial capture on NIP surface is only 8 % of the capture on imprinted MIP surface.

The second method used was the capillary electrophoresis with laser-induced fluorescence (CE-LIF) (**Figure 3 C**). The MP-MIPs with fluorescently labelled bacteria were directly injected into the capillary and the signal of MIPs as well as NIPs was evaluated based on the electrophoretic peak height. As it was found, the detection using CE-LIF is more sensitive in comparison with fluorescence microscopy.

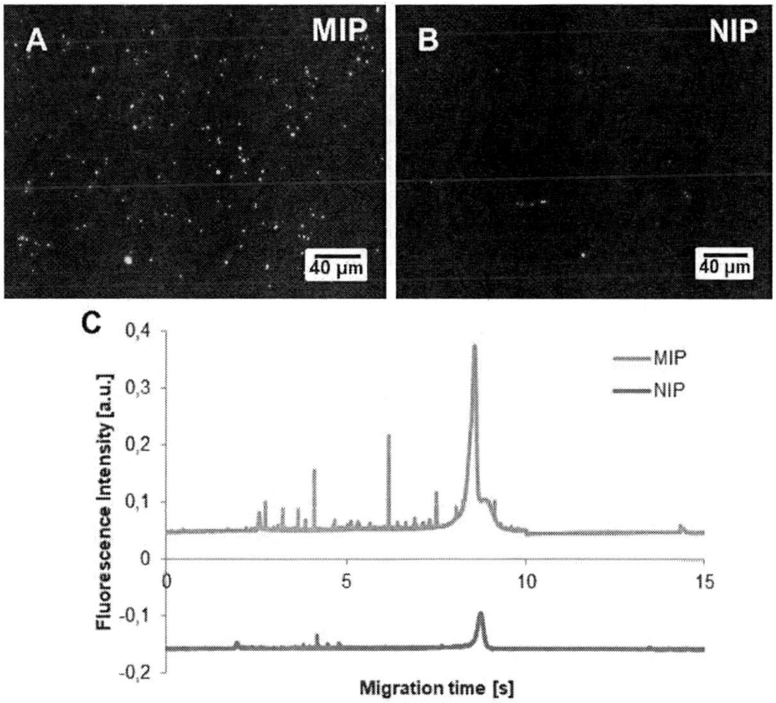

Figure 3 Micrographs of MPs-MIPs (A) and MPs-NIPs (B) with bacteria S. *aureus* (1 × 10^6 CFU·ml^{-1}) stained by fluorescent dye SYTO 9; Electropherograms of MPs-MIPs/NIPs (C)

4. CONCLUSION

In this work, a new alternative method of bacterial detection based on molecular imprinting was developed. The MIPs selective for gram positive bacteria *S. aureus* were prepared on the surface of MPs. Technique of non-covalent molecular imprinting was used in preparation process. Dopamine was used as a functional monomer. Experiment focused on adsorption efficiency of created polymers was carried out. The prepared imprinted polymers exhibited great binding properties under the optimal conditions. Detection of isolated bacteria was performed by fluorescence microscopy and capillary electrophoresis with laser-induced fluorescence detection. These two methods were compared. It was observed that CE-LIF was more sensitive than fluorescence microscopy. By CE-LIF, we were able to detect very low bacterial concentration (1 × 10^2 CFU·ml^{-1}). Moreover, developed method is very simple, specific, low-cost and eco-friendly. The presented technique enables fast and easy bacterial detection and could find potential in food industry. Therefore, future work will be focused on isolation of bacteria from real sample (e.g. milk).

ACKNOWLEDGEMENTS

This work has been supported by the Czech Science Foundation (project No. 16-23647Y) and by the Ministry of Education, Youth and Sports of the Czech Republic under the project CEITEC 2020 (LQ1601).

REFERENCES

[1] MUMTAZ S., WANG S., HUSSAIN S.Z., ABDULLAH M., HUMA Z., IQBAL Z., CRERAN B., ROTELLO V.M., and HUSSAIN I., Dopamine coated Fe3O4 nanoparticles as enzyme mimics for the sensitive detection of bacteria. *Chemical Communications*. 2017. vol. 53, no. 91, pp. 12306-12308.

[2] LAZCKA O., DEL CAMPO F.J., and MUNOZ F.X., Pathogen detection: A perspective of traditional methods and biosensors. Biosensors & Bioelectronics. 2007. vol. 22, no. 7, pp. 1205-1217.

[3] AHMED A., RUSHWORTH J.V., HIRST N.A., and MILLNER P.A., Biosensors for Whole-Cell Bacterial Detection. *Clinical Microbiology Reviews*. 2014. vol. 27, no. 3, pp. 631-646.

[4] CHEN L.X., WANG X.Y., LU W.H., WU X.Q., and LI J.H., Molecular imprinting: perspectives and applications. *Chemical Society Reviews*. 2016. vol. 45, no. 8, pp. 2137-2211.

[5] VASAPOLLO G., DEL SOLE R., MERGOLA L., LAZZOI M.R., SCARDINO A., SCORRANO S., and MELE G., Molecularly Imprinted Polymers: Present and Future Prospective. *International Journal of Molecular Sciences*. 2011. vol. 12, no. 9, pp. 5908-5945.

[6] JIANG J.H., ZHU L.P., ZHU L.J., ZHU B.K., and XU Y.Y., Surface Characteristics of a Self-Polymerized Dopamine Coating Deposited on Hydrophobic Polymer Films. *Langmuir*. 2011. vol. 27, no. 23, pp. 14180-14187.

[7] IQBAL Z., LAI E.P.C., and AVIS T.J., Development of polymer-modified magnetic nanoparticles and quantum dots for Escherichia coli binding test. *Microchimica Acta*. 2012. vol. 176, no. 1-2, pp. 193-200.

[8] PAN J.M., XU L.C., DAI J.D., LI X.X., HANG H., HUO P.W., LI C.X., and YAN Y.S., Magnetic molecularly imprinted polymers based on attapulgite/Fe3O4 particles for the selective recognition of 2,4-dichlorophenol. *Chemical Engineering Journal*. 2011. vol. 174, no. 1, pp. 68-75.

[9] BOLE A.L. and MANESIOTIS P., Advanced Materials for the Recognition and Capture of Whole Cells and Microorganisms. *Advanced Materials*. 2016. vol. 28, no. 27, pp. 5349-5366.

[10] ZHOU W.H., LU C.H., GUO X.C., CHEN F.R., YANG H.H., and WANG X.R., Mussel-inspired molecularly imprinted polymer coating superparamagnetic nanoparticles for protein recognition. *Journal of Materials Chemistry*. 2010. vol. 20, no. 5, pp. 880-883.

DETECTION OF PHYTOESTROGENS BY CAPILLARY ELECTROPHORESIS IN COMBINATION WITH ISOLATION BY MAGNETIC PARTICLES MODIFIED BY MOLECULARLY IMPRINTED POLYMER

Jaroslava ZELNICKOVA[1], Marcela VLCNOVSKA[1], Lucia ZIBEKOVA[1], Kristyna ZEMANKOVA[1], Romana BACOVA[1], Martina KOLACKOVA[1], Marketa VACULOVICOVA[1]

[1]*Department of Chemistry and Biochemistry, Mendel University in Brno, Brno, Czech Republic, EU,* zemankova.kristyna@seznam.cz

Abstract

Phytoestrogens are assumed to play an important role in the prevention of tumor or heart disease, and osteoporosis. Soy food is the most significant source of these compounds present as acetylglucosides, malonylglucosides and/or methyl ethers. The most commonly used methods for phytoestrogen determination include chromatographic or electrophoretic separation techniques.

In this work, the benefits of coupling of capillary electrophoresis with molecularly imprinted polymers for analysis of phytoestrogens are shown. Polydopamin imprinted layer formed on the surface of magnetic particles enables efficient extraction/isolation of model compounds (genistein and biochanin A) from milk sample and subsequent microcolumn separation with absorbance detection enables to distinguish nonspecifically bound interferents.

Keywords: Molecularly imprinted polymers, genistein, biochanin A, micellar electrokinetic chromatography

1. INTRODUCTION

Molecularly imprinted polymers (MIPs) are stable polymers with molecular recognition abilities, provided by the presence of a template during their synthesis and are excellent materials with high selectivity for sample preparation in bioanalytical methods [1]. The use of MIPs has received much attention, not only due to high selectivity, but also for other advantages, such as the low cost of synthesis (compared to antibodies), high mechanical and chemical stability. Thus, MIPs have promising applications in recognition of amino acids and protein [2], important molecules in food [3], pesticides [4], drugs, folic acid and others substances. In more recent applications, magnetic molecularly imprinted polymers (Mag-MIPs) have been used for the recognition of antibiotics in food and environmental samples [2].

Dopamine (DA), (3, 4-dihydroxyphenyl) ethylamine, is considered an important neurotransmitter and significant member ofcatecholamine family. It plays many valuable roles in the functioning of the central nervous system (CNS), hormonal, and renal systems [5]. In addition to biologically significant properties, dopamine also has properties suitable for use in MIPs such as easy polymerization, ability of sorption on almost all inorganic and organic surfaces including superhydrophobic, etc. [6].

Phytoestrogens (PE) and their derivatives have received special attention due to their high concentrations in food, especially in vegetables and soy products. Some of the leading phytoestrogens for cancer treatment or prevention are genistein, coumestrol and glyceollin. Studies imply that countries with high consumption of soy products have a lower incidence of breast cancer in women, this suggests that phytoestrogen intake plays a substantial role in the susceptibility to this cancer type [7]. Soy PE intake may reduce the risk of cardiovascular disease and others chronic diseases such as osteoporosis [8]. Major classes of phytoestrogens include flavonoids, isoflavonoids, lignans and stilbenoids [9].

Selected phytoestrogens, genistein and biochanin A, belong to the group isoflavonoids. In plants, isoflavones are present in a biologically inactive form as glucoside conjugates. In the intestine, by the action of intestinal bacteria, they are hydrolyzed to active forms named aglycones. Taking into account their structural similarity to the natural endogenous hormone 17-estradiol, isoflavones demonstrate the ability of binding to estrogen receptors and consequently the low estrogenic activity. Genistein and biochanin A are considered to be important constituents of animal and human food, as they have a serious influence on human health [10]. Biochanin A and genistein possess anticancer, antioxidant and antiosteoporosis effect [11].

2. MATERIALS AND METHODS

2.1. Chemicals

Dopamine hydrochloride, Trizma® base, phytoestrogen standards (genistein and biochanin A), Sodium dodecyl sulfate and acetic acid were obtained from Sigma-Aldrich (St. Louis, MO, USA) in ACS purity. Sodium tetraborate decahydrate and magnetic particles Dynabeads™ MyOne™ Silane were purchased from Thermo Fisher Scientific (Waltham, MA, USA). Ethanol and methyl alcohol were obtained from Penta (Prague, CZ).

2.2. Capillary Electrophoresis (CE)

Quantification of phytoestrogens, genistein and biochanin A, was performed by CE instrument 7100 (Agilent Technologies, Germany) with absorbance detection at wavelength of 254 nm. Fused silica capillary with an internal diameter of 75 µm, with the total length of 64.5 cm and an effective length of 56 cm was used. The sample was introduced hydrodynamically by 35 mbar for 3 s and a separation voltage of 12 kV was applied. A background electrolyte (BGE) was composed of 30 mM sodium borate buffer, 20 mM sodium dodecyl sulfate (SDS) containing 5 % (v/v) ethanol at pH 9.6. Prior to the analysis, the capillary was washed for 60 seconds using BGE.

2.3. Preparation of Mag-MIPs

Mag-MIPs (**Figure 1**) were prepared from magnetic particles (50 µL), which ware first washed by water (200 µL) for three times. For preparation of MIPs, 400 µL of genistein (1 mg/ml) and 200 µL of 20 Mm TRIS (pH 10) were added. For preparation of non-imprinted polymers (NIPs), 400 µL of 80 % ethyl alcohol and 200 µL of 20 Mm TRIS (pH 10) were added. The mixture was stirred for 1 hour at temperature 45 °C. 100 µL of dopamine (17.5 mg/mL) was added. The mixture was stirred overnight. Then supernatant was removed, MIP/NIP were washed by methanol (200 µL) for three times.

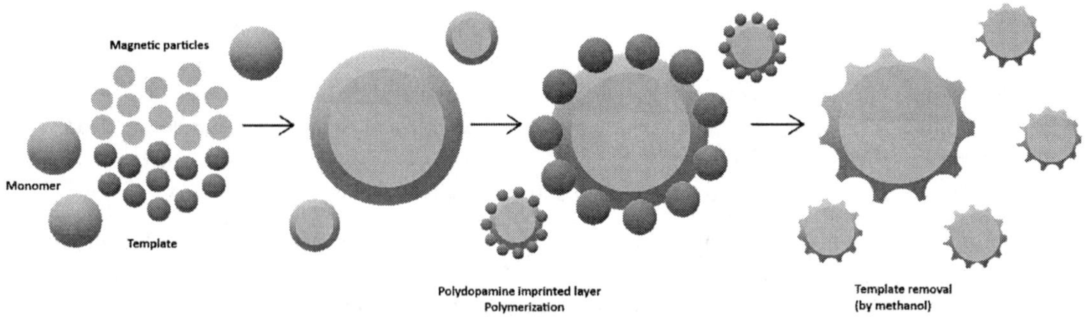

Figure 1 Scheme of preparation of Mag-MIPs

2.4. Sample Preparation

After mag-MIPs/NIPs washing by methanol, genistein (200 µL) was added to mag-MIPs/NIPs at concentrations of 0.125, 0.6, 0.3, 0.015 and 0.0075 mg/ml. Next, biochanin A (0.125 mg/ml), cow milk, milk spiked with genistein (0.125 mg/mL) and soya milk were used as samples. The mixture was stirred for 2 hours at temperature 45 °C. After supernatant removal, mag-MIPs/NIPs were washed with methanol (200 µL) and samples were analyzed by CE.

3. RESULTS AND DISSCUSION

3.1. Detection of phytoestrogens by CE

Genistein extracted from the sample by mag-MIPs was released from the surface by methanol and analyzed by CE with absorbance detection at 254 nm. Standard solution of genistein was added to mag-MIP/NIP at different concentration and a calibration curve was constructed for quantification (**Figure 2**). The difference in signal obtained by mag-MIP and mag-NIP demonstrates that MIP surface is able to recognize the target (template) [12]. Linear dependence in genistein concentration was observed exhibiting the coefficient of determination R^2 = 0.9329.

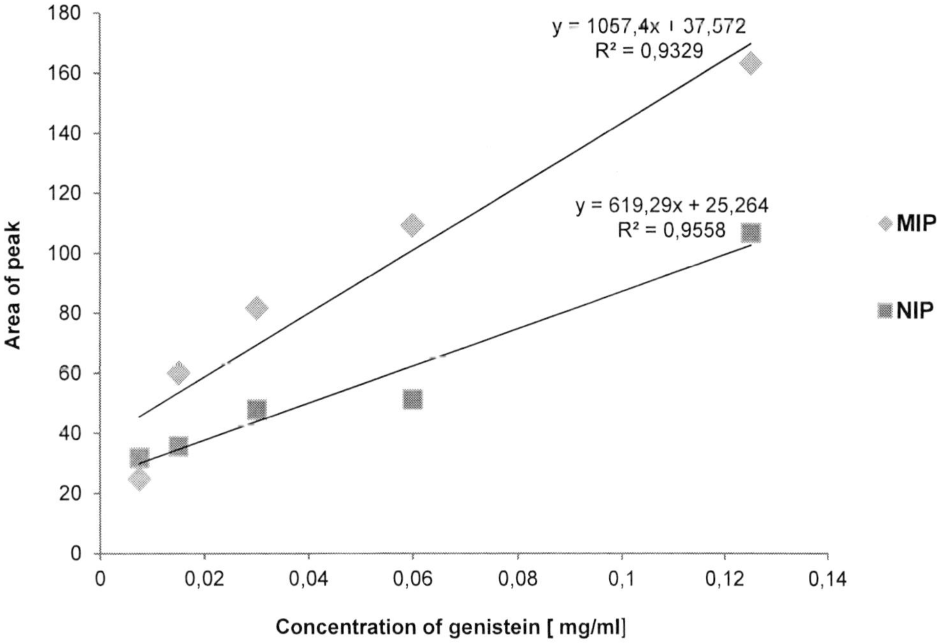

Figure 2 Calibration curve of genistein. CE analysis: BGE - 30 mM sodium borate buffer, 20 mM SDS, 5 % (v/v) ethanol, pH 9.6; detection - 254 nm; hydrodynamic injection - 35 mbar for 3 s; separation voltage - 12 kV

3.2. Real sample analysis

Determination of PEs in real samples is an important goal and therefore, cow milk and soya milk were used as model real samples to test the suggested method. At first, the cow milk was analyzed by CE (pink trace) to demonstrate that not only a number of proteins are present but, also the signal of genistein is very low (marked by asterix). Therefore, the sample preparation using mag-MIPs lead to significant improvement (orange trace) and to identify the genistein signal, the milk sample was spiked by genistein standard solution (blue trace). For comparison, crude soya milk (green trace) was analyzed as well as the soya milk sample purified by mag-MIPs (purple trace). The concentration of PEs (daidzein, glycitein, daidzin etc.) in soya milk is significantly higher compared to cow milk [13] and therefore no spiking was necessary.

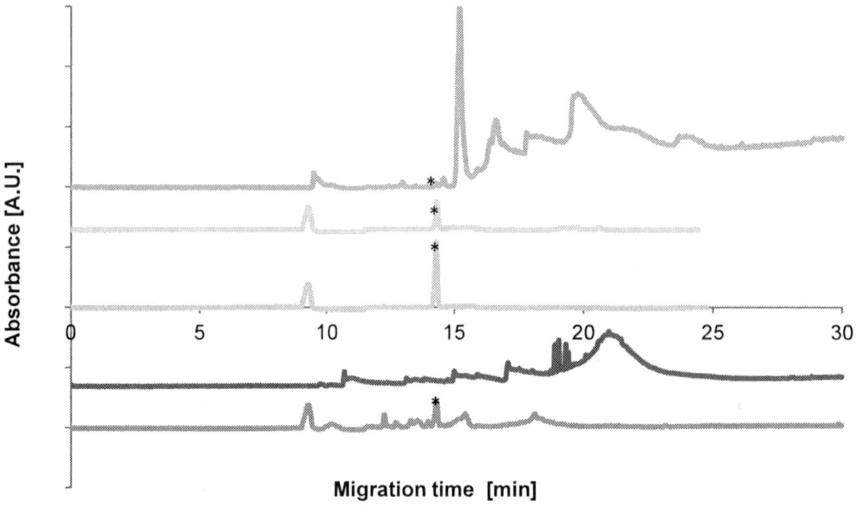

Figure 3 CE real samples (cow and soya milk) purified by mag-MIP. BGE - 30 mM sodium borate buffer, 20 mM SDS, 5 % (v/v) ethanol, pH 9.6; detection - 254 nm; hydrodynamic injection - 35 mbar for 3 s; separation voltage - 12 kV

4. CONCLUSION

In this work, sample purification method using a combination of MIP technology and magnetic particles was developed followed by capillary electrophoretic analysis. Mainly, benefits of simple preparation of polydopamine MIPs combined with effectivity of magnetic particle-based analyte extraction are demonstrated as well as powerful CE-UV/Vis detection. Successful determination of PEs in real samples as cow and soya milk was demonstrated. For comparison, soya milk contains the higher concentration of PEs than cow milk [13]. At first, CE analysis of milk samples was carried out and presence of numerous proteins is evident in electropherograms, however the signal of genistein is very low. The treatment of sample using mag-MIPs caused an increase in the genistein signal.

ACKNOWLEDGEMENTS

The research was financially supported by grant IGA no. TP 4/2017.

REFERENCES

[1] GAMA M.R. and BOTTOLI C.B.G., Molecularly imprinted polymers for bioanalytical sample preparation. Journal of Chromatography B-Analytical Technologies in the Biomedical and Life Sciences, Vol. 1043, No., 2017, pp. 107-121.

[2] UZURIAGA-SANCHEZ R.J., KHAN S., WONG A., PICASSO G., PIVIDORI M.I., and SOTOMAYOR M.D.T., Magnetically separable polymer (Mag-MIP) for selective analysis of biotin in food samples. Food Chemistry, Vol. 190, No., 2016, pp. 460-467.

[3] SUREDA A., SILVA A.S., SANCHEZ-MACHADO D.I., LOPEZ-CERVANTES J., DAGLIA M., NABAVI S.F., and NABAVI S.M., Hypotensive effects of genistein: From chemistry to medicine. Chemico-Biological Interactions, Vol. 268, No., 2017, pp. 37-46.

[4] VILELA M.L.B., WILLINGHAM E., BUCKLEY J., LIU B.C., AGRAS K., SHIROYANAGI Y., and BASKIN L.S., Endocrine Disruptors and Hypospadias: Role of Genistein and the Fungicide Vinclozolin. Urology, Vol. 70, No. 3, 2007, pp. 618-621.

[5] ZAIDI S.A., Development of molecular imprinted polymers based strategies for the determination of Dopamine. Sensors and Actuators B-Chemical, Vol. 265, No., 2018, pp. 488-497.

[6] LIU Y.L., AI K.L., and LU L.H., Polydopamine and Its Derivative Materials: Synthesis and Promising Applications in Energy, Environmental, and Biomedical Fields. Chemical Reviews, Vol. 114, No. 9, 2014, pp. 5057-5115.

[7] MONTES-GRAJALES D., MARTINEZ-ROMERO E., and OLIVERO-VERBEL J., Phytoestrogens and mycoestrogens interacting with breast cancer proteins. Steroids, Vol. 134, No., 2018, pp. 9-15.

[8] CEPEDA S.B., SANDOVAL M.J., RAUSCHEMBERGER M.B., and MASSHEIMER V.L., Beneficial role of the phytoestrogen genistein on vascular calcification. Journal of Nutritional Biochemistry, Vol. 50, No., 2017, pp. 26-37.

[9] RAHEJA S., GIRDHAR A., LATHER V., and PANDITA D., Biochanin A: A phytoestrogen with therapeutic potential. Trends in Food Science & Technology, Vol. 79, No., 2018, pp. 55-66.

[10] CHRZANOWSKA A.M., POLIWODA A., and WIECZOREK P.P., Surface molecularly imprinted silica for selective solid-phase extraction of biochanin A, daidzein and genistein from urine samples. Journal of Chromatography A, Vol. 1392, No., 2015, pp. 1-9.

[11] MA F.-Y., LUO M., ZHAO C.-J., LI C.-Y., WANG W., GU C.-B., WEI Z.-F., ZU Y.-G., and FU Y.-J., Simple and efficient preparation of biochanin A and genistein from Dalbergia odorifera T. Chen leaves using macroporous resin followed by flash chromatography. Separation and Purification Technology, Vol. 120, No., 2013, pp. 310-318.

[12] YE H.L., CHEN X.F., and FENG Z.F., Preparations of magnetic molecularly imprinted polymer for selective recognition and determination of 4-methylimidazole in soft beverage by high performance liquid chromatography. Adsorption Science & Technology, Vol. 35, No. 1-2, 2017, pp. 37-54.

[13] TSANGALIS D., WILCOX G., SHAH N.P., and STOJANOVSKA L., Bioavailability of isoflavone phytoestrogens in postmenopausal women consuming soya milk fermented with probiotic bifidobacteria. British Journal of Nutrition, Vol. 93, No. 6, 2005, pp. 867-877.

COMPARISON OF PDMS MICROTOPOGRAPHICAL PLATES AND PLANAR NANOFIBER SCAFFOLDS FOR IN VITRO CULTIVATION OF CARDIAC CELLS

Josef SKOPALIK[1], Jan SLAVIK[2], Vratislav CMIEL[1], Katerina POLAKOVA[4], Jaroslav PRUCHA[3], Eva GABRIELOVA[4], Larisa BAIAZITOVA[1], Ondrej SVOBODA[1,2], Ivo PROVAZNIK[1]

[1]Department of Biomedical Engineering, Faculty of Electrical Engineering and Communication, Brno University of Technology, Brno, Czech Republic, EU, j.skopalik@gmail.com
[2]Central Institute of Technology (CEITEC), Brno University of Technology, Brno, Czech Republic, EU
[3]Department of Health Care Disciplines and Population Protection, Faculty of Biomedical Engineering, Czech Technical University in Prague, Prague, Czech Republic, EU
[4]Department of Medical Chemistry and Biochemistry, Faculty of Medicine, Palacky University; Olomouc, Czech Republic, EU
[5]Regional Centre of Advanced Technologies and Materials, Department of Physical Chemistry, Experimental Physics and Analytical Chemistry, Faculty of Science Palacky University Olomouc, Czech Republic, EU

Abstract

PDMS microchambers and microchannels were evaluated in many studies of the specific *in vitro* cultivation. Modern microtopographical methods bring new possibilities how to prepare transparent planar blocks with isotropic lamellas. On the other hand, the planar nanofiber sheets are another type of modern culture substrate. Neonatal myocytes and the cardiac muscle cell lineage HL-1 were used for tests of the cell adherence, cell survival and cell orientation on pure or modified polydimethylsiloxane (PDMS) or polycaprolactone (PCL). PDMS with fibronectin coating was shown to be the material with the best cell adherence, viability and isotropic cell orientation from all tested variants.

Keywords: Nanofibers, microlithographical methods, *in vitro* models, tissue engineering, cardiomyocytes

1. INTRODUCTION

The transplantation of cardiomyocytes in the form of a "cell suspension" did not lead to good results in previous animal experiments. To develop successful cardiovascular tissue regeneration methods and also to develop an *in vitro* model mimicking the real cardiac tissue, the preparation of large-scale planar compact cell sheets are needed. The new technologies should be introduced to improve the preparation of the myocardial tissue *in vitro* and to cultivate the typical myocardial structure with an isotropic orientation of cells and with numerous intercellular connections.

The previous experiments on cardiomyocytes carried out over the last 20 years have shown that the microstructure and coating of surfaces of *in vitro* chambers are the key factors which can control the cellular adherence, cellular orientation and cellular connection, and enhance the long-term confluency and maturation of cardiomyocytes. For example, PDMS is a non-toxic material, which was used in many *in vitro* systems as a planar surface, mostly for endothelial cells. Its surface microstructure can be modified to obtain a model mimicking the basal structure of extracellular scaffolds in the real tissue *in vivo*. On the other hand, the PCL prepared by the electrospinning method is another versatile way how to produce a biocompatible material simulating the structure of the extracellular matrix *in vivo* [1, 2]. The purpose of the work presented here was to provide a quantitave comparison of these substrates and their different coatings in terms of the cardiac cell adhesion and survival.

2. METHODS

2.1. Fabrication of microtopographical plates

Creating of PDMS layer needs some solid polymer material as a master substrate. After several tests, silicon (prepared from 6333097 nanopowder - Sigma Aldrich) was chosen as the best master substrate. All steps of the fabrication are illustrated in **Figure 1**. Silicon was deposited from hexamethyldisilazane (HMDS) vapours in N_2 gas at 100 °C for 60 s (for silicon substrate hydrophobicity to increase the photoresist adhesion). Photoresist S1813 (DOW Electronic) was spin-coated on silicon at 1800 RPM for 60 s (final creation of 3 µm thick layer of the photoresist). The phototresist was heated at 115 °C for 60 s. The pattern was exposed to direct writing by DWL at a power of 4.8 mW. The lines obtained were 3 µm wide and distanced 10 µm one from another (or two alternative variants with distanced 4 µm and 30 µm). The photoresist was developed in AZ 726MIF (AZ Electronic Materials USA Corp.). The master silicon with the patterned photoresist was rinsed in DI water and dried under N_2 gas. The master was coated by HMDS vapours in N_2 gas at 100 °C, again. PDMS was prepared by a mixture of silicone elastomer and a silicone elastomer curing agent (Dow Corning Europe S.A.). PDMS elastomers were well mixed and degassed in a vacuum chamber to remove bubbles, poured on HMDS-treated master and cured at 80 °C for 90 min. The PDMS substrate was then peeled off and cut to the shape desired. The final PDMS substrates had lamellas 3 µm tall and 3 µm wide, separated by 4, 10 or 20 µm (**Figure 2**). The final PDMS culture plates had geometrical size of one well in a standard 24-well plate. Each variant of the PDMS plate (variants 3 µm - 4 µm, 3 µm - 10 µm, 3 µm - 20 µm) was also prepared in a subvariant with additional fibronectin coating (F1141, Sigma-Aldrich) or collagen coating (Collagen I Rat, Invitrogen).

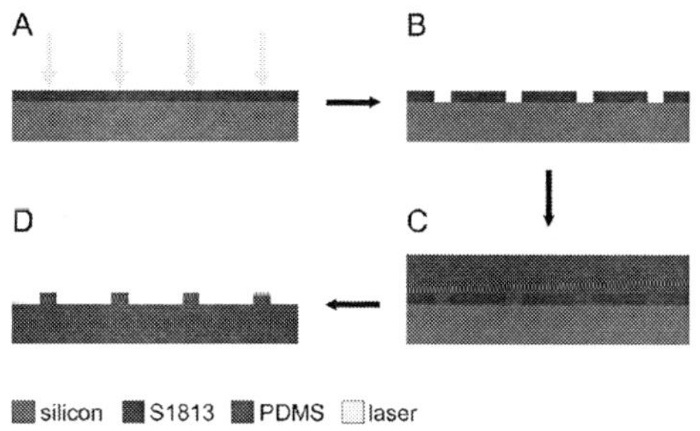

Figure 1 Fabrication process of PDMS substrate with lamellas, (A) silicon substrate coated by S1813 and exposed to laser, (B) development of photoresist, (C) pouring PDMS on master, (D) peeling off PDMS

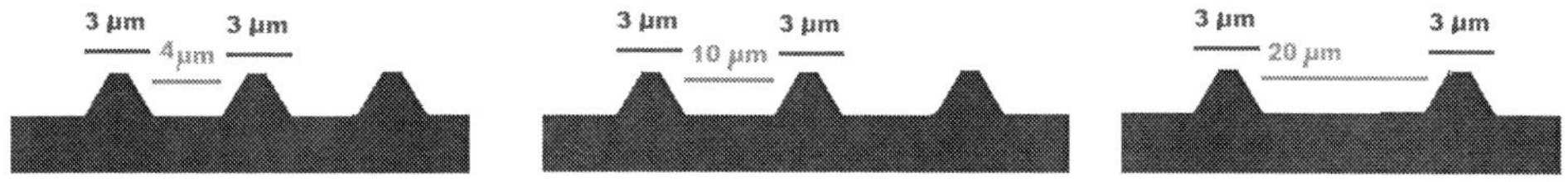

Figure 2 Final shape of PDMS surface prepared for our testing of cell cultivation.

2.2. Fabrication of nanofibers sheets

PCL (Mn = 80 kDa, Sigma-Aldrich) were dissolved in the HFIP for 24 hrs at room temperature and sonicated. The aligned PCL nanofiber yarns were prepared by an enhanced wet–dry electrospinning process - fiber forming technology based on drawing a submicron fiber from a solution or melt. A collection electrode was used for depositing the prepared fibers and preparation of parallel fiber structure with a significant isotropicity nature. Electron microscopy images (by Hitachi SU6600) revealed the structure, where a significant amount of parallel fibres are parallel with one main direction (**Figure 3**). The final nanofiber sheets were adjusted to the size of one well in a standard 24-well plate.

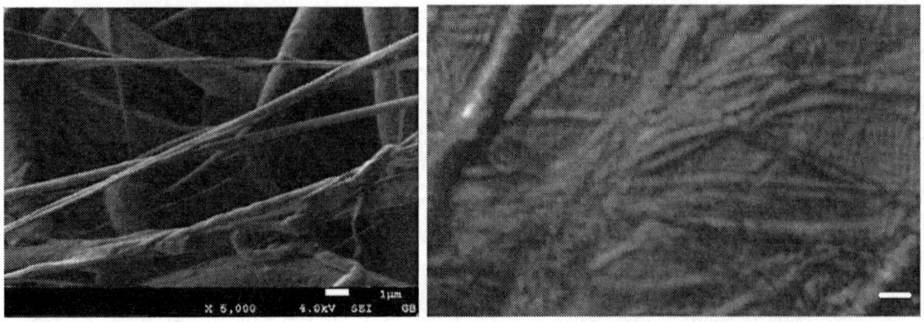

Figure 3 SEM image for control of significant isotropic structure of fibers orientation (scale bar = 1 μm) and illustrative image of isotropy of nanofibers from light microscope (scale bar = 10 μm)

The nanofibers planar plate was also prepared in a subvariant with an additional incubation in the fibronectin solution (F1141, Sigma-Aldrich) or collagen solution (Collagen I Rat, Invitrogen) for 2 hours at 37°C.

2.3. Neonatal myocardial cell preparation and test of culture plates

Neonatal rat cardiomyocytes were isolated from Wistar neonatal rats aged 2-5 days. Hearts were removed and washed in a salt solution containing 20 mM HEPES, 120 mM NaCl, 1 mM NaH_2PO_4, 5.5 mM glucose, 5.4 mM KCl, and 0.8 mM $MgSO_4$ (pH 7.3-7.4). Cardiomyocytes were isolated by the collagenase digestion, after which the cells were resuspended in a medium containing Iscove's modified Dulbecco's medium (IMDM) (Sigma-Aldrich) and Medium 199 (Sigma-Aldrich) (4:1) supplemented with horse serum (10 %) and fetal calf serum (5 %). The separation of non-myocardial cells (heart fibroblasts) and selecting the non-adhesive cells (cardiomyocytes) were done in accordance with protocol presented in [3].

Cardiomyocytes were resupended in the culture medium IMDM and transferred (in form of 100 μL drops) to all types of culture PDMS and PCL plates, which were prepared as described above (Chapters 1.1. and 1.2). Cells were seeded at density 5×10^4 cells/cm^2. The cells were incubated under 21 % O_2 and 5 % CO_2 at 37 °C. The cultivation medium was removed after 48 hrs and replaced by a medium containing IMDM and Medium 199 (4:1) with penicillin. The ratio of the adhered cell and non-adhered cells was quantified using Leica Dmi8 (20 field of view from each sample). The viability of the cells was quantified by using a Calcein viability kit (Biotium Ltd.) and Leica TCS SP8X. Statistical

2.4. HL-1 caridac cell preparation and test of culture plates

HL-1 cells were obtained from Sigma-Aldrich in cryoconserved form. The cells were defrosted, washed and cultivated on a plastic petri dish with fibronectin coating in accordance with the manufacturer´s instruction. After 3 passages, the cells were washed and resuspendend in Claycomb medium (JRH Biosciences) supplemented with 10 % fetal bovine serum (Sigma/Aldrich) and 2 mM L-glutamine (Invitrogen).

HL-1 cells were transferred (by 100-uL drops) to all types of culture microplates, which were prepared as described above (Chapters 1.1. and 1.2). The drops were counted to provide a density 5×10^4 cells/cm^2. The cells were incubated under 21 % O_2 and 5 % CO_2 at 37 °C. The cultivation medium was removed after 48 hrs and replaced by fresh DMEM medium with 5 % FBS (Sigma-Aldrich). The ratio of adhering / non-adhering cells was counted by using Leica Dmi8 or Leica TCS SP8X. The viability and function of the cells were quantified by using a Calcein viability kit (Biotium Ltd.)

2.5. Cell orientation and average isotropy index

The cell cultures on different substrates were recorded by Leica TCS SP8X (fluorescence mode 500/520 nm) and all images of 200 randomly selected cells in each culture was used for the calculation of the "average isotropy index" (AII). This index was computed by Matlab utility (mathematical methods described on example in **Figure 4**).

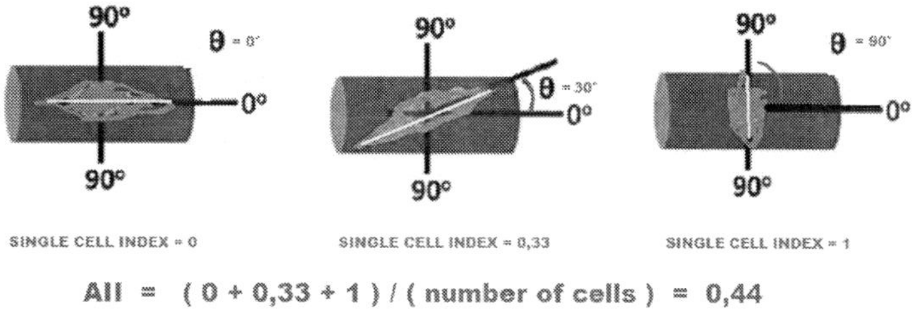

Figure 4 Description of software methods for calculating average isotropy index (AII) for certain set of three cells. Red line (direction 0°) highlights the basal direction of lamellas or main direction of sets of nanofibers.

3. RESULTS

The image of all PDMS substrates displayed a high ratio of adherent cells 24 hrs after seeding. Their viability was higher than 90 % in all variants. The detailed statistical analysis of the effectivity of the cell adherence and AII is summarized in **Table 1**. The most effective adherence and production of an isotropic cell culture was found in substrates "3-10 + fibronectin coating" and "3 µm - 10 µm + collagen coating". For illustrative image of two types of cells in different subvariants of PDMS with coating see **Figures 5 and 6**.

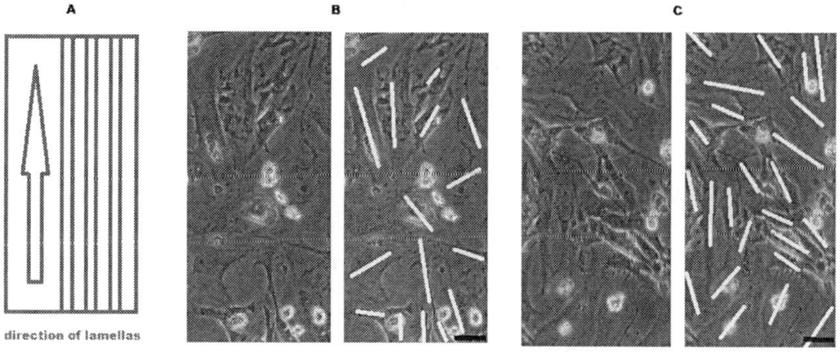

Figure 5 Neonatal myocyte adhesion and orientation on PDMS plate; (A) Direction of lamellas, (B) myocytes on PDMS (variant 3-10) with fibronectin coating (left: wide field microphoto, right: microphoto with marks of main axis of cells), (C) myocytes on collagen (variant 3-10)

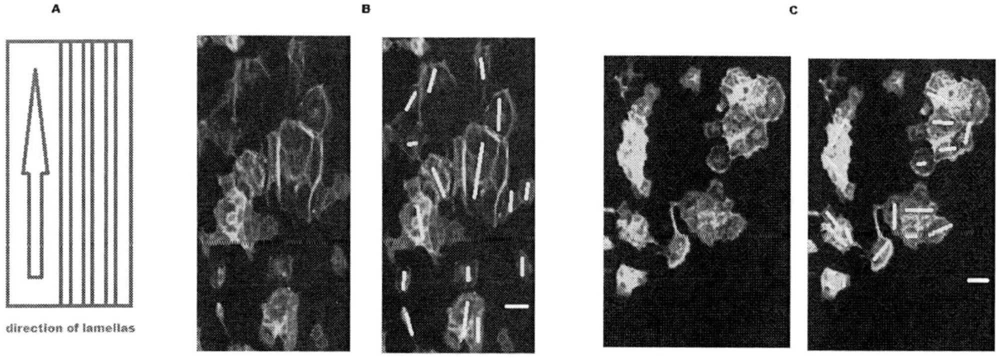

Figure 6 HL-1 myocyte adhesion and orientation on PDMS plate; (A) Direction of lamellas, (B) myocytes on PDMS (variant 3-10) with fibronectin coating (left: wide field microphoto, right: microphoto with marks of main axis of cells), (C) myocytes on control PDMS plate (plate without lamellas - anisotropic orientation of cells)

Table 1 Quantification of cell adhesion and cell isotropic orientation on different PDMS microstructures. Average from 3 independent cell samples.

Profile of plates (lamellas, their distance, their coating)	HL-1 cells		Neonatal myocytes	
	Cell adherence (%)	AII (rel. unit)	Cell adherence (%)	AII (rel. unit)
Triangle basis 3 μm Distance 4 μm	35 ± 5	0.36	12 ± 4	0.52
Triangle basis 3 μm Distance 10 μm	44 ± 8	0.24	18 ± 5	0.33
Triangle basis 3 μm Distance 20 μm	21 ± 6	0.44	14 ± 5	0.43
Triangle basis 3 μm Distance 4 μm + fibronectin	59 ± 16	0.19	36 ± 8	0.16
Triangle basis 3 μm Distance 10 μm + fibronectin	74 ± 15	0.17	38 ± 15	0.14
Triangle basis 3 μm Distance 20 μm + fibronectin	62 ± 20	0.28	37 ± 11	0.24
Triangle basis 3 μm Distance 4 μm + collagen	42 ± 16	0.34	52 ± 8	0.35
Triangle basis 3 μm Distance 10 μm + collagen	40 ± 12	0.31	67 ± 21	0.23
Triangle basis 3 μm Distance 20 μm + collagen	39 ± 11	0.48	59 ± 18	0.38

The microscopic image of all nanofiber substrates displayed a very low ratio of adherent cells 24 hrs after seeding. Their viability was higher than 90 % in all variants. A detailed statistical analysis of effectivity of cell adherence and AII index is summarized in **Table 2**.

Table 2 Quantification of cell adhesion and cell isotropic orientation on different nanofibers

Subvariant of nanofibers	HL-1 cells		Neonatal myocytes	
	Cell adherence (%)	AII (rel. unit)	Cell adherence (%)	AII (rel. unit)
Unmodified	5 ± 2	0.45	2 ± 1	0.52
Fibronectin add.	14 ± 8	0.19	8 ± 3	0.34
Collagen add.	8 ± 1	0.33	4 ± 3	0.29

4. CONCLUSION

The results show that developing of ideal substrates, which would cause the 100 % adhesion and 100 % isotropy of cardiomyocytes after seeding, is still a big challenge. PDMS with fibronectin or collagen coating exerted a significantly higher ratio of adherent cardiomyocytes (both HL-1 and neonatal cells) after seeding, higher than pure PDMS and higher than PCL nanofiber structures. Cardiac cells used in our experiments had a very low adherence to the PCL system compared to data in some recent communications [4,5], where another cells (stem cells or fibroblasts) were tested on similar PCL fibers. Geometrical variants of PDMS lamellas show a significant impact on the isotropy of final cultures of cardiomyocytes, the best variant for effective unification of cell direction in cultures being variant "3-10 + fibronectin" (triangle basis 3 μm, distance 20 μm, fibronectin coating). The positive results will be used for next additional optimising of surfaces created from PDMS and PCL nanofibers in our lab and could accelerate the development of new devices for cardiovascular medicine in near future.

ACKNOWLEDGEMENTS

The work was supported by the project by the project FEKT-S-17-4487 and by the project MPO TRIO FV20422 (Czech Ministry of Industry).

REFERENCES

[1] SILL, Travis; VON RECUM, Horst A. Electrospinning: applications in drug delivery and tissue engineering. Biomaterials, 2008, 29.13: 1989-2006.

[2] RIM, Nae Gyune; SHIN, Choongsoo S.; SHIN, Heungsoo. Current approaches to electrospun nanofibers for tissue engineering. Biomedical materials, 2013, 8.1: 014102.

[3] PSOTOVÁ, Jitka, et al. Chemoprotective effect of plant phenolics against anthracycline-induced toxicity on rat cardiomyocytes. Part III. Apigenin, baicalelin, kaempherol, luteolin and quercetin. Phytotherapy Research: An International Journal Devoted to Pharmacological and Toxicological Evaluation of Natural Product Derivatives, 2004, 18.7: 516-521.

[4] TATU, Carmen Sofica, et al. Adhesion and Secretory Profile of Mesenchymal Stem Cells Upon Contact with Some Biomaterials. REVISTA DE CHIMIE, 2017, 68.9: 2079-2082.

[5] ASADIAN, Mahtab, et. al. Effects of pre- and post- electrospinning plasma treatments on electrospun PCL nanofibers to improve cell interactions. Journal of Physics: Conference Series (Vol. 841, No. 1, p. 012018). IOP Publishing.

PLASMA TREATMENT IN SURFACE MODIFICATION OF SILICA DIOXIDE NANOFIBERS

[1,2,*]Hana TOMÁNKOVÁ, [1,3]Miroslava RYSOVÁ, [1]Adam PAZOUREK

*[1]Institute for Nanomaterials, Novel Technologies and Innovation (CXi), TUL, Liberec, Czech Republic, EU, *hana.tomankova@tul.cz*
[2]Institute of Physiology, First Faculty of Medicine, Charles University, Prague, Czech Republic, EU
[3]Institute of New Technologies and Applied Informatics, FM, TUL, Liberec, Czech Republic, EU

Abstract

In this work, silica dioxide nanofibers (SiNFs) were prepared by sol-gel method followed by electrospinning to be further functionalized by silanization. To improve the chemical modification, plasma treatment of various parameters was applied and the effect on the surface properties before and after plasma activation was studied. The increase of the amine groups accessible on the surface after the combined physico-chemical modification, together with the improved hydrophilicity of the so-modified inorganic nanomats, were demonstrated. Finally, the optimal parameters of the SiNFs plasma activation have been applied and their biocompatibility, as well as the time stability of the physical modification, have been proven.

Keywords: Silica dioxide nanofibers, silanization, plasma treatment

1. INTRODUCTION

Due to their unique physical and chemical properties, especially biocompatibility with low imunogenicity and eventual biodegradability, silica dioxide nanostructures are promising candidates for biomedical applications such as treatment of infection associated surgery [1], tissue engineering [2], or pharmaceutical drug-delivery vehicles [3], [4]. Over the past 10 years, research studies have focused on silica nanofibers (SiNFs) which may be an excellent material for regenerative medicine, especially for wound healing [5], [6]. The potential in biomedicine is often conditioned by a certain surface modification of a carrier material. Different methods to modify surface of the nanofibers for biological application are known. They can be of physical (e.g. plasma or UV modification), chemical (via chemical agents), mechanical or biological character and they can significantly influence wettability, functionality, stability and compatibility towards living cells. Concerning drug carrier applications, reactive surface groups are especially needed for attachment of bioactive molecules. When reinforcing them into nanofibrous matrices, not only dispersion, but also strong covalent attachment to the matrix is required [7]. Primary amine groups are widely used to enhance binding potential of the material, silanization being one of the well-known method for their chemical incorporation [8], [9]. At the same time, SiNFs are known for their hydrophobicity diminishing the efficiency of their secondary chemical treatment. In paralel, plasma treatment is a technique applied not only in the industrial fields [10] but also in the biomedical research to increase the hydrophilicity of materials [11]. The advantage of this process is also that it may enhance cell adhesion and biocompatibility of the material [12] while not affecting its properties due to low depth of ion penetration [13].

In order to improve the efficiency of previously applied simple silanization of SiNFs, we realized the combination of both types of modifications, plasma as the physical and silanization as the chemical one. Different parameters of such treatment were tested (time for plasma modification using different atmospheres) as well as its impact on the morphology of the nanomaterial studied. We also proved no changes in biocompatibility of silylated SiNFs that had been plasma treated before. Finally, time stability of plasma pre-treatment was studied.

2. MATERIALS AND METHODS

2.1. Chemicals

Fluorescein isothiocyanate (FITC, fluorescent label), tetraethyl orthosilicate (TEOS, silica fibers precursor), (3-aminopropyl)triethoxysilane (APTES, the coupling agent) were purchased from Sigma-Aldrich (St. Louis, MO, USA). The remaining chemicals, all of analytical reagent grade, were supplied by Penta (Prague, Czech Republic). Nitrogen gas was provided by Linde Gas, a.s., Czech Republic.

2.2. Preparation of nanofibers

The silica nano-fibrous sheets of surface density of 470 g/m^2 were electrospun under stable conditions, at 22°C (air-conditioned space) using the NanoSpider device (NS 1WS500U, Elmarco Ltd.). The distance between an electrode and a collector was 175 mm. Spinning solution was prepared by sol-gel method using TEOS as a precursor, voltage of 70kV was applied. Finally, electro-spun sheets were thermally treated (180°C) to ensure their prolonged stability.

2.3. Physical modification - Plasma treatment

In order to physically modify the surface, plasma treatment with various parameters was used in this study. Three types of discharge were tested: the Gliding Arc (GA) with 25sccm of air in nozzle and 50mm/s speed of treatment (SurfaceTreat a.s.), the corona discharge (KO) with 800W power generator and 16.7mm/s speed of treatment (CXi, TUL), and the low pressure microwave plasma (MW) with radiofrequency of 2.45 GHz and working pressure of 100 Pa (CXi, TUL). The treatment time was 1, 2 or 5 minutes. Besides the air or nitrogen gas was applied as a working atmosphere.

2.4. Chemical modification - Silanisation

Silica nanofi-brous mats (pristine or plasma modified) were cut into 5 x 5 cm sheets and immersed into 3 % (v/v) APTES solution containing 4 % of water in ethanol (v/v). The pH was adjusted to 5.3 using acetic acid. After 2-hour silanization of nanofibers at a laboratory temperature under continual shaking, the samples were washed by the solution containing 4 % water in ethanol (v/v). Finally, the nanofibers were dried at 110°C/30min. Control samples were prepared evenly but immersed in the mixture water/ethanol (4/96) with pH 5.3 only, without APTES.

2.5. Quantification of amine groups

Quantification of primary amine groups was realised according to the work [14] with some modifications. At first, the 3.2 mM stock solution of FITC in ethanol was prepared. Subsequently, silica nanofibers were immersed into 25 fold diluted (130 µM) solution of FITC and incubated in dark overnight under gentle shaking. After removing all unreacted FITC from the nanofibers by thorough washing with ethanol, 0.2M NaOH was added and the samples were incubated in dark under vigorous shaking until the nanofibers fully dissolved. Fluorescence of prepared solution was measured at F_{ex}: 485/20nm, F_{em}: 528/20nm using Multi-Mode Microplate Reader (BioTek Instruments, USA).

2.6. Morphology characterization

Scanning electron microscopy, SEM (Vega3 SB, TESCAN Ltd.), was used to analyse silica nanofibers prior and after the modification process. Samples were coated with 5nm Au/Pd using a sputter coater equipment (SC7620 Mini Sputter Coater, Quorum Technologies Ltd.). Average fiber diameters and standard deviations (SD) were determined from at least 50 randomly chosen measurements of SEM images using VegaTC software.

2.7. Water contact angle measurement

To study the hydrophilicity/hydrophobicity rate, the contact angle was measured at different positions on the samples at room temperature using a Kruss Drop Shape Analyzer DS4. A total volume of 2 μL of distilled water was dropped on the surface of a dry membrane surface, and the average values of the contact angle were calculated.

2.8. Cytotoxicity testing

Biocompatibility of the functionalized nanofibers was evaluated *in vitro* by the „direct contact" method. Human keratinocytes HaCaT were seeded into 24-well plate (40.000 cells) and cultivated for 48 hours in DMEM medium supplemented with stable glutamine dipeptide alanyl-glutamine and 10 % FBS. Nanofibrous discs of 6 mm in a diameter were placed on the cell layer and covered with a fresh medium. After 24 hours of exposition, the cell viability was evaluated by MTT method. Resultant viability was expressed as a portion of viable cells compared to the cell control (CC). Tested samples were also compared to negative (NC) and positive (PC) controls and untreated silica nanofibers prior and after silanization. Plasma treated samples were tested 15 days after the treatment was performed. Each sample was tested in triplicate.

3. RESULTS AND DISCUSSION

Among the different discharges tested, microwave plasma promoted the highest level of surface activation for further chemical functionalization. Under the conditions tested, this type of plasma discharge allowed an increase of grafted primary amine groups by 22±3 % in comparison to silanization of pristine SiNFs, while the corona and gliding arc offered an increase only by 12±2 % and 4±1 %, respectively (**Figure 1**). For further surface pre-treatment, microwave plasma was tested with or without a specific atmosphere of nitrogen gas. The specific atmosphere did not provide any benefit in activating the surface for further silanization. Not even an increasing treatment time reaching 5 min offered any special improvement in incorporation of amines via the coupling agent reaction. The results of the effect of various time and atmospheres are summarized in **Figure 2a**, where an increase of primary amines is represented by percentage vis-a-vis the functional groups grafted on pristine SiNFs (column labeled „APTES"). The effect of microwave plasma pre-treatment on the silanization was similar using both atmospheres, which is why a hypothesis regarding possible functional group grafting provided via nitrogen gas atmosphere [15] under used conditions was rejected. The functional impact ranged between 46 % and 51 % increasement in both cases. Nonetheless, the smallest values were associated to 1-minute plasma exposure which, regarding economic aspects, were the reason to choose a 2-minute activation procedure for further research. Herewith, based on gathered knowledge, the activation time did not exceed 5 minutes [16].

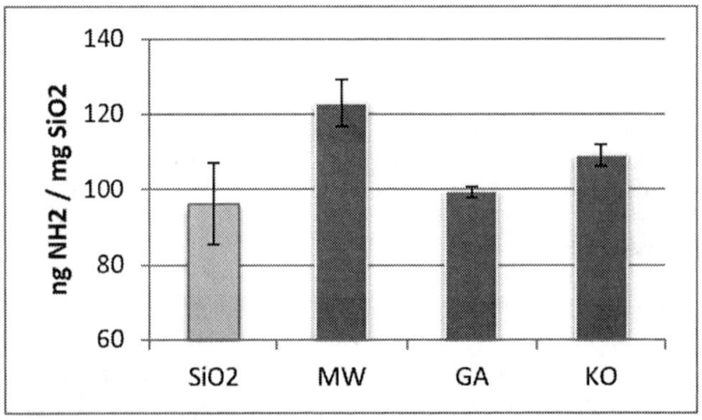

Figure 1 Effect of plasma treatment (with various discharge) on silanization of SiNFs: SiO2-pristine SiNFs, MW-microwave plasma treated, GA-Gliding Arc treated and KO-corona discharge treated SiNFs

Stability of plasma activation on SiNFs was followed during one week. Each activated SiNFs sample was chemically modified after a specific time period ranging from 5 minutes to 7 days. Surface NH_2 quantification showed that 10 % decrease in activation occurred only within the first hours, while the plasma effect on the surface remained stable within the next days (**Figure 2b**). These results are in a good correlation with previous studies dealing with long-term stability of plasma-grafted functional groups [16]. The consequent morphology study of the samples showed that none of the surface modifications (neither via plasma, nor via silanization nor via combination of both) did not generate major mechanical damages on the tested substrate, only unchanged surface morphology was observed, no differences could be seen in means of fibre morphology (scanning electron micrographs are summarized on **Figure 3**). It can be suggested that the material crystallinity was largely unaffected by the applied conditions of plasma activation. The effect of plasma modification on hydrophilisation of the substrate was explored by measuring the water contact angle. Without treatment, the contact angle reached approximately 119.3°, and decreased to 27° after 1-minute plasma exposure (**Figure 4a**), and complete wetting was observed after all longer exposure times. Improved hydrophilicity is supposed to have a positive effect on further silanization.

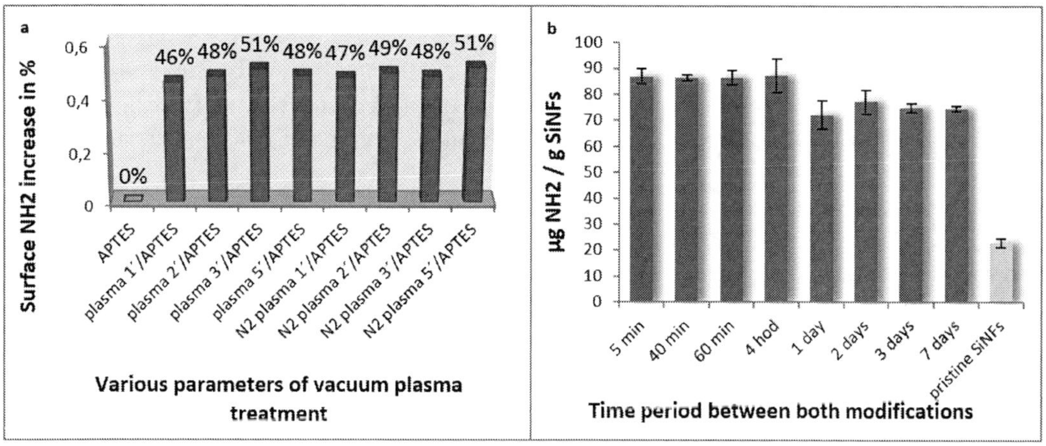

Figure 2 a) Effect of MW plasma treatment (with various time and/or atmosphere) on silanization of SiNFs; **b)** Temporal stability of plasma activated SiNFs via their posttreatment by silanization

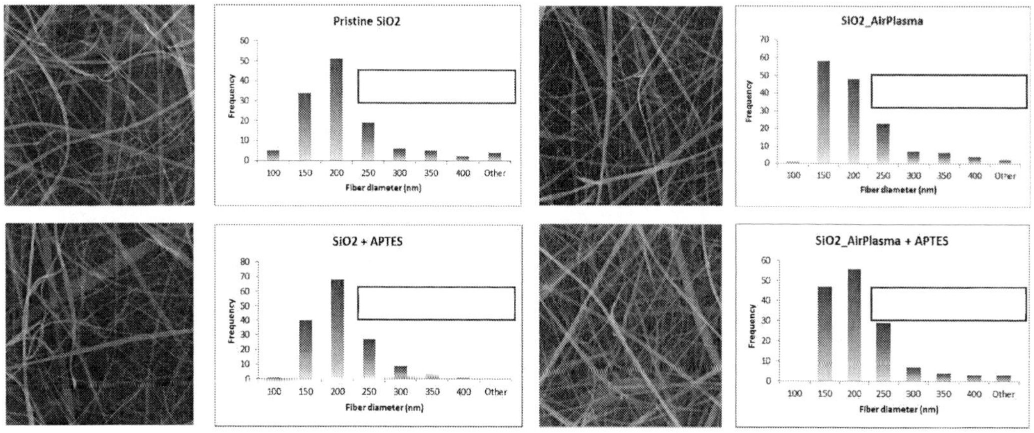

Figure 3 SEM analysis (magnification 10000x) - comparison of pristine SiNFs (SiO_2) with the physically and/or chemically modified nanosheets

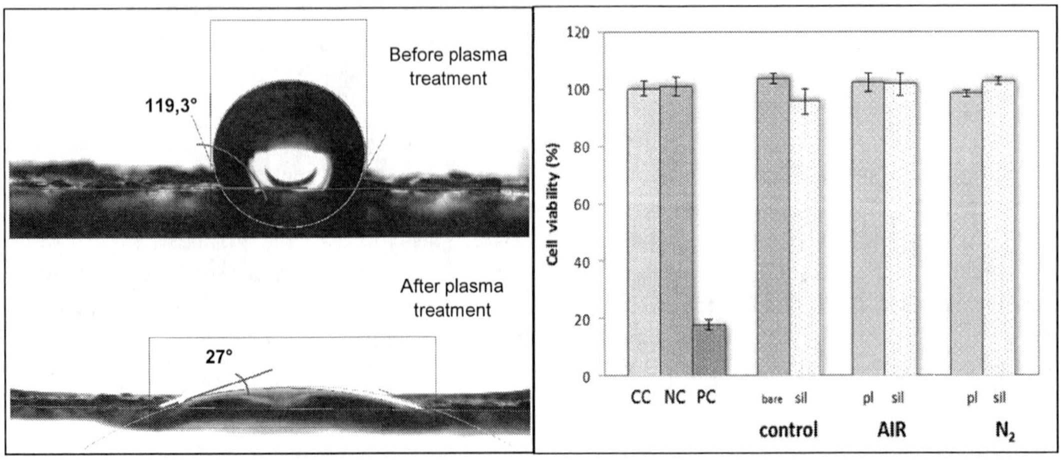

Figure 4 a) Contact angle of pristine and plasma treated SiNFs; **b)** Viability of the HaCaT cells obtained after 24hours exposure to the tested samples. (CC, NC, PC- cell, negative and positive controls; bare NF control - untreated silica NFs; pl- plasma treated, sil - samples silanized after plasma treatment).

Finally, all samples tested in a contact with the HaCaT cell line were proven to be biocompatible, as shown on **Figure 4b**. Viability of all the tested samples exceeded 90 % of the CC. According to these results, plasma treated and silanized silica nanofibers are declared cytocompatible to the Hacat cells according to the EN ISO 10993:5. Exposure to untreated silica nanofibers (NF control) and some plasma treated and silanized samples led to increased viability exceeding 100 % of CC. The best results were obtained by treatment in the air atmosphere, where viability after the plasma treatment was 102.1±3.4 % and 101.4±3.8 % after the subsequent silanization.

4. CONCLUSION

In this study, we showed the effects of plasma treatment on a following chemical modification; silanization of silica nanofibers. We demonstrated that among the discharges tested, microwave plasma provided the best results in improvement of further chemical modification of the surface. At the same time, plasma significantly improved hydrophilicity of the nanofibers. Biocompatibility of silanized SiNFs was unchanged after plasma treatment, and time stability of the the plasma-treated nanofibrous sheets was proven as well. Plasma modification followed by silane coupling agent reaction is thus the most promising method for surface functionalization, not only for industrial polymers but also for engineered materials in biomedicine.

ACKNOWLEDGEMENTS

Research reported in this paper was supported in part by the* TAČR PROJECT *TH02020786 realized at the Institute for Nanomaterials, Novel Technologies and Innovation of the Technical University of Liberec, and by the Ministry of Education, Youth and Sports of the Czech Republic and the European Union - European Structural and Investment Funds in the frames of Operational Programme Research, Development and Education - project Hybrid Materials for Hierarchical Structures (HyHi, Reg. No. CZ.02.1.01/0.0/0.0/16_019/0000843)

REFERENCES

[1] ECHEZARRETA-LOPEZ, MM., DE MIGUEL, T., QUINTERO, F., POU, J. and LANDIN, M. Antibacterial properties of laser spinning glass nanofibers. *International Journal of Pharmaceutics*. 2014. vol. 477, pp. 113-124

[2] KARUNAKARAN, G., SURIYAPRABHA, R., RAJENDRAN, V. and KANNAN, N. Effect of contact angle, zeta potential and particles size on the in vitro studies of Al2O3 and SiO2 nanoparticles. *IET Nanobiotechnol.* 2015. vol. 9, pp. 27-34

[3] MENG, L.-Y., JIANG, W., PIAO, W. and MENG, W. Effect of bio-template on the properties of SiO2/Al2O3 composites for drug delivery. *Journal of Industrial and Engineering Chemistry.* 2016. vol. 37, pp. 14-17

[4] YAN, S., HU, F., HONG, X. and SHUAI, Q. Water-soluble ferrocene complexes (WFCs) functionalized silica nanospheres for WFC delivery in HepG2 tumor therapy. *Materials Science and Engineering: C.* 2018, vol. 90, pp. 397-406

[5] XUE, J., XIE, J., LIU, W. and XIA, Y. Electrospun Nanofibers: New Concepts, Materials, and Applications. *Acc. Chem. Res.* 2017, vol. 50, pp. 1976-1987

[6] TOMANKOVA, H., ROTKOVA, J. and RYSOVA, M. Inorganic Silica Nanofibers in a Role of Bioactive Compound Carrier. 7th International Conference on Nanomaterials - Research and Application. 2015

[7] BERTÓTI, I., MOHAI, I., MOHAI, M. and SZEPVOLGYI, J. Surface modification of multi-wall carbon nanotubes by nitrogen attachment. *Diamond and Related Materials.* 2011, vol. 20, pp. 965-968

[8] LUNG, C. Y. K. and MATINLINNA, J. P. Aspects of silane coupling agents and surface conditioning in dentistry: An overview. *Dental Materials.* 2012, vol. 28, pp. 467-477

[9] MATINLINNA, J. P. and VALLITTU, P. K. Bonding of resin composites to etchable ceramic surfaces - an insight review of the chemical aspects on surface conditioning. *Journal of Oral Rehabilitation.* 2007, vol. 34, pp. 622-630

[10] JIMENEZ, M., HAMZE, H., ALLION, A., RONSE, G., DELAPLACE, G. and TRAISNEL, M. Antifouling Stainless Steel Surface: Competition between Roughness and Surface Energy. Materials Science Forum. Available from 2012 at https://www.scientific.net/MSF.706-709.2523

[11] CHERUTHAZHEKATT, S., CERNAK, M., SLAVICEK, P. and HAVEL, J. Gas plasmas and plasma modified materials in medicine. *Journal of Applied Biomedicine.* 2010, vol. 8, pp. 55-66

[12] DORRAKI, N., SAFA, N. N., JAHANFAR, M., GHOMI, H. and RANAEI-SIADAT, S.-O. Surface modification of chitosan/PEO nanofibers by air dielectric barrier discharge plasma for acetylcholinesterase immobilization. *Applied Surface Science.* 2015, vol. 349, pp. 940-947

[13] TAACA, K. L. M. and VASQUEZ, M. R. Hemocompatibility and cytocompatibility of pristine and plasma-treated silver-zeolite-chitosan composites. *Applied Surface Science.* 2018, vol. 432, pp. 324-331

[14] RITTER, H. and BRUHWILER, D. Accessibility of Amino Groups in Postsynthetically Modified Mesoporous Silica. *J. Phys. Chem. C.* 2009, vol. 113, pp. 10667-10674

[15] SAKA, C. Overview on the Surface Functionalization Mechanism and Determination of Surface Functional Groups of Plasma Treated Carbon Nanotubes. *Critical Reviews in Analytical Chemistry.* 2018, vol. 48, pp. 1-14

[16] MERENDA, A., LIGNERIS, E., DES SEARS, K., CHAFFRAIX, T., MAGNIEZ, K., CORNU, D., SCHUTZ, J.A. and DUMEE, L.F. Assessing the temporal stability of surface functional groups introduced by plasma treatments on the outer shells of carbon nanotubes. *Scientific Reports.* 2016, vol. 6, pp. 31565

ORGANOSILICON PLASMA POLYMERS BASED ON TRIMETHYLSILYL ACETATE MONOMER

Štěpánka KELAROVÁ, Vojtěch HOMOLA, Vilma BURŠÍKOVÁ

Department of Physical Electronics of Masaryk University, Brno, Czech Republic, EU
bittstep@gmail.com, vojhom@outlook.cz, vilmab@physics.muni.cz,

Abstract

Plasma-polymerized organosilicon coatings have been playing an important role in many research studies due to their wide range of applications. These materials can be used in industry as transparent wear resistant layers, protective hard a-Si:C:H or SiO2-like films, corrosion protection coatings, barrier films etc. Plasma-polymerization using organosilicon precursors has been investigated lately for development of unique thin film materials for bioapplications as well, including surface coating of surgical and dental implants to improve their biocompatibility. In the present work, low pressure RF capacitively coupled discharge was used to deposit thin solid films suitable for medical application from mixture of trimethylsilyl acetate (TMSA) and oxygen. Since resistance to liquid environment is the critical parameter that determines usability of the material for bioapplications, the deposited TMSA plasma polymers were exposed to water for 48 hours. The degree of stability varied with deposition parameters as well as other properties of organosilicon plasma polymers. This study summarizes changes of chemical structure and mechanical properties of resulting coatings in dependence on the ratio of TMSA and oxygen flow rates during the deposition process. Results of water stability tests included in present study proved that it is possible to prevent delamination of the coating during exposition to the aqueous environment by appropriate choice of discharge parameters.

Keywords: Plasma polymers, trimethylsilyl acetate, PECVD, FTIR, microindentation, confocal microscopy, ellipsometry

1. INTRODUCTION

Thin solid films based on organosilicon monomer prepared by PECVD technology have been investigated intensively for many industrial applications such as corrosion protection coatings, transparent wear resistant layers, protective hard and antiscratch coatings for plastics, barrier films for pharmaceutical packaging etc. [1]. Organosilicon plasma polymers are very perspective for medical applications as well. It is possible to create organosilicon thin films suitable for protection of medical implants [2,3], organosilicon antibacterial coatings doped by silver nanoparticles [4] or materials enhancing osseointegration for dental implants [5]. Properties of plasma polymers (chemical composition, mechanical properties, stability in different environments etc.) are directly dependent on discharge parameters. These parameters (e.g. flow rates of the monomer and carrier gases, gas pressure, bias voltage etc.) can be varied to adapt resulting coatings for a particular purpose [1-7].

This study deals with characterization of organosilicon coatings deposited in RF glow discharge from gaseous trimethylsilyl acetate (TMSA) and oxygen. The aim of present work is to obtain water-resistant coatings suitable for bioapplications containing C=O functional groups which are presented in TMSA itself [8]. Changes in mechanical and chemical properties in dependence on the ratio of TMSA and O_2 flow rates were investigated as well as stability in water for 48 hours at 37 °C.

2. EXPERIMENTAL DETAILS

Organosilicon thin films were prepared in RF glow capacitively-coupled discharges at low pressures (20÷ 40 Pa) from mixture of TMSA ($CH_3CO_2Si(CH_3)_3$) and oxygen in a parallel plate reactor. The bottom electrode served as the substrate holder and it was coupled to RF generator (13.56 MHz) via a blocking capacitor. Silicon substrates were placed on the bottom electrode, the RF voltage of which was superimposed with a negative DC self-bias. The supplied power was kept at 50 W for all depositions and flow rates of TMSA and O_2 were changed. Changes of the flow rates are represented by the O_2 flow to TMSA flow rate ratio *R*.

Properties of the prepared films were studied by several characterization methods. The chemical composition of the resulting coatings was investigated by Fourier transform infrared spectroscopy (FTIR) using spectrometer Bruker Vertex 80v, in range from 370 cm^{-1} to 7500 cm^{-1} with 500 scans and resolution of 8 cm^{-1}. Thickness of each coating was measured using profilometer Bruker DektaXT immediately after deposition. Measurements of Martens hardness, elastic modulus and indentation hardness were realized using Fisherscope H100C microindentor. Stability of TMSA-based coatings in water for 48 hours at 37 °C was examined by observation of changes in surface structure using confocal laser microscope LEXT OLS4000 3D. The thickness loss caused by exposition to the liquid environment was determined by non-destructive ellipsometric method. All ellipsometric data were measured by Jobin Yvon UVISEL equipment at angle of incidence 65° in the spectral range from 1 eV to 5.5 eV. Thicknesses of plasma polymers were determined from fitting these data in program newAD by using PJDOS dispersion model [9]. Analyzed plasma polymers were immersed in liquid immediately after deposition.

3. RESULTS AND DISCUSSION

Chemical composition of TMSA plasma polymers was determined from baseline-corrected absorbance normalized to the thickness of the film in MIR spectral range (**Figure 1**) according to the available litarature [10-13]. Measured IR spectra (**Figure 1**) showed strong absorptions from 1035 cm^{-1} to ~1200 cm^{-1} in all studied TMSA-based thin films. These absorption peaks are related to the Si-O-C stretching vibrations in a continuous random network [10]. Structure of plasma polymers created using ratio of O_2/TMSA flow rates $R \geq 3$ is very similar to coatings based on siloxane chains [11-13]. Presence of Si-O-Si stretching vibrations typical for siloxane units [11-13] in this spectral range can not be excluded. However according to the chemical structure of TMSA monomer [8], Si−O−C presented in skeleton of original monomer are assumed to be predominant. Absorption peak at ~1265 cm^{-1} was identified as deformation mode of CH_3 in Si−CH_3 group [11,12]. Medium absorptions presented at lower wavenumbers (spectral range from 750 cm^{-1} to 950 cm^{-1}) including peaks at 800 cm^{-1}, 840 cm^{-1} and 890 cm^{-1} are probably related to CH_3 rocking in Si−CH_3, to asymmetric stretching of Si−C/Si−O bond and CH_3 rocking mode in $Si(CH_3)_{x=2,3}$ groups [11-13]. Deformation modes of Si−O−Si may influence absorptions at 800 cm^{-1} and 890 cm^{-1} as well [12]. Content of Si−CH_3 in analyzed samples was confirmed by presence of absorption peaks typical for CH_3 stretching modes at 2880 cm^{-1} and 2960 cm^{-1} [10-13]. Symmetric and asymmetric stretching modes of CH_2 were determined in MIR spectra as well [10-12]. Since values of normalized absorbance in region of wavenumbers near 1400 cm^{-1} related to Si−CH_2−Si [11,13] can be considered negligible, significant amount of carbon is presumably bonded in hydrocarbon chains. According to the IR spectra (**Figure 1**), TMSA-based plasma polymers contain C=O functional groups characterized by absorption at 1720 cm^{-1} [10], however this peak has relatively low intensity in comparison with CH_x vibrations. All investigated TMSA coatings include amount of hydrogen bonded to silicon which has characteristic peaks in region 2000-2450 cm^{-1} [10,14]. The absorption band spread in region above 3050 cm^{-1} is related to OH groups [10-13].

Summarizing main trends observable in IR spectra, it is possible to conclude that amount of carbon bonded in Si-CH_3 and hydrocarbon chains significantly increased with decreasing ratio of O_2/TMSA flow rates *R* (increasing amount of TMSA monomer) used during deposition process. On the contrary, reaching the values of $R \geq 3$ amount of CH_x bonds rapidly decreased. Their absorptions were overlapped by spread peak of OH groups (**Figure 1**). These OH groups can be associated with silanols (other peaks at 920 cm^{-1} and 1190 cm^{-1}) [12] and partially bonded in adsorbed water molecules (OH stretching mode at ~3400 cm^{-1} and weak peak of bending mode at 1630 cm^{-1}) [11]. At the same time, absorptions in region from 1035 cm^{-1} to ~1200 cm^{-1}

related to Si−O−C/Si−O−Si network became more predominant (**Figure 1**). Changes in spectral region inluding Si-H bonds were observed as well: spread absorption around 2190 cm^{-1} transformed into narrow peak at 2343 cm^{-1} usually presented hexamethyldisiloxane coatings [14].

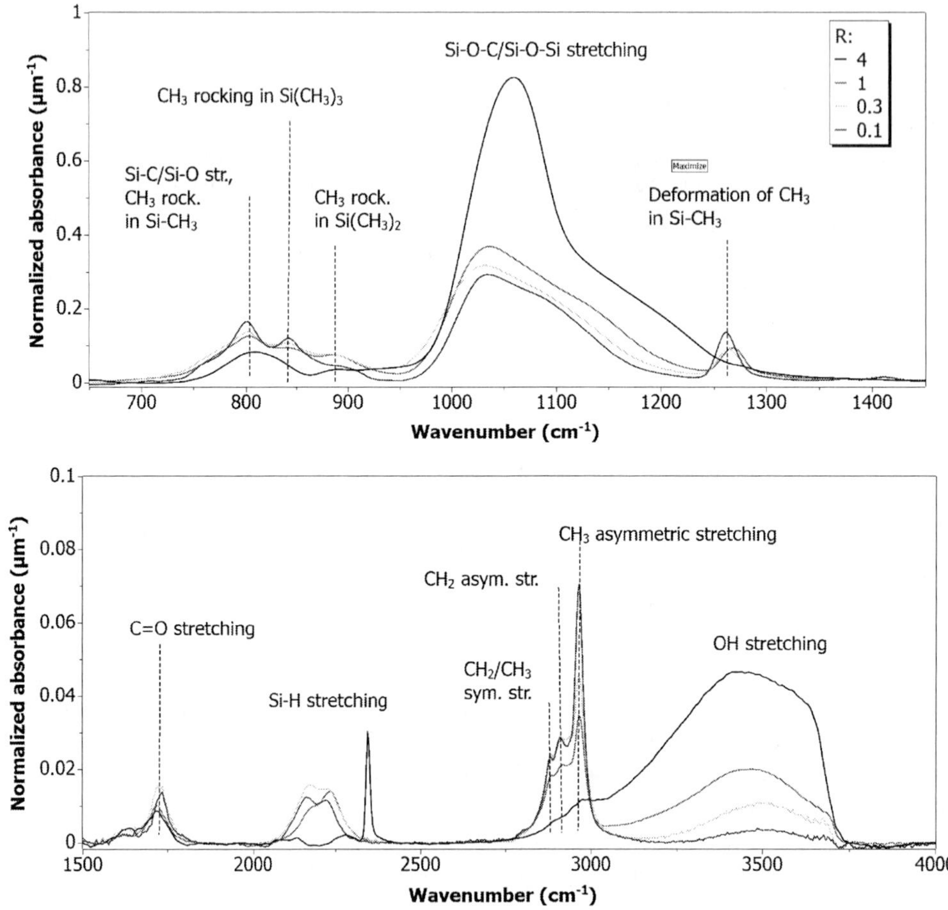

Figure 1 Infrared spectra of several TMSA plasma polymers prepared using different ratio R of O_2 and TMSA flow rates

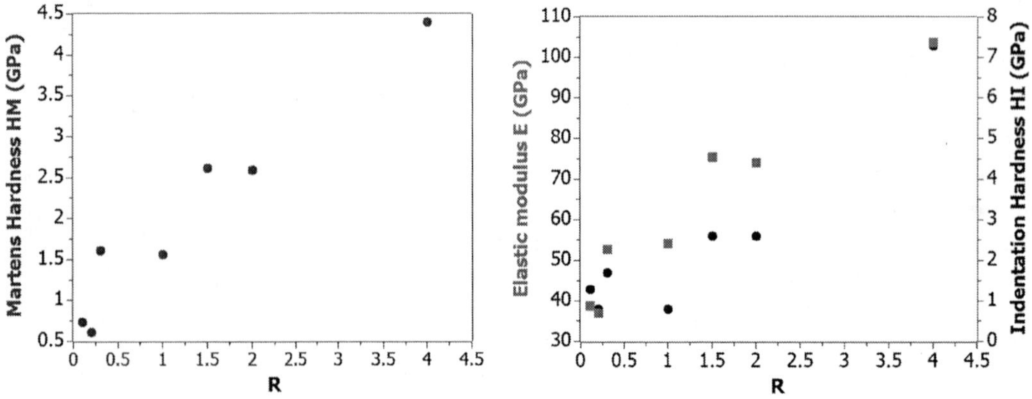

Figure 2 Mechanical properties of TMSA coatings in dependence on parameter R

Changes in chemical structure of deposited coatings described using IR spectroscopy had significant impact on mechanical properties of TMSA thin films. Trends of Martens hardness *HM*, elastic modulus *E* and indentation hardness *HI* in dependence on the parameter *R* are shown in **Figure 2**. All dependences shown in **Figure 2** have an increasing character with increasing ratio *R*.

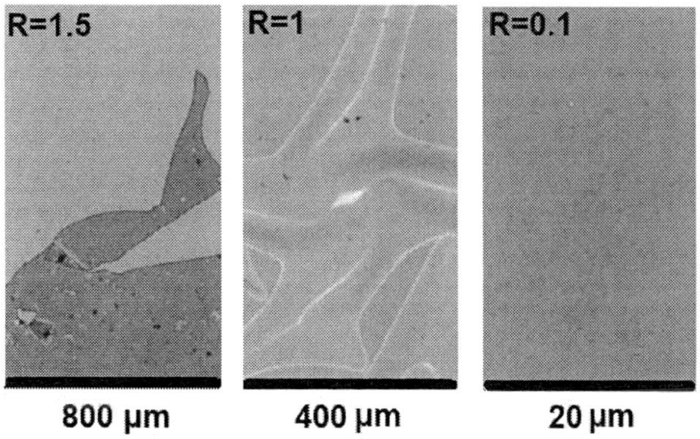

Figure 3 Pictures of chosen TMSA coatings after immersion in water taken by confocal microscope

The stability of TMSA-based plasma polymers in water for 48 h varied with parameter *R* as well as mechanical properties. Thin films prepared by using $R > 1$ were almost completely delaminated (**Figure 3**), while coatings deposited using high amount of TMSA ($R < 1$) were considered to be stable. Surfaces of these plasma polymers after immersion in water stayed smooth without defects and thickness losses determined by ellipsometry were at low values of ~3 %.

4. CONCLUSION

Properties of TMSA-based plasma polymers prepared in RF glow discharge were investigated by several characterization methods. Chemical structure of deposited coatings is directly linked to settings of flow rates of TMSA monomer and oxygen during deposition process. Changes in chemistry of plasma-polymerized TMSA influences mechanical properties and stability in water. Results of water stability tests included in present study proved that it is possible to prevent delamination of the coating during exposition to the aqueous environment by appropriate choice of TMSA and O_2 flow rates. Water-resistant TMSA coatings have a great potential for bioapplications, therefore our further research will be oriented in this direction.

ACKNOWLEDGEMENTS

This research has been supported by the project CZ.1.05/2.1.00/03.0086 funded by European Regional Development Fund and project LO1411 (NPU I) funded by MEYS of Czech Republic and by the project Brno Ph.D. Talent.

REFERENCES

[1] ZAJÍČKOVÁ, L., BURŠÍKOVÁ, V., KUČEROVÁ, Z., FRANCLOVÁ, J., SŤAHEL, P., PEŘINA, V., MACKOVÁ, A. Organosilicon thin films deposited by plasma enhanced CVD: Thermal changes of chemical structure and mechanical properties. *Journal of Physics and Chemistry of Solids.* 2007. vol. 68, no. 5-6, pp. 1255-1259

[2] VOGEL, K., WESTPHAL, N., SALZ, D., THIEL, K., WITTIG, L., CIACCHI, L.C., GRUNWALD, I. Dental implants coated with a durable and antibacterial film. *Surface Innovations.* 2015. vol. 3, no. 1, pp. 27-38

[3] BATORY, D., JEDRZEJCZAK, A., KACZOROWSKI, W., KOLODZIEJCZYK, L., BURNAT, B. The effect of Si incorporation on the corrosion resistance of a-C:H:SiOx coatings. *Diamond & Related Materials.* 2016. vol. 67, pp. 1-7

[4] SAULOU, C., DESPAX, B., RAYNAUD, P., ZANNA, S., SEYEUX, A., MARCUS, P., AUDINOT, J.N., MERCIER-BONIN, M. Plasma-Mediated Nanosilver-Organosilicon Composite Films Deposited on Stainless Steel: Synthesis, Surface Characterization, and Evaluation of Anti-Adhesive and Anti-Microbial Properties on the Model Yeast Saccharomyces cerevisiae. *Plasma Process. Polym.* 2012. vol. 9, no. 3, pp. 324-338

[5] HYAKAWA, T., YOSHINARI, M., NEMOTO, K. Characterization and protein-adsorption behavior of deposited organic thin film onto titanium by plasma polymerization with hexamethyldisiloxane. *Biomaterials.* 2004. vol. 25, no. 1, pp. 119-127

[6] WRÓBEL, A. Aging process in plasma-polymerized organosilicon thin films. *Journal of Macromolecular Science: Part A - Chemistry.* 1985. vol. 22, no. 8, pp. 1089-1100

[7] BLANCHARD, N.E., NAIK, V. V., GEUE, T., KAHLE, O., HEGEMANN, D., HEUBERGER, M. Response of plasma-polymerized hexamethyldisiloxane films to aqueous environments. *Langmuir.* 2015. vol 31, no. 47, pp. 12944-12953

[8] BALL, I.K., HARRISON, P.G., TORR, A. FTIR study of thermally induced transformations of trimethylsilylmethyl acetate in the temperature range 623-813 K. *Journal of Organometallic Chemistry.* 2000. vol. 599, no. 2, pp. 185-194

[9] FRANTA, D., NEČAS, D., ZAJÍČKOVÁ, L. Models of dielectric response in disordered solids. *Optics Express.* 2007. vol. 15, no. 24, pp. 16230-16244

[10] CECH, V., ZEMEK, J., PERINA, V. Chemistry of Plasma-Polymerized Vinyltriethoxysilane Controlled by Deposition Conditions. *Plasma Process. Polym.* 2008. vol. 5, no. 8, pp. 745-752

[11] TSANKOV, D., RADEVA, E., HINRICHS, K., RÖSELER, A., KORTE, E.H. Infrared spectroscopic ellipsometry and atomic force microscopy study of plasma polymerized hexamethyldisiloxane layers post-treated by NH3 plasma. *Thin solid films.* 2005. vol. 476, no. 1, pp. 174-180

[12] CASSERLY, T.B., GLEASON K.K. Chemical Vapor Deposition of Organosilicon Thin Films from Methylmethoxysilanes. *Plasma Process. Polym.* 2005. vol. 2, no. 9, pp. 679-687

[13] Borvon, G., Goullet, A., Mellhaoui, X., Charrouf, N., Granier, A. Electrical properties of low-dielectric-constant films prepared by PECVD in O_2/CH_4/HMDSO. *Materials Science in Semiconductor Processing.* 2003. vol 5, no. 2-3, pp. 279-284

[14] MURESAN, M.G., CHARVÁTOVÁ CAMPBELL, A., ONDRAČKA, P., BURŠÍKOVÁ, V., PEŘINA, V., POLCAR, T., REUTER, S., HAMMER, M.U., VALTR, M., ZAJÍČKOVÁ, L. Protective double-layer coatings prepared by plasma enhanced chemical vapor deposition on tool steel. *Surface & Coatings Technology.* 2015. vol. 272, pp. 229-238

PREPARATION OF ANTIBACTERIAL ELECTROSPUN SCAFFOLD FOR SKIN CELLS CULTURE

[1]Emi GOVORCIN BAJSIC, [1]Matea VECERIC,[2] Emilija ZDRAVEVA, [2]Budimir MIJOVIC, [3]Tamara HOLJEVAC GRGURIC, [4]Massimo UJCIC, [5]Mirna TOMINAC TRCIN, [6]Igor SLIVAC

[1]University of Zagreb, Faculty of Chemical Engineering and Technology, Zagreb, Croatia, EU egovor@fkit.hr
[2]University of Zagreb, Faculty of Textile Technology, Zagreb, Croatia, EU, budimir.mijovic@ttf.hr
[3]University of Zagreb, Faculty of Metallurgy, Sisak, Croatia, EU, tholjev@simet.hr
[4]Institute of Chemical Processes Fundamentals of the Czech Academy of Science, Prague, Czech Republic, EU, ujcic.massimo@gmail.com
[5]University Hospital Centre Sestre Milosrdnice, Tissue bank at University Department of Traumatology, University of Zagreb, Zagreb, Croatia, EU, mirna.tomtrcin@gmail.com
[6]University of Zagreb, Faculty of Food Technology and Biotechnology, Zagreb, Croatia, EU islivac@pbf.hr

Abstract

In this work the function and application of titanium dioxide as a filler in a composite system polycaprolactone/titanium dioxide (PCL/TiO_2) was examined. Titanium dioxide was applied by ultrasonic bath on already electrospun PCL fibrous scaffold treated and non-treated with NaOH. A procedure of surface modification of the electrospun PCL fibrous scaffold was made to enhance the interaction of the surface with the TiO_2 particles. The surface modification was performed using NaOH for the formation of carboxyl groups on the fibers' surfaces. The water contact angle was measured by goniometer to prove the change from hydrophobic to hydrophilic polymer surface. SEM was used to study the morphology structure of the electrospun PCL fibrous scaffold before and after NaOH treatment and introduction of TiO_2. The content of TiO_2 on the electrospun PCL fibrous scaffold was determined by TGA. After NaOH treatment the surface of the electrospun PCL fibrous scaffolds changed from hydrophobic to hydrophilic. SEM micrographs show that with the sonification of 30 min homogeneous TiO_2 particles distribution was obtained, while after sonification of 60 min, the TiO_2 particles tend to agglomerate. The modification of the scaffold surface with NaOH enhances the adhesion of the TiO_2 filler. TG analysis show that longer treatment of the electrospun PCL fibrous scaffolds in the ultrasonic bath gives lower thermal stability. The time of 30 minutes in the ultrasonic bath is optimal to provide sufficient amount of the TiO_2 particles on the electrospun PCL fibrous scaffold.

Keywords: Electrospinning, NaOH, surface modification, titanium dioxide

1. INTRODUCTION

The tendency of human organism, tissue and organs to get injured and the flaws of human tissue caused by innate defects are everyday problems that physicians encounter. Treatment usually focuses on tissue transplantation from one site to another in the same patient (autograft) or from one person, the donor to the other (alograft or transplant). Although these treatments are revolutionary and save human lives, they face many problems. This is why tissue engineering has been developed with the aim of repairing and regenerating damaged tissue by developing biological replacements that restore, maintain, or improve tissue function. Tissue engineering uses different natural and artificial biomaterials as 3D scaffolds where the cells previously taken from the patient's body are planted. 3D scaffold structure is important for imaging specific microgeometry of tissue. In order to obtain scaffold of the exact 3D microgeometry and topography, electrospinning technique is used in combination with 3D printing. First, 3D printing technique is used to obtain collector of specific topography, and then they are used in the electrospinning procedure for fiber collecting. Electrospinning enables the production of very fine fibers of precisely specified properties. This way, certain micro-topography scaffolds are created which is important for linking the cells and the development of new tissue [1]. Recently, biodegradable fiber polymers are used as matrices for the cultivation and growth of different tissue cells. Specifically, mats obtained by electrospinning of poly (ε-caprolactone) (PCL) exhibit good potential as skin regeneration tissue carriers as they do not produce harmful degradation products such as poly (lactic acid) (PLA) or poly (lacto-co-glycolic acid) (PLGA). In addition, PCL has good mechanical properties and prolonged degradation time [2,3]. Poly (ε-caprolactone) (PCL) is a partially crystalline aliphatic polyester with a melting point at 60 ° C. The low melting point makes it easier to process. PCL is hydrophobic due to the presence of nonpolar methylene groups. PCL is the most widely used polyester in tissue engineering. In this paper, PCL scaffolds were electrospun on the collector of parallel channel geometry [4]. Subsequently, TiO_2 was applied from aqueous solution, to neat PCL scaffold and to scaffold previously treated with NaOH, using ultrasonic bath. The aim of this work is to investigate how pre-treatment of the surface and the time in the ultrasonic bath affects the adhesion of TiO_2 to the electrospun PCL fibrous scaffold in order to use the material for fibroblast cell growth and regeneration of the skin tissue.

2. EXPERIMENTAL

2.1. Materials

In this work PCL (Polycaprolactone 440744-500G) average molecular mass *Mn* of 70,000-90,000 g/mol by GPC, Mw/Mn<2, dencity 1.145 g/ml at 25°C) was supplied by Sigma-Aldrich, Germany. As solvents glacial acetic acid and acetone from Sigma-Aldrich, Germany were used.

2.2. Preparation of polymer solutions

Polymer solution of 18 % PCL with 8:2 ratio of glacial acetic acid and acetone was made. It was mixed by magnetic stirer at the temperature of 50 °C.

2.3. Electrospinning procedure

Electrospinning of PCL solutions was made on electrospinning machine, NT-ESS-300, NTSEE Co. Ltd. South Korea. The conditions for electrospinning were: electric voltage: 15 kV, distance from the needle end to collector 18 cm, speed of solution flow 1 mL/h and total time of 4 hours. Syringe is filled with 4 mL of PCL solution. It is then connected to the pump where flow conditions are regulated. One end of the source is connected to the needle and the other to the collector for the dispersion of the solution. 3D printed collector is set to distance of 18 cm from the syringe with the needle and the electrospinning is performed for 4 hours. After that time the source is unplugged and electrospun scaffold is used for further analysis.

2.4. NaOH treatment

Suface of samples is treated before applying TiO_2 with NaOH to change the surface energy and lower down the contact angle with water to make hydrophillic suface of electrospun PCL fiber from originally hydrophobic surface. If the surface is hydrophillic it is considered that it will react better with applied polar particles of TiO_2.The solution of NaOH was 1 N and the samples were treated in it for 1 hour then washed with distilled water until pH neutral.

2.5. Application of TiO_2

TiO_2 was applied on the electrospun PCL fibrous scaffold from 1 % water solution afterwards by ultrasonic bath device. In total 4 (2 without and 2 treated with NaOH) samples were held in the ultrasonic bath for 30 and 60 minutes to make comparison and decide the best technique of applying the highest amount of equally dispersed TiO_2 particles on the electrospun PCL fibrous scaffold without agglomerations.

2.6. Experimental part

Water contact angle were determined by measuring on a contact angle goniometer DataPhysics OCA 20 Instruments GmbH by *Sessile drop* method. The water contact angle was measured for non-modified and PCL modified by NaOH. The volume of the drop was from 1.000 µL to 5.000 µL. On every sample 3 drops were ejected and after 20 seconds the contact angle was measured.

For determination of the fiber's morphology samples are scanned by electron microscope SEM QUANTA 250, FEI. The magnification was 50, 1000, 2000 or 3000 times on different parts of the samples. The thermal stability was measured by TGA analyzer Q500 from TA Instruments. Samples of about 10 mg were analyzed in a nitrogen atmosphere with heat speed of 10°C/min, in a temperature range from 25 °C to 600 °C.

3. RESULTS AND DISCUSION

3.1. Water contact angle measurement

The water contact angle was measured before and after NaOH treatment. The interaction of water and the surface of the sample is detected from the contact angle data. Contact angle measurement plays an important role in a characterization of the surface of materials. It is a measurement of moisturizing the solid with the liquid. A drop is spilled on a solid surface and the balanced shape is awaited. Contact angle is formed from the forces of the periphery of a drop and a solid surface. Surface condition is determined concerning it's polarity, homogeneity and roughness. Hidrophobicity of the electrospun PCL fibrous scaffold could be changed by NaOH treatment. To confirm the change from hydrophobic to hydrophilic water contact angle measurement for treated and non-treated electrospun PCL fibrous scaffold is made. It is well-known that the electrospun PCL fibrous scaffold has a water contact angle of around 80°. The water contact angle measurements are shown in **Figure 1**. For non-treated electrospun PCL fibrous scaffold contact angles on two different position of the scaffold were obtained around 90° and 125.8 ± 5.6°, which clearly show hydrophobicity (**Figure 1a and 1b**). On the other hand the electrospun PCL fibrous scaffold treated with NaOH shows extreme hydrophilicity with water contact angle of about 0° which proves that the surface of the PCL became polar. Unfortunately, it was not possible to take contact angle images of a drop for the electrospun PCL fibrous scaffold treated with NaOH because the material instantly soaked the drop of water. Generally, NaOH treatment is made to lower the polymer surface hydrophobicity and to enhance the interaction between the polymer and TiO_2.

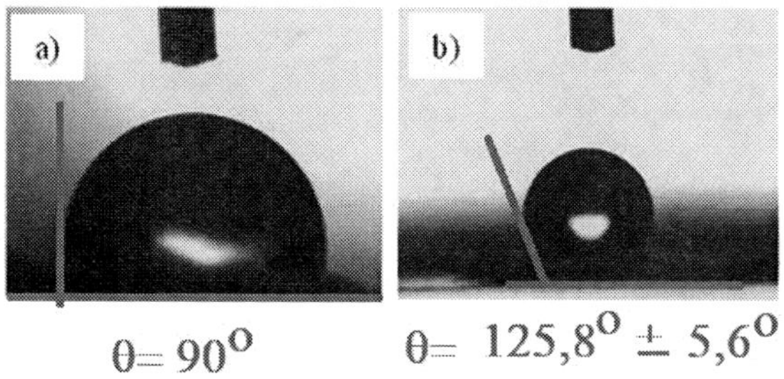

Figure 1 Water contact angle measurements for the electrospun neat PCL scaffold on the different position of the scaffold before treatment with NaOH (Figures 1a and 1b)

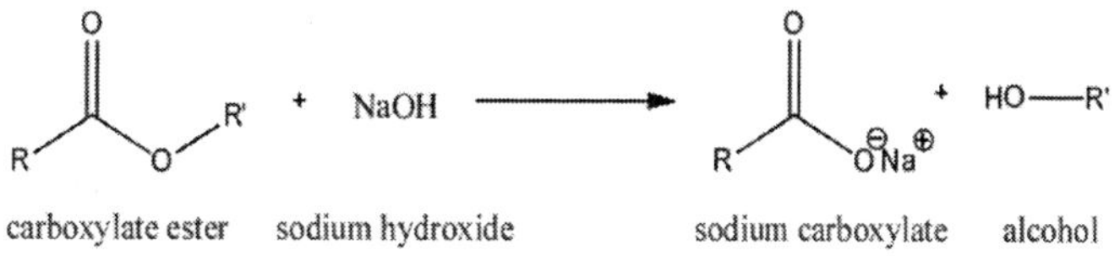

Figure 2 Sodium hydroxide treatment of PCL structure

NaOH hydrolyzes ester bonds of the PCL which leads to carboxyl functional group formation on the electrospun PCL. Carboxyl groups make the surface less hydrophobic as they lower the contact angle value.

Because of the NaOH surface treatment the polymer becomes nano-rough which is enlarging the specific surface that can react with the particles of TiO_2 (**Figure 2**) [5].

3.2. Determination of morphology by scanning electron microscopy (SEM)

Figure 3 shows the structure of the electrospun PCL fibrous scaffold before NaOH treatment and application of TiO_2. Porous, fiber-like structure (**Figure 3a**) that follows geometry of the electrospinning collector (**Figure 3b**) can be seen. Except for fibers, there are also beads formations of the PCL.

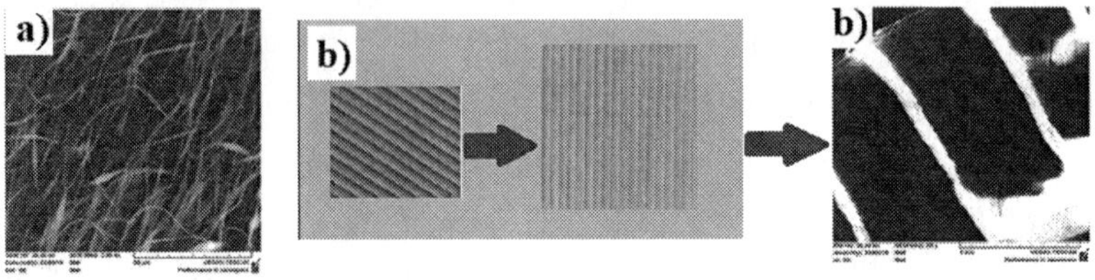

Figure 3 SEM images of the electrospun: a) PCL neat fibrous scaffolds, b) PCL neat fibrous scaffolds on the 3D printed collector, before NaOH treatment and application of TiO_2

The electrospun PCL scaffolds consist of random oriented and intertwined fibers with specific inter pore structure. The surface of the fibers doesn't have any irregularities. The fibers diameters were about 1 µm. The electrospun PCL fibrous scaffold will be used to grow skin cells so it is necessary to provide double porosity of the electrospun fiber carrier by micro pores big enough for skin cells penetration and nano pores small enough to enhance adhesion of the skin cell receptors, transfer of nutrients and oxygen needed for cells growth [6]. For antibacterial properties of the carrier, the TiO_2 is added. **Figure 4a** shows homogeneous TiO_2 particle distribution inside of the electrospun PCL fibrous scaffold after 30 min in the ultrasonic bath.

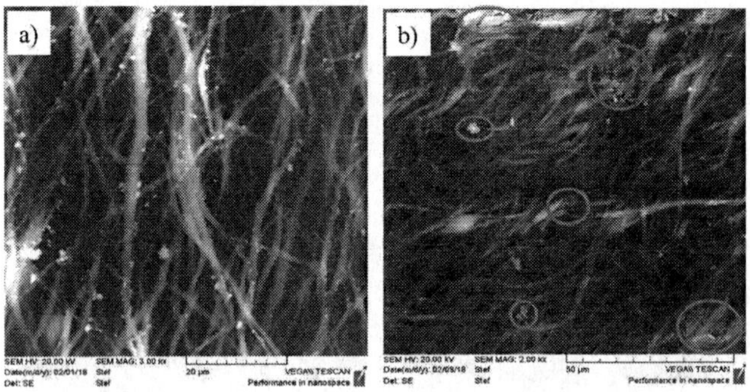

Figure 4 SEM micrographs of the electrospun PCL fibrous scaffold with TiO_2 after a) 30 min and b) 60 min of sonification

After 60 minutes of sonification large quantity of TiO_2 inside of the electrospun PCL fibrous scaffold was obtained, but the particles distribution was not uniform and the particles tend to agglomerate as seen in **Figure 4b**. The modification with NaOH changed the smoothness of the surface. The SEM micrograph in **Figure 5** shows that the geometry of the material changed. In spite of small changes of the surface morphology it is evident in **Figure 5a** that the TiO_2 particles after 30 minutes in an ultrasonic bath got integrated in large agglomerates. After 60 minutes, **Figure 5b**, agglomeration is not so noticeable as without NaOH treatment and distribution of the filler particles is even along the whole surface and between the pores. There is an indication of possible penetration of the filler particles inside of the pores of fiber.

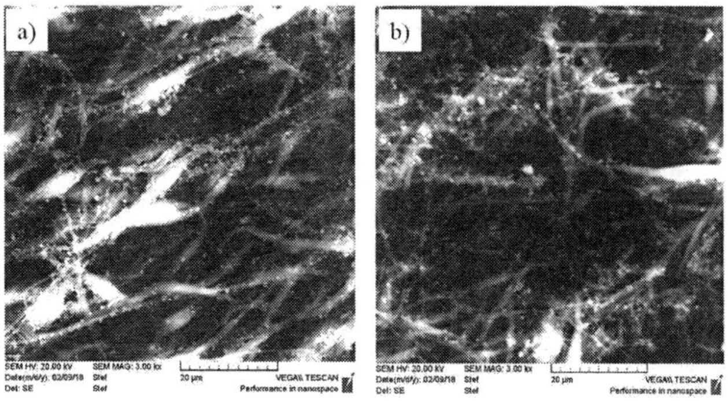

Figure 5 SEM micrographs of the electrospun PCL fibrous scaffold treated with NaOH and with TiO_2 after a) 30 min and b) 60 min of sonification

The mapping micrographs clearly show where the TiO_2 is placed inside the electrospun PCL fibrous scaffolds. The red color marks indicate the titan dioxide particles, thus it is proven that NaOH treatment of the polymer surface enhances the interactions between the polymer and the titanium dioxide (**Figure 6**).

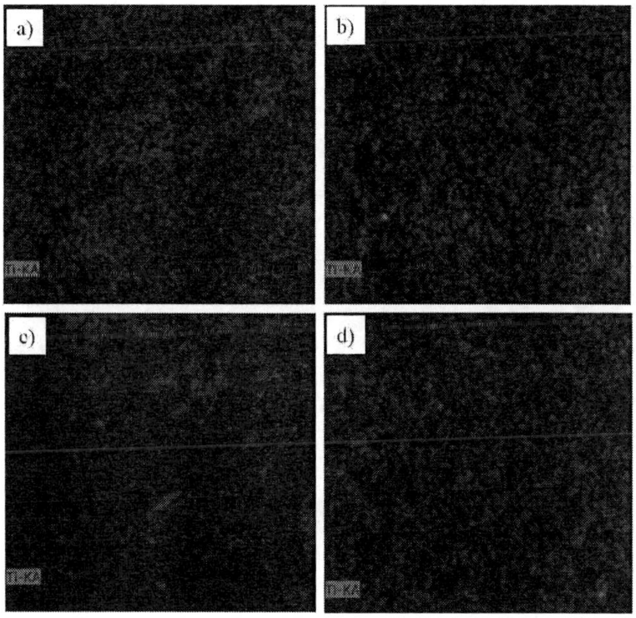

Figure 6 Mapping TiO_2: a) 30 min without NaOH, b) 30 min with NaOH; c) 60 min without NaOH, d) 60 min with NaOH

3.3. Results of thermogravimetric analysis

The effect of the TiO_2 as well as the surface treatment with NaOH on the thermal stability of the electrospun PCL fibrous scaffold was examined by TGA. From the **Table 1** we can see that the addition of TiO_2 in the electrospun PCL fibrous scaffold enhances thermal stability of the scaffold because the initial temperature of degradation increased. Opposite from that, with longer time of ultrasound treatment the degradation starts at lower temperature which means that the time of 60 min is too long for keeping the electrospun PCL fibrous scaffold in the ultrasonic bath. The initial temperature of degradation is also lower for the samples treated with NaOH which affects negatively on the thermal stability of the fibrous scaffold. Another important parameter is the percentage of the residue which is related to the weight of TiO_2 implemented in the electrospun PCL fibrous scaffold. By surface modification with NaOH the weight of the TiO_2 in the PCL fibrous scaffold increases significantly as it is shown in **Table 1**. Also the longer time of treatment enhances the percentage of TiO_2.

Table 1 TGA weight residue and start and end degradation temperatures of the PCL/TiO_2 scaffolds

Sample	Residue at 600°C (mg)	T_i (°C)	T_f (°C)
PCL	0	365.5	426.9
PCL/TiO_2 30 min.	0.67	384.0	452.6
PCL/TiO_2 60 min.	0.79	367.6	426.5
PCL(NaOH)/TiO_2 30 min.	0.80	372.6	432.1
PCL(NaOH)/TiO_2 60 min.	0.88	363.0	427.2

4. CONCLUSION

In this work a possibility to introduce TiO_2 to electrospun PCL fibrous scaffolds by ultrasonic bath method was examined. Further NaOH surface modification of the electrospun PCL fibrous scaffold was made to enhance the interaction with the TiO_2 filler particles. After NaOH treatment the surface of the electrospun PCL fibrous scaffolds changed from hydrophobic to hydrophilic. SEM micrographs show that with the sonification of 30 min homogeneous TiO_2 particles distribution was obtained, while after sonification of 60 min, the TiO_2 particles tend to agglomerate. The modification of the scaffold surface with NaOH enhances the adhesion of the TiO_2 filler. TG analysis show that longer treatment of the electrospun PCL fibrous scaffolds in the ultrasonic bath gives lower thermal stability. The time of 30 minutes in the ultrasonic bath is optimal to provide sufficient amount of the TiO_2 particles on the electrospun PCL fibrous scaffold.

ACKNOWLEDGEMENTS

This work has been fully supported by Croatian Science Foundation under the project IP-2016-06-6878, Custom Tailored Fibrous Scaffold Prototype for Tissue Cells Culture via Combined Electrospinning, COMBOELECTROSPUN.

REFERENCES

[1] HIREMATH, Nitilaksha and BHAT, Gajanan. Melt Blown Polymeric Nanofibers for Medical Applications- An Overview. *Nanoscience and Technology*, 2015. vol. 2, no. 1, pp. 1-9.

[2] YOSHIMOTO, H., SHINA, Y. M., TERAI, H. and VACANTI, J. P. A biodegradable nanofiber scaffold by electrospinning and its potential for bone tissue engineering. *Biomaterials*, 2003, vol. 24, no.12, pp. 2077-2082.

[3] CHEN, Jinglu, CHU, Benjamin and HSIAO, Benjamin S. Mineralization of hydroxyapatite in electrospun nanofibrous poly(L-lactic acid) scaffolds. *Journal of Biomedical Materials Research Part A,* 2006, vol. 79A, 307-317.

[4] LEE, Kun-Hong, KIM, H. Y., KHIL, M. S, RA, Y. M. and LEE, D. R. Characterization of nano-structured poly(1-caprolactone) nonwoven mats via electrospinning. *Polymer*, 2003, vol. 44, no. 4, pp. 1287-1294.

[5] LAM, Christopher XF, TEOH, SWEE, Hin and HUTMACHER, Dietmar W. Comparison of degradation of PCL and PCL-TCP scaffolds in alkaline medium. *Polymer International*, 2007, vol. 56, no. 6, pp. 718-728.

[6] ZAISS, Sascha, BROWN, Toby D, REICHERT, Johannes C and BERNER, Arne. Poly (ε-caprolactone) Scaffolds Fabricated by Melt electrospinning for Bone Tissue Engineering, *Materials*, 2016, vol. 9, no. 4, 232-247.

Ag AND Pd NANOSTRUCTURES: FROM MATERIAL PROPERTIES TO BIOLOGICAL APPLICATIONS

Markéta KAIMLOVÁ, Markéta PIŠLOVÁ, Václav ŠVORČÍK, Jakub SIEGEL

Department of Solid State Engineering, University of Chemistry and Technology Prague, Prague, Czech Republic, EU, polivkoa@vscht.cz

Abstract

Excellent workability of polymers brings them a considerable potential to be used as matrices for metal nanostructures broadly applicable in medicine. For instance, biologically active metal/polymer composites could be effectively prepared by biocompatible polymer coating by thin metal nanolayers. Together with the fact that biological properties of such prepared composites could be improved with increasing metal surface area, we nowadays have a great opportunities for the preparation strongly effective materials. In this study, we prepared Ag and Pd nanolayers (NLs) on the surface of polyethylene naphthalate by means of cathode sputtering. Such NLs were transformed into the form of nanoislands (NIs) by low temperature post-deposition annealing. Subsequently, biological properties of commonly used NLs were compared to novel ones (NIs) with considerably higher specific surface area. Surprisingly, increase of both surface roughness and specific surface area caused by NIs formation did not support improved biological response.

Keywords: Silver, palladium, nanostructures, sputtering, annealing, biological properties

1. INTRODUCTION

In particular, long-term stability and excellent workability make polymers suitable matrices for nanostructure-based composites, especially those containing metal components. The combination of polymeric materials with metal nanostructures often brings an advantage of excellent mechanical, optical, catalytic, electrical or magnetic properties, which are significantly different unlike the bulk state of material [1]. Thus, the dominating mechanical characteristics of polymeric matrices and antimicrobial potency of metal nanostructures may significantly improve such composite properties in the specific area of use. Especially silver nanostructures, entering 3rd decade of prosperous usage in polymeric composites, exhibit excellent electrical, optical and magnetic properties, of which the most promising is undoubtedly their antibacterial effect, widely used in medicine [2]. Composite medical devices, however, must fulfil the criteria of biocompatibility and antibacterial effects. Modern nanotechnology nowadays opens up new possibilities to improve these properties and minimizes the risk of side-effect due to the massive application of these materials [3].

The improvement of biological properties can be accomplished by several methods. Modification of surface morphology and roughness has already been referred by several studies [4-6]. The alteration of these parameters might significantly improve biocompatibility of material promoting cell adhesion and proliferation [6]. To add the antibacterial efficacy to polymeric materials, one can prepare metal nanolayers on their surface. Nevertheless, one can increase the antibacterially active surface area using nanoislands as default structures. Possible way is thermally induced transformation of nanolayers into island-like, which goes hand in hand with the alteration of resulting surface morphology and roughness. Thereafter, the increase of metal specific surface area might lead to the improved antibacterial response. The formation of island-like structures by low temperature annealing of nanolayers has already been described for silver [4], and palladium [5]. Thus, the appropriately performed surface modification enables a direct control over the biological properties of composite materials.

In this work we studied the influence of the surface morphology on resulting biological properties. Metal nanolayers (common Ag and novel Pd) were sputtered on the surface of polyethylene naphthalate.

Subsequently, the samples were annealed and originally continuous layers were transformed into the form of nanoislands with considerably increased specific surface area. Surface properties of each sample were characterized by X-ray photoelectron spectroscopy (XPS) and atomic force microscopy (AFM). Evaluation of biological suitability of prepared coatings was performed by their mutual antibacterial and cytotoxicity testing.

2. EXPERIMENTAL

2.1. Materials, apparatus and procedures

Polyethylene naphthalate foil (PEN, thickness of 50 μm, Goodfellow Ltd., UK) was used as a substrate. PEN samples (Ø 2 cm) were deposited by metals on BAL-TEC SputterCoater SCD 050 device in the range of sputtering times of 10-200 s using 99.999 % pure Ag and Pd targets (Goodfellow Ltd., UK). The deposition parameters were: 20 °C, current of 15 mA and total Ar pressure of 5 Pa (99.99 % purity). Samples annealing was conducted in a Binder thermostat in air atmosphere at 250 °C for 1 h. Then the samples were cooled down and stored under laboratory conditions.

2.2. Analytical methods

Effective thickness of metal layers was determined by AFM scratch test [7] on glass substrate by AFM VEECO CP II device. The series of the samples of various metal thicknesses (sputtering times of 10-200 s) were acquired, from which 20 nm thick ones (and corresponding annealed alternatives) were chosen as representatives. Variations of the thickness did not exceed 5 %.

Compositional changes induced by the deposition and annealing process were studied by XPS. The atomic concentrations of silver Ag(3d), palladium Pd(3d), carbon C(1s) and oxygen O(1s) were measured using Omicron Nanotechnology ESCAProbeP spectrometer. The electron take-off angle was set to 0° which means typical access depth of 8-10 atomic layers [8]. The X-ray source was monochromated at 1486.7 eV (step size of 0.05 eV). The spectra were evaluated by CasaXPS software.

Surface morphology and roughness of the samples were studied on AFM VEECO CP II device working in tapping mode to minimize a potential damage of the sample surface. Bruker Antimony-doped Silicon probe CONT20A-CP attached to a flexible micro-cantilever was used near its resonant frequency of 300 kHz. The scans were measured at the line scanning rate of 0.5 Hz. Surface roughness, characterized by the mean roughness value (R_a), represents the arithmetic average of the deviation from the center plane of a sample.

2.3. Biological assays

Antibacterial potency of prepared coatings was investigated by drop plate method [9] using Gram-negative (G^-) *Escherichia coli* (DBM 3138) and Gram-positive (G^+) *Staphylococcus epidermidis* (DBM 3179). Tested and control samples (physiological solution, PS) in triplicates were immersed in 2 ml of PS, and inoculated with 1.1×10^4 of colony forming units (CFU) per 1 ml of *E. coli* and 2.2×10^4 of CFU per 1 ml of *S. epidermidis*. The samples were incubated under both static (laboratory table) and dynamic (shaking at 130 rpm) conditions at 24 °C for 3 and 24 h. Afterwards, the aliquots of 25 μl from each sample were placed on agar plates (LB for *E. coli*, PCA for *S. epidermidis*) in 10-fold repetitions. After overnight incubation, CFUs were counted by Lucia Image 4.8 software. These experiments were accomplished under sterile conditions.

Cytotoxicity was examined by WST-1 assay [10] using mouse embryonic fibroblasts (L929). The samples were sterilized by UV irradiation, inserted into 12-well plates (VWR, USA, Ø 2.14 cm), and seeded with 30,000 cells per well in triplicates in 1 ml of Minimum Essential Medium (MEM) supplemented with 10 % Fetal Bovine Serum. The cells were incubated for 24, 48 and 72 h (cultivation conditions: 37 °C, 5 % CO_2, and 95 % humidity). The culture medium was replaced with fresh phenol red free FluoroBrite DMEM (480 μl) supplemented with 20 μl of WST-1 reagent per well. After 2 h of incubation, the medium was aliquoted into 96-microtiter plates. Then, the absorbance was measured at 450 nm (reference 630 nm) using UV-Vis spectrometer BioRad. Cells cultivated on standard tissue culture polystyrene (TCPS) served as controls.

3. RESULTS AND DISCUSSION

3.1. Preparation and surface characterization

Atomic concentrations of elements (at. %) were studied by XPS analysis (**Table 1**). Determined concentrations of C and O corresponded well to the PEN stoichiometry with mild distortion caused by hydrocarbon impurities adsorbed from air [11]. Detected concentrations of metals (Ag, Pd) were given by sputtering process. One can see that annealing of pristine PEN led to insignificant differences in its chemical composition. Compared to as-sputtered samples annealing caused a diffusion and aggregation of metals which decreased the detected concentrations of Ag and Pd in case of Ag NIs/PEN and Pd NIs/PEN, respectively. Simultaneously, element concentrations originating from the underlying polymer (C and O) were increased due to PEN substrate become partially uncovered. This phenomenon has already been studied in detail by determining the rate of surface ablation (XPS) and metal release (ICP-MS) [5]. This study uncovered that compositional changes after annealing were caused by the coalescence of metal into separate nanoislands with partial embedding of these clusters into polymer interior. Principally, the ultrathin (ones of nm) polymer overlay covered metal islands reaching almost to their tops. This phenomenon being known as "curtain effect". Considerably higher concentration of O in Ag NIs/PEN after annealing was presumably caused by higher propensity of Ag to oxidation process.

Table 1 Atomic concentrations of C(1s), O(1s), Ag(3d) and Pd (3d) measured by XPS

Sample	Atomic concentrations of elements (at. %)			
	C	O	Ag	Pd
Pristine PEN	72.2	27.8	-	-
Annealed PEN	73.1	26.9	-	-
Ag NLs/PEN	58.4	12.0	29.6	-
Ag NIs/PEN	61.5	30.1	8.4	-
Pd NLs/PEN	73.5	1.2	-	25.3
Pd NIs/PEN	65.6	18.3	-	16.1

Because the surface morphology and roughness has a great impact on resulting biological properties of materials, AFM analysis was involved in this study. In **Figure 1**, one can see that the surfaces of pristine PEN and as-sputtered samples of both Ag and Pd NLs/PEN were mildly corrugated and similar values of R_a (**Table 2**) were measured. Compared to as-sputtered samples (see **Figure 1a**), however, noticeable alteration of the surface morphology of both metal-coated samples can be observed after the annealing (**Figure 1b**). The samples´ surface was completely rearranged during annealing process; thermal accumulation in the metal layers resulted in an alteration of the amorphous phase of PEN (T_g^{PEN} = 120 °C), which led to the significant increase of the R_a of annealed samples (**Table 2**). The values of surface roughness R_a increased two orders of magnitude compared to as-sputtered samples. Isolated island-like structures (NIs) homogeneously distributed over the surface of PEN are apparent from AFM scans of both Ag and Pd/PEN samples. This phenomenon is in accordance with the results of XPS analysis (see **Table 1**), which revealed decreased values of metal concentrations in case of both Ag and Pd after annealing at the expense of the concentrations of C and O originating from PEN substrate. Furthermore, the size and shape of metal nanoislands is effectively controllable by the thickness of metal coating preceding the annealing process [5].

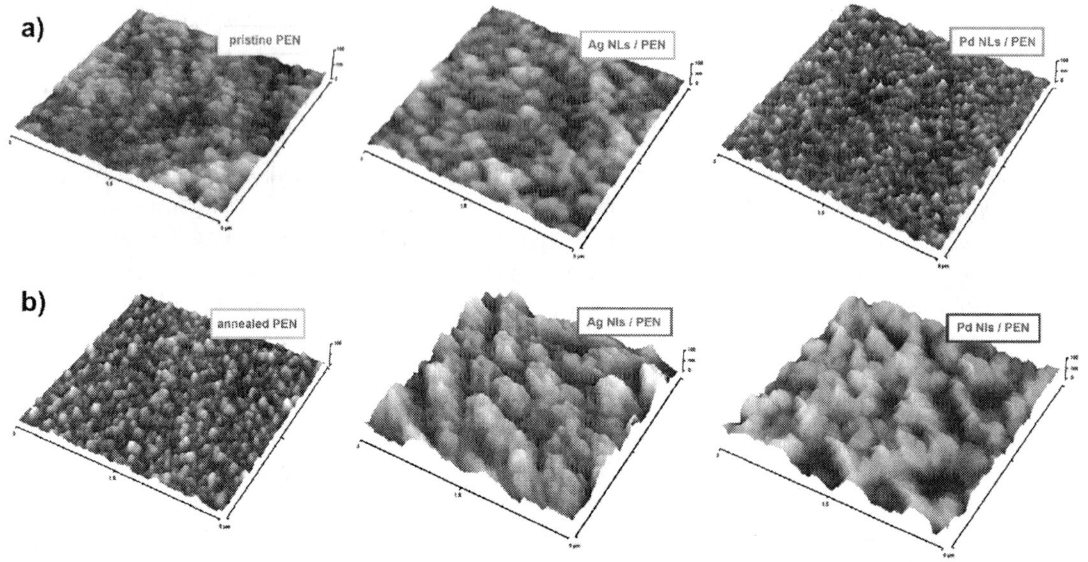

Figure 1 AFM scans of the samples a) before and b) after annealing

Table 2 Mean surface roughness (R_a) of the samples measured by AFM

Samples	Pristine PEN	Annealed PEN	Ag NLs/PEN	Ag NIs/PEN	Pd NLs/PEN	Pd NIs/PEN
R_a **(nm)**	4.4	4.7	4.8	151.6	4.7	148.1

3.2. Biological response

Antibacterial effects of the samples are shown in **Figures 2a,b** for *E. coli*, and *S. epidermidis*, respectively. Missing columns corresponds to no countable amount of CFU; total inhibition, respectively. Generally, pristine and annealed PEN exhibited no antibacterial effects, except an insignificant one for annealed sample after 3 h of dynamic incubation with *E. coli* (**Figure 2a**), presumably caused by a noticeable surface roughness of this substrate (see **Figure 1** AFM). Due to this fact, together with dynamic shaking at 130 rpm, bacterial colonies were mildly inconformed during their adaptation period to surrounding environment [12]. Generally, increased antibacterial effects after longer incubation time (24 h) occurred for all metal/polymer composites in case of both bacterial strains (**Figures 2a,b**) in all incubation conditions. These effects might be caused by i) increased concentrations of released metals into PS after longer incubation period (ICP-MS results were published elsewhere [5]), and ii) longer contact of bacteria with the antibacterially-active surface of the composites. When comparing the results of both metal forms, NLs and NIs ones, one can see that NLs/PEN samples exhibited similar or higher antibacterial response compared to NIs. It is in good accordance with the results of XPS (see **Table 1**) and mechanism of "curtain" effect, mentioned above. This phenomenon precluded to effectively increase the antibacterial response by the alteration of the surface morphology (increase of specific surface area of metal). When comparing the results for Ag and Pd samples, more significant antibacterial response was determined for Pd ones, in case of both types of composites. These results suggest that Pd in its diverse nanostructure forms is able to compete with commonly used Ag in its medical applications, as more antibacterial effective alternative. The comparison of the results presented in **Figures 2a** and **2b** showed generally increased antibacterial effects against *S. epidermidis* (**Figure 2b**), which was more sensitive to both metals. The most testifying are the results for both types of Pd composites, which revealed no countable amount of CFU in all conditions. The differences in the antibacterial effects against used bacterial strains might be explained by more facile penetration of positively charged noble metals particles through the cell walls of G^+ bacteria [13].

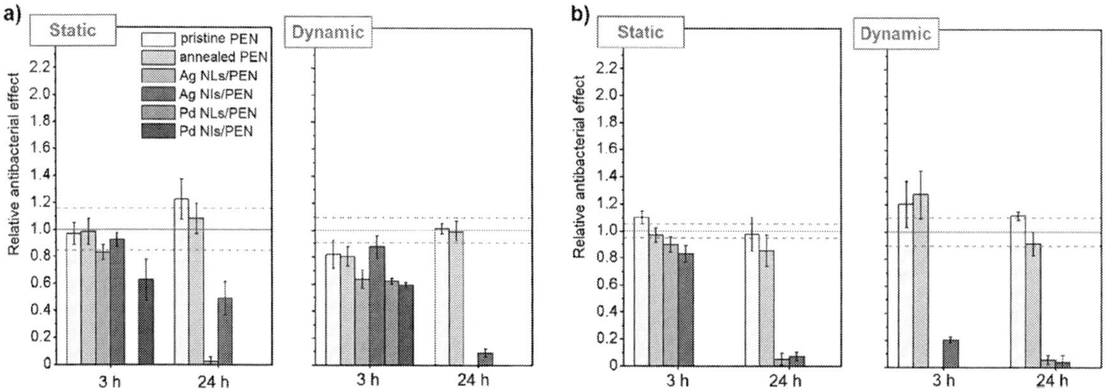

Figure 2 Relative antibacterial effect (CFU for examined sample divided by CFU for control sample) of the samples against bacterial colonies of a) *E. coli* and b) *S. epidermidis* measured in static and dynamic mode. Gray line represents a reference level (number of CFU in physiological solution) for corresponding bacterial strains together with its uncertainty (dash line).

Accessible literature [6] reports about the influence of the surface roughness on biocompatibility. Another studies [14,15], however, suggest significant cytotoxicity of metal nanostructures, especially silver ones. For these reasons, cytotoxicity assay (**Figure 3**) was performed to determine cytotoxic effects of both investigated metal nanostructure forms based on Ag and Pd.

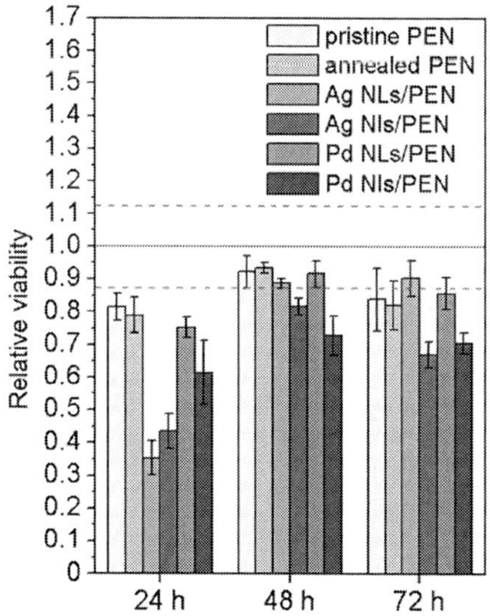

Figure 3 Relative viability (absorbance for examined sample divided by absorbance for control sample), of the samples against L929 model cell line. Gray line represents a reference level (absorbance value for TCPS) together with its uncertainty (dash line).

One can see that L929 cell viability on PEN substrate was not influenced by annealing; relative viability is comparable to control samples after longer cultivation times (48 and 72 h). Slight decrease of relative viability observed for pristine and annealed PEN samples, alike NLs and NIs samples of both metals, after 24 h of cultivation was presumably caused by a difficult adaptation of L929 cell culture on the samples´ surface. Generally, **Figure 3** revealed low cytotoxic effects of all tested samples with minor response of NLs samples both Ag and Pd ones. Interestingly, cytotoxicity of both metals NLs samples, alike NIs ones, after longer cultivation times (48 and 72 h) was comparable; both metals demonstrated comparable cell-conformity. More significant cytotoxicity of annealed samples of both metals (NIs/PEN) compared to as-sputtered ones (NLs/PEN) was presumably caused by high sensitivity of L929 cell line to rougher surfaces [12], rather than by cytotoxic effects of metals themselves. This finding was supported the results of XPS analysis (see **Table 1**), together with "curtain" effect [5], which caused decreased amount of the metals able to evoke cytotoxicity.

4. CONCLUSION

This work presents the comprehensive study of biological effects of various kinds of metal nanostructure-based coatings of biocompatible polymer. Commonly used nanostructured Ag was confronted with Pd one. Surprisingly, it was found that the alteration of the surface morphology and roughness did not improve neither antibacterial effects of all nanostructured metal/PEN composites, due to the influence of „curtain" effect, nor biocompatibility, because of high sensitivity of L929 cell line to rougher surfaces. Both types of Ag and Pd coatings exhibited insignificant cytotoxic effects, which, together with a higher antibacterial response of Pd-coated samples, points to its potential applicability in health-care industry.

ACKNOWLEDGEMENTS

Financial support from specific university research (MSMT No. 21-SVV/2018) and GACR project No. 17-10907S are gratefully acknowledged.

REFERENCES

[1] JORDAN, J., JACOB, K.I., TANNENBAUM, R., SHARAF, M.A. and JASIUK, I. Experimental trends in polymer nanocomposites-a review. *Mater. Sci. Eng. A*. 2005. vol. 393, no. 1, pp. 1-11.

[2] MARAMBIO-JONES, C. and HOEK, E.M.V. A review of the antibacterial effects of silver nanomaterials and potential implications for human health and the environment. *J. Nanoparticle Res.* 2010. vol. 12, no. 5, pp. 1531-1551.

[3] LEAPER, D.J. Silver dressings: their role in wound management. *Int. Wound J.* 2006. vol. 3, no. 4, pp. 282-294.

[4] SIEGEL, J., POLIVKOVA, M., STASZEK, M., KOLAROVA, K., RIMPELOVA, S. and SVORCIK, V. Nanostructured silver coatings on polyimide and their antibacterial response. *Mater. Lett.* 2015. vol. 145, no. 1, pp. 89-90.

[5] POLIVKOVA, M., VALOVA, M., SIEGEL, J., RIMPELOVA, S., HUBACEK, T., LYUTAKOV, O. and SVORCIK, V. Antibacterial properties of palladium nanostructures sputtered on polyethylene naphthalate. *RSC Adv.* 2015. vol. 5, no. 90, pp. 73767-73774.

[6] REBOLLAR, E., FRISCHAUF, I., OLBRICH, M., PETERBAUER, T., HERING, S., PREINER, J., HINTERDORFER, P., ROMANIN, C. and Heitz, J. Proliferation of aligned mammalian cells on laser-nanostructured polystyrene. *Biomaterials.* 2008. vol. 29, no. 12, pp. 1796-1806.

[7] STEINMANN, P.A., TARDY, Y. and HINTERMANN, H.E. Adhesion testing by the scratch test method: The influence of intrinsic and extrinsic parameters on the load. *Thin Solid Films.* 1987. vol. 154, no. 1-2, pp. 333-349.

[8] TYLER, B.J., CASTNER, D.G. and RATNER, B.D. Regularization: A stable and accurate method for generating depth profiles from angle-dependent XPS data. *Surf. Interface Anal.* 1989. vol. 14, no. 8, pp. 443-450.

[9] CHEN, CH.Y. and CHIANG, CH.L. How to optimize the drop plate method for enumerating bacteria. *Matter. Lett.* 2001. vol. 44, no. 2, pp. 121-129.

[10] PESKIN, A.V. and WINTERBOURN, C.C. A microtiter plate assay for superoxide dismutase using a water-soluble tetrazolium salt (WST-1). *Clin. Chim. Acta.* 2000. vol. 293, no. 1-2, pp. 157-166.

[11] LEWIS, K.B. and RATNER, B.D. Observation of surface rearrangement of polymers using ESCA. *J. Colloid Interf. Sci.* 1993. vol. 159, no. 1, pp. 77-85.

[12] TAYLOR, R.L., VERRAN, J. and LEES, G.C. The influence of substratum topography on bacterial adhesion to polymethyl methacrylate. *J. Mater. Sci. Mater. Med.* 1998. vol. 9, no. 1, pp. 17-22.

[13] JUNG, W.K., KOO, H.C., KIM, K.W., SHIN, S., Kim, S.H. and PARK, Y.H. Antibacterial activity and mechanism of action of the silver ion in Staphylococcus aureus and Escherichia coli. *Appl. Environ. Microbiol.* 2008. vol. 74, no. 7, pp. 2171-2178.

[14] ZHANG, T., WANG, L., CHEN, Q. and CHEN, C.Y. Cytotoxic potential of silver nanoparticles. *Yonsei Med. J.* 2014. vol. 55, no. 2, pp. 283-291.

[15] KIM, M.J. and SHIN, S. Toxic effects of silver nanoparticles and nanowires on erythrocyte rheology. *Food Chem. Toxicol.* 2014. vol. 67, no. 1, pp. 80-86.

ELECTROSPINNING AND BIOCOMPATIBILITY OF POLYMER-CERAMIC NANOFIBERS FOR TISSUE ENGINEERING

Eva ŠŤASTNÁ, Klára ČÁSTKOVÁ

Ceitec - Central European Institute of Technology, Brno, Czech Republic, EU, Eva.Jindrova@ceitec.vutbr.cz

Abstract

Polycaprolactone is known as biocompatible material for long time and it is considered as possible material for bioscaffolds. Since natural structure of hard tissues (bones) and most of soft tissues (skin, neural tissue) is fibrous, there are attempts to mimic such a structure. Polycaprolactone (PCL) nanofibers with diameter of approximately 180 nm were prepared by electrospinning method. For better biocompatibility and bioactivity, a composite fibrous structure was prepared. Hydroxyapatite nanoparticles were synthetized via precipitation reaction and further hydrothermal treatment. The ceramic particles were added to modified polycaprolactone precursor and the solution was electrospun. The prepared fibers and as well as dried original precursor were in-vitro tested for cytotoxicity by direct contact test. Mouse fibroblast from L929 line were used for cell cultivation. It was found out that fibrous structure and presence of ceramic particles have a positive influence on cell activity and proliferation - the growth rate of the cells was significantly higher compared to bulk polymer precursor.

Keywords: Nanofibers, electrospinning, biocompatibility, polycaprolactone

1. INTRODUCTION

Regenerative medicine requires new materials and structures for faster and gentler treating of large wounds. Nowadays, allo- or autografts are the most common way how to cure extensive skin injuries and large bone injuries are usually treated with bioinert metallic (e.g. titanium) or ceramic (alumina or zirconia) implants. More convenient method would be implantation of a scaffold that would support natural cell activity on the damaged cells and that would be replaced by new tissue after some time without any residues in the patient's body. There are several attempts to prepare such a scaffold - bulk implants, porous materials and fibrous structures. Considering natural structure of the human tissues, fibrous scaffolds are very promising for regenerative medicine. [1,2,3]

There are several ways of preparation of fibrous structures, e. g. electrospinning, self-assembling and phase separation. [4] The electrospinning is a versatile technique suitable for preparation of polymer, ceramic and composite fibers with range of the diameter from 100 nm to 100 μm. The final morphology of the fibers is influenced by electrospinning parameters (mainly by accelerating voltage) and composition of the precursor. Ceramic fibers can be also modified by subsequent heat treatment. The versatility of the process opens wide field for research in biomaterials - modification of prepared structure with aim to influence the activity of the cells is still an open chapter. [2,3,5]

2. EXPERIMENTAL

2.1. Materials and methods

Polycaprolactone with molecular weight of 40 000 g/mol (Sigma Aldrich, Germany) was used as initial material for preparation of the nanofibers. Polycaprolactone was dissolved in mixture of acetic and formic acid in ratio 3:1 to obtan 10 wt% polycaprolactone solution. To prepare ceramic-polymer composite structure, hydroxyapatite nanoparticles were added to the precursor. The hydroxyapatite ultrafine powder was prepared according to precipitation reaction (1). The product was then hydrothermally treated in a reactor (Ultraclave

Milestone, Italy) at the temperature of 250 °C for 5 hours. The powder was added to 6 wt% solution of polycaprolactone in mixture of acetic acid and formic acid in ratio 3:1. The concentration of the hydroxyapatite in the solution was 4.5 wt%. To obtain a homogenous precursor, the ceramic particles were dispersed by ultrasound probe for 3 minutes.

$$Ca(NO_3)_2 + (NH_4)_2HPO_4 \longrightarrow Ca_5PO_4)_3OH + NH_4NO_3 \quad (1)$$

The precursors were electrospun by 4SPIN (Contipro, The Czech Republic). Parameters of electrospinning were the same for both solution and they are listed in **Table 1**. For better morphology and uniformity of the fibers, warm air (35 °C) was blown at rate of 24 $l.s^{-1}$ around the emitter.

Table 1 Electrospinning parameters of polycaprolactone based precursors

Accelerating voltage (kV)	Emitter-collector distance (mm)	Feeding rate ($ml.hod^{-1}$)	Size of the emitter needle (ga)
25	100	28	19

The as-spun fibers were analyzed by scanning electron microscope (Verios, FEI, USA) - structural was done. After structural analysis, cytotoxicity tests were provided on the both type of fibers. The fibers were deposited on a circle slide glass in a thin layer to guarantee transparency of the samples that is crucial for cytotoxicity tests. The test was done on five samples of each type always for three days. Mouse fibroblasts from L929 line were used for the testing. A photo was taken every 30 s on seven fields of vision. The results were compared with data obtained from the baseline sample (blank sample) and samples with continuous layer of the precursors.

3. RESULTS

3.1. Preparation of the ceramic nanopowder

Structure analysis of the hydroxyapatite powder was done before and after the hydrothermal treatment. As-synthetized ceramics was in the form of agglomerated flake-like particles. The hydrothermal treatment resulted in change of morphology of the particles - the shape was changed in needle-like one. The needles had diameter of approximately 30 nm and they were about 600 nm long which was promising for their further incorporation into the nanofibers. The effect of the hydrothermal treatment is visible in **Figure 1**.

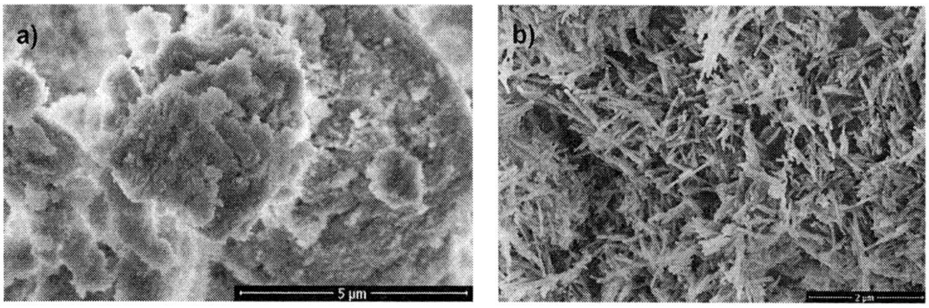

Figure 1 a) Highly agglomerated hydroxyapatite powder before the hydrothermal treatment, b) the nanoparticles with needle-like shape after the hydrothermal treatment

3.2. Preparation of the fibers

Uniform and smooth nanofibers with diameter of approximately 180 nm were prepared by electrospinning in hot air of the polycaprolactone precursor (see **Figure 2a**). Fibers with incorporated hydroxyapatite particles were not as uniform as pore polycaprolactone ones. The ceramic-polymer fibers had visible defects in the structure and the ceramic particles penetrated through the polymer out of the fibers as can be seen in **Figure 2b**. These spots were identified as the places that would enhance proliferation of the cells.

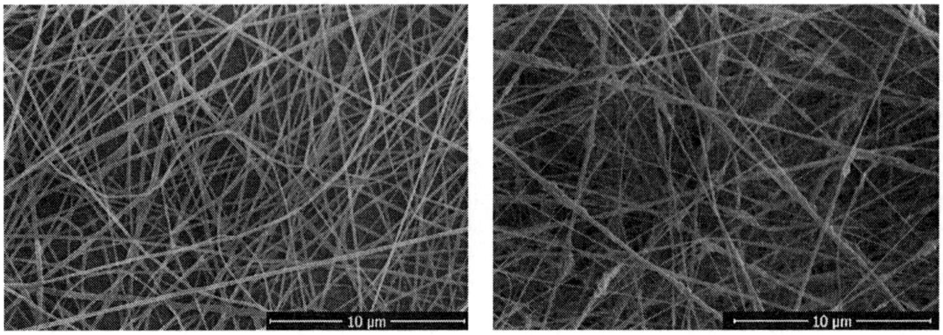

Figure 2 a) As-spun polycaprolactone nanofibers,
b) as-spun composite nanofibers with hydroxyapatite particles

3.3. Cytotoxicity tests

Test of the cells proliferation confirmed beneficial effect of the fibrous structure and presence of the ceramic compound on the cells activity. As can be seen in **Figure 3**, polycaprolactone as bulk material is biocompatible but it is not bioactive. Addition of the ceramic particles enhanced the biological properties, but the proliferation was still slower than on the baseline sample. However, on the electrospun ceramic-polymer nanofibers the proliferation was the fastest in comparison to all tested samples. Specific growth rate was calculated by measuring the numbers of the cells attached to the sample (see **Figure 4**). With exception of layered samples, the specific growth rate increased until 48 hours. The highest specific growth rate after 48 hours had polymer-ceramic fibrous sample. The specific growth rate then decreased that could be caused by formation of cells monolayer on the top of the sample (see **Figure 5**), so the cells did not have enough space for further proliferation. The high specific growth rate of the ceramic-polymer sample after 48 hours of cultivation shows accelerated cells activity on the fibrous samples at the beginning of the cells cultivation that indicates good bioactivity properties.

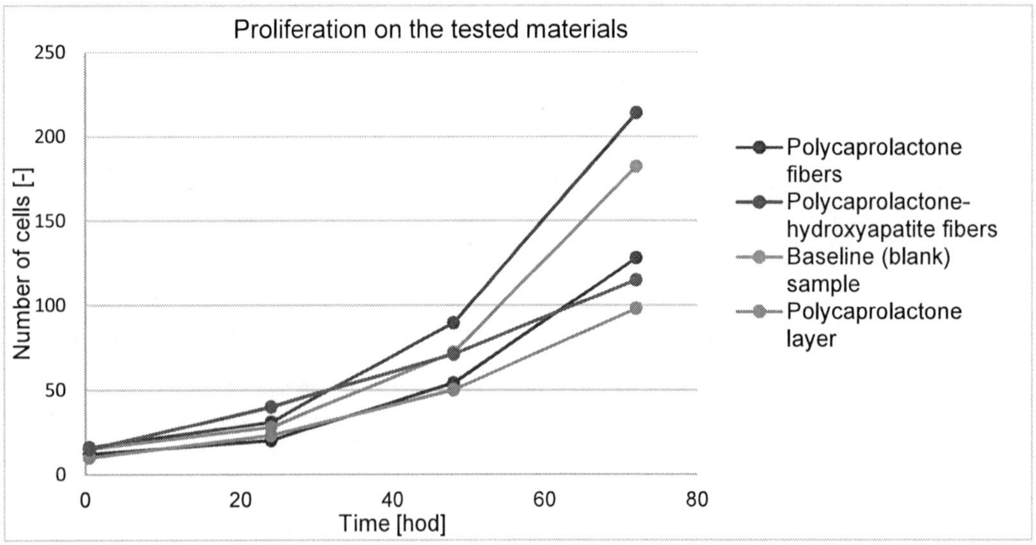

Figure 3 Proliferation of the cells on the tested materials

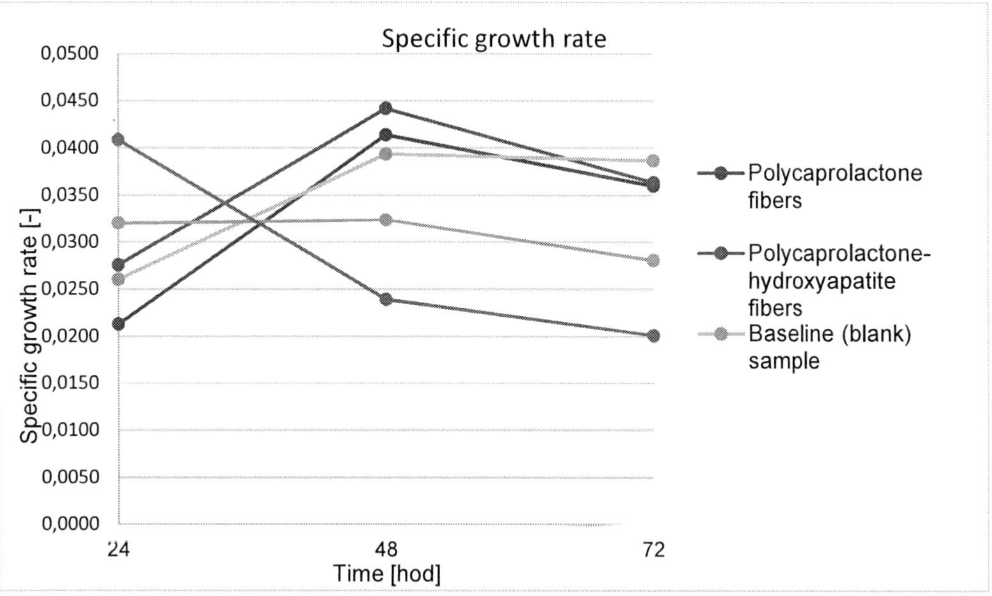

Figure 4 Specific growth rate of the cells on the tested materials

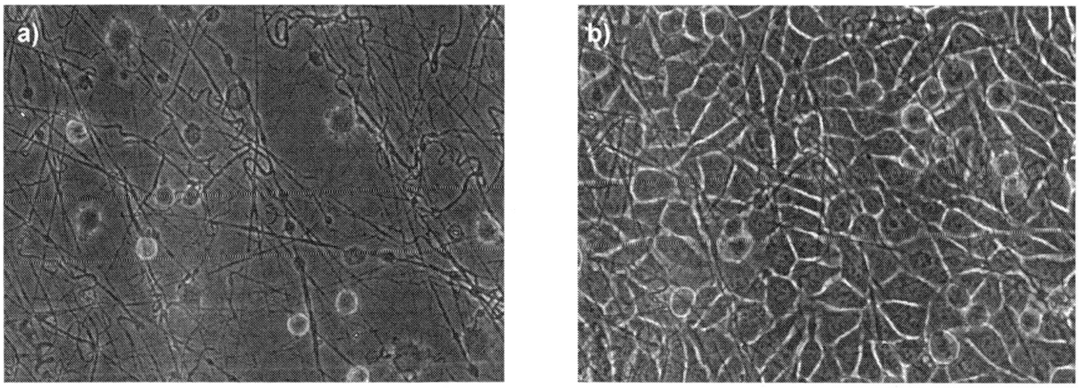

Figure 5 Polycaprolactone-hydroxyapatite fibrous structure with cell culture a) at the beginning of the test, b) after 3 days of testing

4. CONCLUSION

The experiments proved positive effect of the fibrous structure and ceramic phase in the composite on activity of the cells. While ceramic particles enhance proliferation of the cell culture, the polymer compound provides mechanical properties important for manipulation with the material which is beneficial compared to usual ceramic scaffolds. The future research will concern to better surface characteristics (e.g. wettability) of the composite fibers to reach even faster proliferation of the cells.

ACKNOWLEDGEMENTS

The work was financially supported by the junior grant STI-J-18-5564. Part of the work was carried out with the support of CEITEC Nano Research Infrastructure (ID LM2015041, MEYS CR, 2016-2019), CEITEC Brno University of Technology. Part of the work was carried out at the University of South Bohemia, Faculty of Fisheries and Protection of Waters, Institute of Complex Systems.

REFERENCES

[1] BLACK, Cameron R. M., Vitali GORIAINOV, David GIBBS, Janos KANCZLER, Rahul S. TARE a Richard O. C. OREFFO. Bone Tissue Engineering: the materials side. *Current Molecular Biology Reports*. 2015, **1**(3), 132-140. DOI: 10.1007/s40610-015-0022-2. ISSN 2198-6428. Available from: http://link.springer.com/10.1007/s40610-015-0022-2

[2] PINA, Sandra, Joaquim M. OLIVEIRA a Rui L. REIS. Natural-Based Nanocomposites for Bone Tissue Engineering and Regenerative Medicine: A Review. *Advanced Materials*. 2015, **27**(7), 1143-1169. DOI: 10.1002/adma.201403354. ISSN 09359648. Available from: http://doi.wiley.com/10.1002/adma.201403354

[3] PINA, Sandra, Joaquim M. OLIVEIRA a Rui L. REIS. Natural-Based Nanocomposites for Bone Tissue Engineering and Regenerative Medicine: A Review. *Advanced Materials*. 2015, **27**(7), 1143-1169. DOI: 10.1002/adma.201403354. ISSN 09359648. Available from: http://doi.wiley.com/10.1002/adma.201403354

[4] ENGEL, Elisabeth, Alexandra MICHIARDI, Melba NAVARRO, Damien LACROIX a Josep A. PLANELL. Nanotechnology in regenerative medicine: the materials side. *Trends in Biotechnology*. 2008, **26**(1), 39-47. DOI: 10.1016/j.tibtech.2007.10.005. ISSN 01677799. Available from: http://linkinghub.elsevier.com/retrieve/pii/S0167779907002892

[5] LANNUTTI, J., D. RENEKER, T. MA, D. TOMASKO a D. FARSON. Electrospinning for tissue engineering scaffolds. *Materials Science and Engineering: C*. 2007, **27**(3), 504-509. DOI: 10.1016/j.msec.2006.05.019. ISSN 09284931. Available from: http://linkinghub.elsevier.com/retrieve/pii/S0928493106001421

THE EFFECT OF POLY-E-CAPROLACTONE NANOFIBERS WITH PLATELETS ON CELL PROLIFERATION AND VIABILITY

[1,2,3]Věra SOVKOVÁ, [1,2,3]Karolína VOCETKOVÁ, [2,3]Evžen AMLER

[1]*Department of Tissue Engineering, Institute of Experimental Medicine of the Czech Academy of Sciences, Prague, Czech Republic, EU, vera.sovkova@iem.cas.cz*
[2]*2nd Faculty of Medicine, Charles University, Prague, Czech Republic, EU, karolina.vocetkova@iem.cas.cz*
[3]*University Centre for Energy Efficient Buildings CTU, Prague, Czech Republic, EU, karolina.vocetkova@cvut.cz*

Abstract

Platelets are effective source of growth factors and other bioactive molecules. They can be combined with an appropriate scaffold where the mechanical properties of the scaffold and bioactive properties of platelets modulate cell behavior in the site of the injury. In this study, we compared different concentrations of platelets adhered on PCL nanofibers and evaluated their effect on mouse 3T3 fibroblasts. Platelets in maximum concentration significantly stimulated the viability when it is compared with control. Proliferation of fibroblasts was significantly higher than in the control group cultured on the plain scaffolds. Cells were well- spread on the scaffolds with adhered platelets, whereas on the plain scaffolds, the cells created clusters. The effect of platelets was dose-dependent. It is obvious, that the combination of bioactive substances and PLC nanofibres significantly stimulates cells growth and metabolic activity of the tested cell type.

Keywords: Platelets, nanofibers, cell culture

1. INTRODUCTION

Proper wound dressings should enhance cell adhesion and stimulate cell proliferation and viability. Nowadays, lot of efforts is being made to create biocompatible and biodegradable scaffolds which mimic the natural microenviroment of the tissue of interest. This scaffold serves as a physical support for cell adhesion and it is subsequently degraded by the cells for the formation of the new extracellular matrix where they will grow and cooperate. Scaffold can be enriched by bioactive substances which affect cellular fate.

Platelets are small anucleated fragments derived from megakaryocytes and contain various growth factors and bioactive molecules such as PDGF, VEGF, EGF and IGF [1]. These mediators can influence cell behavior and are involved in the early stage of wound healing and regenerative processes. Several types of platelet products including platelet rich plasma, platelet lysate, and platelet rich fibrin can be fabricated. Platelet rich plasma is defined as an autologous platelet concentrate of suspended in a small amount of plasma [2]. It is a cost-effective and easy way of delivering the bioactive substances.

Electrospinning is a unique technique to create polymeric nanofibers. The fibers have nanoscale diameter and thus can mimic the microarchitecture of extracellular matrix. They are highly porous with interconnected pores, enabling the exchange of gases and nutrients [3]. Moreover, compared to micrometer-sized surfaces, nanofibers have a higher rate of protein adsorption support an adhesion of proteins [4].

Polycaprolactone is a biodegradable, biocompatible, linear, aliphatic polyester, which is widely used in tissue engineering applications of bone [5], cartilage [6] and heart and vessels [7]-[9]. It was approved by the FDA as sewing material and material for drug delivery [10].

This study was focused on comparison of the viability and proliferation of fibroblast cell line cultured on composite PCL nanofibers scaffolds enriched with different concentrations of platelet.

2. MATERIALS AND METHODS

2.1. Scaffold fabrication

Scaffolds were prepared from poly-ε-caprolactone by needleless electrospinning method. 24 % (w/v) solution of PCL (45 000 Da, Sigma Aldrich, MO, USA) was resuspended in chlorophorm:ethanol (ratio 9:1 v/v). Samples were electrospun by Nanospider NS500 (Elmarco, Czech Republic) using wire needleless electrode. The applied voltage range was 80-100 kV. The temperature was 24 ± 2°C and relative humidity 55 ± 10 %. The distance between the wire electrode and the collector was 20 cm.

2.2. Composite scaffold preparations

Samples with diameter of 0.6 mm were punched out from the electrospun mat. Subsequently, they were sterilized using 70 % ethanol for 30 minute and washed thrice with PBS. Human deleucotized platelet concentrate in plasma solution derived from apheresis was purchased from Blood Transfusion Service (UHKT, Czech Republic). Three concentrations of platelets were prepared (TAD max - maximum concentration, $915*10^9$ platelets/L; TAD 1/2 - half the maximum concentration, $457.5*10^9$ platelets/L; TAD ¼ - quarter the maximum concentration, $228.75*10^9$ platelets/L). Each concentration was adhered on the sterile nanofibrous scaffolds. Samples were incubated for 2 hours at 22°C and then were rinsed with PBS. The plain scaffolds served as a control.

2.3. Cell culture and seeding

Mouse 3T3-A21 cell line was purchased from Sigma-Aldrich, Munich, Germany. Cells were cultured in Dulbecco's modified Eagle's medium supplemented with 10 % FBS and treated with penicillin/streptomycin (100 IU/mL, 100 µg/mL). Cells were kept in a humidified incubator (37°C, 10 % CO_2 and 80-90 % RH) and medium was changed every 3-4 days. Subconfluent cells were washed with PBS containing 0.02 % (w/v) ethylenediaminetetraacetic acid (EDTA) and treated with trypsin/EDTA solution. The detached cells were counted using a Bürker chamber. The scaffolds were seeded with fibroblasts with a density 8,850 cells/cm^2. All groups were supplemented with 5 % FBS.

2.4. Cell metabolic activity measurement

The metabolic activity of the fibroblast was determined using the MTS assay (CellTiter96® AQueous One Solution Assay; Promega, Madison, WI, USA). On the experimental days, the scaffolds were transferred to the new wells. To each scaffold, 20 µL of MTS solution within 100 µl fresh medium were added. After 2 hours of incubation, the absorbance of the metabolized solution (100 µL) was detected at 490 nm using Tecan Infinite® PRO 200 series reader (Tecan, Männedorf, SUI). The background absorbance (690 nm) was subtracted from the measured data, as well as the absorbance of the plain media.

2.5. Cell proliferation

To determine the cell proliferation, a fluorescence-based kit (Quant- iT™ PicoGreen® dsDNA Assay Kit; Invitrogen, Thermo Fisher Scientific, Waltham, MA, USA) was used. The samples used for the MTS assay were transferred to 500 µL of cell lysis buffer (10 mmol L−1 Tris, 1 mmol L−1 EDTA, 0.2 % v/v Triton X-100). Then, three freeze/ thaw cycles were done to facilitate of the release DNA from the samples. Between each step, the samples were thoroughly vortexed. Fluorescence intensity was detected using Tecan Infinite® PRO 200 series reader Tecan (λex = 485 nm, λem = 528 nm) and the DNA content was determined using calibration curve of standards from the kit.

2.6. Confocal microscopy visualization

Samples were fixed with frozen methanol and subsequently washed with PBS and stained with DiOC (to 1 µg/mL in PBS, 30 minute) to visualize the cellular membranes and propidium iodide (5 µl/mL) to visualize the cell nuclei. Between the incubations the samples were rinsed with PBS. A Zeiss LSM 510 DUO confocal microscope was used for imaging (λex = 488 nm, λem = 501 nm for DiOC6; λex = 536 nm,λem = 617 nm for propidium iodide).

2.7. Scanning electron microscopy visualization

Samples were fixed with 2.5 % glutaraldehyde ((Sigma-Aldrich, MO, USA) for 2 hours in 4°C. Afterwards the samples were dehydrated with the ethanol changes (35 %, 48 %, 70 %, 96 % and 100 %). Hexamethyldisilazane (Sigma-Aldrich, MO, USA) was used to remove any residual water from the samples. The samples were sputter-coated with gold (app. 30 nm) in the Quorum Q150R (Quorum Technologies, United Kingdom) and visualized using Vega 3 SBU microscope (Tescan, Czech Republic).

2.8. Statistical analysis

Quantitative data are presented as mean ± standard deviation. The data were evaluated using SigmaStat 3.5 software. Statistically significant differences between the groups with normal distribution were evaluated with ANOVA method and Tukey test for post hoc analysis. The data with abnormal distribution were tested with Kruskal-Wallis test and Dunn's method for post hoc analysis. The statistically significant level was accepted as 5 %

3. RESULTS

Metabolic activity of the seeded cells was evaluated using MTS metabolic assay on day 1, 3, 8, 10 and 14 (**Figure 1**). PCL nanofibers functionalized with the maximum concentration of platelets (TAD max) exhibited a significantly higher viability of cells in comparison to the control group (PCL). On day 1 and 3, there was no significant difference between the groups. From day 8 up to the day 14, it was obvious that with increasing concentration of platelets, the metabolic activity of the cells increased. Proliferation of the cells were measured by PicoGreen assay (**Figure 2**). The results showed increase of DNA on all the scaffolds functionalized with platelets from day 3. All groups with platelets exhibited gradually increasing number of cells in a dose-independent manner. On the control group, the DNA content was very low and oscillated during the experiment. It is obvious that the scaffolds functionalized with platelets stimulated cell proliferation in vitro. Results from both assays are in concordance with the results of Vocetkova et. al., where platelets in additive solution adhered on PCL scaffold were used and seeded with dermal cells. Fibroblast exhibited better proliferation and metabolic activity when seeded on the functionalized scaffolds [3]. Diaz-Gomez et. al. cultured human adipose-derived mesenchymal stem cells on PCL nanofibers functionalized with platelets. These functionalized scaffolds increased cell adhesion and proliferation [11].

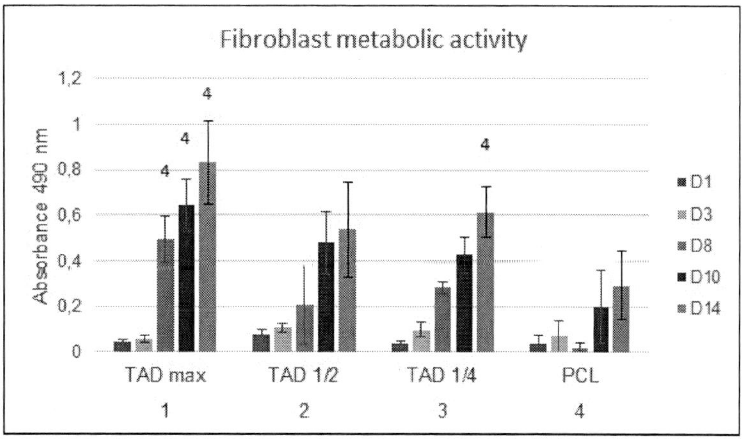

Figure 1 Metabolic activity of 3T3 fibroblast

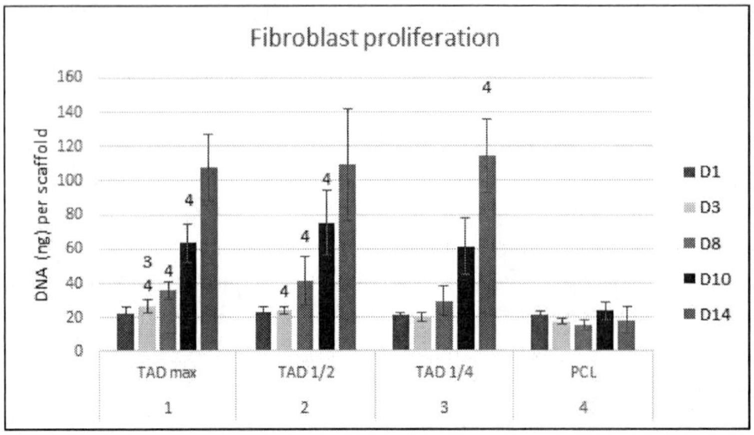

Figure 2 Proliferation of fibroblast

Confocal microscopy was used to visualized the cells seeded on the scaffolds (**Figure 3**). Cells seeded on the functionalized scaffolds were well spread and did not form the clusters as on the PCL control scaffolds. These clusters indicated unfavorable cultering conditions.

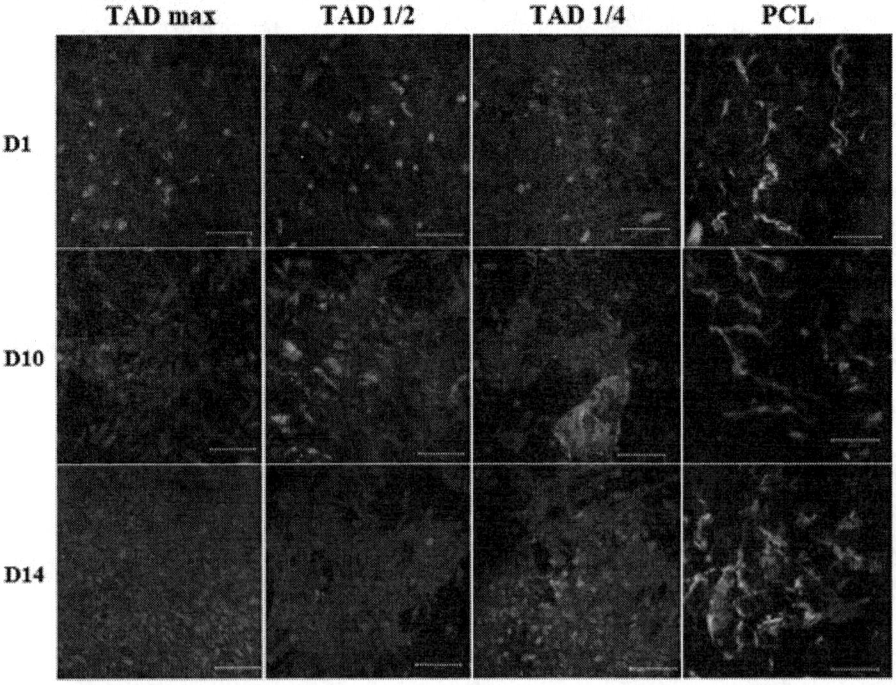

Figure 3 Confocal microscopy visualization of fibroblasts seeded on the scaffolds (magnification 200x, scale bar 100 μm)

To visualize the surface of the nanofibers with platelets and cells, scanning electron microscopy was used (**Figure 4**). The fibrin network was visible in all the samples with platelets. Lesczak et al. tested different types of surfaces to observe blood compatibility. They found that platelet/leukocyte complexes were formed and specific markers for platelet activation were expressed on PCL surfaces [12].

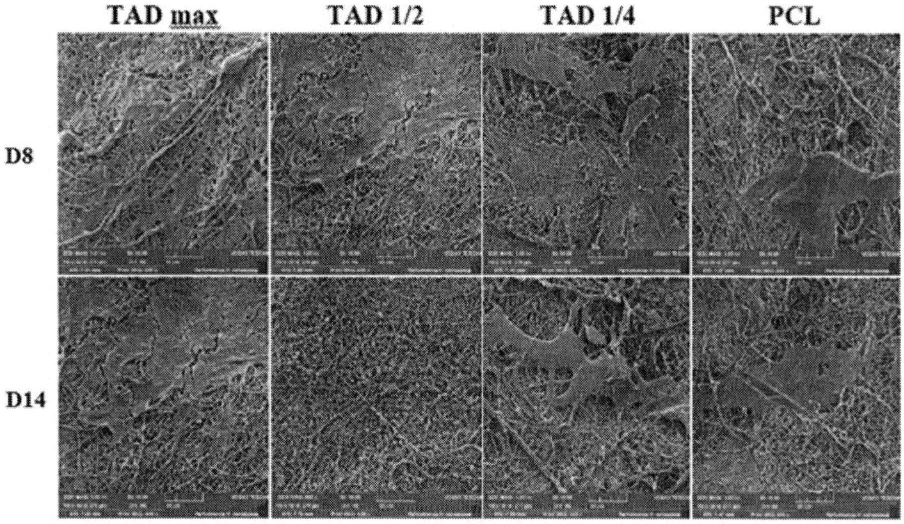

Figure 4 SEM visualization of fibroblasts on scaffolds (magnification 1000x, scale bar 50 μm)

4. CONCLUSION

Nanofibrous scaffold from poly-ε-caprolactone is, thanks to its properties, a suitable material for tissue engineering applications. The scaffold can be combined with bioactive molecules which affect cellular fate. We have previously tested the effect of platelets in additive solution, adhered on PCL nanofibers prepared by centrifugal spinning, on MG-63 cell lines. In the current study, we used pooled platelets in plasma adhered on PCL nanofibers prepared by electrospinning method. As a model for *in vitro* testing, 3T3 fibroblasts were used. It was observed that platelets and plasma had synergistic effect and positively influenced cell proliferation and viability. Our results clearly demonstrated that all concentrations of platelets in plasma are sufficient to ensure cells growth.

ACKNOWLEDGEMENTS

This study has been supported by the Research Program NPU I:LO1508, Prague Competitiveness CZ.2.16/3.1.00/21528, Internal Grant Agency of the Ministry of Health of the Czech Republic No 17-32285A, Ministry of Industry and Trade within TRIO program grant no FV30311.

REFERENCES

[1] FITCH-TEWFIK Jennifer L. and FLAUMENHAFT Robert, *Platelet granule exocytosis: a comparison with chromaffin cells.*: *Frontiers in endocrinology*. 2013. vol. 4 p. 77.

[2] FOSTER Timothy E., PUSKAS Brian L., MANDELBAUM Bert R., GERHARDT Michael B., and RODEO Scott A., *Platelet-Rich Plasma*: *The American Journal of Sports Medicine*. 2009. vol. 37, no. 11, pp. 2259-2272.

[3] VOCETKOVA Karolina, BUZGO Matej, SOVKOVA Vera, BEZDEKOVA Dagmar, KNEPPO Peter, and AMLER Evzen, *Nanofibrous polycaprolactone scaffolds with adhered platelets stimulate proliferation of skin cells*: *Cell Proliferation*. 2016. vol. 49, no. 5,.

[4] DAHLIN Rebecca L., KASPER, Kurtis F., and MIKOS, Antonios G., *Polymeric nanofibers in tissue engineering.*: *Tissue engineering. Part B, Reviews*. 2011. vol. 17, no. 5, pp. 349-64.

[5] SHAHREZAEE Mostafa, SALEHI Majid, KESHTKARI Sara, ORYAN Ahmad, KAMALI Amir, and SHEKARCHI Babak, *In vitro and in vivo investigation of PLA/PCL scaffold coated with metformin-loaded gelatin nanocarriers in regeneration of critical-sized bone defects*: *Nanomedicine: Nanotechnology, Biology and Medicine*. 2018. .

[6] FU, Na, LIAO, Jinfeng, LIN, Shiyu, SUN, Ke, TIAN, Taoran, ZHU, Bofeng, AND LIN, Yunfeng, *PCL-PEG-PCL film promotes cartilage regeneration in vivo*: *Cell Proliferation*. 2016. vol. 49, no. 6, pp. 729-739.

[7] BERTRAM, Ulf, STEINER, Dominik, POPPITZ, Benjamin, DIPPOLD, Dirk, KÖHN, Katrin, BEIER, Justus P., DETSCH, Rainer, BOCCACCINI, Aldo R., SCHUBERT, Dirk W., HORCH, Raymund E., AND ARKUDAS, Andreas, *Vascular Tissue Engineering: Effects of Integrating Collagen into a PCL Based Nanofiber Material*: *BioMed Research International*. 2017. vol. 2017 pp. 1-11.

[8] HASAN, Anwarul, SOLIMAN, Sherif, EL HAJJ, Fatima, TSENG, Yuan-Tsan, YALCIN, Huseyin C., AND MAREI, Hany Elsayed, *Fabrication and In Vitro Characterization of a Tissue Engineered PCL-PLLA Heart Valve*: *Scientific Reports*. 2018. vol. 8, no. 1, p. 8187.

[9] TALLAWI, M., DIPPOLD, D., RAI, R., D'ATRI, D., ROETHER, J. A., SCHUBERT, D. W., ROSELLINI, E., ENGEL, F. B., AND BOCCACCINI, A. R., *Novel PGS/PCL electrospun fiber mats with patterned topographical features for cardiac patch applications*: *Materials Science and Engineering: C*. 2016. vol. 69 pp. 569-576.

[10] MALIKMAMMADOV, Elbay, TANIR, Tugba Endogan, KIZILTAY, Aysel, HASIRCI, Vasif, AND HASIRCI, Nesrin, *PCL and PCL-based materials in biomedical applications*: *Journal of Biomaterials Science, Polymer Edition*. 2018. vol. 29, no. 7-9, pp. 863-893.

[11] DIAZ-GOMEZ, Luis, ALVAREZ-LORENZO, Carmen, CONCHEIRO, Angel, SILVA, Maite, DOMINGUEZ, Fernando, SHEIKH, Faheem A., CANTU, Travis, DESAI, Raj, GARCIA, Vanessa L., AND MACOSSAY, Javier, *Biodegradable electrospun nanofibers coated with platelet-rich plasma for cell adhesion and proliferation.*: *Materials science & engineering. C, Materials for biological applications*. 2014. vol. 40 pp. 180-8.

[12] LESZCZAK, Victoria AND POPAT, Ketul C., *Improved in vitro blood compatibility of polycaprolactone nanowire surfaces.*: *ACS applied materials & interfaces*. 2014. vol. 6, no. 18, pp. 15913-24.

INCORPORATING ANTIBACTERIAL CLAY MINERALS INTO NANOFIBROUS LAYERS BY ELECTROSPINNING

[1]Lenka BARDOŇOVÁ, [2]Adéla KOTZIANOVÁ, [3]Kateřina MAMULOVÁ KUTLÁKOVÁ, [3]Sylva HOLEŠOVÁ, [2]Jan KLEMEŠ

[1]VSB - Technical University of Ostrava, Faculty of Metallurgy and Materials Engineering, Ostrava, Czech Republic, EU, lenka.bardonova.st@vsb.cz
[2]CONTIPRO a.s., Dolní Dobrouč, Czech Republic, EU, adela.kotzianova@contipro.com, jan.klemes@contipro.com
[3]Nanotechnology Centre, VSB - Technical University of Ostrava, Ostrava, Czech Republic, EU, katerina.mamulova.kutlakova@vsb.cz, sylva.holesova@vsb.cz

Abstract

The aim of this study was to describe electrospinning method for preparation of self-supporting homogenous nanofibrous layers with a presence of pristine clay minerals and clay minerals containing antibacterial agent chlorhexidine acetate in their interlayer space. One clay and two polymers were used. Vermiculite was used as obtained with size fraction < 40 µm. Chlorhexidine/vermiculite was prepared through the intercalation process. Nanofibers were made of hydrophobic polymers, polyurethane (PU) and polycaprolactone (PCL), to gain water stable and durable layers. Polymer solutions for electrospinning contained 2, 5 and 8 wt. % (according to the total weight of the solution) of clay or chlorhexidine/clay. These suspensions were homogenized and immediately spun using 4SPIN LAB. Self-supporting homogenous fibrous layers were prepared under the same electrospinning conditions as neat polymers and were further analysed. Morphology was characterized using scanning electron microscopy (SEM) and presence of clay minerals in the layer was confirmed by digital microscopy and EDX mapping. From SEM images, diameter of the fibres was evaluated. Fibre diameter decreased after adding the clay and was ranging from 600 nm to 1200 nm. Clay particles were present both in fibres and on the surface. Antibacterial chlorhexidine was found in the vermiculite matrix as well as separately in the fibres (result of imperfect intercalation).

Keywords: Clay, nanofibers, electrospinning, chlorhexidine, antibacterial

1. INTRODUCTION

Electrospinning is a method using an electrostatic field/high voltage to fabricate micro- and nanoscaled fibres. This method is constantly evolving and gains attention in both scientific community and industry. It is versatile method which can find novel applications in diverse fields - bioengineering, biomedicine, sensors, filtration or electronics. Electrospun fibres have unique properties such as high specific surface area and large number of smaller pores. There is a large variety of polymer materials which could be used to prepare nanofibers, from natural and to synthetic ones. There is a possibility to prepare simple layers made of a single polymer as well as a composite, which is composed of e.g. several polymers with some organic or inorganic components [1,2]. In this study two polymers were used - polyurethane (PU) and polycaprolactone (PCL). Both are hydrophobic, water stable and biodegradable and are often used for bone tissue engineering, cardiac grafts, wound dressing or for engineering of blood vessel substitutes [1]. Clay minerals are hydrated aluminium phyllosilicates with layered structure. They may undergo process called intercalation, which allows ion exchange with functional molecules or particles. Vermiculite belongs to 2:1 group, its basic structure consists of a magnesium octahedral sheet between two silica tetrahedral sheets. Interlayer space is filled with hydrated cations or water [3]. Since the interlayer space can be easily modified, vermiculite has a great potential in the fields of drug carriers [4-6], tissue engineering and regenerative medicine [7]. In several studies, nanocomposite chlorhexidine acetate/vermiculite was prepared through intercalation process and characterized. Chlorhexidine (CH) was stable in clay matrix (outflow <5 %) in aqueous solution with conditions close to those in human body [6]. Composite had very good antibacterial effect against *Escherichia faecalis*, *Escherichia coli* and *Staphylococcus aureus*, and showed poor results against *Pseudomonas aeruginosa* [4-6]. Vermiculite was also tested for toxicity on mammal model organism (mice). Neither local nor systemic reaction was observed, therefore toxicity can be excluded [5]. Studies focused on electrospinning of polymers with added clay often investigate the effect of clay content on morphology and processability of prepared nanofibers [8, 9, 10]. Nanofibers with added clay had enhanced filtrating ability [8], increased thermal stability [9,11], improved tensile strength [10, 11] or chemical stability [12]. Wide range of polymers was used with clay such as montmorillonite and hydroxyapatite. After intercalating drugs into the interlayer space nanofiber mats obtained antibacterial properties [13,14] with drug sustained release activity [13]. Prepared materials have a potential in wound healing applications [12], drug delivery systems [13,14], as an antioxidant product in food and pharmaceutical industries [15] or filtration [16,17].

The aim of this study was to find optimal conditions for preparation of self-supporting homogenous nanofibrous layers with a presence of pristine vermiculite and composite chlorhexidine acetate/vermiculite since there is no similar research. Further to characterize morphology of prepared samples using different methods (digital microscopy, scanning electron microscopy, EDX mapping). Prepared materials might have a great potential as inexpensive drug delivery system.

2. MATERIALS AND METHODS

2.1. Preparing solutions for electrospinning

Polycaprolactone (PCL) and chloroform were purchased from Sigma Aldrich. Polyurethane (PU), *N,N*-dimethylformamide (DMF) and tetrahydrofuran (THF) were purchased from BASF Polyurathanes GmbH, Lachner and Honeywell, respectively. Vermiculite from Brazil was used with particle size fraction < 40 μm. Composite chlorhexidine acetate/vermiculite (1:1) was prepared through intercalation process. Both clay samples were provided by Nanotechnology Centre (VSB) and used as received. Two basic polymer solutions were prepared. First one was the 10 wt. % PU dissolved in DMF and THF (1:1). PU was added under constant stirring and solution was stirred until homogenous. The second one was the 10 wt. % PCL dissolved in DMF and chloroform (2:8). PU was added under constant stirring and solution was stirred until homogenous. Suspensions for electrospinning contained 2, 5 and 8 wt. % of vermiculite (Ver) or chlorhexidine/vermiculite (CH/Ver). The weight percent of added clay was with respect to the weight of the final suspension. Each suspension was stirred first in hand and then placed into stand with coolant (water and ice) and homogenized for 5 min by homogenizer IKA® T25 digital Ultra Turrax (5600 rpm, 15 s on, 5 s pause). Prepared suspension was immediately spun.

2.2. Electrospinning conditions

Electrospinning was performed using the 4SPIN LAB device (Contipro, Czech Republic). Two moving needles were used as an emitter and rotating cylinder of width 10 cm covered by substrate (aluminium foil) was used as a collector. The collector rotating speed was 1000 rpm. A distance between the emitter and the collector was set to 18 cm. Each sample have been spun for 120 min, but after 60 min the suspension was changed for newly homogenized one. Feed rate was 20 μl/min for all suspensions except 8 wt. % of CH/Ver in PCL where feed rate had to be increased to 30 μl/min. Applied voltage varied depending on polymer - 25 kV was used for the PU suspensions and 20 kV for the PCL suspension. Ambient conditions were approximately same for all spinning processes - relative humidity was between 30-43 % and temperature was between 23-28 °C.

2.3. Characterization of materials

A digital microscopy was used to confirm the presence of vermiculite in prepared nanofibrous layers, as vermiculite has yellow-brown to green-brown colour and thus was easily distinguishable from polymer. Images were taken in transmission using the digital microscope VHX Multi Scan Keyence with Keyence ZS-260 lens (magnitude 200x-2000x). All samples were removed from the substrate and placed onto a glass microscope slide. No other treatment of samples was performed. Morphology was characterized by scanning electron microscopy (Zeiss Ultra Plus, Carl ZEISS, applied voltage 3,5 kV). Prior analysis, all samples were coated with a thin layer of chrome. From SEM images (1000 x), diameter of fibres was evaluated using ImageJ 1.48 (average of 30 fibres). Presence of vermiculite and chlorhexidine in the layer was also confirmed by energy-dispersive X-ray spectroscopy mapping (EDX mapping, Zeiss Ultra Plus, Oxford X-MAXN 80). The measurement was conducted at a magnification of 1800, the accelerating voltage was set to 10 kV and the energy range was between 0-10 keV. The intensity of the characteristic X-ray peaks was imaged - for vermiculite (aluminium K-series - 1.557, 1.487, 1.486, magnesium K-series - 1.302, 1.253, silica K-series - 1.836, 1.740, 1.739) and chlorhexidine (chlorine K-series 2.816, 2.622, 2.620).

3. RESULTS AND DISCUSSION

Self-supporting homogenous nanofibrous layers were successfully prepared from above-mentioned suspensions via electrospinning. For clarity, the notation *wt.%Clay_polymer* will be used in the following text, e.g. 2Ver_PU refers to 2 wt.% of vermiculite in polyurethane or 8CH/Ver_PCL refers to 8 wt.% of chlorhexidine/vermiculite in polycaprolactone. Control samples were also prepared (further marked as PU and PCL). Both SEM imaging and EDX mapping were used strictly for qualitative analysis of the composite, to characterize the morphology and whether the clay or CH/clay additive was incorporated into the polymer matrix.

3.1. Digital microscopy

The **Figure 1** shows the differences between 2, 5 and 8 wt.% Ver in PCL (a-c) - the increasing clay content in the nanofibrous layers is evident. All other samples followed the same trend. Some agglomerates in **Figure 1c** exceed the size of 100 μm even though vermiculite size fraction was under 40 μm. This trend was visible especially in samples with PCL. Suspensions with PCL were more viscous, so the viscosity probably affects creation of agglomerates or aggregates.

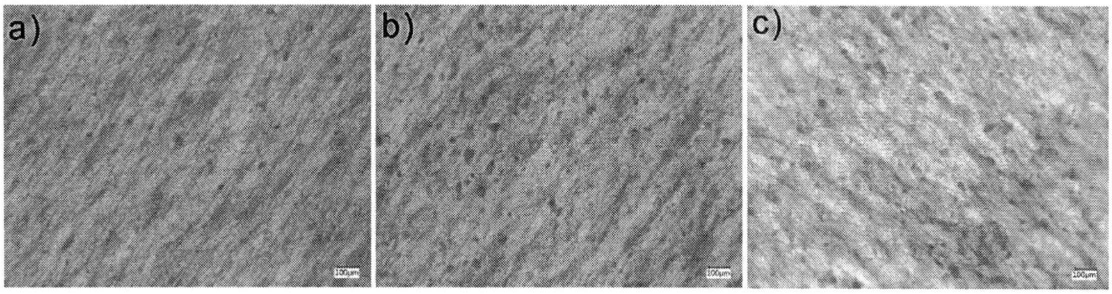

Figure 1 Digital microscopy images of PCL nanofibers with vermiculite (magnification 200x): a) 2Ver_PCL, b) 5Ver_PCL, c) 8Ver_PCL.

3.2. Morphology of the fibres

Neat polymer nanofibers were smooth and continuous with inhomogeneous diameter which was (1416 ± 522) nm for PCL and (1629 ± 311) for PU. The **Figure 2** shows selected SEM images (magnification 5000x) of 5Ver_PCL (a), 5CH/Ver_PCL (b), 5Ver_PU (c), and 5CH/Ver_PU (d). Clay particles were probably incorporated into the PCL fibres as there are not many visible particles on their surface. PU fibres are thinner than PCL fibres and there are more clay particles visible on the surface of the fibres. Diameter of clay/PU fibres is also more homogeneous compared to clay/PCL fibres. The **Figure 3** shows a mean diameter of the fibres with a standard deviation. Diameter of the fibres decreased after adding the clay. The highest diameter was observed at samples with 2 wt. % of clay, then diameter decreased at samples with 8 wt. %, and samples with 5 wt. % of clay had the lowest diameter. But samples with Ver in PCL did not meet this correlation, the highest diameter was at 5Ver_PCL and lowest at 8Ver_PCL. It was observed that electrospun fibres were thicker the more clay they contained [8]. This thickening was caused by higher viscosity of electrospun solution - fibres haven't stretched so much. There is also another factor that affects the final diameter of the fibres. Vermiculite is a clay with nonzero layer charge, so by adding the clay suspension had higher conductivity and therefore was more stretched in the electric field [10]. In this case both principles seemed to combine, suspension with 2 and 5 wt.% stretched more thanks to higher conductivity but suspension with 8 wt.% of clay were too viscose to stretch.

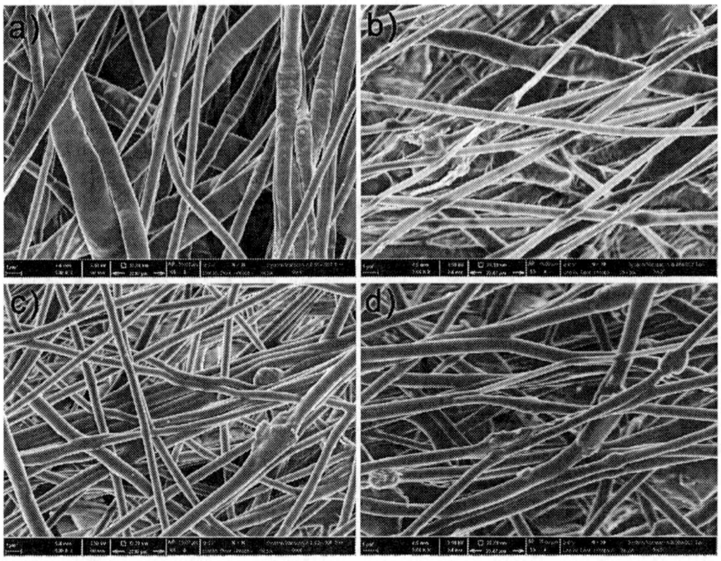

Figure 2 SEM images of PCL or PU nanofibers with vermiculite or chlorhexidine/vermiculite (magnification 5000x): a) 5Ver_PCL, b) 5CH/Ver_PCL, c) 5Ver_PU, d) 5CH/Ver_PU.

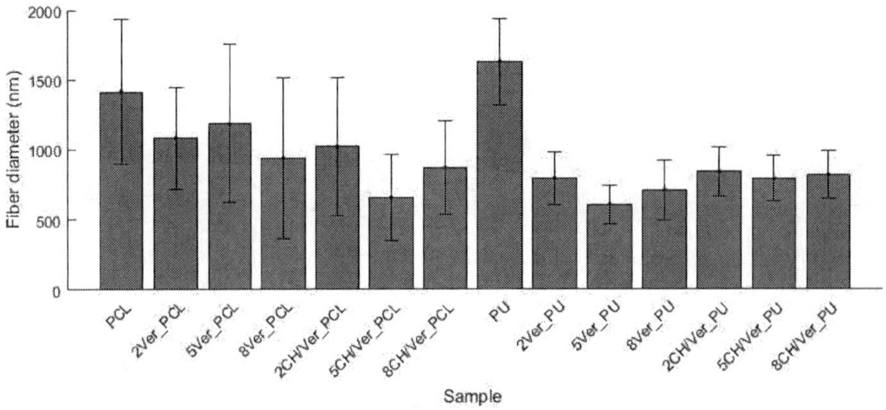

Figure 3 Mean fiber diameter (30 fibres) with standard deviation.

3.3. EDX mapping

In EDX maps (**Figure 4**), major elements contained in vermiculite are blue-coloured (silicon, aluminium and magnesium) and chlorine (recognizable element for chlorhexidine) is red-coloured. Clay particles were mostly concentrated along the fibres and some bigger particles (size exceeding fibre diameter multiple times) were between the fibres. Chlorhexidine was not always situated with vermiculite, there were many spots, where only chlorhexidine or only vermiculite could be found. This means that the intercalation process was not perfect and not all the chlorhexidine was intercalated into interlayer space of vermiculite. These findings fit with results of Holešová et. al. [6] that even the highest used chlorhexidine concentration above CEC value (cation exchange capacity) did not lead to full chlorhexidine intercalation into vermiculite matrix and chlorhexidine is also anchored on clay surface.

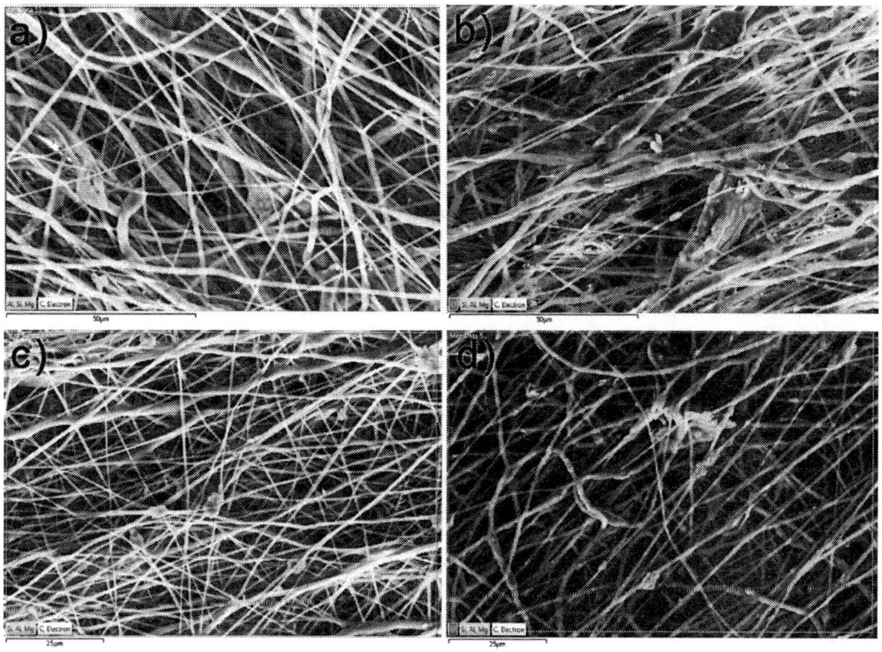

Figure 4 EDX maps, blue coloured: Si, Al, Mg, red coloured: Cl: a) 5Ver_PCL, b) 5CH/Ver_PCL, c) 5Ver_PU, d) 5CH/Ver_PU

4. CONCLUSION

Nanofibrous layers with different weight concentration of vermiculite or chlorhexidine/vermiculite were successfully prepared under the same electrospinning conditions as neat polymers. Fibre diameter decreased after adding the clay and was 600-1200 nm. Clay content in the layer increased with concentration and clay particles were incorporated into the polymer fibres. Antibacterial chlorhexidine was found in the vermiculite matrix as well as separately in the fibres (as a result of imperfect intercalation). Prepared CH/Ver/polymer layers seem to be promising, relatively inexpensive, and easy-to-produce material in the field of antibacterial applications.

ACKNOWLEDGEMENTS

This paper was created in cooperation with Contipro a.s. and in the frame of the SP2018/71 project while financially supported by the Ministry of Education, Youth and Sports of the Czech Republic.

REFERENCES

[1] BHARDWAJ, N., KUNDU, S. C. Electrospinning: A fascinating fiber fabrication technique. *Biotechnology Advances*. 2010. vol. 28, no. 3, pp. 325-347.

[2] HAIDER, A., HAIDER, S., KANG, I. A comprehensive review summarizing the effect of electrospinning parameters and potential applications of nanofibers in biomedical and biotechnology. *Arabian Journal of Chemistry*. 2015. Open access available from: DOI: 10.1016/j.arabjc.2015.11.015.

[3] WEISS, Z., KUŽVART, M. *Jílové minerály: Jejich nanostruktura a využití*. 1st ed. Praha: Karolinum, 2005. p. 281.

[4] HOLEŠOVÁ, S., VALÁŠKOVÁ, M., PLEVOVÁ, E., PAZDZIORA, E., MATĚJOVÁ, K. Preparation of novel organovermiculites with antibacterial activity using chlorhexidine diacetate. *Journal of Colloid and Interface Science*. 2010. vol. 342, pp. 593-597.

[5] HOLEŠOVÁ, S., ŠTEMBÍREK, J., BARTOŠOVÁ, L., PRAŽANOVÁ, G., VALÁŠKOVÁ, M., SAMLÍKOVÁ, M., PAZDZIORA, E. Antibacterial efficiency of vermiculite/chlorhexidine nanocomposites and results of the in vivo test of harmlessness of vermiculite. *Materials Science and Egineering C*. 2014. vol. 42, pp. 466-473.

[6] SAMLÍKOVÁ, M., HOLEŠOVÁ, S., HUNDÁKOVÁ, M., PAZDZIORA, E., JANKOVIČ, Ľ., VALÁŠKOVÁ, M. Preparation of antibacterial chlorhexidine/vermiculite and release study. *International Journal of Mineral Processing*. 2017. vol. 159, pp. 1-6.

[7] MOUSA, M., EVANS, N. D., OREFFO, R.O.C, DAWSON, J.I. Clay nanoparticles for regenerative medicine and biomaterial design: A review of clay bioactivity. *Biomaterials*. 2018. vol. 159, pp. 204-214.

[8] ISARANKURA NA AYUTHAYA, S., TANPICHAI, S., SANGKHUN, W., WOOTTHIKANOKKHAN J. Effect of clay content on morphology and processability of electrospun keratin/poly(lactic acid) nanofiber. *International Journal of Biological Macromolecules*. 2016. vol. 85, pp. 585-595.

[9] PARK, J. H, LEE, H. W., CHAE, D. K., WEONTAE, O., YUN, J. D., DENG, Y., YEUM, J. H. Electrospinning and characterization of poly(vinyl alcohol)/chitosan oligosaccharide/clay nanocomposite nanofibers in aqueous solutions. *Colloid Polym Sci*. 2009. vol. 287, pp. 943-950.

[10] KIM, S. W., HAN, S. O., SIM, I N., CHEON, J. Y., PARK, W. H. Fabrication and Characterization of Cellulose Acetate/Montmorillonite Composite Nanofibers by Electrospinning. *Journal of Nanomaterials*. vol. 2015, Article ID 275230.

[11] ALMUHAMED, S., BONNE, M., KHENOUSSI, N., BRENDLE, J., SCHACHER, L., LEBEAU, B., ADOLPHE, D. C. Electrospinning composite nanofibers of polyacrylonitrile/synthetic Na-montmorillonite. *Journal of Industrial and Engineering Chemistry*. 2016. vol. 25, pp. 146-152.

[12] HEYDARY, A. H., KARAMIAN, E., POORAZIZI, E., KHANDAN, A., HEYDARIPOUR, J. A Novel Nano-Fiber of Iranian Gum Tragacanth-Polyvinyl Alcohol/Nanoclay Composite for Wound Healing Applications. *Procedia Materials Science*. 2015. vol. 11, pp. 176-182.

[13] SAHA, K., BUTOLA, B. S., JOSHI, M. Drug-Loaded Polyurethane/Clay Nanocomposite Nanofibers for Topical Drug-Delivery Application. *Journal of Applied Polymer Science*. 2014. vol. 131, no. 10.

[14] RAMÍREZ-AGUDELO, R., SCHUERMANN, K., GALA-GARCÍA, A., MONTEIRO, A. P. F., PINZÓN-GARCÍA, A. D., CORTÉS, M. E., SINISTERRA, R. D. Hybrid nanofibers based on poly-caprolactone/gelatin/hydroxyapatite nanoparticles-loaded Doxycycline: Effective anti-tumoral and antibacterial activity. Materials Science and Engineering: C. 2018. vol. 83, pp. 25-34.

[15] SANGEETHA, K., ALSHARANI, F. A., VIJAYALAKSHMI, K., ANIL, S., SUDHA, P.N. Synthesis and Antioxidant ability of Nanofiber mat prepared using Polyvinyl alcohol and Montmorillonite clay. *Journal of Pharmacy*. 2017. vol. 7, pp 7-12.

[16] ZHU, Y., CHEN, D. Novel clay-based nanofibrous membranes for effective oil/water emulsion separation. *Ceramics International*. 2017. vol. 43, pp. 9465-9471.

[17] PRINCE, J.A., SINGH, G., RANA, D., MATSUURA, T., ANBHARASI, V., SHANMUGASUNDARAM, T.S. Preparation and characterization of highly hydrophobic poly(vinylidene fluoride) - Clay nanocomposite nanofiber membranes (PVDF-clay NNMs) for desalination using direct contact membrane distillation. *Journal of Membrane Science*. 2012. vol. 397-398, pp. 80-86.

ZINC PHOSPHATE NANOPARTICLES PREPARATION AND THEIR ANTIMICROBIAL ACTIVITY

[1,2]Pavel KOPEL, [1,2]Natalia CERNEI, [1]Milica GAGIC, [3]Pavel HORKY, [1,2]Kristyna SMERKOVA, [1,2]Vojtech ADAM

[1]Department of Chemistry and Biochemistry, Faculty of Agronomy, Mendel University in Brno, Brno, Czech Republic, EU, vojtech.adam@mendelu.cz
[2]Central European Institute of Technology, Brno University of Technology, Brno, Czech Republic, EU
[3]Department of Animal Nutrition and Forage Production, Faculty of Agronomy, Mendel University in Brno, Brno, Czech Republic, EU

Abstract

Zinc phosphate particles were prepared by reaction of zinc nitrate with hydrogen phosphate, diphosphate and triphosphate, and characterized by Dynamic Light scattering (DLS) and Scanning Electron Microscopy (SEM). Antibacterial effect of ZnO nanoparticles and zinc phosphates ones were compared. Bacterial strains of (*Escherichia coli* NCTC 13216, *Staphylococcus aureus* NCTC 8511, and Methicillin-resistant *S. aureus* CCM 7110 were used for the testing. Oxidative stress was determined the GSH/GSSG ratio by using High-pressure liquid chromatography with twelve-channel CoulArray electrochemical detector (HPLC-ED). It has been demonstrated the zinc phosphate particles shown better antibacterial properties and smaller toxicity, than the conventional ZnO.

Keywords: Zinc phosphate, Zinc oxide, Antimicrobial properties

1. INTRODUCTION

Zinc is major element, which a responsible for physiological function in the body of animals or human, such as reproduction [1], replication and transcription [2], immune system, protection from bacterial endotoxins and antibody production [3]. On the other hand, Zn treatment of antibacterial activity have positive results in the concentration (2000 - 2500 mg/kg) during 14 days and these high doses, however, carry a high risk of large amounts of zinc in the food chain [4]. Alternative forms of Zn, such as ZnO, Zn phosphate or ZnS have a best potential and widely used in agriculture in the form of animal feed supplement [5], biomedical applications, due to their low toxicity and biodegradability [6]. Nowadays, Zn nanoparticles may be used how alternative as antibiotics for animals, due as their antibacterial properties [7]. Aim of our study was to prepared of Zn NPs and their applications on *Escherichia coli* NCTC 13216, *Staphylococcus aureus* NCTC 8511, and *Methicillin-resistant S. aureus* CCM 7110. Morphology of Zn nanoparticles were characterized by scanning electron microscope.

2. METHODOLOGICAL BASES AND EXPERIMENTAL PART

2.1. Chemicals

Zinc nitrate, $(NH_4)_2HPO_4$, Na_2HPO_4, $Na_4P_2O_7$, $Na_5P_3O_{10}$, GSH, GSSG, ZnO, methanol, trifluoroacetic acid were obtained from Sigma Aldrich (St. Louis, Missouri, USA).

2.2. Determination of reduced and oxidised glutathione and selenium

Reduced and oxidised glutathiones were determined using HPLC-ED. Experimental conditions were adopted from [8].

2.3. Dynamic light scattering and Scanning electron microscopic (SEM) characterization of Zn NPs

The average particle size and size distribution were determined by quasielastic laser light scattering with a Malvern Zetasizer (NANO-ZS, Malvern Instruments Ltd., Worcestershire, UK). Structures of particles were characterized by electron microscope MIRA 3 XMU (Tescan, a.s., Brno, Czech Republic). The SEM was fitted with Everhart-Thronley type of SE detector, high speed YAG scintillator based BSE detector, panchromatic CL Detector. For automated acquisition of selected areas, a TESCAN proprietary software tool called Image Snapper was used.

2.4. Powder X-ray analysis

Powder X-ray analysis of Zn NPs was carried out on D8 Advance ECO (Bruker, AXS GmbH, Karlsruhe, Germany). Bragg-Brentano geometry, CuKα radiation (λ = 1.54178 Å), the range of 2θ = 4-60° and room temperature were used.

2.5. Preparation of microparticles

Zn-A. Zinc nitrate dissolved in water was heated to 60 °C and $(NH_4)_2HPO_4$ was added with stirring. White precipitate was formed.

Zn-B. Zinc phosphate was prepared similarly, only Na_2HPO_4 was used instead of ammonium salt.

Zn-C. Solution of $Na_4P_2O_7$ was added to stirred solution of zinc nitrate. White precipitate was obtained, filtered and dried.

Zn-D. Zinc nitrate was mixed with $Na_5P_3O_{10}$ solution. White precipitate was immediately formed. The suspension was filtered and product dried.

Zn-AC. Zinc nitrate dissolved in water was heated to 60 °C and solution of citric acid and $(NH_4)_2HPO_4$ was added with stirring. White precipitate was formed.

2.6. Determination of antibacterial properties by method of growth curves

The method is based on measuring of the absorbance using the apparatus Multiskan EX (Thermo Fisher Scientific, Germany) and subsequent analysis in the form of growth curves. Bacteria were cultivated in GTY medium for 24 h with shaking and were diluted with GTY medium using Specord spectrophotometer 210 (Analytik, Jena, Germany) at a wavelength of 600 nm to absorbance 0.1. On the microplate, these cultures were mixed with various concentrations of NPs and a control.

3. RESULTS AND DISCUSSION

Aim of our study was preparation of zinc phosphates under different reaction conditions and their physico-chemical study. For preparation of zinc phosphate sodium as well as ammonium salt were used. From powder XRD it follows that using ammonium salt only hopeite $Zn_3(PO_4)_2{\cdot}4H_2O$ was obtained (see **Figure 1**). When sodium salt was used under the same conditions than 90 % of product is hopeite but 10 % is zinc phosphate dehydrate. For Zn-C preparation diphosphate was applied and product was identified as $Zn_2P_2O_7{\cdot}5H_2O$. More complicated situation is for Zn-D. XRD analysis shows on a mixture of three products. The most abundant component (60 %) was as expected zinc triphosphate $Zn_2HP_3O_{10}{\cdot}6H_2O$ but there is 30 % of $Zn_2P_2O_7{\cdot}5H_2O$ and 10 % of $ZnHPO_4{\cdot}3H_2O$. We have also tried addition of citric acid to the reaction mixtures. For the product Zn-AC $Zn_3(PO_4)_2{\cdot}4H_2O$, $Zn_3(PO_4)_2{\cdot}4H_2O$ and $ZnHPO_4{\cdot}H_2O$ (89, 8 and 3 %) were obtained. Addition of citric acid (Zn-DC) to reaction mixture containing triphosphate leads to nearly complete decomposition of triphosphate and $Zn_2P_2O_7{\cdot}5H_2O$ (97 %) with only minor part of $Zn_2HP_3O_{10}{\cdot}6H_2O$ resulted.

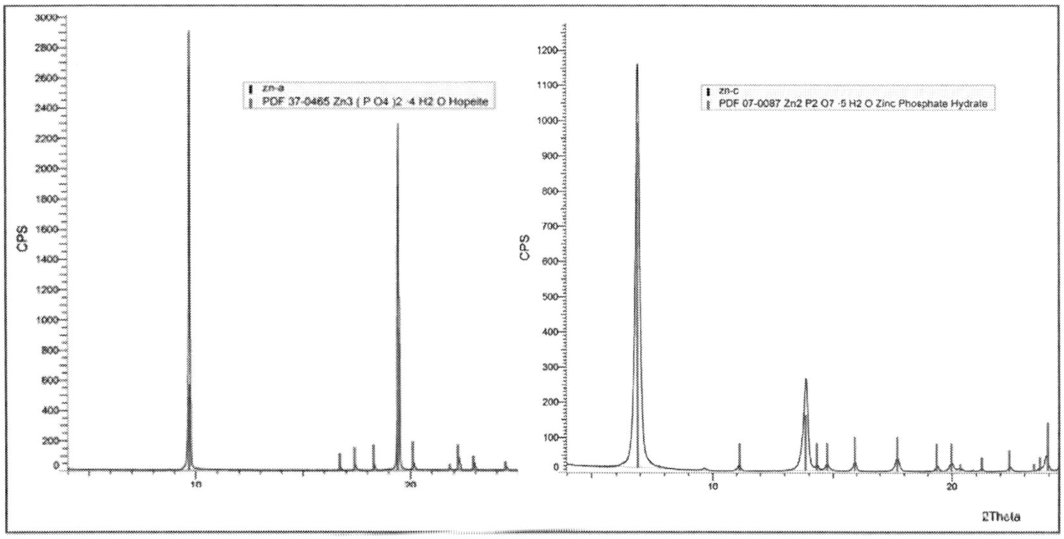

Figure 1 XRD spectra of Zn-A (upper part) and Zn-C. (CPS = counts per second)

DLS analysis of obtained materials show on particles in the range of 50 - 520 nm as well as some particles are much bigger probably due to the formation of aggregates. The size and shape were confirmed by SEM analysis (see **Figure 2**). For comparison zinc oxide nanoparticles and Zn-D are presented. ZnA and ZnB particles had spherical shape with the average diameter 477 and 520 nm, whereas ZnC formed small aggregates with average diameter 452 nm. Zn-D nanoparticles are comparable with those of ZnO nanoparticles with diameters from 50 to 100 nm. It can be clearly seen that nanoparticles were formed as well as their crystalline nature that was also proved by XRD analysis.

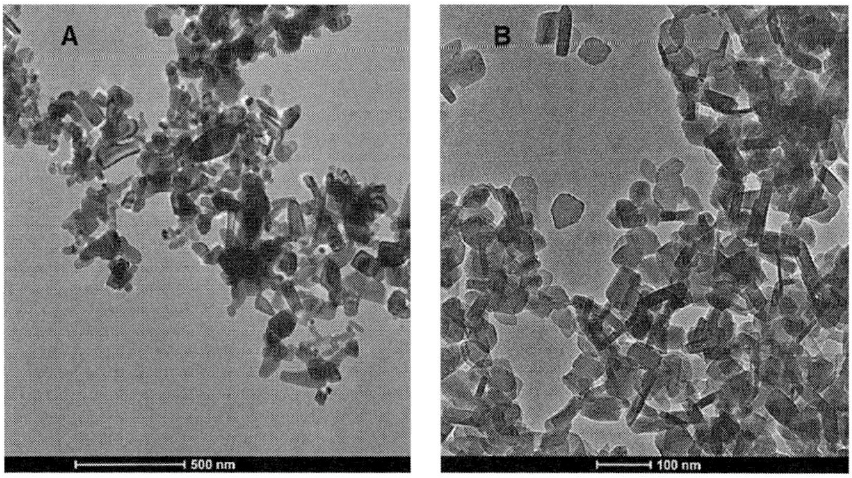

Figure 2 SEM micrographs of ZnO-N (A) and Zn-D (B)

The antibacterial activity of ZnNPs, ZnO and ZnO-N after 24 h was confirmed by the method of the growth curves. For the comparison ZnO particles and nanoparticles were used. From the growth curves inhibition curves (**Figure 3**) were obtained showing on either similar or better antibacterial properties of prepared zinc phosphates. From the literature, it is known that mostly higher antibacterial properties can be expected for lower sized nanoparticles and the activity depends on shape and solubility of nanoparticles. Interestingly, it was found for *E. coli* that with higher concentrations of nanoparticles inhibition of bacteria is lower. It can be explained by recovery of bacteria or by aggregation of nanoparticles in solution leading to lower active surface for interaction with microbial cell. Thus, better activity against *E. coli* show zinc phosphates Zn-A and Zn-B.

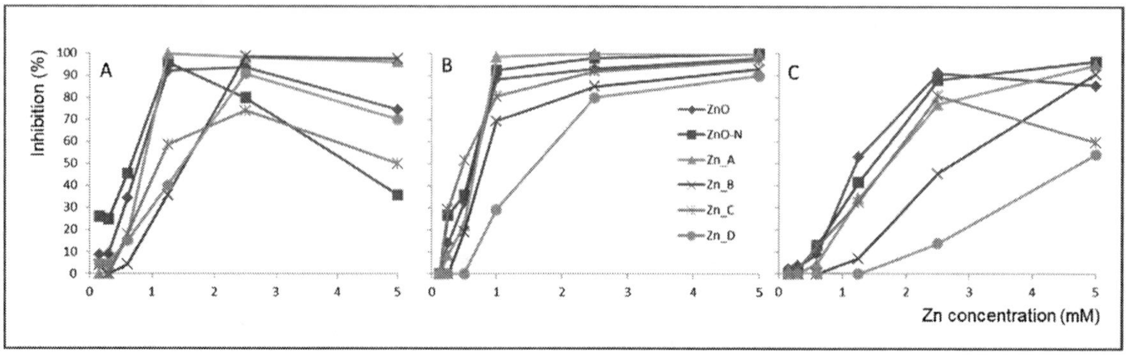

Figure 3 Inhibition curves obtained for (A) E. coli, (B) S. aureus and (C) MRSA obtained under different concentrations of zinc phosphates (Zn-A, Zn-B, Zn-C, Zn-D) and zinc oxides (ZnO, ZnO-N)

Levels of GSH were obtained by analysis of blood, kidney and liver samples of Laboratory male rats of the outbreed strain *Wistar albino* were selected as model animals. The rats were divided into four groups of ten rats each by phosphates or zinc oxides. Four groups of rats were fed with phosphate-based zinc nanoparticles (ZnA, ZnB, ZnC, ZnD) in the dose of 2000 mg Zn/kg diet. All groups of rats had monodiet including 2.7 mg/kg of zinc. The experiment duration was 28 days. The animals had access to feed and drinking water ad *libitum*. From animals were putted to death the blood, kidney and liver samples were collected for subsequent analyses of GSH and GSSG using HPLC-ED. The levels were related to total protein concentration. From the data it was apparent that GSH concentration significantly increases up to concentration 13 ±2 nmol GSH/g protein after zinc treatment in liver. The applied zinc nanoparticles have different effects on the GSH levels in the blood and kidneys and values of GSH are usually slightly higher in comparison with control. In the case of GSSG no significant difference between the different types of zinc phosphate nanoparticles was observed.

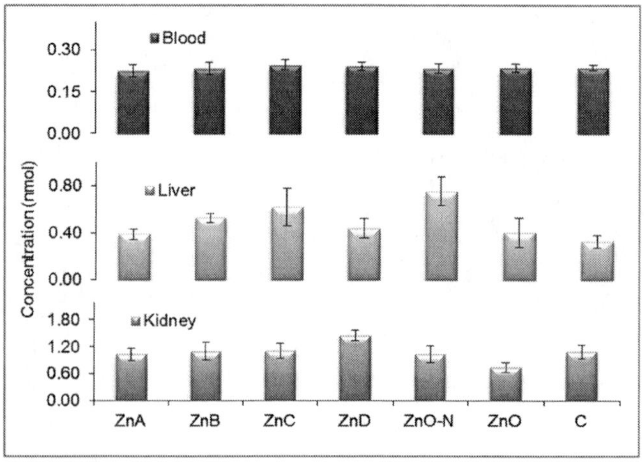

Figure 4 Levels of GSH on blood, liver and kidney. Results are related to the total protein content.

3. CONCLUSION

In this study, zinc phosphate particles were prepared and characterized by SEM and powder X-ray analysis. The particles were tested for their antibacterial properties and it was found that they inhibit grow of bacteria better than zinc oxide which is usually used in veterinary practice. Lowering of zinc content is promising for application in agriculture.

ACKNOWLEDGEMENTS

The financial support from NAZV QK1720349 is highly acknowledged. This work has been supported by the Czech Science Foundation (project No. 16-23647Y) and by the Ministry of Education, Youth and Sports of the Czech Republic under the project CEITEC 2020 (LQ1601).

REFERENCES

[1] FAVIER, A.E. The role of zinc in reproduction. Hormonal mechanisms. *Biol Trace Elem Res* 1992, *32*, 363-382.

[2] WU, F.Y.H.; WU, C.W. Zinc in DNA replication and transcription. *Annual Review of Nutrition* 1987, *7*, 251-272.

[3] HOJYO, S.; FUKADA, T. Roles of zinc signaling in the immune system. *Journal of Immunology Research* 2016, *2016*, 6762343.

[4] MILTON, A.; JOHNSON, M.S. Food chain transfer of zinc within the ecosystems of old and modern metalliferous mine wastes. *Environ Technol* 2002, *23*, 525-536

[5] SWAIN, P.S.; RAO, S.B.N.; RAJENDRAN, D.; DOMINIC, G.; SELVARAJU, S. Nano zinc, an alternative to conventional zinc as animal feed supplement: A review. *Animal Nutrition* 2016, *2*, 134-141.

[6] ZHANG, Y.; NAYAK, T.R.; HONG, H.; CAI, W. Biomedical applications of zinc oxide nanomaterials. *Current molecular medicine* 2013, *13*, 1633-1645.

[7] LIU, Y.; HE, L.; MUSTAPHA, A.; LI, H.; HU, Z.Q.; LIN, M. Antibacterial activities of zinc oxide nanoparticles against escherichia coli o157:H7. *J Appl Microbiol* 2009, *107*, 1193-1201.

[8] URBANKOVA, L.; HORKY, P.; SKLADANKA, J.; PRIBILOVA, M.; SMOLIKOVA, V.; NEVRKLA, P.; CERNEI, N.; LACKOVA, Z.; HEDBAVNY, J.; RIDOSKOVA, A., *et al.* Antioxidant status of rats' blood and liver affected by sodium selenite and selenium nanoparticles. *PeerJ* 2018, *6*.

THE EFFECT OF TEMPERATURE TREATMENT TiO_2 NANOPARTICLES ON ANTIBACTERIAL PROPERTIES

[1]Zuzana BYTESNIKOVA, [1]Vendula VALECKOVA, [1,2]Pavel SVEC, [1,2]Lukas RICHTERA, [1,2]Kristyna SMERKOVA, [3]Petr VITEK, [1,2]Vojtech ADAM

[1]Department of Chemistry and Biochemistry, Mendel University in Brno, Brno, Czech Republic, EU
[2]Central European Institute of Technology, Brno University of Technology, Brno, Czech Republic, EU
[3]Global Change Research Institute of the Czech Academy of Sciences, Brno, Czech Republic, EU

Abstract

The synthesis of TiO_2 nanoparticles (NPs) under various temperature treatments was described and TiO_2 NPs was subsequently tested as an antibacterial agent. Scanning electron microscopy (SEM), Raman spectroscopy, Dynamic light scattering (DLS) were used to confirm structure of TiO_2 NPs and detect differences between individual batches treated with different temperature. Antibacterial properties were tested on *Escherichia coli* (E. coli). TiO_2 NPs as photocatalyst was incubated with bacterial cells under ambient light. Changes in temperature treatment can affect diameter size and crystal structure of TiO_2 NPs as well as its antibacterial properties.

Keywords: Titanium dioxide, pathogenic bacteria, nanoparticles, antibacterial properties, *E.coli*, nanomaterial

1. INTRODUCTION

Bacterial infections are source of serious diseases and can be risk for human health. Antibiotics are relatively cheap, high-yield, widely used solution of bacterial infections. But there have been increasing number of resistant bacteria and even super bacteria, which are resistant to almost all spectrum of antibiotics [1]. Antibiotics have been widely used in livestock production and veterinary medicine to increase production [2-6]. But antibiotics are water-soluble and 30-90 % antibiotics can be excluded in environment [7]. Nanomaterials are perspective solution of this global issue, because the way of interaction with bacteria is different in compare with antibiotics. Antibiotics can diffuse and need penetrate bacterial cell wall in contrast with nanomaterials which act with direct contact with bacterial cells [7-9]. Titanium dioxide (TiO_2) nanoparticles (NPs) is widely known antibacterial agent which was even approved by the American Food and Drug Administration (FDA) for use in human food and materials for food-contact [10]. TiO_2 shows photocatalytic activity and after illumination with light can kill bacteria with contact. After illumination hydroxyl radicals (OH) and reactive oxygen species (ROS) are generated onto TiO_2 and react with bacteria by oxidizing the polyunsaturated phospholipid component of the cell membrane [11-13]. OH is more effective for E.coli inactivation than common disinfectants (e.g. ozone, chlorine or chlorine dioxide) [14]. There are many factors which can affect antibacterial properties of TiO_2 e.g. crystal phase or size [15]. It is general opinion that anatase TiO_2 NPs are more toxic than rutile NPs by inducing stronger oxidative stress [15]. It is also widely known that smaller particles are more toxic. In this present work effect of applied temperature in synthesis of TiO_2 on size, crystal phase and antibacterial properties was evaluated.

2. MATERIAL AND METHODS

2.1. Preparation of TiO_2

Titanium isopropoxide (IV) (Sigma-Aldrich, 97 %) was diluted in isopropyl alcohol (Lachner, 99.7 %). TiO_2 NPs were precipitated on dropwise addition of alkaline Milli-Q water (pH 8). The as-prepared precipitate was washed with Milli-Q water triplicate and heated at 200 °C, 300 °C, and 500 °C [16].

2.2. Scanning Electron Microscopy

The structures of the TiO_2 NPs were characterized by scanning electron microscopy (SEM). For documentation of the nanoparticles structure, a MIRA3 LMU (Tescan, Brno, Czech Republic) was used. This model is equipped with a high brightness Schottky field emitter for low noise imaging at fast scanning rates. The SEM was fitted with In-Beam SE detector. For automated acquisition of selected areas a TESCAN proprietary software tool called Image Snapper (Tescan, Brno, Czech Republic) was used. The software enabled automatic acquisition of selected areas with defined resolution. An accelerating voltage of 15 kV gave satisfactory results regarding maximum throughput.

2.3. Dynamic light scattering

Particle size was measured using the Zetasizer Nano ZS instrument (Malvern Instrument Ltd, UK). The parameters of particle size measurements were follows: refraction index of the dispersive phase of 3.00 and 1.333 for the dispersive environment, adsorption coefficient 10−3, temperature 25 °C, equilibration time 120 s, measurement angle of 173° backscatter. For measurement, disposable cuvettes type ZEN 0040, were used, containing 50 µL of sample.

2.4. Raman spectroscopy

The samples of TiO_2 obtained at three different temperatures were characterized by Raman spectroscopy. Measurements were performed on an InVia Reflex Raman microspectrometer (Renishaw, Wotton-under-Edge, UK) equipped by the 514.5 nm line of an argon laser for excitation. A Leica microscope equipped with a standard 50x objective were used. Scans of 10 s were accumulated 5 times and the laser power was typically set to 0.2 mW at source. Resulting spectra were baseline-corrected in WiRE 3.4 oftware (Renishaw, Wotton-under-Edge, UK).

2.5. Cultivation of bacteria strains

Bacterial culture Escherichia coli NCTC 13216 was cultivated in Muller-Hinton broth (MHB; Oxoid, Hampshire, UK) overnight at 37 °C and 150 rpm.

2.6. Minimal inhibitory concentration

The bacterial culture was diluted with physiological solution to a concentration ~1 × 10^6 CFU/mL. The bacterial concentration was measured by optical density at 600 nm (OD_{600}). Then, the bacterial culture was diluted 100 times. The 75 µL of the bacterial suspension was added into a 96-well microplate and mixed with TiO_2 nanoparticles in ratio 1:1, with total volume 150 µL. The TiO_2 nanoparticles concentration range was from 0.5 mg/mL to 0.063 mg/mL. In case of control, bacteria were diluted only with MHB. The microplate was irradiated with visible light for 30 minutes and placed in Plate Shaker-Thermostat (Biosan, Riga, Latvia) for 16 hours at 37 °C and 800 rpm.

3. RESULTS

The SEM micrographs (**Figure 1**) confirmed nanostructure of TiO_2 and showed structural differences between samples treated at various temperature conditions. It can be easily seen that NPs prepared at 200 °C (A) are smaller than NPs prepared at 500 °C (C). Average size of TiO_2 NPs at different temperature treatment was confirmed by DLS and results are displayed in **Table 1.** There can be observed trend between increasing temperature during synthesis and increasing size of NPs.

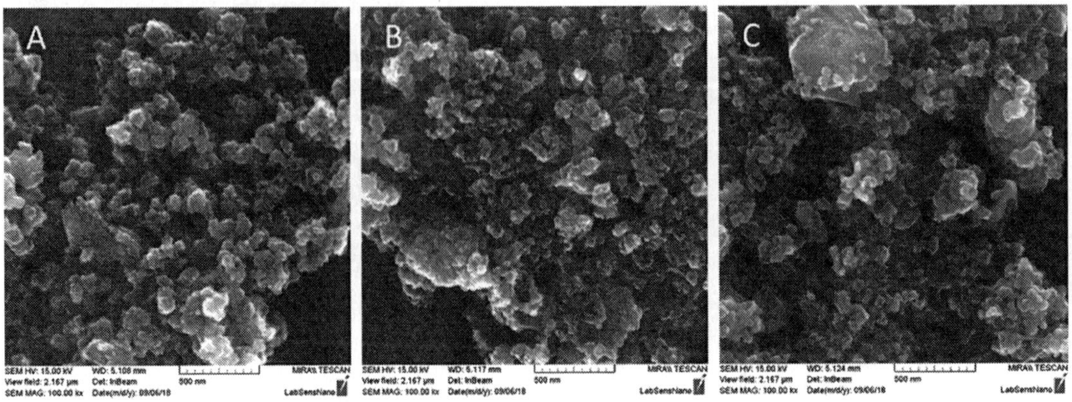

Figure 1 Images of TiO_2 NPs under different heating conditions (A) 200 °C, (B) 300°C, (C) 500°C by using SEM

Table 1 Size parameters of TiO_2 NPs under different temperature treatment by using DLS

Treating temperature of TiO_2 (°C)	Average particle size (nm)
200	92
300	129
500	846

Raman spectroscopy of the three samples synthesized at 200, 300 and 500 °C, respectively, revealed the anatase type of TiO_2 structure. Temperature-dependent spectral changes were observed. The shift of wavenumber position of the band corresponding to Eg(1) phonon mode towards lower value was observed with increasing temperature (**Figure 2**). Namely, at 200°C it was detected at 148.7 cm^{-1}, whereas at 300°C and 500°C the detected position was 147 and 143.7 cm^{-1}, respectively. According to Choi et al. [17], the shift in wavenumber position of the Raman band likely reflects the differences in particle sizes. Simultaneously, the band widths decreased with an increased temperature; for FWHM values corresponding to the band of Eg(1) phonon mode see the **Table 2**. It is interpreted to be a result of increasing crystallinity of anatase with higher temperature.

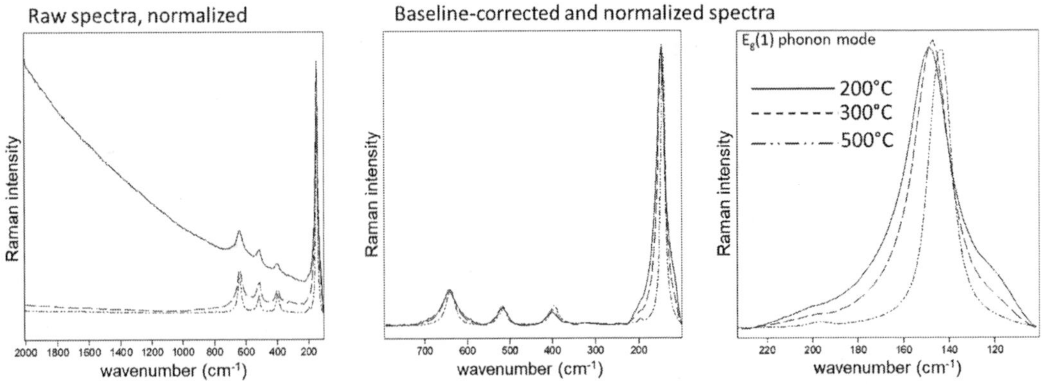

Figure 2 Raman spectra of TiO2 NPs under different heating conditions

Table 2 The position and FWHM of the Eg mode in the TiO_2 NPs

Treating temperature of TiO_2 (°C)	Position of the Eg(1)	FWHM Eg(1)
200	148.7	25.1
300	147.0	19.3
500	143.7	11.6

FWHM, full width at half maximum.

Antibacterial properties of three samples of TiO_2 were tested on *E.coli* as a model organism. The best results exhibited sample prepared at 300 °C in the highest concentration (**Figure 3**). The reason of the best efficiency of TiO_2 NPs treated at 300 °C is still small size and possible presence of large amount organic residues in sample treated at 200 °C, which can difficult the adsorbate- adsorbent interaction [16]. Differences between other samples and concentrations were not so significant although all samples exhibited antibacterial activity. Each experiment was repeated three times. The student t-test was used to evaluate statistically significant differences ($p < 0.05$) between experimental group and control group. The null hypothesis that the frequency was not different between different samples was rejected at a value less than or equal to 0.05. Error bars such as standard error were used.

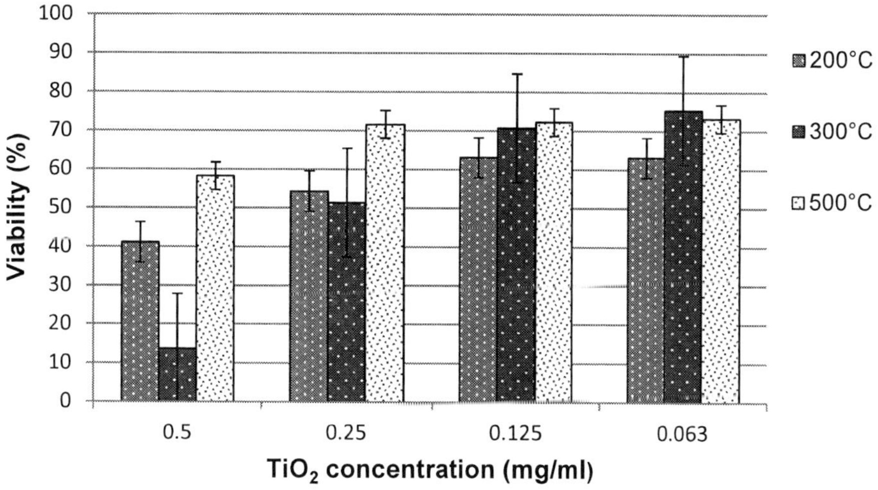

Figure 3 Minimal inhibitory concentration results of TiO_2 NPs after 30 minutes ambient light irradiation

4. CONCLUSION

In this study TiO_2 NPs were synthesized under different temperature treatment. The size of NPs increasedwith increase of heating temperature. The anatase type of TiO_2 structure was detected in all samples. The best antibacterial properties exhibited sample prepared at 300 °C in concentration 0.5 mg/ml against *E.coli* as a model organism.

ACKNOWLEDGEMENTS

This work was also supported by the Ministry of Education, Youth and Sports of CR within the National Sustainability Program I (NPU I), grant number LO1415 and by the Internal Grant Agency of Mendel University in Brno (IP 20/2017).

REFERENCES

[1] Arya, S.C. and N. Agarwal, *Response to "New Delhi Metallo-beta-lactamase (NDM-1): an emerging Threat Among Enterobacteriaceae".* Journal of the Formosan Medical Association, 2010. 109(12): p. 921-922.

[2] Bonnefoy, C., et al., *Temporal and spatial analyses of temperature in a French wine-producing area: the Loire Valley.* International Journal of Climatology, 2013. 33(8): p. 1849-1862.

[3] Carvalho, M.B., et al., *The response of Mucor plumbeus to pentachlorophenol: A toxicoproteomics study.* Journal of Proteomics, 2013. 78: p. 159-171.

[4] Cavagna, A.J., et al., *Water column distribution and carbon isotopic signal of cholesterol, brassicasterol and particulate organic carbon in the Atlantic sector of the Southern Ocean.* Biogeosciences, 2013. 10(4): p. 2787-2801.

[5] Detilleux, J., et al., *Structural equation models to estimate risk of infection and tolerance to bovine mastitis.* Genetics Selection Evolution, 2013. 45.

[6] Froidmont, E., et al., *Association between age at first calving, year and season of first calving and milk production in Holstein cows.* Animal, 2013. 7(4): p. 665-672.

[7] Gallagher, I., G.S. Koch, and F. Planchon, *A profile decomposition approach to the L-t(infinity)(L-x(3)) Navier-Stokes regularity criterion.* Mathematische Annalen, 2013. 355(4): p. 1527-1559.

[8] Gauthier-Rouviere, C., et al., *Flotillin microdomains stabilize Cadherins at cell-cell junctions.* Molecular Biology of the Cell, 2013. 24.

[9] Wang, L.L., C. Hu, and L.Q. Shao, *The antimicrobial activity of nanoparticles: present situation and prospects for the future.* International Journal of Nanomedicine, 2017. 12: p. 1227-1249.

[10] Gutierrez-Arcelus, M., et al., *Passive and active DNA methylation and the interplay with genetic variation in gene regulation.* Elife, 2013. 2.

[11] Cho, M., et al., *Linear correlation between inactivation of E-coli and OH radical concentration in TiO2 photocatalytic disinfection.* Water Research, 2004. 38(4): p. 1069-1077.

[12] Zheng, H., et al., *Bactericidal mode of titanium dioxide photocatalysis.* Journal of Photochemistry and Photobiology a-Chemistry, 2000. 130(2-3): p. 163-170.

[13] Pleskov, Y.V., *A. Fujishima, K. Hashimoto, and T. Watanabe, TiO2 photocatalysis: Fundamentals and applications, Tokyo: Bks, Inc., 1999.* Russian Journal of Electrochemistry, 1999. 35(10): p. 1137-1138.

[14] Wist, J., et al., *Evaluation of photocatalytic disinfection of crude water for drinking-water production.* Journal of Photochemistry and Photobiology a-Chemistry, 2002. 147(3): p. 241-246.

[15] Lin, X.C., et al., *Toxicity of TiO2 Nanoparticles to Escherichia coli: Effects of Particle Size, Crystal Phase and Water Chemistry.* Plos One, 2014. 9(10).

[16] Viana, M.M., V.F. Soares, and N.D.S. Mohallem, *Synthesis and characterization of TiO2 nanoparticles.* Ceramics International, 2010. 36(7): p. 2047-2053.

[17] Choi, H.C., Y.M. Jung, and S.B. Kim, *Size effects in the Raman spectra of TiO2 nanoparticles.* Vibrational Spectroscopy, 2005. 37(1): p. 33-38.

NUMERICAL MODEL OF THE MECHANICAL BEHAVIOR OF COATED MATERIALS IN THE PAIR FRICTION «HEAD - ACETABULUM CUP» OF HIP JOINT

[1]Galina EREMINA, [2]Alexey SMOLIN

[1]Institute of Strength Physics and Materials Science SB RAS, Tomsk, Russian Federation, anikeeva@ispms.ru
[2]National Research Tomsk State University, Tomsk, Russian Federation, asmolin@ispms.ru

Abstract

Wear in a friction pair of components of the endoprosthesis has a significant impact on its operational life. The structure of the surface layers of the contacting elements plays a key role in the wear process. To improve the tribological characteristics of metal endoprostheses, hardening nanostructured coatings are used. In this case, titanium is usually used as a metal, and titanium nitride (TiN) is used as a coating. In this paper, an approach is proposed for multiscale modeling of the friction pair of the endoprosthesis "head-acetabulum cup" based on the method of movable cellular automata. The models of materials involved in the friction pair were verified by simulating the nanoindentation process. At the micro level, the process of friction of two rough surfaces is modeled. The results of modeling at the micro level were used in the macromodel of the friction unit as the tribological characteristics of the material surface of contacting pairs after one loading cycle. To test the performance of the micro and macro models, test calculations were performed with a single cycle loading.

Keywords: Nanostructured coatings, indentation, pair friction, simulation, movable cellular automata method

1. INTRODUCTION

For the treatment of pathologies of the hip joint in the modern world, endprosthetics are widely used. Wear in a friction pair of components of the endoprosthesis has a significant impact on its operational life. The structure of the surface layers of the contacting elements plays a key role in the wear process. To improve the tribological characteristics of metal endoprosthesis (EP), hardening nanostructured coatings are used. In this case, titanium is usually used as a metal, and titanium nitride (TiN) is used as a coating. The structure of the coating is determined by the modes of its application. Promising is the coating by powder immersion reaction assisted coating nitriding (PIRAC). Testing of EP has several stages - these are preclinical and clinical trials. Clinical trials are carried out through the installation of an EP in a living human body. Preclinical studies of the mechanical behavior of the EP can be divided into experimental and theoretical. Experimental studies are tests using a technological unit that simulates the dynamic loading experienced by EP. Theoretical studies of the mechanical behavior of the EP using computer simulation make it possible to study the mechanical behavior of EP taking into account the influence of various factors on it. Therefore, in preclinical studies, computer modeling is used to predict the mechanical behavior of endoprostheses. In the numerical study of wear in a friction pair by the finite element method, modeling is used without explicit account of wear particles, first of all this is attributed to high computational costs and the complexity of explicit modeling of material failure. Most modern papers [1-2] concerning the theoretical study of wear in the friction pair of the endoprosthesis are based on the method for calculating the wear proposed in [3]. The surface roughness is taken into account implicitly [4]. In the case of a numerical study of hip joint endoprostheses with modified surface layers or coatings [5], structural features in multilevel modeling are taken into account using the homogenization method (asymptotic averaging), i.e. features of the structure are taken into account implicitly, and are reflected in the integral characteristics of materials [6-7]. However, this approach may give an incomplete pattern of the EP failure, including the improper distribution of stress-strain state in the bone - endoprosthesis system. Therefore, it is promising to develop numerical models of the mechanical behavior of the friction of the endoprosthesis with an explicit account for wear.

The purpose of this work is to develop a multi-level numerical model of the mechanical behavior of the node of EP "head-acetabulum cup", taking into account such geometric parameters of the TiN coating (PIRAC) as the coating thickness, its roughness and its mechanical characteristics. To achieve the goal, three tasks were set. The first task was to verify the models of materials of the friction unit; for this, a numerical model of the nanoindentation process of the coating - substrate system was built, then numerical calculations were performed, the processing results of which were compared with the data of experiments. It is known that during friction a large contribution to wear is made by the surface roughness. To reduce the computational costs, it was proposed to take into account the tribological characteristics and surface wear, in the macro model, through the coefficient of friction obtained by friction of the contacting surfaces at the micro level. Therefore, the second task was to create a numerical model of friction of materials of the contacting pair at the micro level and to conduct test calculations at the micro level. Wear in the coating was clearly taken into account through the value of the thickness of the layer in which the material was mixed during friction at the micro level. The third task was to create a numerical model of the friction node "head-acetabulum cup" at the macro level using data on the tribological characteristics of materials in the contacting elements obtained from the micromodel and carrying out numerical calculations.

2. METHOD OF MOVABLE CELLULAR AUTOMATA

The method of movable cellular automata [9] is a new efficient numerical method in particle mechanics that is different from methods in the traditional continuum mechanics. Namely, to construct a many-particle interaction model, the expression for the force acting on a discrete element *i* from the surrounding N_i particles should be written in the form (1):

$$\vec{F}_i = \sum_{j=1}^{N_i} \vec{F}_{ij}^{pair} + \vec{F}_i^{\Omega} \quad (1)$$

This force is represented as a superposition of the pair components $\vec{F}_{ij}^{pair}$ dependent on the spatial position or displacement of the automaton *i* relative to the neighbor *j* and the volume-dependent component $\vec{F}_i^{\Omega}$ related to collective effects of the surroundings.

Within the frame of MCA, it is assumed that any material is composed by a certain amount of elementary objects of finite size (automata) which interact among each other and can move from one location to another, thereby simulating a real deformation process. The automaton motion is governed by the Newton-Euler equations (2):

$$\begin{cases} m_i \dfrac{d^2\vec{R}_i}{dt^2} = \sum_{j=1}^{N_i} \vec{F}_{ij}^{pair} + \vec{F}_i^{\Omega}, \\ \hat{J}_i \dfrac{d\vec{\omega}_i}{dt} = \sum_{j=1}^{N_i} \vec{M}_{ij}, \end{cases} \quad (2)$$

where $\vec{R}_i$, $\vec{\omega}_i$, m_i and $\hat{J}_i$ are the location vector, rotation vector, mass and moment of inertia of *i*th automaton respectively, $\vec{M}_{ij} = q_{ij}(\vec{n}_{ij} \times \vec{F}_{ij}^{pair}) + \vec{K}_{ij}$, here q_{ij} is the distance from the center of *i*th automaton to the point of its interaction ("contact") with *j*th automaton, $\vec{n}_{ij} = (\vec{R}_j - \vec{R}_i)/r_{ij}$ is the unit vector directed from the center of *i*th automaton to the *j*th one and r_{ij} is the distance between automata centers, $\vec{K}_{ij}$ is the torque caused by relative rotation of automata in the pair.

The total force acting on automaton *i* can be represented as a sum of explicitly defined normal $\vec{F}_{ij}^{n}$ and tangential (shear) $\vec{F}_{ij}^{\tau}$ components (3):

$$\vec{F}_i = \sum_{j=1}^{N_i} \left(\vec{F}_{ij}^{pair} - AP_i S_{ij} \vec{n}_{ij} \right) = \sum_{j=1}^{N_i} \left[\left(F_{ij}^{pair,n}(h_{ij}) - AP_j S_{ij} \right) \vec{n}_{ij} + F_{ij}^{pair,\tau}\left(\vec{l}_{ij}^{\,shear} \right) \vec{t}_{ij} \right] = \sum_{j=1}^{N_i} \left(\vec{F}_{ij}^{n} + \vec{F}_{ij}^{\tau} \right) \tag{3}$$

where $F_{ij}^{pair,n}$ and $F_{ij}^{pair,\tau}$ are the normal and tangential pair interaction forces depending respectively on the automata overlap hij and their relative tangential displacement $\vec{l}_{ij}^{\,shear}$ calculated with taking into account rotation of the both automata. Note, that although the last expression of equation (3) formally corresponds to the form of element interaction in conventional discrete element models, it differs fundamentally from them in many-particle central interaction of the automata.

Using homogenization procedure for stress tensor in a particle described in [10], the expression for components of the average stress tensor in automaton i takes the form (4):

$$\bar{\sigma}_{\alpha\beta}^{i} = \frac{1}{V_i} \sum_{j=1}^{N_i} q_{ij} n_{ij,\alpha} F_{ij,\beta}\,, \tag{4}$$

where α and β denote the axes *X, Y, Z* of the laboratory coordinate system, V_i is the current volume of automaton *i*, $n_{ij,\alpha}$ is the α-component of unit vector $\vec{n}_{ij}$ and $F_{ij,\beta}$ is β-component of the total force acting at the point of "contact" between automata *i* and *j*.

The equations of motion (2) for the system of movable cellular automata are numerically integrated with the use of velocity Verlet algorithm modified by introducing a predictor for estimation of $\bar{\sigma}_{\alpha\beta}^{i}$ at the current time step.

3. CALCULATIONS

For verification of materials models, a nanoindentation model was built (**Figure 1, a**) of the "coating-substrate" system. The indenter was defined as a rigid non-deformable diamond. The movement of the indenter was set through the velocity in the vertical direction. Next, a micromodel of the friction process was created for the contacting surfaces of the EP taking into account the average height of the surface layer roughness, depending on the mode of coating on the substrate. Since the knowledge of the tribological characteristics of the coating surface after loading cycles necessary for use in the macromodel, the substrate and the transition layer were not taken into account in the micromodels, the contacting blocks with roughness were located on bases that were a fixed base and a piston (**Figure 1, b**), for visibility contacting blocks are shown in different colors. The piston movement was set through the velocities. At the first stage, the velocity was applied in the vertical direction to contact the friction surfaces, then the speed was set in the horizontal direction to simulate friction. The macromodel consisted of acetabulum cup with an outer diameter of 52 mm and an EP head with a diameter of 28 mm (**Figure 1, c**). At the first stage, the head of the EP moved in the direction of the acetabulum cup and the contact of the surfaces, then one cycle of rotational movement along one direction was set using speed.

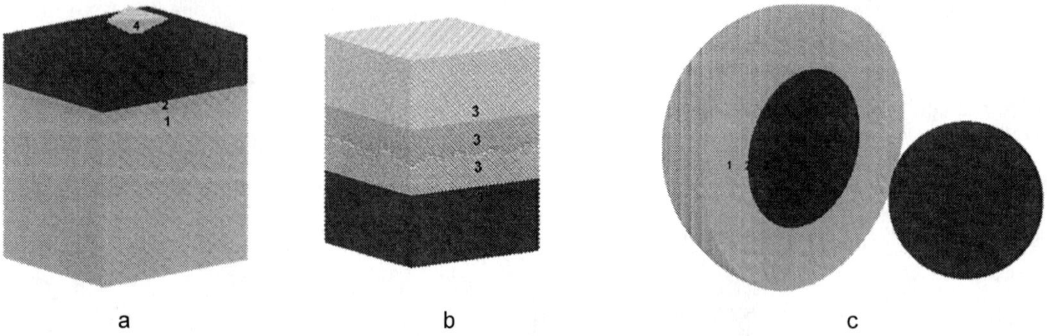

Figure 1 The numerical models, presented in the form of a package of automaton a-nanoindentation, b-friction, c-friction unit "head-acetabulum cup" of hip joint implant (numbers indicate the model materials: 1-titanium, 2-interface, 3-coating, 4- diamond)

According to literature data [11], the following parameter values were chosen for calculating Ti6Al4V samples: ρ = 4420 kg / m^3, G = 41 GPa, K = 92 GPa, E = 107, $\sigma_{y0.02}$ = 0.99 GPa, σ_y = 1.07 GPa and ε_b = 0.10. The geometrical features of the TiN coating and its mechanical properties are determined by the deposition regimes during PIRAC [8] formation. So, at a deposition temperature of 700 °C and a processing time of 48 hours of coating on a titanium substrate, it has a thickness of 1.3 µm with an average roughness height of 0.15 µm (1 mode) and an elastic modulus, and the elastic modulus values were E_1 = 258 GPa; at a deposition temperature of 800 °C and a processing time of 4 hours of coating on a titanium substrate, it has a thickness of 1.4 µm with an average roughness height of 0.132 µm (mode 2) and an elastic modulus of E_2 = 258 GPa; at a deposition temperature of 900 °C and a processing time of 2 hours of coating on a titanium substrate, it has a thickness of 1.5 µm with an average roughness height of 0.265 µm (3 mode) and an elastic modulus of E_3 = 321 GPa. In accordance with this, the following values of the coating parameters were set: ρ = 5220 kg / m^3, G1 = 104 GPa, G_2 = 104 GPa, G_3 = 129, K_1 = 173Gpa, K_2 = 173GPa, K_3 = 205 GPa. The data on the yield stress σ_y = 7.5 GPa, σ_b = 8.5 GPa, ε_b = 0.075 were obtained using the inverse analysis of the indentation curve [12].

4. RESULTS AND DISCUSSION

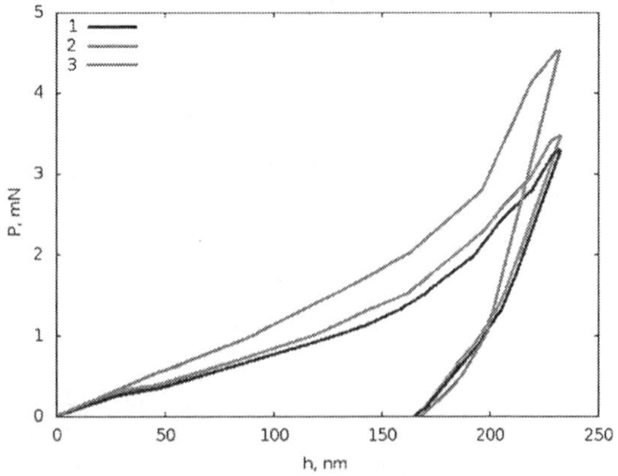

Figure 2 Load-displacement curve P(h) for different application modes (numbers indicate the coating application modes)

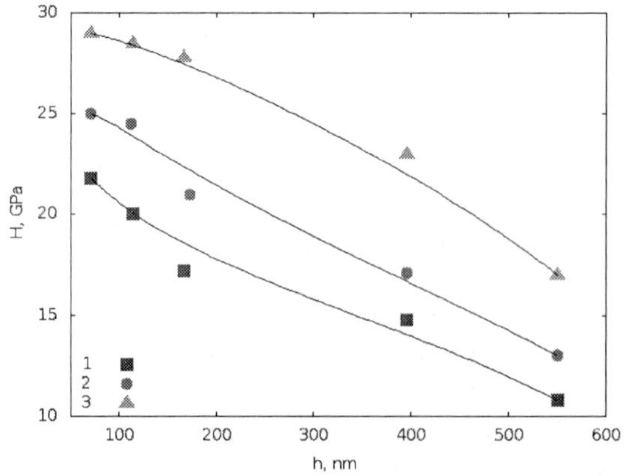

Figure 3 Dependence of the values of hardness on the depth of penetration (numbers indicate the coating application modes)

According to the results of processing the simulation data by the Oliver - Farr method [13], the dependences of the hardness of the systems under consideration on the depth of penetration of the indenter were obtained (**Figure 2**) for three types of coatings obtained with a different method of application.

From the dependence shown in **Figure 3**, it can be seen that the hardest system is the composition of the substrate with the coating applied in the third mode. The results obtained correspond to the data published in [8], which in turn suggests that the material models are developed correctly and can be used to create a multi-level numerical model of the friction node of the "head-acetabulum cup" of EP.

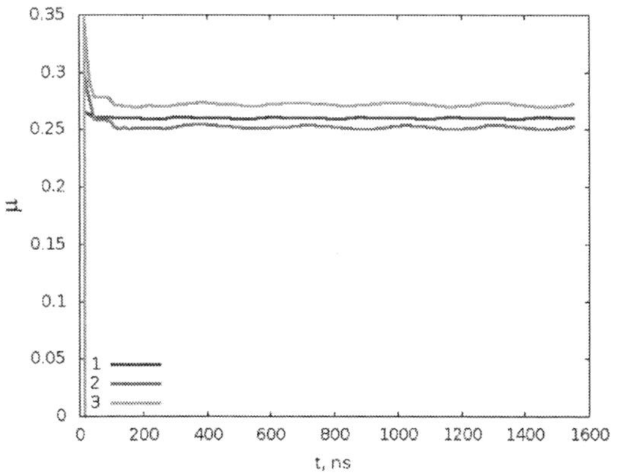

Figure 4 Dependence of friction coefficient on the calculation time when modeling the process of friction at the micro level (numbers indicate the coating application modes)

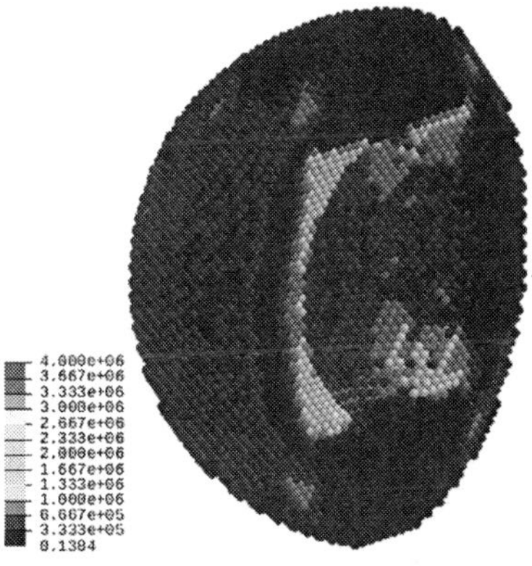

Figure 5 The stress intensity distribution in the friction unit of the EP with a coating obtained by the 3 application mode

According to the results of test calculations the mechanical behavior of two contacting surfaces during friction at the micro level, the values of the friction coefficient and the thickness of the layer in which mixing occurred after one cycle were obtained. Further, obtained values were used in the macromodel of the friction node of the EP to describe the tribological properties of the contacting surfaces after one cycle, thus it was proposed to take into account the wear in the macromodel explicitly. Then, the motion of the EP head in the acetabulum cup was simulated and the stress-strain state of the friction unit was analyzed (**Figure 5**). According to the results of simulation 1 of the friction unit loading cycle, taking into account the data on the tribological state of the surfaces of the contacting elements obtained from the simulation results of the friction at the micro level, it was found that the highest stress value was observed in a friction pair with a coating applied at mode 3, and the minimum value for friction pairs obtained in the second mode of application.

5. CONCLUSION

This article presents a numerical multilevel model of the friction unit of friction of the EP of hip joint. For verification of models of EP materials, numerical nanoindentation experiments were performed. It is shown that the use of friction results for the contacting surfaces of the EP at the micro level in the macromodels of the friction unit hip joint is promising, since it allows one to explicitly take into account the wear in the macromodels through the value of the friction coefficient and the wear of the coating obtained in the micromodels. In the future, it is planned to carry out calculations for the multi-cycle loading of the friction unit of the EP using the proposed multilevel modeling approach.

ACKNOWLEDGEMENTS

The reported study was funded by RFBR according to the research project № 18-38-00323.

REFERENCES

[1] ZHANG, Teng, HARRISON, Noel M., McDonnell, P.F., MCHUGH, Peter E., LEEN, Sean B. A finite element methodology for wear-fatigue analysis for modular hip implants. Tribology International. 2013. vol. 65, pp. 113-127.

[2] KRUGER, Karen M., TIKEKAR, Nishant M., HEINER, Anneliese D., BAER, Thomas E., LANNUTTI, John J., CALLAGHAN, John J., and BROWN, Thomas D. A Novel Formulation for Scratch-Based Wear Modeling in Total Hip Arthroplasty. Computer methods in biomechanics and biomedical engineering. 2014. vol. 17. no. 11, pp. 1227-1236.

[3] MAXIAN, Tina A, BROWN, Thomas D, PEDERSEN, Douglas R, MCKELLOP, Harry A, Lu B, CALLAGHAN, Jhon J. Finite element analysis of acetabular wear. Validation, and backing and fixation effects. Clin Orthop Relat Res 1997. vol. 344, pp. 111-117

[4] ASHKANFAR, Ariyan, LANGTON, David J., JOYCE Thomas J. A large taper mismatch is one of the key factors behind high wear rates and failure at the taper junction of total hip replacements: A finite element wear analysis. Journal of the Mechanical Behavior of Biomedical Materials. 2017. vol. 69, pp. 257-266.

[5] ANTONIALLI, Armando.Í.S., BOLFARINI, Claudemiro. Numerical evaluation of reduction of stress shielding in laser coated hip prostheses. Materials Research. 2011, vol. 14 no.3, pp. 331-334.

[6] OSHKOUR, Azim A., ABU OSMAN, Noor A., YAU, Yat H., TARLOCHAN, Faris, ABAS, Wab W. Design of new generation femoral prostheses using functionally graded materials: a finite element analysis. Proceedings of the Institution of Mechanical Engineers H: Journal of Engineering in Medicine. vol. 227, no. 1, pp. 3-17.

[7] TSUKINO, Makoto, TAKANO, Naoki, MICHEL, Adrien and HAÏAT, Guillaume Multiscale stress analysis of trabecular bone around acetabular cup implant by finite element mesh superposition method Mechanical Engineering Letters. 2015. vol. 1, pp. 15-00354.

[8] BONELLO, Thelma, WILSON, Avelar-Batista J.C., HOUSDEN, Jonathan, GUTMANAS, Elazar Y., GOTMAN, Irene, MATTHEWS Allan, LEYLAND, Adrian, CASSAR, Glenn. Evaluating the effects of PIRAC nitrogen-diffusion treatments on the mechanical performance of Ti-6Al-4V alloy. Materials Science & Engineering. 2014. vol. 619, pp. 300-311.

[9] SMOLIN, Alexey Yu., EREMINA, Galina, M., SERGEEV Valery V., SHILKO, Evgeny V. Three-Dimensional Movable Cellular Automata Simulation of Elastoplastic Deformation and Fracture of Coatings in Contact Interaction with a Rigid Indenter. Physical Mesomechanics. 2014. vol. 17, no. 4, pp. 292-303.

[10] CUNDALL, Peter A., STRACK, Otto D.L. A discrete numerical model for granular assemblies. Géotechnique. 1979. vol. 29, no. 1, pp. 47-65.

[11] Material Datasheet: TIMETAL 6-4s, Titanium Metals Corporation, Denver,Colorado, United States of America, 2000.

[12] GIANNAKPOULOS, Antonios E., SURESH Subra. Determination of elastoplasic properties by instrumented sharp indentation. Scripta Materialia. 1999. vol. 40, no.10, pp. 1191-1198.

[13] OLIVER, Warren, PHARR, George M. An improved technique for determining hardness and elastic modulus using load and displacement sensing indentation experiments. Journal of Materials Research.1992. vol. 7, no. 6, pp.1564-1583.

EFFECT OF COLD ATMOSPHERIC PRESSURE PLASMA GENERATED BY DIFFUSE COPLANAR SURFACE BARRIER DISCHARGE ON POLY(2-OXAZOLINE) THIN FILMS

Petra ŠRÁMKOVÁ[1], Jakub KELAR[1], Richard KRUMPOLEC[1], Zlata TUČEKOVÁ, Monika STUPAVSKÁ[1], Juraj KRONEK[2], Dušan KOVÁČIK[1], Mirko ČERNÁK[1]

[1]*CEPLANT, Department of Physical Electronics, Faculty of Science, Masaryk University, Brno, Czech Republic, EU, petra.sramkova@mail.muni.cz*
[2]*Polymer Institute, Slovak Academy of Sciences, Bratislava, Slovak Republic, EU*

Abstract

Prevention of the non-specific adhesion of bacteria, cells or proteins to a solid surface is essential for sustaining the functionality and stability of medical implants. Poly(2-oxazolines) (POx) are a class of synthetic polymers, which among other features (e.g., biocompatibility, water solubility) possess anti-biofouling properties and accordingly, they are very promising for this purpose. In our contribution, we investigate the effect of cold atmospheric pressure plasma generated by diffuse coplanar surface barrier discharge (DCSBD) on thin POx-based films. Our main objective is to adjust the appropriate conditions of the plasma treatment to achieve the cross-linking of the POx layer, while the anti-biofouling properties stay maintained. For this purpose, POx-based statistical copolymer was deposited on plasma activated silicon wafers by spin-coating. Three various concentrations (2, 5, 10 wt%) of the POx solution resulted in three different thickness of the layer. Achieved POx layers were exposed to cold DCSBD plasma and chemical, and morphological changes were characterized by means of X-ray photoelectron spectroscopy (XPS), atomic force microscopy (AFM), Fourier-transform infrared spectroscopy (FTIR) and profilometry. Very similar values of roughness before and after treatment prove the minor effect of DCSBD plasma on the morphology of POx layer; however, FTIR spectra indicate the undesired etching of the POx layer during the higher exposure times. Despite that the cross-linking of POx layer was not confirmed by available methods, these results provide a good starting point for better understanding of specific processes and further adjusting the suitable conditions of plasma treatment for achieving stable POx coatings.

Keywords: Poly(2-oxazolines), coating, anti-biofouling, DCSBD, atmospheric pressure plasma

1. INTRODUCTION

Due to the biocompatibility, synthetic versatility and potential anti-biofouling properties, nowadays poly(2-oxazolines) (POx) hold a leading position as a valuable alternative to the gold standard polyethylene glycol (PEG) in the biomedical area [1]. Till now, there are a plethora of examples for POx conjugates, micelles, modified surfaces, or hydrogels in connection with their possible applications as therapeutic, imaging and protein/cell repellent agents [2]. Another important application field is surface modifications of implantable medical devices, where biocompatibility and anti-biofouling properties are important to ensure their long-term stability and functionality. POx coatings, which are very promising for this purpose, are usually prepared by challenging multi-step grafting method [3], which includes the surface pre-treatment and functionalization. In addition, these procedures often employ harmful catalysts or solvents. In contrast to this complex procedure, a recently published plasma polymerization technique for POx-based surface coatings introduced more facile, effective and solvent-free method for the modification of various materials [4]. In that case, the POx layer was deposited in one-step by low-pressure plasma generated in the capacitively coupled bell-chamber reactor. Coatings with a thickness of tens nanometers were non-cytotoxic to human dermal fibroblasts, and biofilm adhesion significantly decreased from >99 % to 9.2 % [5]. However, the current applications of plasma in industry or medicine require the more economically friendly and simpler approach; therefore, the low-pressure work regimes are often substituted by atmospheric pressure mode. Diffuse coplanar surface barrier discharge (DCSBD) [6] is a source of cold atmospheric pressure plasma which has a potential to be involved in many industrial or medical applications. Until now, DCSBD has been successfully used for surface activation [7], hydrophilic treatment [8] and adhesion enhancement [9] of various materials, as well as for bacteria and fungi decontamination [10] and plant seeds treatment [11]. In this study, we employed DCSBD for treatment of POx thin films deposited on the model substrate (silicon wafers) with the aim to achieve the homogeneous POx layer with anti-biofouling properties. Compared to multistep grafting or low-pressure plasma polymerization, this process should be faster, easier, and economically friendly. Our main objective in this work is to monitor the chemical and morphological changes in the thin POx film after the plasma treatment by means of X-ray photoelectron spectroscopy (XPS), atomic force microscopy (AFM), Fourier-transform infrared spectroscopy (FTIR) and profilometry.

2. EXPERIMENTAL SECTION

2.1. Materials and plasma specification

The silicon wafers with dimensions 1x1 cm^2 were used as the model surface for all experiments. They were ultrasonically cleaned in isopropanol, acetone, and water for 15 min and dried in the nitrogen flow prior to use. A statistical copolymer poly(2-methyl-2-oxazoline)-*stat*-(2-butenyl-2-oxazoline) (POx) with the degree of polymerization equal to 100 containing 5 % of butenyl groups in the polymer backbone (POx) was synthesized by the living cationic ring-opening copolymerization and characterized as previously described [12]. Polymer films were prepared by spin-coating the POx aqueous solutions (150, 250 and 350 µL) with different concentrations (2, 5 and 10 wt.%) on plasma activated silicon wafers (2000 rpm, 2 min) using a spin-coater WS-650 (Laurell Technologies).

For plasma pre-treatment of silicon wafers and treatment of POx thin films, a diffuse coplanar surface barrier discharge has been used. During the experiments DCSBD was operated in ambient air at atmospheric pressure and it was supplied by sinusoidal voltage (20 kV peak-to-peak, 15 kHz) at input power 400 W. The distance between the treated silicon wafers or POx thin films and DCSBD ceramics was 0.35 mm. For pre-treatment of the wafers, time 10 s was selected to achieve sufficient wettability. POx thin films were treated for 3, 10 and 30 s.

2.2. Surface analyses

The thickness of spin-coated POx films was measured by DektakXT stylus surface profiler (Bruker). Before spin-coating, a small piece of tape was applied on the edge of silicon wafers, which was then removed before the thickness measurements. The step height between the film and wafer was evaluated on three independent wafers. Resulting thickness was determined from 5-10 spots on every wafer and expressed as an average ± standard deviation. Atomic force microscopy (AFM) measured by Ntegra prima (NT-MDT) microscope in semi-contact mode was carried out to measure the root mean square (RMS) surface roughness of the POx films before and after plasma treatment.

Chemical analysis of POx thin films was performed by X-Ray photoelectron spectroscopy (XPS) and Fourier transform infrared spectroscopy (FTIR). The XPS measurements were done on the ESCALAB 250Xi (ThermoFisher Scientific). The system is equipped with 500 mm Rowland circle monochromator with microfocused Al Kα X-Ray source. An X-ray beam with 200 W power (650 microns spot size) was used. The survey spectra were acquired with pass energy of 50 eV and resolution of 1 eV. High-resolution scans were acquired with pass energy of 20 eV and resolution of 0.1 eV. In order to compensate the charges on the surface electron flood gun was used. Spectra were referenced to the hydrocarbon type C1s component set at a binding energy of 284.8 eV. The spectra calibration, processing, and fitting routines were done using Avantage software. Infra-red spectra were measured by Bruker Vertex 80V spectrometer. The samples were measured in transmission mode. All measurements were made at a vacuum of 2.51 hPa to suppress undesirable effects caused by the presence of ambient humidity and CO_2. Each spectrum was scanned 100 times and evaluated by Bruker OPUS software (version 6.5).

3. RESULTS AND DISCUSSION

Statistical copolymer poly(2-methyl-2-oxazoline)-*stat*-(2-butenyl-2-oxazoline) (POx) was synthesized by cationic ring-opening polymerization and characterized by methods as previously described [12]. The resulting molar mass of the copolymer was 8,300 g mol^{-1}, which approximately corresponded to the required degree of polymerization equal to 100, and the fraction of 2-butenyl-2-oxazoline was 5 % what corresponded to the percentage in the feed. As the recent studies indicate, such water-soluble POx (co)polymers should possess good biocompatibility and anti-biofouling properties [13,14], therefore we have chosen this material as suitable for anti-biofouling surfaces. POx thin layer was deposited by spin-coating on silicon wafers. The thickness of the layer was controlled by the different concentration of the aqueous solution (2, 5 and 10 wt.%). The volume of applied solution (150, 250 and 350 μL) during spin-coating did not influence the thickness of the layer; hence the resulting values listed in **Table 1** were determined as an average from 6-8 samples disregarding the used volume (2-3 samples from each used volume). Samples with deposited POx layer were after that treated by cold atmospheric pressure plasma generated by DCSBD. POx samples with the lowest thickness (POx 2wt%) were excluded from the further discussion due to a nonuniform character of the film and possible removal of the thin layer by the plasma etching process. Plasma treatment often leads to morphological changes in the surface layer of many materials. Therefore, we evaluated the roughness before and after plasma treatment by AFM. Resulting values for three different plasma exposure times - 3, 10, and 30 s are summarized in **Table 1** and selected AFM images are depicted in **Figure 1**.

Table 1 The values of POx film thickness and RMS roughness before and after plasma treatment

Sample	thickness (nm)	RMS roughness (nm)			
Plasma treatment time (s)		0	3	10	30
Si wafer		0.15			
POx 2wt%	46.6 ± 20.6	-	-	-	-
POx 5wt%	167.56 ± 55.9	0.25	0.19	0.20	1.85
POx 10wt%	361.93 ± 74.1	0.26	0.22	0.24	0.19

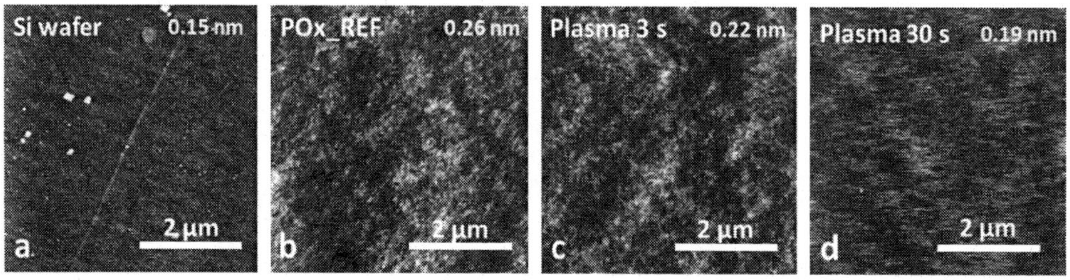

Figure 1 Selected AFM images of a) cleaned Si wafer, wafer with deposited 10wt.% POx layer b) before plasma treatment, c) after 3 s and d) after 30 s of plasma treatment. Resulting RMS roughness values are inserted in the AFM scans

Surface morphology is evidently different after the deposition of POx thin film on the silicon wafer (**Figure 1 a, b**), along with the roughness, which increased from 0.15 nm (wafer) to 0.25 or 0.26 nm for 5 and

10 wt.% solution, respectively. However, after the plasma application, only negligible changes occurred. Except the sample POx 5wt% treated for 30 s, all layers were slightly smoother compared to untreated POx film, which indicates the minor effect of gentle DCSBD plasma on the morphology of POx film.

With the aim to reveal the chemical changes in the POx layer after the plasma treatment, X-Ray photoelectron spectroscopy (XPS) and Fourier transform infrared spectroscopy (FTIR) were employed. In general, plasma modification takes place on the upper layer of material, which represents approximately several tens of nm. XPS analysis is suitable for monitoring the surface, which was in direct contact with the plasma, while FTIR spectra provide the view into the deeper layer of studied material. Accordingly, the possible chemical changes can be detected. Nevertheless, the depth of penetration of the IR beam into the sample depends on many factors (crystal, the angle of incidence, sample's refractive index, the wavelength of IR beam), and thickness of the layer limits such measurements. In our case, samples prepared from 2 wt% and 5 wt% POx solution with a thickness in the range 46.6 ± 20.6 and 167.56 ± 55.9, respectively, were under the detection limit of FTIR measurements. Already before the plasma treatment, the intensity of characteristic signals was very low (data not shown). For this reason, we report here only the results from samples POx 10wt% before and after plasma treatment (**Figure 2**). FTIR spectra of POx demonstrate the presence of typical amide carboxyl peak (1645 cm^{-1}), which corresponds to carbonyl stretching of the amide backbone of the POx chain. Another signal in the range 1420-1478 cm^{-1} is assigned to CH_x symmetrical deformations, and small signal around 1260 cm^{-1} appertains to C-N stretching [15]. The characteristics signals of the silicon wafer are in the range 500-1100 cm^{-1}; hence they do not overlap with POx peaks. FTIR spectra in **Figure 2** demonstrate the decreasing intensity of POx typical absorption bands with the duration of plasma treatment. Plasma irradiation probably causes the etching of the POx layer and thus the removing of the upper layer of POx film leading to decreasing of the film thickness. Accordingly, the longer treatment times (10-30 s) are not suitable for achieving stable POx coatings. Formation of the new chemical bonds, which could confirm the cross-linking, was not detected by FTIR spectroscopy.

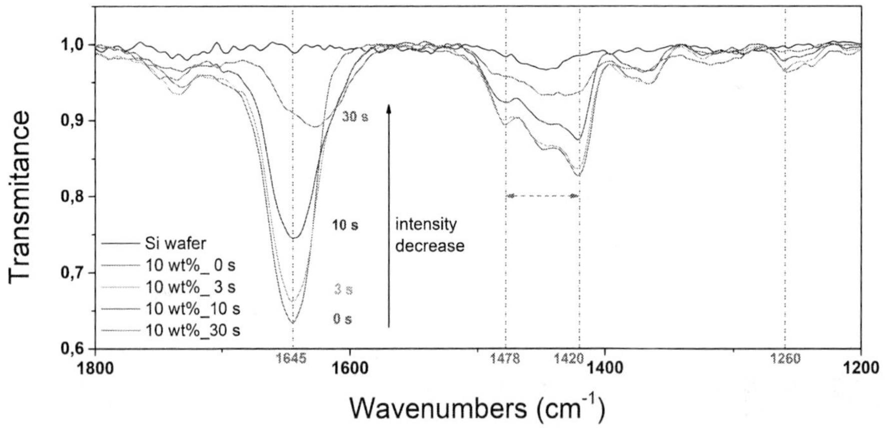

Figure 2 FTIR spectra of Si wafer and POx 10wt% thin film before and after the plasma treatment

Chemical composition acquired by XPS measurement is listed in **Table 2**. Average N/C and O/C ratio, as an indicator of surface oxidation, slightly increased due to the plasma treatment. The influence of the layer thickness was in all cases negligible, whereas with the increasing treatment time the new oxygen groups were formed and O/C ratio increased (e.g. POx 10wt% - 0.24, 0.29 and 0.46 for treatment time 3, 10 and 30 s, respectively). Additionally, a detailed analysis of the high-resolution C1s peak was performed. The C1s peak of untreated POx layer was deconvoluted into three components (**Figure 3**, POx_REF), which were attributed to C-C/C-H (284.8 eV), C-N (285.8 eV) and N-C=O (287.7 eV) bonds [14]. Air plasma is a great source of reactive oxygen and nitrogen species, therefore also in our case plasma treatment induced the formation of additional C=O / N-C=O (287.7 eV) groups and new C-O (286.1 eV) and O-C=O (289.3 eV) functionalities (**Figure 3**, 0 days). In general, detection of the plasma-induced cross-linking is a serious challenge. Besides UV radiation, the plasma electrons interacting with the substrate can initiate a formation of radicals in the polymer chains and starting the cross-linking. However, also the chain scission can simultaneously occur. In our case, the most pronounced increase was achieved in a number of C-O and N-C=O/C=O groups, while the percentage of C-C and C-N bonds decreased. OH radicals produced in air plasma can abstract the hydrogen from polymer backbone and formed alkyl radicals, which initiate the incorporation of O-containing groups onto the polymer surface. On the other hand, there is an assumption that reduction of the percentage of C-C bond and forming of new C-O can indicate the disruption of the polymer chain and following creation of ether (C-O-C) cross-linking junctions.

Table 2 Concentration of elements (at%) and relative area of C1s chemical bonds (%) before (POx REF) and immediately after plasma treatment (POx 5wt%, POx 10wt%)

Sample	plasma treatment time (s)	N1s	C1s	O1s	N/C	O/C	C-C	C-N	C=O	C-O	COO	N-C=O
POx REF	0	14.2	69.7	16.1	0.20	0.23	38.1	48.6	0	0	0	13.3
POx 5wt%	3	14.5	66.5	19.0	0.22	0.29	27.1	22.4	3.3	22.8	2.3	22.1
	10	14.7	40.6	44.7	0.36	1.10	18.6	20.2	8.6	30.5	4.5	17.5
	30	11.2	70.6	18.2	0.16	0.26	13.1	15.4	0	35.4	2.2	33.9
POx 10wt%	3	13.7	69.5	16.8	0.20	0.24	21.9	20.3	5.8	24.9	0.8	26.4
	10	16.7	64.5	18.8	0.26	0.29	29.3	24.7	4.2	22.6	1.5	17.8
	30	17.5	56.4	26.1	0.31	0.46	22.5	18.0	8.4	24.3	3.3	23.6

Ageing of POx layer in ambient air was monitored during the first 3 days and it is depicted in **Figure 3** for. the selected sample POx 10wt% treated by DCSBD plasma for 30 s. Considerable increasing of C-O groups after three days could involve the post-oxidation. However, concerning the O/C ratio, no markedly changes occurred, values almost in all cases stayed maintained (data not shown). In summary, no considerable changes after three days were revealed. POx film exposed to plasma remained chemically stable while stored in air.

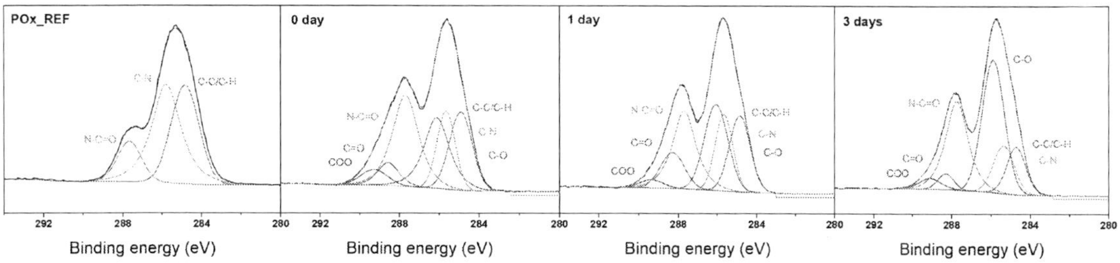

Figure 3 Deconvolution of high-resolution C1s spectra of initial POx 10wt% film and its ageing (0, 1, 3 days) after 30 s of plasma treatment

4. CONCLUSION

The influence of cold atmospheric pressure plasma generated by diffuse coplanar surface barrier discharge on thin poly(2-oxazoline) films was evaluated during the different exposure time (3, 10 and 30 s). Very similar values of roughness before and after plasma treatment proved the minor effect of DCSBD plasma on the morphology of POx layer. However, FTIR spectra indicated the possible etching of the POx layer and thus the removing of the upper layer of POx film in the case of higher exposure time (10 and 30 s). Air plasma is great source of reactive oxygen and nitrogen species, thus induced the oxidative reactions in the POx layer, which was revealed by the XPS. To detect the plasma-induced cross-linking is not trivial by using the FTIR and XPS, and in the future experiments, the additional measurement techniques as MALDI or SIMS could be utilized. Accordingly, using the inert gases instead of ambient air could enhance the cross-linking of the film, and it will be subject of further experiments.

ACKNOWLEDGEMENTS

This work was supported by Operational Programme Research, Development And Education - Project „Postdoc@MUNI" (no. CZ.02.2.69/0.0/0.0/16_027/0008360).

REFERENCES

[1] KONRADI, R., ACIKGOZ, C. and TEXTOR, M. Polyoxazolines for nonfouling surface coatings - a direct comparison to the gold standard PEG. *Macromol. Rapid Commun.* 2012. vol. 33, no. 19, pp. 1663-1676.

[2] LORSON, T., LÜBTOW, M.M., WEGENER, E., BOROVA, S., NAHM, D., JORDAN, R., SOKOLSKI-PAPKOV, M., KABANOV, A.V. and LUXENHOFER, R. Poly(2-oxazoline)s based materials: A comprehensive and critical update. *Biomaterials.* 2018. vol. 178, pp 204-280.

[3] MORGESE G. and BENETTI, E. M. Polyoxazoline biointerfaces by surface grafting. *Eur. Polym. J.* 2017. vol. 88, pp. 470-485.

[4] RAMIASA, M. N., CAVALLARO, A. A., MIERCZYNSKA, A., CHRISTO, S. N., GLEADLE, J. M., HAYBALL, J. D. and VASILEV, K. Plasma polymerised polyoxazoline thin films for biomedical applications. *Chem. Commun.* 2015. vol. 51, no. 20, pp. 4279-4282.

[5] CAVALLARO, A. A., MACGREGOR-RAMIASA, M. N. and VASILEV, K. Antibiofouling Properties of Plasma-Deposited Oxazoline-Based Thin Films. *ACS Appl. Mater. Interfaces.* 2016. vol. 8, no. 10, pp. 6354-6362.

[6] ČERNÁK, M., KOVÁČIK, D., RÁHEĽ, J., SŤAHEL, P., ZAHORANOVÁ, A., KUBINCOVÁ, J., TÓTH, A. and ČERNÁKOVÁ, Ľ. Generation of a high-density highly non-equilibrium air plasma for high-speed large-area flat surface processing. *Plasma Phys. Control. Fusion.* 2011. vol. 53, no. 12.

[7] KELAR, J., SHEKARGOFTAR, M., KRUMPOLEC, R. and HOMOLA, T. Activation of polycarbonate (PC) surfaces by atmospheric pressure plasma in ambient air. *Polym. Test.* 2018. vol. 67, no. February, pp. 428-434.

[8] HANUSOVÁ, J., KOVÁČIK, D., STUPAVSKÁ, M. and ČERNÁK, M. Atmospheric pressure plasma treatment of polyamide-12 foils. *Open Chem.* 2015. vol 13, pp. 382-388.

[9] BUČEK, A., BRABLEC, A., KOVÁČIK, D., SŤAHEL, P. and ČERNÁK, M. Glass bond adhesive strength improvement by DCSBD atmospheric-pressure plasma treatment. *Int. J. Adhes. Adhes.* 2017. vol. 78, pp. 1-3.

[10] TUČEKOVÁ, Z., KOVAL'OVÁ, Z., ZAHORANOVÁ, A., MACHALA, Z. and ČERNÁK, M. Inactivation of Escherichia coli on PTFE surfaces by diffuse coplanar surface barrier discharge. *Eur. Phys. J. Appl. Phys.* 2016. vol. 75, no. 2, p. 24711: 6p.

[11] ZAHORANOVÁ, A., HENSELOVÁ, M., HUDECOVÁ, D., KALIŇÁKOVÁ, B., KOVÁČIK, D., MEDVECKÁ, V. and ČERNÁK, M. Effect of Cold Atmospheric Pressure Plasma on the Wheat Seedlings Vigor and on the Inactivation of Microorganisms on the Seeds Surface. *Plasma Chem. Plasma Process.* 2016. vol. 36, no. 2, pp. 397-414.

[12] ŠRÁMKOVÁ, P., ZAHORANOVÁ, A., KRONEKOVÁ, Z., ŠIŠKOVÁ, A. and KRONEK, J. Poly(2-oxazoline) hydrogels by photoinduced thiol-ene 'click' reaction using different dithiol crosslinkers. *J. Polym. Res.* 2017. vol. 24, no. 5, p. 82: 13p.

[13] KRONEK, J., KRONEKOVÁ, Z., LUSTOŇ, J., PAULOVIČOVÁ, E., PAULOVIČOVÁ, L. and MENDREK, B. In vitro bio-immunological and cytotoxicity studies of poly(2-oxazolines). *J. Mater. Sci. Mater. Med.* 2011. vol. 22, no. 7, pp. 1725-1734.

[14] KONRADI, R., PIDHATIKA, B., MÜHLEBACH, A. and TEXTOR, M. Poly-2-methyl-2-oxazoline: A peptide-like polymer for protein-repellent surfaces. *Langmuir.* 2008.vol. 24, no. 3, pp. 613-616.

[15] MUMTAZ, F., CHEN, C., ZHU, H., PAN, C. and WANG, Y. Controlled protein adsorption on PMOXA / PAA based coatings by thermally induced immobilization (SI). *Appl. Surf. Sci.* 2018. vol. 439, pp. 148-159.

IN VIVO ANTIBACTERIAL STUDY OF ZINC PHOSPHATE-BASED NANOPARTICLES

Silvia KOCIOVA[1], Vendula VALECKOVA[1], Zuzana BYTESNIKOVA[1], Pavel HORKY[2], Pavel KOPEL[1,3], Vojtech ADAM[1,3], Kristyna SMERKOVA[1,3]

[1]*Department of Chemistry and Biochemistry, Mendel University in Brno, Brno, Czech Republic, EU, xkociova@mendelu.cz*
[2]*Department of Animal Nutrition and Forage Production, Mendel University in Brno, Brno, Czech Republic, EU*
[3]*Central European Institute of Technology, Brno University of Technology, Brno, Czech Republic, EU*

Abstract

In animal nutrition, the nanotechnology is mostly used for the preparation of nanominerals, especially trace minerals, due to their low bioavailability. These nanominerals also showed advantageous effects at lower doses compared to conventional minerals. In addition, nanominerals can improve digestive efficacy, immunity and overall performance in livestock. In that case, we suggested and synthesized four types of zinc phosphate-based nanoparticles (ZnNPs), which achieved a significant antibacterial activity. After successful *in vitro* testing, we realized also *in vivo* study. We investigated the effect of zinc nanoparticles on rats after oral administration. Among others, the influence on total aerobic bacteria and coliforms in rat feces at day 10 or day 30 of treatment was monitored. In general, it has been observed a decrease of total aerobic and coliform bacteria population compared to untreated control group.

Keywords: Antibiotics, antimicrobial activity, zinc nanoparticles, nanominerals, rats

1. INTRODUCTION

Nanotechnology is emerging technology which has massive potential to global revolutionize agriculture and livestock sectors [1]. With regard to animals, its major applications are in administration of drugs and probiotics, nutrients and supplements, diagnosis and treatment, and also they are useful in hormonal immune-sensors in the management of reproduction. In animal nutrition, the main use of nanotechnology is in preparations of nanominerals, particularly trace minerals, due to their low bioavailability [1,2]. Nanominerals are stable at higher temperature and pressure [3] and can be lightly taken up by gastrointestinal tract and consequently utilized in the animal system [4]. In addition, minerals as nanoparticles increase the absorption of trace minerals and reduce intestinal mineral antagonism, which consequently reduces minerals excretion and thus pollution of the environment. Nano-minerals can be used in animal and poultry feed for effective nutrient intake to make better use of feed and other dietary supplements. Studies also indicate that nutrition by nanoparticles also improved digestion, immunity and performance of livestock [1,5]

The third most globally produced metal nanomaterial is nano zinc, particularly zinc oxide, which is one of the most studied representative of this type of particles [6]. The uniqueness of zinc is precisely the fact that it is the second most numerous trace element in animal body, but it cannot be stored in the body, so its regular intake in food is needed [7]. The reason for this sudden increase in demand for this type of nanoparticles is due to their better immuno-modulatory effect, but mainly antibacterial effects as the conventional Zn, or more precisely ZnO. Zn-based nanoparticles (ZnNPs) exhibited bactericidal effects against both, gram-positive and gram-negative bacteria [8, 9]. Within pathogenic bacteria, the species, which are responsible for the bacterial causes of severe secretory diarrhoea, especially in pre- and post-weaned piglets, are still dominant and still can be a significant cause of death. There are many different pathogens and factors that can cause diarrhoea so outbreaks should be investigated thoroughly. However, it is important to keep in mind that the use of nano Zn as an antimicrobial and immune agent depends on its dosage [7, 10-12].

So far, the main source of zinc for feed supplementation is used in the form of its inorganic salts, such as zinc sulphate ($ZnSO_4$), zinc chloride ($ZnCl_2$) and zinc oxide (ZnO). However, bioavailability of zinc from inorganic sources has been found to be relatively low, and therefore it is important to looking for the other sources with higher bioavailability [13]. In this case, the major objective of this work was to synthetize zinc-based nanoparticles, specifically four types of zinc-phosphate nanoparticles, with the antimicrobial activity. These zinc-phosphate NPs (ZnNPs) were prepared by synthesis of zinc nitrate with hydrogen phosphate (ZnA and ZnB), diphosphate (ZnC) or triphosphate (ZnD). First of all, ZnNPs were charecterized by Dynamic light scattering (DLS) and Transmission electron microscopy (TEM). After that, these ZnNPs were examined for their antibacterial activity *in vitro*, using three different methods to determine their antibacterial effects. Because of the successful results from that part of the project, in next work we focused on the main purpose of the study - the *in vivo* experiments. The current study focuses on the effect of in-house prepared ZnNPs on rats after oral administration.

2. MATERIAL AND METHODS

2.1. Particle characterization

ZnNPs were prepared by reaction of zinc nitrate with hydrogen phosphate (ZnA and ZnB), diphosphate (ZnC) or triphosphate (ZnD). The particle size and distribution were measured by dynamic light scattering, DLS, on a Malvern Zetasizer (NANO-ZS, Malvern Instruments Ltd., Worcestershire, UK). For visualization of synthetized particles, the transmission electron microscopy, TEM, were used. The images were taken with Tecnai F20 microscope (FEI, Eindhoven, Netherlands) at appropriate magnifications.

2.2. Analyses of *in vitro* antibacterial activity

The *in vitro* antimicrobial effect of synthetized ZnNPs and also with ZnO and commercial ZnO based NPs (ZnO-N) was determined and evaluated on three bacterial cultures – *Staphylococcus aureus*, *Escherichia coli*, methicillin-resistant *S. aureus* (MRSA).

For counting of bacterial colonies, bacteria were diluted in Muller-Hinton broth to a concentration $\sim 1\times10^8$ CFU/ml, and further diluted in tenfold steps. The cultures were mixed with samples to a final concentration 5 mM, in ratio 9:1. After 2 h of incubation at 37 °C, 100 µl of each suspension was applied on agar plates. After 24 h of incubation, the colonies were counted.

Bacterial cultures at concentration $\sim 1\times10^6$ CFU/ml were used for the growth curves method. In 96-well microplate, these diluted cultures were mixed with samples in maximal concentration 5 mM in ratio 1:1, where total volume in one single well was 200 µl. The growth curves were carried out by measuring the optical density at 600 nm during 24 h (Multiskan EX; Thermo Fisher 144 Scientific, Bremen, Germany).

For Live/dead assay, the MRSA culture ($\sim 1\times10^8$ CFU/ml) was incubated with samples in ration 9:1 for 24 h, and the individual suspensions were then centrifuged and washed by 0.85 % NaCl solution. Fluorescent dyes, SYTO9 and propidium iodide (PI), were used for observation under fluorescence microscope (Olympus IX71 inverted fluorescence microscope – Olympus, Tokyo, Japan).

2.3. *In vivo* antibacterial testing

For this experiment, a male rat of the *Wistar albino* strain was selected as a model animal. The rats were divided into 7 groups - four groups of rats were fed by Zn-phosphate NPs (ZnA, ZnB, ZnC and ZnD), the fifth was fed by ZnO, the sixth by commercial zinc nanoparticles (ZnO-N). The last group served as a control group (C) without a zinc supplementation in the diet. Each of them were administrated in a dose of 2000 mg Zn/kg diet. Feed and water were available *ad libidum*. The rats were treated with ZnNPs and ZnO during 30 days.

The experiment was carried out in the facility of the Institute of Animal Nutrition and Forage Production at Mendel University in Brno. Throughout the experiment, the microclimatic conditions in the laboratory were observed and controlled.

2.4. Counting of total aerobic bacteria and coliforms

The samples of rat's feces were collected at day 10, 20 and 30 of treatment. They were homogenized with sterile physiological solution 1:9 w/v. Then, the homogenate was furthermore serially diluted in tenfold steps. Subsequently, 1 ml of selected diluted suspension was places into empty Petri dishes and then potting by Plate count agar (PCA) and MacConkey agar (MCA) in duplicate. Total counts of aerobic bacteria from PCA and counts of coliforms from MCA were determined after 24 h of incubation in 37 °C. The results are expressed as CFU/g feces.

3. RESULTS AND DISCUSION

Four different precursors were used for ZnNPs synthesis, hydrogen phosphate for ZnA and ZnB, diphosphate for ZnC and triphosphate for ZnD, and the characterization of these particles was determined by DLS and TEM. By these two analyses, we find out that the ZnA and ZnB particles were in spherical shape with average diameter about 470 and 525 nm, respectively. Whereas ZnC and ZnD had irregular shape, and created small aggregates. The average diameter was then determined as 448 (ZnC) and 1030 (ZnD) nm. The different shapes are shown in **Figure 1A**. After the characterization of in-house prepared ZnNPs, the antibacterial activity against three bacterial species was investigated, and also the difference between individual particles connected with their structure was found out.

A

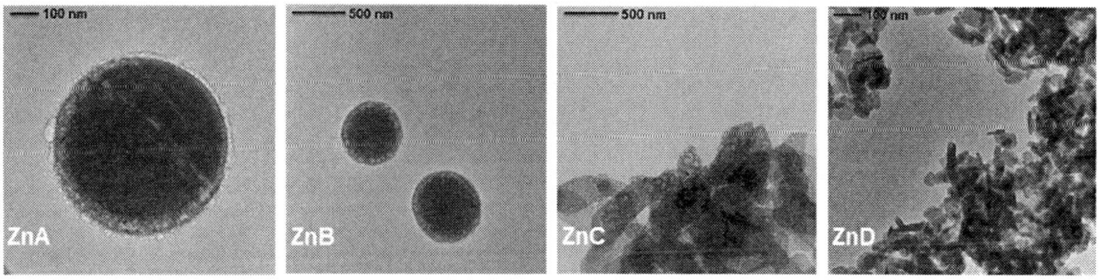

B

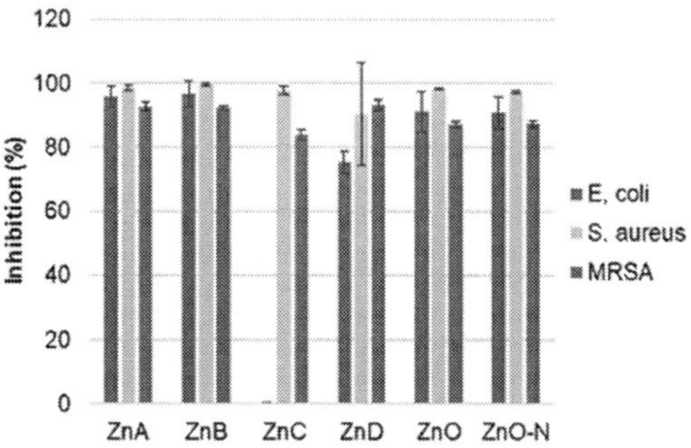

C

Sample	Live/dead	Dead
Control	474	10
ZnA	19	1
ZnB	56	2
ZnC	422	61
ZnD	369	43
ZnO	29	5
ZnO-N	62	12

Figure 1 (A) TEM images of synthetized ZnNPs, (B) inhibition effect of particles on three bacterial strains by spread-plate method, (C) the average amount of live and dead bacteria (SYTO9) and dead bacteria (PI) during microscope assay

Three methods - counting of bacterial colonies, growth curves and Live/dead assay - have been employed to determine their antimicrobial effects, and their antibacterial activity was assessed. These analyses were performed with synthetized ZnNPs and also with ZnO and commercial ZnO based NPs (ZnO-N) for comparison. By spread-plate method, the amount of viable bacteria after 5 mM Zn treatment and 24 h incubation was found out. Overview of % inhibition effects of ZnNPs and ZnO are shown in **Figure 1B**, where almost all types of Zn caused more than 90 % inhibition of bacterial growth.

A significant trend of antibacterial activity was found out by measuring the optical density during 24 h. From these growth curves the minimal inhibition concentrations (MIC) were determined, and ranged from 1.25 mM for *S. aureus* and *E.coli* to 2.5 mM for MRSA type. It is known, that MRSA phenotype confer different resistance characteristics present in SCCmec genotype cassette, which are probably responsible for this lower efficacy [14]. In various MRSA isolates, a metal resistance genes were also found [15].

The bacteria viability was analyzed by the fluorescence microscopy. The individual images were taken on three fields of sight with a 20 µm approximation. From the resulting bacterial population, the average was calculated and the results are shown in the table in **Figure 1C**. From the given values, similar differences can be observed as are seen, in the physical characterization of the particles. While nanoparticles ZnA and ZnB treatment led to inhibition of bacterial growth, ZnC and ZnD did not reduce the bacterial growth, but they caused the increase of the dead cells.

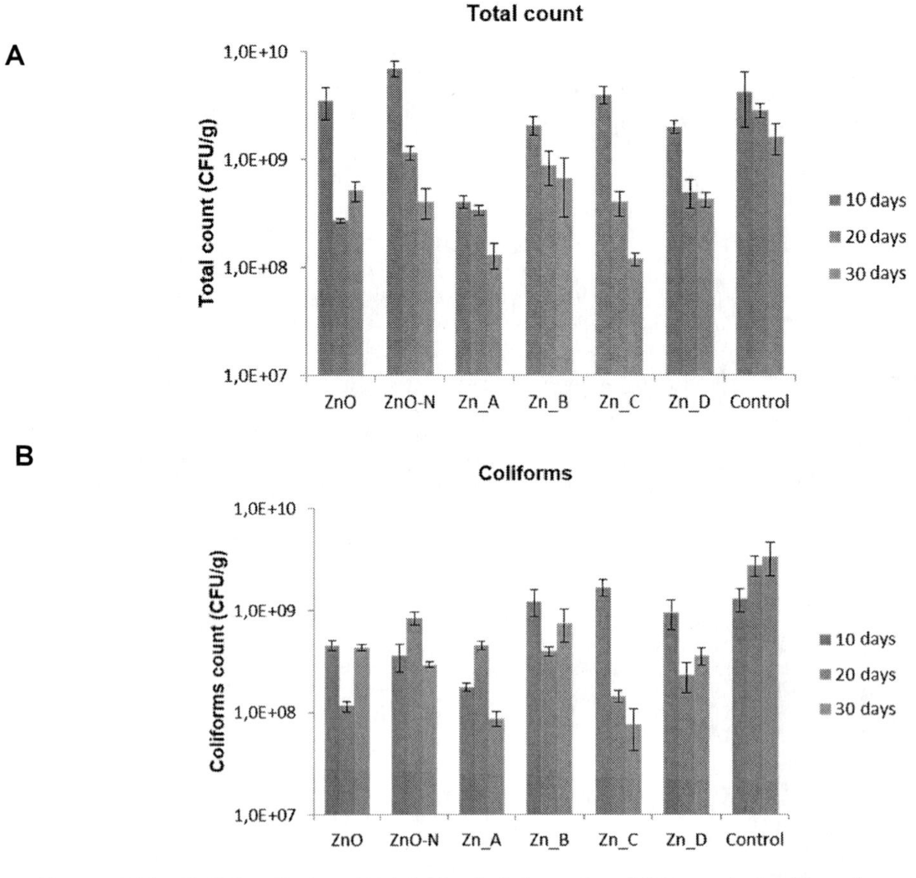

Figure 2 The ZnNPs effect on (A) total bacterial count and (B) count of coliform bacteria in rat feces

After promising and successful tests, the effects of these ZnNPs were subsequently monitored in rats, specifically their effect on rats gut microbiota. There were investigated the effect of zinc nanoparticles on rats after oral administration (**Figure 2**). In general, a decrease in the bacterial population was observed compared to the untreated control group with the duration of the experiment.

The decrease in both, the total aerobic bacteria and coliforms in rats feces was monitored, especially at day 30 of the treatment. After 10 and 20 days, few changes can be observed, but after 30 days there were apparent decrease of bacterial populations. Especially in case of nanoparticles types ZnA and ZnC the decrease of coliforms were significant (**Figure 2B**). Currently, these bacterial populations can cause digestive problems, if they are pathogens and their abundance is high in the intestines. They also cause problems with immunity and overall performance.

4. CONCLUSION

This work was concerned with testing the antimicrobial activity of zinc nanoparticles. At the beginning, their antibacterial effects were characterized *in vitro* by three testing methods against three bacterial strains. Zn-based nanoparticles have shown sufficient antimicrobial activity, which was the key point for their testing throughout the whole animal organism, the rats. The zinc nanoparticles effect on intestinal microbiota has been evaluated during the 30 day period.

ACKNOWLEDGEMENTS

The study was financially supported by Ministry of Agriculture of the Czech Republic (QK1720349 "Nanoparticles zinc as an alternative to antibiotics in pigs"). Financial support from ERDF "Multidisciplinary research to increase application potential of nanomaterials in agricultural practice" (No. CZ.02.1.01/0.0/0.0/16_025/0007314) is gratefully acknowledged as well.

REFERENCES

[1] GOPI, M., et al., Role of Nanoparticles in Animal and Poultry Nutrition: Modes of Action and Applications in Formulating Feed Additives and Food Processing. International Journal of Pharmacology, 2017. 13(7): p. 724-731.

[2] JOZWIK, A., et al., The Effect of Different Levels of Cu, Zn and Mn Nanoparticles in Hen Turkey Diet on the Activity of Aminopeptidases. Molecules, 2018. 23(5).

[3] STOIMENOV, P.K., et al., Metal Oxide Nanoparticles as Bactericidal Agents. Langmuir, 2002. 18(17): p. 6679-6686.

[4] SWAIN, P.S., et al., Preparation and effects of nano mineral particle feeding in livestock: A review. Veterinary World, 2015. **8**(7): p. 888-891.

[5] M. PEZ-ALONSO, Trace Minerals and Livestock: Not Too Much Not Too Little. ISRN Veterinary Science, 2012. 2012: p. 18.

[6] PICCINNO, F., et al., Industrial production quantities and uses of ten engineered nanomaterials in Europe and the world. Journal of Nanoparticle Research, 2012. 14(9): p. 1109.

[7] SWAIN, P.S., et al., Nano zinc, an alternative to conventional zinc as animal feed supplement: A review. Animal Nutrition, 2016. 2(3): p. 134-141.

[8] PADMAVATHY, N. and R. Vijayaraghavan, Enhanced bioactivity of ZnO nanoparticles-an antimicrobial study. Science and Technology of Advanced Materials, 2008. 9(3): p. 035004.

[9] AZAM, A., et al., Antimicrobial activity of metal oxide nanoparticles against Gram-positive and Gram-negative bacteria: a comparative study. International Journal of Nanomedicine, 2012. 7: p. 6003-6009.

[10] HONGFU, Y.B.Z., Effects of Nano-ZnO on growth performance and diarrhea rate in weaning piglets. China Feed, 2008. 1: p. 008.

[11] MISHRA, A., et al., Growth performance and serum biochemical parameters as affected by nano zinc supplementation in layer chicks. J. Anim. Nutr., 2014. 31: p. 384-388.

[12] SAHOO, A., et al., Serum biochemical indices of broiler birds fed on inorganic, organic and nano zinc supplemented diets. Int. J. Recent Sci. Res., 2014. 5(11): p. 2078-2081.

[13] UNIYAL, S., et al., Comparative efficacy of zinc supplementation from different sources on nutrient digestibility, hemato-biochemistry and anti-oxidant activity in guinea pigs. Livestock Science, 2017. 204: p. 59-64.

[14] KAITO, C., et al., Transcription and Translation Products of the Cytolysin Gene psm-mec on the Mobile Genetic Element SCCmec Regulate Staphylococcus aureus Virulence. Plos Pathogens, 2011. 7(2).

[15] CAVACO, L.M., et al., Cloning and Occurrence of czrC, a Gene Conferring Cadmium and Zinc Resistance in Methicillin-Resistant Staphylococcus aureus CC398 Isolates. Antimicrobial Agents and Chemotherapy, 2010. 54(9): p. 3605-3608.

RELEASE OF SILVER NANOPARTICLES FROM POLYSACCHARIDE FILMS BASED ON METHYLCELLULOSE

Markéta PIŠLOVÁ, Tereza JAČKOVÁ, Dominik FAJSTAVR, Kateřina KOLÁŘOVÁ, Václav ŠVORČÍK

Department of Solid State Engineering, University of Chemistry and Technology Prague, Czech Republic, EU, pislovam@vscht.cz

Abstract

The present work is focused on preparation and characterization polysaccharide films based on methylcellulose. Films were doped with silver nanoparticles (AgNPs). AgNPs were prepared and stabilized by reduction of silver nitrate. Polyethylene glycol and glycerol were used as additives for better material properties of methylcellulose films. Properties of these films were analyzed with a number of analytical techniques. Wettability and water absorption of films were evaluated. Concentration of the AgNPs released into the solution during dissolution was studied by atomic absorption spectroscopy. The presence of AgNPs in solutions and solid films was monitored by the ultraviolet-visible spectroscopy. Solid films were dissolved at different temperatures and release of AgNPs from the film was observed during the dissolution time by the ultraviolet-visible spectroscopy. The presence of AgNPs was confirmed both in the solid films and in the solutions by the above mentioned methods. It was found that additives changed materials properties. Our research was aiming on use of these films in medicine as a new type of wound dressing with antibacterial properties or as a drug delivery system.

Keywords: Polysaccharides films, silver nanoparticles, methylcellulose

1. INTRODUCTION

Cellulose (β-1,4-glucopyranose) is the most abundant polysaccharides on the Earth. It belongs among a wide group of natural polysaccharides with excelent biological properties. Natural polysaccharides have excellent biological properties, such as non-toxicity, biocompatibility and biodegradability [1,2]. Their main advantages are good mechanical properties, easy preparation and low cost. Biological properties could be changed by the chemical or physical modification of their surface. Surface properties of the materials are inevitably bonded with tissue engineering, especially for polymers when improvement of the cell adhesion to the surface, the cell growth and uniformity are studied [3].

Cellulose is commonly used in health care like wound dressing or as a pharmaceutical excipients. Nowadays i tis also used as a laxative.[4]. Cellulose is further used in the paper, textile, cosmetic, food and chemical industries. Nowadays, cellulose has a lot of derivatives e.g. organic esters - celulose acetate or propionate, inorganic esters - nitrocelulose, celulose ethers - methylcellulose, hydroxypropylmethyl celulose, carboxymethyl celulose, etc. Every derivatives have special properties like solubility. This work is focused on methylcellulose. It is hydrophilic white powder which dissolves only in cold water. Dissolved powder makes a clear viscous solution or gel [5,6,7]

This work is focused on the preparation of polysaccharide films based on methylcellulose. The prepared films were doped with silver nitrate yielding AgNPs embedded in the polysaccharide matrix [8,9]. The influence of addition of glycerol or polyethylene glycol on the material properties was studied. Wettability and water absorption of the films were evaluated. The UV-Vis absorption of prepared solution and dissolved films was measured. Release of AgNPs was studied by UV-Vis, too.

2. EXPERIMANTAL

2.1. Material

Methylcellulose (MC) and silver nitrate ($AgNO_3$) were obtained from Sigma-Aldrich. Sodium citrate was purchased from Penta. Polyethylene glycol 400 (PEG 400) was obtained from Fluka Analytical and glycerol was purchased from Lach-Ner and distilled water. PEG 400 and glycerol were used plasticizers.

2.2. Preparation of polysaccharide films

Polysaccharide films were prepared via process shown in **Figure 1**. Chronologically: methylcellulose (2 g) was stirred in 200 mL distilled water. The solution was constantly stirred and heated at 80 °C [5,6]. After 45 min, suspension of MC removed from stirring and heating device. Suspension was cooled down for 20 min. The cooled-down suspension of MC was then placed in an ice bath. During the cooling down, 10 mL of sodium citrate (0.085 g of sodium citrate was dissolved in 10 mL of water) was slowly dropped to the suspension a then 10 mL of $AgNO_3$ was added (0.085 g of $AgNO_3$ was dissolved in 10 mL of water), too. This mixture was stirred for 30 min. After that, 4 mL of PEG 400 (or glycerol) were dropped to suspension. Suspension was still stirred. Then the solution was cooled down to the laboratory temperature and aliquots of 10 mL were poured into round silicone molds, 5 cm in diameter. The samples were dried for 17 h at 60 °C. Finally, we got round homogenous polysaccharide films. The rest of solutions was used to measurements as well. The other films were prepared by the same method under the condition of no PEG 400 or glycerol. These solutions and films were used to as control samples.

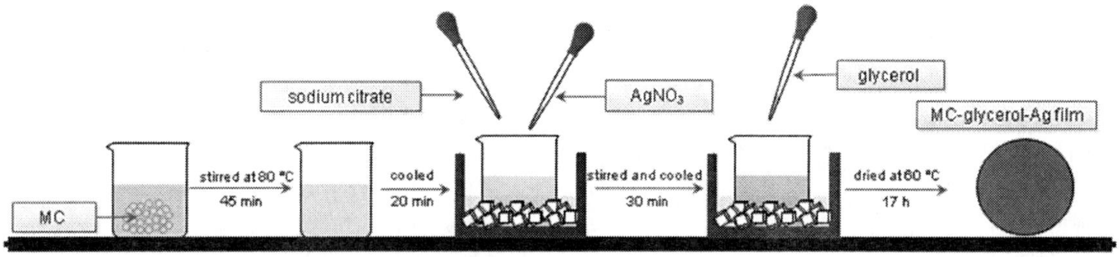

Figure 1 Schema of the preparation of polysaccharide films with AgNPs

2.3. Characterization methods

The presence and concentration of AgNPs in solutions and on the surface of the prepared films were studied by Ultraviolet-visible spectroscopy (UV-Vis, LAMBDA 25, PerkinElmer, USA). Wettability of the prepared films was evaluated by Surface Energy Evalution System Advex Instruments, Brno). Water absorption of the films was evaluated gravimetrically (RADWAG MAX 60, Poland). The release of AgNPs was observed at 2 different temperatures. Solid films were dissolved in ice bath 4°C and heated at 37°C. These solutions with release AgNPs were measured by UV-Vis. Concentration of release AgNPs from films was determined by atomic absorption spectroscopy (AAS).

3. RESULTS AND DISCUSSION

Three types of polysaccharide films based on methylcellulose were prepared. All these films contained AgNPs (**Figure 2**). First type of film was pristine, hereby referred as **MC-Ag**, consisting only of methylcellulose and AgNPs. The second type of film is here referred as **MC-glycerol-Ag**, consisting of metylcellulose film which was modified by plasiticizer glycerol during the preparation and also AgNPs. The third type is here referred as **MC-PEG 400-Ag** and it is aslo methylcellulose film which was modified by plasiticizer PEG 400 during the preparation. Then films with plasticizers and AgNPs were prepared, too. Films that contain AgNPs have orange-brown of gray-green colour in solid state. The colour change indicates the presence of AgNPs, on the other hand polysaccharide films without AgNPs are pure /citace/.

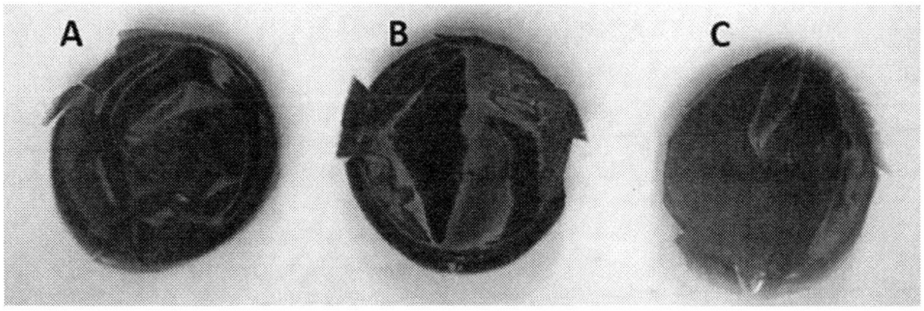

Figure 2 Prepared films: A - MC-Ag, B - MC-glycerol-Ag, C - MC-PEG 400-Ag

3.1. Ultraviolet-visible spectroscopy (UV-Vis) of solution

AgNPs were confirmed by UV-Vis spectroscopy in all three solutions that contained AgNPs. AgNPs have characteristic peak at wavelength around 400 nm [8]. **Figure 3** shows that the highest absorbance had solution **MC-Ag** that was used to prepare **MC-Ag** film. Very similar peak with a little bit lower absorbance was observed for the solution **MC-glycerol-Ag** film. The lowest absorbance had solution **MC-PEG 400-Ag**.

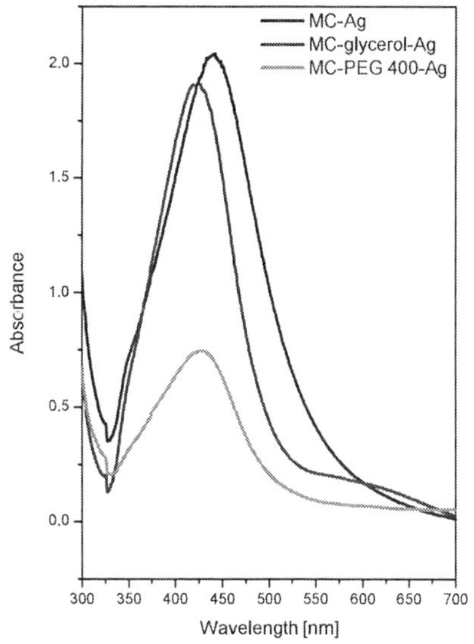

Figure 3 UV-Vis spectra of solutions for preparation MC films: A - MC-Ag, B - MC-glycerol-Ag, C - MC-PEG 400-Ag

3.1. Characterization of solid films

3.1.1. Wettability and water absorption

Wettability and water absorption are very important properties of materials which could be as a wound cover. The values of the contact angle and water absorption determined wettability of the films. It can subsequently be used as an indicator of the ability to absorb exudate from injury. These properties are necessary to design wound healing material. The measurements of wettability and water absorption were done only with films that contained AgNPs. From **Figure 4** is apparent that contact angles decreased with addition of plasticizers. The lowest wettability value and the highest contact angle value had film **MC-Ag**. On the other hand the lowest contact angle was measured at film **MC-PEG 400-Ag**. The film **MC-Ag,** which had the highest value of contact angle, had also the highest value of water absorption. The lowest value of water absorption had film **MC-glycerol-Ag**. We found out that the addition of plasticizers significantly reduced the wettability and water absorption. Plasticizers also change material properties like elasticity. The films which only based on MC were stroger than the others film with additives.

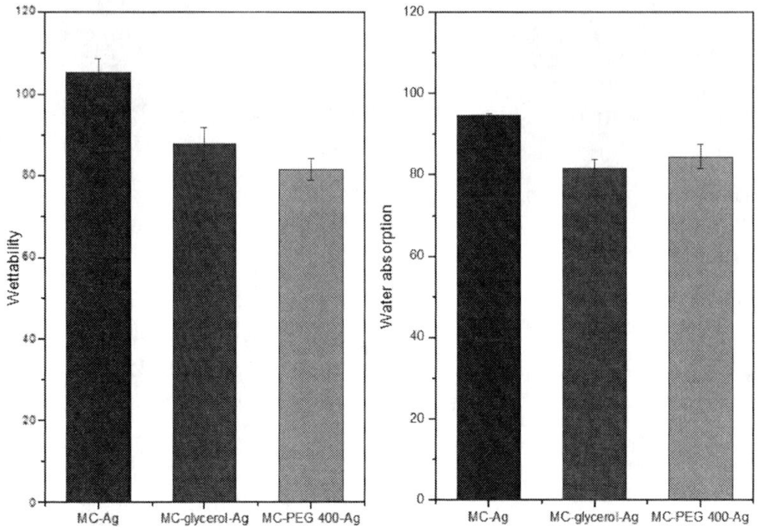

Figure 4 Wettability and water absorption of the prepared films

3.2. Characteristics of dissolved films

Due to the natural properties of methylcelullose, the solubility of prepared films and release of AgNPs were tested at distilled water. We chose two different temperatures: 4 and 37°C. The temperature of 4°C were chosen for termo was chosen due to the thermoreversible properties of the methylcellulose. And the temperature of 37°C as the temperature as the temperature simulating the warm human body. We dissolved a quarter of the prepared film in conditions and then ultraviolet-visible spectroscopy (UV-Vis) was measured. The concentration of AgNPs of the dissolved films was also determined by atomic absorption spectroscopy (AAS).

3.2.1. Atomic absorption spectroscopy (AAS)

Atomic absorption spectroscopy was used to examine concentration of AgNPs that were released from the films in cold distilled water (4°C). During 3 days whole films were dissolved and all AgNPs were in the solution. **Table 1** shows that after 3 days the highest amount of AgNPs contained film **MC-Ag.** The lowest amount of AgNPs was measured at film **MC-PEG 400-Ag**. In this case, it is possible that between MC and PEG 400 or glycerol were created H bonds and thus did not completely reduce $AgNO_3$ to AgNPs as was the case with Mc-Ag. The concentration was measured with experimental error ±5 %.

Table 1 Concentration of released AgNPs studied by AAS

Sample	Ag (mg/L)
MC-Ag	99.3
MC-glycerol-Ag	88.1
MC-PEG 400-Ag	82.1

3.2.3. Release of AgNPs from polysaccharide films

The release of AgNPs from prepared films were studied by UV-Vis spectroscopy. Two different temperatures were used during the measurements. From **Figure 5** show that the AgNPs were most released from the film **MC-Ag**. It is possible that hydrogen bond between AgNPs and MC were weaker and AgNPs were easier released from the film. It was observed that temperature of solution influenced the released of AgNPs. AgNPs released from the films more and more rapidly at 37°C than at 4°C.

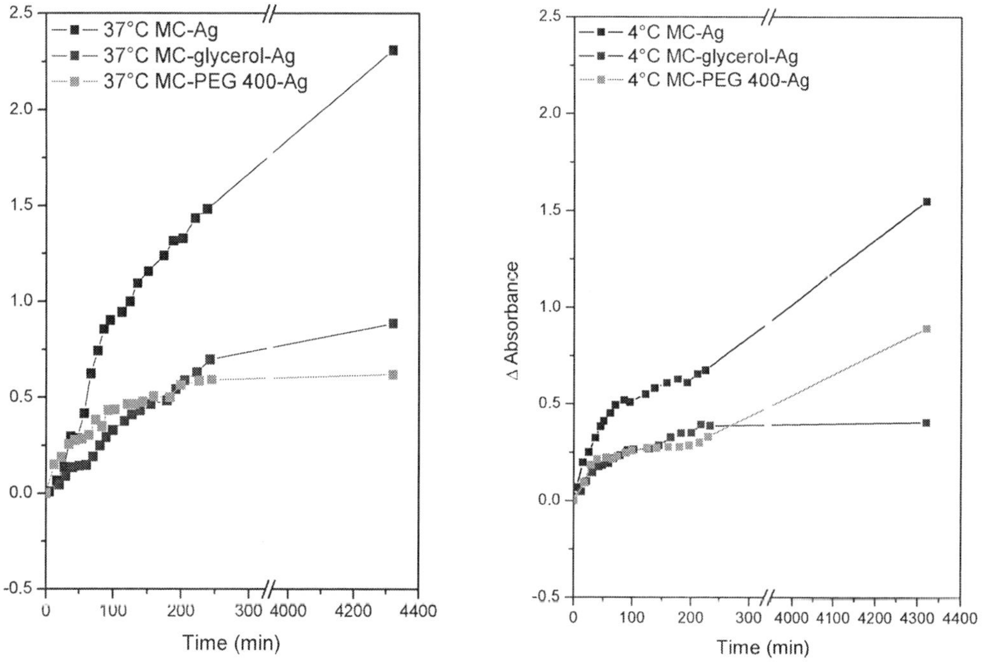

Figure 5 Differences in absorbance of released AgNPs from films: MC-Ag, MC-glycerol-Ag, MC-PEG 400-Ag

CONCLUSION

Three polysaccharide films based on methylcellulose were prepared. The presence of AgNPs was confirmed by UV-Vis and AAS. Prepared solutions were measured by UV-Vis before forming films. The highest concentration of AgNPs had **MC-Ag** films. It was found out that polysaccharides films containing plasticizers (glycerol, PEG 400) reduced wettability and water absorption. Release of AgNPs from polysacharide films were observed. It was found that the AgNPs were most released from the film **MC-Ag**. It was observed that temperature of solution influenced the released of AgNPs. AgNPs released from the films more and more rapidly at 37°C than at 4°C.

ACKNOWLEDGEMENTS

This work was supported by the Grant Agency of CR (project No. 17-00885S) and by the specific university research (MSMT No 21/2018).

REFERENCES

[1] DUTTA, P. K., DUTTA, J., TRIPATHI, V. S. Chitin and chitosan: Chemistry, properties and applications. *Journal of scientific and industrial research*, 2004, vol. 63, no. 1, pp. 20-31.

[2] SHEN, X., SHAMSHINA, J. L., BERTON, P., GURAU, G., ROGERS, R. D. Hydrogels based on cellulose and chitin: fabrication, properties, and applications. *Green Chemistry*, 2016, vol. 18, no. 1, pp. 53-75

[3] SLEPIČKA, P., SIEGEL, J., LYUTAKOV, O., ŠVORČÍK, V. Nanostructuring of Polymer Surface Stimulated by Laser Beam for Electronics and Tissue Engineering. *Chemické* listy, 2012, vol. 106, no. 10, pp. 875-883.

[4] HAMILTON, J.W., WAGNER, J., BURDICK, B. B., BASS, P. Clinical evaluation of methylcellulose as a bulk laxative, *Digestive Diseases and Sciences,* 1988, vol. 33, no. 8,pp.993-998.

[5] NASATTO, P. L., PIGNON, F., SILVEIRS, J. L., DUARTE, M. E. R., NOSEDA, M. D., RINAUDO, M. Methylcellulose, a cellulose derivative with original physical properties and extended applications. *Polymers*, 2015, vol. *7, no.* 5, pp. 777-803.

[6] CHALABALA, M. *Technologie léků*. 2nd ed. Praha:Galén, 2001. p. 408.

[7] THIELKING, H., SCHMIDT, M. Cellulose ethers. *Ullmann's encyclopedia of industrial chemistry*, 2000.

[8] PIŠLOVÁ, M., KOLÁŘOVÁ, K., VOSMANSKÁ, V., KVÍTEK, O., ŠVORČÍK, V. Preparation of Polysaccharide Films Based on Chitosan and Cellulose, *Chemické listy*, 2015, vol. 109, no. 12, pp. 942-945.

[9] DARROUDI, M., AHMAD, M. B., ZAMIRI, R., ZAK, A. K., ABDULLAH, A. H., IBRAHIM, N. A. Time-dependent effect in green synthesis of silver nanoparticles. *International journal of nanomedicine*, 2011, vol. *6*, pp. 677-681.

GREEN SYNTHESIZED SILVER NANOPARTICLES DERIVED FROM AN EXTRACT OF THE BETULA PENDULA TREE

[1]Šárka HRADILOVÁ, [2]Aleš PANÁČEK, [2]Radek ZBOŘIL

[1]RCPTM, Faculty of Science, Palacky University, Olomouc, Czech Republic, EU, sarka.hradilova@upol.cz
[2]RCPTM, Department of Physical Chemistry, Faculty of Science, Palacky University, Olomouc, Czech Republic, EU

Abstract

This is a first report for the synthesis of silver nanoparticles using *branch* extract *of Betula pendula* - one of the most common trees in Europe, China and southwest Asia. The present study is dealing with ecofriendly green synthesis of silver nanoparticles (AgNPs) using the aqueous extract of branches from *B. pendula* tree. The reaction mixture contained only the extract (10 or 25 % of the total volume of the reaction mixture) from the branches and the source of silver ion without the addition of any catalyst, surfactant or template. The bio-synthesis was carried out at room temperature for 24 hours. The described method is perfectly green, simple, coast-effective and eco-friendly. Reducing substance from birch branches are purely natural and can be obtained simply without damaging the tree. The AgNPs were characterized by UV-vis., DLS and TEM. Surface plasmon spectra for AgNPs are centered at 430-436 nm. The synthesized AgNPs were found to be spherical in shape with size 22.7 ± 6.9 nm and 25.2 ± 7.5 nm, respectively for a reaction mixture containing 10 or 25 % of the extract. Biomolecules from branches acted as a reducing agent as well as stabilizer. Common green synthesis uses "green" parts of plants, especially leaves or fruits. It should be noted that in countries with alternating seasons, leaves are not available throughout the year. Branches, on the other hand, are available all year round and their composition is constant regardless of the season.

Keywords: Green synthesis, *Betula pendula*, silver nanoparticles, eco-friendly

1. INTRODUCTION

Nanoparticles (NPs) of noble metals are used in a wide range of applications from electronics to biology and medicine. Nanoparticles thanks to its "nano" size exhibit properties entirely different from the bulk material. This is essentially a high surface to volume ratio which increases with decreasing size of the nanoparticles. The large free surface is particularly advantageous for catalytic purposes [1,2], as well as for enhanced Raman spectroscopy (SERS) [3,4]. Among the nanoparticles of noble metals silver occupies an important place due to its unique antibacterial properties [5-7]. With increasing demand for silver nanoparticles is also growing effort to prepare them in a way that is effective, but also inexpensive. Chemical preparation methods of nanoparticles with different size and shape are very well known. These methods often utilize strong reducing agents e.g. sodium borohydride [8], run at elevated temperatures, and often are used toxic or hazardous substances [9,10] or high pH [11]. On the other hand, so-called green syntheses as defined by Sharma [12] are based on selection of suitable solvent medium, they are environmentally friendly and reducing substances comes from nature. The main sources of reducing agents are plants [13-23], fungi [24] and microorganisms [25,26]. Reducing agent figures often as a stabilizer and capping agent [20,23,27] too. Because green synthesis take place under mild conditions, sources of reducing agents comes from nature and the additional chemicals are not necessary, this procedure is not only environmentally friendly, but also inexpensive. As already mentioned, frequently used parts of plants are, e.g. leaves [14,19,25,26,28], roots [27,29], seeds [18] or fruits [23]. In areas where there is a change of seasons it is not possible to obtain the leaves or fruits all year round. In contrast, bark, or parts of trees are available year-round. Yet there have only minimal information on the use of the bark

[16] for the preparation of nanoparticles. Data on the use of branches or whole wood extract are altogether missing. Our aim was to test using branches extract to bio-reduction of silver nanoparticles. As a suitable source of reduction substances appears to be silver birch tree, which is wide-spread throughout Europe, China and southwest Asia. Birch tree (*B. pendula*, Linné, 1753) is a genus of deciduous trees or shrubs of the family *betulaceae*. It is a genus of mild to sub polar zone of the northern hemisphere, widespread mainly in the Eurasian region. *B. pendula* also known under the name of silver birch or warty birch is a small to medium-size tree with white peeling bark. Birch is typical for rapid growth; the thin ends of the branches can be easily removed without damaging the tree. The water content and composition of wood slightly varies depending on the season [30], but if dried branches are used, the water content does not play any role.

2. MATERIALS AND METHODS

2.1. Chemicals

All chemicals/reagents used in the study were of analytical grade and were purchased from Sigma-Aldrich, and they were used without further purification. All solutions were prepared with Milli Q water.

2.2. Synthesis of AgNPs (silver nanoparticles)

Accurate concentration of 10mM silver nitrate was prepared by dissolving $AgNO_3$ in 25 ml of Milli Q water. A fresh solution was prepared before each experiment. Birch tree (*Betula pendula*) branches were collected from the tree in Olomouc town, Czech Republic during the beginning of March, 2017, at a time when the tree does not have leaves yet. The branches were cut into pieces of about 2 cm, rinsed with distilled water thrice followed by Milli Q water to remove the dust and other contaminants and then dried at 60°C for 24 hours to remove the moisture. Weight of dried branches compared to fresh halved. Then 10 g of dried birch branches were weighed, mixed with 100 ml of Milli Q water and boiled in a 250 ml Erlenmeyer flask under continuous stirring for 10 minutes. After cooling the extract was filtered using Whatman No.1 filter paper and stored at 4°C for further use. We have observed how the concentration of the reducing agent will affect the formation of AgNPs. Two different volume of branch extract (10 ml or 25 ml) was added to $AgNO_3$ solution (final concentration of $AgNO_3$ was 1 mM, total volume 100 ml) for bio reduction process at room temperature and observed for change in color. The final solution containing bio-synthesized AgNPs was centrifuged at 10 000 rpm for 15 min. The precipitate was washed 3 times to remove unreacted silver ions, kept in petri dish and left in the oven at 60°C for 24 hrs. Dry AgNPs were scrapped and used for characterization.

2.3. Characterization of AgNPs

The bio reduction of Ag^+ ions in solutions was monitored by measuring a spectrum of the reaction medium. The UV-VIS spectral analysis of the sample was done by using UV−VIS absorption spectroscopy with the Specord S 600 spectrophotometer (Analytik Jena AG). The average sizes and size distributions of the prepared noble metal NPs were determined by dynamic light scattering (DLS) using the Zetasizer Nano ZS (Malvern). The nanodimensions of the synthesized silver NPs were confirmed by transmission electron microscopy using the JEM-2010 (Jeol). Zeta potential was measured using the Zetasizer Nano ZS (Malvern). Fourier transform infrared (FTIR) spectral measurements were carried out to identify the potential biomolecules in *Betula p.* branch extract which is responsible for reducing and capping the biogenic silver nanoparticles. FTIR absorption spectra were measured on an FTIR spectrometer iS5 Thermo Nicolet employing attenuated total reflection (ATR) on a ZnSe crystal, using 64 scans with the resolution of 2 cm^{-1}. Air-dried samples were placed directly on a ZnSe crystal and slightly pressed in order to remove any residual air containing humidity and CO_2.

3. RESULTS AND DISCUSSION

The addition of birch (*Betula pendula*) branch extract to the silver nitrate solution resulted in the change in color from light yellowish to dark yellow on incubation at room temperature (24 °C). Color change of the reaction mixture containing 10 % and 25 % extract immediately after the addition of the extract, and after 24 hrs. are shown in **Figure 1**. Silver particles exhibit dark yellow color in the solution because of the excitation of surface plasmon vibrations caused by the silver nanoparticles as described elsewhere [31]. The color gradually darkened with increasing time. The increasing concentration of silver nanoparticles was confirmed by UV spectrophotometry (**Figure 2**). The presence of the surface plasmon provides information about the size and the morphology of nanoparticles [31-33]. The larger volume of the extract of the reaction mixture showed higher reduction efficiency as seen in the UV-VIS spectra, as well as larger size of the nanoparticles and high yield. Nanoparticles reduced 10 % extract exhibited plasmon with a maximum at 430 nm, 25 % at 436 nm respectively. Bio synthesis of silver nanoparticles was complete after 24 hours as can be seen from the spectra (**Figure 2**), location of maxima and peak height did not change since the twenty fourth hour.

Figure 1 Reaction mixtures Immediately after addition of birch extract (left), reaction mixture after 24 hrs. (right)

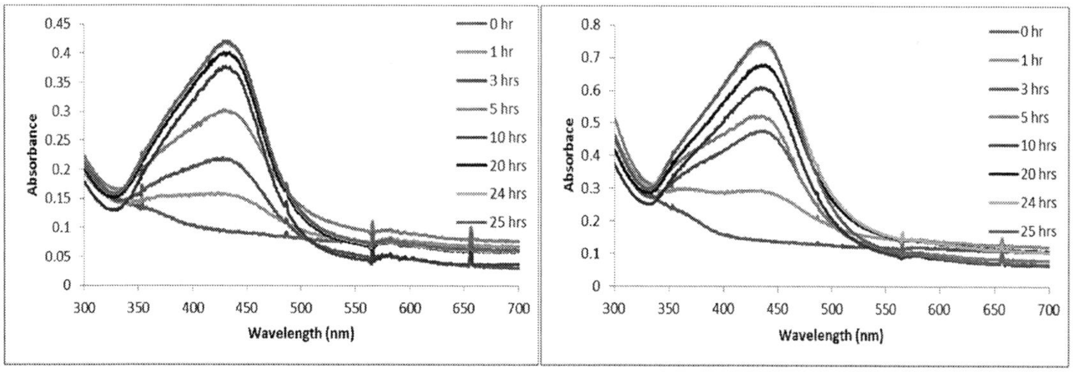

Figure 2 UV-VIS. Absorption spectra recorded in time horizon from 0 to 25 hours. Mixture containing 10 % (v/v) of extract (left), and 25 % (v/v) of extract (right).

The size and morphology of bio synthesized silver nanoparticles were examined using TEM (**Figure 3**) and DLS. The obtained data are summarized in **Table 1**. Size obtained from DLS was somewhat larger than TEM data. It is evident that the biosynthesized nanoparticles are more or less spherical in shape and their size variances in the interval between 7 and 37 nm. The nanoparticles prepared with 25 % extract showed more negative values of zeta potential, it can therefore be assumed that the stability of the dispersion is higher.

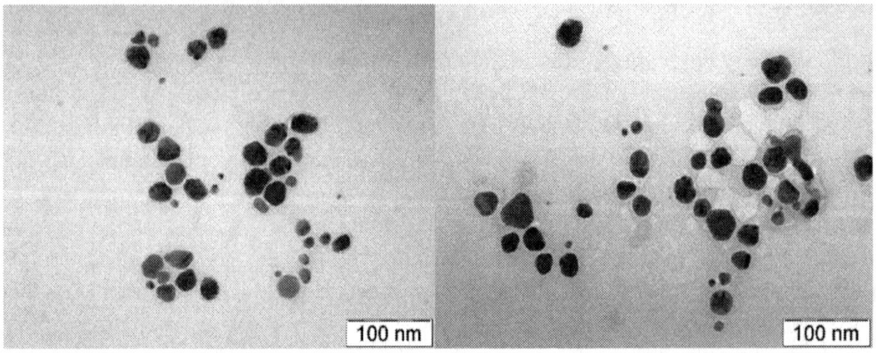

Figure 3 TEM images of silver NPs. 10 % extract (left), 25 % extract (right)

Table 1

	Size TEM	Size DLS	Zeta potential	Absorption peak
10 % extract	22.7 ± 6.9 nm	32.14 ± 0.08 nm	-22 ± 0.86 mV	430 nm
25 % extract	25.2 ± 7.5 nm	51.45 ± 1.49 nm	-24 ± 0.63 mV	436 nm

Fourier transform infrared (FTIR) spectral measurements were carried out to identify the potential biomolecules in birch tree branches extract which is responsible for reducing and capping the biogenic silver nanoparticles. Typical FTIR spectrum of *Betula pendula* branch extract is shown on **Figure 4**.

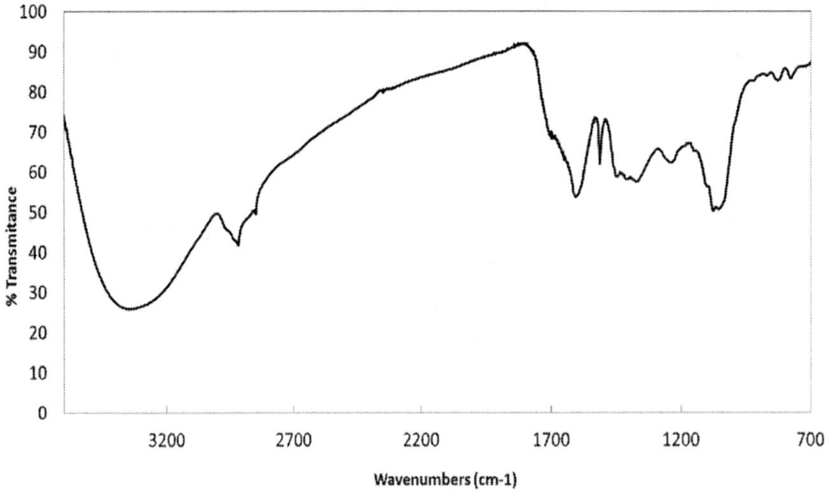

Figure 4 FTIR spectrum of *Betula pendula* branch extract

There are several characteristic bands. The finger prints region lied between wavenumbers ranging from 700 and 1600 cm^{-1} and typical stretching of chemical groups was observed at 2800 to 3500 cm^{-1}. The absorption bands at 1605 cm^{-1} and 1515 cm^{-1} are characteristic of phenyl ring skeletal vibrations of lignin macromolecules [34,35]. The broad band at 3350 cm^{-1} was due to presence of hydroxyls. The band 2917 cm^{-1} was contributed to methyl and methylene groups. In the finger print region band located at 1327 cm^{-1} was characteristic for the presence of methyl groups and 1076 cm^{-1} for alcohol groups. These bands originated from the functional groups present in the various tree metabolites of the branches extract. The functional groups such as -OH present in extract might be responsible for bio reduction of Ag^+ to AgNPs. *Betula p.* is rich in tannins (an astringent, bitter plant polyphenolic compounds that binds to and precipitates proteins) and other (poly)-phenols. So the capping agents might be tannins or phenols and it's -OH groups might be involved in bio reduction of silver nitrate. These results also correlate with the literature that had already been reported [22,27].

3. CONCLUSION

This is a first report for the synthesis of silver nanoparticles using *branch* extract *of Betula pendula. B. pendula* branch extract was applied for bio reduction of silver ions and environmentally friendly green synthesis of spherical silver nanoparticles. Preparation of AgNPs was carried at room temperature for 24 hours and the reaction mixture was composed of only the extract from the branches and the source of silver ion. The described method is absolutely green, simple, coast-effective and eco-friendly. Biogenic silver nanoparticles are stable due to the layer of bioactive substances adsorbed on their surface and can be used for the same purposes as nanoparticles prepared by conventional techniques. Branches of a tree as a source of reducing agents are available for all seasons unlike the leaves or fruits, the composition of the branches is more less constant regardless of the season and the end parts of the branches used for the reduction can be obtained simply without greater damage to the tree.

ACKNOWLEDGEMENTS

The authors gratefully acknowledge the support provided by project LO1305 of the Ministry of Education, Youth and Sports of the Czech Republic.

REFERENCES

[1] HERVÉS, P., PÉREZ-LORENZO, M., LIZ-MARZÁN, L. M., DZUBIELLA, J., LU, Y. and BALLAUFF, M. Catalysis by metallic nanoparticles in aqueous solution: model reactions. *Chemical Society Reviews* 2012, 41, pp. 5577-5587.

[2] SUCHOMEL, P., PANACEK, A., PRUCEK, R., ZBORIL, R., and KVITEK, L. Synthesis of Small Silver Nanoparticles and Their Catalytic Activity in 4- Nitrophenol Reduction. In NANOCON 2013: *Conference Proceedings.* Brno: Tanger Ltd., 2015, pp. 225-231.

[3] KVITEK, L., PRUCEK, R., PANACEK, A., NOVOTNY, R., HRBAC, J., and ZBORIL, R. The influence of complexing agent concentration on particle size in the process of SERS active silver colloid synthesis. *Journal of Materials Chemistry* 2005, 15, 1099-1105.

[4] HE, L. L., LIN, M. S., Li, H., and KIM, N. J. Surface-enhanced Raman spectroscopy coupled with dendritic silver nanosubstrate for detection of restricted antibiotics. *Journal of Raman Spectroscopy* 2010, 41, pp. 739-744.

[5] FRANCI, G., FALANGA, A., GALDIERO, S., PALOMBA, L., RAI, M., MORELLI, G., and GALDIERO, M. Silver Nanoparticles as Potential Antibacterial Agents. *Molecules* 2015, 20, pp. 8856-8874.

[6] LI, W. R., XIE, X. B., SHI, Q. S., ZENG, H. Y., OU-YANG, Y. S., and CHEN, Y. Ben Antibacterial activity and mechanism of silver nanoparticles on Escherichia coli. *Applied Microbiology and Biotechnology* 2010, 85, pp. 1115-1122.

[7] SCENIHR Nanosilver: safety, health and environmental effects and role in antimicrobial resistance. 2013, pp. 1-102.

[8] POLTE, J., TUAEV, X., WUITHSCHICK, M., FISCHER, A., THUENEMANN, A. F., RADEMANN, K., KRAEHNERT, R., and EMMERLING, F. Formation Mechanism of Colloidal Silver Nanoparticles : Analogies and Differences to the Growth of Gold Nanoparticles. *ACS Nano* 2012 *6* (7), pp. 5791-5802.

[9] BOCA, S. C. and ASTILEAN, S. Detoxification of gold nanorods by conjugation with thiolated poly(ethylene glycol) and their assessment as SERS-active carriers of Raman tags. *Nanotechnology* 2010, 21, 235601.

[10] ALI, M. R. K., SNYDER, B., and EL-SAYED, M. Synthesis and optical properties of small Au nanorods using a seedless growth technique. *Langmuir* 2012, 28, pp. 9807-9815.

[11] PANÁČEK, A., KVÍTEK, L., PRUCEK, R., KOLÁŘ, M., VEČEŘOVÁ, R., PIZÚROVÁ, N., SHARMA, V. K., NEVĚČNÁ, T., and ZBOŘIL, R. Silver colloid nanoparticles: Synthesis, characterization, and their antibacterial activity. *Journal of Physical Chemistry B* 2006, 110, pp. 16248-16253.

[12] SHARMA, V. K., YNGARD, R. A., and LIN, Y. Silver nanoparticles: Green synthesis and their antimicrobial activities. *Advances in Colloid and Interface Science* 2009, 145, pp. 83-96.

[13] BORASE, H. P., SALUNKE, B. K., SALUNKHE, R. B., PATIL, C. D., HALLSWORTH, J. E., KIM, B. S., and PATIL, S. V. Plant extract: A promising biomatrix for ecofriendly, controlled synthesis of silver nanoparticles. *Applied Biochemistry and Biotechnology* 2014, 173, pp. 1-29.

[14] DEVARAJ, P., KUMARI, P., AARTI, C., and RENGANATHAN, A. Synthesis and characterization of silver nanoparticles using cannonball leaves and their cytotoxic activity against MCF-7 cell line. *Journal of Nanotechnology* 2013, Article ID 598328, 5 pages

[15] GHAFFARI-MOGHADDAM, M. and HADI-DABANLOU, R. Plant mediated green synthesis and antibacterial activity of silver nanoparticles using Crataegus douglasii fruit extract. *Journal of Industrial and Engineering Chemistry* 2014, 20, pp. 739-744.

[16] IRAVANI, S. and ZOLFAGHARI, B. Green synthesis of silver nanoparticles using *Pinus eldarica* bark extract. *BioMed Research International* 2013, Article ID 639725, 5 pages

[17] KHATOON, N., AHMAD, R., and SARDAR, M. Robust and fluorescent silver nanoparticles using *Artemisia annua*: Biosynthesis, characterization and antibacterial activity. *Biochemical Engineering Journal* 2015, pp. 91-97.

[18] MUNIYAPPAN, N. and NAGARAJAN, N. S. Green synthesis, characterisation and biological studies of silver nanoparticles using *Dalbergia sissoides* seeds and ecofriendly reduction of 4-nitrophenol. *International Journal of Biomedical Nanoscience and Nanotechnology* 2014, Vol.3, No.3, pp.262-277.

[19] NAIK, L. S., MARX, K. P., SREE, P., CH, P., and RAMANA, V. Green synthesis of silver nanoparticles using Strawberry leaf extract (*Arbutus unedo*) and evaluation of its antimicrobial activity-a Novel study. 2013, 3, pp. 47-50.

[20] PARALIKAR, P. Biogenic Synthesis of Silver Nanoparticles Using Leaves Extract of *Epiphyllum Oxypetalum* and its Antibacterial Activity. *Austin J Biotechnol Bioeng.* 2014, 1, pp. 1-5.

[21] POURMORTAZAVI, S. M., TAGHDIRI, M., MAKARI, V., and RAHIMI-NASRABADI, M. Procedure optimization for green synthesis of silver nanoparticles by aqueous extract of *Eucalyptus oleosa. Spectrochimica Acta Part A: Molecular and Biomolecular Spectroscopy* 2015, 136, pp. 1249-1254.

[22] RAJA, K., SARAVANAKUMAR, A., and VIJAYAKUMAR, R. Efficient synthesis of silver nanoparticles from *Prosopis juliflora* leaf extract and its antimicrobial activity using sewage. *Spectrochimica acta. Part A, Molecular and biomolecular spectroscopy* 2012, 97, pp. 490-494.

[23] RAMESH, P. S., KOKILA, T., and GEETHA, D. Plant mediated green synthesis and antibacterial activity of silver nanoparticles using *Emblica officinalis* fruit extract. *Spectrochimica acta. Part A, Molecular and biomolecular spectroscopy* 2015, 142, pp. 339-343.

[24] MISHRA, A., TRIPATHY, S. K., and YUN, S. Fungus mediated synthesis of gold nanoparticles and their conjugation with genomic DNA isolated from *Escherichia coli* and *Staphylococcus aureus. Process Biochemistry* 2012, 47, pp. 701-711.

[25] REDDY, A. S., CHEN, C. Y., CHEN, C. C., JEAN, J. S., CHEN, H. R., TSENG, M. J., FAN, C. W., and WANG, J. C. Biological synthesis of gold and silver nanoparticles mediated by the bacteria *Bacillus subtilis. Journal of nanoscience and nanotechnology* 2010, 10, pp. 6567-6574.

[26] RAJESHKUMAR, S., PONNANIKAJAMIDEEN, M., MALARKODI, C., MALINI, M., and ANNADURAI, G. Microbe-mediated synthesis of antimicrobial semiconductor nanoparticles by marine bacteria. *Journal of Nanostructure in Chemistry* 2014, 4, 96, 7 pages.

[27] RATHI SRE, P. R., REKA, M., POOVAZHAGI, R., ARUL KUMAR, M., and MURUGESAN, K. Antibacterial and cytotoxic effect of biologically synthesized silver nanoparticles using aqueous root extract of *Erythrina indica* lam. *Spectrochimica Acta Part A: Molecular and Biomolecular Spectroscopy* 2015, 135, pp. 1137-1144.

[28] JOSEPH, S. and MATHEW, B. Microwave assisted facile green synthesis of silver and gold nanocatalysts using the leaf extract of *Aerva lanata. Spectrochimica Acta Part A: Molecular and Biomolecular Spectroscopy* 2015, 136, pp. 1371-1379.

[29] NARAGINTI, S. and SIVAKUMAR, A. Eco-friendly synthesis of silver and gold nanoparticles with enhanced bactericidal activity and study of silver catalyzed reduction of 4-nitrophenol. *Spectrochimica Acta - Part A: Molecular and Biomolecular Spectroscopy* 2014, 128, pp. 357-362.

[30] GIBBS, R. D. Studies of Wood: Ii. On The Water Content of Certain Canadian Trees And On Changes In The Water-Gas System During Seasoning And Flotation. *Canadian Journal of Research* 1935, 12, pp. 727-760.

[31] MULVANEY, P. Surface Plasmon Spectroscopy of Nanosized Metal Particles. *Langmuir* 1996, 12, pp. 788-800.

[32] KELLY, K. L., CORONADO, E., ZHAO, L. L., and SCHATZ, G. C. The Optical Properties of Metal Nanoparticles: The Influence of Size, Shape, and Dielectric Environment. *The Journal of Physical Chemistry B* 2003, 107, pp. 668-677.

[33] LINK, S. and EL-SAYED, M. A. Optical properties and ultrafast dynamics of metallic nanocrystals. *Annual review of physical chemistry* 2003, 54, pp. 331-66.

[34] CASAS, A., ALONSO, M. V., OLIET, M., ROJO, E., and RODRÍGUEZ, F. FTIR analysis of lignin regenerated from *Pinus radiata* and *Eucalyptus globulus* woods dissolved in imidazolium-based ionic liquids. *Journal of Chemical Technology & Biotechnology* 2012, 87, pp. 472-480.

[35] YANG, H., YAN, R., CHEN, H., LEE, D. H., and ZHENG, C. Characteristics of hemicellulose, cellulose and lignin pyrolysis. *Fuel* 2007, 86, pp. 1781-1788.

SESSION D
MONITORING AND TOXICITY OF NANOMATERIALS

Chairman

Ing. Jan TOPINKA, DrSc. Institute of Experimental of the CAS, Prague, Czech Republic, EU

BIOMARKERS OF OXIDATIVE STRESS AND INFLAMMATION IN RESEARCHERS EXPOSED TO NANOPARTICLES BY INHALATION DURING THE HANDLING OF NANOCOMPOSITES

Daniela PELCLOVÁ[1], Vladimír ŽDÍMAL[2], Jaroslav SCHWARZ[2], Martin KOMARC[3], Štěpánka VLČKOVÁ[1], Zdenka FENCLOVÁ[1], Lucie LISCHKOVÁ[1], Štěpánka DVOŘÁČKOVÁ[4], Andrea RÖSSNEROVÁ[5], Pavel RÖSSNER, Jr.[5]

[1]*Charles University in Prague and General University Hospital in Prague, First Faculty of Medicine, Department of Occupational Medicine, Prague, Czech Republic, EU,*
daniela.pelclova@lf1.cuni.cz, stepanka.vlckova@vfn.cz, zdenka.fenclova@lf1.cuni.cz, lucie.lischkova@vfn.cz
[2]*Institute of Chemical Process Fundamentals of the CAS, Prague, Czech Republic, EU,*
zdimal@icpf.cas.cz, schwarz@icpf.cas.cz
[3]*Charles University in Prague and General University Hospital in Prague, First Faculty of Medicine, Institute of Biophysics and Informatics, Prague, Czech Republic, EU, komarc@ftvs.cuni.cz*
[4] *Technical University in Liberec, Faculty of Mechanical Engineering, Department of Machining and Assembly, Liberec, Czech Republic, EU, stepanka.dvorackova@tul.cz*
[5]*Institute of Experimental Medicine, Czech Academy of Sciences, Prague, Czech Republic, EU,*
andrea.rossnerova@iem.cas.cz, rossner.pavel@iem.cas.cz, fatima.elzeinova@iem.cas.cz

Abstract

At present, little is known about the health effects in the workers processing nanocomposites. In our study, 20 researchers (41.8±11.4 y/o), handling nanocomposites for 17.8±10.0 years were examined pre-shift and post-shift, together with 21 controls (42.7±11.5 y/o). Biomarkers of oxidative stress derived from lipids, nucleic acids, proteins and markers of inflammation were analyzed in the exhaled breath condensate (EBC). Aerosol exposure was monitored during three nanoparticle generation operations: smelting, welding and nanocomposite machining. Mass concentrations during these operations ranged from 0.120 to 1.840 mg/m^3, and median particle number concentrations from 4.8x10^4 to 5.4x10^5 particles/cm^3. Nanoparticles accounted for 40 to 95 % of particles, with Fe and Mn prevailing. Significant elevations were already seen in most oxidative stress markers and in several inflammation markers in the pre-shift samples relative to the controls. Significant associations were found between working in nanocomposite synthesis and the majority of EBC biomarkers. Chronic bronchitis was more frequent in researchers. A minor, but significant post-shift decrease of lung function parameters was found. We conclude that workers in nanocomposite synthesis may be at risk of developing airway disorders with time. From all the markers analyzed in EBC, the following markers were most robust and could be recommended for preventive examinations: 8-hydroxy-2-deoxyguanosine (8-OHdG) and 5-hydroxymethyl uracil (5-OHMeU) from nucleic acids; o-tyrosine (o-Tyr) and 3-nitrotyrosine (3-NOTyr) from proteins; and malondialdehyde and aldehydes C6-C13 from lipids. Among the markers of inflammation, tumor necrosis factor (TNF) and leukotriene B4 appeared to be the most useful.

Keywords: Nanocomposites, exhaled breath condensate, oxidative stress, inflammation, spirometry

1. INTRODUCTION

Thousands of workers are engaged in the research and commercial scale production of nano-enabled composites. However, among tens of thousands of papers focusing on the research of the toxicity of nanoparticles found in PubMed, those concerning workers, i.e. the most exposed subjects, is very limited and will be mentioned in this study. The safety of nanoparticles for humans is the main goal of all *in vitro* and *in vivo* testing [1-6]. Experimental studies especially highlight respiratory health impairment. A common feature of many aerosols in the workplace is their ability to generate reactive oxygen species, which induce oxidative damage to biomolecules, leading to the activation of redox signaling pathways [7,8]. Elevated markers of oxidative stress in the biological fluids pro-inflammatory markers, such as tumor necrosis factor (TNF), pro-inflammatory leukotrienes (LTs), antioxidant enzymes and cardiovascular markers could give a signal of impairment. Lipid mediators, known to be immediately generated in response to tissue injury, might be strong candidates for priming this particle-triggered change of the inflammatory balance [9-10]. The results from the few publications focusing on exposed humans agree with the experimental studies [11-19]. Accordingly, our studies in 2012 and 2013, showed an elevation of markers of oxidative stress and inflammation in the exhaled breath condensate (EBC) [20] both in the production workers, and office employees, visiting the workshops for about 15 min a day [21-24]. In this article we focus on workers producing and handling nanocomposites as new materials with better technical characteristics, especially hardness and resistance. This study started in September 2015 and continues with a yearly follow-up of the group of researchers and matched controls.

2. MATERIALS AND METHODS

2.1. Working processes

Researchers usually work in two workshops: welding and smelting of mixtures containing nano-additives in workshop 1, and machining of the finished nanocomposite in workshop 2. All tasks occur simultaneously in both workshops and last on average 2.5 hours. On the day of medical examination in September 2016, in workshop 1, 11 out of a total of 20 researchers performed welding on metal surfaces on mild steel (content in wt%): Fe, 97.39; Mn, 1.70; C, 0.24; Si, 0.6; P, 0.035; S, 0.035) and smelting at 760°C. In workshop 2, the remaining 9 researchers performed the machining of surfaces of nanocomposite blocks of geopolymers and epoxide resins with nanoSiO_2 fillers. No respiratory protection was used. A detailed description of the working process was given in our last article [25].

2.2. Subjects

EBC samples were collected in 20 nanocomposite researchers (15 men, 5 women, one smoker, 19 non-smokers, with mean age 41.8±11.4 years; exposure duration 17.8±10.0 years). For simplicity, we refer to examinations of the researchers as pre-shift and post-shift, even though they were in fact post-exposure task measurements. The remainder of their total 8-hr shift was spent in their offices. There were 21 controls (15 men, 6 women; two smokers, 19 non-smokers, mean age 42.7±11.5 years), working as office employees in the same town.

2.3. Aerosol measurement

Nano-aerosol monitoring was performed with several aerosol spectrometers, including a Scanning Mobility Particle Sizer (TSI SMPS 3936L, USA), and an Aerodynamic Particle Sizer (TSI APS 3321, USA), covering the size range of aerosol particles from 6 nm up to 20 µm. Additionally, an Ultrafine Condensation Particle Counter (TSI UCPC 3025, USA) was used to measure the total particle number concentration (3 nm -~1 µm), as well as three Optical Particle Sizers (TSI OPS 3330, USA) used to measure the number size distribution in the range of 300 nm-10 µm. Area and time integrated samplings were conducted using a Berner Low Pressure Impactor (BLPI, HAUKE GmbH, Austria) to sample aerosol particles onto 10 stages corresponding to their aerodynamic diameter covering the 25 nm - 13.6 µm size range. These samples were then analyzed by gravimetry, ion chromatography and Scanning Electron Microscope (SEM) (Tescan Indusem, Czech Republic) equipped by Energy-Dispersive X-Ray Spectroscopy (EDS) (XFlash detector 5010, Bruker, Germany) to analyze the elemental composition of size-resolved aerosol fractions, as previously described in detail [25].

2.4. Biological samples

EBC samples were collected using Ecoscreen Turbo (DECCS, Jaeger, Germany) All samples were immediately spiked with deuterium labelled standards, immediately frozen and stored at -80 °C for subsequent processing [20]. TNF was analyzed by a matrix-assisted laser desorption ionization time-of-flight (MALDI-TOF) and analyses of low-molecular biomarkers LTs were performed using liquid chromatography-electrospray ionization tandem mass spectrometry (LC-ESI-MS/MS), after solid-phase extraction (SPE), and MS/MS detection were used [22,25].

2.5. Statistics

Basic descriptive statistics were calculated using Microsoft Excel 2013. Following tests were used for statistical comparisons of the pre-shift and post-shift markers (both within workers and between workers and controls): T-test, Mann-Whitney U test, paired sample t-test and the Wilcoxon signed-rank test. The bivariate relationship between the variables under study was assessed using the Spearman correlation coefficient. A multiple regression analysis was used to predict the markers in EBC by a set of predictors. All analyses were conducted using SPSS version 22.0 (SPSS, Inc., Chicago, Illinois, USA).

3. RESULTS AND DISCUSSION

Exposure data concerning aerosol measurements were provided in detail in our previous manuscript [25]. The average total mass concentration in the workshops was 0.120 mg/m^3 during smelting, 0.804 mg/m^3 during machining, and 1.840 mg/m^3 during welding. The total median particle number concentration was 1.3 x 10^5 particles/cm^3 (particles/cm^3) during welding, 4.8 x 10^4 particles/cm^3 during smelting and 5.4 x 10^5 particles/cm^3 during machining. The highest proportion of particles smaller than 100 nm in diameter was found during smelting (95 %), followed by machining (61 %), and welding (40 %).

The group characteristics did not differ significantly, including smoking (5 % smokers among workers with 10 pack-years, 9.5 % in controls with average 14 pack-years). Chronic bronchitis and dyspnea, associated with minor limitations in ordinary physical activity, were only found in the exposed researchers (20.0 % and 15 %, respectively). A cough was 5 times more frequent than in the controls. A significant decrease in forced expiratory volume in 1 s (FEV1) and FEV1/ forced vital capacity (FVC) was found in the researchers post-shift ($p<0.05$).

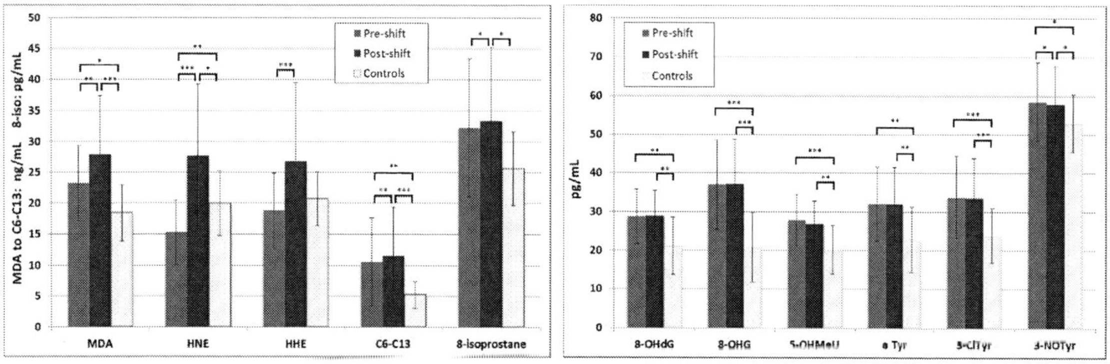

Figure 1 Markers of oxidation of lipids (left) and markers of oxidation of nucleic acids and proteins (right) MDA=malondialdehyde, HNE=4-hydroxy-trans-nonenal, HHE=4-hydroxy-trans-hexenal, C6-13=aldehydes C6-C13, 8-isoprostane=8-isoProstaglandin F2α, 8-OHdG=8-hydroxy-2-deoxyguanosine, 8-OHG=8-hydroxyguanosine, 5-OHMeU=5-hydroxymethyl uracil, o-Tyr=o-tyrosine, 3-ClTyr=3-chlorotyrosine, 3-NOTyr=3-nitrotyrosine *$p<0.05$, **$p<0.01$, ***$p<0.001$

The majority of the pre-shift and post-shift markers of oxidative stress and the three markers of inflammation were already elevated pre-shift, as shown in **Figure 1** and **Figure 2**. Additionally, all markers of oxidation of lipids reacted by post-shift elevation (**Figure 1**, left). However, no difference was seen between the markers of inflammation and markers of oxidative stress in the subgroups of the 11 researchers working in workshop 1, and 9 researchers from workshop 2. The only exception was 8-isoProstaglandin F2α (8-isoprostane), produced by free-radical lipid peroxidation of arachidonic acid, which was higher both pre-shift and post-shift in the researchers from workshop 1 ($p<0.05$) [25].

Both experimental and epidemiological studies have indicated that chronic inflammation is involved in, and plays a critical role in, several chronic diseases, including respiratory and cardiovascular diseases, and lung tumorigenesis.

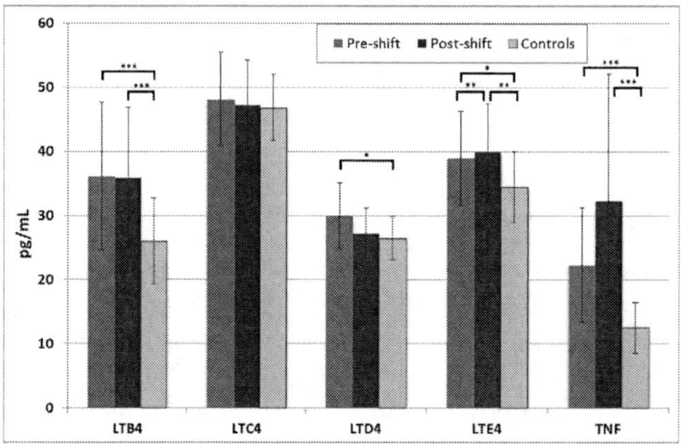

Figure 2 Markers of inflammation

TNF=tumor necrosis factor, LT=leukotriene B4, C4, D4, E4, *$p<0.05$, **$p<0.01$, ***$p<0.001$

3. CONCLUSION

Handling nanocomposites is very complex and may lead to exposure nanoparticles in the workplace aerosol during welding, smelting of the metallic products at hot temperatures and machining and grinding of geopolymers epoxide resins with nanocomponents [26-27]. Since inhalation remains the major exposure pathway to nanoparticles during nanocomposite manufacturing, the respiratory system is the main portal of entry and the primary target organ of concern [28-30]. The initial minor effects on the obstructive parameters of lung function have been observed. EBC markers of oxidative stress and inflammation were already elevated in the pre-shift samples, therefore they may be useful in monitoring the chronic effect. The markers of oxidation of lipids may reflect the early effect of the shift. EBC is one of the few means of non-invasive monitoring of individuals exposed to nanoparticles. These markers could be further tested for their potential use in preventive examinations of the workers in parallel with spirometry.

ACKNOWLEDGEMENTS

GACR 18-02079S, Progres Q25 and Q29 of the Charles University, and MEYS CR LO1508.

REFERENCES

[1] SCHULTE, Paul A., KUEMPEL, Eileen D., DREW, Nathan M. Characterizing risk assessments for the development of occupational exposure limits for engineered nanomaterials. *Regulatory Toxicology and Pharmacology.* 2018, vol. 95, pp. 207-219.

[2] IAVICOLI, Ivo, LESO, Verucska, SCHULTE, Paul A. Biomarkers of susceptibility. State of the art and implications for occupational exposure to engineered nanomaterials. *Toxicology and Applied Pharmacology.* 2016. vol. 299, pp.112-124.

[3] ROSSNEROVA, Andrea; POKORNA, Michaela, SVECOVA, Vlasta, SRAM, Radim J., TOPINKA, Jan, ZÖLZER, Friedo, ROSSNER, Pavel, Jr. Adaptation of the human population to the environment: Current knowledge, clues from Czech cytogenetic and "omics" biomonitoring studies and possible mechanisms. *Mutation Research.* 2017, vol. 773, pp. 188-203.

[4] LIOU, Saou-Hsing, TSAI, Candace S., PELCLOVA, Daniela, Schubauer-Berigan, Mary K., Schulte, Paul A. Assessing the first wave of epidemiological studies of nanomaterial workers. *Journal of Nanoparticle Research.* 2015, vol. 17: 413.

[5] BELLO, Dhimiter, WARDLE, Brian L., ZHANG, Jie, YAMAMOTO, Namiko, SANTEUFEMIO, Christopher, HALLOCK, Marilyn, VIRJI, M. Abbas. Characterization of exposures to nanoscale particles and fibers during solid core drilling of hybrid carbon nanotube advanced composites. *International Journal of Occupational and Environmental Health* 2010. vol.16, no. 4, pp. 434-450.

[6] BRADNA, Pavel, ONDRACKOVA, Lucie, ZDIMAL, Vladimir, NAVRATIL, Tomas, PELCLOVA, Daniela. Detection of nanoparticles released at finishing of dental composite materials. *Monatshefte für Chemie/Chemical Monthly.* 2017, vol. 148, no. 3, pp. 531-537.

[7] RUNA, Sabiha, HUSSEY, Michael, PAYNE, Christine K. Nanoparticle-Cell Interactions: Relevance for Public Health. *Journal of Physical Chemistry B.* 2018. vol. 122, no. 3, pp. 1009-1016.

[8] BECK-SPEIER, Ingrid, KARG, Erwin, BEHRENDT, Heidrun, STOEGER, Tobias, ALESSANDRINI, Francesca. Ultrafine particles affect the balance of endogenous pro- and anti-inflammatory lipid mediators in the lung: in-vitro and in-vivo studies. *Particle and Fibre Toxicology.* 2012. vol. 9: 27.

[9] PETERS-GOLDEN, Marc. Expanding roles for leukotrienes in airway inflammation. *Current Allergy and Asthma Reports.* 2008, vol. 8, no. 4, pp. 367-373.

[10] BARNES, Peter, J. Targeting cytokines to treat asthma and chronic obstructive pulmonary disease. *Nature Reviews Immunology.* 2018, vol. 18, no. 7, pp. 454-466.

[11] LIOU, Saou-Hsing, TSOU, Tsui-Chun, WANG, Shu-Li, LI, Lih-Ann, CHIANG, Hung-Che, LI, Wan-Fen, LIN, Pin-Pin, LAI, Ching-Huang, LEE, Hui-Ling, LIN, Ming-Hsiu, HSU, Jin-Huei, CHEN, Chiou-Rong, SHIH, Tung-Sheng, LIAO, Hui-Yi, CHUNG, Yu-The. Epidemiological study of health hazards among workers handling engineered nanomaterials. *Journal of Nanoparticle Research.* 2012, vol. 14: pp. 878.

[12] LIAO, Hui-Yi, CHUNG, Yu-Teh, LAI, Ching-Huang, WANG, Shu-Li, CHIANG, Hung-Che, LI, Lih-Ann,TSOU, Tsui-Chun, LI, Wan-Fen, LEE, Hui-Ling, WU, Wei-Te, LIN, Ming-Hsiu, HSU, Jin-Huei, HO, Jiune-Jye, CHEN, Chiou-Jong, SHIH, Tung-Sheng, LIN, Chin-Chi, LIOU, Saou-Hsing. Six-month follow-up study of health markers of nanomaterials among workers handling engineered nanomaterials. *Nanotoxicology.* 2014. Suppl 1, pp. 100-110.

[13] LEE, Jong Seong, CHOI, Young Chul, SHIN, Jae Hoon, LEE, Ji Hyun, LEE, Yurim, PARK, So Young, BAEK, Jin Ee, PARK, Jung Duck, AHN, Kangho, YU, Il Je. Health surveillance study of workers who manufacture multi-walled carbon nanotubes. *Nanotoxicology.* 2015. vol. 9, no. 6, pp. 802-811.

[14] LIOU, Saou-Hsing, WU, Wei-Te, LIAO, Hui-Yi, CHEN, Chao-Yu, TSAI, Cheng-Yen, JUNG, Wei-Ting, LEE, Hui-Ling. Global DNA methylation and oxidative stress biomarkers in workers exposed to metal oxide nanoparticles. *Journal of Hazardous Materials.* 2017. vol. 331, pp. 329-335.

[15] KURJANE, Natalja, ZVAGULE, Tija, RESTE, Jelena, MARTINSONE, Zanna, PAVLOVSKA, Ilona, MARTINSONE, Inese, VANADZINS, Ivars. The effect of different workplace nanoparticles on the immune systems of employees. *Journal of Nanoparticle Research.* 2017, vol. 19, no. 9, pp. 320.

[16] ZHAO, Lin, ZHU, Yifang, CHEN, Zhangjian, XU, Huadong, ZHOU, Jingwen, TANG, Shichuan, XU, Zhizhen, KONG, Fanling, LI, Xinwei, ZHANG, Yifei, LI, Xianzuo, ZHANG, Ji, JIA, Guang. Cardiopulmonary effects induced by occupational exposure to titanium dioxide nanoparticles. *Nanotoxicology.* 2018, vol. 12, no. 2, pp. 169-184,

[17] FOREST, Valérie, VERGNON, Jean-Michel, GUIBERT, Cyril, BITOUNIS, Dimitrios, LECLERC, Lara, SARRY, Gwendoline, POURCHEZ, Jérémie. Metal load assessment in patient pulmonary lavages: towards a comprehensive mineralogical analysis including the nano-sized fraction. *Nanotoxicology* 2017. vol. 11, no. 9-10, pp. 1211-1224.

[18] PIRELA, Sandra Vanessa, MARTIN, John, BELLO, Dhimiter, DEMOKRITOU, Philip. Nanoparticle exposures from nano-enabled toner-based printing equipment and human health: State of science and future research needs. *Critical Reviews in Toxicology.* 2017, vol. 47, no. 8, pp. 683-709.

[19] BOONRUKSA, Pongsit, BELLO, Dhimiter, ZHANG, Jinde, ISAACS, Jacqueline A., MEAD, Joey L., WOSKIE, Susan R. Exposures to nanoparticles and fibers during injection molding and recycling of carbon nanotube reinforced polycarbonate composites. *Journal of Exposure Science and Environmental* Epidemiology. 2017, vol. 27, pp. 379-390.

[20] KUBÁŇ, Petr, FORET, František. Exhaled breath condensate: determination of non-volatile compounds and their potential for clinical diagnosis and monitoring. *Analytica Chimica Acta.* 2013. vol. 805, pp. 1-18.

[21] PELCLOVA, Daniela, ZDIMAL, Vladimir, KACER, Petr, FENCLOVA, Zdenka, VLCKOVA, Stepanka, KOMARC, Martin, NAVRATIL, Tomas, SCHWARZ, Jaroslav, ZIKOVA, Nadezda, MAKES, Otakar, SYSLOVA, Kamila, BELACEK, Jaroslav, ZAKHAROV, Sergey. Leukotrienes in exhaled breath condensate and fractional exhaled nitric oxide in workers exposed to TiO2 nanoparticles. *Journal of Breath Research.* 2016. vol. 10, no. 3: 036004.

[22] PELCLOVA, Daniela, ZDIMAL, Vladimir, FENCLOVA, Zdenka, VLCKOVA, Stepanka, TURCI, Francesco, CORAZZARI, Ingrid, KACER, Petr, SCHWARZ, Jaroslav, ZIKOVA Nadezda, MAKES Otakar, SYSLOVA, Kamila, KOMARC, Martin, BELACEK, Jaroslav, NAVRATIL, Tomas, MACHAJOVA, Michaela, ZAKHAROV, Sergey. Markers of oxidative damage of nucleic acids and proteins among workers exposed to TiO2 (nano) particles. *Occupational and Environmental Medicine.* 2016, vol. 73, no. 2, pp.110-8.

[23] PELCLOVA, Daniela, ZDIMAL, Vladimir, KACER, Petr, VLCKOVA, Stepanka, FENCLOVA, Zdenka, NAVRATIL, Tomas, KOMARC, Martin, SCHWARZ, Jaroslav, ZIKOVA, Nadezda, MAKES, Otakar, ZAKHAROV, Sergey. Markers of nucleic acids and proteins oxidation among office workers exposed to air pollutants including (nano)TiO2 particles. *Neuroendocrinology Letters.* 2016, vol. 37(Suppl1), pp. 13-16.

[24] PELCLOVA, Daniela, ZDIMAL, Vladimir, KACER, Petr, ZIKOVA, Nadezda, KOMARC, Martin, FENCLOVA, Zdenka, VLCKOVA, Stepanka, SCHWARZ, Jaroslav, MAKEŠ, Otakar, SYSLOVA, Kamila, NAVRATIL, Tomas, TURCI, Francesco, CORAZZARI, Ingrid, ZAKHAROV, Sergey, BELLO, Dhimiter. Markers of lipid oxidative damage in the exhaled breath condensate of nano TiO_2 production workers. *Nanotoxicology.* 2017. vol. 11, no. 1, pp. 52-63.

[25] PELCLOVA, Daniela, ZDIMAL, Vladimir, SCHWARZ, Jaroslav, DVORACKOVA, Stepanka, KOMARC, Martin, ONDRACEK, Jakub, KOSTEJN, Martin, KACER, Petr, VLCKOVA, Stepanka, FENCLOVA, Zdenka, POPOV, Aleksey, LISCHKOVA, Lucie, ZAKHAROV, Sergey, BELLO, Dhimiter. Markers of Oxidative Stress in the Exhaled Breath Condensate of Workers Handling Nanocomposites. *Nanomaterials (Basel).* 2018, vol. 8, no. 8.

[26] HUANG, Yue-Wern, CAMBRE, Melissa LEE, Han-Jung. The toxicity of nanoparticles depends on multiple molecular and physicochemical mechanisms. *International Journal of Molecular Sciences.* 2017, vol. 18, no. 12: 2702.

[27] KUKA, Stanislav, HURBANKOVA, Marta, DRLICKOVA, Martina, BASKA, Tibor, HUDECKOVA, Henrieta, TATARKOVA, Zuzana. Nanomaterials - a new and former public health issue. The case of Slovakia. *Central European Journal of Public Health.* 2016, vol. 24, no. 4, pp. 308-313.

[28] WOOD, Alice M., TAN, See Ling, STOCKLEY, Robert A. Chronic obstructive pulmonary disease: towards pharmacogenetics. *Genome Medicine.* 2009, vol 1, no. 11: 112.

[29] ZHANG, Rong, DAI, Yufei, ZHANG, Xiao, NIU, Yong, MENG, Tao, LI, Yuanyuan, DUAN, Huawei, BIN, Ping, YE, Meng, JIA, Xiaowei, SHEN, Meili, YU, Shanfa, YANG, Xiaofa, GAO, Weimin, ZHENG, Yuxin. Reduced pulmonary function and increased pro-inflammatory cytokines in nanoscale carbon black-exposed workers. *Particle and Fibre Toxicology.* 2014, vol. 11: 73.

[30] KHATRI, Madhu, BELLO, Dhimiter, MARTIN, John, BELLO, Anila, GORE, Rebecca, DEMOKRITOU, Philip, GAINES, Peter. Chronic upper airway inflammation and oxidative stress in photocopier operators: Mechanistic insights. *NanoImpact.* 2017. vol. 5, pp. 133-145.

EXPOSURE TO COPPER OXIDE NANOPARTICLES: BIOLOGICAL IMPACTS

Pavel RÖSSNER, Jr.

Institute of Experimental Medicine, The Czech Academy of Sciences, Prague, Czech Republic, EU,
pavel.rossner@iem.cas.cz

Abstract

Copper (Cu) is a biogenic element that plays an important role in many biological processes. Cu is also used in various industrial products and its nanosized form has antibacterial and antiviral properties. Cu nanoparticles (NPs) are further applied in industry, chemistry and electronic devices. When compared with other metal oxide NPs, Cu NPs are less commonly used. However, the potential of copper oxide NPs (CuO NPs) to induce cytotoxicity and DNA damage *in vitro* was higher than that of other tested NPs. Despite this fact, studies on CuO NPs toxicity are relatively scarce and their effects on human health have never been investigated. *In vitro* tests in model cell lines showed the ability of CuO NPs to cause oxidative stress, but induction of apoptosis and cell cycle arrest were also observed. In animal studies, inflammatory responses in lungs were most common, but some studies revealed damage to other organs, including the brain. Considering toxicity of CuO NPs and the fact that these NPs are routinely manufactured and used in various products, more research is needed to elucidate mechanisms of their negative health effects in humans.

Keywords: CuO nanoparticles, toxicity, experimental systems

1. INTRODUCTION

As other metal-based nanoparticles, copper oxide nanoparticles (CuO NPs) may enter body either by inhalation, dermal or ingestion route. For human exposure, the inhalation route is usually the most significant one, particularly for occupational exposures. In the body, nanoparticles (NPs) may reach lungs, where they can be deposited, but also other organs, including the central nervous system. Their presence usually causes oxidative stress or inflammatory responses.

Copper (Cu) is a biogenic element with many biological functions. Cu is necessary for the proper growth, development and maintenance of bones, connective tissues and various organs [1]. It is involved in the stimulation of immune system in the protection against infection, in repair of injuries and healing. It plays an important role as a cofactor in some enzymes (e.g. superoxide dismutase, cytochrome oxidase), has roles in cell signaling and cell proliferation. In mollusks and some arthropods Cu is responsible for oxygen transport. Also, its metabolism is critical for tumor progression [2]. Nanosized Cu has antimicrobial and antiviral properties and is therefore used is face masks, wound dressings and socks [3]. It is

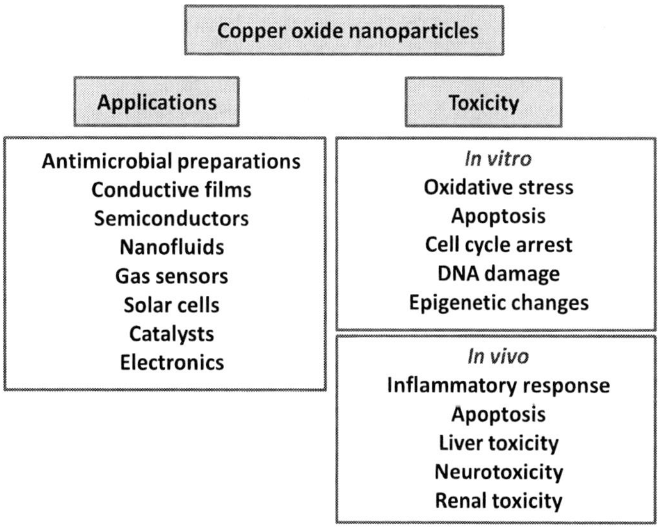

Figure 1 Overview of application and toxic effects of copper oxide nanoparticles

further applied in conductive films, lubrication, nanofluids, gas sensors, solar cells, lithium batteries and catalysts. Cu-based nanostructures are also used in various electronic devices, e.g. inkjet printers [4] (**Figure 1**). Unlike some other nanomaterials, Cu-based materials are relatively cheap. The global production of CuO nanoparticles (NPs) was 570 tons in 2014 and it is estimated to increase to 1600 tons in 2025 [5]. In comparison with other metal-based NPs (e.g. Ag or Zn), Cu-based NPs are less commonly used. Similar to other metal-based NPs, toxicity of Cu-based NPs depends on solubility of the material. In general, solubility depends on pH, dissolved organic carbon content and water hardness [3]. Acidic pH increases solubility; similarly, organic material may coat and disperse NPs thus supporting solubility.

2. TOXICITY OF CuO NANOPARTICLES IN VITRO

The production of CuO NPs is relatively low when compared with other metal oxide NPs. Nevertheless, CuO NPs properties and behavior in biological systems make them an important subject for toxicological investigation. Theoretical mechanisms of CuO NPs toxicity include (1) release of Cu ions from the surface of NPs, (2) oxidative damage mediated by Fenton-type reactions [reactive oxygen species (ROS) generation] induced by particles and (3) cell death associated with autophagy caused by CuO NP exposure. Autophagy is manifested by formation of autophagosomes that engulf and degrade defective proteins and organelles thus protecting the cells from damaged structures and helping the cells to survive. However, autophagy may also lead to cell death that is independent from apoptosis.

Although metal oxide NPs (e.g. Cu, Zn, Ti, Fe, Si) share similarities in their biological effects often mediated by ROS production, CuO NPs have been shown to have the highest potential to induce cytotoxicity and DNA damage in vitro [6]. This seems to be related to the ability of CuO NPs to better overwhelm antioxidant defenses of the organism. Toxicity of metal-based NPs may be caused by both ions and the particles themselves. These effects depend on the ability of NPs to release ions in the aqueous environment. CuO NPs have been shown to release Cu ions in culture medium but not in water [7], but the results on negative biological effects of the released ions are conflicting (reviewed in [6]): Cu ions accounted for part of the cytotoxicity of CuO NP in A549 [7] and BEAS-2B cells; however, other authors did not observe any contribution of Cu ions to toxic effects of CuO NPs in HEp-2 cells and A549 cells concluding that particles themselves are mostly responsible for toxicity of CuO NPs.

Toxicity of CuO NPs has been investigated in various cell lines in several studies; most of them observed oxidative stress-related response (for overview of in vitro effects see **Figure 1**). Thus, damage to mitochondria leading to increased production of ROS has been observed in A549 cells exposed to CuO NPs suggesting induction of oxidative damage [8]. CuO NPs have been further shown to induce lipid peroxidation, trigger intracellular signaling resulting in oxidative stress response and induction of apoptosis. In another study, CuO NPs generated ROS that subsequently induced the expression of p38 and p53 indicating induction of DNA damage in A549 cells [9]. CuO NPs have been further shown to cause oxidative stress induction measured by increased levels of 8-isoprostane and glutathione levels in human airway epithelial cells. Oxidative stress may induce autophagy; there are reports indicating that CuO NPs exposure is linked to this process in A549 cells [7]. On the other hand, CuO NPs may cause autophagic dysfunction by affecting lysosomal enzymes or disrupting vesicle trafficking [10].

Other processes potentially mediating toxic effects of CuO NPs have also been investigated. Apoptosis as a mechanism induced by CuO NPs was described (reviewed in [10]). The response is initiated by ROS generation resulting in destruction of mitochondrial membrane. This process results in activation of p53 which increases Bax/Bcl2 ratio. CuO NPs cause cell cycle arrest and DNA damage that was manifested by γ-H2AX formation [11]. CuO NPs may also induce IL-8 production which may consequently cause activation of NF-κB pathway [11]. CuO NPs may be associated with selective damage to cellular membranes because they release some amino acids, mostly glutamate, and cellular K^+ [10]. CuO NPs have been shown to bind to proteins via hydrogen bonds [12]. This may result in changes of protein phosphorylation and ubiquitination which plays an important role in cell signaling.

CuO NPs may affect gene expression through epigenetic mechanisms, specifically by histone modification [13]. They upregulate the expression of DNA repair proteins and directly interact with structural elements of the cell (e.g. cytoskeleton). This may disrupt mitotic spindle and cause aberrant cell division [10]. CuO NPs further directly interact with DNA causing replication arrest possibly leading to genome instability.

Molecular mechanisms of CuO NPs cytotoxicity in A549 cells were assessed using whole genome gene expression analysis. Exposure to CuO NPs caused upregulation of the expression of 648 genes and downregulation of the expression of 562 genes [7]. These genes included e.g. those involved in nucleic acids metabolic processes, response to stress, cell cycle, mitosis, cytokinesis, chromosome segregation, cellular component organization and morphogenesis. Genes encoding heat shock proteins or DNA damage-inducible genes were in the category of response to stress. This suggests that CuO NP stimulate protein denaturation and induce cell cycle arrest in the G1 and G2 phases. On the other hand, p53 protein does not seem to play a major role in the response to CuO NPs treatment. Cu ions alone induced expression of superoxide dismutase 2 and genes encoding metallothionein isomers. Proteins encoded by these genes help to protect the cells against oxidative stress induced e.g. by the presence of Cu ions. Thus, in A549 cells CuO NPs seem to damage both mitochondria and DNA; Cu ions contribute to the mitochondrial damage and lead to cell cycle arrest.

3. TOXICITY OF CuO NANOPARTICLES IN VIVO

Although in vitro tests in model cell lines provide information on cellular mechanisms of CuO NPs exposure and give general hints on their toxicity, experiments in laboratory animals represent a more relevant approach to assess the impact of CuO NPs exposure on human health. Despite this fact, such animal studies are relative scarce. Most of them were conducted in rodents and focused on pulmonary and neurotoxic effects, but also on changes in other organs associated with CuO NPs exposure (**Figure 1**).

Effects of various ways of pulmonary delivery of CuO NPs were investigated in both rats and mice (reviewed in [1]). Most of the studies reported inflammatory responses. Thus, lung instillation of CuO NPs <50 nm induced acute and chronic inflammation in rats. In another study, intratracheal instillation of CuO NPs induced neoplastic lesions [14]. A decrease in levels of antioxidants following intratracheal instillation was detected by another group [6]. Yokohira et al. studied CuO NPs toxicity in F344 male rats after intratracheal instillation and observed numerous negative effects including acute cell death, neutrophil infiltration or edema [15]. Cho et al. reported that intratracheal instillation of CuO NPs caused inflammation in lungs of rats both after 24 h and 4 weeks postinstillation [16]. Gosens et al. focused on pulmonary toxicity of CuO NPs in rats after a short-term inhalation exposure [17]. The animals were exposed to five doses of CuO NPs for five days, 6 hours/day and examined either 1 day or 22 days after exposure. The exposure resulted in inflammation and cytotoxicity detected 24 hours after the treatment. Histopathological examination of the lungs revealed alveolitis, bronchiolitis, vacuolation of the respiratory epithelium and emphysema of the lung. These effects almost completely resolved after a 3-week post exposure period. Thus, clearance of NPs is an important factor affecting their toxicity. Hirano et al. reported that intratracheally-instilled CuO NP had a clearance half-time of 37 h [18].

So far, there is only one study that focused on analysis of whole genome transcriptional changes after inhalation of CuO NPs [19]. The authors exposed rats for 6 hours to two doses (3.3 mg/m^3 and 13.2 mg/m^3) of CuO NPs and analyzed transcriptomics response 1 day post-exposure and after a 22 day recovery period. The exposure resulted in deregulation of about 1000 genes in the high-dose group and about 200 genes in the low-dose group. After the recovery period, the number of deregulated genes dropped to about 20. The main processes affected by the exposure included cell proliferation/survival and inflammation. Interestingly, no oxidative stress related pathways were affected. The authors observed proliferation of alveolar epithelial cells and detected upregulation of CCl2 (monocyte chemoattractant protein 1) both on the level of mRNA and protein and upregulation of ECT2 oncoprotein. Finally, no aberrant DNA methylation of inflammation-related genes was observed.

Intranasal instillation of CuO NP caused pulmonary inflammation, induction of apoptosis, collagen accumulation and expression of the progressive fibrosis marker α-SMA in C57BL/6 mice [21]. The animals were exposed to 3 doses of CuO NPs and lung tissue was collected on day 7, 14, 21 and 28. Interestingly, in another nasal instillation study necrosis of hepatocytes in the liver of exposed mice was observed [1]. In these animals, renal glomerulus and olfactory bulb were also affected. Intraperitoneal injection of CuO NPs resulted in increased frequency of micronuclei and induction of oxidative DNA damage in mice [21] indicating genotoxic effects of these NPs. Whole-body inhalation was accompanied by increased levels of inflammation markers in mice. The effects were persistent even 3 weeks post-exposure [1].

Although few investigations have been performed, the current data indicate that CuO NPs are neurotoxic [22]. In animal studies, they altered function of blood-brain barrier, damaged neurons and caused brain edema. Oxidative damage and neuronal apoptosis was also observed [22]. Significant toxic effects of CuO NPs of various sizes and concentrations were detected in dorsal root ganglion neurons of rats [1]. Neurotoxicity of Cu NPs following intraperitoneal, intraveneous, intracarotid and intracerebroventricular administration was detected in mice and rats [1]. Importantly, nasal instillation of CuO NPs affected brain tissue of experimental mice [23]. The exposure caused various pathological changes in neurons, including chromatin congregation and mitochondria shrinkage in olfactory cells, or increase of endoplasmic reticulum and disassociation of ER ribosomes in hippocampus. No change in antioxidant enzymes expression was observed, although malondialdehyde levels significantly increased.

Based on the tests in crustaceans, algae and fish, CuO NPs have been classified as toxic to aquatic organisms [3]. The effect of CuO NPs exposure on changes of expression of selected genes was studied in marine bivalve *Mytilus edulis* [24]. The mussels were exposed to CuO NPs for 24 h and expression of genes involved in immune response, antioxidant activities, cell metabolism, cell transport and cytoskeleton was assessed. The authors also compared the effect of Cu ions with the impact of the particles. The authors observed deregulation of expression of a number of genes, including e.g. β-tubulin, actin, antioxidant enzymes (catalase, SOD) and cell cycle control genes (p53, Gadd45). In general, the effects were more pronounced after treatment with CuO NPs than Cu ions suggesting the toxicity of CuO NPs is not caused solely by Cu ions release.

So far, no study investigated negative health effects of CuO NPs exposure in human populations. Nevertheless, the increased cancer risk among copper smelter workers suggests that Cu NPs exposure is harmful to humans [1]. The only study that used human samples to investigate toxicity of CuO NPs focused on human blood cells [25]. Lymphocytes obtained from blood of healthy volunteers were treated with CuO NPs and several parameters, including cell viability, ROS production, lipid peroxidation, glutathione levels and mitochondrial and lysosomal damaged were analyzed. The results indicated effects on cell viability associated with increased ROS production along with mitochondrial and lysosomal damage.

4. CONCLUSIONS

CuO NPs are used in many industrial products and as antimicrobial agents. Although their production is lower than other metal-based NPs, their toxicity is high. *In vitro* they have been shown to induce oxidative stress, apoptosis and cell cycle arrest. In laboratory animals exposed to CuO NPs inflammation was most commonly detected, but neurotoxicity was also noted. Due to potential negative health effects of exposure in humans, more studies are needed to better characterize mechanisms of CuO NPs toxicity.

ACKNOWLEDGEMENTS

This work was supported by project LO1508 from the Ministry of Education, Youth and Sports of the Czech Republic.

REFERENCES

[1] MAGAYE, R., ZHAO, J., BOWMAN, L. and DING, M. Genotoxicity and carcinogenicity of cobalt-, nickel- and copper-based nanoparticles. *Exp Ther Med.* 2012. vol. 4, pp. 551-561.

[2] SENGUPTA, J., GHOSH, S., DATTA, P., GOMES, A. and GOMES, A. Physiologically important metal nanoparticles and their toxicity. *J Nanosci Nanotechnol.* 2014. vol. 14, pp. 990-1006.

[3] BONDARENKO, O., JUGANSON, K., IVASK, A., KASEMETS, K., MORTIMER, M. and KAHRU, A. Toxicity of Ag, CuO and ZnO nanoparticles to selected environmentally relevant test organisms and mammalian cells in vitro: a critical review. *Arch Toxicol. 2013.* vol. 87, pp. 1181-1200.

[4] RUBILAR, O., RAI, M., TORTELLA, G., DIEZ, M.C., SEABRA, A.B. and DURÁN, N. Biogenic nanoparticles: copper, copper oxides, copper sulphides, complex copper nanostructures and their applications. *Biotechnol Lett.* 2013. vol. 35, pp. 1365-1375.

[5] SU, Y., ZHENG, X., CHEN, Y., LI, M. and LIU, K. Alteration of intracellular protein expressions as a key mechanism of the deterioration of bacterial denitrification caused by copper oxide nanoparticles. *Scientific Reports* [online]. 2015. vol. 5, 15824. Available from http://doi.org/10.1038/srep15824.

[6] AHAMED, M., AKHTAR, M.J., ALHADLAQ, H.A. and ALROKAYAN, S.A. Assessment of the lung toxicity of copper oxide nanoparticles: current status. *Nanomed.* 2015. vol. 10, pp. 2365-2377.

[7] HANAGATA, N., ZHUANG, F., CONNOLLY, S., LI, J., OGAWA, N. and XU, M. Molecular Responses of Human Lung Epithelial Cells to the Toxicity of Copper Oxide Nanoparticles Inferred from Whole Genome Expression Analysis. *ACS Nano.* 2011. vol. 5, pp. 9326-9338.

[8] KARLSSON, H.L., CRONHOLM, P., GUSTAFSSON, J. and MÖLLER, L. Copper Oxide Nanoparticles Are Highly Toxic: A Comparison between Metal Oxide Nanoparticles and Carbon Nanotubes. *Chem Res Toxicol.* 2008. vol. 21, pp. 1726-1732.

[9] WANG, Z., LI, N., ZHAO, J., WHITE, J.C., QU, P. and XING, B. CuO Nanoparticle Interaction with Human Epithelial Cells: Cellular Uptake, Location, Export, and Genotoxicity. *Chem Res Toxicol.* 2012. vol. 25, pp. 1512-1521.

[10] CHIBBER, S. and SHANKER, R. Can CuO nanoparticles lead to epigenetic regulation of antioxidant enzyme system?: CuO NPs can result in epigenetic regulation: a hypothetical view. *J Appl Toxicol.* 2017. vol. 37, pp. 84-91.

[11] KUNG, M.L., HSIEH, S.L., WU, C.C., CHU, T.H., LIN, Y.C., YEH, B.W. and HSIEH, S. Enhanced reactive oxygen species overexpression by CuO nanoparticles in poorly differentiated hepatocellular carcinoma cells. *Nanoscale.* 2015. vol. 7, pp. 1820-1829.

[12] RABBANI, G., KHAN, M.J., AHMAD, A., MASKAT, M.Y. and KHAN, R.H. Effect of copper oxide nanoparticles on the conformation and activity of β-galactosidase. *Colloids Surf B Biointerfaces.* 2014. vol. 123, pp. 96-105.

[13] CHENG, T.F., CHOUDHURI, S. and MULDOON-JACOBS, K. Epigenetic targets of some toxicologically relevant metals: a review of the literature. *J Appl Toxicol.* 2012. vol. 32, pp. 643-53.

[14] YOKOHIRA, M., HASHIMOTO, N., YAMAKAWA, K., SUZUKI, S., SAOO, K., KUNO, T. and IMAIDA, K. Lung Carcinogenic Bioassay of CuO and TiO2 Nanoparticles with Intratracheal Instillation Using F344 Male Rats. *J Toxicol Pathol.* 2009. vol. 22, pp. 71-78.

[15] YOKOHIRA, M., KUNO, T., YAMAKAWA, K., HOSOKAWA, K., MATSUDA, Y., HASHIMOTO, N., SUZUKI, S., SAOO, K. and IMAIDA, K.: Lung Toxicity of 16 Fine Particles on Intratracheal Instillation in a Bioassay Model Using F344 Male Rats. *Toxicol Pathol.* 2008. vol. 36, pp. 620-631.

[16] CHO, W.S., DUFFIN, R., POLAND, C.A., HOWIE, S.E.M., MACNEE, W., BRADLEY, M., MEGSON, I.L. and DONALDSON, K. Metal oxide nanoparticles induce unique inflammatory footprints in the lung: important implications for nanoparticle testing. *Environ Health Perspect.* 2010. vol. 118, pp. 1699-1706.

[17] GOSENS, I., CASSEE, F.R., ZANELLA, M., MANODORI, L., BRUNELLI, A., COSTA, A.L., BOKKERS, B.G.H., DE JONG, W.H., BROWN, D., HRISTOZOV, D. and STONE, V. Organ burden and pulmonary toxicity of nano-sized copper (II) oxide particles after short-term inhalation exposure. *Nanotoxicology.* 2016. vol. 10, pp. 1084-1095.

[18] HIRANO, S., EBIHARA, H., SAKAI, S., KODAMA, N. and SUZUKI, K.T. Pulmonary clearance and toxicity of intratracheally instilled cupric oxide in rats. *Arch Toxicol.* 1993. vol. 67, pp. 312-317.

[19] COSTA, P.M., GOSENS, I., WILLIAMS, A., FARCAL, L., PANTANO, D., BROWN, D.M., STONE, V., CASSEE, F.R., HALAPPANAVAR, S. and FADEEL, B. Transcriptional profiling reveals gene expression changes associated with inflammation and cell proliferation following short-term inhalation exposure to copper oxide nanoparticles. *J Appl Toxicol.* 2018. vol. 38, pp. 385-397.

[20] LAI, X., ZHAO, H., ZHANG, Y., GUO, K., XU, Y., CHEN, S. and ZHANG, J. Intranasal Delivery of Copper Oxide Nanoparticles Induces Pulmonary Toxicity and Fibrosis in C57BL/6 mice. *Scientific Reports* [online]. 2018. vol. 8, 4499. Available from http://doi.org/10.1038/s41598-018-22556-7.

[21] SONG, M.F., LI, Y.S., KASAI, H. and KAWAI, K. Metal nanoparticle-induced micronuclei and oxidative DNA damage in mice. *J Clin Biochem Nutr.* 2012. vol. 50, pp. 211-216.

[22] MIGLIORE, L., UBOLDI, C., DI BUCCHIANICO, S. and COPPEDÈ, F. Nanomaterials and neurodegeneration: Nanomaterials and Neurodegeneration. *Environ Mol Mutagen.* 2015. vol. 56, pp. 149-170.

[23] LIU, Y., GAO, Y., LIU, Y., LI, B., CHEN, C. and WU, G. Oxidative stress and acute changes in murine brain tissues after nasal instillation of copper particles with different sizes. *J Nanosci Nanotechnol.* 2014. vol. 14, pp. 4534-4540.

[24] CHÂTEL, A., LIÈVRE, C., BARRICK, A., BRUNEAU, M. and MOUNEYRAC, C. Transcriptomic approach: A promising tool for rapid screening nanomaterial-mediated toxicity in the marine bivalve Mytilus edulis -Application to copper oxide nanoparticles. *Comp Biochem Physiol Part C Toxicol Pharmacol.* 2018. vol. 205, pp. 26-33.

[25] ASSADIAN, E., ZAREI, M.H., GILANI, A.G., FARSHIN, M., DEGAMPANAH, H. and POURAHMAD, J. Toxicity of Copper Oxide (CuO) Nanoparticles on Human Blood Lymphocytes. *Biol Trace Elem Res.* 2018. vol. 184, pp. 350-357.

DNA STRAND BREAKS AND OXIDATIVE DAMAGE IN LYMPHOCYTES OF WORKERS HANDLING NANOCOMPOSITE MATERIALS

[1]Božena NOVOTNÁ, [2]Daniela PELCLOVÁ, [1]Pavel RÖSSNER, Jr., [1]Táňa BRZICOVÁ, [1]Andrea RÖSSNEROVÁ, [2]Lucie LISCHKOVÁ, [3]Štěpánka DVOŘÁČKOVÁ

[1]*Institute of Experimental Medicine, Czech Academy of Sciences, Prague, Czech Republic, EU,* *bozena.novotna@iem.cas.cz, pavel.rossner@iem.cas.cz, tana.brzicova@iem.cas.cz, andrea.rossnerova@iem.cas.cz*

[2]*Charles University in Prague and General University Hospital in Prague, Prague, Czech Republic, EU, daniela@pelclova.cz, Lucie.Lischkova@vfn.cz*

[3]*Technical University in Liberec, Liberec, Czech Republic, EU,* stepanka.dvorackova@tul.cz

Abstract

Many experimental studies have demonstrated the ability of nanoparticles to induce DNA damage and oxidative stress in a number of organs, probably due to the formation of reactive oxygen species. Our knowledge, however, is still limited, especially as regards the persistence of genotoxic effects after long-term exposure to nanoparticles and their health risks. We decided, therefore, to analyze DNA damage in lymphocytes of workers long-term exposed to nanoparticles due to processing of nanocomposite materials.

Exposed group involved 20 employees examined pre-shift and post-shift, samples from 20 controls were taken in the morning. Lymphocytes were isolated from heparinized blood on a ficoll gradient and frozen until further processing. An alkaline comet assay with enzymes of excision DNA repair (Fpg and Endo III) was used to detect DNA damage.

Within the exposed group, the values of total DNA damage in post-shift samples slightly exceed those in pre-shift ones, but the difference was not statistically significant (% tail DNA 4.47 ± 1.43 vs 3.95 ± 1.23, respectively). Compared to controls, however, the workers showed significantly higher levels of genotoxic damage, both in terms of DNA breakage (% tail DNA 3.08 ± 1.04 vs 1.63 ± 0.50, respectively; $p<0.0001$), and oxidized bases (% tail DNA 0.88 ± 0.41 vs. 0.57 ± 0.25, respectively; $p <0.01$). Our preliminary results thus indicate that long term occupational exposure to nanoparticles is associated with an increased risk of DNA damage.

Keywords: Genotoxicity, human, comet assay, nanoparticles, occupational exposure

1. INTRODUCTION

The current period is characterized by an extensive development of nanotechnologies and nanomaterials, and toxicologists thus face a number of questions about the possible health risks of long-term exposure to nanoparticles (NP). The harmful biological effects of NP are usually associated with induction of reactive oxygen species (ROS). Overproduction of ROS may induce oxidative stress and subsequent failure of physiological cell functions resulting in DNA damage, disturbance of cell signaling, protein and lipid alterations, cytotoxicity, apoptosis and carcinogenesis [1-2]. Such effects have been described in a number of *in vitro* and animal *in vivo* studies [3-5]. Nevertheless, it is difficult to extrapolate simply these results to judge the effects in human and so far only few studies reported the human response to NP [6-8]. Namely the workers handling nanomaterials represent a population potentially at risk of chronic effects due to day-to-day exposure at the workplace.

Alkaline version of single-cell gel electrophoresis (comet assay) with an analog of mammalian OGG1-formamidopyrimidine DNA glycosylase (Fpg) and endonuclease III (Endo III) allows the detection of DNA strand breaks, alkali-labile sites, transient gaps arising during base excision repair, as well as a broad spectrum of oxidized purines and pyrimidines [9]. Recently, this approach has been approved as a reliable test for genotoxicity of nanomaterials [10-11].

In our study we therefore used comet assay with the aim to assess DNA damage in lymphocytes of workers long-term exposed to nanoparticles during manufacturing and processing nanocomposite materials.

2. METHODS

2.1. Study groups

The exposed group involved 20 nanocomposite-synthesizing and processing workers (13 men and 7 women), the control group consisted of 20 volunteers (13 men and 7 women) from the same town, not employed in this plant nor occupationally exposed to dust or other health risks. All participants of the study completed questionnaire on personal and occupational history, medical treatments, dietary habits, smoking habits, and alcohol intake. History of tuberculosis, myocarditis, congenital heart disease, lung cancer, and recent fever and/or common cold symptoms were the criteria for exclusion from the study.

The workers used personal protective equipment for welding (welding helmets, leather gloves and leather aprons) and smelting (gloves). No respiratory protection was used during any of the procedures.

2.2. Exposure

The workers were exposed to nanoparticles (NP) during three different operations: welding on metal surfaces, smelting of mixtures containing nanoadditives, and machining of the finished nanocomposite. Iron, manganese and silicon represented the highest proportion of elements in the workplace aerosols generated during synthesis and post-processing of nanocomposites and associated operations. Detailed information about the chemical and physical characterization of nanoparticles and aerosols such as the concentration, size distribution and elemental composition has been already reported [12].

2.3. Blood collection and processing

Workers provided blood samples before and after aerosol exposure in the workshop during September 2017. For simplicity, these examinations are referred to as pre-shift and post-shift, despite that the workers spent the remainder of their total 8-hour shift in the office. The controls were withdrawn only once at approximately the same time of day as workers. The pre-shift samples were taken to study the subacute/chronic effects resulting from previous exposures, comparison of the pre-shift and post-shift samples was performed to assess the acute effect of exposure during the shift.

Lymphocytes were isolated from the whole heparinized blood by density gradient centrifugation over Ficoll-Paque PLUS (Sigma, Germany) and washed with phosphate buffered saline (PBS). Then, the cells were diluted with freezing medium as described earlier [13] and stored at -80 °C until the further processing.

2.4. DNA damage assessment

DNA damage was analyzed using an alkaline version of the comet assay [14-15]. The cells were quickly thawed in a 37 °C water bath, and the viability of cells was estimated by trypan blue exclusion. The number of trypan blue-positive cells did not exceed 15 % per sample. Four slides were prepared per sample (for details see [13]). Two slides per sample were treated with 45 μl of formamido-pyrimidine-glycosylase (FPG) and endonuclease III (ENDO III) in a 1:1 mixture (final concentration of both enzymes 2.5 μl/ml; Sigma-Aldrich, Germany) for 1 h at 37°C. In parallel, two slides were treated with the same volume of buffer used for the dilution of enzymes (0.1M KCl, 4mM EDTA, 2.5mM HEPES, 2 % BSA, all Sigma-Aldrich, USA). Subsequently, the slides were equilibrated for 40 min in alkaline buffer (0.3M NaOH, 1mM EDTA, pH 13) to allow the DNA to unwind. Electrophoresis was performed in fresh alkaline buffer (30 min, 1.2 V/cm, 300 mA). Finally, the slides were neutralized in 0.4M Tris (pH 7.5), stained with 0.005 % ethidium bromide (Sigma-Aldrich, Germany) for 7 min, washed with distilled water (7 min), fixed in methanol (15 min), dried at room temperature and stored.

Before analysis, the slides were rehydrated in distilled water, and images were captured with a CCD-13008 camera (VDS, Vosskuhler, Germany) attached to a BX51 fluorescence microscope (Olympus, Japan). The extent of DNA migration was quantified using Lucia Comet Assay 7.00 software (Laboratory Imaging, Prague, Czech Republic), and the results were expressed as the percentage of DNA in the tail (Tail DNA %). Both total DNA damage (with enzymes) and DNA strand breaks (DNA-SB; without enzymes) were measured in 100 randomly selected cells per sample. Finally, each sample was characterized by two medians from the measured data, i.e., by medians calculated from (a) values of total DNA damage and (b) values of DNA-SB, and the level of oxidative DNA damage was assessed as the difference between these medians.

2.5. Statistical analysis

An unpaired two-tailed Student's t-test was used to analyze the significance of differences between the groups.

3. RESULTS AND DISCUSSION

The basic information about the groups is summarized in **Table 1**. We found no significant differences in age, gender, body mass index, and alcohol consumption between the analyzed groups (all $p > 0.05$). On the other hand, 20 % of exposed subjects used vitamins, while no one in the control group. This probably reflects the outcomes of raising awareness of workers on potential health risks of nanomaterials.

Table 1 Characterization of the exposed and control subjects. SD - standard deviation; BMI - body mass index.

	Exposed Mean ± SD (min - max)	**Controls** Mean ± SD (min - max)
Age (years)	38.6 ± 10.7 (23 - 64)	39.8 ±7.1 (27 - 55)
BMI (kg/m^2)	24.6 ± 4.4 (17.5 - 34.3)	26.2 ± 4.4 (20.1 - 36.6)
Smoking	1 + 2 ex	3
Vitamin intake (C, E)	2 + 2 occasionally	0

On average, the workers were exposed to aerosol for 7 years, with the usual length of stay in the workshop around 2 hours a day (**Table 2**). As concerns the total length of exposure, there were big differences between individual workers. Nevertheless, the workers No 9 and 10 with the longest working age (18 and 21 years, respectively) exhibited the highest levels of DNA damage within the exposed group (% Tail DNA 6.26 and 5.87, respectively), although their average daily exposure was relatively short - 1 and 1.5 hours, respectively.

Table 2 The length of exposure of workers handling nanomaterials. SD - standard deviation.

Length of exposure	**Mean ± SD (min - max)**
Total (years)	7.3 ± 4.8 (2.5 - 21)
Common working day (hours/day)	2.1 ± 1.4 (0.5 - 6)
Current - i.e. on the monitoring day (hours)	3.4 ± 0.9 (2.5 - 6)

The average values of total DNA damage in the exposed group appeared to be slightly higher in post-shift samples than those in samples obtained prior to beginning of the shift, however this difference was not statistically significant (**Figure 1**). In contrast, pre-shift as well as post-shift values of DNA damage in lymphocytes of exposed workers considerably exceeded the levels detected in control subjects.

Figure 1 documents that both DNA-SB and oxidized lesions were significantly increased in exposed group compared to controls. The previous study from the same workplace demonstrated in the exhaled breath condensate (EBC) of exposed subjects an increase of markers of nucleic acid oxidation over the control values already in the pre-shift samples, and no further significant elevation was observed in post-shift EBC samples [12]. This fully corresponds with our present finding.

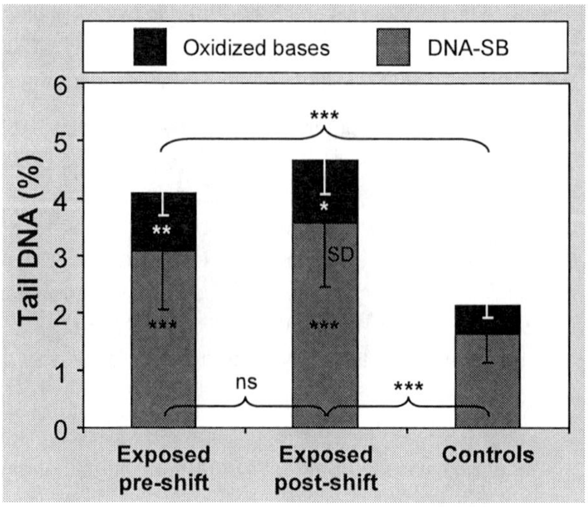

Figure 1 DNA damage in lymphocytes of workers long-term exposed to nanoparticles and control subjects. DNA-SB - DNA strand breaks, SD - standard deviation, $*p < 0.05$, $**p < 0.01$, $***p < 0.001$.

Depression of antioxidant enzymes (superoxide dismutase and glutathione peroxidase) and increased expression of cardiovascular markers (fibrinogen and intercellular adhesion molecule) were found among workers handling nanomaterials [16]. The same author subsequently suggested that exposure to metal oxide NP may lead to global methylation, DNA oxidative damage, and lipid peroxidation [17]. Chronic upper airway inflammation and systemic oxidative stress were documented in photocopier operators chronically exposed to nanoparticles [18]. A cross-sectional study performed in a nano-TiO_2 manufacturing plant in China described an association between the occupational exposure to nano-TiO_2 and the markers of lung damage, cardiovascular disease, oxidative stress and inflammation [19].

Our results thus support the recent data on harmful biological effects associated with long-term occupational exposure to nanoparticles and illustrate the need for further studies aimed at a detailed analysis of the mechanism of genotoxic effects and their health consequences.

4. CONCLUSION

Long-term inhalation exposure of workers handling nanocomposite material led to considerable increase of DNA damage. The levels of both DNA-SB and oxidized lesions in peripheral blood lymphocytes of exposed subjects significantly exceeded those detected in control persons.

Comet assay provided a useful tool for monitoring potential harmful effects of nanoparticles on DNA.

ACKNOWLEDGEMENTS

This work was supported by project LO1508 from the Ministry of Education, Youth and Sports of the Czech Republic and project 18-02079S from the Grant Agency of the Czech Republic.

REFERENCES

[1] DU, Hairong, ZHU, Xiaoling, FAN, Chuangang, XU, Song, WANG, Youjie and ZHOU, Yikai. Oxidative damage and OGG1expression induced by a combined effect of titanium dioxide nanoparticlesand lead acetate in human hepatocytes. *Environmental Toxicology*. 2012. vol. 27, no. 10, pp 590-597.

[2] FU, Peter P, Xia, Qingsu, HWANG, Huey M, Ray, PARESH C. and YU, Hongtao. Mechanisms of nanotoxicity:generation of reactive oxygen species. *Journal of Food and Drug Analysis*. 2014. vol. 22, no. 1, pp 64-75.

[3] MANKE, Amruta, WANG, Living and ROJANASAKUL, Yon. Mechanisms of nanoparticle-induced oxidative stress and toxicity. *Biomed Research International*. 2013. vol 2013, Article ID 942916, pp 1-15.

[4] MAGDOLENOVA, Zuzana, COLLINS, Andrew, KUMAR, Ashutosh, DHAWAN, Alok, STONE, Vicki and DUSINSKA, Maria. Mechanisms of genotoxicity. A review of in vitro and in vivo studies with engineered nanoparticles. *Nanotoxicology*. 2014. vol 8, no.3, pp 233-278.

[5] KHANNA, Puja, ONG, Cynthia, BAY, Boon H. and BAEG, Gyeong H. Nanotoxicity: An interplay of oxidative stress, inflammation and cell death. *Nanomaterials*. 2015. vol 5, no. 2, pp 1163-1180.

[6] LEWINSKI, Nastassja, GRACZYK, Halshka and RIEDIKER, Michael. Human inhalation exposure to iron oxide particles. *BioNanoMaterials*. 2013. vol. 14, no. 1-2, pp 5-23.

[7] SCHULTE, Paul A., IAVICOLI, Ivo, RANTANEN, Jorma H., DAHMANN, Dirk, IAVICOLI, Sergio, PIPKE, Rüdiger, GUSEVA CANU, Irina, BOCCUNI, Fabio, RICCI, Maximo, POLCI, Maria L., SABBIONI, Enrico, PIETROIUSTI, Antonio and MANTOVANI, Elvio. Assessing the protection of the nanomaterial workforce. *Nanotoxicology*. 2016. vol. 10, no. 7, pp 1013-1019.

[8] IAVICOLI, Ivo, LESO, Veruscka and SCHULTE, Paul A. Biomarkers of susceptibility. State of the art and implications for occupational exposure to engineered nanomaterials. *Toxicology and Applied Pharmacology*. 2016. vol. 299, pp 112-124.

[9] COLLINS, Andrew R. The comet assay for DNA damage and repair: principles, applications, and limitations. *Molecular Biotechnology*. 2004. vol. 26, no. 3, pp 249-261.

[10] KARLSSON, Hanna L. The comet assay in nanotoxicology research. *Analytical and Bioanalytical Chemistry*. 2010. vol. 398, no. 2, pp 651-666.

[11] MAGDOLENOVA, Zuzana, LORENZO, Yolanda, COLLINS, Andrew and DUSINSKA, Maria. Can standard genotoxicity tests be applied to nanoparticles? *Journal of Toxicology and Environmental Health, Part A*. 2012. vol. 75, no. 13-15, pp 800-806.

[12] PELCLOVA, Daniela, ZDIMAL, Vladimir, SCHWARZ, Jaroslav, DVORACKOVA, Stepanka, KOMARC, Martin, ONDRACEK, Jakub, KOSTEJN, Martin, KACER, Petr, VLCKOVA, Stepanka, FENCLOVA, Zdenka, POPOV, Alexey, LISCHKOVA, Lucie, ZAKHAROV, Sergey and BELLO, Dhimiter. Markers of Oxidative Stress in the Exhaled Breath Condensate of Workers Handling Nanocomposites. *Nanomaterials*. 2018. vol. 8, pp 611-629.

[13] NOVOTNA, Bozena, Topinka, Jan, Solansky, Ivo, Chvatalova, Irena, Lnenickova, Zdena and Sram, Radim J. Impact of air pollution and genotype variability on DNA damage in Prague policemen. *Toxicology Letters*. 2007. vol. 172, no. 1-2, pp 37-47.

[14] SINGH, Narendra P., McCOY Michael T., TICE, Raymond R. and SCHNEIDER, Edward I. A simple technique for quantitation of low levels of DNA damage in individual cells. *Experimental Cell Research*. 1988. vol. 175, no. 1, pp 184-91.

[15] TICE, Raymond R, AGURELL. Eva, ANDERSON, Diana, BURLINSON, Brian, HARTMANN, Andreas, KOBAYASHI, Hiroshi, Miyamae, Youichi, Rojas, Emilio, Ryu Jae C. and Sasaki, Yu F. Single cell gel/comet assay: guidelines for in vitro and in vivo genetic toxicology testing. *Environmental and Molecular Mutagenesis*. 2000. vol. 35, no. 3, pp 206-221.

[16] LIOU, Saou H., TSOU, Tsui C., WANG, Shu L., LI, Lih A., CHIANG, Hung C., LI, Wan F., LIN, Pin P., LAI, Ching H., LEE, Hui L., LIN, Ming H., HSU, Jin H., CHEN, Chiou R., SHIH, Tung S., LIAO, Hui Y. and CHUNG, Yu T. Epidemiological study of health hazards among workers handling engineered nanomaterials. *Journal of Nanoparticle Research*. 2012. vol. 14, pp 1-15.

[17] LIOU, Saou H., WU, Wei T., Hui-Yi LIAO, Chao Y. C., TSAI, Cheng Y., JUNG, Wei T. and LEE, Hui L. Global DNA methylation and oxidative stress biomarkers in workersexposed to metal oxide nanoparticles. *Journal of Hazardous Materials*. 2017. vol. 331, pp 329-335.

[18] KHATRI, Madhu, BELLO, Dhimiter, MARTIN, John, BELLO, Anita, GORE, Rebecca, DEMOKRITOU, Philip and GAINES, Peter. Chronic upper airway inflammation and oxidative stress in photocopier operators: Mechanistic insights. *NanoImpact*. 2017. vol. 5, pp 133-145.

[19] ZHAO, Lin, ZHU, Yifang, CHEN, Zhangjian, XU, Huadong, ZHOU Jingwen, TANG, Shichuan, XU, Zhizhen, KONG, Fanling, LI, Xinwei, ZHANG, Yifei, LI, Xianzuo, ZHANG, Ji and JIA, Guang. Cardiopulmonary effects induced by occupational exposure to titanium dioxide nanoparticles. NANOTOXICOLOGY. 2018. vol. 12, no. 2, pp 169-184.

CYTOTOXICITY OF NANOMATERIALS APPLICABLE IN RESTORATION AND CONSERVATION

[1,2]Táňa BRZICOVÁ, [3]Monika REMZOVÁ, [3]Radek ŽOUŽELKA, [3]Jiří RATHOUSKÝ, [1]Kristýna VRBOVÁ, [1]Pavel RŐSSNER, [1]Jan TOPINKA

[1]*Institute of Experimental Medicine, CAS, Czech Republic, EU,*
tana.brzicova@iem.cas.cz, kristyna.vrbova@iem.cas.cz, prossner@biomed.cas.cz, jan.topinka@iem.cas.cz
[2]*VSB - Technical University of Ostrava, Ostrava, Czech Republic, EU, tana.brzicova@vsb.cz*
[3]*J. Heyrovsky Institute of Physical Chemistry, CAS, Prague, Czech Republic, EU,*
monika.remzova@jh-inst.cas.cz, radek.zouzelka@jh-inst.cas.cz, jiri.rathousky@jh-inst.cas.cz

Abstract

Nanotechnologies represent new promising approach for conservation and preservation of cultural heritage. Nanomaterials (NMs) have been shown to improve efficacy, durability and quality of the restoration due to their unique properties that can be achieved and tuned by controlling particle physico-chemical characteristics. The same characteristics, however, affect the way how NMs interact with biological systems and may potentially underlie their enhanced or unexpected toxicity. In this pilot study, we compared the toxic potential of representatives of three of the most common oxide materials applicable in restoration: TiO_2 (standard and purified P25, a mixture of prevailing anatase with rutile crystalline modifications), SiO_2 (bare A200, and R805, R9200 as coated forms of A200), and ZnO. Using two *in vitro* cytotoxicity assays, WST-1 and LDH, evaluating metabolic activity and cell membrane integrity, respectively, we preliminary ranked the tested substances according to their cytotoxic potential, which may be used for their prioritization for further testing and applications. After 24h exposure, a dose-dependent decrease in cell viability was only detected in ZnO NPs and uncoated silica (A200). Hydrophobic coated silicas (R805 and R9200) and TiO_2 NPs (purified and unpurified P25) did not exhibit cytotoxic effects up to the highest tested concentration of 250 µg/mL. Toxicological data related to the physico-chemical characteristics will be applicable in developing both more efficient and safer nano-based products for restoration and conservation.

Keywords: Nanotechnology, cytotoxicity, restoration, work of art, safe-by-design

1. INTRODUCTION

Traditional techniques for preservation of cultural and heritage works of art (e.g. frescoes, oil-paints, sculptures) using organic solvents and polymers exhibit several limitations and drawbacks, such as promotion of chemical degradation or modification of permanent color. Moreover, exposure to organic solvents has been associated with various adverse health effects. Reducing toxicity might be a crucial factor for successful introduction of novel materials in restoration and conservation practices. NMs have been shown to improve efficacy, durability and quality of the restoration due to their unique properties that can be achieved and tuned by controlling particle chemistry, size, shape, crystal structure, surface, and other characteristics during their production [1]. The same characteristics, however, affect the way how NMs interact with biological systems and may potentially underlie their enhanced or unexpected toxicity. The ideal NM should exhibit low toxicity while maintaining desirable technical properties. By the controlled synthesis of NMs, individual characteristics can be designed in or out and affect both toxicity and performance. Controlling toxicity at an early stage of material or technology development is a core principle of so called safe-by-design approach. It is aimed at designing products or processes that bear an intrinsically low risk potential, instead of confining this potential by application of protective measures [2]. Designing out the hazard represents one of the main safe-by-design strategies, and in NMs can be achieved by e.g. NP doping, surface passivation, reduction of photo-catalytic efficiency, formation of composites, or surface functionalization (reviewed by [2]). Simple *in vitro* assays able to examine basic toxicological endpoints offer a possibility to perform a preliminary screening evaluation of a wide range of NMs that may offer new or enhanced properties applicable in restoration practice.

In the present work, we used representatives of three of the most common oxide NMs: TiO_2 (standard and purified P25, a mixture of prevailing anatase with rutile crystalline modifications), SiO_2 (bare A200 and R805, R9200 as coated forms of A200), and ZnO (see **Table 1**). All tested samples are relevant for applications in restoration practice. P25 offers a solution to limit degradation of materials caused by pollutants (e.g. volatile organic compounds, NOx, SO_2) and pathogens due to its excellent photocatalytic properties in combination with high stability, availability and low-cost [3]. ZnO NPs exhibit antimicrobial and antifungal properties, resistance to moisture, UV, and leaching. Similarly to nano-TiO_2, nano-ZnO is relatively cheap and available [4]. Silica NPs may be especially beneficial in consolidants, where they reduce crack formation, prevent dimensional instability owing to swelling and shrinkage induced by changes in temperature and relative humidity [5]. Silica NPs have been reported to increase cohesion, compression and flexural resistances, and to reduce the porosity after setting, and overall to increase the durability [6].

In this pilot study, we compared the toxic potential of the NP samples using two *in vitro* cytotoxicity assays WST-1 and LDH, evaluating metabolic activity and cell membrane integrity, respectively, as diverse markers of cell viability.

2. METHODS

2.1. NMs

Five commercially available metal oxide NPs were tested. Their key characteristics of the tested NMs are reported in **Table 1**. All samples were provided in a powder form. NM dispersions were prepared by probe sonication according to the protocol published by Jensen et al. [7]. Before cell exposure, the NP dispersions were gradually diluted in the cell culture medium containing 1 % fetal bovine serum (FBS).

Table 1 Physico-chemical characterization of the tested NMs

Label	Manufacturer	Commercial name	Surface modification	Surface area BET* (m^2/g)	Diameter* (nm)
A200	Evonik Industries	Aerosil® SiO_2 A200	-	220	12
A805	Evonik Industries	Aerosil® SiO_2 R805	$-CH_3(CH_2)_7$	125-175	12
A9200	Evonik Industries		$-CH_3$	150-190	12
P25	Evonik Industries	Aeroxide® TiO_2 P25	-	35-65	25
P25 purified	Evonik Industries	Aeroxide® TiO_2 P25	Purification	35-65	25
ZnO	Bochemie		-	90-110	12

2.2. Cytototoxicity assays

A549 cells (Human Lung Adenocarcinoma, ATCC No CCL-185™) were cultivated in MEM GlutaMAX™ Supplement medium (Gibco) with 10 % fetal bovine serum (FBS; Sigma-Aldrich) at 37°C in a humidified atmosphere containing 5 % CO_2. For cytotoxicity assays, the cells were seeded in 96-well microplates at a concentration of 7,500 cells per well and incubated overnight to get attached. Before exposure, the cells were washed with phosphate saline buffer. Serial dilutions of NMs in the cell culture medium supplemented with 1 % FBS were added to the cells in the volume of 100 µL per well.

WST-1 assay: After the exposure period, the cell culture medium was removed and cells were rinsed twice with PBS. The Cell Proliferation Reagent WST-1 (Roche Diagnostics) was mixed with MEM with 1 % FBS in a ratio of 1:10 and delivered to the test wells in a volume of 120 µl per well. The plate was incubated at 37 °C for 1h. To prevent interference of NPs adsorbed on the plate plastic with absorbance reading, 100 µL of supernatants from each well were transferred to a new plate. Absorbance was measured at 450 nm using a SpectraMax® M5 Plate Reader (Molecular Device). To determine the viability, background values of wells without cells were subtracted. Negative control (NC) cells were defined to be 100 % viable and the viability was expressed as a percentage of viability of NC.

LDH assay: At the end of the 24h incubation, 50 µL of the supernatant from each well was transferred into a new 96-well microtiter plate for the measurement of released LDH activity ($LDH_{supernatants}$). The rest of the medium was discarded. The survived cells were washed with PBS and lysed with 100 µL of 1 % Triton X-100 (Sigma-Aldrich) at 37°C for 30 min. Fifty µL of the supernatant from each well with lysed cells was transferred into a new plate for the measurement of LDH activity of survived cells ($LDH_{lysates}$). Fifty µL of the reaction mixture containing the detection dye and the catalyst from the Cytotoxicity Detection Kit (LDH) (Roche Diagnostics) was added to both the $LDH_{supernatants}$ and the $LDH_{lysates}$. Plates were incubated in the dark for 15 min. After adding 25 µL of 10 nM HCl to each well to terminate the reaction, the absorbance was measured at 490 nm using the SpectraMax® M5 Plate Reader. Background values of wells without cells were subtracted. The percentage of cell viability was calculated as ($LDH_{lysates}/LDH_{lysates}$ + $LDH_{supernatants}$) × 100. The NC represented 100 % viability and the results were expressed as a percentage of NC viability.

2.3. Statistical analysis

Three independent experiments were performed for each NP sample and both cytotoxicity assays. Data are presented as mean ± standard deviation (SD). To determine significance, an analysis of variance (ANOVA) followed by a Dunnett's test was performed (The Prism 5 program, GraphPad Software, San Diego, CA, USA). Statistical significance was established at p-level <0.05. IC50 values (concentrations that inhibited cell survival by 50 %) were calculated using four-parameter log-logistic models in the drc package [8] in the statistical software R (version 3.4.0.).

3. RESULTS AND DISCUSSION

NMs offer numerous benefits over the conventional chemicals when used in restoration and conservation practices, e.g. high compatibility with historic materials, very high chemical effectiveness due to their increased surface-to-volume ratio and a higher concentration of active substances, ability to penetrate into various and variable porous systems, deeper penetration, antimicrobial properties, self-cleaning, suitable optical properties (such as transparence), etc. [9]. The direct contact of the restorater with the applied subsances and the potential to their release after application emphasize the importance of their safety.

In the present study, we evaluated cytotoxicity as a basic toxicological endpoint of six NMs applicable in restoration and conservation practice that represent three most often used metal oxides. *In vitro* cytotoxicity assays represent a rapid screening tool applicable to preliminary ranking the tested substances and to their prioritization for further testing and applications. Since inhalation represents the main route of exposure to NPs during consolidation processes, A549 cell line was used as an *in vitro* model of human lung epithelial cells. We carried out two independent methods, WST-1 and LDH assays evaluating cell metabolic activity and cell membrane integrity, respectively. This combination allows better interpretation of the results and minimizes the risk that results would be affected by interference of NPs with the test system.

After 24h exposure, a dose-dependent decrease in cell viability was only detected in ZnO NPs and uncoated silica (A200). Hydrophobic coated silicas (R805 and R9200) and TiO_2 NPs (purified and unpurified P25) did not exhibit cytotoxic effects up to the highest tested concentration of 250 µg/mL (**Figure 1**).

Zinc is a vital component, essential for cellular metabolism and cell proliferation. However, high concentrations of Zn^{2+} can cause cell death by breakdown of the mitochondrial membrane potential [10]. It has been postulated that internalized soluble metal oxide NPs dissolve in the acidic lysosomal fluid releasing ions [11]. To address the contribution of Zn^{2+} ions to the observed cytotoxicity of ZnO NPs, the cells were treated with $ZnCl_2$ as an ionic form of zinc and tested using the WST-1 assay that exhibited higher sensitivity in detection of ZnO cytotoxicity than the LDH assay. The results showed that $ZnCl_2$ had significantly higher toxic effects than ZnO NPs when the exposure concentrations were adjusted to provide the same dose of elemental Zn (**Figure 1E**, IC_{50} values in **Table 2**). The lower cytotoxic potential of ZnO NPs in comparison with $ZnCl_2$ may be explained by a gradual release of Zn^{2+} from the internalized NPs contrarily to complete dissolution of $ZnCl_2$ in the cell culture medium. Lower sensitivity of the LDH assay may be explained by a limited period of the released LDH present in the cell culture medium. At the selected time point (24h) the enzyme may be still kept inside the dying cells during programmed cell death. Contrarily, LDH can already be degraded after 24h if the cells die early after the exposure. The discrepancy between the results obtained using LDH and WST-1 assay shows the importance of employing cytotoxicity assays with different endpoints to avoid underestimation of the results.

A200, R805, and R9200 represent the same type of silica NPs differing only in their surface properties. According to an extensive literature review [12], silica NPs induce significant cytotoxic effects at or above the concentration of 25 µg/mL, which is in concordance with our results for uncoated SiO_2 NPs (**Figure 1C,D**). Based on the obtained results, organosilane coating rendering R805 (SiO_2-octyl), and R9200 (SiO_2-methyl) hydrophobic surface significantly reduced cytotoxic effects of the NPs towards the A549 cells (**Figure 1C,D**). The mechanism of reduction of cytotoxicity of silica NPs by the hydrophobic coating was not investigated in this study. However, it has been shown that reactive silanol groups (Si-OH) on the surface of bare silica are involved in ROS generation inside cells, as well as in cell membrane damage, predominantly mediated by hydrogen bonding [13]. Surface modifications reducing the abundance of surface silanol groups may thus decrease reactivity and related toxicity of silica NPs.

No cytotoxicity of P25 was detected in A549 cells in our study. However, in the available literature, results on cytotoxic effects of TiO_2 NPs (including P25) are very inconsistent. Experimental conditions, e.g. exposure to UV light may significantly affects the results [14]. The contaminants of P25 (Al_2O_3, SiO_2, Fe_2O_3, HCl; the content of all molecules other than TiO_2 based on ignited material is reported to be below 1 % (wt) did not decrease viability, thus the purification did not have any detectable effects on cytotoxicity of P25.

The tested NMs were ranked according to their cytotoxic potential expressed as IC50 (concentration causing 50 % decrease in cell viability) (**Table 2**). In general, soluble metal-oxide NPs, such as ZnO, are considered more toxic than those exhibiting limited solubility (e.g. TiO_2). Manipulation of NM surface chemistry in an easy and effective way to adjust both biological and performace (e.g. consolidating) properties.

Cell death is a basic toxicological endpoint, however, absence of cytotoxic effects alone does not guarantee safety of the tested NPs. Other toxic effects, such as genotoxicity (DNA damage), immunotoxicity, and others, should be taken into account. Nevertheless, *in vitro* cytotoxicity assays provide an effective preliminary tool to screen for low toxic NPs.

Table 2 IC_{50} values (the inhibition concentration that cause 50 % decrease in cell viability) of NPs and $ZnCl_2$ in A549 exposed for 24h. IC_{50} values for ZnO NPs and $ZnCl_2$ were recalculated to the amount of zinc ions (values in brackets). NA - not available (not tested).

Sample	WST-1 assay (µg/mL)	LDH assay (µg/mL)
ZnO	9.4 (7.6 Zn^{2+})	12.6
A200	87.4	88.0
$ZnCl_2$	9.8 (4.7 Zn^{2+})	NA

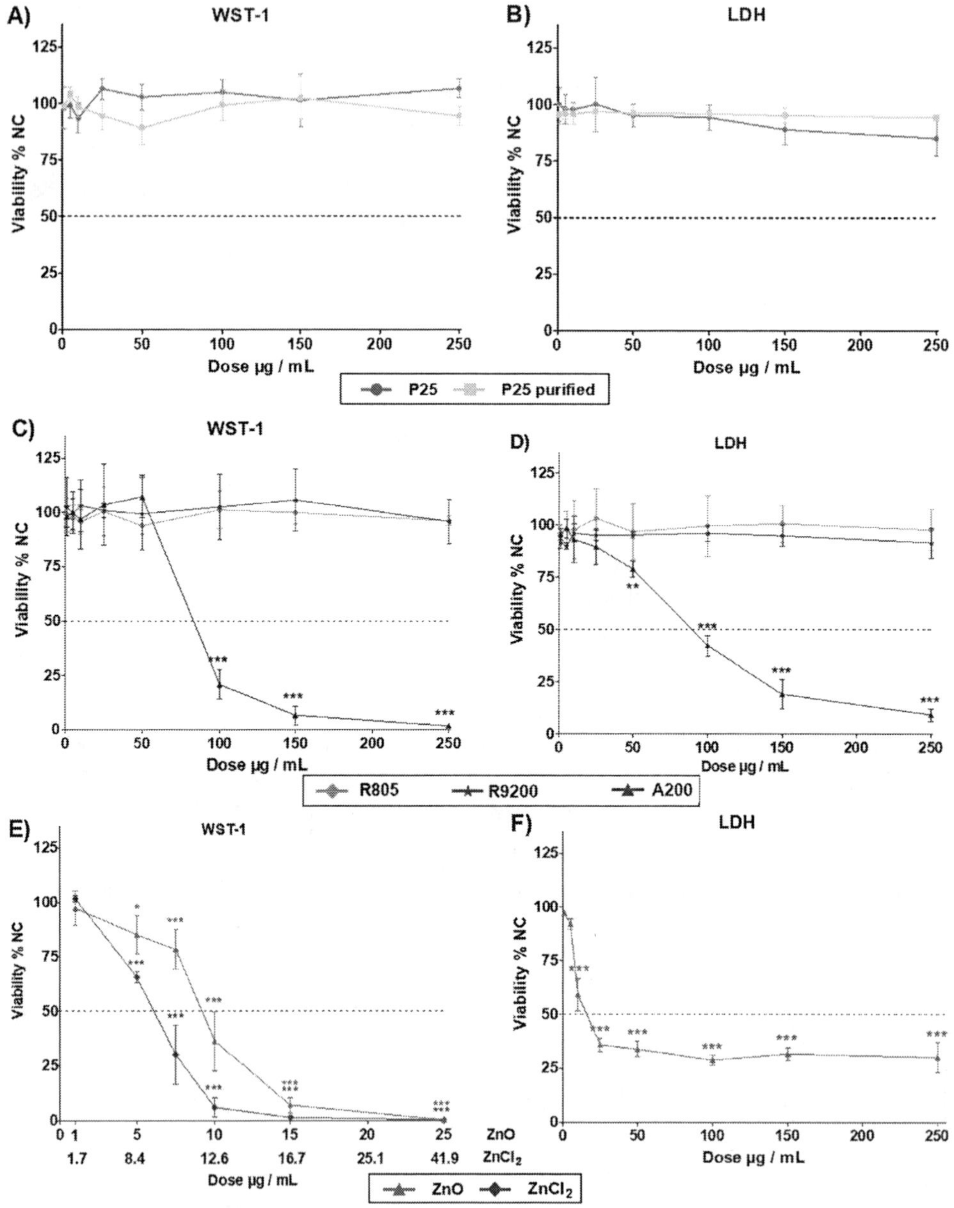

Figure 1 Cytotoxic effects of NPs and zinc salt ($ZnCl_2$) in A549 cells after 24h exposure measured by the WST-1 (A, C, E) and LDH assays (B, D, F). Results are expressed as % of cell viability (N=3). Concentrations of ZnO and $ZnCl_2$ are adjusted to provide the same concentration of molar Zn. Data were analyzed by ANOVA followed by Dunnett's Multiple Comparison. *$p<0.05$, **$p<0.01$ and ***$p<0.001$.

4. CONCLUSION

In this pilot study, we compared three of the most common oxide materials (TiO_2, ZnO a SiO_2) applicable in restauration practice in term of their cytotoxic effects *in vitro* towards the lung epithelial cell line A549 using WST-1 and LDH assays. The results showed that hydrophobic coating prevents cytotoxicity of silica NPs, which may be related to decreased abundance of surface silanol groups and reactivity. ZnO as soluble metal oxide NPs exhibited highest cytotoxicity, probably as a consequence of intracellular dissolution and release of Zn^{2+} ions. Toxicological data will be considered in combination with experimental data evaluating performance of the NPs in restauration processes with the aim to find develop more efficient and safer nano-based products for restoration and conservation.

ACKNOWLEDGEMENTS

This work was supported by the Ministry of Youth, Education and Sports of the Czech Republic (LO1508; LM2015073; CZ.02.1.01/0.0/0.0/16_013/0001821) and by Operational Program Prague-Competitiveness (CZ.2.16/3.1.00/2450).

REFERENCES

[1] BONINI M, BAGLIONI P, and CHELAZZI D. Inorganic NMs: Synthesis and Properties. In *Nanoscience for the Conservation of Works of Art*. 2013. pp. 315-344.

[2] KRAEGELOH, A, SUAREZ-MERINO, B, SLUIJTERS, T, and MICHELETTI, C. Implementation of Safe-by-Design for NM Development and Safe Innovation: Why We Need a Comprehensive Approach. *Nanomaterials*. 2018. vol. 8, no. 4, pp. 239.

[3] GEMMELLARO, P. *Titanium Dioxide nanoparticles in the Field of Conservation of Cultural Heritage*. PhD Thesis. Università degli studi di Catania Facoltà di scienze MM. FF. 2012.

[4] ABBASI, J, SAMANIAN, K, and AFSHARPOR, M. Evaluation of polyvinyl butyral and zinc oxide nano-composite for consolidation of historical woods. *International Journal of Conservation Science*. 2017, vol. 8, no. 2.

[5] BAGLIONI, P, CARRETTI, E, CHELAZZI, D. Nanomaterials in art conservation. *Nature nanotechnology*. 2015. vol. 10, no. 4, pp. 287.

[6] ZORNOZA-INDART, A, and LOPEZ-ARCE, P. Silica nanoparticles (SiO_2): Influence of relative humidity in stone consolidation. *Journal of Cultural Heritage*. 2016. vol. 18, pp. 258-270.

[7] JENSEN, K A, KEMBOUCHE, Y, CHRISTIANSEN, E, JACOBSEN, N R, WALLIN, H, GUIOT, C, SPALLA, O, and WITSCHGER, O. Final protocol for producing suitable manufactured NM exposure media. Web-Report. *The generic NANOGENOTOX dispersion protocol - Standard Operation Procedure (SOP)*. 2011.

[8] RITZ, C, and STREIBIG, J C. Bioassay analysis using R. *Journal of Statistical Software*. 2005, vol. 12, pp. 1-22.

[9] WAKED, A M. Nano materials applications for conservation of cultural heritage. *WIT Transactions on The Built Environment*. 2011. vol. 118, pp. 577-588.

[10] WÄTJEN, W, HAASE, H, BIAGIOLI, M, and BEYERSMANN, D. Induction of apoptosis in mammalian cells by cadmium and zinc. *Environmental Health Perspectives*. 2002. vol. 110, Suppl. 5, pp. 865.

[11] CHO, W, S, DUFFIN, R, THIELBEER, F, BRADLEY, M, MEGSON, I L, MACNEE, W, POLAND, C, A, TRAN, C L, and DONALDSON, K. Zeta potential and solubility to toxic ions as mechanisms of lung inflammation caused by metal/metal oxide nanoparticles. *Toxicological Sciences*. 2012. vol. 126, no. 2, pp. 469-477.

[12] NAPIERSKA, D, THOMASSEN, L C J, LISON, D, MARTENS, J A, and HOET, P H. The nanosilica hazard: Another variable entity. *Particle and Fibre Toxicology*. 2010. vol. 7, no. 1, pp. 39.

[13] FRUIJTIER-PÖLLOTH, C. The toxicological mode of action and the safety of synthetic amorphous silica-A nanostructured material. *Toxicology*, 2012. vol. 294, no. 2-3, pp. 61-79.

[14] CLEMENTE, Z, CASTRO, V, MOURA, M, JONSSON, C M, and FRACETO, L F. Toxicity assessment of TiO2 NPs in zebrafish embryos under different exposure conditions. *Aquatic toxicology*. 2014. vol. 147, pp. 129-139.

MICRONUCLEI LEVELS IN NANOCOMPOSITES PRODUCTION WORKERS: INTERPRETATION OF RESULTS FROM TWO YEARS OF MONITORING

Andrea RÖSSNEROVÁ[1], Daniela PELCLOVÁ[2], Vladimír ŽDÍMAL[3], Pavel RÖSSNER, Jr.[1], Fatima ELZEINOVÁ[1], Kristýna VRBOVÁ[1], Jan TOPINKA[1], Jaroslav SCHWARZ[3], Jakub ONDRÁČEK[3], Martin KOŠTEJN[3], Martin KOMARC[2], Štěpánka VLČKOVÁ[2], Zdenka FENCLOVÁ[2], Lucie LISCHKOVÁ[2], Štěpánka DVOŘÁČKOVÁ[4]

[1]Institute of Experimental Medicine, Czech Academy of Sciences, Prague, Czech Republic, EU, andrea.rossnerova@iem.cas.cz, rossner.pavel@iem.cas.cz, fatima.elzeinova@iem.cas.cz, kristyna.vrbova@iem.cas.cz, jan.topinka@iem.cas.cz

[2]Charles University in Prague and General University Hospital in Prague, Prague, Czech Republic, EU, daniela@pelclova.cz, komarc@ftvs.cuni.cz, Stepanka.Vlckova@vfn.cz, zdenka.fenclova@lf1.cuni.cz, Lucie.Lischkova@vfn.cz

[3]Institute of Chemical Process Fundamentals, Czech Academy of Sciences, Prague, Czech Republic, EU, zdimal@icpf.cas.cz, schwarz@icpf.cas.cz, ondracek@icpf.cz, kostejn@icpf.cz

[4]Technical University in Liberec, Liberec, Czech Republic, EU, stepanka.dvorackova@tul.cz

Abstract

Micronucleus assay, which has been used for almost sixty years, is one of the basic methodological approaches of genetic toxicology. This is also evident from more than 10,500 references found in PubMed database. These studies concern the evaluation of cytotoxicity and genotoxicity of numerous chemicals and in the last decade also nanomaterials (NM) and nanoparticles (NP). Surprisingly, no relevant human studies focused on MN formation following the exposure to NP have been reported.

In our study we analyzed 4x (in September 2016 and 2017; pre-shift and post-shift each year) samples in a group of workers, working for almost 18 years in nanocomposites research, and matched controls. Detail aerosol exposure monitoring of particulate matter (PM) including nano-sized fractions was completed during working shift in a sampling day. The micronucleus assay using Pan-Centromeric Chromosome Paint was applied to recognize, beside the frequency of total micronuclei (MN) in binucleated cells (BNC), also other types of chromosomal damage (losses and breaks), including the centromere positive (CEN+) and centromere negative (CEN-) micronuclei.

The monitoring data showed differences in the risk of exposure to NP related to individual working processes, as well as differences in chemical composition of nano-fraction. Cytogenetic results demonstrated consistently (both years): (i) possible adaptation to long-term exposure of NP (related to total frequency of MN), (ii) short-term (2.5 h) exposure could be a reason for the aberration increase, particularly a chromosomal losses (aneugenic effect).

Keywords: Genotoxicity, human, micronucleus assay, nanoparticles, pan-centromeric FISH

1. INTRODUCTION

Assessment of genotoxicity of nanomaterials (NM) and nanoparticles (NP) is one of the current objectives of genetic toxicology due to the increase of their use during the last decade in many areas of human life. One of the important and frequently used methodological tools of researchers is micronucleus assay which is evident from more than 10,500 references found in PubMed database in the middle of 2018. These *in vitro* and *in vivo* studies concern mainly the evaluation of cytotoxicity and genotoxicity of numerous chemicals, which are discussed in context of deleterious effects for humans. Only 300 of them (majority *in vitro*) concern the toxicity of NM and/or NP. Related to *in vivo* testing of genotoxicity of NP by micronucleus test, about 20 studies in mice (Swiss mice, C57BL/6J mice) or rat models (Wistar rats, Sprague-Dawley rats) are currently found in PubMed databases, but did not show consistent results due to their broad spectrum and various characteristics.

Surprisingly, still no relevant human studies focused on MN formation following the exposure to NP have been reported [1]. Alternatively, it is possible to recognize four studies [2-5] analyzing various cytogenetic markers in welders where the exposure to particulate matter in nano-size range also occurs [6]. A study published already in 1983 analyzed long-term occupational exposure (mean: 19 years) reported no significant differences between the exposed and controls, even though three various cytogenetic methods were used: conventional cytogenetic analysis (CCA), analysis of total MN and sister chromatid exchanges (SCE) [2]. A Mexican study that focused on genetic damage in exfoliated oral mucosa cells also did not observe differences between cases and controls [3] opposite to a comparison of chromosomal alterations induction, including MN, between nasal and buccal cells of welders in another study where a higher sensitivity of the epithelial cells from respiratory system was shown [4]. Additional identification of centromeres and genetics polymorphisms in DNA repair and detoxification genes was performed in another study where the authors reported a significantly higher level of chromosome/genome damage in welders suggesting that the combined analysis of genetic polymorphisms and centromeres in MN may improve the sensitivity of the micronucleus assay in detecting genotoxic effects [5].

Due to this gap in knowledge and potential negative health effects of NP, we analyzed 4x (in September 2016 and 2017; pre-shift and post-shift each year) a group of persons, long-term (years) working in nanocomposites research, and matched controls. Detail aerosol exposure monitoring of particulate matter (PM) including nano-sized fractions was completed during working shift in a sampling day. The micronucleus assay using Pan-Centromeric Chromosome Paint was applied to recognize, beside the frequency of total micronuclei (MN) in binucleated cells (BNC), also origin of chromosomal damage (losses and breaks), including the centromere positive (CEN+) and centromere negative (CEN-) micronuclei.

2. METHODS

2.1. Study groups

A total of 41 and 40 participants were involved in a collection of blood samples in September 2016 and 2017, respectively. The sample set included a group of twenty occupationally exposed nanocomposite processing researchers, working long-term (years) in nanocomposite research, sampled twice each year (pre-shift and post-shift, including processes such as welding mild steel S355J2 and smelting in workshop 1; and machining including the milling and grinding of epoxide resin with nanoSiO_2, and geopolymer nanocomposites in workshop 2) and 21/20 control volunteers living in the same location. Both genders, with a prevalence of men were involved in each subgroup. The researchers spent 2.5 hours during working operations. For simplicity, we refer to these samples as pre-shift and post-shift, even though the remainder of their total 8-h shift was spent in their offices. The exposed subjects did not use personal respiratory protection. More details on basic characteristics are presented in the results and discussion section.

2.2. Aerosol exposure monitoring

The online monitoring during the shift for individual processes and prior to it (background), included two standard aerosol spectrometers Scanning Mobility Particle Sizer (SMPS) (TSI SMPS 3936L, USA) and Aerodynamic Particle Sizer (APS) (TSI APS 3321, USA) which were used to obtain more details in the nanoscale range from 6 nm up. The offline Berner Low Pressure Impactor (BLPI) (HAUKE Gmbh., Austria) was used to sample aerosol particles onto 10 stages up to 13.6 µm including two nanoscales (25-56 nm and 56-100 nm). These samples were consecutively analyzed by gravimetry, ion chromatography [7] and Scanning

Electron Microscope (SEM) (Tescan Indusem, Czech Republic) equipped by Energy-Dispersive X-Ray Spectroscopy (EDS) (XFlash detector 5010, Bruker, Germany) to analyze the elemental composition of size-resolved aerosol fractions.

2.3. Micronucleus assay with Pan-centromeric FISH

Blood samples: Blood cell cultures were incubated at 37°C for 72 h. Cytochalasin B was added to a final concentration of 5 µg/ml after 44 h [8]. Cultures were harvested by centrifugation, treated with hypotonic solution of KCl and fixed with methanol/acetic acid.

Centromeric FISH technique: The protocol performed with FITC labelled Human Pan Centromeric probes was adapted from the manufacturer protocol (Cambio, Cambridge, UK). Stained slides were counterstained with DAPI mixed with Vectashield mounting medium.

Microscopic analysis: One thousand binucleated cells (BNC) were analyzed for each subject under the fluorescence microscope (Axioskop - Zeiss) equipped with filters for DAPI (blue signal) and FITC (green signal). BNC with micronuclei (MN), centromere positive (CEN+) and centromere negative (CEN-), were recorded and analyzed by use of the ISIS software version 5.0 (MetaSystems). See an example including explanation in **Figure 1**.

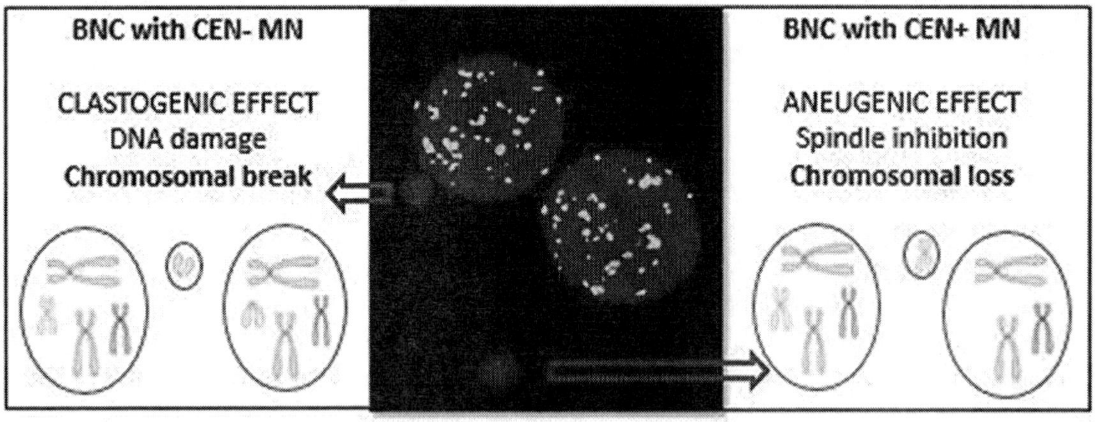

Figure 1 Example of BNC with two MN (both: CEN- and CEN+) with the explanation of their origin (chromosomal break on the left or chromosomal loss on the right)

2.4. Statistical analysis

Basic descriptive statistics [mean, standard deviation (SD), median and range (minimum and maximum)], were calculated using Microsoft Excel 2013. T-test, Mann-Whitney *U* test, paired sample t-test or the Wilcoxon signed-rank test were used for the comparison of the studied parameters (total MN/1000 BNC, CEN+ MN/1000 BNC and CEN- MN/1000 BNC).

3. RESULTS AND DISCUSSION

Main characteristics of the studied groups working in nanocomposite research analyzed repeatedly in September 2016 and 2017 twice per day (pre-shift and post-shift) each year, and matched controls are summarized in **Table 1**. There was no significant difference between the exposed and control groups for any of the selected characteristics and year even though the studied groups were not 100 % identical between years (an overlap of 14 and 11 participants in exposed and controls between years, respectively.

Table 1 Characteristics of the studied groups (format: N or mean±SD (min-max))

	2016		2017	
Characteristics	Exposed	Controls	Exposed	Controls
Number (N)	20	21	20	20
Males/Females (N)	15/5	15/6	13/7	13/7
Age (years)	41.8±11.4 (29-63)	38.7±9.1 (20-55)	38.6±10.7 (23-63)	39.8±7.1 (27-55)
BMI (kg/m²)	28±6.2 (18-42.1)	25±4.7 (18-37.2)	24.6±4.4 (17.5-34.3)	26.2±4.4 (20.1-36.6)
Smoking (N)	1	4	1	3

Two types of exposure data were available for the group of nanocomposite researchers: (i) related to time of exposure (years of exposure, common working day exposure and exposure in monitoring day), (ii) particular data from aerosol exposure monitoring including nano-fraction as described in detail previously [9]. Comparison of these data revealed differences in years of exposure between groups monitored in year 2016 and 2017 (17.8±10 and 7.3±4.8 years, respectively). Also, individual working processes were accompanied by differences in nano-fraction exposure and their chemical composition.

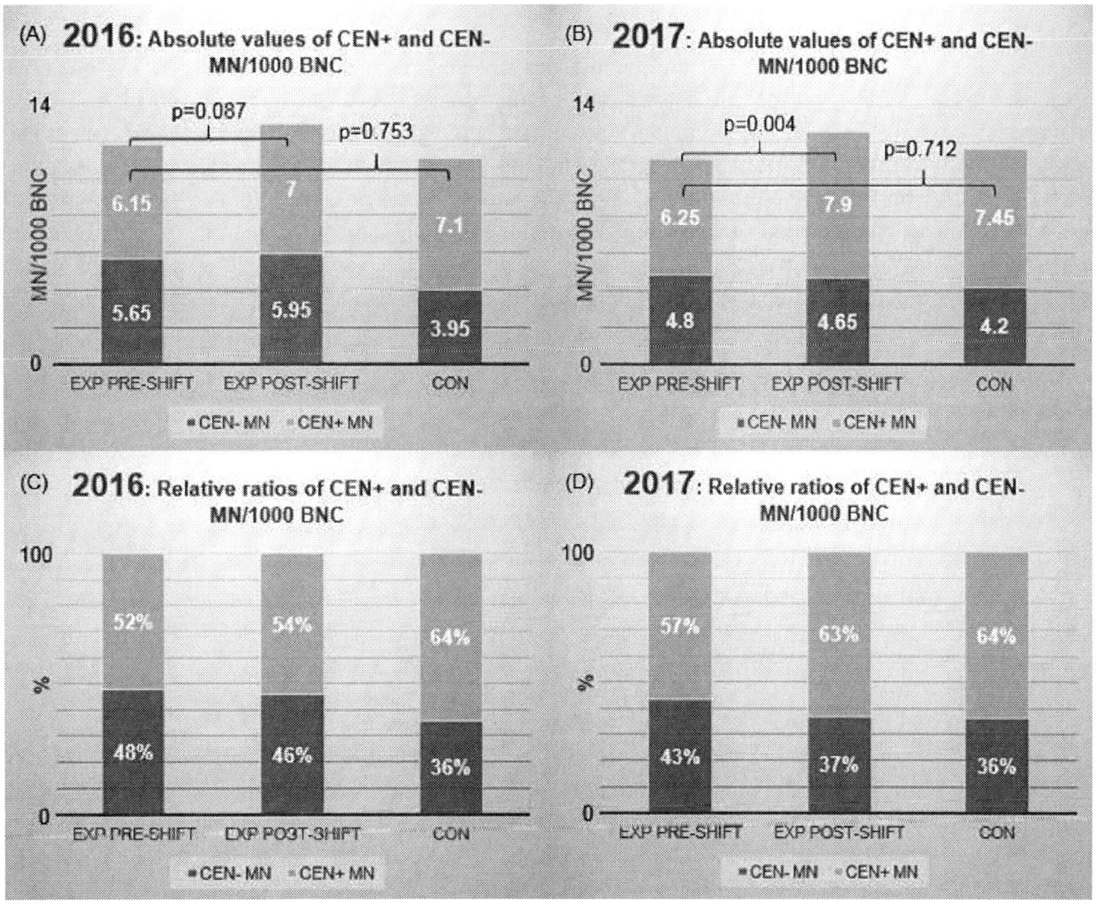

Figure 2 Overview of cytogenetic results (absolute values (A, B) and relative ratios (C, D) of CEN+ and CEN- MN in 1000 BNC) obtained by micronucleus assay using Pan-Centromeric Chromosome Paint in years 2016 and 2017

Summary of most important cytogenetic results (absolute values and relative ratios of CEN+ and CEN- MN in 1000 BNC) obtained by micronucleus assay using Pan-Centromeric Chromosome Paint is shown in **Figure 2**. We already analyzed in detail the sample set followed in year 2016 [10]. These results showed differences in the impact of exposure to NP related to individual working processes as well as differences in chemical composition of nano-fraction. Cytogenetic results demonstrated possible adaptation to long-term (years) exposure to NP (related to total frequency of MN) and generally corresponded with above mentioned studies related to welders [2-5], although this exposure may be responsible for DNA damage pattern changes (increase of chromosomal breaks from 36 % to 48 % - clastogenic effect). However, short-term (daily) exposure could be a reason for the increase of chromosomal losses (aneugenic effect) as well as breaks (clastogenic effect) in the exposed group, depending on the particular type of exposure. New data from year 2017 are generally in agreement with the previous year trend of results (related to both long-term and short-term exposure). Moreover, a trend to increase of DNA damage related to daily exposure (short-term exposure of long-term exposed subjects) has even greater significance (p=0.004) in comparison with the previous year (p = 0.087). This can be explained by a change in participants of the studies during the year and their individual exposure history differences. The exposure history of a group followed in 2017 was shorter and the effect of adaptation was not so strong.

4. CONCLUSION

The principal aim of this study was to fill the gap in human biomonitoring studies [1] and focus on the effects of exposure to NP on DNA damage in occupationally exposed subjects. We also aimed to verify the versatility of our hypotheses on human adaptation to the environment [11]. Obtained cytogenetic results demonstrated consistently (for both years): (i) possible adaptation to long-term exposure of NP (related to total frequency of MN), (ii) short-term (about 2.5-h per shift) daily exposure could be at least a reason for the aberration increase, particularly chromosomal losses (aneugenic effect), although particular type of exposure as well as exposure history play crucial role in resulting effects. As this type of human study is unique, more work is urgently needed for the understanding of the effect of both long-term and short-term exposure to NP on the level of DNA damage.

ACKNOWLEDGEMENTS

This work was supported by the Grant Agency of the Czech Republic #18-02079S and the grant of the Ministry of Education Youth and Sports CR #LO1508.

REFERENCES

[1] GONZALEZ, Laetitia, KIRSCH-VOLDERS, Micheline. Reprint of "Biomonitoring of genotoxic effects for human exposure to nanomaterials: The challenge ahead". *Mutation Research*, 2016, vol. 770, pp. 204-216.

[2] LITTORIN, Margareta, HŐGSTEDT, Benkt, STRŐMBÄCK, Bodil, KARLSSON, Anita, WELINDER, Hans, MITELMAN, Felix, SKERFVING, Staffan. No cytogenetic effects in lymphocytes of stainless steel welders. *Scandinavian Journal of Work, Environment & Health*, 1983, vol. 9, pp. 259-264.

[3] JARA-ETTNGER, Ana C., LÓPEZ-TAVERA, Juan C., ZAVALA-CERNA, Maria G., TORRES-BUGARÍN, Olivia. Genotoxic Evaluation of Mexican Welders Occupationally Exposed to Welding-Fumes Using the Micronucleus Test on Exfoliated Oral Mucosa Cells: A Cross-Sectional, Case-Control Study. *PLoS One*, 2015, vol. 10, pp. 1-12.

[4] WULTSCH, Georg, NERSESYAN, Armen, KUNDI, Michael, JAKSE, Robert, BEHAM, Alfred, WAGNER, Karl-Heinz, KNASMUELLER, Siegfried. The sensitivity of biomarkers for genotoxicity and acute cytotoxicity in nasal and buccal cells of welders. *International Journal of Hygiene and Environmental Health*, 2014, vol. 217, pp. 492-498.

[5] IARMARCOVAI, Gwenaelle, SARI-MINODIER, Irene, ORSIÈRE, Thierry, DE MÉO, Michel, GALLICE, Philippe, BIDEAU, Chantal, INIESTA, Danielle, POMPILI, Jocelyne, BERGÉ-LEFRANC, Jean-Louis, BOTTA, Alain. A combined analysis of XRCC1, XRCC3, GSTM1 and GSTT1 polymorphisms and centromere content of micronuclei in welders. *Mutagenesis*, 2006, vol. 21, pp. 159-165.

[6] DIERSCHKE, Katrin, ISAXON, Christina, ANDERSSON, Ula B.K., ASSARSSON, Eva, AXMON, Anna, STOCKFELT, Leo, GUDMUNDSSON, Anders, JŐNSSON, Bo A.G., KÅREDAL, Monica, LŐNDAHL, Jakob, PAGELS, Joakim, WIERZBICKA, Aneta, BOHGARD, Mats, NIELSEN, Jörn. Acute respiratory effects and biomarkers of inflammation due to welding-derived nanoparticle aggregates. *International Archives of Occupational and Environmental Health*, 2017, vol. 90, pp. 451-463.

[7] TALBOT, Nicholas, KUBELOVA, Lucie, MAKES, Otakar, ONDRACEK, Jakub, CUSACK, Michael, SCHWARZ, Jaroslav, VODICKA, Petr, ZIKOVA, Nadezda, ZDIMAL, Vladimir. Transformations of Aerosol Particles from an Outdoor to Indoor Environment. *Aerosol and Air Quality Research*, 2017, vol. 17, pp. 653-665.

[8] FENECH, Michael, MORLEY, Alexander A. Measurement of micronuclei in lymphocytes. *Mutation Research*, 1985, vol. 147, pp. 29-36.

[9] PELCLOVA, Daniela, ZDIMAL, Vladimir, SCHWARZ, Jaroslav, DVORACKOVA, Stepanka, KOMARC, Martin, ONDRACEK, Jakub, KOSTEJN, Martin, KACER, Petr, VLCKOVA, Stepanka, FENCLOVA, Zdenka, POPOV, Alexey, LISCHKOVA, Lucie, ZAKHAROV, Sergey, BELLO, Dhimiter. Markers of oxidative stress in exhaled breath condensate of workers handling nanocomposites. *Nanomaterials*, 2018, vol. 8, pp. 1-19.

[10] ROSSNEROVA, Andrea, PELCLOVA, Daniela, ZDIMAL, Vladimir, ROSSNER, Pavel Jr., ELZEINOVA, Fatima, VRBOVA, Kristyna, TOPINKA, Jan, SCHWARZ, Jaroslav, ONDRACEK, Jakub, KOSTEJN, Martin, KOMARC, Martin, VLCKOVA, Stepanka, FENCLOVA, Zdenka, DVORACKOVA, Stepanka. The Repeated Cytogenetic Analysis of Subjects Occupationally Exposed to Nanoparticles. Mutagenesis, 2018, submitted.

[11] ROSSNEROVA, Andrea, POKORNA, Michaela, SVECOVA, Vlasta, SRAM, Radim J., TOPINKA, Jan, ZŐLZER, Friedo, ROSSNER, Pavel Jr. Adaptation of human population to the environment: Current knowledge, clues from Czech cytogenetics and "omics" biomonitoring studies and possible mechanisms. Mutation Research, 2017, vol. 773, pp. 188-208.

TOXICITY OF SURFACE-MODIFIED COPPER OXIDE NANOPARTICLES IN A MOUSE MACROPHAGE CELL LINE: INTERPLAY OF PARTICLES, SURFACE COATING AND PARTICLE DISSOLUTION

[1]Tereza ČERVENÁ, [1]Helena LÍBALOVÁ, [2]Pedro M. COSTA, [2]Magnus OLSSON, [2]Lucian FARCAL, [3]Simona ORTELLI, [3]Magda BLOSI, [1]Jan TOPINKA, [3]Anna L. COSTA, [2]Bengt FADEEL

[1]Institute of Experimental Medicine, Academy of Sciences of the Czech Republic, Prague, Czech Republic, EU, tereza.cervena@iem.cas.cz
[2]Division of Molecular Toxicology, Institute of Environmental Medicine, Karolinska Institutet, Stockholm, Sweden, EU
[3]Institute of Science and Technology for Ceramics, National Research Council of Italy, Faenza, Italy, EU

Abstract

Copper oxide nanoparticles (CuO NPs) are used in numerous applications including antimicrobial agents, semiconductors, chemical sensors, catalysts and others. However, their high toxic potential and growing industrial production increases the concern for health and environmental risks. The present study aimed to investigate the toxicity of pristine versus surface-modified CuO NPs in relation to their stability/dissolution in cell culture medium, cellular uptake, cytotoxicity and ROS production in mouse macrophages RAW264.7. Our results showed that NPs modified with different coating agents [citrate (CIT), ascorbate (ASC), polyethyleneimine (PEI) and polyvinylpyrollidone (PVP)] affected the dissolution of the NPs in water but not in cell culture medium. Further, in vitro testing demonstrated that PEI-coated NPs were most cytotoxic while ASC-coated material exerted the lowest cytotoxicity. The anionic coatings CIT and ASC also provided protection against CuO NPs-induced oxidative stress and decreased levels of ROS production. Cellular internalization assessed by TEM was detected in cells exposed to CIT, ASC, PVP-coated and pristine NPs but not PEI-CuO NPs. Intracellular Cu content in cells tended to decrease with time, resulting in reduced cell/total Cu ratios, with the highest reduction noted for cells exposed to uncoated NPs. Overall, the toxicity of the surface-modified CuO NPs could be explained by synergistic interactions between the NPs, their dissolution, and the toxicity of the coating agents.

Keywords: Nanoparticles, surface coating, cytotoxicity, macrophages, oxidative stress

1. INTRODUCTION

Copper oxide nanoparticles (CuO NPs) are used in wide range of applications including antimicrobial agents, wood preservatives, doping materials in semiconductors, chemical sensors, catalysts as well as agents used in biomedicine and many others. However, the growing industrial production of engineered nanomaterials, not only metal oxide nanoparticles, increases the concern for health and environment. Several recent studies demonstrated toxicity in vitro [1,2] and in vivo [3], adverse effect has also been documented in workers exposed to welding fumes containing Cu-based NPs [4].

Although it is generally accepted that toxicity of CuO NPs is attributed to the rapid dissolution in aqueous media and massive release of Cu^{2+} ions [5], recent studies proposed a so-called Trojan horse mechanism by which various metallic nanoparticles are firstly internalized within cell and the subsequent intracellular ion release is elicited by the acidic conditions of the lysosomal cellular compartment [6,7]. Solubility (both extracellular and intracellular) is a function of surface stability of the NPs. Thus, considerable efforts have been devoted to the development of surface modifiers to improve nanoparticle's colloidal properties and enhance their biocompatibility. Surface modification of NPs controls their colloidal stability, surface charge and reactivity and thus may substantially influence their interaction with biomolecules and living organisms.

The present study aimed to investigate the toxicity of pristine versus coated CuO NPs in relation to their stability/dissolution in cell culture medium, cellular uptake, cytotoxicity and ROS production. The surface modifiers were selected to allow for a comparison of anionic, neutral and cationic agents. The molecules ascorbate (ASC) and citrate (CIT) are known for their anti-oxidant properties and are used as reducing agents and stabilizers of NPs. Polyvinylpyrrolidone (PVP) is a bulky, non-toxic neutral polymer used as a dispersing and shape-control agent. Polyethyleneimine (PEI) is a synthetic cationic polymer which provides electrosteric stabilization and enhanced particle cellular uptake.

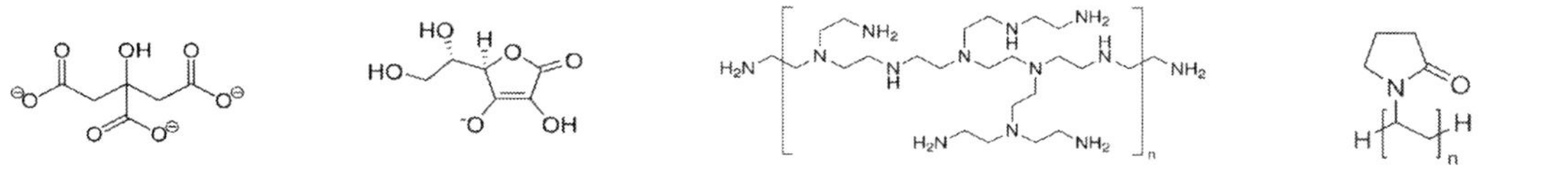

Figure 1 Chemical structure of selected coating agents used in the study

2. MATERIALS AND METHODS

2.1. Nanomaterial characterization

CuO NPS with a primary particle size of 12 nm (± 4 nm) were obtained from PlasmaChem (Germany). Surface of pristine nanoparticles was modified by self-assembly technique, i.e. CuO NPs were dispersed in buffer (PBS, pH 7.4), combined with organic additives at the proportion 10 % m/m relative to CuO and 95h milled with zirconia spheres as grinding media [8]. Particle dispersion in water cell culture medium was achieved through sonication. Dynamic light scattering (DLS) and electrophoretic light scattering (ELS) were used to determine particle hydrodynamic diameter and zeta potential, respectively (Zetasizer nano ZSP ZEN5600, Malvern Instruments). The dissolution of the NPs was estimated as the ratio between Cu^{2+} and CuO determined by suspension ultrafiltration with 10 kDa filters (Millipore), followed by analyzing the filtered solutions through inductively-coupled plasma optical emission spectrometry (ICP-OES) using 5100 Synchronous Vertical Dual View equipment (Agilent).

2.2. RAW264.7 macrophage cell line culture

RAW264.7 cell line was obtained from EACC (UK). Cells were cultured in DMEM (Sigma) supplemented with 10 % FBS, 1 mM sodium pyruvate, penicillin (100 U/mL) and streptomycin (100 μg/mL). Cells were maintained in humidified atmosphere with 5 % CO_2 at 37°C.

2.3. Cytotoxicity assessment

RAW264.7 cells were grown in DMEM (Gibco) supplemented with 10 % FBS, 1mM of sodium pyruvate,100U/ml Penicillin and 100 μg/ml Streptomycin. Cells were treated with 18 different concentrations of CuO NPs ranging from 1.41 to 62.5 μg CuO/ml of and incubated 24h in 37°C. $CuCl_2$ was included as and ionic control. For some experiments, N-acteylcysteine was used for pre-incubation of cells. Upon the treatment, Alamar Blue reagent (Life Technologies) was used to detect cellular viability. Dose-response cytotoxicity data were analyzed by the Benchmark dose method in R software with Proast 38.9 package.

2.4. Intracellular Cu content.

Cells were exposed to the BMD20 dose of each NP to achieve equitoxic conditions. After exposure, supernatants (medium) and cells (adherent cells + residual pellet) were collected separately and digested with

65 % HNO_3 for 24h at RT. Cu was determined through inductively-coupled plasma mass spectrophotometry (ICP-MS) using an iCAP Q apparatus. Results were expressed as ratios of intracellular/total Cu.

2.5. Detection of reactive oxygen species (ROS)

Cells were exposed to 25 µg/mL of CuO NPs for 4h, then 1 µM dihydroethidium was added (15 min, 37°C, dark). Afterwards, cells were collected, washed in PBS and DHE signal was analyzed by flow cytometry using a FACScan (Becton Dickenson).

3. RESULTS

3.1. Nanoparticle characterization

Pristine and modified CuO NPs were tested for their stability and solubility in water and cell culture medium (DMEM). According to DLS data indicating the hydrodynamic size of NP's agglomerates, NPs coated with ionic coating agents CIT, ASC and PEI were better dispersed in water than pristine material while neutral PVP worsened the stability so much that no data could be collected. In DMEM, all CuO NPs were better stabilized comparing to water and showed similar hydrodynamic size. Rate of dissolution in water showed slight differences among CuO NPs. The highest rate of dissolution after 24h exerted ASC-coated (1.99 %) comparing to PVP-coated NPs (0.23 %). However, the dissolution rate in DMEM indicating massive release of Cu^{2+} ions (almost 70 %) was similar for all nanomaterials irrespectively to the surface modification. Zeta potential of CuO NPs in water was consistent with the charge given by the coating agents but in DMEM, all CuO NPs gained the uniform negative charge (overview is presented in **Table 1**).

Table 1 Physicochemical properties of CuO NPs

	d DLS (nm)		**Zeta potential ELS (mV)**		**Dissolution after 24h (Cu^{2+}/Cu mass (%))**	
	MilliQ	DMEM	MilliQ	DMEM	MilliQ	DMEM
Uncoat. CuO	2052.3 ± 347	55.1 ± 16	-9.1 ± 0.4	-8.2 ± 0.4	0.1	67.41
CIT-CuO	313.4 ± 8	37.4 ± 2	-18.0 ± 0.3	-9.7 ± 0.6	1.98	69.19
PVP-CuO	nd	52.9 ± 25	-17.4 ± 0.3	-9.2 ± 0.2	2.08	65.39
PEI-CuO	383.5 ± 22	44.6 ± 14	-8.1 ± 2.3	-9.4 ± 0.8	0.23	66.93
ASC-CuO	109.7 ± 2	72.8 ± 21	+28.3 ± 0.7	-10.1 ± 0.7	1.68	66.01
DMEM		17.4 ± 0.3				

3.2. Cytotoxicity assessment

The cytotoxic response was assessed by Alamar Blue assay and dose-response effect was analysed by benchmark mark dose (BMD) method. BMD_{20} values of modified and pristine CuO NPs were inferred from dose response curves and could be ranked as follows: PEI-coated > uncoated > PVP-coated > ASC-coated > CIT-coated (**Table 2**).

Table 2 BMD_{20} estimates (given as µg/mL and 95 % confidence intervals) in RAW264.7 macrophages.

CuO NPs	**BMD_{20}**	**Coating agent**	**BMD_{20}**
Uncoated NPs	3.43 (2.67 - 4.28)	-	-
CIT-coated NPs	13.2 (11.3 - 15.3)	Citrate	52.5 (49.38 - 64.32)
ASC-coated NPs	14.6 (12.4 - 16.8)	Ascorbate	1.14×10^{19} (ND)
PVP-coated NPs	6.32 (5.43 - 7.27)	Polyvinylpyrrolidone	436 (132.4 - 8805)
PEI-coated NPs	2.06 (1.4 - 2.91)	Polyethyleneimine	13.4 (11.6 - 15.3)
Ionic control ($CuCl_2$)	55.6 (49.2 - 58.9)		

The cytotoxicity of ionic control $CuCl_2$ and coating agents themselves was also tested. $CuCl_2$ exerted considerably lower cytotoxicity and also the cytotoxicity of the coating agents, excepting PEI, was much reduced comparing to respective modified CuO NPs (**Table 2**).

3.3. Cellular internalization

The intracellular Cu content was determined by using ICP-MS after 4h and 24h exposure. Overall, Cu concentrations were consistently higher in medium than in cells regardless to the treatment an exposure time, Cu concentrations tended do decrease with time. After 4h exposure, differences in the internalization of NPs (expressed as ratio Cu in cells/total Cu) were found and ratios could be ranked as follows: uncoated > ASC-coated > PEI-coated > PVP-coated > CIT-coated. Twenty-four hour exposure resulted in decrease of Cu in all samples of NP-exposed cells on a similar level, the highest reduction was noted for uncoated and PVP-coated CuO NPs (**Figure 2**). Interestingly, the absolute concentration of Cu in cells did not correlated with the Cu ratio.

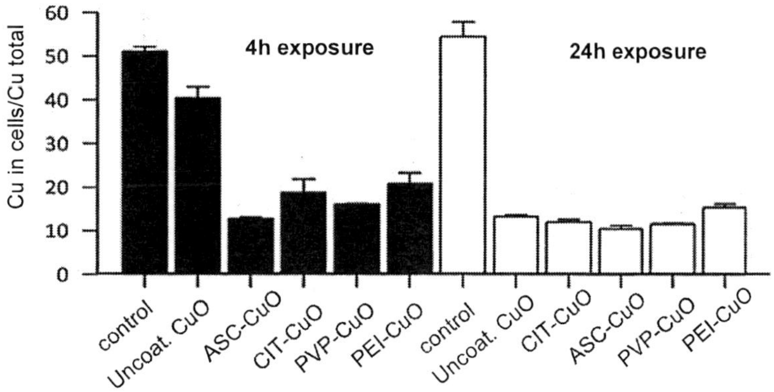

Figure 2 Ratios of Cu between cells and cells + medium (total Cu). Taken and modified from [9].

3.4. Cellular ROS production and the effect of ROS

Intracellular ROS generation in response to coated and uncoated CuO NPs was analysed by flow cytometry using DHE assay. Anionic coating agents clearly lowered levels of ROS when compared to uncoated CuO NPs while PVP-coated NPs produced the opposite effect (**Figure 3**). The rank of NPs according to their pro-oxidant properties is following: PVP > uncoated ~ PEI > CIT > ASC. On the other hand, no significant effect of ROS reduction on cytotoxicity was observed when cells were pre-incubated with 10 mM N-acetylcysteine before the exposure to uncoated CuO NPs (**Figure 4**).

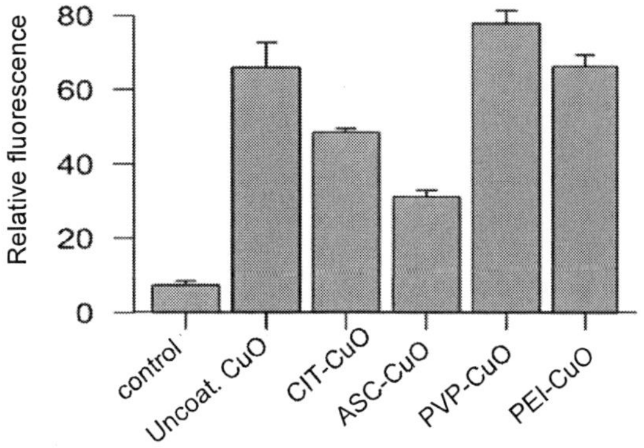

Figure 3 ROS formation after 4h exposure analyzed using redox-sensitive fluorescent probe DHE. Taken and modified from [9].

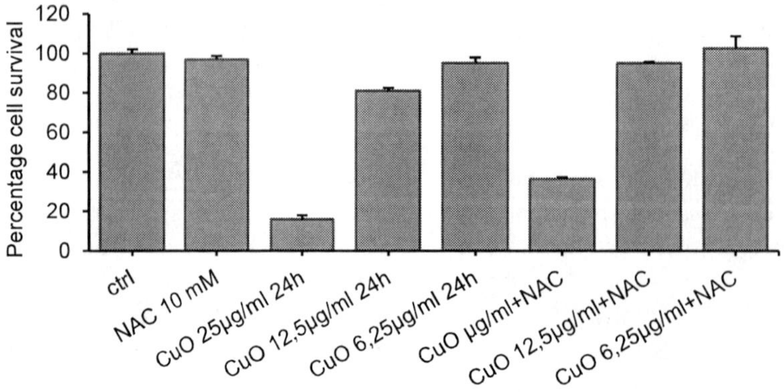

Figure 4 Cytotoxicity of selected toxic concentrations of uncoated CuO NPs after 4h exposure and the effect of NAC pre-treatment. Taken and modified from [9].

4. DISCUSSION

Numerous studies examined the influence of surface coating of diverse nanoparticles in order to prevent their adverse effects towards biological systems and environment [10, 11]. The present study demonstrates that surface coating of CuO NPs significantly affect the NP stability and physicochemical properties thus contributing to their toxicity in RAW264.7 macrophage cell line. Colloidal stability of CuO NPs in water was improved by ionic surface modifiers (CIT, ASC and PEI) but worsened by neutral PVP in comparison with pristine CuO NPs. This can be explained by depletion-flocculation phenomena. Negative ζ potential of uncoated CuO NPs was probably given by PO_4^{3-} from buffer specifically adsorbed on NP surface. Coated NPs diluted in water gained the ζ potential consistent with the charge of the coating agent confirming the preferential interaction with the surface modifier. In contrast, DMEM culture medium with FBS substantially changed properties across all NPs irrespective of the coating agent; all CuO NPs displayed similar Z-potential, mean hydrodynamic diameter and high rate of dissolution. The ability of serum proteins to change the agglomeration or surface chemistry thus reducing the toxicity of nanoparticles was demonstrated in several relevant studies [12, 13]. The cytotoxicity results identified positively charged PEI as the most toxic modifier contributing to the cytotoxicity. PEI alone also exerted a considerable cytotoxic effect as also reported in the study of Wightman 2011. The least cytotoxic were CIT- coated and ASC-coated CuO NPs. In our study, citrate and ascorbate have been shown to reduce pro-oxidant properties of CuO NPs, however, we also demonstrated that the effect of ROS on cytotoxicity was marginal only. Cellular uptake evaluated by transmission electron microscopy (data not shown) indicated complete intracellular dissolution of PEI-coated and uncoated nanoparticles since no particles could be detected after 4h exposure to BMD20. This may imply a Trojan-horse mechanism of toxicity of CuO NPs. Relatively low cytotoxicity of Cu^{2+} ions could further support this hypothesis. Cu concentration in cells measured by ICP-MS decreased after 24h exposure indicating exocytosis or passive release of nanoparticles. On the other hand, intracellular Cu burden did not corelated with the cytotoxicity; the proportion of bioavailable Cu in cells appeared to be more important factor.

5. CONCLUSIONS

Overall, our study suggests that coating agents have a considerable impact on stability of CuO NPs. However, dissolution of the NPs, particularly in cell culture medium, was not affected by different surface modifications. In vitro toxicity testing demonstrated that cytotoxicity of CuO NPs in RAW264.7 was not directly linked to nanoparticle dissolution and Cu burden, nor to cellular ROS production. The interplay between NP cytotoxicity, their dissolution, bioavailability, coating agents toxicity and time could explain the resulting toxic effect in RAW264.7 cells.

ACKNOWLEDGEMENTS

This work was supported by the European Commission through the Seventh Framework Programme (FP7-SUN; grant agreement no. 604305) and the COST Action MODENA (TD1204), and by the Czech Science Foundation (P503/12/G147), and the Ministry of Education, Youth and Sport of the Czech Republic (LO1508).

REFERENCES

[1] KARLSSON Hanna L., CRONHOLM Pontus, GUSTAFSSON Johanna, MOLLER Linnart 2008. Copper oxide nanoparticles are highly toxic: a comparison between metal oxide nanoparticles and carbon nanotubes. Chemical Research in Toxicology. 2008. vol. 21, no. 9, pp. 1726-1732.

[2] DI BUCCHIANICO Sebastiano, FABBRIZI Maria R., MISRA Superb K., VALSAMI-JONES Eugenia, BERHANU Deborah, REIP Paul, BERGAMASCHI Enrico, MIGLIORE Lucia. Multiple cytotoxic and genotoxic effects induced in vitro by differently shaped copper oxide nanomaterials. Mutagenesis. 2013. vol. 28, no. 3, pp. 287-299.

[3] CHO Wan S., DUFFIN Rodger, POLAND Craig A., DUSCHL Albert, OOSTINGH Gertie J., MACNEE William, BRADLEY Mark, MEGSON Ian L., DONALDSON Ken. Differential pro-inflammatory effects of metal oxide nanoparticles and their soluble ions in vitro and in vivo; zinc and copper nanoparticles, but not their ions, recruit eosinophils to the lungs. Nanotoxicology. 2012. vol. 6, no. 1, pp. 22-35.

[4] ANTONINI James M., LEWIS Anthony.B., ROBERTS Jenny R., WHALEY David A. Pulmonary effects of welding fumes: review of worker and experimental animal studies. *American Journal of Industrial Medicine.* 2003. vol. 43, no. 4, pp. 350-360.

[5] STARK Wendelin J. Nanoparticles in biological systems. Agnewande Chemie International Edition. 2011. vol. 50, no. 6, pp. 1242-1258.

[6] CRONHOLM Pontus, KARLSSON Hanna L., HEDBERG Jonas, LOWE Troy A., WINNBERG Linna, ELIHN Karine, WALLINDER Inge O., MOLLER Lennart. Intracellular uptake and toxicity of Ag and CuO nanoparticles: a comparison between nanoparticles and their corresponding metal ions. Small. 2013. vol. 9, no. 7, 970-982.

[7] STRAUCH Bettina M., NIEMAND Rebeca K., WINKELBEINER Nicola L., HARTWIG Andrea. Comparison between micro- and nanosized copper oxide and water soluble copper chloride: interrelationship between intracellular copper concentrations, oxidative stress and DNA damage response in human lung cells. *Particle and Fibre Toxicology.* 2017. vol. 14, no. 1, pp. 28.

[8] ORTELLI Simona, COSTA Anna L., BLOSI Magda, BRUNELLI Andrea, BADETTI Elena, BONETI Alessandro, HRISTOZOV Danail, MARCOMINI Antonio. Colloidal characterization of CuO nanoparticles in biological and environmental media. *Environmental Sciences Nano.* 2017. vol. 4, no. 6, pp. 1264-1272.

[9] LIBALOVA Helena, COSTA Pedro M., OLSSON Magnus, FARCAL Lucian, ORTELLI Simona, BLOSI Magda, TOPINKA Jan, COSTA Anna L., FADEEL Bengt. Toxicity of surface-modified copper oxide nanoparticles in a mouse macrophage cell line: Interplay of particles, surface coating and particle dissolution. *Chemosphere.* 2018. vol. 196, pp. 482-493.

[10] BASTOS V., FERREIRA DE OLIVEIRA José M., CAROLLA Joana, DANIEL DA SILVA Ana L., DUARTEM Iola F., SANTOS Conceição, OLIVEIRA Helena. Coating independent cytotoxicity of citrate- and PEG-coated silver nanoparticles on a human hepatoma cell line. Journal of Environmental Sciences. 2017. vol. 51, pp. 191-201.

[11] GLIGA Anda R., SKOGLUND Sara, WALLINDER Inger O., FADEEL Bengt, KARLSSON Hanna L. Size-dependent cytotoxicity of silver nanoparticles in human lung cells: the role of cellular uptake, agglomeration and Ag release. *Particle and Fibre Toxicoogy.* 2014. vol. 14, no. 11, pp. 11.

[12] KWON Dongwook, PARK Jonghoon, PARK Jaehong, CHOI Seo Y., YOON Tae H. Effects of surface-modifying ligands on the colloidal stability of ZnO nanoparticle dispersions in in vitro cytotoxicity test media. *International Journal of Nanomedicine.* 2014. vol 9, no. 2, pp. 57-65.

[13] MURDOCK Richard C., BRAYDICH-STOLLE Laura, SCHRAND Amanda M., SCHLAGER John J., HUSSAIN Saber M. Characterization of nanomaterial dispersion in solution prior to in vitro exposure using dynamic light scattering technique. *Toxicological Sciences.* 2008. vol. 101, no. 2, pp. 239-253.

CELLULAR UPTAKE OF GRAPHENE ACID BY HEL AND HELA CELLS STUDIED BY UV RAMAN SPECTROSCOPY

Anna BALZEROVÁ, Kateřina POLÁKOVÁ, Tomáš MALINA, Jan BELZA, Václav RANC, Radek ZBOŘIL

Regional Centre of Advanced Technologies and Materials, Department of Physical Chemistry, Faculty of Science, Palacký University in Olomouc, Olomouc, Czech Republic, EU, anna.balzerova@upol.cz

Abstract

The last few years have witnessed rapid development of biological and medical applications of graphene-based materials in drug delivery, biosensing, and bioimaging. Graphene acid is a novel graphene-based material with many interesting properties, including colloidal stability and biocompatibility [1]. However, mechanisms of its cellular uptake are not yet fully understood, and their study is crucial for further applications of the material. Study of a cell - nanomaterial interactions of the graphene-based nano materials is commonly performed using fluorescence imaging in a combination with respective fluorescent tags. However, labelling could potentially influence the material key properties, and new ways for label-free imaging are thus demanded. Here we developed a methodology for a label-free imaging of a cellular uptake of graphene acid by cancer (HeLa) and non-cancerous (HEL) cell lines based on Raman microscopy equipped with UV excitation laser.

Keywords: Graphene acid, Raman, imaging, cells

1. INTRODUCTION

Graphene is a two-dimensional allotrope of carbon [2]. Due to its excellent properties, such as strength, elasticity, high thermal conductivity, high electron mobility, and tuneable band gap, graphene become one of the most studied materials in last few years [2,3]. Its range of applications can be significantly extended by a surface functionalization using numerous functional groups which further enable to modulate its electronic, magnetic, and optical properties [1]. Graphene acid represents one of the strong candidates. Carboxylic groups on the surface of graphene acid significantly improve mechanical and electronic properties, sorption capacity, the dispersibility of the material in an aqueous environment and improve the stability of resulting colloidal solutions [4]. One of strong applications of graphene acid, derived from its high biocompatibility and colloidal stability, lies in the drug delivery and gene transport. However, its interactions with cells have not been fully described and understood.

Interactions of nanomaterials with cells are usually studied using fluorescent microscopy, flow cytometry and established toxicological tests [5-7]. However, there is a requirement for a utilisation of fluorescent labels in cases, where non-fluorescent materials are being studied. This can lead, because of unwanted interactions of the anchored fluorescent labels, to biased results [8,9]. The issue presents a scientific challenge worth to pursuit. Molecular spectroscopy, particularly Raman microscopy, presents an interesting alternative. Its potential has been earlier demonstrated in the imaging of graphene oxide in cell cytoplasm. However, the study was performed without an accent on a description of possible uptake mechanisms or effects on various cell lines [10,11].

Here we present an application of Raman spectroscopy for the examination of cellular uptake of novel material - graphene acid, into the different cell lines. The study was performed on two cell lines - human healthy lung fibroblasts (HEL) and human cervical cancer cells (Hela), with a variable time of uptake: 4 hours and 24 hours. We demonstrated that the UV Raman microscopy can be successfully applied in the study of the cellular uptake behaviour of graphene acid, and we have described differences of the nanomaterial uptake between cancerous and healthy cell lines.

2. EXPERIMENTAL SECTION

2.1. Cell Culture

For analysis we used adherent HEL 12469 (human healthy lung fibroblasts) and HeLa (human cervical cancer) cell lines. HEL cells were cultivated in EMEM - Minimum Essential Medium (Sigma Aldrich) enriched by L-Glutamine and Non-essential amino acids (NEAA) and HeLa cells were cultivated in low glucose DMEM medium (Life Technologies); both at 37 °C under a 5 % CO_2 atmosphere.

2.2. Cell viability assessment

Cell viability assay was performed using BD FACSVerse flow cytometer (BD Biosciences, USA). 10,000 cells were seeded per well in a 96-well plate. Cells were incubated with various concentrations of GA (10, 25, 50, 75 and 100 µg/ml) for 24 h. After 24 h, supernatant was collected, and cells were gently washed with PBS solution (0.1 M, 7.4 pH). Then cells were detached with trypsin (0.25 % in EDTA, Sigma Aldrich), resuspended in 100 µl of culture media and added to supernatant. Viability of cells was determined by calcein-AM fluorescent probe. Cells were incubated with 2 µl of calcein-AM, diluted in DMSO (50 µM). Viable cells had active intracellular esterases that catalyzed non-fluorescent calcein-AM to highly fluorescent green calcein. The fluorescence signal of calcein was measured by flow cytometry (exc. 488/em. 527) and viability of cells was established.

2.3. Preparation of cells for Raman measurements

The cells were fixed by a simple procedure for Raman analysis. Briefly, CaF_2 substrate was put to 24 well plate, 200 µl of a cell suspension (5000 cells/slide) were added, only on the substrate area. After 24 hours cells become adhered, the medium was replaced with one containing graphene acid, prepared according to protocol published earlier by our group [1]. The concentration of the nanomaterial was set to 50 µg/ml and incubation time was 4 and 24 hours, respectively. After incubation, the cell medium on the surface of the substrate was gently washed twice with PBS buffer (Phosphate-buffered saline). Then, the substrate was carefully removed and placed on the mount. After that, 200 µl of 4 % paraformaldehyde was added and incubated for 10 minutes. Finally, cells were washed twice with PBS and once with ultrapure water and kept drying in the fridge.

2.4. Apparatus

Raman spectra of the cells were acquired using a Thermo Scientific DXR Raman Microscope (Thermo Electron, USA) equipped with laser operating at the 455 nm wavelength. The laser power on sample was set up to 3 mW and the measured range was 350-3400 cm^{-1}. The exposition time was set to 2 s with 8 expositions per spectrum. Sixteen spectra were averaged to obtain one experimental data point. Cell viability assay was performed using BD FACSVerse flow cytometer (BD Biosciences, USA).

3. RESULTS AND DISCUSSION

First, basal cytotoxicity of the graphene acid had been studied to tune its concentration before a consecutive Raman imaging. The toxicity was evaluated using a flow cytometry to avoid possible false positive or negative outcomes, which can generally occur, when a standard MTT test is applied [12,13]. The concentrations were evaluated in the range from 10 to 100 mg/L. **Figure 1** shows that only an unimportant effect of applied nanomaterial is observed. The viability decreased from 100 % for a control group to 88 % for HEL cells lines and 90 % for Hela, respectively. These results are comparable with a cytotoxicity of graphene oxide up to a concentration of 50 mg/L. Published results indicate that graphene oxide starts to be substantially toxic at higher concentration with viability ranging from 90 % for 50 mg/L to 50 % for 100 mg/L [14,15]. Concenration of 50 mg/L of the graphene acid was selected for further experiments.

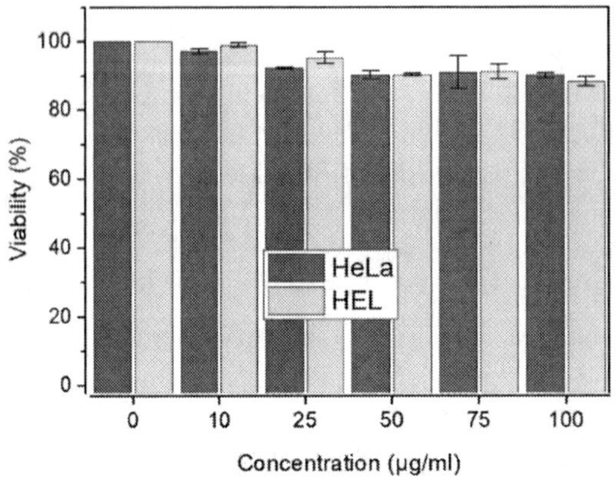

Figure 1 Cell viability of HeLa and HEL cells after 24 h of incubation with different concentrations of GA. Viability of cells was normalized to control untreated cells (100 %).

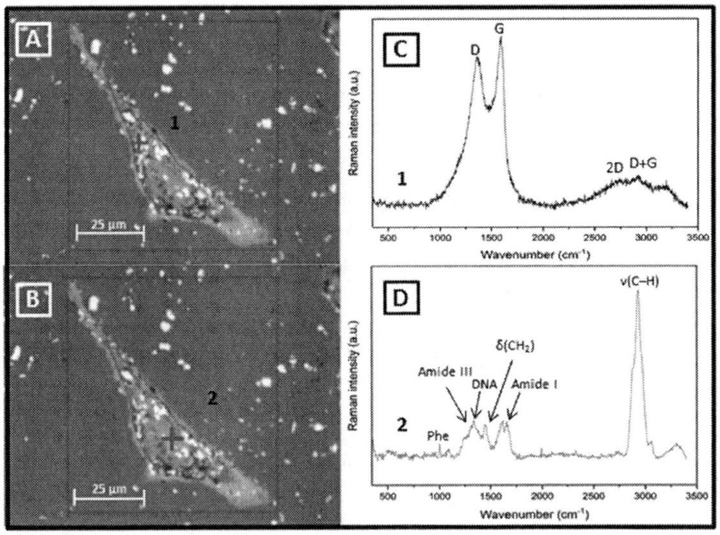

Figure 2 A), B) microscopic images of a HeLa cell.
C), D) Raman spectra obtained at different spots marked in A), B) in the HeLa cell.

Second, Raman spectra were acquired at several points on fixed cells to evaluate spectral characteristics of cells, as well as spectra characteristics of the internalized nanomaterial. **Figures 2A** and **2B** show microscopic images of the fixes HeLa cell. Figures contain two marks, labelled 1 and 2, respectively. Mark 1 corresponds to the nanomaterial (visible as light spots) and mark 2 corresponds to the cell, particularly nucleus. Raman spectrum shown in **Figure 2C** represents the mark 1, and contains two spectral bands located at 1360 and 1595 cm^{-1}, assigned to D and G spectral bands of graphene acid [11,16]. The D/G ratio of 0.8 confirms sp3 hybridisation in the graphene lattice, indicating a presence of functional carboxyl groups on its surface. **Figure 2D** represents a mark 2 and shows Raman spectrum of the nucleus with several characteristic spectral bands. The most intense band is located at 2930 cm^{-1}. This can be interpreted as C-H vibrations (v(C-H)) of -

CH_2 and $-CH_3$ groups presented in proteins, lipids or nucleic acids [17]. Additional Raman bands were assigned to phenylalanine at 1005 cm^{-1}, DNA backbone (O-P-O stretching) at 1092 cm^{-1}, amide III band at 1313 cm^{-1}, DNA at 1340 cm^{-1}, CH_2 bending mode found in proteins at 1450 cm^{-1}, and lipid C=C bond and amide I band of proteins at 1660 cm^{-1} [18,19].

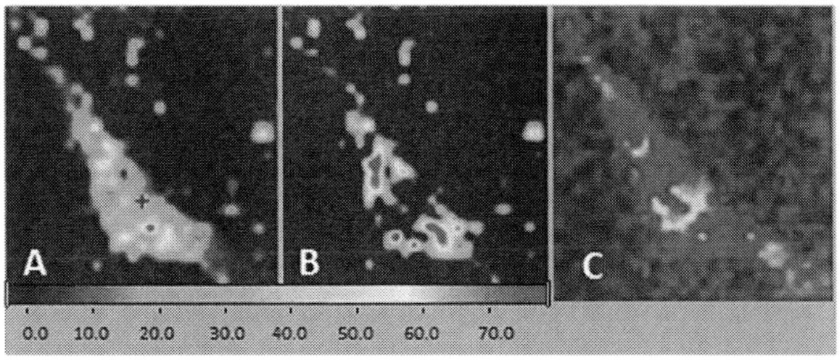

Figure 3 Raman maps of HeLa cell after the incubation with GA for 24 hours demonstrating area of: (A) cell, (B) GA, (C) image from multivariate curve resolution analysis of GA internalized in cell

Raman images obtained by the scanning of Hela cells after a 24-hour incubation with graphene acid is shown in **Figure 3**. **Figure 3A** was obtained by the univariate analysis performed at spectral band located at 2930 cm^{-1} (ν(C-H)) and reflects the spatial distribution of above mentioned compounds inside the cell. Importantly, weak Raman signal was observed also outside of the scanned cell. This phenomenon is caused by the used cell medium, which contains proteins with similar spectral characteristics. The **Figure 3B** is a result of univariate analysis performed on the spectral band localized at 1595 cm^{-1}, which is characteristic for the present nanomaterial. Most of the graphene acid is localized inside the cell, particularly in the cytoplasm. However, some particles remained outside the cell. Finally, the multivariate curve resolution was performed to visualize both signals, the cell and the nanomaterial. The result is shown in **Figure 3C**, where red colour represents the graphene acid, and the green represents the cellular structure. The resulting image confirms the uptake of the graphene acid inside the HeLa cell. Interestingly, the nanomaterial is not present in the cell nucleus, which can be caused by a combination of its negative charge and the size [20].

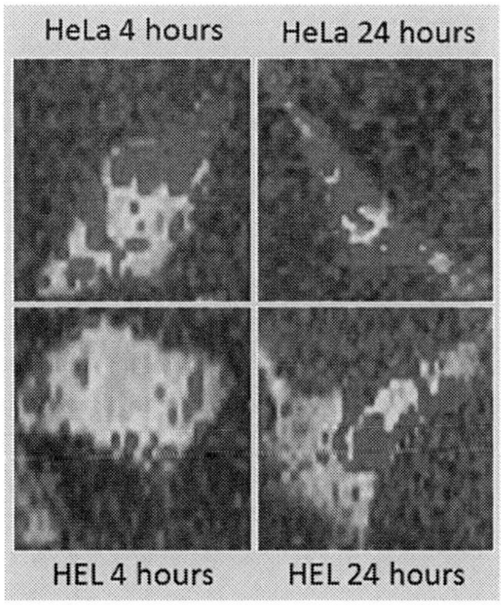

Figure 4 Raman mapping of GA internalization in the HeLa cells and HEL cells after incubation for 4 hours and 24 hours

Next, effect of the incubation time and effects of the nanomaterial on various cell lines were studied. Two cell lines were selected for the study, namely human healthy lung fibroblasts (HEL) and cancer cells (HeLa). Both lines were incubated with a total concentration of 50 mg/L of graphene acid for 4 and 24 hours. Resulting data are shown in **Figure 4** with considerable differences between healthy and cancer cell lines. The Hela cells incubated for 4 hours contained noticeably higher amount of the internalized nanomaterial, compared to normal cell lines. This can be possibly caused by the altered metabolism, characteristic for HeLa cells [21]. Nonetheless, the nanomaterial was found in the cytoplasm of both cell lines, indicating a possible exploitation of the nanomaterial in a drug delivery.

4. CONCLUSION

In summary, the new methodology for a study of the cellular uptake of graphene acid, based on the Raman microscopy, equipped with UV laser, was developed. Cellular uptake was studied on two different types of cells - healthy human lung fibroblasts (HEL) and human cervical cancer cells (HeLa). It was found that the nanoflakes of the graphene acid are easily internalized into cancer and non-cancerous cells, with a preference towards the studied cancer cell line.

ACKNOWLEDGEMENTS

The authors gratefully acknowledge support from project NPU LO1305 of the Ministry of Education, Youth and Sports of the Czech Republic.

REFERENCES

[1] BAKANDRITSOS, A., PYKAL, M., BŁOŃSKI, P., JAKUBEC, P., CHRONOPOULOS, D. D., POLÁKOVÁ, K., GEORGAKILAS, V., ČÉPE, K., TOMANEC, O., RANC, V., BOURLINOS, A. B., ZBOŘIL, R. and OTYEPKA, M., "Cyanographene and Graphene Acid: Emerging Derivatives Enabling High-Yield and Selective Functionalization of Graphene," *ACS Nano*, vol. 11, no. 3, pp. 2982-2991, Mar. 2017.

[2] DASARI, B. L., NOURI, J. M., BRABAZON, D. and NAHER, S., "Graphene and derivatives - Synthesis techniques, properties and their energy applications," *Energy*, vol. 140, pp. 766-778, 2017.

[3] ZHONG, Y., ZHEN, Z. and ZHU, H., "Graphene: Fundamental research and potential applications," *FlatChem*, vol. 4, pp. 20-32, 2017.

[4] PUMERA, M. and SOFER, Z., "Towards stoichiometric analogues of graphene: Graphane, fluorographene, graphol, graphene acid and others," *Chem. Soc. Rev.*, vol. 46, no. 15, pp. 4450-4463, 2017.

[5] ZUCKER, R. M., DANIEL, K. M., MASSARO, E. J., KARAFAS, S. J., DEGN, L. L. and BOYES, W. K., "Detection of silver nanoparticles in cells by flow cytometry using light scatter and far-red fluorescence.pdf," *J. Int. Soc. Adv. Cytom.*, vol. 83, no. 10, pp. 962-972, 2013.

[6] WANG, C., CHENG, L. and LIU, Z., "Drug delivery with upconversion nanoparticles for multi-functional targeted cancer cell imaging and therapy," *Biomaterials*, vol. 32, no. 4, pp. 1110-1120, 2011.

[7] LU, J., LIONG, M., LI, Z., ZINK, J. I. and TAMANOI, F., "Biocompatibility, biodistribution, and drug-delivery efficiency of mesoporous silica nanoparticles for cancer therapy in animals," *Small*, vol. 6, no. 16, pp. 1794-1805, 2010.

[8] JENSEN, E. C. "Use of Fluorescent Probes: Their Effect on Cell Biology and Limitations," *Anat. Rec. Adv. Integr. Anat. Evol. Biol.*, vol. 295, no. 12, pp. 2031-2036, Dec. 2012.

[9] SAHOO, H., "Fluorescent labeling techniques in biomolecules: a flashback," *RSC Adv.*, vol. 2, no. 18, p. 7017, 2012.

[10] ELIÁŠOVÁ SOHOVÁ, M., BODÍK, M., SIFFALOVIC, P., BUGÁROVÁ, N., LABUDOVÁ, M., ZAŤOVIČOVÁ, M., HIANIK, T., OMASTOVÁ, M., MAJKOVÁ, E., JERGEL, M. and PASTOREKOVÁ, S., "Label-free tracking of nanosized graphene oxide cellular uptake by confocal Raman microscopy," *Analyst*, vol. 143, no. 15, pp. 3686-3692, 2018.

[11] HUANG, J., ZONG, C., SHEN, H., LIU, M., CHEN, B., REN, B. and ZHANG, Z., "Mechanism of cellular uptake of graphene oxide studied by surface-enhanced Raman spectroscopy," *Small*, vol. 8, no. 16, pp. 2577-2584, 2012.

[12] BELYANSKAYA, L., MANSER, P., SPOHN, P., BRUININK, A. and WICK, P., "The reliability and limits of the MTT reduction assay for carbon nanotubes-cell interaction," *Carbon N. Y.*, vol. 45, no. 13, pp. 2643-2648, Nov. 2007.

[13] CASEY, A., HERZOG, E., DAVOREN, M., LYNG, F. M., BYRNE, H. J. and CHAMBERS, G., "Spectroscopic analysis confirms the interactions between single walled carbon nanotubes and various dyes commonly used to assess cytotoxicity," *Carbon N. Y.*, vol. 45, no. 7, pp. 1425-1432, 2007.

[14] PERUZYNSKA, M., CENDROWSKI, K., BARYLAK, M., TKACZ, M., PIOTROWSKA, K., KURZAWSKI, M., MIJOWSKA, E. and DROZDZIK, M., "Comparative in vitro study of single and four layer graphene oxide nanoflakes - Cytotoxicity and cellular uptake," *Toxicol. Vitr.*, vol. 41, pp. 205-213, 2017.

[15] KARKI, N., TIWARI, H., PAL, M., CHAURASIA, A., BAL, R., JOSHI, P. and SAHOO, N. G., "Functionalized graphene oxides for drug loading, release and delivery of poorly water soluble anticancer drug: A comparative study," *Colloids Surfaces B Biointerfaces*, vol. 169, no. May, pp. 265-272, Sep. 2018.

[16] MALARD, L. M., PIMENTA, M. A., DRESSELHAUS, G. and DRESSELHAUS, M. S., "Raman spectroscopy in graphene," *Phys. Rep.*, vol. 473, no. 5-6, pp. 51-87, 2009.

[17] VINET, L. and ZHEDANOV, A., "A 'missing' family of classical orthogonal polynomials," Antimicrob. Agents *Chemother.*, vol. 58, no. 12, pp. 7250-7257, Nov. 2010.

[18] NOTINGHER, I., "Raman spectroscopy cell-based biosensors," *Sensors*, vol. 7, no. 8, pp. 1343-1358, 2007.

[19] SU, L., CHEN, Y., ZHANG, G. N., WANG, L. H., SHEN, A. G., ZHOU, X. D., WANG, X. H. and HU, J. M., "In vivo and in situ monitoring of the nitric oxide stimulus response of single cancer cells by Raman spectroscopy," *Laser Phys. Lett.*, vol. 10, no. 4, p. 045608, Apr. 2013.

[20] ZHANG, D., WEI, L., ZHONG, M., XIAO, L., LI, H. W. and WANG, J., "The morphology and surface charge-dependent cellular uptake efficiency of upconversion nanostructures revealed by single-particle optical microscopy," *Chem. Sci.*, vol. 9, no. 23, pp. 5260-5269, 2018.

[21] CHAVES, N. L., ESTRELA-LOPIS, I., BÖTTNER, J., LOPES, C., GUIDO, B. C., DE SOUSA, A. R. and BÁO, S. N., "Exploring cellular uptake of iron oxide nanoparticles associated with rhodium citrate in breast cancer cells," *Int. J. Nanomedicine*, vol. 12, pp. 5511-5523, 2017.

ANALYSIS OF THE WAY TO INCREASE SAFETY IN TERMS OF POSSIBLE INHALATION OF NANO AND MICROPARTICLES DURING GUN SHOOTING

[1]Petra ROUPCOVA, [1]Stanislav LICHOROBIEC, [1,2]Karel KLOUDA, [3]Lucie BEDNÁŘOVÁ, [1]Lucie GEMBALOVÁ

[1]VSB - Technical University of Ostrava, Ostrava, Czech Republic, EU, petra.roupcova@vsb.cz; stanislav.lichorobiec@vsb.cz; lucie.gembalova@vsb.cz
[2]Research Institute for Occupational Health and Safety, Praha, EU, klouda@vubp-praha.cz
[3]National Institute of Nuclear, Chemical and Biological Protection, Milin, Czech Republic, EU, bednarova@sujchbo.cz

Abstract

A firearm shot is a complicated physico-chemical process in where the by-product of the shot is a shot-fire product that contains both metallic and non-metallic micro and nanoparticles. The main aim of the experiment was to verify the efficiency of respirators against gunshot residues from pistol shooting on an indoor shooting range and testing the phytotoxicity of sediment particles taken from the shooting range floor. The experiments were focused on measuring their concentration of an observer (range 20-1000 nm) after gunshots residues (GSR) from pistol. The various types of respirators detected particles penetration. The efficiency was calculated from the measured values especially their FIT factor and S (%) - efficiency of protective equipment. The most efficient respirator was considered REFIL 651 (FIT factor = 93, S = 98.9. The samples from the shot fills were further tested on their phytotoxicity - a component of environmental toxicity. The test of phytotoxicity was conducted from specific places at a distance of 1 m and 3m from the boxes. The root growth inhibition *IC* (%) for 1 m was 65.84, for 2 m 75.08. The toxic effect of sediment particles might be considered high. The measurement proved the growth of nano- and microparticles by 3 digits in comparison to the standard in a normal atmosphere. The results should inspire the shooting range operators to extend the use of the personal protective equipment to respirators, along with hearing protectors and goggles and lastly to threat the deposit as a dangerous waste.

Keywords: Concentration of nanoparticles and micro particles, gunshot residue, respirators, FIT factor, phytotoxicity

1. INTRODUCTION

The paper follows up on the results of previous research measurements of aerosol micro- and nanoparticles of GSR when shooting at an indoor shooting range. This article enriches the tests of efficiency of respirators against the residues. The main aim of the experiment was to verify the efficiency of the five chosen respirators against GSR from pistol shooting on an indoor shooting range (FIT factor and S- efficiency of protective equipment) and also testing the phytotoxicity of sediment particles taken from the shooting range floor and lastly to recommend some safety measures for the operators. Detailed measurements during shooting with a pistol, submachine gun and a shotgun provided comprehensive data on the amount of nanoparticles, their weight and distribution [1,2]. The analysis of the samples established metallic particles including their distribution, always in relation to the type of firearm and ammunition [3]. The results showed the presence of nanoparticles of shot- fire products and therefore an increased risk (toxicity) for the shooters in areas with atmosphere polluted by GSR, either in gaseous or solid form. In order to prove the reduction of inhaled micro- and nanoparticles, an experiment was conducted, during which the observer were equipped with respirators and the concentration of nanoparticles was measured in front of and behind the respirator while shooting a pistol, which is the most frequently used firearm on the shooting range. At the same time samples of sediment particles were taken from the shooting range floor, to be tested for phytotoxicity - germination and root elongation test. The tests were conducted on seeds of white mustard (*Sinapis alba* L.).

2. EXPERIMENTAL PART

The "Patriot" indoor shooting range is located in Ostrava in a basement. The area available for shooting has internal dimensions of 33 x 4.5 m. The shooting post has dimensions of 25 x 4.4 m with 4 shooting boxes 1.1 m wide, see. Space for spectators (observers, instructors) is located behind the shooting zone, including a table, bench and cabinet. **The shooting** was conducted with a pistol: CZ 75 D Compact. **Ammunition**: 9 mm LUGER, SELLIER&BELLOT a.s. Lead bullet, nitrocellulose smokeless powder Vektan Ba-9 0.34 g. **Tested respirators:** Triosin T-500 (FFP -3), REFIL 641 (FFP-2), REFIL 651 (FFP-3), REFIL 731 (FFP-2), REFIL 711 (FFP-1). The marking in brackets specifies, as per EN 149, the penetration limit of the filter (Filtering Face Piece): FFP 3 (the lowest penetration), then FFP 2 followed by FFP 1 (the highest penetration). Measuring instruments: for measurement of carbon oxide concentration: GAS Detector MSA Orion Plus. For measurement and comparison of concentrations of nanoparticles in front of and behind a respirator PORTACOUNT Pro+ Fit Tester 8038 from TSI, measuring range 20-1000 nm. Protection factor, so-called FIT factor:

$$F = \frac{c_0}{c_1} \qquad (1)$$

c_0 - concentration of pollutants outside the respirator, c_1 - concentration beneath the respirator, i.e. concentration entering the respiratory organ. FIT factor is a measure of efficiency of the protection of respiratory organs. The higher the *F* value, the safer the protective equipment for the user. Even though the *F* factor specifies how many times the particles concentration is reduced compared to the environment, sometimes it is more practical to use the expression "efficiency of protective equipment", *S* (%):

$$S = 100 - \frac{100}{F} \qquad (2)$$

The tested person was an observer. He received the tested respirators, which were fitted with a hole for a sensor of particles concentration measurement. The shooter shot from the pistol at a target from the first box in intervals of app. 15 seconds. The observer stood behind the first box in the spectators' zone, app. 1.5 m from the shooting pistol. During a 7-minute cycle the observer performed, in one-minute intervals, 1. normal breathing, 2. deep breathing, 3. head from side to side, 4. head up and down, 5. speaking including grimaces, 6. bending at the waist, 7. normal breathing (exercises performed as per OSHA 29 CFR 1910.124, see **Figure 1**). The first measurement took place with the shooter with a unique respirator TRIOSIN T-500, but without exercises. The measurements for the REFIL 641 and 651 respirators were taken for the observer. For the duration of the experiment the shooter shot in regular intervals, one shot every 15 to 20 sec.

Figure 1 Course of exercises by the observer [L. Frishansova VUBP-Praha v.v.i.]

3. RESULTS AND DISCUSSION OF EFFICIENCY OF RESPIRATION

The results were processed in the table and graphics; the FIT factor and the efficiency of individual respirators was determined. The measured values of concentration of micro- and nanoparticles in the observers' area and behind the respirator tested by them are listed in **Table 1** together with the determined (calculated) FIT factors and the efficiency for the specific type of respirator. For clarity and quick orientation the number of particles in the observer's space is shown in graphics (**Figure 2**) for the individual type of experiment, and the FIT factor value as well (**Figure 3**).

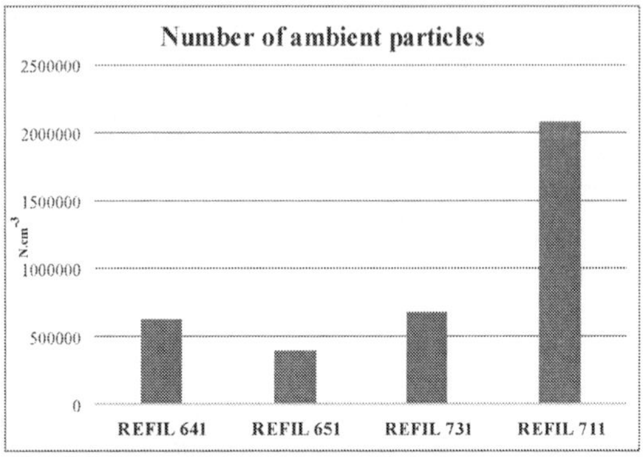

Figure 2 Graphical illustration of the number of particles in the area during individual tests in relation to the respirator type (observer)

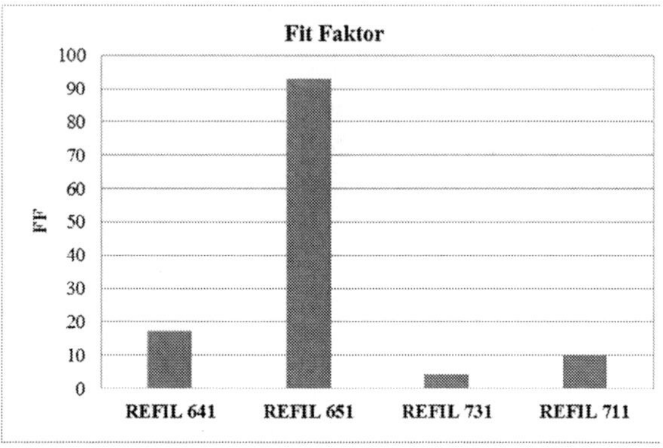

Figure 3 Graphical illustration of the F value in relation to the respirator type (observer)

The shooter was practically inhaling the particle values of the normal natural background. The *FIT* factor values are one of two measurements, without any exercises, so they do not comply with the OSHA 29 CFR 1910.134 standard in relation to the respirator quality (FFP value) and the *F* values, for the observer *F* is higher for the FFP1 respirator (REFIL 711) as compared to FFP2 (REFIL 731). The impact of different factors for the same types of respirators (**Table 1**) can be attributed to the effect of the respirator sealing [2] on the face of the observer and the shooter. We were interested in the effect of testing exercises on the *F* values and the respirator efficiency. The dependency of our experiments cannot be scientifically proved. The statistical reduction of the FIT factor and the efficiency is shown for exercise (bending at the waist).

Table 1 Overview of *FIT* factors and efficiency (*S*) of respirator in relation to the exercises

Test exercise**	REFIL 641		REFIL 651		REFIL 731		REFIL 711	
	F	S (%)	F	S (%)	F	S (%)	F	S (%)
1	28	96.4	261	99.6	7	85.7	8.3	87.9
2	49	97.9	183	99.4	2	50.0	9.7	89.7
3	20	95.0	109	99.1	5	80.0	12	91.6
4	20	95.0	117	99.1	6	83.3	12	91.6
5	19	94.7	67	98.5	4	75.0	7.6	86.8
6	7	85.7	51	98.0	3	66.6	13	92.3
7	15	93.3	77	98.7	3	66.6 or 60.6	11	90.9
Σ	17	94.1	93	98.9	4	75.0	10	90.0

4. RESULTS AND DISCUSSION ON TOXICITY IN THE ENVIRONMENT OF THE INDOOR SHOOTING RANGE DURING INTENSIVE SHOOTING

Along with an abnormal amount of micro and nanoparticles from the growth from 10^3 (background) to 10^6 during full shooting utilization of the shooting range we also measured the concentration of carbon oxide of app. 90 ppm, which is an allowable value for short-term intervention of professional firefighters. Along with carbon oxide, based on the study of the pyrolysis of nitrocellulose powder, other substances can be expected. In the quoted study [4] the mass spectrometer identified 26 pollutants: CO, NO, N_2O, CH_4, CO_2 and H_2O specified as primary, and acetylene, formaldehyde, acetonitrile, HCN, propene and ethylene oxide as secondary. These pollutants are not allied products for health. Establishing the composition and contents of metallic particles identified from volumes from atmospheres and identified by the EDAX system, including traction differentiation of particle sizes, were for shooting as follows Pb>Sb>Cu>Zn>Sn>Cd. Even though the weight of Pb particles in cm^3 was the greatest, particles with dimensions of 750-10000 nm prevailed. The weight of most nanoparticles up to 100nm was measured for Zn and Cu [3]. Lead poisoning has been known to mankind since antiquity; during the industrial revolution from the 18th century the situation deteriorated. Lead has no known biological function in organisms and when it enters the body it has a severe impact on health, which can be irreversible [5]:

- effect on nervous system, central as well as peripheral nervous system,
- impact on hemopoietin systems causes inhibition of vital enzymes (ALAD, ALA, enzyme catalyzing ferro chelatase - inserted Fe blocks the production of HEME (hemoglobin),
- nephrotoxicity - dysfunction of proximal porphyrin,
- cardiovascular effects,
- reproductive effects.

Lead is probably the most studied heavy metal. The primary location of lead deposits in the human body is the bones. The main mechanism of toxicity is the formation of ROS (formation of radicals •OH, •O2-, H_2O_2, H-O-O•). Depletion of reserves of antioxidants creates oxidative stress, which represents an imbalance between the production of free radicals and therefore the ability of biological systems to detoxify between products or to repair any damages. Free radicals will trigger a chain reaction, which leads to peroxidation of lipids, disturbance of cell oxidation of proteins and oxidation of nucleic acids (DNA, RNA) leading to cancer. Toxicity of so-called ion mechanism is caused by the lead's capacity to replace other ions Cu^{2+}, Mg^{2+}, Fe^{2+} and Na^+. This affects the basic biological processes in the body. The ion mechanism contributes to neurological deficit; after the calcium ion exchange the blood-brain barrier can be breached. Lead damages immature astroglial cells. The above-mentioned principle of ROS nanoparticles toxicity and oxidative stress applies also to so-called essential metals, which play an irreplaceable role in the bio organism. Zinc is necessary for the function of different mammal enzymes and it is present virtually in all plant and animal tissues [6]. The inhalation of vapors or fine dust of metallic Zinc and ZnO causes "brass founder's fever" (fever caused by metals and their oxides). It manifests itself as lethargy, headache, chest pain, irritating cough and it is accompanied by fever and temperatures up to 59 °C. The same applies to exposure to vapors of copper or its dust aerosols. Nano zinc and nano copper (and their oxides) are used for various industrial purposes (catalysts, polymer composites) and in healthcare. There are a number of research publications on the toxicity of metals and their oxides [7,8]. With results in vivo, as well as in vitro, including the test on human cells (liver, lung) and the phytotoxicity [9,10]. In essence, a certain level of genotoxicity based on ROS formation is always identified. Zinc oxide is in an ambivalent situation, because it has anti-bacterial properties and is applied as a composite of packaging of food [11] and its genotoxicity is used against carcinogenic cells [12,13]. The documented neuro-toxic effect of copper nanoparticles in the atmosphere can be viewed as an increased risk, by penetration through the skin and nose to soma sensory neurons in spinal ganglia [14].

4.1. Sediment toxicity on the indoor shooting range floor

The GSR are absorbed on different materials (clothing, skin, hair) and also sedimentation on surfaces within the shooting range. Phytotoxicity is an integral part of the environmental toxicity. Therefore the deposited particles were swept from the shooting range floor after a Czech Police shooting practice session and the surface particle collection and a germination and root growth test were conducted on *Sinapis alba* L. (white mustard) seeds according to relevant methodologies of determination of waste eco-toxicity. Introduced dust load into Petri dishes, pipetted 7mL of diluent, 20 seeds for germination stored in Petri dish with diluent at given temperature (20±2 °C), readout executed after 72 hours of germination out of reach of light. The ecotoxicity tests according to OECD Guidelines 208/1984, which is used as a standard test in Europe. Readout of keys according to relationship:

$$IC = \frac{(L_c - L_v)}{L_c} \cdot 100 \quad (3)$$

IC is root growth inhibition (%), L_c arithmetic average of root length (mm), L_v arithmetic average of root length in tested media (mm).

The test of phytotoxicity was conducted from specific places with the area of an A4 paper sheet (210 x 297 mm), at a distance of 1 m and 3 m from the boxes. **Table 2** shows that the weight of the deposit of GSR grew on the shooting position with the distance from the shot. Also the toxic effect of sediment particles (see inhibition value) can be considered high.

Table 2 Weight gradient of the deposit in the shooting post and the box, including the specified inhibition of root growth

Distance from the deposit collection place (m)	Deposit collection point	Weight of deposit from area of A4 size (g)	Recalculation of deposit to m² (g)	Root growth inhibition *IC* (%)
1	Box 1	0.302	4.832	65.84
	Box 3	0.426	6.716	
3	Box 1	0.602	9.632	75.08
	Box 3	0.403	6.448	
0	Box 1	0.125	2.000	67.92
	Box 3	0.249	3.984	

5. CONCLUSION

The measurement showed the growth of nano- and microparticles by 3 digits in comparison to the standard in a normal atmosphere. This growth was caused by GSR particles, which, after partial chemical identification, can be included among toxic pollutants. The toxicity of the atmosphere on the shooting range is increased also by the high content of carbon monoxide. The results of our tests showed that the application of normal respirators substantially reduces the contents of inhaled particles. The measurement results should inspire the shooting range operators to extend the use of the personal protective equipment to respirators, along with hearing protectors and goggles. Especially for the instructors, who are exposed to shooting for extended periods. We can also conclude that the shooting results of our shooter were not affected by the use of the respirators. The shooter had no negative experience during shooting. The phytotoxicity showed the high toxicity of the deposit from GSR from the shooting range floor. The shooting range operators should remember this fact and treat the deposit as dangerous waste.

ACKNOWLEDGEMENTS

This article has been supported within the Student Grant Competition 'Development of application and preparation of new carbon nanomaterials and their modification with polymers and other selected materials'. Project number SP2018/137.

REFERENCES

[1] VYORAL, P., *Shot fume dispersion problems and issues.* Diploma thesis, Tomas Bata univesity, Zlin, 2013. (in Czech)

[2] SLABOTINSKY, J., BRADKA, S. Real protective effect of respirator. *The Science for Popolation Protection,* 2014, vol. 1, no. 2, pp. 1-8. (in Czech)

[3] FRISHANSOVA, L., KLOUDA, K., BERNATIKOVÁ, S., LACH, K., LICHOROBIEC, S. The quantity, distribution and dispersion and analyses of nano- and microparticles from gunshot residues. In: Proceedings of the *XVII. international conference of Occupational Health and Safety 2017.* Sepetna: SPBI, 2017, pp. 18-23. (In Czech)

[4] CROPEK, Donald M., KEMME, Patricia A., DAY, Jean M. Pyrolytic decomposition studies of AA2, a double-base propellant. *US Army Corps of Engineer.* 2001.

[5] NOVAKOVA, Z. *Toxicity of nanoparticles of lead.* Dissertation thesis, Masaryk University, Brno, 2015. (in Czech)

[6] BENCKO, V., CIKRT M., LENER, J. *Toxic metals in the living and working environment of humans.* Grada Publishing, Prague, 1995, p. 236. (In Czech)

[7] SEABRA, A. B., DURAN, N. Nanotoxicology of Metal Oxide Nanoparticles. *Metal.* 2015, vol. *5*, pp. 934-975.

[8] KUMAR, V., KUMARI, A., GULERIA, P., YADAV, S. K. Evaluating the toxicity of selected types of nanochemicals. *Reviews Environmental contamination and Toxicology.* 2012, vol. 215, pp. 39-121.

[9] LIN, D., XING., B. Phytotoxicity of nanoparticles: Inhibition of seed germination and root growth. *Environmental Pollution*, 2007, vol. 150, no. 2, pp. 243-250.

[10] DIMKPA, CH. O. MCLEAN J., E., LATTA, D. E., MANANGON, E., BRITT, D. W., JOHNSON W., P., BOYANOV, M. I., ANDERSON, A. J. CuO and ZnO nanoparticles: phytotoxicity, metal speciation, and induction of oxidative stress in sand-grown wheat. J. Nanopart Res. 2012, vol. 14, no. 9, pp. 1-15.

[11] ESPITA, P. J. P., SOARES, N. F. F., COIMBRA J. S. R., ANDRADE, N. J., ANDRADE, N. J., CRUZ, R. S., MEDEIROS, E. Zinc oxide nanoparticles: synthesis, antimicrobial activity and food packaging applications. *Food Bioprocess Technol,* 2012, vol. 5, pp. 1447-1464.

[12] NAIR, S., SASIDHARAN, A., DIVYARANI, V., MENON, D. *Role of size scale of ZnO nanoparticles and microparticles on toxicity toward bacteria and osteoblast cancer cells. Journals of Materials Science. Materials in Medicine*, 2009, vol. 20, pp. 235-241.

[13] PREMANATHAN, M., KARTHIKEYAN, K., JEYASUBRAMANIAN, K., MANIVANNAN, G. Selective toxicity of ZnO nanoparticles toward Gram-positive bacteria and cancer cells by apoptosis through lipid peroxidation. *Nanomedicine:Nanotechnology, Biology and Medicine*, 2011, vol. 7, no. 2, pp. 184-192.

[14] PRABHU, Banavalu M., ALI, Syed F., MURDOCK, Richard C., HUSSAIN, Saber M., SRIVATSAN, M. Copper nanoparticles exert size and concentration dependent toxicity on somatosensory neurons of rat. *Nanotoxicology.* 2010, vol. 4, no. 2, pp. 150-160.

MONTMORILLONITE STABILITY AND ITS EFFECTS ON THE ENVIRONMENT

Klára DROBÍKOVÁ[1, 2], Michaela TOKARČÍKOVÁ[1], Kateřina MAMULOVÁ KUTLÁKOVÁ[1], Oldřich MOTYKA[1,3], Jana SEIDLEROVÁ[1]

[1]*Nanotechnology Centre, VSB - Technical University of Ostrava, Ostrava, Czech Republic, EU*
[2]*IT4Innovations, VSB - Technical University of Ostrava, Ostrava, Czech Republic, EU*
[3]*ENET centre, VSB - Technical University of Ostrava, Ostrava, Czech Republic, EU*

Abstract

Montmorillonite is one of the most important groups of phyllosilicates and is an excellent sorbent. For the facilitation of the post-sorption sorbent separation from the medium (e.g.) waste water, it is convenient to use its magnetically modified form. Magnetically modified montmorillonite composites can be prepared using a simple microwave synthesis; the efficiency of this preparation method has already been proven. The aim of this study was to evaluate the effect of the microwave radiation during the preparation of its magnetic form on the structure and stability of montmorillonite. Water extracts were prepared by the batch method using deionized water. The leachate was prepared for 24 hours in different ratios of solid and liquid phase (1:10 000, 1:1 000, 1:500 and 1:100) from the montmorillonite and the magnetically modified montmorillonite. Concentrations of Al, Ca, K, Mg, Na, Si and Fe were determined using atomic emission spectrometry with inductively coupled plasma, pH of the filtrate was also determined. The results of the experiments showed that Al and Ti do not leach under the observed conditions. The assumption that the ion exchange elements (K, Na, Ca, Mg) leach to the solution was confirmed. Their amount is dependent on the leaching time and on the solid/liquid phase ratio. Silicon, a part of the fundamental layer structure also leached, however, in an amount not affecting the montmorillonite structure. The process of the microwave-assisted magnetic modification of the montmorillonite was proved not to affect the stability of the material.

Keywords: Montmorillonite, magnetic form, leaching, microwave treatment

1. INTRODUCTION

Heavy metal pollution is one of the most important environmental problems today. Various industries (e.g., mining, metallurgy, surface finishing, iron and steel production, metal surface treating, etc.) produce and discharge wastes containing numerous heavy metals and organic compounds. Such a human activity brings about serious environmental pollution, threatening both human health and ecosystems. The metals cannot be degraded to harmless products and hence persist in the environment indefinitely. Both the selected metals and organic compound have a negative effect on human health [1]. Pollution is commonly categorized according to the character of the pollutant as inorganic or organic; however, in reality, both categories of pollutants are present in various ratios.

One of the most commonly applied methods for the water environment is sorption. A whole range of sorption methods are utilized in both laboratory and field and they are usually divided according to the principal, environment (gas, liquid, solid) or the characteristics of the pollutant. Frequently applied sorbent is activated carbon which, however, is a rather costly material [2,3]. Therefore, many studies were focused on the cheaper sorbent alternatives such as sheet materials - phyllosilicates [4-8], slag [9,10], ashes [11] and, finally, biomaterials, mainly various waste by-products (sawdust, orange peel, rice husks etc.) [12,13]. New sorbents are being designed continuously in order to enhance the sorption efficacy, decrease the cost of the cleaning technology and improve the functionality of the material. Introduction of new types of sorbents based on "smart" materials combines the sorption and (e.g.) catalytic characteristics of the material, and, thus, simultaneous removal of more pollutants.

In the case of the liquid environment, it is necessary to separate the sorbent from the cleansed liquid; for this purpose, filtration is used. Application of a sorbent with magnetic characteristics would allow the realization of the aforementioned separation using a magnetic field. Magnetically modified sheet materials or bio-sorbents can both be used for this purpose.

Bentonite was modified with ferrofluids by Mockovciakova et al. [4]. Sorption experiments were focused on the removal of low concentrations of Cd^{2+} and Ni^{2+} ions. Other authors describe sorption of heavy metal ions on magnetically modified zeolites [4, 5], bentonites [6,7] and montmorillonite [10]. Moreover, new composites designed by combining two or more sorbents such as magnetic nano-sorbent from the orange peel waste simultaneously precipitated with Fe_3O_2 nanoparticles and either hydroxyapatite nanoparticles [14] or zeolite [15] were successfully used for the removal of Cd^{2+} a Zn^{2+} from liquid solutions.

2. MATERIALS AND METHODS

For the purpose of this study, Montmorillonite (Mt) from the Ivančice (Czech Republic) deposit was selected. Based on the chemical composition, the crystallochemical formula of this montmorillonite was determined [16]:

$$(Si_{3.99}Al_{0.01})(Al_{1.32}Fe(III)_{0.16}Mg_{0.50}Ti_{0.02})O_{10}(OH)_2(Ca_{0.16}K_{0.04}Na_{0.12}) \quad (1)$$

Raw clay material was dried, milled, and sieved. Mt particles having size ≤ 40 μm were used for the experiments.

Chemical composition of Mt was determined using an energy dispersive fluorescence spectrometer (XRFS), SPECTRO XLAB, Spectro. For this measurement, samples in powder form were pressed into tablets using wax as a binder. Al, Ca, K, Mg, Na, Si and Fe were determined, after acid digestion, by Atomic emission spectrometry with inductively coupled plasma (AES-ICP), SPECTRO CIROS VISION, Spectro.

Two sample types were used and compared: (1) the native Mt leached in deionized water (sample denoted as M) and (2) the magnetically modified Mt prepared using microwave radiation (sample denoted as MMW). Suspension of the $FeSO_4$ in deionized water was treated in a microwave oven for 10 min at power 700 W, 2450 MHz. After cooling, the Mt was added into suspension. After 24 h, the Mt with Fe_xO_y was separated by filtration (pore size 0.23 μm) and dried. Both M and MMW were leached in deionized water using the batch method. The leachates were prepared for 24 hours in 1:10 000, 1:1 000, 1:500 and 1:100 ratios of solid and liquid phase from M and MMW. In the filtrate, concentrations of Al, Ca, K, Mg, Na, Si and Fe were determined using AES-ICP, the pH value was determined as well.

To determine the phases and structure of prepared samples, X-ray powder diffraction analysis (XRPD), BRUKER AXS was used. X-ray powder diffraction patterns were recorded under Co Kα irradiation (λ = 1.789 Å) using the Bruker D8 Advance diffractometer equipped with a fast position sensitive detector VÅNTEC 1. Samples in powder form were pressed in a rotational holder. Reflection mode was used for all measurements. Phase composition was evaluated using database PDF 2 Release 2004 (International Centre for Diffraction Data).

3. RESULTS AND DISCUSSION

The chemical composition of Mt, determined in the dried sample, revealed the high amount of Si and Al (**Table 1**).

XRPD patterns of native Mt, M and MMW shows (**Figure 1**) that the process of magnetization did not change the structure of the native Mt because the width is same. Therefore, the Fe_xO_y particles were bound on the Mt surface. The preliminary study proved, that FexOy particles prepared using the above-mention procedure are a mixture of γ-Fe_2O_3 a Fe_3O_4.

Table 1 The chemical composition of the Montmorillonite, elemental content is express as oxide

Parameter	Content (wt. %)	Parameter	Content (wt. %)	Parameter	Content (wt. %)
Na_2O	0.093±0.047	SiO_2	57.47±4.03	TiO_2	0.296±0.030
MgO	3.07±0.62	K_2O	0.925±0.084	Fe_2O_3	3.40±0.14
Al_2O_3	13.9±0.84	CaO	2.04±0.63	LOI	17.75±07

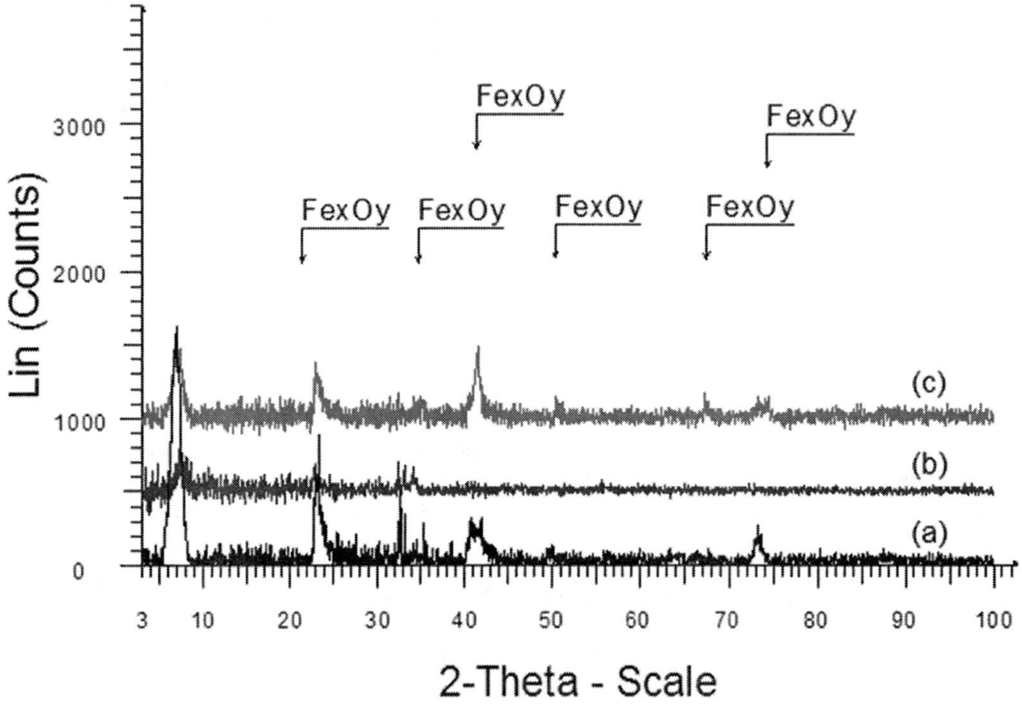

Figure 1 XRPD patterns of native Mt (a), M (b) and MMW (c) with detected Fe_xO_y phases

Iron content in Mt a MMW is presented in **Table 2**. In the magnetically modified montmorillonite, iron was found mainly in the form of FeO, Fe_2O_3 and Fe_3O_4 [13]. Magnetically modified Mt treated by microwaves contained 8.5 wt % of the total iron which is 3.5 times more than in the original Mt (2.4 wt. %). Content of ferrous oxide was six times higher in the case of MMW.

Table 2 Iron content in Mt and MMW

Parameter	Mt (wt. %)	MMW (wt. %)
Total Fe	2.4±0.1	8.5±0.4
FeO	0.24±0.03	1.49±0.06

The pH values of the leachate provide important information on the two-way interaction between the solid and liquid phase. Elements present in the solid phase may dissolve into the extraction solution (deionized water) and alter the pH, as shown in the **Figure 2**. Microwave treatment of the material led to a shift of the leachate pH towards more basic values (around 7.5), while the pH was unaffected by the particular ratio of the solid and liquid phase. The leachate pH from the native material was 6.5-7.2 which reflects the inhomogeneity of the Mt samples.

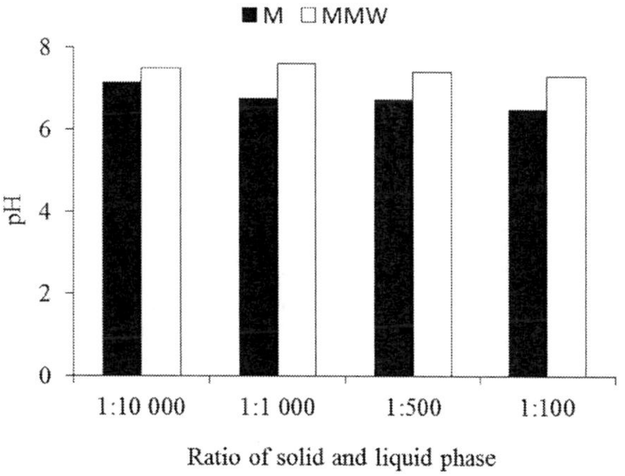

Figure 2 pH of extract solution prepared from M and MMW

Leached amounts of the observed elements in the extracts are presented in the **Figures 3 - 5**. The amounts of the leached Al and Fe were below the detection limit of the used method (< 0.01 mg/L and (< 0.005 mg/L, respectively). Iron oxides are, thus, bound strongly to the Mt surface and the material is stable even after the 24-hours in contact with the water environment.

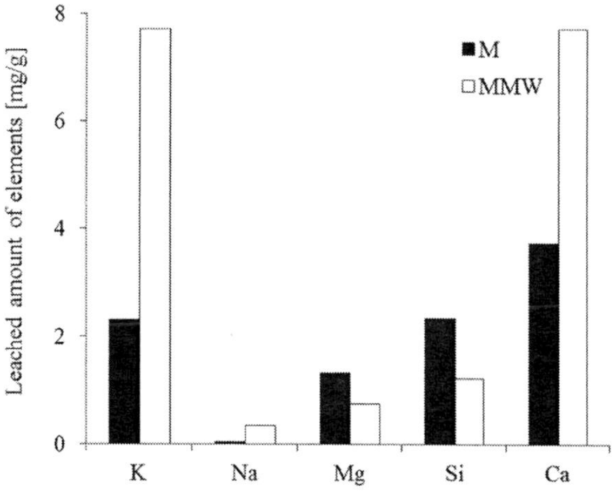

Figure 3 Leached amount of elements from M and MMW - ratio 1:10 000

Relatively highest amount of the leached elements was observed when the ratio of solid and liquid phase was 1 : 10 000 (**Figure 3**). From the MMt, K and Ca were the elements leached the most while Na, Mg and Si were leached slightly more from M. In the rest of the solid: liquid phase ratios, the same trend in the leached amount of Si was observed - this amount was also very similar in both of the samples. Hence, it can be concluded that the preparation of the magnetically modified did not alter the structure of the montmorillonite. The leached amounts of the ion-exchange elements (Na, K, Ca and Mg) differed between the materials. As expected, more K, Mg and Ca was leached from M than from MMW. This was, however, not observed in the case of Na, which was leached more from MMt than from M suggesting an alteration of the Na stability in Mt structure caused by microwave radiation. In all the elements, their amount in the leachate was affected by the solid: liquid phase ratio.

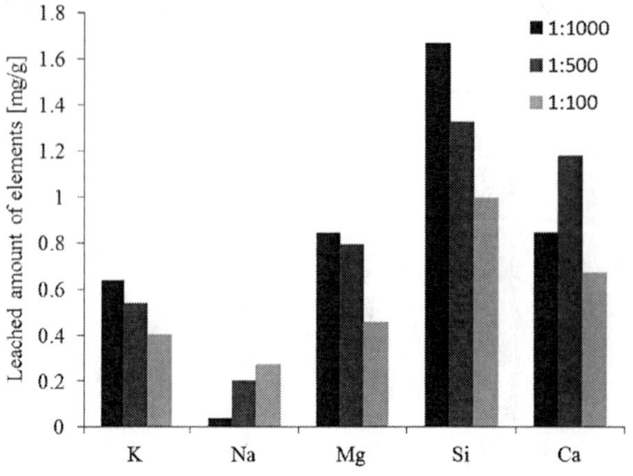

Figure 4 Leached amount of elements from M

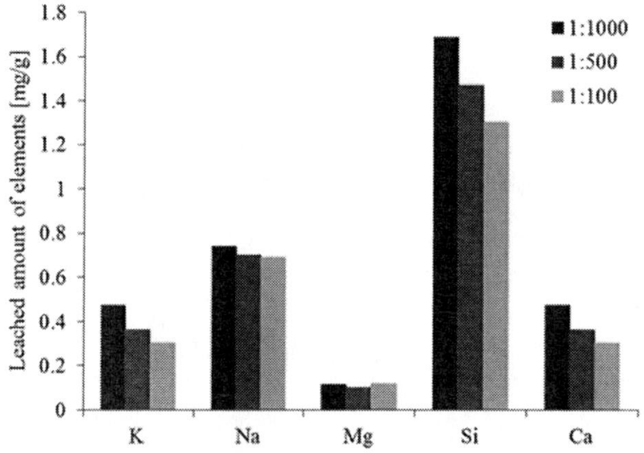

Figure 5 Leached amount of elements from MMW

4. CONCLUSION

Magnetically modified montmorillonite was prepared by a simple method using microwave radiation. The original Ivančice montmorillonite and its modified form were subjected to 24 hour leaching in deionized water applying various ratios of solid and liquid phase. It was proved that the magnetic iron oxides are bound to the Mt strongly and are, most probably, located on its surface. Microwave radiation was found not to affect the leaching of the elements of the basic structure of the montmorillonite (Al and Si). However, it seems to affect the leaching of the ion-exchange elements. Solid: liquid phase ratio affected the amount of observed elements in the leachate too.

ACKNOWLEDGEMENTS

Funding: This research has been funded by the LQ1602 „National Programme of Sustainability II - IT4Innovations excellence in science“ and LO1404 „Sustainable Development of ENET Center“.

REFERENCES

[1] WANG, Jianlong and CHEN, Can. Biosorbents for heavy metals removal and their future. *Biotechnology advances.* 2009. vol. 27, pp. 195-226.

[2] KADIRVELU, Krishnagounder, THAMARAISELVI, Kaliannan and NAMASIVAYAM, Chinnaiya. Removal of heavy metals from industrial wastewaters by adsorption onto activated carbon prepared from an agricultural solid waste. *Bioresource technology*. 2001. vol. 76, pp. 63-65.

[3] MOUNI, Lotfi, MERABET, Djoudi, BOUZAZA, Abdelkrim and BELKHIRI, Lazhar. Adsorption of Pb(II) from aqueous solutions using activated carbon developed from Apricot stone. *Desalination*. 2011. vol. 276, pp. 148-153.

[4] MOCKOVČIAKOVÁ, Annamária, OROLÍNOVÁ, Zuzana and ŠKVARLA, Jiří. Enhancement of the bentonite sorption properties. Journal of Hazardous Materials. 2010. vol. 180, pp. 274-281.

[5] BOURLINOS, Athanasios B., ZBORIL, Radek and PETRIDIS, Dimitrios. A simple route towards magnetically modified zeolites. Microporous and Mesoporous Materials. 2003. vol. 58, pp. 155-162.

[6] MATIK, Marek, VÁCLAVÍKOVÁ, Miroslava, HREDZÁK, Slavomír, LOVÁS Michal and JAKABSKÝ, Štefan. Possibility of Modification of Zeolites by Iron Oxides and its Utilization for Removal of Pb(II) from Water Solutions. Acta Montanistica Slovaca. 2004. vol. 9, pp. 418-422.

[7] OLIVEIRA, Luiz C.A., RIOS, Rachel V.R.A, FABRIS, José D., SAPAG, Karim, GARG, Vijayendra K. and LAGO, Rochel M. Clay-iron oxide magnetic composites for the adsorption of contaminants in water. Applied Clay Science. 2003. vol. 22, pp. 169-177.

[8] VEREŠ Ján and OROLÍNOVÁ, Zuzana. Study of the treated and magnetically modified bentonite as possible sorbents of heavy metals. Acta Montanistica Slovaca. 2009. vol. 14, pp. 152-155.

[9] GUPTA, Vinod K. Equilibrium Uptake, Sorption Dynamics, Process Development, and Column Operations for the Removal of Copper and Nickel from Aqueous Solution and Wastewater Using Activated Slag, a Low-Cost Adsorbent. Industrial & Engineering Chemistry Research. 1998. vol. 37, pp. 192-202.

[10] DIMITROVA, Sonia V. and MEHANDGIEV, Dimitar R. Lead removal from aqueous solutions by granulated blast-furnace slag. Water Research. 1998. vol. 32, pp. 3289-3292.

[11] CALACE, Nicoletta, NARDI, Elisa, PETRONIO, Bianca M. and PIETROLETTI, Marco. Adsorption of phenols by papermill sludges. Environmental Pollution. 2002. vol. 118, pp. 315-319.

[12] AZIZ, Abdellah, OUALIA, Mohand S., ELANDALOUSSI, El H., L. C. DE MENORVAL, Louis C. and LINDHEIMER, Marc. Chemically modified olive stone: A low-cost sorbent for heavy metals and basic dyes removal from aqueous solutions. Journal of Hazardous Materials. 2009. vol. 163, pp. 441-44717.

[13] GUPTA, Vinod Kumar and NAYAK, Arunima. Cadmium removal and recovery from aqueous solutions by novel adsorbents prepared from orange peel and Fe_2O_3 nanoparticles. Chemical Engineering Journal. 2012. vol. 180, pp. 81-90.

[14] FENG, Yuan, GONG, Ji-Lai, ZENG, Guang-Ming, NIU, Qiu-Ya, ZHANG, Hui-Ying, NIU, Cheng-Gang, J. H. DENG, Jiu-Hua and YAN, Ming. Adsorption of Cd (II) and Zn (II) from aqueous solutions using magnetic hydroxyapatite nanoparticles as adsorbents. Chemical Engineering Journal. 2010. vol. 162, pp. 487-494.

[15] NAH, In Wook, HWANG, Kyung-Yub, JEON, Choong and CHOI, Hong Bok. Removal of Pb ion from water by magnetically modified zeolite. Minerals Engineering. 2006. vol. 19, pp. 1452-1455.

[16] VALÁŠKOVÁ, Marta, HUNDÁKOVÁ, Marianna, MAMULOVÁ KUTLÁKOVÁ, Kateřina, SEIDLEROVÁ, Jana, ČAPKOVÁ, Pavla, PAZDZIORA, Erich, MATĚJOVÁ, Kateřina, HEŘMÁNEK, Martin, KLEMM, Volker and RAFAJA, David. Preparation and characterization of antibacterial silver/vermiculites and silver/montmorillonites. Geochimica et Cosmochimica Acta. 2010. vol. 74, pp. 6287-6300.

[17] LIN, Su-hsia and JUANG, Ruey-shin. Heavy metal removal from water by sorption using surfactant-modified montmorillonite. Journal of Hazardous Materials. 2002. vol. 92, pp. 315-32.

PLANT RESPONSE TO NANO METAL OXIDES AND THEIR COMPOSITES: CHLOROPHYLL CONTENT AS A STRESS INDICATOR

[1,2,3]Oldřich MOTYKA, [2,3]Kristína ŠTRBOVÁ, [2]Inga ZINCOVSCAIA

[1]*Nanotechnology centre, VSB - Technical University of Ostrava, Ostrava, Czech Republic, EU,* oldrich.motyka@vsb.cz
[2]*Joint Institute for Nuclear research, Dubna, Russian Federation*
[3]*ENET centre, VSB - Technical University of Ostrava, Ostrava, Czech Republic, EU*

Abstract

Interaction between engineered nanoparticles and plants is still a widely unexplored area. While there is a growing body of research on the toxicity of nanoparticles to plants, their stress-inducing capabilities are less known. Determination of chlorophyll content in plant leaves is a relatively easy way of real-time assessment of the physiological state of the plant. Cauliflower (*Brassica oleracea* convar. Botrytis) seedlings were subjected to the TiO_2 nanoparticles either by being planted in the nano-TiO_2 enriched soil (1g/kg) or by exposure of its leaves to nano-TiO_2 suspension (1g/L). Chlorophyll content in the shoots (mmol/m^2) was measured both before and following the exposure in set interval; shoots were then harvested and prepared for Ti content analysis. While no significant decrease of chlorophyll content was observed in the unexposed plants, chlorophyll content significantly decreased after four days in the seedlings growing in the nanoparticle-enriched soil and after twelve days in the seedlings subjected to the suspension. All the seedlings appeared vital, yet the decrease in chlorophyll content indicates nanoparticle-induced stress that may seriously affect the yield from cultivated crops when these are exposed to nano-metal oxides or their composites as well as raise health concerns due to their bioaccumulation in the plant tissues.

Keywords: Nano metal oxides, chlorophyll, exposure, stress

1. INTRODUCTION

Rapid development of nanoparticles (NPs) production and manufacturing results in increase their release to all parts of the environment. Since the plants are essential constituent of the ecosystems - standing at the beginning of the food chain - investigation of their interaction with nanoparticles (NPs) is important to avoid to potential ecological risk as well as negative effect on human health [1]. NPs are known to negatively affect plant growth, cell structure, as well as the plant physiology (reduction in seed germination, suppression of root elongation, reduction of biomass and leaf production) [2] biochemical functions and, finally, can cause the plant death [3]. Regarding the impact of metal oxide nanoparticles (including nano-TiO_2, one of the most frequently used nanomaterials) on photosynthesis, both positive and negative effects were reported [4] [5] [6].

Since chlorophyll (namely chlorophyll a) is the most sensitive to photodegradation from all the other photosynthetic pigments, its content in shoots is a valuable indicator of the impact the exposure to nanoparticles has on the plant [3]. Several studies on photosynthetic organisms proved that before any detrimental effect on post-exposure viability or growth was observed, the chlorophyll levels already decreased noticeably [4] [5]. Chlorophyll content was also found in plants when no signs of toxicity were apparent using the conventional toxicological methods [6]. In addition, non-destructive methods using a method based on radiation transmittance allow measurements in vivo and monitoring of the response of an individual plant over the course of the experiment [7]

The aim of this pilot study was to assess whether, in plants exposed to metal oxide (TiO_2) nanoparticles either via root system or leaves, the chlorophyll content determined *in vivo* -- can be used for NP-exposure-induced physiological stress.

2. MATERIALS AND METHODS

Store-bought young cauliflower plants (*Brassica oleracea* convar. Botrytis, n = 9) were transferred to the laboratory and, after three days of acclimatization, divided into three equally sized groups. The first group of plants was planted in pots containing a mixture of sowing and propagation soil substrate (AGRO) and nano-TiO_2 powder (1 g/kg) prepared by mixing both components together in a rotational shaker for 24 hours. The second group was planted in the same, yet uncontaminated, substrate and their leaves were sprayed, biweekly, with 5 ml of 1g/L water suspension of nano-TiO_2. The last group was only planted to the substrate and subjected to no treatment (control). During the experiment, all plants were watered with tap water when necessary (dry substrate).

TiO_2 NPs for the experiments were prepared from titanyl sulfate (Precheza a.s.) precursor containing 102 g of TiO_2 per 1 dm^3 of suspension. The suspension was hydrolysed and, consequently continuously stirred for 5 hours at 100°C. The solid phase was then separated by decantation, washed several times with distilled water until the conductivity of the filtrate was lower than 100 $\mu S \cdot cm^{-1}$, and dried at 105 °C for 24 h.

The substrates used - both enriched and original - were leached and the leachate was analysed for Ti content. Atomic emission spectroscopy with inductively coupled plasma (ISP-AES) showed no titanium leaching from the original substrate (<0.01 mg/L) and 0.028 mg/L leaching from the contaminated substrate. The chlorophyll content measurement was performed *in vivo* using MC-100 Chlorophyll Concentration Meter (Apogee) in ten replications per measurement. All the leaves available for the measurement of the plant were targeted. The measurement was performed immediately after planting, after 24 hours, 4 days and 12 days from the planting. Measured chlorophyll content was expressed in $\mu mol\ m^{-2}$.

After the exposure period, the above-ground parts of plants were harvested and dried to constant weight at 45 °C; due to the small yield, the samples were merged according to the treatment, homogenized and pulverized. After thermal digestion in a mix of chemicals ($HF+HNO_3+H_2O_2$), the samples were analysed for the content of Ti (as well as Ca, K, Mg and Na, essential mineral elements) using ICP-AES (SPECTRO CIROS VISION, Spectro).

Statistical analysis of the measured data as well as their graphical representation using the R [8] One-way analysis of variance (ANOVA) was used to assess the differences in chlorophyll content among the treatments, post hoc Tukey HSD test was used for the pair-wise comparison and identification of the particular differences.

3. RESULTS AND DISCUSSION

The resulting content of the selected elements in the shoots (after 12 days of experiment) showed marginal difference in the content of essential elements among the treatments (**Figure 1**). However, the titanium content was significantly higher in the above-ground cauliflower biomass of the plants subjected to the nano-TiO_2 exposure via leaves (in suspension) - while the difference between the control plants and plants grown on NP-enriched soil was negligible.

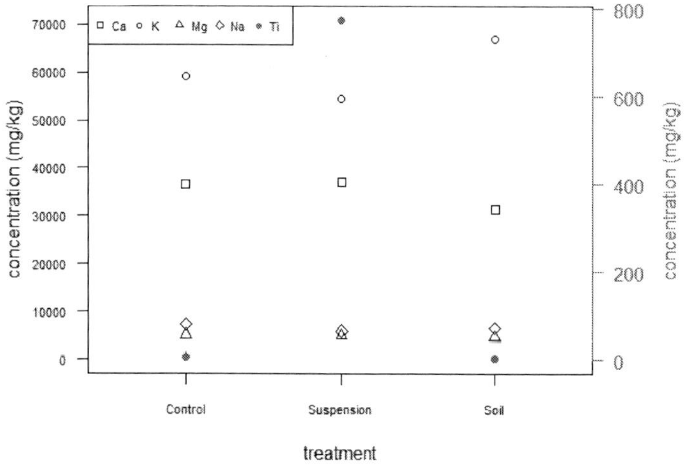

Figure 1 Content of selected elements in the plants depending on the treatment

In the control plants, a slight decrease in chlorophyll content was observed at the end of the experimental period, however, it was not found to be significant by ANOVA (**Figure 2**). It is, indeed, conceivable that the stress of replanting had some effect on the experimental plants - expressed in chlorophyll decrease - nevertheless, the experimental period was too short to fully address this issue.

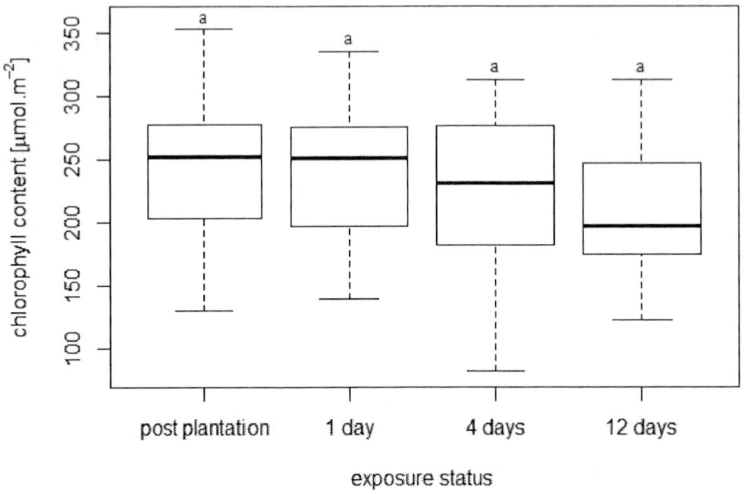

Figure 2 Chlorophyll content in the unexposed (control) plants. Identical letters above the boxplots denote the non-significant difference between the measurements.

In the plants subjected to the nano-TiO_2 suspension sprayed on their leaves, the decrease in chlorophyll was much more prominent around the 12th day (**Figure 3**). ANOVA confirmed this decrease as significant ($p = 2.02 \cdot 10^{-9}$), Tukey HSD test specified that the difference is between the 12-day measurements and all the other measurements while the rest of the measurements did not significantly differ from each other.

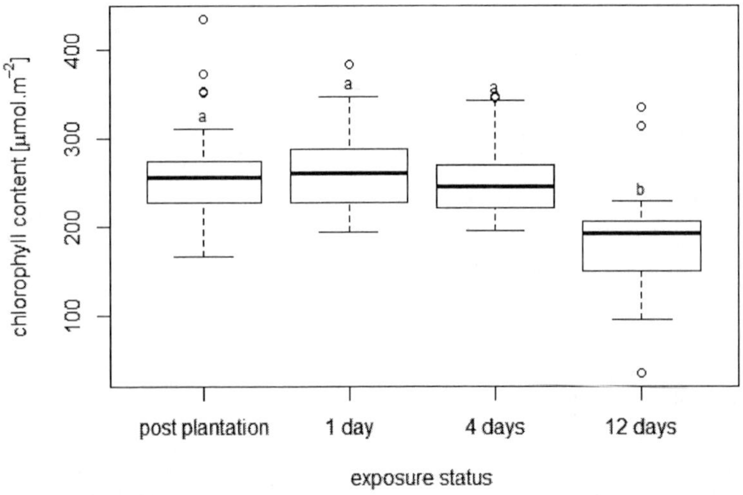

Figure 3 Chlorophyll content in the plants exposed to the 1g/L nano-TiO_2 suspension. Identical letters above the boxplots denote the non-significant difference between the measurements.

Chlorophyll content in the plants grown in the substrate enriched with nano-TiO_2 showed the most distinct decrease - this time, it was as early as at the 4-day mark. Further, a steep decrease in chlorophyll content was observed after 12 days from replanting. Both of these decreases were found to be significant by ANOVA ($p = 3.62 \cdot 10^{-11}$), Tukey HSD test further confirmed significant difference not only between the particular decreases and post-plantation state but also between each of the decreases.

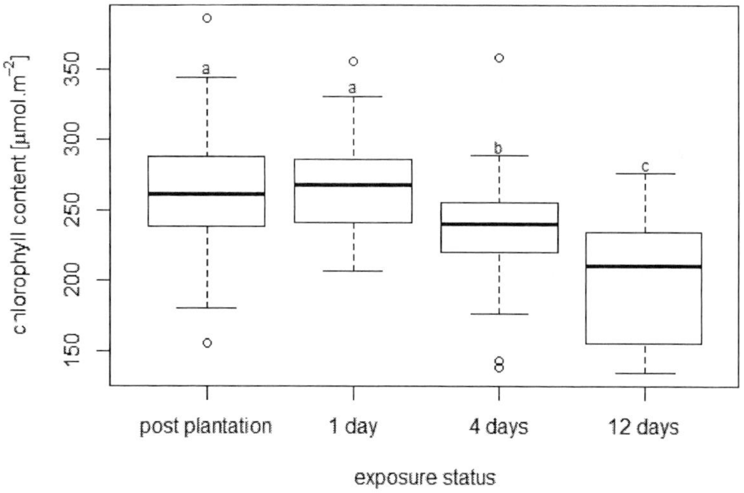

Figure 4 Chlorophyll content in the plants planted on in the nano-TiO_2 - enriched soil. Identical letters above the boxplots denote the non-significant difference between the measurements.

It is apparent that both treatments considerably affected the chlorophyll content in the assessed plants while the exposure via roots (growing on the artificially contaminated substrate) proved to have a more damaging impact on the chlorophyll content than the exposure via leaves. In addition, chlorophyll content - measured using the chosen non-invasive method - was confirmed to be an indicator of the stress induced by the nano-TiO_2 exposure suitable for a real-life assessment of the physiological state of the exposed plant even in such a short period of exposure time.

The significant decrease in chlorophyll content in the plants grown on the NP-enriched soil (even in comparison with plants exposed to nano-TiO_2 via leaves) in combination with the fact that no titanium was found in the above-ground parts of these plants is noteworthy. It indicates that while the plants absorbed enough nanoparticles to negatively affect their photosynthesis, the translocation of the accumulated NPs was low to none. This is of great importance since modelling of environmental concentrations of engineered TiO_2 nanoparticles suggests the soils to be the greatest source of nanoparticles available to plants [9]. It is also consistent with the findings of Asli and Neumann [10] who found that the nanoparticles enter the root hairs of maize (*Zea mays* L.) and adhere to their cell walls, which causes the physical clogging of the root system leading to reduced water and nutrient uptake. Due to the environmentally rather unreasonable concentration of NPs in the substrate, this may be easily the case, however, the specific mechanism of the nano-TiO_2 penetration and accumulation in the roots of cauliflower is yet to be uncovered.

4. CONCLUSION

Nanoparticles of TiO_2 were found to affect the chlorophyll content in the cauliflower plants regardless of whether they have been exposed via roots or leaves. Leaf exposure lead to accumulation of the titanium in the shoots, while root exposure affected the plant physiology both quicker and more severely. On the other hand, transport of the nano-TiO_2 from the roots to the above-ground parts of the plants was not confirmed; hence, the alteration of the plant physiology must take place immediately in the plant root system. Chlorophyll content measured using non-invasive method was proved to be a valuable indicator of the plant stress levels following the exposure to the TiO_2 nanoparticles in both assessed ways of exposure.

ACKNOWLEDGEMENTS

This paper was created in the frame of the SP2018/71 and project while financially supported by the Ministry of Education, Youth and Sports of the Czech Republic, and by the project ID No. LO1404 „Sustainable Development of ENET Center".

REFERENCES

[1] YANG, Jie, CAO, Weidong and RUI, Yukui, Interactions between nanoparticles and plants: Phytotoxicity and defense mechanisms. *Journal of Plant Interactions*. 2017. vol. 12, pp. 158-169.

[2] LEE, Chang W., MAHENDRA, Shaily, ZODROW, Katherine, LI, Dong, TSAI, Yu-Chang, BRAAM, Janet and ALVAREZ, Pedro J. J. Developmental phytotoxicity of metal oxide nanoparticles to *Arabidopsis thaliana*. *Environmental Toxicology and Chemistry*. 2010. vol. 29, pp. 669-675.

[3] RICO, C.M., PERALTA-VIDEA, J.R. and GARDEA-TORRESDEY, J.L. Chemistry, biochemistry of nanoparticles and their role in antioxidant defense system in plants. In: SIDDIQUI, Manzer H., AL-WHAIBI H. Mohammed and MOHAMMAD Firoz, eds. *Nanotechnology and plant sciences*. New York, NY: Springer Berlin Heidelberg, 2015, pp. 1-17.

[4] MINGYU, Su, XIAO, Wu, CHAO, Liu, CHUNGXIANG, Qu, SIAQING, Liu, LIANG, Chen, HAO, Huang and FASHUI, Hong. Promotion of energy transfer and oxygen evolution in spinach photosystem II by nano-anatase TiO_2. *Biological Trace Elemenent Research*. 2007, vol. 119, pp. 183-192.

[5] LEI, Zheng, MINGYU, Su, CHAO Liu, LIANG, Chen, HAO, Huang, XIAO, Wu, XIAOQING, Liu, FAN, Yang, FENGQING, Gao and FASHUI Hong. Effects of nanoanatase TiO_2 on photosynthesis of spinach chloroplasts under different light illumination. *Biological Trace Element Research*. 2007. vol. 119, pp. 68-76.

[6] GAO, Jianguo, XU, Gendi, QIAN, Huanhuan, LIU, Peng, ZHAO, Ping and HU, Yong. Effects of nano-TiO_2 on photosynthetic characteristics of *Ulmus elongata* seedlings. *Environmental Pollution*. 2013. vol. 176, pp. 63-70.

[7] WEI, Chenxi, ZHANG, Yanbo., GUO Jing., HAN, Bing., YANG, Xu and YUAN, Junlin. Effects of silica nanoparticles on growth and photosynthetic pigment contents of *Scenedesmus obliquus. Journal of Environmental Sciences* 2010. vol. 22, pp. 155-160.

[8] URSACHE-OPRISAN, Manuela, FOCANICI, Ecaterina, CREANGA, Dorina and CALTUN, Ovidiuet al () Sunflower chlorophyll levels after magnetic nanoparticle supply. *African Journal of Biotechnology*. 2011. vol.10, pp. 7092-7098.

[9] RICO Cyren M., HONG Jie, MORALES Maria I., ZHAO, Lijuan, BARRIOS, Ana C., ZHANG, Jian-Ying, PERALTA-VIDEA, Jose R. and GARDEA-TORRESDEY, Jorge, L. Effect of cerium oxide nanoparticles on rice: a study involving the antioxidant defense system and in vivo fluorescence imaging. *Environmental Science and Technology*. 2013. vol. 47, pp. 5635-5642.

[10] PARRY, Christopher, BLONQUIST Jr., Mark, J. and BUGBEE, Bruce. *In situ* measurement of leaf chlorophyll concentration: analysis of the optical/absolute relationship. *Plant, Cell and Environment*. 2014. vol. 37, pp. 2508-2520.

[11] R Core Team. R: A language and environment for statistical computing. R Foundation for Statistical Computing, Vienna, Austria. 2018, URL https://www.R-project.org/

[12] GOTTSCHALK, Fadri, SONDERER, Tobias, SCHOLZ, Roland, W. and NOWACK, Bernd. Modeled environmental concentrations of engineered nanomaterials (TiO_2, ZnO, Ag, CNT, fullerenes) for different regions. *Environmental Science and Technology*. 2009. vol. 43, pp. 9216-9222.

[13] ASLI, Sare and NEUMANN, Peter M. Colloidal suspensions of clay or titanium dioxide nanoparticles can inhibit leaf growth and transpiration via physical effects on root water transport. *Plant, Cell and Environment*. 2009. vol. 32, pp. 577-584.

SESSION E
ADVANCED METHODS OF PREPARATION AND CHARACTERIZATION OF NANOMATERIALS

Chairmen

Ing. Pavel JELINEK, Ph.D. Institute of Physics of the CAS, Prague, Czech Republic, EU

Mgr. Petr KLAPETEK, Ph.D. Czech Metrology Institute, Brno, Czech Republic, EU

STABILIZATION OF CADMIUM SULFIDE NANOPARTICLES USING BLOCK-COPOLYMERS

Bernhard SCHUMMER[1], Benedikt SOCHOR[1], Ivan D. POPOV[2], Stefan GERTH[3]

[1]*University Würzburg, Department for Physics and Astronomy, Chair for X-Ray Microscopy, Würzburg, Germany, EU, bernhard.schummer@physik.uni-wuerzburg.de, benedikt.sochor@physik.uni-wuerzburg.de*
[2]*Institute of Solid State Chemistry, Ural Branch of the Russian Academy of Sciences, Ekaterinburg, Russian Federation, idpopov65@gmail.com*
[3]*Fraunhofer Institute for integrated circuits IIS, Development Center X-Ray Technology EZRT, Fürth, Germany, EU, stefan.gerth@iis.fraunhofer.de*

Abstract

This study investigated the stabilization of cadmium sulfide (CdS) nanoparticles in aqueous solutions of the triblock copolymer Pluronic P123. The concentration of the added cadmium sulfide varied between 0.002 and 0.02 M. For the synthesis, sodium sulfide and cadmium chloride were both dissolved in aqueous P123 solutions with concentrations of 23 and 30 weight percent (wtp) of polymers. Both solutions were cooled to low temperatures and mixed in a heated sample cell. The stabilization of CdS nanoparticles was successful for both P123 concentrations. The stabilized nanoparticle-polymer solutions were investigated using ultraviolet-visible spectroscopy (UV-Vis) and small- and wide-angle-x-ray-scattering (SAXS/ WAXS). In the absorption spectra measured by UV-Vis spectroscopy, no shift of the absorption edge, depending of the CdS concentration in solution is visible indicating a bulk like behavior. SAXS showed, that the distance between the Pluronic micelles is not changing in the lyotropic liquid crystal (LC) phase by adding CdS, but a significant rise of the scattering intensity was visible. We assume, that the nanoparticles are located in the micellar core. As proved by WAXS, in some cases even crystalline nanoparticles could be produced. This sample system consisting of polymeric micelles and nanoparticles can be seen as a model for studying the formation of nanoparticles because the crystallization process of the nanoparticles was enlarged form a timescale of nanoseconds to minutes.

Keywords: Pluronic, P123, cadmium sulfide, nanoparticles, SAXS, WAXS, UV-Vis, kinematics, stabilization

1. INTRODUCTION

Cadmium sulfide nanoparticles were studied, because of their size-dependent optical and electronic properties. As an II-VI semiconductor it has a wide band gap in the range from $2.45\ eV$ as bulk material to $3.25\ eV$ as a $3\ nm$ small particle [1]. They can be used in catalysis [2], the purification of water [3], as biological labels [4] or as materials for solar cells [5]. One of the most interesting effect of CdS is the size quantization effect or quantum size effects (QSE). At first it was calculated by Burs [6] and also shown by Schmidt et al. [7]. The calculation is based on the lowest Eigenstates of an electron-hole pair, also called Wannier exciton. The lowest 1S exciton of CdS is about 60 Å in diameter [6]. Lowering the diameter of the nanoparticles from 100 to 40 Å, a shift of $0.43\ eV$ towards higher energies will accrue. Thus, the absorbance will be different for different sized nanoparticles. Since CdS is very hydrophobic, it is not possible to synthesize nanoparticles in aqueous solutions without added stabilizers. Various ways to synthesize CdS nanoparticles and quantum dots can be found in literature [4], [8], [9]. A. Rempel et. al. found a way to synthesize CdS nanoparticles in water with EDTA disodium salt ($C_{10}H_{14}N_2O_8Na_2$, Na_2H_2edta) [10]. They found that the nanoparticle size distribution could be described with a normal distribution function with a maximum at $2.6\ nm$ surrounded by an organic shell of around $5\ nm$.

In our opinion, another promising way to stabilize CdS are polymers, which form micelles with hydrophobic cores. With this method, the properties of the polymers can be used - e.g. the formation of liquid crystals - to modify the structuring of the nanoparticles. A suitable candidate is Pluronic P123, because it is already well investigated and has the required behavior [11]-[14]. At low temperatures (5°C), P123 is present as unimers, elongated chains in an aqueous solution. Reaching the critical mizellization temperature, unimers will form to micelles. In this manner P123 is a promising system to stabilize CdS.

2. SAMPLES AND METHODS

2.1. Samples

Pluronic P123 was purchased from Sigma Aldrich (CAS Number: 9003-11-6) and used as received without any further purification. P123 (PEO_{19}-PPO_{69}-PEO_{19}) is a triblock polyether copolymer consisting of polyethyleneoxid (PEO) and polypropyleneoxide (PPO). Fourier transform infrared studies show that both PEO and PPO chains undergo a dehydration while increasing the temperature. The degree of dehydration is lower for PEO than for PPO [15].

CdS was synthesized with the method of A. Rempel et.al. [10]. There cadmium chloride ($CdCl_2$) and sodium sulfide (Na_2S), each in an aqueous solution with EDTA, have to be mixed to produce nanoparticles. Both were purchased from Sigma Aldrich and used as received (CAS-Numbers: 10108-64-2 and 1313-84-4). Instead of EDTA we used P123 as a stabilizer. Thus, two solutions were prepared. One with water, P123 and $CdCl_2$ (Sol_C) and one with water, P123 and Na_2S (Sol_S). When those two solutions were mixed at the right temperatures and the right concentrations, stable solutions can be realized. We used two concentrations 23 and 30 wtp of P123 and seven concentrations 2, 4, 6, 8, 10, 15 and 20 mM of $CdCl_2$ and Na_2S.

2.2. Methods

In principle, the mixing and stabilization process consists of two steps: cooling the reactants (Sol_C) and (Sol_S) and mixing them in a tempered sample cell. Thus P123 at 30 wtp is forming a liquid crystal, which is gel like, the reactant temperature has to be below the critical crystallization temperature. The reactants were filled in syringes and cooled in a tempered metal block. The liquids were lead through short pips into a Y-connector where they were mixed and the mixed product was lead into a tempered cell after the mixing. This flow through cell was also the sample cell for the SAXS and WAXS measurements and is suitable for vacuum. Before the mixing, the solvents were cooled to 5°C and tempered to 30°C after the mixing to have a defined start for the stabilization of the nanoparticles. For uniformity, the 23 wtp solutions were treated the same way. For the UV-Vis-spectroscopy measurements the sample were premixed in a glass container before transferring them in the designated sample holders.

Small angle x-ray scattering (SAXS) and wide angle x-ray scattering (WAXS) measurements were accomplished at an in-house self-build machine, at the Chair of X-ray Microscopy (University Würzburg), which is optimized for low contrast measurements. The machine is equipped with a rotating anode micro focus x-ray source with a copper target, a paralyzing mirror and slits for a determinable and variable beam shape with a high flux. The sample can be tempered between 2 °C and 70 °C. The sample detector distance (SDD) is variable between 5cm and 3.5 m with a resulting Q-range of 0.005 to 5 $Å^{-1}$. The detector is a Dectris EIGER 1M. The complete system operates in vacuum to minimize scattering and absorption of air. All data sets were reduced with Nika a macro for Igor Pro [16].

Optical absorption spectra were recorded by a FS-5 spectrofluorometer from Edinburgh Instruments, at the Institute of Solid State Chemistry (Ural Branch of the Russian Academy of Sciences, Eekaterinburg) in the UV and visible ranges. The device is equipped with continuous Xe-lamp (450 W) and silicon photodiode for transmission measurements. Quartz cuvettes with a optical light paths of 0.5 mm were used. The reference sample for measurements was an aqueous solution of P123.

3. RESULTS

To choose the right temperatures and conditions for the stabilization and mixing process, pure aqueous solutions of P123 with a concentration of 30 wtp were precisely examined. At 10 °C unimers start to form micelles, which can be seen in the **Figure 1**. The scattering curve of unimers can be described as an exponential decay. At 17°C the micelles can be seen in the scattering curve with a strong form factor. Heating the solution above 20 °C a liquid crystal is formed, which can be seen by the peaks of the red curve. The solution forms a gel. Above 37 °C the slope changes again to a less ordered system. The black curve has no distinct peaks due to the peak broadening. Also the gels start to liquefy again.

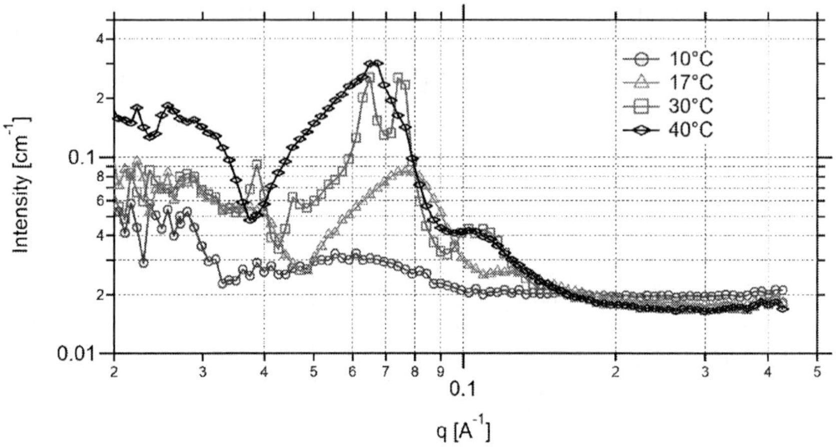

Figure 1 SAXS measurements of Pluronic P123 with a concentration of 30 wtp at different temperatures. Each measurement was integrated for 2h.

The two reactants solutions (see 2.1) were cooled down to 5 °C and mixed in a heated capillary at 30 °C. CdS was stabilized with this method, proved in **Figure 2**. On the left side (**Figure 2a**) various CdS concentrations are shown starting from 0.002M on the right up to 0.02M on the left. The gels have a light yellow color at low concentrations changing to orange for higher concentrations, which are stable for days. In **Figure 2b** shows pure CdS in water mixed in a capillary. An orange separated phase arises due to the high hydrophobicity of CdS.

Figure 2 Exaples of stabilized CdS nanoparticles from 0.002M rigth and 0.02M left (a). The capilliry of the sample cell is shown in (b-f), where (b) is filled CdS in pure water, and (c-f) shows the stabilization of CdS with P123 rigth after the mixing of both solutions (c) until five minutes after the mixing (f).

Figures 2c-f depict the formation process of a 0.005M CdS solution at different times. Until a stable color settles, 5 minutes passed. No phase separation is visible. P123 with an CdS concertation of 0.005 and 0.01M were analyzed with SAXS and WAXS. The SAXS scattering pattern are shown in **Figure 3**. The peaks corresponding to the LC structure for pure P123 and with added CdS do not shift (see **Table 1**). The peaks were fitted with a gaussian function. By adding CdS to P123, a significant higher scattering intensity is visible. Aqueous solutions of P123 with a concentration of 23 wtp were also used to produce CdS nanoparticles and measured with SAXS. Those samples show the same color as CdS in a 30 wtp solution of P123, but 23 wtp do not form lyotropic LC. Those samples were able to stabilize CdS, but have been still liquid. To study the optical properties and thus to draw conclusions about the nanoparticle sizes UV-Vis Spectroscopy measurements of different CdS concentrations were conducted (see **Figure 4a**). The absorbance increases with raising the concentration of CdS. The region between 450 and 500 nm is interesting to determine the bandgap of the CdS nanoparticles. For this purpose an analysis of the absorption spectra was carried out using current concepts of semiconductor physics developed for bulk materials [17]. According to Tauc theory the spectra indicate a direct optical transition [18]. Using the spectral curves, the absorption edge or band gap is specified as the intersection of the first linear segment of the absorption curve with the x-axis [18]. The resulting bandgap does not change within the error with changing concentration of CdS (see **Figure 4b**). For both concentrations of CdS also WAXS measurements were conducted, to characterize the crystallinity of the produced CdS nanoparticles. For 0.005M, no reflexes were visible. However, for a concentration of 0.01M CdS stabilized in 30wtp P123 very narrow single reflexes are visible and can be assigned to circles (see **Figure 5**). The reflexes are marked with white rings.

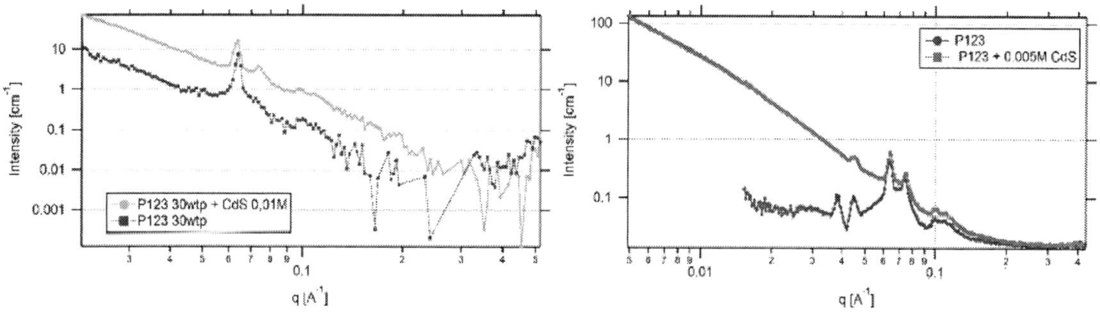

Figure 3 SAXS measurements of pure P123 and with added CdS with concentrations of 0.005M on the left and 0.01M on the right. Note that on the left side water was not subtracted, where water is subtracted from both data sets in the right graph.

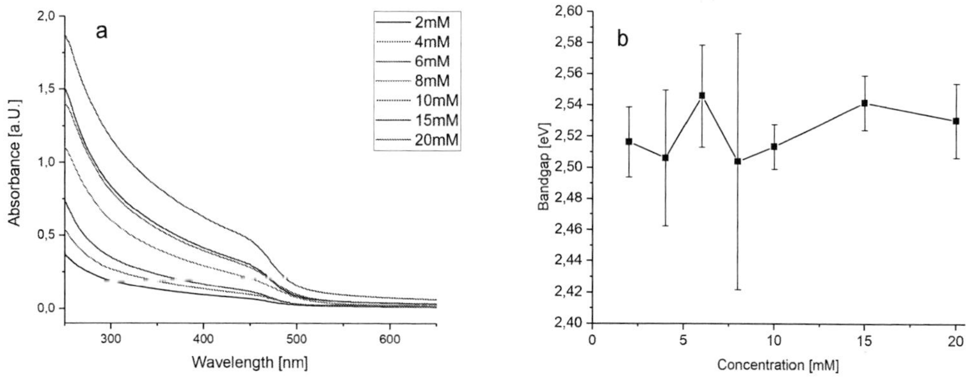

Figure 4 UV-Vis measurments of 30 wtp P123 and different concentrations of CdS on the left (a) and the calculated bandgaps with Tauc theory [20] on the right (b).

Table 1 Positions of the structure peaks of pure P123 with a concentration of 30 wtp and added Cds

Composition	1. Peak [1/Å]	2. Peak [1/Å]
P123	0.063730 ± 0.000046	0.07402 ± 0.00027
P123 + CdS	0.063980 ± 0.000016	0.07492 ± 0.00023

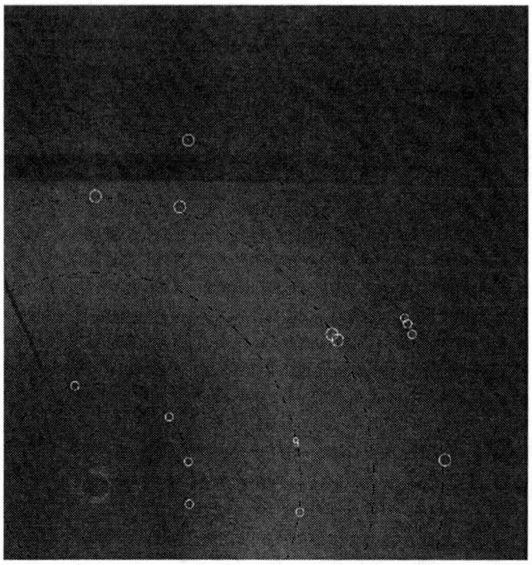

Figure 5 WAXS measurement of 0.01M CdS stabililized with 30wtp P123. Reflexes, which are located on rings were marked with white rings

4. DISCUSSION

Having in mind that CdS and PPO of P123 are distinct hydrophobic (see 2.1), we assume, that the stabilization process is driven by a hydrophobic-hydrophobic interaction. The SAXS data for pure 30wtp P123 and for added CdS show, that the inter micellar distance, represented by the peaks, does not change. This is an indication that there was no change in the size of the micelles. The succinct raise of the intensity is caused by an amplification of the scattering contrast due to the presence of CdS. According to this interpretation, the CdS could be found either in the micelle or between the micelles stabilized by the lyotropic LC. Since the liquid 23wtp P123 solutions were able to stabilize CdS and no precipitation of CdS was visible, the latter option must be dismissed. Considering that the shell of the micelle is usually more hydrated then the core, we assume that the CdS is stabilized inside the micellar core. Furthermore, this confirms the hydrophobic-hydrophobic interaction. It was not possible to determine the size of the CdS nanoparticles with SAXS. In this regard, UV-Vis measurements were conducted to draw conclusions about the size of the nanoparticles. The measurements showed that CdS nanoparticles display the same behavior as bulk CdS, with a bandgap of $E_G = 2.45\ eV$. By varying the concentration of CdS, no correlation to the quantum size effect could be determined. The reason for this could be that the number of nanoparticles larger than $60Å$ in diameter is much higher than for particles with smaller diameters showing a quantum size effect. This indicates a broad size distribution. However, note that the mixing procedure of UV-Vis and SAXS was not identical, due to experimental requirements. WAXS analyzes the crystallinity of the CdS nanoparticles. The optical absorption spectra prompt the formation of a CdS phase. WAXS measurements of 0.01M CdS in 30wtp P123 showed sharp single crystal reflexes. Although the exact crystal structure cannot be determined out of the date, it clearly shows that the crystallinity is temperature dependent. By controlling the temperature, it is possible to influence the micellization and formation of LC. Thus, it is possible to control the kinematics of the crystallization process of cadmium sulfide. Additionally the timescale for the crystallization is enlarged from the timescale of nanoseconds [19] towards minutes, which is a deceleration of seven orders of magnitude in time.

5. CONCLUSION

It is possible to stabilize hydrophobic CdS nanoparticles with Pluronic P123. The structure of 30wtp P123´s lyotropic LCs does not change upon adding CdS, but it's scattering intensity rises significantly. Also CdS could be stabilized with a liquid P123 solution, proving that the nanoparticles are located inside the micelles. Due to the lower hydration level of core compared to the shell, the nanoparticles should be stabilized by the core, which confirms a hydrophobic-hydrophobic interaction. UV-Vis measurements yield mostly nanoparticles bigger than 60Å, because of a bulk-like behavior of the nanoparticle's bandgap. With WAXS it was proved, that it is possible to produce crystalline CdS nanoparticles. First investigations of the kinematical process show an enlargement of the crystallization timescales form nanoseconds to minutes.

ACKNOWLEDGEMENTS

We want to thank Robert Luxenhofer for the possibility to use the lab at the chair of chemical technologies and material synthesis (University Würzburg) and Michael Lübtow for the support in the lab. We also thank Dectris Ltd for supporting this work. This work was partially carried out in accordance with the scientific and research plans and state assignment if the ISSC UrB RAS (no. AAAA-A16-116122810208-8).

REFERENCES

[1] R. ROSSETTI, J. L. ELLISON, J. M. GIBSON, and L. E. BRUS, "Size effects in the excited electronic states of small colloidal CdS crystallites," *J. Chem. Phys.*, vol. 80, no. 9, pp. 4464-4469, 1984.

[2] J. C. COLMENARES, R. LUQUE, J. M. CAMPELO, F. COLMENARES, Z. KARPIŃSKI, and A. A. ROMERO, "Nanostructured photocatalysts and their applications in the photocatalytic transformation of lignocellulosic biomass: An overview," *Materials (Basel).*, vol. 2, no. 4, pp. 2228-2258, 2009.

[3] H. WEIB, A. FERNANDEZ, and H. KISCH, "Electronic Semiconductor - Support Interaction-A Novel Effect in Semiconductor," *Angew. Chemie - Int. Ed.*, vol. 40, no. 20, pp. 3825-3827, 2001.

[4] M. BRUCHEZ Jr., "Semiconductor Nanocrystals as Fluorescent Biological Labels," *Science (80-.).*, vol. 281, no. 5385, pp. 2013-2016, 1998.

[5] P. SHARMA, S. BROWN, G. WALTER, S. SANTRA, and B. MOUDGIL, "Nanoparticles for bioimaging," *Adv. Colloid Interface Sci.*, vol. 123-126, no. SPEC. ISS., pp. 471-485, 2006.

[6] L. E. BRUS, "Electron-electron and electron-hole interactions in small semiconductor crystallites: The size dependence of the lowest excited electronic state," *J. Chem. Phys.*, vol. 80, no. 9, pp. 4403-4409, 1984.

[7] H. M. SCHMIDT, U. KOCH, S. BARAL, W. KUNATH, K. WEISS, and E. DIEMAN, "Photochemistry of colloidal semiconductors. onset of light absorption as a function of size of extremely small cjs particles *," *Chem. Phys. Lett.*, vol. 124, no. 6, pp. 0-3, 1986.

[8] C. B. MURRAY, D. J. NORRIS, and M. G. BAWENDI, "Synthesis and Characterization of Nearly Monodisperse CdE (E = S, Se, Te) Semiconductor Nanocrystallites," *J. Am. Chem. Soc.*, vol. 115, no. 19, pp. 8706-8715, 1993.

[9] Y. YIN and A. P. ALIVISATOS, "Colloidal nanocrystal synthesis and the organic-inorganic interface," *Nature*, vol. 437, no. 7059, pp. 664-670, 2005.

[10] A. A. REMPEL', N. S. KOZHEVNIKOVA, and S. V. REMPEL', "Structure of cadmium sulfide nanoparticle micelle in aqueous solutions," *Russ. Chem. Bull.*, vol. 62, no. 2, pp. 398-402, Dec. 2013.

[11] P. ALEXANDRIDIS and T. A. HATTON, "Poly(ethylene oxide)poly(propylene oxide)poly(ethylene oxide) block copolymer surfactants in aqueous solutions and at interfaces: thermodynamics, structure, dynamics, and modeling," *Colloids Surfaces A Physicochem. Eng. Asp.*, vol. 90, pp. 1-46, 1995.

[12] K. MORTENSEN, "Structural studies of aqueous solutions of PEO - PPO - PEO triblock copolymers, their micellar aggregates and mesophases; a small-angle neutron scattering study," *J. Phys. Condens. Matter*, vol. 8, no. 25A, pp. A103-A124, Jun. 1996.

[13] Y. SASANUMA, "Conformational analysis of poly (propylene oxide) and its model compound 1, 2-dimethoxypropane," *Macromolecules*, vol. 28, no. 25, pp. 8629-8638, 1995.

[14] Z. ZHOU and B. CHU, "Light-Scattering Study on the Association Behavior of Triblock Polymers of Ethylene Oxide and Propylene Oxide in Aqueous Solution Light-Scattering Measurements," *J. Colloid Interface Sci.*, vol. 126, no. 1, 1988.

[15] C. GUO, H.-Z. LIU, and J.-Y. CHEN, "A Fourier transform infrared study of the phase transition in aqueous solutions of Ethylene oxide-propylene oxide triblock copolymer," *Colloid Polym. Sci.*, vol. 277, no. 4, pp. 376-381, Apr. 1999.

[16] J. ILAVSKY, "Nika: Software for two-dimensional data reduction," *J. Appl. Crystallogr.*, vol. 45, no. 2, pp. 324-328, 2012.

[17] K. SEEGER, *Semiconductor Physics*. Springer, Berlin, Heidelberg, 2004.

[18] J. TAUC, "Absorption edge and internal electric fields in amorphous semiconductors," *Mat. Res. Bull.*, vol. 5, no. 4, pp. 721-730, 1970.

[19] A. SCHIENER *et al.*, "In situ investigation of two-step nucleation and growth of CdS nanoparticles from solution," *Nanoscale*, vol. 7, no. 26, pp. 11328-11333, 2015.

MICROSTRUCTURE REFINEMENT IN MARTENSITIC CREEP RESISTANT STEEL VIA SEVERE PLASTIC DEFORMATION

Petr KRAL[1*], Jiri DVORAK[1], Vaclav SKLENICKA[1], Yoichi TAKIZAWAA[2,3,4], Takahiro MASUDA[2,3], Zenji HORITA[2,3], Kveta KUCHAROVA[1], Marie KVAPILOVA[1], Marie SVOBODOVA[5]

[1]Institute of Physics of Materials, Academy of Sciences of the Czech Republic, Brno, Czech Republic, EU
[2]Department of Materials Science and Engineering, Faculty of Engineering, Kyushu University, Fukuoka, Japan
[3]WPI, International Institute for Carbon-Neutral Energy Research (WPI-I2CNER), Kyushu University, Fukuoka, Japan
[4]Technology Department, Nagano Forging Co., Ltd, Nagan, Japan
[5]UJP PRAHA a.s., Prague, Czech Republic, EU

Abstract

In this study, the effect of severe plastic deformation on grain size refinement and mechanical properties of a coarse-grained steel was investigated. The experimental material used in the present investigation was advanced tungsten modified 9 %Cr P92 steel. The coarse-grained state of P92 steel was deformed at room temperature by high-pressure sliding or high-pressure torsion. The microstructure was investigated using Tescan Lyra 3 scanning electron microscope equipped with an electron-back scatter unit. Severe plastic deformation at room temperature led to grain size reduction down to sub-microscopic level. The grain size and hardness exhibited significant changes to the equivalent strain of about 10, and slight changes between the equivalent strains of 10 - 20. Equivalent plastic strains higher than 20 only led to insignificant changes in the mean grain size and hardness. The creep behaviour of the ultrafine-grained state exhibited the minimum creep rate of about two orders of magnitude greater when compared to the coarse-grained P92 steel.

Keywords: Steel, ultra-fine grained materials, creep, electron back scatter diffraction

1. INTRODUCTION

Tempered martensitic 9-12 % chromium steels are widely used for high-temperature components of steam power plants. The creep resistance of these steels is particularly given by the microstructure containing free dislocations and low energy boundaries, such as low angle grain boundaries and martensitic lath boundaries pinned by precipitates [1,2]. Their microstructure stability is significantly reduced during long-term operation by recovery of tempered martensite, coarsening and/or dissolving of primary precipitates and precipitation of new secondary phases.

Bulk ultra-fine grained and nanostructured materials can advantageously be produced by the methods of severe plastic deformation (SPD) [3-6]. The most attractive method is high pressure torsion (HPT) refining the structure of metallic materials to the ultra-fine (<1 μm) or even nano (<100 nm) scale [7-10]. The grain size of HPT-processed materials is so small that new low-angle grain boundaries are not formed in the grain interior. Thus the grain boundaries limit the mean free dislocation path and the dislocations easily recover at high-angle boundaries. Recent works [11,12] showed that grain size reduction to the microscopic level and high number of high-angle grain boundaries (HAGB) increase the contribution of boundary-mediated processes to the total creep deformation. Also, creep resistance was found to be reduced by high density of mobile dislocations in P91 steel after ageing for 1 week at 973 K and subsequent deformation by 1 ECAP pass at an elevated temperature [13]. The aim of this work is to study the influence of the application of severe plastic deformation on microstructure and creep behaviour of ultrafine-grained P92 steel.

2. EXPERIMENTAL MATERIAL AND PROCEDURES

The used experimental material was advanced tungsten modified 9 %Cr P92 steel. The chemical composition and heat treatment of the as-received state are given elsewhere [14]. Discs of 30 mm diameter and 1.1 mm thickness and sheets with dimensions of 10x100x1.1 mm^3 were cut from as-received P92 steel. Further, the discs were processed by 1 rotation HPT at room temperature under the pressure of 6 GPa and rotation speed of 0.1 mm per second. The sheets were processed by high-pressure sliding (HPS) at room temperature with the sliding distance of 15 mm under the pressure of 4 GPa. The value of von Mises equivalent strain during HPT is proportional to the radius of the disc according to the following equation [5]:

$$\varepsilon_{eq} = 2\pi rN/\sqrt{3}t \quad (1)$$

where r is the distance from the disc centre, N is the number of turns and t is the thickness of disc. The value of von Mises equivalent strain during HPS process was estimated using the following equation [15]:

$$\varepsilon_{eq} = x/\sqrt{3}t \quad (2)$$

where x is the sliding distance of plunger with respect to anvil and t is the thickness of the sample.

Constant load tensile creep tests were conducted at 873 K and 150 MPa. The testing was performed in protective argon atmosphere using flat specimens with the gauge length of 10 mm and cross section of 3 x 1 mm^2. The centre of the gauge length of tensile HPT-processed specimen was situated 7.5 mm from the central part of the disc. The UFG creep specimen was thus manufactured from the disc region with the equivalent strain of about 20-30. The gauge length of the HPS tensile specimen was taken from the central part of the sheet. The equivalent strain of the HPS specimen was about 7.9.

Microstructure investigations were performed using scanning electron microscope Tescan Lyra 3 equipped with NordlysNano EBSD detector operating at the accelerating voltage of 20 kV with specimen tilted at 70°. The electron back scatter diffraction (EBSD) analyses of UFG P92 steel were performed in the metallographic section perpendicular to the rotation axis of the disc and in the section parallel to the anvil. The step size of 25 nm for the EBSD maps was selected with respect to the inter-boundary spacing. The EBSD data were analyzed using HKL Channel 5 software developed by Oxford Instruments. The high-angle grain boundaries (HAGBs) were characterized as boundaries with misorientation angle $\theta > 15°$ and low-angle grain boundaries (LAGBs) as boundaries with $\theta < 15°$. Standard intercepts counting of LAGBs and HAGBs were carried out along tests lines. The value of the inter-boundary spacing was determined as the arithmetic average of the results in two perpendicular directions (horizontal, vertical).

3. EXPERIMENTAL RESULTS AND DISCUSSION

3.1. Microstructure after severe plastic deformation

Figure 1 shows the microstructure of P92 after 1 HPT revolution at room temperature in various positions from the center of the HPT disc. The microstructure acquired ~ 0.5 mm from the centre of HPT disc contained about 55 % of HAGBs (**Figures 1a and 1b**). The equivalent strain of about 1.6 caused reduction of the mean grain size down to 0.57 µm. However the microstructure was not homogeneous and contained the mixture of fine and large grains exceeding 1 µm. The further increase in plastic strain up to 23 led to additional reduction of grain size down to about 0.15 µm. The microstructure acquired ~ 7 mm from the centre of HPT disc exhibited nearly random distribution of misorientation angles (**Figures 1c and 1d**). The results demonstrate that the microstructure homogeneity increases with increasing value of the plastic strain occurring during HPT. Nevertheless, the microstructure was not fully homogeneous even after ε_{eq}=23 (r = 7 mm).

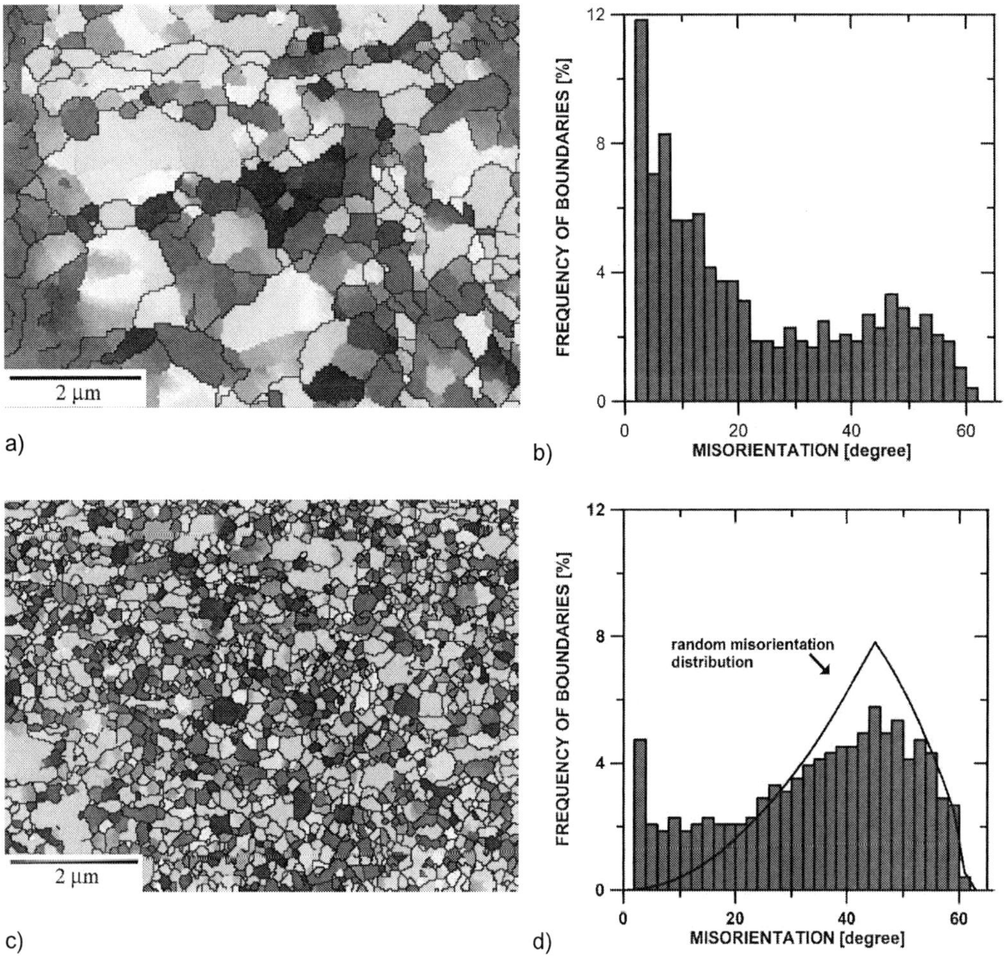

Figure 1 Microstructure and misorientation distribution of P92 steel after HPT deformation a, b) r = 0.5 mm (ε_{eq}~1.6), c, d) r = 7 mm (ε_{eq}~23)

Figure 2 shows the microstructure of the specimen processed by 1 HPS pass taken from the rear (**Figures 2a and 2b**) and middle (**Figures 2c and 2d**) parts of the sheet. Microstructure in the middle part of the sheet contained grains with the mean size of about 0.3 µm. In the microstructure about ~ 77 % of HAGBs was estimated. EBSD analyses revealed that the microstructure of the HPS-processed sheet was not fully homogeneous. The microstructure observed in the edges was different in comparison with the microstructure situated in the middle part.

The different microstructure at the edge can be the consequence of the material flow - excess material at the front part and lack of material at the rear part [15]. The results demonstrate that plastic strain of about 1.6 led to the formation of ultra-fine grained microstructure with high number of HAGBs. It means that the boundaries of tempered martensite in the as-received state with the preferential misorientations near Σ 3 (111/60°) [16] were replaced by random HAGBs. The further increase in plastic strain led to the formation of more or less homogeneous structure with almost random grain misorientation distribution. Similar microstructures have been found after large strains in regions of microstructure refinement saturation [17].

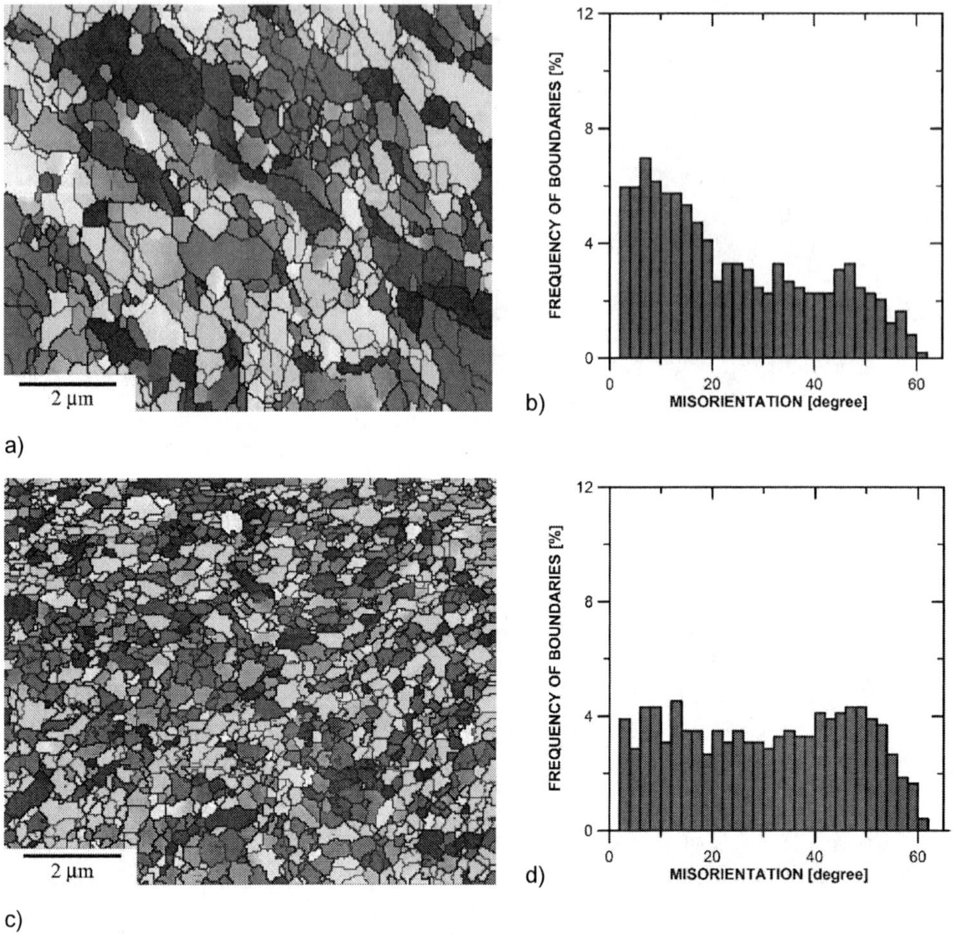

Figure 2 Microstructure and misorientation distribution of P92 steel after HPS deformation a,b) rear part, c,b) middle part (ε_{eq}~7.9)

3.2. Mechanical properties of P92 steel after severe plastic deformation

Figure 3 shows the dependence of average hardness on equivalent strain measured after HPT and HPS. The results show that hardness changes significantly to the strain of about 10. The further increase in plastic strain to 10 - 20 only caused slight additional increase in hardness. Equivalent plastic strains higher than 20 led to insignificant changes in hardness. The results also show hardness saturation at large strains. Thus, present observation is consistent with the results published in other works [17,18]. The microstructure refinement saturation often begins at strains between 10 and 30. It means that the grain size and HAGBs frequency do not change with further deformation. The comparison of creep curves acquired for as-received state of P92 steel and specimens processed by HPT and HPS are shown in **Figure 3b**. Inspection of **Figure 3b** revealed that the minimum creep rate becomes faster and the time to fracture becomes shorter with increasing plastic strain introduced into the microstructure during HPT and HPS. The faster creep rate observed in HPT and HPS processed specimens is related to the grain refinement and can be explained by boundary-mediated processes such as enhanced storage and recovery of dislocations at HAGBs and grain-boundary sliding [11,12]. The slight differences in creep behaviour found in HPT and HPS processed specimens could be caused by larger grains found in HPS processed specimen. Thus the lower frequency of HAGBs measured in HPS processed microstructure is available for boundary-mediated processes.

a) 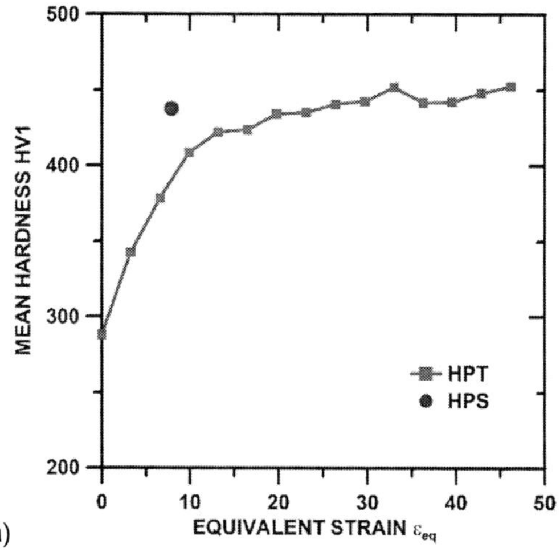

b) 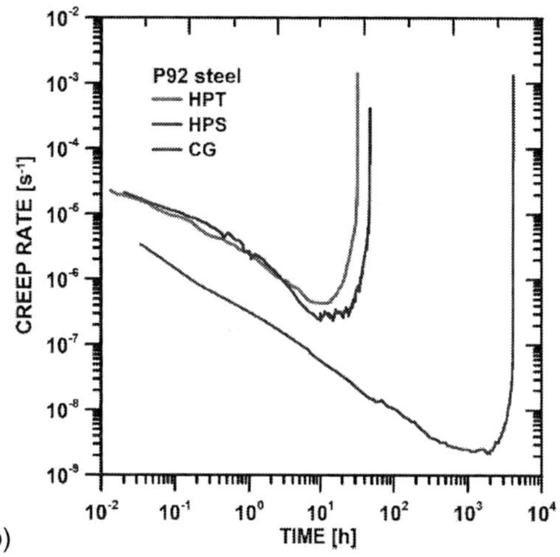

Figure 3 Mechanical properties of P92 steel processed by HPT and HPS deformation
a) hardness at RT, b) creep behaviour at 873 K

4. CONCLUSIONS

Microstructure investigations showed that HPT and HPS at room temperature reduced grain size down to submicrometer level. P92 steel processed by HPT and HPS exhibited higher hardness at room temperature and faster minimum creep rate in comparison with coarse-grained P92 state. It is suggested that the change of mechanical properties in UFG steel is influenced by HAGBs. They strengthen UFG state at room temperature serving as the obstacles to dislocation glide but may soften the UFG material through boundary mediated processes such as enhanced storage and recovery of dislocations at HAGBs or grain-boundary sliding at elevated temperatures.

ACKNOWLEDGEMENTS

We acknowledge the financial support provided by the Czech Science Foundation under the project no. 16-09518S. This work was also supported by a Grant-in-Aid from the MEXT, Japan, for Scientific Research (S) (No.26220909). The HPT process was carried out in the International Research Center on Giant Straining for Advanced Materials (IRC-GSAM) at Kyushu University, Japan.

REFERENCES

[1] ABE, F., KERN,T-U., Viswanathan, R. *Creep-resistant steels*, Woodhead Publishing: Sawston, Cambridge, UK, 2008.

[2] ORLOVA, A., BURSIK, J., KUCHAROVA, K., SKLENICKA, V. Microstructural development during high temperature creep of 9 % Cr steel. *Mater. Sci. Eng. A*, 1998, vol. 245, pp. 39-48, https://doi.org/10.1016/S0921-5093(97)00708-9.

[3] LUKÁČ, P., KOCICH, R. GREGER, M., PADALKA, O., SZARAZ, Z. Microstructure of AZ31 and AZ61 Mg alloys prepared by rolling and ECAP. *Kovove Materialy*, 2007, vol. 45, pp. 115-120.

[4] KOCICH, R., GREGER, M., MACHÁČKOVÁ, A. Finite element investigation of influence of selected factors on ECAP process. *Metal 2010: 19th international metallurgical and materials conference*, Roznov pod Radhostem, Czech Republic, 2010, pp. 166-171.

[5] VALIEV, R.Z., ISLAMGALIEV, R.K., ALEXANDROV, I.V. Bulk nanostructured materials from severe plastic materials. *Prog. Mater. Sci.*, 2000, vol. 45, pp. 103-189, https://doi.org/10.1016/S0079-6425(99)00007-9.

[6] KOCICH, R., KUNČICKÁ, L., KRÁL, P., MACHÁČKOVÁ, A. Sub-structure and mechanical properties of twist channel angular pressed aluminium. *Mater. Character.*, 2016, vol. 119, pp. 75-83,

[7] FURUKAWA, M., HORITA, Z., NEMOTO, M., LANGDON, T.G. The use of severe plastic deformation for microstructural control. *Mater. Sci. Eng. A,* 2002, vol. 324, pp. 82-89, https://doi.org/10.1016/S0921-5093(01)01288-6.

[8] KUNČICKÁ, L., KOCICH, R. Comprehensive characterisation of a newly developed Mg-Dy-Al-Zn-Zr alloy structure. *Metals*, 2018, vol. 8. pp. 73. https://doi.org/10.3390/met8010073.

[9] TAKIZAWA, Y., KAJITA, T., KRAL, P., MASUDA, T., WATANABE, K., YUMOTO, M., OTAGIRI, Y., SKLENICKA, V., HORITA, Z. Superplasticity of Inconel 718 after processing by high-pressure sliding (HPS). *Mater. Sci. Eng. A*, 2017, vol. 682, pp. 603-612, http://dx.doi.org/10.1016/j.msea.2016.11.081.

[10] KUNČICKÁ, L., LOWE, T.C., DAVIS, C.F., KOCICH, R., POHLUDKA, M. Synthesis of an Al/Al2O3 composite by severe plastic deformation. *Mater. Sci. Eng. A*, 2015, vol. 646, pp. 234-241. doi:10.1016/j.msea.2015.08.075.

[11] BLUM, W., LI, Y.J., ZHANG, Y., WANG, J.T. Deformation resistance in the transition from coarse-grained to ultrafine-grained Cu by severe plastic deformation up to 24 passes of ECAP. *Mater. Sci. Eng. A*, 2011, vol. 528, pp. 8621-8627, https://doi.org/10.1016/j.msea.2011.08.010.

[12] KRAL, P., DVORAK, J., BLUM, W., KUDRYAVTSEV, E., ZHEREBTSOV, S., SALISHCHEV, G., KVAPILOVA, M., SKLENICKA, V. Creep study of mechanisms involved in low-temperature superplasticity of UFG Ti-6Al-4V processed by SPD. *Mater. Character.,* 2016, vol. 116, pp. 84-90, http://dx.doi.org/10.1016/j.matchar.2016.04.007.

[13] KOSTKA, A., TAK, T-G., EGGELER, G. On the effect of equal-channel angular pressing on creep of tempered martensite ferritic steels. *Mater. Sci. Eng. A*, 2008, vol. 481-482, pp. 723-726, https://doi.org/10.1016/j.msea. 2007.02.154.

[14] KRAL, P., DVORAK, J., SKLENICKA, V., MASUDA, T., HORITA, Z., KUCHAROVA, K., KVAPILOVA, M., SVOBODOVA, M. The Effect of Ultrafine-Grained Microstructure on Creep Behaviour of 9 % Cr Steel. Materials 11, 2018, Issue 5, No. 787, pp. 117-118, doi:10.3390/ma11050787.

[15] FUJIOKA, T., HORITA, Z., Development of High-Pressure Sliding Process for Microstructural Refinement of Rectangular Metallic Sheets. *Materials Transactions*, 2009, vol. 50, No. 4, pp. 930 - 933.

[16] SKLENICKA, V., KUCHAROVA, K., KRAL, P., KVAPILOVA, M., SVOBODOVA, M., CMAKAL, J. The effect of hot bending and thermal ageing on creep and microstructure evolution in thick-walled P92 steel pipe. *Mater. Sci. Eng. A,* 2015, vol. 644, pp. 297-309.

[17] PIPPAN, R., SCHERIAU, S., TAYLOR, A., HAFOK, M., HOHENWARTER, A., BACHMAIER, A. Saturation of Fragmentation during Severe Plastic Deformation. *Annu. Rev. Mater. Res.*, 2010, vol. 40, pp. 319-343.

[18] PIPPAN, R., WETSCHER, F., HAFOK, M., VORHAUER, A., SABIROV, I. The Limits of Refinement by Severe Plastic Deformation. Adv. Eng. Mater., 2006, vol. 8, No. 11, pp. 1046-1056.

LOW TEMPERATURE SYNTHESIS OF SILVER NANOPARTICLES

Libor KVÍTEK, Adéla BÁRTOVÁ, Robert PRUCEK, Aleš PANÁČEK, Jana SOUKUPOVÁ, Radek ZBOŘIL

Palacký University in Olomouc, RCPTM, Olomouc, Czech Republic, EU, libor.kvitek@upol.cz

Abstract

The influence of lower temperature on size and polydispersity of silver nanoparticles prepared by reduction of the silver(I) diamine complex cation by sodium borohydride was studied. The preparation was carried out in ethylene glycol and ethanol in the presence of stabilizing agents. Silver nanoparticles were characterized by UV-Vis spectroscopy, dynamic light scattering (DLS) and transmission electron microscopy (TEM). The best results were achieved in ethanol. The smallest nanoparticles were prepared at the reaction temperature of -5 °C (3,4±1,7 nm). With decreasing temperature, the average size of silver nanoparticles increased up to 7.2±2.9 nm (-25 °C).

Keywords: Silver, nanoparticles, low temperature, preparation

1. INTRODUCTION

Present period of intense progress of nanotechnology brings a new life to materials, which have been used for centuries - materials, which enabled establishment of a significant branch of physical chemistry - the colloid chemistry. Colloid metal particles or currently more often labelled as nanoparticles always played an important role in technology, especially in the process of glass and ceramics dyeing [1], where gold colloids were of primary importance. The last quarter of the 20th century was characteristics with the renaissance of interest in metal colloid particles, especially due to the discovery of surface enhanced Raman scattering (SERS) [2], which initiated rapid development of the novel, highly sensitive spectroscopic technique enabling studies of single molecule [3]. The interest in silver nanoparticles (Ag NPs) as antibacterial agent is also worth mentioning as growing resistance of bacteria against antibiotics proves to be a key obstacle for the medicinal practice in the future [4]. Silver is probably the most frequently studied metal in the area of nanomaterial chemistry from the point of view of both unique physicochemical properties and area of application. However, the task of the preparation of silver nanoparticles with desired properties is far away from being completely solved.

Contemporary intense development of „wet processes“ commonly used in the Ag NPs preparation is closely connected with its SERS application. This area was pioneered by Creighton, Blatchford a Albrecht who prepared the silver particles using the reduction of $AgNO_3$ with a strong reducing agent ($Na[BH_4]$) in ice cooled aqueous medium [5]. The resulting nanoparticles were in the diameter from 5 - 20 nm with the absorption maximum („surface plasmon“) located at approx. 400 nm. The second, from already well-established reduction processes, is known as Lee and Meisel's process which uses weaker reducing agent, the citrate anion [6]. The citrate anion has additional function to stabilize silver nanoparticles in the course of their formation [7]. The reduction process is completed after boiling the reaction mixture for one hour; the prepared nanoparticles are significantly bigger and of higher polydispersity than nanoparticles prepared using borohydride as the reduction agent. Their diameters range from 30 to 120 nm, their surface plasmon is located at approx. 420 nm and the stability of the dispersion is limited, when compared to the borohydride process. Therefore, borohydride reduction method offers more reliable fabrication of small and stable Ag NPs. However, reproducibility of this process is lower and strongly dependent on the good laboratory experience. The aspect of reproducibility can be positively influenced using appropriate stabilizers. However, the addition of a modifier in the course of the reduction can result in a certain increase in the particle diameter (when compared to the original process) [8]. Therefore, the idea of the lower temperature application could represent a solution how to increase reproducibility of the reduction process with the result of production of particles under 10 nm in diameter. Unfortunately, conducting of the reduction in aqueous environment is limited with the freezing point of water. Lowering the reaction temperature under 0°C could be achieved only with anti-freezing mixture of water with alcohols. This mixture and also pure alcohols are often used as non-aqueous environment for Ag NPs production. The alcohols are weak reduction agents. Therefore, they can be used directly for production of Ag NPs without utilization of any other reduction agent, especially under elevated temperature. Polyols are used in this way for preparation of the Ag NPs. The stability of the nanoparticles in non-aqueous environment is low and suitable stabilizers should be used for prevention of the aggregation. Typically, Ag particles prepared in ethyleneglycol at 180°C are due to aggregation bigger than 1 μm. This particle diameter, however, steeply decrease to 50-80 nm when polyvinylenepyrrolidone (PVP) is used as polymeric stabilizer [9]. This size could be additionally lowered down to 10 nm when ammonia was used as a complexing agent [10]. Also simple alcohols can be used directly as reducing agent. Typically, ethanol can reduce Ag^+ directly to Ag NPs and on the contrary to the polyols the reduction process can be conducted at laboratory temperature. of course, utilization of the stabilizers as PVP is necessary also in this case [11,12]. Nanoparticles with diameters below 10 nm can be produced by following the same preparative process when some specific stabilizers are applied. Ag NPs in diameter range from 5 up to 100 nm can be produced via solvothermal synthesis in ethanol and under the support of dodecanethiol stabilization (unfortunately, this compound strongly inhibited the surface activity of the Ag NPs) [13]. Application of stabilizers from the group of non-ionic surfactants (Brij, AOT, Tween) is connected with synthesis of Ag NPs with size under 10 nm (6-10 nm) performed under laboratory temperature [14], and the application of latex microparticles as stabilizers in combination with microwave synthesis resulting in fabrication of Ag NPs in the diameter range 5-10 nm (note: the system exhibit high polydispersity) [15]. The smallest Ag NPs were prepared using aminosilanes as stabilizers - the diameter range was 1-3 nm under elevated temperature (30-60°C). Several studies were also conducted in ethanol environment using e.g. hydrazine as a reducing agent to produce stable Ag NPs (stabilized by PVP) but the obtained particles were in the diameter range 50-70 nm [16]. Also studies employing photons as initiator of the reduction process were published - γ-radiation proved to be more effective (Ag NPs fabricated were in the diameter range 15-35 nm) [17] than visible light (diameter in the range 20-100 nm) [18].

In spite of the number of the above mentioned studies, none of them was aimed on utilization of temperature lower than 0°C as a key condition for a reproducible production of truly small (under 10 nm) Ag NPs. Therefore, the study of the influence of the temperature under the freezing point of water became the main task of the presented study.

2. EXPERIMENTAL

Silver nanoparticles were prepared via the reduction of silver(I) diamine complex cation $[Ag(NH_3)_2]^+$ using sodium borohydride as a reducing agent. 50 % v/v ethanol was used as reaction medium to have possibility conduct the reaction down to -25 °C. 5 mL of 5 mmol/L aqueous solution of $AgNO_3$ (p.a., Sigma Aldrich) was mixed in a baker with 5 mL of 25 mmol/L ethanolic solution of ammonia (p.a., Sigma Aldrich) and 2.5 mL of 1 % w/w aqueous solution of polyacrylic acid (M_r=1,200, Sigma Aldrich). The reaction mixture was cooled down to reaction temperature in cryostat after adding additional ethanol (7.5 mL, UV purity, Sigma Aldrich) and water (4 mL). After the adjustment of the temperature, the reaction was started through the injection of 1 mL of precooled 5 mmol/L aqueous solution of $NaBH_4$ (p.a., Sigma Aldrich) using injection rate 0.2 mL/min. Although the reaction rate decreased with decreasing temperature, it was anyway completed after 10 min from the injection of the reducing agent which was indicated by stable yellow color of the reaction mixture. The reaction mixture was stirred using external stirrer with glass propeller the whole reaction time. All used solutions were prepared with deionized water (18 MΩ·cm, Millipore).

After completing of reaction the reaction mixture was taken out from cryostat and was kept freely on the air to reheat it to the room temperature. Then the prepared dispersion of nanoparticles was characterized using UV-vis spectroscopy (Specord S 600, Analytik Jena AG, Germany) in 1 cm cuvette after 5 times diluting by water, dynamic light scattering (DLS) on Zetasizer Nano ZS (Malvern Co., UK), and transmission electron microscopy (TEM) on JEM 2010 (Jeol, Japan) instrument.

Catalytic activity of the prepared nanoparticles was evaluated using a model reaction based on the reduction of methylene blue (MB) by $NaBH_4$. The reaction course was monitored via the UV-vis spectroscopy thanks to the observation of the intensive absorption peak of the oxidized MB form at the wavelength of 664 nm. The catalytic experiments were performed directly in the spectroscopic cuvette under laboratory temperature. 1 mL of 10 μmol/L aqueous solution of MB in ammoniacal buffer (pH=10) was pipetted to the cuvette with 10 μL of the prepared nanoparticle suspension and 1 mL of water. The reaction was started through a quick injection of 1 mL of 0.5 mmol/L aqueous solution of $NaBH_4$. During reaction course, spectra of the MB in the interval 360-760 nm were acquired in the regular time intervals. The evaluation of the kinetic data was performed using values of maxima absorbance MB from the recorded spectra using approximation of the 1. order kinetic equation.

3. RESULTS AND DISCUSSION

The used preparation method for Ag NPs is based on the previously mentioned Creighton, Blatchford a Albrecht borohydride method in combination with Tollens procedure where silver(I) diamine complex cation is used as silver source instead simple silver cation. Additionally, polymeric stabilizer (polyacrylic acid) was used as adjuvant in reaction mixture to avoid a problem produced Ag NPs aggregation in mixed environment water-ethanol (1:1). Thanks to low freezing point of this solvents mixture the preparation procedure could be proceeded at temperatures down to freezing point of pure water. In our experiments we have tested three reaction temperatures: -5, -15 and -25 °C. Lowering of the reaction temperature was accompanied with decreasing of the reaction rate of the studied redox reaction. The first change of colour of the reaction mixture (it is colorless) was observed immediately after mixing of reaction component at temperature -5 °C. At -15 °C the first colour change was observed approximately after 30 s from the point of mixing of reaction components and at -25 °C this time was prolonged to about 60 sec. This change in reaction rate is probably a reason for change of diameter of the emerged Ag NPs as was observed primarily from DLS measurements (see **Table 1**) and consequently also from TEM pictures (**Figure 1**). The real size distribution of diameter of the prepared Ag NPs was evaluated from these TEM micrographs and their graphical representations are presented on **Figure 2**. Average values of diameters of the prepared Ag NPs at different reaction temperatures evaluated from the size distribution are presented with values of polydispersity in **Table 1**. The extraordinary difference between average values of Ag NPs diameters determined form DLS and from TEM are clearly visible in **Table 1**. The abnormally big diameters determined from DLS are very probably connected with presence of polymeric stabilizer which can interconnect prepared Ag NPs to weakly bonded bigger objects which can be stable only in liquid environment and not on dry microscopic grid. The existence of this bigger object is clearly visible also from the UV-vis spectra at which two absorption maxima are observed. The first located below 400 nm is connected with localized surface plasmon of individual particles while emergence of second absorption maxima at about 450 nm is caused by existence of weakly interconnected nanoparticles. Therefore, the real diameters can be evaluated only from TEM micrographs. And from them a clear dependence on the reaction temperature is visible. Lowering of reaction temperature results in growing of average diameters of the prepared Ag NPs. The reason of this dependency is clearly connected with the change of the reaction rate in initiation part of reaction curve. The reduction procedure for preparation of metal nanoparticles is typical example of the autocatalytic reaction. The reaction curve for this type of reaction has a typical S shape. At the start of reaction nuclei of new metal phase are formed by slow homogeneous mechanism. However, when first stable nuclei are formed the second reaction phase of quick catalytic growth of the particles is rapidly getting on. Homogeneous mechanism formation of stable nuclei is strongly temperature dependent due to high activation energy. Therefore, with lowering temperature less nuclei are formed at the initiation period and final diameter of Ag NPs is bigger due to large amount of silver ions which are left for reduction on individual nuclei during quick catalytic step. It is also important to note improvement of polydispersity of the prepared disperse systems of Ag NPS with lowering of reaction temperature which is also very probably caused by the slowdown of the reaction rate in the initiation step of the whole autocatalytic reaction. Due to high reaction rate in catalytic step of the reaction mechanism silver cations are quickly exhausted and first initiation step is suppressed by this way. When new nuclei are not formed during reaction steps after initiation period polydispersity of the emerging system is improved.

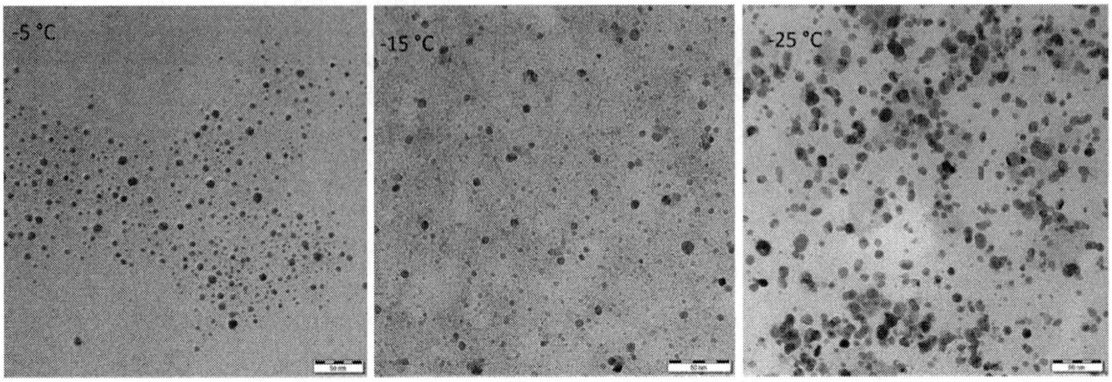

Figure 1 TEM micrographs of Ag NPs prepared by reduction $Ag[NH_3]_2^+$ by borohydride in the presence of PAA as stabilizer at different temperatures below freezing point of water

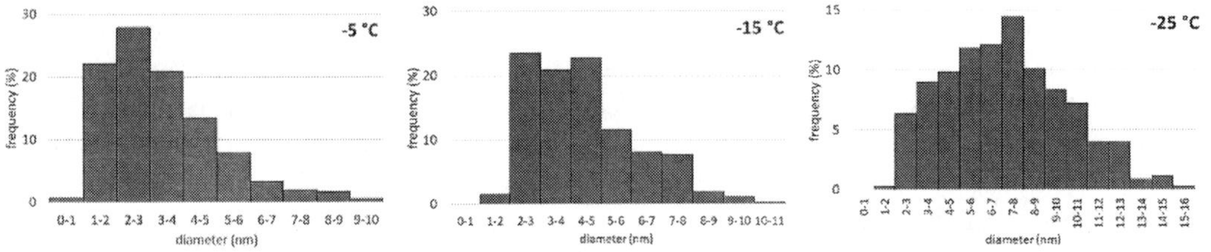

Figure 2 Distribution diagrams determined from TEM for Ag NPs prepared by reduction $Ag[NH_3]_2^+$ by borohydride in the presence of PAA as stabilizer at different temperatures below freezing point of water

Very small silver nanoparticles prepared via the newly proposed method can have promising properties, especially thanks to its perspective catalytic activity. As it can be concluded from numerous published catalytic studies, aimed on the evaluation of catalytic activity of the metal nanoparticles, the smaller the nanoparticles are, the high catalytic activity can be expected. The simple catalytic experiments using model reaction of reduction methylene blue (MB) with borohydride were conducted to confirm this conclusion also in the case of so small nanoparticles as were prepared in this study. In course of the reaction, the spectrum of MB was taken at certain time intervals and after the reaction end the values of the absorbance at maximum for MB were evaluated and these obtained kinetic data (**Figure 3**) were processed using approximation of kinetics of the first order (big excess of borohydride in comparison with MB). The reference measurement of the used model reaction was made without Ag NPs as the catalyst - in this case the apparent rate constant was 0.0035 s^{-1}, which is significantly lower (minimally one order) in comparison with the lowest apparent rate constant for the slowest reaction catalyzed with the prepared Ag NPs (see values of k_{aoo} in **Table 1**). The obtained kinetic data confirm growing catalytic activity of the metal nanoparticles with decreasing of their diameter. In our case, this conclusion is valid for the Ag NPs with diameter 3.4 nm, which one approximatively contents 10^3 Ag atoms.

Therefore, for Ag NPs, containing minimally 10^3 silver atoms, the main rule about the dependence of catalytic activity of Ag NPs on their diameter is fulfilled.

Table 1 The values of average diameters and maxima of absorbance from UV-vis spectra for Ag NPs prepared by reduction $Ag[NH_3]_2^+$ by borohydride in the presence of PAA as stabilizer at different temperatures below freezing point of water

T [°C]	TEM		DLS		UV-vis		k_{app} [s^{-1}]
	diameter [nm]	polydispersity	diameter [nm]	polydispersity	1st peak [nm]	2nd peak [nm]	
- 5	3.4±1.7	0.250	115	0.189	342.5	451.5	0.1236
-15	4.4±1.8	0.167	59	0.083	344.0	445.0	0.0541
-25	7.2±2.9	0.162	50	0.022	344.5	446.0	0.0415

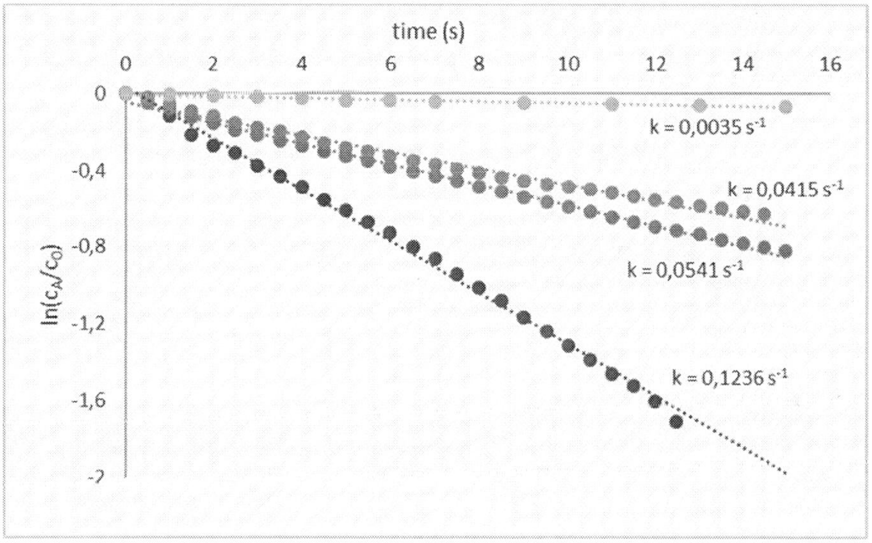

Figure 3 Kinetic curves for the reduction of methylene blue with borohydride catalyzed with the Ag NPs prepared at different reaction temperatures via the reduction of $Ag[NH_3]_2^+$ with borohydride in the presence of PAA as stabilizer

4. CONCLUSION

The reported study confirmed a possibility to tailor of the diameter and polydispersity of the silver nanoparticles prepared via a reduction method due to the change of reaction temperature, i.e. decrease of the reaction temperature below the freezing point of water. The average diameter of the prepared Ag NPs increased with the decrease of the reaction temperature and simultaneously polydispersity is improved due to the slowed down reaction rate in initiation step of nuclei formation. The smallest prepared nanoparticles with average diameter 3.4 nm are the best redox catalyst as proved with the used model reaction of reduction methylene blue by borohydride, which is in accordance with other studies aimed at evaluation of influence of the nanocatalyst size on its catalytic activity.

ACKNOWLEDGEMENTS

The authors are grateful for the support by the projects LO1305 of the Ministry of Education, Youth and Sports of the Czech Republic.

REFERENCES

[1] BOBIN, O., SCHVOERER, M., NEY, C., RAMMAH, M., PANNEQUIN, B., PLATAMONE, E. C., DAOULATLI, A. and GAYRAUD, R. P. The role of copper and silver in the colouration of metallic luster decorations. *Color Research and Application*. 2003. vol. 28, no. 5, pp. 352-359.

[2] FLEISCHMANN, M., HENDRA, P. J. and MCQUILLAN, A. J. Raman-Spectra of Pyridine Adsorbed at a Silver Electrode. *Chemical Physics Letters*. 1974. vol. 26, no. 2, pp. 163-166.

[3] NIE, S. M. and EMERY, S. R. Probing single molecules and single nanoparticles by surface-enhanced Raman scattering. *Science*. 1997. vol. 275, no. 5303, pp. 1102-1106.

[4] PANACEK, A., SMEKALOVA, M., VECEROVA, R., BOGDANOVA, K., RODEROVA, M., KOLAR, M., KILIANOVA, M., HRADILOVA, S., FRONING, J. P., HAVRDOVA, M., PRUCEK, R., ZBORIL, R. and KVITEK, L. Silver nanoparticles strongly enhance and restore bactericidal activity of inactive antibiotics against multiresistant Enterobacteriaceae. *Colloids and Surfaces B-Biointerfaces*. 2016. vol. 142, no. pp. 392-399.

[5] CREIGHTON, J. A., BLATCHFORD, C. G. and ALBRECHT, M. G. Plasma Resonance Enhancement of Raman-Scattering by Pyridine Adsorbed on Silver or Gold Sol Particles of Size Comparable to the Excitation Wavelength. *Journal of the Chemical Society-Faraday Transactions Ii*. 1979. vol. 75, no. pp. 790-798.

[6] LEE, P. C. and MEISEL, D. Adsorption and Surface-Enhanced Raman of Dyes on Silver and Gold Sols. *Journal of Physical Chemistry*. 1982. vol. 86, no. 17, pp. 3391-3395.

[7] PILLAI, Z. S. and KAMAT, P. V. What factors control the size and shape of silver nanoparticles in the citrate ion reduction method? *Journal of Physical Chemistry B*. 2004. vol. 108, no. 3, pp. 945-951.

[8] PEDROZA-TOSCANO, M. A., LOPEZ-CUENCA, S., RABELERO-VELASCO, M., MORENO-MEDRANO, E. D., MENDIZABAL-RUIZ, A. P. and SALAZAR-PENA, R. Silver Nanoparticles Obtained by Semicontinuous Chemical Reduction Using Carboxymethyl Cellulose as a Stabilizing Agent and Its Antibacterial Capacity. *Journal of Nanomaterials*. 2017. vol. no.

[9] MOUDIR, N., MOULAI-MOSTEFA, N. and BOUKENNOUS, Y. Silver Micro- and Nano-Particles Obtained Using Different Glycols as Reducing Agents and Measurement of Their Conductivity. *Chemical Industry & Chemical Engineering Quarterly*. 2016. vol. 22, no. 2, pp. 227-234.

[10] ZHAO, T., SUN, R., YU, S. H., ZHANG, Z. J., ZHOU, L. M., HUANG, H. T. and DU, R. X. Size-controlled preparation of silver nanoparticles by a modified polyol method. *Colloids and Surfaces a-Physicochemical and Engineering Aspects*. 2010. vol. 366, no. 1-3, pp. 197-202.

[11] AMRAOUI, M., DAOUDI, C. and REMRAM, M. Preparation and Characterization of Silver Nanospheroids: Theoretical and Experimental Approaches. *Photonics Letters of Poland*. 2017. vol. 9, no. 2, pp. 63-65.

[12] WU, W. J., WU, M. Z., SUN, Z. Q., LI, G., MA, Y. Q., LIU, X. S., WANG, X. F. and CHEN, X. S. Morphology controllable synthesis of silver nanoparticles: Optical properties study and SERS application. *Journal of Alloys and Compounds*. 2013. vol. 579, no. pp. 117-123.

[13] TIAN, C. T., MAO, B. D., WANG, E. B., KANG, Z. H., SONG, Y. L., WANG, C. L., LI, S. H. and XU, L. One-step, size-controllable synthesis of stable Ag nanoparticles. *Nanotechnology*. 2007. vol. 18, no. 28.

[14] LIZ-MARZAN, L. M. and I, LADOTOURINO. Reduction and stabilization of silver nanoparticles in ethanol by nonionic surfactants. *Langmuir*. 1996. vol. 12, no. 15, pp. 3585-3589.

[15] IDER, M., ABDERRAFI, K., EDDAHBI, A., OUASKIT, S. and KASSIBA, A. Rapid Synthesis of Silver Nanoparticles by Microwave-Polyol Method with the Assistance of Latex Copolymer. *Journal of Cluster Science*. 2017. vol. 28, no. 3, pp. 1025-1040.

[16] SHAO, W. F., LI, G., ZHU, P. L., ZHANG, Y., OUYANG, Q. L., SUN, R., CHEN, C. H. and WONG, C. P. Facile synthesis of low temperature sintering Ag nanopaticles for printed flexible electronics. *Journal of Materials Science-Materials in Electronics*. 2018. vol. 29, no. 6, pp. 4432-4440.

[17] LIU, F. K., HSU, Y. C., TSAI, M. H. and CHU, T. C. Using gamma-irradiation to synthesize Ag nanoparticles. *Materials Letters*. 2007. vol. 61, no. 11-12, pp. 2402-2405.

[18] SHAO, K. and YAO, J. N. Preparation of silver nanoparticles via a non-template method. *Materials Letters*. 2006. vol. 60, no. 29-30, pp. 3826-3829.

SYNTHESIS OF ULTRA SMALL SILVER NANOPARTICLES USING Sn^{2+}/Sn^{4+} REDOX COUPLE AS STRONG REDUCTANT

Lucie HOCHVALDOVÁ, Libor KVÍTEK, Aleš PANÁČEK, Robert PRUCEK, Radek ZBOŘIL

RCPTM, Palacky University in Olomouc, Olomouc, Czech Republic, EU, lucie.hochvaldova01@upol.cz

Abstract

Application potential of colloidal particles and nanoparticles is related to their size, shape and other physical and chemical properties. Those properties can be influenced by various ways of synthesis, while the presence of reducing agent and modifiers (polymers, surfactants) are the most vital ones. One of the strongest reducing agents is tin with the redox couple Sn^{II}/Sn^{IV} and standard potential -0.384 V. This redox system allows to change redox potential just by addition of hydrogen peroxide to the tin(II) solution, which causes its partial oxidation to tin (IV) and changes the ratio between its oxidized and reduced form. Those changes effect redox potential (which is measured by potentiometer) and influent the tailored silver nanoparticles preparation. Antibacterial activity of prepared nanoparticles, with an average size between 3 to 50 nm, were tested against pathogenic and resistant bacteria, while the smallest nanoparticles have showed high antibacterial activity with minimal inhibition concentration (MIC) in units of mg/L.

Keywords: Silver nanoparticles, synthesis, redox potential, stabilization, antibacterial activity

1. INTRODUCTION

The beginning of the 21st century is the era of nanotechnologies. [1] It is already known that many properties are changing significantly with the size decrease to nanometre scale, different shape and polydispersity of the nanoparticles. Not every size of the nanoparticles is optimal for a specific use. For instance, antibacterial [2] and catalytic activity [3] is increasing with decreasing size of the particles. [4] On the other hand, small nanoparticles are not as effective at analytical applications in biosensors, which are using interesting optical properties of nanoparticles. [5] Another reason why we need different sizes of nanoparticles is Raman spectroscopy, where each laser wavelength requires a different size for ideal enhancement of analyte´s Raman signal. [6]

Thanks to different size requirements of the silver nanoparticles, a big part of research is focused on tailored methods for silver particle preparation. The most commonly used methods for metal nanoparticles synthesis are those based on reduction of soluble salts. [7] Generation of nanoparticles is associated with creation of new phase in originally homogenous system, which is related to saturation. Saturation depends on the speed of the reaction which is directly connected to the value of redox potentials. [8] Therefore, the change of redox potentials can be used to influence reaction and creation of silver nanoparticles. Unfortunately, with common reducing agents - $NaBH_4$ [9], citrate [10], hydrazine [11], hydroxylamine [12] or sugars [13] redox potential cannot be simply changed. On the other hand, with redox system based on redox pair Sn^{2+}/Sn^{4+} redox potential can be simply changed by the change of ratio of both redox form of tin.

In this paper the influence of redox potential on the size of silver nanoparticles has been studied. Defining the change of redox potential of commonly used reducing agents is difficult, which is why reduction system based on tin(II) chloride has been used. By change of the ratio tin(II)/tin(IV) ions in the system, redox potential has been changed, which influenced the reduction of silver ions. First, the synthesis of small nanoparticles prepared by reduction of silver ions by tin(II) chloride has been optimized. The antibacterial activity of the system has been tested and eventually removal of the access of tin compounds from the system has been studied.

2. EXPERIMENTAL SECTION

Silver nitrate $AgNO_3$ (Fagron), polyacrylic acid (PAA) Mr 8000 (45 % w/v, Aldrich), ammonia NH_3 (28-30 %, Sigma Aldrich), tin(II) chloride $SnCl_2$ (Sigma Aldrich), hydrogen peroxide H_2O_2 (30 %, Penta) and sodium hydroxide NaOH (Lach: Ner) were used without any additional purification. Deionized water (18 MΩ·cm, Millipore) was used to prepare all solutions.

Zetasizer Nano ZS (Malvern, UK) and transmission electron microscope (TEM) JEM 2010 (JEOL, Japan, 160 kV) have been used to determine size and polydispersity of prepared nanoparticles. All UV/VIS spectra have been measured on spectrophotometer Specord S 600 (Analytik Jena AG, Germany). Concentration of silver and tin in the dispersion has been measured by atomic absorption spectroscopy (contrAA300, Analytik Jena, Germany), which allowed us to recalculate obtained absorption of the sample on its concentration. Calibration solution with concentrations of 10, 20, 30 and 40 mg/l of tin have been used for tin and concentrations of 0.25; 0.5; 1; 1.5 a 2 mg/l of silver have been used for silver. Nanoparticles have been prepared with continuous steering on electromagnetic stirrer and redox potential with pH of the solution has been measured by pH 2700 instrument (Geotech Instruments, USA). Centrifugation has been done on centrifuge (Eppendorf Centrifuge 5702) and for redispergation of silver nanoparticles has been used ultrasound generator CL-334 20100097 (Q Sonica sonificators).

3. RESULTS AND DISCUSSION

3.1. Preparation of silver nanoparticles

Colloidal dispersion of silver nanoparticles was prepared by modified Tollens method using tin(II) chloride as a reducing agent. Stock solutions were in a particular order and amount (shown in **Table 1**) added into baker, which was cooled with ice and placed on electromagnetic stirrer. Total amount of the reaction mixture always ended up to 25 ml, while distilled water was used to fill up the missing volume.

Table 1 The amount of the stock solutions (ml) used for preparation of 1-5 samples of silver nanoparticles

Stock solution	Concentration (mol/l)	1	2	3	4	5
$AgNO_3$	0.005	5	5	5	5	5
PAA 8000	0.008	5	5	5	5	5
NH_3	0.05	5	5	5	5	5
$SnCl_2$	0.25	1	1	1	1	1
H_2O_2	0.05	0	1	1.5	2	3

Diamine-silver complex $[Ag(NH_3)_2]^+$, prepared by reaction of silver nitrate with ammonia, was reduced by tin(II) chloride (which was pre-modified by addition of hydrogen peroxide **(Table 1)**. Peroxide was added to the tin(II) chloride solution to partially oxidize tin(II) ions to tin(IV), which affected the ratio of oxidised and reduced form and influenced redox potential of the system and consequently the reduction of silver ions by itself.

After addition of tin(II) chloride to the reaction mixture, the colour of dispersion has changed which indicated reduction of silver ions to silver nanoparticles. System prepared in this way was not stable by itself, particles started to aggregate and sediment in a while. To ensure the stability of the whole system polyacrylic acid (PAA Mr 8000) has been added to the solution.

3.2. Redox potential of the system

Redox potentials of each dispersion were measured by potentiometer. The values of the potential have changed with the change of the ratio tin(II) and tin(IV), which has been caused by addition of hydrogen peroxide to the tin(II) solution. **(Table 1)** More hydrogen peroxide was added the higher amount of tin(II) was oxidised to tin(IV), which has led to the increase in redox potential **(Table 2)** and decrease of its reduction strength.

The size of the prepared silver nanoparticles was determined from TEM images **(Figure 1)**. Three of the images for each sample were evaluated via program Gwyddion which in invert value mode allows to measure the size of the nanoparticles automatically. If the contrast of the nanoparticles was not good enough to ensure correct recording of the size, nanoparticles were measured manually, while all the nanoparticles presented on the image were measured. Minimally 200 particles were accounted in determination of the distribution diagram of the prepared nanoparticles.

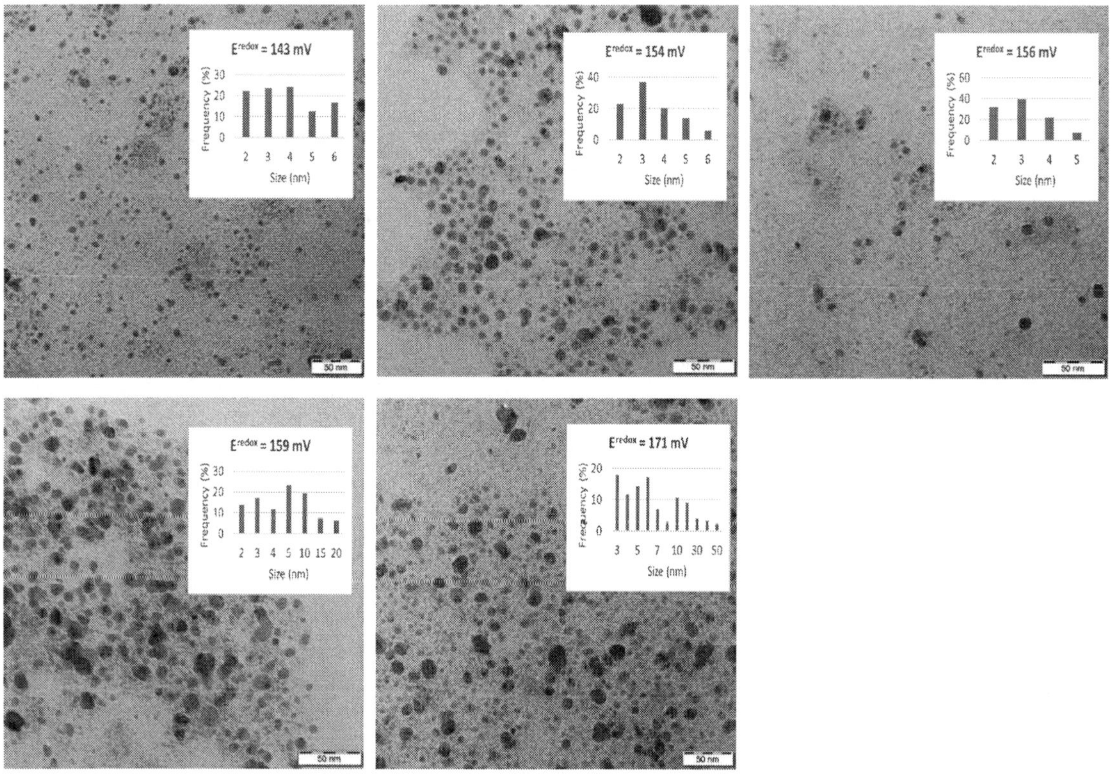

Figure 1 TEM images of silver nanoparticles prepared in systems with different redox potential (distribution diagrams of the prepared nanoparticles are presented in insets)

Mean diameter ($\overline{d}$) of the particles, corrected sample standard deviations (σ), and polydispersity (PDI) were calculated from size distributions presented on **Figure 1** and all the values are shown in **Table 2.**

Table 2 Average size of the nanoparticles and their polydispersity for systems prepared at different redox potentials

Sample	Redox potential (mV)	Size (nm)	Standard deviation (nm)	PDI
1	143	3.8	1.4	0.13
2	154	3.3	1.2	0.14
3	156	2.9	1.0	0.12
4	159	3.7/13.6 (bimodal)	1.2/10.9	0.10/0.65
5	171	4.9/23.5 (bimodal)	1.4/23.9	0.09/1.03

As you can see in **Table 2**, small addition of hydrogen peroxide to the reducing agent does not oxidise huge amount of tin(II), so in this case tin keeps its strong reducing properties and forms small nanoparticles with sizes in the range from 2 to 6 nm. As higher amounts of hydrogen peroxide are added, higher amount of tin(II) is oxidised to tin(IV) and tin is losing its reducing properties and bigger particles are formed. After addition 2 or 3 ml of hydrogen peroxide system becomes polydisperse (bimodal) and next to the small particles (smaller than 10 nm), big nanoparticles in 10ths of nm are formed. This change of the size caused by different addition of hydrogen peroxide proves the dependency of particle size on redox potential of the system, which is controlled by different addition of hydrogen peroxide. **(Table 2)**

3.3. Tin removal

The last part of this research was focused on tin removal from the system. Previously described silver nanoparticles dispersion has contained gel-like precipitate of oxide-hydroxide tin compounds. More tin (IV) was present in the system, the higher amount of precipitate has been formed. At the same time the concentration of the reducing agent (tin(II) chloride) was 10 times higher than concentration of silver therefore many methods have been tested to ensure proper tin removal. The best way how to remove tin from the system has been alkalization of the system which causes aggregation of silver nanoparticles and at the same time soluble tin salts are formed. **(Reaction scheme 1-3)**

$$SnCl_2(aq) + 2\ NaOH\ (aq) \rightarrow SnO \cdot H_2O\ (s) + 2\ NaCl\ (aq) \quad (1)$$

$$SnO \cdot H_2O\ (s) + NaOH\ (aq) \rightarrow NaSn(OH)_3\ (aq) \quad (2)$$

$$SnCl_4(aq) + 4NaOH\ (aq) \rightarrow Sn(OH)_4\ (aq) + 4NaCl\ (aq) \quad (3)$$

The system has been alkalized by addition of NaOH until its final concentration reached concentration 0.24 mol/l. This concentration has been established within partial study focused on influence NaOH concentration on tin compounds separation from silver nanoparticles. After the precipitation, the solid part has been separated from the liquid via centrifuge (5 min, 4,400 revolutions per minute). Afterwards the liquid part has been removed and the solid one has been diluted in the same volume of distilled water with PAA and sonicated for 1 minute with Q Sonica sonicators (E=2.062J). Nanoparticles purified this way stays thanks to PAA stable and their size stays unchanged. Concentration of tin in the system after alkalization has been determined by atomic absorption spectroscopy. The result has shown that only one alkalization is not sufficient enough to remove all tin compounds from the system and it needs to be repeated. After the second alkalization the concentration of the tin has been lower than detection limit of the device. Furthermore, with each alkalization there has been higher losses of silver and system becomes less stable. In table below are shown percentage decreases (%) of tin and silver **(Table 3)**. From these data is apparent that 2 alkalizations have not caused big decrease in silver concentration and more than 97 % of original amount of tin has been removed.

Table 3 Percentage representation of tin and silver content in silver nanoparticles dispersion after individual steps of alkalization

Sample	Tin			Silver			
	1 x alkalization		2x alkalization	1 x alkalization		2x alkalization	
	Sn (%)	σ (%)	Sn (%)	Ag (%)	σ (%)	Ag (%)	σ (%)
1	20.9	9.2	≤ 2.8	97.7	1.2	92.5	3.7
2	18.1	6.7		98.1	1.9	91.7	1.0
3	15.4	2.2		97.4	2.9	95.0	0.0
4	12.9	3.3		98.5	1.0	91.1	3.7
5	8.0	1.5		95.7	3.6	61.9	0.8

3.4. Biological activity of the prepared silver nanoparticles

Antibacterial activity has been tested by microdilution method, which establish minimum inhibitory concentration (MIC). MIC determines minimal concentration, which inhibits the growth of the bacteria. The testing has been carried out on microtiter plates, while all samples have been diluted with 100 μl MH broth by geometric series from 2 to 128 times. Bacterial strain shown in **Table 4** have been injected (concentration 10^6 ml^{-1} CFU (colony forming units)) and incubated for 24 hours (37 °C) and then tested for their antibacterial activity.

Table 4 MIC of silver nanoparticles against gram-positive and gram-negative bacteria

Bacteria	MIC (mg/l)				
	0 ml H_2O_2 (3.8 nm)	1 ml H_2O_2 (3.3 nm)	1,5 ml H_2O_2 (2.9 nm)	2 ml H_2O_2 (3.7/13.6 nm)	3 ml H_2O_2 (4.9/23.5 nm)
Enterococcus faecalis CCM 4224	6.75	3.38	3.38	6.75	13.5
Staphylococcus aureus CCM 3953	1.69	1.69	1.69	3.38	6.75
Escherichia coli CCM 3954	1.69	1.69	0.84	3.38	3.38
Pseudomonas aeruginosa CCM 3955	3.38	3.38	3.38	6.75	13.5

Table 4 provides evidence that silver nanoparticles synthesized with smaller addition of hydrogen peroxide (0-1.5 ml) have higher antibacterial activity than those synthesized with two or more millilitres. This phenomenon is related to the size of the nanoparticles, while for small addition of hydrogen peroxide only small particles were present, which have caused an increase of their antibacterial activity. On the other hand, for higher addition of hydrogen peroxide, bigger particles in tenths of nanometres have been formed as well, which eventually caused slight decrease in antibacterial activity of the system. Values for gram-positive (Enterococcus faecalis, Staphylococcus aureus) and gram-negative (Escherichia coli, Pseudomonas aeruginosa) bacteria are similar, which means silver nanoparticles have on them the same effect.

4. CONCLUSION

The synthesis of silver nanoparticles prepared by modified Tollens method using reduction of diamine-silver complex by tin compounds was studied in this paper. Reduction systems containing tin(II) chloride as a reducing agent allows easy modification of redox potential. Hydrogen peroxide was used to oxidise tin(II) to tin(IV), which changed the ratio of oxidised and reduced form. Additionally, it caused the change of redox potential, which allowed us to study its influence on the size of prepared silver nanoparticles. With a small additions of hydrogen peroxide were prepared small particles with size ranging from 2 to 6 nm. Whereas with addition of two or more millilitres were prepared bigger particles with bimodal distribution. In this case system has contained small particles which were smaller than 6 nm and also bigger particles with a size ranging in tens of nanometres. Antibacterial activity of prepared nanoparticles has been tested against pathogenic and resistant bacteria, while the smallest nanoparticles (prepared by addition of 0-1.5 ml H_2O_2) have shown high antibacterial activity against gram-positive and gram-negative bacteria with minimal inhibitory concentration (MIC) in units of mg/L.

The second aim of this work was to find a method for tin compounds removal (reducing agent) from the system. Many methods were tested, but the most efficient one was alkalization of the system by concentrated sodium hydroxide, which has led to the precipitation of silver nanoparticle´s and dissolution of tin compounds. The liquid phase was easily separated after centrifugation, and after twice repeated alkalization more than 97 % of tin was removed from silver nanoparticle dispersions.

ACKNOWLEDGEMENTS

The authors are grateful for the support by the projects LO1305 of the Ministry of Education, Youth and Sports of the Czech Republic.

REFERENCES

[1] HULLA, J. E., SAHU, S. C. & HAYES, A. W. Nanotechnology: History and future. *Hum. Exp. Toxicol.* 34, 1318-1321 (2015).

[2] PANÁČEK, A., KVÍTEK, L., PRUCEK, R. Silver colloid nanoparticles: Synthesis, characterization, and their antibacterial activity. *J. Phys. Chem. B* 110, 16248-16253 (2006).

[3] PANÁČEK, A., PRUCEK, R., HRBÁČ, J., Polyacrylate-assisted size control of silver nanoparticles and their catalytic activity. *Chem. Mater.* 26, 1332-1339 (2014).

[4] AGNIHOTRI, S., MUKHERJI, S. & MUKHERJI, S. Size-controlled silver nanoparticles synthesized over the range 5-100 nm using the same protocol and their antibacterial efficacy. *RSC Adv.* 4, 3974-3983 (2014).

[5] SISTANI, P.,SOFIMARYO, L., MASOUDI, Z.R., A penicillin biosensor by using silver nanoparticles. *Int. J. Electrochem. Sci.* 9, 6201-6212 (2014).

[6] TSUJI T., IRYO K, W. N. Preparation of silver nanoparticles by laser ablation in solution: influence of laser wavelength on particle size,. *Appl. Surf. Sci.* 202 (1), 80-85 (2002).

[7] RIOUX, D. & MEUNIER, M. Seeded Growth Synthesis of Composition and Size-Controlled Gold-Silver Alloy Nanoparticles. *J. Phys. Chem. C* 119, 13160-13168 (2015).

[8] ZONG, R., WANG, X., SHI, S. & ZHU, Y. Kinetically controlled seed-mediated growth of narrow dispersed silver nanoparticles up to 120 nm: secondary nucleation, size focusing, and Ostwald ripening. *Phys. Chem. Chem. Phys.* 16, 4236 (2014).

[9] CREIGHTON, J. A., BLATCHFORD, C. G. & ALBRECHT, M. G. Plasma Resonance Enhancement of Raman-Scattering by Pyridine Adsorbed on Silver or Gold Sol Particles of Size Comparable to the Excitation Wavelength. *J. Chem. Soc. Trans. Ii* 75, 790-798 (1979).

[10] LEE, P. C. & MEISEL, D. Adsorption and surface-enhanced Raman of dyes on silver and gold sols. J. Phys. Chem. v. 86, p. 3391. *J. Phys. Chem.* 86, 3391-3395 (1982).

[11] NICKEL, U., MANSYREFF, K. & SCHNEIDER, S. Production of monodisperse silver colloids by reduction with hydrazine: The effect of chloride and aggregation on SER(R)S signal intensity. *J. Raman Spectrosc.* 35, 101-110 (2004).

[12] LEOPOLD, N. & LENDL, B. A New Method for Fast Preparation of Highly Surface-Enhanced Raman Scattering (SERS) Active Silver Colloids at Room Temperature by Reduction of Silver Nitrate with Hydroxylamine Hydrochloride. *J. Phys. Chem. B* 107, 5723-5727 (2003).

[13] KVÍTEK, L., PRUCEK, R., PANÁČEK, A., The influence of complexing agent concentration on particle size in the process of SERS active silver colloid synthesis. *J. Mater. Chem.* 15, 1099-1105 (2005).

A NONSTATIONARITY OF SILVER HYDROSOLS FORMATION AND CHANGES OF NANOPARTICLES SHAPE AFTER «FREEZING»

Ekaterina KONONOVA, Irina MIKHALENKO, Viktor YAGODOVSKII

Peoples' Friendship University of Russia, Moscow, Russian Federation
kononova_ea@pfur.ru , mikhalenko_ii@pfur.ru

Abstract

Among the other elements the dispersed silver holds a special place because of its protective biocidal effect and specific physicochemical characteristics depending on the particle size and shape. Absorption spectra of silver hydrosols were monitored in processes of Ag nanoparticles (Ag-NPs) formation due to reducing reaction of Ag^+ ions by tannin with $AgNO_3$ or Ag_2SO_4 as precursors. The dependence of the optical absorption on the wavelength in the long-wavelength range of plasmon spectra has been analyzed in framework of Mie theory by an original method. Several parameters may be estimated, such as the volume fraction of NPs (NV), the effective concentration of conduction electrons (Ne), and the damping factor of the metal electron plasma vibrations (γ). The changes in γ are the most interesting because γ includes surfaces' defects of metallic nanoparticles. It was found that in case of $AgNO_3$ precursor the increase of NV leads to γ decrease with oscillations for both parameters when the reaction $Ag^+_n \rightarrow Ag^0_n$ proceeds. Low temperature treatment of the stable silver particles (10 nm in diameter) undergo aggregation (rearrangement) because silver NPs surface becomes defective. At 77 K changes in NV and γ may be explained by tunnel mechanism of surface defects formation. These defects include mobile small surface clusters $H^+(H_2O)_n$ and $OH^-(H_2O)_n$. We compare the influences of precursor and Ag-NPs concentration on NV and γ parameters before and after freezing.

Keywords: Silver nanoparticles, freezing of Ag hydrosols, plasmon spectra, surface defects

1. INTRODUCTION

Metal nanoparticles (NPs) of 1-20 nm range attracted significant attention of researchers because of the wide range of applications and extraordinary optical and electronic properties [1-3]. Localized plasmons are the collective oscillations of free electrons in metal nanoparticles. The optical properties of silver clusters and NPs largely depend on their sizes [4,5]. In [6,7], it has been found that the peak width Δ of the surface plasmon spectrum linearly depends on 1/r (r is the radius of a Ag-NPs) and that the peak shifts to the lower frequencies with a decrease in the particle size. The dependence of the optical density on the wavelength in the long-wavelength range (see **Figure 1a**) has been analyzed in terms of the Mie theory in [8,9] and its representation corresponded by equation (1).

$$\frac{l}{2{,}3Abs} = \frac{\theta_2}{\theta_1} + \frac{1}{\theta_1}\left(\frac{\lambda_{max}^2}{\lambda} - \lambda\right)^2 \ (1), \quad \theta_1 = \frac{9\pi NVc\, m_0^3}{\sigma} \ (2), \quad \theta_2 = \frac{e^4\lambda_{max}^4 N_e^2}{4\pi c^2 m_e^2 \sigma^2} \ (3).$$

The parameters θ_1 (cm) and θ_2 (cm^2) defined by formulas (2) and (3) are used to estimate the volume fraction *NV of* metal particles, the effective concentration N_e *of* conduction electrons, and the damping factor γ of the metal electron plasma vibrations. The latter factor is related to the particle conductivity (formulas 4).

$$\sigma = \frac{e^2 N_e}{m_e \gamma} \ (4), \quad \lambda_{max}^2 = \frac{(2\pi c)^2 m_e(\varepsilon_0 + 2m_0^2)}{4\pi N_e e^2} \ (5), \qquad \gamma = \gamma_{ee} + \gamma_{ef} + \gamma_{ed} = \widetilde{\gamma_0} + n_d\sigma_S V_F \ (6), \ \widetilde{\gamma_0} = \gamma_{ee} + \gamma_{ef}$$

The wavelength at the absorption peak λ_{max} is related to *Ne* by (formula 5), where ε_0 is the λ-independent dielectric constant of silver; *e* and m_e are the charge and effective mass of electron, respectively; *c* is the velocity of light; m_0 is the refractive index of the medium. The coefficient γ determines the half-width of the absorption band and includes the coefficients of conduction electrons scattering by electrons (γ_{ee}), by phonons (γ_{ef}), and by crystal lattice defects (γ_{ed}) (formula 6), where $\widetilde{\gamma}_0$ denotes the constant at given temperature, n_d the defect concentration in the near-surface layer of a metallic particle, σ_S the effective section of electron scatter by defects, and V_F the electron speed at the Fermi level.

2. FIGURES, SCHEMA, TABLES

Silver hydrosols (0.001 and 0.002 wt.%) were prepared by reduction of $AgNO_3$ (sol_1, sol_2) or Ag_2SO_4 (sol_3) (reagent grade) with tannin ($C_{76}H_{52}O_{46}$, high-purity grade, Germany) as reducing agent and stabilizer in the presence of a sodium tetra borate buffer (reagent grade, pH 9.0). Sols which were synthesized this way showed high stability and reprodusibility of absorption spectra during two years. The absorption spectra were recorded at room temperature from 300 to 750 nm on an SK-101 spectroscopic complex (Russia) at 25°C before and after cryotreatments, with an accuracy of optical density determination Abs ± 0.002. Freezing was performed by holding a sol (4 ml) in a low-temperature bath for 15-120 s. The baths were cooled with mixtures of liquid nitrogen with pentane (142 K), toluene (178 K), octane (217 K), and butanol (230 K). Dewar flask and special thin-walled test tubes with a ~1.5 mm interval gap were used. This excluded gradients across the layer of the colloidal solution to be cooled. After freezing a sample was rapidly (1 min) heated up to room temperature and its absorption spectrum was recorded. The optical path was 1 cm, water was in reference channel. It was preliminarily checked that freezing treatment has no effect on the state of tannin/buffer solution.

Figure 1b demonstrates the spectral controlling synthesis of Ag hydrosol ($AgNO_3$). This routine procedure is applied by many researchers. The rate of Ag-NPs formation in different conditions can be defined. The particularity of our approach consists in the analysis of Ag-NPs absorption curve that permits to clarify the changes in the states of forming Ag-NPs.

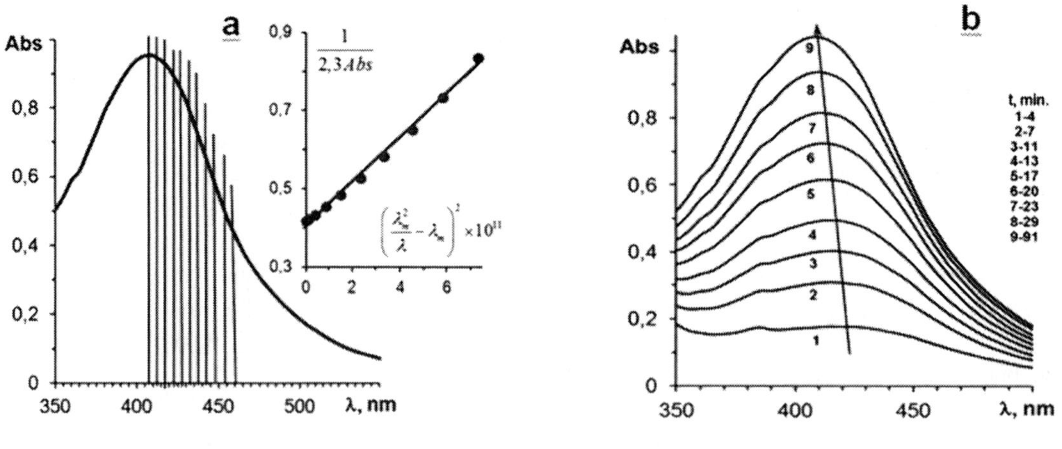

Figure 1 Absorption curve analysis (a) and AgNPs absorption spectra in Ag sol_1 synthesis (b)

It is seen from **Figure 2** that the volume fraction of the dispersed phase NV and γ coefficient of plasma-resonance vibrations vary during the synthesis of hydrosols. The time dependencies of NV and γ are antisymbatic: the growth of NV is accompanied by a decrease in γ, that is, a decrease in the defectiveness of the surface of the emerging particles (healing of biographical defects). On the kinetic curves we can see a nonstationary period with oscillations for both parameters NV and γ. Nonstationary phenomena plays an important role in chemistry and self sustained oscillations are known for different heterogeneous catalytic reactions [10,11]. The appearance of nonstationarity mode in Ag NPs formation processes is a very interesting result. To explain the parameters of NV and γ oscillations we considered the auto catalytic way of ions-metal NPs complex formation in the schema of reduction reaction $Ag^+ \rightarrow$ Ag-NPs.

1. $Ag_V^+ \rightleftarrows Ag_S^+$ (Adsorption of Ag+-ions)

2. $red_V \rightleftarrows red - s$ (Adsorption of reducing agent)

3. $Ag_S^+ + (Ag_n^0)^* \rightleftarrows Ag_S^+ - Ag_n^0$ (Formation of the complex Ag+-NPs)

4. $Ag_S^+ - Ag_n^0 + red - s \rightleftarrows Ag^0 - Ag_n^0 - ox \longrightarrow Ag_{n+1}^0 + ox$ (Reduction step: the interaction of the adsorbed reducer and a complex to form a product AgOn+1)

5. $Ag_S^+ - Ag_n^0 + Ag_S^+ \rightleftarrows Ag_S^+ - Ag_n^0 - Ag_S^+$ (Inhibiting of complex Ag-ion on the second adsorption stage).

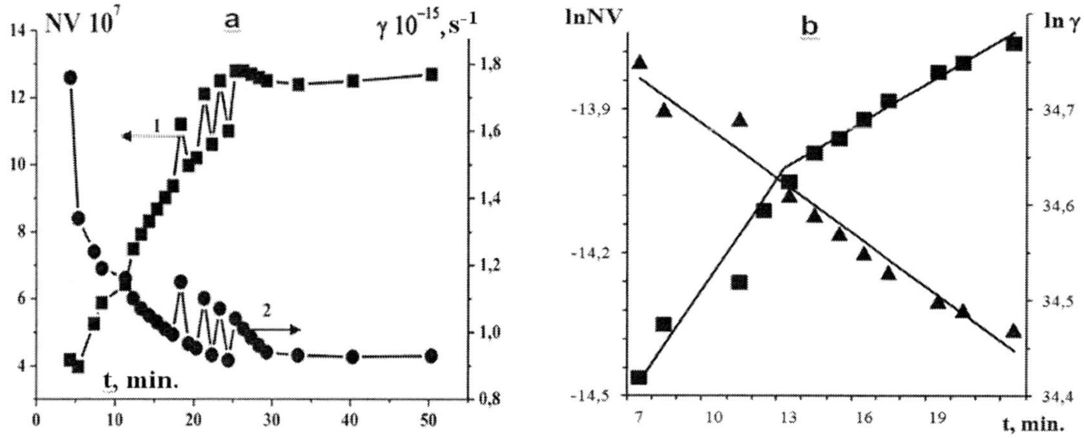

Figure 2 Nonlinear time dependences of Ag-NPs parameters in silver hydrosol ($AgNO_3$) synthesis processes with NV and γ self oscillations (**a**) and its linear plots in first kinetic order coordinates (**b**)

The kinetic of formation of Ag-NPs corresponds to first order. The rate constant values K_{NV} and $K\gamma$ were calculated from semilogarithmic plots (**Figure 2b**). Only for *NV-time* dependence two periods of AgNPs formation were observed with diminishing in 2 times of K_{NV} as it can be seen from data in the **Table 1**.

Table 1 Constant rates (min.$^{-1}$) of optical parameters changes for Ag hydrosols formation

Ag sol ($AgNO_3$) 0.001 % wt.			Ag sol (Ag_2SO_4) 0.002 % wt.	
k_γ	k_{NV} (1)	k_{NV} (2)	k_γ	k_{NV}
0.02	0.06	0.03	0.01	0.06

Figure 3a *illustrates the spectra monitoring in situ for synthesis of Ag-hydrosol with* Ag_2SO_4 precursor (sol*3).* The unusual shape of the spectrum line and red shift of Abs_{max} are associated with the formation of unstable aggregated Ag-NPs and clusters Ag^+_n (sharp peaks below 380 nm). After mixing and dispersion stabilization we see the shift of peak maximum in a short wave area and typical plasmon spectra of isolated stable AgNPs of 8-10 nm in diameter that appearance due to disintegration of unstable agglomerates and incorporation of unstable Ag^+_n in NPs. As in the case of Ag-sol ($AgNO_3$) at the initial stage (up to 20 minutes) the parameters NV and γ change also antisymbiatically and in the final stage we see its synchronous decrease due to disaggregation of initially formed associated AgNPs with weak oscillations (**Figure 3b**).

The effect of freezing treatment was studied for two *Ag-hydrosol obtained from* $AgNO_3$. Sol_1 was a freshly prepared sample and the sol_2 was the sol_1 stored for a year. It was found preliminarily that low temperature treatment (LTT) did not influence the state of tannin/buffer system: UV spectra were reproducible after LTT. The spectra of sols had one absorption maximum at about 416 nm. According to literature data and our TEM images, the Ag particles diameters with plasmon spectra maximum 410-430 nm is about 8-10 nm.

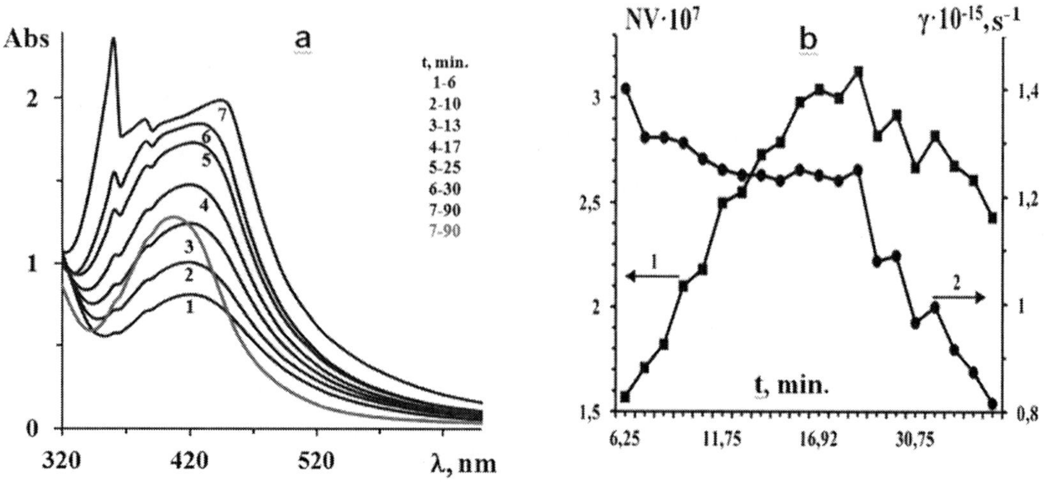

Figure 3 Absorption spectra of Ag-sol$_3$ formation (a) and time dependences of Ag-NPs parameters volume fraction of metallic particles *NV* and damping factor of electron plasma vibrations γ (b)

For example, spectra obtained before and after silver hydrosol treatment at 77K for 20 s are shown in **Figure 4**. We found that the absorption maximum did not shift as the temperature decreased, while the optical density at the maximum Abs$_m$ lowered (**Table 2**). However, Abs$_m$ values for T=77 K fell out of this dependence. Absorption peak at 500-550 nm appeared at T < 182 K and related to the formation of a new fraction of larger Ag particles and/or nonspherical anisotropic NPs and their aggregation [12]. Holding sols subjected to low-temperature treatment at room temperature for 48 h additionally decreased Abs$_m$ and increased absorption in the long-wave region, that is, changes in degree of dispersion also occurred after LTT.

Table 2 Influence of cryotreatment temperature (duration 20 s) on γ (s^{-1}) and NV of Ag hydrosols

		Sample 1				Sample 2			
№	T, K	Abs$_m$	ΔAbs*, %	$\gamma\cdot10^{-15}$,	NV·10^6	Abs$_m$	ΔAbs*, %	$\gamma\cdot10^{-15}$,	NV·10^6
1	298	0.94	100	0.72	1.09	1.01	100	0.50	0.79
2	230	0.91	85	0.70	1.02	0.95	94	0.54	0.82
3	207	0.77	82	0.74	0.92	0.74	73	0.63	0.80
4	182	0.66	70	0.81	0.87	0.72	71	0.64	0.74
5	142	0.50	53	0.93	0.75	0.52	51	0.72	0.70
6	77	0.86	92	0.71	0.99	0.63	62	0.64	0.68

The hexagonal ice structure with a disordered arrangement of protons transforms into a denser structure with an ordered arrangement of H^+ as the temperature decreases to 77 K [13]. Explanation of this effect is discussed in detail in [14] for gold hydrosols by means of tunnel mechanism for "cryodefects" formation from positively charged surface complexes $X^+(H_2O)_n$-Au (or Ag), e.g., $H^+(H_2O)_n$ and $Au^{+\delta}(H_2O)_n$ (n = 1-100).

To exclude the concentration influence, we have studied the effect of freezing at 77 K of Ag-sol$_3$ in terms of different Ag-NPs concentration in the dispersions. (**Table 3**, **Figure 4b**).

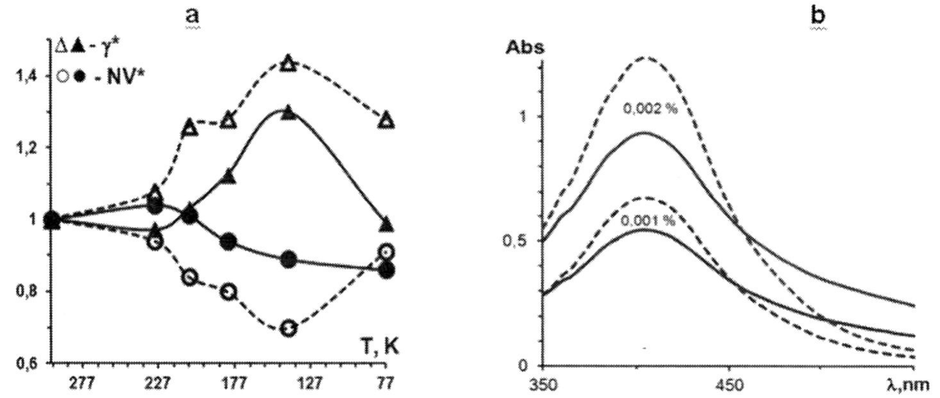

Figure 4 a) Influence of freezing temperature on the relative parameters γ*, and NV* calculated from the spectra of freshly (dashed line) and stored for a year Ag ($AgNO_3$) hydrosols; **b)** Absorption spectra of Ag (Ag_2SO_4) hydrosol before (dashed line) and after freezing at 77 K. Freezing duration is 20 s.

Table 3 Effect of Ag sol concentration in Abs_{max} and $\gamma\times10^{-15}$ (s^{-1}) after freezing Ag-sol_3 (77 K, 20 s)

C, wt. %	**0.001**	**0.0012**	**0.0014**	**0.0016**	**0.0018**	**0.002**
Abs_m before freezing	0.675	0.776	0.910	0.995	1.062	1.237
Abs_m after freezing	0.543	0.591	0.674	0.757	0.794	0.934
ΔAbs_m, %	**–19**	**–24**	**–26**	**–24**	**–25**	**–24**
γ before freezing	1.02	1.03	1.02	0.993	1.04	1.01
γ after freezing	1.3	1.31	1.28	1.23	1.3	1.3
γΔ, %	**+27**	**+27**	**+26**	**+24**	**+25**	**+28**

On the **Figure 4b** it is shown that the absorption in maximum Abs_m after freezing decreases. However, the relative change of *ΔAbs_m* in this environment remains constant (~25 %). It was found that the freezing of dilute Ag-$sols_3$ does not lead to aggregation of the particles, but the rise of the surface defects is sufficient (~27 **%). Thus freezing effect in relative changes of parameters does not depend on the concentration of the** silver hydrosol.

3. CONCLUSIONS

Our investigation of the kinetics of Ag-hydrosols formation and the effect of sols freezing revealed a changes of optical parameters: a decrease in the effective concentration of unbound electrons and an increase in the imperfections of particles. These leads to formation of small ensembles (floccules) or aggregates depending on LTT duration. Flocculated or aggregated Ag-NPs are formed, respectively, upon short or long expositions. The low activation energy of the cryoaggregation of NPs is dictated by the tunnel mechanism of the formation of "cryodefects" from positively charged surface complexes $X^+(H_2O)_n$-Ag, e.g., $H^+(H_2O)_n$ and $Ag^{+\delta}(H_2O)_n$ (n = 1-100). It is determined that defects on Ag-NPs surface increases with decreasing of temperature of freezing due to its cryoaggregation. **Relative effect of LTT does not depend on the Ag hydrosol concentration.**

REFERENCES

[1] Leila Gharibshahi, Elias Saion, Elham Gharibshahi, Abdul Halim Shaari and Khamirul Amin Matori. Influence of Poly (vinylpyrrolidone) concentration on properties of silver nanoparticles manufactured by modified thermal treatment method. *Materials*. 2017. V. 10. № 4. p. 402; DOI:10.3390/ma10040402

[2] Yu-xi Zhang and Yu-hua Wang. Nonlinear optical properties of metal nanoparticles. (Review Article) RSC Adv., 201. V. 7. pp. 45129-45144; DOI: 10.1039/C7RA07551K

[3] Lufsyi Mahmudin, Edi Suharyadi, Agung Bambang Setio Utomo, Kamsul Abraha. Influence of stabilizing agent and synthesis temperature on the optical properties of silver nanoparticles as active materials in surface plasmon resonance (SPR) biosensor. *Journal of Modern Physics.* 2015, V.6, pp. 1071-1076.

[4] A. V. Sokolov. Optical Properties of Metals. Gos. Izd Fiz.mat. Lit.. Moscow. 1961, p. 464.

[5] A.D. Pomogailo, A.S. Rozenberg and I.E. Uflyand. Metal Nanoparticles in Polymers, Moscow: Khimiya. 2000. p. 337.

[6] L. Rodriguez-Sanchez, M.L. Blanko and M.A. Lopez-Quintela. Electrochemical synthesis of silver nanoparticles. *Journal Phys. Chem.* B, 2000, V. 104, p. 9683.

[7] H.J. Freund. Clusters and islands on oxides: from catalysis via electronics and magnetism to optics. *Surface Science*. 2002. V. 500. p. 271.

[8] W. T. Doyle. Absorption of light by colloids in alkali halide crystals. *Physical Review.*1958. V. 111 № 4. p. 1067.

[9] K. A. Sal'sedo, V. V. Tsvetkov, and V. D. Yagodovskii. Adsorption from solutions on the surface of silver particles. I. Spectroscopic Determination of the Parameters of Small Silver Particles. *Zh. Fiz. Khim.* 1989. V. 63. №12. p. 3295.

[10] I. Prigogine, I. Stengers. Order out of chaos. Man's new dialogue with nature. Heinemann. London. 1984. p. 431.

[11] G.F.Fromet, K.C.Waugh. Reaction kinetics and the development and operation of catalytic processes. 1st Ed. Elsever Science, 2001, V.133, p. 664.

[12] B. M. Sergeev, M. V. Kiryukhin, F. N. Bakhov, and V. G. Sergeev. Photochemical synthesis of silver nanoparticles in the water solutions of polycarboxylic acids. The effect of polymer matrix on the nanoparticle size. *Vestnik. Mosk. Univ*. Moscow: Khimiya. 2001. V. 42. № 5. p. 308.

[13] A. F. Wells, Structural Inorganic Chemistry. Clarendon, Oxford. 1984. p. 1375. https://books.google.ru/books?hl=ru&lr=&id=uR77AAAAQBAJ&oi=fnd&pg=PP2&ots=6w77u6GN3i&sig=Z6Z77MEtBCi4UHLzuhfVbR3aPU0&redir_esc=y#v=onepage&q&f=false

[14] E. A. Kononova, I. I. Mikhalenko and V. D. Yagodovskii. Changes in the state of silver sol particles after low-temperature treatments. Russian Journal of Physical Chemistry A . 2008. V. 82. pp. 669-674. DOI: 10.1007/s11504-008-4028-9.

AN EFFECTIVE TOOL TO CONTROL SPATIAL RESOLUTION COLLOID LITHOGRAPHY

[1,2]Elena MILIUTINA, [1,2]Olga GUSELNIKOVA, [1]Roman ELASHNIKOV, [1]Vasilii BURTSEV, [1]Vaclav ŠVORČÍK, [1,2]Oleksiy LYUTAKOV

[1]VŠCHT - University of Chemistry and Technology, Prague, Czech Republic, EU, lyutakoo@vscht.cz
[2]Tomsk Polytechnic University, Research School of Chemistry and Applied Biomedical Sciences, Tomsk, Russian Federation

Abstract

Polymer-based colloid lithography represents the very effective approach for creation of a large-scale template for subsequent surface modification. Utilization of this approach, however, is significantly restricted by the ability of used micro/nanospheres to create only limited number of symmetrical structures. In this work, we propose the treatment of polystyrene (PS) microspheres in the solvent vapor, before subsequent surface grafting with diazonium salts as an alternative method for the creation of structures with required geometry and symmetry. Application of vapor treatment allows the effective modification of PS microspheres screened area and. As a result, the subsequent grafting procedure leads to the creation of ordered micropores array with tunable size and height. Combination of different solvents with slight heating was studied and found to be suitable for further creation of different surface grafting pattern. The success of surface grafting through the treated PS microsphere mask was verified using Raman spectroscopy and AFM microscopy techniques.

Keywords: Colloid lithography; polystyrene microspheres; vapor annealing; periodical structures

1. INTRODUCTION

The physicochemical methods of surface modification changing surface morphology or chemistry provide a unique way of creation of new materials with functional surfaces [1-4]. As examples, the introduction of superhydrophilicity/phobicity, water or oil adhesion and repellence, specific chemical recognition events, externally triggered or self-related smart surface response, as well as many other useful surface singularities can be mentioned [5-10]. Among the common techniques for the introduction of specific surface properties different masks-based approaches, where the material is area-selective chemically modified are of considerable importance [11-13]. In this field even, the colloid lithography can be considered as one of the more common approach due to its simplicity and reliability [13-16]. In particular, deposited by spin- or deep-coating polymer latex microspheres tends to spontaneously form the ordered array on the samples surface, effectively creating the mask for further surface pattering [17,18] Thus, this simple lithographic approach has been commonly applied for the preparation of ordered microporous structures via the surface modification through the polymer microspheres mask and subsequent surface modification or electrodeposition of metals [19-23].

In this paper, we propose an alternative approach, enhancing the range of surface patterns accessible by the common polystyrene (PS) colloid lithography. Vapor annealing of closed packed polymer microspheres array deposited on the gold substrate is used. Deformation and conglutination of microspheres depending on the applied solvent and partial pressure is expected, leading to the effective modification of uncovered by PS microspheres gold surface, available for further grafting by a diazonium salt.

2. EXPERIMENTAL

2.1. Materials

Gold targets for metals deposition (purity of metals 4 N) were purchased from Safina. Deionized water, methanol (reagent grade, ≥ 99.8 %), acetone (97 %), 4-nitrobenzenediazonium tosylate, toluene (99 %) were purchased from Sigma-Aldrich and used without further purification. Colloid suspension of polystyrene (PS) microspheres (PS microspheres size - 500 nm) was supplied by Alfa Aesar.

2.2. Samples preparation

Thin gold films (thickness 20 nm) were deposited onto glass surface (Thermo Scientific), covered by adhesive titanium layer (thickness ca 10 nm), using thermal evaporation method. PS suspension was drop deposited and slowly evaporated on Au surface, under the careful control of temperature (40°C). After deposition of PS colloid mask, samples were annealed in (i) toluene (≈0.01 and ≈0.03 bar partial pressure) or (ii) ethanol vapors (≈0.02 and ≈0.05 bar partial pressure) during 10 min. The 4-nitrobenzenediazonium tosylate (ADT-NO_2) salt was grafted electrochemically (4 mM freshly prepared water solutions, -1.0 V, 7 min) through the annealed PS microspheres mask. After grafting, the PS mask was removed by rinsing in the toluene under sonication for 10 min. Then the metal substrates were sequentially rinsed under sonication with deionized water, ethanol, and acetone for 10 min and dried in a desiccator for 3 h.

2.3. Measurement techniques

For the characterization of the sample surface, the atomic peak force microscopy (AFM) technique was used. Surface mapping was performed with Icon (Bruker) set-up on the areas of 3×3 μm^2 Surface conductivity characterization was performed in Peak Tuna AFM mode, using the Pt/Pd coated tip with constant voltage (3.0 V). Raman spectra were measured on ProRaman-L (Laser power 30 mW) Raman spectrometer with 785 nm excitation wavelengths. Spectra were measured 30 times, each of them with 3 s accumulation time.

3. RESULTS AND DISCUSSION

The array of PS latex microspheres was deposited on the Au surface and a closely spaced periodical structure with hexagonal symmetry was formed, according the previous experiments. The samples were then annealed in the vapors of organic solvents (ethanol or toluene) with the aim to initiate PS microspheres swelling and partial conglutination. After the probing of several organic solvents, we chose for further experiments the toluene and ethanol, which conserve the PS microspheres periodicity and do not create periodical structure irregularities. It was previously expected that vapor-initiated conglutination of PS microspheres could lead to changes in the tuning of PS microspheres screened surface area. In the next step, the electrochemical surface grafting by ADT-NO_2, through the pristine or vapor annealed PS masks, followed by sequential removing of microspheres was applied.

First, the grafting of organic moities to tjhe gold surface was verified using the Raman spectroscopy. Results are presented in the **Figure 1** and it is evident that after the grafting procedure, the Raman bands, typical for C_6H_4-NO_2 becomes apparent. In particular, the Raman spectra show the presence of following characteristic peaks: 482 cm^{-1} (C-N bending); 637 cm^{-1} (NO_2 def vib, in-plane Ar_{C-C}); 795 cm^{-1} (Ar_{C-H} out of plane def); 830 cm^{-1} (NO_2 scissors bending); 868 cm^{-1} (Ar_{C-H} out of plane def); 1046 cm^{-1}, 1075 cm^{-1} (Ar_{C-H} in plane bends); 1334 cm^{-1}, 1352 cm^{-1} (NO_2 symmetric stretch); 1520 cm^{-1}, 1554 cm^{-1} (Ar_{C-C} ring deformation). Additionally, it is evident that Raman bands intensities are a strong function of previous vapor treatment of PS microspheres. The greatest intensities were observed in the case of nontreated PS microspheres mask, which indicate the attachment of largest amount of organic moieties. In turn, grafting through the PS mask annealed in the ethanol vapor leads to a peak intensity decrease, which is proportional to partial solvent pressure. Even more pronounced decrease of Raman bands intensity is also observed for the grafting through PS microspheres array annealed in toluene. It is evident that at higher toluene partial pressure the Raman bands become barely noticeable. So, we can conclude that the highest amount of organic moieties was grafted in the case of nontreated PS microsphere, while the annealing in the ethanol leads to a slight decrease in the amount of grafted organic moieties, and the annealing with toluene significantly screens the Au surface against grafting.

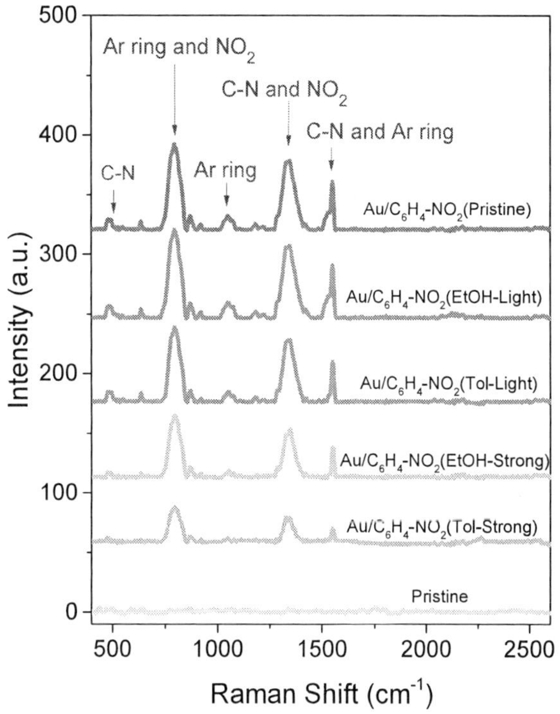

Figure 1 Raman spectroscopy measurements performed after the ADT-NO2 grafting through the pristine and vapor annealed PS colloid mask

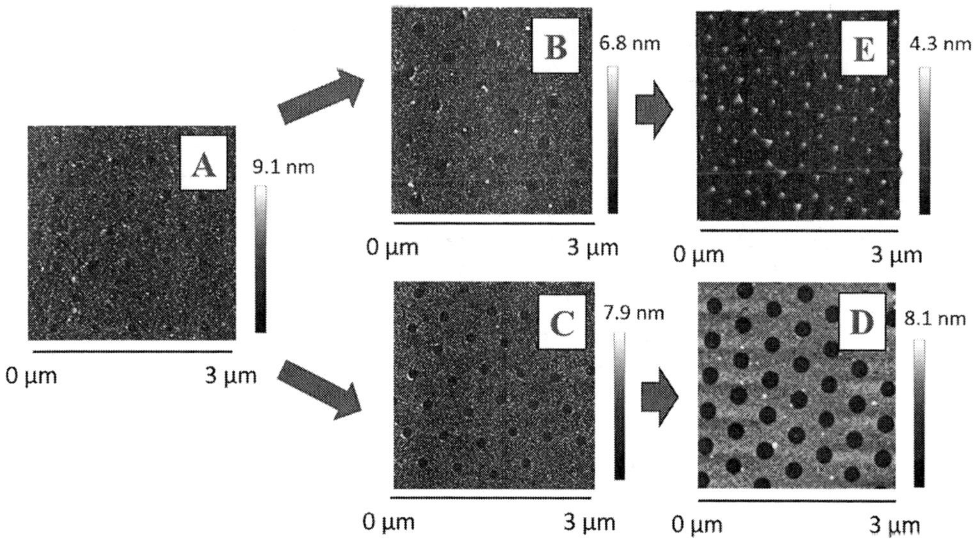

Figure 2 AFM characterization of grafted through the PS microspheres masks gold surface, measured after the removing of PS microspheres: A - ADT-NO_2 grafting through the pristine PS mask, B - ADT-NO_2 grafting through the annealed (ethanol, partial pressure ≈0.02 bar) PS mask, C - ADT-NO_2 grafting through the annealed (ethanol, partial pressure ≈0.05 bar) PS mask, D - ADT-NO_2 grafting through the annealed (toluol, partial pressure ≈0.01 bar) PS mask, E - ADT-NO_2 grafting through the annealed (toluol, ≈0.03 bar) PS mask.

The different residual surface patterns, determined by the surface regions "available" for ADT-NO2 grafting, were created after the PS microspheres removal, according the AFM results (**Figure 2**), in dependence on the applied vapor annealing procedure. In particular, the strong vapor annealing conditions (i.e. utilization of high solvent pressure and/or solvent with high affinity to PS microspheres) lad to almost full surface blocking and the grafting through such "strongly-annealed" structure produced the columnar-like pattern (as is evident in thre **Figure 2E**). Oppositely, the softer annealing conditions resulted in the slight increase of the area screened by PS microspheres and the pores, remained on the surface have larger size compared to the case of nontreated PS (**Figures 2B, 2C**).

Generalizing the AFM investigations, it can be concluded that the formation of bi-dimensional well-ordered pores with a regular round shape and size determined by the previous vapor annealing procedure was observed. In the case of nonannealed PS microspheres, the diameter of the pores was found to be 0.10 µm, and it increases with increasing ethanol vapor pressure (**Figure 2D**). The high surface blocking in the case of toluene vapors utilization leads to the grafting of organic layers in a columnar form. The thickness of the grafted organic layer was also estimated from z-axis scale bars on the presented AFM scans and as can be expected the thickness decreases with increasing impact of organic vapor.

4. CONCLUSION

In this work, we propose the vapor annealing of ordered PS microspheres array, followed by the diazonium surface modification as an alternative possibility to effective tune the PS microspheres screened surface area and altering of the size of created surface chemical and morphological pattern. The grafting through the polymer colloid mask procedure results in the creation of ordered micropores array with tunable size or, moreover, columnar array. Combination of ethanol and toluene solvents with slight heating (40° C) was founded to be the more suitable for PS microspheres blocked surface area.

ACKNOWLEDGEMENTS

This work was supported by the Czech Science Foundation under the projects 18-26170S and Tomsk Polytechnic University (VIU-RSCABS-196/2018).

REFERENCES

[1] BELANGER, Daniel, PINSON, Jean. Electrografting: a powerful method for surface modification. Chem. Soc. Rev. 2011. 40, pp. 3995-4048.

[2] PINSON, Jean, PODVORICA, Fetah. Attachment of organic layers to conductive or semiconductive surfaces by reduction of diazonium salts. Chem. Soc. Rev. 2005. 34, pp. 429-439.

[3] LYUTAKOV, Oleksiy, HUTTEL, Ivan, SIEGEL, Jakub, ŠVORČÍK, Vaclav. Regular surface grating on doped polymer induced by laser scanning. Appl. Phys. Lett. 2009. 95, 173103.

[4] GUSELNIKOVA, Olga, POSTNIKOV, Pavel, KALACHYOVA, Yevgeniya, KOLSKA, Zdenka, LIBANSKY, Milan, ZIMA, Jiri, ŠVORČÍK, Vaclav, LYUTAKOV, Oleksiy. Large-Scale, Ultrasensitive, Highly Reproducible and Reusable Smart SERS Platform Based on PNIPAm-Grafted Gold Grating. CHEMNANOMAT 2017. 3, 2, pp. 135-144.

[5] UEDA, Erica, LEVKIN, Pavel A. Emerging Applications of Superhydrophilic-Superhydrophobic Micropatterns. Adv. Mater. 2013. 25, pp. 1234-1247.

[6] ZHANG, Xi, SHI, Feng, NIU, Jia, JIANG, Yugui, WANG, Zhigiang. Superhydrophobic surfaces: from structural control to functional application. J. Mater. Chem. 2008. 18, pp. 621-633.

[7] ZHANG, Junji, MA, Wenjing, HE, Xiao-Peng, TIAN, He. Taking Orders from Light: Photo-Switchable Working/Inactive Smart Surfaces for Protein and Cell Adhesion. ACS Appl. Mater. Interfaces. 2017. 9, pp. 8498-8507.

[8] GUSELNIKOVA, Olga, POSTNIKOV, Pavel, SAJDL, Petr, ELASHNIKOV, Roman, ŠVORČÍK, Vaclav, LYUTAKOV, Oleksiy. Functional and Switchable Amphiphilic PMMA Surface Prepared by 3D Selective Modification. Adv. Mater. Interfaces. 2018. 5, 1701182

[9] ELASHNIKOV, Roman, TRELIN, Andrej, OTTA, J., FITL, P., MARES, D., JERABEK, V., ŠVORČÍK, Vaclav, LYUTAKOV, Oleksiy. Laser patterning of transparent polymers assisted by plasmon excitation. SOFT MATTER 2018. 14, 23, pp. 4860-4865.

[10] Slepicka, Petr, Siegel, Jakub, Lyutakov, Oleksiy, Kasalkova, Nikola, Kolska, Zdenka, Bacakova, Lucie, Svorcik, Vaclav. Polymer nanostructures for bioapplications induced by laser treatment. Biotechnology Adv. 2018. 36, 3, pp. 839-855.

[11] LEDERHOSE, Paul, ABT, Doris, WELLE, Alexander, MÜLLER, Rouven, BARNER-KOWOLLIK, Christopher, BLINCO, James P. Exploiting λ-Orthogonal Photoligation for Layered Surface Patterning. Chem. Eur. J. 2018. 24, pp. 576-580.

[12] FORNER-CUENCA, A.; MANZI-OREZZOLI, V.; KRISTIANSEN, P. M.; GUBLER, L.; SCHMIDT, T. J.; BOILLAT, P. Mask-assisted electron radiation grafting for localized through-volume modification of porous substrates: influence of electron energy on spatial resolution. Radiat. Phys. Chem. 2017. 135, pp. 133-141.

[13] WANG, Yuli, LAI, Hsuan-Hong, BACHMAN, Mark, SIMS, Christopher E.; LI, G. P.; ALLBRITTON, Nancy L. Covalent micropatterning of poly (dimethylsiloxane) by photografting through a mask. Anal. Chem. 2005. 77, pp. 7539-7546.

[14] YANG, Seung-Man, JANG, Se Gyu, CHOI, Dae-Geun, KIM, Sarah, YU, Hyung Kyun. Nanomachining by colloidal lithography. Small 2006. 2, pp. 458-475.

[15] HOOGENBOOM, J. P.; VOSSEN, D. L. J.; FAIVRE-MOSKALENKO, C.; DOGTEROM, M.; BLAADEREN, A. V. Patterning surfaces with colloidal particles using optical tweezers. Appl. Phys. Lett. 2002. 80, 4828-4830.

[16] LIN, Meng-Hsien; CHEN, Chi-Fan, SHIU, Hung-Wei, CHEN, Chia-Hao, GWO, Shangjr. Multilength-scale chemical patterning of self-assembled monolayers by spatially controlled plasma exposure: nanometer to centimeter range. J. Amer. Chem. Soc. 2009. 131, pp. 10984-10991.

[17] GOSECKA, Monika, GRIFFETE, Nebewia, MANGENEY, Claire, CHEHIMI, Mohamed M, SLOMKOWSKI, Stanislaw, BASINSKA, Teresa. Preparation and optical properties of novel bioactive photonic crystals obtained from core-shell poly (styrene/α-tert-butoxy-ω-vinylbenzyl-polyglycidol) microspheres. Colloid Polym. Sci. 2011. 289, 1511-1518.

[18] NGUYEN, Van-Quynh, SCHAMING, Delphine, TRAN, Dai Lam, LACROIX, Jean-Christophere. Ordered Nanoporous Thin Films by Nanosphere Lithography and Diazonium Electroreduction: Simple Elaboration of Ultra-Micro-Electrode Arrays. ChemElectroChem. 2016. 3, pp. 2264-2269.

[19] ZHANG, Junhu, LI, Yunfeng, ZHANG, Xuemin, YANG, Bai. Colloidal self-assembly meets nanofabrication: From two-dimensional colloidal crystals to nanostructure arrays. Adv. Mater. 2010. 22, pp. 4249-4269.

[20] BARTLETT, Philip N.; BIRKIN, Peter R.; GHANEM, Mohamed A. Electrochemical deposition of macroporous platinum, palladium and cobalt films using polystyrene latex sphere templates. Chem. Commun. 2000. pp. 1671-1672.

[21] BARTLETT, P. N.; BAUMBERG, J. J.; BIRKIN, P. R.; GHANEM, M. A.; NETTI, M. C. Highly Ordered Macroporous Gold and Platinum Films Formed by Electrochemical Deposition through Templates Assembled from Submicron Diameter Monodisperse Polystyrene Spheres. Chem. Mater. 2002. 14, pp. 2199-2208.

[22] HU, Jin, ABDELSALAM, Mamdouh, BARTLETT, Philip, COLE, Robin, SUGAWARA, Yoshihiro, BAUMBERG, Jeremy, MAHAJAN, Sumeet, DENUAULT, Guy. Electrodeposition of highly ordered macroporous iridium oxide through self-assembled colloidal templates. J. Mater. Chem. 2009. 19, pp. 3855-3858.

[23] CAO, Bingqiang, CAI, Weiping, SUN, Fengqiang, LI, Yue, LEI, Yong, ZHANG, Lide. Fabrication of large-scale zinc oxide ordered pore arrays with controllable morphology. Chem. Commun. 2004, pp. 1604-1605.

SURFACE CHARACTERIZATION OF TITANIUM AFTER ARGON SPUTTER CLEANING

Tomas HORAZDOVSKY, Zdenek TOLDE

Faculty of Mechanical Engineering, Czech Technical University in Prague, Czech Republic, EU

Abstract

Surface integrity plays a key role in the surface modifications and coating processes that provide high and long-term performance of the treated components. Argon ion bombardment is widely used in many surface modification processes and for surface pre-treatment. The effect of argon sputter cleaning on the residual stress, hardness and surface morphology of commercially pure titanium is investigated in this work. Titanium samples were mechanically polished and then etched by argon ions. The thickness of the sputtered layer was measured by a quartz thickness monitor. The hardness was investigated by nanoindentation, the residual stress was measured by X-ray diffraction, and the surface morphology was monitored by atomic force microscopy. Removal of the surface layer by argon ion bombardment caused a decrease in residual stress and a decrease in hardness. Uneven sputtering of grains resulted in increased surface roughness. The observed changes in surface properties increased with increasing thickness of the removed layer.

Keywords: Ion etching, morphology, residual stress, hardness

1. INTRODUCTION

Titanium and its alloys are used in many engineering applications in the aerospace, automotive, chemical and biomedical industries, due to their high strength, low density and good corrosion resistance in many environments [1-4]. However, a significant disadvantage of titanium alloys is the high coefficient of friction and wear [5]. Advances and innovations in surface engineering have significantly enhanced the performance of commercially available titanium materials. Surface properties can be improved by advanced methods using accelerated ions, such as magnetron sputtering, plasma-assisted chemical vapour deposition, ion implantation and ion beam assisted deposition [6-9]. Surface integrity plays a key role in the surface modifications and coating processes that provide high and long-term performance of the treated components. A major aspect of the structural bonding of coating materials to titanium is the surface pre-treatment of the substrates. The pre-deposition treatment of the substrate has a significant influence on the durability of the adhesion bonds. Ion sputter cleaning in the pre-deposition phase of the coating process is often used to improve the adhesion of coatings [10,11]. Ion etching of the surface leads to the removal of impurities after mechanical polishing and the removal of contaminants from the air. The conditions of the sputter cleaning process (e.g. ion current density, processing time, incidence angle, etc.) influence the amount of material that is removed. Different thickness of the removed layer can have a significant effect on the surface properties of titanium and on the quality of its surface treatment.

This work presents the effect of argon sputter cleaning on the internal stress, hardness and surface morphology of commercially pure titanium grade 2. Hardness was investigated by nanoindentation, residual stress was measured by X-ray diffraction (XRD), and surface morphology was monitored by atomic force microscopy (AFM).

2. EXPERIMENTAL

The substrates were made of commercially pure titanium grade 2 in the form of a cylinder 14 mm in diameter and 3 mm in height. The samples were cut from a titanium bar, and were then ground with a series of waterproof abrasive papers. Final polishing was performed with diamond paste. Before the sputter cleaning process, the samples were ultrasonically degreased in acetone for 20 minutes, and in isopropyl alcohol, also for 20 minutes.

The samples were placed on a sample holder mounted on a rotary manipulator in a vacuum chamber, and were sputter cleaned with 700 eV argon ions to etch and remove the surface layer. The ion beam current density was about 90 μA·cm^{-2}. The sputtering rate and the thickness of the removed layer were measured by a quartz crystal monitor located in the vacuum chamber. An experimental arrangement of the sputter cleaning process is presented in **Figure 1**. The sputter cleaned samples were prepared in two groups. The thickness of the removed layer was 1 μm for the first group and 3 μm for the second group.

The nanoindentation method in continuous measurement mode (CMX) was applied to obtain depth profiles of the surface hardness [8]. The CMX indentation function was prescribed by a quasistatic force (P_{qstat}) in the range from 2 to 10000 mN. The dynamic actuation force (P_{dyn}) was prescribed in the range from 3 to 202.9 mN at a frequency of 200 Hz. For the sputtered sample, average values were obtained from 16 indents in a 4 x 4 matrix with a separation step of 5 μm. The residual stress was investigated by XRD analysis. A PANalytical X'Pert PRO horizontal powder diffractometer was used with a Co anode (λ= 0.178901 nm), a Göbel mirror in the primary beam and a parallel plate collimator in a diffracted beam. The residual stresses were evaluated using the $sin\psi^2$ method. The sputter cleaned surface was analysed by an atomic force microscope (AFM). The AFM was used in tapping mode.

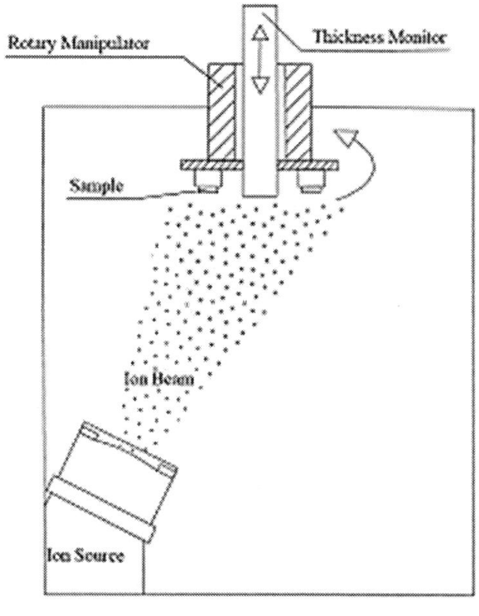

Figure 1 Experimental arrangement of the sputter cleaning process

3. RESULTS AND DISCUSSION

Figure 2 shows the evolution of the macroscopic residual stress with the thickness of the surface layer removed by argon sputter cleaning. The residual stress after polishing was found to be approximately 190 MPa. Argon sputter cleaning reduced the residual stress to approximately 124 MPa for a removed layer 1 μm in thickness, and to approximately 116 MPa for a removed layer 3 μm in thickness. There are two causes for the reduction in stress, both of which act simultaneously. In the first case, the residual stress is reduced by removing the layer in which the stress was induced by polishing. In the second case, the pressure stress induced by argon sputter cleaning counteracts the stress caused by polishing.

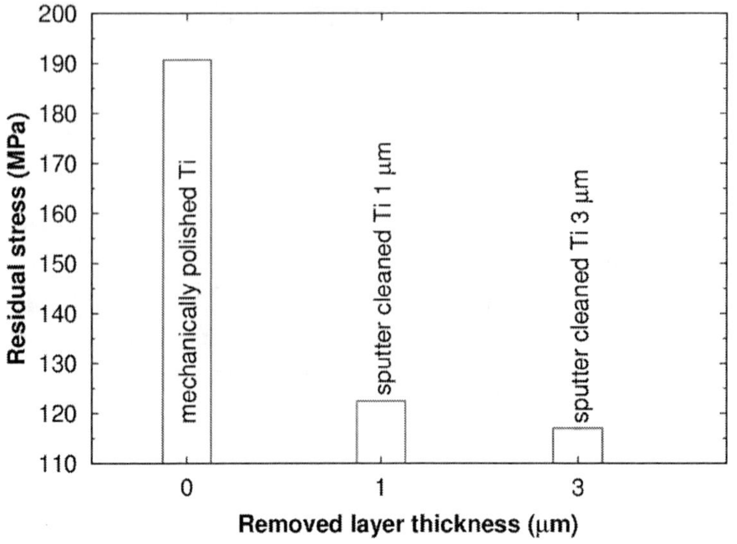

Figure 2 Macroscopic residual stress vs. thickness of surface layer removed by argon sputter cleaning

Figure 3 shows the hardness profiles measured on the polished sample and on the samples sputter claned by argon ions. It can be seen that the course of the surface hardness is not constant. Locally increased surface hardness is probably caused by internal stress after mechanical polishing. The hardness, with the maximum in the near surface region, decreases towards the depth of the sample. The results show that the surface hardness decreases with increasing thickness of the layer removed by argon sputter cleaning. A very flat profile with an almost constant course of hardness was measured on the sample with the removed layer 3 μm in thickness. The results presented here indicate that a reduction in internal stress leads to lower surface hardness values. This phenomenon has also been observed in the strained titanium lattice after ion implantation [8].

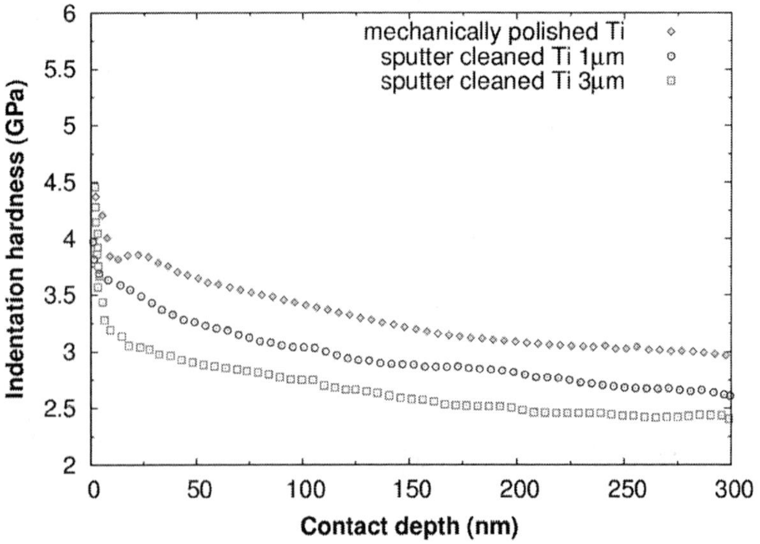

Figure 3 Hardness profiles of mechanically polished titanium and ion sputter cleaned titanium with different thicknesses of the removed surface layer

Representative AFM surface morphology scans are shown in **Figure 4**. The mechanically polished surface is relatively smooth, without major protrusions. However, the surface appears to be wavy. The typical morphology with exposed grain boundaries is observed on the surface sputter cleaned by argon ions (**Figure 4b**). Uneven sputtering of grains with different orientations leads to increasing roughness. The roughness (Ra) increased from 0.006 μm for a mechanically polished sample to 0.135 μm for a sputter cleaned sample with a removed layer 3 μm in thickness. The results show that sputter cleaning of a titanium substrate leads to increased surface roughness and reduced residual stress. These factors play a significant role in the adhesion of coatings [10-12].

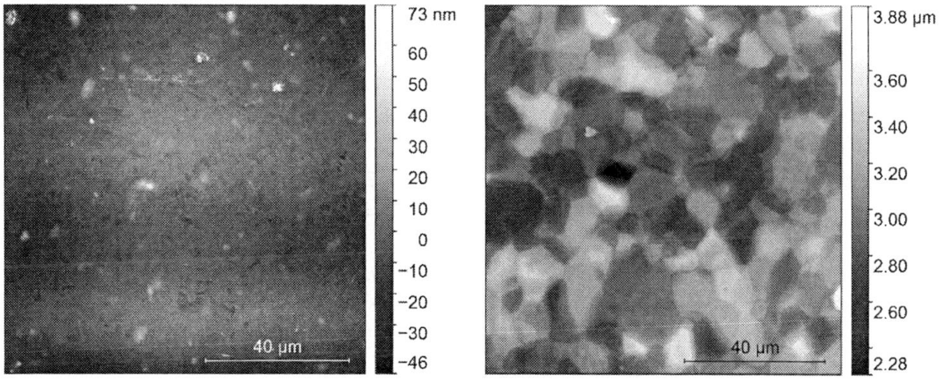

Figure 4 Representative AFM images of a mechanically polished surface (a), and an argon sputter cleaned surface (removed layer 3μm in thickness) (b)

4. CONCLUSION

The influence of the argon sputter cleaning process on the residual stress, the hardness and the surface morphology of commercially pure titanium has been demonstrated. Argon sputter cleaning reduces the residual stress from approximately 190 MPa for a mechanically polished sample to approximately 124 MPa for a removed layer 1 μm in thickness, and to approximately 116 MPa for a removed layer 3 μm in thickness. The reduction in residual stress had an impact on surface hardness. It has been demonstrated that the surface hardness decreases with increasing thickness of the removed layer. Uneven sputtering of grains with a different orientation led to increasing roughness after sputter cleaning.

ACKNOWLEDGEMENTS

This research has been supported by Ministry of Education, Youth and Sports of the Czech Republic project SGS18/173/OHK2/3T/12 of the Czech Technical University in Prague.

REFERENCES

[1] PETERS M., KUMPFERT J., WARD C.H., LEYENS C. Titanium alloys for aerospace applications, Advanced Engineering Materials, Vol. 5, 2003, 419-427.

[2] PODZIMEK S., TOMKA M., NEMETH T., HIMMLOVA L., MATUCHA P., PROCHAZKOVA J. Influence of metals on cytokines production in connection with successful implantation therapy in dentistry, Neuroendocrinology Letters, Vol. 31, No. 5, 2010, pp. 657-662

[3] FOJT J, JOSKA L, MALEK J, SEFL V. Corrosion behavior of Ti-39Nb alloy for dentistry. Materials Science and Engineering C, Vol. 56, 2015, pp.532-537.

[4] VLCAK, P. Evidence of a defect structure in high fluence nitrogen ion implanted titanium and its effect on corrosion, Results in Physics, Vol. 11, 2018, pp. 274-277.

[5] POHRELYUK I.M., SHEYKIN S.E., PADGURSKAS J., LAVRYS S.M. Wear resistance of two-phase titanium alloy after deformation-diffusion treatment, Tribology International, Vol. 127, 2018, pp. 404-411.

[6] CHODUN R., NOWAKOWSKA-LANGIER K., WICHER B., OKRASA S., MINIKAYEV R., ZDUNEK K. Reactive sputtering of titanium compounds using the magnetron system with a grounded cathode, Thin Solid Films, Vol. 640, 2017, pp. 73-80.

[7] MOVASSAGH-ALANAGH F., ABDOLLAH-ZADEH A., ALIOFKHAZRAEI M., ABEDI M. Improving the wear and corrosion resistance of Ti-6Al-4V alloy by deposition of TiSiN nanocomposite coating with pulsed-DC PACVD, Wear, Vol. 390-391, 2017, pp. 93-103.

[8] VLCAK P., DRAHOKOUPIL J., VERTAT P., SEPITKA J., DUCHON J. Hardness response to the stability of a Ti(+N) solid solution in an annealed TiN/Ti(+N)/Ti mixture layer formed by nitrogen ion implantation into titanium. Journal of Alloys and Compounds, Vol. 746, 2018, pp. 490-495.

[9] VLCAK P., JIRKA I. Protective sliding carbon-based nanolayers prepared by argon or nitrogen ion-beam assisted deposition on Ti6Al4V alloy, Journal of Nanomaterials, Vol. 2016, Art. No. 1697090, 2016, pp. 1-9, Doi:10.1155/2016/1697090.

[10] UTRERA O.H., ABUNDIZ-CISNEROS N., SANGINES R., DILIEGROS-GODINES C.J., MACHORRO R. Cleaning level of the target before deposition by reactive direct current magnetron sputtering, Thin Solid Films, Vol. 646, 2018, pp. 98-104.

[11] VLCAK P. The effect of ion irradiation and elevated temperature on the microstructure and the properties of C/W/C/B multilayer coating, Applied Surface Science, Vol. 365, 2016, pp. 306-313.

[12] TOLDE Z., STARY V., CVRCEK L., VANDROVCOVA M., REMSA J., DANIS S., KRCIL J., BACAKOVA L., SPATENKA P. Growth of a TiNb adhesion interlayer for bioactive coatings, Materials Science and Engineering C, Vol. 80, 2017, pp. 652-658.

SYNTHESIS AND CHARACTERIZATION OF NANOPARTICLES USING MICROFLUIDICS

Aliye Hazal KOYUNCU, Ondřej KAŠPAR, Pavel ULBRICH, Viola TOKÁROVÁ

University of Chemistry and Technology Prague (UCT Prague), Prague, Czech Republic, EU,
tokarovv@vscht.cz

Abstract

In recent years, nanoparticles have high importance in various areas such as biomedical, environmental or pharmaceutical applications due to their chemical and physical properties. In all these applications, maintaining control over the particle size and shape is of high significance due to their size and shape-dependent properties. The most abundant method of synthesis of nanoparticles is a batch process because of its fast and easy reaction setup. However, many types of nanoparticles are very sensitive to reaction parameters which directly affect the size and shape of the final particles and the process is hardly reproducible.

In this work, we present nanoparticle synthesis using microfluidic devices. Compared to a batch process, microfluidics offers a high control over the reaction process by easy control of the system parameters (e.g. mixing and flow rates of reactants and reaction time) with the ability to make an operative change of parameters during the synthesis. Specifically, the synthesis of silica and silver nanoparticles using a droplet generation microfluidic device will be presented. The size and morphology of final nanoparticles will be analyzed and compared to a traditional batch method.

Keywords: Nanoparticles, microfluidics, silver, flow-focusing

1. INTRODUCTION

Nanoparticles have high importance in various areas such as bio-pharmaceutical, tissue-engineering, gene, and drug delivery, biomedicine and environmental technology due to their unique structural, chemical and physical properties [1,2]. Especially, biomedical applications like cancer treatment or bactericidal applications of nanoparticles have been used extensively [2,3]. For all these applications, controlling the particle size and producing a unique size and shape of particles are significant due to their size and shape-depended properties. Traditionally batch synthesis methods are the most preferred method for particles production [4,5]. However, in the batch methods, it is often hard to control bulk shearing forces, affecting the nucleation and consequently size and shape of the final particles. Besides, traditional methods are not flexible or efficient, and obtained particle size distributions are usually wide [6]. Therefore, there is an increasing demand for better synthesis methods to produce particles with a monodispersed size and defined morphology suitable for high-end applications [7].

Microfluidics is nowadays more than ever employed as a promising method for nanoparticle synthesis overcoming limits of batch-wise synthesis [8]. Firstly, large surface area to volume ratio of microchannels helps to increase mass and heat transfer in the system. It increases the efficiency of the reaction in a smaller volume compared to a high-volume batch method. Secondly, microchips are more suitable to work in harsh conditions in comparison to the batch reactors in regards to rapid temperature or pressure changes while using toxic and/or explosive materials [9]. Moreover, very fast reactions need to be effectively controlled in the batch, by mixing conditions or immediate parameter changes in the system. In those cases, microfluidic devices allow to control the parameters and to screen the system more flexible than batch reactor [6],[8] and it can provide higher reproducibility of final products [9].

In this paper, the synthesis of silver nanoparticles using the microfluidic device is discussed. The structure of the paper follows the experimental description of silver nanoparticle production by both microfluidic and batch methods, evaluation of the results obtained by characterization techniques, critical discussion of the suitability of both methods for efficient production of nanoparticles, and finally a conclusion.

2. MATERIALS AND METHODS

2.1. Chemicals

The following chemicals were used: silver nitrate ($AgNO_3$, Sigma, purity > 99 %), trisodium citrate dihydrate ($C_6H_5Na_3O_7.2H_2O$, Penta, purity > 99 %), tannic acid ($C_{76}H_{52}O_{46}$, Sigma Aldrich), Rhodamine B (Aldrich), polydimethylsiloxane (PDMS, Sylgard 184, Dow Corning), silicone elastomer (Sylgard 184, Dow Corning), mineral oil (Sigma), diethyl ether ($C_4H_{10}O$, Penta). Demineralized water (Aqual 25, conductivity ~ 0.07 µS/cm) was used for all reactions and treatment processes where noted.

3. EXPERIMENTAL PART

3.1. Batch process

The synthesis of silver nanoparticles in the batch was done by the precipitation of silver nitrate as the silver source according to Ranoszek-Soliwoda et al. [10]. An aqueous solution of silver nitrate (0.9625 mM, 95 mL), trisodium citrate dihydrate (0.07 M, 4.03 mL) and tannic acid (0.0015 M, 0.60 mL) were prepared using demineralized water. The molar ratio of the silver nitrate, trisodium citrate dihydrate, tannic acid was calculated to be 1.0:3.1:0.1. Silver nitrate solution was placed in a 250 ml glass round bottom flask, and the mixture of silver nitrate solution with the tannic acid solution was added dropwise into the flask at room temperature under continuous stirring at 500 rpm. When the addition was completed, stirring was kept for 15 more minutes. Samples were collected every minute for the first 5 minutes and every 5 minutes for the rest of the synthesis process. The samples were analysed by a dynamic light scattering (DLS, Malvern Zetasizer NanoZS) and spectrophotometer (Tecan Infinite 200 PRO Series). Additionally, nanoparticles were visually characterized by transmission electron microscopy (TEM, Jeol JEM-1010).

3.2. Microfluidic chip design

The microchip (**Figure 1A**) was designed to have three inputs, i.e. two for the disperse phases (1) and (2) and one for the continuous phase (3), the flow-focusing junction (4), where the droplet generation occurs, and one output (5). The basic principle using the microfluidic setup for nanoparticle preparation is based on a droplet formation where all reactants are mixed. The two immiscible fluids meet at the flow focusing point (**Figure 1B**) to generate droplets (aqueous solution) surrounded by the continuous phase (mineral oil). Each input was connected by a polytetrafluoroethylene (PTFE) capillary to a syringe (Hamilton) containing a reagent. A computer controlled linear pump (neMESYS) was used to precisely control the flow rates of the reagents and the continuous phase.

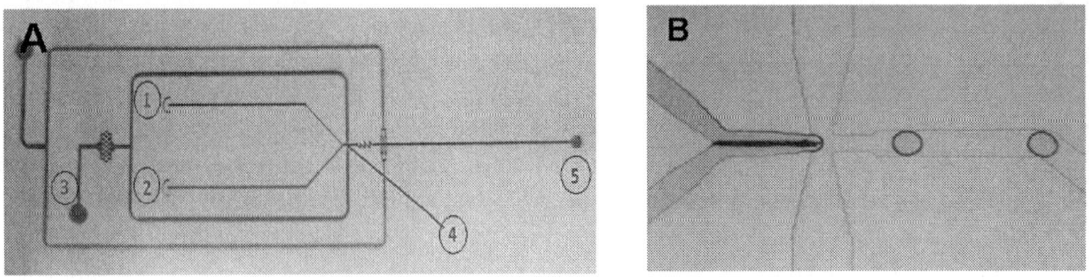

Figure 1 Microfluidic chip: A) silicon wafer with the chip design; B) the detailed image of the flow focusing part for droplet generation

3.3. Fabrication of PDMS microfluidic chips

The microchips were prepared using a soft lithography method using polydimethylsiloxane (PDMS) and a silicon wafer serving as a master mould. The PDMS was prepared by mixing the curing agent and the silicone elastomer (1:5 w:w) and poured over the silicon wafer placed in a Petri dish. The polymer mixture was degassed and heat-treated at 75 °C for 22 minutes. The crosslinked PDMS was peeled off from the wafer and holes were punched for the channel inlets and outlets. The bottom part of the chip was prepared using a PDMS mixture of 1:14 w:w ratio (curing agent to silicone elastomer) poured inside an empty Petri dish and put in the oven at 75 °C for 22 minutes. The surface of the PDMS layer remained sticky, and the upper part of the chip with channels was placed on top of it. The resulting chip made of the two combined parts was left in the oven at 75°C overnight to ensure strong bonding. The width of the nozzle orifice and outlet microchannel was 46 µm and 70 µm, respectively with a uniform height of 20 µm.

3.4. Microfluidic process

Silver nanoparticles were synthesized by the reduction reaction of silver nitrate using trisodium citrate dihydrate solution as a reduction agent. Briefly, silver nitrate (SN, $AgNO_3$) solution (1 mL and 0.92 mM), trisodium citrate dihydrate (SC, $C_6H_5Na_3O_7.2H_2O$) solution (1 mL and 5.72 mM) and tannic acid (TA, $C_{76}H_{52}O_{46}$) solution (1 mL and 0.185 mM) were prepared with demineralized water. The molar ratio of all reactants (1.0:3.1:0.1) was kept the same as in the batch process. 1 ml of TA solution was added into 1 ml SC solution as a stabilizing agent. They were mixed, and 1 ml of the mixture solution was taken as one of the dispersed phases. The both disperse phases - silver nitrate and trisodium citrate dihydrate with tannic acid solutions were pre-filtered before use. 1 ml SN solution was transferred into a 1 ml volume glass syringe. Then, 1 ml of SC and TA solution was transferred to the other glass syringe (Hamilton) of 1 ml volume. Finally, 2 ml of a mineral oil (as continuous phase) was transferred into a 2.5 ml volume glass syringe, and they were all connected to the microchip via PTFE (Adtech, 4 mm diameter) tubings. The flow rates were set for each phase and experimental setup individually by the linear pump software. The final product was collected into sample tubes pre-filled with 0.5 mL demineralized water to quench the reaction and particles growth. Then the W/O emulsion was separated by centrifugation for 3 minutes at 13.4 rpm (Eppendorf, MiniSpin) to oil and aqueous phases. After discarding of the oil phase, samples were further cleaned by the extraction with diethyl ether to remove remaining traces of the oil. **Table 1** summarizes studied experimental conditions for all samples.

Table 1 Experimental compositions and conditions of all studied samples of silver nanoparticle synthesis

Sample	C_{SN} [mM]	C_{SC} [mM]	C_{TA} [mM]	Flow Rate 1 Q_{SN} [µl/h]	Flow Rate 2 $Q_{SC\ and\ TA\ mixture}$ [µl/h]	Total Flow Rate $Q_{SC\ TA}+Q_{SN}$ [µl/h]	Flow Rate 3 $Q_{mineral\ oil}$ [µl/h]
1	0.92	5.72	0.185	8	8	16	60
2	0.92	5.72	0.185	8	8	16	80
3	0.92	5.72	0.185	16	16	32	60
4	0.92	5.72	0.185	16	16	32	80

4. RESULTS

4.1. Comparison of batch and microfluidic methods of silver nanoparticles production

Two different methods were used to produce silver nanoparticles using a consistent molar ratio of reagents (1.0:3.1:0.1) in the systems as described in the Experimental Section. The dynamic behaviour of the batch synthesis was examined using dynamic light scattering (DLS) and opposite to the expectations a linear growth of particles in time was not observed. Additionally, the samples were relatively polydisperse. On the other hand, in microfluidic production process, DLS measurements of the final product show less polydisperse populations than in the case of batch production (**Figure 2A**), except for one case, the product which was produced at 16 µl/h flow rates for both dispersed phases and 60 µl/h flow rate of the continuous phase (sample 3 from **Table 1**). The polydispersity in the final product is probably caused by the presence of many particle nuclei in some samples as it will be shown in section 4.3. All DLS data are shown in the result **Table 2**.

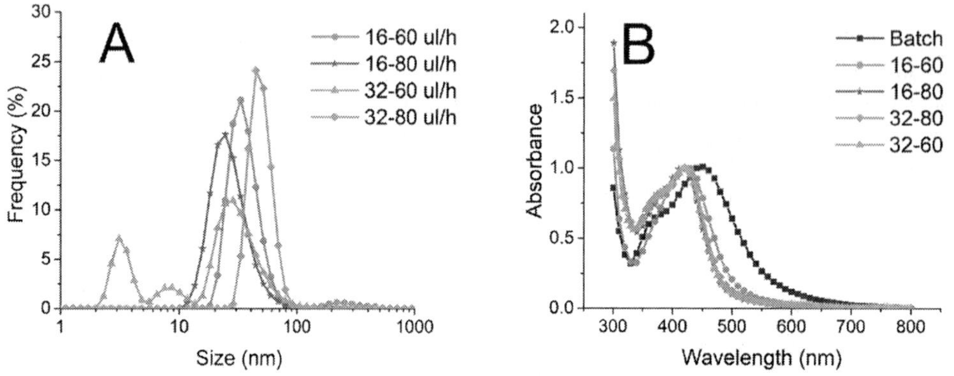

Figure 2 (A) Volume size distribution graphs of Ag nanoparticle samples produced by microfluidics with various flow rates; (B) UV-Vis spectra of Ag NPs prepared by both methods.

Figure 2B shows UV-Vis spectra for all prepared samples. The batch sample showed relatively constant spectra after the first 5 minutes with the maximal intensity at 440 nm. The higher wavelength of the maxima and broader peak are both indicators of larger particles and the presence of agglomerates. In contrast, results from microfluidic preparation show a much narrower spectra with a sharp peak with the absorbance maxima at 420 nm.

4.2. Image analysis of particles from TEM images

The transmission electron microscope (TEM) was used to observe particle structure, morphology, and size. The results show that at the end of the microfluidic process, the final particles are obtained mostly spherical and relatively monodisperse. Additionally, it also proves that there is no agglomeration in the final product. The batch product shows more polydisperse and larger particles having more irregular shape compared to microfluidic products.

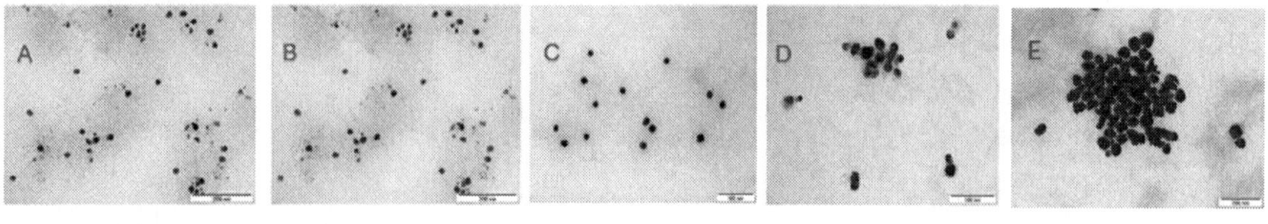

Figure 3 TEM pictures of final silver nanoparticles from microfluidic (A-D) and batch (E) proces: A) sample 1 B) sample 2, C) sample 3, D) sample 4 (all microfluidic samples correspond with the designation in **Table 10**), E) sample of the batch process collected at 15 min.

4.3. Effect of flow rate on particle size

The average sizes and their standard deviations for particles prepared using different flow rates of dispersed and continuous phases are shown in **Table 2**. The results of DLS are defined based on the volumetric size distributions. The results show that increased flow rate of the dispersed phase and fixed continuous phase lead to increased values of the final particle size measured by both DLS and image analysis of TEM images. By increasing the flow rate of the dispersed phase from 16 to 32 µl/h, the size of the particles increased from 27.7 nm to 46.1 nm. The results of the batch process shows the biggest size values of the silver nanoparticles.

Table 2 Influence of the flow rates of dispersed and continuous phases on particle size distribution for microfluidic experiments and comparison with the batch process.

Experiment No	$Q_{disperse}$ [µl/h]	$Q_{continuous}$ [µl/h]	Particle Size DLS (nm)	Particle Size TEM (nm)
1	16	60	32.8 ± 7.9	16.3 ± 9.0
2	16	80	27.7 ± 8.3	12.8 ± 9.9
3	32	60	35.7 ± 5.6	20.7 ± 14.4
4	32	80	46.1 ± 9.1	22.3 ± 8.2
Batch		-	65.2 ± 26.5	42.4 ± 29.1

5. CONCLUSION

In this work, silver nanoparticle synthesis is carried out by two different methods, batch and microfluidic. The nanoparticles are prepared using trisodium citrate dihydrate as a reducing agent, tannic acid as a stabilizer and finally silver nitrate as a metal reductant. The experimental results presents that both methods were successful in nanoparticle production. However, the microfluidic method provides better control over the system parameters via internal settings. In comparison to batch process, microfluidic process results in narrower particle size distribution and more spherical morphology. Additionally, silver solution changes colour to yellow-brown during the synthesis. Thus, due to the susceptibility of silver nanoparticles to light absorption, the particle size and shape could be measured during the process and their final properties could be controlled. Finally, both methods were compared successfully and the important parameters of the syntheses in relation to the resulting particle qualities were identified.

ACKNOWLEDGEMENTS

This paper is based upon work supported by the Czech Science Foundation (project GACR 17-11851Y).

REFERENCES

[1] Y. Deng, C. Deng, D. Yang, C. Wang, S. Fu, and X. Zhang, "Preparation, characterization and application of magnetic silica nanoparticle functionalized multi-walled carbon nanotubes," *Chem. Commun.*, no. 44, p. 5548, 2005.

[2] C.-H. Weng, C.-C. Huang, C.-S. Yeh, H.-Y. Lei, and G.-B. Lee, "Synthesis of hexagonal gold nanoparticles using a microfluidic reaction system," *J. Micromechanics Microengineering*, vol. 18, no. 3, p. 035019, Mar. 2008.

[3] J. R. Morones *et al.*, "The bactericidal effect of silver nanoparticles," *Nanotechnology*, vol. 16, no. 10, pp. 2346-2353, Oct. 2005.

[4] J. B. Wacker, I. Lignos, V. K. Parashar, and M. A. M. Gijs, "Controlled synthesis of fluorescent silica nanoparticles inside microfluidic droplets," *Lab Chip*, vol. 12, no. 17, p. 3111, 2012.

[5] S. S. Eun, Y. Sunhye, C. H. Hee, and C. Soonja, "Fully crosslinked poly(styrene-co-divinylbenzene) microspheres by precipitation polymerization and their superior thermal properties," *J. Polym. Sci. Part A Polym. Chem.*, vol. 42, no. 4, pp. 835-845.

[6] G. A. Patil, M. L. Bari, B. A. Bhanvase, V. Ganvir, S. Mishra, and S. H. Sonawane, "Continuous synthesis of functional silver nanoparticles using microreactor: Effect of surfactant and process parameters," *Chem. Eng. Process. Process Intensif.*, vol. 62, pp. 69-77, Dec. 2012.

[7] J. Wang *et al.*, "Droplet Microfluidics for the Production of Microparticles and Nanoparticles," *Micromachines*, vol. 8, no. 1, p. 22, Jan. 2017.

[8] C.-X. Zhao, L. He, S. Z. Qiao, and A. P. J. Middelberg, "Nanoparticle synthesis in microreactors," *Chem. Eng. Sci.*, vol. 66, no. 7, pp. 1463-1479, 2011.

[9] N. Hao, Y. Nie, and J. X. J. Zhang, "Microfluidic synthesis of functional inorganic micro-/nanoparticles and applications in biomedical engineering," *Int. Mater. Rev.*, pp. 1-27, 2018.

[10] K. Ranoszek-Soliwoda *et al.*, "The role of tannic acid and sodium citrate in the synthesis of silver nanoparticles," *J. Nanoparticle Res.*, vol. 19, no. 8, p. 273, Aug. 2017.

LASER-GENERATED NANOPARTICLES TO CHANGE PHYSICAL PROPERTIES OF SOLIDS, LIQUIDS AND GASES

Alfio TORRISI[1], Mariapompea CUTRONEO[1], Giovanni CECCIO[1], Antonino CANNAVO'[1], Pavel HORÁK[1], Lorenzo TORRISI[2], Jiří VACIK[1]

[1]*Nuclear Physics Institute of the CAS, Řež, Czech Republic, EU, torrisi@ujf.cas.cz*
[2]*Department of Mathematical and Computer Science, Physical Sciences and Earth Sciences, University of Messina, Italy, EU*

Abstract

Synthesis of nanoparticles was possible employing a Nd:YAG pulsed laser at fundamental harmonic. The production of nanoparticles in water depends mainly on the laser parameters (pulse duration, energy, wavelength), the irradiation conditions (focal spot, repetition rate, irradiation time) and the medium where the ablation occurs (solid target, water, solution concentration). The nanoparticles can be introduced in solids, liquids or gases to change many physical characteristics. The optical properties of polymers and solutions, the wetting ability of liquids, the electron density of laser-generated plasma, represent some examples that can be controlled by the concentration of metallic nanoparticles (Au, Ag, Ti, Cu). Some bio-medical applications will be presented and discussed.

Keywords: Nanoparticles, laser, plasma, bio-medical applications

1. INTRODUCTION

Laser interaction with solid targets in liquids represents a promising physical method to produce nanoparticles (NPs) without chemical agents. This technique allows to control the size, structure, concentration and morphology of the NPs by monitoring of the laser parameters (i.e., wavelength, fluence, spot size, pulse width, repetition rate and irradiation time), the irradiation conditions (focus, incidence angle, ...) and the target properties (composition, thickness, ...) [1]. Laser ablation induces physical modification of matter, due to photo-thermal and photo-chemical effects, and molecular fragmentation from solid into nano-particulates when the irradiation occurs in liquids instead in vacuum [2,3].

The pulsed laser ablation (PLA) process [4] in liquids involves different steps [5,6]: the laser beam is focused on the surface of the target, which is plunged into the liquid, and it is partially absorbed producing a direct photoionization on the target surface. A detachment of the material from the target occurs by thermal and Coulomb mechanisms with a consequent cavitation bubble effect, due to the energy released by the plasma plume to the surrounding liquids at the plume-liquid interface, ending with the bubble collapse. In the end, the plasma plume formation and its confinement at the metal-liquid interface induce the enhancement of temperature and pressure, a localised emission of atoms and electrons occurs and their strong interaction, due to the not possible expansion in the dense liquid, induces NPs production and a liquid solution generation [7].

The solutions, containing metallic nanoparticles at different concentrations, significantly modify the original physical and chemical properties of the liquid, such as the surface tension, density, viscosity, thermal and electric conduction [8,9]. The localised laser pulse induces target heating and the water causes fast cooling resulting in preferential crystallisation and grain growth of NPs [10].

The produced NPs generally have a spherical shape, a composition depending on the irradiated target and a size distribution depending on the used laser parameters and irradiation conditions. A special attention is given to metallic NPs of Au, Ti, Ag and C. The produced NPs in water can be employed to be inserted in solids, liquids and gases to modify their physical and chemical properties.

Surface Plasmon Resonance (SPR) absorption of the incident light, producing high absorption bands at particular wavelengths, is one of the main optical properties of metallic nanoparticles. It consists of collective oscillations of the conduction electrons excited by the electromagnetic field of the incident light. Thus, a metallic nanoparticle can be described as quasi-free sphere conduction electrons, moving on its surface. The light of the electromagnetic field exerts a force on moving conduction electrons of the surface, which induces an electric dipole when a nanoparticle is illuminated [11]. This dipole, in turn, generates an electric field opposite to that of the incident light, which forces electrons to return to their equilibrium position. After removing the external electric field, the electrons oscillate with a so-called "plasmon frequency", which is proportional to the square root of their density [11]. The surface plasmon oscillations are affected by various factors such as the composition, the size and shape of the nanoparticles. Its role plays also the surrounding medium and the electric field of the other close nanoparticles. Thanks to their peculiar properties, the Au nanoparticles are finding many potential applications for different scientific fields, from physics to chemistry, from material sciences to microelectronics and bio-medicine [1,7].

Introducing Au-NPs in polymer, it is possible to change their optical, mechanical, electrical and thermal properties, as reported in literature [1,8]. Depending on the NPs concentration, a transparent polymer can be modified in a very absorbent material, an insulator in a conductive medium, an elastic material in a very strong one [12]. The employment of biocompatible nanoparticles at low concentration in water, physiological solutions and blood determines a different wetting ability, liquid density and viscosity, electrical and thermal conductivity and a different color. For example, the different concentration of Au-NPs in water changes the color solution, the optical absorbance at specific bands in the visible region depends on the NPs size and produces a change in the solution electrical conductivity [8]. The production of ionized gas using pulsed and high intensity lasers irradiating of solids produces plasmas, which expand at high velocity in vacuum. If the irradiated target contains NPs, it is possible to modify the plasma properties. For example, the use of Au-NPs increases the electron density of the plasma and its temperature. Moreover, the electric field generated at the target surface and responsible of the ion acceleration in plasma can be increased, as reported in literature [13]. The high-energy absorption due to NPs enhances the local temperature of the light illumined medium containing metallic nanoparticles. Moreover, metallic NPs in biological tissues can be efficiently employed for photo thermal therapy induced by lamps or lasers operating at specific wavelengths [14].

In this paper, it will be described a physical method to generate nanoparticles, mainly based on gold, using a high intensity pulsed laser. Successively, three applications will be considered for employment of the gold nanoparticles in solids, liquids and gases.

2. EXPERIMENTAL SETUP

Metallic NPs, based on Au, Ag, Cu and Ti, were prepared using the "top-down" method by laser ablation in water [15,16]. A Nd:YAG laser operating at 1064 nm wavelength, with 3 ns pulse duration, 200 mJ pulse energy, 1 mm^2 focused spot and 10 Hz pulse repetition rate was employed. A metallic target having a surface of 1 cm^2 and 1 mm thickness was placed on a bottom of a glass tube filled with 10-30 ml of distilled water. The laser light was focused onto the metallic target with a 500 mm focal length convergent lens to obtain 1 mm of the spot diameter. An exposition time of the order of 10-30 min with repetitive laser shots was used. The ablation yield depends on the irradiated material, generally concentration of solution of the order of 0.1-10 mg/ml is prepared. A scheme of the experimental setup is shown in **Figure 1a** [2].

Figure 1b reports on typical curves of ablation yield for different targets (Au, Ag, Cu, Ti) as a function of the laser fluence (energy pulse per surface unit, J/cm^2) [17]. One can see that for the same value of fluence, the ablation yield is higher for Au and lower for Ti. The increment of the ablation yield means that the high concentration of NPs in solution can be obtained for relatively low laser fluence and irradiation times. The ablation yield has been measured using the weight loss with a micro-scale of a little target irradiated for long time (30 min) at the 10 Hz repetition rate. The NP structure was obtained by irradiation the solid target in water, placed in about 5 mm depth with respect to the free water surface. After the laser irradiation, in order to avoid effect of the NP coalescence, 1-5 ml surfactant (sodium citrate) was added to the solution.

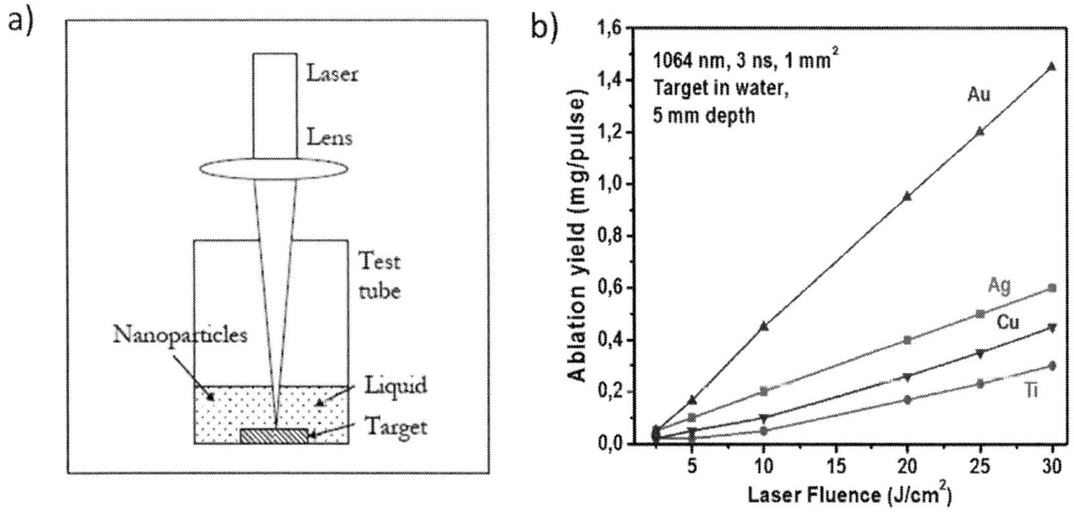

Figure 1 a) Scheme of the experimental set-up used for the laser generation of nanoparticles in water and b) measurements of ablation yield vs. laser fluence for different metallic targets

The NPs production was investigated using the SEM and TEM microscopy. The EDX analysis were performed using detection of the characteristic X-rays emitted from the nanoparticles in order to verify the purity of the produced nanoparticles and their composition. Their size distribution was plotted in order to obtain information on the average size and on its indetermination by TEM image acquisition, considering a large number of nanoparticles. Using an image containing about 100 NPs or more, the single diameters were considered, employing an appropriate imaging software, as reported in [3].

Liquid solutions containing NPs were characterized using the near-UV-VIS-near-IR wavelengths produced by different lasers and lamps in order to measure the solution absorbance. The wetting ability of the solution was also measured by the contact angle of 1 ml liquid drop placed on different polished substrates. The contact angle was measured using an appropriate optical microscope. Polymers with NPs were obtained by embedding the NPs into the melted polymers. The optical properties of such polymers were measured using the transmission and absorption coefficients in thin films. The electrical conductivity of the polymers with the embedded NPs were measured using a Keithley modules for I-V measurements in DC [18]. Measurements of plasma generation by Nd:YAG laser, operating at 1064 nm in single pulse (3 ns), were generated in high-vacuum conditions. The ion velocity generated in such gas, as well as the properties of NPs employed in laser irradiated target were measured by time-of-flight (TOF) technique, employing ion collectors attached to a fast storage oscilloscope, as reported in literature [19].

3. RESULTS

Figure 2 shows a SEM (a) image of metallic NPs of Au deposited on a carbon substrate on the stab. The average dimension of such spherical NPs is about 10 nm. Their composition was measured using the EDX analysis incited by the 20 keV electron beam. A typical EDX investigation reporting on the Au-NPs purity is presented in **Figure 2b**; TEM microscopy gives an average diameter of 10 nm (c) [17].

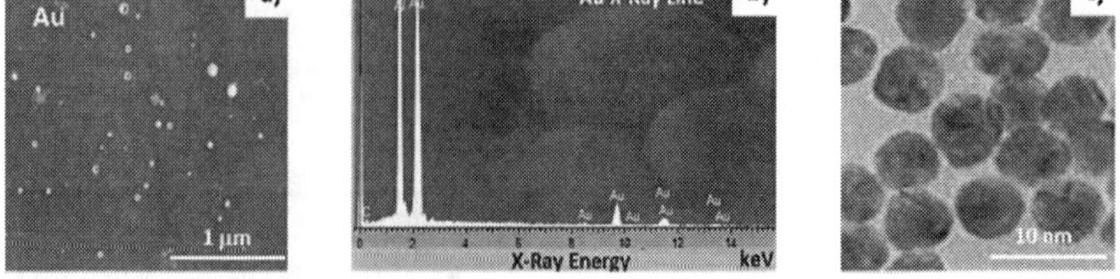

Figure 2 a) SEM images of Au-NPs on carbon surface, b) typical EDX analysis and c) TEM image.

3.1. Applications to liquids

The liquid containing NPs can be analyzed by optical spectroscopy to determine its absorbance in the UV-Vis-IR wavelength regions, to obtain information about the characteristics (shape and size) of the embedded NPs. **Figure 3a** reports on an example of Au-NPs generated in water and analyzed in absorbance vs. wavelength [1]. The optical density of the liquid contained in a cuvette shows an absorption peak at specific wavelength ranges, increasing with the size of the nanoparticles. For nanoparticles of about 20 nm in diameter the SPR peak is localized at about 525 nm, while for NPs with 100 nm in diameter the peaks shifts to about 580 nm. **Figure 3b** reports on the different coloration of the solution, which turns from rose to red and then violet with increasing of the Au-NPs concentration.

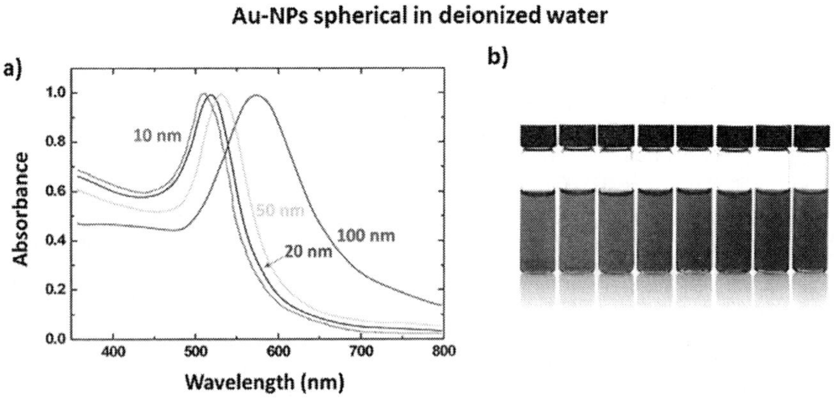

Figure 3 a) Measurements of absorbance in water containing Au-NPs vs. wavelength and diameter of the NPs, and b) different colorization of the solutions with increasing the NPs concentration [20]

Figure 4a-f reports on the typical examples of the contact angle measurements of water without and with Au-NPs deposited as a drop on different polished substrates (polyethylene-PE (a, b), Ti (c, d) and Si (e, f) [8]. The solution had an Au-NPs concentration of 0.3 mg/ml. The contact angle θ was measured by the resulting angle between the forces applied to the solid-gas interface (γ_{SG}), solid-liquid interface (γ_{SL}) and liquid-gas interface (γ_{LG}), as reported in the equation given in the scheme of **Figure 4g** [8]. One can see that the wetting ability of the solution increases using the gold NPs, as it is demonstrated by the lower contact angle in the presence of NPs. The contact angle in fact decreases from 94° to 85° in PE, from 91° to 76° in Ti and from 67° to 42° in Si. It means that the Au-NPs induce a decreasing of the cohesion forces of the liquid solution molecules and enhancing the adhesion forces with the solid surface. This result was obtained also for other substrates [8], such as Cu, Ag, Al, polymethyl methacrylate (PMMA), SiO_2 and glass, demonstrating that the increased wetting ability of the solution containing Au-NPs represents a general behavior. This result is important for applications in different fields, such as in bio-medicine, where to improve the wetting ability of biomaterials with liquids and biological tissues Au-NPs can be injected in the liquid or deposited on the surface of the biomaterials.

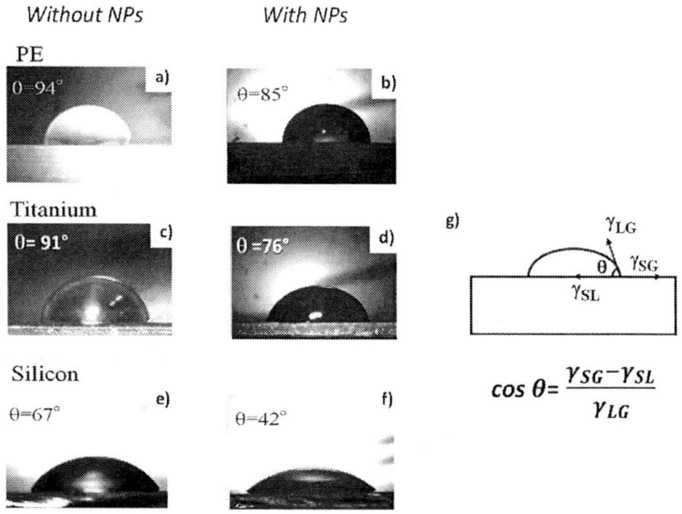

$$\cos\theta = \frac{\gamma_{SG} - \gamma_{SL}}{\gamma_{LG}}$$

Figure 4 Examples of contact angle measurements of water (a, c, e) and water containing Au-NPs (right) deposited as a drop on different substrates of b) PE, d) TI and f) Si. In g) a scheme is depicted a relation between contact angle and surface tension is depicted

The employment of metallic NPs both in the liquid and on the solid surfaces increases their hydrophilic trend, as it is reported in literature [8] This behavior can improve the cell growth on the implanted prosthesis stimulating the tissue integration and other physiological processes that prolong the prosthesis life, as indicated in the **Figure 5**.

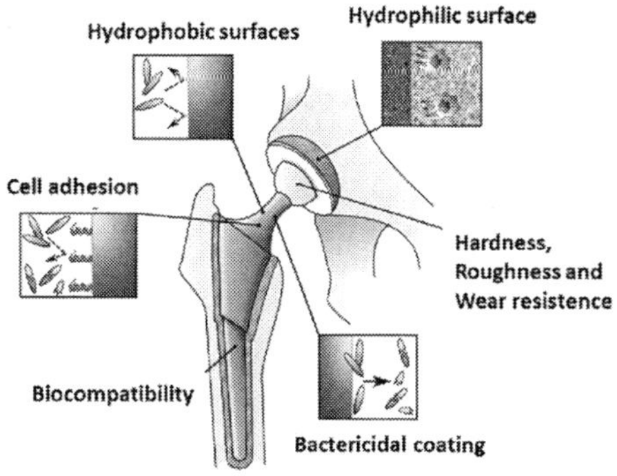

Figure 5 Scheme of the prosthesis implantation in biological environment with the use of NPs at its surface [21]

3.2. Applications to solids

The introduction of NPs in polymers changes their physical and chemical properties. In particular, the optical properties can be modified drastically also by employing low NPs concentrations. **Figure 6a** [17] shows the measured absorbance vs. wavelength for polyethylene pure and containing uniformly embedded graphene nanoparticles at different concentration up to 12 ml solution, and containing also Au-NPs at a concentration of 1 % in weight. One can see that the absorbance in the visible range increases significantly, for example at 500 nm from 50 % in pure PE increases up to 98 % in PE with graphene oxide water dispersion NPs at 0.4 wt%

concentration. In presence of Au-NPs with about 10 nm in diameter, the absorbance shows an evident SPR resonant absorption peak centered at about 510 nm. **Figure 6b** shows another example of PE absorption coefficient vs. wavelength [22]. The polymer is of high density ultra-high-molecular-weight-poly-ethylene) (UHMWPE) in which carbon nanotubes (CNT) are embedded during the melting phase at different concentrations from zero up to 1 % in weight.

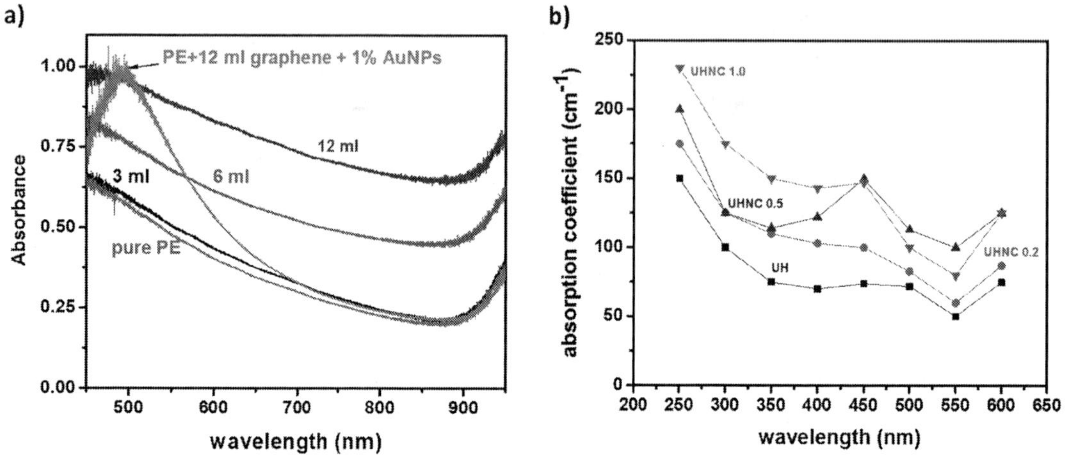

Figure 6 a) Absorbance vs wavelength for polyethylene containing graphene and Au-NPs, and b) absorption coefficient in UHMWPE vs. wavelength containing the CNT nanoparticles

A special application due to such polymer modification can be found in the field of laser welding of polymeric materials. Two polymers, one as pure and transparent to the laser wavelength and the other rich in C-NPs content, black and strongly absorbent, can be welded at their interface using the technique of known as Laser Transmission Welding (LTD). The laser pulse is transmitted by the first polymer foil and absorbed to the surface of the under-placed second polymer. The laser energy released at the interface of the two pressed foils produces local heating and polymer melting inducing a resistant welding of the two polymeric foils, as was already reported in [22]. **Figure 7a** reports on a scheme of the used experimental set-up for the LTD technique in polymers, and **Figure 7b** a photo of the produced polymeric welding.

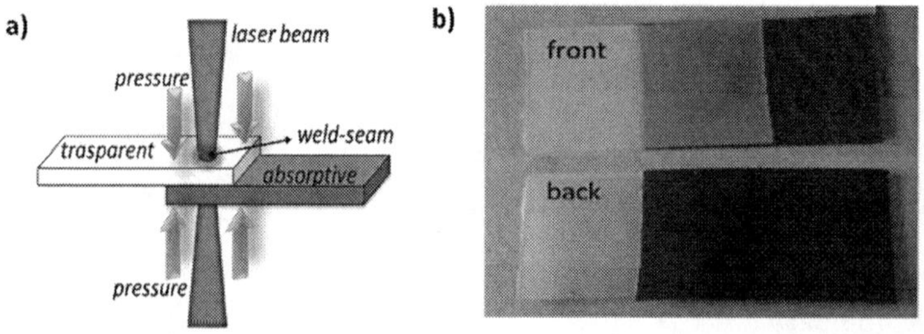

Figure 7 a) Scheme of the LTD laser welding technique, and b) photo of the welded polymers

3.3. Applications to gases

Employment of minimum quantities of PE, of the order of 0.1 % in weight, changes the color of the polymer from white and transparent to black strongly absorbent material. The use of metallic nanoparticles can be extend also to gas phases in order to modify, significantly, their properties. Many applications concern the modifications of the laser-generating plasma properties. A plasma represents an ionized gas at high temperature and density, which, in absence of charge equilibrium, shows high electric field responsible to high ion acceleration along preferential directions. Nano- and sub-nano laser pulses at high intensity (>10^{10} W/cm^2) irradiating a solid material (target) placed in vacuum, generate a non-equilibrium plasma which characteristics of temperature, density, photon, electron and ion emission, etc., can be measured with different on line techniques (TOF, optical spectroscopy, Langmuir probes, ion collectors, semiconductor detectors, mass quadrupole spectrometry, plastic detectors, etc.). A typical scheme of the experimental set-up used to generate plasma is shown in **Figure 8a**.

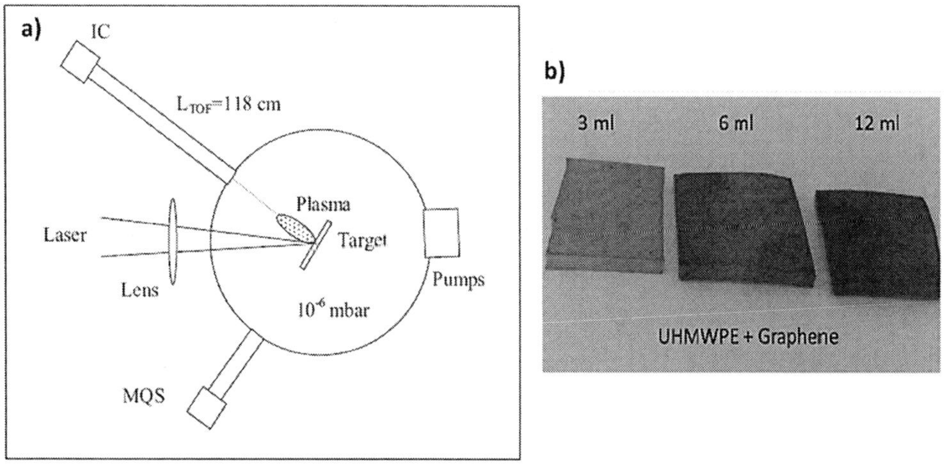

Figure 8 a) Scheme of the set-up for plasma produced by laser, and b) photo of PE with different C-NPs content

A long tube is employed to detect the emitted ions with an ion-collector (IC) used in TOF approach, i.e., by knowing the detector-target distance and measuring the time of arrival of the ions with a fast storage oscilloscope an energy of ions can be found. The irradiated targets are placed in the centrum of the vacuum chamber and consists of polymers without and with embedded NPs [17]. **Figure 8b** shows the photo of some samples, in which different concentrations of graphene nanoparticles are embedded. The presence of NPs modifies the plasma temperature and electron density, depending strongly on the atomic number and concentration of the used NPs species. For example, the use of Au-NPs enhances the effective atomic number of the target and permits to generate a plasma with a greater electron density and a great charge not-uniformity in the plasma [23]. Typical results can be observed by the TOF spectra acquired during the laser-plasma generation measuring the velocity of the emitted ions. Using a Nd:YAG laser at 3 ns pulse duration and 10^{10} W/cm^2 intensity operating in single pulse at 1064 nm wavelength, plasmas from polymer targets were generated. **Figure 9** shows an example of the results obtained by embedding graphene NPs and Au-NPs in UHMWPE [17]. Their presence increases the number of emitted particles, as was observed from the higher ion yield (mV), and the higher ion velocity, as was observed from the lower TOF time (μs) necessary to reach the IC detector 118 cm distant form the target. This confirmed high kinetic energy of the emitted ions (protons and carbons) achieving 200 eV/charge state [24].

Using high laser intensity (>10^{18} W/cm^2) and brief pulses (tens of femtoseconds), it is possible to accelerate ions in target-normal-sheat-acceleration (TNSA) regime at about 10 MeV per charge state. This allows to use the energetic ions for many applications, from nuclear physics (to produce nuclear reactions) to chemistry (to induce chemical reaction), from microelectronics (to implant ions in substrates) to medicine (to use the ions for radiotherapy), and others [24].

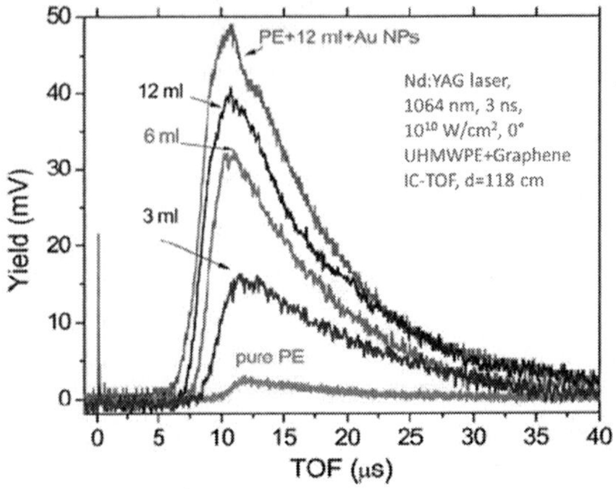

Figure 9 TOF spectra of the ions emitted form a laser-generated plasma form pure polymer and polymers containing C-NPs and added Au-NPs

4. CONCLUSIONS

The production of nanoparticles from the laser ablation in water technique permits to generate metallic and not metallic nanoparticles with narrow size distribution, generally below 100 nm in diameter. The repetitive laser pulse irradiation and the long irradiation time permit to control the solution concentration and the use of surfactants, such as sodium citrate, avoid the nanoparticle coalescence.

These nanoparticles are employed more and more in different scientific fields to modify the properties of liquids, solids and gases. Their composition, shape, size and concentration permit to modify the physical and chemical properties of many materials, such as biological solutions, polymers and plasma gases. The paper presents results about the liquid color and absorbance variation due to SPR effects, the optical properties of polymers in which the NPs are embedded uniformly during the melting phase, and the use of nanoparticles in laser irradiated targets to generate plasma with high temperature, density and energetic particles. Moreover, inhere only few applications of NPs were shown, but, obviously, many other examples can be given, from the preparation of special materials at high mechanical resistance and low density, to the use of special sensors, of chemical and pharmaceutical species and the use of NPs to increase the medical imaging contrast and the enhancement of absorbed dose in radiotherapy.

In this contest, the particular interest is devoted to Au-NPs due to their high atomic number and atomic weight, their easy synthesis and stability, their exhibition of excellent biocompatibility and chemical passivation, their high mass absorption coefficient for X-rays, and high particle stopping powers. Due to the high electron density, Au-NPs significantly enhance the absorption of X-rays in the medium where they are embedded, in dependence of the used concentration. In the same manner, their introduction in biological liquid, soft or hard tissues enhances the electron and ion stopping powers, i.e., the energy released by the incident corpuscular radiation. At microscopic level, Au particles of dimensions below 5-10 nm can be filtered by the kidney and spleen and transferred to all organs. Such nanoparticles may diffuse not only in the extracellular liquids, but can also cross the cell membrane, diffuse in the intracellular liquid arriving in the nucleus and permitting their bonding with the DNA. These effects are of special interest in the case of tumor cells because radiotherapy permits high release of radiation energies sufficient to damage the DNA and the entire cell. Thus, it means to increase the absorbed dose and to destroy more specifically tumor sites containing the gold solution during radiotherapy expositions [14]. Moreover, the SPR effect can be employed to localize heating during photo-thermal therapy induced by lamps or laser light [3].

ACKNOWLEDGEMENTS

This research was supported by University of Messina Research & Mobility 2016 Project (project code RES_AND_MOB_2016_TORRISI) n. 74893496 and by Grant Agency for the Czech Republic, project GACR "Center of Excellence", N. P108/12/G108.

REFERENCES

[1] TORRISI, Lorenzo, CUTRONEO, Mariapompea, SILIPIGNI, Letteria, BARRECA, Francesco, FAZIO, Barbara, RESTUCCIA, Nancy and KOVACIK, Lubomir. Gold nanoparticles produced by laser ablation in water and in graphene oxide suspension. *Philosophical Magazine.* 2018. vol. 98, iss.24, pp. 2205-220. Available from: DOI: 10.1080/14786435.2018.1478147.

[2] TORRISI, Lorenzo and SCOLARO, Cristina. Nanoparticles improving the wetting ability of biological liquids. *Journal of Thermodinamics & Catalysis.* 2017. vol. 8, iss. 1, pp. 1000184, 1-6. Available from: DOI: 10.4172/2157-7544.1000184.

[3] TORRISI, Lorenzo, RESTUCCIA, Nancy and PATERNITI, Irene. Gold Nanoparticles by Laser Ablation for X-Ray Imaging and protontherapy Improvements. *Recent Patents on Nanotechnology.* 2018. vol. 12, iss.1, pp. 59-69. Available from: DOI: 0.2174/1872210511666170609093433.

[4] YANG, Guowei, Laser ablation in liquids: Applications in the synthesis of nanocrystals. *Progress in Materials Science.* vol. 52, iss. 4, pp. 648-698, 2007. Available from: DOI: 10.1016/j.pmatsci.2006.10.016.

[5] AMENDOLA, Vincenzo and MENEGHETTI, Moreno. What controls the composition and the structure of nanomaterials generated by laser ablation in liquid solution? *Physical Chemistry Chemical Physics.* 2013. vol. 15, iss. 9, pp. 3027-3046. Available from: DOI: 10.1039/C2CP42895D.

[6] TACCOGNA, Francesco, DELL'AGLIO, Marcella, RUTIGLIANO, Maria, VALENZA, Gabriele and DE GIACOMO, Alessandro. On the growth mechanism of nanoparticles in plasma during pulsed laser ablation in liquids. *Plasma Sources Sciences and Technology.* 2017. vol. 26, iss. 4, pp. 045002, 1-10. Available from: DOI: 10.1088/1361-6595/aa595b.

[7] TORRISI, Lorenzo and SCOLARO, Cristina. Blood Wettability of Haemocompatible Carbon-based Materials. *Journal of Advanced Chemical Engineering.* 2017. vol. 7, iss.2, pp. 1000179, 1-6. Available from: DOI: 10.4172/2090-4568.1000179.

[8] TORRISI, Lorenzo, SCOLARO, Cristina and RESTUCCIA, Nancy. Wetting ability of biological liquids in presence of metallic nanoparticles. *Journal of Materials Science: Materials in Medicine.* 2017. vol. 28, iss. 63, pp. 1-12. Available from: DOI: 10.1007/s10856-017-5871-1.

[9] TORRISI, Lorenzo, RESTUCCIA, Nancy, CUZZOCREA, Salvatore, PATERNITI, Irene, LELO, Isidora, PERGOLIZZI, Stefano, CUTRONEO, Mariapompea and KOVACIK, Lubomir. Laser-produced Au nanoparticles as X-ray contrast agents for diagnostic imaging. *Gold Bullettin.* 2017. vol. 50, iss.1, pp 51-60. Available from: DOI: 10.1007/s13404-017-0195-y.

[10] JEON, Jin-Woo, YOON, Sangwoo, CHOI, HaeWoon, KIM, Joohan, FARSON, Dave and CHO, Sung-Hak. The Effect of Laser Pulse Widths on Laser-Ag Nanoparticle Interaction: Femto- to Nanosecond Lasers. *Applied Sciences.* 2018. vol. 8, iss. 1, pp. 1-13. Available from: DOI: 10.3390/app8010112.

[11] GARCIA, Miguel Ángel. Surface plasmons in metallic nanoparticles: fundamentals and applications. *Journal of Physics D: Applied Physics.* 2011. vol. 44, iss. 28, pp. 283001, 1-22. Available from: DOI: stacks.iop.org/JPhysD/45/389501.

[12] GALTIERI, Giuseppe, VISCO, Anna Maria, NOCITA, Davide, TORRISI, Lorenzo, CECCIO, Giovanni and SCOLARO, Cristina. Polyethylene laser welding based on optical absorption variations. *Journal of Instrumentation.* 2016. vol. 11, pp. C04013 1-9. Available from: DOI: 10.1088/1748-0221/11/04/C04013.

[13] TORRISI, Lorenzo, CALCAGNO, Lucia, CUTRONEO, Mariapompea, BADZIAK, Jan, ROSINSKI, Marcin, ZARAS-SZYDLOWSKA, Agnieszka and TORRISI, Alfio, Nanostructured targets for TNSA laser ion acceleration. *Nukleonika.* 2016. vol. 61, iss. 2, pp.103-108. Available from: DOI: 10.1515/nuka-2016-0018.

[14] TORRISI, Lorenzo. Evaluation of the Radiotherapy and Protontherapy improvement using gold nanoparticles, Gold Bullettin. 2017. vol. 50, iss.4., Available from: DOI:10.1007/s13404-017-0216-x.

[15] TIWARI, Dhermendra, BEHARI, Jitendra and SEN Prasenjit. Time and dose dependent antimicrobial potential of Ag nanoparticles synthesized by top-down approach. 2008. *Current Science.* vol. 95, iss.5, pp. 647-655. Available from: http://www.currentscience.ac.in/Downloads/download_pdf.php?titleid=id_095_05_0647_0655_0.

[16] GARRIGUE, Patrick, DELVILLE, Marie-Hélène, LABRUGERE, Christine, CLOULET, Eric, KULESZALL, Pawel J. Kuleszall, MORAND, Jean Pierre, and KUHN, Alexander.Top−Down Approach for the Preparation of Colloidal Carbon Nanoparticles. *Chemistry of material.* 2004. vol. 16, iss. 16, pp. 2984-2986. Available from: DOI: 10.1021/cm049685i.

[17] TORRISI, Lorenzo, CECCIO, Giovanni, RESTUCCIA, Nancy, MESSINA, Elena, GUCCIARDI, Pietro Giuseppe and CUTRONEO, Mariapompea. Laser-generated plasmas by graphene nanoplatelets embedded into polyethylene. *Laser and Particle Beams.* 2017. vol.35, iss. 2 pp. 294-303. Available from: DOI: 10.1017/S0263034617000179.

[18] Keithley, "I-V Systems." [Online]. Available: https://www.tek.com/datasheet/system-93-i-v-system.

[19] WORYNA, Eugeniusz, PARYS, Piotr, WOLOWSKI Jerzy and MROZ, Waldemar. Corpuscular diagnostics and processing methods applied in investigations of laser-produced plasma as a source of highly ionized ions. *Laser and Particle beams.* 1996. vol. 14, iss. 3, pp. 293-321. Available from: DOI: 10.1017/S0263034600010053.

[20] NanoComposix, "Gold Nanoparticles in Solution." [Online]. Available: https://nanocomposix.com/collections/gold-spheres.

[21] GALLO, Jiri, HOLINKA, Martin And MOUCHA, Calin S. Antibacterial surface treatment for orthopaedic implants. *International Journal of Molecular Sciences.* 2014. vol. 15, iss. 8, pp. 13849-13880. Available from: DOI: 10.3390/ijms150813849

[22] VISCO, Anna Maria, TORRISI, Lorenzo, GALTIERI, Giuseppe and SCOLARO, Cristina, Effect of the filler amount on the optical absorption properties and the surface features of polymeric joints based on biomedical UHMWPE welded by a Nd:YAG Laser. *Journal of Thermoplastic Composite Materials.* 2017. vol. 30, iss. 12, pp. 1675-1692. Available from: DOI: 10.1177/0892705716662515.

[23] TORRISI, Lorenzo, CALCAGNO, Lucia, GIULIETTI, Danilo, CUTRONEO, Mariapompea, ZIMBONE, Massimo and SKALA, Jiří. Laser irradiations of advanced targets promoting absorption resonance for ion acceleration in TNSA Regime. *Nuclear Instruments and Methods B.* 2015. vol.355, pp. 221-226. Available from: DOI: 10.1016/j.nimb.2015.01.019.

[24] TORRISI, Lorenzo, CUTRONEO, Mariapompea, CECCIO, Giovanni, CANNAVO', Antonino, BATANI, Dimitri, BOUTOUX, Guillaume, JAKUBOWSKA, Katarzyna and DUCRET, Jean-Eric. Near monochromatic 20 Me V proton acceleration using fs laser irradiating Au foils in target normal sheath acceleration régime. *Physics of Plasmas.* 2016. vol. 23, iss.4, pp. 043102, 1-6. Available from: DOI: 10.1063/1.4945637.

STUDY OF EFFICIENCY OF HALLOYSITE NANOTUBES DISPERGATION IN POLY(VINYL ACETATE) MATRIX

Přemysl MENČÍK, Radek PŘIKRYL, Veronika MELČOVÁ

Brno University of Technology, Brno, Czech Republic, EU, mencik@vutbr.cz

Abstract

Nanocomposite system formed by poly(vinyl acetate) (PVAc) matrix and halloysite nanotubes (HNT) filler was prepared in three different ways. The first method was kneading the composite melt in laboratory kneader. The second one was direct emulsion polymerization of vinyl acetate in the presence of HNT. The third method of composite preparation was to disperse HNT in toluene solution of PVAc in planetary ball mill. The filler content was 10 wt. % in all composites. The samples from all three preparation methods were subjected to the observation of nanoparticles dispergation efficiency within the bulk material using SEM. The best level of particles dispersion was achieved using the ball mill. The particles were effectively dispersed, forming the agglomerates in the range from below one micrometer down to single nanotubes.

Keywords: Nanocomposite, halloysite, dispergation, ball mill

1. INTRODUCTION

The usage of nanometric size filler is keeping its supreme position thanks to the specific properties of the output nanocomposite. Nanotubular clay mineral Halloysite (HNT) is composed of aluminosilicate naturally structured in layers. It has significant reinforcing impact on polymer composites. It has large influence to tensile strength of polymer composites even in small contain - less than 7 % [1]. The unique structure of HNT allows its usage as filler for nanocomposites for medical application. For example as a drug deliverer [2] or reinforcing material for tissue replacements with very low cytotoxicity [3].

A sufficient nanofiller dispersion and preparation of nanocomposites remains a large problem till present time. There exist a large number of nanocomposites preparation methods. The simplest method is to disperse nanofiller in polymer melt mechanically, most often using twin screw extruder [4]. The less common way how to disperse filler can be by high energy planetary ball mill. The dispersion of the filler in matrix can be than performed while dry phase [5] or in the polymer solution [6]. Other ways to disperse the particles can be performed by ultrasonication of particles in the polymer solution [7], or radical grafting of the polymer matrix directly to particle surface. It is about the matrix polymerization with the filler directly present [8]. In this paper three methods of nanocomposite preparation will be introduced: To disperse filler in the polymer melt by laboratory kneader, direct polymerization of poly(vinyl acetate) with the HNT present and to disperse the HNT filler in polymer solution using planetary ball mill.

The sample preparation by laboratory kneader introduces a low-volume composite sample preparation by mixing of the matrix and the filler in the melt. It also simulates the most common way of composite production - twin screw extrusion. In the laboratory kneader the particles are dispersed in matrix by shear rate. It is caused by mixing segments affecting the polymer melt directly and nanofiller particles indirectly.

During the vinyl acetate polymerization with presence of HNT, better level of nanofiller dispersion was assumed, than during preparation in the polymer melt. Low molecular vinyl acetate has a better diffusion to nanofiller agglomerates than high molecular PVAc. Moreover, the HNT has a certain amount of hydroxyl groups on its surface which causes it partly more polar. That is why the particles are sufficiently dispersed in the polar solvent - the water.

The HNT filler was to disperse in toluene PVAc solution using the planetary ball mill, considering its high milling efficiency. The particles are dispersed in high viscous polymer matrix solution by the effect of high speed movement of milling segments. At the same time the particles agglomerates are grinded by impaction of the milling segments to the wall of the milling jar and their rolling alongside the milling jar wall. The combination of those effects provides a highly effective mixing method.

2. MATHERIALS AND METHODES

2.1. Used material

Poly(vinyl acetate) Vinnex® 2525, kindly provided by Wacker. Halloysite nanotubes, poly(vinyl alcohol), dodecyl sulfate sodium salt, benzoyl peroxide and ammonium iron(II) sulfate hexahydrate purchased by Sigma Aldrich. Sodium pyrophosphate decahydrate and toluene was obtained from Penta s.r.o.

2.2. Preparation of testing samples

HNT in the amount of 10 wt % was used to prepare all of the PVAc based composites.

The first of testing composite samples had been prepared in melt on laboratory kneader Plasti-Corder BRABENDER with temperature of 140°C and 45 RPM of sigmoid kneading segments. The PVAc had been brought to kneader and plasticized by kneading for 1 minute. Then HNT filler was added and sample had been mixed for 11 minutes. After that the melt had been removed from the kneader and leaved to cool down to the laboratory temperature.

The second sample was prepared by polymerization of vinyl acetate with presence of HNT filler using following procedure: An emulgator containing poly(vinyl alcohol) and dodecyl sulfate sodium salt dispersed in water was prepared in reaction container. The initiator benzoyl peroxide, catalyzer ammonium iron(II) sulfate hexahydrate and the free ion catcher sodium pyrophosphate decahydrate were added to the emulgator under the nitrogen atmosphere. After complete dissolution of all parts, vinyl acetate and HNT filler were added. The mixture had been heated for two hours and stirred intensively, then cooled down. PVAc composite precipitate was done by freezing. Catalyzers and dispersants were separated and removed in the form of a green precipitate. The required white composite was dried in the dry box at 40°C to the constant weight value.

The third composite sample was prepared by milling of PVAc solution with HNT filler in planetary ball mill. Commercial PVAc was dissolved in toluene to get 10 % solution and placed to steel milling jar containing the milling balls together with HNT filler. The milling balls weight was calculated as 10 times the weight of the polymer mass. Two steel milling jars was placed into Retch planetary mill and the milling was repeated in 6 cycles, each for 30 minutes. 20 minutes breaks were set in between the milling cycles to let the milling jars to cool down. At the end of milling process the output composite solution was precipitated to water and filtered. Then it was dried in the drying box to constant weight.

2.3. Scanning electron microscopy

Samples for the scanning electron microscopy were first cryogenically broken under liquid nitrogen to create fracture area. Then they were covered by thin Pt/Pd layer using Chemical Vapor Deposition method in argon atmosphere. For the particle distribution analysis, scanning electron microscope JEOL 7500F with resolution of 1 nm and acceleration voltage of 10 kV was then used.

3. RESULTS AND DISCUSION

Photography **Figure 1** is showing the fracture area of composite sample prepared on laboratory kneader. On picture on the left a large agglomerate of undispersed particles can be seen. Particle agglomerates in range of size of several tens of micrometers was observed on the photographs. A detail view of matrix - halloysite particles agglomerates interface can be seen on the figure left. Nanotubes on the edges of the agglomerates are relatively well soaked with matrix. This is the indication of sufficient adhesion of untreated HNT to PVAc matrix. The presence of large particle agglomerates is then not caused by low interfacial compatibility in the composite, but by insufficient ability of the kneader to disperse the filler. Still, even individual nanotubes was observed in the sample.

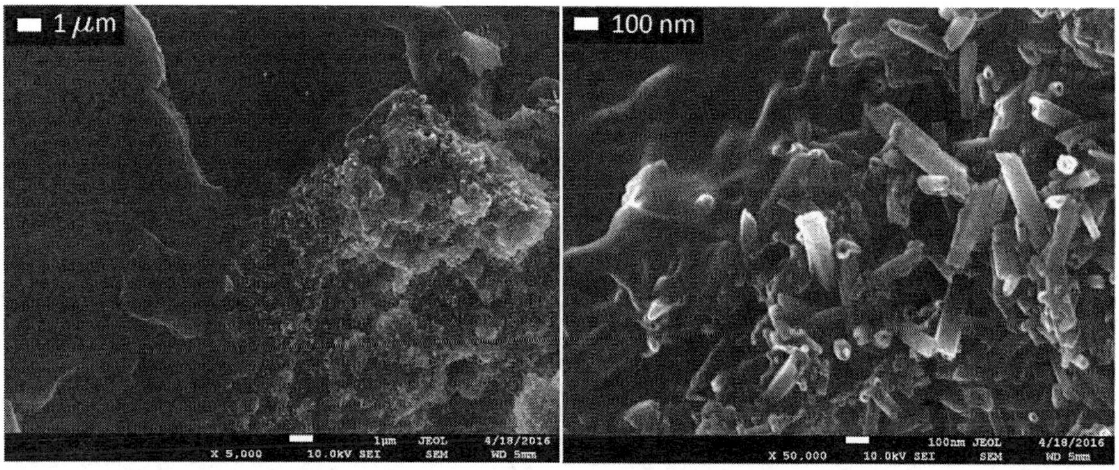

Figure 1 SEM photograph of the PVAc - HNT samples prepared by melt kneading

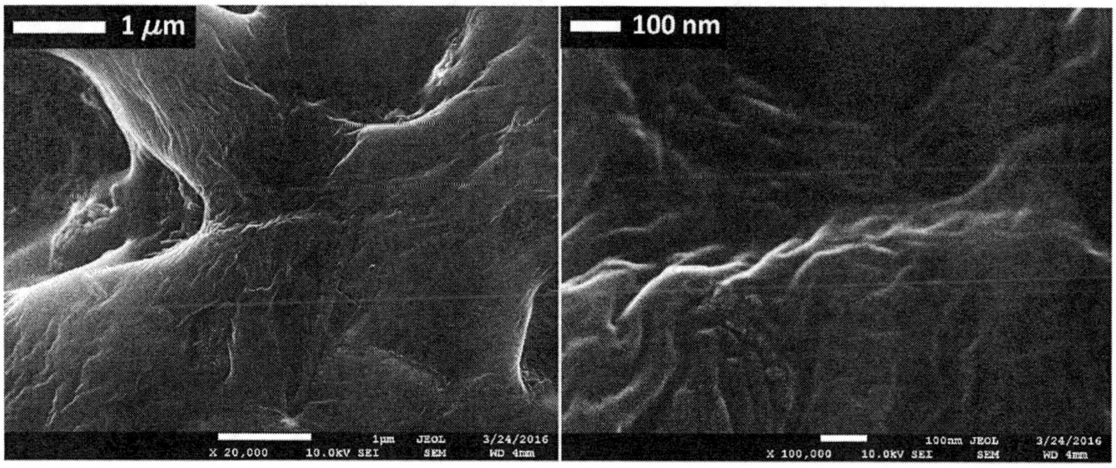

Figure 2 SEM photograph of the PVAc - HNT samples prepared by direct polymerization

The fracture area of composite sample prepared by emulsion polymerization of PVAc with HNT present is displayed on picture **Figure 2**. On a picture on the left sponge structure of the sample can be seen. It is due to the method of preparation. This state of sample makes the nano reinforcement observation more complicated. The particle agglomerate densely covered with PVAc matrix is visible on the upper and the left side of the photograph. A large, vertical needle shaped agglomerate of size of several micrometers can be seen at the picture's front. This indicates a lower compatibility of prepared PVAc matrix to the nanofiller. A detail of HNT and PVAc matrix interface is shown on the picture on the right. The particles of the filler are covered by the matrix but with worst soaking. The interface adhesion is not ideal then. That can be related to good vinyl acetate diffusion to the particle surface in the agglomerate, but with bad wettability of the individual particles by the vinyl acetate solution. That's why the PVAc is then separated from the particle surface during polymerization process. Incorrectly dispersed agglomerates of nanofiller are also appearing during this composite sample preparation method, even if these agglomerates are significantly smaller than those prepared in the laboratory kneader.

The picture **Figure 3** is showing a surface structure of cryogenically broken composite sample, prepared in planetary ball mill. The SEM analysis did not show significant agglomerate of HNT particles. The largest particles agglomerates found is shown on the picture on the left, where the agglomerates are composed only by a few nanotubes. A detailed view of matrix and nanofiller interface is depicted on the picture on the right. The agglomerate found is well covered with polymer matrix. Processing of composites in the planetary ball mill has no negative impact on interfacial adhesion then.

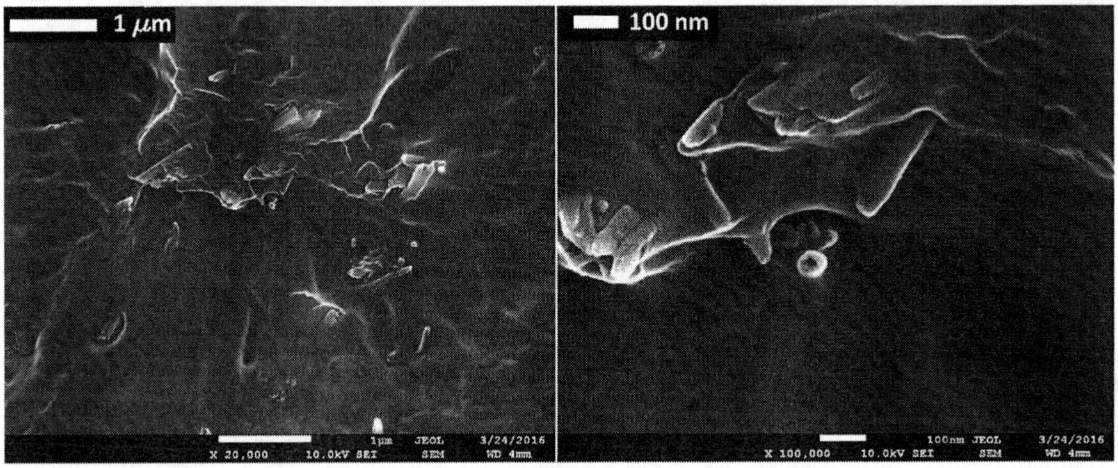

Figure 3 SEM photograph of the PVAc - HNT samples prepared from toluene solution, dispersed by ball mill

4. CONCLUSION

In this paper three methods of nanocomposites preparation were compared and level of particle dispersion in matrix was observed. Composite preparation by kneading in polymer melt is the least time consuming method. Unfortunately, large number of insufficiently dispersed particles is appearing in the composite structure. This method is not very suitable for nanocomposite preparation then.

Composite preparation using direct polymerization with presence of the filler was favorable because of its low demands of laboratory equipment. It was medium time consuming in comparison with other methods. It also produces an average affectivity of dispersion, but the composite structure is strongly porous. In theory good results could be reached by combination of both methods mentioned, meaning directly polymerized composite is then fabricated by extrusion.

Ball mill method of filler dispersion in polymer solution provides the highest level of HNT dispersion compared with previously discussed composites preparation methods. This method allowed dispersing the filler in polymer almost to the level of individual nanotubes. Due to the high time requirement for milling, this method of composite preparation remains highly ineffective.

ACKNOWLEDGEMENTS

This work was supported by the Materials Research Centre is supported by the Ministry of Education, Czech Republic (project No. LO1211).

REFERENCES

[1] GAAZ, Tayser, Abu SULONG, Abdul KADHUM, Ahmed AL-AMIERY, Mohamed NASSIR and Ahed JAAZ. The Impact of Halloysite on the Thermo-Mechanical Properties of Polymer Composites. *Molecules* [online]. 2017, vol. 22 iss 5, pp. 838-841 [viewed 2018-09-25]. Available from DOI: 10.3390/molecules22050838.

[2] TULLY, Joshua, Rawil FAKHRULLIN and Yuri LVOV. Halloysite Clay Nanotube Composites with Sustained Release of Chemicals. *NATO SCI PEACE SEC C* [online]. 2015, pp. 87-118 [viewed. 2018-09-25]. Available from DOI: 10.1007/978-94-017-9921-8_5.

[3] LIU, Mingxian, Rui HE, Jing YANG, Zheru LONG, Biao HUANG, Yongwang LIU and Changren ZHOU. Polysaccharide-halloysite nanotube composites for biomedical applications: a review. *Clay Miner.* [online]. 2016, Vol. 51 iss. 03, pp. 457-467 [viewed 2018-09-25]. Available from DOI: 10.1180/claymin.2016.051.3.02.

[4] KELNAR, Ivan and Jaroslav KRATOCHVÍL. Effect of platy and tubular nanoclays on behaviour of biodegradable PCL/PLA blend and related microfibrillar composites. *AIP Publ.* [online]. 2016, Vol. 1736 iss. 1, [viewed 2018-09-25]. Available from DOI: 10.1063/1.4949587.

[5] GORRASI, Giuliana, Valentina SENATORE, Giovanni VIGLIOTTA, Sandra BELVISO and Rachele PUCCIARIELLO. PET-halloysite nanotubes composites for packaging application: Preparation, characterization and analysis of physical properties. *Eur Polym J.* [online]. 2014, Vol. 61, pp. 145-156 [viewed 2018-09-25]. Available from DOI: 10.1016/j.eurpolymj.2014.10.004.

[6] MUNKHBAYAR, B., Md. J. NINE, Jinseong JEOUN, Munkhjargal BAT-ERDENE, Hanshik CHUNG and Hyomin JEONG. Influence of dry and wet ball milling on dispersion characteristics of the multi-walled carbon nanotubes in aqueous solution with and without surfactant. *Powder Technol.* [online]. 2013, Vol. 234, pp. 132-140 [viewed. 2018-09-25]. Available from DOI: 10.1016/j.powtec.2012.09.045.

[7] BIDSORKHI, Hossein Cheraghi, Hossein ADELNIA, Raheleh HEIDAR POUR and Mohammad SOHEILMOGHADDAM. Preparation and characterization of ethylene-vinyl acetate/halloysite nanotube nanocomposites. *J. Mater. Sci.* [online]. 2015, Vol. 50, iss. 8, pp. 3237-3245 [viewed. 2018-09-25]. Available from DOI: 10.1007/s10853-015-8891-6.

[8] LI, Cuiping, Jiguang LIU, Xiaozhong QU, Baochun GUO a Zhenzhong YANG. Polymer-modified halloysite composite nanotubes. *J Appl Polym Sci.* [online]. 2008, Vol. 110, iss. 6, pp. 3638-3646 [viewed 2018-09-25]. Available from DOI: 10.1002/app.28879.

PRODUCTION AND CHARACTERIZATION OF MICRO-SIZE PORES FOR ION TRACK ETCHING APPLICATIONS

Antonino CANNAVO'[1], Vladimir HAVRANEK[1], Vasily LAVRENTIEV[1], Lorenzo TORRISI[2], Mariapompea CUTRONEO[1], Giovanni CECCIO[1], Alfio TORRISI[1], Pavel HORAK[1], Jiri VACIK[1]

[1]*Nuclear Physics Institute of the CAS, Husinec Řež, Czech Republic, EU, cannavo@ujf.cas.cz*
[2]*Dipartimento di Scienze Fisiche - MIFT, Università di Messina, Messina, Italy, EU*

Abstract

For many years the applications of ion track etch materials have increased considerably, like charged particles detection, molecular identification with nanopores, ion track filters, magnetic studies with nanowires and so on. Over the materials generally used as track detector, the Poly-Allyl-Diglycol Carbonate (PADC), offers many advantages, like its nearly 100 % detection efficiency for charged particle, a high resistance to harsh environment, the lowest detection threshold, a high abrasion resistance and a low production costs. All of these properties have made it particularly attractive material, even if due to its brittleness, obtaining a thin film (less than 500 μm) is still a challenge.

In this work, PADC foils have been exposed to α-particles emitted by a thin radioactive source of ^{241}Am and to C ions from the Tandetron 4130 MC accelerator. The latent tracks generated in the polymer have been developed using a standard etching procedure in 6.25 NaOH solution. The dependence of the ion tracks' geometry on the ion beam energy and fluence has been evaluated combining the information obtained through a semiautomatic computer script that selects the etched ion tracks according to their diameter and mean grey value and nanometric resolution images by atomic force microscopy.

Keywords: PADC polymer, AFM, OM, Image analysis

1. INTRODUCTION

In addition to the conventional application of the ion-track polymeric materials in detection of charged particles [1] and neutrons [2], the controlled fabrication of nanopores with specific geometries and characteristics have opened in the last decades the way to other fields of research and technology, e.g., to biosensors [3], nuclear membranes for filtration [4] to templates for nanostructure preparation [5], various industrial applications [6], filters for X-rays imaging [7], and so on [8].

One of the most important advantages of such a technology is that the shape and density of the etched tracks are two independent parameters that can be tuned according to the specific criteria demanded by the applications. In fact, several aspects do concur to the development of pores - mainly the irradiation conditions and the etching procedure that can be optimized to achieve a certain purpose.

The majority of the works present now in literature [9,10] is focused on the characterization and application of ion tracks in polyethylene terephthalate (PET). This is mainly due to the advantageous properties of PET, such as the high stability in acids, biological compatibility, mechanical hardness and high etch ratio. Nevertheless, the poly-allyl-diglycol carbonate (PADC) has also unique characteristics, e.g., the highly sensitive to ionizing particles [11] that allow to create the track-etch foils for special purpose. In fact, the experimental conditions used in this paper, namely the light elements (α, C) with relatively low energy (1 MeV - 10 MeV), would not be enough to damage effectively PET and other similar polymers. However, the wide utilization of PADC is still limited because of its brittleness that makes the manufacturing of thin foils (with a thickness < 100 μm) a challenge. In this paper are reported first results obtained by study of etched ion tracks generated in PADC and measured by a standard optical microscopy (OM) and by an atomic force microscopy (AFM).

2. EXPERIMETAL SETUP

The PADC sheets, from Page Moudlings (Pershore), of density 1.32 g/cm^3 and thickness 750 μm, have been cut to a rectangular shape (15 mm × 25 mm).

The irradiation by α-particles and carbon ions have been carried out at the CANAM infrastructure of the Nuclear Physics Institute of CAS in Řež. The irradiations by α-particles have been performed with a thin ^{241}Am radioactive source, having a main α-peak cantered at 5.486 MeV. To change the energy of the emitted α-particles, the pressure inside the vacuum chamber has been varied in the range between 430 mbar and 20 mbar, allowing to investigate the energy range between 3.5 MeV and 5.5 MeV, respectively. The energy loss by α-particles due to the variation of the pressure has been taken into account through the SRIM code. In the simulations, the pressure has been modelled by a thin absorbing layer (of air) in front of the PADC sample whose thickness (x) is given by the residual pressure inside the chamber (P_{ch}) through the equation:

$$x = \frac{P_{ch}}{P_0} D$$

where P_0 is the pressure for standard operation condition, while D is the distance between the α-source and the PADC sample, i.e., 45 mm. For each α-particle energy, the exposition time has been kept constant (20 minutes) in order to have a constant particles density hitting the surface of the PADC sample, namely 10^4 particles/cm^2. Conversely, the exposition to carbon ions have been performed along the Ion Microprobe beamline (+10°) of the Tandetron 4130 MC. The accelerated ions have been focused and scanned on the PADC foil in a surface of 1 mm^2 by a focusing quadrupole with 1 μm lateral resolution. This methodology has allowed to keep a low ions density on the PADC surface (3×10^5 particles/cm^2) in order to reduce the pores overlapping. The analysis has been repeated at different ions energy, from 0.9 MeV up to 10 MeV using 3$^+$ and 4$^+$ charge ion state.

Before the exposition of the PADC foils to ion beams, the energy peak and the ion current have been tested using a conventional solid state detector (ORTEC Ultra series) connected to a standard spectrometric chain. As expected, and confirmed also from the SRIM simulations, the spectra related to α-particles present a widening of the energy distribution when the energy decreases. In practice, the lower energies have been obtained with the higher pressure values inside the experimental chamber. This makes an increase of the α-particles straggling that results from their interaction with the residual air molecules. However, the broadening of the energy distribution does not affect the measurements with PADC, because even if the FWHM of the particles' energy rises from 38 keV to 83 keV, the difference is still lower than the minimal energy distribution that can be observed by the analysis of the etched ion tracks in PADC using the proposed technique. On the other hand, the C-ion beam has not showed any significant change of the energy resolution (it remained below 20 keV).

After irradiations, all PADC samples have been chemically etched in 6.25 N solution of NaOH. After 2 h etching at a constant temperature of 70 ± 2 °C, the samples have been washed in distilled water, dried in air and then analysed by Optical Microscopy (OM) and Atomic Force Microscopy (AFM).

The analysis with OM have been performed using an Optika DM microscope with a CCD camera interfaced with a computer through Optika Vision Lite 2.0 software. A modular NTEGRA scanning probe microscope (NT-MDT Ltd.) has been used for the AFM study.

3. RESULTS AND DISCUSSION

In **Figure 1**, two typical pictures obtained by OM are depicted. The pictures refer to the PADC foils irradiated by 5 MeV α-particles (**Figure 1a**) and C-ions (**Figure 1b**). The pores are visible as black circles in the bright background. Their regular shape is due to the fact that the ions have hit the PADC foil orthogonally to the surface with a divergence in both cases < 5°. Each picture has a resolution of 2048 x 1536 pixel corresponding to a sample surface of 237.6 μm × 178.2 μm.

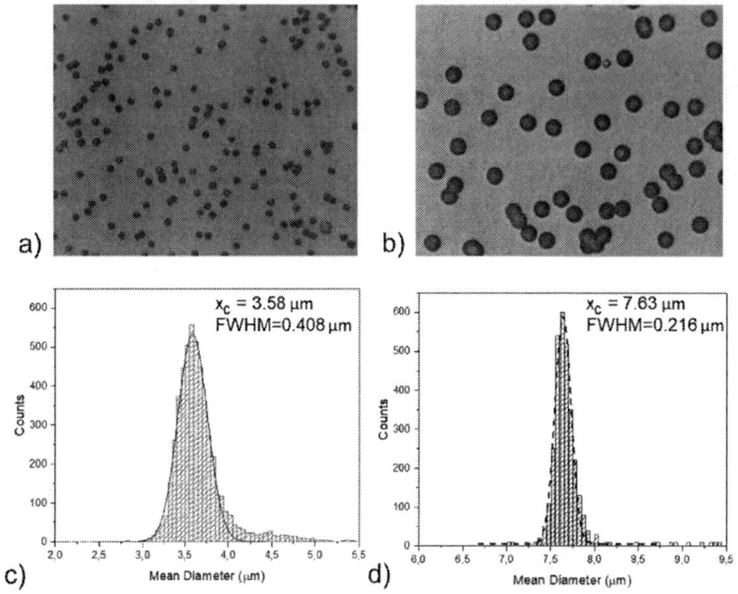

Figure 1 OM pictures of PADC foils irradiated by 5 MeV α-particles (a) and C-ions (b), and evaluation of the corresponding mean diameter (c) and (d), respectively

In order to improve the statistic of the pores (concerning the accuracy of their diameters), the sample surface has been scanned by optical microscopy and 20 pictures have been made and analysed for each sample. The pores identification has been done through an automatic script written under ImageJ environment. The selection rules of pores have been included in the script in order to select pores based on the size and shape to limit the presence of artefacts in the data analysis. These artefacts are due to several aspects, such as the overlapping of two or more pores or to a local damage of the polymer. From the counting of the total number of pores a mean density of 2.126×10^5 particles/cm^2 and 1.653×10^5 particles/cm^2 has been evaluated for α-particles and C-ions, respectively. These values are in good agreement with the ion fluences expected from the imposed experimental conditions, testifying the good efficiency of the method. Finally, standard software for the data analysis has been used to collect data from the different pictures and the pores have been classified based on area, mean gray value, major and minor axis length, position of centers, and so on. An example is depicted in the **Figure1c** and **Figure 1d** that show the evaluation of the mean diameter of pores for the samples irradiated by 5 MeV α-particles and 5 MeV C-ions, respectively. From the Gaussian fit of these data, a mean diameter of 3.58 ± 0.20 μm has been evaluated for 5 MeV α-particles, while for 5 MeV C-ions the mean diameter is 7.64 ± 0.11 μm.

Another significant example is depicted in **Figure 2**, showing how it is possible to distinguish in the same foil the pores generated by α-particles with different energies. For this purpose, a PADC foil has been irradiated twice with α-particles of 1.01 MeV and 5.48 MeV. The difference in energy determines the difference in conformation in the developed pores. In fact, the higher energy results in a mean diameter of 3.73 ± 0.19 μm, while the lower energy produces pores with a mean diameter of 4.87 ± 0.25 μm. The pores also differ in the gray tonality (clear and dark) that reflects the different depth of the etched tracks. This aspect becomes clearer when the pore's diameter is given as a function of its mean gray value, as it is showed in **Figure 2b**. In this picture, at least two different regions can be distinguished: the first one is placed in the lower left corner that means a darker gray tonality, and it is related to the 5.48 MeV α-particles; the second area, caused by the 1.01 MeV α-particles, is positioned in the central part of the graph and has an elongate shape due to the higher broadening in the diameter distribution.

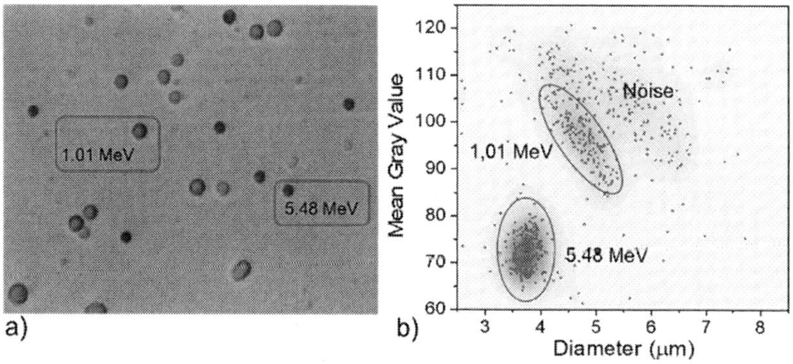

Figure 2 OM picture of a PADC sample irradiated by 1.01 MeV and 5.48 MeV α-particles, and diameter vs mean gray value of the detected pores

Following the guidelines of the OM results on the possible distribution in pores '*diameter vs. energy*', a detailed analysis of the single pores has been performed through AFM. **Figure 3a** shows the AFM picture of one of the pores of the sample already shown in **Figure 1a**.

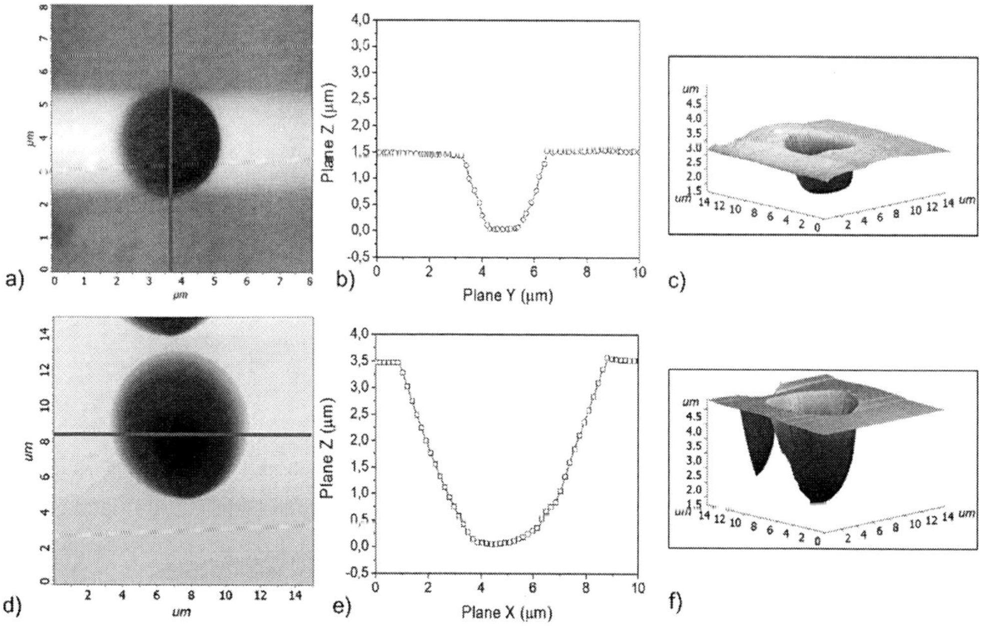

Figure 3 AFM pictures of a single pore in a PADC sample irradiated by 5 MeV α-particles a) and C-ions d), their corresponding profile analysis b) and e), and their 3D reconstructions c) and f), respectively

The profiling analysis of such a pore, **Figure 3b**, reveals a depth of 1.655 μm and confirms a diameter of 3.61 μm. The 3D reconstruction of this pore, given in **Figure 3c**, shows its conical shape. For comparison, also a pore generated by 5 MeV C-ions has been studied and the corresponding results are depicted in **Figures 3d-f**. In this case the pore has both a higher depth (3.49 μm) and a greater diameter (7.61 μm) with respect to the previous case. The same chemical etching has been done for both samples, so as expected, also in this case the pore has a conical shape, as it is showed in its 3D reconstruction in **Figure 3f**.

Combining the results carried out from OM and AFM, finally, the graphs in **Figure 4** can be obtained. Here, the dependence of the pore diameter on the particles' energy (**Figure 4a**) and the depth of the pores (**Figure 4b**) is showed. The higher energy loss of the C ions leads to a higher damage of the polymers, and this determines their faster etching rate. For this reason, at the same energy, pores related to C ions (square) have a bigger diameter and a greater depth than α-particles (circle).

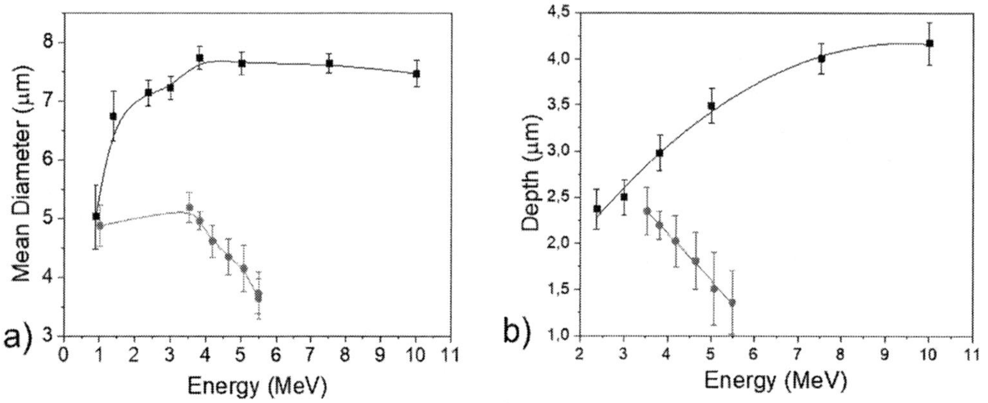

Figure 4 Dependence of the diameter and depth of the pores on energy of the α-particle (circle) and C-ions (square)

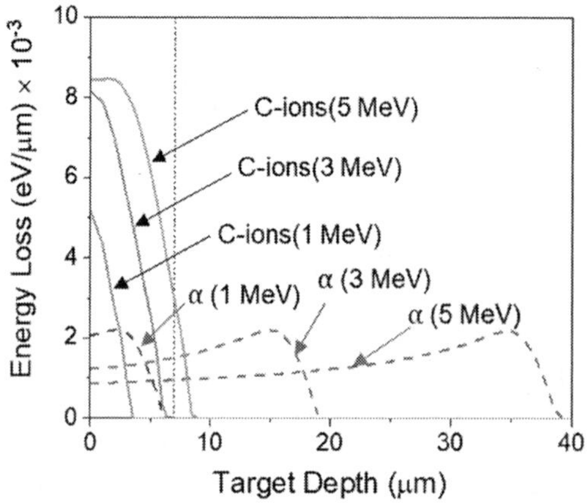

Figure 5 Energy loss for α-particles and C energy with an energy of 1 MeV, 3 MeV and 5 MeV

It should be also noted that both pictures show different trends. In the case of α-particles the mean diameter decreases quickly with the reduction of the energy, but in the case of C-ions the reduction occurs slowly. Concerning the '*depth vs energy*' dependence, the trend is totally different: when the energy increases, the pores become more/less deep in the case of C-ions/α-particles, respectively. Again, the reason of such behaviour is in the different energy loss and particles' range of the two ion species inside the PADC material.

As can been seen in **Figure 5**, α-particles have a higher range in PADC than C-ions. Consequently, α-particles lose their energy in a wider region of the sample and the Bragg peak is shifted into a deeper point of the PADC sample. For example, the range of 3 MeV α-particles in PADC, evaluated by SRIM simulation, is 18.5 μm, while for C-ions it is 5.62 μm. Considering a track etching rate of 3.5 μm/h [12], after 2 h etching the PADC foils has been etched 7 μm along the ion track path (vertical dashed line of **Figure 5**).

This means that the whole region, involved in the energy loss of C ions, can be etched out, while for α-particles the etching has not been enough to reach the Bragg peak. If the energy of the particles increases (e.g., 5 MeV) the situation for the α-particles is even worse, because the particle range becomes higher (38.2 μm) and the energy lost in the etched foil is lower. On the other hand, almost all the energy loss by the Carbon ions is included in the 7 μm etched region.

4. CONCLUSION

Encouraged by the unique properties of PADC, the characterization of the pores, etched ion tracks generated by the exposition of this plastic material to two ion species (α-particles and C ions), has been presented in the paper. The study has taken an advantage from the combination of OM and AFM study of the etched foils. Using the first technique, a large surface of the sample has been analysed. This 2D analysis has allowed to determine the '*diameter vs energy*' dependence. The latter has been used to perform the 3D AFM analysis of single pores. In such a way, a general trend has been deduced to describe also the '*depth vs energy*' dependence. OM has also permitted to distinguish between pores generated by the same type of ion (α-particles), but with two different energies (1.01 MeV and 5.48 MeV). For this purpose, the inspected pores have been classified according to the diameter and their mean gray value. The obtained data have clearly showed that the pores can be divided into distinct areas specific to given energy of ions. It should be noted that the analysis described in this paper represents a mandatory step before the introduction of PADC material to ion track-etch applications. The paper presents only preliminary results that however will be improved in the next planned experiments that will be performed using different ions (fission fragments, backscattered ions, results of nuclear reactions induced by thermal neutrons), and also with different etching parameters.

ACKNOWLEDGEMENTS

The authors acknowledge the financial support from the Grant Agency of Czech Republic under the contract No. 18-07619S. The measurements were carried out at the 'CANAM' infrastructure supported by MEYS in projects No. LM2015056.

REFERENCES

[1] MALINOWSKA, A., et al. Characterization of solid state nuclear track detectors of the polyallyl-diglycol-carbonate (CR-39/PM-355) type for light charged particle spectroscopy. *Review of Scientific Instruments.* 2014. vol. 85, no.°12, p. 123505.

[2] PUGLIESI, R., et al. Characteristics of the solid state nuclear detector CR-39 for neutron radiography purposes. *Applied radiation and isotopes.* 1999. vol. 50, no.°2, pp. 375-380.

[3] GARCÍA-ARELLANO, H., et al. Dependence of yield of nuclear track-biosensors on track radius and analyte concentration. *Nuclear Instruments and Methods in Physics Research Section B: Beam Interactions with Materials and Atoms.* 2018. vol. 420, pp 69-75.

[4] WANG, P. et al. Ultrafast ion sieving using nanoporous polymeric membranes. *Nature communications.* 2018. vol. 9, no.°1, p. 569.

[5] MAURER, F. et al. Field emission of copper nanowires grown in polymer ion-track membranes. *Nuclear Instruments and Methods in Physics Research Section B: Beam Interactions with Materials and Atoms.* 2006. vol. 245, no.°1, pp. 337-341.

[6] APEL, P. Track etching technique in membrane technology. *Radiation Measurements*. 2001. vol. 34, no. 1-6, pp. 559-566.

[7] MITROFANOV, A.V. et al. Ion track filters in imaging X-ray astronomy. *Nuclear Instruments and Methods in Physics Research Section B: Beam Interactions with Materials and Atoms*. 2006. vol. 245, no.°1, pp. 332-336.

[8] PRICE, P. B. Recent applications of nuclear tracks in solids. *Radiation Measurements*. 2008. vol. 43, pp.S13-S25.

[9] APEL, P.Y. et al. Accurate characterization of single track-etched, conical nanopores. *Physical Chemistry Chemical Physics*. 2014. vol. 16, no.°29, pp. 15214-15223.

[10] FINK, D. et al. Coupled chemical reactions in dynamic nanometric confinement: IV. Ion transmission spectrometric analysis of nanofluidic behavior and membrane formation during track etching in polymers. *Radiation Effects and Defects in Solids*. 2015. vol. 170, no.°3, pp. 155-174.

[11] BALESTRA, S. et al. Bulk etch rate measurements and calibrations of plastic nuclear track detectors. *Nuclear Instruments and Methods in Physics Research Section B: Beam Interactions with Materials and Atoms*. 2007. vol. 254, no.°2, pp. 254-258.

[12] LOUNIS, Z. et al. Track etch parameters in CR-39 detectors for proton and alpha particles of different energies. *Nuclear Instruments and Methods in Physics Research Section B: Beam Interactions with Materials and Atoms*. 2001. vol. 179, no.°4, pp. 543-550.

SOLID STATE SYNTHESIS OF IRON(III) OXIDE POLYMORPHS FROM PRUSSIAN BLUE WITH DIFFERENT MORPHOLOGY

Claudia APARICIO*, Jiří TUČEK, Jan FILIP, Libor MACHALA

Regional Centre of Advanced Technologies and Materials (RCPTM), Department of Experimental Physics, Faculty of Science, Palacky University, Olomouc, Czech Republic, EU,
claudia.aparicio@upol.cz, jiri.tucek@upol.cz, jan.filip@upol.cz, libor.machala@upol.cz

Abstract

Two Prussian blue (PB) samples with different morphologies (spheres, mean size 160 nm; cubes, mean size 1.48 μm) and variable content of potassium (K-free and K-bearing) were used as a precursor material for the preparation of iron(III) oxides by solid state thermal decomposition method. A mixture of iron(III) oxide polymorphs (α-Fe_2O_3, β-Fe_2O_3, and γ-Fe_2O_3 nanoparticles) or pure maghemite (γ-Fe_2O_3 nanoparticles) were obtained by thermal decomposition of the cubic (K-bearing) or spherical (K-free) PB particles, respectively, at 350 °C in air. The particle's morphology of the starting material (i.e. PB) was mostly retained after the thermal decomposition. The PB samples and the as-formed iron(III) oxides were characterized by using X-ray powder diffraction (XRD), ^{57}Fe Mössbauer spectroscopy, and scanning electron microscopy (SEM). A high amount (50 wt%) of β-Fe_2O_3 polymorph was produced when K-bearing PB was used, simultaneously cubic clusters of maghemite were formed. A single phase (maghemite) clustered nanoparticles with spherical morphology were obtained when we used a K-free PB. The maghemite nanoparticles (4 nm) have proven to be superparamagnetic.

Keywords: Thermal decomposition, Prussian blue, iron oxides

1. INTRODUCTION

The control of size and morphology has been one of the main goals in the synthesis of nanoparticles. A variation of size and shape of nanoparticles leads to different optical, thermal, electrical, magnetic, and even catalytic properties of the materials [1,2]. Nanoparticles with defined shapes are usually synthesized by colloidal or thermal decomposition methods [1], using a suitable surfactant [3] or ultrasound during the synthesis [4]. It is also known that particle size have a direct effect on the decomposition temperature [1,5], also the morphology of particles might have a similar effect.

Iron(II) hexacyanoferrate ($Fe_4^{+3}[Fe^{2+}(CN)_6]_3{\cdot}xH_2O$), also known as Prussian blue (PB) has been used as a precursor for iron oxides, prepared via solid state decomposition approach [6-8]. PB with different sizes and morphologies were used in those studies, including hollow particles [9], where in most of the cases the morphology and size of the precursor was maintained after the formation of iron oxides [8,9]. Generally, particles size [6] of a precursor and even the thickness of a layer [10] on the crucible affects the temperature of thermal decomposition and phase composition of iron oxide products as well. Another factor, which might influence the thermal behavior of the material and the formation of thermal decomposition products, is the particle's morphology of the precursor. The goal of the present work is to prepare iron oxides from Prussian blue, with emphasis on the influence of particle morphology on thermal decomposition of PB, and how it affects the shape of the resulting iron oxide particles.

2. MATERIALS AND METHODS

2.1. Synthesis of Prussian blue (PB)

Spherical particles of PB were prepared using the synthesis described by Shen et al. [3]. Polyvinylpyrrolidone (7.5 g, PVP, K-30, Mw = 40 000, Fluka) was dissolved in hydrochloric acid (100 ml, HCl, 0.01 M). Then, potassium hexacyanoferrate(III) (225 mg, $K_3[Fe(CN)_6]$, Lachema) was slowly added to the former solution under stirring. The resultant solution was refluxed at 80 °C for 2 hours, and afterwards centrifuged. The bright blue precipitate was dried in vacuum for 20 h, then powdered and labeled as PB-S.

Cubic particles of PB were synthesized using a sonochemical synthesis described by Wu et al. [4]. Potassium hexacyanoferrate(II) (423 mg, $K_4[Fe(CN)_3]\cdot 3H_2O$, Penta) was dissolved in hydrochloric acid (100 ml, HCl, 0.1 M). Then, the solution was immersed in an ultrasound bath at 40 °C for 4 hours. After that, the cooled solution was centrifuged. The dark blue precipitate was dried in air, powdered and labeled as PB-C.

2.2. Preparation of Iron(III) Oxides

The samples PB-S and PB-C (26 mg) were heated in a muffle furnace (Linn Hightherm Gmbh LM 112.07) from room temperature up to 350 °C in air (heating rate 5 °C/min); and then kept at 350 °C for 1 hour. The samples PB-C and PB-S, initially blue, changed color after heating to reddish brown color and dark brown color, respectively. The samples obtained after the heating were labeled as PBC350 and PBS350.

2.3. Samples Characterization

Simultaneous thermogravimetric analysis (TGA) and differential scanning calorimetry (DSC) were performed on PB-C and PB-S samples (10 mg of sample in an open alumina crucible) in air from 35 °C to 900 °C (heating rate 5 °C/min). The equipment used was a thermal analyzer STA 449 C Jupiter from Netzsch. After heating, samples were cooling down to room temperature under nitrogen.

X-ray powder diffraction (XRD) patterns were recorded with a PANalytical X'Pert PRO MPD diffractometer (Co-Kα radiation) in Bragg-Brentano geometry, equipped with an X'Celerator detector and programmable divergence and diffracted beam anti-scatter slits. Phase identification and Rietveld refinement were performed using the HighScore Plus software (PANalytical) in conjunction with PDF-4+ and ICSD databases.

Transmission ^{57}Fe Mössbauer spectroscopy was carried out in a constant acceleration mode using a $^{57}Co(Rh)$ source at room temperature (RT) and low temperatures (20 and 5 K). Spectra were folded and fitted using CONFIT2000 and Mosswinn software; the spectrometer was calibrated with a metallic α-iron foil.

Scanning electron microscopy (SEM) images of samples were obtained using a scanning electron microscope Hitachi SU-6600, with 5 kV and 15 kV setting. Complementary, energy dispersive X-ray analysis (EDS) was carried out as well on samples PB-C and PB-S.

3. RESULTS AND DISCUSSION

3.1. Prussian blue: Cubic and Spherical Shapes

XRD patterns and Mössbauer spectra confirm the presence of Prussian blue in both PB-S and PB-C samples (**Figure 1a, 1b**). From Mössbauer spectroscopy, the ratio Fe^{2+}/Fe^{3+} for samples PB-C (0.73) and PB-S (0.74) are very close to value 0.75 for stoichiometric Prussian blue. From SEM images (**Figure 1c**) is observed that particle sizes follow a normal distribution, with average sizes of 1.48 μm for cubic particles (PB-C) and 160 nm for spherical particles (PB-S). Moreover, EDS results (**Figure 1d**) show that both samples contain potassium in addition to the expected Fe, C, and N (N is overlapped by C contribution). PB-S sample contains only a negligible amount of K, where PB-C has a considerable amount of potassium.

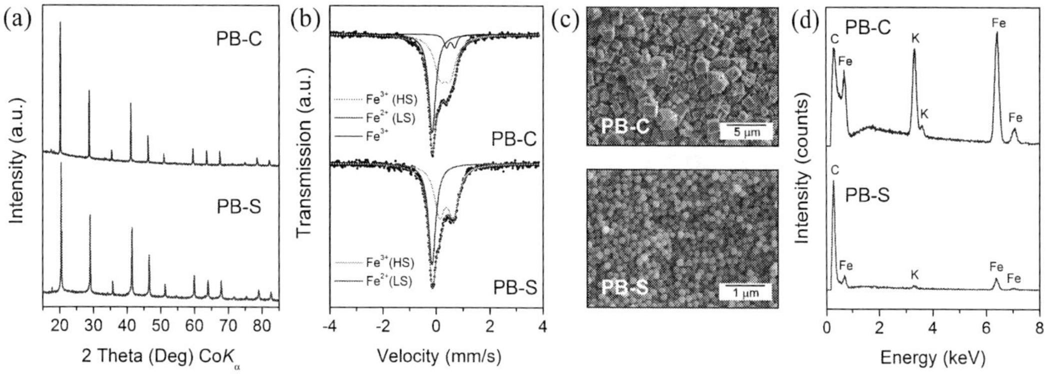

Figure 1 XRD patterns (a), RT Mössbauer spectra (b), SEM images (c), and EDS spectra (d) of cubic (PB-C) and spherical (PB-S) particles of Prussian blue.

3.2. Iron Oxide Formation after Heating of PB in Air

Above 250 °C, PB starts to decompose in air and the decomposition finishes after the total release of crystal water and cyanide groups from the lattice. There are visible differences in the TGA/DSC curves (**Figure 2**) for PB-S and PB-S. To explain the differences in thermal decomposition of PB, two structural models are used:

a) PB of spherical shape and without K in sample PB-S, $Fe_4[Fe(CN)_6]_3 \cdot xH_2O$ (space group *Pm*3*m*) with non-random distribution of vacancies [12];

b) PB of cubic shape and containing K in sample PB-C, $Fe_4[Fe(CN)_6]_3 \cdot [K^+_h \cdot OH^-_h \cdot mH_2O]$ (space group *Fm*3*m*), in this model we have coordinated water in a lattice with random vacancies of $[Fe^{2+}(CN)_6]^-$ units [11].

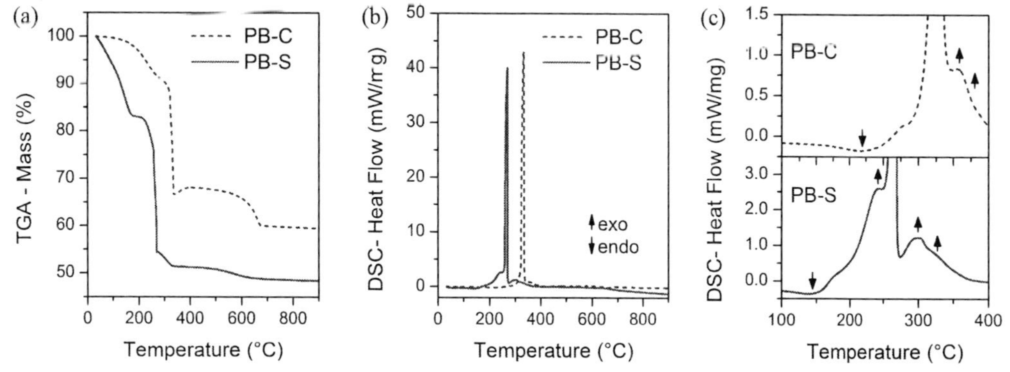

Figure 2 TG (a) and DSC (b) curves for PB-S (solid line) and PB-C (dashed line) samples. A detailed view of DSC curves (c), where the arrows represent endothermic (↓) and exothermic (↑) effects.

The differences in total mass loss (**Figure 2a**), for PB-S (-51.7 %) and PB-C (-40.6 %) are due to the different amount of water in the samples (9.5 in PB-S and 4.6 in PB-C) and the K presence in PB-C sample, where is expected the formation of KOCN and KNO_3 in addition to iron(III) oxides. Further heating above 400 °C leads to formation of hematite and potassium ferrite ($KFe_{11}O_{17}$), potassium ferrite was formed by the reaction between KNO_3 and Fe_2O_3. The position of exothermic peak on DSC curve (**Figure 2b**) is different for spherical particles (269 °C) in comparison with cubic particles (334 °C). Such a difference was observed previously by other authors in DSC curves of PB samples (K-free and K-bearing) with similar particles size (60 - 80 nm), where the main exothermal effect was reported at 260 °C for K-free PB and 278 °C for K-bearing

PB [12]. However, TGA performed on PB with bigger particles (150 - 200 nm) gave an exothermal effect slightly below 300 °C [7].

The proposed decomposition mechanisms are restricted only up to 400 °C (right after PB decomposition). Using the first model for PB-S, the equation (1) could explain the PB decomposition with the theoretical mass loss of 45.7 %, value just slightly smaller than the experimental one (48.6 %).

$$Fe_4[Fe(CN)_6]_3 \cdot 9.5\,H_2O + 5.25\,O_2 \rightarrow 3.5\,Fe_2O_3 + 9(CN)_2 + 9.5\,H_2O \quad (1)$$

After PB-C dehydration and release of cyanogen gas at 330 °C, iron(III) oxides and potassium cyanate (KOCN) are formed. In TG curve, between 335 and 400 °C (**Figure 2a**), the increment in mass (+1.5 %) accompanied by a small exothermal effect at 360 °C (**Figure 2c**) is explained by the decomposition of KOCN into KNO_3 and CO_2. Equation (2) summarizes the PB-C decomposition mechanism, where experimental (31.9 %) and theoretical (32.0 %) mass loss agree very well.

$$Fe_4[Fe(CN)_6]_3 \cdot K^+_{1.3}OH^-_{1.3} \cdot 4.6\,H_2O + 8.175\,O_2 \rightarrow 3.5\,Fe_2O_3 + 1.3\,KNO_3 + \ldots$$
$$\ldots + 8.35(CN)_2 + 5.25\,H_2O + 1.3\,CO_2 \quad (2)$$

3.3. Phase Composition and Morphology of the Iron(III) Oxide Samples

Despite the identical thermal treatment, the samples PBC350 and PBS350 differ significantly in phase composition. Sample PBC350 contained hematite (12 wt%), maghemite (38 wt%) and β-Fe2O3 (50 wt%), while only maghemite was found in sample PBS350 (**Figure 3a, 3b**). The RT Mössbauer spectra of samples PBC350 and PBS350 (**Figure 3c, 3d**) are similar; the Mössbauer parameters for both samples are listed in **Table 1**. In sample PBC350, the values of sub-spectral areas reveal that the dominant phase is β-Fe_2O_3 (58.7 %), followed by maghemite (34.2 %), and hematite (7.2 %).

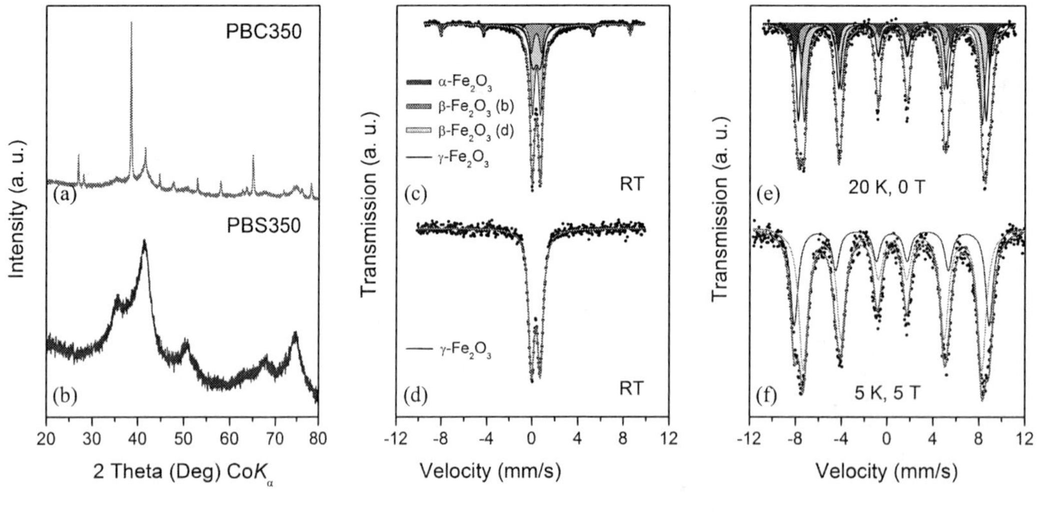

Figure 3 XRD patterns of the samples PBC350 (a) and PBS350 (b).
Mössbauer spectra for samples PBC350 (c, e) and PBS350 (d, f).

Low temperature Mössbauer spectra confirmed the presence of maghemite in both samples (**Figures 3e, 3f**). The ratio of ferric ions in O and T sites ($[Fe^{3+}]_O/[Fe^{3+}]_T$), in Mössbauer spectrum of PBS350 (**Figure 3f**), was determined as 1.67, which correspond to stoichiometric maghemite with vacancies only in octahedral sites. Using the relationship $\theta_c = arcsin[sqrt(6r/(4+3r))]$ (where: $r = A_{2,5}/A_{1,6}$; $A_{i,j}$ line intensity of line *i* or *j* of a sextet) it was calculated the canting angle of spins for each sub-lattice of maghemite: $\theta_{c,A}$ = 44.6° (tetrahedral site) and $\theta_{c,B}$ = 59.0° (octahedral site) [13].

Table 1 Mössbauer parameters of samples PBC350 and PBS350. IS: isomer shift (± 0.01), QS: quadrupole splitting or quadrupole shift (± 0.01), W: line width (± 0.01), B_{hf}: magnetic hyperfine field or effective magnetic field[a] (± 0.1), Area (± 1.0), SP: superparamagnetism, WF: weak-ferromagnetism.

Sample	T (K)	Phase	IS (mm/s)	QS (mm/s)	W (mm/s)	B_{hf} (T)	Area (%)
PBC350	298	β-Fe_2O_3 b	0.35	0.90	0.41	-	15
		β-Fe_2O_3 d	0.37	0.70	0.41	-	44
		γ-Fe_2O_3 SP	0.35	0.65	0.75	-	34
		α-Fe_2O_3 WF	0.37	-0.20	0.25	51.4	7
PBS350	298	γ-Fe_2O_3 SP	0.35	0.77	0.59	-	100
PBC350	20	β-Fe_2O_3 b	0.51	0.64	0.53	51.5	13
		β-Fe_2O_3 d	0.48	0.07	0.53	48.2	39
		α-Fe_2O_3 WF	0.50	-0.20	0.41	52.8	13
		γ-Fe_2O_3	0.44	-0.05	0.48	50.8	35
PBS350	5	γ-Fe_2O_3 O	0.47	0.00	0.91	48.3[a]	62.5
	(5 T)	γ-Fe_2O_3 T	0.40	0.00	0.89	52.6[a]	37.5

The morphology of the iron(III) oxide samples vary depending on the PB precursor morphology, as can be observed from SEM images (**Figure 4**). When the particles of PB are spherical, the maghemite form spherical clusters (**Figure 4a**) with a smaller size than the PB-S precursor (116 nm). In the case of cubic PB particles, the maghemite forms cubic clusters with similar sizes as the PB-C precursor particles (**Figure 4b**). The preservation of the morphology during thermal transformation was addressed in previous works [7-9]. Maghemite clusters - cubic and spherical - are formed by small nanoparticles (4 nm, based on XRD). Because of their reduced sizes, each particle behaves as a single magnetic domain, and at room temperature they exhibit superparamagnetism, as confirmed by Mössbauer spectroscopy.

The β-Fe_2O_3 particles in PBC350, have a mean size of 70 nm (**Figure 4c**), similar to the estimated crystallite size (68 nm) from XRD. The β-Fe_2O_3 particles grow on the cubic cluster surfaces where the surface energy permits the polymorph transformation from γ- to β- ferric oxide.

Figure 4 SEM images of the samples PBS350 (a) and PBC350 (b, c) showing the clusters of γ-Fe_2O_3 (b) and β-Fe_2O_3 nanoparticles (c)

4. CONCLUSION

The Prussian blue with two different morphologies (cubic and spherical) decomposes above 250 °C in air. At the beginning of the thermal decomposition of both PB samples (with and without K), after the initial dehydration and release of CN^- units, amorphous iron oxides are first formed, afterwards when the temperature increases above 350 °C the amorphous iron oxides crystallize into γ-Fe_2O_3 or a mixture of α-Fe_2O_3, β-Fe_2O_3,and γ-Fe_2O_3 polymorphs. The maghemite (γ-Fe_2O_3) is in the form of clusters-forming nanoparticles (4 nm) retaining the original shape of the original precursor. Both maghemite clusters behave as superparamagnets at room temperature, making then interesting for potential applications. Finally, the differences in the TG and DSC curves for PB-C and PB-S are more likely due to a combined effect of the presence of potassium and particle sizes and not because of the particle's morphology.

ACKNOWLEDGEMENTS

The authors gratefully acknowledge the support by the project LO1305 of the Ministry of Education, Youth and Sports of the Czech Republic).

REFERENCES

[1] BURDA, Clemens, CHEN, Xiaobo, NARAYANAN, Radha and EL-SAYED, Mostafa A. Chemistry and properties of nanocrystals of different shapes. *Chemical Reviews*. 2005. 105, pp. 1025-102.

[2] XIA, Younan, XIONG, Yujie, LIM, Byungkwon and SKRABALAK, Sara E. Shape-Controlled Synthesis of Metal Nanocrystals: Simple Chemistry Meets Complex Physics? *Angewandte Chemie International Edition*. 2009. 48, pp. 60-103.

[3] SHEN, Xiaoping, WU, Shikui, LIU, Yan, WANG, Kun, XU, Zheng and LIU, Wei. Morphology Synthesis and Properties of Well-Defined Prussian Blue Nanocrystals by a Facile Solution Approach. *Journal of Colloid and Interface Science*. 2009. 329, pp. 188-95.

[4] WU, Xinglong, CAO, Minhua, HU, Changwen and HE, Xiaoyan. Sonochemical Synthesis of Prussian Blue Nanocubes from a Single-Source Precursor. *Crystal Growth & Design*. 2006. 6, pp. 26-28.

[5] REYES-LABARTA, Juan A. and MARCILLA Antonio. Kinetic Study of the Decompositions Involved in the Thermal Degradation of Commercial Azodicarbonamide. *Journal of Applied Polymer Science*. 2007. 107, pp. 339-346.

[6] MACHALA, Libor, ZOPPELLARO, Giorgio, TUČEK, Jiří, ŠAFÁŘOVÁ, Klára, MARUŠÁK, Zdeněk, FILIP, Jan, PECHOUŠEK, Jiří and ZBORIL, Radek. Thermal Decomposition of Prussian Blue Microcrystals and Nanocrystals Iron(III) Oxide Polymorphism Control Through Reactant Particle Size. *RSC Advances*. 2013. 3, pp. 19591-19599.

[7] ZHANG, Lei, BIN WU, Hao, XU, Rong and LOU, Xiong Wen (David). Porous Fe_2O_3 Nanocubes Derived from MOFS for Highly Reversible Lithium Storage. *CrystEngComm*. 2013. 15, pp. 9332-9335.

[8] ZAKARIA, Mohamed B., HU, Ming, HAYASHI, Naoaki, TSUJIMOTO, Yoshihiro, ISHIHARA, Shinsuke, IMURA, Masataka, SUZUKI, Norihiro, HUANG, Yu-Yuan, SAKKA, Yoshio, ARIGA, Katsuhiko, WU, Kevin C.-W. and YAMAUCHI, Yusuke. Thermal Conversion of Hollow Prussian Blue Nanoparticles into Nanoporous Iron Oxides with Crystallized Hematite Phase. *European Journal of Inorganic Chemistry*. 2014. 2014, pp. 1137-1141.

[9] ZHANG, Lei, WU, Hao Bin and LOU, Xiong Wen (David). Metal-Organic-Frameworks-Derived General Formation of Hollow Structures with High Complexity. *Journal of the American Chemical Society*. 2013. 135, pp. 10664-10672.

[10] HERMANEK, Martin and ZBORIL, Radek. Polymorphous Exhibitions of Iron(III) Oxide During Isothermal Oxidative Decompositions of Iron Salts: A Key Role of the Powder Layer Thickness. *Chemistry of Materials*.2008. 20, pp. 5284-5295.

[11] BUENO, Paulo R., FERREIRA, Fabio F., GIMÉNEZ-ROMERO, David, SETTI, Grazielle O., Setti, FARIA, Ronaldo C., GABRIELLI, Claude, PERROT, Hubert, GARCIA-JAREÑO, Juan J. and VICENTE, Francisco. Synchrotron Structural Characterization of Electrochemically Synthetized Hexacyanoferrates Containing K^+: A Revisited Analysis of Eletrochemical Redox. *The Journal of Physical Chemistry C*. 2008. 112, pp. 13264-13271.

[12] SAMAIN, Louise, GRANDJEAN, Fernande, LONG, Gary J., MARTINETTO, Pauline, BORDET, Pierre and STRIVAY, David. Relationship Between the Synthesis of Prussian Blue Pigments, Their Color, Physical Properties, and Their Behavior in Paint Layers. *The Journal of Physical Chemistry C*. 2013. 117, pp. 9693-9712.

[13] TUČEK, Jiří, ZBORIL, Radek and PETRIDIS, Dimitris. Maghemite Nanoparticles by View of Mössbauer Spectroscopy. *Journal of Nanoscience and Nanotechnology*. 2006. 6, pp. 926-947.

SIMULATION OF ELECTRON INTERACTION WITH LIQUID WATER AND PROCESSES RELATED TO SUB-NANOSECOND ELECTRICAL BREAKDOWN

Petr BÍLEK, Zdeněk BONAVENTURA

Department of Physical Electronics, Faculty of Science,
Masaryk University, Brno, Czech Republic, EU,
petrbilek248@gmail.com

Abstract

Sub-nanosecond electrical breakdown in dielectric liquids is of vital interest for applications, e.g. in high-voltage insulation, high-current switching and electric field-cell interactions. Liquid dielectrics in strong nonuniform electric fields are under influence of electrostrictive force that tends to move the fluid into the regions with higher electric field. If the voltage rise is fast enough(nanoseconds), the liquid does not have enough time to be set into motion in order to reduce the internal stress. In this case the electrostrictive force induces significant stress in the bulk of the liquid which is manifested as a negative pressure. At certain threshold, the negative pressure causes cavitation ruptures of the fluid. Subsequently, free electrons can be produced by emission from the surface inside the cavity and accelerated to energies exceeding the energy for ionization of water and thus, contribute to electrical breakdown of water.

We present numerical model, which gives us basic ingredients necessary to study interaction of accelerated electrons with liquid water during sub-nanosecond electrical breakdown. We aim to determine geometrical conditions in electrode vicinity needed for electron multiplication and subsequent build up of space charge, for which we propose to use Monte Carlo software Geant4-DNA. Using of proper simulation techniques could enable us to explain some experimentally observed properties of sub-nanosecond electrical breakdown in liquid water.

Keywords: Liquid water, sub-nanosecond electrical breakdown, cavitation, negative pressure, Geant4-DNA

1. INTRODUCTION

While electrical breakdown in gases seems to be already well explained [1], the challenge stays to set light to electrical breakdown in dielectric fluids such as deionized water. The breakdown in liquids depends on the rise time of the applied high voltage pulse and for „long" high-voltage pulse (~0.1 - 100 μs) occurs through gas bubbles. In this case it is not very different from the usual electrical breakdown in gases, and it has been well studied [2]. Gas bubbles creation in sub-nanoseconds is impossible, since the Joule heating has not enough time to provide energy for evaporation and the expansion of the bubbles, as was pointed out in [3]. Experiments [4, 5, 6, 7] performed in liquid water in presence of nanosecond voltage in a point to plane geometry have measured formation of streamer-like branching and were able to associate characteristic emission spectra with the most important phases of the discharge. The emission spectra showed a broad-band continuum evolving during the first expansion of the discharge and collapse, followed by H/O atomic lines occurring together with continuum emission during the second expansion and collapse [4]. Detailed study of discharge emission dynamics has been performed by [7].

The following text will be devoted to liquid water breakdown with nanosecond rise time, which can be used for wide range of applications, e.g. in high-voltage insulation, high-current switching [8] and sub-nanosecond electric pulse - cell interactions [9].

2. MOTIVATION

Nowadays, accepted scenario of nanosecond electrical breakdown in dielectric liquids was described by [3] and consists of the following steps (see **Figure 1**):

1) At the beginning, the strong inhomogeneous electric field in the vicinity of the needle electrode creates a region saturated by cavitation nanopores.
2) In the nanopores of sufficient sizes, the primary electrons are accelerated by electric field to energies exceeding the ionization potential of water molecule.
3) Many ionization collisions form electron avalanche and build-up of space charge.
4) After neutralization of electrons at the electrode, positive charged liquid forms a virtual needle electrode, and electrostrictive conditions for the appearance of the next set of nanocavities are fulfilled.

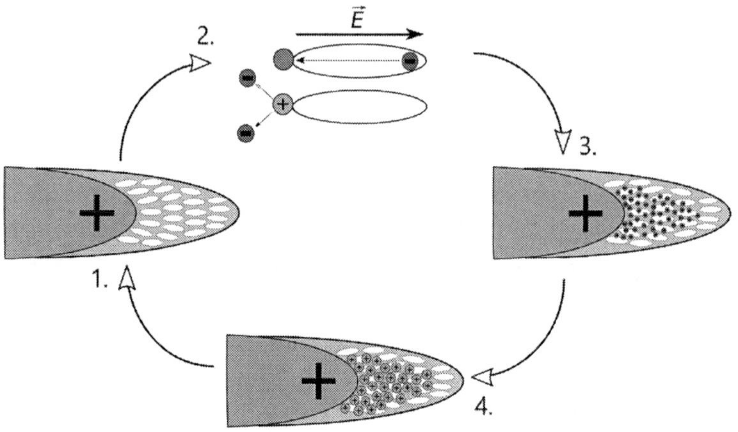

Figure 1 Accepted scenario of nanosecond electrical breakdown in dielectric liquids

The aim of our work is to formulate and develop complex numerical model to study nanosecond electrical breakdown in deionized liquid water. We believe that the model could explain experimentally observed features of the phenomenon such as spatio-temporal characteristics of the discharge and emission spectra. Presented model can be divided into three main stages(submodels):

A) Model of dielectric liquid under influence of electrostrictive forces.
B) Model of cavitation ruptures grow.
C) Model of electron interaction with liquid water containing nanopores.

3. RESULTS

Each of the previous submodels has different spatio-temporal scale and thus requires different physical approach. In what follows, we give short description of each of the submodels (A-C).

3.1. Model of dielectric liquid under influence of electrostrictive forces

Liquid dielectric in strong inhomogeneous electric field is under influence of electrostrictive force, which can cause negative pressure acting on the fluid and initiate discontinuity in the fluid. Experimental data show that the critical negative pressure for water rupture and start of nanometer size cavitation is about 30 MPa.

We have developed hydrodynamic model to study the dynamics of water in a pulsed strongly inhomogeneous electric field in the approximation of compressible flow described by equation of continuity for mass and momentum [10,11]. The hydrodynamic model allows us to find time dependence for pressure and velocity in the liquid for given electrode radius and voltage pulse. In the **Figure 2** we present results of our hydrodynamic model for electrode radius 100 µm and voltage amplitude 54 kV. We have pointed out that in case of long rise time of voltage pulse(left part of the **Figure 2**), the hydrodynamic pressure compensates the electrostrictive pressure and the total pressure (red curve) does not exceed critical negative pressure for the water rupture. When we shorten rise time up to few nanoseconds(right part of the **Figure 2**), electrostrictive pressure prevails the hydrodynamic and the total pressure exceeds the critical pressure, which is necessary for start of nanometer-size cavitation. This model enables us to identify conditions favourable for nanocavitation generation [12].

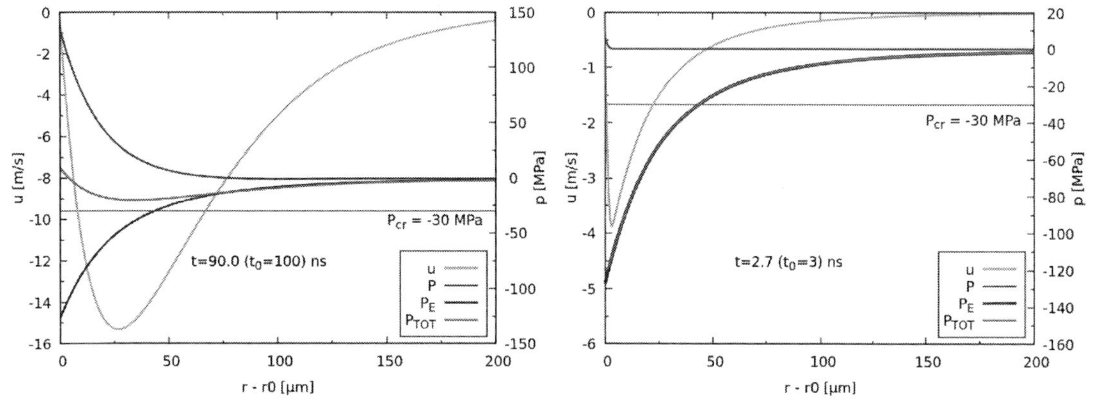

Figure 2 Hydrostatic pressure P, electrostrictive pressure P_E, total pressure P_{tot}, and velocity of fluid u as a function of distance from the electrode at: t_0 = 100 ns (on the left), t_0 = 3 ns (on the right). Horizontal line indicates critical pressure P_{cr} = −30 MPa. Negative values of the velocity corresponds to the motion of the fluid towards the electrode.

3.2. MODEL OF CAVITATION RUPTURES GROW

The previous model confirmed existence of nanopores in the presence of very steep high voltage pulse. Model of cavitation ruptures grow presented in this section aims to reveal grow of these nanocavities. Forces averaged over the surface of the nanocavity give rise to pressure which consists of electrostrictive pressure, surface tension and hydrodynamics pressure:

$$P_{\text{av}} = (3/4)(\alpha - 1)\varepsilon_0\varepsilon E^2 - k_\sigma\sigma/R - p,$$

Nanocavities will collapse if the pressure is negative, on the other hand they will expand if the pressure is positive. Dynamics of the expanding nanocavity is driven by the balance for the kinetic energy of expanding fluid and by the work of pressure forces [4, 13]. Radius of the microcavity as a function of time can be described by the following equation:

$$\frac{d}{dt}\left(R^3\left(\frac{dR}{dt}\right)^2\right) = \frac{4}{\rho}R\frac{dR}{dt}\left(\frac{3}{8}R(\alpha-1)\varepsilon_0\varepsilon E^2 - \frac{pR}{2} - k_\sigma\sigma\right).$$

Solution of this equation allows us to find time dependence of cavity radius at the end of the pulse rise time and determine energy that electron gains inside the cavity. This is important, in order to identify volume in the vicinity of the electrode where ionization of water is possible [3]. To illustrate this, we present **Figure 3** with a numerical solution of the previous equation for these parameters: electrode radius (r_0= 100 µm), water permittivity (ε = 81), α = 1.5, voltage pulse amplitude (U_0 = 54 kV), rise time (t_0 = 3 ns), water surface tension (σ = 0.072N/m) and iniatial nanocavity radius (R_0 = 2 nm).

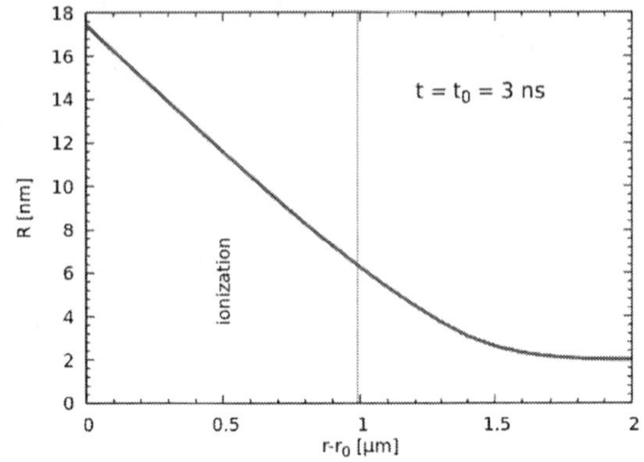

(a) Radius of expanded micropore as a function of distance from the electrode at the end of the voltage rise time. Vertical line separates regions where ionization of water is and is not possible.

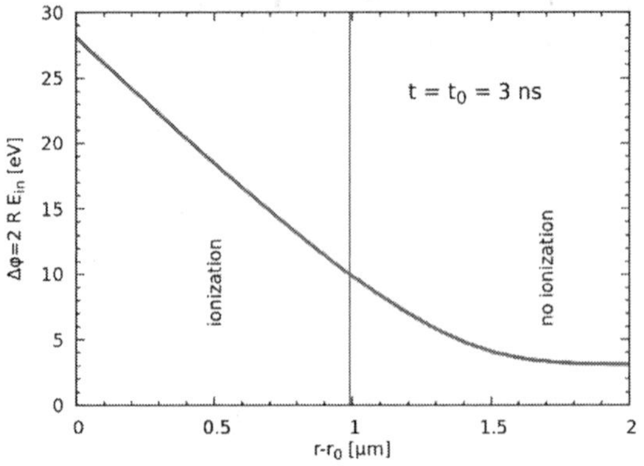

(b) Energy gained by an electron when crossing a cavity as a function of distance from the electrode at the end of the voltage rise time.

Figure 3 Development of nanopores in inhomogeneous pulsed electric field

3.3. MODEL OF ELECTRON INTERACTION WITH LIQUID WATER CONTAINING NANOPORES

Having geometrical parameters (ionization region and cavity radius as a function of distance from electrode) enables us to investigate electron propagation on the nanoscopic scale. For this reason we decided to use Geant4 toolkit (GEometry ANd Tracking) [15,16]. Geant4 is a toolkit for the simulation of the passage of particles through matter and it is developed since 1994. Its areas of application include high energy, nuclear and accelerator physics, as well as studies in medical and space science. Recent release contains a new set of physics models - Geant4-DNA extension for modelling of low-energy electron transport in liquid water (for energies lower than 10keV) [15,16].

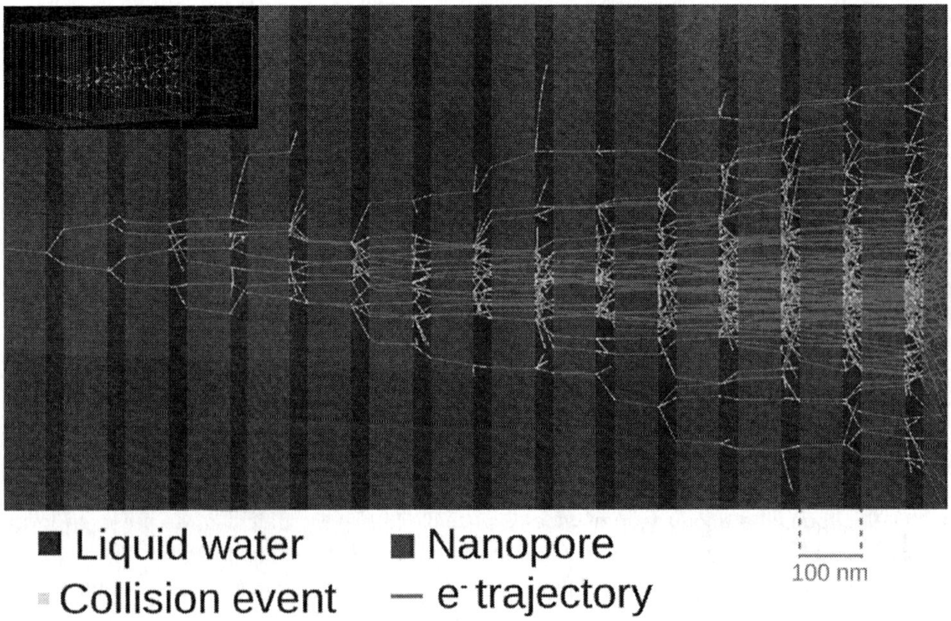

Figure 4 Electron avalanche on the set of nanocavity/liquid layers simulated using Geant4-DNA

Within Geant4-DNA we had to define: geometry as a set of liquid/nanocavity layers, source of primary electrons, electric field based on applied electric pulse and finally, we chose physical models (processes and cross sections). We are able to simulate development of electron avalanche originating from one seed electron accelerated on the set of layers composed from cavities and liquid-water layers as it is shown in the **Figure 4**. It seems that the main source of electron energy loss comes from vibrational excitations of water molecules. Geant4-DNA gives us way how to compute electron energy distribution function after avalanche formation, with which is connected light emission during the sub-nanosecond electrical breakdown.

4. OPEN PROBLEMS AND CONCLUSION

Within the conclusion we should mention some open problems which we have to face: Implementation of scattering distribution for vibrational excitations of electron on water molecule in Geant4-DNA is still missing and the collision has currently only forward scattering. Further, we should try to answer questions as: What is the origin of primary electrons during the electric breakdown? and What exactly happens after electron avalanche formation? With the aim to quantitatively describe build-up of space charge after the electron avalanche formation.

Fluid dynamics with electrostrictive forces(model A), nanopore-grow model(model B) and Geant4-DNA Monte Carlo code with appropriate processes and cross sections(model C) could enable us to describe whole scenario of the electrical breakdown in liquid water. The presented conjunction of models gives us basic tools necessary to study sub-nanosecond electrical breakdown in liquid water. Microscopic view on the electron population could help us to explain experimentally observed emission spectra.

ACKNOWLEDGEMENTS

This contribution is funded by the Czech Science Foundation grant no. 18-04676S.

REFERENCES

[1] RAIZER, Yuri P. *Gas Discharge Physics.* Springer-Verlag Berlin Heidelberg, 1991. p. 449.

[2] USHAKOV, Vasily Y., KLIMKIN, V. F. and KOROBEYNIKOV S. M. *Impulse Breakdown of Liquids.* Springer-Verlag Berlin Heidelberg, *2007.* p. 398.

[3] SHNEIDER, Mikhail N, PEKKER, Mikhail. Liquid Dielectrics in an Inhomogeneous Pulsed Electric Field. *IOP Publishing,* 2016.

[4] ŠIMEK, Milan, PONGRÁC, Branislav, BABICKÝ, Václav, ČLUPEK, Martin and LUKEŠ, Petr. Luminous phase of nanosecond discharge in deionized water: morfology, propagation velocity and optical emission. *Plasma Sources Science and Technology.* 2017. vol. 26, no. 7, pp. 9.

[5] MARINOV, Ilya L., GUAITELLA Olivier, ROUSSEAU, Antoine, STARIKOVSKAIA Svetlana M. Succesive Nanosecond Discharge in Water. *IEEE Transactions on Plasma Science.* 2011. 392672.

[6] SEEPERSAD Y., PEKKER Mikhail, SHNEIDER Mikhail N., FRIDMAN A. And DOBRYNIN D. Investigation of positive and negative modes of nanosecond pulsed discharge in water and electrostriction model of initiation. *J. Phys. D.: Appl. Phys.* 2013. 48424012.

[7] STARIKOVSKIY, A., YANG A., CHO Y. I., FRIDMAN A. Non-equlibrium plasma in liquid water: dynamics of generation and quenching. *Plasma Sources Science and Technology.* 2011. 20024003.

[8] JOSHI, Ravindra P., KOLB, Jurgen F., XIAO Shu, SCHOENBACH Karl H. Aspects of Plasma in Water: Streamer Physics and Applications. *Plasma Processes and Polymers.* 2009. vol. 6, no. 11, pp. 763-777.

[9] SCHOENBACH, K. H. et al. Biological Effects of Intense Subnanosecond Electrical pulses. *IEEE Transactions on Plasma Science.* 2006. 1-4244-0019-8, pp. 573-576.

[10] SHNEIDER, Mikhail N, PEKKER, Mikhail. Liquid Dielectrics in an Inhomogeneous Pulsed Electric Field. *Physical Review E.* 2013. vol. 87, no. 4, 043004.

[11] SHNEIDER, Mikhail N, PEKKER, Mikhail. Cavitation in dielectric fluid in inhomogeneous pulsed electric field. *Journal of Applied Physics.* 2013. vol. 114, 214906.

[12] ŠLAPANSKÁ, M., BÍLEK, P., HNILICA, J., BONAVENTURA, Z. Study of cavitation in liquid water under the action of inhomogeneous pulsed electric fields: application to sub-nanosecond electrical breakdown. *45th Conference on Plasma Physics.* Žofín Palace, Prague.

[13] BÍLEK, P., ŠLAPANSKÁ, M., HODER, T., ŠIMEK, M., BONAVENTURA, Z. Simulation of electron interactions with liquid water and processes related to sub-nanosecond electrical breakdown. *45th Conference on Plasma Physics.* Žofín Palace, Prague.

[14] SHNEIDER, Mikhail N, PEKKER, Mikhail. Pre-breakdown processes in a dielectric fluid in inhomogeneous pulsed electric fields. *Journal of Applied Physics.* 2015. vol. 117, 224902.

[15] BERNAL M.A. et al. Track structure modeling in liquid water: A review of the Geant4-DNA very low energy extension of the Geant4 Monte Carlo simulation toolkit. *Physica Medica.* 2015, vol. 31, no. 8, pp. 861 - 874.

[16] INCERTI, S. et al. GEANT4 very low energy cross section models with experimental data in water. *Medical Physics.* 2010. vol. 37, no. 9.

SORPTION OF ORGANIC DYE METHYLENE BLUE TO HUMIC SUBSTANCES

Jan RYBÁRIK, Martina KLUČAKOVÁ

BUT - Brno University of technology, Brno, Czech Republic, EU,
xcrybarik@fch.vutbr.cz

Abstract

Humic substances represent a wide range of organic compounds generated by gradual decomposition of plant residues and dead organisms in our ecosystems. They are among the most widespread compounds in the world that contain a huge amount of carbon in its structure. Is contains a lot of functional groups and mainly carboxyl -COOH group. We decided to use humic acid methylation to investigate the effect of the carboxyl group on our experiments.

Humic substances are able to bind various pollutants from Industrial production, agriculture and other human activities (like heavy metals, pesticides, herbicides, dyes, etc.). The interaction possibilities can be different: ion exchange, hydrogen bridge formation, coulombic interactions, hydrogen bridges, van der Waals interactions, hydrophobic interactions, absorption, etc.

The purpose of this work is to measure the decrease of selected initial concentrations of the selected organic dye probe (methylene blue) with a sample of lignitic humic acid and a methylated sample of that humic acid. The obtained humic acid sample will undergo a desorption experiment under the same time conditions. The output of the work will also monitor the ratio between the amount of adsorbed and desorbed methylene blue.

The results can provide more details about the stability dye-humic complexes and the mobility of dye in systems containing humic substances.

Keywords: Humic substances, humic acid, methylation, methylated humic acid, adsorption, desorption

1. INTRODUCTION

As was mentioned, humic acid represents a huge group of carbon contained compounds, which can be find all over the world (water, soil, air). Their structure isn't universal, but it changing with a place of finding and with an origin of the sample. Generally it is a supramolecular structure which contains a huge amount of carbon in its structure. It contains as well wide range of functional groups (carboxyl, phenolic, alcohol, ether, nitrogen, etc.) [1].

In nature humic acids are able to bind various substances from soil - we call this effect soil self-cleaning. It can interact with some herbicides, pesticides, heavy metals, organic dyes, medicines, etc. An adsorption experiments to humic acids can be found in some previous work. The adsorption of methylene blue dye on the pillared clay with immobilized humic acid is 2.6 times more effective, then the pillared clay without humic acid. The maximum dye adsorption on pillared clay with immobilized humic acid was observed at a pH of 5.0-7.0 (it was 95.2 - 99.2 %) [2].

The removal of methylene blue from aqueous solution by peat (with humic substances) was analyzed. Various initial concentrations of methylene blue and temperature were monitored. The absolute amount of adsorbed methylene blue increased with a higher initial dye concentration and rising temperature. It was described as endothermic reaction in which the methylene blue diffuses into the pores after saturation of the outer bonding sites [3].

Humic acid-coated Fe_3O_4 nanoparticles as magnetic adsorbents were use. Nanoparticles possessed much higher adsorbed amount of humic acid than Fe_3O_4 nanoparticles and humic acid powder. The time when 50 %

of the methylene blue was adsorbed was 7 min and maximum adsorbed amount of methylene blue was 0.291 mmol/g. [4]

We decided to use humic acid methylation to investigate the effect of the carboxyl group on our experiments. By methylation, the carboxyl groups are blocked and interactions with pollutants are avoided at these active sites. Then it is possible to distinguish them from the other comparisons with humic acids without modification. Diffusions experiments with humic, methylated humic acid and methylene blue with methylene blue bring us inspiration. [5], [6], [7], [8].

For comparison of successful adsorption we done desorption experiments as well.

2. EXPERIMENTAL PART

Our humic acids sample was prepared by extraction from finely granulated South Moravian lignite (Mikulčice, Czech Republic). 60 g of lignite was mixed with 1 dm^3 of concentrated 0.1 M HCl, the suspension was left for 1 hour on the shaker. The decalcified lignite was washed with sufficient distilled water to remove chloride ions. Prepared suspension of lignite was transferred to a plastic container with extraction agent (0.1M sodium hydroxide and 0.1M anhydrous pyrophosphate). The bottle with the extraction agent was left on the shaker for 12 hours.

The suspension was centrifuged and the supernatant (solution of humic acid) was filtered, this filtrate was acidified with hydrochloric acid up to a pH = 2. After two hours, all humic acid had been precipitated, then centrifuged from solution. Solid part (humic acids) were placed into dialysis membrane (MWCO: 3,500; Spectra/Por®, Spectrum laboratories, Inc.) and were immersed in distilled water. The distilled water was regularly replaced with a test for the presence of chloride ions (with $AgNO_3$), after 10 time change it was lyophilized (Virtis benchtop K (Bio Trade s.r.o).

Half of the sample of prepared humic acid was subjected to selective methylation. For 1 g of humic acids were added with 4 cm^3 of chloroform and 2 cm^3 of methanol. Subsequently, 4 cm^3 of a 2M solution of trimethylsilyl diazomethane in hexane (TMS-DM) was added. The mixture was stirred for 2 hours on a mixer in the hood. An additional 0.75 cm^3 TMS-N21 was added after. It was necessary to evaporate the unreacted reagent, the samples were dried overnight in the digester at room temperature.

From the modified acids obtained, it was necessary to evaporate the unreacted reagent. The samples were dried overnight in the digester at room temperature. The picture of obtain humic acid and methylated humic acid can be seen at **Figure 1**.

Figure 1 Picture of color differences (left - humic acid, right - methylated humic acid)

Extracted humic acid and selectively methyleneic acid was subjected to analysis of the FTIR Nicolet iS50 spectrophotometer using the ATR method to determine the spectrum of humic acids and to verify the successful methylation of the prepared sample of methylated humic acids.

For adsorb experiments we used 100 mg of humic acid (HK) and methylated humic acid (mHK), each dissolved in 20 mL of methylene blue of different concentration. The overview of inicial methylen bule concentrations used for is shown in **Table 1**. The diferences in absorbance were meassured on spektrofotomeher UV-VIS HITACHI U-3900H, up to 288 hours.

Table 1 Initial methylene blue dye concentration used for adsorption experiments

100 mg humic acid with methylen blue	0.1 g/L	0.075 g/L	0.05 g/L	0.025 g/L	0.01 g/L
100 mg methylate humic acid with methylen blue	0.1 g/L	0.075 g/L	0.05 g/L	0.025 g/L	0.01 g/L

After 288 hours, the experiment was stopped. The solid fraction of humic acid/methylated humic acid with adsorbed methylene blue was centrifuged and dryed on the petri dish at room temperature. Then we done desorption expetiments. 10 of obtained samples were desolved each in 10 mL of destilated water. It was observed as well for 288 hours to obtain information about amout of desorbet methylen blue dye. All experiment were realised on laboratory temperature with magnetic stirrer (250 rpm) in vial flask. Picure of dryed samples after adsorption, can be seen on **Figure 2**.

Figure 2 Prepared samples of humic acids after adsorption of methylene blue

3. RESULTS AND DISSCUSION

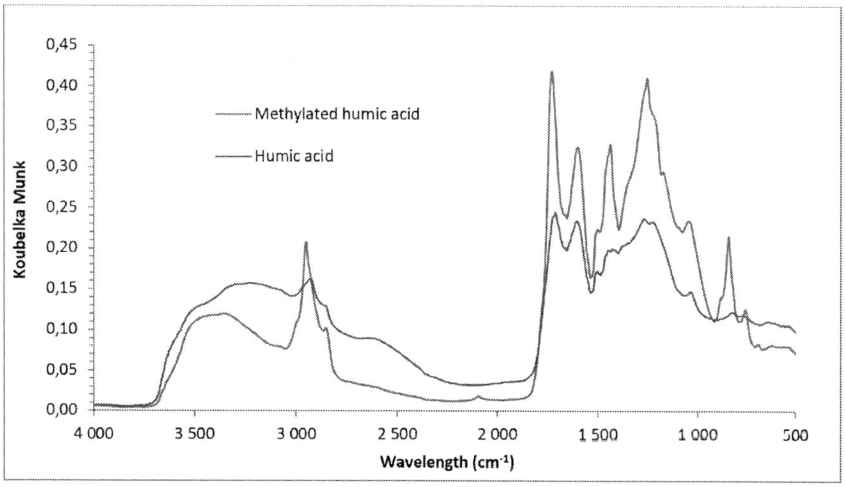

Figure 3 FTIR spectrum -humid acids (blue curve), methylated humic acid (red curve)

Due to FT-IR spectroscopy was confirmed successful methylation of prepared methylated humic acid. On the FT-IR spectra (**Figure 3**) it can be seen, the methylation of functional groups resulted in changes of FT-IR spectra characteristic for esters. The intensity of the broad peak at around 3500 cm^{-1} was decreased by methylation and the shallow band at 2580 cm^{-1} was completely absent. The absorption band corresponding to C=O stretching was shifted to higher wavenumbers after methylation (from ~1720 cm^{-1} to ~1735 cm^{-1}) and its intensity was increased. The observed changes in spectra caused by methylation corresponded with results published previously. [9], [10].

Adsorption experiments with humic acid was observed via measuring of absorption of methylene blue at peak 665 nm. As can be seen on **Figure 3**, the initial concentration of methylene decreased from 0.002 g of methylene blue (initial concentration) up to 0.001992 g of methylene blue in adsorption to humic acid during 288 hours. Same tendency was observed for all adsorption experiments with humic acid only with different initial concentrations, which are mentioned in **Table 2**.

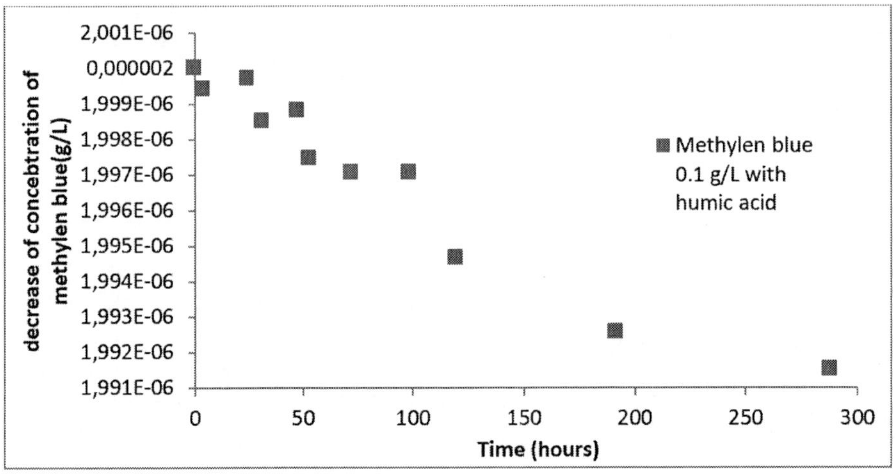

Figure 4 Decreasing of initial concentration of methylene blue with 100 mg of humic acid

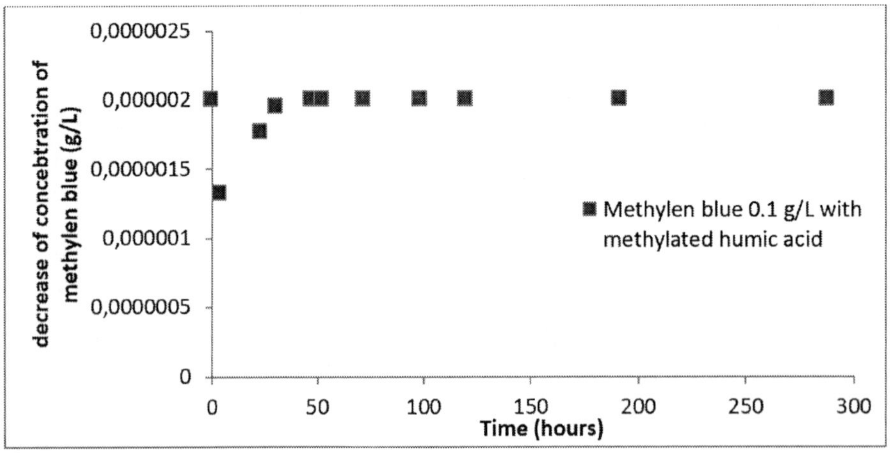

Figure 5 Decreasing of initial concentration of methylene blue with 100 mg with methylated humic acid

Adsorption experiments with selectively methylated humic acid were observed as well. The initial concentration of methylene blue dye had decreased in first 40 hours, but then it increased almost up to the initial concentration of the methylene blue dye. This tendency can be seen on **Figure 5** for initial concentration 0.1 g/L.

As was expected, a lignitc humic acids were able to adsorbed methylene blue dye. The highest adsorption we obtained with initial concentration 0.01 g/L of methylene blue - in this case the concentration of adsorbed dye rise up to 7.3 %. In case of methylated humic acid, we obtain very low decrement of adsorption, it was able to bind only 0.3 % of dye in case of initial concentration 0.025 and 0.001 g/L. All results can be seen on **Table 2.**

Table 2 Values of adsorbed methylene blue dye into humic acid and methylated humic acids in %

Initial concentration of methylene blue	0.1 g/L	0.075 g/L	0.05 g/L	0.025 g/L	0.01 g/L
Adsorption info 100 mg of humic acid	0.4 %	0.76 %	0.74 %	1.9 %	7.3 %
Adsorption info 100 mg of methylated humic acid	0.1 %	0.1 %	0.1 %	0.3 %	0.3 %

Desorption experiments have provided information on how stable is the adsorption of methylene blue into humic acid and methylated humic acid as well. For 288 hours, we observed concentration of the dye in vial flasks via spectrophotometry. From obtained results is clear that methylene blue dye adsorbed in the first part of experiment partly desorbed. For initial concentration of 0.05 g/L, desorption reached the value 54 % for humic acid and 66.9 % for selectively methylated humic acid. All values are mentioned in the **Table 3.**

Table 3 Table with values of the amount of desorbed methylene blue dye from humic acid in %

Initial concentration of methylene blue	0.1 g/L	0.075 g/L	0.05 g/L	0.025 g/L	0.01 g/L
Desorption from humic acids	34.2 %	31.1 %	54.0 %	46.9 %	24.45 %
Desorption from methylated humic acids	18.4 %	50.0 %	28.6 %	66.7 %	60.0 %

4. CONCLUSION

We prepared sample of humic acid from lignite. Half of prepared humic acid were selectively methylated to block carboxyl groups in its structure. Visually there was a color different of samples and as well due to FT-IR spectroscopy was confirmed successful methylation.

Then we continued with adsorption experiments - humic acid was able to bind up to 7.3 % of dye, methylated humic acid was able to bind maximally 0.3 % of dye. From results is clear that carboxyl groups are mainly responsible for binding with a dye. Because different of binding was big.

Desorption experiment had shown that binding into its structure is week. In case of humic acids it desorbed up to 54 % for concentration 0.01 g/L of dye and in case of methylated humic acid it was 66.7 % for initial concentration 0.025 g/L.

All this experiment bring us interesting data. For future experiment we would like to repeat experiments with same dye and run experiment for longer time. Then retaliate experiment with another type of dye.

ACKNOWLEDGEMENTS

Materials Research Centre at FCH BUT-Sustainability and Development, REG LO1211, with financial support from National Programme for Sustainability I(Ministry of Education, Youth and Sports).

REFERENCES

[1] RYBÁRIK, J. *Strukturní a termodynamické aspekty interakcí biopolymerů s organickými ionty.* Brno: Brno university of technology, Faculty of chemistry, 2017. Ú 1-64.

[2] VINOD, V. P. a T. S. ANIRUDHAN. *Adsorption Behaviour of Basic Dyes on the Humic Acid Immobilized Pillared Clay. SpringerLink [online].* 2003, 2003(150), 193-217 [cit. 2018-26-09]. DOI: 10.1023/A:1026145631713.

[3] FERNANDES, A.N., C.A.P. ALMEIDA, C.T.B. MENEZES, N.A. DEBACHER a M.M.D. SIERRA. Removal of methylene blue from aqueous solution by peat. Journal of Hazardous Materials [online]. 2007, Vol. 144(1-2), p 412-419 [cit. 2018-26-09]. DOI: 10.1016/j.jhazmat.2006.10.053.

[4] ZHANG, Xian, Panyue ZHANG, Zhen WU, Ling ZHANG, Guangming ZENG a Chunjiao ZHOU. Adsorption of methylene blue onto humic acid-coated Fe3O4 nanoparticles [online]. [cit. 2018-26-09]. DOI: 10.1016/j.colsurfa.2012.12.056.

[5] SEDLÁČEK, P.; SMILEK, J.; KLUČÁKOVÁ, M. How Interactions with Polyelectrolytes Affect Mobility of Low Molecular Ions - Results from diffusion cells. *Reactive and Functional Polymers 2013.* Vol. 73, No. 11, p. 1500-1509.

[6] SEDLÁČEK, P.; SMILEK, J.; KLUČÁKOVÁ, M. How the Interactions with Humic Acids Affect the Mobility of Ionic Dyes in Hydrogels - 2. Non-stationary Diffusion Experiments. *Reactive and Functional Polymers 2014.* Vol. 75, No. 1, p. 41-50.

[7] SMILEK, J.; SEDLÁČEK, P.; KALINA M.; KLUČÁKOVÁ, M. On the role of humic acids' carboxyl groups in the binding of charged organic compounds. *Chemosphere 2015.* Vol. 138, p. 503-510.

[8] SMILEK, J.; SEDLÁČEK, P.; LAŠTŮVKOVÁ, M.; KALINA, M.; KLUČÁKOVÁ, M. Transport of Organic compounds Through Porous Systems Containing Humic Acids. *Bulletin of Environmental Contamination and Toxicology 2017,* Vol. 3, No. 98, p. 373-377.

[9] KLUČÁKOVÁ, M.; KALINA, M.; SEDLÁČEK, P., GRASSET, L. Reactivity and Transport Mapping of Cu(II) Ions in Humic Hydrogels. Journal of Soil and Sediments 2014, Vol. 14, No. 2, p.368-376.

[10] KLUČÁKOVÁ M., KALINA M. Diffusivity of Cu(II) ions in humic gels - influence of reactive functional groups of humic acids. *Colloids and Surfaces, A: Physiochemical and Engineering Aspects* 2015, vol. 483, pp. 162-170.

DETERMINATION OF THICKNESS REFINEMENT USING STEM DETECTOR SEGMENTS

Radim SKOUPÝ[1], Vladislav KRZYŽÁNEK[2]

[1]Institute of Scientific Instruments of the CAS, Brno, Czech Republic, EU, ras@isibrno.cz
[2]Institute of Scientific Instruments of the CAS, Brno, Czech Republic, EU, krzyzanek@isibrno.cz

Abstract

Quantitative STEM imaging together with Monte Carlo simulations of electron scattering in solids can bring interesting results about properties of many thin samples. It is possible to determine thickness of a sample, to calculate mass of particles and measure mass per length/area. Appropriate calibration is one of the crucial parts of the method. Even a small error or inaccuracy in detector response to electron beam - either blanked or full - brings significant error into thickness determination. This problem can be overcome by parallel STEM imaging in more segments of the detector. Comparing more segments gives a possibility to use a signal from different segments for different thicknesses of a sample. Accuracy of individual parts of the detector depends on the captured signal quantity. It is desirable to use such a STEM detector segment that provides the greatest signal change to a unit of thickness. To demonstrate the usage, we used a sample of Latex nanospheres placed on thin carbon lacey film; diameter of the nanospheres was around 600 nm in order to compare the results from different detector segments. Thanks to the known thickness of the sample (calculated from its geometrical shape), it is possible to estimate the optimal acquisition settings and post processing steps with the known and the true state of the sample.

Keywords: Quantitative STEM, thickness determination, detector segments, Monte Carlo simulation

1. INTRODUCTION

The field of electron microscopy offers many types of imaging techniques. In general, it is possible to divide imaging techniques into two groups by dimensionality of provided information. The first group gives 3D information about the investigated sample. It can be focused ion beam milling (FIB-SEM), serial block face imaging (SBF-SEM) or volume scope (deconvolution of images captured at different acceleration voltages) in the case of scanning electron microscopes (SEM). In the case of transmission electron microscopy (TEM) it is possible to use tomography or single particle analysis workflows. The second group is imaging in 2D either conventional imaging in SEM using of variate detectors and signals or planar projection of a sample in TEM and STEM. Quantitative scanning transmission electron microscopy (STEM) imaging brings a way how to get more quantitative information about the thickness, such as a thickness of the sample at each pixel from a single image (other available information may be a mass per length of filamentous structures, mass per area of sheets, or mass of particles). The main assumptions for this technique are 1) known geometry of the detection system that usually contains the geometry and position of the detector, the sample holder and the pole piece of the final demagnifying lens 2) density and composition of investigated samples. As clear from principle of STEM imaging, the samples are mounted on TEM grids. It depends on the type of the sample if the supporting film is used. Provided that it is necessary e.g. for imaging of individual nanoparticles, it is desirable to use a film as thin as possible with sufficient endurance to the electron beam. Correlative quantitative imaging in SEM and STEM is described in [1].

2. MATERIAL AND METHODS

Method based on workflow described by Volkenandt et al in [2] was used in this study. The main assumption is a ***linear response of a detector to intensity of incident electron beam***. It is necessary to calibrate the detector system response on incident electron beam over the whole range of used probe currents. We used commercially available STEM III detector mounted on Magellan 400L SEM (both FEI, Czech Republic). Images were taken by the original internal scanning unit of the microscope with a dwell time of 5 µs and image size of 1024 × 882 pixels. Five amplifier lines were used for simultaneous acquisition of four independent signals: bright field (BF) - one channel, dark field 1 (DF1) - one channel, dark field 3 (DF3) - one channel, and high angle annular dark field (HAADF) - two channels. Dark field 2 and dark field 4 were not used in this study. Other SEM settings are as follows: acceleration voltage 30 kV, probe current 25 pA, working distance (WD) 4 mm, field free mode, pixel size approximately 1 nm. Quantitative imaging in immerse high resolution mode (described in [3]) was not used for its uncertainty and difficulty.

For Monte Carlo (MC) simulation of signals captured by the STEM detector, it is essential to use correct geometrical parameters of the used detector together with working distance of a sample in view of system pole piece - sample - detector. Those characteristics of our system are shown in the **Table 1**. The MC simulation was performed in Casino software [4] with appropriate settings (beam energy 30 keV, 100,000 electrons per point, total and partial cross section taken from Elsepa database, supporting thin carbon layer was taken into account). Captured electron microscopy images were calibrated to full and blanked electron beam by the equation (1)

$$I_{NORM} = \frac{I_{SIG} - I_{DARK}}{I_{FULL} - I_{DARK}} \tag{1}$$

where:

I_{SIG} - the pixel value

I_{DARK} - the mean value of dark image

I_{FULL} - the mean value of full electron beam on the detector.

Table 1 Radius and angle of individual detector segments of STEM III detector in working distance of 4 mm. The segments used in this study are highlighted by gray color.

Segment of the detector	**Radius (mm)**		**Angle (mrad)**	
BF	0.00	0.46	0.0	32.5
DF1	0.53	0.79	37.8	52.0
DF2	0.80	1.03	57.3	73.7
DF3	1.11	1.45	79.1	103.1
DF4	1.52	2.21	108.4	156.3
HAADF	2.28	10.00	161.5	650.3

Used calibration images are shown in **Figure 1.** Segments DF3 and HAADF are only visible in our SEM when the retractable STEM detector is manually shifted. When those images were taken, the detector was centred according to the electron beam. Notice the dark spot in the middle of the BF segment of the detector - this response inhomogeneity of our detector has clearly visible impact on the captured BF image. Slightly scattered electrons passing through the edge of spherical nanoparticle are landing on more sensitive part of the BF segment and creating the bright rim of the particle. It is recognisable in final determination of thickness, where influence of this inhomogeneity is visible in wrong estimation of thickness of supporting carbon layer (3 nm in thick).

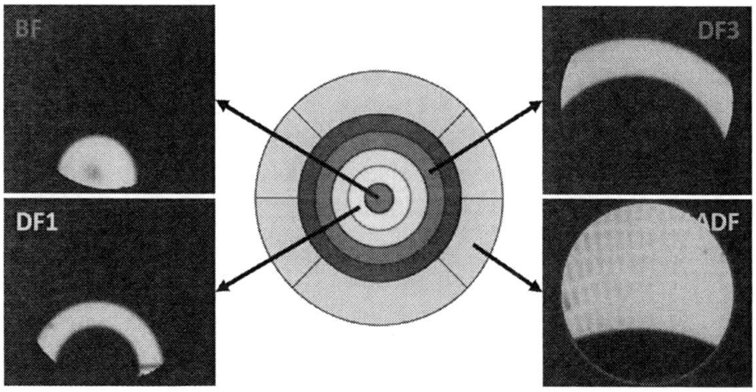

Figure 1 Calibration images of used detector segments with schematic drawing of their arrangement in used STEM detector

To compare individual detector segments and their accuracy we used - analogically to our previous work with a BSE signal [5] - a sample with a thickness known in each point - Latex nanosphere with a nominal diameter of 616 nm (S130-6, Agar Scientific, United Kingdom). We measured diameter of 583 nm in the case of the shown particle. Further processed images are shown in **Figure 2.** Even with the naked eye, there is low contrast change of DF3 image visible in direction from the edge of the particle to the centre. For MC simulation, the used density of latex was 1.055 kg/m^3. Results of MC simulation are shown in **Figure 3 right**.

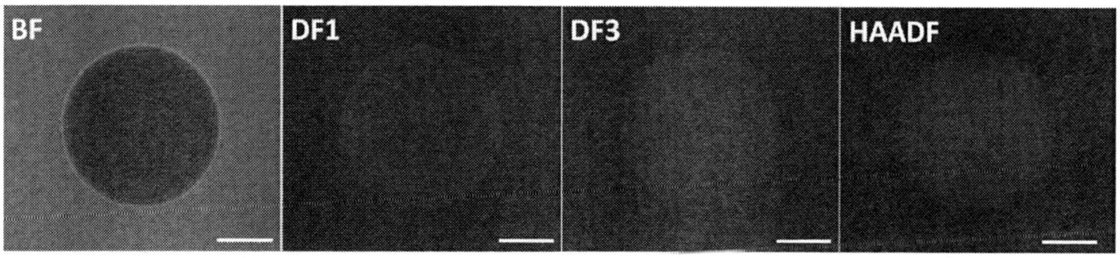

Figure 2 Latex nanosphere captured simultaneously in BF, DF1, DF3 and HAADF. Bar 200nm.

For noise limitation and easier comparison of the results, the center of nanoparticle is detected and then rotation averaging is performed (shown in **Figure 3**). In this step a distance of each individual pixel to center of the sphere is calculated. Then the pixels with distance in chosen range are averaged. The advantage of this method is the absence of an interpolation error. However, a small amount of pixels in the middle of the particle causes significantly higher noise in this area.

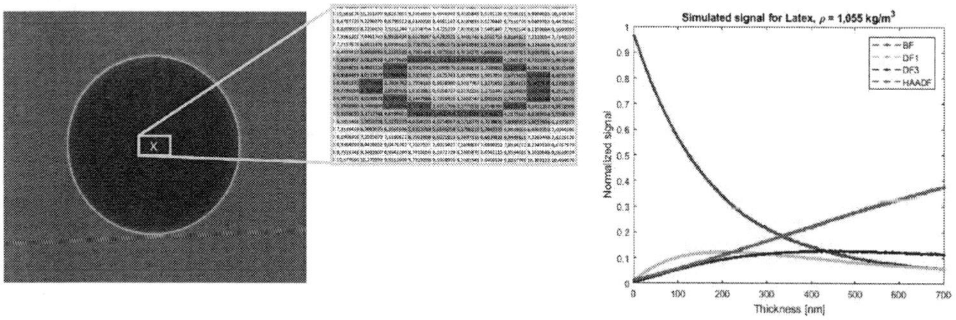

Figure 3 Left: Calculation of rotationally averaged signal. Detected sphere nanoparticle is green, control of integrity of the particle (red lines) and detected center (yellow cross). Right: Monte Carlo simulation for Latex.

3. RESULTS AND DISCUSSION

Our results show high differences of accuracy in estimated thickness by individual detector segments. Captured data are shown in **Figure 4 left** and final correspond thickness profiles are shown in **Figure 4 right**. The highest accuracy we found in the case of HAADF segment, where the estimated thickness corresponds with the theoretical one in the whole measured thickness range. The BF segment brings similar sensitivity as HAADF does, except for the center of the sphere and the edge area of the sphere including the thin support carbon layer. This inaccuracy is caused by reduction of sensitivity in the center of the BF segment of the used STEM detector (as mentioned above and shown in **Figure 1**).

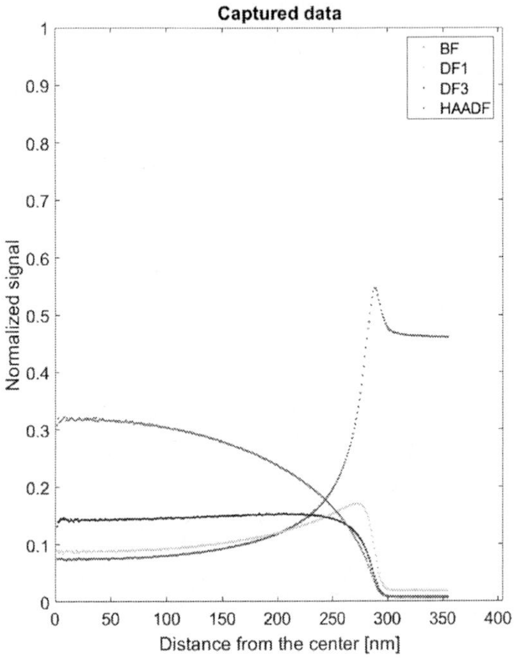

Figure 4 Captured data after rotationally averaging (left). Thickness of the sample estimated by quantitative STEM imaging in four different detector segments (right).

Results of DF1 and DF3 segment imaging have not been satisfying within our experiment so far. It is for further examinations, it is highly probable the calibration of both segments may be problematic. In the case of DF1 (green) there is dependency similar to what was expected clearly visible but moved about 100 nm lower in thickness. The straight line indicates the thickness over simulated range (values of the points where captured signal is higher than the highest simulated signal are removed). It brings the assumption, that the active size of those detector segments is higher than we used for MC simulation and our information about properties of used STEM detector is not accurate. The DF3 segment is not the appropriate choice for thickness estimation of our test sample, because of flat course at most of the captured data profile. Low change of signal due to varying thickness brings low sensitivity in thickness determination. In general, all segments DF1-DF4 have very fine angular range and thus their sensitivity for measured samples is very low.

4. CONCLUSION

We demonstrated the ability of quantitative low voltage STEM imaging to precise determination of local thickness in wide range from 0 to approximately 600 nm in the case of latex (polystyrene) sample. The proof was done at the sample of latex spherical nanoparticles that have relatively low density. Particles are composed of light elements - carbon and hydrogen. In the case of samples containing higher atomic elements, the penetration depth of the electron beam is much lower. Unfortunately, not all used segments gave us the same results. We observed calibration errors which were devaluating measurements in DF1 and DF3 segments. Compared to that, BF and HAADF segments bring accurate results which correspond with the reality of the sample. Nowadays other experiments are underway with the aim to observe origin of partly inaccurately calibration. We recommend to process similar comparative experiment after any change in imaging process of a microscope.

ACKNOWLEDGEMENTS

The research was supported by Czech Science Foundation (project GA17-15451S), Ministry of Education, Youth and Sports of the Czech Republic (project LO1212). The research infrastructure was funded by Ministry of Education, Youth and Sports of the Czech Republic and European Commission (project CZ.1.05/2.1.00/01.0017).

REFERENCES

[1] SUN, Cheng, MÜLLER, Erich, MEFFERT, Matthias and GERTHSEN, Dagmar. On the Progress of Scanning Transmission Electron Microscopy (STEM) Imaging in a Scanning Electron Microscope. *Microscopy and Microanalysis*. 2018. vol. 24, no. 2, pp. 99-106. DOI: 10.1017/S1431927618000181.

[2] VOLKENANDT, Tobias, MÜLLER, Erich and GERTHSEN, Dagmar. Sample Thickness Determination by Scanning Transmission Electron Microscopy at Low Electron Energies. *Microscopy and Microanalysis*. 2014. vol. 20, no. 1, pp. 111-123. DOI: 10.1017/S1431927613013913.

[3] WALKER, Christopher.G.H., KONVALINA, Ivo, MIKA, Filip, FRANK, Luděk a MÜLLEROVÁ, Ilona. Quantitative comparison of simulated and measured signals in the STEM mode of a SEM. *Nuclear Instruments and Methods in Physics Research Section B: Beam Interactions with Materials and Atoms*. 2018. vol. 415, pp. 17-24. DOI: 10.1016/j.nimb.2017.10.034.

[4] DEMERS, Hendrix, POIRIER-DEMERS, Nicolas, COUTURE, Alexandre Réal, JOLY, Dany, GUILMAIN, Marc DE JONGE, Niels and DROUIN, Dominique. Three-dimensional electron microscopy simulation with the CASINO Monte Carlo software. *Scanning*. 2011. vol. 33, no. 3, pp. 135-146. DOI: 10.1002/sca.20262.

[5] KRZYZANEK, Vladislav, SPORENBERG, Nora, KELLER, Ulrike, GUDDORF, Jessica, REICHELT, Rudolf and SCHÖNHOFF, Monika. Polyelectrolyte multilayer capsules: nanostructure and visualisation of nanopores in the wall. *Soft Matter*. 2011. vol. 7, no. 15, pp. 7034-7041 DOI: 10.1039/c1sm05406f.

INVESTIGATION OF ATMOSPHERIC PRESSURE PLASMA SURFACE MODIFIED ALUMINIUM OXIDE NANOPOWDER BY THERMAL DESORPTION SPECTROSCOPY

[1]Ján DUGÁČEK, [1]Lubomír PROKEŠ, [1]Pavel SŤAHEL, [1]Jozef RÁHEĽ, [1]Martina ILČÍKOVÁ

[1]CEPLANT - R&D Centre For Low-cost Plasma and Nanotechnology Surface Modifications, Masaryk University, Brno, Czech Republic, EU, 373920@mail.muni.cz

Abstract

Aluminium oxide nanopowder is used for sintering into ceramic materials with attractive mechanical properties. The simplest and most common sintering methods are wet methods during which the nanopowder is dispersed in a liquid. High surface energy increases the quality of the sintered material. Atmospheric pressure plasma is known improve its wetting properties and the quality of ceramic sintered using the electrophoretic method. However, polarisation abnormalities during the deposition suggested an addition of negatively charged species on the powder's surface during the treatment, probably NO_X molecules. To verify this hypothesis, we have used the Thermal Desorption Spectroscopy on the treated and the untreated nanopowder to study the changes that occurred during the treatment. We have found that a large number of additional NO, NO_2, CO_2 and hydroxyl were attached to the surface, at least some of them were grafted rather than adsorbed. This provided additional evidence that the abnormalities were caused by NO_X molecules on the surface.

Keywords: Alumina, sintering, electrophoretic method, plasma treatment, Thermal Desorption Spectroscopy

1. INTRODUCTION

Aluminium oxide is a material with many useful properties, such as high hardness, high wear resistance, corrosion resistance, thermodynamic stability and can also have good transparency. It has a lot of application in cutting tools, optical objects, implants etc. [1]

Alumina ceramic's properties are improved with smaller grain size of the material it's sintered from, changing significantly as the grains reach the nanoscale. Such materials are made of nanopowders via sintering. These methods also allow mixing more types of nanopowders together to create nanocomposite materials [2].

Electrophoretic deposition is a cost-effective and versatile sintering method. This method requires a high concentration suspension of the nanopowder in a liquid, otherwise particles deposit at different speeds [3]. Hydrophilicity of the nanopowder helps ensure good mixing of the powder in the liquid and thus increase the quality of the sintered material.

Tomáš Morávek and colleagues have verified that plasma treatment in DCSBD reduces the surface roughness and slightly improves the density and grain size of electrophoretically sintered coatings [4]. XPS measurements had shown an increased amount of C-O bonds and AlOOH bonds. However, while investigating the behaviour of plasma treated and as-received aluminium oxide powders during the electrophoretic deposition, it was discovered that the main difference between the processes is a change of the polarization of the electrode where the deposit was created. The plasma treated powders created the deposit on the positive electrode while the deposit created from the suspension containing non-treated powder was formed on the negative electrode. To be possible for this to happen, some negatively charged species had to be adsorbed on the powder's surface during the treatment. By using different analytical methods, it was concluded that one of possible causes is an adsorption of NO_X molecules.

In order to investigate if there were any NO_X gases, we have analysed the result using Thermal Desorption Spectroscopy (TDS) [5] on alumina nanopowders before and after treatment with a Diffuse Coplanar Surface Barrier Discarge (DCSBD) [6]. We have compared the results to data obtained from FTIR.

2. EXPERIMENTAL

We have used high-purity (99.99 %) Taimicon TM-DAR nanopowder from company Taimei chemicals with grain size 0.1 μm. According to the manufacturer, the powder's surface is 14.5 m^2/g.

We have used a DCSBD discharge with area 196 x 78 mm, containing two electrodes of interlocking comb pattern with 30 stripes placed 1 mm apart from each other (**Figure 1**). We have placed the nanopowder on the discharge's surface with homogeneous width. The experiment was performed in air at atmospheric pressure. The discharge was fed by 500 W power with frequency 30 kHz. The process lasted for 30 seconds.

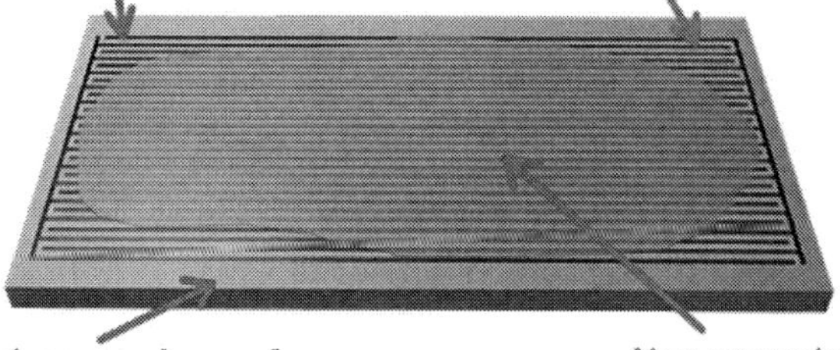

Figure 1 Using a DCSBD to process the nanopowder

The Thermal Desorption Spectroscopy measurement was performed on a custom-built device (**Figure 2**). About 300 mg of the nanopowder was placed in a metal vessel and inserted into a vacuum chamber evacuated by a Pfeiffer D-35614 Asslar turbomolecular pump, at high vacuum at pressure around $5 \cdot 10^{-5}$ Pa. Using a Clasic laboratory fournace, the powder was heated by 10K/min until temperature 1000 °C and kept at 1000 °C for 30 min. Mass spectra of gases desorbed during the reactions caused by increasing temperature were measured using a Prisma 80 quadrupole mass spectrometer from Pfeiffer with range from 1 to 80 amu. Results were computed and visualised using a custom program developed with Qt.

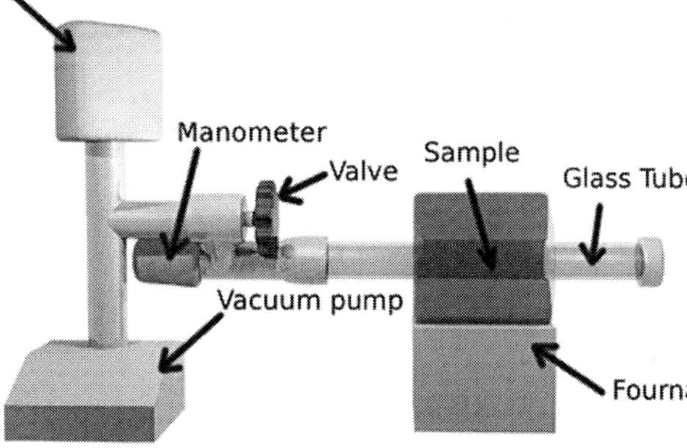

Figure 2 The apparatus used for TDS

3. RESULTS AND DISCUSSION

First, we have measured thermal desorption spectra of treated and untreated nanopowder at temperatures up to 1000 °C. The data were rendered into a heatmap where every point's colour indicates the amount of desorbed gas of mass determined by its x position at temperature determined by its y position (**Figure 3**).

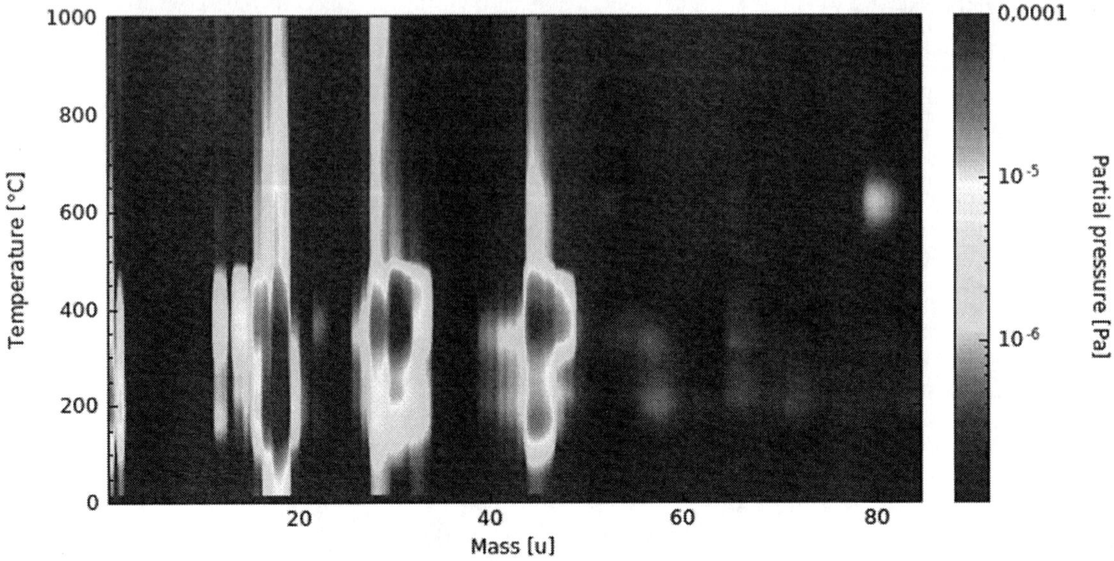

Figure 3 Thermal Desorption heatmap of plasma-treated alumina nanopowder

We can see a large desorption peak of water around temperature 200 °C. There are also desorption peaks of other materials at that temperature, especially oxygen, nitrogen and carbon dioxide. There is also a desorption maximum around temperature 380 °C, especially for oxygen, nitrogen, nitrogen dioxide and carbon dioxide. The data from untreated powder are in the following heatmap: Then, we treated the nanopowder with DCSBD plasma for 30 s and measured its effects on the spectra (**Figure 4**).

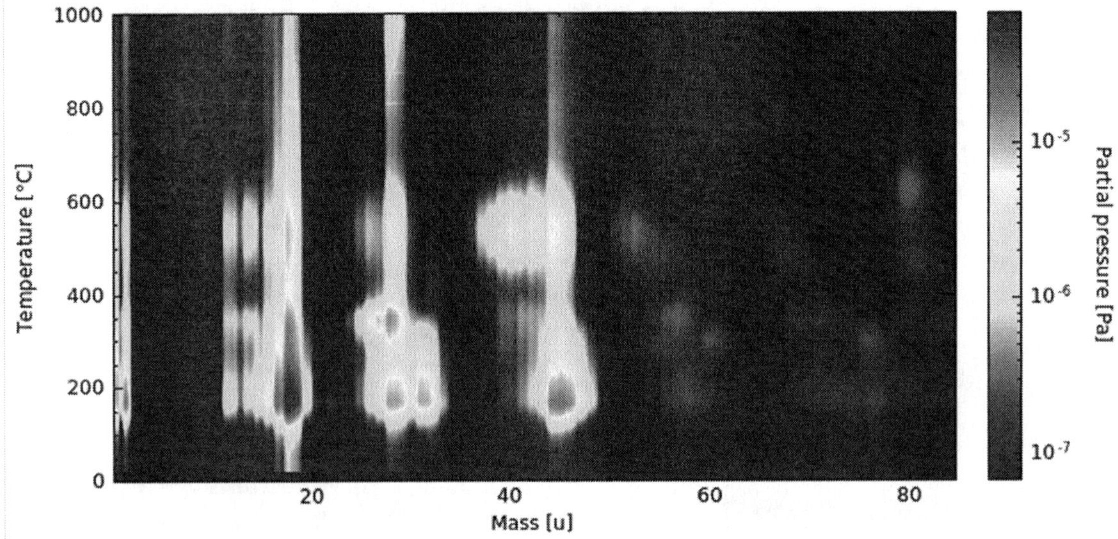

Figure 4 Thermal Desorption heatmap of untreated alumina nanopowder

Besides the large desorption peak around temperature 200 °C, we can see a smaller group of desorption peaks of oxygen and nitrogen around temperature 400 °C. The series of small peaks released at temperature around 500 °C appeared also in experiments without the nanopowder. Therefore, the plasma treatment greatly affected the peak appearing at temperature 380 °C. Graph at **Figure 5** shows this release more closely to allow to determine which gases were released and their relative amounts compared to the release from untreated sample.

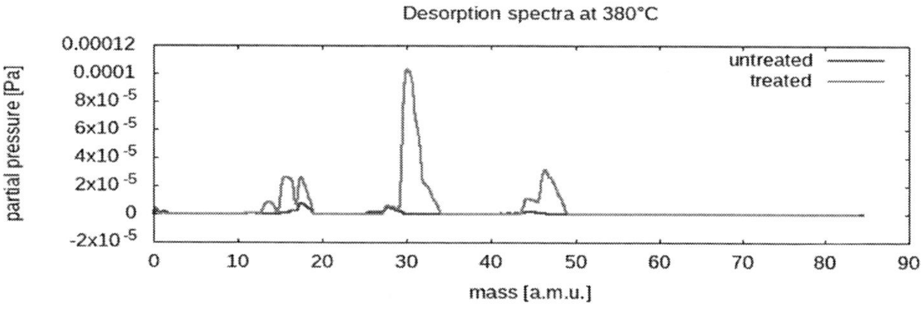

Figure 5 TDS spectrum at temperature 380 °C

In the spectra of the untreated nanopowder, we can see mostly atmospheric gases and water vapour, which may originate from the vacuum chamber's walls. In the spectra of the plasma-treated nanopowder, there is a large additional release at masses 14, 16, 17, 18, 30 32, 44 and 46 u (**Table 1**).

Table 1 Probable species corresponding to the desorbed masses

Mass	Probable species
14	N (fragment of N_2, NO or NO_2)
16	O (fragment of O_2, NO, NO_2 or H_2O)
17	OH
18	H_2O
30	NO
32	O_2
44	CO_2
46	NO_2

It is unlikely that species at masses 30 and 44 were ethane and propane respectively because of the absence of fragments that are typical for hydrocarbons. The massive releases at masses 30 and 46 corresponding to nitrogen oxide and nitrogen dioxide suggest that the sample indeed contains NO_X as hypothetised.

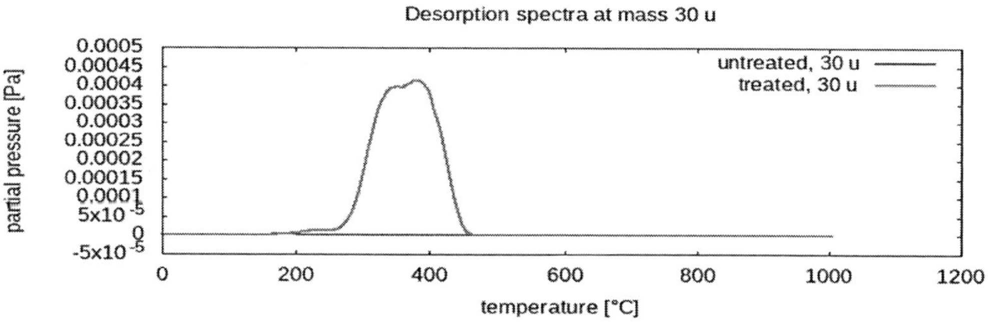

Figure 6 TDS trends of species at mass 30 u

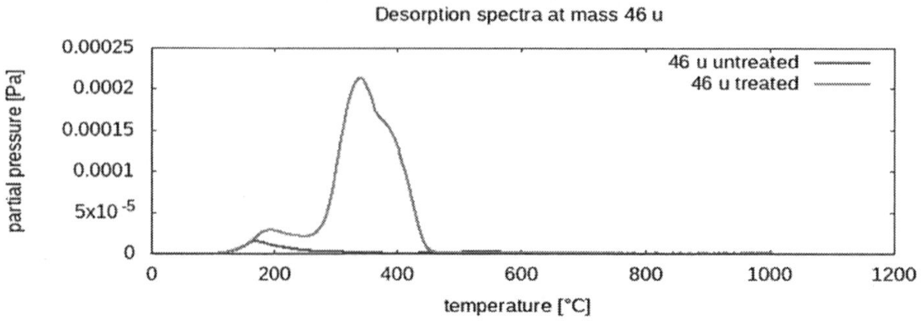

Figure 7 TDS trends of species at mass 46 u, corresponding to nitrogen dioxide

The trends at masses 30 u and 46 u (**Figures 6 and 7**) give strong evidence that the plasma-treated samples contain much larger amounts of NO_X. We can also observe an additional peak in the data from the treated sample that did not appear in the data from the untreated sample. Therefore, the nitrogen oxides are also bonded differently in the treated sample.

To determine if these new species were grafted to the surface or adsorbed from the surrounding atmosphere, we have used a FTIR measurement. We were able to identify some of the detected species and determine how did their amount change during the plasma treatment (**Table 2**).

Table 2 Changes in concentrations of species detected by FTIR

Species	Change
hydroxyl	substantial increase
adsorbed H_2O	removed
possibly NO_X	increase
CH_X	removed
Al_2O_3	unchanged

The replacement of adsorbed water with hydroxyl groups suggests that the treatment grafted new functional groups to the nanopowder's surface. Unfortunately, we could not determine if the NO_X was grafted or adsorbed to the surface.

While both TDS and FTIR measurements confirm that the amount of NO_X increased during the treatment, neither allowed us to determine if it was grafted or adsorbed. This might be possible to decide by modelling the peaks and determining their activation energy and decide if they were grafted or adsorbed, but it is a challenging task because the peaks appear to be composed of two individual reactions culminating at similar temperatures.

4. CONCLUSION

We have measured the Thermal Desorption Spectra of untreated alumina nanopowder and of alumina nanopowder treated by DCSBD plasma. The results show that the chemical properties of the nanopowder surface were significantly altered. The plasma modified surface contained additional functional groups, either able to capture or consisting of additional oxygen, hydroxyl, nitrogen monoxide, nitrogen dioxide and carbon dioxide. This provides additional evidence to an earlier hypothesis that nitrogen oxides were added to the nanopowder's surface during the treatment. Further analysis has shown that some of the species were grafted at the surface, but we were not able to determine how were the nitrogen oxides attached to the surface. Additional modelling of the thermal desorption process might determine the nature of the attachment.

ACKNOWLEDGEMENTS

The work was financially supported by the Czech Science Foundation, contract No. GA17-05620S, project CZ.1.05/2.1.00/03.0086 funded by European Regional Development Fund and project LO1411 (NPU I) funded by Ministry of Education, Youth and Sports of Czech Republic.

REFERENCES

[1] ESKANDARI A., AMIN ZARE A., SADRNEZHAAD S. K. and REZAVI HESABI Z. Compaction response and sintering behaviour of alumina nanoparticle: Effect of agglomeration and particle coordination. Effect of spark plasma sintering of alumina nanopowder on the mechanical properties. 2nd International Conference on Ultrafine Grained and Nanostructured Materials Center of Excellence For High Performance Materials School of Metallurgy and Materials Engineering, University College of Engineering, University of Tehran, 2009

[2] CHAIM R., LEVIN M., SHLAYER A and ESTOURNES C., Sintering and densification of nanocrystalline ceramic oxide powders: a review. Advances in Applied Ceramics: Structural, Functional, 2008, vol. 107, pp. 159-169.

[3] BESRA L., Liu M. A review on fundamentals and applications of electrophoretic deposition (EPD). Progress in Materials Science, 2007 vol. 52, no. 1, pp. 1-61

[4] DRDLÍK D., MORÁVEK T., RÁHEĽ J., STUPAVSKÁ M., CIHLAŘ J., DRDLÍKOVÁ K., MACA K., Electrophoretic deposition of plasma activated sub-micron alumina powder, Ceramics International, 2018, vol. 44, n° 8, pages 9789-9793

[5] ZÁBRANSKÝ L., BURŠÍKOVÁ V., DANIEL J., SOUČEK P., VAŠINA P., DUGÁČEK J., SŤAHEL P., CAHA O., BURŠÍK J., PEŘINA V. Comparative analysis of thermal stability of two different nc-TiC/a-C:H coatings, Surface and Coatings Technology, 2015, vol 267, pp. 32-39

[6] BUČEK A., BRABLEC A., KOVÁČIK D., SŤAHEL P. and ČERNÁK M. Glass bond adhesive strength improvement by DCSBD atmospheric-pressure plasma treatment, International Journal of Adhesion and Adhesives, 2017, vol. 78, pp 1-3Z

MODIFICATION OF VARIOUS POLYMER SURFACES USING ATMOSPHERIC PRESSURE REDUCING PLASMA

Zdeňka DOUBKOVÁ, Zlata TUČEKOVÁ, Jakub KELAR, Richard KRUMPOLEC, Miroslav ZEMÁNEK

R&D Centre for Low-Cost Plasma and Nanotechnology Surface Modifications, Masaryk University, Brno, Czech Republic, EU, z.doubkova@mail.muni.cz

Abstract

This paper presents the results of surface modification of polymers induced by atmospheric pressure non-isothermal plasma generated in reducing atmosphere. The technology based on diffuse coplanar surface barrier discharge (DCSBD) was used with the aim to achieve highly hydrophobic surfaces. One rigid (polymethylmethacrylate-PMMA) and three flexible polymers (PMMA, polytetrafluoroethylene-PTFE and polyethylene terephthalate-PET) were modified by plasma. The changes in wetting phenomena and surface free energy were observed depending on plasma treatment conditions. The most significant modification occurred on flexible PMMA and PTFE. Using pure hydrogen plasma, we were able to achieve an increase of water contact angle up to ~115° from the original value 78° (untreated PMMA). On the contrary, the water contact angle on PTFE decreased to ~91° from the value 109° (untreated PTFE). Observed etching process on polymer surfaces and the changes in surface roughness were detected by a scanning electron microscopy (SEM) and atomic force microscopy (AFM).

Keywords: Non-isothermal plasma, reduction atmosphere, PMMA, PET, PTFE

1. INTRODUCTION

Plasma treatment of polymer surfaces is nowadays a highly exploited technique used to induce changes in surface wettability, for adhesion improvement, surface functionalization, sterilization, cleaning and etching. Plasma processing is widely applied in the industry e.g. for the production of functional coatings, in microelectronics, the modification of packaging materials as well as biomaterials.

The non-isothermal atmospheric pressure plasma sources provide the solution for efficient improvement of surface properties of low adhesive, chemically inert and heat sensitive polymeric materials such as polymethylmethacrylate-PMMA, polyethylene terephthalate-PET and polytetrafluoroethylene-PTFE.

Plasma-induced etching and surface structuring of polymers can result in the change of its wettability and surface morphology without the undesirable modification of the bulk and optical properties. The nano- and micro-scale surface structuring can result from the short and longer plasma exposures, respectively [1,2].

Plasma etching is widely studied for the fabrication of superhydrophobic and superoleophobic surfaces on polymers [3] with subsequent self-cleaning properties [4]. Ellinas et al. and Tserepi et al. [5,6] studied attachment of cells on highly hydrophobic surfaces for different biological applications. For this process, the low-pressure plasma generated mostly in Ar, He, O_2 and fluorocarbons containing gases is used [7,8].

Dry etching of different polymeric materials by plasma generated in various gas mixtures at atmospheric pressure was recently reported by authors of [9-11]. Reducing plasma can also be used for the improvement of the wettability and adhesion properties of chemically highly inert materials, such as PTFE [12,13].

In this paper, the diffuse coplanar surface barrier discharge (DCSBD), capable of meeting the industrial production requirements in reducing atmosphere, was used for surface etching and structuring at atmospheric pressure. The effect of high energetic plasma, with the electron density of 1.3×10^{16} cm^{-3} and mean electron temperature of approx. 19×10^3 K [14], generated in pure hydrogen was tested on one rigid (PMMA) and three flexible (PMMA, PTFE, PET) polymeric substrates.

2. EXPERIMENTAL

Three flexible polymeric samples and one rigid polymer sheet were used in this work. Flexible PMMA foil (0.05 mm, DENZ BIO-MEDICAL GmbH, Austria), PET foil (0.25 mm) and PTFE foil (0.25 mm both from GOODFELLOW CAMBRIDGE Ltd., England) were used as samples. Moreover, 2-mm thick Acrylic sheet (CHO CHEN Ind. Co., Ltd, Taiwan) was also tested.

The polymers were treated by plasma generated by DCSBD plasma source. Single DCSBD unit generated a thin plasma layer on area of 20×8 cm^2 when powered by a high voltage generator VF700 (LIFETECH s.r.o., Czech Republic) with the frequency of 15 kHz and input power of 400 W. The samples were treated at the distance of 0.2 mm from the DSCBD ceramics in a dynamic regime. The schematics of DCSBD, as well as a picture of plasma generated in pure hydrogen, can be seen in **Figure 1**.

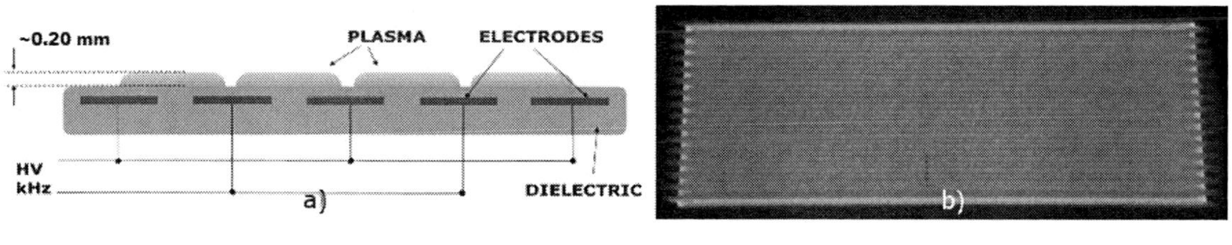

Figure 1 a) Schematics of DCSBD plasma unit and b) a picture of DCSBD plasma in pure hydrogen [15]

The DSCBD plasma source was placed in a sealed reactor chamber of volume approximately 2 l. The gas flow of pure hydrogen (99.998 %, MESSER TECHNOCAS s.r.o., Czech Republic) was kept at 1 l/min during the plasma treatment.

The changes in water contact angle (WCA) and surface free energy (SFE) were measured using surface energy evaluation system See System (ADVEX INSTRUMENTS s.r.o., Czech Republic). The WCA value was estimated as an average of contact angle measurement of 10-15 water droplets with a volume of 1 µl. The SFE value was measured using the average contact angles of three different liquids: water, ethylene glycol and diiodomethane (SIGMA-ALDRICH, Inc., USA), estimated in analogy with the WCA measurement. The surface free energy was then calculated according to the Owens-Wendt regression method [16].

The morphological changes of polymer surfaces after the plasma treatment were measured using Scanning Electron Microscope Mira3 (TESCAN s.r.o., Czech Republic) with an accelerating voltage of 5 kV using secondary electron emission detector. To prevent charging of the polymer surface, the samples were coated with 10 nm of Au/Pd composite layer. The change of surface roughness was measured using Atomic Force Microscope NTEGRA Prima (NT-MDT, Russia) in a semi-contact mode.

3. RESULTS AND DISCUSSION

Figure 2 shows the results of contact angle measurement (a) and surface energy analysis (b) of plasma-treated polymers after 5, 10 and 30-min modification in pure hydrogen DCSBD plasma.

The contact angle analysis revealed the increase of WCA on both PMMA samples and PET surface. The maximum value of WCA on both PMMA substrates was achieved already after 10 minutes of treatment by reducing plasma. The increase of WCA was rather steep for the PMMA foil from initial 78.6° to 109.2° after 5 minutes and to 114.8° after the 10 minutes of treatment. The slighter increase of WCA on PMMA sheet can be related to the thickness and rigidity of the substrate. The observed PMMA changes after DCSBD plasma treatment are similar to the results of chemical fluorination of PMMA presented by [17] and the treatment of PMMA by fluorocarbon plasma at low-pressure [8].

The WCA on PET foil increased from initial 81.4° to the measured maximum 97.0° after 30 minutes of the reduction plasma treatment. The slower increase of WCA and the previous work of Krumpolec et al. [10] is indicating, that the further change in contact angle can be achieved by the plasma etching of the substrate surface after extended treatment time.

The surface energy of PMMA and PET surfaces decreased because of a decrease of polar component of SFE. In both cases, the disperse component of SFE mostly preserved its value. It is apparent that a decrease of SFE by reducing plasma is due to the substantial decrease of the polar component and the mechanical structuring of the substrate surface observed by microscopy, which will be described in more detail later.

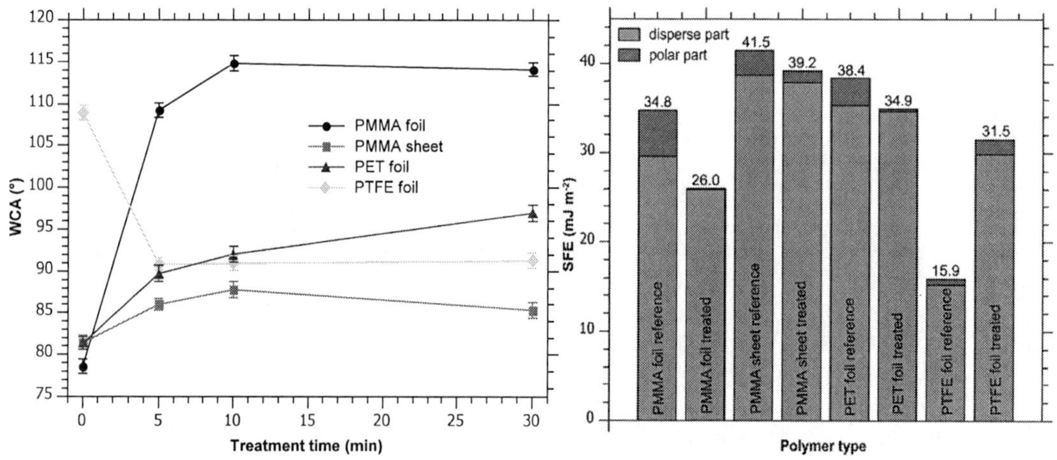

a) b)

Figure 2 The WCA (a) and SFE values (b) measured on plasma-treated polymer surfaces after 30 minutes in hydrogen DCSBD plasma and the comparison with reference values

On the other hand, the wettability of PTFE showed different results. A decrease of WCA from reference value 108.9° to 90.8° was observed after 5 minutes exposure in hydrogen plasma. No further decrease of WCA with increasing treatment time was observed.

Similar WCA and SFE results were observed by Jie-Rong and Wakida [12] who modified PTFE in low-temperature low-pressure hydrogen and methane plasma for several minutes. Authors related the result of an increase of the polar and disperse component to the plasma induced changes of chemical composition on the surface. The electron spectroscopy for chemical analysis revealed a decrease of fluorine concentration and an increase of carbon, oxygen and nitrogen intensity on the PTFE surface. However, authors didn't analyse the changes in surface morphology.

Figure 3 and **Table 1** present the roughness data measured on PMMA samples. The AFM analysis revealed similar tendencies in the roughness increase. The measured roughness for both substrates increased approx. 10 times after 10 min treatment in hydrogen DCSBD plasma. The maximum average roughness after the 30 min treatment achieved similar values for flexible PMMA and PMMA sheet. However, the structures created on the surface have different density, shape and diameter of the peaks. As revealed, the shape of the structures significantly influences the surface wettability.

Table 1 The average roughness of flexible and rigid PMMA after the DCSBD plasma treatment in pure hydrogen (area 5×5 μm^{-2})

Treatment time (min)	PMMA foil		PMMA sheet	
	Roughness (nm)	Δ (nm)	Roughness (nm)	Δ (nm)
0	2.8	0.1	1.1	0.2
5	20.1	0.3	7.1	3.2
10	22.7	4.5	11.6	3.0
30	65.8	1.4	59.1	3.9

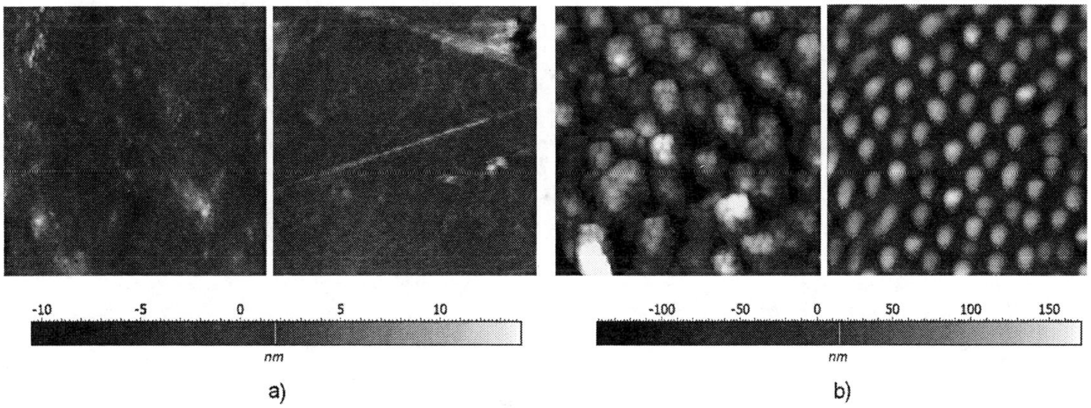

Figure 3 The AFM images (from left to right) of flexible and rigid PMMA references (a) and after 30 min (b) of DCSBD plasma treatment in pure hydrogen (area 5×5 μm^{-2})

SEM images show the morphology of plasma-treated polymers in **Figure 4**. A time-dependent formation of pillar and dot like structures on both PMMA and PET foils was observed. This type of morphology was reviewed in [18] and associated with bactericidal properties of naturally occurring, self-cleaning superhydrophobic surfaces. Coen et al. [1] observed a similar effect on PMMA foil treated by low-pressure helium plasma. As reported, small pillar structures (called worm-like chains) appeared after short treatment times. The thicker grooves were formed after the further (longer) treatment with small pillars preserved on top. Simultaneously, the authors linked this similar time-dependent change of the structure with kind of melting process involving chain scission and etching effect.

The SEM microscopy of PTFE surface revealed the arise of root-like nanoscale lines on the surface. These structures had no effect toward the improvement of surface hydrophobicity. In [1], the root-like structure grew more visibly after several minutes of helium plasma treatment. This effect was explained as the etching of amorphous part of the polymer. Then the appearing root-like structures can be composed of a backbone of crystallised polymer chains surrounded by the amorphous part. This assumption can be verified by long-lasting etching by atmospheric pressure hydrogen plasma treatment in the future.

4. CONCULUSION

The effect of atmospheric pressure DCSBD plasma generated in pure hydrogen was studied on one rigid substrate and three different polymeric foils. The change of the wetting phenomena towards hydrophobic properties and structuring of the surface was confirmed for both PMMA samples and PET foil. The contact angle analysis performed on these substrates revealed a substantial decrease of polar component of surface energy. The microscopic analyses showed the change of morphology and formation of submicron structures on both PMMA and nanoscale structures on PET substrate. On the other hand, hydrogen plasma led to the increase of contact angle and improvement of wettability of PTFE foil with an insignificant change of substrate topography and morphology.

The atmospheric pressure low-temperature DCSBD plasma in pure hydrogen is presenting a novel approach to structuring the polymeric surfaces on a submicron scale. At the same time, it can improve the wettability and adhesion of the chemically inert polymer like PTFE without the need for low-pressure or time-consuming chemical processing of the substrate.

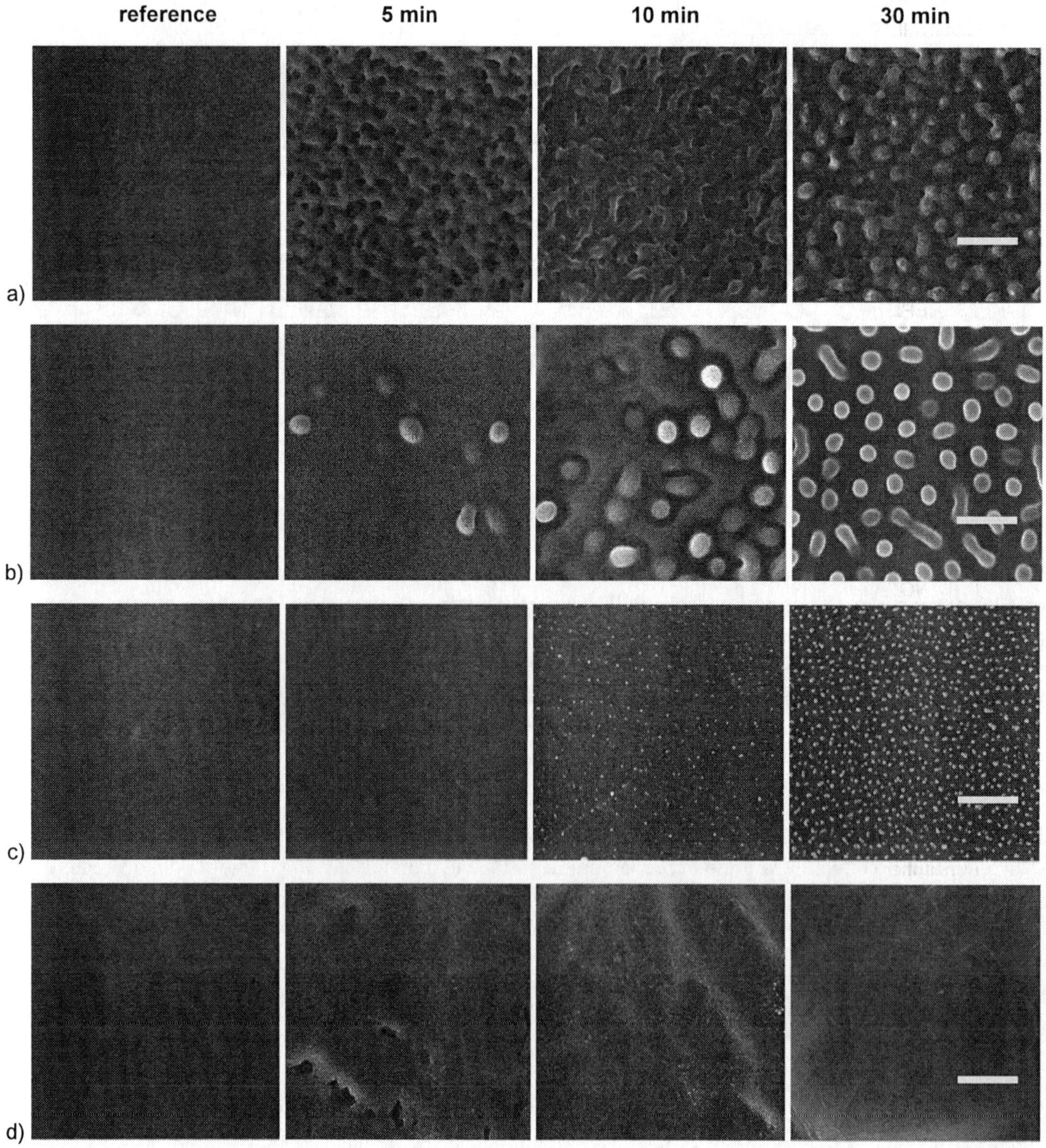

Figure 4 SEM micrographs of the PMMA foil (a), PMMA sheet (b), PET (c) and PTFE foil (d) (from left to right) reference sample and the samples after 5, 10 and 30 min treatment by DCSBD plasma generated in pure hydrogen (scale bar 1 μm)

ACKNOWLEDGEMENTS

The authors would like to acknowledge financial support from the project LO1411 (NPU I), funded by Ministry of Education, Youth and Sports of the Czech Republic and from the project TJ01000327, funded by Technology Agency of the Czech Republic.

REFERENCES

[1] COLLAUD COEN, M. et al. Modification of the micro- and nanotopography of several polymers by plasma treatments. *Applied Surface Science*. 2003. vol. 207, no. 1-4, pp. 276-286.

[2] GOGOLIDES, E. et al. Controlling roughness: from etching to nanotexturing and plasma-directed organization on organic and inorganic materials. *Journal of Physics D: Applied Physics*. 2011. vol. 44, no. 17, pp. 174021.

[3] BORMASHENKO, E. et al. Robust technique allowing manufacturing superoleophobic surfaces. *Applied Surface Science*. 2013. vol. 270, pp. 98-103.

[4] WATSON, Gregory S. et al. A gecko skin micro/nano structure - a low adhesion, superhydrophobic, anti-wetting, self-cleaning, biocompatible, antibacterial surface, *Acta Biomaterialia*. 2015. vol. 21, pp. 109-122.

[5] ELLINAS, K. et al. Three-dimensional plasma micro-nanotextured cyclo-olefin-polymer surfaces for biomolecule immobilization and environmentally stable superhydrophobic and superoleophobic behavior. *Chemical Engineering Journal*. 2016. vol. 300, pp. 394-403.

[6] TSEREPI, A. et al. Plasma Nanotextured Polymeric Surfaces for Controlling Cell Attachment and Proliferation: A Short Review. *Plasma Chemistry and Plasma Processing*. 2016. vol. 36, no. 1, pp. 107-120.

[7] REN, H.-X., et al. Conventional route to scalable morphology-controlled regular structures and their superhydrophobic/hydrophilic properties for biochips application. *Lab Chip*. 2009. vol. 9, no.15, pp. 2140-2144.

[8] HER, E.K. et al. Superhydrophobic transparent surface of nanostructured poly(methyl methacrylate) enhanced by a hydrolysis reaction. *Plasma Processes and Polymers*. 2013. vol. 10, no. 5, pp. 481-488.

[9] FRICKE, K.et al. High rate etching of polymers by means of an atmospheric pressure plasma jet. *Plasma Processes and Polymers*, 2011. vol. 8, no. 1, pp. 51-58.

[10] KRUMPOLEC, R. et al. Roll-to-roll surface etching of polymers using hydrogen plasma at atmospheric pressure. In *NANOCON 2017 - 9th International Conference on Nanomaterials - Research and Application*. Brno, 2017. pp. 283-288.

[11] KRUMPOLEC, R. et al. Atmospheric pressure plasma etching of silicon dioxide using diffuse coplanar surface barrier discharge generated in pure hydrogen, *Surface and Coatings Technology*. 2017. vol. 309, pp. 301-308.

[12] JIE-RONG, C. and WAKIDA, T. Studies on the surface free energy and surface structure of PTFE film treated with low temperature plasma. *Journal of Applied Polymer Science*. 1997. vol. 63, no. 13, pp. 1733-1739.

[13] SARANI, A. et al. Surface modification of PTFE using an atmospheric pressure plasma jet in argon and argon+CO2. *Surface and Coatings Technology*. 2012. vol. 206, no. 8-9, pp. 2226-2232.

[14] PRYSIAZHNYI, V. et al. Generation of Large-Area Highly-Nonequilibrium Plasma in Pure Hydrogen at Atmospheric Pressure, *Contribution to Plasma Physics*. 2014, vol. 54, pp. 138-144.

[15] KELAR, J. et al. Nanostracturing of polymethyl methacrylate surface with use of higly etching plasma. In *PLASTKO 2018, Conference proceedings*. Zlín, 2018. pp. 6.

[16] OWENS, D.K. and WENDT, R.C. Estimation of the surface free energy of polymers. *Journal of Applied Polymer Science*. 1969. vol. 13, no. 8, pp. 1741-1747.

[17] TOKUDA, K. et al. Simple method for lowering poly(methyl methacrylate) surface energy with fluorination. *Polymer Journal*. 2015, vol. 47, no. 1, pp. 66-70.

[18] ELBOURNE, A. et al. Nano-structured antimicrobial surfaces: From nature to synthetic analogues. *Journal of Colloid and Interface Science*. 2017. vol. 508, pp. 603-616.

MODIFICATION OF ELECTROMAGNETIC PROPERTY OF COPPER COATED MILIFE FABRICS

Mohanapriya VENKATARAMAN, Kai YANG, Aravin Prince PERIYASAMY, Xiaoman XIONG, Jiri MILITKY, Rajesh MISHRA

Department of Material Engineering, Faculty of Textile Engineering, Technical University of Liberec, Liberec, Czech Republic, EU, mohana.prasad@gmail.com

Abstract

In this study, milife fabrics (polyester nonwoven fabrics) were used. Three common silanes including octyltriethoxysilane, phenyltriethoxysilane and tetraethoxysilane have been prepared by sol-gel technique to treat the milife fabrics with and without copper coating. The main aim of the study is to protect the conductive copper layer of the milife fabric by the silanization proces. Surface morphology have been measured by scanning electron microscope to prove the deposition of silane, surface roughness was measured by confocal microscope and electromagnetic shielding effectiveness (EMI) have been tested to prove the change in EMI values with respect to silane coating. From the results, it was obvious that EMI of the copper coated milife treated with phenyltriethoxy silane ranging from 40 to 50 dB increases about 80 % of the values of copper coated milife compared to other two silane coatings. In addition, SEM showed that deposition of silane on the fabric surface depending on the different types of silane used for the sol-gel coating.

Keywords: Electromagnetic, silane, coating, milife fabrics, copper, modification

1. INTRODUCTION

Sol-gel coating is one of the most important emerging technology in textile industry due to its many advantages such as highly effective, adjustable coating thickness, good coating durability, easy combination with various functions and less environmental impacts. Typically, metal oxides starting from a colloidal solution (sol) that acts as the precursor for an integrated network (or gel) of either discrete particles or network polymers are used for the fabrication of materials. Typical precursors are metal alkoxides and metal salts (such as chlorides, nitrates and acetates), which undergo various forms of hydrolysis and polycondensation reactions. Both in liquid and solid phase contained in the formation of a gel-like diphasic system are evolved by the 'sol' in this chemical procedure whose morphologies range from discrete particles to continuous polymer networks. A significant amount of fluid may need to be removed initially for the gel-like properties to be recognized in the case of the colloid, due to slow volume fraction of particles (or particle density) which can be accomplished through a number of ways. The method which is simple is to allow time for sedimentation to occur and then pour off the remaining liquid. A drying process is required in the removal of the remaining liquid (solvent) phase which is typically accompanied by a significant amount of shrinkage and densification. The distribution of porosity in the gel determines the rate at which the solvent can be removed. During this phase of processing, changes imposed upon the structural template will strongly influence the microstructure of the final component [1-4]. In a single precursor component system, interconnected nanoscale porosity can be designed to the final material and hence a high surface area, it depends upon the precursors used and the solvent employed [4,5-7]. For thin films and coatings, the process of sol-gel is accepted as a technology. An alternative to chemical vapour deposition, sputtering, and plasma spray is the sol-gel process. It is shown that sol-gel thin films to be technically sound alternatives and to be commercially viable. For over 30 years, the technology of sol-gel thin films has been used and the process is quite simple. With a solvent and water, a solution containing the desired oxide precursor is prepared and is applied to a substrate by spinning, dipping or draining. Even though most needed to be calcined and densified with heating [8-16], coatings can be at room temperature. As a method to deposit the functional materials to different materials, sol-gel coating was widely used in textile field by many researchers [10,14,15,17]. One of the great advantages of this technique is the possibility of producing transparent, porous solids at a low processing temperature. In order to develop microporous matrices which are excellent hosts for organic dopants, organically modified silica (organosilica) can be processed through the sol-gel method [17-20].

2. EXPERIMENTAL PROCEDURE

2.1. Materials

Milife fabric (100 % polyester nonwoven fabrics) with or without copper coating have been used for this study. Three types of precursors, octyltriethoxysilane (OCTYL), phenyltriehoxysilane (PHENYL), and tetraethoxysilane (ORTHO) were selected for silane coating. All the silanes were purchased from Sigma-Aldrich, USA. Triacetoxyvinyl silane and nitric acid were used as catalyst, ethanol and deionized water was used for the preparation of sol solutions. The sample description is given in **Table 1**.

Table 1 Sample description of nonwoven fabrics

Type of fabric	Sample code	Fabric conditions	Thickness (mm)		Fabric density ($g.m^{-3}$)	
			Mean	*CV%*	*Mean*	*CV%*
Milife	A	Copper coated	0.110	06.42	38.88	1.68
	B	Without Copper coated	0.072	11.62	11.12	4.33

2.1.1. Silica sol-gel synthesis and coating

The silica sols were prepared by mixing different precursors with catalyst and solvent, the mixing can be done with proper proportion of silane, catalyst, water and solvent by 32:1:100:320 molar ratios respectively. Many studies could be conducted to optimize this proportions. The mixture was stirred until a clear solution was obtained when the crystals of the catalyst was dissolved. Later, the deionized water was added to the above mixture by using the syringe pump at the speed of 1 $mL.h^{-1}$. Triacetoxyvinyl silane can be used as catalyst in terms of OCTYL and PHENYL coating (it has the dual functions like a catalyst, certainly it will be the third precursor) because it slowly releases the acetic acid during the hydrolysis reaction and ensures to avoid the rapid hydrolysis, whereas HNO_3 has been used for preparing the sol with **Tetraethoxy silane.** The solution was continuously stirred for 24 hours at room temperature to allow the complete hydrolysis of precursors. The fabrics were dried in atmospheric condition and then cured at 110 °C for 10 minutes. Before measurement, the coated fabric was left in atmospheric condition for 24-48 hours to ensure complete stabilizing of the silica matrices.

2.2. Methods

The surface morphology of control and sol-gel coated fabrics were observed by using TS5130 Vega-Tescan scanning electron microscope (SEM). Laser scanning confocal microscope (LSCM analysis of copper coated fabric, Olympus OLS 3100 model laser scanning confocal microscope was employed to measure surface properties of control and coated fabrics. Surface roughness and other related properties were calculated according to the ISO 4287:1997 standard. Infrared thermography measurement was conducted by using thermal camera. A vertical hot plate maintained at constant temperature around 40°C was used as heat source, the specimen was flatly attached onto the hot plate with the help of a frame tool, and after 2 min thermal images were taken by thermal camera at a distance of 40 cm from the hot plate. The room temperature was kept at 25 ± 2°C. A thermal image reveals the amount of radiation emitted by the heat plate through the fabric. The thermal insulation value can be computed via the following equation (1);

$$I = \left(\frac{B - F}{B - R} \right) \tag{1}$$

where *B*- Instrument temperature (i.e. 40 °C), *F*- Fabric temperature, *R*- Room temperature.

The electromagnetic (EM) shielding effectiveness (SE) of the samples was measured according to ASTM D4935 -10 standard for the planar materials. Surface resistance and volume resistance of the samples have been measured by using concentric electrodes (applied pressure 2.3kPa) under a 100V direct current (DC) power supply, according to standard ATSM D257 -07. This standard is intended for insulating samples and it was chosen due to the wide sample set containing insulating, semi-conductive and conductive samples allowing a comparison of the results.

3. RESULTS AND DISCUSSION

3.1. Surface morphology of sol-gel coated fabric

The surface of coated and uncoated fabric samples was observed using SEM. **Figure 1** shows the surface morphologies of sol-gel coated fabrics with different silane coating. It is apparent from the micrographs that a thin film was formed on the coated fabric surface and the film formation is significantly depending on the type of silane used for the sol-gel coating. In fact, each fiber is covered with a continuous layer of sols, these layer formation on the fiber surface confirmed that the copper coated fabric has been protected through sol-gel coating techniques.

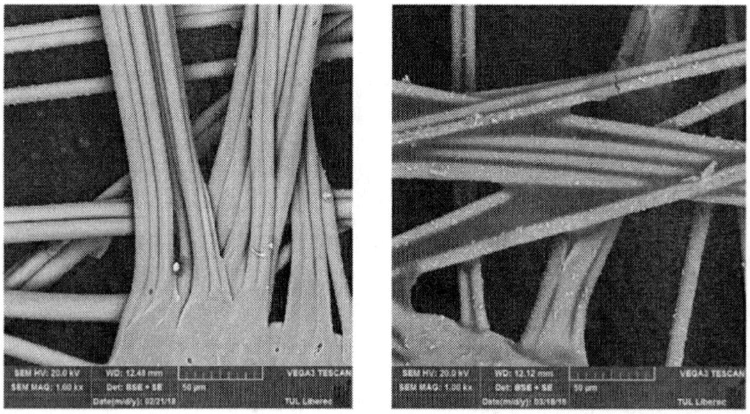

(a) (b)

Figure 1 Impact of silanization on the copper coated milife fabric;
(a) before sol-gel coating, (b) after sol-gel coating (PhTES)

From the figure, it can be seen that the sol-gel coating made significant changes in the surface by creating the roughness, also the surface roughness depends on the type of silane used. In this study, sols prepared with phenyl and ortho based silane creates more roughness than the sols prepared with octyl alone, which is due to the chemical groups presented in the silane and its structure. However, octyl having the flexible long alkyl chain groups provided the significantly less roughness. Meanwhile, phenyl has benzene groups which make more roughness on the network, therefore, it produces more roughness as well as it reduces the physical properties of coated materials. **Table 2** shows the roughness characteristic of the fabric before and after coating. In overall, numerical values showed the drastic changes in the peak height (Rp), depth valley (Rv) and the addition of both parameters (Rz). As a result, the highest numerical value indicates the more roughness, and thus significant changes in the surface roughness.

Table 2 Surface roughness characteristics of milife fabrics

Sample	Type of fabric	Max. Peak height (Rp) (µm)	Max. Valley depth (Rv) (µm)	Max. Depth (Rz) Rz=Rp+Rv (µm)
Control	Milife	0.92	0.44	1.36
	Milife (CU-coated)	1.22	1.18	2.4
Octyl	Milife	1.1	1.65	2.75
	Milife (CU-coated)	0.65	1.62	2.27
Phenyl	Milife	1.45	1.68	3.13
	Milife (CU-coated)	3.21	2.54	5.75
Ortho	Milife	0.65	1.28	1.93
	Milife (CU-coated)	3.8	4.21	8.01

3.2. Electromagnetic shielding effectiveness

Figures 2 to 5 shows the dependence of the EM SE in dB on the whole measured frequency range for all the samples. **Figures 2 and 4** shows the SE of milife samples coated with copper. It is found that the control sample exhibits a certain SE performance and the values gradually decrease with the increasing frequency. The SE of milife samples containing copper coated with phenyltriethoxysilane silane roughly doubles that of control samples, and its downward trend with increasing frequency slows down. Oppositely, with increasing frequency, SE has a downward trend for both the octyltriethoxysilane-coated milife samples containing copper and the tetraehoxysilane-coated milife samples, SE of the former gets around 0 dB at 1.5 GHz. In **Figures 3 and 5**, the reflection of milife samples coated with copper have closer values at the beginning. Only the reflection of samples treated with octyl silane gets increased with increasing frequency.

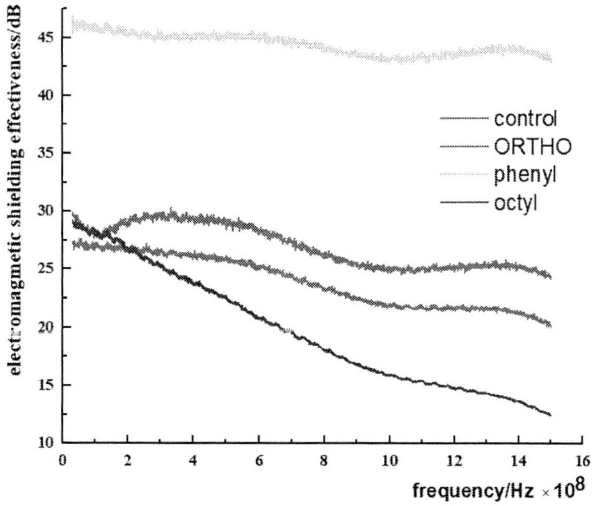

Figure 2 EMI of Milife coated with copper (total electromagnetic shielding effectiveness)

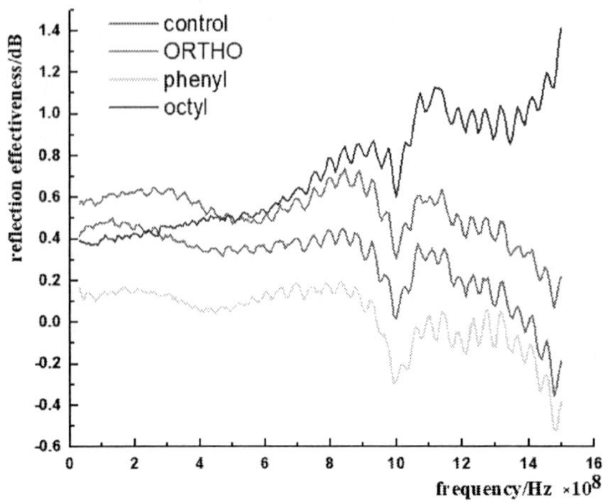

Figure 3 EMI of Milife coated with copper (reflection shielding effectiveness)

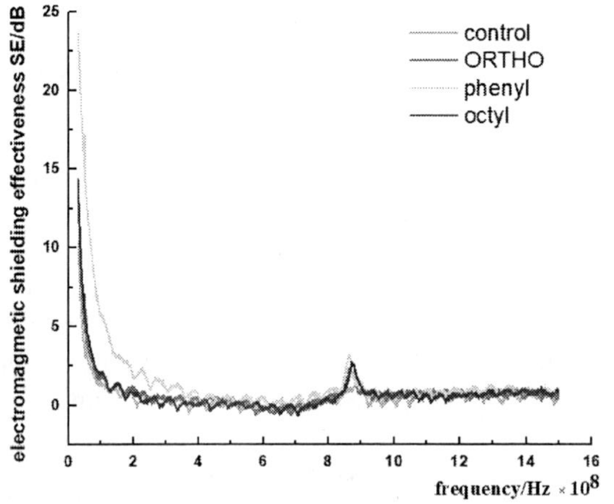

Figure 4 EMI of Milife (total electromagnetic shielding effectiveness)

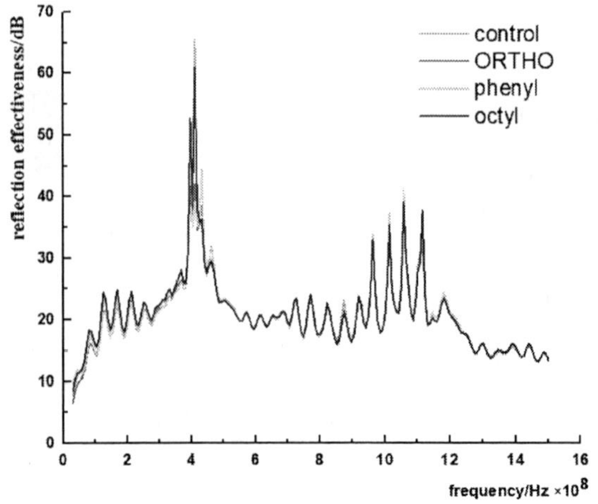

Figure 5 EMI of Milife (reflection shielding effectiveness)

4. CONCLUSION

From this study, it can be concluded that the conductive copper layer of the milife fabric can be protected by silanization process. The results from surface morphology and surface proved the deposition of silane coating and the roughness factor of the different silane coating had an adverse effect on the copper coated milife fabrics. Electromagnetic shielding effectiveness (EMI) of phenyltriethoxy silane coated fabrics have the highest value of total shielding effectiveness ranging from 40 to 50 dB increases about 80 % of the values of copper coated milife compared to other two silane coatings.

ACKNOWLEDGEMENTS

This work was supported by the Ministry of Education, Youth and Sports of the Czech Republic and the European Union - European Structural and Investment Funds in the frames of Operational Programme Research, Development and Education - project Hybrid Materials for Hierarchical Structures (HyHi, Reg. No. CZ.02.1.01/0.0/0.0/16_019/0000843).

REFERENCES

[1] ISMAIL W.N.W. *Sol-gel technology for innovative fabric finishing-A Review.* Journal of Sol-Gel Science and Technology, 2016, vol. 78, pp. 698-707. doi: 10.1007/s10971-016-4027-y

[2] TOMSIC, B., SIMONCIC, B., Orel, B. *Sol-gel coating of cellulose fibres with antimicrobial and repellent properties.* Journal of Sol-Gel Science and Technology, 2008, vol. 47, pp.44-57, doi: 10.1007/s10971-008-1732-1.

[3] Chemistry S *Sol-Gel Synthesis.* In: Garche J (ed) Encycl. Electrochem. Power Sources. Elsevier, Amsterdam, 2011, pp. 613-624.

[4] LI, F., XING, Y., DING, X. *Silica xerogel coating on the surface of natural and synthetic fabrics.* Surface Coatings and Technology, 2008, vol. 202, pp. 4721-4727. doi: 10.1016/j.surfcoat.2008.04.048.

[5] KOWALCZYK, D., BRZEZIŃSKI, S., KAMIŃSKA, I. *Multifunctional nanocoating finishing of polyester/cotton woven fabric by the sol-gel method.* Textile Research Journal, 2017, 4051751769397. doi: 10.1177/0040517517693979

[6] AVNIR, D., LEVY, D., REISFELD, R. *The nature of the silica cage as reflected by spectral changes and enhanced photostability of trapped Rhodamine* 6G. Journal of Physical Chemistry, 1984, vol. 88, pp. 5956-5959, doi: 10.1021/j150668a042

[7] MAHLTIG, B., AUDENAERT, F., BÖTTCHER, H. *Hydrophobic silica sol coatings on textiles-the influence of solvent and sol concentration.* Journal of Sol-Gel Science and Technology, 2005, vol. 34, pp. 103-109. doi: 10.1007/s10971-005-1321-5.

[8] WU, L.Y.L., ZHAO, Q., HUANG, H., LIM, R.J. *Sol-gel based photochromic coating for solar responsive smart window.* Surface Coatings Technology, 2017, vol. 320, pp. 601-607. doi: 10.1016/j.surfcoat.2016.10.074.

[9] CHEN, Z., WU, L.Y.L. *Scratch damage resistance of silica-based sol-gel coatings on polymeric substrates.* In: Tribology of Polymer Nanocomposites Friction and Wear of Bulk Material Coatings Second Ed. Elsevier, Oxford, 2013, pp. 467-511.

[10] YI, G.C., YATSUI, T., OHTSU, M., GROARKE, R. *ZnO Nanorods and their Heterostructures for Electrical and Optical Nanodevice Applications.* In: Reference Module in Material Science and Materials Engineering Elsevier, 2016, pp 335-374.

[11] HOU, L., SCHMIDT, H. *Photochromic properties of a silylated spirooxazine in sol gel coatings.* Materials Letters, 1996, vol. 27, pp. 215-218, doi: 10.1016/0167-577X(95)00287-1.

[12] TOMONAGA, H., MORIMOTO, T. (2001) *Photochromic coatings including silver halide microcrystals via sol-gel process.* Thin Solid Films, 2001, vol. 392, pp.355-360, doi: 10.1016/S0040-6090(01)01057-4.

[13] HU, J.L., LU, J. *Memory polymer coatings for smart textiles.* In: Active Coatings for Smart Textiles. Elsevier, 2016, pp. 11-34.

[14] HU, J.L. *Introduction to active coatings for smart textiles.* In: Active Coatings for Smart Textiles. Elsevier, 2016, pp. 1-7.

[15] WANG, C., YIN, Y. *Functional modification of fiber surface via sol-gel technology.* In: Active Coatings for Smart Textiles. Elsevier, 2016, pp. 301-328.

[16] SAHOO, P., DAS, S.K., PAULO DAVIM, J. *Surface Finish Coatings.* In: Comprehensive Materials Finishing. Elsevier, Oxford, 2016, pp. 38-55.

[17] RANI, A., GOEL, G. *Coated and laminated textiles.* In: Asian Dyer. Elsevier, 2006, pp. 67-75.

[18] MEIROWITZ, R. *Microencapsulation technology for coating and lamination of textiles.* In: Smart Textile Coatings Laminates. Elsevier, 2010, pp. 125-154.

[19] SHIM, E. *Bonding requirements in coating and laminating of textiles.* In: Joining Textiles. Woodhead Publishing, 2013, pp. 309-351.

[20] GUGLIELMI, M., ZENEZINI, S. *The Thickness of Sol-Gel Silica Coating Obtained by dipping.* Journal of Non-Crystalline Solids, 1990, vol.121 pp. 303-309, doi: https://doi.org/10.1016/0022-3093(90)90148-F.

STUDY OF LITHIUM ENCAPSULATION IN POROUS MEMBRANE USING ION AND NEUTRON BEAMS

Giovanni CECCIO, Antonino CANNAVO', Pavel HORAK, Alfio TORRISI, Ivo TOMANDL, Vladimir HNATOWICZ, Jiri VACIK

Nuclear Physics Institute, Academy of Sciences of the Czech Republic, Řež, Czech Republic, EU,
ceccio@ujf.cas.cz

Abstract

Ion track-etched membranes are porous systems obtained by etching of the latent ion tracks using a suitable etchant solution. In this work, control of the pores' spatial profiles and dimensions in PET polymers was achieved by varying etching temperature and etching time. For determination of the pores' shape, Ion Transmission Spectroscopy technique was employed. In this method, alterations of the energy loss spectra of the transmitted ions reflect alterations in the material density of the porous foils, as well as alterations of their thickness. Simulation code, developed by the team, allowed the tomographic study of the ion track 3D geometry and its evolution during chemical etching. From the doping of porous membranes with lithium-based solution and its analysis by Thermal Neutron Depth Profiling method, the ability of porous PET membranes to encapsulate nano-sized material was also inspected. The study is important for various applications, e.g., for catalysis, active agents, biosensors, etc.

Keywords: Latent ion tracks, etching, doping, Ion Transmission Spectroscopy, Thermal Neutron Depth Profiling

1. INTRODUCTION

The energetic ions, passing through polymers, may induce latent ion tracks with supposedly cylindrical shape. The transformation of the latent ion tracks into pores via the process of the chemical etching [1] provokes a high interest because by changing the etching conditions a different shape of the pores can be obtained [2]. A promising approach to obtain information about the micro-channels (pores) in a non-destructive way is to measure the energy loss of the mono-energetic charged particles passing through the thin foil of the analyzed material [3]. With this method, the alteration of the energy loss reflects the change in the density and thickness of the thin film. If the non-etched film is uniform with no pores, it is possible to observe the Full Energy Peak (FEP). By etching the material, it is possible to observe the change in the position of the FEP. If there are also passing-through channels, in addition to the REP will also be present in the energy spectrum the Reduced Energy Peak (REP). The shape of the spectra between the REP and FEP gives information about the shape of the micro-channels.

Obviously, the formed porous membranes can be used to store the nano(micro)-particles. To clarify this possibility, one-side etched membranes were doped with the lithium solution. The solution was analyzed by Thermal Neutron Depth Profiling (TNDP), nondestructive nuclear analytical technique [4,5]. The possibility to use membranes as storage of the micro-particles or nano-particles can found application in many fields of research and technology, spanning the micro-electronics to biological studies [6].

2. EXPERIMENTAL SETUP

Polyethylene terephthalate (PET) thin film, 18 μm in thickness, was irradiated with the Xe ions with energy 1.2 MeV/n (157 MeV) and fluence of 10^6 Xe/cm^2. These ions passing though the PET create latent tracks that can be visualized after chemical etching. The chemical etching was performed using the NaOH solution of 9 M concentration. Important parameters for the chemical etching are the temperature and the time of etching. We performed both double-side and one-side etching at different temperatures for the fixed time, or at the fixed temperature (50 °C, 65 °C or 75 °C) for different time (ranging from 5 to 60 minutes). The one-side and double-side etched membranes were analyzed using the Ion Transmission Spectroscopy (ITS) thermal Neutron Depth Profiling (TNDP). To perform the ITS study, a thin ^{241}Am α-source was employed to provide quasi-mono-energetic ions. The transmitted spectra were analyzed by the Monte Carlo tomography codes developed at the NPI Rez for the laboratory purposes. The codes calculate the energy loss spectra of the α-particles by numerical integration. The resulting spectrum is obtained as a sum of all particular spectra, taking into account the density of the pores as an input parameter. The one-side etched PET samples were doped using the 5 M LiCl water solution for 3 hours and dry in air for 48 hours in the controlled environment (at RT and lower relative humidity). To analyze the encapsulation of lithium inside the sample, a series of the TNDP measurements was performed and examined. The TNDP analysis was carried out at the CANAM infrastructure, on the LWR15 research reactor (operated by the Research Center Rez). The samples were irradiated with a thermal neutron beam of the 10^7 n_{th} $cm^{-2}s^{-1}$ intensity, provided by a short mirror neutron guide system. Thermal neutrons induced an intensive, isotropic nuclear reaction ^{6}Li(n,α)T, producing two reaction products, α(2.051 MeV) and T(2.734 MeV), emitted in the opposite directions. Even though the natural abundance of the ^{6}Li isotope is only about 7.59 %, the high cross section (940 barns) assures (under the intense neutron beam bombardment) a fast data collection. The Li depth profiling was evaluated from the energy spectra using the LIBOR code developed by our team [7]. By this way, it is possible to get an estimation of encapsulation of Li in the pores and the size (length) of the (incomplete) pores (dips).

3. RESULTS

Successive energy spectra of the 5.486 MeV alpha-particles, transmitted through the PET foil etched gradually at the fixed temperature of 65°C for different (increasing) time, are shown in **Figure 1**. Before etching the spectrum consists of only one peak, the full energy loss peak (FEP).

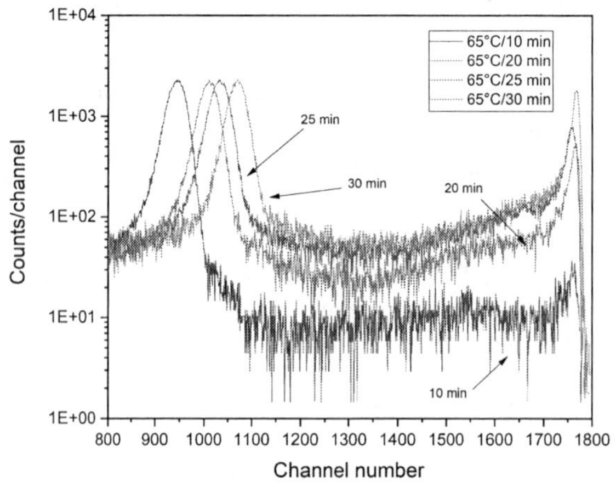

Figure 1 Series of the ITS spectra of alpha particle transmitted through the PET foil etched in the 9 M NaOH solution for the indicated times at 65°C

With the etching, additional counts begin to grow at the higher energy side, gradually up to the FEP. Consequently, by increasing the etching time, also the ratio between the FEP and REP counts is growing. An interesting feature of the spectra, seen in **Figure 1,** is a gradual shift of the REP towards the higher energy. This is due to the step-like bulk etching (i.e., lowering of the film's thickness). Using the Monte Carlo simulation, the shape of the emerging pore and the etched-out thickness of the foil can be calculated for all etching times.

A typical result of the simulation (with the overlapping ITS experimental spectrum) is shown in **Figure 2a.** From the given simulation follows that the shape of the pores is cylindrical with a diameter of 5 μm. Concerning the etching of the foil bulk in the 9 M LiCl solution, for etching temperature of 65 °C an etching rate of 117 nm per minute was determined, see **Figure 2b.**

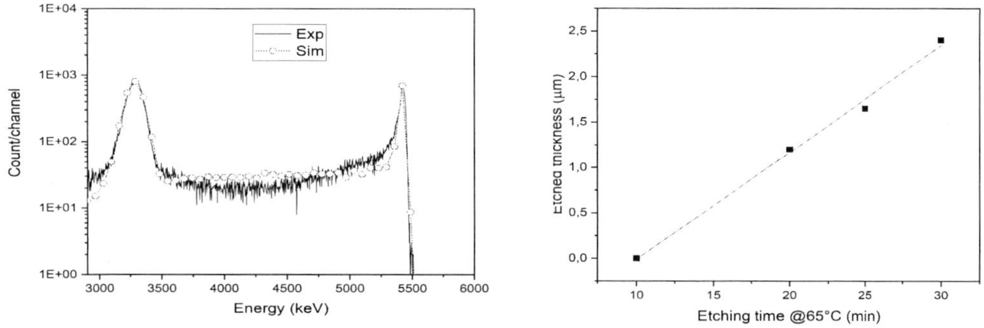

Figure 2 (left-a) Comparison of the ITS spectrum (measured on the PET foil etched for 30 minutes at 65 °C) with a Monte-Carlo simulation; (right-b) bulk etched-out thickness calculated by the Monte-Carlo simulation considering the REP position

Similar results can be seen in **Figure 3**, where successive energy spectra for the PET foils etched for the fixed etching time and at different temperatures, are shown. From the **Figure 3** it is clear that the condition of the 55°C/15 minutes etching step is close to the threshold for the creation of the pass-through pores.

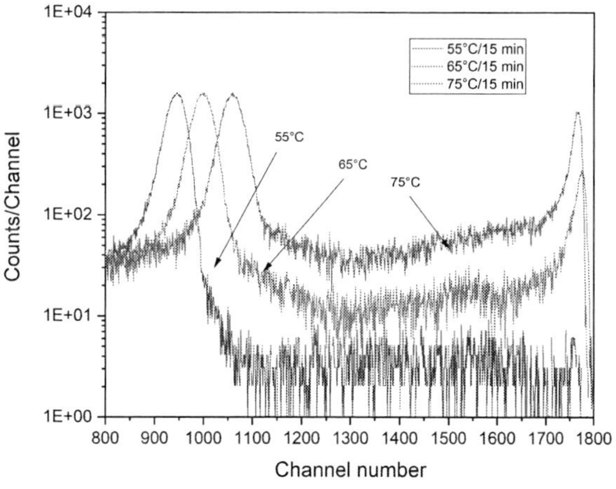

Figure 3 Series of the ITS spectra for α-particles transmitted through the PET foils etched isochronally for the fixed time of 15 minutes and at different temperatures

Using these data, a set of foils with incomplete pores (dips) was produced. This was achieved by one-side etching in the 9 M NaOH water solution with the fixed temperature of 50 °C and different times between 15 and 60 minutes. These foils were then doped only on the etched side with the 5 M LiCl water solutions for 3 hours, and then left to dry in controlled environment (at RT and lower relative humidity) for 48 hours. After the preparation, the samples were analyzed using the TNDP technique. The measurement was performed with the non-etched side of the foil faced towards the solid-state detector (Canberra PIPS). The **Figure 4** shows the TNDP spectrum obtained for the sample etched time 45 minutes. In the spectrum, both alpha (α_{2051}) and triton (T_{2734}) peaks are depicted, together with an additional ($T_{opposite}$) peak that corresponds to tritons from the opposite (i.e., etched) side of the sample. This part is dominant, indicating that most of Li (43 %) was attached to the etched surface zone, a part of it in-diffused towards the surface zone of the non-etched side (37 %) and a smaller part was gathered inside the pores (20 %). Using the LIBOR code, the distribution of the Li atoms in the one-side etched foil was evaluated (**Figure 5**). One can see that Li was concentrated mainly on both (etched and non-etched) sides (or near both surfaces) of the foil with a certain amount resisting also inside the pores. Surprisingly, in the center part of the foil in the pores a certain elevation in the Li concentration is indicated. This is not clear, it can be related to the effect of free radicals created in the pores during etching, where the Li ions are trapped and reside there.

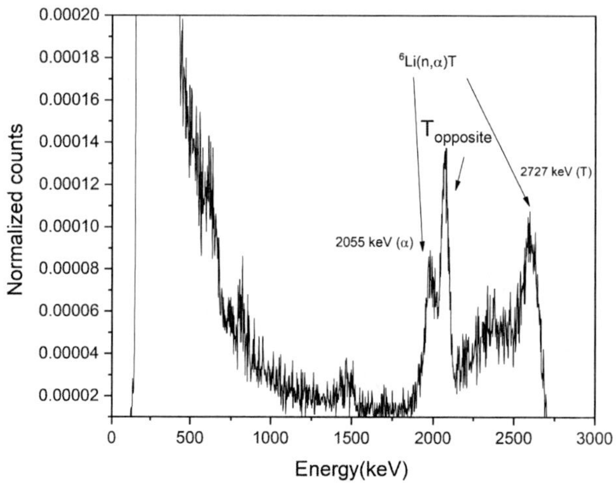

Figure 4 The TNDP spectrum measured for the PET foil etched at 50°C for 45 minutes and doped with the 9 M LiCl water solution

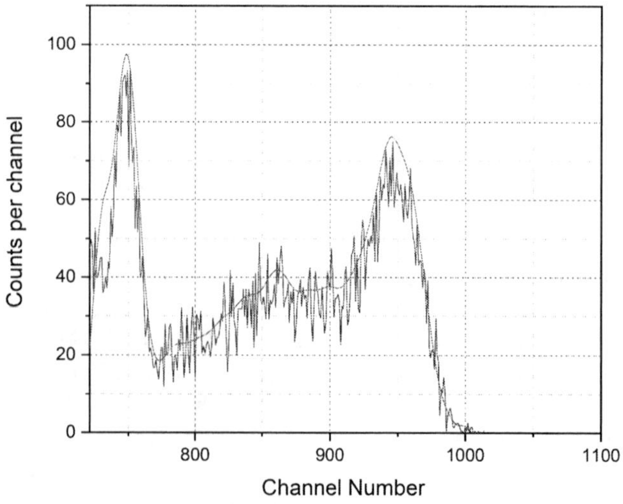

Figure 5 Simulation fitting performed by the LIBOR code on the TNDP spectrum shown in Figure 4

The **Figure 6** reports on the distribution of Li in the samples, etched on both-sides isothermally at 50 °C for different time, and analyzed by TNDP. One can see that the concentration of Li in PET, etched for 15 and 30 minutes, is the highest on both surfaces (and negligible in the pores). This can be assigned to the role of just emerging (incomplete) pores (dips), where Li can gather and reside. Surprisingly, however, the concentration on both sides is not even, and also the Li depth distribution is different. This fact is not clear, but it might be assigned to unequal etching process proceeding on both sides due to creation of slightly different radiation defects caused by slowing down Xe ions penetrating through the foil. The foils etched for longer time (45 - 60 minutes) show the higher incorporation of Li inside the pores, and, to a certain surprise, a decline (loss) of Li on the surface. This effect can lead to a desirable process of the Li encapsulation in the pores.

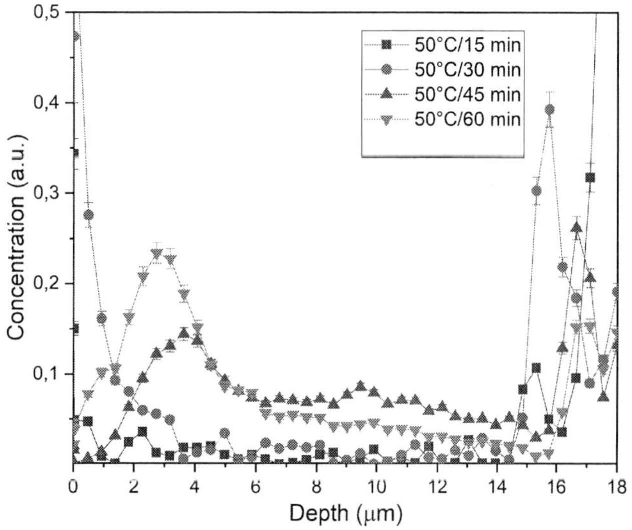

Figure 6 Depth profiles of Li in the PET foil, etched in 9 M NaOH obtained by the LIBOR code

4. CONCLUSIONS

The PET foils irradiated with swift heavy ions (157 MeV Xe, 10^6 Xe/cm^2) were isochronally or isothermally etched in the 9 M NaOH water solution under a control regime (at 50 - 75°C for 5 - 60 minutes) and inspected by the Ion Transmission Spectroscopy (ITS). Partially, they were also doped with Li (in 5 M LiCl water solution for 3 hours, subsequently dried for 48 hours) and analyzed by noninvasive Thermal Neutron Depth Profiling (TNDP).

The ion transmission study showed how the pores' shape can be evolved (from the initially incomplete to the finally fully opened 3D form of a certain parameters) during the gradually etching process. In the present experiment, the pores could be developed into a cylindrical form with a large diameter of 5 μm within the high etching rate procedure. The study of the Li distribution in the PET porous foils pointed out on an interesting behavior of the Li incorporation into pores in their different developing stage. One could see that Li resides mainly on the surface of the foils after etching for the first several tens of minutes (for the etching temperature 50°C), for longer etching time, however, Li penetrates into the pores massively and forms there a stock of Li-based matter (obviously micro- and nano-particles). Surprisingly, on the surface of the etched foils a negligible amount of Li remains only. This process is interesting for a possible, desirable encapsulation of Li in the pores of thin polymers. An important result is that there exists the possibility to manipulate the quantity and the position of the incorporated Li by changing of the etching parameters.

Further information on study of etching process, control of the pore shape and filling of the pores with different materials (including Li-based nanoparticles) for possible future applications will be collected in next experiments.

ACKNOWLEDGEMENTS

The authors acknowledge the financial support from the Grant Agency of Czech Republic under the contract No. 18-07619S. The measurements were carried out at the 'CANAM' and 'Reactors LVR-15 and LR-0' infrastructures supported by MEYS in projects No. LM2015056 and LM2015074, respectively.

REFERENCES

[1] APEL, P. *Track etching technique in membrane technology* Radiation Measurement 34, 2001, p. 559-566.

[2] MO, D., LIU, J.D., DUAN, J.L., YAO, H.J., LATIF, H., CAO, D.L., CHEN, Y.H., ZHANG, S.X., ZHAI, P.F. and LIU, J. *Fabrication of different pore shapes by multi-step etching technique in ion-irradiated PET membranes* Nuclear Instruments and Methods in Physics Research B 333, 2014, p. 58-63

[3] VACIK, J., HAVRANEK, V., HNATOWICZ, V., LAVRENTIEV, V., HORAK, P., FINK, D. And APEL, P. *Tomographic study of ion tracks by ion energy loss spectroscopy* AIP Conference Proceedings 1525, 2013, p. 663

[4] VACIK, J., CERVENA, J., HNATOWICZ, V., SVORCIK, V., KOBAYASHI, Y., FINK, D. and KLETT, R. *Damage profile examination on ion irradiated peek by ^{6}Li doping and neutron depth profiling technique* Nuclear Instruments and Methods in Physics Research B 141, 1998, p. 216-222

[5] VACIK, J., CERVENA, J., HNATOWICZ, FINK, D. and STRAUSS, P. *Study of LiOH etching of polyethyleneterephtalate irradiated with 11.4 MeV/amu Pb ions by neutron depth profiling and alpha particle transmission* Nuclear Instruments and Methods in Physics Research B 146, 1998, p. 509-512

[6] KIM, M.Y., LI, D.J., PHAM, L.K., WONG, B.G. and HUI, E.E. *Microfabrication of high-resolution porous membranes for cell culture* Journal of Membrane Science 452, 2014, p. 460-469

[7] HNATOWICZ, V., VACIK, J. and FINK, D. *Deconvolution of charged particle spectra from neutron depth profiling using Simplex method* Review of Scientific Instruments 81, 2010,

NANOPLATINUM PREPARATION BY IRRADIATION METHODS IN MICELLAR SYSTEMS

Rostislav SILBER, Patrik BECK, Mojmír ČAMRA

Czech Technical University in Prague, Czech Republic, Tesla-Mikroel, Prague, Czech Republic, EU,
Silber@fjfi.cvut.cz

Abstract

This work deals with the preparation of platinum nanoparticles using ionizing radiation - accelerated electrons with energy 4.5 MeV produced by linear accelerator Linac-4-1200 Tesla VT Mikroel. Aqueous solutions of platinum chloride with Triton X-100 as a reducing and stabilizing agent were used. Solutions were irradiated with doses up to 20 kGy in 20 ml glass vials. Solutions containing platinum ions were colored from yellow to black color after irradiation, depending of the dose. This points platinum ions reduction and nanoplatinum formation. Irradiated solutions were monitored by UV/VIS spectrometry, photon cross-correlation spectroscopy to determine particles size and distribution, transmission electron microscopy and X-ray diffraction analysis to provide information on the forms of prepared nanoparticles. The effects of $PtCl_4$ and Triton X-100 concentrations and also dose deposited energy were studied.

Keywords: Nanoplatinum preparation, electron irradiation, Triton X-100, micelles

1. INTRODUCTION

Nanoparticles have properties which differ from those of larger particles of the same substance as effects related to particles surface increase in significance. In recent years, many methods have been developed to prepare particles with specific sizes, morphologies and composition. Metal nanoparticles in particular are valued for their many and varied applications.

Nanoplatinum usually occurs in the form of suspensions or colloids. Platinum nanoparticles have different shapes - spherical, rod-shaped, cubic or tetrahedral. Nanoparticle sizes range from 2 to 100 nm, and its size strongly depends on the method of preparation. Although nanoparticles are more or less the subject of research, they are used in the areas of catalysis, medicine and in the production of new materials of specific properties. This allows for its unique catalytic, magnetic and optical properties [1].

Regarding the catalytic properties NP are used for the reactions, such as for example hydrogen oxidation, methanol and ethanol oxidation, formic acid oxidation and oxygen reduction reaction. Regarding the catalytic properties, platinum is highly stable and has very good electrocatalytic activity in fuel cell reactions. Platinum catalysts also play a major role in chemical detection. These include, in particular, the use of platinum for the detection of carbon monoxide and oxides of nitrogen, oil refining and many others [1]. The optical properties of nanoparticle-based nanomaterials can be monitored using Raman spectroscopy. Unlike nanoparticles of other metals have a plasma peak in the visible spectrum; nanoplatinum particles have a plasma peak in the ultraviolet region around 215 nm [2].

The goal of this work is to mention the basic irradiation methods - notably electron beam, gamma radiation and UV photolysis - in which nanoplatinum can be generated from aqueous solutions of PtC_4 in nonionic surfactants especially in Triton X-100 and to describe some of the physicochemical properties of the particles prepared. The most notable of these properties are the effects of the initial concentration of PtC_4 and of the surfactant used on the subsequent concentration of nanoparticles prepared.

2. RADIOLYTIC PREPARATION METHODS

During radiolysis of water conventional species such as H^+, OH^- and H_2O_2, as well as several species more reactive than regular chemical compounds, form [3] and are shown in equation 1. Species thus formed from water compose a majority of the products of ionizing radiation in solutions.

$$H_2O \rightarrow e_{aq}^-, H^+; {}^{\bullet}OH; H_2; H_2O_2; H^+; OH^- \quad (1)$$

For the reduction of silver ions several of these products are useful. Perhaps the best reducing agent due to its high standard redox potential is the hydrated electron (e_{aq}^-) with $E^0(e_{aq}^-/e^-) = -2.9$ V. These solvated electrons are thermalized and surrounded by oriented water molecules and are highly reactive. A second, weaker reducing agent - which gains significance with decreasing pH values - is the $H^{\bullet}$ radical which forms according to the reaction in equation 5. It has a standard redox potential $E^0(H^{\bullet}/H^+)$ of -2.3 V.

$$e_{aq}^- + H^+ \rightarrow H^{\cdot\bullet} \quad (2)$$

The third highly reactive product is the hydroxyl radical ${}^{\bullet}OH$, which is a strong oxidizing agent. To be able to produce a reducing environment, the influence of ${}^{\bullet}OH$ radicals must be reduced To do this, a radical scavenger is chosen to transform the oxidizing ${}^{\bullet}OH$ radicals into reducing - if weak - radicals of the scavenger. The general scavenging equation is equation 6. Common scavengers include alcohols (equation 7), such as methanol, ethanol and propan-2-ol and format (equation 8).

$$OH + RH \rightarrow R^{\bullet} + H_2O \quad (3)$$

$$OH + RR'CHOH \rightarrow RR'COH + H_2O \quad (4)$$

$$OH + HCOO^- \rightarrow COO^- + H_2O \quad (5)$$

After this modification, the solvated electrons or various reducing radicals act to reduce the platinum ions.

Use ionizing radiation to induce redox reactions has the advantage of a relatively simple relation between the dose and dose rate of irradiation and those which produce nanoparticle. Changes in the dose rate can modify the size; changes in the dose up to a certain level increase the concentration [4,5].

3. STABILIZERS

The high surface energy of platinum nanoparticles, which opens up so many new applications, also causes them to be unstable and have a tendency to form aggregates. To prevent aggregation, nanoplatinum must be stabilized in some way. Common methods include dispersing the particles in a gel] or adding a stabilizing substance which binds to the molecule. A simple separation and stabilization method which relies on physical rather than chemical bonds is the use of surface active substances, which stabilize the nanoparticles within micelles.

Nonionic surfactants generally consist of a long polyoxide-ether chain, most commonly a polyethylene glycol either with an aliphatic and/or aromatic tail. The two surfactants used for our experiments are Triton X-100 and Brij L4 (also known under another trade name as Brij 30) and in the experiments described in this paper Triton X-100 was used. Triton X-100 is one of the trade names for 4-(1,1,3,3-Tetramethy(lbutyl)phenylpolyethylene glycol (see **Figure 1**), where the number of ethylene glycol groups averages between 9 and 10. In an aqueous solution with a pH of less than 8, it takes on the form of a symmetrical 95 700 u particle with a radius of gyration of 2.93 nm. Its critical micellar concentration is 0.2 - 0.9 $mmol.dm^{-3}$ [6], which, given its average molecular weight of 625 $g.mol^{-1}$, is equal to about 0:013 ± 0:056 w%. Brij L4 is the trade name used to refer to Tetraethylene glycol dodecyl ether [80] (see **Figure 2**). In aqueous solutions it is considerably less soluble than Triton X-100, due to its shorter ethylene glycol chain and is thus prone to forming emulsions rather than colloids. Its critical micellar concentration is about 10 $mg.dm^{-3}$ [7,8].

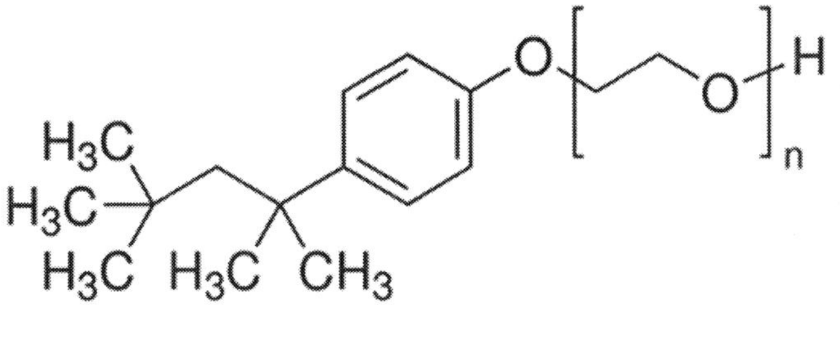

Figure 1 Triton X-100, n = 9-10

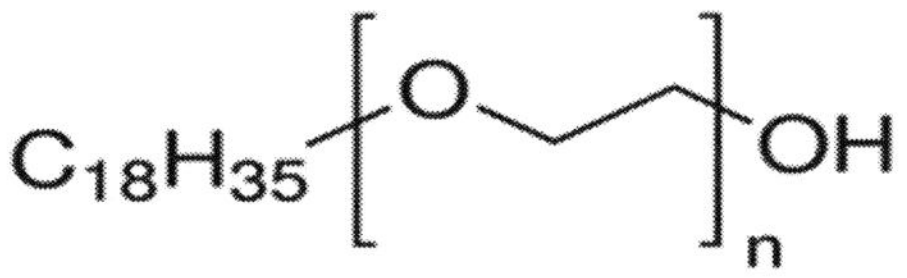

Figure 2 Brij L4, n = 4

Nonionic polyethylene glycol type surfactants were chosen not only for their suitable stabilizing character, but also for their radiation properties. They can act as radical scavengers, forming water and a radical on the first carbon (see equation 12) much like the diols described by Billany et al. [8]. The radical is then free to reduce silver ions in solution or adsorbed on the surface of a silver nanoparticle (n = 1, respectively n > 1 in equation 13). These reactions are analogous to the reaction with glycol by Soroushian et al. [9].

$$R\text{-}OH_2CCH_2OH + OH^{\cdot\bullet} \rightarrow R\text{-}OH_2CC'HOH + H_2O \quad (6)$$

$$R\text{-}OH_2CC'HOH + Pt^{+} R \rightarrow {}^{\bullet}OH2CCHO + Pt^{0} \quad (7)$$

Where R = $C_{12}H_{25}\text{-}(OCH_2CH_2)_3$ for Brij or t-Oct-$C_6H_4\text{-}(OCH_2CH_2)x$ and x = 8 - 9 for Triton X-100.

4. EXPERIMENTAL

4.1. Chemicals

Triton X-100 and $PtCl_4$ were purchased from Merck. All chemicals purchased were analytical grade and were used without further purification.

4.2. Sample Preparation

Nanoplatinum in aqueous solution of surfactant was prepared at the desired concentration (2 wt.% for Triton X-100) by adding the liquid surfactant to water and continuously mixing it until a homogenous solution was formed in the case of Triton X-100. It usually took about one hour to prepare the desired solution. Solid $PtCl_4$ was then dissolved in the continuously stirred solution. Samples were stored in the dark to shield them from the effects of UV light present in natural sunlight.

4.3. Electron Irradiation

A 10 mL sample of the prepared solution was then transferred into 10 or 20 mL ampoules, which were secured with parafilm. The ampoules were placed on a tray on the conveyor and irradiated with 4.5 MeV electrons from a linear accelerator with doses ranging from 1-32 kGy. The samples were placed on the tray all at once and irradiated until a sample was to be taken then irradiation resumed. If the jump between doses increased, then the dose rate was adjusted accordingly. An identical sample containing all the components except the platinum chloride was also irradiated at 24 kGy. This was then used as the standard for taking base readings

4.4. Instrumentation and Sample Evaluation

UV-Vis Spectrometry. Absorbance at certain peaks serves to describe the concentration of individual species of platinum nanoparticles. To obtain Pt spectra, the measurements were performed on a Cary 100 Conc double beam spectrometer from Varian, Inc. (measurement range 190 - 900 nm). It uses a tungsten halogen lamp for visible light and a deuterium arc as the ultraviolet source and signals are captured on a R928 photomultiplier tube. Analysis was done in the Cary WinUV control program. Samples were measured in fused silica cuvettes. Solutions were diluted with water to the concentration necessary to give an absorption in the 0 - 1 range, where the Beer-Lambert law is valid. Standards were used at the same dilution

Photon Cross-correlation Spectroscopy. (PCCS) in the configuration supplied by Sympatec - their NANOPHOX machine in combination with the WINDOX 5 software were used in the measurements. The NANOPHOX system uses a He-Ne laser with a wavelength of 632.8 nm with an adjustable position. Samples are kept in a thermostatic bath. In photon correlation spectroscopy the scattering of a beam of light on a particle is measured. Such scattering is used to detect the Brownian motion of particles which depends on temperature, the size and shape of the particle, as well as on other factors such as viscosity.

Transmission electron microscopy was used to obtain sample images that could serve to compare the particles size of the sample and also to determine the shape of nanoparticles. Nanoparticle images were taken with the JEOL JEM 3010 transmission electron microscope, which includes so-called edox (for electron dispersion spectroscopy). High concentration samples for this electron microscope were diluted with distilled water and ethanol. They were then homogenized by ultrasound. Samples were plated on a copper plate, on the carbon-coated side and plates were placed in a holder, which was then inserted into an electron microscope and then pumping of the vacuum was started. In addition, cathode heating was switched on at 2000 ° C. After the cathode was irradiated, it was possible to monitor the sample on the electron microscope shield. This image was digitized and transferred to the computer by cameras. The images were processed by Digital Micrograph.

Rtg. structural analysis. The measurement was carried out on a Rigaku Miniflex 600 X-ray diffractometer. The x-ray source of this device is a copper X-ray tube with a nominal output of 1 kW, cooled by water from an external cooling circuit. For detection a NaI(Tl) scintillation detector with a beryllium window on a moving arm was used. The diffraction measurements were performed in a pseudo-continuous mode in the range of 10-80°. The sample was placed in the center of the goniometer in the stationary holder (recessed plate). The measured samples were processed using the PDXL 2 program.

Figure 3 Appearance of irradiated samples. The samples are sorted from the back to the front according to the increasing concentration: 1 mM $PtCl_4$, 2 mM, 5 mM and 10 mM and from left to right according to the increasing dose: 0, 4, 8, 12, 16 and 24 kGy

5. RESULTS AND DISCUSSION

Appearance of irradiated samples is shown in the **Figure 3** Solutions containing platinum ions were after irradiation colored from yellow to black colour, depending of the dose. This points platinum ions reduction and nanoplatinum formation.

5.1. UV-VIS Spectrometry

UV/VIS spectra obtained by electron irradiation are in the **Figure 4**.

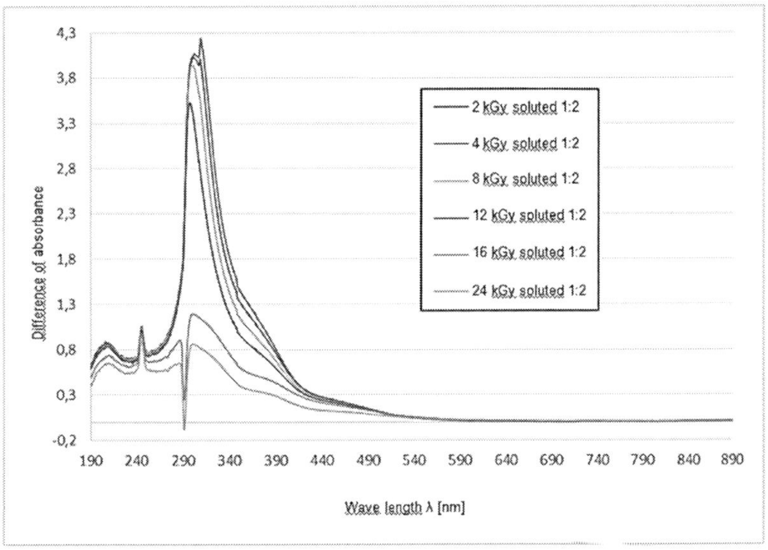

Figure 4 UV/VIS spektra for 10 mM $PtCl_4$ solution (2 wt. % Triton X-100) irradiated by the doses 2-24 kGy

5.2. Photon Cross-correlation Spectroscopy

Some results obtained by PCCS are listed in **Table 1**. It concerns samples prepared from 1 mM $PtCl_4$ solution in 2 wt.% Triton X-100 irradiated by the dose 2, 4, 6 and 8 kGy were measured.t

Table 1 Particle size of samples 1 mM $PtCl_4$ in 2 wt.% Triton X-100 irradiated by the dose 2, 4, 6 and 8 kGy

Dose (kGy	Particle Size (nm)	kcps
2	4.3	562
4	5.3	1435
6	4.6	680
8	4.0	716

5.3. Transmission electron microscopy 7

Some results obtained by TEM are in the **Figure 5**. It concerns samples prepared from 0.5 wt. % Triton X-100 + 1 mM $PtCl_4$

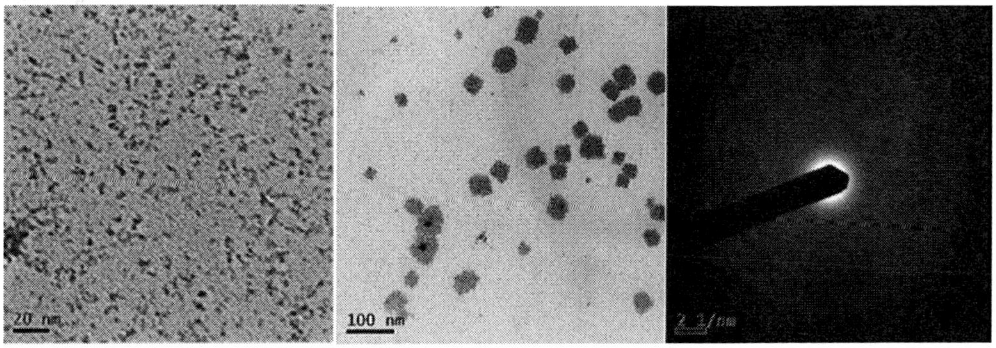

Figure 5 Samples irradiated with a dose of 4 kGy - particle size 4.0 nm and with a dose of 6 kGy - particle size 4.6 nm. Aqueous solution of 0.5 wt. % Triton X-100 + 1 mM $PtCl_4$

5.4. Rentgen diffraction analysis

The resulting diffractogram is shown in the **Figure 6**. A sample containing 2 mM $PtCl_4$ in 2 wt. % Triton X-100 irradiated with a dose of 8 kGy was used. The figure shows the sample diffractogram (red line), the membrane diffractogram (green) and the resulting diffraction pattern of the sample and the membrane. According to the available database, platinum has diffraction peaks around the diffraction angles of 39.76°, 26.24° and 67.46°. In the **Figure 6** the occurrence of these peaks indicates the presence of reduced platinum Pt^0.

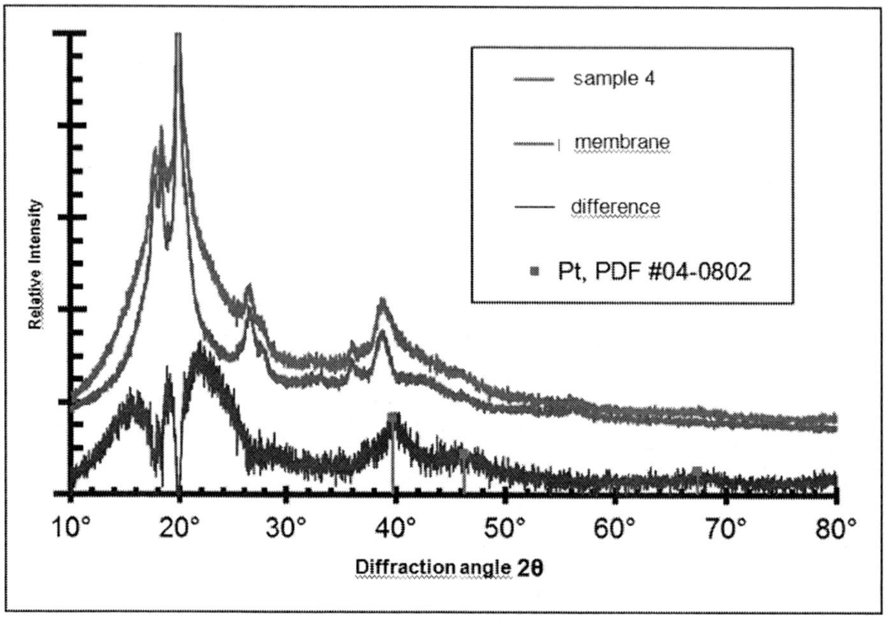

Figure 6 The resulting diffractogram. Sample 2 mM $PtCl_4$ in 2 wt.% Triton X-100, 8 kGy

6. CONCLUSION

The irradiation of samples with accelerated electron changed the original colour of the solution from yellow to black. This indicates the formation of nanoplatinum.

The results of UV / VIS spectrophotometric measurements indicate that the nanoplantinum may be formed relative to the plasma peak around the wavelength of 215 nm. This peak, however, is not 100 % demonstrable due to the degradation of the UV / VIS spectrum in the UV area by the presence of Triton X-100.

Results of transmission electron microscopy and photon cross correlation spectrometry indicate the platinum nanoparticles occurrence with size within the range of several nanometers. In addition to these nanoparticles, there were also larger particles likely to occur due to the aggregation of nanoparticles in larger units. These large particles, however, occupy a minority share. X-ray diffraction analysis results the presence of platinum diffraction peaks.

These results demonstrate the possibility of radiation preparation of nanoplatinum in associative colloids represented by Triton X-100. For more accurate nanoplatinum identification, spectrum degradation in UV / VIS spectrophotometry by Triton X-100 is to be eliminated.

ACKNOWLEDGEMENTS

This work was supported by grant of Ministry of Industry and Trade TRIO III FV30139.

REFERENCES

[1] CHEN, AICHENG; HOLT-HINDLE, PETER. Platinum-based nanostructured materials: synthesis, properties, and applications. Chemical reviews, 06/2010, 110 (6)3

[2] STEPANOV, A.L.; GOLUBEV, A.N. and NIKITIN S.I. (2013) Synthesis and Applications of Platinum Nanoparticles: A Review in Nanotechnology Vol. 2: Synthesis and Characterization, pp. 173-199. Studium Press. ISBN 1626990026

[3] BAXENDALE J.H., BUSI F. (editors) The Study of Fast Processes and Transient Species by Electron Pulse Radiolysis 1982. ISBN

[4] NAGHAVI K., SAION E., REZAEE K., YUNUSW.M.M. Influence of dose on particle size of colloidal silver nanoparticles synthesized by gamma radiation Radiation Physics and Chemistry. 2010, 79 (12): 1203-1208

[5] REMITA H., LAMPRE I., MOSTAFAVI M., BALANZAT E., BOUFFARD S. Comparative study of metal clusters induced in aqueous solutions by - rays, electron or C6+ ion beam irradiation Radiation Physics and Chemistry.2005, 72 (5): 575-586

[6] BHAITI S.M., MOHAN C. Detergents - A guide to the properties and uses of detergents in biological systems 2007 [80] BrijR L4 - Safety Data Sheet Sigma Aldrich. 2012

[7] ZHU H., AITKEN M.D. Surfactant-Enhanced Desorption and Biodegradation of Polycyclic Aromatic Hydrocarbons in Contaminated Soil Environmental Science and Technology. 2010, 44 (19): 7260-7265

[8] BILLANY M.R., KHATIB K., GORDON M., SUGDEN J.K. Alcohols and ethanolamines as hydroxyl radical scavengers International Journal of Pharmaceutics.1996, 137: 143-147

[9] SOROUSHIAN B., LAMPRE I., BELLONI J., MOSTAFAVI M. Radiolysis of silver ion solutions in ethylene glycol: solvated electron and radical scavenging yields Radiation Physics and Chemistry. 2005, 72 (2-3): 111-118

MONITORING OF RELEASED PARTICLES BY HEATING RF POLYMER FOR CONVERSION TO CARBON UNDER INERT ATMOSPHERE

Kristýna HRABOVÁ, Ján POLÁČIK, Vladimír ADAMEC, Barbora SCHÜLLEROVÁ, Ladislav ŠNAJDÁREK, Jiří POSPÍŠIL

Brno University of Technology, Brno, Czech Republic, EU

Abstract

Monitoring and counting of the fine particles released during the temperature rise of the resorcinol-formaldehyde polymer suitable for conversion to carbon aerogel are discussed in the poster. Carbon aerogel is produced by heat treatment under inert atmosphere. The presented approach combines the thermogravimetric analysis with detailed monitoring of the size distribution of fine particles produced. Thermogravimetric analysis (TGA) allows the exact temperature influence of the sample and identifies its weight loss. Aerosol stream leaving TGA enters a Scanning Mobility Particle Sizer (SMPS) where the particle size fractions are separated. A number of particles in the particles size fractions are identified by the condensation particle counter (CPC). Long-term use of Carbon Aerogels results in degradation of these materials and release of fine particles, which could be harmful to health.

Keywords: Carbon, aerogel, thermogravimetrics, aerosol monitoring, fine particles

1. INTRODUCTION

Aerogel is a solid substance with the lowest known density. One cubic meter of the latest and lightest variation of this material weighs mere 1.9 grams. It is also called „solid smoke" as up to 99.8 % of its volume is air. Aerogel is the only material with porosity over 95 % and a very wide pore distribution. The pores are open, i.e. gases or liquids can penetrate the material with minimal limitations. Aerogels have a low coefficient of thermal conductivity ranging from λ= 0.015 to 0.020 W/m.K [1,2]. Aerogels are used for catalytic reactions, production of microfiltration membranes, adsorbents etc. Thanks to high surface area, they can absorb toxic and oil substances found in water. Substances with high surface area, thermal stability and the ability to adsorb heavy metal ions from aqueous solutions are used as adsorbent materials. Aerogel is made by thermal removal of liquid from the substance. They are prepared through sol-gel polycondensation of suitable organic monomers such as resorcinol and formaldehyde producing organic gel. The solid structure formed in this way is filled with liquid phase which is replaced with air through drying. During this process, organic aerogel is created, and it subsequently becomes carbonaceous aerogel through pyrolysis, which allows a wide application spectrum as adsorbents, catalyst carriers or super-condensers. During its production, like during other processes subject to heat treatment, ultrafine particles are released including PM_1 particles. Types of aerogels Oxides (Quarz, Titania, Zirkonia, Mixed oxides), Polymers (Resorcinol, Melamine-Formaldehyde), Carbon (pyrolysed Polymer), cellulose, starch, alginates, and almost everything that can be gelled [3,4,5].

2. SAMPLES AND MATERIALS

To produce carbon-aerogel in our experiment we heated a sample of resorcinol-formaldehyde polymer in nitrogen internal atmosphere to 850 °C. During this process we monitored the weight loss of the sample using thermogravimetric analysis. At the same time, we detected emerging fine particles using SMPS (Scanning Mobility Particle Sizer). For comparison, we used a similar approach for a sample of pure carbon.

Carbon aerogels are a unique group of artificially prepared carbonaceous materials, whose main component is air (95 vol.%) dispersed in their solid structure. Carbon aerogels are highly porous amorphous carbon materials that are derived from organic aerogels. They consist of interconnected clusters of carbon nanoparticles with diameters ranging from 1 to 20 nm. To transform organic aerogels to their carbon form, a pyrolysis or carbonization process in an inert gas atmosphere (nitrogen, argon, or helium) at temperatures above 600 °C is necessary. Carbon can be included in many inorganic nanostructures to produce nanocomposites with considerably better electronic, chemical, thermal, and mechanical properties. Our sample was compressed from expanded graphite with carbon purity of 99.85 %. The sample was compressed into tablets [6,7].

3. EXPERIMENTAL SETUP

3.1. Thermogravimetric analysis (TGA)

The laboratory procedure which was developed combines thermogravimetric analysis with detailed monitoring of the size distribution of the fine particles produced. Thermogravimetric analysis (TGA) allows monitoring the exact temperature influence of a small fuel sample according to the desired schedule. TGA also influences the composition of the atmosphere flowing around the sample. During the measurement, the entire module is inserted into a gas-tight laboratory furnace with controlled heating rate. In our case, the atmosphere was composed of pure nitrogen. During the experiment, TGA monitors the weight of the heated sample and identifies its weight loss. The result of the measurement is a TGA curve showing the weight change in dependence on the temperature of the sample. From a subsequent analysis of the graph drawings, it is possible to obtain information on the processes taking place in the sample such as endothermic and exothermic reactions, thermal decomposition of a particular substance etc. Gaseous substances released from the sample monitored in a TGA instrument are cooled to the ambient temperature and they leave through a hose attached to an SMPS device, where the number and size of fine particles are monitored.

3.2. Scanning Mobility Particle Sizer (SMPS)

Based on the electrical mobility diameter (D_{EM}), we use the SMPS method (Scanning Mobility Particle Sizer) to measure the size of particles and their concentration. Such an apparatus consists of an impactor for separation of particles larger than PM_{10}, and a neutralizer for electric charge redistribution, a DMA and a CPC. DMA - Differential Mobility Analyzer is used for particle classification based on electric mobility. It consists of two cylinder-shaped electrodes. The particles fly into the interspace of the electrodes and based on the height of the voltage and the particle charge, the SMA device releases monodisperse aerosol. CPC - Condensation Particle Counter is a device for detecting the number of monodisperse particles. The particles are too small to be detected using the optical method. That is why the particles first go through saturated butanol vapours. Then they are cooled, and they grow to the size of over 1 μm, which allows their optical laser detection (**Figure 1**).

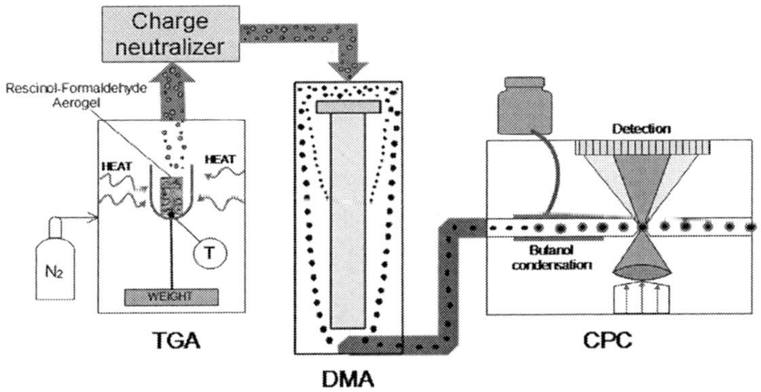

Figure 1 Experimental setup - on the left: Thermogravimetric analysis; on the right - an SMPS device

The SMPS device provides the number of fine particles and their size distribution in one-minute intervals. In this experiment, we detect particles whose size ranges from 18nm to 533nm. The minute particle distributions are shown in the colour diagram with temperature curves, sample weight and derivation (DTG) for a clearer display.

4. RESULTS

Samples of resorcinol-formaldehyde polymer and pure carbon differed in weight, but they were heated under the same conditions of inert nitrogen atmosphere at the speed of 15 Kelvin per minute to the temperature of 850 °C as shown in the **Table 1**.

Table 1 Properties of the samples examined and their weight change

	Aerogel	Carbon
Original sample weight	93.47 mg	584.74 mg
Weight after heating	40.04 mg	565.86 mg
Residual mass	42.84 %	96.77 %
Temperature of the largest weight loss	273 °C	200 °C
Temperature increase	15 °C / min	
Maximum temperature / endurance	850 °C / 4 hours	

As **Figure 2** shows, the change in the carbon sample is minimal, only to 96.8 % of its original weight. This proves that in higher temperatures in inert atmosphere pure carbon compressed from expanded graphite is stable. The DTG curves do not show a significant weight loss, which confirms that under these conditions there are no significant exothermic or endothermic reactions taking place in the graphite sample. We suppose that the slight weight loss is mainly caused by the release of moisture bound in the carbon sample.

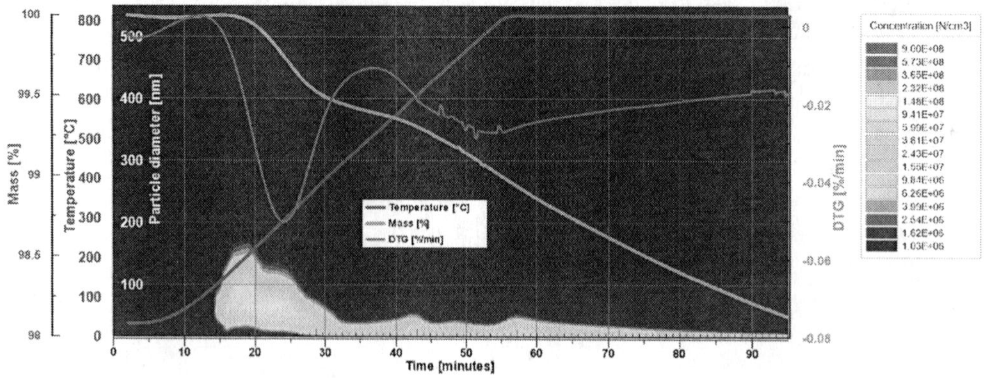

Figure 2 Diagram of fine particle production in time with temperature, weight loss and DTG for carbon curves

The production of fine particles was also minimal compared with the resorcinol-formaldehyde polymer and it already started at 110 °C. Mainly small size particles of up to 200 nanometres are produced in the nucleation and condensation modes. The maximum particle concentration was approximately 1.00E+08 particles per cm^3.

For the resorcinol-formaldehyde polymer, the weight fell as low as 40 % of the original weight after the sample was heated. A significant weight reduction occurred at around 100 °C. However, the most considerable weight loss occurred between 300 °C and 600 °C.

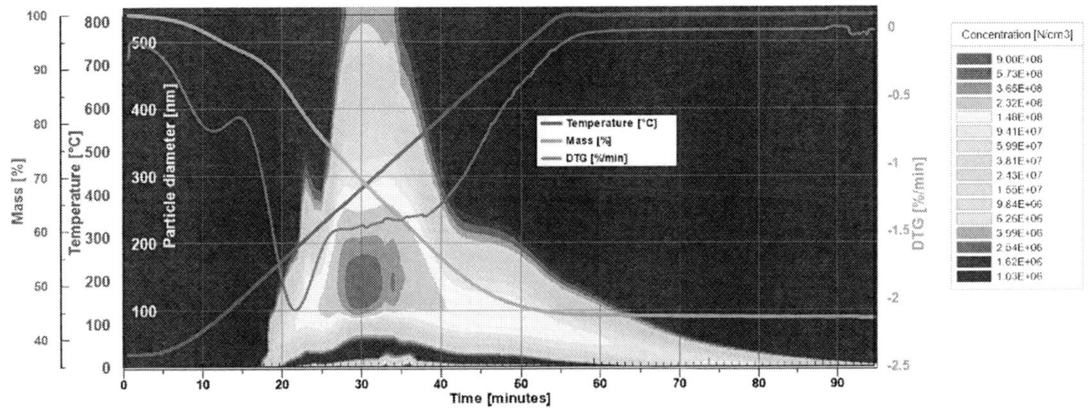

Figure 3 Diagram of fine particle production in time with temperature, weight loss and DTG for resorcinol-formaldehyde polymer curves

The production of fine particles monitored using the SMPS method was the most significant at temperatures ranging from 200 to 800 °C. The highest concentration of fine particles was detected at around 400 °C, specifically up to 9.00E+08 particles per cm^3.

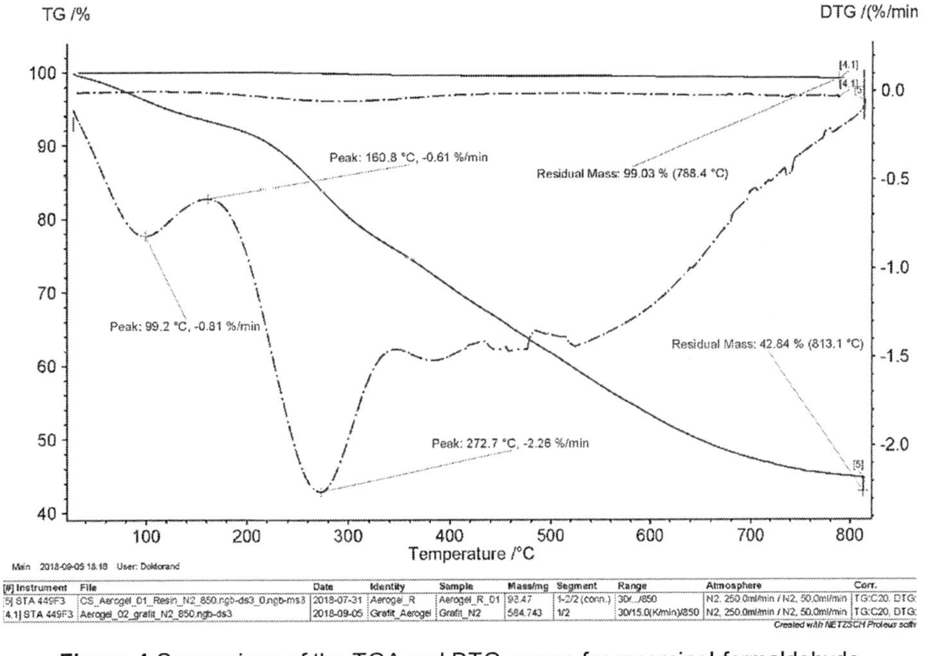

Figure 4 Comparison of the TGA and DTG curves for resorcinol-formaldehyde polymer (blue) and carbon (purple)

CONCLUSION

In this experiment, we monitored the behaviour of a heat-affected sample of resorcinol-formaldehyde polymer, which is the input product in the production of aerogel, in comparison with pure carbon. The new SMPS particle detection method was used, and weight loss was monitored using TGA analysis. We discovered that the production of ultrafine particles is higher for RF than for pure carbon and the particles also differ in size. The RF sample shows residual weight of 43 % after heat treatment, which is a considerable difference in comparison with pure carbon (96 %). Fine particles produced during combustion and heat treatment of various products are a current issue. Their presence in the atmosphere may have a serious impact on human health, especially the respiratory system.

REFERENCES

[1] HUANG, X.; YIN, Z. Y.; WU, S. X.; QI, X. Y.; HE, Q. Y.; ZHANG, Q. C.; YAN, Q. Y.; BOEY, F.; ZHANG, H. Graphene-Based Materials: Synthesis, Characterization, Properties, and Applications Small 2011, 7, 1876- 1902 DOI: 10.1002/smll.201002009

[2] AL-MUHTASEB, S. and RITTER, J. (2003), Preparation and Properties of Resorcinol-Formaldehyde Organic and Carbon Gels. Adv. Mater., 15: 101-114. doi:10.1002/adma.200390020

[3] Organic and Carbon Aerogels: What is an Organic Aerogel?. Aerogel.org [online]. 2008 [cit. 2018-09-19]. Dostupné z: http://www.aerogel.org/?p=71

[4] RATKE, L. Aerogels: Structure, properties and applications [online]. Köln: Institut für Materialphysik im Weltraum, 2011 [cit. 2018-09-18]. Dostupné z: https://elib.dlr.de/72455/1/Aerogel-Applications-2011_k.pdf

[5] SERP, P. and FIGUEIREDO, J. L., Carbon materials for catalysis. Hoboken, N.J.: John Wiley, c2009. ISBN 9780470178850.

[6] CHANDRASEKARAN, S., CAMPBELL, P., BAUMANN, T., WORSLEY, M. (2017). Carbon aerogel evolution: Allotrope, graphene-inspired, and 3D-printed aerogels. Journal of Materials Research, 32(22), 4166-4185. doi:10.1557/jmr.2017.411

[7] ALEKI, H. and HÜSING, N. Current status, opportunities and challenges in catalytic and photocatalytic applications of aerogels: Environmental protection aspects. Applied Catalysis B: Environmental [online]. Elsevier B.V, 2018, 221, 530-555 [cit. 2018-09-20]. DOI: 10.1016/j.apcatb.2017.08.012. ISSN 0926-3373.

PHASE COMPOSITION OF CHOSEN Mg-BASED MATERIALS DURING HYDROGEN SORPTION

Jiří ČERMÁK[1, 2], Lubomír KRÁL[1], Pavla ROUPCOVÁ[1, 3]

[1]Institute of Physics of Materials of the CAS, Brno, Czech Republic, EU
[2]CEITEC-Institute of Physics of Materials of the CAS, Brno, Czech Republic, EU
[3]CEITEC-Brno University of Technology, FME, IMSE, Brno, Czech Republic, EU

Abstract

Phase transformation during hydrogen sorption was investigated in ten chosen magnesium-based hydrogen storage (HS) materials. Chemical composition of the materials consisted of Mg, as a principal hydrogen-binding element, additive X and amorphous carbon (CB), as an anti-sticking component. In order to assess the effect of X itself upon the structure, values of concentration of both X and CB were fixed to about 12 wt. %. The influence of X = Mg_2Si, Mg_2Ge, $Mg_{17}Al_{12}$, Mg_5Ga_2, NaCl, LiCl, NaF, LiF and two combinations Ni+$Mg_{17}Al_{12}$ and Ni+Mg_2Si upon the changes in phase composition was tested. Phase content in HS materials was observed (i) after the intensive ball milling (BM), (ii) after the BM followed by hydrogen charging at 623 K and (iii) after the BM and one hydrogen charging/discharging cycle (C/D) at temperature 623 K. The study was carried out by SEM and XRD. It was found that, the C/D is approximately structurally reversible for X = Mg_2Ge, $Mg_{17}Al_{12}$, NaF and LiF. However, additives X = $Mg_{17}Al_{12}$ and NaF decompose already during the BM. In alloys with combination of Ni with $Mg_{17}Al_{12}$, new phases Ni_mAl_n are formed. Phase composition changed during C/D for X = Mg_2Si Mg_5Ga_2 and Ni+Mg_2Si due to equilibration of phases composition. Observed structure changes of HS materials with chloride ionic additives NaCl and LiCl are, most likely caused by the relatively strong affinity between Mg and Cl. Hydrogen storage capacity of all studied alloys was 6.0 ± 0.3 wt. % H_2.

Keywords: Hydrogen storage, Mg alloys, carbon black

1. INTRODUCTION

Mg-based alloys are promising materials for HS applications [1,2]. The principal HS phase in alloys of this type is MgH_2 hydride. This phase shows high HS capacity (7.7 wt. % H_2) and excellent cycling behavior. The substantial drawback is its high thermodynamic stability, which implies high operation temperatures (starting from approximately 573 K). Much research work was devoted to decrease the high stability of MgH_2 [see, e.g., in 3], i.e. to increase hydrogen pressure in equilibrium, p_{eq}, above the Mg hydride. This would shift working temperatures of HS reservoirs down to desired values (to some 400 K). However, this proved to be an extraordinary difficult task. After a considerable research effort, only moderate success was reached in this respect. Instead of improvement of the hydrogen sorption dynamics, rather an acceleration of sorption process stands in the focus of research groups in the last years. of course, the two aspects are closely interconnected, but the smoothing the kinetic paths of the process seem to be easier task then to affect effectively the driving forces. The main ways how to achieve the target seem to be the catalysis of partial sorption steps [4] and nanosizing [5].

In the present work, structure of ball-milled (BM) HS blend of Mg with chosen additives during the C/D cycle was investigated. Five types of additives were tested: (i) Mg intermetallics formed by two elements from 13th group (Mg_2Ge, Mg_2Si), (ii) by two elements of 14th group (Mg_5Ga_2, $Mg_{17}Al_{12}$), (iii) additives formed by two combination Ni+$Mg_{17}Al_{12}$ and Ni+Mg_2Si (iv) two simple fluorides with Na^+ and Li^+ (NaF, LiF), and (v) two chlorides with the same kations (NaCl, LiCl). Expectation of beneficial effect of the intermetallics with elements from 13th and 14th group upon the C/D kinetics was based on results published in [6], synergy of those elements and Ni can be expected with regard to results from [7] and improvement of C/D kinetic by anions F^-, Cl^- and/or cations Na^+, Li^+ may be expected from results published in [8-10]. Sufficient nanosizing of the BM blend was assured by introduction of carbon black (CB) [11].

2. EXPERIMENTAL

Samples were made from pure components Mg(purity: 3N8), Ge(5N), Si(6N), Ga(4N), Al(5N), Ni(3N), LiF, LiCl, NaF, NaCl (all halides >2N) and CB (2N). Intermetallics Mg_2Si and Mg_2Ge were prepared by mechanical alloying in form of powders. Compounds $Mg_{17}Al_{12}$ and Mg_5Ga_2 were induction melted in pure Ar(4N) and grinded into form of splinters and halides and CB were purchased as fine powders. All the additives were mixed with splinters of Mg and ball milled (BM) in hydrogen atmosphere (6N) using Fritsch Pulverisette6 ball-mill. Two HS samples were prepared with admixture of fine powder of Ni. The mass ratio of the milling balls to the milled blend was about 60 and the milling cycle - 10 min milling/50 min cooling - was repeated 90 times. Hydrogen charging under pressure p = 2.5 MPa and discharging was carried out using Sieverts-type gas sorption analyzer PCT-Pro Setaram Instrumentation at constant temperature 623 K. All manipulations of the milled blend were done in the glove box in protective Ar atmosphere. Nominal chemical composition of ball milled batches is listed in **Table1**.

Table 1 Experimental samples

Sample label	Nominal composition of batch (wt. %)	Chemical composition (wt. %)
Si	76.36 Mg - 10.25 Mg_2Si - 13.39 CB	81.79 Mg - 4.82 Si - 13.39 C
Ge	75.28 Mg - 12.61 Mg_2Ge - 12.11 CB	80.32 Mg - 7.57 Ge 12.11 C
Al	76.58 Mg - 10.40 $Mg_{17}Al_{12}$ - 13.02 CB	82.48 Mg - 4.5 Al - 13.02 C
Ga	81.61 Mg - 10.91 Mg_5Ga_2 - 7.48 CB	86.72 Mg - 5.80 Ga -7.48 C
NiSi	71.87 Mg - 3.26 Ni - 12.47 Mg_2Si - 12.40 CB	79.76 Mg - 3.26 Ni - 4.58 Si - 12.40 C
NiAl	72.37 Mg - 2.53 Ni - 12.84 $Mg_{17}Al_{12}$ - 12.31 CB	79.82 Mg - 2.53 Ni - 5.39 Al - 12.31 C
LiF	74.04 Mg - 13.18 LiF - 12.78 CB	74.04 Mg - 3.35 Li - 9.65 F 12.78 C
NaF	74.81 Mg - 12.41 NaF - 12.78 CB	74.81 Mg - 6.81 Na - 5.60 F - 12.78 C
LiCl	75.94 Mg - 11.57 LiCl - 12.49 CB	75,94 Mg - 1.89 Li - 9.68 Cl - 12.49 C
NaCl	74.61 Mg - 12.99 NaCl -12.40 CB	74.60 Mg - 5.20 Na -7.80 CL - 12.40 C

Phase composition of samples was obtained by XRD with the help of the X'Pert Pro MPD device using CoKa radiation and the results were interpreted (Rietveld analysis) with the HighScore Plus SW and ICSD databases. Accuracy of phase composition was about 1.5 wt. %. Morphology of milled samples, average chemical composition and chemical maps were observed by SEM TESCAN LYRA3 equipped with X-max80 EDS in the area approximately 300 × 500 μm containing about 10^2 grains. Accuracy of average concentration of substitution elements (except for Li) was within 1.5 wt%.

Due to considerable uncertainty in estimation of content of amorphous phase (AM) by XRD, c_{AM}, the value of c_{AM} was fixed to known concentration of CB, c_{CB}, and, due to similar reasons, for the carbon concentration, c_C, in SEM analyses, the known value $c_{AM} = c_C$ was taken (see **Table 1**). Concentration of hydrogen was measured by sorption experiment (c_H = 6.0 ± 0.3 wt. % H_2 in all cases). For Li concentration, c_{Li}, in LiF-containing samples the value $c_{Li} = 0.36537 \times c_F$ and for, c_{Li}, in LiCl-containing samples the value $c_{Li} = 0.1957 \times c_{Cl}$ was taken, where concentrations c_F and c_{Cl} were obtained by SEM.

3. RESULTS AND DISCUSSION

3.1. Morphology

Morphology of all samples after the BM, its change after C and after the D steps follows very similar evolution scheme. It can be illustrated by the sample Ge. The structure of Ge after BM is shown in **Figure 1a**, where relatively smooth grains of size typically between 5 and 50 μm can be seen. These grains composed principally of solid solution (Mg) transform during C to hydride phase and their appearance becomes fragmented - **Figure 1b**. After D, when the hydrided phase(s) transformed back to hydride-free components, the fragmented structure persists and, moreover, additional transversal cracks of great grains appear - **Figure 1c**.

a b c

Figure 1 Typical morphology of samples after BM (a), C (b) and D (c). Sample Ge.

Fine bright particles of size about 1 μm (for Ge, see in **Figure 1a**) were identified as impurities with traces of Ti, Al, W and Fe. Concentration of these elements, was lower than 1 wt. %, and therefore it was neglected in SEM analyses.

3.2. XRD measurements

XRD patterns were measured for each sample in BM, C and D states. The obtained data could be classified qualitatively into two considerably different groups. The first typical behavior can be described by the scheme: c_{AM} in BM ≅ c_{AM} in C ≅ c_{AM} in D with $c_{AM} \cong c_{CB}$ (c_{AM} stands for concentration of amorphous phase from SEM analysis, c_{CB} is known concentration of carbon black). This type of behavior was observed in all alloys except for chloride-containing samples. The samples modified by chlorides (LiCl and NaCl) on the other hand, formed the other group where the measured content of AM was very high, especially in BM and D. The typical XRD patterns are exemplified in **Figure 2** for Mg_2Ge and NaCl. No carbides were detected in any sample; the components, which content was below 1 % were neglected.

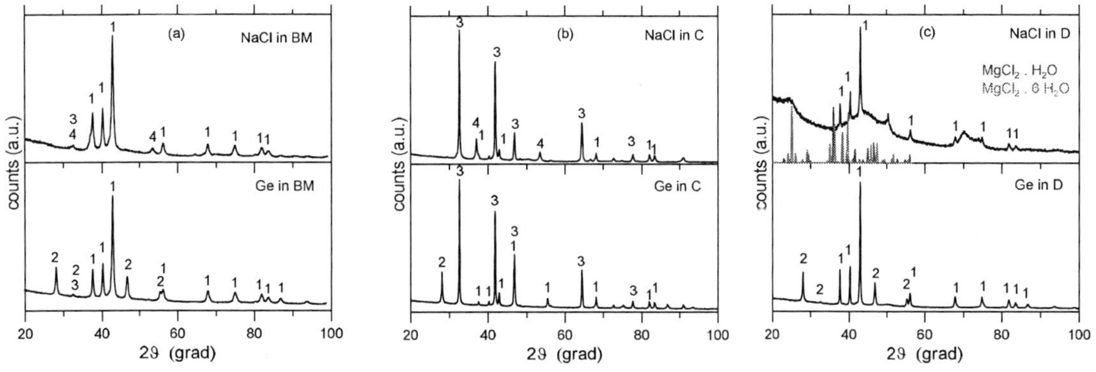

Figure 2 Two types of XRD patterns (exemplified by Ge and NaCl) in states BM (a), C (b) and D (c). 1-(Mg), 2-Mg_2Ge, 3-MgH_2, 4-NaCl. Patterns in color in (c) - magnesium chlorides [13].

3.3. Observed structure transformations - survey of results and discussion

Sample Si. After BM (phases: 79(Mg)-4Mg_2Si-2Si-2αMgH_2-13AM; elements: 84Mg-3Si-13C), Si was present both as regularly dispersed fine particles of Mg_2Si and as scattered elemental Si in the form of particles of size up to 1 μm. Separate XRD analysis of ball milled blend Mg+Si showed that the BM produced Mg_2Si compound only up to about 21 %. This explained observed differences between nominal and measured phase composition in Si sample. Presence of αMgH_2 is a consequence of BM in H_2. Chemical analysis agreed well with the nominal values (**Table** 1) within 3 wt.% (2×1.5 %). Carbon was distributed regularly. After C (78αMgH_2-9 Mg_2Si-13AM; 78Mg-3Si-13C-6H), low-pressure alpha hydride phase appeared, all elemental Si reacted with Mg and formed particles Mg_2Si (size up to 2 μm). In D state (76(Mg)-11 Mg_2Si-13AM; 84Mg-3Si-13C), particles Mg_2Si (size up to 2 μm) persisted.

Sample Ge. After BM (75(Mg)-12Mg_2Ge-13CB; 81Mg-7Ge-12AM), Mg_2Ge particles (~5 μm) were observed, C was uniformly distributed, chemical concentration agreed with nominal composition. After C (70αMgH_2-11Mg_2Ge-7(Mg)-12AM; 78Mg-5Ge-11C-6H), Mg_2Ge particles become less distinctly confined. After D (75(Mg)-13Mg_2Ge-12AMB; 83Mg-5Ge-12C), the original state was restored.

Sample Al. After BM (83(Mg)-3$Mg_{1.95}Al_{0.05}$-1MgH_2-13AM; 82Mg-5Al-13C), no $Mg_{17}Al_{12}$ phase was detected. It seems that it was decayed during BM and the products entered (Mg) phase. Al is distributed approximately uniformly; slightly higher Al concentration can be detected in $Mg_{1.95}Al_{0.05}$ particles (size between 1 and 10 μm). After C (74αMgH_2-13(Mg)-13AM; 79Mg-3Al-12C-6H), low-pressure hydride αMgH_2 appeared, presence of (Mg) solid solution is a consequence of incomplete charging. After D (86(Mg)-1MgH_2-13AM; 82Mg-5Al-13C), the state observed before after BM was restored.

Sample Ga. After BM (86(Mg-14AM; 86Mg-7Ga-7C), all components were uniformly scattered. Mg_5Ga_2 phase decomposed during BM and the decay products entered the AM and partly also the (Mg). After C (86βMgH_2-4Mg_5Ga_2-3Mg_2Ga-7AM; 85Mg-2Ga-7C-6H), one of high pressure phases [12] was formed (β). Loci (~1μ) with significantly higher concentration of Ga were observed, which were identified as two Mg_mGa_n phases. They precipitated from AM. After D (85((Mg)-8Mg_5Ga_2-7AM; 88Mg-5Ga-7C), Mg_5Ga_2 particles persisted.

Sample NiSi. After BM (77(Mg)-4Mg_2Si-1Ni-2Si-5αMgH_2-12AM; 80Mg-4Ni-4Si-12C), Si particles (~2 μm) that anti-coincided with Mg were observed. Mg_2Si not completely formed (similarly as in sample Si above). After C (76αMgH_2-1(Mg)-10Mg_2Si-1Ni-12AM, 77Mg-2Ni-3Si-12C-6H), Formation of Mg_2Si phase completed (particles ~2-3 μm). After D (74(Mg)-13Mg_2Si-1$Mg_2NiH_{3.8}$-12AM; 81Mg-3Ni-4Si-12C), particles of Mg_2Si were restored, trace amount of ternary Mg-Ni hydride (particles ~3 μm) was a residuum after the C.

Sample NiAl. After BM (85(Mg)-1$Ni_{0.89}Al_1$-1Ni-1αMgH_2-12AM; 80Mg-3Ni-5Al-12C), the $Mg_{17}Al_{12}$ decayed and the residuals entered (Mg). After C (82αMgH_2-1(Mg)-5N_2Al_3-2AM; 77Mg-2Ni-3Al-12C-6H), Mg was mostly bound in MgH_2, Sporadic Mg particles were also observed. Al was scattered uniformly; no localized phase particles composed of Ni, Al and Mg were observed. Diffuse coincidence between Al and Ni seems to be more pronounced than that between Al and Mg. After D (83(Mg)-4Ni_2Al_3-1NiAl-12AM; 80Mg-3Ni-5Al-12C), new-formed phases Ni_mAl_n replaced completely originally introduced phase $Mg_{17}Al_{12}$ in hydrogen-free sample.

Sample LiF. After BM (75(Mg)-12LiF-13AM; 75Mg-3Li-9F-13C), Small particles LiF (~3 μm) were observed on the surface of great Mg grains. C was scattered uniformly. After C (83βMgH_2-2LiF-2(Mg)-13AM; 68Mg-4Li-10F-12C-6H), the identity of LiF phase remained even after the hydrogen charging. One of high pressure phases of MgH_2 (β phase) was formed by BM. (Mg) phase was a consequence of incomplete charging. After D (75(Mg)-11LiF-1MgH_2-13AM; 73Mg-4Li-10F-13C), the original state observed after BM was almost exactly restored.

Sample NaF. After BM (80(Mg)-7$NaMgF_{1.17}H_{1.83}$-13AM; 75Mg-6Na-6F-13C), NaF decomposed during BM and the decay products entered the (Mg) phase. A small fraction of ternary hydride was formed. All chemical components are scattered uniformly. After C (84βMgH_2-1$NaMgH_3$-2(Mg)-13AM; 74Mg-5Na-3F-12C-6H), high pressure hydride βMgH_2 was formed, ternary hydride changed its composition and F remained solved in major phases. After D (80(Mg)-7$NaMgF_{1.17}H_{1.83}$-13AM; 79Mg-4Na-4F-13C), the original state was approached, observed after BM.

Sample LiCl. After BM (73(Mg)-11LiCl-4βMgH_2-12AM; 76Mg-2Li-10Cl-12C), both phase and chemical composition agree approximatelly with nominal values; high pressure βMgH_2 was formed. After C (72αMgH_2-15Li-1(Mg)-12AM; 67Mg-2Li-12Cl-12C-6H), LiCl was found completely decayed. Cl entered hydride phase and Li precipitated in elemental form. After D (2(Mg)-98AM; 73Mg-2Li-12Cl-12C), AM phase grew significantly. It was most likely due to interaction of Cl with Mg [13] that produced finally extremely fine (almost amorphous) phase. Positions of chlorides lines are plotted in **Figure 2c.** In case of fine structure they may broaden and superimpose over LiCl pattern. In results, they may be involved in overall AM phase.

Sample NaCl. After BM (77(Mg)-8NaCl-3αMgH_2-12AM; 75Mg-5Na-8Cl-12C), NaCl phase was found partly decomposed; the decay products entered (Mg) phase. After C (74αMgH_2-9NaCl-5Mg-12AM; 70Mg-5Na-7Cl-12C-6H), NaCl phase remained also after hydrogen charging. After D (10(Mg)-90AM; 74Mg-5Na-9Cl-12C), NaCl phase decayed and the transformation products entered the phase that behaves as AM. AM contains also prevailing fraction of Mg. Most likely, this is a result of fragmentation of MgH_2 into very fine particles during interaction of Mg with Cl - reasoning is similar as in the case of sample LiCl.

3.4. Observed structure transformations - summary

Chemical composition of all samples in states BM and D are reasonably equal to elements nominal concentrations in milling batches. In sample Ge, the added phase Mg_2Ge is unaffected by C/D cycle. In samples Si and NiSi the phase Mg_2Si behaves similarly to Mg_2Ge - its lower content in the BM state is caused by its incomplete formation in the beginning of the C/D cycle. Phase $Mg_{17}Al_{12}$ in samples Al and NiAl decayed already during BM and does not reform during D. Moreover, Al from decayed $Mg_{17}Al_{12}$ forms very stable phases Ni_mAl_n in sample NiAl. Added phases in samples Ga and LiF decay during BM, but during D, they are (partly) reformed. In samples NaF, LiCl and NaCl the added phases decay during BM, but they do not restore in D state. In samples LiCl and NaCl, significantly high fraction of AM was observed after the C/D cycle. Hydrogen charging of samples with halides and with Mg_5Ga_2 led to a formation of high pressure phase of Mg hydride (βMgH_2).

4. CONCLUSION

Different structure evolution schemes were observed in Mg-based HS mixtures modified by chosen admixtures that could be effective in HS behavior. It was found that composition modified by combination of Ni and $Mg_{17}Al_{12}$, forms Ni_mAl_n phase(s), which decreased their beneficial catalytic effect. On the other hand, admixture of halides (NaF, NaCl, LiF and LiCl), especially chlorides that support formation of AM, may be good components of HS Mg-based alloys. Sorption experiments that may support these conclusions are running.

ACKNOWLEDGEMENTS

This work was supported by project of Czech Science Foundation No. 17-21683S.

REFERENCES

[1] HUOT, J. Metal Hydrides. In: HIRSCHER, Michael ed. *Handbook of hydrogen storage* Weinheim: WILEY-VCH Verlag GmbH &Co. KGaA, 2010, pp. 81-110.

[2] CRIVELO, J.-C., DENYS, R.V., DORNHEIM, M., FELDERHOFF, M., GRANT, D.M., HUOT, J., JENSEN, T.R., deJONGH, P., LATROCHE, M., WALKER, G.S., WEBB, C.J. and YARTYS, V.A. Mg-based compounds for hydrogen storage. *Applied Phys. A.* 2016. vol.85, pp 122-139.

[3] WANG, H., LIN, H.J., CAI, W.T., OUYANG, L.Y. and ZHU, M. Tuning kinetics and thermodynamics of hydrogen storage in light metal element based systems-A review of recent progress. *J. Alloys Compd.* 2016. vol. 658, pp.280-300.

[4] Li, Q., LIN, Q., CHOU, K-C. and JIANG, L. A study on the hydriding-dehydriding kinetics of $Mg_{1.9}Al_{0.1}Ni$. *J. Mater Sci.* 2001. vol. 39, pp.61-65.

[5] YHU, Y., LIU, Z., YANG, Y., GU, H., LI, L. and CAI, M. Hydrogen storage properties of Mg-Ni-C system hydrogen storage materials prepared by hydriding combustion synthesis and mechanical milling. *Int. J. Hydrogen energy.* 2010. vol. 35, pp.6350-6355.

[6] CERMAK, J. and KRAL, L. Alloying of Mg/Mg_2Ni eutectic by chosen non-hydride forming elements: Relation between segregation of the third element and hydride storage capacity. *J. Power Sources.* 2012. vol. 197, pp.116-120.

[7] CERMAK, J. and KRAL, L. Hydrogen diffusion in Mg-H and Mg-Ni-H alloys. *Acta Mater.* 2008. Vol. 56, pp.2677-2686.

[8] VAJO, J.,J., MERTENS, F., AHN, C.,C., BOWMAN jr., R.,C. and FULTZ, B. Altering hydrogen storage properties by hydride destabilization through alloy formation: LiH and MgH_2 destabilized with Si. *J. Phys. Chem. B.* 2004. vol. 108, pp.13977-13983.

[9] LIU, H., WU, C., ZHOU, H., CHEN, T.,LIU, Y., WANG, X., DONG, Z., GE, H., Li, S. and YAN, M. Synergistically thermodynamic and kinetic tailoring of the hydrogen desorption properties of MgH_2 by co-addition of AlH_3 and CeF_3. *RSC Advances.* 2015. vol. 5, pp.22091-22096.

[10] LYCI, G. and SURENDRA, K., S. Structural stability of metal hydrides, alanates and borohydrates of alkali and alkali- earth elements: A review. *Int. J. hydrogen energy.* 2010. vol. 35, pp.5454-5470.

[11] CERMAK, J. and KRAL, L. Hydrogen absorption in carbon allotropes and talc. *Kovove Mater.-Metallic Mater.* 2018, vol. 56, pp.75-80.

[12] VAJEESTON, P., RAVINDRAN, P., HAUBACK, B.C., FJELLVAG, H., KJEKSHUS, A., FURUSETH, S. and HANFLAND, M. *Phys. Rev.B.* 2006. vol. 73. p.224102.

[13] HUANG, Q., LU, G., WANG, J and YU, J. Thermal decomposition mechanism of $MgCl_2.6H_2O$ and $MgCl_2.H_2O$. *J. Analyt.Appl. Pyrolysis.* 2011. vol. 91, pp. 159-164.

ON-LINE MICROFLUIDIC SERS MESUREMENRS ON THE BASE OF SURFACE PLASMON-POLARITON EXCITATION

[1]Vasilii BURTSEV, [1,2]Elena MILIUTINA, [1,2]Olga GUSELNIKOVA, [1]Vaclav ŠVORČÍK, [1,2]Oleksiy LYUTAKOV

[1]University of Chemistry and Technology, Prague, Czech Republic, EU, lyutakoo@vscht.cz
[2]Tomsk Polytechnic University, Research School of Chemistry and Applied Biomedical Sciences, Tomsk, Russian Federation

Abstract

There are many methods that allow detection of complex analytes such as soil contaminants and explosives, which can be dispersed in the various of liquid and solid. Nevertheless, it remains a challenge to create a lab-on-a-chip (LoC) platform for on-line detection of small analytes volumes (below the few femtoliters), with high sensitivity, specificity and reliability. It is expected that the combination of microfluidics and plasmonic will create a highly sensitive and affordable lab-on-a-chip platform for the solving of above mentioned challenge. The proposed approach will utilize the unique advantages of microfluidics that explore and uses unusual behavior of microscale-restricted liquids and surface-enhanced Raman analytical method, which is able to recognize the single molecule of the targeted analyte. In this work, we demonstrated the model of a microfluidic micromixer incorporating with a plasmon active substrate. The design and realization of proposed experimental concept include the solving of following tasks: (i) utilization of Comsol software for mathematical simulation of possible processes in microfluidic mixer; (ii) preparation of microfluidic platform using 3D printing; (iii) application of excimer UV laser large-scale patterning and local gold deposition for creation of plasmonic active area. Our LoC platform allows detection of model analyte (R6G) at concentrations down to 10 fM (10^{-15} $mol\cdot L^{-1}$). Proposed unique features are the key to new scientific experiments and innovations in the field of lab-on-a-chip devices and analytical approaches.

Keywords: Lab-on-a-chip, SERS, R6G, 3D printing, microfluidics system

1. INTRODUCTION

In recent years, many advances in science are associated with a reduction in the size of various analytical devices and the improvement of their characteristics [1-4]. One of the more promising trends in this area is microfluidic-based analytical platforms. Utilization of microfluidic allows managing the small volumes of liquids (of the order of micro-and nanoliter). Moreover, the microfluidics made it possible to develop a new class of devices - so-called lab-on-a-chip complex arrangement (LoC) [5-7]. The use of LoC modules in medicine, biology, pharmaceuticals, industry and other fields opens the new opportunities for significantly reducing the cost, complexity and timing of various analyzes and scientific research. Development of LoC can be closely attributed to the utilization of surface enhanced Raman spectroscopy (SERS). The SERS phenomenon is the result of localized surface plasmon resonance (LSPR). LSPR arises due to the excitation of a collective oscillation of electrons inside a metallic nanostructure induced by incident light, which leads to a huge optical amplification of the near field near the surface of metal nanostructures [8, 9]. It has been scientifically proven that the surface geometry of metallic nanostructures (size, shape, periodicity) [10-12] affects the intensity and distribution of the electromagnetic field on SERS substrates. This in its own way leads to the creation of a "hot spot" effect. Such a structure can be a nanoscale lattice.

So, SERS can be considered as a fundamental method for chemical analysis of small materials amounts, since it can be used to detect analytes adsorbed on metal structures at low concentrations [13-16]. The first priority to create the effective SERS-based component of microfluidic device is to increase the sensitivity of the

SERS signal, which can be implemented in a combination of Raman scattering and microfluidics. In this paper, we propose novel technique for the fabrication of periodic metal grille inside 3D microfluidic channel using femtosecond laser processing. The resulting 3D microfluidic SERS chip was tested using the model analyte (R6G). The results show that the microchips developed by us can work as 3D microfluidic SERS chip for super-sensitive online measurements with high performance.

2. EXPERIMENTAL

2.1. Materials

3D microfluidic model was formed with polymer Clear resin (purchased from Formlabs, USA). Isopropyl alcohol (≥98.5 %), deionized water, Rhodamine 6G (R6G) were purchased from Sigma-Aldrich and used without further purification. The target for Au deposition (purity of Au was 4 N) was purchased from Safina.

2.2. Sample preparation

The 3D model was made using 3D printing (the model is shown in **Figure 1**). Model for printing was performed in the program COMSOL Multiphysics. After the model printing the created samples were irradiated by UV-source for 30 min and dried at 24° C for 2 h.

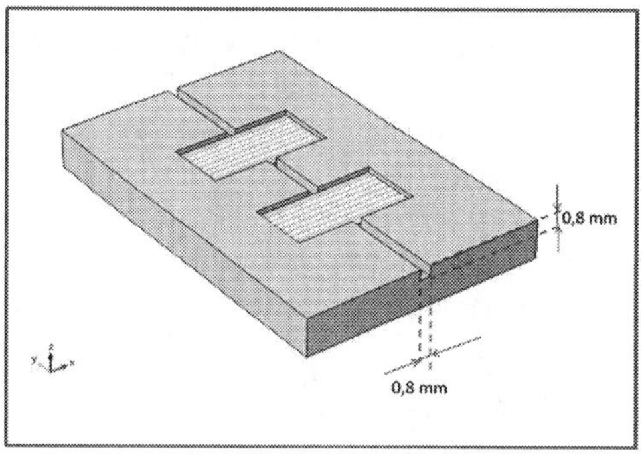

Figure 1 The 3D model of microfluidic chip, used for further 3D printing and deposition of surface plasmon-polariton areas

Grating preparation

Polymer films (Solution of epoxy resin, photoresist Su-8, purchased from Microchem, Germany) were spin-coated (1000 rpm, 10 min) from a solution onto 3D microfluidic SERS chip surface. The prepared samples were dried at 60°C for 24 h, irradiated by UV-source for 30 min and annealed at 90°C for 2 h. Pattering procedure of Su-8 film was performed using the excimer laser irradiation (KrF excimer laser, COMPexPro 50F, Coherent, Inc., wavelength 248 nm, pulse duration 20-40 ns, repetition rate 10 Hz) [15]. The laser beam was polarized linearly with a cube of a UV-grade fused silica with an active polarization layer. The samples were irradiated with 3000 laser pulses with laser fluencies above 9 mJ cm^{-2}. The angle of laser beam incidence with respect to the sample surface normal was 45° and the periodic surface structures were created on the Su-8 surface with 1×2 cm^2 patterned area size. The Au was deposited onto a patterned surface by vacuum sputtering (DC Ar plasma, gas purity of 99.995 %, a gas pressure of 4 Pa, a discharge power of 7.5 W, sputtering time 200 s, resulted thicknesses 25 nm). The deposition of Au was accomplished from Au target (purity of 99.99 %, provided by Safina, Czech Republic).

Surface measurement techniques

For characterization of sample surface, the peak force AFM technique was applied. Surface mapping was performed with Icon (Bruker) set-up on the areas of 8×8 μm^2.

Microfluidic SERS investigations- concentration dependence

In the 3D microfluidic SERS chip was to filed water solutions of R6G in different concentrations 10^{-6} M, 10^{-8} M, 10^{-10} M, 10^{-12} M, 10^{-14} M and the SERS response was on-line collected from the gold grating surface. Raman scattering was measured on portable ProRaman-L spectrometer (Laser power 20 mW) Raman spectrometer with 785 nm excitation wavelengths. Spectra were measured 120 times, each of them with 3 s accumulation time.

3. RESULTS AND DISCUSSION

The morphology of excimer-created gold grating was measured using the AFM technique and typical periodical surface pattern is presented in the **Figure 2A**. As is evident the surface topography represents a well ordered grating array, with sinusoidal shape. As was proven in our previous work such geometry of metal surface is favorable for the excitation and propagation of so-called surface plasmon-polariton. At the other word, photon entrapping the metal grating surface will lead to the appearance of plasmon waves, which focused the light energy near the surface and allow the efficiently exciter the SERS phenomenon. To evaluate the SERS performance of nanostructure the typical Raman analyte, R6G was used. In particular, the SERS spectra were acquired in the microfluidic chip at ambient temperature, using a customized microscopic Raman spectrometer equipped with a 780 nm excitation laser and a spectrometer grating having 500 grooves per mm. During the test, the R6G solutions of varied concentration were loaded into a syringe and then injected into the microfluidic microchannel with SERS active area using a syringe pump. (**Figure 2B**) **Figure 2C** shows the typical SERS spectra of R6G solutions of with different concentrations (from 10^{-6} to 10^{-14} mol·L^{-1}). Characteristic R6G SERS bands can be clearly distinguished in the **Figure 2C**: 605, 771, 1185, 1310, 1355, 1505, 1602, and 1645 cm $^{-1}$ are identified [17]. As can be expected the intensities of the peaks of Raman scattering gradually decrease with the decreasing of R6G concentration The SERS signal becomes almost unnoticeable at a concentration below 10^{-14} mol·L^{-1}.

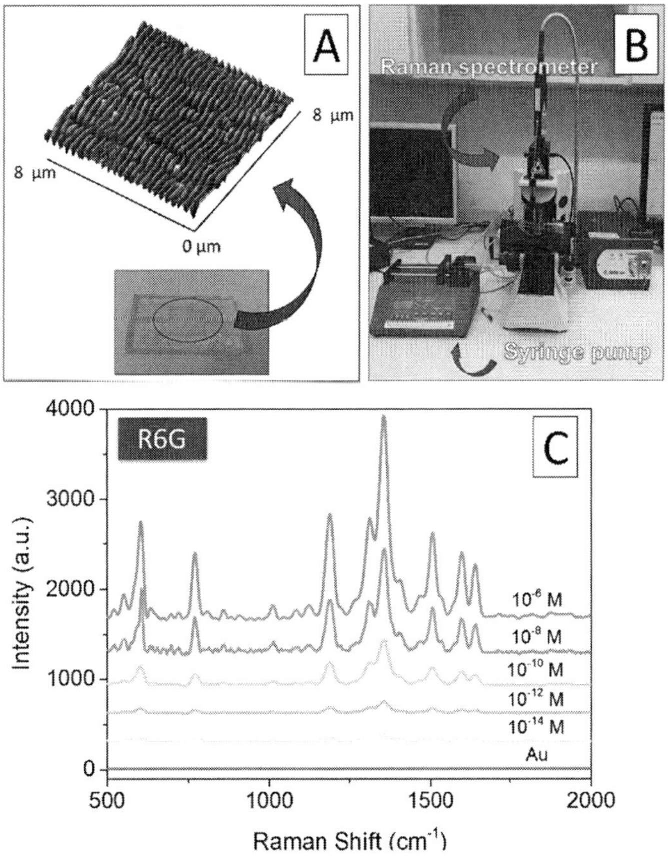

Figure 2 A) The photo of SERS-active microfluidic chip and the AFM-characterized periodic metal grating surface; B) the experimental setup used to perform the on-line SERS measurements in the microchannel; C) Raman spectra of R6G water solutions with varying concentrations, measured using proposed microfluidic SERS-active chip.

The enhancement factor (EF) of the SERS signal was determined by calculating the analytical gain using the following equation: $(EF) = (I_{sers} * N_{ref})/(I_{ref} * N_{sers})$ [18] (10^{-6} M concentration of the

R6G solution was used). The I_{SERS} and I_{ref} are the Raman intensities from the sensor and a reference measurement from a bulk sample; N_{SERS} and N_{ref} are the number of molecules contributing to the corresponding Raman intensity, respectively. In our case, calculations were made for a microfluidic SERS chip with and without a grid. Since the molecule numbers are the same in our case, the Raman intensities were used directly for SERS EF calculations. To determine the EF parameter, a peak of 1355 cm^{-1} was chosen and the EF was found to be equal 10^8. So, in our case, the microfluidic SERS chip with a lattice shows a pronounced enhancement of SERS in the combination with the possibility to analyze the small samples volume.

4. CONCLUSION

In this paper, we proposed a novel technique for the fabrication of plasmon-active SERS component inside 3D microfluidic channel using femtosecond laser surface pattering, followed by the deposition of thin metal film. An experimental model of a microfluidic channel was made using 3D printing. The created structures were applied for real-time surface-enhanced Raman spectroscopy (SERS) measurements of small samples volumes. Created microfluidic SERS platform allows detection of model analyte (R6G) at concentrations up to 10 fM (10^{-15} mol·L^{-1}). Proposed unique features are the key to new scientific experiments and innovations in the field of lab-on-a-chip devices and analytical approaches.

ACKNOWLEDGEMENTS

This work was supported by the Grant Agency of Health Ministry 15-33459A

REFERENCES

[1] FOUDEH, Amir, DIDAR, Tohid F., VERES, Teodor, TABRIZIAN, Maryam. Microfluidic Designs and Techniques Using Lab-on-a-Chip Devices for Pathogen Detection for Point-of-Care Diagnostics. *Lab on a Chip*. 2012. Issue 12, pp. 3249-3266.

[2] LAWANSTIEND, Duangtip, GATEMALA, Harchana, NOOTCHANAT, Supeera, EAKASIT, Sanong, WONGRAVEE, Kanet, SRISA-ART, Monpichar. Microfluidic approach for in situ synthesis of nanoporous silver microstructures as on-chip SERS substrates. *Sensors and Actuators B: Chemical*. 2018. 270, pp. 466-474.

[3] XIE, Jun, MIAO, Yunan, SHIH, Jason, TAI, Yu-Chong, LEE, Terry D. Microfluidic Platform for Liquid Chromatography−Tandem Mass Spectrometry Analyses of Complex Peptide Mixtures. *Anal. Chem*. 2005, 77, pp. 6947−6953.

[4] ASHOK, Praveen C., SINGH, Gajendra P., RENDALL, Helen A., KRAUSS, Thomas F., DHOLAKIA, Kishan. Waveguide confined Raman spectroscopy for microfluidic interrogation. *Lab on a Chip*. 2011. 11(7), pp. 1262-1270.

[5] HAEBERLE, Stefan, ZENGERLE, Roland. Microfluidic platforms for lab-on-a-chip applications. *Lab on a Chip*. 2007. 7(9), pp. 1094-1110.

[6] MARK, Daniel, HAEBERIE, Stefan, ROTH, Gunter, VON STEFEN, Felix, ZENGERIE, Roland. Microfluidic lab-on-a-chip platforms: requirements, characteristics and applications. *Chem. Soc. Rev*. 2010. 39(3), pp. 1153-1182.

[7] QUANG, Ly Xuan, LIM, Chaesung, SEONG, Gi Hun, CHOO, Jaebum, DO, Ki Jun, YOO, Seoung-Kyo. A portable surface-enhanced Raman scattering sensor integrated with a lab-on-a-chip for field analysis. *Lab Chip*. 2008. 8(12), pp. 2214-2219.

[8] ZHANG, Xiao-Yang, HU, Anming, ZHANG, Tong, LEI, Wei, XUE Xiao-Jun, ZHOU, Yunhong, DULEY, Walt W. Self-Assembly of Large-Scale and Ultrathin Silver Nanoplate Films with Tunable Plasmon Resonance Properties. *ACS Nano*. 2011. 5(11), pp. 9082-9092.

[9] ELASNIKOV, R., TRELIN, A., OTTA, J., FITL, P., MARES, D., JERABEK, V., SVORCIK Vaclav, LYUTAKOV Oleksiy. Laser patterning of transparent polymers assisted by plasmon excitation. *Soft Matter*. 2018. 14, 23, pp. 4860-4865.

[10] LEE, J. N., PARK, C., WHITESIDES, G. M. Solvent Compatibility of Poly(dimethylsiloxane)-Based Microfluidic Devices. *Anal. Chem*. 2003, 75, 6544−6554

[11] GUSELNIKOVA, O., POSTNIKOV, P., ERZINA, M., KALACHYOVA, Y., SVORCIK, V., LYUTAKOV, O. Pretreatment-free selective and reproducible SERS-based detection of heavy metal ions on DTPA functionalized plasmonic platform. *Sensors and Actuators B-Chemical.* 2017. 253, pp. 830-838.

[12] GUSELNIKOVA, Olga, KALACHYOVA, Yevgeniya, HROBONOVA, Karolina, TRUSOVA, Marina, BAREK, Jiri, POSTNIKOV, Pavel, SVORCIK Vaclav, LYUTAKOV Oleksiy. SERS platform for detection of lipids and disease markers prepared using modification of plasmonic-active gold gratings by lipophilic moieties. *Sensors and Actuators B-Chemical.* 2018. 265, pp. 182-192.

[13] PARATORE, Federico, ZEIDMAN KALMAN, Tal, ROSENFELD, Tally, KAIGALA, Govind V., BERCOVICI, Moran. Isotachophoresis-Based Surface Immunoassay. *Anal. Chem.* 2017. 89, 14, pp. 7373-7381.

[14] ZHOU, Qitao, KIM, Taesung, Review of microfluidic approaches for surface-enhanced Raman scattering. *Sensors and Actuators B: Chemical.* 2016. 227, pp. 504-514.

[15] KALACHYOVA, Yevgeniya, MARES, David, JERABEK, Vitezlav, ULBRICH, Pavel, LAPCAK, Ladislav, SVORCIK, Vaclav, LYUTAKOV, Oleksiy. Ultrasensitive and reproducible SERS platform of coupled Ag grating with multibranched Au nanoparticles. *Physical chemistry chemical physics.* 2017. 19, 22, pp. 14761-14769.

[16] GUSELNIKOVA, Olga, POSTNIKOV, Pavel, KALACHYOVA, Yevgeniya, KOLSKA, Zdenka, LIBANSKY, Milan, ZIMA, Jiri, SVORCIK, Vaclav, LYUTAKOV, Oleksiy. Large-Scale, Ultrasensitive, Highly Reproducible and Reusable Smart SERS Platform Based on PNIPAm-Grafted Gold Grating. *Chemnanomat.* 2017. 3, 2, pp. 135-144.

[17] KIM, Woong, GUO, Mingquan, YANG, Peidong, WANG, Daojing. Microfabricated monolithic multinozzle emitters for nanoelectrospray mass spectrometry. *Anal. Chem.* 2007. 79(10), pp. 3703-3707.

[18] BAI, Shi, SERIEN, Daniela, HU, Anming, SUGIOKA, Koji. 3D Microfluidic Surface-Enhanced Raman Spectroscopy (SERS) Chips Fabricated by All-Femtosecond-Laser-Processing for Real-Time Sensing of Toxic Substances. *Advanced Functional Materials.* 2018. Volume 28, Issue 23, 1706262.

INSTRUMENTATION FOR STUDY OF NANOMATERIALS IN NPI REZ

(New laboratory for material study in Nuclear Physics Institute in Rez)

Václav BEJŠOVEC, Antonino CÁNNAVO, Giovanni CECCIO, Vladimír HNATOWICZ, Pavel HORÁK, Vasily LAVRENTIEV, Anna MACKOVÁ, Ivo TOMANDL, Alfio TORRISI, Jiří VACÍK

Nuclear Physics Institute (NPI) of Academy of Sciences of the Czech Republic, Rez, phorak@ujf.cas.cz

Abstract

Nano-sized materials become irreplaceable component of a number of devices for every aspect of human life. The development of new materials and deepening of the current knowledge require a set of specialized techniques - deposition methods for preparation/modification of the materials and analytical tools for proper understanding of their properties. A thoroughly equipped research centers become the requirement for the advance and development not only in nano-sized field.

The Center of Accelerators and Nuclear Analytical Methods (CANAM) in the Nuclear Physics Institute (NPI) comprises a unique set of techniques for the synthesis or modification of nanostructured materials and systems, and their characterization using ion beam, neutron beam and microscopy imaging techniques. The methods are used for investigation of a broad range of nano-sized materials and structures based on metal oxides, nitrides, carbides, carbon-based materials (polymers, fullerenes, graphenes, etc.) and nano-laminate composites (MAX phases).

These materials can be prepared at NPI using ion beam sputtering, physical vapor deposition and molecular beam epitaxy. Based on the deposition method and parameters, the samples can be tuned to possess specific properties, e.g., composition, thickness (nm - μm), surface roughness, optical and electrical properties, etc.

Various nuclear analytical methods are applied for the sample characterization. RBS, RBS-channeling, PIXE, PIGE, micro-beam analyses and Transmission Spectroscopy are accomplished at the Tandetron 4130MC accelerator, and additionally the Neutron Depth Profiling (NDP) and Prompt Gamma Neutron Activation (PGNA) analyses are performed at an external neutron beam from the LVR-15 research reactor.

The multimode AFM facility provides further surface related information, magnetic/electrical properties with nano-metric precision, nano-indentation, etc.

Keywords: LEIF, MBE, AFM, PVD, Ion Beam Analysis, Ion scattering, NDP, Microbeam

1. INTRODUCTION

Nanostructured materials with a characteristic length scale of which is below 100 nm, have specific properties deviating from those of bulk ones. This deviation results from the reduced size and/or dimensionality, as well as from the numerous interfaces between adjacent crystallites or layers. The synthesis of materials and/or structures with new properties is an emerging interdisciplinary field based on solid state physics, chemistry and materials science [1]. The research in the new laboratory is mostly focused on preparation and characterization of two-dimensional thin films and nano-laminate structures (MAX phases) exhibiting nanoscale features. The structures are based on combinations of inorganic materials, metals, synthetic polymers and carbon allotropes. Here, only the MAX phases are mentioned in details. These composites have a layered structure of the hexagonal carbides and nitrides that were discovered in the 1960s. They exhibit unusual chemical, physical, electrical, and mechanical properties under various conditions [2]. The MAX phases are bridging the gap between metals and ceramics. For this reason they have become an object of intensive research. They exhibit a high potential applicability in different technologies, such as high efficiency engines, or damage-tolerant systems. Despite of the rather long lasting research, there remain open scientific questions to answer, as well as technological hurdles to overcome. In our laboratory the first MAX phase specimens have been already prepared and characterized by various methods. The techniques of the laboratory are briefly described in the following paragraph.

2. INSTRUMENTATION AVAILABLE AT THE NPI

2.1. Low energy ion facility (LEIF)

The Low Energy Ion Facility (LEIF) established recently in the NPI is a modular system developed for the fundamental research in the field of materials science. The system (**Figure 1**) is divided into two lines with the first line consisting of ion source, ion optics chamber, ion beam diagnostics and a sputtering chamber. After the sputtering chamber, a 1T separation magnet bends the ion beam to 90° and directs it through ion diagnostics unit and quadrupole magnetic optics into the implantation/analytical chamber.

The fundamental element of the system is the multicusp ion source capable of producing uniform, high density plasma with a high gas and electrical efficiency. A filament (based on the thoriated tungsten wire) is surrounded by eight permanent magnets to produce a beam of single and multiple charged ions with the current of tens of nA up to hundreds of μA. In the anode area a small vessel is placed in order to generate ion species not only from gases but also from solid compounds. The maximum voltage applied on the ion source is 35 kV, thus generating the ion beam with the energy up to 35 keV for single charged ions.

The ion beam extracted from the source by the movable extraction electrode is shaped by the Einzel lense system and directed to the sputtering chamber, where the Ion Beam Sputtering (IBS) is utilized for the production of thin layers with thicknesses up to hundreds of nm. In another regime, the beam can be directed through the first chamber to the bending separation magnet, where only ions with specific A/Z ration are selected. The ions are used for irradiation, implantation of samples within the implantation/analytical chamber. The magnet is designed to separate ion species with the energy of 35 keV and A/Z ratios up to 200, so that the implantation within a broad spectrum of ions to the ion fluencies from 1e14 - 1e17 ions/cm^2 is possible.

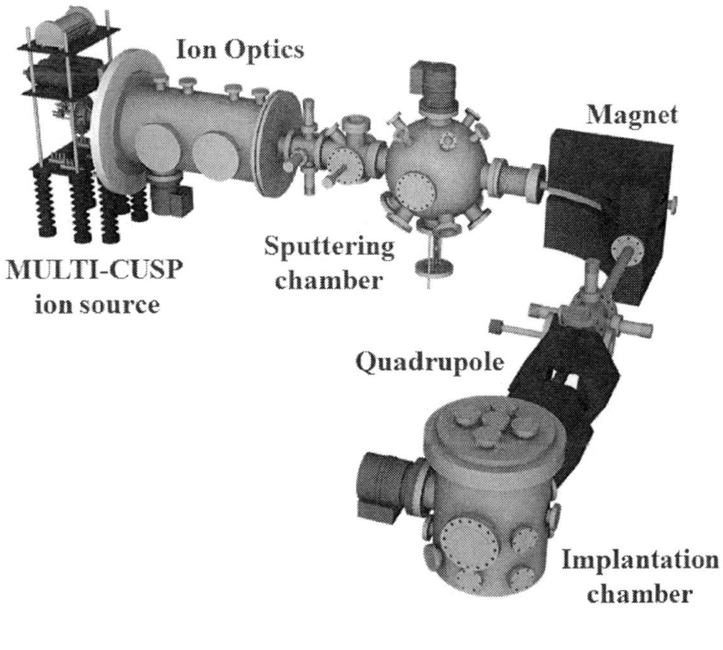

Figure 1 The schematics of the Low Energy Ion Facility (LEIF)

2.2. MBE

Molecular beam epitaxy (MBE) is an Ultra-High-Vacuum (UHV) device for production of epitaxial structures with high purity, uniformity and monolayer control.

The physical working-principle of MBE is based on the interaction between molecular or atomic beams (provided by evaporation in the effusion cells) and a heated crystalline substrate. Despite the conceptual simplicity, this technique needs particular technological requirements to be fulfilled. The Ultra-High-Vacuum operation, i.e., $p = 10^{-9}$ Pa, is a mandatory condition because it guarantees the highest purity of growing material, permitting the use of electron diffraction probes as diagnostic tool. The abruptness, variable growth temperature (20 - 500 °C) and the low growth rate (1 μm/h) makes possible to control of interfaces and doping profiles.

The MBE system installed in NPI is equipped with 2 electron guns and a Knudsen cell as a molecular beam sources. Pyrolytic boron nitride (PBN) is chosen for the crucibles which gives a low rate of gas evolution and chemical stability up to 1400 °C. Careful variation of the temperatures of the cells via PID controllers permits the control of the intensity of the flux of every component or dopant of better than 1 %. An additional ion gun is present for Ion Beam Assisted Deposition (IBAD). A special UHV positioning system allows rotation and 3D manipulation of the substrate to improve the growth homogeneity. Tantalum shutters permit the control of composition and doping of the growing structure.

The in-situ characterization tool is the RHEED (reflection high energy electron diffraction) system. The oscillation of the RHEED signal exactly corresponds to the time needed to grow a monolayer and the diffraction pattern on the RHEED window gives direct indication over the state of the surface. The main deposition chamber can be connected to an additional analytical chamber for ex-situ sample analyses by, e.g., atomic force microscopy, van der Pauw method and low energy electron diffraction.

2.3. PVD-C-plus facility

Fabrication of metal-organic hybrid nanomaterials is presently one of the crucial topics of modern nanoscience promising new fundamental fields and advanced applications. We contribute to this topic using our PVD system (physical vapor deposition, PVD) designed for the controlled fabrication of the Me-C60 hetero-nanostructures (we call this system as "PVD-C-plus" facility). The PVD system can also be applied for fabrication of pure C60 or other C-nanostructures, as well as for fabrication of pure metals and silicon. Variety of hybrid nanostructures combining the components from the former (i.e., C-materials) and the latter (metals and Si) groups can be also produced under this PVD facility.

Operation of the PVD system implies a proper adjustment of setup parameters, which allow to design rather complex nanostructures. Deposition of thin films (up to 200 nm in thickness) of the materials occurs in a high vacuum (up to 10^{-6} mbar) on nonvolatile solid substrates kept at the temperature ranging from RT to 500 °C. Deposition of thin films can be realized through thermal evaporation of the materials using two kinds of sources, namely mini-effusion cell (MEC) and electron gun (EG). The MEC source is used for evaporation of C60, as well as the EG can be used for evaporation of metals and Si. Parameters of deposition influencing the film composition, nanostructure and thickness can be tuned precisely with the digital controllers driving the efficiency of the deposition process with high precision. Design of the hybrid nanostructures also implies the proper use of electrical shutters and digital quartz thickness monitor, which are equipped inside of the vacuum chamber of the PVD system.

2.4. NTEGRA Aura AFM system

The main instrument able to perform a comprehensive test of our deposited films or other suitable samples is the commercial atomic force microscope (AFM) called as NTEGRA Aura (producer is NT-MDT). The NTEGRA Aura is an advanced, powerful and complex AFM system possessing a modular design and including all functional opportunities of the AFM method known at present. The modular structure allows one to upgrade the system continuously and to realize a specific design of each experiment that greatly improve a research potential of this system.

The main feature of the NTEGRA Aura is a possibility to realize the AFM experiment in vacuum or in controllable gas atmosphere (N_2, O_2, Ar, etc.). Our AFM system also includes other important opportunities, such as heating of the sample (up to 120 °C) and application of the external magnetic fields with in-plane and out-of-plane orientations of the magnetic vectors. Depending on the sample origin and the required task, we can apply contact, semi-contact (tapping) and non-contact scanning modes, which can yield a unique 3D pictures of the surface nanostructure (with resolution of 1 nm), unique information about the local mechanical (elasticity, friction), electrical or magnetic properties of the sample surface (with resolution of 10 nm). The research importance of the AFM system are related also with the opportunity to perform spectroscopic analysis in a certain point of the sample surface in order to clarify origin of some selected nano-object. The scientific opportunities of our AFM system is greatly enhanced in scanning tunneling microscopy (STM) mode, which allows to study the surface under atomic resolution. The latter mode, however, requires the samples with special (atomically-flat and clean) surfaces and with rather good conductivity. The digital PX-controller provides a high functionally of all AFM modes with negligible electrical noise.

2.5. ITS

Ion Transmission Spectroscopy (ITS) [3] is a specific method of non-destructive examination of inhomogenities in thin films (e.g., presence of micro-channels or pores). ITS is based on measuring of energy loss of the mono-energetic charged particles passing through the film. The mono-energetic beam of alpha particles from radioactive ^{241}Am source or from Tandetron 4130MC accelerator are used. The transmitted particles are detected by the Si detector connected to a standard spectrometric chain. The energy spectra are evaluated off-line by a computer code (developed at the NPI Rez for laboratory purposes). The energy spectra of particles reflect not only the presence of inhomogenities, but also their size and form.

2.6. NDP/PGAA

The Neutron Depth Profiling (NDP) [4] is a nondestructive nuclear analytical technique based on measurement of energies of reaction products from nuclear reactions of certain light isotopes with thermal neutrons (E_n = 0.025 eV). The emission of reaction products is isotropic and the detector can be placed at any angle with respect to the sample surface. In the Nuclear Physics Institute at Rez, a 5.6 m long neutron guide was installed at the LVR-15 nuclear research reactor (the reactor is operated by Research Center Rez). The neutron guide provides a thermal neutron beam of 10^7 $n \cdot cm^2 \cdot s^{-1}$ intensity and Cadmium Ratio (R_{Cd}) = 10^5 with a beam dimension of 4 mm (height) x 80 mm (width). A classical reaction that can be studied by NDP is $^6Li(n,\alpha)^3H$. In this reaction, thermal neutrons induce an intensive, isotropic nuclear reaction (n, α) on the ^{6}Li isotope producing two reaction products, α (2.055 MeV) and ^{3}H (2.727 MeV), that are emitted from the sample in opposite directions. Even though the natural abundance of the ^{6}Li isotope is only about 7.5 %, the high cross section (940 barns) assures (under the intense neutron beam bombardment) a fast data collection. This reaction is investigated for important application, such as thin Li ion batteries, important and actual research topic.

2.7. RBS, NRA, ERDA and RBS/Channeling

The Rutherford Back-Scattering (RBS) [5] is a non-destructive analytical method based on elastic scattering of impacting particles. Protons, alpha particles or heavier ions (produced by the Tandetron MC 4130 accelerator) with energy from few keV to tens of MeV are penetrating a sample where these species are scattered and subsequently detected by the detection system. In NPI the solid state detector with energy resolution of 15 keV is placed at the angle of 170°. From the obtained energetic spectrum of scattered particles and intensity of individual components it is possible to identify presence of specific elements, as well as its concentration and spatial distribution.

The RBS technique in its most common setup, i.e., alpha particles with energy of 2000 keV, is more suitable for the detection of heavier elements within the sample composed of lighter elements. In this situation it is possible to detect elements on an areal concentration level of 10^{12} - 10^{13} at/cm^2. The accessible depth resolution is ca 10 nm. To analyze the concentration of lighter elements, a modification of the method is necessary. For elements, such as C or O, an ion beam of specific energy can be used so the impacting particles penetrate the atomic core, resonance occurs and the significantly increased cross-section allows the estimation of these elements even within heavier matrix. This analysis is called Nuclear Reaction Analysis, NRA. Other possibility is a method of Elastic Recoil Detection Analysis, ERDA, where instead of scattered particles the particle recoiled from the samples are detected.

A special variant of the RBS technique is RBS/Channeling [5]. In this mode the collimated beam of particles irradiates the sample in the direction of one of the crystal planes. Most of the impacting particles are in this situation funneled into channels formed by the organized rows of atoms. Channeling particles do not get into the contact with atomic cores and thus the probability of scattering is much lower. Presence of other elements, interstitials or defects will be observed as significant increase of the intensity of recoiled particles. RBS/Channeling is thus an importing method providing unique information about the crystal structure.

2.8. PIXE/PIGE

In contrast with the RBS technique, the Proton Induced X-ray/Γ-ray emission methods are based on inelastic interaction between incident proton and the sample material (experimental set-up is depicted in the **Figure 2a**). In case of PIXE, the characteristic X-ray radiation occurs as a result of the interaction between incident proton and electrons in atomic shell. The particle can eject electron from the inner shell of the atoms that is replaced by the electron from the outer shell. The energy difference between these two levels is ejected in the form of X-ray radiation. Under feasible conditions, i.e., heavier elements within the light matrix, the detection limit of the PIXE method can be below 1 ppm. The PIXE method is an important technique for analysis of trace elements within the biological samples (Pb, As, Hg, Cd, etc.). It is also a valuable tool in the analysis of the composition of aerosols in the air created as a result of burning of fossil fuels (analysis of traces of toxic elements).

The Proton Induced Γ-ray emission is a versatile, non-destructive analytical method based on the (p,γ) reaction. In this case, the incident proton causes excitation of the atomic core of the sample element. In a short time, the core returns to the basic level and the energetic difference is emitted as γ-ray. The energy and the number of photons is detected by the HPGe detector and from the spectrum elements in the sample can be identified. The concentration of the elements is measured with the help of the standard sample. For protons with energies from 1 to 3 MeV, the best sensitivities was found for Li, B, F, Na, and Al. The highest cross sections are for light isotopes (A<30), which can be determined with a sensitivity of 10 μg. g^{-1} or less.

2.9. Microbeam

One of the beam lines connected to the Tandetron MC 4130 accelerator in NPI (**Figure 2b**) is used for the analysis and modifications of a sample at small scale. The ion beam from the accelerator passes through a lens, i.e., a combination of magnetic quadrupoles with alternated polarities that focuses the beam to a spot as small as hundreds of nm in diameter. Standard IBA techniques (mainly PIXE, RBS, and NRA) are used to characterize the irradiated object. By raster-scanning the beam over the sample surface, 2D or 3D distribution of elements can be determined with nm depth resolution and lateral resolution limited by the size of the beam spot. The mapped area is from ca 10x10 μm up to 1x1 mm in size.

When samples cannot be measured in vacuum (due to their size, or presence of volatile components), a setup with an external ion beam can be used. Such arrangement is often applied for archaeological artefact or artworks. The ion beam is extracted from the chamber/beam line into air through a thin window. The sample is surrounded by an array of detectors (X-ray detectors for soft and hard X-rays etc.) obtaining the information on the sample composition and concentration. The IBA methods can be employed on the external microbeam, either individually or in the PIXE-PIGE-RBS combination.

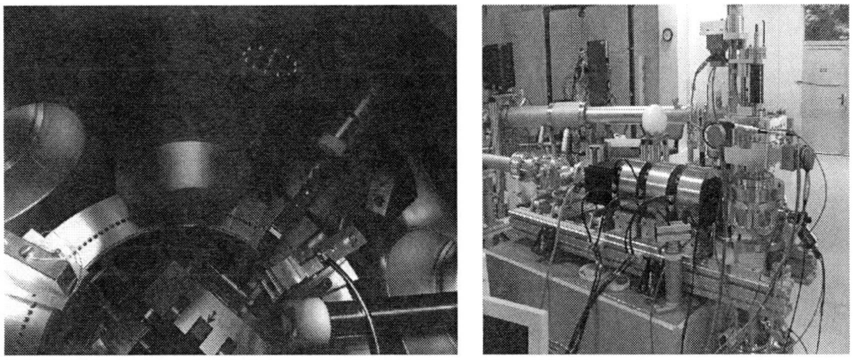

Figure 2 Instrumentation in NPI; 2a - Analytical chamber with multiple detectors for PIXE, PIGE and RBS measurement, 2b - The microbeam system

3. CONCLUSION

A laboratory complex has been developed at NPI SCR Rez, which is endowed with various standard deposition and analytical techniques for preparation and characterization of new prospective nano- and micro-structured systems. The uniqueness of the laboratory lies in combination of common techniques with diagnostic methods making use of charged particle and neutron beams. Most important are the devices for vacuum deposition, deposition by ion sputtering, ion implantation and by molecular beam epitaxy (MBE). Full spectrum of the IBA methods together with Neutron Depth Profiling (NDP) and Ion Transmission Spectroscopy (ITS) is used for specimen characterization. Complementary information on the structure, composition and properties of specimens are obtained with Atomic Force Microscopy (AFM) and using other techniques (FTIR, Raman, XRD, nano-indentation, SEM, TEM, HRTEM, etc.) available at collaborating organizations.

ACKNOWLEDGEMENTS

The authors acknowledge financial support from the Grant Agency of the Czech Republic (grant project No. P108/12/G108). The work was carried out at the CANAM infrastructure of the NPI ASCR Rez.

REFERENCES

[1] H. Gleiter, Acta Materialia, Volume 48, Issue 1, 1 January 2000, Pages 1-29

[2] Barsoum, M.W. (2006) "Physical Properties of the MAX Phases" in Encyclopedia of Materials Science and Technology, K. H. J. Buschow (eds.). Elsevier, Amsterdam.

[3] J.Vacik, J.Cervena, V.Hnatowicz, D.Fink, Y.Kobayashi, K.Hirata, P.Y.Apel and P.Strauss, Study of latent and etched tracks by a charge particle transmission technique, Radiation Measurements 31 (1999) 81-84.

[4] D. Fink, Neutron depth profiling, *Hahn-Meitner-Institute, Report* HMI-B 539 (1996).

[5] Y. Wang, M. Nastasi, Handbook of Modern Ion-Beam Materials Analysis, Materials Research Society, Pittsburgh, 2009.

CHARACTERIZATION OF OLD CONCRETE FOR NEW USE: GRID NANOINDENTATION AND SCANNING ELECTRON MICROSCOPY

Zdeněk PROŠEK, Pavel TESÁREK, Vladimír HRBEK

Czech Technical University in Prague, Faculty of Civil Engineering,
Prague, Czech Republic, EU,
zdenek.prosek@fsv.cvut.cz, pavel.tesarek@fsv.cvut.cz, vladimir.hrbek@fsv.cvut.cz

Abstract

Concrete waste after further processing can be used in the form of a micronized powder in which the non-hydrated portions are reactivated. For this purpose, nanoindentation and scanning electron microscopy with X-ray spectroscopy (SEM-EDS) are combined to determine properties of the concrete sample. The approximate composition of old concrete is calculated from SEM BSE image analysis. The nanoindentation technique over two main material levels enabled evaluation of the indentation modulus of these phases due to the spectral deconvolution of measured data. All these techniques serve for determining the activity of old concrete and characterization shape and textures of activation micronized old concrete powder.

Keywords: Old concrete, nanoindentation, SEM analysis, micronized powder

1. INTRODUCTION

Construction creates the high amount of waste from all sectors. Construction waste is almost 33 % of the 2.5 billion tons of waste generated in the EU [1]. The old concrete is taken to the recycling line. Materials from recycling facilities are most often used as coarse aggregates in new concrete, but only coarse fractions are processed [2]. In the case of a very fine fraction (grain size below 1 mm), there is still no ideal solution to allow its use on a broader scale. Several possible ways of application of this material are examined in cement composites for example as thermally activated binder replacement [3], as mixtures for cement production [4] or as a part of geopolymers based binders [5]. Due to high demands on energy consumption and CO2 production the most sophisticated and environmentally friendly method is when the very fine fraction of recycled concrete is used in its raw form.

For this reason, the paper deals with the possibility of mechanical activation of concrete recyclate with high-speed milling. Mechanical activation of the old concrete occurs when hydrated cement paste is milled, and this reveals still non-hydrated grains of clinker. The amount of non-hydrated clinker is directly dependent on the type of cement and the age of the old concrete. In general, the amount of non-hydrated clinker in the cement matrix can be estimated between 10 to 20 % [6]. Up to 20 % of old concrete can be reactivated and can form binder in the future concrete and the remaining amount will be inert and can form a filler. Micronized milling process can reduce not only the amount of cement but also the amount of natural filler in future concrete [7]. Since up to approximately from 80 to 90 % of the total volume of old concrete consists of inert particles, inert particles characteristics significantly affect the performance of use of old concrete in a new production. Shape and texture of fine inert particles have the most significant impact on the resulting properties of fresh and hardened concrete. The inert particles not only make concrete economical by occupying more volume but also impart volume stability and increase durability [8]. Shape and texture of particles have a significant effect on the performance of fresh concrete. Concrete mixture with well-shaped, rounded, and smooth particles require less amount of cement paste on the excellent workability of fresh mixtures than mixtures with flat, elongated, angular, and rough particles. As a result, well-shaped particles will have less durability caused by the paste such as heat generation, porosity, and drying shrinkage [9].

2. EXPERIMENTAL

For the experiment on mechanical activation of old concrete was used recycled concrete made of concrete railway sleepers. The used old concrete is a material that has already undergone a recycling process (steel reinforcement was removed, and the concrete was pre-crushed). For our purposes, we used a fraction of 0-4mm that does not have a practical use for reuse in new concrete. Concrete sleepers were PB2 and SB8, which was 50 years old.

In the first step, the experiment examined concrete fractions of 0-4 mm. For this purpose, a combination of methods: nanoindentation for determination of micromechanical properties and electron microscopy with EDS detector was used. In the second step, fractions 0-4 were mechanically activated using the micro-milled process. The fine milling of the material was delivered by the company Lavaris Ltd. (procedure protected by patents) using the high-speed milling technique. Subsequently, the micronized concrete powder was subjected to image analysis using an electron microscope. The image analysis provided information on the shape, activation, and suitability of micronized concrete powder for further use.

2.1. Sample preparation

For the investigation, the sample of old concrete (fraction 0-4 mm) and micronized concrete powder were embedded in epoxy resin. Parts of specimens were cut (approx. thickness 15 mm), and silica-carbon papers and diamond suspensions were used for grinding and polishing of the specimen surface. Polished sections were prepared on a Tegramin from Struers. The samples were polished in multiple steps to achieve the best surface quality of the samples. This technique ensured adequate surface roughness of the sample for both indentation and SEM investigation. The coordination mark was inscribed on the polished surface, serving as the origin for further orientation. The mark enabled the positioning of both investigation techniques in the same area of the sample.

2.2. NanoIndentation

The technique of nano-indentation (Ti 750 series, Hysitron Inc.) was applied on the specimen to evaluate elastic mechanical properties (Young's modulus and hardness) of the material. For the purpose of this study, the displacement driven load function with "loading" and "unloading" segments lasting 5 seconds each with an in-between 60 seconds of "holding" time segment was adopted. The maximum prescribed indentation depth was selected 150 nm. The measurement setting meets the grid-indentation specifications of the heterogeneous composite with further application of spectral deconvolution on obtained data [10-14].

The measurement was performed on two crucial levels of the composite (cementitious matrix and aggregate-matrix inter-phase zone) to achieve efficient investigation of each material phase. The first level grid (located in the matrix) was designed for detailed evaluation of phases present in cement paste. The grid consisted of 25 by 25 indents separated by 2 µm (covering area approx. 17.26 %). Grid applied on inter-phase zone location was constructed of 21 by 21 indents in pattern with 5 µm spacing (covering 2.81 % of total area of the grid).

2.3. Electron microscopy

The investigation of the samples consisted of two main techniques, scanning electron microscopy combined with Energy Spectrometer (SEM-EDS) and backscattered electron images with image analysis (SEM BSE-IA). As a result, with nanoindentation, a combination of micromechanical properties, phases chemical composition and material structure enabled the proper description of the material.

The scanning electron microscopy (SEM) investigation was performed on FEG SEM Merlin ZEISS on polished specimens coated with a thin layer of carbon necessary to ensure proper conductivity of the surface. The FEG SEM Merlin ZEISS scanning electron microscope is located in the Laboratory of Electron Microscopy and Microanalysis at the University Center of Energy Efficient Buildings. Qualitative and quantitative analysis of the chemical composition of the samples was performed using an X-ray microanalysis, namely an Energy Spectrometer (EDS) by Oxford Instruments. The point ID method was used in the first step, where only selected areas (areas near indents) were tested. Five microanalyses were performed to determine the weight and atomic representation of individual elements in percent for each phase. The resulting phase was calculated by using stoichiometry. The working distance of the electron microscope was set to 8.5 mm, the accelerating voltage was set to 15 kV, and the current was set to 1 nA. This set up provides a good signal.

Backscattered electron images (SEM-BSE) was performed on FEG SEM Merlin ZEISS with same set up as in the previous case. SEM-BSE has energy comparable to that of the primary electron beam. They come out of greater depth (tens of millimeters in diameter) and thus provide information on local material changes. The information is a material contrast (each phase has its shade of gray). Their resolution is 50-200 nm, which is sufficient in our case. When combining SEM-BSE and image analysis (IA), we get not only the phase contrast but also the percentage representation of the individual phases and the character of the individual grains. The calculation of phase volumetric representations in the viewed field of each phase was performed by Phase Analysis 1.1 (software). Determination of structure and character of individual particles was performed by ImageJ-Analyze particles (software). SEM-BSE IA was made from 64 images at 1k× magnification and 1024×768 px resolution. Such a number of images covers an area of 2.05 mm x 2.05 mm, i.e., 4.2 mm^2.

Structure and character of individual particles were determined by a set of parameters describing the shape of the particle. These shape descriptors were circularity, aspect ratio, roundness, and ferret angle. The circularity (Circ.) describes how particles similar to a circle with a value of 1.0 indicating a perfect circle and value of near to 0.0 indicates an increasingly elongated shape. The value of circularity is calculated by equation (1):

$$Circ. = 4 \cdot \pi \frac{A}{P^2} \tag{1}$$

where:

Circ. - the circularity (-)
A - the area of grain (μm^2)
P - the perimeter (The length of the outside boundary of the grain) (μm).

The aspect ratio (2) is defined as:

$$AR = \frac{MA}{MI} \tag{2}$$

where:

AR. - the aspect ratio of the particle's fitted ellipses (-)
MA - the major axis (the length of the longest axis best fitting ellipse to grain) (μm)
MI - the minor axis (the length of the shortest axis best fitting ellipse to grain) (μm).

The roundness (3) is also a useful parameter for describing the similarity degree of a particle to a circle. Unlike circularity, roundness calculates with rough edges. The roundness is a shape factor that has a maximum value of 1 for a circle and lower values for shapes having a higher ratio of perimeter to area, longer or thinner shapes, or objects having rough edges. For the image analysis used in this experiment, Roundness is defined as:

$$Round. = 4 \frac{A}{\pi \cdot MA^2} \tag{3}$$

where:

Round. - the roundness (-)
A - the area of grain (μm^2)
MA - the major axis (the length of the longest axis best fitting ellipse to grain) (μm).

Feret angle is the angle between the longest side of the grain and the shortest side of the grain.

3. RESULTS AND DISCUSSION

Figure 1 shows SEM images in BSE mode of old concrete and micronized old concrete with the same magnification. BSE image shows white for non-hydrated clinker, light gray for the cement matrix, gray for quartz (aggregate) and dark gray or black for pore, crack or epoxy resin. Also, there are three other phases in the cement matrix, which have different micromechanical properties, namely low-density C-S-H gel, high-density C-S-H gel, and portlandite. Micromechanical properties namely reduced the modulus of individual phases were 19.2 ± 3.5 GPa for low-density C-S-H gel, 39.8 ± 6.0 GPa for high-density C-S-H gel, 58.6 ± 6.3 GPa for portlandite, 79.8 ± 3.1 GPa for quartz and 118.4 ± 16.9 GPa for non-hydrated clinker. **Table 1** shows the result of SEM BSE IA. The results show approximately 5.6 % clinker in old concrete. Micronized old concrete had 2.5 % of clinker, but if percentage value was calculated without epoxy resin, the amount was still the same as for old concrete, namely 5.1 %. This fact, however, does not apply in the case of a cement matrix, where the quantity has been reduced by up to 20 %. This fact was most likely due to the low mechanical properties and milling of the cement matrix into dust particles that were not captured by the filter. Also, it can be seen in **Figure 1** that mechanical re-activation of the already hydrated clinker has occurred partly by their uncover.

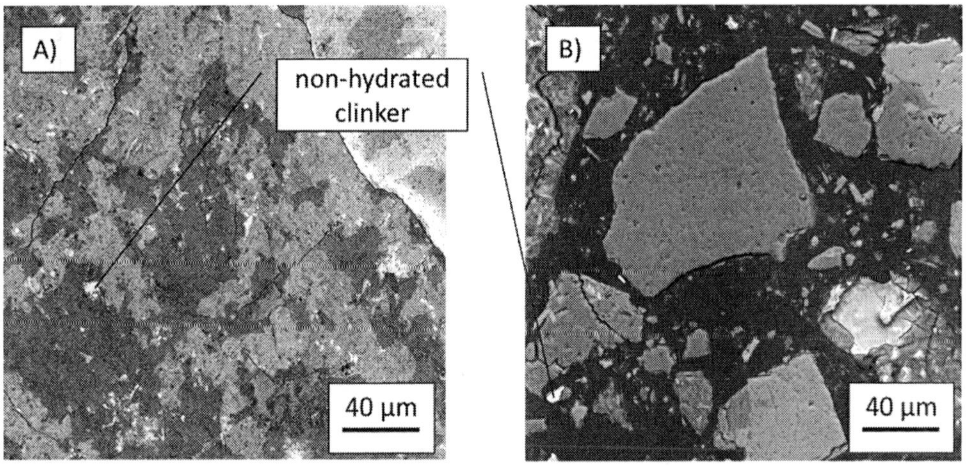

Figure 1 BSE images of microstructure, magnification 1k× a) old concrete, b) micronized old concrete

Table 1 Results of phase volumetric representations in viewed field

Material	Clinker (%)	Cement matrix (%)	Quartz (%)	Pore / Epoxy (%)
Old Concrete	5.6 ± 0.5	48.1 ± 2.1	45.6 ± 1.5	0.7 ± 0.1
Micronized Old Concrete	2.5 ± 0.2	6.6 ± 0.9	41.8 ± 3.5	49.1 ± 3.5

Figure 2 shows a grain characterization procedure where the individual grains are replaced by a fitting ellipse describing their shape. The individual phases were not separated by the micro-milled process so that the shape characterization was performed on all grains. Results in the form of shape descriptors are shown in **Table 2**. During the analysis, more than 91,000 grains were examined. Value of circularity was equal to 0.73. This value determines that the grains are slightly elliptical than the circular shape. The value of roundness was equal to 0.57, so although grains are elliptical, the perimeter is very uneven due to rough edges. Aspect ratio was equal to 2.03, so grains had a slightly elongated shape. The Feret angle was 96.3°, the value is very close to 90°, and therefore characterization of shape grain can be used by fitting ellipse technique.

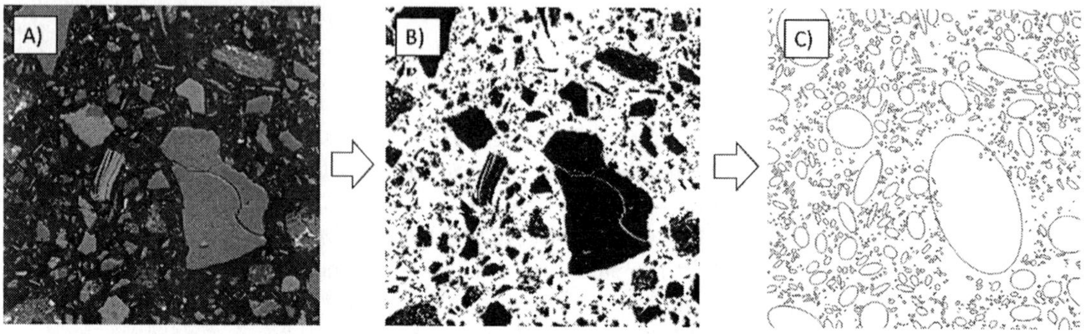

Figure 2 Image analysis of BSE images: A) BSE image, B) marking grains, C) fitting an ellipse to grain

Table 2 Summary of a parameter describing the shape of grains

Material	Count (pc.)	Circularity (-)	Roundness (-)	Feret Angle (°)	Aspect ratio (-)
Micronized Old Concrete	91115	0.73 ± 0.13	0.57 ± 0.11	96.3 ± 12.5	2.03 ± 0.3

4. CONCLUSION

This work deals with the use of high-speed milling technique for recycling of old concrete and the direct characterization of the old concrete and micronized old concrete. Two different experimental methods were used for this purpose, and the results show that:

- Tested old concrete contained approximately 5 % of non-hydrated clinkers that were uncovered by the high-speed grinding process and reactivated;
- SEM-BSE images showed the grains of non-hydrated clinker were uncovered only partially;
- The high-speed milling process had a negative effect on the amount of old cement matrix, where the amount was reduced by 20 % due to low mechanical properties;
- Grains of micronized old concrete has an elongated ellipse shape with rough edges.

In the future, the work will focus on other different types of concrete waste, such as monolithic structures, concretes for supporting structures and concretes for transport infrastructure. The concrete will have different compositions depending on their properties as well as varying amounts of non-hydrated clinker and the strength of the individual phases.

ACKNOWLEDGEMENTS

This was financially supported by the Czech Science Foundation research projects 17-06771S and by the Czech Technical University in Prague project SGS16/201/OHK1/3T/11.

REFERENCES

[1] Waste statistics. In: EUROSTAT. Date: 04-04-17 [update]. [viewed 2018-09-27]. Available from https://ec.europa.eu.

[2] XIAO, Jianzhuang, LI, Wengui, FAN, Yuhui, HUANG, Xiao. An overview of study on recycled aggregate concrete in China (1996-2011), *Construction and Building Materials.* 2012. vol 31, pp. 364-383.

[3] SHUI, Zhonghe, XUAN, Dongxing, WAN, Huiwen, CAO, Beibei. Rehydration reactivity of recycled mortar from concrete waste experienced to thermal treatment, *Construction and Building Materials.* 2008. vol. 22, pp. 1723-1729.

[4] SCHOON, Joris, HEYDEN, Luc Van der, ELOY, Pierre, GAIGNEUX, Eric M., BUYSSER, Klaartje De, DRIESSCHE, Isabel Van, BELIE, Nele De. Waste fibrecement: An interesting alternative raw material for a sustainable Portland clinker production, *Construction and Building Materials,* 2012. vol. 36, pp. 391-403.

[5] AHMARI, Saeed, REN, Xin, TOUFIGH, Vahab, ZHANG Lianyang. Production of geopolymeric binder from blended waste concrete powder and fly ash, *Construction and Building Materials.* 2012. vol. 35, pp. 718-729.

[6] TOPIČ, Jaroslav, PROŠEK, Zdeněk, FLÁDR, Josef, TESÁREK, Pavel. Vliv jemnosti recyklované betonové moučky na vývin hydratačního tepla a vliv jejího množství na mechanickofyzikální vlastnosti cementové pasty, *Wate Forum.* 2018. vol. 2, pp. 268-274.

[7] PROŠEK, Zdeněk, TOPIČ, Jaroslav, ĎUREJE, Jakub, TREJBAL, Jan. Srovnání vlivu mikromletého betonu a mramorové moučky na mechanické vlastnosti cementových past, *Waste Forum.* 2018. vol. 2, pp. 262 - 267.

[8] KWAN, Albert KH; MORA, C. F.; CHAN, H. C. Particle shape analysis of coarse aggregate using digital image processing. *Cement and Concrete Research.* 1999. vol. 29. no. 9, pp. 1403-1410.

[9] ERDOGAN, S. T., et al. Three-dimensional shape analysis of coarse aggregates: New techniques for and preliminary results on several different coarse aggregates and reference rocks. *Cement and Concrete Research,* 2006. vol. 36, no. 9, pp. 1619-1627.

[10] OLIVER, Warren C., PHARR, George M. An improved technique for determining hardness and elastic modulus using load and displacement sensing indentation experiments. *Journal of materials research.* 1992. vol. 7, no. 6, pp. 1564-1583.

[11] PHARR, G. M., BOLSHAKOV, A. Understanding nanoindentation unloading curves. *Journal of Materials Research.* 2002. vol. 17, no. 10, pp. 2660-2671.

[12] OLIVER, Warren C., PHARR, Georges M. Measurement of hardness and elastic modulus by instrumented indentation: Advances in understanding and refinements to methodology. Journal of materials research. 2004. vol. 19, no.1, pp. 3-20.

[13] CONSTANTINIDES, Georgios, et al. Grid indentation analysis of composite microstructure and mechanics: Principles and validation. *Materials Science and Engineering: A.* 2006. vol. 430, no.1-2, pp. 189-202.

[14] ULM, Franz-Josef, et al. Statistical indentation techniques for hydrated nanocomposites: concrete, bone, and shale. *Journal of the American Ceramic Society.* 2007. vol. 90, no. 9, pp. 2677-2692.

SPONSORS / EXHIBITORS

OPTIK INSTRUMENTS

www.brukeroptics.cz

NenoVision

www.nenovision.com

MOLECULAR SPECTROSCOPY
NICOLET CZ
www.nicoletcz.cz

www.nicoletcz.cz

RAITH
NANOFABRICATION

www.raith.com

TESCAN
PERFORMANCE IN NANOSPACE

www.tescan.com

ZEISS

www.zeiss.com

PROFESSIONAL CONTACT ORGANIZATION
FOR THE SAFETY AND RISKS OF NANOMATERIALS IN THE ENVIRONMENT

www.tul.cz

ANAMET
ANALYTICAL & MEASURING & TESTING

www.anamet.cz

ABL&E
JASCO

www.ablelab.com

CHROMSPEC
SPOL. S R.O.

www.chromspec.cz

mtm.eu

www.mericitechnikamorava.cz

AMERICAN ELEMENTS
THE MATERIALS SCIENCE COMPANY ®

www.americanelements.com

MEDIA PARTNERS

Envi Web
www.enviweb.cz

chemosensors

Nanoscale
Editor-in-chief
Chunli Bai

Nanoscale Advances
Editor-in-chief
Chunli Bai
OPEN ACCESS

Environmental Science Nano
Editor-in-chief
Peter Vikesland

nano tech 2018
International Nanotechnology Exhibition & Conference

nanomaterials
an open access journal by MDPI

Nanotechnology Congress 2018
October 11-13, 2018 | Dubai, UAE

The International NanoScience Community
www.nanopaprika.eu
"Spicy world of NanoScience"

condensed matter

HAPTIC